2nd International Tropical Renewable Energy Conference (i-TREC 2017)

IOP Conference Series: Earth and Environmental Science
Volume 105

Bali, Indonesia
3 - 4 October 2017

Editors:

Filbert H. Juwono
Eny Kusrini
Cindy Rianti Priadi
Eko Adhi Setiawan

ISBN: 978-1-5108-5686-8
ISSN: 1755-1307

Printed from e-media with permission by:

Curran Associates, Inc.
57 Morehouse Lane
Red Hook, NY 12571

Some format issues inherent in the e-media version may also appear in this print version.

Printed by Curran Associates, Inc. (2018)

For permission requests, please contact the Institute of Physics
at the address below.

Institute of Physics
Dirac House, Temple Back
Bristol BS1 6BE UK

Phone: 44 1 17 929 7481
Fax: 44 1 17 920 0979

techtracking@iop.org

Additional copies of this publication are available from:

Curran Associates, Inc.
57 Morehouse Lane
Red Hook, NY 12571 USA
Phone: 845-758-0400
Fax: 845-758-2633
Email: curran@proceedings.com
Web: www.proceedings.com

2nd International Tropical Renewable Energy Conference (i-TREC 2017)

IOP Conference Series: Earth and Environmental Science
Volume 105

Bali, Indonesia
3 - 4 October 2017

IOP
Institute of Physics

Table of contents

Volume 105

2nd international Tropical Renewable Energy Conference (i-TREC) 2017

3–4 October 2017, Bali, Indonesia

Accepted papers received: 18 December 2017
Published online: 25 January 2018

Preface

Papers

IOP Conf. Series: Earth and Environmental Science **105** (2017) 011001 doi:10.1088/1755-1315/105/1/011001

Preface

The 2nd international Tropical Renewable Energy Conference (i-TREC) 2017 has been successfully held on 3 – 4 October, 2017 in the beautiful island of Bali, Indonesia. The 2nd i-TREC 2017 was proudly organised by Faculty of Engineering, Universitas Indonesia.

This conference aimed to promote research in the field of development of tropical renewable energy, to facilitate an exchange of new ideas in these fields, and to create a dialogue between scientists, professors, industries, and practitioners. The main theme of the 2nd i-TREC 2017 was "Towards Tropical Renewable Energy Innovation and Technology Integration."

Six keynote speakers and three invited speakers participated in this event. In addition, there were 168 presenters presenting their papers which have passed through peer-review process. Among those, 126 papers were published in this issue of IOP Conference Series: Earth and Environmental Science. The authors came from various countries, such as Australia, Brunei Darussalam, Brazil, France, Indonesia, Japan, and Malaysia.

The committee thanks our sponsors, Kantor Riset Pengelolaan Produk Riset & Inovasi, Universitas Indonesia (main sponsor), PT PLN (platinum sponsor), PT Wijaya Karya (bronze sponsor), and PT Luas Birus Utama. We also thank all participants of the 2nd i-TREC 2017 that have made this conference a successful event. Finally, we hope that the research results of this conference can have contribution in the field of tropical renewable energy.

The Editors of the 2nd international Tropical Renewable Energy Conference (i-TREC) 2017

Filbert H. Juwono, PhD.
Eny Kusrini, PhD.
Dr. Cindy Rianti Priadi.
Dr.-Ing. Eko Adhi Setiawan.

IOP Conf. Series: Earth and Environmental Science **105** (2017) 011002 doi:10.1088/1755-1315/105/1/011002

List of committees

Advisory Board

Dr.-Ing Eko Adhi Setiawan (Universitas Indonesia)
Prof. Dr. rer.nat. Rosari Saleh (Universitas Indonesia)
Prof. Dr. Ir. Dedi Priadi, DEA (Universitas Indonesia)
Worajit Setthapun, Ph.D. (Chiang Mai Rajhabat University)
Prof. Mohd Ikmar Nizam Bin Hj Mohamad Isa (Universiti Malaysia Terengganu)
Prof. Dr. Kenneth Baldwin (Australian National University)
Prof. Yanjun Dai (Shanghai Jiao Tong University, Shanghai)

Scientific Committee

Prof. Dr. Heri Hermansyah (Universitas Indonesia)
Prof. Dr. Akhmad Herman Yuwono (Universitas Indonesia)
Dr.-Ing.Nasruddin (Universitas Indonesia)
Prof.Dr. Adi Surjosatyo (Universitas Indonesia)
Prof. Anne Zulfia, Ph.D. (Universitas Indonesia)
Prof.Dr.-Ing. Nandy Putra (Universitas Indonesia)
M. Ali Berawi, Ph.D. (Universitas Indonesia)
Dr. Ir. Muhammad Asvial, M.Eng. (Universitas Indonesia)

Organizing Committee

General Chair : Eny Kusrini, Ph.D. (Universitas Indonesia)
General Secretary : Dr. Cindy Priadi (Universitas Indonesia)
Publication Chair : Dr. Filbert H. Juwono (Universitas Indonesia)
Technical Committee : Nofrijon Sofyan, Ph.D. (Universitas Indonesia)
Design : Dita Trisnawan, S.T., M.Arch., STD (Universitas Indonesia)
Website : Dr. Muhammad Arif Budiyanto (Universitas Indonesia)

IOP Conf. Series: Earth and Environmental Science **105** (2017) 011003 doi:10.1088/1755-1315/105/1/011003

Conference Photographs

The opening of The 2nd I-TREC 2017

Group Photo of the Participants

IOP Conf. Series: Earth and Environmental Science **105** (2017) 011003 doi:10.1088/1755-1315/105/1/011003

Keynote Speaker Presentation

Speech by Conference Chair

Group Photo at the Gala Dinner

IOP Conf. Series: Earth and Environmental Science **105** (2017) 011004 doi:10.1088/1755-1315/105/1/011004

Sponsor or Funding Acknowledgements

We would like to thank our sponsors, who have supported the 2nd international Tropical Renewable Energy Conference (i-TREC) 2017.

Main Sponsor

UNIVERSITAS INDONESIA

KANTOR PENGELOLAAN PRODUK RISET & INOVASI

Platinum sponsor

PLN

Bronze sponsor

WIKA

Supported by

STARBORN

Peer review statement

All papers published in this volume of *IOP Conference Series: Earth and Environmental Science* have been peer reviewed through processes administered by the proceedings Editors. Reviews were conducted by expert referees to the professional and scientific standards expected of a proceedings journal published by IOP Publishing.

IOP Conf. Series: Earth and Environmental Science **105** (2017) 012001 doi:10.1088/1755-1315/105/1/012001

Preliminary study of natural zeolite from Bayah for solar powered cooling application

Djubaedah[1,2], **Wulandari**[1,3], **Nasruddin**[4,a]

[1]Doctoral Program in Mechanical Engineering Department, Universitas Indonesia, Depok, Indonesia.

[2]Agency for the Assessment and Application of Technology (BPPT), Serpong, Indonesia.

[3]State University of Jakarta, Jakarta, Indonesia.

[4]Mechanical Engineering Department, Universitas Indonesia, Depok, Indonesia.

[a]E-mail: nasruddin@eng.ui.ac.id

Abstract. This paper presents the first attempt to investigate of natural zeolite for cooling adsorption application. Representative samples were collected from Bayah-Banten, Indonesia. As known that zeolites can be used as an adsorbent. Unfortunately, natural zeolite has many limitation, among them contains a lot of impurities. To improve the characteristic of natural zeolite, activation and modification should be conducted beforehand. Activation was conducted by reducing grain size (1-2mm), washed by aquadest and heated using microwave. While the chemical activation is done through acidification by adding HCl solution. Finally, the activated natural zeolite was calcined gradually initiated at room temperature to 150°C for 2 hours and then heating continued from 150°C to 300°C for 4 hours. The final zeolite activated product was characterized by X-Ray Diffraction (XRD), Brunauer–Emmett–Teller (BET) and Transmission Electron Microscopy (TEM). The result showed that the morphology of the natural zeolte was cubic shave as observed by TEM and the surface area of activated natural zeolite is geater than the raw material.

1. Introduction

The commercial buildings are synonymous with high energy consumption, about 40% - 50% of energy consumption is due to HVAC usage[1]. The usage of unfriendly refrigerant on HVAC systems also causes ozone depletion [2]. Currently, worldwide energy policy encourage an efforts of renewable energy utilization and energy conservation activities in all sectors [3, 4]. Nowadays, alternative air-conditioning units that are derived from renewable energies, present a promising solution [5, 6]. Solar energy is one of the renewable energy sources which suitable to be applied for air-conditioning system [5]. Adsorption cooling system which can be carried-on by low-grade heat sources, such as solar energy or waste heat. Adsorption is a physical or chemical reaction process between a solid (adsorbent) and a gas (refrigerant). Generally in the adsorption process of cooling machine, the physical reaction between the adsorbent and refrigerant occurred through the connections of Van der Waals force [7].

Figure 1 shows the ideal cycle of the cooling adsorption process that occurs: A-B is the preheating process (isoteric), B-C is the desorption process where the refrigerant contained in the adsorbent will be issued along with the increased in the temperature on the adsorbent, then after saturation occurs the

C-D process is precooling isoteric) and subsequently is the D-A evaporation process (adsorption). The adsorbent temperature continues to fall and the liquid refrigerant evaporates and extracts heat from the evaporator which creates a cooling effect [8].

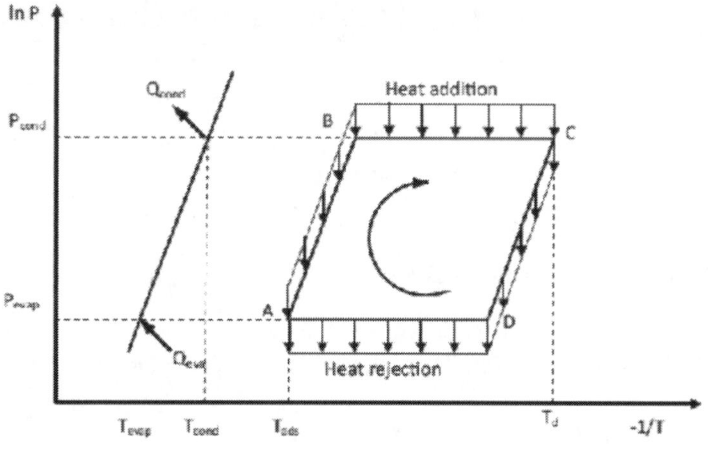

Figure 1. Ideal adsorption cooling cycle [8].

On the cooling of the adsorption system, there are some of the main components that must be available i.e. heat source, adsorber, condenser and evaporator. The selection of pairs of adsorbents should consider the temperature of the heat source, the desired characteristics of the cooling system, the working properties of the adsorbent pairs (the physical, chemical and thermodynamic properties of the substances), the price, the availability and the environmental impact [8].

Ammonia, methanol, and water are have relatively larger latent calorific values (1368, 1160, and 2258 kJ/kg respectively) and small specific volumes (in units of 10^{-3} m^3/kg) so they are the most commonly used as a refrigerants. Water is the most thermally stable refrigerant, followed by methanol, and ammonia. Water deficiency is not applicable for cooling requirements below 0°C [5]. A good adsorbent material should have a large pore (ranging between 600 m^2/g or more) to absorb a large amount of refrigerant, but this causes the thermal conductivity to be low, thus limiting the performance of the cooling system [9].

The most commonly used adsorbents pairs today are zeolite - water, silica gel - water, activated carbon - methanol and activated carbon with ammonia. The isothermal adsorption of Zeolite is highly dependent on non-linear pressure, and thus makes Zeolite a suitable choice for solar cooling application [5]. Silica gel almost meets all the criteria required in the selection of adsorbents, besides it often experience the phenomenon of decreasing adsorption and weathering capacity [10, 11].

Carbons Active - methanol are operate at low regeneration temperatures <120°C, has a large adsorption capacity, low adsorption heat and low freezing point. The activated carbon pairs - ammonia require a regeneration temperature above 150°C, but has lower adsorption capacity, has a very strong odor and corrosive [12, 13]. For zeolite - water pair, regeneration temperature above 200°C and adsorption evaporation temperature of up to 70°C or more, they are remains stable at high temperatures as well as latent heat from water higher than methanol, so zeolite - water pairing is more appropriate to apply to room air conditioning [5] because solidified water temperatures can resist clotting processes while the disadvantage that are not too high cooling capacity and low working pressure can inhibit mass transfer [13]. In a study conducted by Anyanwu and Ogueke [14] can be concluded that for the activated carbon pairs - ammonia and activated carbon - methanol is very congested for ice manufacture and Frozen food pickling applications. As for zeolite – water pair is more suitable for air conditioning application of adsorption system because the lowest water temperature is 1°C and its high latent heat is suitable for producing cold water [7].

Zeolite is a nonmetallic mineral material with a structure of alumina-silicate having pores and skeletons in molecular dimensions [15]. There are four processes as a description of the formation of zeolites, namely the process of sedimentation of volcanic dust on the lake environment that is alkaline, alteration process, diagenesis process and hydrothermal process [16]. One of the uniqueness of the zeolite is to have an area surface and acidity that is easy to be modified. To improve the character of natural zeolite such as impurity and its crystallinity that is not good, activation and modification is the common method to be conducted beforehand. Activation process is intended to modify the properties of zeolites, such as surface area and acidity. The increased surface area and acidity will caused the catalytic activity of the zeolite to increased [17] . The objective of this research is utilizing natural zeolite from Indonesia as an adsorbent for cooling application while the novelty is improving a low grade natural zeolite from Indonesia become an effective cooling adsorbent.

2. Material and Methods

2.1. Material

Natural Zeolite (ZA) was supplied by PT. Trasindo Citra Utama which is obtained from Bayah, Province of Banten-Indonesia. The zeolite was crushed into small size (1-2mm) and sifted using a micro-size screen to remove fine particle and then weighed the weight of the zeolite to be used. Other materials such as Aquadest and hydrochloric acid (HCl) used for activating of zeolite.

2.2. Treatment of zeolite

This treatment was started by washing of natural zeolite by Aquadest. Washing done 3 times to remove undesirable materials and then filtered it. The washed zeolite dried using microwave oven. the use of microwave oven is to shorten the drying time as did by A. Arafat when he prepared Zeolite Y and ZSM-5 in microwave oven [18]. Before heating, zeolite is weighed first to know the amount of water depreciation during the drying process. The washed zeolite is placed in a clean cup then heated in the microwave with full heating for 20 minutes. Every 1 minute the zeolite is removed and weighed. This warming is done continuously until the shrinkage of weight from zeolite does not happen again. This can be interpreted that the water content in the zeolite is almost nonexistent. This sample was coded as ZAH. After the zeolite is dried, the next step is the activation of zeolite. In general, The use of Acid in the activation process of zeolite is to remove impurities that block the pores [19] and to increased Si/Al ratio [20]. The acid used in this study is hydrochloric acid (HCl).

About 15g of natural zeolite sample in 40.5 ml HCl – 500 ml Aquadest solution. The suspension was stirred in 1000 ml beaker glass using a magnetic stirrer type IKA C-MAG at a rate of 2 Mot for 24 h and room temperature. Subsequently, the suspension was filtered and washed with distilled water. The wet modified material was dried gradually using Themolyne furnace starting from room temperature - 150°C for 2 hours. Then calcined at a temperature of 150 °C - 300 °C at a rate of 10C / min for 4 hours. This sample was coded as ZAAH.

2.3. Characterization

The raw natural zeolites (ZA), heater natural zeolite (ZAH) and activated natural zeolites (ZAAH) were characterized by a variety of techniques describe below:

2.3.1. BET

The surface area and porous properties are the most important characteristics of porous materials because it is strongly related to the performance in many applications. The surface parameters of the prepared zeolite samples were measured by Micromeritics ASAP 2020 series. The determination was based on the measurements of the adsorption isotherms of nitrogen at 77°K. All samples were outgassed for 4 h at 350°C. The surface area was estimated from the BET equation in the relative pressure range of between 0.05 and 0.3, over five adsorption points. The micropore volume was calculated from tplot Harkins and Jura equation. Corresponding mesopore volume was determined by subtracting the micropore volume from the volume of liquid nitrogen adsorbed at relative pressure of 0.99. Pore size distributions were determined using non-local density functional theory for slit-like pore geometry.

2.3.2. XRD
The crystallinity phase of natural zeolite was analyzed by X-Ray Diffractometer (XRD) at PT. Nanotech Hebal Indonesia, Balai Inkubator Teknologi BPPT- Serpong, Indonesia.

2.3.3. TEM
The technique of Transmission Electron Microscopy was conducted to observe the structural characterization or morphologies of natural zeolite. TEM analysis was conducted at Universitas Indonesia.

3. Result and Discussion

In previous study, Khaidir et al. [21] reported have modified natural zeolite from Bayah and use it as a molecular sieves material on bioethanol dehydration. The result showed that the modified natural zeolite by acidification can increased the properties of natural zeolite. Therefore, in the present paper, effect of hydrochloric acid for the activation of natural zeolite from Bayah was investigated, and their properties were characterized. The results described in the following section. From BET testing conducted on the samples obtained the results of surface area, pore size and pore volume as shown in the table 1.

Table 1. Surface area, pore volume and average pore size of natural zeolite from Bayah

Sample of Zeolite	ZA	ZAH	ZAAH
BET Surface Area (m^2/g)	70.03	72.02	81.5
t-Plot micropore volume (cm^3/g)	0.03	0.03	0.03
Adsorption average pore width (4V/A by BET) (nm)	5.7	5.8	5.8
BJH Adsorption average pore diameter (4V/A) (nm)	16.4	14.9	15.1
BJH Desorption average pore diameter (4V/A) (nm)	12.3	12.3	13.9

X-Ray Diffraction (XRD) pattern is shown by the figure 2. There are two strongest peaks available i.e. peak at 2Θ as describe in the table 2.

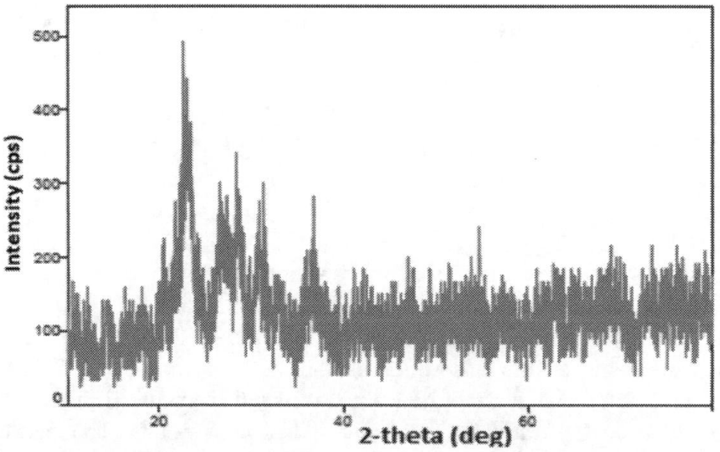

Figure 2. XRD pattern of natural zeolite from Bayah

Table 2. Strongest Peak List of XRD spectrum of natural zeolite from Bayah

No.	2-theta(deg)	d(ang.)	Height(cps)	FWHM(deg)	Int. I(cps deg)	Int. W(deg)	Size(ang.)
1	23.19(5)	3.832(8)	188(40)	1.41(9)	422(28)	2.2(6)	60(4)
2	26.92(6)	3.309(8)	94(28)	3.5(4)	574(94)	6(3)	24(3)

Razzak et al. have also conducted research on the characterization of zeolites from Bayah [22]. In their study which compared the XRD peaks with the ICDD 49-0924 and concludes that the natural zeolites from Bayah belong to a mineral mordenite group. It is different with that done by Supandi in his research [23] who reported that natural zeolite from Bayah belong to be a mineral clinoptilolite group and crystal structure of natural zeolite from Bayah may contains two crystal systems, there are crystal monoclinic with space group C2/m and crystal orthorhombic with space group of CmCm. This difference may occur due to the different mineral content of each mining site.

The technique of Transmission Electron Microscopy (TEM) was employed for the structural characterization of the natural zeolite. Figure 3 is illustrated a bright field TEM image corresponding to natural zeolite from Bayah, in which the presence of different phases can be observed. The result indicates the presence of the clinoptilolite phase and TEM image shows the cubic morphology of the material, which agrees with the XRD result to know more details about morphology and mineral content of natural zeolite, it is better to do SEM and EDX test.

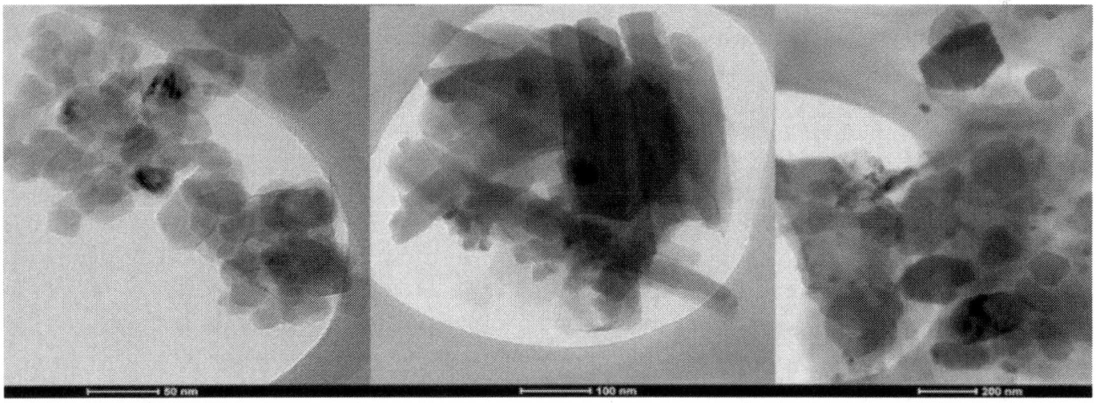

Figure 3. The morphology of natural zeolite from Bayah using TEM examination.

4. Conclusion

Based on the results presented at above, it can be taken conclusions that through acidification treatment followed by calcination of natural zeolite from Bayah, it can be increased a surface area but the surface area obtained is still too small to be used as an adsorbent material to be applied for cooling. So another modifications are needed which can be improve the zeolite properties better.

5. Acknowledgment

The authors gratefully acknowledge the financial support that provide by JICA Project for AUN/SEED-Net under Contract No: UICRC1601-UICRC1701. Acknowledgment is also directed to Mr. Jody Arya Laksmono for sample preparation, and Mr. Toni Toha Apandi, Director PT. Transindo Citra Utama for supplying zeolites.

References

[1] Pérez-Lombard, L., J. Ortiz, and C. Pout, 2008. A review on buildings energy consumption information. *Energy and buildings*. **40**(3) 394-98.

[2] Hu, P., J.-J. Yao, and Z.-S. Chen, 2009. Analysis for composite zeolite/foam aluminum–water mass recovery adsorption refrigeration system driven by engine exhaust heat. *Energy Conversion and Management*, **50**(2) 255-61.

[3] Ellabban, O., H. Abu-Rub, and F. Blaabjerg, 2014. Renewable energy resources: Current status, future prospects and their enabling technology. *Renewable and Sustainable Energy Reviews*, **39** 748-64.

[4] Dincer, I., 2000.Renewable energy and sustainable development: a crucial review. *Renewable and Sustainable Energy Reviews*, **4**(2) 157-75.

[5] Fernandes, M.S., et al., 2014. Review and future trends of solar adsorption refrigeration systems. *Renewable and Sustainable Energy Reviews*, **39** 102-23.

[6] Gugulothu, R., et al., 2015. A Review on Solar Powered Air Conditioning System. *Procedia Earth and Planetary Science,* **11** 361-67.

[7] Choudhury, B., P. Chatterjee, and J. Sarkar, 2010. Review paper on solar-powered air-conditioning through adsorption route. *Renewable and sustainable energy reviews*, **14**(8) 2189-95.

[8] Younes, M.M., et al., A review on adsorbent-adsorbate pairs for cooling applications. *Applied Thermal Engineering*, 2017. **114** 394-414.

[9] Tatlıer, M., B. Tantekin-Ersolmaz, and A. Erdem-Şenatalar, 1999. A novel approach to enhance heat and mass transfer in adsorption heat pumps using the zeolite–water pair. *Microporous and Mesoporous Materials*, **27**(1) 1-10.

[10] Alghoul, M., et al., 2007. Advances on multi-purpose solar adsorption systems for domestic refrigeration and water heating. *Applied thermal engineering*, **27**(5) 813-22.

[11] Wang, D., et al., 2012. Investigation of adsorption performance deterioration in silica gel–water adsorption refrigeration. *Energy Conversion and Management*, **58** 157-62.

[12] Wu, J., J. Cheng, and Q. Wang, 1997.Study on a new solid absorption refrigeration pair: active carbon fiber—methanol.

[13] Wang, D., et al., 2010. A review on adsorption refrigeration technology and adsorption deterioration in physical adsorption systems. *Renewable and Sustainable Energy Reviews*, **14**(1) 344-53.

[14] Anyanwu, E. and N. Ogueke, 2005. Thermodynamic design procedure for solid adsorption solar refrigerator. *Renewable energy*, **30**(1) 81-96.

[15] Jacobs, P., et al., 2001. Introduction to zeolite science and practice. *Elsevier*. **137**.

[16] Kusdarto, K., 2008.Potency of zeolite in Indonesia. *Jurnal Zeolit Indonesia*, **7**(2) 78-87

[17] Lestari, D.Y. 2010. Kajian modifikasi dan karakterisasi zeolit alam dari berbagai negara. in Prosiding Seminar Nasional Kimia dan Pendidikan Kimia.

[18] Arafat, A., et al., 1993.Microwave preparation of zeolite Y and ZSM-5. *Zeolites*, **13**(3) 162-65.

[19] Pardoyo, P., 2005.Pengaruh perlakuan asam dan kalium permanganat terhadap karakter zeolit tasik. *Jurnal Kimia Sains dan Aplikasi*, **8**(1).

[20] Ertan, A. and F. Çakicioğlu-Özkan, 2005. CO_2 and N_2 adsorption on the acid (HCl, HNO_3, H_2SO_4 and H_3PO_4) treated zeolites. *Adsorption*, **11** 151-56.

[21] Khaidir, K., D. Setyaningsih, and H. Haerudin, 2009. Modification of natural zeolite as molecular sieve material on bioethanol dehydration. *Jurnal Zeolit Indonesia*, **8**(2) 97-105.

[22] Razzak, M.T., T. Las, and P. Priyambodo, 2013. The Characterization of Indonesian's Natural Zeolite For Water Filtration System. *Jurnal Kimia valensi*, **3**(2).

[23] Suminta, S., 2006. *Characterization of natural zeolite by X-Ray diffractometer*. Jurnal Zeolit Indonesia, **5**(2) 52-68.

Process optimization of microalgae cultivation in a bubble-column photobioreactor

Y Muharam[1,2], Dianursanti[1] and A S Wirya[1]

[1] Department of Chemical Engineering, University of Indonesia, Depok 16424, Indonesia

[2]E-mail: muharam@che.ui.ac.id

Abstract. A process optimization was performed in this study to obtain the values of the geometry and process parameters that provide an optimum average microalgae concentration in an internally illuminated bubble-column photobioreactor. A phenomenological model of the photobioreactor is used to simulate the process in the reactor. The model considers gas-phase mass balance and liquid-phase mass balances accompanied by the rates of CO_2 and nutrient intakes in liquid phase. The process optimization gives the values of the small and large cylinder diameters, the culture height, the inlet air flowrate, the CO_2 mole fraction in the inlet air and the initial microalgae concentration that promotes the microalgae growth to 0.773 g/L from 0.2 g/L for less than four days.

1. Introduction

As one of the third-generation renewable energy sources that does not compete with the needs of other industries, microalgae can be a solution in sustainable production. Oil contained in microalgae is amazing, reaching 40-85% of the dry weight [1]. That is much higher when compared to oil palm, which is only about 20%. Microalgae can be processed into several derivates of bioenergy products such as biodiesel, bioethanol, biobutanol, straight vegetable oil and green fuels.

Mass production of microalgae will be efficient if reactors used for microalga cultivation are designed as good as possible. This is achieved when the design considers process and geometry parameters. A design method considering those parameters is simulation using theoretical or phenomenological models. By simulation, the process is optimized so that the best performance of the reactor is reached.

Several studies on the modelling of microalgae cultivation reactors have been reported. Seo et al. modeled a bubble-column photobioreactor with computational fluid dynamics to learn mixing in the reactor [2]. The microalgae growth rate was not introduced into the model. Pegallapati and Nirmalakhandan developed the model of an internally illuminated bubble-column photobioreactor that has been validated with the experimental data for 16-day cultivation [3]. They assumed that gas phase and bulk culture medium were perfectly mixed. Sforza et al. modeled a flat-plate photobioreactor that assumed as a perfectly stirred-tank reactor without recycle [4]. The material balance was modeled at macroscopic level. Modeling of an airlift-raceway photobioreactor has been performed by Kethesan and Nirmalakhandan [5]. The model is macroscopic level and includes CO_2 transfer, biomass growth, light effect and nitrogen consumption. Nauha and Alopaeus combined computational fluid dynamics and compartmental models to predict mixing, mass transfer characteristics and light effect in a photobioreactor [6]. The models have not been validated with experimental data. Zhang et al. modeled a flat-plate photobioreactor describing algae growth, the consumptions of CO_2, phosphate, nitrate, and ammonium by *Chlorella kessleri* cultured in waste water for 21 days [7]. Recently, Muharam et al. developed the axisymmetric phenomenological model of an internally illuminated bubble-column photobioreactor for microalgae cultivation [8]. The model considers mixing in gas phase and bulk culture medium using dispersion approach and has been validated using the experimental data by Pegallapati and Nirmalakhandan [3].

The objective of the present study is to obtain an optimum process for *Nannochloropsis salina* microalgae cultivation in an internally illuminated bubble-column photobioreactor through simulation. The internal illumination is chosen because it provides more homogeneous intensity to the angular direction of the reactor. The axisymmetric phenomenological model that has been developed by Muharam et al. [8] was used.

2. Methodology

As described in [8], the internally illuminated bubble-column photobioreactor model comprises a gas-phase mass balance and liquid-phase mass balances. The uptakes of CO_2, phosphate, nitrate and sulphate take place in liquid phase. The photobioreactor is illustrated in Figure 1. Two cylinders are arranged concentrically. The microalgae culture is placed in the annular space. A light with constant intensity is situated inside the small cylinder. The sparger holes to spray CO_2-containing air locate at the bottom of the cylinder.

Process optimization is to obtain the values of decision variables in terms of geometry and process parameters that provide the optimum value of an objective function, which in this case is the average microalgae concentration in the photobioreactor. Therefore, the culture height as well as the small and large cylinder diameters as the geometry parameters vary. The inlet air flowrate and pressure, as well as the CO_2 mole fraction in the inlet air and the initial microalgae concentration as the process parameters vary. Nutrient concentrations do not vary.

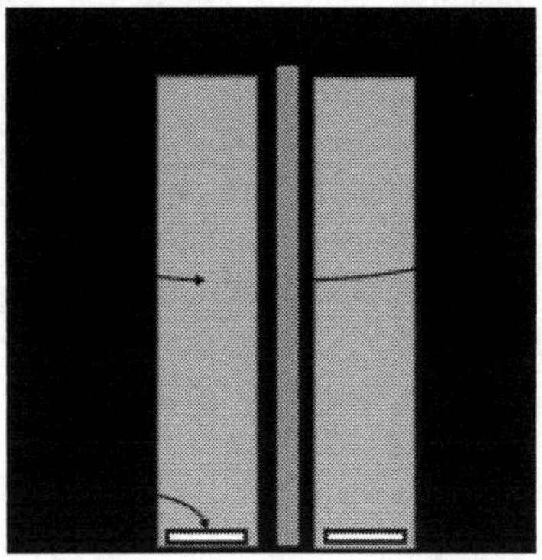

Figure 1. Model of internally illuminated bubble column photobioreactor.

The gas-phase CO_2 mass balance is as follows:

$$\frac{\partial C_{CO_2,G}}{\partial t} - \left(\frac{1}{r}\frac{\partial}{\partial r}\left(rD_{r,G}\frac{\partial C_{CO_2,G}}{\partial r} \right) + \frac{\partial}{\partial z}\left(D_{z,G}\frac{\partial C_{CO_2,G}}{\partial z} \right) \right) + \left(u_r \frac{\partial C_{CO_2,G}}{\partial r} + u_z \frac{\partial C_{CO_2,G}}{\partial z} \right) = -R_{CO_2,GL} \quad (1)$$

where $C_{CO_2,G}$ is the gas-phase CO_2 concentration, u_r is the r-component velocity, and u_z is the z-component velocity.

The mass balances of CO_2, microalgae and nutrients in liquid phase are described by:

$$\frac{\partial C_{j,\mathrm{L}}}{\partial t}-\left(\frac{1}{r}\frac{\partial}{\partial r}\left(rD_{r,\mathrm{L}}\frac{\partial C_{j,\mathrm{L}}}{\partial r}\right)+\frac{\partial}{\partial z}\left(D_{z,\mathrm{L}}\frac{\partial C_{j,\mathrm{L}}}{\partial z}\right)\right)=R_{j,\mathrm{L}} \quad (2)$$

where $C_{j,\mathrm{L}}$ is the liquid-phase j-component concentration, and $R_{j,\mathrm{L}}$ is the j-component reaction rates.

The gas-phase axial dispersion coefficient, $D_{z,\mathrm{G}}$ and the gas-phase radial dispersion coefficient, $D_{r,\mathrm{G}}$ are in accordance with Joshi [9]:

$$D_{z,\mathrm{G}} = 50D_R^{3/2}\left(\frac{u_x}{\varepsilon_\mathrm{G}}\right)^3 \quad (3)$$

and

$$D_{r,\mathrm{G}} = D_{z,\mathrm{G}} \quad (4)$$

The liquid-phase axial dispersion coefficient, $D_{z,\mathrm{L}}$ and the liquid-phase radial dispersion coefficient, $D_{r,\mathrm{L}}$ are in accordance with Krishna et al. [10]:

$$D_{z,\mathrm{L}} = 0.0632\,\mathrm{lg}^{0.5}D_\mathrm{R}^{1.5}\left(\frac{u_x^3}{g\upsilon_\mathrm{water}}\right)^{0.125} \quad (5)$$

and

$$D_{r,\mathrm{L}} = 0.01D_{z,\mathrm{L}} \quad (6)$$

The gas-liquid mass transfer, $R_{\mathrm{CO_2},\mathrm{GL}}$ is formulated by

$$R_{\mathrm{CO_2},\mathrm{GL}} = k_\mathrm{L}a_{\mathrm{CO_2}}\left(C_{\mathrm{CO_2}}^{*} - C_{\mathrm{CO_2},\mathrm{L}}\right) \quad (7)$$

where $k_\mathrm{L}a_{\mathrm{CO_2}}$ is correlated to the liquid-phase diffusion coefficients of CO_2 and O_2, $d_{\mathrm{CO_2}}$ and $d_{\mathrm{O_2}}$ by

$$k_\mathrm{L}a_{\mathrm{CO_2}} = k_\mathrm{L}a_{\mathrm{O_2}}\sqrt{\frac{d_{\mathrm{CO_2}}}{d_{\mathrm{O_2}}}} \quad (8)$$

as reported by Grima et al. [11].

The CO_2 concentration in liquid phase that is in equilibrium with the one in gas phase, $C_{\mathrm{CO_2}}^{*}$ is related to the CO_2 mole fraction, $y_{\mathrm{CO_2}}$ in gas phase by

$$C_{\mathrm{CO_2}}^{*} = \frac{py_{\mathrm{CO_2}}}{RTH_{\mathrm{CO_2}}} \quad (9)$$

where $H_{\mathrm{CO_2}}$ the Henry's dimensionless constant of CO_2 in water.

The microalgae growth rate is expressed by

$$R_\mathrm{algae} = \mu C_\mathrm{algae} - k_\mathrm{d}C_\mathrm{algae} \quad (10)$$

where μ is the specific growth rate of microalgae as the function of CO_2 and nutrient intakes, as well as temperature and light intensity, and k_d is the death rate of microalgae.

3. Results and discussion

The simulation results are shown in figure 2 to figure 7. The simulation was carried out with the small cylinder diameter of 5 cm, the large cylinder diameter of 30 cm, the culture height of 60 cm, the inlet air flowrate of 1 L/min, the inlet air pressure of 1 atm, the inlet CO_2 mole fraction of 0.02, the initial microalgae concentration of 0.2 g/L, and the inlet and ambient temperatures of 30 °C. When a process or geometry parameter varies, all other parameters do not change.

Figure 2 shows the relationship between the average microalgae concentration in the reactor and the large cylinder diameter. From the figure, it can be seen that the increase in the large cylinder diameter leads to the decrease in the average microalgae concentration, but the change is not significant. This indicates that the agitation by rising gas bubbles provides good mixing in the axial and radial direction of the reactor. The effect is significant if the gas flowrate is very low, which is within the regime of less than 0.5 L/min. In this regime, the increase in the large cylinder diameter decreases significantly the average microalgae concentration. This is because the light penetration inside the reactor is obstructed by microalga. The culture that is away from the light source receives low-intensity light.

The microalgae concentration increases almost linearly during the first few days, then tends to be constant after reaching a maximum value. In this steady state, the growth rate and the death rate from overcrowding are comparable. The time needed to achieve the steady state depends on the large cylinder diameter. It takes longer time for the large cylinder with larger diameter to reach the maximum value.

Figure 3 exhibits the effect of the small cylinder diameter on the average microalgae concentration. The simulation was performed with the same values of the parameters set above, except that the large cylinder diameter was set to 50 cm. As shown in the figure, the highest maximum concentration occurs in the reactor with the small cylinder diameter being 5 cm.

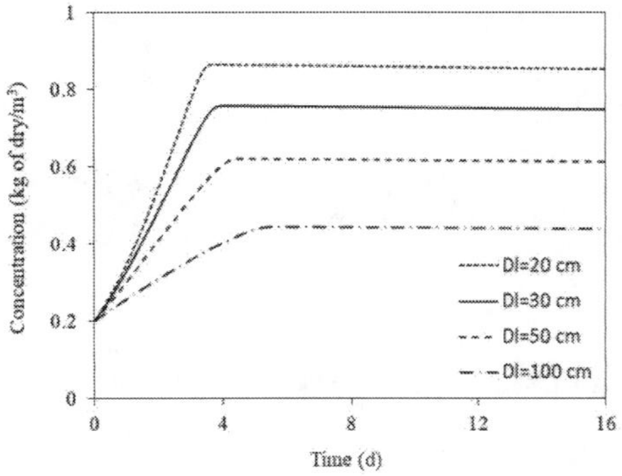

Figure 2. The average microalgae concentration vs the large cylinder diameter.

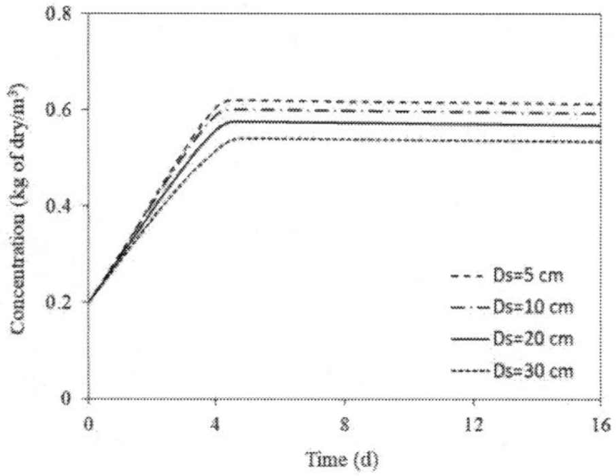

Figure 3. The average microalgae concentration vs the small cylinder diameter.

The change in the culture height does not affect significantly on the average microalgae concentration as shown in figure 4. This happens when the bubble velocity can be preserved along the reactor. Indeed, the model does not consider the change in the bubble velocity when bubbles rise because the culture height is not very high. In the case of very high culture height, the bubble velocity decreases as a result of part of bubble mass moves into liquid phase along its way to axial direction.

Figure 5 depicts the microalgae growth as the function of the initial concentration. The initial concentration of 0.05 g/L results in the steady-state concentration of 0.68 kg/m^3 or 10-fold increase. The initial concentration of 0.2 kg/m^3 promotes the concentration to 0.75 kg/m^3 at steady state or fourfold increase. With constraints such as self-shielding factor, the ratio of the maximum concentration to the initial concentration for higher initial concentration is smaller than for the lower one. The ratio for higher initial concentration are certainly smaller, but their steady state is achieved faster than lower one, and more microalgae are yielded.

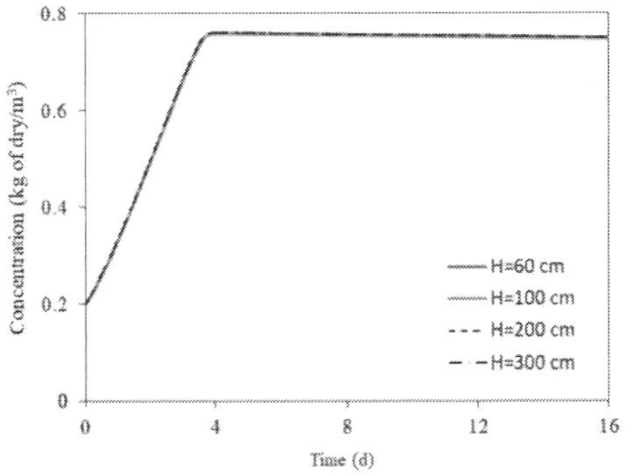

Figure 4. The average microalgae concentration vs the culture height.

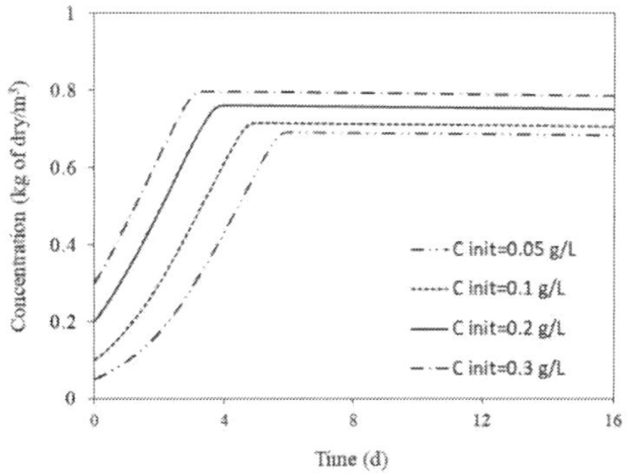

Figure 5. The average microalgae concentration vs the initial microalgae concentration.

Figure 6 shows the effect of the inlet air flowrate on the average microalgae concentration in the reactor. The inlet flowrate varies within the range that complies the superficial velocity requirement for homogeneous mixing in bubble-column reactors. It can be seen from the figure that the change in the inlet flowrate does not significantly affect the average microalgae concentration. For low flowrate regime, the effect is clear, as happened for 0.01 L/min.

Figure 7 displays the effect of the CO_2 mole fraction in the inlet air. The increase in the CO_2 mole fraction leads to the increase in the CO_2 intake by microalgae. Thus, the microalgae concentration goes up. However, due to the limitation of CO_2 solubility in water, the maximum CO_2 mole fraction in the inlet flow is 0.05.

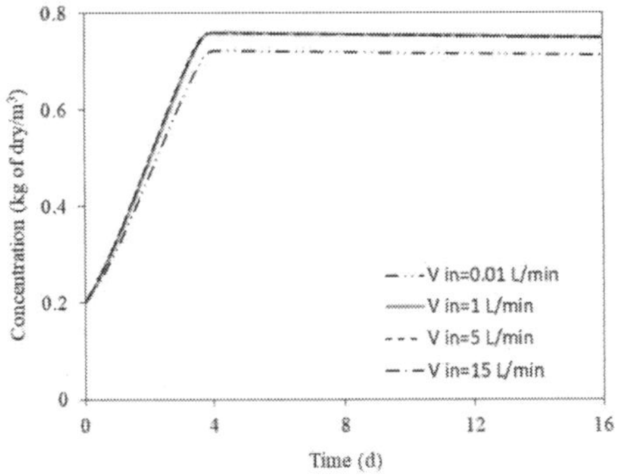

Figure 6. The average microalgae concentration vs the inlet flowrate.

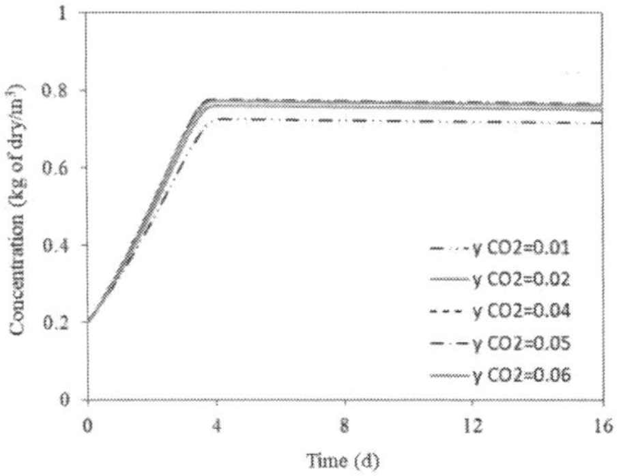

Figure 7. The average microalgae concentration vs the CO_2 mole fraction.

From the above simulation results, the values of the geometry and process parameters giving the optimum microalgae growth are listed in table 1. These values provide the maximum average microalgae concentration of 0.773 g/L achieved in less than four days as shown in figure 8. This information is important to design an internally illuminated bubble-column photobioreactor with a desired production capacity. For high capacity, the culture height may be increased to several meters. In order to anticipate the reduction of CO_2 content in rising bubbles, some spargers may be installed at certain elevations. If one reactor column is not sufficient, identical reactor columns may be arranged in parallel and treated equally to meet the desired capacity.

Table 1. Optimization result of the bubble-column photobioreactor.

Parameters	Value
Large cylinder diameter	30 cm
Small cylinder diameter	5 cm
Inlet air flowrate	1 L/min
CO_2 mole fraction in inlet air	0.05
Initial microalgae concentration	0.2 g/L
Inlet and ambient temperatures	30 °C

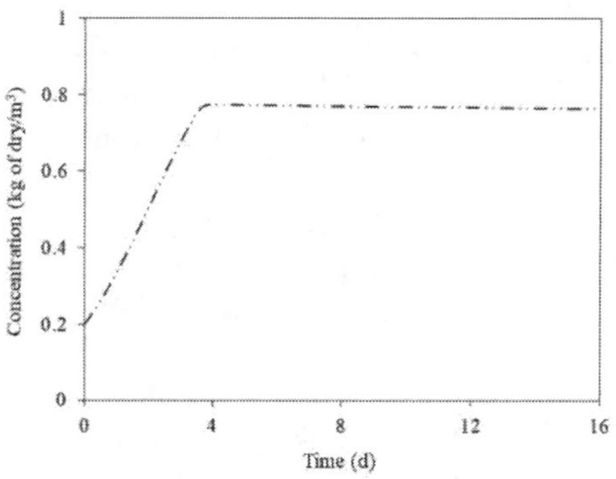

Figure 8. The optimum average microalgae concentration.

4. Conclusion

The process optimization gives the values of the geometry and process parameters as listed in table 1 with the optimum average microalgae concentration being 0.773 g/L achieved in less than four days.

5. Acknowledgement

We express our gratitude to the University of Indonesia, which funded this research through the scheme of Penelitian Unggulan Perguruan Tinggi No 2769/UN2.R3.1/HKP.05.00/2017.

6. References

[1] Borowitzka M 1996 Algae as food *Microbiology of Fermented Foods* ed B J B Wood (London: Blackie Academic & Professional) pp 585-602
[2] Seo I, Lee I, Huang H, Hong S, Bitog J, Kwon K, Lee C, Kim Z, Cuello J 2012 Biosyst. Eng. **113** (13) 229
[3] Pegallapati A K, Nirmalakhandan N 2012 Bioresource Technol. **124** 137
[4] Sforza E, Enzo M, Bertucco A 2014 Chem. Eng. Res. Des. **92**(6) 1153
[5] Ketheesan B, Nirmalakhandan N 2013 Bioresour. Technol. **136** 689
[6] Nauha E K, Alopaeus V 2013 Chem. Eng. J. **229** 550
[7] Zhang X, Gong X, Chen Y 2015 J. Ind. Microbiol. Biotechnol. **23**(1) 691
[8] Muharam Y, Dianursanti, Pramadana A B, Wirya A S 2017 Chem. Engineer. Trans. **56** 1555
[9] Joshi J B 1980 Trans. Inst. Chem. Engs. **58** 228
[10] Krishna R, Urseanu R, van Baten J M, Ellenberger J 1999 Chem. Eng. Sci. **54** 4903
[11] Grima E M, Perez J A S, Camacho F I G, Medina A R 1993 J. Chem. Technol. Biotechnol. **56**(4) 329

Performance prediction of vaccine carrier using adsorption process and 13x/$c_a cl_2$ composite zeolite as adsorbent

Nasruddin[1,a], Q H Alius[1], Djubaedah[1,2], A Taufan[3], R G Gurky[1], and A P Arsyad[1]

[1]Departement of Mechanical Engineeering, Universitas Indonesia, Depok – Indonesia
[2]Agency for the Assessment and Application of Technology (BPPT), Serpong, Indonesia.
[3]Development Center for Appropriate Technology, Indonesian Institute of Sciences, Subang -Indonesia

[a]E-mail: nasruddin@eng.ui.ac.id

Abstract. Vaccine is one of the biggest problem that happen in Indonesia. Almost 75% vaccine is frozen on shipment process from Province to healthcare in district. Vaccine usually is transferred using a vaccine box whose the temperature has to be maintained at 2-8°C. The newest technology of vaccine box uses adsorption cooling and solar energy to keep the cold chain. This technology is separated into 3 modules which are cooling module (contain zeolite and water), vaccine box, and solar heater. The cooling module using adsorption as the cooling process, where the adsorption makes a pressure drop and causes the decrease of the temperature. The decrease temperature will be used as source of the cooling in the vaccine box. After the zeolite in cooling module reach the saturation because of adsorption process, it has to be heated up using solar cooker as the media. The solar cooker collects the energy from the sun using the principle of greenhouse effect. This study aimed to analyze if the design of vaccine box using adsorption cooling system and 13x/$c_a cl_2$ composite zeolite as the adsorbent is suitable or not for holding the optimum temperature for vaccines. From simulation using MATLAB, an integrated technical computing including computation, visualization, and programming, the vaccine box could reach temperature 5°C and can be maintained for 12 hours with COP is 0.146. For the desorption process, it needs heat 150 KW and 23 minutes to complete the process.

1. Introduction

Vaccine, as described on The Basic Concept of Vaccination by Pharmaceutical Research and Manufacturers of America (PhRMA), is a biological preparation that enhance immunity against a disease by stimulating the production of antibodies [1]. Vaccines are administered in liquid form, either by injection, by oral, or by intranasal routes. This method which also known as immunization is proven to be the most cost-effective public health intervention diseases such as polio, measles, hepatitis B, tetanus, pertussis, diphtheria, and childhood tuberculosis. WHO guidelines and vaccine labels state that the use of hepatitis B, diphtheria– tetanus–pertussis, diphtheria–tetanus, and tetanus toxoid vaccines must be stored at temperatures of 2°C–8 °C to ensure the quality of vaccine. It must not be frozen because it can lead a loss of potency and compromised protective immunogenicity in recipients.

Unfortunately, there still some obstacles faced by health worker in Indonesia to distribute vaccine to reach rural area due the lack of transportation access and source of electricity. A study that monitored Hepatitis B vaccine in Indonesia state that 75% of vaccine shipments were being frozen, mostly during the transport from provincial to district warehouse.

The laboratory research that the newest technology of vaccine cold box using cool water packs [3] to prevent freezing during transportation and the result is with the extreme ambient temperature (43°C) with the temperature inside cool box around 20°C. In while there are technology that use a thermoelectric [2] as cooling system and the temperature could reach -10°C and range of the temperature is 2°C-8°C. However from all the technology need electricity to used and not suitable for remoted area.

Adsorption cooling system in vaccine cooling module believed can give a better performance since this technology has simple control system, low maintenance, and also this technology doesn't need electricity at all. Using zeolite as the adsorbent also give an advantages since the adsorbent can be heated using solar energy so it can applied in rural area. This study aimed to analyze if the design of vaccine box using adsorption cooling system and $13x/c_acl_2$ composite zeolite as the adsorbent is suitable or not for holding the optimum temperature for vaccines.

2. Methodology and Concept Design

2.1. Adsorption and Desorption Process

On thermocycle of adsorption process cooling consist of two process which is adsorption and desorption process. The working principle of the adsorption cooling can be described by using clayperon diagram as shown in figure 1.

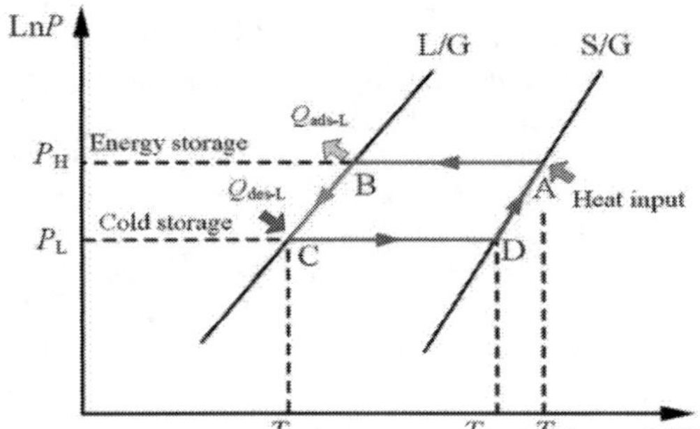

Figure 1. Clayperon schematic diagram [11]

The process of cooling happened at line C – D. These cooling process actives when the pressure below one atmospheric and the valve between two chambers is opened. Unsaturated adsorbent will adsorb the refrigerant on the other side. The result is the chamber 2 will decrease the temperature swiftly because the decreasing of pressure (Pressure drop). The temperature on chamber 2 depends many variable such as mass of adsorbent, mass of water, isotherm of adsorbent and etc. At line A-D is the process when adsorbent reach saturated condition and the cooling process not run effectively anymore. Therefore the line C-D-A is what called as adsorption process, while at line A-B-C is the process of desorption. Line A-B is when the heat from the heat source input into saturated adsorbent. The heat input continuously until the water inside of the adsorbent transform into vapor. Because of the difference the pressure and temperature between the chamber 1 and 2 the vapor will move into chamber where the temperature and pressure are low. At B-C line is process when the vapor became water because of condensation.

2.2. System Description

The vaccine box separated into 3 module which are cooling module, cool box, and solar cooker. Each module have difference function. The cooling module consist of two chambers and manual valve as shown by Figure 2. These two chambers contains zeolite and water for adsorption cooling. These two chambers made of stainless steel 304 to prevent from deformation. The water will adsorp by zeolite through pipe that connected to valve. Cooling box has a function to hold the temperature between 2-8°C while shipment.

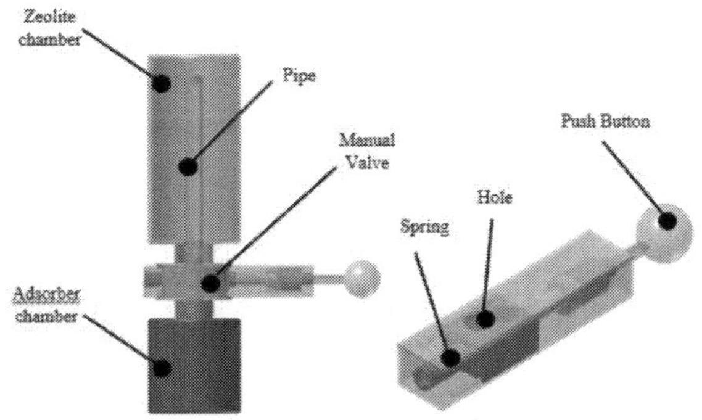

Figure 2. Parts of cooling module and valve

Cooling module will put in the vaccine storage. The cool box specification is on the table 1.

Table 1. Specification of cooling box

Vaccine Storage Capacity	1.6 liter
Approximation External Dimension	400 mm X 400 mm X 300 mm
Vaccine Storage Dimension	10 mm X 10 mm X 10 mm
Insulation Thickness	30 mm
Insulation Material	Polyurethane Foam
Vaccine Storage Material	High Impact Polystyrene (HIPS)
External Surface Material	High Density Polyethylene (HDPE)

2.3. Analyze Performance of Cooling Module

This Analyze the performance of this cooling box is using MATHLAB as the software for simulation. There are several assumptions that used in this simulation

1. Pressure and temperature are homogenous in adsorbed
2. The process of adsorption and desorption are fully isobaric
3. The refrigerant is adsorbed homogenously
4. There are no differences of pressure between two chambers.

There are several variable that used in simulation. The variable is shown in table 2.

Table 2. Variable and value used in simulation

C_a J/KgK	C_{ss} J/KgK	$C_{w,v}$ J/KgK	C_{air} J/KgK	C_w J/KgK	$U_{e/c}$ W/K	U_b W/K	ΔH J/Kg	M_w Kg	M_{air} Kg	M_a Kg	$M_{e/c}$ Kg	E_a J/mol	R_p m	L J/Kg	D_{s0} m²/s	R J/molK
836	520	1850	1005	4721.8	0.08	0.38	3800000		0.2	2.3	0.606	28035	0.0001	2500000	3.92x 10^{-4}	8.314

The equation based on energy balance in every part in cooling module and cool box. There are equations on the adsorbed chamber, chamber 2, and the cool box. The chamber 2 can be used as evaporator while adsorption process and can be used as condenser in desorption process. The non-equilibrium adsorption of zeolite water kinetics [1] is defined by formula [1]:

$$\frac{dX}{dt} = \frac{15}{R_p^2} D_{s0}\, exp\left(-\frac{E_a}{RT_b}\right)(X_{eq} - X) \tag{1}$$

Where X_{eq} is the equilibrium capacity of the adsorber and can be defined by formula [2]

$$X_{eq} = {}^{1}\!/_{(C_0 + C_1 T_b + C_2 T_b^2 + C_3 T_3^3)} \tag{2}$$

Where $C_{i=0-3}$ is the numerical constant for composite zeolite. Where C_1, C_2, C_3 for absorption is 381.4, -3.463, 0.01008, and -9.153 x 10^{-6} while for desorption 400.8, -3.532, 0.01002, and -8.908 x 10^{-6}. θ is the flag refers to process absorption and desorption.

If the θ have valued 1 then simulation at adsorption phase, while if θ is 0 then simulation at desorption phase. θ is the flag refers to process adosrption and desorption. If the θ have valued 1 then simulation at adsorption phase, while if θ is 0 then simulation at desorption phase.

2.3.1 Energy Balance at Adsorber Chamber

$$\frac{dT_b}{d\tau}\left[M_a\left(C_a + C_{p,w}X\right)\right] = M_a\,\Delta H\,\frac{dw}{d\tau} - (1-\delta)U_b A_b\left(T_b - T_{amb}\right) - \theta\, M_a C_{w,v}\left(T_b - T_e\right)\frac{dw}{d\tau} + Q_h \tag{3}$$

Q_h is heat that input to adsorber chamber for desorption process while ΔH is the isotheric adsortion at adsorption process. On the left part of the equation is heat transfer inside the adsorber and case of the chamber while and the other side is the heat transfer between adsorber chamber to environment, adosorber chamber to evaporator, adsorption heat and heat source.

2.3.2 Energy Balance at Adsorber Chamber as Evaporator

$$\frac{dT_e}{d\tau}\left[M_{e,w}\,C_{p,w} + C_{ss}\,M_e\right] = \delta\left[-L\,M_a\,\frac{dX}{d\tau} - U_e A_e\left(T_e - T_{room}\right)\right] \tag{4}$$

Where L is evaporation latent heat of transfer. The equation on left part represents the heat transfer inside the evaporator included the refrigerant and the Chamber. The right side represent the heat transfer between evaporator latent heat to adsorbed and evaporator to environment.

2.3.3 Energy Balance at Adsorbed Chamber as Condenser

$$\left[M_{e,w}\,C_{p,w} + C_{ss}\,M_c\right]\frac{dT_c}{d\tau} = (1-\delta)\left[-L\,M_a\,\frac{dX}{d\tau} - C_{wv}\,M_a\,\frac{dX}{d\tau}\left(T_c - T_b\right) - UA\left(T_c - T_{amb}\right)\right] \tag{5}$$

The left side of the formula is the heat transfer inside the chamber. The other side contains 3 terms of heat transfer. There are heat transfers between condensers with latent heat of water's condensation, condenser to environment, and condenser to bed.

2.3.4 Energy Balance at Cooling Box

$$C_{air}\, M_{air}\, \frac{dT_r}{d\tau} = Q_l + U_{cb}A_{cb}\, (T_{amb} - T_r) + \; U_eA_e\, (T_e - T_r) \tag{6}$$

Where Q_l is the internal heat loads that are comes from the vaccine itself. Right side of the equation represents of heat transfer between the evaporator to cool box and also cool box to environment.

2.3.5 Performance

The performance of Vaccine box can be determined by cooling capacity of the adsorbent. Cooling capacity shows the difference temperature between starting point of adsorption process and end point when the adsorbent on saturated condition. To determine the performance can use equation

$$Q_r = \int_0^{\tau} (C_{air}\, M_{air}) * \,(T_{r,end} - T_{r,start})d\tau \tag{7}$$

$$COP = {}^{Q_r}\!/_{Q_h} \tag{8}$$

3. Result and Discussion

From the model of cold box and cooling module that have designed, the performance could be obtained by the Simulation of MATHLAB. The simulation begin with assumption of the internal heat load of vaccine is 20.16 Watt based on capacity of vaccine storage. The internal heat load will affect temperature of vaccine storage room. Adsorbent begin adsorption with capacity 0.08 kgw/kgad where the maximum capacity for adsorption is 0.8 kgw/kgad. For desorption phase, author makes assumption that the solar collector will gain 150 watt constantly from solar sun. From figure 5, the temperature suddenly decrease when the adsorption process begin and vaccine storage room temperature reach 6°C - 5°C in 2 minutes while the temperature of evaporator reach 3°C. After that temperature tend to decrease because of adsorption still running. The temperature of vaccine storage will constantly below 5°C as long for 12 hours but for adsorption process it only takes 2 minutes and after that the zeolite will completely saturated. The adsorption process still continuous as long the zeolite isn't heated up. However, adsorbent still adsorb but definitely doesn't affect significantly to cooling capacity. The temperature of adsorbent increase because of isotheric heat comes from adsorption process. In figure 5, the capacities of adsorbent reach the maximum capacity in 2 minutes. Range of time adsorption process is short because of the mass of adsorbent is 2.3 kg and while the mass of refrigerant is only 1 kg. The simulation result of cooling capacity that can occur when adsorption process is 1043.7 kJ.

For desorption process the solar irradiation is based on the location of depok which is 106.818610 longitude and -6.4 longitudinal with assumption constantly at 150 Watt. Based on the power that given on solar collector. From figure 7 and 6 the desorption process occurs for 23 minutes until all the water changed phase into vapor. The beginning capacity of adsorbent is 0.7 kg/kg and reaches 0.045 kg/kg. From figure 9, the graphic tend to decrease linearly between time and fraction of adsorbent.

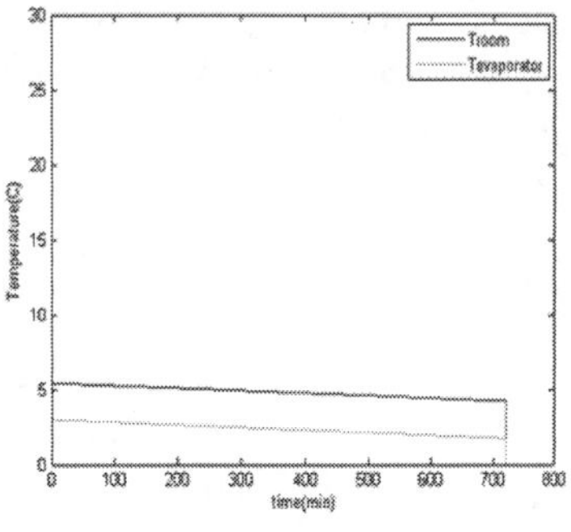

Figure 3. Temperatur of vaccine storage (adsorption)

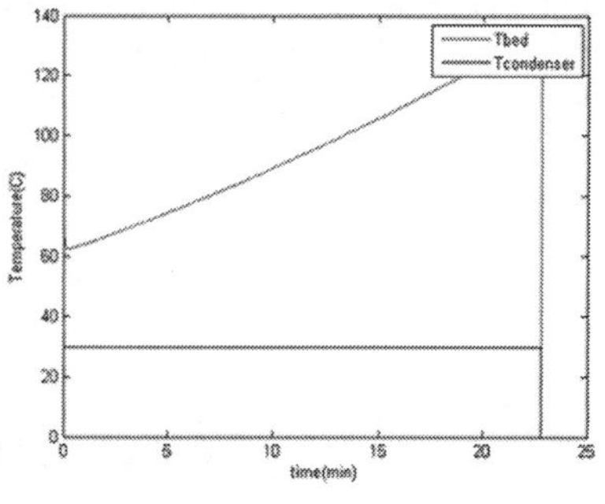

Figure 4. Temperatur of vaccine storage (desorption phase)

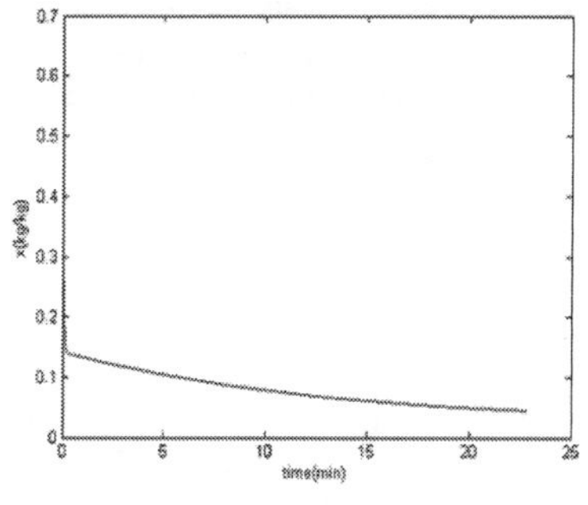

Figure 5. Capacity Vs Time at Desorption Phase

4. Conclusion

The simulation of these vaccine box is designed for shipment of vaccine from province to healt center in each district of remoted area in Indonesia. The aim of this study is to get an information the design is suitable or not for holding the optimum temperature for vaccines. The design of these vaccine box is very mobile with mass 15 Kg included the cooling module and cool box. From the simulation the cooling capacity that produced by cooling module is 1043.7 Kj or 24 Watt with the internal load of vaccine is 20.16 watt. As the result the temperature of vaccine storage is below than 5°C with evaporator temperature is 3°C. Adsorption process takes 6 second to decrease the temperature from ambient temperature to 5°C. However, vaccine box can hold the temperature for 24 hours without any removal the vaccine in vaccine storage. After the vaccine is stored to healt centre in district, it can be used again by heating up the adsorbent chamber. The adsorbent chamber is heated up for 24 minutes until the fraction of the chamber is decrease from 0.7Kg/Kg to 0.045 Kg/Kg. As the result COP of this design is 0.146.

References

[1] E I B and S A A. 2015, Perfomance Evaluation of a Solar Powered Solid Adsorption Refrigerator under Trophical Humid Climate. *European Journal ofEngineering and Technology*, **3** (4).

[2] Putra N 2009. Design, manufacturing and testing of a portable vaccine carrier box employing thermoelectric module and heat pipe, *Journal of Medical Engineering & Technology*, **33**(3), 232-37.

[3] Kartoglu, U., et al., *Use of cool water packs to prevent freezing during vaccine transportation at the country level.* PDA J Pharm Sci Technol, 2009. **63**(1) 11-26.

[4] Miyazaki T A-S 2009 A New Cycle Time Allocation for Enhancing the Performance of Two-bed Adsorption Chillers. *International Journal of Refrigeration,* **32** 864-53.

[5] Wang Kai V E. 2011. New Opportunities for Solar Adsorption Refrigeration. *American Society of Heating, Refrigerating and Air conditioning Engineers Journal Technical Feature:*, 14-24R.

[6] L.W. Wang. 2009. A review on adsorption working pairs for refrigeration. *Renewable and Sustainable Energy Reviews*, 518-34.

[7] WHO 2013. *Immunization systems and technologies for tomorrow;Innovative passive cooling options* . PATH Publication.

[8] Verde M, Cortes L, Corberan J M, Sapienza A, Vasta S, Retuccia G 2010. Modelling of an Adsorption System Driven by Engine Waste Heat for Truck Cabin A/C. Performance Estimation for a Standard Driving Cycle. *Applied Thermal Engineering,* **30** 1511-22.

[9] Ghassan T M, Bourhan T M and Mustafa M J 2012. Experimental Study of a Solar Adsorption Refrigeration. *Smart Grid and Renewable Energy*, 126-32.
[10] Ghassan, Bourhan and Mustafa 2012. Streamlining Immunization Logistics in The Province of Central Java and Yogyakarta, Indonesia. *A Collaborative Project of The Ministry of Health, Republic of Indonesia and PATH*. April 2005-March 2006.
[11] Chan K C, Chao Y H 2012 Performance predictions for a new zeolite 13X/CaCl2 composite adsorbent. *International Journal of Heat and Mass Transfer*, 3214-24.
[12] Hannun R M and Lafta Z A 2016 Modelling of Solar Adsorbtion Refrigeration System in Nasiriya City. *Journal of Solar Energy Research* **11** 13-21.

Design, development and performance prediction of solar heater for regeneration of adsorbent chamber

Nasruddin[1,a], Djubaedah[1,2], R G Gurky[1], Q H Alius[1], and A P Arsyad[1]

[1]Department of Mechanical Engineering, Universitas Indonesia, Depok, Indonesia.
[2] Agency for the Assessment and Application of Technology (BPPT), Serpong, Indonesia.

[a] E-mail: nasruddin@eng.ui.ac.id

Abstract. Adsorption cooling is the newest technology that can be used in the distribution of vaccines. The principle of adsorption cooling is applicable to make a new technology called portable vaccine cooler, using zeolite as the absorbent and heat energy to keep cold chain of vaccine which needs to be stored in temperature of 2°C to 8°C. There are many ways to heat the zeolite up. Instead of using fire that induces environmental pollution to heat the zeolite up, the authors choose to use solar energy as the source of the desorption process. The aim of this study is to support cooling module and get the best results of two types of solar heaters to be applied on cooling box. In this paper, performance prediction of two solar heater are simulated using ANSYS workbench, a platform which integrate simulation technologies and parametric CAD system. The result shows that the solar heater type 1 is better to use because it can heat the absorbent chamber up at the temperature of 109°C in an hour running, whereas the absorbent only needs the temperature around 70°C-80°C to make the process of desorption running perfectly.

1. Introduction

Vaccine has a huge effect to prevent some of the most horrible diseases in history (such as Hepatitis B, diphtheria, polio, etc.). Over the years, vaccines have put an end to myriad cases of diseases and saved millions of lives. With the large number of vaccines needed and they will increase linearly with total population, scientists spend a lot of time to developing the availability of vaccines, especially in Indonesia's rural areas. Indonesia is an archipelago state and this brings a challenge to distribute the vaccines from one place to another in Indonesia. WHO has published a bulletin which shows that freezing temperatures were recorded in 75% of baseline shipments in Indonesia. The highest rates of freezing occurred during transport from a province to a district, storage in district-level ice-lined refrigerators, and storage in refrigerators in health centers [1].

WHO states that vaccines need to be stored in temperature of 2°C to 8°C, this will enable flexible vaccine distribution strategies that could reduce vaccine freezing and costs, and increase capacity. There are various advanced technologies used to maintain the temperature of vaccines in the course of distribution process, such as implementation of thermoelectric, phase changing material, solar direct-drive refrigerators, and many others.

Adsorption cooling is the newest technology that can be used in the distribution of vaccine. The principle of this device is the refrigerant is adsorbed and desorbed during the process, analogues to the principle of absorption device. However, because the adsorbent is stationary (a solid), adsorption is built differently than desorption system. Usage of stationary adsorbents leads to intermittent operation of adsorption system. Continuous cooling requires two or more adsorbents, in which adsorption and desorption process occur alternately. [2]

The principle of adsorption cooling is applicable to make a new technology called portable vaccine cooler, using zeolite as the absorbent and heat energy to keep the chain cold. This technology is separated into 3 modules which are cooling module (contain zeolite and water), vaccine box, and heater. The cooling module uses adsorption as the cooling process using zeolite as the adsorbent to make a pressure drop and causes the decreasing temperature. There are many ways to heat the zeolite up. Instead of using fire that induces environmental pollution to heat the zeolite up, the authors choose to use solar energy as the source of desorption process using solar heater.

There is various type of solar heater that has been developed over the world. Generally, solar heater classified into three types; parabolic, box-type, and panel type solar heater. Two types of solar heater discuss in this paper have been developed with portable, simple and compact design as it needs to get the rural area to support cooling module in portable vaccine cooler.

The aim of this study is to support cooling module and get the best results of two types of solar heaters to be applied on vaccine cooler in Indonesia. In this paper, performance prediction of two solar heater are simulated using ANSYS workbench, a platform which integrate simulation technologies and parametric CAD system. As the results, one of those solar heaters uses the principle of greenhouse effect can work for an hour to reach the temperature of 109°C at the portable vaccine cooler adsorbent chamber. The other type of solar heater uses reflective material which is capable of focusing incident rays onto a focal area and it can work for an hour to reach the temperature of 45.7°C – 65.6°C. Those solar heaters could be applied in Indonesia and offer the following novel and user-friendly features: (1) Heat the zeolite at any time convenient to the user (2) Fast Heating (3) Affordable cost (4) environmentally friendly.

2. Methodology and Materials

2.1. Solar Heater Type 1

This is a type of solar heater which uses a multi-faceted reflective panel using transparent medium cover to create greenhouse effect. The direct sunrays will be reflected into the solar heater and heat up the dark painted portable vaccine cooler. Panel cooker as a combination of box-type and parabolic cooker is however, not quite as robust as the box-type cooker, and it cannot heat as much load at once as the parabolic cooker.

Overall thermal efficiency of the Solar Heater was calculated by following equation [3]:

$$\eta_u = \frac{M_f C_f \Delta T_f}{I_{av} A_C \Delta t} \quad (1)$$

where η_u is overall thermal efficiency (%), M_f is mass of cooking fluid (kg), C_f is specific heat of cooking fluid (j/kg.K), ΔT_f is difference between the maximum and ambient air temperature, I_{av} is average solar intensity (W / m^2), A_C is the aperture area (m^2) of the cooker, and Δt is time required to achieve the maximum temperature of the cooking fluids (s).

The Cooking Power of the solar heater was calculated by the following equation [4]:

$$P = \frac{T_{w_2} - T_{w_1}}{t} m_w C_{pw} \quad (2)$$

where P is heating power (W), T_{w_2} is final water temperature (°C), T_{w_1} is initial water temperature (°C), t is time (s), m_w is mass of water (kg), C_{pw} is water heat capacity (4.168 kJ/kg.K).

2.1.1 Design of Solar Heater Type 1

Figure 1 shows the parts of the solar heater type 1. This solar heater uses PYD aluminum foil to reflecting the sunrays and plastic as the lid to make the sunrays trapped inside. The lid can be opened and closed using a zipper. The hole located on the base of the solar heater type 1 is an adsorbent chamber mounting hole.

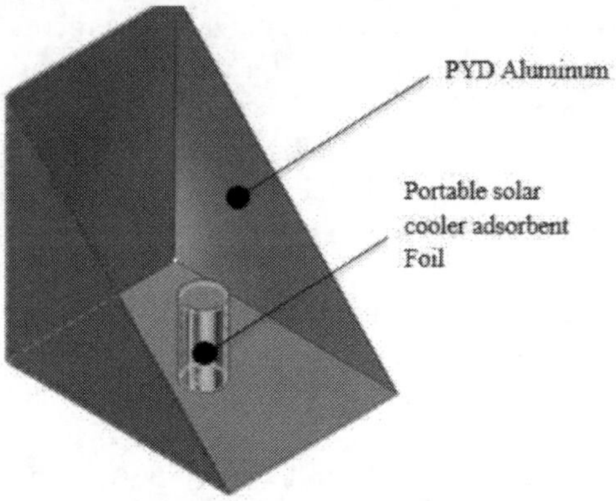

Figure 1.Parts of solar heater type 1

Table 1. Specifications of solar heater type 1

The height of the solar heater type 1	490 mm
The width of the solar heater type 1	370 mm
The length of the solar heater type 1	490 mm
Adsorbent chamber mounting hole diameter	76.2 mm
Solar heater type 1 internal material	PYD Aluminum Foil
Portable vaccine cooler adsorbent	Zeolite
Solar heater type 1 lid material	Plastic

2.2. Solar Heater Type 2

This type of solar heater is made by reflective material made from solar zenith with conical frustum shape. This system is a conical lens capable of focusing incident rays onto a focal area. When this solar heater is directed to sunlight, incident rays on the reflective internal surface are reflected towards the focal area at the base. Design of solar heater type 2 is shown by figure 2 as below:

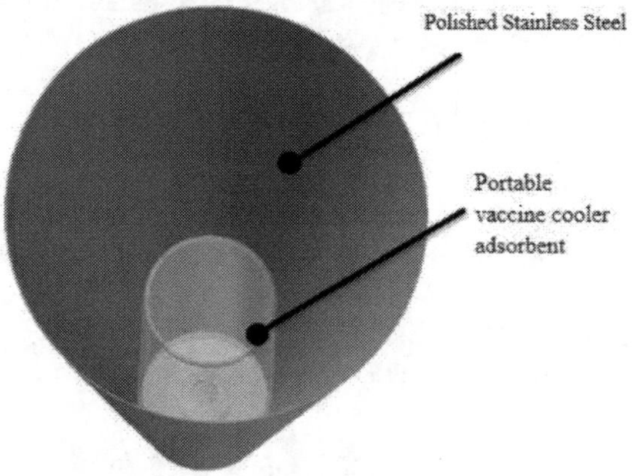

Figure 2. Parts of solar heater type 2

The formula derivation of geometric concentration of this type is expressed below [5]:

$$C = \frac{A}{a} = \frac{\pi R_1^2}{\pi R_2^2} = \frac{R_1^2}{R_2^2} \quad (3)$$

Where A=area of the large end; a=area of the small end; R_1=radius of opening; R_2=radius of the focus and $\Delta C'CB'$, where $C'B'C = \theta$, where θ is the apex angle of cone:

$$R_1 = R_2(2\cos\theta + 1) \tag{4}$$

From equation (3) and (4) we can obtain:

$$\text{Then } C = (2\cos\theta + 1)^2 \tag{5}$$

Equation (3) represents the average concentration at CC' in (figure 1)

Concentration systems are approximately classified into: low concentration range (C<10), medium concentration range (10<C<100), and high concentration range (C>100) [3].

With an apex angle of 30°, the average geometric concentration calculated by equation 5 is 7.46. The concentrated sunrays at the focal area used to heat the zeolite up to keep the cold chain.

Details of calculated solar heater type 2 dimensions are stated in table 2.

Table 2. Details of calculated solar heater type 2

No	Dimension	Value	Unit
1	R_1	11.98	cm
2	R_2	3.81	cm
3	L	46.28	cm
4	l	31.56	cm
5	H	44.7	cm
6	h	30.48	cm
7	A	450.65	cm
8	a	45.58	cm
9	C	7.46	-
10	E	300	watt
11	θ	30°	Degree
12	Φ	93.17°	Degree

The appropriate dimension for the conical frustum solar heater can be calculated as below [5]:
The length of cone side L is represented by An from $\Delta Ann'$

$$L = \frac{R_1}{\sin\frac{\theta}{2}} \tag{6}$$

Where $\left(\cos\theta = 1 - 2sin^2\frac{\theta}{2}; R1 = 11.98;\ \theta = 30°\right)$

$$L = R_1\left[\frac{2}{(1-\cos\theta)}\right]^{0.5}$$

The length of the reflecting part l is represented by AC in ΔABC where (R_2=3.81)

$$l = (R_1 - R_2)\left[\frac{2}{(1-\cos\theta)}\right]^{0.5} \tag{7}$$

The height of the reflective part h is represented by nn' by $\Delta Ann'$

$$\frac{R_1}{H} = \tan\frac{\theta}{2}, where\ \left(\tan\frac{\theta}{2} = \sqrt{\frac{1-\cos\theta}{1+\cos\theta}}\right) \tag{8}$$

$$\therefore H = (R_1)\left[\frac{1+\cos\theta}{1-\cos\theta}\right]^{0.5}$$

The height of the reflective part h is represented by BC. In ΔABC we can obtain

$$\frac{BA}{BC} = \frac{R_1 - R_2}{h} = \tan\frac{\theta}{2} \tag{9}$$

$$\therefore h = (R_1 - R_2)\left[\frac{1+\cos\theta}{1-\cos\theta}\right]^{0.5}$$

IOP Conf. Series: Earth and Environmental Science **105** (2017) 012004 doi:10.1088/1755-1315/105/1/012004

The expanded angle Φ of the complete cone is the angle at the apex when the frustum is undone From (equation (6)) in (Equation (10)):

$$\Phi = \frac{2\pi R_1}{L} \tag{10}$$

$$\Phi = 180[2(1-\cos\theta)]^{0.5} = 180[2(1-\cos 30°)]^{0.5} = 93.17°$$

3. Results and Discussion

The simulation is assumed running for an hour in Indonesia condition, at Depok coordinate, 6.4025° S, 106.7942° E. The simulation is running with assumption of the time is 12:00 pm, the weather is sunny, hot, and not windy with irradiation constant of $700 watt/m^2$ (Department of Mechanical Engineering Universitas Indonesia, 2015). The solar thermal will affect the adsorbent chamber through the solar heaters with each of its ways. The first solar heater type 1 will make the solar thermal trapped inside and use the solar thermal as the source to heat the adsorbent chamber. Meanwhile, the solar heater type 2 reflects sun rays onto a focal area and uses the solar thermal as the source to heat the adsorbent chamber. With the low concentration range of the solar heater type 2, the authors predict that the solar heater type 2 will not be able to heat the adsorbent chamber up in an hour in the temperature up to100°C.

Figure 3 . The simulation result of solar heater type 1 (running for an hour)

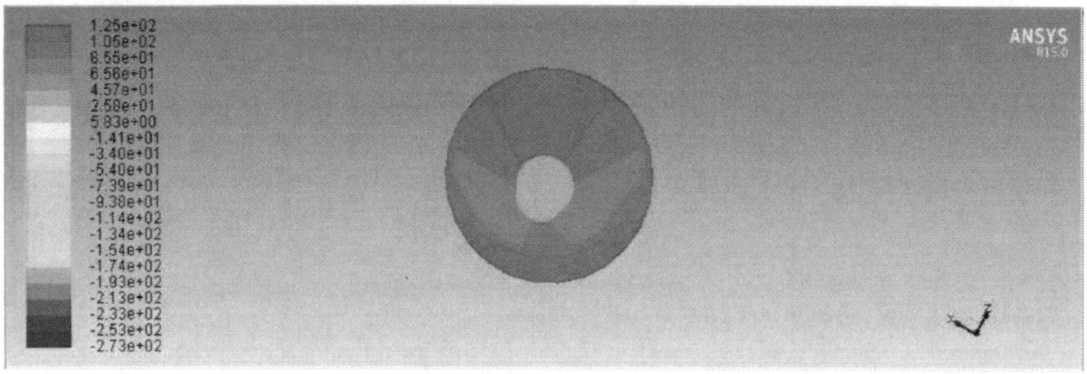

Figure 4. The simulation result of solar heater type 2

In figure 3, the solar heater type 1 wall temperature is heated up to the temperature of $262°C - 491°C$. The wall succeeds to reflect the sun rays and make it trapped inside, so the adsorbent chamber can be heated up to temperature of $109°C$ in an hour of running simulation. In figure 2, the solar heater type 2 reflective material is heated up to the temperature of $85.5°C - 125°C$, but the reflective material cannot make the adsorbent chamber be heated up like the solar heater type 1 does. The solar heater type 2 can only make the adsorbent chamber be heated up to temperature of $45.7°C - 65.6°C$ in an hour of running simulation. Just as the authors' prediction before the simulation is started, the solar heater type 2 is not able to heat the adsorbent chamber up in the temperature up too 100°C because of the low concentration range of the solar heater to focusing the sun rays onto its focal area.

As a reference, in another research by N. L. Panwar [6] found that maximum temperature of water was 94.85°C, whereas its exergy output was in the range, 7.37-46.46 W. This experiment analyzed a domestic size parabolic solar cooker in actual use which is running from 10:00 to 13:30 solar time. Compare to this experiment, solar heater type 1 has the closest result than the solar heater type 2 [6].

Table 3. Details of comparison between solar heater type 1, type 2, and reference

	Type 1	Type 2	Reference
Maximum wall temperature (°C)	491	125	-
Maximum adsorbent temperature (°C)	109	65.6	94.85

4. Conclusion

The solar heaters are made in order that they can be taken to rural areas with the portable vaccine cooling easily. The design of the solar heaters are very mobile because they can be folded and be carried easily with compact design, standard dimension. The solar heater type 1 simulation result shows that the solar heater type 1 can heat the adsorbent chamber up at the temperature of 109°C in an hour running; now in fact we only need the temperature around 70^0-80^0 C to make the process of desorption running perfectly. This means that the solar heater type one can be used to keep the cold chain. This simulation result is different from the solar heater type 2 simulation result where the solar heater type 2 can only heat the adsorbent chamber up at the temperature of 45.7°C – 65.6°C in an hour running. This means it needs more time to heat the adsorbent chamber up at the temperature of 70-80°C. So, it can be concluded that it is better to use solar heater type 1 because it can heat the adsorbent chamber up faster than the solar heater type 2.

5. Acknowledgment

The authors gratefully acknowledge the financial support that provide by KEMENRISTEKDIKTI under INSINAS project 2016.

References:

[1] Nelson C M, Wibisono H, Purwanto H, Mansyur I, Moniaga V, Widjaya A 2004, Hepatitis B vaccine freezing in the Indonesian cold chain: evidence and solution. *Bulletin of the World Health Organization* **82** 99-105.

[2] Gwadera M and Kupiec K Adsorption 2011 Cooling as an effective method of waste heat utilization. *Biblioteka Cyrowa Politechniki Krakoskiej*, 62-69.

[3] El-Sebaii A A and Ibrahim A 2005 Experimental testing of a box solar cooker using the standard procedure of cooking power. *Renewable energy*, **30** 1861-71.

[4] Kundapur A and Sudhir C V 2009 Proposal for new world standard for testing solar cookers. *Journal of engineering science and technology*, **4** 227-81.

[5] Sharaf E 2002 A new design for an economical, highly efficient, conical solar cooker.

[6] Panwar N L, Kaushik S C, and Kothera Surendra 2012. Experimental investigation of energy and exergy efficiencies of domestic size parabolic dish solar cooker Cooling as an effective method of waste heat. *J. Renewable Sustainable Energy* **4** 023111.

[7] Duffie J A, Beckman W, *Solar Engineering of Thermal Processes*, Chapter 7: introduction through Section 7.2. 322-27.

[8] Fairey et al 2002 *Solar Oven Development an Testing: Final Report, Florida Solar Energy Center*, FSEC 26-58-876.

[9] Akoy E O M and Ahmed A I A 2015 Design, construction and performance evaluation of solar cookers. *Journal of Agricultural Science and Engineering* **1**(2) 75-82.

IOP Conf. Series: Earth and Environmental Science **105** (2017) 012005 doi:10.1088/1755-1315/105/1/012005

Heat transfer coefficient of near boiling single phase flow with propane in horizontal circular micro channel

Sentot Novianto[1], Raldi Koestoer[1], Agus S Pamitran[1]
[1]Department of Mechanical Engineering, Faculty of Engineering, Universitas Indonesia, Kampus UI Depok 16424, Indonesia

Email: sentot.novianto41@ui.ac.id, koestoer@eng.ui.ac.id, pamitran@eng.ui.ac.id

Abstract. Experimental study of heat transfer coefficient of single phase has conducted successfully. Technology to reduce energy consumption is important to be investigated, moreover with implementation of renewable energy. The study used natural refrigerant of propane as working fluid with heating process. The aim of the present research is to study near boiling single phase heat transfer coefficient on the micro channel of 500 μm diameter and 0.5 meter length. Variable of research are heat flux of 1 to 15 kW/m^2, mass flux of 297 to 1102 $kg/m^2.s$, and test section inlet temperature of 21 to 26°C. The experimental results showed single phase laminar-turbulence flow with Reynolds number less than 10000. Heat transfer coefficient of near boiling single phase flow is significantly affected by Reynolds number and Prantl number. Using the present experimental data, new correlation of Nusselt number is developed as a function of Reynolds number and Prantl number.

1. Introduction

Technology to reduce energy consumption is important to be investigated, moreover with implementation of renewable energy. Many of renewable energy system use heat exchanger as cooling or evaporating units. Discussion about environmental issue is more intensively, mainly to reduce the using of refrigerant as working fluid in cooling system. Propane is one alternative refrigerant to replace conventional refrigerants. Propane has zero ODP (Ozone Depletion Potential) and less than four GWP (Global Warming Potential) [1]. [2] proposed single phase heat transfer with conventional channel. The correlation has range limited for smooth channel with ratio D/L higher than 60, turbulence flow, and Pr between $0.7 \leq Pr \leq 160$. [3] proposed a modified Petukhov correlation. The correlation has developed as function of Reynolds, Prantl, friction factor with laminar flow. [4] found heat transfer coefficient in microchannel was higher than that in conventional channel for both laminar and turbulence flow. [5] found heat transfer coefficient in microchannel is higher for laminar flow. A new correlation has proposed for laminar correlation.

[6] studied the single-phase forced-flow convection of water or methanol flowing through micro-channels. A modified correlation of heat transfer coefficient has been proposed with a factor of 0.023 to 0.00805. [7] explained a wide range empirical Nusselt-type correlation for channel flow; these were generally obtained using hydraulic diameters of 2 cm or more. Some author has studied heat transfer i.e. [8] reviewed on the convective heat transfer on micro-channels. In many cases the experimental data of the Nusselt number in micro-channels disagree with the conventional theory but they also appear to be inconsistent with one other. [9] have investigated fully developed liquid and vapor flow with hydraulic diameters varying from 69.5 to 304.7 μm. R 134a liquid and vapor were used as testing fluids. During the experiments, the Reynolds numbers were varied between 112 and 9180. The goal of the present research is to study single phase heat transfer coefficient on the microchannel 500 μm diameter and length 0.5 meter. The new correlation can be used to determine heat transfer coefficient on design of heat exchanger with microchannel.

2. Methodology

2.1. Experiment set up

The main observation is on the test section heated by electrical heater. Figure 1 describes the experimental apparatus. The test section is a horizontal tube with diameter of 500μm and length of 0.5 m.

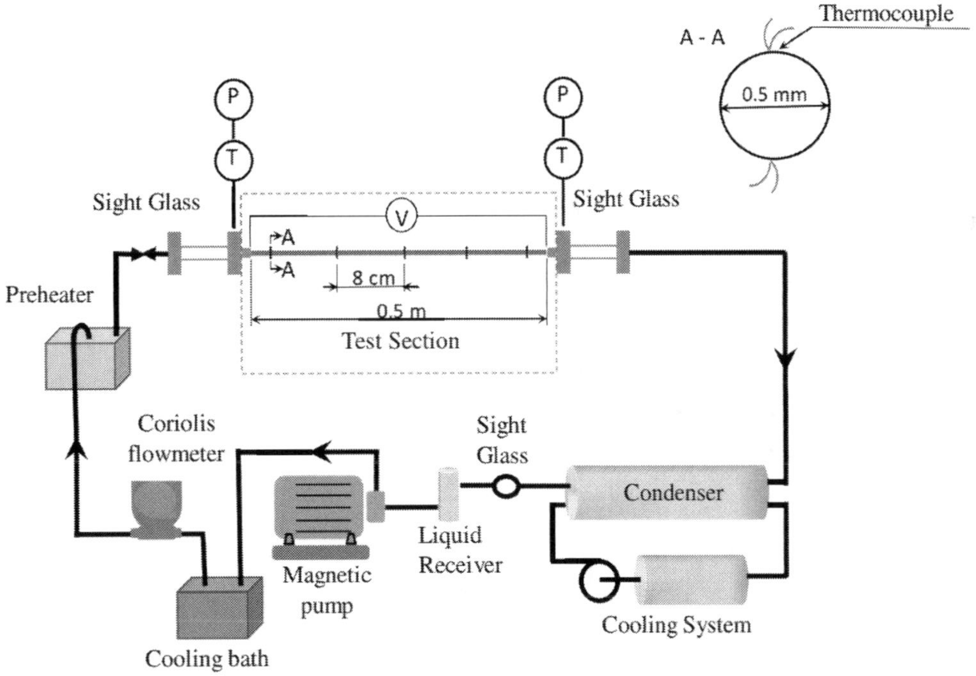

Figure 1. Experimental apparatus

Temperature of refrigerant flowed in the test section were measured by attached K-type thermocouples at the top and bottom of the test section, as shown in figure 1. There are five temperature measurement locations on the test section. Moreover, inserted thermocouple and pressure transmitter are installed at the inlet and outlet of the test section. A condensing unit is used to condense the evaporated refrigerant. After the condensation process, the liquid refrigerant is pumped by a magnetic pump. A cooling bath is placed after the magnetic pump to maintain the working fluid in liquid phase condition. A Coriolis flow meter is used for measuring flow rate of the refrigerant. Before the working fluid enters to the test section, it is flowed in a preheater to adjust the inlet temperature of working fluid. There are two sight glasses at the inlet and outlet of the test section for showing the phase of working fluids.

2.2. Data Reduction

[10] introduced confinement number (Co) as a ratio of capillary length and hydraulic diameter. Kew and Cornwell defined that the micro channel flow occurred when Co > 0.5. The average calculated Co for the present experimental data is 2.38. Based on the classification proposed by Kew and Cornwell, it can be classified as a micro channel flow. The experimental uncertainty is shown in table 1.

Table 1. Experimental uncertainties

Variable	Uncertainty
Average Temperature (°C)	0.46
Mass flow (%)	0.05
Heat (%)	1

The heat from electrical heater is given to the test section can be calculated as

$$q_{given} = V.I \tag{1}$$

Where V, and I represent voltage and current, respectively.

The equation for calculating heat received to the fluid working is given as

$$q_{received} = \dot{m}C_p(T_{out} - T_{in}) \tag{2}$$

where $\dot{m}, C_p, T_{out}$ and T_{in} represent flow rate, specific heat, outlet temperature and inlet temperature, respectively. Heat flux is given as

$$\ddot{q} = \frac{q_{received}}{A_{surface}} \text{ , with } A_{surface} = \pi DL \tag{3}$$

where D, and L represent inner diameter and length test section, respectively.

Table 2. Existing single phase heat transfer coefficient correlation

Author	Heat transfer	Range
(Dittus and Boelter, 1930)	$Nu = 0.023Re^{0.8}Pr^{n}$	$Re > 10000$
(Gnielinski, 1975)	$Nu = \frac{\left(\frac{f}{8}\right)(Re-1000)Pr}{1+12.7\left(\frac{f}{8}\right)^{0.5}(Pr^{\frac{2}{3}}-1)}$	$3 \times 10^{3} < Re < 5 \times 10^{6}$
(Wu and Little, 1984)	$Nu = 0.00222Pr^{0.4}Re^{1.08}$	$Re > 3000$
(Choi et al., 1991)	$Nu = 0.000972Re^{1.17}Pr^{1/3}$	$Re < 2000$
(Wang and Peng, 1994)	$Nu = 0.00805Re^{0.8}Pr^{1/3}$	$Re > 1500$

3. Result and discussion

The importance parameter used in the data reduction and analyze are Reynolds number (Re), and Nusselt number (Nu). During the present experiment, Reynolds number was varied between 1522 and 5549. Liquid phase were flowing on the inlet and outlet of the test section. Figure 2 showed the effect of Reynolds number on Nusselt number. The red trend line shows increasing Reynolds number results in increasing of Nusselt number.

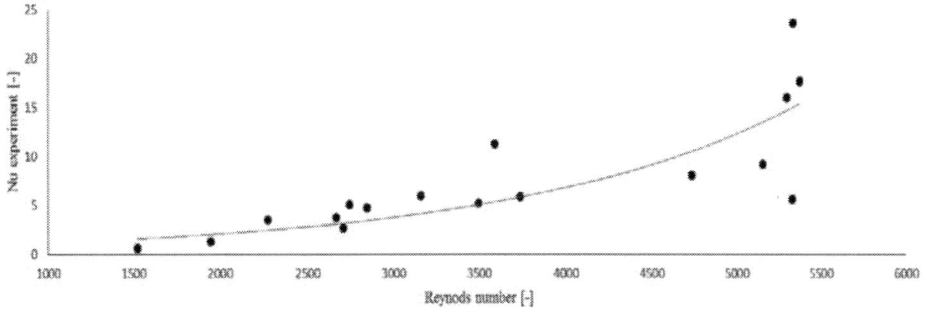

Figure 2. Effect of Reynolds number on Nusselt number

Reynolds number on the microchannel is predominately affected by diameter. The experimental Reynolds number experiment was less than 5600. Laminar, transition, and turbulence flow is observed on the experiment.

Liquid temperature on regime of convective flow was less than wall temperature. The increasing of heat flux caused increasing of liquid temperature and wall temperature. But more increasing heat flux, made the different of liquid temperature and wall temperature is lower, therefore the heat transfer coefficient is higher.

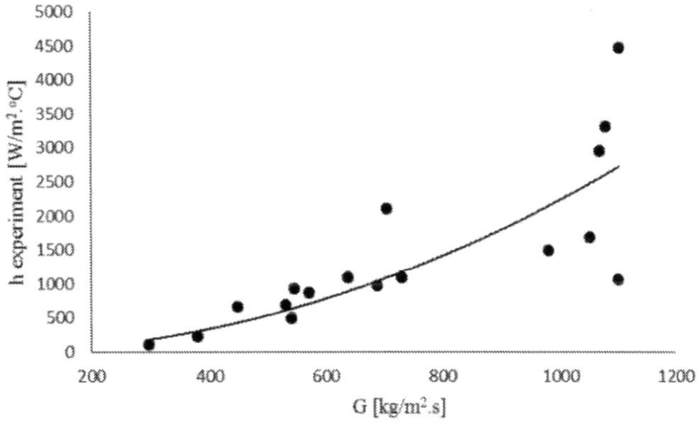

Figure 3. Effect of mass flux on experimental heat transfer coefficient.

Figure 3 showed effect of mass flux on experimental heat transfer coefficient. Increasing mass flux results in increasing heat transfer coefficient. The same results were found by [11].

The predicted Nusselt number were calculated with [2] equation. The equation is given as

$$Nu = 0.023Re^{0.8}Pr^{0.4} \quad (4)$$

Where Nu, and Pr represent Nusselt number, and Prantl number, respectively.

3.1. Development of a new Nusselt number with microchannel

Generally, the Nusselt number can be expressed as a function of Reynolds number and Prantl number.

$$Nu = C.Re^{m}Pr^{n} \quad (5)$$

Dittus-Boetler used m = 0.8 and n = 0.4 for heating process. With present experimental data, a new correlation is proposed with C of 0.00789. The news proposed Nusselt number is given as

$$Nu = 0.00789Re^{0.8}Pr^{0.4} \quad (6)$$

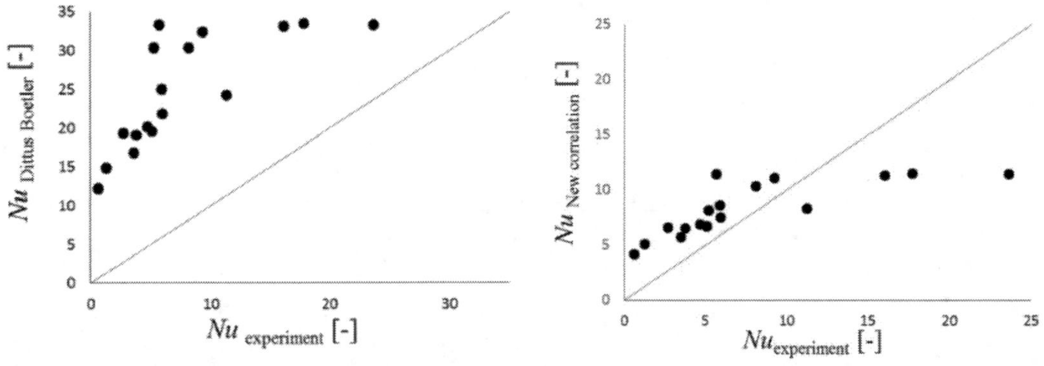

Figure 4. Comparison of experimental Nusselt number and calculated Nusselt number with Dittus-Boetler and comparison of experimental Nusselt number and calculated Nusselt number with New Correlation

Figure 4 showed comparison of experimental Nusselt number and calculated Nusselt number with Dittus-Boetler. The predicted Nusselt number with Dittus-Boetler was higher than the experimental Nusselt number. The mean deviation and absolute deviation of the Nusselt Dittus Boetler are both 245%. Figure 4 also showed comparison of experimental Nusselt number and calculated Nusselt number with a new correlation. The mean deviation and absolute deviation of the new correlation are 18% and 42%, respectively.

3.2. Comparison heat transfer coefficient with some existing correlation

Predicted heat transfer coefficient with [3] shows high mean deviation 195%. Gnielinski correlation is function of friction factor, Prantl number and Reynolds number, with range of Reynolds number of $3 \times 10^3 < Re < 5 \times 10^6$. Predicted heat transfer coefficient with [4] shows high mean deviation 240%. Wu and Little correlation is function of Prantl number and Reynolds number, with Reynolds number higher than 3000. Predicted heat transfer coefficient with [5] shows high mean deviation 297%. Choi correlation is function of Prantl number and Reynolds number, with Reynolds number lower than 2000. Predicted heat transfer coefficient with [6] shows high mean deviation 13%. Wang and Peng is function of Prantl number and Reynolds number, with Reynolds number higher than 1500.

4. Conclusion

Experimental study of heat transfer coefficient of single phase flow near boiling has conducted successfully. Characteristic of heat transfer is affected by Reynolds number and Prantl number. Increasing Reynolds number results increasing heat transfer coefficient. Heat flux is predominately affected heat transfer coefficient of single phase in microchannel. Increasing of heat flux caused the different of liquid temperature and wall temperature become lower. Heat transfer coefficient will be higher when the different of liquid temperature and wall temperature become lower. New correlation of heat transfer coefficient in micro channel has been developed based on Dittus-Boetler correlation. The new correlation can be used to determine heat transfer coefficient on design of heat exchanger with microchannel.

5. Acknowledgement

This study is funded by Research grant of KOPIT 2016 Universitas Indonesia.

6. References

[1] Choi, K I., Pamitran A., Oh J T. & Saito K. 2009. Pressure drop and heat transfer during two-phase flow vaporization of propane in horizontal smooth minichannels. *International Journal of Refrigeration* **32** 837-45.
[2] Dittus F. & Boelter L. 1930. University of California publications on engineering. *University of California publications in Engineering,* **2** 371.
[3] Gnielinski, V. 1975. New equations for heat and mass transfer in the turbulent flow in pipes and channels. *NASA STI/recon technical report A,* **75** 8-16.
[4] Wu, P. & Little W. 1984. Measurement of the heat transfer characteristics of gas flow in fine channel heat exchangers used for microminiature refrigerators. *Cryogenics,* **24** 415-20.
[5] Choi S., Barron R F. & Warrington R. Fluid flow and heat transfer in microtubes. ASME DSC, 1991. 123-34.
[6] Wang B. & Peng X. 1994. Experimental investigation on liquid forced-convection heat transfer through microchannels. *International Journal of Heat and Mass Transfer,* **37** 73-82.
[7] Adams T., Abdel Khalik S., Jeter S. & Qureshi Z. 1998. An experimental investigation of single-phase forced convection in microchannels. *International Journal of Heat and Mass Transfer,* **41** 851-57.
[8] Morini G L. 2004. Single-phase convective heat transfer in microchannels: a review of experimental results. *International Journal of Thermal Sciences,* **43** 631-51.
[9] Hrnjak P. & Tu X. 2007. Single phase pressure drop in microchannels. *International Journal of Heat and Fluid Flow,* **28** 2-14.
[10] Kew P A. & Cornwell K. 1997. Correlations for the prediction of boiling heat transfer in small-diameter channels. *Applied Thermal Engineering,* **17** 705-715.
[11] Mcneil D A., Raeisi A H., Kew P A. & Hamed R S. 2013. Flow boiling heat-transfer in micro to macro transition flows. *International Journal of Heat and Mass Transfer,* **65** 289-307.

Effect of physical and chemical modification on surface area of low-grade bauxite

E Kusrini[1], A Rahman[2], Aji A Bumi[1], C S Utami[1], E A Prasetyanto[3]

[1]Department of Chemical Engineering, Faculty of Engineering, Universitas Indonesia, Kampus Baru UI Depok, 16424, Indonesia
[2] Department of Chemistry, Faculty of Mathematics & Natural Sciences, Universitas Negeri Jakarta, Indonesia
[3]Faculty of Medicine, Atma Jaya Catholic University of Indonesia, Jl. Pluit Raya 2, Jakarta, 14440 Indonesia

E-mail: ekusrini@che.ui.ac.id

Abstract. In this research, the effect of physical and chemical modifications of low-grade bauxite (LB) with particle size of 200 mesh were studied. Here, we demonstrate chemically technique to synthesize low-grade bauxite/CeO_2 composites using $Ce(NO_3)_3.6H_2O$ with weight variations of 0.5, 1 and 2 g. The effect of $Ce(NO_3)_3.6H_2O$ concentrations on BET surface area of prepared low-grade bauxite/$CeO_{2(0.5)}$ (ALB0.5), low-grade bauxite/$CeO_{2(1)}$ (ALB1), and low-grade bauxite/$CeO_{2(2)}$ (ALB2) composites were observed. The results showed the BET surface area of LB is 59.85 m^2/g. After acid activation with 1M HNO_3, activated low-grade bauxite (ALB) has BET surface area of 101.6 m^2/g. By chemically modification with $Ce(NO_3)_3.6H_2O$ in concentration ranging from 0.5; 1 and 2 g, the BET surface areas were reduced in the respectively order of 94.5, 87.9, and 85.5 m^2/g for ALB0.5, ALB1, and ALB2. With the increase of the $Ce(NO_3)_3.6H_2O$ concentration, the BET surface area reduced compare to those found in the ALB. The acid activation treatment has significant effect compare to the chemical modification using $Ce(NO_3)_3.6H_2O$.

1. Introduction

The bauxite exploitation increased significantly due to high demand of alumina and aluminum for various applications [1]. This exploitation process generated huge quantities of mining wastes such as low-grade bauxite. Therefore, there is an increasing interest in the utilization of low-grade bauxite waste. For this purpose, further application of low-grade bauxite is needed to be explored.

As our previous report [2], low-grade bauxite (LB) was collected from Bintan islands, Indonesia, consists essentially of SiO_2, Al_2O_3, and Fe_2O_3, which makes it possible to use it as source materials for some industrial applications such as synthetic zeolites, adsorbent and catalyst support. Low-grade bauxite with alumina content of~44% was used for the present study. It contained a small amount of rare-earth elements including Y_2O_3 (0.0052%) and La_2O_3 (0.0041%). Mineral composition was determined using X-Ray Fluorescence with a proprietary software based on data reconciliation including multiple inputs such as chemical analysis, LOI and so on. Detail mineral composition of the low-grade bauxite used in this report is presented in table 1.

Table 1. Chemical composition of low-grade bauxite (mass fraction, %)

Oxide	% Composition
SiO_2	24.09
Al_2O_3	44.11
Fe_2O_3	9.77
LOI	20.05
TiO_2	0.912

The mineral composition of low-grade bauxite was determined by XRD analysis. The mineral composition of LB is listed in table 2. It is found that kaolinite, and gibbsite are the major components. Other minerals such as cristobalite, hematite, anatase, muscovite and quartz are also present in low quantity. The combination of chemical and mineral composition of the LB make it as a suitable candidate for catalyst support.

Table 2. XRD analysis of low-grade bauxite [2]

Sample	Clay Mineral (%)				Other Minerals (%)					
	Ectite	Illite	Kaolinite	Calcite	Quartz	Cristobalite	Gibbsite	Hematite	Muscovite	Anatase
Low-grade bauxite	*	*	42	*	2	8	30	7	4	7

Cerium oxide has been reported as good catalyst for various application and/or catalyst supporter [3]. As we know that, CeO_2 can provide favorable catalytic activity and provide a large area of chemical surface absorption and improve reducibility [4]. The addition of CeO_2 will also increase the number of Ce^{3+} ions that provide redox-active sites and also provide a large surface area for physical and chemical adsorption [5-7]. Therefore, in this research, we used $Ce(NO_3)_3.6H_2O$ for the modification of low-grade bauxite (LB) to increase the BET surface area.

In this study, we explored the effect of physical and chemical treatments on the surface area properties of low-grade bauxite (LB), activated low-grade bauxite (ALB). Also, we prepared composites of low-grade bauxite/$CeO_{2(0.5)}$ (ALB0.5), low-grade bauxite/$CeO_{2(1)}$ (ALB1), and low-grade bauxite/$CeO_{2(2)}$ (ALB2).

2. Experimental

2.1. Materials

Low-grade bauxite was obtained from Madong located in Bintan island, Indonesia. Other chemicals and materials were used in this study without purification.

2.2. Characterization

Mechanical treatment of low-grade bauxite was done in a laboratory scale using *Planetary ball mill* (NOAH NQM-4). Two stainless steel balls of 30 mm diameter with weight of 111 g and six stainless steel balls of 20 mm with weight of 32.5 g were used in the milling experiment. The stirrer speed used was 660 rotations per min (11 Hz) during 120 min.

The surface area of each powder sample was determined using a BET analyzer (model: ASAP 2020 V4.02 (V4.02 E), Micromeritics, US) with analysis adsorptive N_2, equilibration interval of 5 second, no low-pressure dose and analysis bath temperatures at -195.858; -195.822; -195.791; -

195.794 and -195.814°C (77K). Prior to BET analysis, the sample was subjected to degassing under high vacuum at 350°C for 4 hours.

The characteristics of the LB has been evaluated using XRD, XRF and BET analysis. Whereas, the characteristics of ALB and the prepared composites have been analyzed only using BET analyzer.

2.3. Preparation of ALB

Low-grade bauxite (LB) was milled to obtain the fine powder and sieved to obtain a 200 mesh (75 μm) particles size. The sieved low-grade bauxite powder using *Humboldt Motorized Siever Shaker* and then was kept in sealed plastic container. Low-grade bauxites (200 mesh) were washed with distilled water several times to remove any other impurities and allowed to dry at 110°C for 6 hours. Then sample was activated with 1 M HNO_3 solution in ratio of acid: solid of 20 mL/g, followed was stirred continuously for 8 hours at 80°C, then the powders were filtrated and washed with distilled water several times until reached the pH at 7. The obtained solid was transferred to an oven at 110°C for 24 hours. Then sample was transferred to a furnace at 500° C for 3 hours to perform calcination process. For low-grade bauxite (LB), and activated low-grade bauxite (ALB) were stored in desiccator to further characterization.

2.4. Preparation of ALB/CeO_2 composites

The ALB composites are prepared according to the method reported by Kusrini *et al.* [7,8]. The obtained activated low-grade bauxite (ALB) samples were chemically modified using $Ce(NO_3)_3.6H_2O$. The weight of $Ce(NO_3)_3.6H_2O$ was varied at 0.5, 1 and 2 g, and NH_3 solution (1M) were added drop-wisely into the sample solution for 30 min and stirred slowly with a magnetic stirrer, and the pH was adjusted to 9. After mixing for one hour, all samples were dried for 12 hours at 110°C, and then calcined in the furnace at 500°C for 5 hours. All samples of low-grade bauxite/$CeO_{2(0.5)}$ (ALB0.5), low-grade bauxite/$CeO_{2(1)}$ (ALB1), low-grade bauxite/$CeO_{2(2)}$ (ALB2) were stored in a desiccator to further characterization.

3. Results and Discussion

This study presents the properties of Indonesian low-grade bauxite sample after physical treatment (ball mill) and chemical modifications using $Ce(NO_3)_3.6H_2O$. After the treatment, samples were subjected to a surface area and porosity characterization.

The N_2 adsorption isotherms at 77K of activated and treated sample are shown in figure 1, and the properties including BET surface area, nanoparticle size, surface from external and microporous calculated from t-plot are summarized in table.1. As seen from Figure 1, the adsorption isotherm of N_2 at 77K for all samples showed type II of sorption isotherms, typical pattern for nonporous material [8].

BET results show good correlation between the mill treatment and surface area of the samples, ; Also the material after chemical modifications have significant difference compare to the raw material. The various $Ce(NO_3)_3.6H_2O$ weight ratio was ranging from 0.5 to 2 g resulted in different BET surface area, external surface and average particle size. Increasing of adsorbed N_2 to the samples indicates the increasing of surface area and diminished bauxite particle size. Activation process by milling and acid treatment using 1M HNO_3 increased surface area of low-grade bauxite (LB) significantly up to almost twice in activated low bauxite (ALB) (see table 3). BET surface area of LB has sharply increased after acid activation using 1M HNO_3 (101.60 m^2/g), but after chemical modifications using $Ce(NO_3)_3.6H_2O$ was reduced eventually. Increasing concentration of $Ce(NO_3)_3.6H_2O$ was ranging from 0.5 to 2 g, the BET specific surface area reduced, namely 94.523; 87.925 and 85.467 m^2/g, respectively. The increasing amount of $Ce(NO_3)_3.6H_2O$ used in chemical treatment causes a decrease in the number of adsorbed N_2 which indicates decreasing surface area and increasing particle size. A quite similar was observed for modification of graphite waste using $Ce(NO_3)_3.6H_2O$ to produce graphite/CeO_2 composites [7,8]. The largest surface area that is found in graphite/CeO_2 composite with variation concentration of $Ce(NO_3)_3.6H_2O$ of 2 g is equal to 26.82 m^2/g [7].

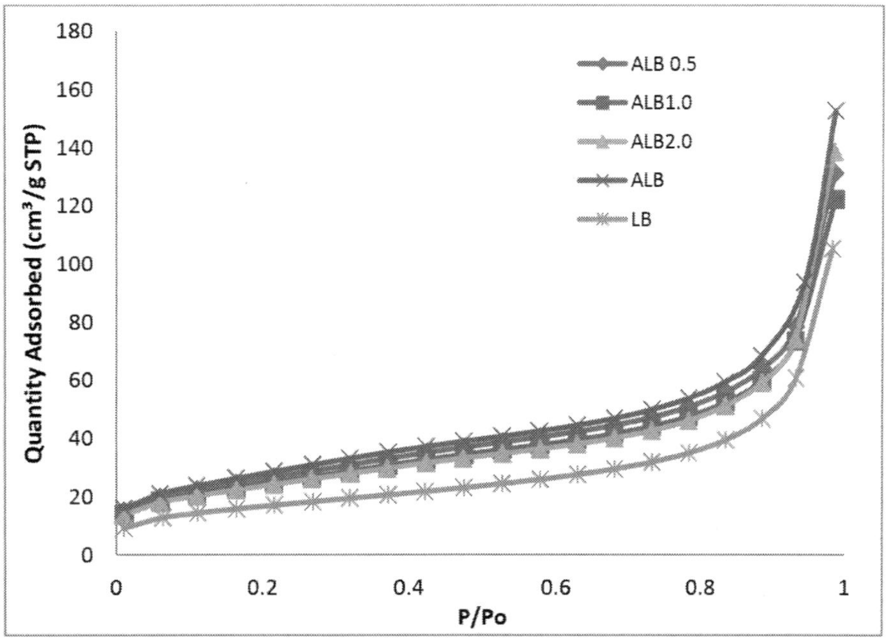

Figure 1. Isotherm adsorption using N_2 gas of quantity adsorbed vs relative pressure (P/Po)

Tabel 3. BET surface area characterization for all samples

	LB	ALB	ALB 0.5	ALB 1.0	ALB 2.0
S_{BET}	59.849	101.602	94.524	87.925	85.467
S_{ext} (m^2/g)	57.625	104.852	95.583	87.914	85.016
S_{micro} (m^2/g)	2.225	-	-	0.0107	0.451
Average Size (nm)	100.251	59.054	63.476	68.240	70.203

In LB contained t-plot small micropore of 2.225 m^2/g, whereas for ALB1 and ALB2 are 0.0107 and 0.451 m^2/g, respectively. A t-plot micropore area for ALB and ALB0.5 are not present. It is confirmed that ALB and ALB0.5 are non-porous. The t-plot external surface area of ALB after milled process and acid activation is the highest (104.852 m^2/g). If we compare both t-plot micropore area and t-plot external surface area for all samples, confirmed that all samples are non-porous. Because no micropore is present in all samples. t-plot external surface area is contributed from the average particle size of samples. The LB has average particle size of 100.25 nm. The acid activation using 1 M HNO_3 has significant effect on the BET surface area as well as t-plot external surface area. We observed all samples are non-porous because increasing surface area is contributed from decreasing average particles size. From the pattern of isotherm adsorbed N_2 and the calculated particle size after chemical treatment, it can be assumed that core shell of CeO_2@ bauxite was formed, where the core is bauxite and CeO_2 coated it. Increasing amounts of $Ce(NO_3)_3.6H_2O$ increases the particle size of CeO_2@bauxite formed, indicating CeO_2 growth on outer side of bauxite particles.

4. Conclusion

The physical and chemical treatments were used to modify the surface properties of low-grade bauxite. With increasing $Ce(NO_3)_3.6H_2O$ concentration, the BET surface area reduced compare to those found in the activated low-grade bauxite (ALB).

5. Acknowledgments

The research was supported by the Hibah Publikasi Internasional Terindeks untuk Tugas Akhir Mahasiswa (Hibah PITTA) No. 774/UN2.R3.1/HKP.05.00/2017 funded by Universitas Indonesia.

6. References

[1] P Smith, 2009,"The processing of high silica bauxites review of existing and potential processes", Hydrometallurgy **98** 162-176.

[2] Kusrini E, Usman A, Wicaksono W, Rahman A, Gunawan C 2017(a), *J Mater Environ Sci*, Under Review

[3] L.Vivier, D.Duprez, 2010, Chem Sus Chem **3** 654-678

[4] Jamalzadeh, Z., Haghighi, M., Asgari, N., 2013. *Front. Environ. Sci. Eng.* **7** 365–381.

[5] Z. Abbasi, M. Haghighi, E. Fatehifar, S. Saedy, 2011, J. Hazard. Mater. **186** 1445–1454.

[6] E Kusrini, C S Utami, Nasruddin, A Usman, A S Kurniawan, 2017(b), Synthesis and Characterization of Graphite/CeO_2 Composites via Mechanically-Thermal-Chemically Methods, IoP: *Mat. Eng. Sci.*

[7] E Kusrini, Nasruddin, C S Utami, AA Bawono, A Usman, *2017(c)*, Surface Modification on Graphite Waste for CO_2 Adsorption Capacity, Inter. *J. Tech,* Submitted.

[8] S Lowell, S J E, Shields, M A, Thomas, and M Thommes, 2004, Characterization of Porous Solids and Powders: Surface Area, Pore Size and Density, *Kluwer Academic Publishers, Dordrecht, The Netherlands*

Product distribution from catalytic conversion of biomass over $B_2O_3/\gamma\text{-}Al_2O_3$ catalyst

Setiadi[1], **Ervandy Haryoprawironoto**[1],

[1]Department of Chemical Engineering, Chemical Engineering Study Program Universitas Indonesia, Kampus Baru UI, Depok, West Java, Indonesia

E-mail: setiadi@che.ui.ac.id or ervandyharyo@gmail.com

Abstract. The catalytic conversion of biomass has been studied by using a conventional fixed-bed reactor. The mixed catalyst is prepared by impregnation method followed by calcination of the catalyst. The $B_2O_3/\gamma\text{-}Al_2O_3$ catalyst composition used in the biomass catalytic conversion process is varied to study the effect on its product distribution. In addition, catalyst variations were performed to test the catalytic effect in producing the target products, so that the optimum catalyst composition to maximize the target products is known. The catalyst composition of B_2O_3 in the mixed catalyst used was 0%, 5%, 10%, 20%, and 25% wt. The use of $B_2O_3/\gamma\text{-}Al_2O_3$ mixed catalyst can convert the small oxygenate compounds produced at the thermal decomposition step of biomass into aromatic hydrocarbon compounds. In addition, other groups of compounds were also observed with different product distributions where the differences were affected by the composition of the mixed catalyst used. Experimental results show that addition of B_2O_3 content up to 20% wt gives the highest catalytic effect with the main product being aromatic hydrocarbons. The decrease in catalytic effect occurs with the addition of 25% wt B_2O_3 content.

1. Introduction

In the future, many see the potential of biomass as an alternative to fossil fuels that can produce similar petroleum products. Various hydrocarbon compounds can be produced from biomass through biological, biochemical, and thermochemical processes [1]. Pyrolysis technology is one of the most promising thermochemical routes in biomass conversion to meet interest in the production of bio-fuels and chemicals [2]. Biomass pyrolysis will produce products consisting of bio-oil, char, and gas. Bio-oil is a product of biomass pyrolysis which has great potential to be utilized. However, the bio-oils are of poor quality. They are thermally unstable, degrade with time, acidic, have a low heating value, and are not compatible with existing petroleum-derived oils [3]. Bio-oils must be catalytically upgraded if they are to be used as a conventional liquid transportation fuel [4].

The catalytic conversion of biomass is one of the proven ways that can be used to improve the quality of bio-oil. Basically, this process involves the biomass thermal decomposition stage as occurs in conventional pyrolysis process, but with the presence of catalyst. Catalyst addition to this process led the biomass pyrolysis vapor consisting of small oxygenate compounds enter the pores of the catalyst in which they are converted to CO_2, CO_2, water, coke, and aromatic hydrocarbon compounds. The conversion of catalytic biomass into aromatic compounds has been successfully performed using zeolite catalysts [5,6]. Introduction of zeolite catalysts into the pyrolysis process can convert oxygenated compounds generated from pyrolysis into aromatics [7]. The zeolite catalyst has an advantage in performing a shape-selective reaction where the operating conditions and proper selection of the catalyst can maximize the production of aromatic hydrocarbons. However, zeolite catalyst is relatively expensive.

Therefore, in this work to convert biomass into aromatic hydrocarbon compounds mixed catalyst of $B_2O_3/\gamma\text{-}Al_2O_3$ are used which is readily available and more cost-effective. The $B_2O_3/\gamma\text{-}Al_2O_3$ catalyst system is an interesting catalyst and is rarely being researched. B_2O_3 is an acidic solid while $\gamma\text{-}Al_2O_3$ is a solid that can act as an ionic conductor. B_2O_3 has a small particle size as $\gamma\text{-}Al_2O_3$ has a large enough surface area (± 250 m^2/g), so that the solid particles B_2O_3 can be well dispersed on the solid surface $\gamma\text{-}Al_2O_3$. Sato [8] reported that the addition of B_2O_3 may form a BO_4 specimen which can add the amount of bronsted acid to the catalyst that plays a role in the formation of aromatic hydrocarbon compounds. Five variations of B_2O_3 addition in $B_2O_3/\gamma\text{-}Al_2O_3$ mixed catalyst used in this work were 0%, 5%, 10%, 20%, and 25% wt. The composition of the catalytic conversion product of biomass will be analyzed to determine the effect of B_2O_3 percentage in the mixed catalyst on the catalytic effect in forming the target product.

2. Experimental

2.1. Materials

The type of biomass used as a feed in this experiment is a corncob. Corn cobs must be grounded to a particle size of 1-3 mm by a grinding machine. Corncob is then dried by oven at 60°C for 5 hours to reduce its water content to <10% of total mass. For a typical run, 1 gram of corn cob and 3 gram of catalyst-quartz sand mixture was used.

The catalyst material used is a solid containing boron oxide and $\gamma\text{-}Al_2O_3$. Boron oxide is obtained from boric acid (H_3BO_3) which acts as the precursor of the oxide. Boric acid will decompose into water vapor and metaboric acid (HBO_2) at a temperature of 170°C. Further heating at 300°C will decompose HBO_2 into water vapor and microcrystalline boron oxide (B_2O_3) solids.

The reaction mechanism of boric acid formation is as follows:

$$H_3BO_{3(s)} \rightarrow HBO_{2(s)} + H_2O_{(g)} \quad (1)$$

$$2HBO_{2(s)} \rightarrow B_2O_{3(s)} + H_2O_{(g)} \quad (2)$$

The catalyst used in each catalytic conversion process with different composition variations is as much as 1 gram. The catalyst used cannot be excessive so as not to cover the flow of N_2 gas which will cause the blockage of flow in the tube. However, if the amount of catalyst used is too small it will cause the catalytic conversion process to be less than optimal.

The preparation of the catalyst is carried out by the impregnation method which steps can be described as follows: preparing the boric acid solution by dissolving the solids into the deionized water maintained at 75°C to completely dissolved solids. The solution is used to impregnate the $\gamma\text{-}Al_2O_3$ solid by adding $\gamma\text{-}Al_2O_3$ solids to the boric acid solution. While stirred and maintained at 75°C so that the water solvent was all evaporated and dried, a solid was obtained which was a mixture of borate solid acid- $\gamma\text{-}Al_2O_3$. To obtain a mixture of $B_2O_3/\gamma\text{-}Al_2O_3$, the solids were calcined under atmospheric conditions respectively at a temperature of 300°C and 600°C for 2 hours each.

2.2. Methods

The biomass catalytic conversion was performed in a stainless-steel reactor tube. The tube reactor has a length of 300 mm and a diameter of 12 mm. The effluent gas product of biomass catalytic conversion was at the bottom end of the reactor tube. This open-ended tube was successively placed from the bottom with some quartz wool, a mixture of 3 grams of quartz catalyst sand, and 1 gram of corn cobs. Quartz sand was mixed with active catalyst to help uniform heat in all parts of the catalyst. Biomass feeds are placed over a mixture of sand-quartz catalysts as fixed beds. Therefore, the non-catalytic fast pyrolysis took place first and then the primary product vapors were upgraded when they passed through the catalyst layer.

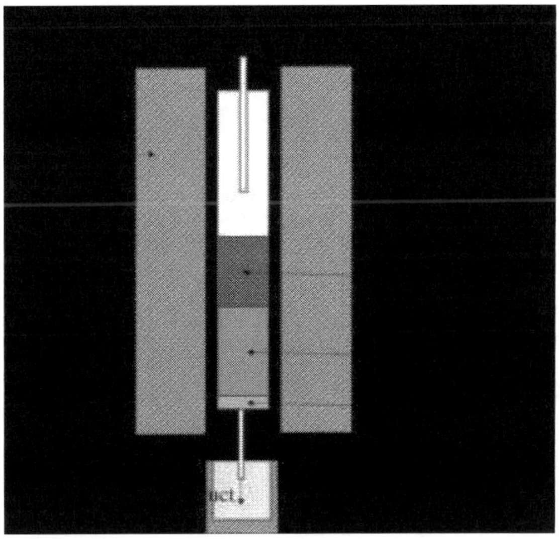

Figure 1. Schematic diagram of experimental apparatus for biomass catalytic conversion

Biomass catalytic conversion was performed at 500°C and 1 atm pressure. N2 gas with a flow rate of 30 ml/min was used as an inert pyrolysis gas. The N_2 gas will remove the air present in the reactor tube before the catalytic conversion reaction takes place. N_2 gas will also force the pyrolysis gas vapor to move downward and enter the pores of the catalyst in which they are upgraded. Schematic diagram of experimental apparatus for biomass catalytic conversion was shown in Figure 1.

The gas product from catalytic conversion was captured by a system called cold gas absorption. Cold gas absorption is a system in which there is a container containing a solution of ethanol pro analyst and is provided with a cooling jacket to cool the solution. The gas product was streamed into the cold gas absorption system and will be in direct contact with the cold ethanol pro analyst solution so that it undergoes quenching. The product gas was then dissolved together with the solvent. The chemical compositions of the products were analyzed, and their relative contents were determined and discussed

3. Results and Discussion

Pure γ-Al_2O_3 and its mixtures with 5%, 10%, 20%, and 25% B_2O_3 additions were used as bed materials for catalytic conversion of corncobs. Figure 2 shows the product distribution resulting from catalytic conversion of corncobs comprising groups of compounds such as aromatics, acids, furans, phenols, and others. The result shows that the addition of B_2O_3 to the catalyst mixture caused the aromatic percentage of the product to increase to its maximum value as the percentage of B_2O_3 increased from 0% to 20%. On the use of 20% B_2O_3, relative aromatic content was obtained with a value of 49.32%, increased from 28.46% with pure γ-Al_2O_3 as a catalyst. When the percentage of B_2O_3 in the mixed catalyst exceeded 20%, the relative aromatic content decreased as shown in the study where a relative aromatic content of 9.27% was obtained with 25% B_2O_3 as catalyst.

Pyrolysis of corncobs without the use of a catalyst will produce large molecules of oxygenates. These oxygenates cannot enter the micro-pores of the B_2O_3 catalyst in which pyrolysis vapor upgrading occurs. As these oxygenates are unable to enter the catalyst micro-pores, this oxygenate molecule undergoes polymerization and forms a large molecule that we know as coke on the surface of the catalyst. It causes the catalyst to be deactivated due to the formation of coke that cover the catalyst surface and block the entrances of the catalyst pores. γ-Al_2O_3 is a mesoporous catalyst which has strong cracking characteristic and can convert the large-molecule oxygenates into small-molecule oxygenates [9]. The combination of γ-Al_2O_3 and B_2O_3 catalysts will make pyrolysis vapors containing a lot of large molecules of oxygenate being converted into small molecules of oxygenate first and then enter the micropores of B_2O_3 catalyst to be converted into aromatics.

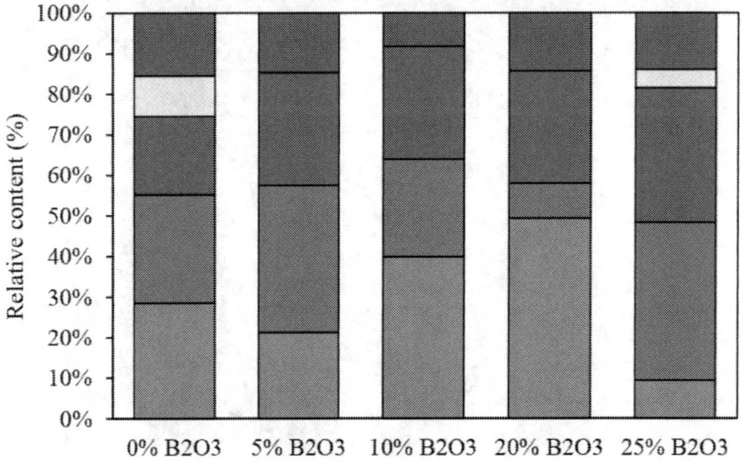

Figure 2. Product distribution from catalytic conversion of biomass over five different composition of $B_2O_3/\gamma\text{-}Al_2O_3$ catalyst: 0% B_2O_3, 5% B_2O_3, 10% B_2O_3, 20% B_2O_3, and 25% B_2O_3. Key—Aromatics: green, acids: red, furans: purple, phenols: white, and others: blue.

Figure 2 also shows that different percentages of B_2O_3 in the catalyst mixtures will result in difference of aromatic composition in the product. Experimental result show that B_2O_3 is a type of catalyst that can convert small molecules of oxygenate to aromatic hydrocarbon compounds in which the percentage of aromatic hydrocarbons increases with the addition of B_2O_3 percentages in the catalyst mixtures. However, this increase reaches its maximum value with a 20% B_2O_3 catalyst and begins to decrease as the catalyst adds past 20%. So, it can be said that 20% B_2O_3 in $B_2O_3/\gamma\text{-}Al_2O_3$ catalyst mixture is the best catalyst system to convert biomass to aromatic hydrocarbon.

The addition of B_2O_3 should increase the catalytic effect of the catalyst mixtures in converting biomass to aromatic, since the amount of bronsted acid sites will also increase linearly. However, the results of the research showed a decrease in the percentage of aromatic products with the addition of B_2O_3 percentage above 20%. The addition of B_2O_3 can lead to the formation of a layer that covers $\gamma\text{-}Al_2O_3$ which causes the surface area of $\gamma\text{-}Al_2O_3$ to decrease further. The reduced surface area of the catalyst causes the less contact area between $\gamma\text{-}Al_2O_3$ and pyrolysis vapors. This leads to the ability of $\gamma\text{-}Al_2O_3$ in cracking large molecules of oxygenate into smaller ones to decrease so that fewer oxygenate molecules can be converted to aromatic by B_2O_3.

Figure 3 shows the effect of the percentage of B_2O_3 on the catalyst mixtures used in the catalytic conversion of corncob to the distribution of aromatic products present in the product. Benzene does not form on pure $\gamma\text{-}Al_2O_3$ catalyst and reaches its maximum value at 5% B_2O_3 and then decreases with the addition of B_2O_3. The addition of B_2O_3 increases toluene production and reached its maximum value at 20% B_2O_3 where toluene was a major component in the aromatic group. Xylene increased with the addition of B_2O_3 in the range of 5-20%. The content of indene and naphthalene in the product was reduced by the addition of $\gamma\text{-}Al_2O_3$. It can be inferred from the experimental results that B_2O_3 catalyst mixtures were selective on the formation of benzene, toluene and xylene compounds in which there were only small amounts of indene and naphthalene (less than 4%). Monocyclic aromatic reactions with oxygenate fragments produce naphthalene, indene, and their derivatives [10]. Naphthalene is a type of polycyclic aromatic hydrocarbons (PAHs) whose relative contents are worth noting because PAHs were a potent toxin and had a high carcinogenicity. However, the resulting PAHs in catalytic biomass catalysts with $B_2O_3/\gamma\text{-}Al_2O_3$ catalysts were much lower than in the ZSM-5 catalyst [11].

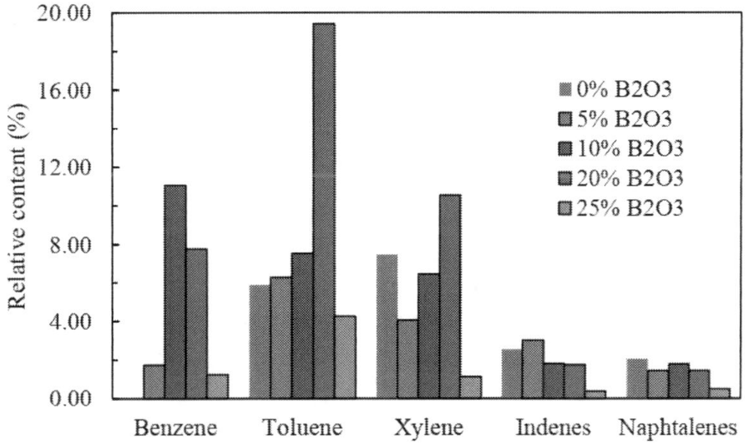

Figure 3. Relative content of aromatics from catalytic conversion of biomass over five different composition of $B_2O_3/\gamma\text{-}Al_2O_3$ catalyst: 0% B_2O_3, 5% B_2O_3, 10% B_2O_3, 20% B_2O_3, and 25% B_2O_3.

The effect of B_2O_3 percentage on catalytic conversion of biomass to relative content from oxygenated species is shown in figure 4. The experimental results show that relative content in most oxygenated species is decreasing with the addition of B_2O_3 percentage. Biomass pyrolysis will produce vapour containing thermally stable oxygenate molecules such as phenols, furans, ketones, acids, aldehydes and others. Thermally stable molecular oxygenate is an intermediate in the production of aromatic.

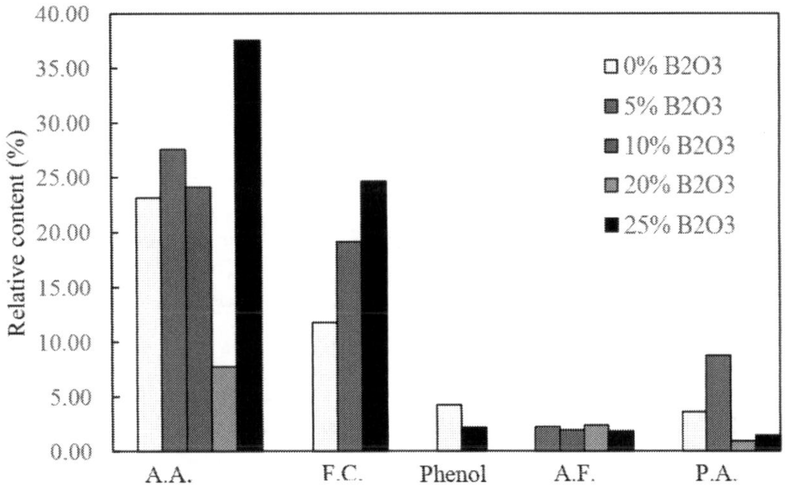

Figure 4. Relative content of oxygenated species from catalytic conversion of biomass over five different composition of $B_2O_3/\gamma\text{-}Al_2O_3$ catalyst: 0% B_2O_3, 5% B_2O_3, 10% B_2O_3, 20% B_2O_3, and 25% B_2O_3.

The results showed that increasing percentage of B_2O_3 in the catalyst mixtures could remove more organic oxygenate compounds in upgraded pyrolysis vapour. These oxygenate molecules was converted to hydrocarbons in the B_2O_3 micro-pores and remove the oxygen as water, CO and CO_2 through deoxygenation reactions. The 20% percentage of B_2O_3 can dramatically decrease the oxygenate content such as acetic acid and propionate acid and remove furancarboxaldehyde and phenol. It has been previously discussed that the 20% percentage of B_2O_3 in the catalyst mixtures is the best catalyst system that provides maximum aromatic production, so most of the oxygenate molecules have been converted to aromatic while relative content of oxygenated organic compounds decreased.

4. Conclusions

The catalytic conversion of corncobs with pure γ-Al_2O_3 catalyst and its mixtures with the addition of B_2O_3 was performed on a fixed bed catalytic reactor. The γ-Al_2O_3 and B_2O_3 catalyst mixtures can be used to produce aromatic hydrocarbon compounds. The maximum aromatic yield of 49.32% can be obtained with 20% B_2O_3/80% γ-Al_2O_3. The addition of B_2O_3 to the catalyst mixture B_2O_3/γ-Al_2O_3 can increase the selectivity of the formation of aromatic compounds such as benzene, toluene, and xylene which reaches its maximum value at 20% B_2O_3.

References

[1] Demirbas F 2009 *Energy 2009* **86** S151-S161
[2] Zhang L, Xu C, and Champagne P 2010 *Energy Convers. Manag.* **51** 969-982
[3] Mohan D, Pittman C Jr, and Steele P 2006 *Energy Fuel* **20** 848
[4] Horne P and Williams P 1996 *Fuel* **75** 1043
[5] Torren R, Geoffrey A, William C, and George W 2009 *Top Catal* **52** 241–252
[6] Bo Zhang, Zhao P, Xiao B, Kuan D, and Zu W 2015 *Fuel Processing Technology* **138** 430-434
[7] Carlson T, Vispute T, Huber G 2008 *ChemSusChem* **1** 397
[8] Sato S, Masakatsu K, and Toshiaki S 1995 *Journal of Molecular Catalysis* 171-177
[9] Zhang H, Xiao R, Jin B, Xiao G, and Chen R 2013 *Bioresource Technology* **140** 256-262
[10] Carlson T, Jae J, and Huber G 2009 *ChemCatChem* **1** 107-110
[11] Torren R, Geoffrey A, William C, and George W 2009 *Top Catal* **52** 244

Sustainable infrastructure transportation to improve society wellbeing in Karawang

D T Winata[1] **and A D Rarasati**[1]
[1]Civil Engineering Department, Faculty of Engineering, Center for Sustainable Infrastructure Development, Faculty of Engineering, Universitas Indonesia, Kampus UI Depok, Indonesia 16424

E-mail: ayomi@eng.ui.ac.id

Abstract. Transportation infrastructure design must follow sustainability concept that requires stability in economic, social and environment, then sustains in the future. Therefore, this research aims to identify time travel, fare level, percentage spending on transport, waiting time, and availability of alternatives mode as sustainable transportation indicators that could be applied in rural and urban area at Karawang Regency, as developing district correspond to society welfare as a social theme variable in sustainability development indicators. This research also identifies existing commutes characteristics and benchmark of the sustainable transportation indicators to improve society wellbeing. This research applies quantitative methods by conducting survey to collect data from respondents. The instruments to obtain data in Karawang Regency are questionnaire form, and then collaborated with structured-interview. Spearman's rank correlation is employed as the quantitative method in order to conduct non-directional and directional hypothesis test. In conclusion, time travel, spending on transport, and fare level variables representing mobility, equity, and affordability have significant influence to experience wellbeing, and then they have negative correlation to social welfare as well. For that reason, time travel and fare level must be reduced to improve society wellbeing.

1. Introduction

Infrastructures are structures and facilities that create system to support society daily activities especially social and economy sector in urban city. There are seven types of infrastructures related with society needs in a city that one of them is transportation infrastructures. Meanwhile human needs are getting complex from time to time, therefore infrastructure have to be developed in order to solve with new problems and fulfil society needs. On the other hand, infrastructure project development will have lot of constraints for example time limitation, budget constraint, and social impact.

Availability of infrastructure aim is to facilitate people with their daily activity to achieve efficiency and improve productivity. In order to achieve efficiency, infrastructure development and plan are needed in the future. The target of infrastructure improvement is to alter their existing and new infrastructure to more sustainable forms. Characteristics of sustainable infrastructure are eco-friendly, low budget construction and maintenance, influence social and culture in good way and durable.

Indonesia as a developing country still faces lots of problem and the most common problem is inequalities in between regions in social welfare and infrastructure development sector. According to the Organisation for Economic Co-operation and Development (OECD) report in 2013, insufficient in quantity and inadequate in quality, transport infrastructure are serious problems in regional development. Most of regency and city roads that account for nearly 80% of the network, are in bad condition [1]. Inequalities produce different solution to solve problem depending on area development [2]. Therefore, development in a region must consider sustainable infrastructure solutions. It needs good analysis to pick the right solution adjusting with problems in society.

As a result, the researcher selected as Karawang Regency as research study case of developing region. Karawang Regency where the regency is located near Jakarta as the capital city of Indonesia, then Karawang becomes regency that supports Jakarta in logistic and agriculture stock [3]. Karawang Regency has the mixture of industrial area, agriculture area, and moreover, the transportation infrastructure development in Karawang is still growing. Therefore, the researcher wants to find the applicable sustainable transportation indicators that could be used to measure transportation condition in Karawang by correlating and assessing impact to society wellbeing as social development goals. Moreover, this research will determine how to improve society wellbeing using selected sustainable transportation indicators. In addition, the research also analyzes commutes characteristics in rural and urban area in Karawang according the selected sustainable transportation indicators.

2. Literature review

Happy Planet Index is one of multidimensional indicators of sustainability development for a country. Since it can be calculated based on experience of wellbeing, then it means improving experience wellbeing very important to represent achieve social goals of sustainable development [4]. Likewise HPI, sustainable transport infrastructures have their own specific indicators measuring impact to social sector, and then it has to bring good social impact especially social wellbeing. In order to develop transportation infrastructure, specific parameter must be determined in the beginning to accommodate society needs [5]. Moreover, the parameter must be measurable and bring good influence to society wellbeing in Karawang based on sustainable concept goals. Eventually, when all the goals are achieved, it will directly increase customer loyalty in order to get long-term financial performance and make a sustained investment [6].

Satisfaction with travel can be described by daily work commute activities. Ettema, et al. [7] and Olsson, et al. [8] argued that travelling satisfactory is important for experience wellbeing because it makes people possible to do various activities. According to Olsson, et al. that discussed about relationship between daily travel and subjective well-being in Sweden, characteristics of commutes could be described in travel time, number of travel modes, and transportation frequency time. Furthermore, subjective wellbeing may increase and having relationship to travel cost and spending transport in household. Due to Olsson, et al., defines commutes characteristics from 5 characteristics of commutes affecting travelling satisfactory and important for people's life satisfaction, and then the commutes characteristics is same with sustainable transportation indicators detail by Gundmundsson [9], Dobranskyte-Niskota [10], Friman & Fellesson [6], and Litman [11], therefore, the researcher selects sustainable transportation indicators that have similarities.

As a result, the researcher selects 5 transportation variables that represent commutes characteristics affecting travelling satisfactory and people's life satisfaction or subjective wellbeing. The variables are travel time or speed, number of alternatives travel modes, transportation frequency from waiting time, travel cost (fare level) and spending transport in household. Those variables also the same with sustainable transportation indicators, therefore it may represent sustainability of transportation infrastructure as well. The first following variable is travel time that same variables to measure mobility goals in sustainable transportation infrastructure [8]. The second is travel cost represents fare level as affordability variable. The third is transportation frequency time can be measured by waiting time in order to rate how efficient the transportation performance. Lastly, the travel cost is indicated by using fare level and spending on transport for passenger.

3. Research methods

The method during data processing and obtaining output data use quantitative methods. The aim of quantitative method is to count data in an attempt of observation. Then the data is in number forms and analyze by using statistical method. Data will be collected by cross-sectional survey. Questionnaire and structured interview were used as an instrument to collect data from respondents who use transport infrastructure frequently. The survey was conducted at Lamping Village in Pangkalan Sub-district, Tanjungpakis Village in Sub-district Pakis Jaya, Teluk Jambe Timur, and surrounding West Karawang Sub-district. The respondent's age is limited as productive age between 15 to 64 years old. The total respondents are 344 respondents, with 63 from rural area (Lamping Village and Tanjungpakis Village) and 281 from urban area (West-Karawang Sub-district and Teluk Jambe Timur).

The questionnaires are used to obtain general information about respondents relate with age, gender, home address, highest education level, monthly income respondent's household, details occupation and how often respondent do the traveling activities. After that, the questionnaire records destination (office or school) address, traveling duration time, daily transportation cost, many options of mode that people have and duration the respondents have to wait for public transport and traffic. Lastly, the questionnaire asks about individual wellbeing according to Gallup [12]. The wellbeing statements are related to supportive social, financial, community, physical, and motivation to achieve purpose of life. Then, the wellbeing will be scored in into 4 points scale base on Likert scale, ranging response from strongly disagree (1) to strongly agree (4).

From the questionnaire, it takes data processing to change to variables going to be applied. Firstly, travel time is same with average speed on trip, therefore it is produced from travelling duration divided by trip distance getting from measuring distance of home address to destination address. Travel time unit is minutes over kilometre. Fare level is how much money that spend on each kilometres travels. Secondly, fare level unit is price over distance where this research uses Rupiah as a currency price and kilometre as a distance unit. Amount of money spent on transport are gather from daily cost from each transportation mode that used in commute trip to destination. Afterwards, fare level is total cost divided by the distance. Thirdly, Percentage spending on transport from household income comes from cost proportion that every family spends on daily travel activity in a month. Transport cost is multiplied by two in order to assume commuting trip, thus it produces daily cost transport each respondent. Daily cost transport will be multiplied by how many days in a week that the respondent goes to work or school and it makes into a month by multiplying four. Final cost of transport is a total spend in a month. It will be divided by household monthly income and converted into percentage. Fourthly, waiting time is sum of total duration each respondent waits in exchange modes. Respondents who fills waiting time are considered as public transportation users. It happens because if the respondents use their own private vehicle, they do not have to wait.

The following step is checking the obtained transport variables samples either well distributed or not, with a purpose determine parametric or non-parametric test. The data distribution is checked by using skewness and graph frequency distribution. The conclusion of travel time or speed, number of alternatives travel modes, transportation frequency from waiting time, travel cost (fare level) and spending transport variables are not distributed well and use non-parametric test. The tests are Mann-Whitney U test and Spearman's Rank Correlation Test. Mann-Whitney U test is used to test either any different in between urban and rural respondent's characteristics. Spearman's Rank wants to identify correlation and benchmark the selected sustainable transport indicators to society wellbeing. Lastly, the researcher does generalization of commutes characteristics using mean, median and inter-quartile range (IQR) as statistical descriptive.

4. Results

Firstly, the researcher assess questionnaire with validity using Product-Moment Correlation and reliability test using Cronbach's Alpha. The validity test result shows the lowest computed r-xy is 0.57 and it is bigger than r-critical with 0.0903 for significance level 0.05 (5%) or the confidence level 95%

with one-tail degree of freedom (df) is 342. Therefore, the questionnaire passed validity test and collect variables accurately with valid answers. The reliability test using Cronbach's Alpha, the result of internal consistency reliability (α) is 0.7015. The reliability of the five statements that going to measure society wellbeing could be accepted because the alpha (α) is bigger than 0.7 categorized in acceptable criteria.

Secondly, the researcher finds that the variables in not well distributed data or not because the lowest skewness value is not zero with skewness value of 0.8904 for alternatives mode. For that reason, the researcher conducted a Mann-Whitney U Test to check whether there are any difference characteristics in urban and rural area in Karawang. The Mann-Whitney U calculation results are presented in table 1.

Table 1. Mann-Whitney test characteristic urban and rural area.

	Urban (281)	Rural (63)	Conclusion
Time Travel			
U-value	2668.5	3183.5	
z-value	0.684		Not different
p-value	0.248		Not significant
Spending on Transport			
U-value	8097.5	9605.5	
z-value	1.052		Not different
p-value	0.144		Not significant
Fare Level			
U-value	1513	4333	
z-value	3.74		Different
p-value	0.000 9		Significant
Waiting Time			
U-value	3839.5	13863.5	
z-val ue	7.024		Different
p-value	0.0001		Significant
Alternatives Mode			
U-value	262.5	1507.5	
z-value	8.752		Different
p-value	0.00001		Significant
Wellbeing Score			
U-value	8330.5		
z-value	0.729		Not Different
p-value	0.252		Not significant

Based on table 1, it shows some different commutes characteristic in urban and rural area in Karawang. The conclusion of different will be obtained if the z-value is higher than z-critical with 1.65 and the p-value do not pass 0.05. Eventually, there are differences in fare level, waiting time, and alternative mode they had between urban and rural people in Karawang. The following step is assessing correlation to society wellbeing, and if the characteristic is not different, the computation results do not need to split up. The result of spearman's rank correlation as non-parametric test is presented in table 2.

Based on table 2, the correlated sustainable transportation variables to society wellbeing are time travel, fare level and spending on transport. It can be categorized as correlated variables if the absolute r-value is bigger than 0.1147 as the r-critical. Moreover, if the r-value is lower than 0 or it is in negative form, it shows that the behaviour of variable is negative correlation. Negative correlation means when the time travel, fare level and spending on transport decrease, and then the society wellbeing is improved. It also similar with the meaning of time travel, fare level and spending on transport is inversely proportional to society wellbeing.

Table 2. Correlation to society wellbeing.

	Urban Area	Rural Area
Fare Level		
R-value	-0.173	-0.3216
P-value	0.0045	0.19
Conclusion correlation to wellbeing	Significant correlated (negative)	No Correlation
Waiting Time		
R-value	-0.00726	0.1825
P-value	0.222	0.1521
Conclusion correlation to wellbeing	No Correlation	No Correlation
Alternatives Mode		
R-value	-0.0726	0.1905
P-value	0.118	0.134
Conclusion correlation to wellbeing	No Correlation	No Correlation
Time Travel		
R-value	-0.1848	
P-value	0.00163	
Conclusion correlation to wellbeing	Significant correlated (negative)	
Spending on transport		
R-value	-0.184	
P-value	0.0016	
Conclusion correlation to wellbeing	Significant correlated (negative)	

After conducting correlation test, in order to examine existing condition of commutes characteristics refers to time travel, spending on transport, fare level, waiting time, and alternatives mode, the researcher conducted statistical analysis. The researcher figures the existing condition by analyzing through average, median and inter-quartile range (IQR). The description about commutes characteristics in urban and rural area in Karawang is presented in table 3.

Table 3. Commutes characteristics urban and rural in Karawang.

Time Travel (min/km or km/hr)		
Average	4.67 minutes/km (12.85 km/h)	
Median	2.92 minutes/km (20.58 km/h)	
IQR	1.74 - 5.82 minutes/km (10.31-34.48 km/h)	
	Urban	**Rural**
Spending on Transport (%)		
Average	11.02%	
Median	8%	
IQR	2%-16%	
Fare Level (Rp/km)		
Average	Rp 1.575,-	Rp 465,-
Median	Rp 850,-	Rp 293,-
IQR	Rp 365 – Rp 1700	Rp 0 – Rp 650
Waiting Time (Minutes)		
Average	13	3
Median	10	0
IQR	0-20	0
Alternatives Mode (Quantity)		
Average	2	0
Median	1	0
IQR	1-2	0

5. Discussion

Before discussing correlation experience wellbeing to each sustainable transport variables, characteristic society of welfare in Karawang will be analyzed generally a bit. From five statements relating with supportive social, financial, community, physical and motivation to reach life goals, the respondents was asked to rate 1 to 4 for each statement. As a result, the score for social welfare will be in range 5 to 20 for the total score. Then the result for total score affecting individual welfare is 15 for median and the inter-quartile range is 14 to 16. If the total score from 5 to 20 is equalized to rate 1 to 10, then the median is 6.67 and the IQR range is within 6 to 7,33. According to Happy Planet Index Report 2016, the experience wellbeing of Indonesia has score 5.4 over 10. Higher score society wellbeing in Karawang Regency than overall in Indonesia also shows that higher GDP sometimes also increase society wellbeing in a region. Based on Karawang Regency Government report in 2016 [3], Karawang Regency Gross Domestic Product per capita in 2014 is Rp 41.986.072, then Gross Domestic Product per capita in Indosnesia is about Rp. 41.8 million.

For alternatives mode, the average and median in urban area is 2 for average and 1 for median, on the other hand the rural area has no option for alternatives modes. Refers to analysis of obtained data in Karawang Region, available alternatives mode is very limited. No modes answers show there are some respondents prefer to go by walking (25% go to work and 10% go to school) than using vehicle especially in rural area. Moreover, most of respondents select motorcycle as main mode that is often used vehicle. Particularly in rural or villages in Karawang Region, public transportations such as minibus or "angkot" do not reach them. Even though the public transportations reach them, they cross rarely and the number of vehicles is small. The public transportations are only available every 1 to 2 hours. Eventually, people who live in villages depend on their private vehicle or motorcycle. Moreover, due to limited public transportations problems common in every suburb area in Karawang Region, most of people already understand and do not bother their society welfare as long as fuel price for their private vehicle not increasing.

Waiting or delaying time during operational transport also do not cause serious influence to society wellbeing likewise the number of alternatives mode. Since most of people rely upon motorcycle as private vehicle, the waiting time for public transport in city is relatively short. The urban area specifically in West Karawang sub-district, the waiting time average is 13 minutes or in range 0 to 20 minutes, in contrary for rural area, the society only have to spend less time or almost no delaying time in doing travelling. No delaying time in rural area or villages is caused by none use public transport due to no public transportation operated and the traffic very low. In addition, based on researcher observation during survey, the overall traffic in Karawang Region specifically at the city is moderate or the traffic only takes about 5 minutes. Owing to short range waiting time in society, the waiting time cannot be correlated and not significant enough with society wellbeing as well

Time travel in Karawang in general same and no difference, then correlation using statistical analysis presents that time travel transportation significantly influences society wellbeing. The research gets the median of travel time 2.92minutes/km (20.58 km/h) and the inter-quartile range (IQR) is 1.74 - 5.82 minutes/km (10.31-34.48 km/h). The average of time travel in Karawang is 4.67 min/km (12.85 km/h). Furthermore, benchmark in order to improve society wellbeing is negative correlation as well. Negative correlation means the faster people could reach the destination, the more welfare they are as well. Logically, if people go anywhere faster, they could save time to do other activities and achieving their life goals. In addition, faster time will save people energy to do their job or study as student. From the behaviour most of people Karawang's village, they would rather buy gasoline at near station to save time even though it is more expensive than normal price, it could be concluded that faster time is the most important matter in a trip.

Percentage spending on transport expenditure is one of variables to measure equity and fairness in social dimension. The median for percentage spending on transport expenditure is 8%, average is 11.02% and the IQR is 2% to 16% from the income. Almost 50% of samples spend percentage 2% to 16% of income to pay daily transportation cost. Through observations, researcher finds equity and fairness in transportation will seriously alter economy and social welfare in society. It is very influencing for rural and urban society welfare because spending on transport significantly correlated to society wellbeing. The correlation has negative correlation, therefore the benchmark in order to improve society wellbeing, percentage of spending on transport must be decreased. The reason monthly expend that have to spend for transportation must be decreased is in order to fulfil other financial needs.

Fare level measures how much money (Rupiah) people have to pay for certain distance and in this research is a kilometre. Through characteristic analysis, the fare level in rural versus urban and student versus non-student show significant difference. From the correlation test shows fare level brings significant impact only to urban society welfare. Based on descriptive statistical, the researcher calculates the median of fare level in urban area is Rp.850,-/km, then the IQR is Rp.365,-/km to Rp.1700,-/km. The average fare level is Rp.1574,-/km. This much higher than fare level in rural area because in people activities in urban area rely on the transportation mode, while in urban area the people go by walking and not spend much money. Likewise the preliminary researcher assumption, using direction correlation hypothesis test, it supports with the negative correlation assumption. Low fare level is also a substantial variable in transportation infrastructure rising up social welfare. This fact is also supported by result of other research showing price is the most crucial factors that encourage society to use public transportation. They demand for cheaper fare in order to boost up the users.

6. Conclusion

According to finding and result of research conducted in Karawang Region, from 5 selected variables that are by time travel, fare level, percentage spending on transport, waiting time, and availability of alternatives mode representing of mobility, equity, affordability, efficiency and accessibility, only 3 variables has significant impact to society welfare. The most substantial transportation variables for society wellbeing in Karawang are mobility represented by time travel, and then the time travel characteristics is same at urban or rural area in Karawang. Moreover, equity and affordability measured by spending on transport and fare level also influence significantly society wellbeing especially in urban area. In conclusion, time travel, spending on transport and fare level can be used to assess efficiency on transportation for sustainability in economic and social perspective in Karawang.
Sustainability of transportation according to time travel, spending on transport and fare level bring significant impact to society wellbeing. It has been proved using hypothesis test in chapter 4, and then it also supported by direction hypothesis. The directional hypothesis describes that they have negative correlation or monotonic decreasing to social welfare. Negative correlation shows if time travel, spending on transportation and fare level get lower, then society wellbeing will tend to increase symbolizing the social welfare gets better. As a result, in order to improve society wellbeing in Karawang, the benchmarks for time travel, spending on transport and fare are the lesser time travel, spending on transport and fare are better.

In addition, this survey may represent existing commutes characteristics based on selected sustainable transport indicator in Karawang. Due to the data is not well-distributed and the researcher check whether any differences between student versus non-student and urban versus rural area, the researcher will split the description if there is significant difference. The commutes characteristic will explain using average median and IQR range to generalize the performance.

7. Acknowledgement

The authors would like to gratitude Universitas Indonesia through PITTA Grant No. 743/UN2.R3.1/HKP.05.00/2017 for conducting this research and disseminating research results and the Center for Sustainable Infrastructure Development for the in-kind contribution.

8. References

[1] The Organisation for Economic Co-operation and Development 2013 *Structural Policy Country Notes Indonesia*

IOP Conf. Series: Earth and Environmental Science **105** (2017) 012008 doi:10.1088/1755-1315/105/1/012008

[2] Asian Development Bank 2012 *Transport Sector Assessment, Strategy, and Road Map.* Manila

[3] Karawang Regency. 2016. *Rencana Pembangunan Jangka Menengah Daerah*. Karawang. Retrieved from http://karawangkab.go.id/sites/default/files/pdf/Perda Nomor 8 Tahun 2016.pdf

[4] Paul D, Tessa P, Andy D, Melanie K, David P, Aki T and Mat W 2006 *Research on The Relationship Between Well-Being and Sustainable Development* 1–52

[5] Yannis T and Constantinos A 2008 Public transit user satisfaction: Variability and policy implications *Transp. Policy* **15** 260–272

[6] Margareta F and Markus F 2009 Service supply and customer satisfaction in public transportation: The Quality Paradox *Journal of Public Transportation* **12** 57–69

[7] Dick E, Tommy G, Lars E O, Margareta F and Sjef M 2013 The road to happiness: Measuring Dutch car drivers' satisfaction with travel *Transp. Policy* **27** 171–178

[8] Lars E O, Tommy G, Dick E, Margareta F and Satoshi F 2013 Happiness and satisfaction with work commute *Social Indicators Research* **111** 255–63

[9] Henrik G 2001 Indicators and performance measures for Transportation, Environment and Sustainability in North America *Research Notes from NERI* **148**

[10] Agne D-A, A Perujo, J Jesinghaus and P Jensen 2007 Indicators to assess sustainability of transport activities *JRC Scientific and Technical Reports* **54971**

[11] Todd L 2007 Developing indicators for comprehensive and sustainable transport planning *Transportation Research Record: Journal of the Transportation Research Board* **2017** 10–15.

[12] Gallup-Heathways Global Well-Being Index 2014 *2014 Country Well-Being Rankings Report* Retrieved from http://www.well-beingindex.com/2014-global-report

IOP Conf. Series: Earth and Environmental Science **105** (2017) 012009 doi:10.1088/1755-1315/105/1/012009

Effect of anode depth and conductivity of electrolyte solution on energy consumption in plasma electrolysis

T Sukreni, N Saksono and S Bismo[1]
[1]Chemical Engineering Department, Faculty of Engineering, Universitas Indonesia, Depok 16424, Indonesia

E-mail: setijo.bismo@ui.ac.id

Abstract. This research was conducted to investigate the effect of anode depth and conductivity of electrolyte solution on required energy for plasma generation in plasma electrolysis. The experimental parameters were composed of different depth of anodes (0, 15, 25, 45, and 65 mm) and various solution conductivity. The difference in electrolyte solution conductivities is presented by different concentration of Na_2SO_4 (0.01, 0.02, 0.03, and 0.05 M). A batch reactor (diameter 130 mm, height 190 mm) with tungsten electrodes (cathode diameter 6 mm and anode diameter 0.5 mm) was used by applying a continuous cooling system. Current was observed in various voltage (40 – 400 V) for 30 second in each voltage. Energy consumption for vapor formation and discharge plasma found higher at deeper position of anode. At higher conductivity of electrolyte solution, although energy consumption for vapor formation was observed lower, energy consumption for plasma discharged was found higher.

1. Introduction

At a sufficiently high voltage, conventional electrolysis changes spontaneously into plasma electrolysis [1] [2]. The process of plasma formation begins with a normal (conventional) electrolysis characterized by the formation of a gas bubble on the electrode. At this stage the voltage increase directly proportional to the increasing current. After the breakdown voltage, gas bubbles containing hydrogen, oxygen and water vapor form a gas layer that encloses the electrode with the smallest surface area [3]. The phenomenon of the formation of gas sheath can be illustrated using the current and voltage characteristic curves during the plasma electrolysis process [4] where the value of voltage and breakdown current (V_B and I_B) and voltage and discharge current (V_D and I_D) can be determined by using the characteristic curves.

The effectiveness of a process can be seen from the amount of energy consumption during plasma electrolysis. During the process of plasma electrolysis, the formation of the sheath caused by joule heating requires the greatest energy consumption [5]. Some variables such as the anode depth and the conductivity of the solution affect the amount of energy required, such as energy breakdown (E_B) and discharge energy (E_D), during the plasma formation process. Previous studies have shown that the value of E_B was greater as the contact surface of the anode is larger. The immersed anode can cause the energy required for the formation of the sheath around the anode is larger. Additionally, hydrostatic pressure in water causes the difficulty of forming a gas layer on the anode so that a larger

E_B is required [6]. In this experiment the calculation of energy consumption will be carried out with the same anode surface area at the depth variation so that the energy changes that occur are only functions of the anode depth position.

The conductivity of the solution also affects the value of V_B and I_D [7]. The increase in the conductivity of the solution can decrease the V_D [1] and increase the I_D, but does not affect the E_D value [5]. The results of the research have shown that the value of E_B and E_D is lower with increasing concentration of solution [8]. While in experimental results conducted by Ahmed et al. [7] showed that the value of E_B will decrease with increasing the conductivity of the solution, but the value of E_D will increase with increasing the conductivity of the solution. In this research, the calculation of energy consumption is also performed on various conductivity where the conductivity changes with the addition of Na_2SO_4 at various concentrations.

2. Materials and Methods

The circuit of plasma electrolysis reactor can be seen in figure 1. The plasma electrolysis reactor is made of transparent tube with tungsten electrodes, 6 mm diameter of cylindrical cathode and 0.5 mm diameter of anode which is coated by glass (quartz). The voltage source at the reactor electrode comes from the PLN's voltage-controlled electricity using a slide regulator, transformer, and bridge diode. Then the flowing current is observed with multi-meter. Tests conducted to determine the energy consumption of plasma formation in the anode depth position variable and the conductivity of the solution. The change in the conductivity of the electrolyte solution was made by mixing Na_2SO_4 at a concentration of 0.01 M; 0.02 M; 0.03 M; 0.05M into the distilled water. While the variation of anode depth position is at 5 mm, 15 mm, 25 mm, 45 mm and 65 mm.

Measurements are made to the voltage, current, time, temperature and conductivity of the solution from the beginning of the electrolysis to the establishment of stable plasma electrolysis. Based on the data obtained, a curve of voltage-current characteristics is used to find out the values of V_B, I_B, V_D and I_D at various anode depths and the conductivity of the solution and also the amount of energy during the plasma electrolysis process.

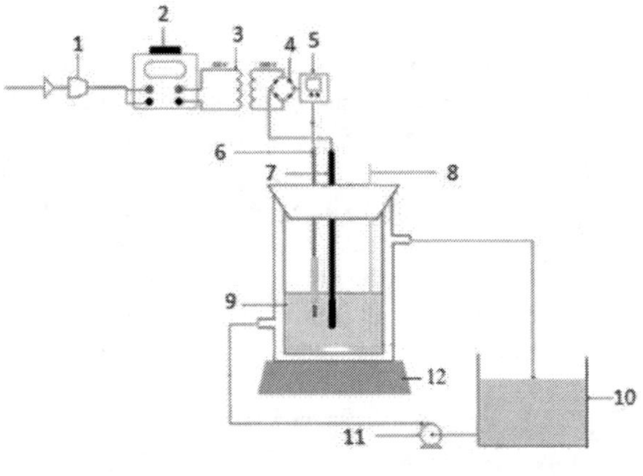

1. MCB	4. Diode bridge	7. Katode	10. Cooler water
2. Slide regulator	5. Multi-meter	8. Thermometer	11. Pump
3. transformator	6. Anode	9. Electrolyte Solution	12. Magnetic stirrer

Figure 1. The circuit schematic of plasma electrolysis reactor.

3. Results and Discussion

3.1. I - V characteristic curve on plasma electrolysis

From the experimental results there is a relationship between the current (I) and the voltage (V) as shown in figure 2. The I-V curve has a characteristic during the plasma electrolysis of which the image can be identified for several zones:

- Ohmic zone: $0 < V < V_B$; the current increases linearly with the added voltage applied in accordance with the ohm law. In this section there is conventional electrolysis in water characterized by gas bubbles around the anode. V_B is referred to as breakdown voltage [9]. The absence of a current line at the beginning of the curve at a very low voltage (0 - 40 V) indicating the absence of current flowing from the electrolysis process that has not yet taken place.
- Transition zone: $V_B < V < V_D$; the current begins to decrease with increasing voltage due to the formation of gas sheets around the anode, which occurs the process of forming and breaking gas bubbles that occur oscillations in the current. When a gas layer is formed on the anode, the current cannot rise again and the current will begin to decrease as the anode and the solution do not come into contact anymore. V_D is called the discharge voltage.
- Plasma zone: $V > V_D$; the current begins to increase as the voltage increases and the light emitted becomes lighter. The colors emitted correspond to the metal ions in the electrolyte solution [2].

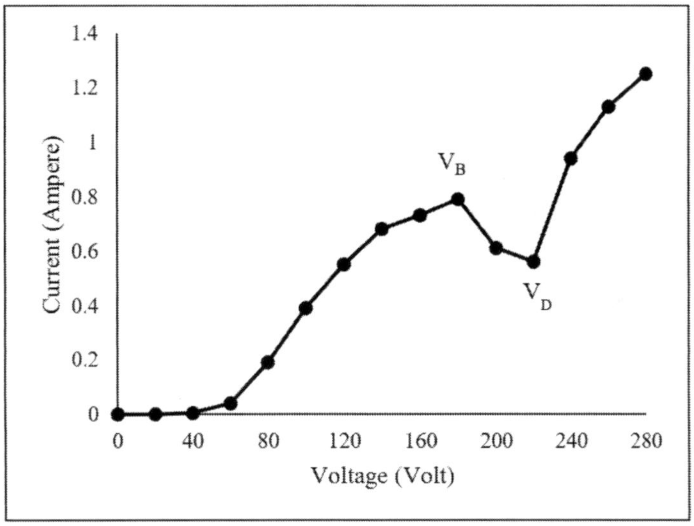

Figure 2. The voltage-current characteristic curve in the 0.02 M Na_2SO_4 electrolyte solution at a depth of 15 mm and a temperature of 65 °C.

3.2. Effect of parameter changes

At variuos experimental parameter it is found that the values of V_B, I_B, V_D and I_D are influenced by the anode depth position and the conductivity of the electrolyte solution, which also affect the energy consumption during the formation of the plasma shell.

*3.2.1. Effect of anode depth position.*The deeper the anode, the higher the energy required to begin the formation of the gas shell (E_B), as shown in Figure 3. Several previous studies have shown similar results [4] [8] despite differences in the extent of contacts of dyed anodes. In the previous experiment the area of anode contact increased with the deepening of the dyed anode causing the energy consumption for the formation of the envelope to become larger. While in this experiment the area of contact is made fixed by mounting the glass sheath on the anode. Thus the increase in energy occurs in the addition of depth occurs due to the growing hydrostatic pressure. The deeper the anode position, the greater the hydrostatic pressure, the higher energy (E_B) required to form the plasma shell.

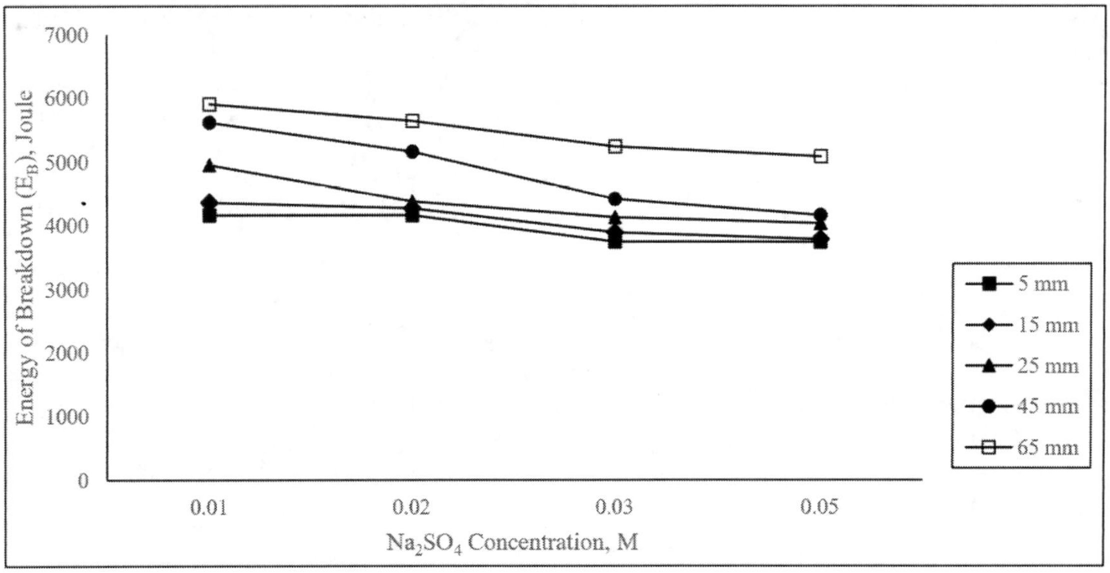

Figure 3. Energy consumption at breakdown voltage (E_B) at temperature 65°C.

3.2.2. Effects of conductivity of electrolyte solution.
Table below shows the values of breakdown voltages and currents (V_B and I_B) as well as the voltage and current discharge (V_D and I_D) on various conductivities using the Na_2SO_4 concentration variations. Changes in the conductivity of the solution affect the breakdown voltage (V_B) on the plasma formation in solution. Figure 4 shows that the higher the conductivity, the V_B required for the formation of the gas shell is lower. At higher conductivity, conduction current increases, ohmic heating increases, requiring a lower breakdown voltage [7]. Conversely, at low conductivity, the resistance in the solution becomes higher so that the conduction current decreases, the ohmic heating decreases, requiring a higher breakdown voltage.

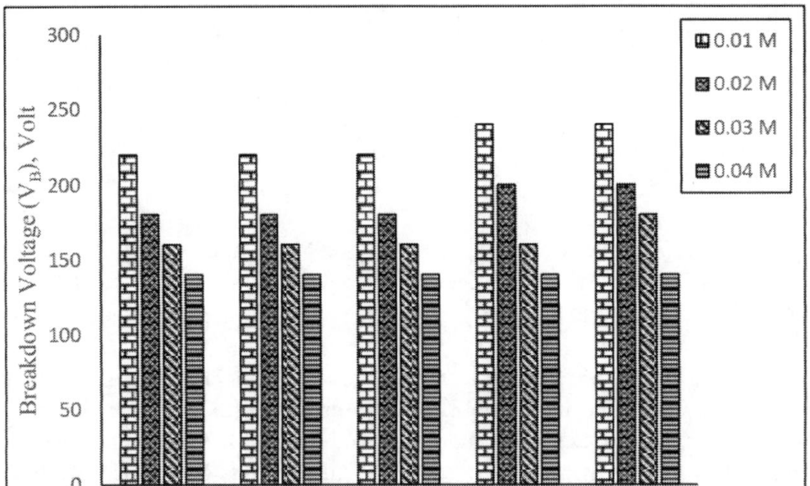

Figure 4. Breakdown voltage (V_B) on the anode depth variation and the conductivity of the solution at 65°C.

Table 1. Effect of conductivity change on V_B, V_D, I_B and I_D values at 25 mm anode depth and temperature 65°C.

Concentration of Na_2SO_4 (M)	Conductivity (mS)	V_B (Volt)	I_B (A)	V_D (Volt)	I_D (A)
0.01	3.48	220	0.75	280	0.46
0.02	7.01	180	0.81	220	0.59
0.03	10.67	160	0.86	200	0.65
0.05	15.13	140	0.96	160	0.83

Changes in conductivity also affect the discharge voltage (V_D) where higher conductivity, discharge occurs at lower voltages (figure 5). However, although the discharge voltage decreases, the discharge energy in plasma formation will increase as the conductivity of the solution increases (figure 6). This is due to the resistivity between the anode and the anolyte. In addition, the high dielectric strength of the solution causes the discharge strength to occur lower in the lower conductivity solution [7]. The higher the conductivity, the greater the power (discharge), which can be seen from the formation of the lighter plasma.

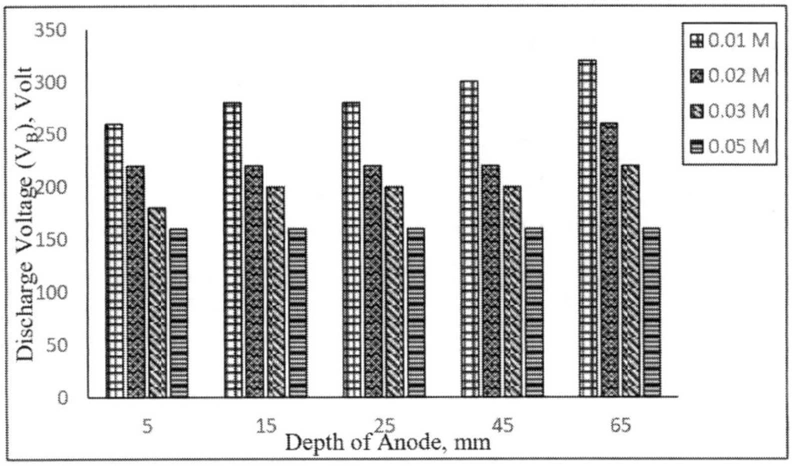

Figure 5. The discharge voltage (V_D) on the anode depth variation and the conductivity of the solution at a temperature of 65°C.

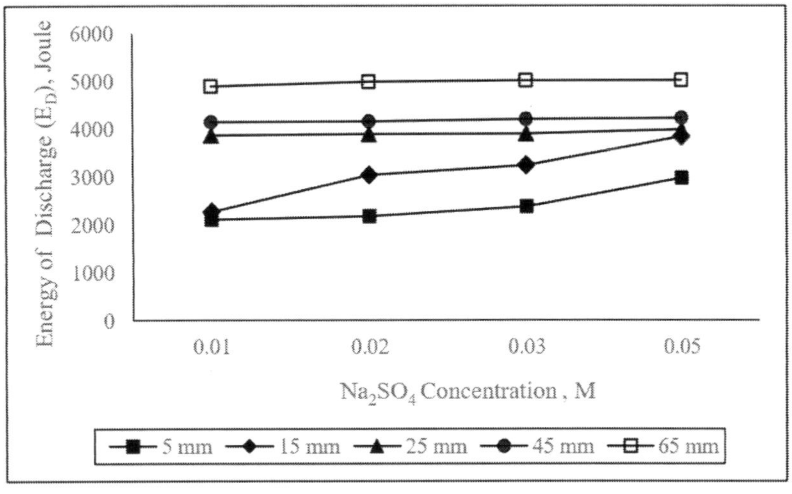

Figure 6. The discharge energy at the anode depth variation and the concentration of the solution at 65°C.

IOP Conf. Series: Earth and Environmental Science **105** (2017) 012009 doi:10.1088/1755-1315/105/1/012009

4. Conclusion

In this study, the effect of anode depth and conductivity of electrolyte solution on required energy for plasma generation in plasma electrolysis was studied. At deeper anode position, the energy required for plasma formation (E_B) and discharge (E_D) were observed higher. Changes in the conductivity of the solution also affect the breakdown and discharge energy on the plasma formation in solution. At large conductivity, the breakdown energy (E_B) was lower, while the discharge energy (E_D) becomes higher.

5. References

[1] Jin X L, Wang X Y, Zhang H M, Xia Q, Wei, D B and Yue J J 2010 influence of solution conductivity on contact glow discharge electrolysis *Plasma Chem. Plasma Process.* **30** 429

[2] Jin X, Wang X, Yue J, Cai Y and Zhang, H 2010 the effect of electrolyte constituents on contact glow discharge electrolysis. *Electrochim. Acta* **56** 925

[3] Sengupta S K and Singh O P 1991 Contact glow discharge electrolysis: a study of its onset and location *J. Electroanal. Chem. Interfacial electrochem.***301** 189

[4] Gupta S K S and Singh R 2016 cathodic contact glow discharge electrolysis: its origin and non-faradaic chemical effects *Plasma Sources Sci. Technol* **26** 15005

[5] Saito G, Nakasugi Y and Akiyama T 2015 generation of solution plasma over a large electrode surface area. *J. Applied Phys.* **118** 023303

[6] Saksono N, Seratri R T, Muthia R and Bismo S 2015 Phenol degradation in wastewater with a contact glow discharge electrolysis reactor using a sodium sulfate *I. J. Tech.* **7** 1153

[7] Ahmed M W, Suresh R, Yang J K, Choi S and Lee H J 2016 effect of water conductivity on the generation of OH radicals in high frequency underwater capillary discharge *Int. J. Renew. Energy Environ. Eng.* **4** 28

[8] Saksono N, Nugraha I, Ibrahim and Febiyanti I A 2015 Hydroxyl radical production on contact glow discharge electrolysis for degradation of linear Alkylbenzene sulfonate *Environ. Prog. Sustain. Energy* **35** 962

[9] Gangal U, Srivastava M and Gupta S K S 2009 mechanism of the breakdown of normal electrolysis and the transition to contact glow discharge electrolysis *J. Electrochem Soc* **156** F131

IOP Conf. Series: Earth and Environmental Science **105** (2017) 012010 doi:10.1088/1755-1315/105/1/012010

Design optimization of hybrid biomass and wind turbine for minapolitan cluster in Domas, Serang, Banten, Indonesia

Erwin Erwin[1,2], Tresna P Soemardi[1], Adi Surjosatyo[1], Julianto Nugroho[1], Kurnia Nugraha[2], Slamet Wiyono[2]

[1] Mechanical Engineering, University of Indonesia, Depok, Indonesia.
[2] Mechanical Engineering, Sultan Ageng Tirtayasa University, Cilegon, Indonesia.

E-mail: erwin@untirta.ac.id

Abstract. This research aims is to design a renewable energy power plant based on the renewable energy potential in Domas village. Domas is a coastline that has a wind speed is has the potential to develop a wind power plant with enough capacity that can be utilized economically; in addition Domas is also a centre of rice and rice husks producer. Economic activity as a farmers and milkfish breeders and this area is set to be minapolitan area of Serang district. The problem is many of the ponds is placed far from electricity, so it requires another alternative as a source of electrical. Aquaculture requires aeration and lighting to increase fish and shrimp production, to keep the oxygen content dissolved in water. Lighting are required starting at 6 pm to 6 am, and the aerators based on the shrimp and fish life cycle are required to be switched on from 5 pm to 6 am. Wind data is obtained from Homer, meteonorm and NOAA databases, which is then processed in Homer for optimization. Potential data of husk is obtained from ministry of energy and mineral resources of the republic of Indonesia (EMR) database for Serang regency in 2016. For 1 hectare of land with 1.5 tons of fish, total electricity needed 2.4 kWh per day for lighting and 13 kWh per-day for aerator, with peak of 0.385 kW and 1.9 kW. By using the Aeolos 500 wind turbine as non-dispatchable resource and biomass generator as a dispatchable resource, a simulation is performed to get the system that produces the lowest LCOE (Levelized Cost of Electricity) and highest renewable energy (REN) factor for wind turbine. The recommended system is 2 kW Biomass Generator, 10 kW Inverter, 10 kW Rectifier, 24 Battery Trojan L18P 360 Ah capacity with 8 units of Aeolos 500 array. With economic LCOE $0.445 / kWh, and 82.2 % REN factor for wind turbine.

1. Introduction

Sources of renewable energy have become an alternative to the supply of electric energy for remote areas. Domas village located in the Serang district within no more than 20 km from Serang city center. The village is near on Java sea beach . Domas village is a Minapolitan Village with milky fish and shrimp as a main aquaculture. This study aims to design and provide a hybrid system of renewable energy plant to powering the fish and shrimp pond, alternative energy sources that will be used is biomass and wind using local potential characteristic.

Recently, research on the fluctuations in energy generated by wind turbines is highly dependent on the intensity of wind turbulence [1] research integration of different renewable energy sources has been done to improve the sustainability of the supply of electric energy [2-3]. The problem that arises is how to manage this energy source [4], optimization [6-7] and determine the size of each generator [8] for efficient and optimum economical and technical value. Research on hybrid biogass and wind turbine simulations based on real data shows more reliable energy production to produce stabilized at 40 MW in Spain [9]. HOMER software that is issued by NREL was become one alternative in the initial study to conduct a feasibility study [5] the integration of renewable energy sources. Sultan wind turbine is design to meet the specific wind potential behaviour in Serang district [11] and in the future research will used as a main wind turbine in this system. In term of gasification, using sand to prevent agglomeration in gasification with coconut shell and wood pellets [12]. using rice husk ash mass effect to sustain pyrolysis process [13]. The development of job creation formation model in renewable energy sector is offered because of the magnitude of renewable energy potential in Indonesia [19]. CFDs are used to analyse and optimize the integrated wind turbine on buildings [20]. This Research focus on developing the optimized of dispatch-able wind power and non-dispatch-able biomass hybrid plant system, that can electrified the aerator and others need in milkfish pond.

2. Method

Hybrid system is more preferred over a single system due to a more stable performance along a year [16]. This research firstly did an investigating of potential renewable energy by collecting data from secondary data such as from Meteonorm database, meteorology station from NOAA website, Homer database for wind speed and ESDM database for biomass potential. This data will be used as an input in optimization designing renewable energy plant using Homer beta 2 Software.

2.1 Wind potential investigation

Annual average wind speed in Domas according to meteonorm 3.18 m/s, also lower than Garissa Kenya [17]. In this research wind data also collected from homer database. Wind velocity data from local meteorological stations are taken in the NOAA database, at an average speed of 3 m/s, in addition to wind speed data also taken from the meteonorm database and the homer database. The village of Domas is located at -5.990 N 106.254 E, in the Meteonorm database has the potential for sunlight and wind throughout the year as shown below.

Table 1. Data of wind potential in Domas village average monthly

Month	Meteonorm	NOAA
January	3.6	2.225806
Febuary	2.6	2.243333
March	3.1	2.080645
April	2.6	1.656667
Mei	2.6	1.877419
June	3.1	1.7
July	3.1	1.682759
August	3.6	1.432258
September	3.6	1.446667
October	3.1	1.325806
November	3.6	1.34
December	3.6	1.43871

2.2. Biomass potential investigation

From ministry of energy and mineral resources, in Serang city, rice husk is a biggest potential of biomass. Domas village with about 50 families, and every family, average have 1 hectare of paddy field, and produce rice 6.15 tons rice per hectare for four times a year, and produce about rice husk 28 tons per year, means every day about 70kg of rice husk for every family. So for biomass have potentially 350 kg daily. In this case biomass will be process to produce biofuel using gasification with max capacity 30 HP, with feeding speed, 30 kg rice husk per hour.

The biomass potential obtained from ministry of energy and mineral resources of the republic of Indonesia (*EMR*) for 2017, from the table indicates considerable potential and is technically entirely unused, potentially equivalent to a medium-sized generator.

Table 2. Biomass potential data obtained from EMR report for Serang Banten area

District	Sources	Industrial Type	Potential (MWe)			Used	
			Global	Technical	Optimized	On grid	Off grid
Serang	Farming	Husk Farming	67.86	4.95	0	0	0
	Municipal Garbage	Municipal Garbage Potential	12.2	0	0	0	0

2.3. Load calculation

To obtain data real electrical energy requirements for a pond then needs to be calculated on the equipment needed and electrical needs, such as aeration tools and magnitude, lighting and other logistic needs.

A pond needs lights for lighting and aerators to keep oxygen levels in the water. Lighting and home appliance are required starting at 6 pm to 6 am, and aerators based on the shrimp and fish life cycle are required to be switched on from 5 pm to 6 am.

Ordinary farmers have up to 2 hectares of ponds, intensive ponds have a distribution capacity of up to 5 tails per m2, and it takes 3 waterwheels per plot (0.5 Ha) the load calculation will use a 2 hectare base with 4 pool plots. Overall average electricity required 2.4 KWh per day for lighting and 13 KWh per day for aerator, with peak of 385 watt and 1.9 Kw respectively.

This system is designed for the whole year where the renewable energy potential condition changes by the time and the energy requirement for the pond changes as needed so that the fish grow optimally, and it is expected that the built system will be able to function throughout the year.

After calculating the duration of the aerator during the life cycle of the fish, and the need for lighting of the pond and home appliance, it can be concluded that the required load is like the figure 3, below.

3. Result

3.1. Plant setup

Because of the unstable wind potential and biomass potential, gasification provides a supply of electricity that can support fluctuating electricity to meet demand load. This is also to reduce the carbon footprint in the plant. In this research, wind energy as a non-dispatchable resource, and gasification serves as a dispatch-able resource. Figure 1 show the system of plant setup for simulation.

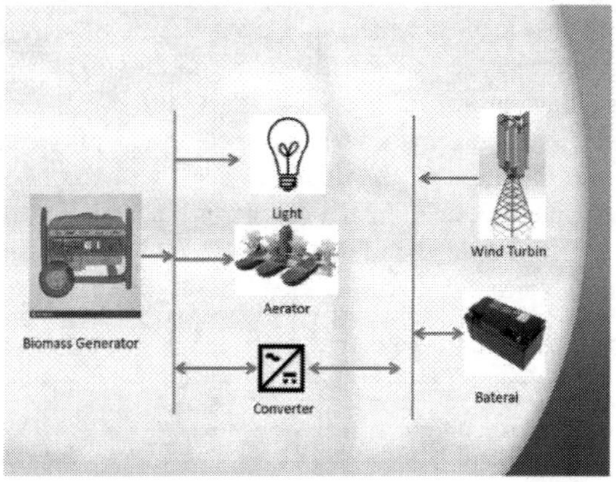

Figure 1. Plant set up in Homer

3.2. *Analysis*

The use of several windmill with specifications of the same capacity to see some of the possibilities of windmill configuration, setup fees, maintenance that eventually affect the LCOE, here there are 2 types of windmills that will be discussed. From result of simulation with homer got result as follows.

From simulation for optimization has 200 working combination of system for each type of wind turbine, and below is table 3 and table 4 for each 5 best combinations.

Table 3. Best combination for Aeolos 500

No	A 500	Label (kW)	L16 P	Converter (kW)	Initial capital	Operating cost ($/yr)	Total NPC	COE ($/kWh)	Renewable fraction	Diesel (L)	Label (hrs)
1	9	2	24	10	$9,900	1,721	$31,894	0.447	0.86	430	705
2	9	2	22	10	$9,700	1,740	$31,948	0.448	0.85	460	770
3	8	2	26	10	$9,600	1,751	$31,984	0.449	0.83	488	786
4	9	2	26	10	$10,100	1,716	$32,031	0.449	0.87	408	657
5	10	2	22	10	$10,200	1,711	$32,074	0.45	0.89	388	663

3.2.1. *Aeolos 500 wind turbine (wind 1)*

Providing the largest power supply compared to other configurations, with the number of turbines 9 units.

Table 4. Best 10 combination for Sunnily

No	Sunily	Label (kW)	L16P	Converter (kW)	Initial capital	Operating cost ($/yr)	Total NPC	COE ($/kWh)	Renewable fraction	Diesel (L)	Label (hrs)
1	12	2	24	10	$11,400	2,393	$41,986	0.589	0.62	931	1,464
2	4	2	5	10	$5,500	2,858	$42,032	0.59	0.22	2,024	4,168
3	12	2	28	10	$11,800	2,372	$42,127	0.591	0.63	882	1,374
4	12	2	22	10	$11,200	2,421	$42,154	0.591	0.62	966	1,548
5	12	2	26	10	$11,600	2,392	$42,181	0.592	0.63	919	1,436

Deficiency, the relative price per unit is expensive compared to other wind turbines and is rare to find in market. 8 units arranged in this 3 x 3 matrix requires an area of 45.9 m x 45.9 m. Providing the largest power supply compared to other configurations, with the number of turbines 8 units.

3.2.2. Sunily wind turbine (wind 2)
Providing a large power supply with a total of 12 units of turbines arranged in a 4 x 3 matrix requires an area of 51 m x 74 m. Due to a number of units more than wind 1 then LCOE (*Levelized Cost Of Electricity*) system is higher than the system that uses wind 1.
3.3 Recommended design
From the result of optimization by Homer, the system that can be recommended is shown in table 5 below.

Table 5. Configuration of renewable energy plant

Renewable Configuration	wind 2	Wind 1
Wind turbine	12 Units	9 Units
Generator 1	2 kW	2 kW
Battery	24 Trojan L16P	24 Trojan L16P
Inverter	10 kW	10 kW
Rectifier	10 kW	10 kW

Table 6. Economic result of renewable energy plant

	Wind 2	Wind 1
Total net present cost	$41,986	$31,894
Levelized cost of energy	$ 0.589/kWh	$ 0.447/kWh
Operating cost	$ 2,393/yr	$ 1,721/yr
REN Factor	0.62	0.86

The cost and price of electricity that arise from the sistem and renewable energy factor for wind turbines shown in table 6.

Figure 2 shows that in some months from August to January the REN factor of the Aeolos 500 wind turbine exceeds 90%. In figure 3, in the months from August to January the REN factor of the Sunily wind turbine is only about 80%.

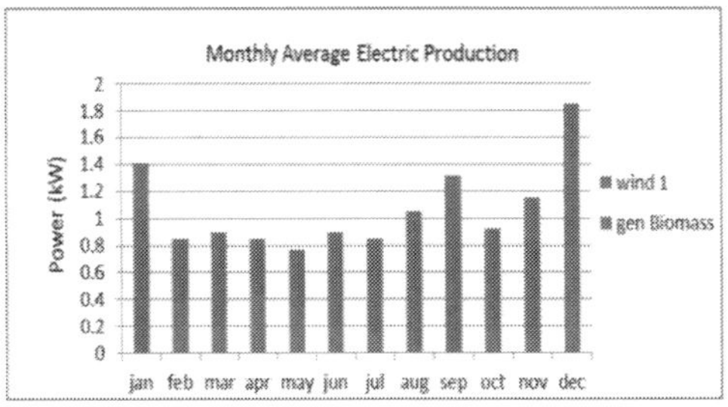

a.

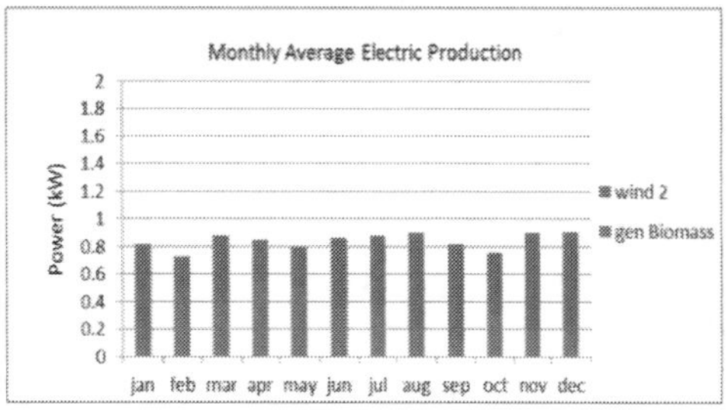

b.

Figure 2. Comparison of energy generated by wind 1 (a) and wind 2 (b) with biogass generator

From the two types of systems with Aeolos 500 (wind 1) yield higher renewable energy factor and lower cost of electricity than systems using Sunily (wind 2).

4. Discussion

Optimization of solar power system has been done in Sleman Java, for aeration of fish pond and managed to get the appropriate system [21]. Evaluation of hybrid plants between solar and wind turbines has also been done for aeration ponds in Thailand [22]. Departing from the success of previous research is the author tries to make a hybrid generator between biogas and windmills for milkfish pond in Domas.

Comparison of several types of wind turbines and REN factor produced indicate that wind turbines in accordance with wind conditions will result in higher renewable energy factor [23]. In this study showed that the Aeolos 500 wind turbines is more suited and more optimum result to wind conditions at Domas, than the Sunily wind turbines. The two best design is feasible according to economic for area isolated from electricity such as milkfish pod in Domas.

For a capacity of 40 MW wind park hybrid with biomass gasification, Pérez-Navarro offers the concept of biomass gasification using syngas storage so that if the combustion process in the gasification reactor produces excessive syngas it can be stored, which can then be used when load needed is increased [9].

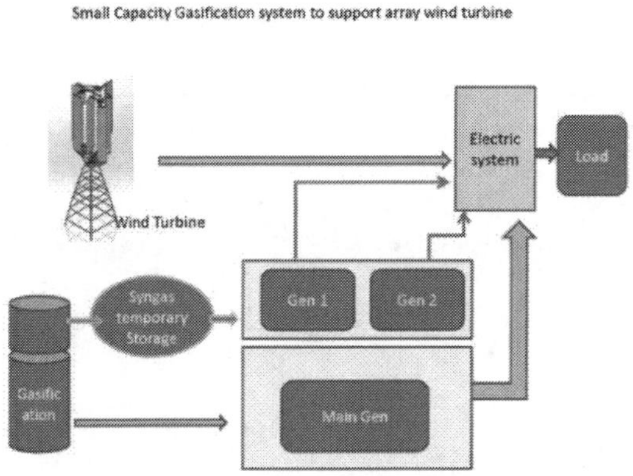

Figure 3. Proposed hybrid system of biomass gasification generator and array Wind Turbine

5. Conclusion

From the study that has been done, the recommended system of new renewable energy generation wind power using several wind turbines are feasible to build and have advantages and disadvantages of each. For overall electricity demand is 2.4 KWh per day for lighting and 13 KWh per day for aerators, with peak of 385watt and 1.9KW respectively.

It can be seen that *Aeolos* 500 in economic analysis, is cheaper and have higher the production rate of electricity than *Sunily* because of to the different power curve owned by each wind turbine. The more characteristic of the power curve at low wind velocity will be higher REN Factor value and will decrease the LCOE (*Levelled Cost Of Electricity*).

REFERENCES

[1] Rosen, A. and Y. Sheinman (1996). "The power fluctuations of a wind turbine." Journal of Wind Engineering and Industrial Aerodynamics **59(1)** 51-68.

[2] Schmidt J, Cancella R, Pereira Jr AO., 2016. An optimal mix of solar PV, wind & hydro power for low-carbon electricity supply in Brazil. *Renewable Energy*. **85** 137

[3] Bhandari B, Lee K-T, Lee CS, Song C-K, Maskey RK, Ahn S-H., 2014. A novel off-grid hybrid power system comprised of solar photovoltaic, wind, and hydro energy sources. *Applied Energy*. **133** 236

[4] Cau G, Cocco D, Petrollese M, Knudsen Kær S, Milan C., 2014. Energy management strategy based on short-term generation scheduling for a renewable microgrid using a hydrogen storage system. *Energy Conversion and Management*. **87** 820

[5] Gang L, Rasul MG, Amanullah MTO, Khan MMK. 2011. Feasibility Study of Stand-Alone PV-Wind-Biomass Hybrid Energy System in Australia. *Power and Energy Engineering Conference* (APPEEC), Asia-Pacific. 1

[6] Osmani A, Zhang J., 2014. Optimal grid design and logistic planning for wind and biomass based renewable electricity supply chains under uncertainties. *Energy*. **70** 514

[7] Schmidt J, Cancella R, Pereira Jr AO., 2016. An optimal mix of solar PV, wind and hydro power for a low-carbon electricity supply in Brazil. *Renewable Energy*. **85** 137

[8] Kolhe ML, Ranaweera KMIU, Gunawardana AGBS., 2015. Techno-economic sizing of off-grid hybrid renewable energy system for rural electrification in Sri Lanka. *Sustainable Energy Technologies and Assessments*. **11** 53

[9] Pérez-Navarro, A., D. Alfonso, et al. (2010). Hybrid biomass-wind power plant for reliable energy generation. *Renewable Energy* **35** 1436

[10] Bhattacharjee S, Acharya S., 2015. PV–wind hybrid power option for a low wind topography. *Energy Conversion and Management*. **89** 942

[11] Erwin, Erny Listijorini, Rina Lusiani, Tresna P Soemardi., 2015. Development of the third darrieus blade of Sultan Wind Turbine for low wind speed. *Applied Mechanics and Materials*. **758** 7

[12] Surjosatyo, A, Muammara, et al, 2017. Prevention Study of Sand Agglomeration on Fluidised Bed Combustor with Co-Combustion Method. *Chemical Engineering Transactions*, **56** 1273

[13] Surjosatyo, A., I. Haq, et al. 2017. "Effect of rice husk ash mass on sustainability pyrolysis zone of fixed bed downdraft gasifier with capacity of 10 kg/hour." *AIP Conference Proceedings* 1826(1): 020009.

[14] Kabalci E. 2013. Design and analysis of a hybrid renewable energy plant with solar and wind power. *Energy Conversion and Management*. **72** 51

[15] Veldhuis AJ, Reinders AHME. 2015.Reviewing the potential and cost-effectiveness of off-grid PV systems in Indonesia on a provincial level. *Renewable and Sustainable Energy Reviews*. **52** 757

[16] Bekele, G. and G. Boneya 2012. "Design of a Photovoltaic-Wind Hybrid Power Generation System for Ethiopian Remote Area." *Energy Procedia* **14**, 1760

[17] Ghaem Sigarchian, S., R. Paleta, et al. 2015. "Feasibility study of using a biogas engine as backup in a decentralized hybrid (PV/wind/battery) power generation system – Case study Kenya. *Energy*. **90** 1830

[18] Meteonorm (2015). Global Meteorological Database for engineers, planners, and education. Rosen, A. and Y. Sheinman 1996. "The power fluctuations of a wind turbine." *Journal of Wind Engineering and Industrial Aerodynamics*. **59** 51

[19] ELFANI, Maria. The Impact of Renewable Energy on Employment in Indonesia. *International Journal of Technology 2011*, [S.l.] **2** 47-55

[20] CHO, Kang-Pyo; JEONG, Seung-Hwan; PERWITA SARI, Dany. Harvesting Wind Energy from Aerodynamic Design for Building Integrated Wind Turbines. *International Journal of Technology,* 2011 [S.1] **2** 189

[21] Prasetyaningsari, I., A. Setiawan, et al. (2013). "Design Optimization of Solar Powered Aeration System for Fish Pond in Sleman Regency, Yogyakarta by HOMER Software." Energy Procedia **32** 90

[22] Nookuea, W., P. E. Campana, et al. (2016). "Evaluation of Solar PV and Wind Alternatives for Self Renewable Energy Supply: Case Study of Shrimp Cultivation." Energy Procedia **88** 462

[23] Yi, Z. and S. Ula (2008). Comparison and evaluation of three main types of wind turbines. 2008 IEEE/PES Transmission and Distribution Conference and Exposition.

IOP Conf. Series: Earth and Environmental Science **105** (2017) 012011 doi:10.1088/1755-1315/105/1/012011

Experimental study of hybrid loop heat pipe using pump assistance for high heat flux system

Iwan Setyawan[1], **Nandy Putra**[1,2], **Imansyah Ibnu Hakim**[1]
[1]Applied Heat Transfer Research Group, Department of Mechanical Engineering, University of Indonesia, 16424 Depok, Jawa Barat, Indonesia

[2]E-mail: nandyputra@eng.ui.ac.id

Abstract. Loop heat pipe is a promising thermal control system, the same as a heat pipe, to realize the efficient release of heat from electric appliances. It is suitable for applications in electronics that have higher power density. Meanwhile, the heat pipe loop (LHP) can also be considered for adoption in the manufacture of solar hot water systems, due to its unique features such as effective thermal conductance and flexible design embodiments. However, the start-up problem of the LHP and transport distance and higher heat release capacity is still existed influencing the operating stability of the device. Base on the problem, this study focuses to carry out work on the developing a novel LHP system in order to provide a robust solution for significant enhance the ability energy transfer. This novel LHP is a conventional LHP that was modified by adding a diaphragm pump to accelerate the fluid transportation (called as hybrid loop heat pipe, HLHP). The pump is installed on the liquid line complete with a reservoir. It will work passively using the wick capillary pressure when there is no sign the occurrence of dry-out. In another hand, the pump was only activated when the evaporator temperature tends increased extremely because of the failure of start-up. The experimental result showed that installed diaphragm pump in LHP modified system was able to prevent the occurrence of dry-out and significantly reduced the evaporator temperature. This study will contribute to energy savings and the utilization of renewable energy.

1. Introduction

Two-phase cooling technology passive systems have been widely used and continue to be developed include heat pipes, loop heat pipe (LHP) and capillary pumped loop (CPL). They are very reliable and have very high thermal conductivity [1]. However the two-phase cooling technology passive system occasionally can no longer meet the cooling needs due to inherent limitations in terms of capillary pumping heat flux, the transport distance and the ability to handle multiple heat sources simultaneously. The above-mentioned problem solving is expected to be accomplished along with the successful testing of the use of mechanical pumps into the exhaust heat system of NASA's Rover Exploration Mars, which sparked the current technical trend with the increasing use of active cooling designs for aerospace applications [2]. This cooling system is known as a hybrid loop cooling system that combines pumping of the active mechanical fluid by pumping the capillary liquid. Thus, a hybrid two-phase loop (hybrid) combines high flux capabilities of mechanical pumps and strong pumping operations from conventional two-phase loops with the simplicity of capillary mechanisms[1, 3]. Moreover, hybrid systems, no longer limited to aerospace applications, but also have been investigated for other applications. Up to now, several experimental studies of hybrid loop cooling systems have been conducted. For example, Park et al [1, 2, 4-6] investigated a loop hybrid system capable of removing the heat of five sources, transient heat flux in the range of heat flux 188 W/cm2 to 250 W/ cm2 and conducted experiment of advanced technology hybrid two-phase loop for thermal management of electronics. Other researchers, Sarraf et al. [3] have demonstrated Hybrid Capillary Pumped Loop for high temperature electronics. Jiang et al. [7] investigated a two-phase loop called pump-assisted capillary loop phase change that is designed to overcome the weaknesses of the temperature oscillations and limited range of heat transfer in the loop heat pipe. Their research carries out the significant advance of the state-of-the-art in hybrid loop heat pipes.

Thus, it is clear that the heat pipe fails to function because the capillary wick is no longer able to restore the liquid to the evaporator, then the evaporator reaches the driest point that causes excess heat and eventually happen dry out [8]. Based on that problem on the heat pipes, hence in this study will be designed a loop heat pipe which is added the pump mechanism which is called hybrid loop heat pipe. The design is slightly different from the design of other researchers. Where in probe the heat pipe loop serves as passive cooling when the heat load and other incoming parameters do not trigger the dry-out and vice versa in case of dry out, the pump will be activated. Thus the heat pipes will be operate on two different conditions.

2. Methodology

2.1. Design and Manufacturing

LHP are used and designed similar to generally of the loop heat pipes. Evaporator of cylinder model was made of copper pipe to the length of 65mm, an inside diameter of 23mm and a wall thickness of 1.2 mm which merges with the capillary wick, having a thickness of 2mm that was made before. The condenser is also made of copper tubing with dimensions of 70 mm long, 23 mm diameter in counter flow model that has two coolant ports, in and out with diameter of 10 mm. For ensuring the availability of liquid will be pumped when necessary, then this design equipped with a reservoir made of pipes with dimensions of 23 mm in diameter and 65 mm length. In this design the liquid line pipe from the pump made separate directly into the evaporator. As for the total length of the vapor line and liguid line, both equal to 402.5 mm. The addition of a diaphragm pump is used to pump the liquid into the evaporator when needed. With the circulation by this hybrid system, will help the circulation of fluid work on the loop. The more optimal circulation can be given by this mechanism because the pumping system make the fluid circulation better and the speed of the fluid returns to the evaporator faster. The HLHP is displayed a simple configuration in figure 1, which consists of evaporator, condenser, reservoir and diaphragm pump.

Figure 2 shows the experimental setup for testing of the HLHP. In this experiment a heater as a heat source is applied and adjusted by using a DC power supply. Absorbed heat will then be transferred to the condenser through capillarity mechanism and subsequently removed in the condenser and then is absorbed by the water supplied from the Circulating Thermostatic Bath (CTB). In this experiment a micro controller is used to control the diaphragm pump. HLHP used 16 thermocouples, type K that were placed on the main parts. Heat was absorbed by the water coolant, which was at a constant temperature of 25°C and had a flow rate of 100 ml/min supplied from the Circulating Thermostatic Bath (CTB). Furthermore, measured data were recorded using a data acquisition system with NI cDAQ-9174 and NI-9213, which was subsequently processed using Lab VIEW software. The HLHP working pressure during the test was measured using a pressure sensor, PSA-C01, and the NI-9203. To analyze the thermal performance of the HLHP properly, the system had to be perfectly sealed with an insulating material to prevent heat loss to the surrounding environment[9]. In this experiment, is done by varying the heat applied to the

IOP Conf. Series: Earth and Environmental Science **105** (2017) 012011 doi:10.1088/1755-1315/105/1/012011

evaporator ie starting from 10W continued 20 W, 30 W, 50 W, 90 W, 140 W, 150 W, 160 W, 170 W, 180 W and 200 W. The thermal resistance R_{th} on the loop can be determined using:

$$R_{th} = \frac{T_e - T_c}{Q_{in}} \quad (1)$$

where, T_e is the evaporator temperature, T_c is the condenser temperature.

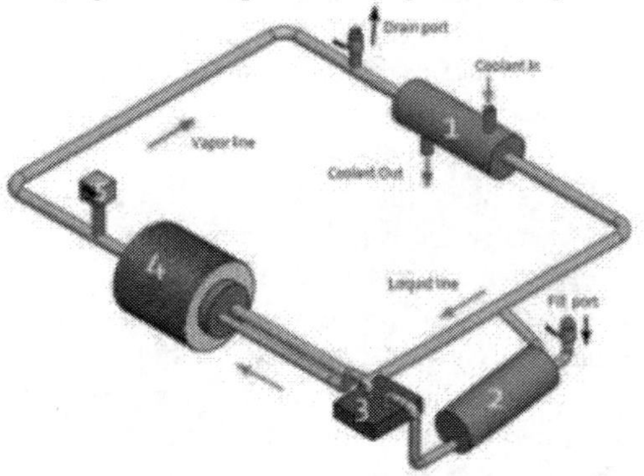

1 : Condenser
2 : Reservoir
3: Diaphragmpump
4 : Evaporator
5 : Pressure sensor

Figure 1. Design of Hybrid loop Heat pipe[9]

2.2. Test Setup

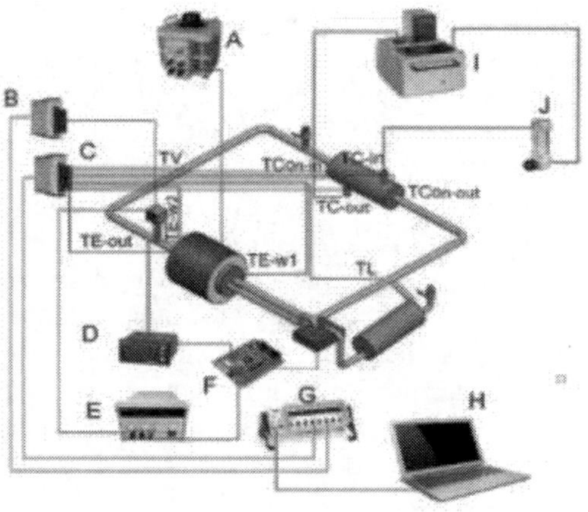

A: AC power supply
B: NI 9203 data acquisition module
C : NI 9213 data acquisition module
D: Temperature controller
E: DC power supply
F : Microcontroller
G: NI Cdaq-9174 chassis
H: Computer
I : Circulating Thermostatic Batch(CTB)
J: Flow meter
H: Computer

Figure 2. Schematic installation of hybrid loop heat pipe.

3. Result and discussion

Figure 3(a) showed temperature distribution of HLHP when the heat load of 10 W is applied to the evaporator. However, the start-up had not been reached. No forward flow in the system was identified,

IOP Conf. Series: Earth and Environmental Science **105** (2017) 012011 doi:10.1088/1755-1315/105/1/012011

where the temperature at the condenser inlet was much lower than the saturation temperature in the evaporator[10, 11]. According to some authors, which are based on their own experimental experience, a start-up at low heat loads is a problematic operating mode for LHP. The problem here is interrelated with the fact that at low heat loads a start-up does not take place i.e there is no circulation of the working fluid According to experimental experience by some researchers, start up on low heat load is a problematic mode of operation for LHP. Where here is related to the fact that start up does not occur, ie there is no working fluid circulation at low heat load[12]. Figure 3(b) showed the start-up of the HLHP at a heat load of 30 W. Because start-up was successfully achieved at a heat load of 30 W, it caused a decrease of the evaporator temperature. This temperature was even lower than for the heat load of 10W, where the start-up was not reached.

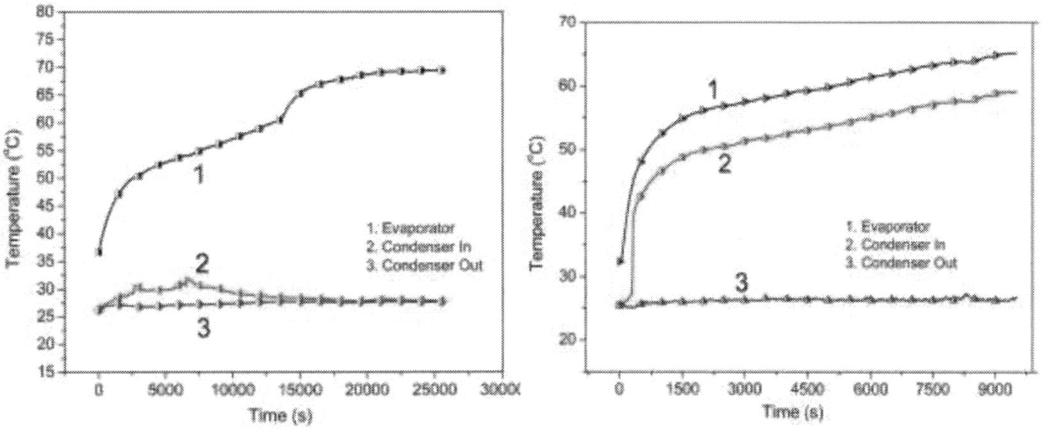

(a) **(b)**

Figure 3. Temperature distribution on the HLHP with heat load

(a). 10 W ; (b). 30 W

As already mentioned earlier that under 30W heat load, HLHP start-up successfully. Furthermore, the heat load was raised to 50W, 90W and 140W. From the test result, it found that 140W is the maximum heat load limit that can be handled by HLHP without the help of the pump. Fig 5. showed evaporator temperature and thermal resistance at various heat loads for working of HLHP passively. In the graph, it can be seen when the heat load 10W is raised to 20W, then the evaporator temperature increases. However, when the heat is raised to 30W, the evaporator temperature is precisely much decreased. Furthermore, by more increasing the heat load it will more increase the evaporator temperature. Similarly, when the heat load is increased to 30 W, the thermal resistance is greatly reduced. Figure 4 confirmed the previous discussion in figure 3, that when the 30W heat load is applied, the HLHP is successful start-up where the two-phase system heat transfer becomes dominant.

Figure 5 shows the results with a heat load of 150 W. Under this condition, more vapor formed in the evaporator chamber, will cause overheating on the surface of evaporator. Therefore, the system did not function properly. Heat was absorbed in the evaporator, and it seems that the condenser was no longer able to handle this heat load, as indicated by the temperature oscillation at the condenser outlet. This temperature oscillation is caused by the vapor carried by the working fluid. When the heat load is applied to the evaporator, the resulting vapor enters the condenser to condense which then blocks the flow inside the condenser [13]. Furthermore, the vapor temperature in the evaporator chamber was superheat caused the temperature drastically increased. This rise continued until the pump was activated when the temperature reached 157°C. At this point, the temperature difference in the evaporator wall and condenser inlet much increased. Actually, this is an early sign of dry-out because the receding liquid in the evaporator [14]. However, once the pump was turned on, the temperature of the evaporator decreased suddenly. Regarding to the pump activating would return successfully to the normal conditions i.e the system worked in two phase.

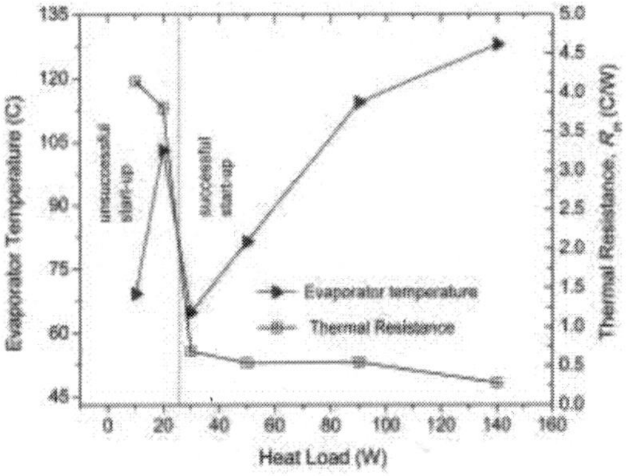

Figure 4. The evaporator temperature and thermal resistance versus heat load on the HLHP

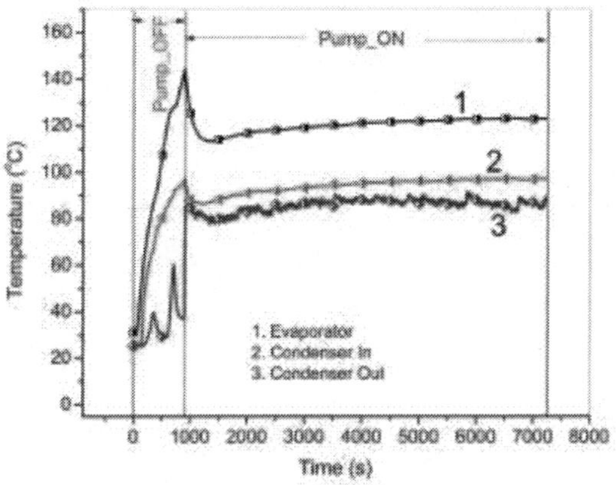

Figure 5. Temperature distribution on the HLHP with heat load 150W

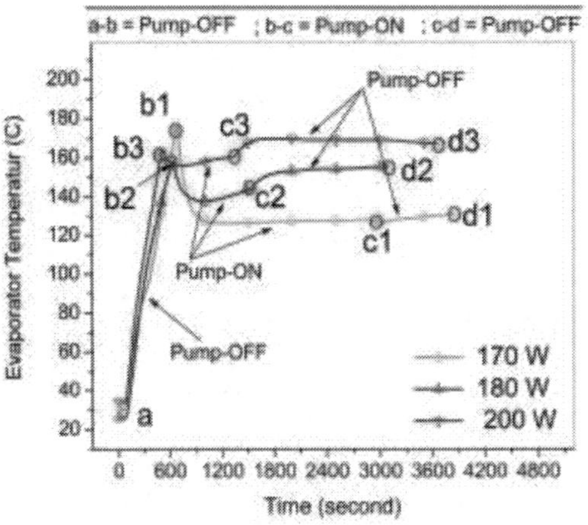

Figure 6. The comparison of the evaporator temperature on the HLHP when the pump deactivated and activated

Figure 6 shows the comparison between HLHP when the diaphragm pump is activated (pump ON) and inactive (pump OFF). In the graph, evaporator temperature can be seen at various heat loads applied to HLHP. Furthermore, it can be seen that with the help of the pump, will successfully prevent the occurrence of dry-out and reduce evaporator temperature significantly, especially for heat load 170W. As the heat load increases, the evaporator temperature will rise as well. Nevertheless, the pump remains successful in preventing dry out. Another result is very interesting from the results of this test is the pump turned out just as a trigger to restore HLHP work well.

4. Conclusion

Design, manufacturing, and testing of the hybrid loop heat pipe (HLHP) with the integration of the diaphragm pump on the system have been carried out. HLHP will run passively until 140 W without dry out occurance. However, when 150 W is applied to the evaporator, there are early signs the occurrence of dry-out. With the integration of the diaphragm pump in the system, HLHP will able to restore normal conditions. The diaphragm pump has been successful to avoid the dry-out and reduced the evaporator temperature. Thus, this HLHP can be considered for use on devices with high heat-flux.

Acknowledgment

The authors would like to thank DRPM Universitas Indonesia, for funding this research through "Hibah PUPT" scheme.

References

[1] Park, C., A. Vallury, and J. Perez, 2006, Advanced hybrid cooling loop technology for high performance thermal management. *4th International Energi Conversion Engineering Conference* 1-8.
[2] Park, C., et al. 2007.Electronics thermal management using advanced hybrid two-phase loop technology. *Thermal Engineering Heat Transfer Summer Conference collocated with the ASME 2007 InterPACK Conference*. **2** 911-16.
[3] Sarraf, D.B. and W.G. Anderson, High Temperature and High Heat Flux Thermal Management for Electronics.
[4] Park, C., et al., 2004, Hybrid Loop Thermal Bus Technology for Vehicle Thermal Management. *DTIC Document*.
[5] Park, C. and M. Crepinsek, 2011. Effect of operational conditions on cooling performance of pump-assisted and capillary-driven two-phase loop. *Journal of Thermophysics and Heat Transfer*, **25**(4) 572-80.
[6] Park, C., A. Vallury, and J. Zuo, 2009.Performance evaluation of a pump-assisted, capillary two-phase cooling loop. *Journal of Thermal Science and Engineering Applications*, **1**(2) 022004.
[7] Jiang, C., et al., 2014. Experimental investigation of pump-assisted capillary phase change loop. *Applied Thermal Engineering*, **71**(1) 581-88.
[8] Vasiliev, L.L., 2005.Heat pipes in modern heat exchangers. *Applied thermal engineering*, **25**(1): p. 1-19.
[9] Setyawan, I., I.I. Hakim, and N. Putra. 2017. Experimental study on a hybrid loop heat pipe. in MATEC Web of Conferences. *EDP Sciences*.
[10] Singh, R., et al., 2007. Novel design of a miniature loop heat pipe evaporator for electronic cooling. *Journal of Heat Transfer*, **129**(10) 1445-52.
[11] Chuang, P.-Y.A., 2003 An improved steady-state model of loop heat pipes based on experimental and theoretical analyses. *ProQuest*.
[12] Maydanik, Y., et al., 2011. Investigation of a compact copper–water loop heap pipe with a flat evaporator. *Applied Thermal Engineering*, **31**(16): p. 3533-3541.
[13] Wang, D., et al., 2015. Operational characteristics of a loop heat pipe with a flat evaporator and two primary biporous wicks. *International Journal of Heat and Mass Transfer*, **89** 33-41.
[14] Crepinsek, M. and C. Park, 2012. Experimental analysis of pump-assisted and capillary-driven dual-evaporators two-phase cooling loop. *Applied Thermal Engineering*, **38** 133-142.

Preparation and characterization of copper oxide catalyst with activated carbon support for ozone decomposition in industrial environment

A S Azhariyah, A Pradyasti, S Bismo[1]

[1] Chemical Engineering Department, Universitas Indonesia, Kampus Baru UI Depok, 16426, Indonesia

E-mail: setijo.bismo@ui.ac.id

Abstract. In this research, ozone decomposition has been synthesized based on copper oxide (CuO_x) with granular activated carbon (GAC) as a support catalyst, being used as ozone decomposer in effluent gas emissions of industries that use ozone. Therefore, catalysts were made as a mask filter to decompose ozone. CuO_x was impregnated to the surface of GAC by using copper carbonate ($CuCO_3$) as precursor and then calcined to release carbon dioxide with temperature of 300 °C for 1 hour. Size of activated carbon and loading percentage of copper oxide to the support were varied to get the optimum value. The quality of a catalyst such as pore diameter and surface area are characterized by BET, the cross-sectional surface of the catalyst and the catalyst elements composition are analyzed by X-ray spectroscopy (SEM-EDX) and X-Ray Fluorescence (XRF), while CuO_x composition and crystal phase are analysed by XRD. Mask filter, which contained catalysts for ozone decomposition, was tested using a fixed bed reactor at room temperature and atmospheric pressure. Results of the catalytic conversion were tested using iodometric method. Activated carbon with smallest diameter (60 - 100 mesh) and highest loading percentage (2 %-wt) showed the highest activity which the ozone conversion to oxygen reached 100%. Amount of CuO_x on the support also determine the efficiency of catalyst due to appropriate amount of CuO_x probably maintain the morphology and crystal phase of the catalyst.

1. Introduction

Needs of catalyst for ozone decomposition are continue to increase. This can be happened because ozone is used in many industries, such as water treatment, antiseptic, air purifier, and food industries. Ozone is used because it is a strong oxidizing agent that can be used in the disinfection process, which can remove bacteria, viruses, germs, improve the quality of water, and preserve foods [1]. However, some industries do not have ozone destructor unit and the excess of ozone is released to the air and have reached 2000 ppm which is exceed the allowed threshold [2]. The allowed threshold for ozone is 0.1 ppm in 8 hours (Recommended Explosure Limits). This excess of ozone is harmful to human, causing respiratory diseases, and leading to death. Ozone decomposition with catalyst is preferred because it can operate at room temperature [3]. Active site of the catalyst has either metal or metal oxide. Metal that can be made as catalyst are Au, Pd, Ag, and Rh, whereas for the metal oxide are Mn, Cu, Cr, Co, Fe, Zn, and Ni. CuO_x is chosen for this research because it has high oxidation-reduction

[1] To whom any correspondence should be addressed

potential values and has been widely used commercially. CuO_x is a metal oxide that is often used as a catalyst in various reserches and has the ability in degradation [4]. But, to get higher conversion of ozone to be oxygen, catalyst needs a support. GAC is used as a support for the catalyst because of its wide specific surface area and its good adsorbent property. With wider surface area, CuO_x is easier to be dispersed on the support.GAC also has oxygen functional group in its structure, so it also can be used as catalyst for ozone decomposition [5].

2. Methodology/Experimental

2.1. *Material*

In this research used granular activated carbon (GAC) as support from coconut shell with three types of GAC size, which was 18-35 mesh, 35-60 mesh, and 60-100 mesh that were obtained from CV Ady Water in Jakarta. Aquades, HCl, NaOH, H_2SO_4, KI, and $Na_2S_2O_4$ and were obtained from CV Sinar Kimia in Tanggerang. $CuCO_3$ was obtained from PD Cipta Bangun Nauli in Bogor. Tetra cloth for mask filter was supplied from CV Cahaya Textile in Bogor. Mask was supplied from houseware store in Asemka.

2.2. *Instrumentation*

The instrument that used to investigate the catalyst activity was a fixed bed reactor (FBR) with an ozone was supplied from a commercial ozone generator. X-Troy Ozonator was used to produce ozone. ASAP 2020 by Micromeritics was used for BET analysis. JEOL JSM 6510L A was used for SEM-EDX analysis. TORONTECH TT-EDXPRT was used for XRF analysis. Empyrean XRD was used for XRD analysis.

2.3. *Procedure*

2.3.1. *Preparation of CuO_x/GAC*

Three types of GAC size was used for investigation, which was 18-35 mesh, 35-60 mesh, and 60-100 mesh and was derived from coconut shell. First, the GAC was pretreated with 1 mol/L HCl and 1 mol/L NaOH both for an hour to remove the impurities or ash content on the surface such as K^+, Na^+, Ca^{2+}, and Mg^{2+}. Then, the treated GAC was impregnated in aqueous $CuCO_3$ solution for 12 hours by agitator with loading percentage of 0-2 %-wt. Loading percentage of CuO_x was calculated using the ratio of the CuO_x mass in the precursor solution to GAC. After impregnated onto the GAC, CuO_x/GAC was calcined at 300 °C for 1 hour to release the CO_2.

2.3.2. *Characterization of CuO_x/GAC*

Specific surface area of GAC was determined by BET method using nitrogen adsorption at the boiling temperature of liquid nitrogen. Surface morphology of CuO_x/GAC were observed by scanning electron microscopy (SEM) and surface composition analysis was carried out by energy dispersive X-ray spectroscopy (EDX) and X-Ray Fluorescence (XRF). The magnitude for SEM was 1,000 times and the voltage was 20 kV. The accelerating voltage for EDX was 20 kV. XRF instrument used soil method. Crystal phase was studied by X-ray diffraction (XRD).

2.3.3. *Catalytic activity test*

Catalytic activity of the sample was investigated using a fixed bed reactor (FBR) whose radius was 10 cm and height was 1 cm. Catalyst with mass of 10 grams was used for investigation. Commercial ozone generator was used to produce ozone from air with concentration of 36 ppm in 1 L of air. Excess of ozone from the reaction was analyzed using iodometric titration method. Amount of ozone was determined by reaction of I^- ion with ozone which produces iodine (I_2) at acidic condition. Acidic condition was fulfilled by addition of H_2SO_4. Then, I_2 was titrated with sodium thiosulfate ($Na_2S_2O_4$) at pH 2 and the volume of $Na_2S_2O_4$ could be used to calculate the volume of ozone which had not yet decompose to oxygen.

2.3.4. *Mask preparation*

Two pieces of tetra cloth with a size of 20 x 12 cm were filled with 10 grams of catalyst and stitched to form 1 cm x 1 cm matrices. Tetra clothes which were used as a mask filter were inserted into a mask which already contained a zipper. Illustration of the mask showed in figure 1, figure 2, and figure 3.

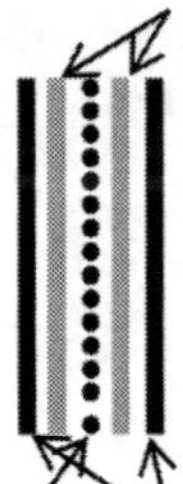

Figure 1. Mask configuration.

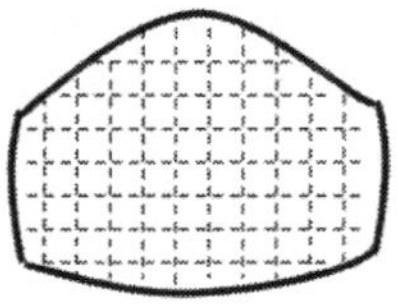

Figure 2. Matrices of tetra clothes.

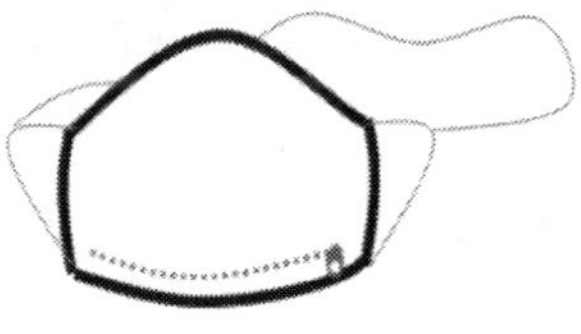

Figure 3. Ozone decomposition mask.

3. Results and Disscussion

3.1. BET Analysis

Specific surface area of GAC was showed in table 1. It showed that there was an increase in specific surface area of the GAC because of the pre-treatment. Therefore, by having a pre-treatment for GAC using strong acid and strong base, impurities would be removed. Aqueous HCl solution were dissolved and Cl^- ion was bonded with any positive charged impurities and H^+ ion bonded with GAC. With the addition of NaOH, H^+ ion which was previously bonded with GAC then bondedwith OH^- ion and formed water, while Na^+ ion was bonded with negative charge impurities. Excess of Na^+ was bonded to GAC.

Table 1. BET characterization of GAC.

Parameter	GAC 18-35 mesh		GAC 35-60 mesh		GAC 60-100 mesh	
	Before pre-treatment	After pre-treatment	Before pre-treatment	After pre-treatment	Before pre-treatment	After pre-treatment
Pore diameter (nm)	2.12	2.13	2.15	2.12	2.20	2.11
Pore volume (cm^3/g)	0.130	0.427	0.157	0.447	0.184	0.478
Surface area (m^2/g)	244.29	801.02	288.92	842.243	333.54	906.21

3.2. SEM-EDX Analysis

Surface morphology and composition of GAC before and after being pre-treated was showed in Figure 4, figure 5, and table 2. Table 2 showed that there were many impurities in the surface of GAC before being pre-treated. The amount of impurities decreased after pre-treatment such as Mg, Fe, Cl, Ca, and K. It can be seen, that the amount of Na is increased because base treatment, where Na^+ was bonded to GAC surface.

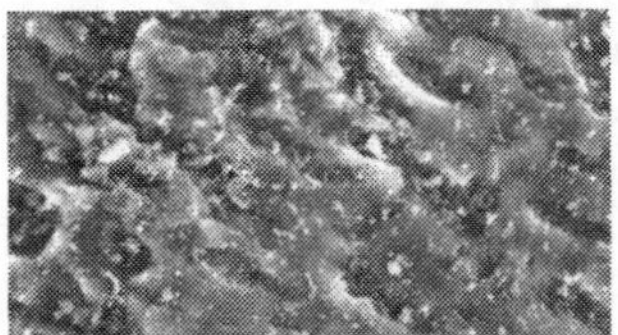

Figure 4. SEM analysis of GAC before pre-treatment.

Figure 5. SEM analysis of GAC after pre-treatment.

Table 2. EDX characterization of GAC before and after pre-treatment.

Elements	Before pre-treatment (%-w)	After pre-treatment (%-w)
C	81.34	78.89
O	9.42	17.53
Na	0.75	2.02
Mg	0.10	0.00
Al	0.00	0.44
Si	0.13	0.00
Zn	0.00	0.49
Cl	1.51	0.00
Ca	0.00	0.63
K	3.39	0.00
Cu	0.00	0.00
Fe	2.56	0.00
Mo	0.31	0.00
P	0.50	0.00

With wider surface area, CuO_x would be easier to be dispersed onto the GAC. Figure 6 and figure 7 showed the surface morphology of CuO_x /GAC for 1-2 %-wt.

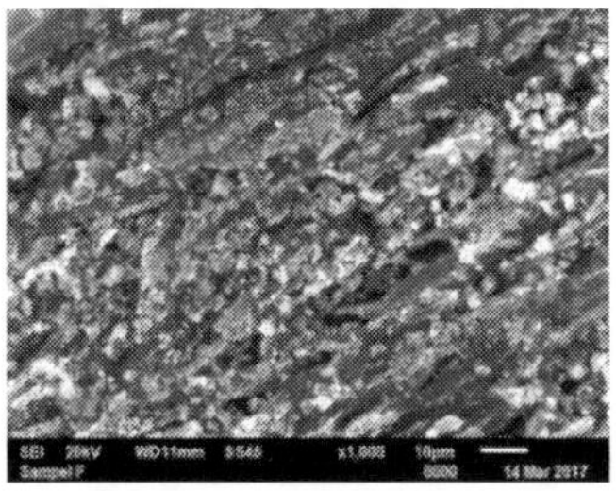

Figure 6. SEM analysis of CuOx/GAC 1%-w.

Figure 7. SEM analysis of CuOx/GAC 2%-w.

3.3. XRD Analysis

Figure 8 and figure 9 consecutively showed the XRD analysis of CuO_x/GAC before and after calcined and table 3 showed the intensity. JCPDS 76-660, JCPDS 65-2309, JCPDS 05-0667 were used as references for $CuCO_3$, CuO, and Cu_2O. It showed that there were a decrease in peak position after calcined process. Calcined decomposed CO_2 from $CuCO_3$ to create CuO_x. From XRD analysis, after calcined, only CuO_x (CuO and Cu_2O) remained in GAC.

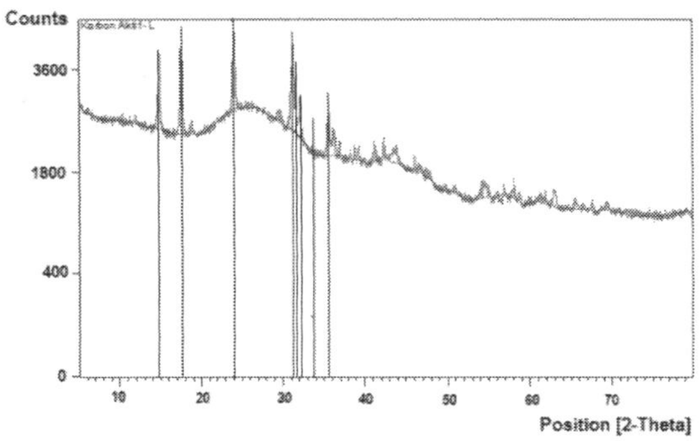

Figure 8. XRD analysis of CuOx/GAC 1%-w.

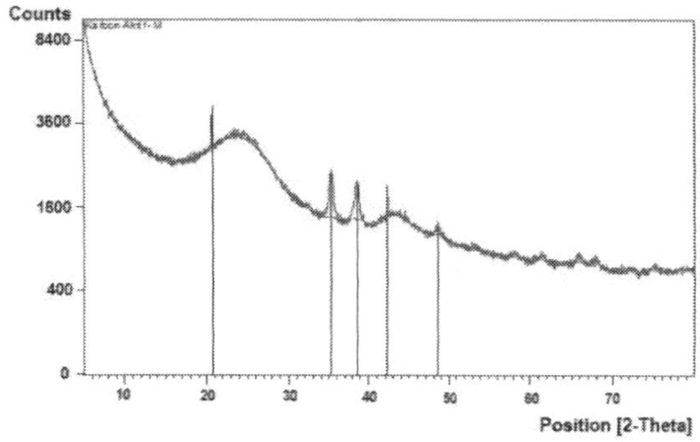

Figure 9. XRD analysis of CuOx/GAC 2%-w.

Table 3. XRD peak list of CuO_X/GAC

Before calcined			After calcined		
Compound	2θ (°)	Rel. Intensity (%)	Compound	2θ (°)	Rel. Intensity (%)
	14.7756	62.29	CuO	35.2256	100.00
	17.5614	86.46		38.6532	79.66
	24.0875	97.95	Cu_2O	42.3340	43.34
$CuCO_3$	31.2850	100.00			
	31.6315	69.99			
	32.1952	38.91			
	35.5834	49.00			

3.4. XRF Analysis

The XRF method was used to see the metal oxide compositions in GAC after impregnation. Table 4 showed the loading perscentage of CuO_x at each size. The loading percentage of all catalysts had a near-real value, except CuOx/GAC 1%-w 60-100 mesh was greater than the desired loading percentage. This could happen because the impregnation process was uneven, so the loading on the GAC was not homogeneous.

Table 4. %-loading CuO_x on GAC surface.

Size (mesh)	%-loading	
	1%	2%
18-35	0.8 %	2.50 %
35-60	0.84 %	2.09 %
60-100	1.66 %	1.90 %

3.5. Catalytic Activity Analysis

Figure 10 showed the conversion of ozone to oxygen for CuO_x/GAC catalyst with size of 18-35 mesh. CuO_x/GAC that had been pre-treated had a capability to have high conversion value in ozone decomposition. From 30 minutes, GAC showed a decrease in the conversion of ozone to oxygen. GAC with size of 18-35 meshand loading percentage of 1-2%-wt had an increase in conversion compared to loading 0%-wt. Figure 10 showed that higher loading of CuO_x would result a higher ozone conversion. With CuO_x/GAC catalyst, conversion of ozone into oxygen occurred not only because of the reaction of ozone with carbon, but also with CuO_x.

Figure 11 showed the conversion of ozone to oxygen for CuO_x/GAC catalyst with size of 35-60 mesh. When CuO_x/GAC had been pre-treated, performance of catalyst increased significantly. Up to 30 minutes, the conversion of ozone for CuO_x/GAC with size of 35-60 mesh and any loading percentage was higher than 90%. This high conversion of ozone also happened with CuO_x/GAC catalyst whose size was 60-100 mesh. This could be seen in figure 12.

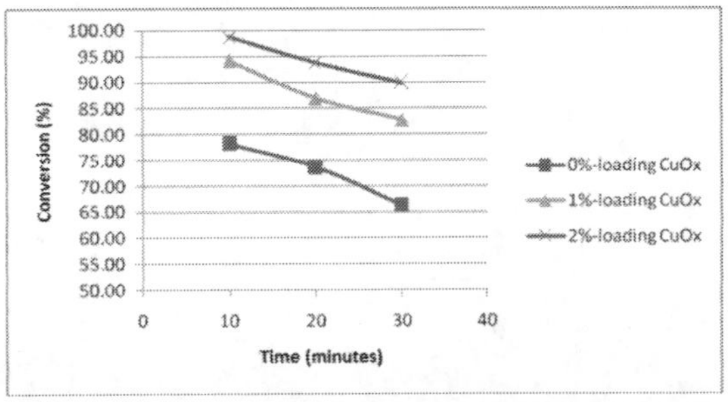

Figure 10. Conversion of ozone as a function of time for some loadings of CuO_x/GAC 18-35 mesh.

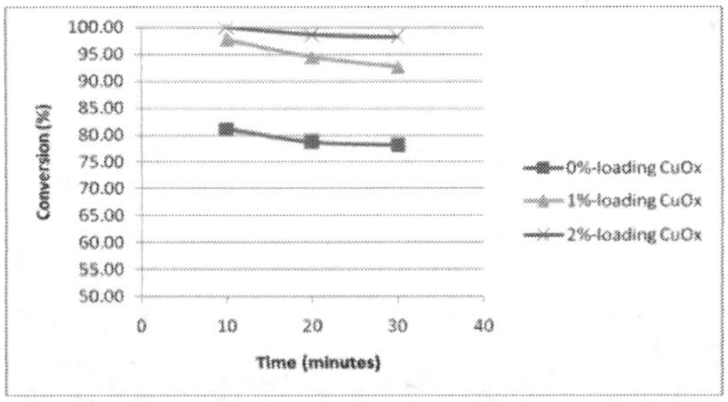

Figure 11. Conversion of ozone as a function of time for some loadings of CuO_x/GAC 35-60 mesh.

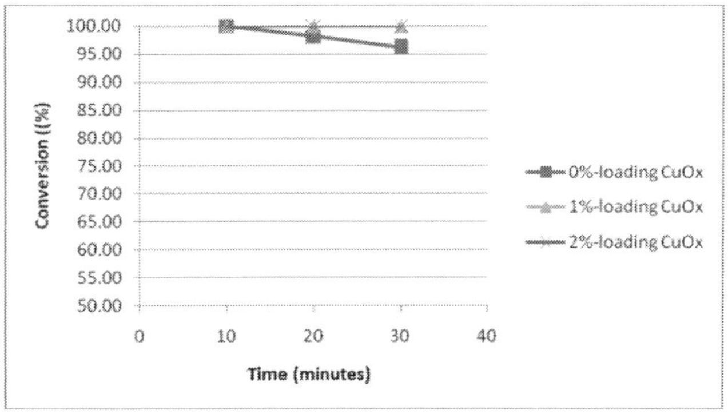

Figure 12. Conversion of ozone decomposition for CuO_x/GAC 60-100 mesh.

The presence of oxide of metal on the catalyst support increasedthe conversion of ozone to oxygen. In this case, the existence of CuO_x improved conversion rate of ozone by converting ozone to oxygen. The existence of CuO_x also extended the lifespan of the GAC. Appropriate amount of CuO_x gave an optimum value of active sites on the catalyst and maintained the morphology and crystal phase of the catalyst. Ozone might interact with CuO_x before reacted with carbon.

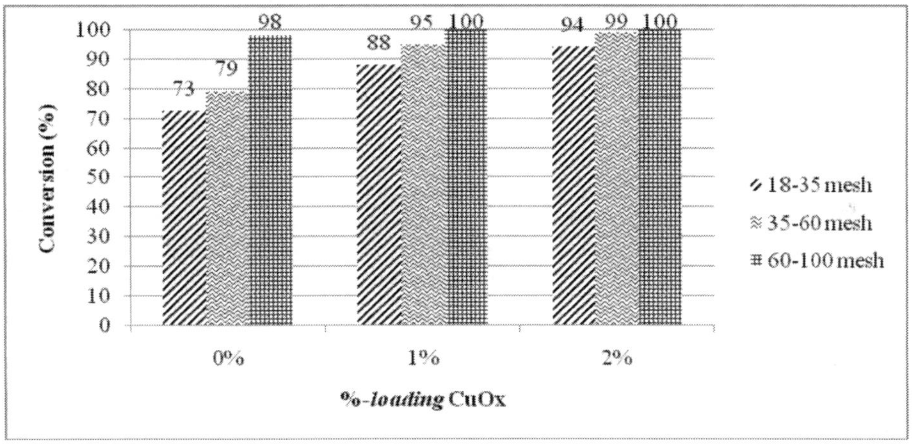

Figure 13. Ozone decomposition or conversion for various mesh of CuO_x/GAC catalysts.

Decomposition of ozone was very good at GAC 35-60 and 60-100 mesh 2%-wt. In addition to loading percentage and pre-treatment, this showed that the size of GAC had a very significant effect to ozone decomposition. Smaller size catalyst had larger surface area and bigger pore volume. Figure 13 showed the relationship of the ozone conversion with the size of the GAC. With the smaller size, the frequency of contact between the GAC ozone was increasing, so the conversion of ozone decomposition was increasing.

Larger surface area was also associated with impurities. The smaller the size of the GAC, the impurities that covered the pore and the surface was less than the bigger size of GAC. Figure 13 also showed the CuO_x influence on ozone decomposition. With bigger loading of CuO_x, the conversion of ozone to oxygen increased because many ozone molecules were converted by CuO_x into oxygen molecules.

3.6. Catalyst Durability
CuOx/GAC catalysts were tested for 8 hours. The results could be seen in figure 14.

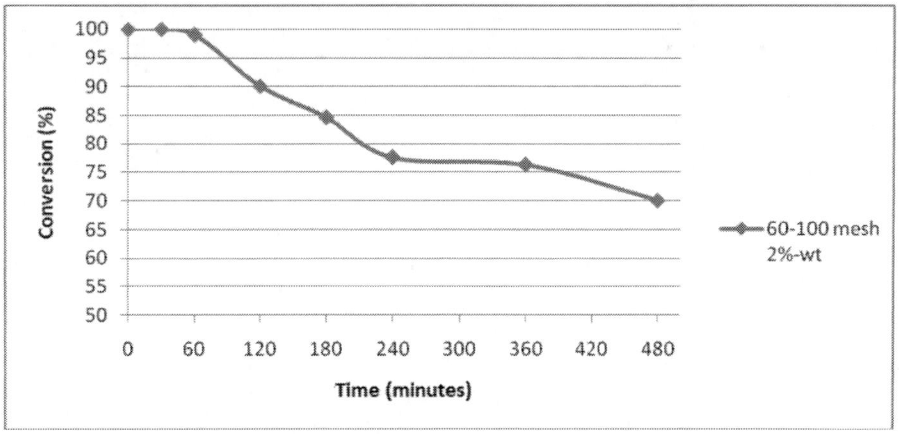

Figure 14. Average conversion of ozone decomposition of GAC.

CuO_x/GAC conversion decreased significantly after 60 minutes. There was a slight decrease in the conversion of CuOx/GAC catalyst from 240 minutes to 360 minutes. It happened because the data were taken discretely, which was tested for 2 days. There was a 15-hour break after 240 minutes before the catalyst was tested again. It means that the catalyst was allowed to regenerate for 15 hours. This break gave time to O_2 and CO_2 to desorb from the catalyst. So, for the next day, the surface of the catalyst would be wider to decompose ozone. The use of mask to decompose ozone is important to our community because it decomposed ozone at least 75.45 ppm for CuO_X/GAC catalyst. Therefore, mask that contained CuO_X/GAC can be used to reduce the ozone concentration to be below the allowed threshold.

4. Conclusion

Decomposition of ozone can be done effectively using CuO_x catalyst with GAC as the support. CuO_x/GAC catalyst with size of 60-100 mesh and loading percentage of 2 %-wt have the highest conversion value being reached 100%. Size and loading percentage of the catalyst had a relationship with the conversion of ozone in which the smaller size of a catalyst based on CuO_x easily dispersed to the support and higher percentage of CuO_x could convert higher ozone to oxygen molecules. CuO_X/GAC 2%-w 60-100 mesh could convert ozone to oxygen with a minimum of 70% within 8 hours.

5. Acknowledgments

The authors acknowledge Universitas Indonesia and PITTA UI-2017 program for the financial support. We are also thankful to Laboratorium Intensifikasi Proses, Departemen Teknik Kimia, Fakultas Teknik Universitas Indonesia; Badan Pengkajian dan Penerapan Teknologi; Badan Tenaga Nuklir Nasional; Laboratorium Pusat Penelitian Fisika-LIPI; and FMIPA Universitas Negeri Jakarta for the instruments to characterize the catalysts.

6. References

[1] O'Donnel C, Tiwari B K, Cullen P J, Rice G 2012 *Ozone in Food Processing* (UK: John Wiley & Sons, Ltd.).
[2] Subrahmanyam C, Bulushev D A, Kiwi-Minsker L 2005 *Applied Catalysis B:Environmental* **61** 98-106.
[3] Sigmund J 2001 *Catalytic Destruction of Ozone: A Cost Effective Approach to Controlling Off-Gas Ozone Emissions* (USA: EIJ Company LLC).
[4] Dhandapani B, Oyama S T 1997 *Applied Chemistry B: Environmental* **11** 129-166.
[5] Wang M, Zhang P, Li J, Jiang C 2014 *Chinese Journal of Catalysis* **35** 335-341.

Preparation of zinc oxide catalyst with activated carbon support for ozone decomposition

A Pradyasti, A S Azhariyah, E F Karamah, and S Bismo[1]
[1] Chemical Engineering Department, Universitas Indonesia, Kampus Baru UI Depok 16424, Indonesia

E-mail: setijo.bismo@ui.ac.id

Abstract. Investigation of catalyst for ozone decomposition was carried out by using zinc oxide (ZnO) catalyst and granular activated carbon (GAC) support. Ozone needs to be decomposed because it is harmful to human and can lead to death. Before GAC was used as a support, GAC was pre-treated using chloride acid (HCl) and sodium hydroxide (NaOH) to remove impurities. ZnO was impregnated onto the surface of GAC by using zinc carbonate ($ZnCO_3$) solution as precursor and then calcined at 300 °C to decompose carbon dioxide (CO_2). Size of GAC and loading percentage of ZnO were varied to get the highest catalytic activity. Size of GAC was varied between 18 – 100 mesh and loading percentage was between 0 – 2%-w. The morphology, composition, and crystal phase were characterized by BET, SEM-EDX, and XRD method. From XRD method, crystal phase of catalyst was changed from $ZnCO_3$ structure to ZnO when calcined with exact temperature. Ozone decomposition was performed at room temperature and atmospheric pressure using fixed bed reactor. ZnO/GAC with smallest size (60 – 100 mesh) and highest loading percentage (2%-w) showed the highest activity which the conversion reached 100% for 30 minutes. ZnO/GAC with smallest size and highest loading percentage had the largest surface area and the most active sites to decompose ozone.

1. Introduction

Need of catalyst for ozone decomposition is continued to increase with the excessive use of ozone in many industries. Ozone is a powerful oxidant and can be used as a disinfectant to remove unwanted chemical compound, bacteria, and even virus [1]. However, some industries do not have ozone destructor unit and the excess of ozone is released to the air, which has reached 2000 ppm. This level exceeds the allowed threshold which is 0.1 ppm for 8 hours [2]. This excess of ozone is harmful to human, causing respiratory diseases, and leading to death.

Catalytic ozone decomposition will decompose ozone to non-lethal compound, which is oxygen. Either metal or metal oxide is used for active sites of catalyst. Metal that can be used for catalyst are platina and palladium, whereas for metal oxide are manganese oxide, cobalt oxide, nickel oxide, silver oxide, and zinc oxide [3]. ZnO is chosen for this investigation because it is cheap and has good catalytic activity [4]. Catalyst support is used to get higher conversion of ozone decomposition. GAC is used as a support because it is a good adsorbent and has large specific surface area. With larger surface area, ZnO is easier to be dispersed onto the support and the amount of ZnO active sites will increase on catalyst support. Moreover, GAC also has oxygen functional groups in its structure which also can be used as catalyst for ozone decomposition [5].

[1] To whom any correspondence should be addressed

2. Methodology/ Experimental

2.1. Samples

All chemicals were analytical grade reagents and used as received without further purification. GAC was obtained from CV Ady Water, Jakarta. Deionized water, HCl, NaOH, H_2SO_4, KI, and $Na_2S_2O_4$ and were obtained from CV Sinar Kimia, Tangerang. $ZnCO_3$ was obtained from CV Inti Periodik Laboratory, Bogor. Tetra cloth for mask filter was supplied from CV Cahaya Textile, Bogor. Mask was supplied from houseware store in Asemka, Jakarta.

2.2. Preparation of ZnO/GAC

Three sizes of GAC were used for this investigation, which were 18 – 35 mesh, 35 – 60 mesh, and 60 – 100 mesh. GAC was pre-treated with 1 mol/L HCl and 1 mol/L NaOH to remove impurities on the surface. Then, the treated GAC was impregnated in $ZnCO_3$ solution for 12 hours with loading percentage of 0 – 2%-w. After impregnated, ZnO/GAC was calcined at 300 °C for 30 minutes to release CO_2.

2.3. Characterization of ZnO/GAC

Specific surface area was determined by BET method using nitrogen adsorption at the boiling temperature of liquid nitrogen. ASAP 2020 by Micromeritics was used for BET instrument. Surface morphology of ZnO/GAC was observed by scanning electron microscopy (SEM) and composition was analyzed by energy dispersive X-ray spectroscopy (EDX). Instrument JEOL JSM 6510L A was used for SEM-EDX. The magnitude for SEM was 1.000x and the voltage was 20 kV. The accelerating voltage for EDX was 20 kV and the energy range for EDX was between 0 – 20 keV. Crystal phase was studied by X-ray diffraction (XRD). Empyrean XRD was used for the instrument within a diffraction interval 2θ of 5 – 80 degrees.

2.4. Catalytic Activity Test

Catalytic activity was investigated using fixed bed reactor which has radius of 10 cm and length of 1 cm. Commercial ozone generator was used to produce ozone from air with concentration of 23.28 mg/hour and flow rate of 2.39 L/minute. Excess of ozone from ozone decomposition reaction was analyzed using iodometric titration method. Amount of ozone was determined by reaction of ozone with iodine ion (I^-) which produces iodine (I_2) at acidic condition. Acidic condition was fulfilled by addition of H2SO4. I2 was titrated with sodium thiosulfate ($Na_2S_2O_4$) and volume of $Na_2S_2O_4$ was used to calculate the volume of ozone which had not been decomposed to oxygen.

3. Results and Discussion

Surface morphology and composition of GAC before and after being pre-treated was showed in figure 1, figure 2, and table 1. Figure 1 and table 1 displayed that there were many impurities on the surface of GAC before pre-treatment. Therefore, by having a pre-treatment using strong acid and strong base solution, impurities were removed because there were ion exchange on catalyst. Aqueous HCl solution were dissolved and chloride ion (Cl^-) was bonded with any positive charged impurities. Meanwhile, hydrogen ion (H^+) was bonded with GAC. With the addition of NaOH, H^+ from strong acid solution then bonded with hydroxide ion (OH^-) and formed water, while sodium ion (Na^+) was bonded with negative charge impurities. From Table 1, it could be seen that the composition of Na, O, and Al were increase. Na and O were on the surface of GAC because the use of NaOH as pre-treatment solution, whereas Al was increase because the use of aluminum foil and aluminum tray when drying catalyst in the oven.

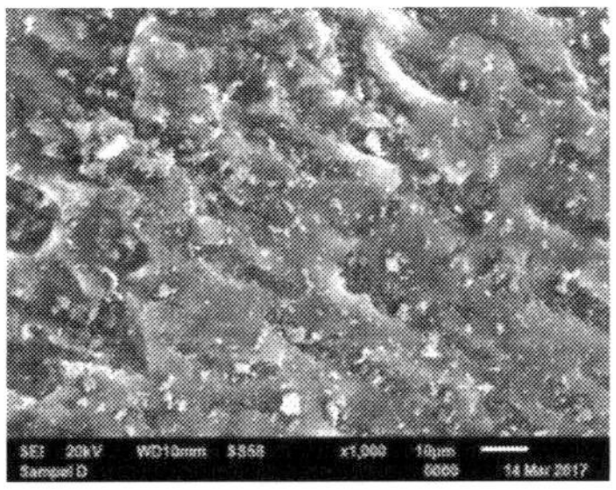

Figure 1. SEM analysis of GAC before pre-treatment.

Figure 2. SEM analysis of GAC after pre-treatment.

TABLE 1. EDX characterization of GAC before and after pre-treatment

Catalyst Composition	Before pre-treatment (%)	After pre-treatment (%)
C	81.34	79.38
O	9.42	17.53
Na	0.75	2.02
Mg	0.10	0.00
Al	0.00	0.44
Si	0.13	0.00
P	0.50	0.00
Cl	1.50	0.00
K	3.39	0.00
Ca	0.00	0.63
Fe	2.56	0.00
Mo	0.31	0.00

Pore volume and surface area of GAC values were showed in table 2. These data were obtained using BET characterization method. It could be seen from the Table that pore volume and surface area were increase after pre-treatment. With larger surface area, ZnO would be easier to be dispersed onto GAC and conversion of ozone decomposition could be increased.

TABLE 2. BET characterization of GAC before and after pre-treatment

Parameter	GAC 18 – 35 mesh		GAC 35 – 60 mesh		GAC 60 – 100 mesh	
	Before	After	Before	After	Before	After
Pore volume (cm^3/g)	0.130	0.427	0.157	0.447	0.184	0.478
Surface area (m^2/g)	244.29	801.02	288.92	842.24	333.54	906.21

Table 3 showed the nominal value of Zn in catalyst after impregnation. Catalysts with 1%-w loading had a lower nominal value than the expected loading. Meanwhile, for 2%-w, all values were higher than the wanted loading. It could be happen because ZnCO3 is an insoluble solid and can make the agitation process in impregnation become uneven to the surface of GAC. The used of electron in SEM-EDX method was focusing in one point of catalyst, so the nominal value could not represent the loading percentage of catalyst.

TABLE 3. Nominal value of ZnO in GAC

Size of GAC (mesh)	Percentage in 1%-w loading (%)	Percentage in 2%-w loading (%)
18 – 35	0.73	3.84
35 – 60	0.05	2.62
60 – 100	0.98	3.50

Figure 3 displayed a peak comparison between ZnCO3/GAC catalyst before calcined and Smithsonite solid ($ZnCO_3$) (JCPDS 8-0449). Smithsonite solid has peaks located at 25°, 32°, 35°, 38°, 42°, 46°, 51°, and 53° with peaks at 25° and 32° have higher intensity than other peaks. From the Figure, it could be seen that catalyst also had a peak at 25° with intensity of 600 counts and peak at 32° with intensity of 100 counts. Both peaks indicated the presence of $ZnCO_3$ structure in catalyst.

From XRD result, catalyst did not show specific peaks like Smithsonite solid. This was happened because GAC is an amorphous solid whose particles are arranged unevenly. This make the length between atoms becomes irregular and will lead a structural aberration that shows unspecific peaks.

Figure 4 showed a peak comparison between ZnO/GAC catalyst after calcined and Zincite solid (ZnO) (JCPDS 36-1451) which also depicted with blue and red. Zincite solid has peaks located at 31°, 34°, 36°, 47°, and 56° with peaks at 31° and 36° have higher intensity than other peaks. From the figure, it could be seen that catalyst also had peaks at 31°, 34°, and 36°, although the intensities were not really high. Peak at 31° had an intensity of 100 counts, peak at 36° had an intensity of 70 counts, and peak at 47° had an intensity of 80 counts. Despite the low intensities, it still could be concluded that there was a ZnO crystal structure in catalyst.

Characterization with XRD method on various size and loading percentage of catalysts concluded in figure 5. Catalysts with 2%-w loading had higher intensity at 31°, 34°, and 36° peaks compared to 1%-w loading. Those three peaks indicated the presence of ZnO in catalysts. With bigger loading of ZnO, X-ray would be frequently hit the ZnO structure thus made the intensity of the specific peaks higher.

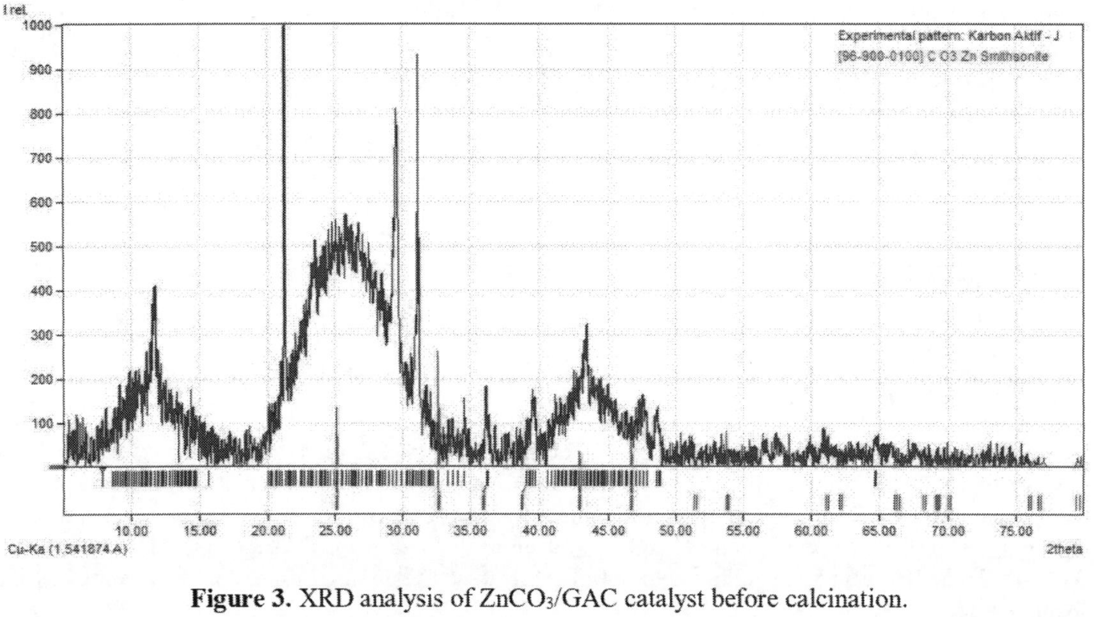

Figure 3. XRD analysis of $ZnCO_3$/GAC catalyst before calcination.

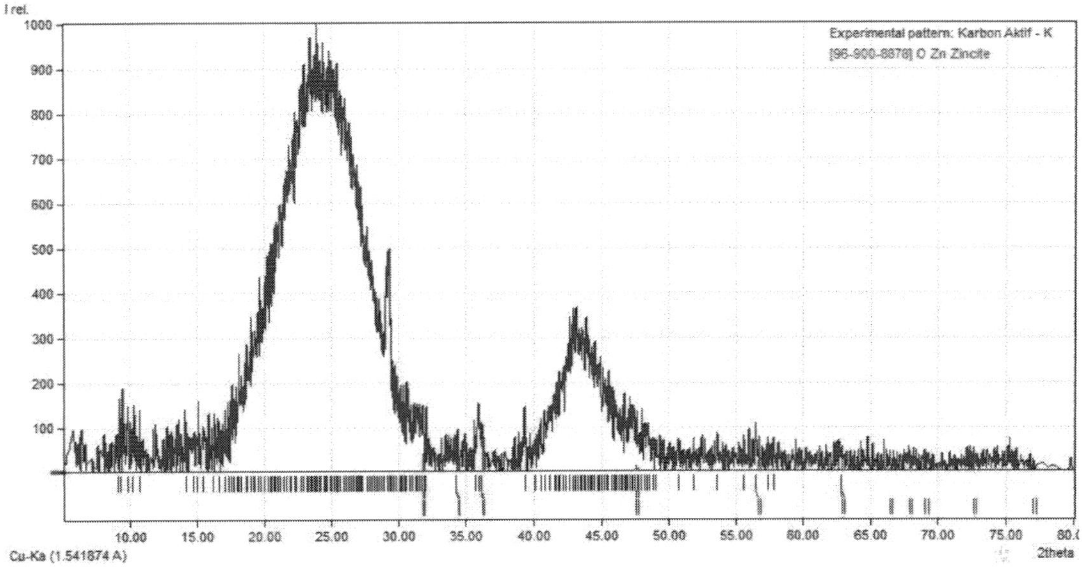

Figure 4. XRD analysis of ZnO_3/GAC catalyst after calcination.

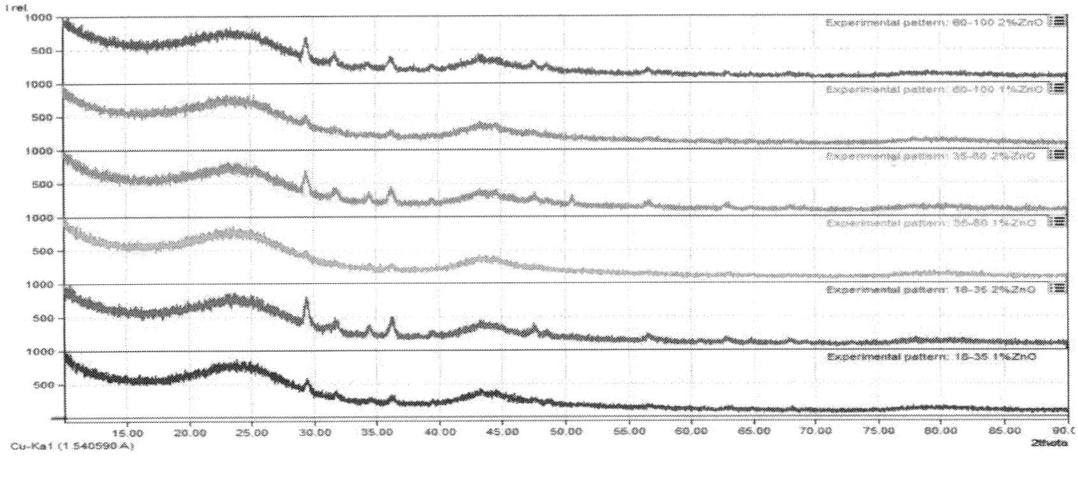

Figure 5. XRD analysis upon various size and loading percentage of catalysts.

Figure 6 and figure 7 showed the conversion of ozone to oxygen for GAC catalysts before and after pre-treatment. Catalysts before pre-treatment had the lowest conversion because impurities were still remained on the surface of GAC. Impurities covered the active sites of catalyst so the surface for ozone decomposition reaction was less than GAC after pre-treatment. Also from both figures, it could be seen that the smallest size of GAC (60 – 100 mesh) had the best catalytic activity compared to any larger sizes. Size of catalyst supports affected the performance of the catalyst because with smaller size used, there were more active sites for ozone to be decomposed to oxygen.

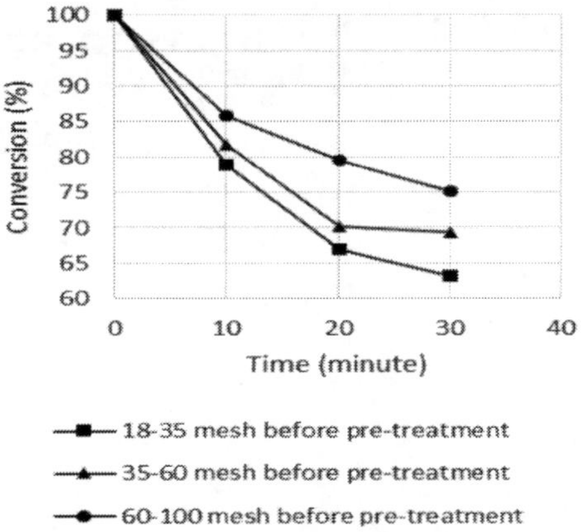

Figure 6. Conversion of ozone decomposition for GAC before pre-treatment.

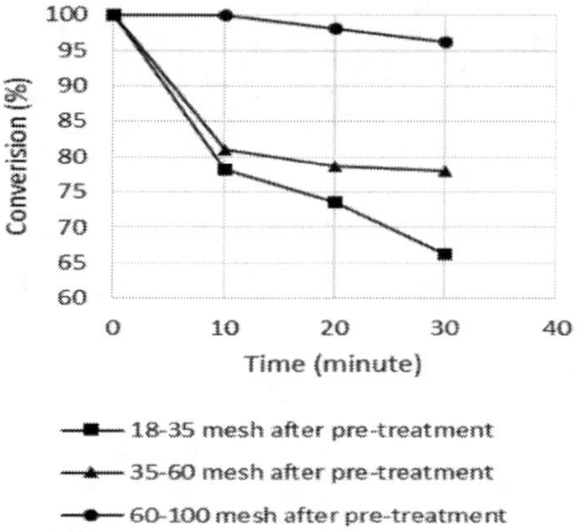

Figure 7. Conversion of ozone decomposition for GAC after pre-treatment.

Figure 8 and figure 9 displayed the conversion of ozone to oxygen for ZnO/GAC catalyst with 1 and 2%-w consecutively. Addition of ZnO in catalyst gave a great impact to the conversion of ozone decomposition. It could be seen from the figure that the conversion was increased significantly when using catalysts with 1 – 2%-w loading and 2%-w loading showed the biggest conversion. Higher conversion after adding ZnO onto the catalysts occurred because the characteristic of ZnO as a donor and acceptor. ZnO had many defect sites like oxygen vacancies [4]. Oxygen vacancies were important parts of ZnO structure because those were the place for ozone to be absorbed and oxygen to be desorbed from catalysts. When ozone molecule was on the surface of catalyst, one oxygen atom from ozone would fill the oxygen vacancies in ZnO and leave the oxygen gas to be released from catalysts [6].

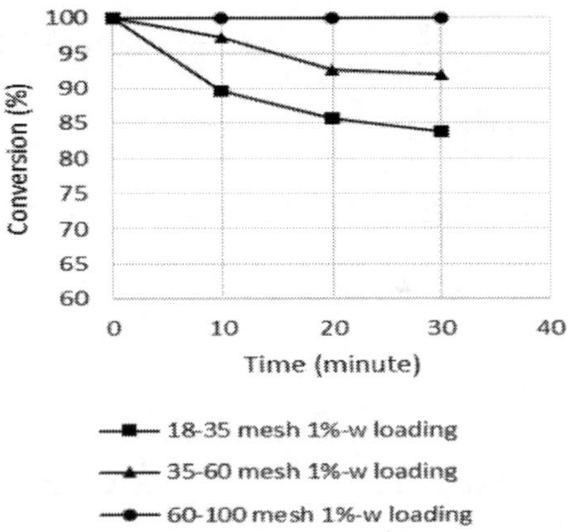

Figure 8. Conversion of ozone decomposition for ZnO/GAC 1%-w loading.

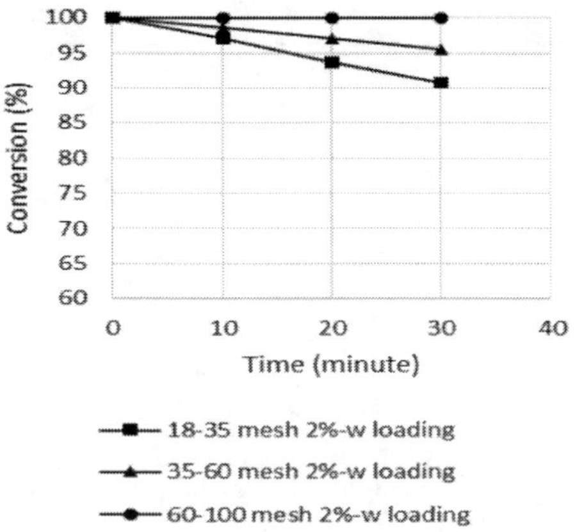

Figure 9. Conversion of ozone decomposition for ZnO/GAC 2%-w loading.

4. Conclusion

Decomposition of ozone could be done effectively using ZnO catalyst with GAC support. ZnO/GAC catalyst had high value of conversion for all sizes and loading percentages. ZnO/GAC catalyst with size of 60 – 100 mesh and loading percentage of 2%-w had the highest conversion value which was reached 100% for 30 minute process.

Size and loading percentage of the catalyst had a relationship with the conversion of ozone in which the smaller size of a catalyst made ZnO easily dispersed to the support and higher percentage of ZnO means more defect sites or oxygen vacancies which could convert more ozone to oxygen molecules.

5. Acknowledgement

The authors acknowledge Universitas Indonesia and PITTA-UI 2017 program for the financial support. We are also thankful to Badan Tenaga Nuklir Nasional and FMIPA Universitas Negeri Jakarta for the instruments to characterize the catalysts.

6. References

[1] Koltunski E and Plumridge J 2005 *Ozone as a Disinfecting Agent in the Reuse of Wastewater* (Duebendorf: Ozonia Ltd.).
[2] Subrahmanyam C, Bulushev D A and Kiwi-Minsker L 2005 *Applied Catalysis B: Environmental* **61** 98-106.
[3] Dhandapani B and Oyama S T 1997 *Applied Chemistry B: Environmental* **11** 129-166.
[4] Milenova K I, Nikolov P M, Kasabova N A and Avramova I A 2014 *Polish Journal of Chemical Technology* **16** 55-59.
[5] Wang M, Zhang P, Li J and Jiang C 2014 *Chinese Journal of Catalysis* **35** 335-341.
[6] Zhu G, Zhu J, Jiang W, Zhang Z, Wang J, Zhu Y and Zhang Q 2017 *Applied Catalysis B: Environmental* **209** 729-737.

IOP Conf. Series: Earth and Environmental Science **105** (2017) 012014 doi:10.1088/1755-1315/105/1/012014

Energy consumption model improvement with principal component analysis method for small village community

Subhan Petrana[1], Eko Adhi Setiawan[2] and M Yunus H A[1]

[1]Department of Electrical Engineering, Universitas Khairun, Ternate, Indonesia
[2]Tropical Renewable Centre, Faculty of Engineering, Universitas Indonesia, Depok, Indonesia

Email : subhanpetrana@gmail.com

Abstract. Daily electrical energy consumption is a key parameter to determine balancing electricity supply and demand particularly in remote area where is not yet being electrified. The absence of daily electrical energy consumption data is a constraint for developing microgrid systems based on renewable energy. The purpose of this study is to improve electrical energy consumption model with adding some variables such as family income and family expenses, the number of family members and the load factor as independent variable, and electrical energy consumption assumption as dependent variable. Principal Component Analysis (PCA) method is used to find the relationship between independent and dependent variables. The electric energy consumption model $Y = (0.69) + (1.77e^{-7}x\ A_1) - (1.54e^{-7}x\ A_2) + (0.037x\ A_3) + (7.45\ x\ A_4)$, with determination factor (R-Square) is 97%.

1. Introduction

Currently, the number of villages throughout Indonesia that have not been electrified reach 2.519 [1]. It is a challenge in development of microgrids system based on renewable energy, especially in village community. Consumption data is important parameter to determine balancing electricity supply and demand particularly in remote area where is not yet being electrified. The purpose of this study is to improve electrical energy consumption model with adding some variables such as family income and family expenses, the number of family members and the load factor as independent variable, and electrical energy consumption assumption as dependent variable.The PCA method is used to predict electric power burden as a basic for planning electrical network [2]. While [3] using PCA to predict electrical load by considering economic condition in the region. Beside PCA, Multiple Regression Method and Grey Models are also used to predict household electrical consumption. Multiple Regression Method is used by [4] to predict electrical energy consumption in Delhi by including the influence of weather change to electrical consumption. The Gray models method by [5] is used to predict electrical consumption for using as regional electrical consumption model in villages.

2. Methods

2.1. Principal Component Analysis (PCA)

PCA method is able to simplify observed variables by reducing its dimensions, by means eliminating the correlation between independent variables through the transformation of original independent variables into new variables that are not correlated to each other or known as free of multicollinearity. These variable will be regression to be seen its influence to dependent variable [3].

2.2. Mathematical Model of PCA

Mathematical model of PCA is as follows [3]:

$$X = \begin{bmatrix} X_{11} & X_{12} & \cdots & X_{1p} \\ X_{21} & X_{22} & \cdots & X_{2p} \\ \vdots & \vdots & \vdots & \vdots \\ X_{n1} & X_{n2} & \cdots & X_{np} \end{bmatrix} = [X_1 \quad X_2 \quad \cdots \quad X_p] \tag{1}$$

where:

$$X_j = \begin{bmatrix} X_{1j} & X_{2j} & \cdots & X_{nj} \end{bmatrix}^T \ ; j = 1,2 \cdots, p$$

the linear matrix combination is as follows:

$$\begin{matrix} PC_1 & = & a_{11}X_1 & + & a_{12}X_2 & + & \cdots & + & a_{1p}X_p \\ PC_2 & = & a_{11}X_1 & + & a_{12}X_2 & + & \cdots & + & a_{1p}X_p \\ \vdots & & \vdots & & \vdots & & \vdots & & \vdots \\ PC_p & = & a_{p1}X_1 & + & a_{p2}X_2 & + & \cdots & + & a_{pp}X_p \end{matrix} \tag{2}$$

to maximize the ratio of PC1 variance to total Variance, with the limit that:

$$a_{i1}^2 + a_{i2}^2 + \cdots + a_{ip}^2 = 1 \tag{3}$$

Principal Component is as follows:

$$PC_i = a_{1i}X_1 + a_{2i}X_2 + \cdots + a_{pi}X_p \quad ; i = 1, 2, \ldots . k \tag{4}$$

and matrix PC is as follow:

$$F = \begin{bmatrix} f_{11} & f_{12} & \cdots & f_{1k} \\ f_{21} & f_{22} & \cdots & f_{2k} \\ \vdots & \vdots & \vdots & \vdots \\ f_{n1} & f_{n2} & \cdots & f_{nk} \end{bmatrix} \tag{5}$$

The equation (1) to (5) are representation of PCA mathematical model.

2.3. PCA Calculation Steps

The steps of PCA calculation system and its explanation is [2]:

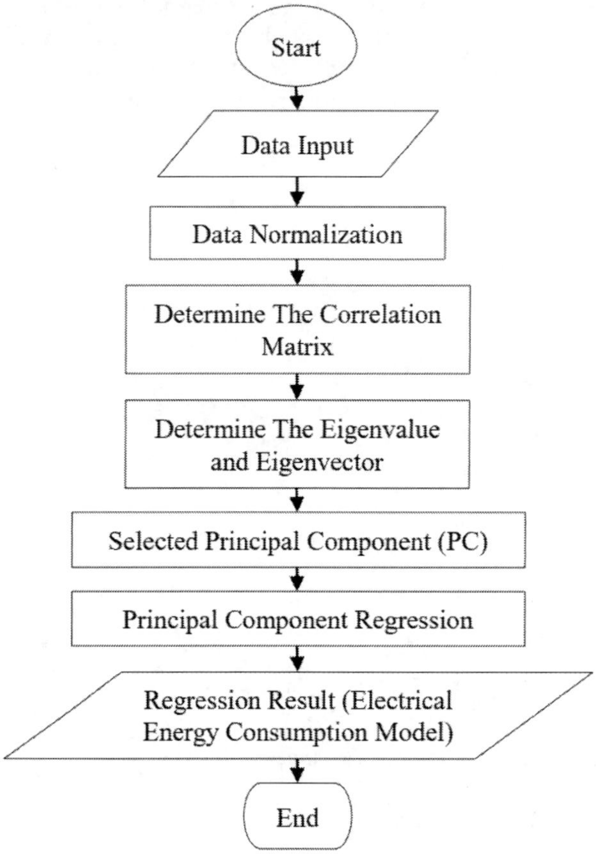

Figure 1. Calculation steps

1. The first steps of PCA calculation is to normalize the initial data as assumed in the matrix below:

$$X = \begin{bmatrix} X_{11} & X_{12} & \cdots & X_{1p} \\ X_{21} & X_{22} & \cdots & X_{2p} \\ \vdots & \vdots & \vdots & \vdots \\ X_{n1} & X_{n2} & \cdots & X_{np} \end{bmatrix} \quad (6)$$

Its matrix size is n x p, where n is number of observation and p represent of five variables. Furthermore, data normalization process is done by procedure that each observed data on a variable is subtracted with mean of its variable and divided with its standard deviation. The equation is as follow:

$$x_{ik}^{1} = {[x_{ik} - \bar{x}_k]}/{S_k} \quad ; i = 1,2,\cdots,n \, ; k = 1,2,\cdots,p \quad (7)$$

Where x_{ik} is observation data, $\bar{x}_k$ is mean, and S_k is deviation standard. The equation of mean and deviation standard is as follow :

$$\bar{x}_k = {1}/{n} \sum_{i=1}^{n} x_{ik} \quad ; S_k = \sqrt{{1}/{n-1} \sum_{n-1}^{n} (x_{ik} - \bar{x}_k)^2} \quad (8)$$

In this equation (8) the variant variable is 1 and mean is 0.

2. After normalizing, the next step is to determine correlation matrix with equation as follows:

$$r_{ij} = \sum_{i=1}^{n} x_{kr}^{1}\ x_{kj}^{1}/n-1 \qquad ; i,j = 1,2,\cdots,p \tag{9}$$

$$R = \begin{bmatrix} r_{11} & r_{ij} & \cdots & r_{1p} \\ r_{i1} & r_{ij} & \cdots & r_{ip} \\ \vdots & \vdots & \vdots & \vdots \\ r_{p1} & r_{pj} & \cdots & r_{pp} \end{bmatrix}$$

3. Next step is the determine of non-negative eigenvalue that uses characteristic equation $|\mathrm{R} - \lambda\mathrm{I}| = 0$ and eigenvector corresponded to the eigenvalues from R matrix:

$$C_i = (C_{1i}, C_{2i}, \cdots, C_{mi})^T \qquad ; i = 1,2,\cdots,m \tag{10}$$

4. Furthermore, M principal components arranged of the eigenvectors using the formula below:

$$PC_i = c_{1i}X_1 + c_{2i}X_2 + \cdots + c_{pi}X_p \qquad ; i = 1,2,\cdots,m \tag{11}$$

Principal components $PC_1, PC \ldots\ldots, PC_m$ are uncorrelated variables to each other.

3. Result and Discussion

3.1. Research Flowchart

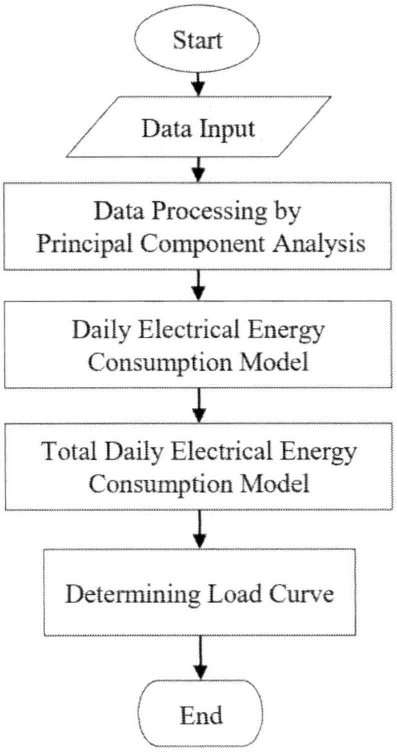

Figure 2. Research flowchart

Figure 2 shows the process of determining the daily electrical energy consumption model using the PCA method. The input data consisted of five variables were not considered before such as family income, family expenses, total family members, load factor is independent variable and electrical energy consumption assumption as dependent variable. To find daily electrical energy consumption model, the input data is processed as shown in figure 1, and HOMER software is applied to form load profile.

3.2. Electrical Energy Consumption Assumption

In this research, electrical energy consumption assumption is determined because the proposed research object is a remote village and it has no electrical energy consumption data. The electrical energy consumption assumption calculation is as follow:

$$\cos\varphi = {}^{P}/_{S} \tag{12}$$

where $\cos\varphi$ is Power Factor, while P, S as real power and apparent power, equation (12) can be written as:

$$P = \cos\varphi \ x\ S \tag{13}$$

so that:

$$P = 0.85\,x\,450\,VA = 382.5\,Watt$$

Based on calculation result above, 382,5 Watt is a power available for each house. While, peak energy consumption is power that used of each house for 24 hours (hr) is determined using formula below:

$$Peak\ Energy\ Consumption = P\,x\,24\,hr \tag{14}$$

$$Peak\ Energy\ Consumption = 382.5\,x\,24\,hr = 9180\,Wh = 9.18\,kWh$$

Electrical energy consumption assumption is determined by use load factor 0.4, 0.6 and 0.8 that grouped based on total family income. its calculation shown in equation below:

$$Electrical\ Consumption\ Assumptions = Daily\ Load\ Peak\ x\ Load\ Factor \tag{15}$$

Thus, based on equation (15) the electrical energy consumption assumption for three type of load factor above are 3.672 kWh, 5.508 kWh and 7.344 kWh.

3.3. Daily Electrical Energy Consumption Model Determination

The correlation matrix (R) is resulting after normalizing process from the initial data using equation (7) to (9), correlation matrix as shown below:

$$R = \begin{bmatrix} 1 & 0.825 & 0.085 & 0.815 & 0.814 \\ 0.825 & 1 & 0.238 & 0.717 & 0.717 \\ 0.085 & 0.238 & 1 & 0.293 & 0.294 \\ 0.815 & 0.717 & 0.293 & 1 & 0.999 \\ 0.814 & 0.717 & 0.294 & 0.999 & 1 \end{bmatrix}$$

The correlation matrix shows that there is a correlation among each variable. For example, if in first column of the matrix elements R21, R41 and R51 have strong correlation among them. The value of their matrix elements 0.825, 0.815 and 0.814. In R31 matrix element, value 0,85 has weak correlation with the three of matrix elements R21, R41 and R51. But, there is no multicollinearity occur because variance inflation for each independent variable are less than 10, as shown table below:

IOP Conf. Series: Earth and Environmental Science **105** (2017) 012014 doi:10.1088/1755-1315/105/1/012014

Table 1. Variance inflation factor (VIF)

Independent variable	Symbol	Variance Inflation
Family Income	A_1	5.383
Family Expenses	A_2	3.439
Total Family	A_3	1.278
Load Factor	A_4	3.507

The next step is to calculate eigenvalue from correlation matrix. The result of the calculation is shown in the table 2.

Table 2. Eigenvalue of correlation matrix

Principal Component	Eigenvalue	Condition Number
1	2.648	1.00
2	0.955	2.77
3	0.281	9.44
4	0.116	22.83

The result of eigenvalue calculating shows that there is no multicollinearity occur, because the condition of number for each eigenvalue are less than 100. Its calculation showing there are four principal component variables that become linier combination which has no correlation each other's. Subsequently, regression of four Principal Components resulting daily electrical energy consumption model with R-square as determinant factor 97 %. The model of daily electrical energy consumption is as follows $Y = (0.69) + (1.77e^{-7}x\, A_1) - (1.54e^{-7}x\, A_2) + (0.037x\, A_3) + (7.45\, x\, A_4)$.

Consumption model as shown the equation above that the largest factor contributed to electrical energy consumption is load factor with the coefficient 7.45. Family income and expenses have small coefficient that are $1.77e^{-7}$ and $1.54e^{-7}$ but they have significant influence because the multiplying factor are in hundreds and million rupiahs. While number of family members with coefficient 0.037 has less influence in the equation above. Based on the model, total daily electrical energy consumption is 22.829 kWh/day.

3.4. Load Profile

Daily electrical energy consumption model that is 22,828 kWh/day is shows in the load curve profile below:

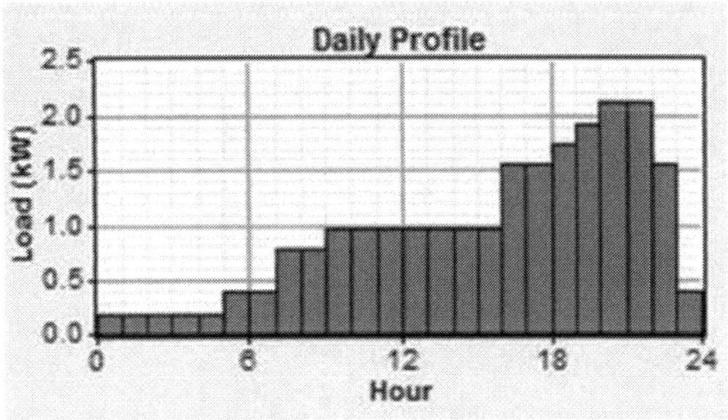

Figure 3. Load Profile

Figure 3 shows a change in the electrical load pattern for one day. Increase electric load occurs in the afternoon to evening, start from 4 pm to 10 pm with power 1.535 kW, 1.727 kW, 1.918 kW, 2.110 kW and 1.535 kW and peak load is 2.110 kW occurs at 8 pm to 9 pm. Increasing electric loads at night are caused by most family member doing their activity at home, such as praying, watching TV, and children learning activity. During the day, the range of load variation is from 0.767 kW to 0.959 kW because most people in the village doing their activity in outside their home.

4. Conclusion

By PCA method, the research successful determine mathematical model of electrical energy consumption $Y = (0.69) + (1.77e^{-7}x\, A_1) - (1.54e^{-7}x\, A_2) + (0.037x\, A_3) + (7.45\, x\, A_4).$ with determination factor 97%, and daily total energy around 22.829 kWh. In calculation process involving several variables such as family income, family expenses, number of family members, load factor, and electrical energy consumption assumption. This result will be used to design microgrid system with optimum configuration.

5. ACKNOWLEDGEMENT

The Authors would like to thank This work supported by "Hibah Penelitian Kerjasama Antara Perguruan Tinggi" Kemenristek Dikti. The Authors would also like to thank for all member in Tropical Renewable Energy Centre (TREC) Universitas Indonesia.

6. References

[1] Badan Pusat Statistik. 2014. Statistik Potensi Desa Indonesia. *Badan Pusat Statistik*, Jakarta-Indonesia.

[2] Jun-long F, Yu X, Yu F, Yang X, Guo-liang L. 2015. Rural Power System Load Forecast Based on Principal Component Analysis. *J Northeast Agric Univ* :67–72. doi:10.1016/S1006-8104(15)30034-9.

[3] Lingli S, Changhong D, Qiushi X, Chu Z. 2012. Load Forecasting Considering The Regional Economic Environment. *Int. Conf. Autom. Control Artif. Intell.* **1** 1310–4.

[4] Nidhi T. 2014 Determinants of the Household Electricity Consumption : A Case Study of Delhi. *Int J Energy Econ Policy* **4** 337–48.

[5] Wu guangwei et al. 2011.The Application of Several Gray Models in Prediction of Electricity Consumption in Rural Areas. *Int. Conf. Bus. Manag. Electron. Inf., IEEE* 2–5. doi:10.1109/ICBMEI.2011.5920938.

IOP Conf. Series: Earth and Environmental Science **105** (2017) 012015 doi:10.1088/1755-1315/105/1/012015

Study on the effects of temperature in the carbothermic reduction laterite ore using palm kernel shell as reducing agent

Risty Hidayanti[1], Sulaksana Permana[1], Ahmad Maksum[1, 2] and Johny Wahyuadi Soedarsono[1, *]

[1]Center of Minerals Processing and Corrosion Research, Department of Metallurgical and Materials Engineering, Universitas Indonesia, Depok 16424, Indonesia

[2]Department of Mechanical Engineering, Politeknik Negeri Jakarta, Depok 16425, Indonesia

*Email: jwsono@metal.ui.ac.id

Abstract. Indonesia has a rich deposit of nickel. However, laterite in Indonesia has not been treated to its full potential. This happens because the refining process of lateritic nickel ore has a high cost, triggered by the amount of energy required and the complexity of the separation process. It needs pre-reduction to make the ore more easily to be reduced and increased the metal content so that it can maximize the nickel refining process and minimize the energy usage. One method of pre-reduction is carbothermic reduction. This research study the effect of temperature variation on the results of the carbothermic reduction of laterite ores using palm kernel shells as the reducing agent. The reduction process was done by heating the # 270 mesh lateritic ore and palm kernel shells with a mass ratio of 1: 4 for 60 minutes in the melting furnace at the temperature variations of 700, 800, 900 and 1000°C. The result of the reduction was then tested using XRF and XRD. Based on the calculation of % recovery, the optimum temperature for reducing the laterite ore with palm kernel shells for 60 minutes was 800°C which the content of NiO as much as 2.68%.

1. Introduction

Indonesia has many laterite ores. Based on the Information from Energy and Mineral Resources in 2010, resources of nickel ore in Indonesia is as much as 263,350,043,000 tonne. But, Indonesia has not processed the nickel ore optimally because the process of purifying laterite is costly. It is triggered by the required energy and complexity in the process of the recovery process [1].

General nickel laterite purification process is categorized into three processes, namely hydrometallurgy, pyrometallurgy, and caron which is a combination of pyrometallurgy and hydrometallurgy [2]. The pyrometallurgical process is still dominantly used in the processing of laterite ore in Indonesia because of its easy-to-use technology and economical production cost. On the other hand, the pyrometallurgical process requires high fuel consumption and high reducing agent, while also produces air pollution [3].

Indonesia has abundant biomass potentials. Biomass is a waste of agricultural or forest residue and other organic materials that can be regulated into fuel and environmentally friendly reducing agents (low emission) through pyrolysis and gasification process.

Reducing gases from pyrolysis and gasification processes includes hydrogen (H_2), carbon monoxide (CO), carbon dioxide (CO_2), nitrogen (N_2), methane (CH_4), and other hydrocarbons is known as syngas. The relative proportion of each component in syngas depends on the operating conditions of gasification, i.e. biomass type, temperature, pressure, catalyst and others [4,5].

One of the natural potentials that can be utilized as reducing agent is palm kernel shell [6] because the reductor gases can be generated in the palm kernel shell gasification process. The application of biomass in metallurgy industry has been widely utilized, including direct reduction of iron ore by biomass char, sintering with wood chips, and production of pig iron by utilizing fine wood as reducing agent[7-9]. However, nickel processing industry in Indonesia currently has not used biomass as their alternative reducing agent. According to the previous study [10] which aimed to optimize the using of potential Indonesian ore as a raw material for producing iron nugget by utilizing palm kernel shell as a reducing agent, this study will analyze the reduction process of laterite ore by using palm kernel shell as a reducing agent.

2. Materials and method

2.1. Materials

The nickel ore sample used in this study was delivered from Halmahera, North Maluku, while palm kernel shell was delivered from Palangkaraya, Central Kalimantan. The experimental tools were ball mill, digital scale, thermocouple, ceramic crucible and sieving machine.

2.2. Experiment

This process took place in the muffle furnace. The laterite ore sample which has been formed to fine powder (# 270) and mixed with palm kernel shell was put into the melting furnace. The process of nickel ore reduction was done in 4 various increasing temperature range, respectively 700°C, 800°C, 900°C, and 1000°C for 1 hour with a mass ratio of 1: 4 (nickel ore: palm kernel shell). After the direct reduction process, XRF and XRD tests were performed to analyze the compounds and nickel content contained in the laterite ore. XRF testing was done at BATAN and XRD testing was done at Department of Metallurgy and Materials Engineering UI Depok. The machine used was XRD-7000 Maxima-X Shimadzu engine.

3. Result and discussion

The palm kernel shell used as a reducing agent had proximate and ultimate composition as seen in table 1.

Table 1. Proximate and Ultimate Analysis of palm kernel shells sample.

Parameter	Content (% adb)
Proximate Analysis	
Moisture in air dried sample	3.70
Ash	2.09
Volatile Matter	74.04
Fixed Carbon	20.17
Ultimate Analysis	
Carbon	49.90
Hydrogen	6.00
Nitrogen	0.28
Oxygen	41.66

Based on result of XRD (figure 1) and XRF analysis (table 2), initial nickel ore could be catagorized as laterite-saprolite nickel ore.

IOP Conf. Series: Earth and Environmental Science **105** (2017) 012015 doi:10.1088/1755-1315/105/1/012015

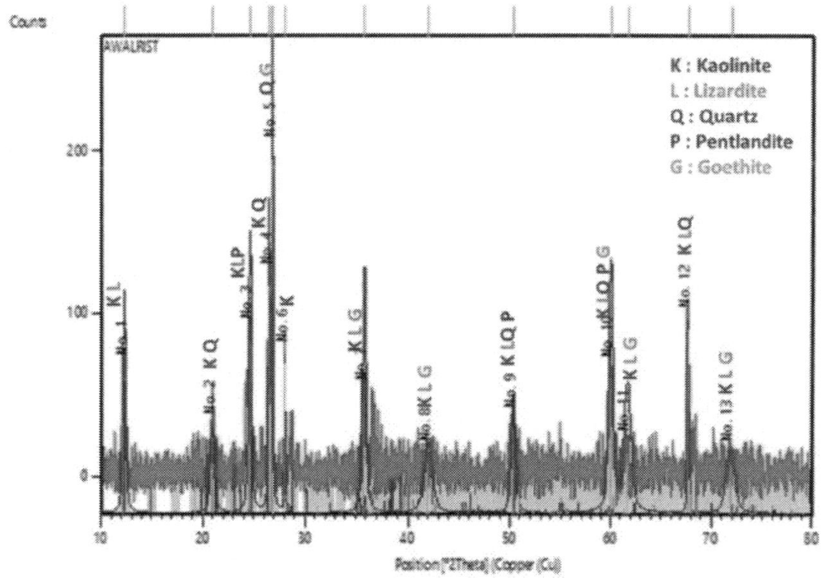

Figure 1. XRD pattern of initial lateritic nickel ore sample.

Table 2. XRF result of initial lateritic nickel ore sample.

Substance	Content (%)
Mg	14.28
Al	2.216
Si	20.12
Mn	0.196
Ni	2.092
Fe	11.06
MgO	23.68
Al_2O_3	4.86
SiO_2	43.05
MnO	0.253
NiO	2.662
Fe_2O_3	14.93

The following graph represents the comparison of the sample XRD test results after carbothermic reduction at all temperatures in figure 2.

IOP Conf. Series: Earth and Environmental Science **105** (2017) 012015 doi:10.1088/1755-1315/105/1/012015

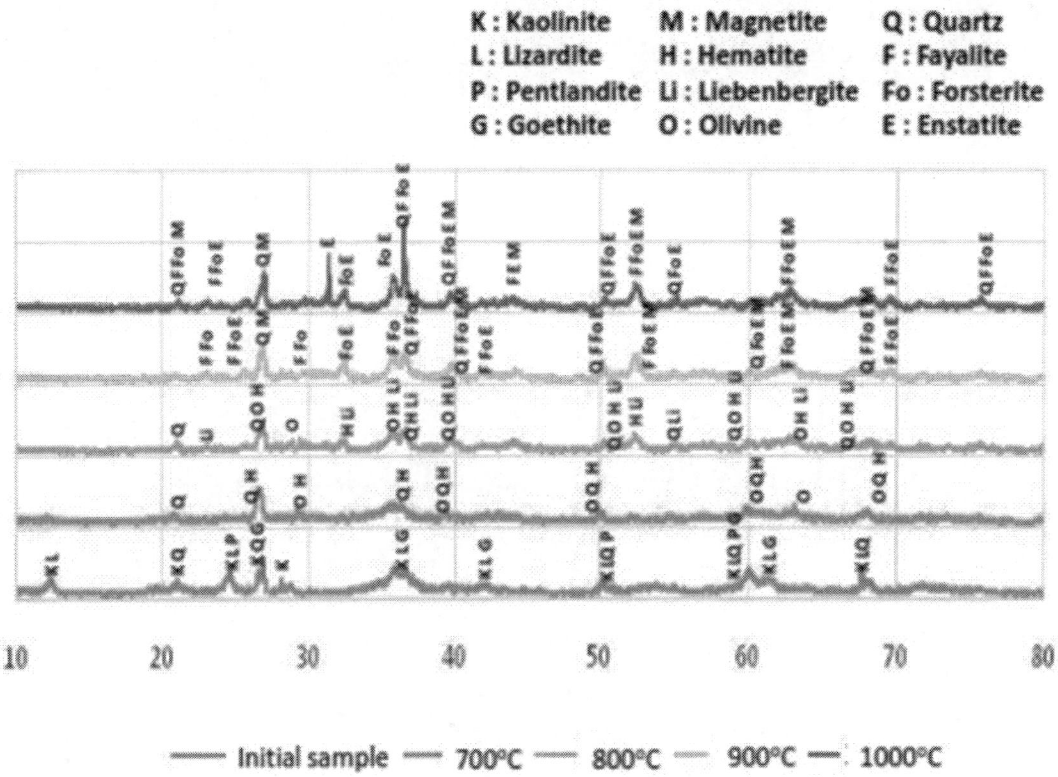

Figure 2. XRD patterns of carbothermic reduction from lateritic nickel ore at all reduction temperature variations.

From the data above, we assumed that during the carbothermic process, there was a change of compounds occured in the temperature range of 700-1000°C. This could be observed among others on the removal and addition of diffraction peaks in samples heated above the temperature of 700°C.

At the temperature of 700°C, the formation of olivine (Mg_2SiO_4) and hematite (Fe_2O_3) compounds could be seen. At the temperature of 800°C, a new diffraction peak was found when compared to the diffraction pattern at 700°C. The diffraction pattern showed the presence of liebenbergite compounds (Ni_2SiO_4). At the temperature of 900°C, a new diffraction peak was assumed showing the presence of forsterite (Mg_2SiO_4) and enstatite ($MgSiO_3$) compounds, in relation to the addition of diffraction peaks which might indicate the presence of magnetite compounds (Fe_3O_4). At the temperature of 1000°C, there were two peaks that could be indicated as the presence of enstatite compounds, in addition to an increase in the intensity of the diffraction peaks at a position value of 2θ 36.5° significantly when compared to the diffraction peaks at 900°C, indicating an increase in the number of forsterite compounds at 1000°C .

This study also conducted XRF testing to determine the amount of oxides and chemical elements contained in the ores that had been reduced. In the results of XRF testing of the lateritic ore used in this study, there were 2 dominant metal oxide compounds: nickel oxide and iron oxide. The amount of each oxide in the lateritic ore could be seen in table 3.

From the XRF test result data in table 3, it was found that the amount of NiO compound increased as the reduction temperature increased. It showed that the highest% NiO was obtained at temperature 900°C. At the temperature of 700°C, there was a decrease in nickel oxide and iron oxide grades when compared to the initial sample. It was occurred as the result of the decomposition process accompanied by oxidation of sulfide ores of the pentlandite ($Fe.Ni_9S_8$) compound encountered in the initial sample. At the temperature of 800°C, NiO and Fe_2O_3 grades increased again and reached the

IOP Conf. Series: Earth and Environmental Science **105** (2017) 012015 doi:10.1088/1755-1315/105/1/012015

maximum value at the temperature of 900°C. It was related to serpentine decomposition process and nickel sulphide decomposition process into nickel oxide. At the temperature of 1000°C, NiO and Fe_2O_3 grades decreased. It happened due to the formation of forsterite and fayalite compounds. When Ni entered and was trapped in the olivine lattice, the nickel would be very difficult to reduce.

Table 3. XRF test results of lateritic nickel ore after carbothermic reduction for 60 minutes at all temperature

Sample	**NiO (%)**	**Fe_2O_3 (%)**
Initial Sample	2.662	14.93
700°C	2.166	14.79
800°C	2.680	15.90
900°C	2.695	16.77
1000°C	2.428	15.85

Table 4. Recovery of laterite ore after carbothermic reduction for 60 minutes at all temperature variations.

Reduction temperature	**Initial weight F (gr)**	**Ni ore. f (%)**	**End weight C (gr)**	**Ni Concentrate c (%)**	**Ni recovery**
700°C	20	2.092	15.586	1.702	63.401
800°C	20	2.092	14.835	2.106	74.674
900°C	20	2.092	13.378	2.118	67.719
1000°C	20	2.092	12.900	1.908	58.826

The calculation of nickel recovery in concentrate based on the XRF testing before and after carbothermic process could be seen in Table 4. The highest recovery value was obtained at the temperature of 800°C. It concluded that between the temperature range of 700-1000°C, the temperature of 800°C was the optimum temperature to obtain nickel concentrate.

In this study, there were no diffraction peaks of NiO compounds in the result of sample characterization using XRD. The reduction of nickel in saprolite ore, which was a silicate-bearing type mineral, was more complicated when compared to the limonite ore. This was related to the olivine formation, where higher reduction conditions were required to reduce olivine [6].

4. Conclusion

The major conclusions drawn from the study were as follow:

a. XRD results of lateritic ores reduced at 700°C, 800°C, 900°C and 1000°C showed that there was no NiO reduction process to Ni. This was due to the mineral type was in the form of silicate-bearing which binded the Ni element so that Ni was difficult to be reduced, but the reduction of hematite (Fe_2O_3) to magnetite (Fe_3O_4) happened. The magnetite compound was detected on the XRD diffraction pattern of laterite ore reduction at the temperature of 800°C.
b. The highest % recovery value of lateritic lateral was obtained on laterite ore which was reduced to the temperature of 800°C, equal to 74.67%.
c. Based on the % recovery, the optimal temperature for reducing laterite ore with oil palm shells was 800°C for 60 minutes resulting in 2.68% NiO content.

5. Acknowledgement

Special thanks to Mr. Agus Budi Prasetyo (Research Centre for Metallurgy, Indonesian Institute of Sciences - LIPI, Indonesia) for helpful supports. The authors also wish to thank the Ministry of Research Technology and Higher Education and Directorate of Research and Community Engagement, Universitas Indonesia, under PUPT Research Grants Contract No. 2747/ UN2.R3.1/HKP05.00/2017 for the financial support of this research.

6. References

[1] Norgate T and Jahanshahi S 2011 Assessing the energy and greenhouse gas footprints of nickel laterite processing *Miner. Eng.* **24** 698–707

[2] Oxley A and Barcza N 2013 Hydro–pyro integration in the processing of nickel laterites *Miner. Eng.* **54** 2–13

[3] Quast K, Connor J N, Skinner W, Robinson D J and Addai-mensah J 2015 Preconcentration strategies in the processing of nickel laterite ores Part 1 : Literature review *Miner. Eng.* **79** 261–8

[4] Dahlquist E 2013 Technologies for Converting Biomass to Useful Energy-Combustion, Gasification, Pyrolysis, Torrefaction and Fermentation ed E Dahlquist (Leiden: CRC Press)

[5] Vassilev S V., Baxter D, Andersen L K, Vassileva C G and Morgan T J 2012 An overview of the organic and inorganic phase composition of biomass *Fuel* **94** 1–33

[6] Moghadam R A, Yusup S, Uemura Y, Chin B L F, Lam H L and Al Shoaibi A 2014 Syngas production from palm kernel shell and polyethylene waste blend in fluidized bed catalytic steam co-gasification process *Energy* **75** 40–4

[7] Matsumura T, Ichida M, Nagasaka T and Kato K 2008 Carbonization behaviour of woody biomass and resulting metallurgical coke properties ISIJ International **48** 572

[8] Zuo H, Hu Z, Zhang J, Li J and Liu Z 2013 Direct reduction of iron ore by biomass char International Journal of Minerals, Metallurgy and Materials **20** 1

[9] Srivastava U, Kawatra S K and Eisele T C 2013 Production of pig iron by utilizing biomass as reducing agent International Journal of Mineral Processing **119** 51–57

[10] Soedarsono J W, Kawigraha A, Sulamet-Ariobimo R D, Johansyah D, Kusuma G D, Suprayogi S, Yosi A, Saputro N L, Sidiq A T, Erwin E and Natanael D 2012 Potential indonesia ores as raw material for producing iron nugget *3rd International Conference on Advances in Materials and Manufacturing Processes, ICAMMP 2012* **652–654** 2529–33

IOP Conf. Series: Earth and Environmental Science **105** (2017) 012016 doi:10.1088/1755-1315/105/1/012016

The influence of palm kernel shell mass ratio as a reducing agent in the lateritic nickel ore carbothermic reduction process

Silmina Adzhani[1], Risty Hidayanti[1], Ahmad Maksum[1, 2], Sulaksana Permana[1] and Johny Wahyuadi Soedarsono[1,*]

[1]Center of Minerals Processing and Corrosion Research, Department of Metallurgical and Materials Engineering, Universitas Indonesia, Depok 16424, Indonesia

[2]Department of Mechanical Engineering, Politeknik Negeri Jakarta, Depok 16425, Indonesia

*Email: jwsono@metal.ui.ac.id

Abstract. Indonesia has many valuable mineral resources, such as lateritic nickel ore. Today, the world demand of lateritic nickel continues to increase. This is an opportunity for Indonesia to develop its potentials in the nickel processing industry. To perform nickel reduction process, reducing agents such as natural gas and coal are needed. In this study, the use of a reductant from palm kernel shell waste as a coal alternative energy in order to reduce the use of fossil fuel which limited availability and cause environmental pollution, being a focus of this research. The purpose of this study is to determine the effect of palm kernel shell as a reductant in lateritic nickel reduction process, using mass ratio variable between mass of nickel ore and reductant. The mass ratio between nickel ore and reducing agent used in this study are 1:1, 1:2, 1:3, and 1:4, with the temperature of reduction in 800°C for 60 minutes. To observe the results of this experiment, the sample characterization was carried out using XRD and XRF. XRD data showed the presence of silica (SiO_2), iron oxide compounds such as maghemite (Fe_2O_3) and magnetite (Fe_3O_4), also compounds from reduction of lizardite such as forsterite (Mg_2SiO_4) and liebenbergite (Ni_2SiO_4). The results of XRF analysis showed improvement of Ni recovery in line with the addition of the mass of reducing agents.

1. Introduction

Today, Nickel is one of the most important elements that are widely used in many industries. Indonesia has a large potential of nickel laterite minerals. Those nickel content is estimated in number up to 1.6 billion tons with an average nickel content of 1.57% which spread around Indonesia such as Sulawesi, Maluku and Papua [1].

Lateritic nickel ore processing is one of the processes that require high energy and high cost production. It is because there are several numbers of processes that should be passed to extract the laterite ore into concentrates, such as rotary-kilns and high-pressure acid leaching (HPAL). In general, nickel laterite ore purification process is categorized into three processes, which are hydrometallurgy, pyrometallurgy and caron process [2,3]. The conventional pyrometallurgy process is still dominantly used in the processing of laterite ore in Indonesia since it is using easy technology and more economical production costs. In other way, conventional pyrometallurgical process usually used coal as a reducing agent which can cause air pollutions [4].

In order to consider environmental issue, the mining and industrial sectors associated with the metal processing are expected to reduce fossil fuel consumption to reduce non-renewable energy consumption and greenhouse gas emissions. Therefore, a process that can reduce the use of coal is needed. The use of biomass energy or waste from natural resources is one alternative that can be done to replace the role of coal as a reducing agent. Palm kernel shell is one of biomass that can be developed as a reducing agent in laterite nickel ore carbothermic process because it has high caloric content, low ash content and low emissions gas production, also the availability of palm plantations in Indonesia is increasing. It is widely expected to produce 7.3 million tons/year of palm oil production's wastes per year. This study was conducted to determine the effect of palm kernel shell as a reducing agent in lateritic nickel ore reduction process..

2. Materials and method

1.1. Materials

Several equipments were in this study including melting furnace, ball mill, ceramic crucible, digital scales, thermocouple, sieving machine, Spectro Xepos METEK XRF machine at BATAN and 7000 Maxima-X Shimadzu XRD machine at Department of Metallurgy and Materials Engineering UI Depok. While materials that were used are lateritic nickel ore from ANTAM field at Pomalaa and palm kernel shell from Kalimantan. First sample were prepared, the lateritic ore were crushed up to # 270 and the particle size of palm kernel shell is in range 1-2 cm.

1.2. Experiment

After sample preparation finished, nickel ore and palm kernel shell were homogenized and mixed into 4 different mass of reducing agent with ratio: 1:1; 1:2; 1:3; 1:4 (nickel ore : palm kernel shell). Then they were placed into ceramic crucible. Roasting process takes place in the muffle furnace at temperature 800°C for 60 minutes. After the reduction process, XRF and XRD test were conducted to analyze the compound and nickel contained in the lateritic ore.

3. Result and discussion

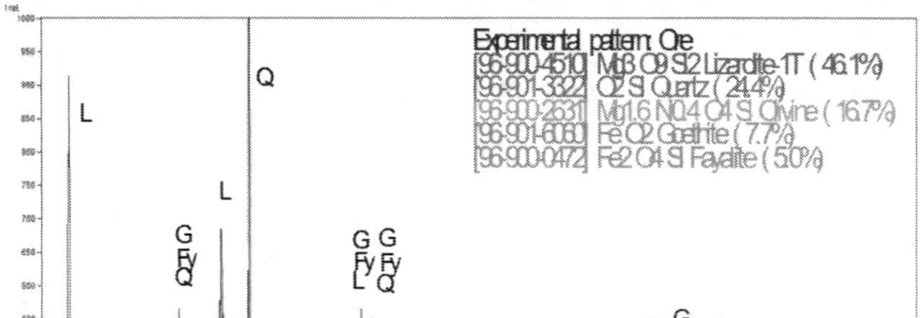

Figure 1. XRD pattern of initial lateritic nickel ore.

To identify the mineral composition of raw materials, proximate analysis were performed for palm kernel shell, while XRD and XRF examination were performed for raw lateritic nickel ore. As the results, figure 1 presents XRD pattern of nickel ore that shows the peaks of goethite, fayalite, lizardite, and quartz which appeared in raw ore.

While the proximate analysis of palm kernel shell as can be seen in table 1. Palm kernel shells are used as reducing agents because of the existing of fixed carbon and CH_4 content in volatile matter which can also use as reducing agents.

Figure 2 presents a comparison of XRD examination result from all of sample with different mass ratio. Semi-qualitative data of laterite ore samples obtained from the examination using X-Ray Diffraction (XRD). Characterization of the initial sample of nickel ore needs to be done in order to know the changes in the compounds form that occurs after the carbothermic reduction at constant temperature 800°C with reduction time in 1 hour. The effect of using different amounts of reducing agents becomes observed in this study.

Figure 2 indicates the different peaks of four samples which have different mass ratio. The variable of mass ratio would be affecting the form of compounds during the carbothermic reduction process determined by different peaks that occurred.

Table 1. Proximate and Ultimate Analysis of palm kernel shells.

Parameter	Content (% adb)
Proximate Analysis	
Moisture in air dried sample	3.70
Ash	2.09
Volatile Matter	74.04
Fixed Carbon	20.17
Ultimate Analysis	
Carbon	49.90
Hydrogen	6.00
Nitrogen	0.28
Oxygen	41.66

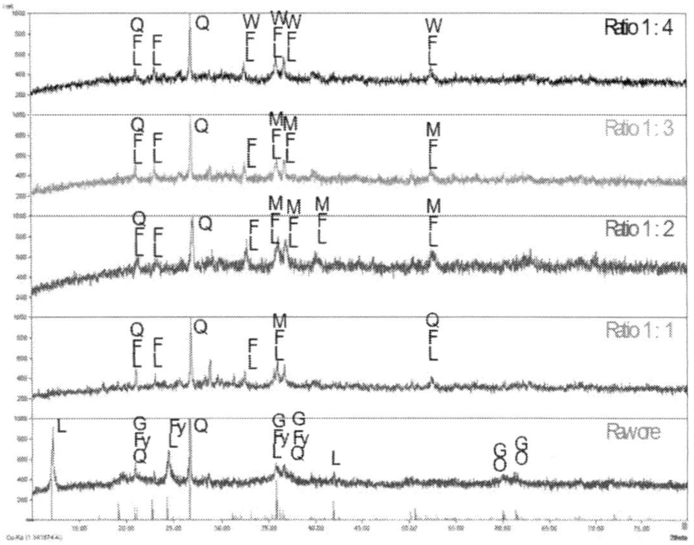

Figure 2. XRD pattern of samples after reduction with different mass ratio (G=ghoethite, Q=quartz, Fy=fayalite, M=maghemite, L=lizardite, F=forsterite, W=wustite).

The change of compound was found differently in every sample with the mass addition of reducing agent from 1:1; 1:2; 1:3; and 1:4. At mass ratio 1:1, the phases of maghemite (Fe_2O_3), forsterite (Mg_2SiO_4), liebenbergite (Ni_2SiO_4), dan quartz (SiO_2) compounds were shown. Furthermore, at mass ratio 1:2 the diffraction pattern shows the presence of diiron-silicate $Fe_2(SiO_4)$ compounds. While at mass ratio 1:3 the presence of Magnetite (Fe_3O_4) compounds was appeared. The magnetite compound that appears in this variable is the results of further reduction of goethite and magnetite compounds. This transformation also occurred in the previous reduction research using other biomass [5,6]. The last variable in this study is the ratio of mass of nickel ore and palm kernel shell of 1:4. Changes occurred in the magnetite compound, where the compound was un-reidentified at results of XRD data above. The presence of new compound, wustite (FeO), at mass ratio 1:4 indicates the result of further reduction of magnetite compounds.

However, there were no peaks diffraction of NiO compunds on the result of sample characterization using XRD. The reduction of nickel in saprolite ore which is a silicate-bearing type mineral is more complicated when compare with limonite ore, this is related to olivine formation, where higher reduction conditions are require to reduce olivine [7].

In this study XRF examination to determine the quantitative data of oxides and chemical elements contained in the ores after reduction process were conducted. As the results of XRF examination, there are two dominant metal oxide compounds, namely nickel oxide and iron oxide, the XRF examination results of each variable can be seen in table 2.

Table 2. XRF test results of lateritic nickel ore after carbothermic reduction at all reduction

Sample	NiO (%)	Fe_2O_3 (%)
Initial Sample	2.66	15.82
Mass ratio 1:1	2.32	15.86
Mass ratio 1:2	2.30	15.41
Mass ratio 1:3	2.72	11.66
Mass ratio 1:4	2.67	16.64

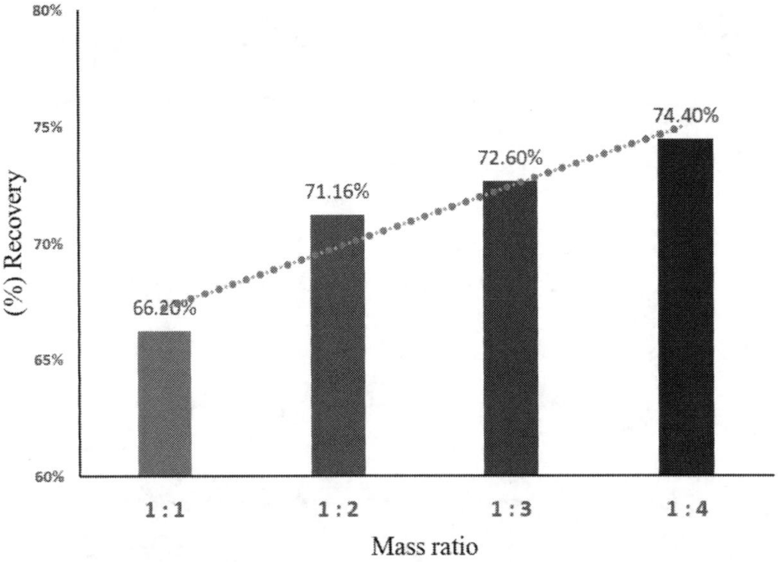

Figure 3. Comparison of % Recovery of samples after reduction with different mass ratio (laterite : palm kernel shell).

The purpose of XRF examination is to know the NiO and Fe_2O_3 content of carbothermic reduction result in each variable, so that we can know the most optimum reduction variable from the process. From the data above, it can be seen that the highest NiO content is obtained in the ratio of 1:3 and the lowest content is in sample 1:1. While for the Fe_2O_3 the highest content is obtained in the ratio 1:4 and the lowest content is in sample 1:3.

By calculating the sample weight and the element content after the reduction process compared with the weight and element content in the initial sample, data %recovery is obtained. The lowest value of nickel recovery was found in the sample 1:1 which is 66.20%, while the highest value of nickel recovery was obtained in the sample 1:4, with a recovery of 74.40%. The value of nickel recovery is projected into figure 3. It can be seen on the trend from the graph, there is an increasing number of recovery value along with the addition of mass of palm kernel shell as a reducing agent. The increase of the reduction degree obtained from the increasing number of reducing agents in this research is similar to the previous research results, both of which are coal [8] and sugarcane bagasse [9] as reducing agent.

4. Conclusion

a. XRD results showed that there is no Ni form appeared, this is due to the form of the mineral type in the form of silicate-bearing which binds the Ni element so that Ni were difficult to be reduced. But the reduction of hematite (Fe2O3) to wustite (FeO) was found. The wustite compound appears on the sample with variable 1:4.
b. The highest Ni % recovery was determined as 74.40% respectively with 80 grams palm kernel shell addition (sample 1:4) for 60 minutes reduction process.
c. The XRF results showed improvement of NiO and FeO in line with the addition of the mass of palm kernel shell. It was observed that n reduction process, the addition of palm kernel shell with various mass ratio does affecting metallization process of Fe and Ni in small amount.

5. Acknowledgement

Special thanks are due to Mr. Agus Budi Prasetyo (Research Centre for Metallurgy, Indonesian Institute of Sciences - LIPI, Indonesia) for helpful supports. The authors also wish to thank the Ministry of Research Technology and Higher Education and Directorate of Research and Community Engagement, Universitas Indonesia, under PUPT Research Grants Contract No. 2748/UN2.R3.1/ HKP05.00/2017 for the financial support of this research.

6. References

[1] U.S. Geological Survey 2017 *Mineral Commodity Summaries 2017* vol 1
[2] Rao M, Li G, Jiang T A O, Luo J U N, Zhang Y and Fan X 2013 Carbothermic Reduction of Nickeliferous Laterite Ores for Nickel Pig Iron Production in China : A Review **65**
[3] Quast K, Connor J N, Skinner W, Robinson D J and Addai-mensah J 2015 Preconcentration strategies in the processing of nickel laterite ores Part 1 : Literature review *Miner. Eng.* **79** 261–8
[4] Domínguez A, Valero A and Valero A 2013 Exergy accounting applied to metallurgical systems : The case of nickel processing *Energy* **62** 37–45
[5] Guo D, Zhu L, Guo S, Cui B, Luo S, Laghari M, Chen Z, Ma C, Zhou Y, Chen J, Xiao B, Hu M and Luo S 2016 Direct reduction of oxidized iron ore pellets using biomass syngas as the reducer *Fuel Process. Technol.* **148** 276–81
[6] Guo D, Hu M, Pu C, Xiao B, Hu Z, Liu S, Wang X and Zhu X 2015 Kinetics and mechanisms of direct reduction of iron ore-biomass composite pellets with hydrogen gas *Int. J. Hydrogen Energy* **40** 4733–40
[7] Forster J, Pickles C A and Elliott R 2016 Microwave carbothermic reduction roasting of a low

grade nickeliferous silicate laterite ore *Miner. Eng.* **88** 18–27
[8] Soedarsono J W, Kawigraha A, Sulamet-ariobimo R D, Asy M A, Yosi A and Putra E M 2013 The Influence of Coal and Reduction Process Parameters in Producing Iron Nugget *Adv. Mater. Res.* **789** 517–21
[9] Soedarsono J W, Arifin C E, Saragi J S, Putra A A, Kawigraha A, Sulamet-Ariobimo R D and Rustandi A 2017 The Effect of Reduction Parameter in Processing Lump Ore with Green Sugarcane Bagasse Reductor in Muffle Furnace *Mater. Sci. Forum* **893** 195–201

IOP Conf. Series: Earth and Environmental Science **105** (2017) 012017 doi:10.1088/1755-1315/105/1/012017

Pyrolysis of activated carbon from coconut shell and its characteristic in the $LiFePO_4/V/C$ composite for lithium ion battery cathode

N Sofyan[1,2], A D Rachmawati[1], A Zulfia[1,2] and A Subhan[3]
[1] Department of Metallurgical and Materials Engineering, Faculty of Engineering, Universitas Indonesia, Depok 16424, Indonesia
[2] Tropical renewable Energy Center, Faculty of Engineering, Universitas Indonesia, Depok 16424, Indonesia
[3] Research Center for Physics, Indonesian Institute of Science (LIPI), Puspiptek Serpong, Cisauk-Banten, Indonesia 15314

E-mail: nofrijon.sofyan@ui.ac.id

Abstract. Most of the time, coconut shells from the coconut farms have not been used but for charcoal purpose. In this work, the charcoal from the coconut shells was converted into an activated carbon and used it for the development of lithium ion battery. The development was begun by firstly synthesizing $LiFePO_4$ (LFP) through a hydrothermal route using stoichiometric amounts of precursors LiOH, $NH_4H_2PO_4$, and $FeSO_4.7H_2O$. The as-synthesized LFP was then mixed with variation of vanadium concentrations and a fix concentration of the carbon pyrolyzed from the coconut shells. X-ray diffraction (XRD) was used to characterize the crystal structure, whereas a scanning electron microscope (SEM) was used to characterize surface morphology of the composite. The characteristic of the composite was further examined an electrochemical impedance spectroscopy (EIS) for the conductivity. The XRD results showed that the $LiFePO_4/V/C$ has been formed successfully with an olivine structure. The SEM results depicted an agglomerate morphology but most of $LiFePO_4/V$ particles have been coated by the carbon. The EIS results showed conductivity values of 1.3387×10^{-2} S/cm, 1.184×10^{-3} S/cm, 1.7241×10^{-3} S/cm, and 6.6423×10^{-4} S/cm for the LFP/C-0V, LFP/C-3V, LFP/C-5V, and LFP/C-7V samples, respectively. The performance test indicated that coconut shell has a great potential as a cheap carbon resource for the development of lithium ion battery cathode.

1. Introduction

Indonesia is one of the largest coconut producers in the world with a plantation area of 2.526 million hectares and a total production of 2.833 million tons in 2016 [1]. In its processing, the commonly used coconut commodity is the inner white meat for copra and oil extraction, while the rest as a shell will be wasted. This coconut shell waste, both in industrial and household scales, is usually discarded, or very few use for its charcoal purposes. At the same time, disposal of this amount of coconut waste would be a problem if a proper waste management and treatment is not performed. On the other hand, this waste could be turned into high value and useful products in many applications. Utilization of the coconut shell wastes as a source of activated carbon, for example, is one of the solutions to minimize the coconut waste, to increase the coconut shell waste value, and at the same time to overcome the aforementioned problems.

Activated carbon is a form of carbon processed to have small particle size but with high pore volume and thus will increase the available surface area used in many applications [2]. The use of coconut shell derived activated carbon as an active anode material to improve the performance of lithium ion battery has been attempted by many investigators [3]. This activated carbon attracts many attentions because of its abundance, cost effective and environment friendly in nature [4]. It has also been known that activated carbon synthesized from coconut shell is considered to have a better performance in comparison to those obtained from other sources because of its mesoporous structure, which will make it more suitable for electrode material [5–7].

Lithium ferro phosphate ($LiFePO_4$) has attracted many attentions since the reversibility of intercalation-deintercalation lithium ion in electrochemical process was observed [8], primarily as a promising candidate for lithium ion battery cathode. Many advantages of this material have been reviewed such as low production cost, environmental friendly and high capacity and stability cycle [9]. However, despite its many advantageous, $LiFePO_4$ also has a drawback in its pure form due to its low electronic conductivity, measured only 10^{-9} S/cm [10]. This low electronic conductivity will result in a low rate capability. Because of that, many efforts have been proposed by many investigators to improve this conductivity, for example through a metal doping [11, 12], carbon coating [9], and co-synthesis with carbon in powder metallurgy method [13].

In this current work, $LiFePO_4$ was synthesized via sol-gel method followed by hydrothermal route. The electrochemical properties of the material were improved by using vanadium as a metal dopant along with the addition of activated carbon pyrolyzed from low cost coconut shell to form a composite of $LiFe_{1-x}V_xPO4/C$. The effects of carbon and vanadium on the structure, morphology and electrochemical performance of the composite used as cathode active material in lithium ion battery are reported and discussed in detail.

2. Experimental Setup

The raw material of coconut was collected from local market in Jakarta. After the white meat was removed from the hard shell and the thin brown rind was pared off, the coconut shell was thoroughly washed with tap water to remove any adhering dirt and dried under room temperature condition. The dried coconut shell was then pyrolyzed at 700 °C and the result in the form of coconut shell charcoal was crushed and sieved into fine carbon fragments. The steps in obtaining the coconut shell charcoal are given schematically in figure 1.

Figure 1. Steps in carbon pyrolysis from coconut shell. Left to right is coconut, coconut shells, chopped and dried coconut shells, and coconut shells after pyrolysis.

The as-prepared coconut fine carbon was then activated by soaking it in the H_3PO_4 0.75 M solution for 7 hours. After the soaking, the mixture was filtered using filter paper and the carbon precipitate was dried in an oven at 105 °C for 10 hours. The dried carbon was further washed with hot distilled water to obtain a pH 7 and was dried again in an oven at 100 °C for 12 hours. The activated carbon was cooled to room temperature and ready for characterization using X-ray diffraction (XRD, ARL OPTX-2050) with Cu Kα radiation (λ = 1.5406 Å).

Lithium ferro phosphate, $LiFePO_4$, was synthesized in according to our previous work [14] and is explained as following. In a separated glass beaker, a stoichiometric amount of the precursors LiOH, $NH_4H_2PO_4$, and $FeSO_4.7H_2O$ (all analytical grade reagents from E-Merck) was dissolved in 30 ml distilled water. After dissolution, the three solutions were mixed into another glass beaker and was stirred until homogenized using a magnetic stirrer for 30 minutes. The homogenous mixed solution was then transferred into a Teflon autoclave and hydrothermally heated at 180°C for 20 hours. The precipitates were separated from the solution and washed with deionized water several times before being dried in an oven at 80°C for 18 hours. Crystal characteristics of the powder were examined using X-ray diffraction (XRD, ARL OPTX-2050) with Cu Kα (λ = 1.5406 Å).

The composites $LiFe_{1-x}V_xPO_4/C$ was prepared by firstly preparing the as-synthesized $LiFePO_4$ powders from the previous step each with variations of 0, 3, 5, and 7 wt.% H_4NO_3V in a ball mill. An amount of 3 wt.% activated carbon was then added into the mixture and the ball-milling process continued until a homogenized mixture was obtained. The mixture was then sintered at 750°C for four hours under an argon environment. The characteristics of the resulting cathode active materials were examined by using X-ray diffraction (XRD, Philips PW3020) with Cu Kα radiation (λ = 1.5406 Å) and a scanning electron microscope (SEM, FEI Quanta-650) to characterize the morphology of the formed composites.

The cathode was prepared using the best active material in terms of crystal structure and homogeneity of the composites from the previous step. The selected material was mixed with poly-vinylidene fluoride (PVDF, MTI) in N-methyl pyrrolidone (NMP, MTI) solvent inside a vacuum mixer. The mixture in the form of paste was then applied onto a current collector of aluminium foil (MTI), dried, hot-rolled and heated in a vacuum oven at 80°C. The conductivity was further measured using an electrochemical impedance spectroscopy (EIS, Hioki LCR 3532-50).

3. Results and Discussion

X-ray diffraction was carried out to confirm the crystal structure of the sample by performing the test using Cu Kα radiation (λ = 1.5406 Å). The diffraction pattern is shown in figure 2. As seen in the figure, inset is the activated carbon from the coconut shell, (a) reference $LiFePO_4$, (b) commercial $LiFePO_4$, (c) as-synthesized $LiFePO_4$ with carbon addition only, (d) 0 wt.% V-doped $LiFePO_4/C$, (e) 3 wt.% V-doped $LiFePO_4/C$, (f) 5 wt.% V-doped $LiFePO_4/C$, and (g) 7 wt.% V-doped $LiFePO_4/C$. For the activated carbon from the coconut shell, as can be seen in the inset, the pyrolyzed sample shows the same diffraction patterns to that of the reference (Nixon et al, 1966) with a structure indexed to the rhombohedral R m space group (JCPDS 74-2328) but with the broaden peaks as an indication of amorphous tendency. The diffractograms show two peaks at around 25° and 43° corresponding to the crystalline reflections from (002) and (101) planes. No typical diffraction peaks of other phases are found in the diffraction patterns.

For the $LiFePO_4$ samples, all of the diffraction peaks of the as-synthesized samples with carbon addition are still in agreement with that of $LiFePO_4$ reference with a structure crystal indexed to the Pnma orthorhombic (JCPDS No.083-2092). This result is in agreement with the result found by others [15, 16]. The same result is true for the synthesized $LiFePO_4$ with carbon addition and various concentrations of up to 7 wt.% vanadium. Some discrepancies exist, however, in terms of impurities and or peaks shifting due to the carbon and or vanadium addition. This is specifically true for the as-synthesized $LiFePO_4$ and the $LiFePO_4$ with carbon addition only. As can be seen in figure 2(c) and (d), some other diffraction peaks are found in the diffractograms. These impurities that are yet to be confirmed could be due to the oxidation process occurred during the sintering process. The impurities peaks, however, are disappeared after vanadium addition at which no typical diffraction peaks of other phases are found in the diffraction patterns.

High crystallinity has been obtained in all of vanadium-doped $LiFePO_4/C$ indicated by strong and sharp diffraction peaks with no typical diffraction patterns of carbon nor vanadium as can be seen in figure 2(e), 2(f), and 2(g). The non-existence of carbon and vanadium diffraction patterns within the materials is an indication that carbon could exists in an amorphous form and that vanadium has substituted and occupied Fe sites in $LiFePO_4$. It could be also that they have been dissolved into $LiFePO_4$ structure to form a solid solution of $LiFe_{1-x}V_xPO_4/C$ composite as expected.

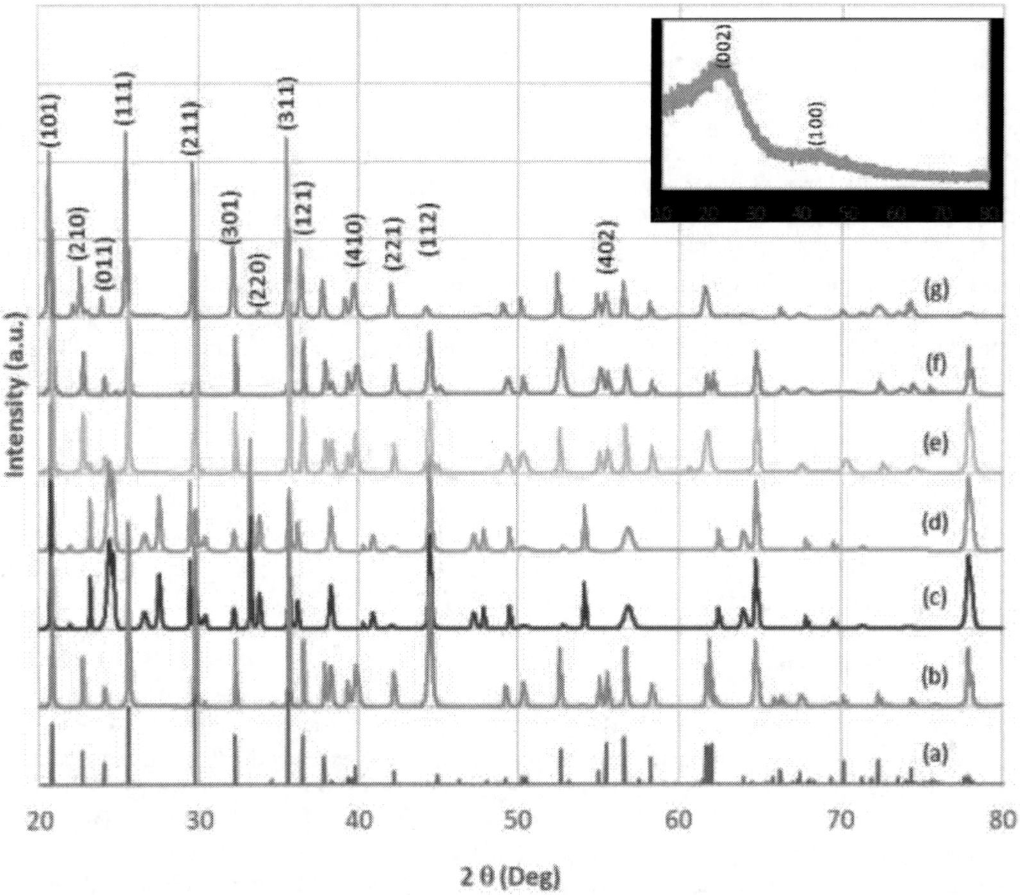

Figure 2. Diffraction patterns of (a) reference $LiFePO_4$, (b) commercial $LiFePO_4$, (c) as-synthesized $LiFePO_4$, (d) $LiFePO_4/C$ at 0 wt.% V, (e) $LiFePO_4/C$ at 3 wt.% V, (f) $LiFePO_4/C$ at 5 wt.% V, and (f) $LiFePO_4/C$ at 5 wt.% V. Inset is the X-ray diffraction pattern of the activated carbon from the coconut shell

Surface morphology of the as-synthesized activated carbon from the coconut shell, $LiFePO_4$, and the $LiFePO_4/V/C$ composites was revealed using an electron microscope. The results from the secondary electron image are given in figure 3. Figure (a) is the activated carbon from the coconut shell, figure 3(b) is the as-synthesized $LiFePO_4$, figure (c) 0 wt.% V-doped $LiFePO_4/C$; (d) 3 wt.% V-doped $LiFePO_4/C$; (e) 5 wt.% V-doped $LiFePO_4/C$; (f) 7 wt.% V-doped $LiFePO_4/C$. There is a difference observed on the surface morphology of the composite between the as-synthesized $LiFePO_4$ and the one with the addition of activated carbon from the coconut shell. In figure 3(a), most of the particles in the as-synthesized activated carbon show the morphology of particles having flake shape and are still agglomerated. The same is true for the as-synthesized $LiFePO_4$, which is the characteristic of $LiFePO_4$ formed through a hydrothermal process [17] as can be seen in figure 3(b). However, the agglomeration tendency is reduced with the addition of carbon. Figure 3(c) shows the morphology of with no activated carbon. The sample has a round shape with a smaller particle than that of the as-synthesized $LiFePO_4$. figure 3(d) is the morphology of $LiFePO_4/C$ at 3 wt.% vanadium, figure 3(e) shows the morphology of $LiFePO_4/C$ at 5 wt.% vanadium, and figure 3(f) shows the morphology of $LiFePO_4/C$ at 7 wt.% vanadium. The morphology reveals that $LiFePO_4/C$ agglomeration tendency is well reduced with the addition of vanadium and thus the particle distributes homogeneously at 7 wt.% vanadium.

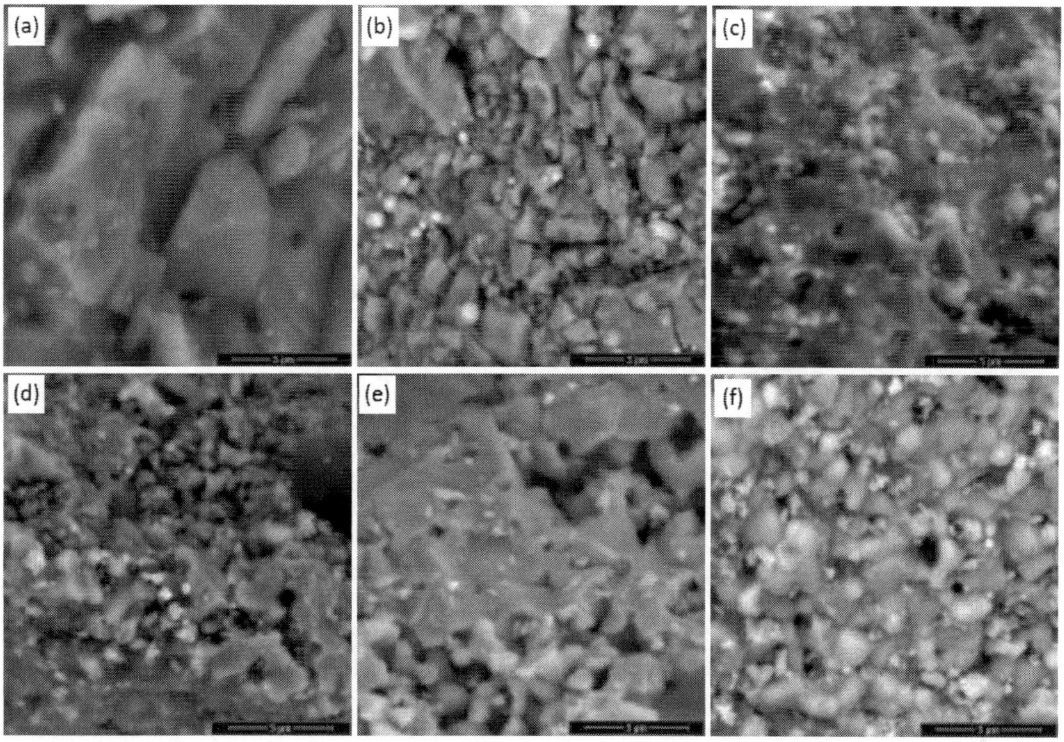

Figure 3. Secondary electron images of (a) activated carbon from the coconut shell; (b) as-synthesized $LiFePO_4$; (c) 0 wt.% V-doped $LiFePO_4/C$; (d) 3 wt.% V-doped $LiFePO_4/C$; (e) 5 wt.% V-doped $LiFePO_4/C$; and (f) 7 wt.% V-doped $LiFePO_4/C$. The bar scale is 5 μm.

Electrical conductivity was performed on the prepared composite cathode sheets using electrical impedance spectroscopy to show the conductivity through a plot diagram depicting the material's capacitance. The results in the forms of Cole plots are shown in figure 4.

As seen in figure 4, the testing was performed on a fix carbon composition and various vanadium concentrations to form $LiFePO_4/C/V$ at 0 wt.% V, 3 wt.% V, 5 wt.% V and 7 wt.% V. Analysis of the Cole-Cole diagrams in figure 4 shows values of $1.3387x10^{-2}$ S/cm, $1.184x10^{-3}$ S/cm, $1.7241x10^{-3}$ S/cm, and $6.6423x10^{-4}$ S/cm for the LFP/C-0V, LFP/C-3V, LFP/C-5V, and LFP/C-7V samples, respectively. When compared to the pure $LiFePO_4$ [10], the carbon coated $LiFePO_4$ shows an increase of conductivity. This is due to the presence of high conductivity carbon that helps intercalation and deintercalation of the Li ion. In this work, however, all V-doped $LiFePO_4/C$ samples shows a lower conductivity compared to that of carbon coated $LiFePO_4$. The uneven presence of vanadium and carbon and the presence of pores can increase resistance by forming a bulk layer that blocks and decreases the electrical conductivity.

Cyclic voltammetry and charge/discharge testing were performed for the two active materials only, i.e. $LiFePO_4/C$ at 3 wt.% V and 7 wt.% V. The resulting curves are given in figure 5. As seen in figure 5, cyclic voltammetry curves shown at the top have reduction-oxidation peaks at 3.59 and 3.20 volts for 3 wt.% V (left) and at 3.59 and 3.05 volts for 7 wt.% V (right), respectively. The charge/discharge testing of the composite was performed at the potential range of 2.5–4.2 V at two capacities, i.e. 0.5C and 4C, and the curves are shown at the bottom in figure 5 for the LiFePO4/C at 3 wt.% V (left) and 7 wt.% V (right). The initial charge and discharge capacities at 0.5C for LiFePO4/C at 7 wt.% V (4.8 mAh g^{-1}) are higher than that of 5 wt.% V (3.2 mAh g^{-1}), however, the charge and discharge capacity of 7 wt.% V drops below 1.2 mAh g^{-1} at 4C whereas for LiFePO4/C at 7 wt.% V, the capacity is still 1.6 mAh g^{-1}. It seems that, as has been mentioned previously, when the vanadium concentration is high, at low cycle, it can accommodate the high capacity. However, at higher cycle, the uneven presence of vanadium and carbon and the presence of pores starts forming a bulk layer that blocks and decreases the electrical conductivity and thus the capacity.

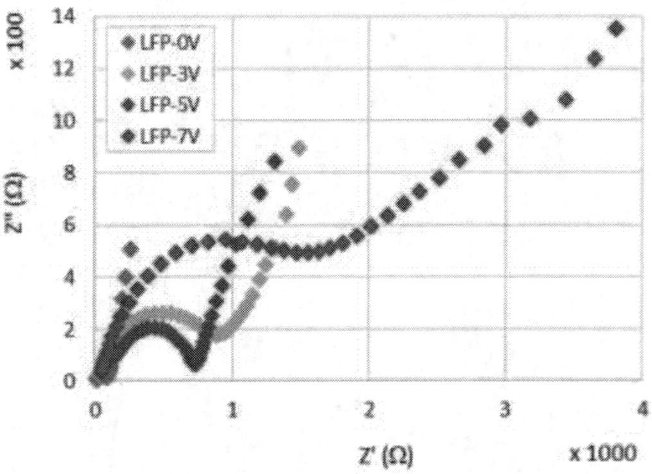

Figure 4. Cole-Cole plots of EIS test results for 0 wt.% V-doped, 3 wt.% V-doped, 5 wt.% V-doped, and 7wt.% V-doped $LiFePO_4/C$

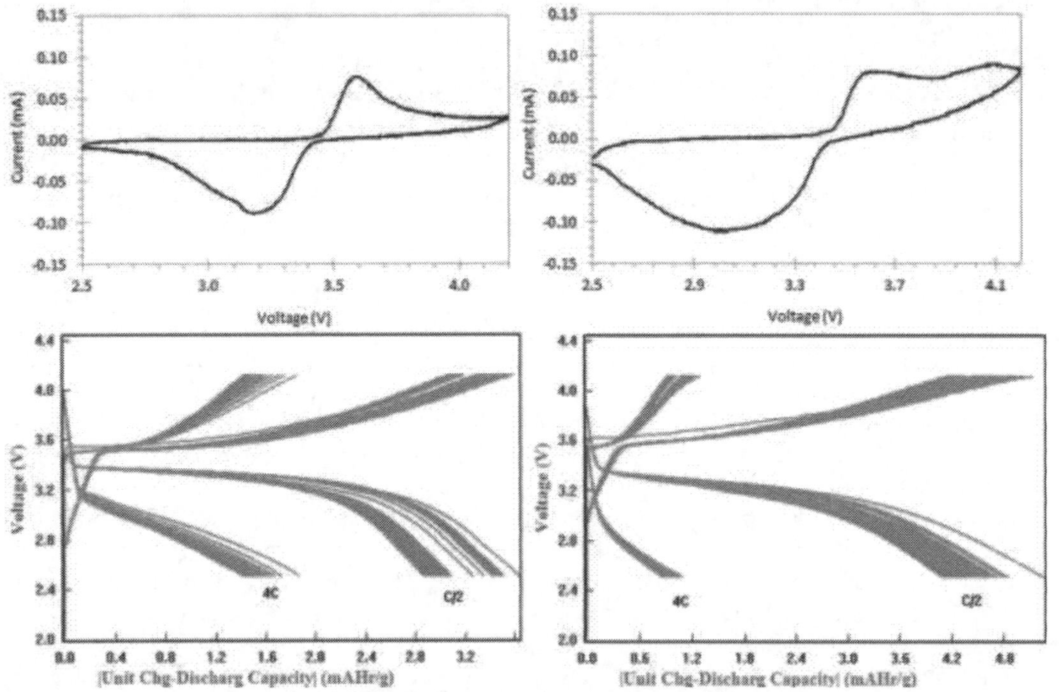

Figure 5. Cyclic voltammetry (top) and charge/discharge (bottom) curves for the $LiFePO_4/V/C$ at 3 wt.% V (left) and 7 wt.% V (right)

4. Conclusion
In this work, the hydrothermal process has resulted in high purity of $LiFePO_4$ and morphologically formed agglomerated small particles containing $LiFePO_4$. The addition of activated carbon from the coconut shell did not affect the purity of the $LiFePO_4$ particles, however, the addition of vanadium shifts the triphylite peaks due to the emergence of the impurity phase. The addition of vanadium also decreases the conductivity because of uneven distribution of vanadium and carbon with a value of $1.3387x10^{-2}$ S/cm, $1.184x10^{-3}$ S/cm, $1.7241x10^{-3}$ S/cm, and $6.6423x10^{-4}$ S/cm for the LFP/C-0V, LFP/C-3V, LFP/C-5V, and LFP/C-7V samples, respectively.

5. Acknowledgement
The authors would like to express their gratitude and appreciation for the funding from the Directorate of Research and Community Services (DRPM), Universitas Indonesia, through Hibah PITTA No. 822/UN2.R3.1/HKP.05.00/2017.

6. References
[1] Subiyantoro M E and Arianto Y (Eds) 2016 *Tree Crop Estate Statistics of Indonesia 2014-2016* (Jakarta: Directorate General of Estate Crops)
[2] Ahmadpour A, Okhovat A, Mahboub M J D 2013 *J. Phys. Chem. Solids* **74** 886
[3] Hwang Y J, Jeong S K, Shin J S, Nahm K S, Stephan A M 2008 *J. Alloys Compd.* **448** 141
[4] Frackowiak E and Béguin F 2002 *Carbon* **40** 1775
[5] Zhou X, Li L, Dong S, Chen X, Han P, Xu H, Yao J, Shang C, Liu Z, and Cui G 2012 *J. Solid State Electrochem.* **16** 877
[6] Xiong W, Liu M, Gan L, Lv Y, Li Y, Yang L, Xu Z, Hao Z, Liu H and Chen L 2011 *J. Power Sources* **196** 10461
[7] Kuratani K, Okuno K, Iwaki T, Kato M, Takeichi N, Miyuki T, Awazu T, Majima M and Sakai T 2011 *J. Power Sources* **196** 10788
[8] Padhi A K, Nanjundaswamy K S and Goodenough J, 1997 *Electrochem. Soc* **144** 1188
[9] Hu L-H, Wu F Y, Lin C T, Khlobystov A N, and Li L J 2013 *Nature Commun* **4** 1687 1
[10] Chung S Y, Bloking J T and Chiang Y M 2002 *Nat. Mater.* **1** 123
[11] Arumugam D, Kalaignan G P and Manisankar P 2008 *J. Solid State Electrochem.* **13** 301
[12] Liu H, Xie J and Wang K 2008 *Solid State Ionics* **179** 1768
[13] Dominko R, Bele M, Gaberscek M, Remskar M, Hanzel D, Pejovnik S and Jamnik J 2005 *J. Electrochem. Soc.* **152** A607
[14] Sofyan N, Putro D Y and Zulfia A 2016 *Int. J. Technol.* **7** 1307
[15] Yang M R and Ke W H, 2008 *J. Electrochem. Soc.* **155** A729
[16] Teng T H, Yang M R, Wu S H 2007 *Solid State Commun.* **142** 389
[17] Shu H, Wang X, Wu Q, Liang Q, Yang X, Yang S, Liu L, Wei Q, Hu B, Zhou M, Chen M, and Zhang L 2012 *J. Electrochem. Soc.* **159** A1904

IOP Conf. Series: Earth and Environmental Science **105** (2017) 012018 doi:10.1088/1755-1315/105/1/012018

Degradation of methylene blue using silver nanoparticles synthesized from *imperata cylindrica* aqueous extract

A A Fairuzi[1], N N Bonnia[1*], R M Akhir[1], M A Abrani[2], H M Akil[3]
[1]School of Physics & Materials Studies, Faculty of Applied Sciences, Universiti Teknologi MARA, 40460 Shah Alam, Selangor, Malaysia
[2]School of Chemistry, Faculty of Applied Sciences, Universiti Teknologi MARA, 40460 Shah Alam, Selangor, Malaysia
[3]School of Materials and Mineral Resources Engineering, Engineering Campus, Universiti Sains Malaysia, 14300 Nibong Tebal, Penang, Malaysia

*E-mail: noornajmi@salam.uitm.edu.my

Abstract. The present study reports about the catalytic degradation of methylene blue (MB) by sodium borohydride ($NaBH_4$) in presence of silver nanoparticles (Ag NPs) as catalyst. Ag NPs was produced from green synthesis method using aqueous *Imperata cylindrica* extract. Formation of Ag NPs from the synthesis was confirmed by using UV-visible spectroscopy with the appearance of surface plasmon (SPR) bands around 427 nm. The agglomeration in microstructure of biosynthesized Ag NPs was revealed by field scanning electron microscope (FESEM). The synthesized of Ag NPs showed five diffraction peaks at 2θ values of 38.62°, 44.51°, 64.84°, 77.68° and 81.96° which corresponds to the (111), (200), (220), (311) and (222) planes of face-centred cubic. Energy dispersive X-ray analysis (EDX) revealed signals at the energy of 3 keV which indicates the presence of elemental silver that contributed to 68.44 weight % of the analysed sample. Upon addition of biosynthesized silver nanoparticles, the degradation of MB increases up to 92.06% within 14 min.

1. Introduction

Nanotechnology is the science of producing and utilizing nano-sized particles [1]. Ag NPs is useful various fields such as catalysis, optics, biomedical, pharmaceutical and sensor technology [2]. Although there are many methods that can be used to synthesize Ag NPs, most of them utilized toxic chemicals and may include the use of enormous energy. This scenario is not economic and cause serious pollution to environment. Synthesis of nanomaterials using biological entities is gaining attention as; biological methods are less expensive, nontoxic and environmentally acceptable "green chemistry" procedures [3]. The use of plant and plant extract in nanoparticles synthesis is advantageous compared to microbial based system because it eliminates the intricate process of maintaining cell cultures. In this study, Ag NPs nanoparticles were synthesized from aqueous extract of *I. Cylindrica* (Figure 1). This perennial grass lived well in Southeast Asia, the Philippines, China, and Japan [4]. *Imperata cylindrica* (L.) Beauv. or cogon grass is a plant in division of *Magnoliophytai*; class *Liliopsida*; order *Cyperales*; family *Poaceace/Gramieane* and genus *Imperata Cirillo*. Study of this plant by Krishnaiah et al. (2009) proved the traditional medicinal value *I. cylindrica* was due to presence of several bioactive constituents such as alkaloids, flavonoid, tannins and cardiac glycoside

[5]. These phytoconstituents is vital in biosynthesis of various metallic nanoparticles as reducing and capping agent.

Figure 1. Photograph of *I. cylindrica* plant.

In this study, the chemical and physical properties of biosynthesized Ag NPs were characterised. Silver nanoparticles as describe in previous study is a good catalyst. As an extension to previous study, catalytic ability of the Ag NPs synthesized from *I. cylindrica* extract was investigated. The investigation was carried out studying degradation of aqueous MB, an organic dye as model.

2. Experimental section.

2.1. Biosynthesis of Ag NPs

The leaves of *I. cylindrica* were collected from specialized area inside the Universiti Teknologi MARA, Shah Alam, Selangor, Malaysia campus. The collected leaves were cleaned thoroughly by distilled water and then oven dried at 60°C for 48 hrs. The dried leaves were finely ground to powder using conventional food processor. Aqueous leaves extract was prepared by dissolving 5 g of the powder in 200 ml distilled water. The mixture was stirred and heated at 60 ^{0}C for 30 min and the leaves residue was filtered out from the extract. Ag NPs was synthesized by adding10 ml of the aqueous extract into 90 ml of 5 mM aqueous $AgNO_3$. The mixture was heated using water bath at 60 ^{0}C for 10 min. Appearance of brown colour marks the formation of Ag NPs.

2.2. Characterization of biosynthesized Ag NPs

The formation of Ag NPs was studied by measuring the absorption spectra of the mixture at a regular time interval using UV-visible double beam spectrophotometer (Perkin Elmer, Lambda 35 UV/Vis System) operated in 700 -300 nm range at a resolution 1 nm. X-ray diffractograms of the samples were recorded by X-ray diffractometer (Rikagu Ultima IV, Japan). FTIR spectrum biosynthesized Ag NPs was recorded in diffuse reflectance mode operated at a resolution 4cm^{-1} at wavelength of 4000 - 400 cm^{-1} (Thermo Scientific, Nicolet 6700). Morphology of the Ag NPs was obtained by field emission scanning microscope (FESEM) SUPRA-40 VP attached with energy dispersive X-ray (EDX) unit.

2.3. Dye degradation study

The experiment was conducted by adding 2 ml of freshly prepared NaBH4 solution (0.2 M) to 50 ml (10 mg l^{-1}) aqueous solution of MB at room temperature (30 ±1°C). To this 1 ml of silver nanocatalyst were added and the reactants were mixed quickly. At a regular interval of time, 0.5 ml of the reaction mixture was withdrawn and diluted in 2.5 ml of distilled water. The absorbance of the mixture was monitored using UV–visible spectrophotometer at regular intervals of time. The degradation was also monitored without silver nanoparticles.

3. Results and discussion

3.1. UV-visible spectroscopic study

UV-visible spectroscopy characterization method was then used to monitor formation of Ag NPs and its properties. The absorption of light by the nanoparticles at different wavelengths provides an indication of particle size, while the breadth of the peaks signifies the particle size distribution [6]. The characteristic SPR band of Ag NPs is in the range of 400-450 nm [7]. Biosynthesized Ag NPs were evaluated at periodically for 60 minutes. Figure 2 shows the intensity UV- visible spectra of Ag NPs increase as the reaction time increase. This indicates the increased in concentration of the Ag NPs as time of reaction increased. The UV-vis spectroscopy showed that the nanoparticles started to form after 15 min the reaction commence, as depicted by the appearance of surface plasmon bands around 427 nm. As the reaction continued, absorbance peaks became narrower and were shifted to shorter wavelength. These condition signifying that the particles became smaller and are getting monodispersed [8].

3.2. X-ray diffraction (XRD) analysis

The powder XRD pattern was recorded for the identification of phases exhibited by the biosynthesized Ag NPs. The synthesized of Ag NPs showed five diffraction peaks (Figure 3) at 2θ values of 38.62°, 44.51°, 64.84°, 77.68° and 81.96° which corresponds to the (111), (200), (220), (311) and (222) planes of face-centred cubic. The peak corresponding to the (111) plane is more intense than that of other planes suggesting the predominant growth of Ag NPs along (111) direction.

3.3. FESEM and EDX analysis

The silver colloid was oven dried at 90°C for 72 hrs to remove water from the synthesized nanoparticles. The dried particles were brought to FESEM analysis. Micrograph (figure 5) of Ag NPs revealed agglomeration of the particles. The average particle size was 31 nm. EDX analysis displayed signal at the energy of 3 keV indicates that the silver has been correctly identified [9]. Elemental silver contributed 68.44 weight % of the analysed sample. A representative EDX spectrum is presented in figure 4.

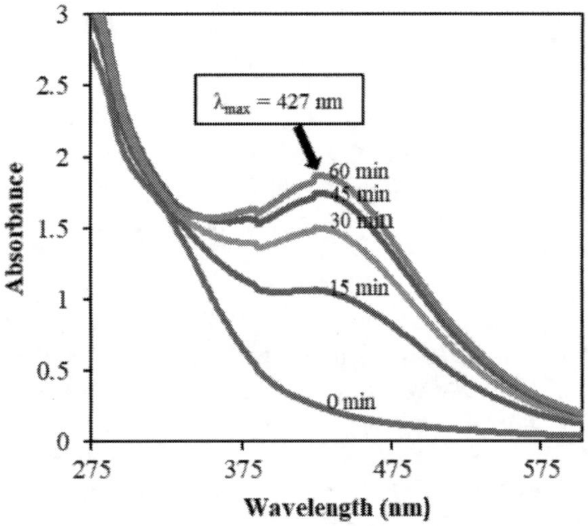

Figure 2. UV- visible spectra of Ag NPs

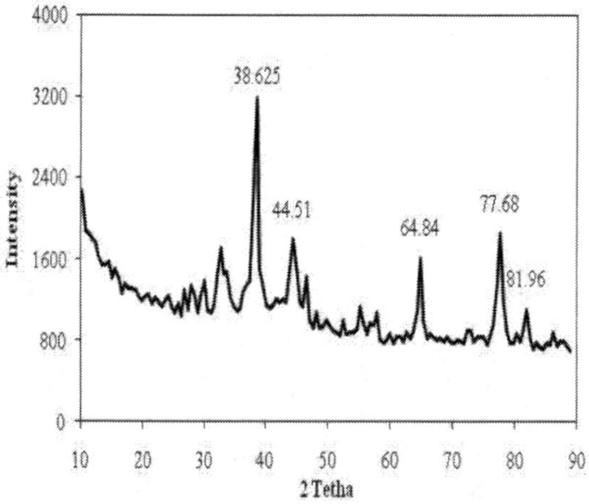

Figure 3. XRD pattern of Ag NPs

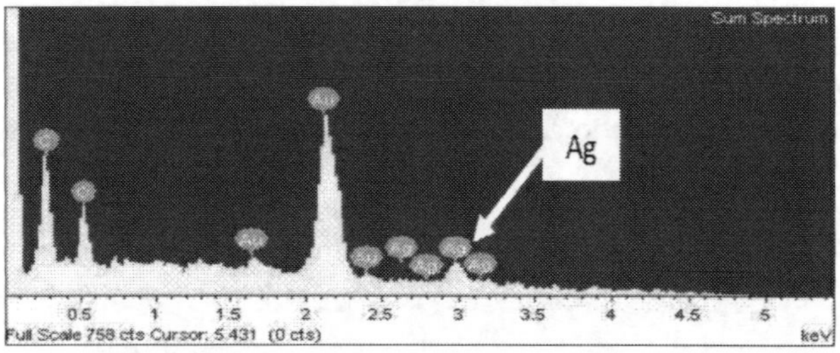

Figure 4. EDX spectrum of AgNPs.

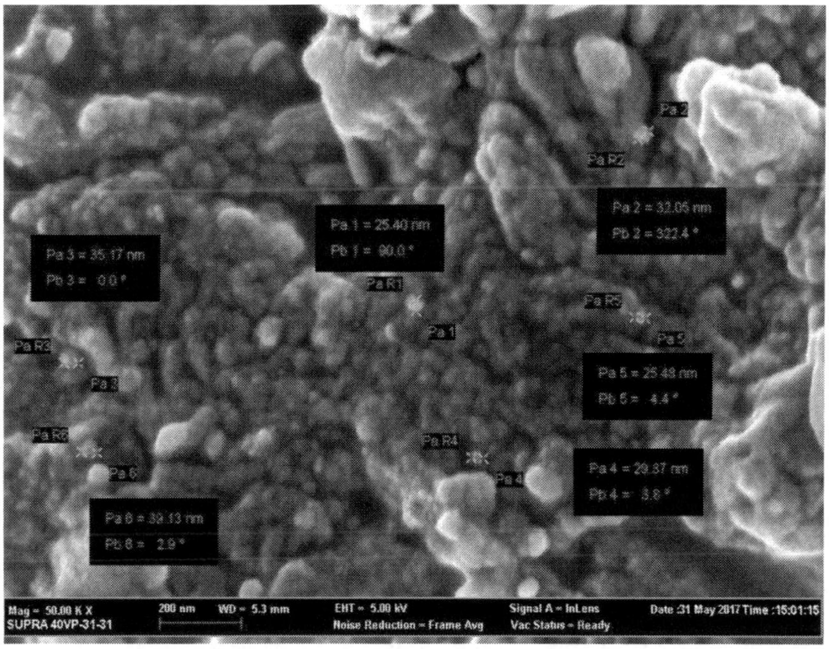

Figure 5. FESEM micrograph of biosynthesized Ag NPs.

3.4. Fourier transform infrared spectroscopy analysis

FTIR analysis on the biosynthesized Ag NPs was carried out to identify the functional groups responsible for the reduction of metal precursors ($AgNO_3$). The spectral (figure 6) analysis of the extract shows IR bands at 3253.70, 2922.30, 1594.00, 1397.53, and 1260.87 cm^{-1}. The bands corresponds to O-H vibration, C-H vibration, C=C groups or from aromatic rings, germinal methyls [7] and Amide III [8].

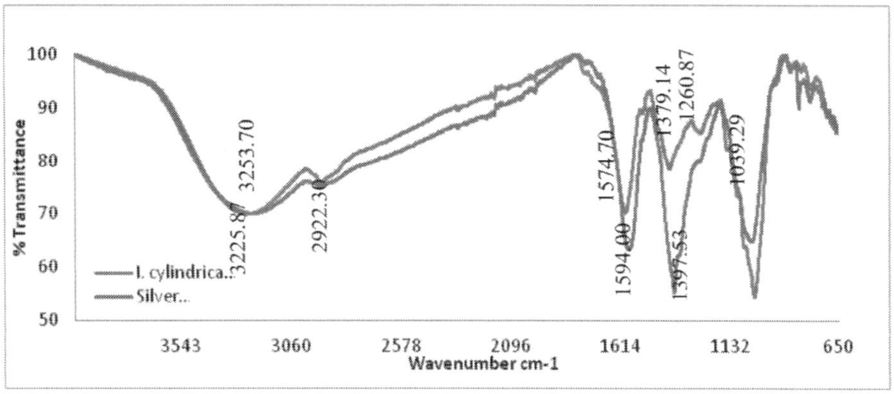

Figure 6. FTIR spectra of *I. cylindrica* extract and biosynthesized AgNPs

3.5. Catalytic activity of Ag NPs

A potential application of biosynthesized Ag NPs was as catalyst in the reduction of aqueous MB to Leuco MB in the presence of $NaBH_4$. Ag NPs act as an electron relay and initiate shifting of electron from BH_4^- ion (donor B_2H_4/BH_4^-) to acceptor (acceptor LMB/MB) and thus causing reduction of the dye. BH_4^- ion simultaneously adsorbed on the surface of NPs and thus electron transfer occurs from BH_4^- ion to the dye through NPs[12] .In aqueous medium, MB shows maximum absorption at 664 nm and a shoulder peak at 615 nm. The reaction was monitored by UV-visible spectroscopy in the wavelength range between 300 and 900 nm at room temperature. The decrease of absorbance at maximum wavelength (664 nm) with time was followed spectrophotochemically (figure 7a and 7b). The intense blue colour of the MB solution faded and become colourless during degradation process. Percentage of degradation (table 1) was calculated as quantitative representation of the degraded dye.

$$Dye\ degradation\,(\%) = [(C_0 - C)/C_0] \times 100 \qquad (1)$$

Where C_0 is the initial concentration of methylene blue solution and C is the concentration of the dye solution after t minutes of reaction. In the absence of biosynthesized nanocatalyst, the reduction process is negligible (dye degradation 1.02-2.61 %). Addition of biosynthesized silver nanoparticles improved the reduction process (dye degradation up to 92.06% within 14 min). Analysis of the kinetic data of degradation reactions reveals pseudo-first order reaction kinetics. The linear plot (figure 8) of ln $[C/C_o]$ versus time supports the kinetic theory, where the k value obtained was 0.137 min^{-1}.

Table 1. Degradation percentage of MB

Time (min)	% Degradation	
	Ag NPs	Control
2	15.39	1.02
4	25.00	2.61
6	34.74	2.55
8	46.84	1.80
10	56.39	2.59
12	86.69	2.53
14	92.06	1.90

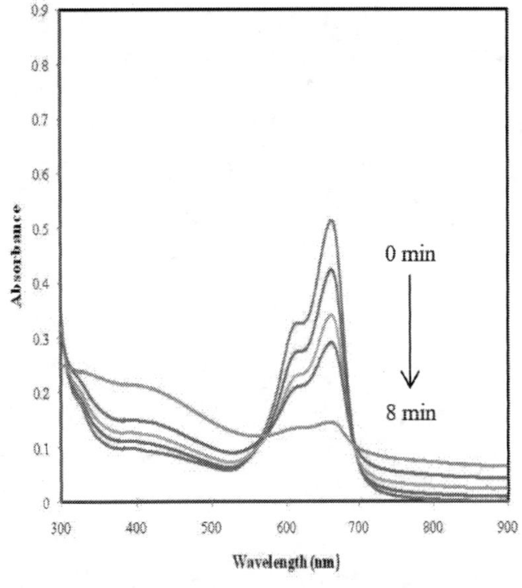

Figure 7a. Absorbance spectra of MB with nanocatalyst

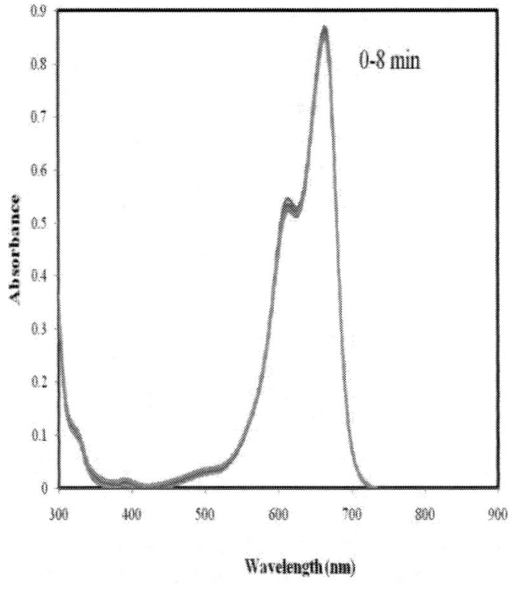

Figure 7b. Absorbance spectra of MB without nanocatalyst (control)

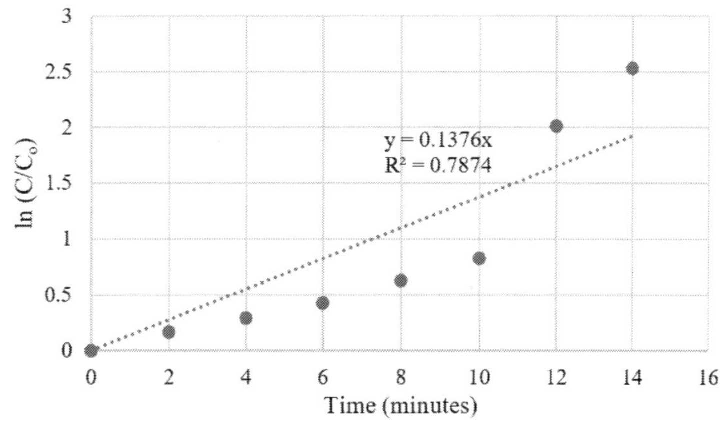

Figure 8. Kinetic studies of MB degradation catalysed by Ag NPs

4. Conclusion

The biosynthesized Ag NPs degrade (presence of $NaBH_4$) 92% MB within 14 min. Catalytic properties of Ag NPs synthesized from aqueous extract of *I. cylindrica* in degrading organic dye was demonstrated.

Acknowledgments
Authors wish to acknowledge Ministry of Higher Education Malaysia for RACE Grant Scheme (600-RMI/RACE 16/6/2(3/2014) and Faculty of Applied Sciences, Universiti Teknologi Mara for research facilities.

References

[1] M. Sardar 2017, Current Biotechnological Applications of Green Synthesized Silver Nanoparticles, Journal of Nanosciences . **2** 1–8.

[2] P. Logeswari, S. Silambarasan, and J. Abraham 2013, Sharif University of Technology Ecofriendly synthesis of silver nanoparticles from commercially available plant powders and their antibacterial properties, *Sci. Iran.* **20** 1049–1054

[4] S. Joseph and B. Mathew, 2015 Microwave-assisted green synthesis of silver nanoparticles and the study on catalytic activity in the degradation of dyes, *J. Mol. Liq.* **204** 184–191,.

[5] G. E. MacDonald, 2004 Cogongrass (Imperata cylindrica) - biology, ecology, and management, *CRC. Crit. Rev. Plant Sci.* **23** 367–380.

[6] D. Krishnaiah, T. Devi, A. Bono, and R. Sarbatly, 2009 Studies on phytochemical constituents of six Malaysian medicinal plants,. **3** 67–72 .

[7] H. O. Edeoga, D. E. Okwu, and B. O. Mbaebie, 2005. Phytochemical constituents of some Nigerian medicinal plants, *African J. Biotechnol.*. **4** 685–688.

[8] M. Sastry, K. Mayyaa, and K. Bandyopadhyay, 1997. pH dependent changes in the optical properties of carboxylic acid derivatized silver colloid particles, *Colloids Surf A*, 221–228.

[9] N. Kumar, R. Bogireddy, H. Anand, K. Kumar, and B. K. Mandal, 2016. Biofabricated silver nanoparticles as green catalyst in the degradation of different textile dyes, *Journal of Environmental Chemical Engineering Biochem. Pharmacol.*, **4** 56–64.

[10] M. Vanaja, 2013. Phytosynthesis of silver nanoparticles by Cissus quadrangularis : influence of physicochemical factors,. 1–17.

[11] G. A. Bhaduri, R. Little, R. B. Khomane, S. U. Lokhande, B. D. Kulkarni, B. G. Mendis, 2013. Green synthesis of silver nanoparticles using sunlight, *J. Photochem. Photobiol. A Chem.*, **258** 1–9.

[12] M. Meena Kumari, J. Jacob, and D. Philip, 2015. Green synthesis and applications of Au-Ag bimetallic nanoparticles, *Spectrochim. Acta - Part A Mol. Biomol. Spectrosc.* **137** 185–192.

Dynamic operation of water gas shift reaction over $Fe_2O_3/Cr_2O_3/CuO$ catalyst in Pd/Al_2O_3 membrane reactor

Yogi Wibisono Budhi[1,a], Dhinny Dwi Putri[1], Afifa Husna[1], Hans Kristian Irawan[1], Manabu Miyamoto[2], Shigeyuki Uemiya[2]
[1]Chemical Engineering Study Program, Institut Teknologi Bandung, Jl. Ganesha 10, Bandung, Indonesia
[2]Department of Materials Science and Technology, Gifu University, 1-1 Yanagido, Gifu 501-1193, Japan

E-mail: Y.Wibisono@che.itb.ac.id

Abstract. Hydrogen has been considered as promising energy carrier that can be produced from renewable resources, such as biomass through gasification. This process results in producer gas containing CO, CO_2, H_2, N_2, and CH_4. The conventional enhancement of hydrogen is typically conducted using several unit operations such as water gas shift reactor (WGSR) and separation unit such as Pressure Swing Adsorption (PSA). Process intensification offers a new method to integrate both WGSR and separation unit into single membrane reactor. This research aimed to investigate the influence of dynamic operation on membrane reactor performance. The steady state fixed bed reactor and membrane reactor were used as base case to judge the performance of dynamic membrane reactor. The water gas shift reaction over $Fe_2O_3/Cr_2O_3/CuO$ was carried out at 350°C and 1 atm by varying the feed composition and gas residence time. The feed composition ratio of H_2O/CO consisted of 2 and 3 on mole basis, while the gas residence times were 1.2 s and 2.3 s. The membrane reactor consists of shell and tube sides made of Pd/Al_2O_3 material with technical specification of 10 mm inner diameter, 20 μm Pd thickness supported by alumina, and 10 cm reactor length. The compositions of the feed gas and products were measured using gas analysers such CO gas detector (Bacharach PCA® 3) and H_2 gas detector (Cosmos XP-3140). The dynamic operation was performed following the square wave perturbation of the feed gas at switching time of 15 s. The experiment results showed that increasing the feed composition ratio and gas residence time increased the conversion of CO and hydrogen production in the fixed bed reactor and membrane reactor. Higher production of hydrogen also improved the recovery of hydrogen in membrane reactor. The use of membrane reactor increased significantly the conversion of CO when compared to fixed bed reactor. Moreover, the dynamic membrane reactor would give much better performance in term of CO conversion and hydrogen recovery. The stability of the Pd/Al_2O_3 membrane reactor was proven for at least 10 h operation.

1. Introduction

Nowadays, many researchers develop sustainable energy resources due to limitation to fossil fuel. Hydrogen is expected and has been widely agreed as an alternative energy instead of fossil fuels because it is clean, environmentally friendly, and can be produced from various resources [1]. The energy density of hydrogen per mass unit is approximately 2.6 times higher than LPG [2]. Hydrogen could also be easily converted to electricity by fuel cell technology [3].

Water Gas Shift Reaction (WGSR) is often used as synthesis path to enhance the production hydrogen. The WGSR typically converts carbon monoxide (CO) by using steam to produce hydrogen and CO_2. The WGSR is an exothermic and reversible reaction, and therefore it is highly influenced by operating conditions. The enhancement of H_2O/CO ratio will increase the CO conversion [4] and prevent the carbon deposition in catalyst [5]. High reaction temperature will increase the reaction rate, while low temperature will increase the conversion. As consequence, the optimum design of WGSR results in two reactors in series, which are High Temperature Shift Converter (HTSC) and the Low Temperature Shift Converter (LTSC). High purity hydrogen can only be obtained through product separation process.

Process intensification opens new way in integrating the HTSC and LTSC, results in the so-called palladium (Pd) membrane reactor. Stankiewicz and Moulijn [6,7] defined process intensification as the development of novel equipment, processing techniques, and methods, when compared to traditional ones, offer substantial improvements in (bio) chemical processes. This technique will make the process more economical and efficient. Pd is a metal that have high permeability and selectivity of hydrogen [8]. Hydrogen diffused through Pd membrane according to solution-diffusion mechanism [9]. Enhancement of mechanical strength and reduction of capital expenditure can be achieved by setting a membrane support, such as alumina (Al_2O_3). The catalyst used in membrane reactor for WGSR is normal catalyst such as iron oxide.

There are many variables that influence the membrane performance, such as pressure difference, reaction temperature, sweep gas velocity, palladium thickness, and H_2O/CO feed ratio [10,11]. The membrane permeability during steady state operation could decrease over time due to concentration polarization and inhibition [12]. Low hydrogen recovery during steady state operation can also be induced by surface deactivation. The dynamic operation can produce higher hydrogen recovery through when certain time scale is applied. A square wave perturbation can be created by modulating the feed flow rate in high and low value at certain switching time [13,14]. According to Hoebink et al. [15], there are three different regimes of unsteady state operation: relaxed steady state (sliding), dynamic, and quasy-steady state.

2. Materials and Methods

2.1. Experimental System

Materials used in this research consist of high purity gases (H_2, N_2, CO), steam, $Fe_2O_3/Cr_2O_3/CuO$ catalyst (90/8/2-%w/w, density 1.12 kg/L, and surface area is about 110 m^2/g) in the range of 60 mesh to 80 mesh, inert for catalyst dilution in the fixed bed reactor, and liquid soap as the carrier of gas volumetric flow measurement. The fixed bed reactor has length (L) of 34.5 cm and inside diameter (ID) of 10 mm. The catalyst bed length inside the reactor is 5 cm. Meanwhile, the membrane reactor (L = 10 cm and ID = 10 mm) consists of Pd and porous Al_2O_3 layer with a thickness of 20 μm and 2 mm, respectively. The catalyst is placed on the shell side of membrane reactor. Feed and sweep gas (N_2, 5 ml/s) are flowed on a co-current configuration. Water was introduced to the reactor inlet through syringe pump then heated by heating coil. The heat needed for reaction was supplied by tube furnace. The reaction products were analysed by CO gas detector (Bacharach PCA® 3) and H_2 gas detector (Cosmos XP-3140). Both main equipment schemes are presented in Figure 1. Steady state operation in membrane reactor was carried out without modulator.

2.2. Experimental Procedure

The catalytic WGSR was conducted in the fixed bed reactor and Pd/Al_2O_3 membrane reactor. Based on Augustine [11], purging the reactor was accomplished by flowing N_2 for 30 minutes. The catalyst was subsequently activated by flowing steam and H_2 (1:1) for an hour at 400°C, followed by second purging using N_2 for 15 minutes and decreasing the temperature to reaction temperature of 350°C. After purging, heating, and activating the catalyst, main experiments were performed according to the specified variations. Data interpretations consist of CO conversion (X_{CO}) and hydrogen recovery (HR).

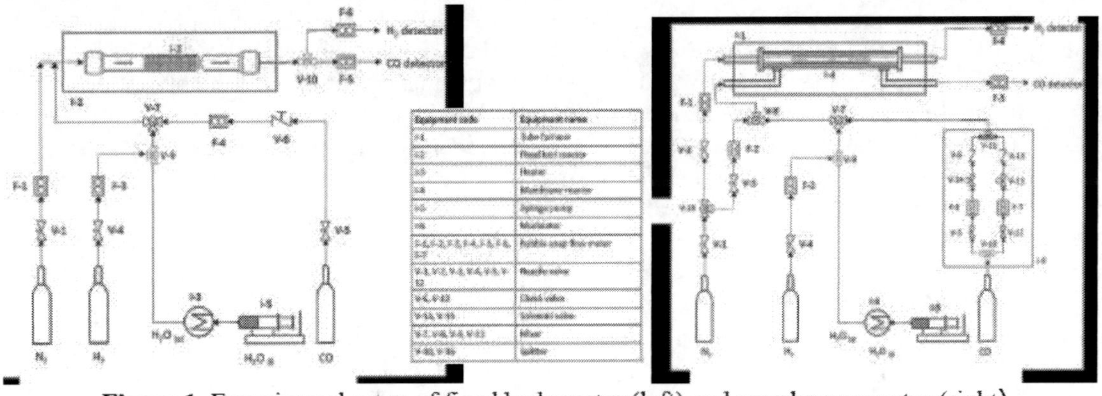

Figure 1. Experimenal setup of fixed bed reactor (left) and membrane reactor (right)

3. Results and Discussion

3.1. WGSR using Fixed Bed Reactor

The conversion of CO at space time of 1.2 s (GHSV 3,130 h^{-1}) achieved 7.5% and 11.5% for H_2O/CO ratio 2 and 3, respectively. Meanwhile, CO conversion is 9.8% and 13.2% for H_2O/CO ratio 2 and 3 at 2.3 s space time. The CO conversion results for all variations are presented in Figure 2. Based on work from Demirel and Azcan [16], enhancement of steam in feed will shift the reaction equilibrium to the product so it increase the production of H_2. As the result, it increase CO conversion. Meanwhile, The higher the space time inside the reactor, the longer time for the reaction to be occurred. Longer reaction time increases the amount of reactants to produce H_2 as the main product. Therefore, experiments with higher space time resulted in higher CO conversion. Fishtik and Datta [17] found the same tendencies.

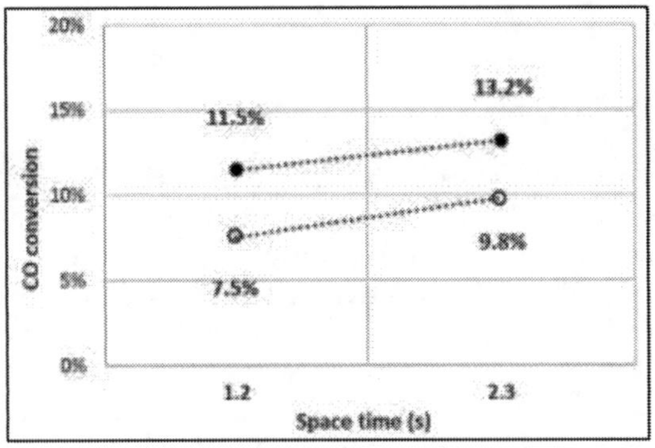

Figure 2. CO conversion in fixed bed reactor with H_2O/CO ratio = 2 (○) and H_2O/CO ratio = 3 (●)

3.2. Performance of Membrane Reactor in Steady State Operation

The performance of membrane reactor indicated in CO conversion and H_2 recovery was investigated at 350°C by varying feed composition and space time. The experiment results are presented in Figure 3. At the same space time (2.3 s), when the feed ratio (H_2O/CO) was 2, the CO conversion reached 47.4% and H_2 recovery was 40.4%. Meanwhile, when feed ratio was 3, the CO conversion reached 55.4% and H_2 recovery was 59.8%. Increasing the steam flow rate will shift the reaction equilibrium toward the product due to thermodynamic aspect. As the result, CO conversion is high. Moreover, hydrogen produced by WGSR in membrane reactor was continuously permeating to the tube side, leading to increased H_2 production significantly. This result showed the same tendency as research conducted by Uemiya et al. [10] and Augustine et al. [11].

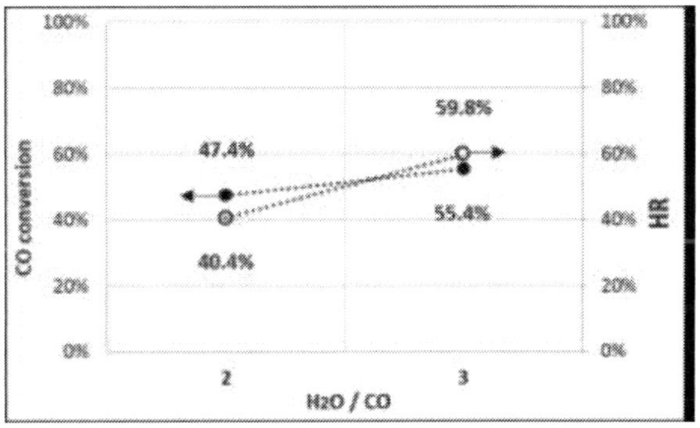

Figure 3. CO conversion (•) and H_2 recovery (○) as function of H_2O/CO feed ratio in space time=2.3s

The experiment was also conducted out for space time of 1.2 s (GHSV 3130 h^{-1}) and 2.3 s (1585 h^{-1}) to evaluate the effect of space time in the membrane reactor. The feed ratio in this experiment remained constant at 3. The experiment results are shown in Figure 4. When the space time is 1.2 s, CO conversion is 50.5%. Meanwhile, H_2 recovery reached 46.4%. Experiment which is held in space time 2.3 s shows higher CO conversion and H_2 recovery. CO conversion and H_2 recovery value respectively are 55.4% and 59.8%. The higher space time, the higher CO conversion and H_2 recovery. It is caused by longer time for reaction to be occurred in space time of 2.3s. As the result, variation in high space time produce high amount of H_2 and those H_2 have a proper chance to permeate through the membrane. Augustine et.al.[11] found the same tendencies.

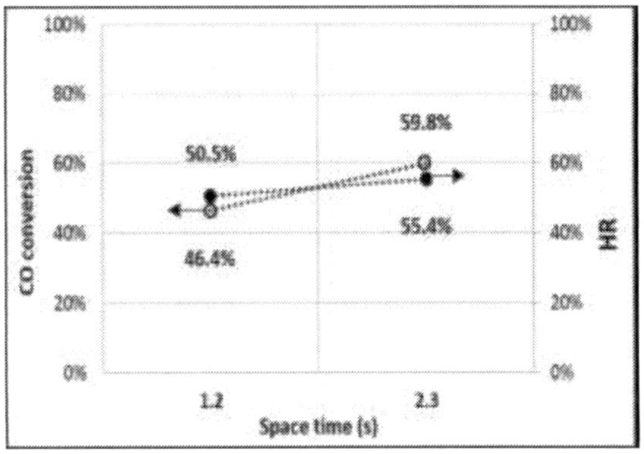

Figure 4. CO conversion (•) and H_2 recovery (○) as feed space time function in feed ratio of 3

3.3. Performance Comparison between Fixed Bed Reactor and Membrane Reactor

Referred to Le Chatelier principle, the equilibrium shifts to the reaction product if the product concentration decreases. The membrane reactor concept may improve H_2 recovery in the WGSR due to hydrogen separation through the membrane. In this work, the WGSR at 1 bar and 350°C showed better performance when using Pd/Al_2O_3 membrane reactor than fixed bed reactor. The comparisons of both results are shown in Figure 5. The conversion of CO at constant space time of 2.3 s achieved 9.8% and 13.2% for H_2O/CO ratio 2 and 3, respectively, at the experiment conducted in fixed bed reactor. Meanwhile, CO conversion is 47.4% and 55.4% for H_2O/CO ratio 2 and 3 at the experiment conducted in membrane reactor. This means that the use of Pd/Al_2O_3 membrane reactor gave much better performance than the use of conventional fixed bed reactor. The data is also supported by experiments with constant feed ratio=3.

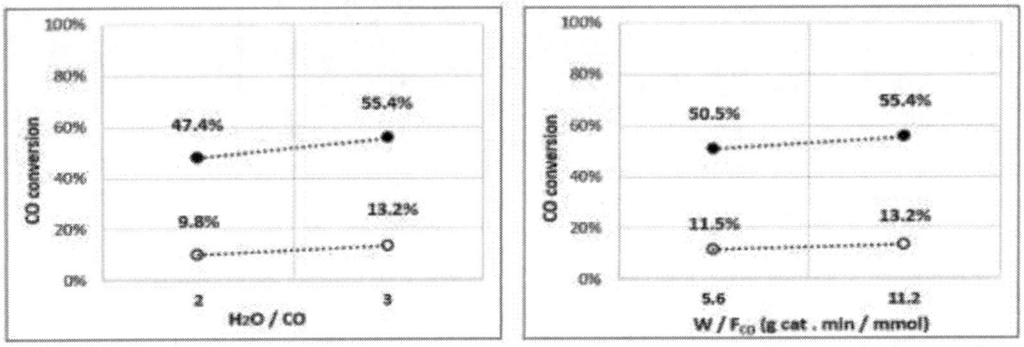

Figure 5. CO conversion as function of H_2O/CO ratio in space time=2.3s (left) and of space time in feed ratio=3 (right) in Pd/Al_2O_3 membrane reactor (●) and fixed bed reactor (○)

3.4. Dynamic Operation in Membrane Reactor

Unsteady state operation or dynamic operation has been recognized as one of successful methods on process intensification to reach better performance. In this study, the dynamic operation was operated by applying a square wave disturbance periodically. The CO molar flow was periodically changed to disturb the membrane reactor, particularly the concentration fluctuation on the membrane surface. This dynamic concentration of CO on the surface of membrane at proper switching time would influence the solution diffusion rate. This dynamic WGSR experiment using Pd/Al_2O_3 membrane reactor was operated at 1 atm and 350°C with H_2O/CO ratio of 2, space time 2.3 s and the switching time was 15 s. Figure 6 shows the dynamic CO conversion and Hydrogen recovery as function of time on stream after stable oscillation. Based on Hoebink et al. [15], the regime of operation formed in this switching time was dynamic, indicating that the system may respond the dynamic perturbation.

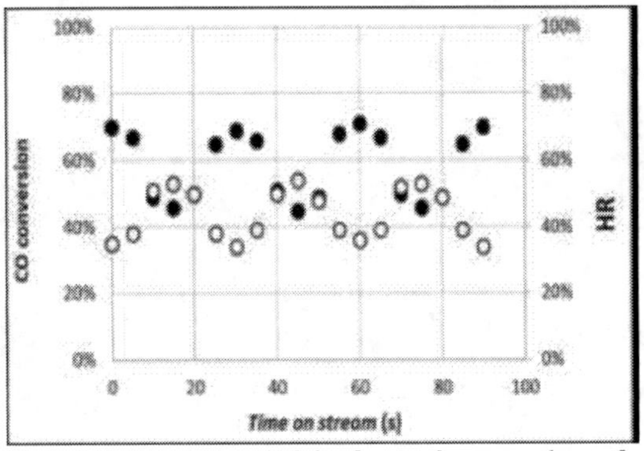

Figure 6. CO conversion (●) and H_2 recovery (○) in dynamic operation of membrane reactor

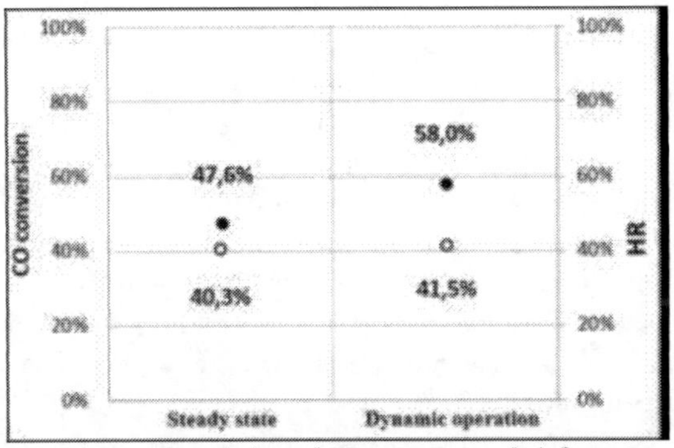

Figure 7. Comparison of CO conversion (●) and H_2 recovery (○) between steady state and dynamic operation of WGSR using Pd/Al_2O_3 membrane reactor

The comparison of membrane reactor performance under steady state and dynamic operation is presented in figure 7. The CO conversion and H_2 recovery resulted from steady state operation were 47.6 % and 40.3%, respectively. Meanwhile, the CO conversion and H_2 recovery resulted from dynamic operation exceeded 58.0% and 41.5%, respectively. The dynamic operation increased the CO conversion and H_2 recovery because it was able to create "sweeping" effect by changing the feed flow rate that minimized the concentration polarization effect (external mass transfer resistance) over the membrane surface.

4. Conclusion

The WGSR to produce hydrogen has been studied in the fixed bed reactor and membrane reactor either under steady state and dynamic operation. In both fixed bed and membrane reactor, increasing H_2O/CO ratio and space time gave rise to the increased CO conversion. In WGSR using Pd/Al_2O_3 membrane reactor over iron oxide catalyst, increasing of H_2O/CO ratio and space time also increased H_2 recovery. The performance of Pd/Al_2O_3 membrane reactor for WGSR has proven to be better than fixed bed reactor for various operation conditions in term of CO conversion and H_2 recovery. The dynamic operation in the membrane reactor improves higher CO conversion and H_2 recovery compared with steady state operation.

5. Acknowledgments

The financial support provided by Ministry of Research, Technology and Higher Education, Republic of Indonesia, through National Competitive Research Grant "Overseas Collaborative Research" no. 009/SP2H/LT/DRPM/IV/2017, is gratefully acknowledged.

6. References

[1] Nakajima T, Kume T, Ikeda Y, Shiraki M, Kurokawa H, Iseki T, Kajitani M, Tanaka H, Hikosaka H, Takagi Y, Ito M 2015 *International Journal of Hydrogen Energy* **40** 11451-11456
[2] Ball M and Wietschel M 2009 *International Journal of Hydrogen Energy* **34** 615-627
[3] Basile A, Companari S, Manzolini G, Iulianelli A, Longo T, Liguori S 2011 *International Journal of Hydrogen Energy* **36** 1531-1539
[4] Reddy G K and Smirniotis P G 2015 *Water Gas Shift Reaction: Research Developments and Applications* (Oxford: Elsevier)
[5] Cornaglia L, Tosti S, Sansovini M, Múnera J, Lombardo E 2013 *Applied Catalysis A: General* **462–463**, 278–286.
[6] Stankiewicz A and Moulijn J A 2000 Chem Eng Process **96** 22-34
[7] Stankiewicz A and Moulijn J A 2004 *Re-engineering the chemical processing plant* (New York: Dekker)
[8] Basile A and Gallucci F 2011 *Membrans for membran reactors: preparation, optimization and selection* (Eindhoven: John Wiley and Sons)
[9] Falco M D, Marrelli L, Iaquaniello G 2011 *Membran Reactors for Hydrogen Production Process* (London: Springer)
[10] Uemiya S, Sato N, Ando H, Kikuchi E 1991 *Ind. and Eng. Chem. Research* **30** 589-591
[11] Augustine A S, 2011 *International Journal of Hydrogen Energy* **36** 5350-5360
[12] Caravella A, Scura F, Barbieri G, Drioli E 2010 *Journal of Physical Chemistry B.* **114** 12264-76
[13] Budhi Y W, Rionaldo H, Padama A A B, Kasai H, Noezar I 2015 *International Journal of Hydrogen Energy* **40** 10081-10089.
[14] Budhi Y W, Noezar I, Aldiansyah F, Kemala P V, Padama A A B, Kasai H, Subagjo 2011 *International Journal of Hydrogen Energy* **36** 15372-15381.
[15] Hoebink J, Nievergeld A, Marin G 1999 *Chem. Eng. Sci.* **54** 4459-4468
[16] Demirel E and Azcan N 2012 *Proc. of the WCECS* **II** (San Francisco)
[17] Fishtik I and Datta R 2002 *Surface Sciences* **512** 229-254

IOP Conf. Series: Earth and Environmental Science **105** (2017) 012021 doi:10.1088/1755-1315/105/1/012021

The effect of pyrogallol antioxidant addition and storage temperature to the change of biodiesel quality during storage period

B H Susanto[1], Lukman[1], H Sutanto[1] and M Nasikin[1]

[1] Bioprosess Engineering Study Program, Department of Chemical Engineering, Faculty of Engineering, Universitas Indonesia, Depok, West Java, Indonesia

Email: bambanghs@che.ui.ac.id

Abstract. Biodiesel is an alkyl ester compound of fatty acids prepared from a source of naturally renewable triglycerides and used as diesel engine fuel, commonly made through the process of esterification or transesterification. In its application and storage, biodiesel is potentially damaged by oxidation reaction due to internal factors (high unsaturated fatty acid content) and external factors (air, heat or light) resulting in changes of the characteristics and quality of biodiesel. In order to maintain the characteristics and quality of biodiesel to conform to established standards, it is necessary to add antioxidants that can inhibit oxidation reaction in biodiesel. In this study, pyrogallol antioxidant is added to the biodiesel of palm oil with various concentrations and storage temperatures. The observed biodiesel parameters during the storage period are those that can represent oxidation such as changes in kinematic viscosity, density, acid value, and iodine value. The results showed that the addition of pyogallol antioxidant can inhibit the oxidation reaction in biodiesel. The use of antioxidant with a concentration of 0.1% at storage temperatures of 30°C and 60°C is known to retain the characteristics and quality of biodiesel during storage from damage caused by oxidation reaction.

1. Introduction

Some developments of alternative energy from renewable resources (bioenergy) namely biodiesel, bioethanol, and biogas. Biodiesel is known to be used as one of the renewable alternative energy sources because it does not produce excess sulfur emissions, biodegradable and has better combustion efficiency than diesel [1].

Although the production of biodiesel has met the established standards, biodiesel can be degraded to other compounds during the storage that results in the change of the biodiesel component so that the product no longer meets the standard. The change occurs due to storage conditions and its use. Biodiesel is easily degraded into compounds such as acids, aldehydes, ketones, esters, peroxides and alcohols [2]. Changes in biodiesel characteristics due to the oxidation process include increased viscosity value, peroxide value, acid value and decreased induction period [3]. Therefore, it is necessary to ensure that biodiesel does not degrade or damage due to oxidation reaction during storage period. One way is to do the addition of antioxidants. Antioxidants are compounds that can resist the rancidity and inhibit oxidation reaction in materials containing fat or oil, even at small concentrations [4].

In this research, the addition of antioxidant is by using antioxidant pyrogallol (PY). Study of antioxidant effect to biodiesel is done by using antioxidant PY with two storage conditions by setting two different temperatures for each sample; temperature 30°C or room temperature and at temperature of 60°C which is kept constant. Biodiesel is synthesized from palm oil. The observed parameters during the storage period that indicate the oxidation of biodiesel are kinematic viscosity, density, acid value, and iodine value. Observations will be carried out once a week for 4 weeks of storage.

2. Material and methods

2.1. Addition of pyrogallol antixodiant to biodiesel

Biodiesel was added with pyrogallol antioxidant. Characteristics and specifications test of biodiesel were conducted which include acid value, iodine value, kinematic viscosity, and 40°C density. This was done to know the initial condition of biodiesel that has been obtained used as a reference.

After the biodiesel characteristics was obtained, the next step is the addition of pyrogallol antioxidant into biodiesel by varying different concentrations of 0.05%, 0.07% and 0.1% of the total volume of the biodiesel sample to be used as the test samples. Pyrogallol solution is made from 4.5 ml methanol mixture, 2 ml ethyl acetate, and 0.03 g of pyrogallol antioxidant. Afterwards, the antioxidant was mixed into all test samples.

2.2. Storage of biodiesel test sample

Biodiesel sample will be stored after antioxidant PY was mixed into biodiesel sample. There were two different storage conditions: at room temperature 30^{0}C and 60^{0}C.

2.3. Oxidation Resistance Test on Each Biodiesel Sample

After PY antioxidant was mixed into the biodiesel then oxidation resistance test would be conducted. The oxidation resistance carried out in this study include acid value test, iodine value test, kinematic viscosity test and density test. Each test was done in every week of storage period. In this study, the storage was evaluated for 4 weeks.

3. Result and discussion

3.1. Palm oil biodiesel product test

Following is the results of initial test of biodiesel used in this research:

Figure 1. Palm oil biodiesel.

Table 1. Results of biodiesel product test.

Parameter	Unit	SNI 04-7182-2012	Value
Acid Value	mg KOH/g	0,6	0,067
Iodine Value	%mass (g-I2/100 g))	Max 115	72,7
Kinematic Viscosity 40°C	mm^2/s (cSt)	2,3 – 6,0	4,46
Density 40°C	kg/m^3	850 – 890	0,8521
Ester Content	% wt	>96,5	98,7
Free Glycerol	% wt	<0,02	0,006
Total Glycerol	% wt	<0,24	0,198

Based on the test results, the biodiesel used is still in an acceptable condition and still in accordance with the specified standard. It is seen that the value of several parameters tested is still in the range of Indonesian standard SNI 04-7182-2012.

3.2. Effect of antioxidant addition to acid value during storage

One indication of oxidation damage to biodiesel is the increase of acid value, which is caused by the decomposition of peroxide compounds from the oxidation reaction. High acidity level in biodiesel is unfavorable because it can cause damage to the vehicle's machine. The higher the acid value then the lower the quality of biodiesel [5]. High acid value is associated with the corrosion of the media [5]. Additionally, it will shorten the life of pumps and filters on machines [6].

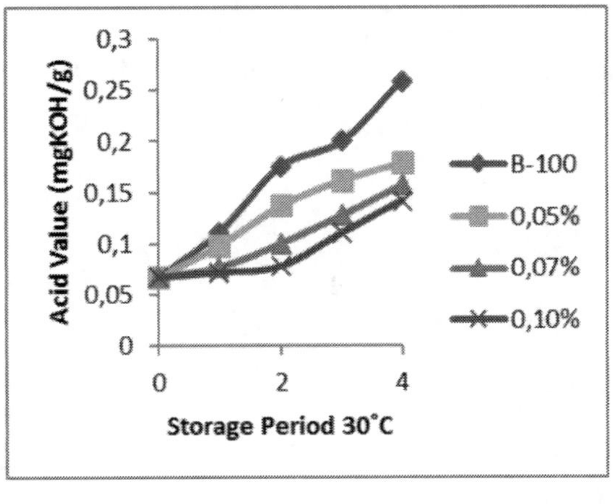

Figure 2. Acid value at 30°C.

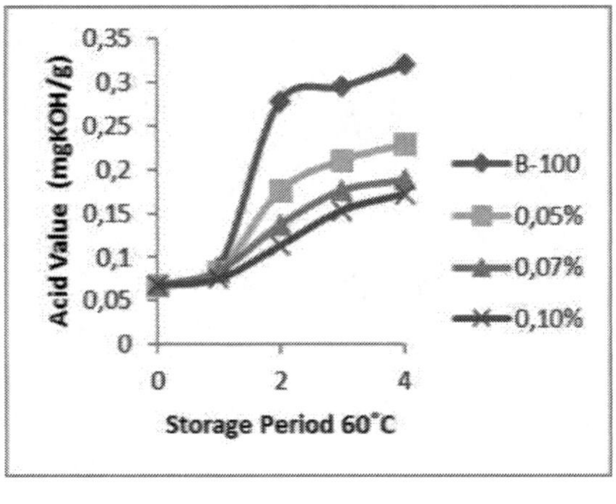

Figure 3. Acid value at 60°C.

According to the graphs, the type of the added antioxidant concentration has an effect on the change of the acid value. The higher the added concentration then the lower the acid value [7]. In this study, the addition of antioxidant at the concentration of 0.1% was able to suppress the lowest acid value, whereas the addition of antioxidant at 0.05% and 0.07% showed no significantly different results in inhibiting the increase of acid values. On the other hand, biodiesel samples without the addition of antioxidant underwent the highest increase of acid values compared to biodiesel samples with the addition of antioxidant. This is caused by peroxide compounds from the oxidation of oxidized unsaturated fatty acids that form aldehyde compounds. Aldehydes will be oxidized further to form carboxylic acids which lead to an increase in the acid values [8].

3.3. Effect of antioxidant addition to iodine value during storage

The effect of antioxidant addition to iodine value can be seen in figure 4 and 5.

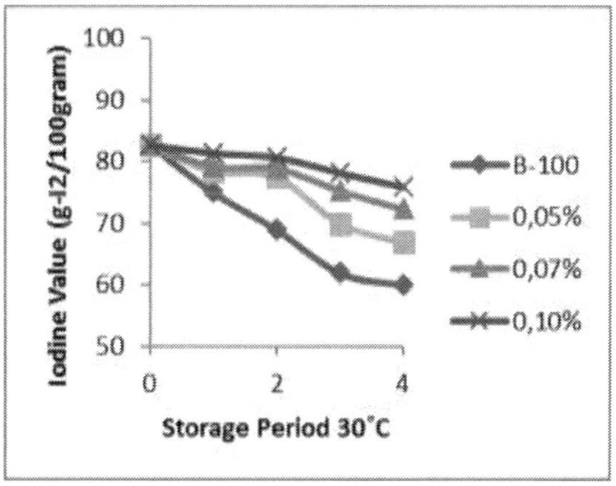

Figure 4. Iodine value at 30^{0}C.

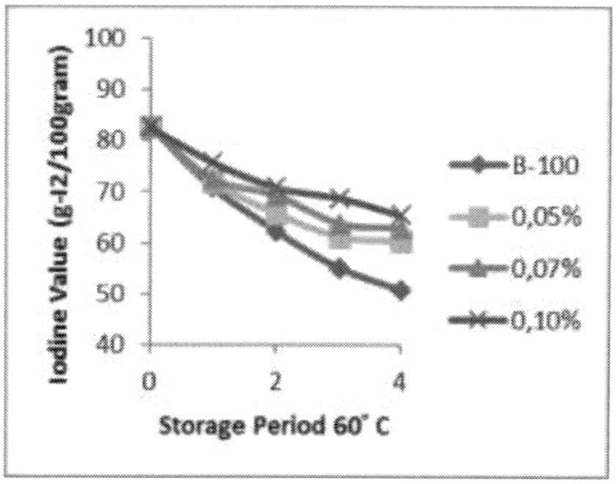

Figure 5. Iodine value at 60^{0}C.

The iodine value is determined by the degree of oil unsaturation. If the level of oil unsaturation is high then the oil will bind the iodine in larger quantities so that the iodine value is higher. High level of oil unsaturation makes oil become more easily to be oxidized. If the oil is easily oxidized, the degree of unsaturation decreases as the double bond is broken and the result is the smaller iodine value.

In this study, it can be seen that the iodine value of each sample decreased. Figure 4 and 5 above shows the effect of addition of pyrogallol antioxidant to the decrease of iodine number from each sample. According to the result, the higher the concentration of antioxidant used then the iodine value is still high. However in the sample that does not use antioxidant, the iodine value shows a significant decrease.

3.4. Effect of antioxidant addition to kinematic viscosity of biodiesel

The effect of pyrogallol antioxidant addition to the value of kinematic viscosity of biodiesel is shown in figure 6 and 7.

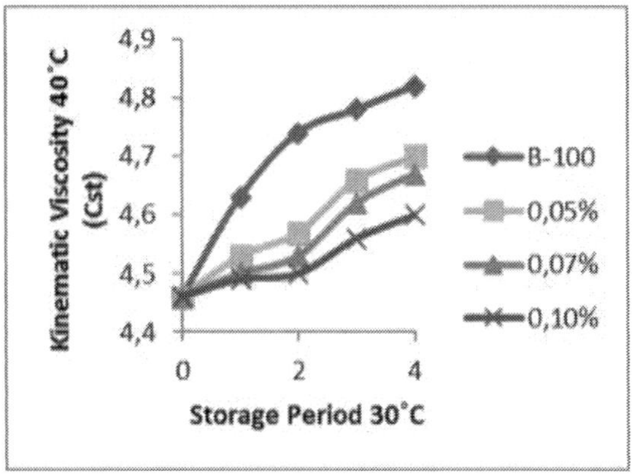

Figure 6. Kinematic viscosity 30^{0}C.

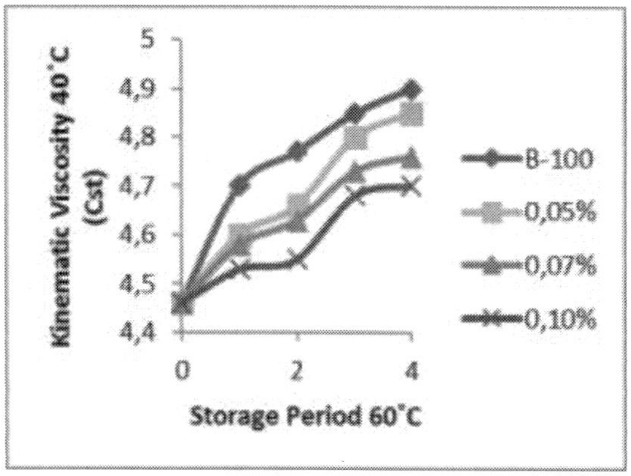

Figure 7. Kinematic viscosity 60^{0}C.

From figures 6 and 7, it can be seen that the kinematic viscosity of all samples during the storage period tends to increase in viscosity. This happens because biodiesel underwent oxidation reactions that trigger the formation of new compounds or polymers with high molecular weight. Khothe [9] stated that the increase in the viscosity of biodiesel during the storage occurs due to the process of oxidation of biodiesel which produces free fatty acid compounds, double bond isomerization, saturation and formation of higher molecular weight compounds.

Kinematic viscosity is a measurement of fluid flow influenced by gravity. The viscosity of oils is expressed by the number of seconds in which the price or number represents a certain volume of oil to flow through a hole of a certain small diameter. A smaller number of seconds means lower viscosity [10].

3.5. Effect of antioxidant addition to the density of biodiesel

The effect of pyrogallol antioxidant addition to the density of biodiesel from palm oil is shown in figure 8.

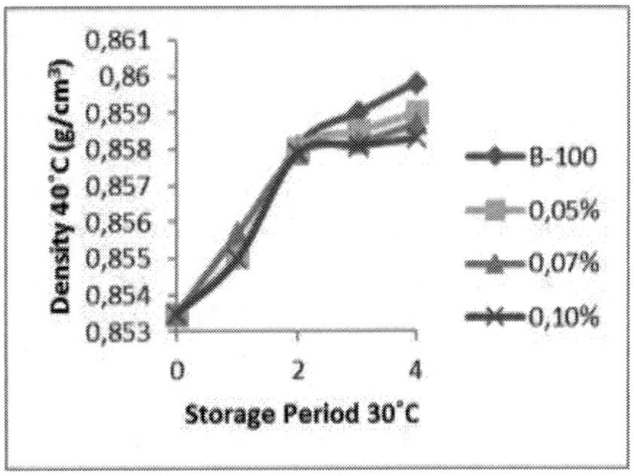

Figure 8. Density of biodiesel 30^{0}C.

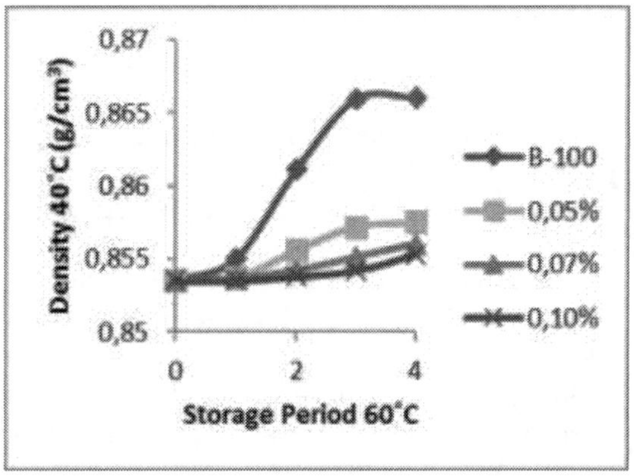

Figure 9. Density of biodiesel 60^{0}C.

According to figure 8, the increase in the density value of each sample at a storage temperature of 30^{0}C is not quite significant. This small change in density value is due to the fact that biodiesel is still relatively stable and there is no oxidation that results in the formation of short chain compounds, so the density values are still likely to remain the same.

On the other hand, in figure 9 with a storage temperature of 60^{0}C primarily in the B-100 sample without the addition of antioxidant, it can be shown that there is an increase in the biodiesel density value. It happens due to the oxidation reactions has begun to occur in the sample during storage at high temperatures. Increased density values occur because of the degradation of the sample resulting in products such as gums and deposits which have high density values or the formation of high molecular weight compounds with a deposit nature [9]. It is also due to the oxidation reaction of biodiesel that produces short chain compounds and other compounds that potentially increase the density of biodiesel. Oxidation of biodiesel will yield short chain compounds that tend to crystallize so that it will decrease the volume and increase the density of biodiesel.

4. Conclusion

The addition of pyrogallol antioxidant in low concentration is able to maintain the quality of biodiesel from the damage of oxidation reaction. During 4 weeks of storage, all test samples underwent the change of several biodiesel quality test parameters. The acid value, kinematic viscosity and density increase while the iodine value decrease.

The quality of biodiesel which was made from palm oil could be maintained by the addition of pyrogallol antioxidant with concentration ranges from 0.05% to 0.1% during 4 weeks of storage. The best quality was achieved by the addition of 0.1% pyrogallol antioxidant.

5. References

[1] Suryani E A, Dadang, Hariyadi, Hanafie H, Reksowardojo I K, Rivai M, Ihsanur M, Suryadarma P, Tjitrosemito S, Soerawidjaja T H, Prawitasari T, Prakoso T, Purnomo W and Hambali 2008 *Jarak Pagar Tanaman Penghasil Biodiesel* (Jakarta: Penebar Swadaya)

[2] Jiayu, Imahara, Hiroaki, Saka and Xin S 2009 Kinetics on oxidation of biodiesel stabilized with antioxidant fuel **II** p 282-286

[3] Agarwal A K and Khurana D A 2013 Long-term storage oxidation stability of Karanja biodiesel with the use of antioxidants *Fuel Processing Technol.* **106** 447-452

[4] Matz S A 1984 *Snack Food Technology* (Westport, Conn: AVI Publishing Co)

[5] Martin, Schober, Mittelbach S 2004 The influence of antioxidants on the oxidation stability of biodiesel *J. of the American Oil Chemists Society* **80**(8) 817-823

[6] Soerawidjaya T H 2005 Studi kebijakan penggunaan biodiesel di Indonesia *Kajian kebijakan dan kumpulan artikel penelitian biodiesel. Kementrian Riset dan Teknologi RI-MAKSI IPB*

[7] Fessenden R J, Fessenden J S 1992 *Kimia Organik Edisi Ketiga, a.b. Pudjatmaka* (Jakarta: Gramedia)

[8] Gordon M H 1990 The Mechanism of Antioxidant Action in Vitro. In B. J. F. Hudson (Ed.) *Food Antioxidant* pp 1-18 (Netherlands: Springer)

[9] Khothe 2006 Dependence of biodiesel fuel properties on structure of fatty acid alkyl esters *Fuel Processing Technol* **47**(2) 1059-70

[10] O'Malley P J 2002 The reaction profile for hydrogen atom transfer from phenol to peroxyl free radicals *Chemical Physics Letters* **363**(3) 318-322

IOP Conf. Series: Earth and Environmental Science **105** (2017) 012022 doi:10.1088/1755-1315/105/1/012022

Techno-economic feasibility of flare gas utilization using adsorbed natural gas

Yuswan Muharam[1,3], Fiqi Giffari[1] and Mirza Mahendra[2]

[1] Department of Chemical Engineering, University of Indonesia, Depok 16424, Indonesia
[2] Indonesia Ministry of Energy and Natural Resources

E-mail: muharam@gmail.com

Abstract. Adsorbed natural gas (ANG) is a natural gas storage technology, which is potential to transport natural gas in small and medium quantity. The objective of this study is to determine the techno-economic feasibility of ANG technology for flare gas to consumers around flare gas sources. The method is process simulation, economic calculation and optimization with variables being ANG selling price, flare gas price, and the percentage of capital financing by the government. The process simulation shows that the ANG product is in the range of 2.07 MMSCFD – 2.97 MMSCFD. The economic calculation results in the interest rate of return of less than 10%. With optimization, the IRR increases to more than 20%.

1. Introduction

The supply of 587 MMSCFD flare gas in Indonesia is potential to be utilized. However, due to its low flowrate and remote location, the appropriate gas transportation technology is required to utilize flare gas [1]. Compressed Natural Gas (CNG) is one of the technologies used to distribute natural gas in small to medium quantities and distances of less than 965 km. However, CNG has a deficiency in terms of safety due to relatively high operating pressure (> 200 bar). This disadvantage of CNG leads to the development of adsorbed natural gas (ANG) [2, 3]. This technology does not require liquefaction process such as liquid natural gas (LNG) or compression process such as CNG. In ANG technology, natural gas is deposited on a porous material (adsorbent) as adsorbed phase at the pressure of 3.5-4.0 MPa. The energy density of ANG at the pressure of 500 psi (~ 3.4 Mpa) is proportional to that of CNG at 2400 psi (~ 16.5Mpa) pressure [4].

ANG technology continues to grow up in line with the development of research on adsorbents for methane. It is reported in 2013 that the adsorbent of type HKUST-1 has the storage capacity of 267 V(STP)/V at 10 MPa [5]. With this advantage, ANG technology is considered to be feasible as a flare gas transportation technology that has a relatively small capacity. Therefore, the recent study is to determine the techno-economic feasibility of ANG technology for flare gas utilization to consumers around flare gas sources.

[3] To whom any correspondence should be addressed.

2. Methodology

The study uses flare gases from three natural gas fields with the specification as shown in table 1.

Table 1. Detailed specification of flare gas from three natural gas fields.

Composition	Field X	Field Y	Field Z
	Mole fraction		
C_1	0.49886	0.75060	0.80752
C_2	0.03943	0.08900	0.00498
C_3	0.04001	0.08450	0.00499
iC_4	0.01398	0.01920	0.00498
nC_4	0.02027	0.02250	0.00279
iC_5	0.01098	0.00730	0.00158
nC_5	0.00991	0.00490	0.00132
C_6+	0.02528	0.00820	0.00267
N_2	0.00227	0.00800	0.01733
CO_2	0.32692	0.00580	0.15004
H_2S	0.01209	0.00000	0.00180
Location	Onshore	Offshore	Offshore
Pressure (psig)	77.00	71.00	75.00
Temperature (°F)	117.00	108.00	105.00
Flowrate (MMSCFD)	4.80	2.50	3.77

Figure 1, figure 2 and figure 3 show the process flow diagram of Field X, Field Y and Field Z. Field X requires compression, gas sweetening and fractionation facilities to process the flare gas. Field Y needs compression and fractionation facilities. Field Z involves compression facilities and gas sweetening facilities in the plant. The process simulation results in mass balance, energy consumption, and equipment sizes, which in turn were used to estimate the capital and operating costs. The cashflow was calculated using the following assumption: project duration = 10 years, discount factor = 10%, operating time = 330 days/year, flare gas price = 2.7 US$/MMBTU, energy price = flare gas price, ANG selling price = 7 US$/MMBTU, condensate selling price = 40 US$/BBL, LPG selling price = 300 US$/ton, and HKUST-1 price = 6.5 US$/kg.

The output of the cashflow calculation is interest rate of return (IRR), net present value (NPV), period of time (POT) and profitability index (PI). If the economic indicators indicate that the project is not feasible, then feasibility optimization is performed by: setting the ANG selling price in the range of 8-10 US$/MMBTU, based on the existing range of CNG selling price for the industrial sector in Indonesia; setting flare gas price ranging from 0.45 US$/MMBTU to 2.7 US$/MMBTU, based on the government regulation and the average selling price of flare gas in other countries; alternating the financing scheme using public private partnership where the government capital financing share is in the range of 0-100%, based on the history of government cooperation with private parties ever undertaken.

3. Results and discussion

The results of the process simulation on Field X, Field Y, and Field Z exhibits in table 2. It can be seen in the table that the ANG production in Field Z is the highest, followed by Field X and Field Y. If viewed from the flare gas flowrates, Field Z yields 3.77 MMSCFD flare gas, lower than 4.80 MMSCFD from Field X. The higher ANG from Field Z if compared to that from Field X is because the methane fraction in flare gas from Field Z is twice of that from Field X. The energy consumption is directly proportional to the flare gas flowrate. The higher flare gas flowrate consumes higher energy.

The capital cost, as shown in table 3 is insignificantly effected to the flare gas flowrate. This is because the flowrate in the three fields is almost the same. Capital cost is heavily influenced by infrastructure. Field Y, although its flare gas flowrate is the lowest, but its capital cost is the highest. This is because the distance between Field Y and the mainland is the farthest among the three fields that requires the capital costs for the flowline and the gathering facility.

The operating cost, as exhibited in table 4 is directly proportional to the flare gas flowrate. Field X needs the highest capital cost because its flare gas is the highest.

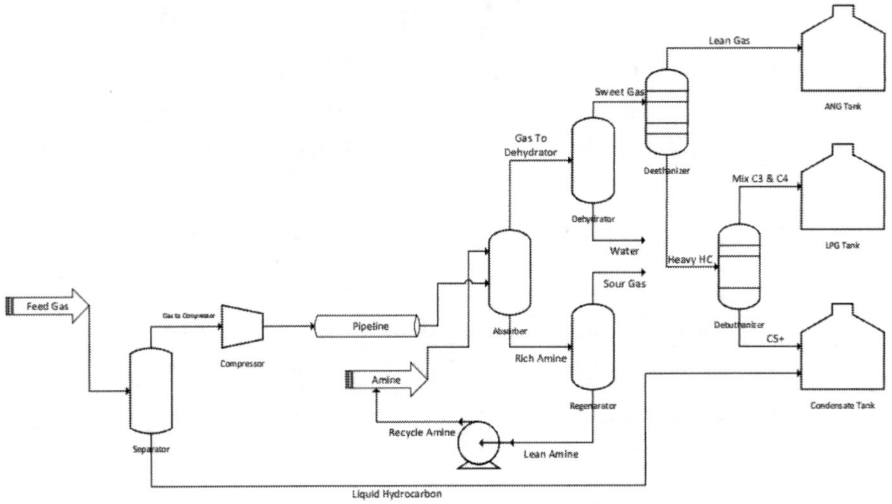

Figure 1. Process flow diagram of Field X.

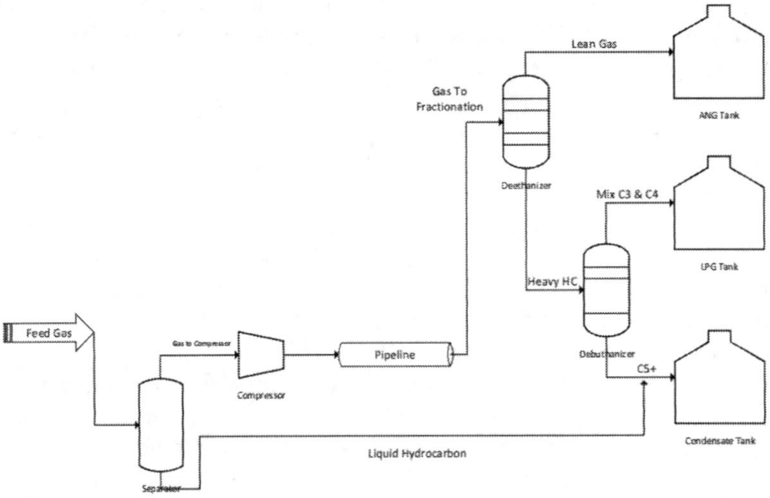

Figure 2. Process flow diagram of Field Y.

Table 5 shows that Field Y has the best economy than the others although the flare gas flowrate is the lowest and the capital cost is the highest. This is because the flare gas from Field Y possesses the lowest impurities (CO_2 and H_2S) resulting in the highest conversion ratio among the three fields. It appears from the table that the economics of the Field X, Field Y, and Field Z are not feasible. Therefore, it is necessary to optimize the projects of the three fields.

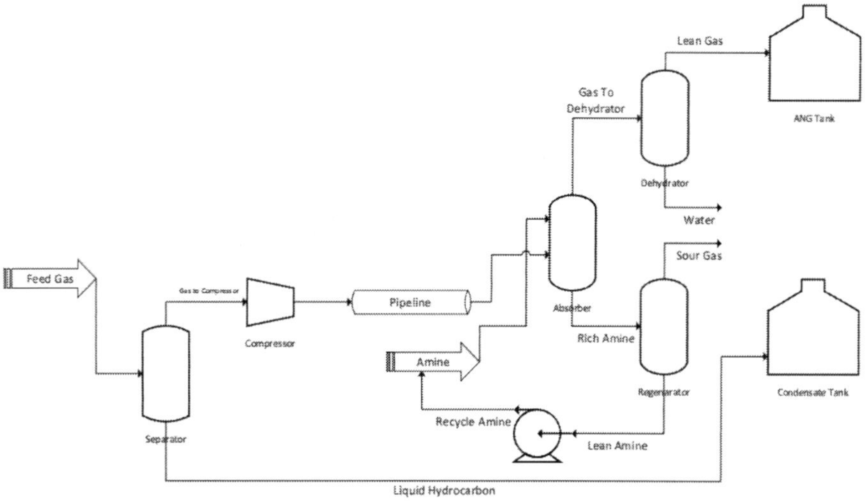

Figure 3. Process flow diagram of Field Z.

Optimization is carried out by calculating hundreds of possible scenarios. Table 6, table 7 and table 8 exhibit only five scenario results including the best scenarios. Optimization result for Field X, as seen in table 6, shows that without the Government financing Field X will be feasible with IRR being higher than 20% if the flare gas price is set at 1 US$/MMBTU and the ANG selling price is 9 US$/MMBTU, whereas with the Government financing, it will be feasible with IRR more than 20% if the government bears 70% of the capital cost and the flare gas price is set at 2 US$/MMBTU. As for Field Y, as seen table 7, with the Government financing, Field Y will be feasible with IRR greater than 20% if the government bears 40% of the capital cost, the flare gas price is 2.5 US$/MMBTU, and the ANG selling price is 10 US$/MMBTU. As to Field Z, as seen in table 8, with the Government financing, Field Z will be feasible with IRR higher than 20% if the government endures 70% of the capital cost, the flare gas price is set at 2.7 US$/MMBTU, and the ANG selling price set to 9 US$/MMBTU.

Table 2. Mass balance and energy consumption.

Field	**X**	**Y**	**Z**
Feed (MMSCFD)	4.80	2.50	3.77
Product			
Condensate (BBL/day)	88.76	46.47	0.00
LPG (ton/day)	17.89	18.38	0.00
ANG (MMSCFD)	2.33	2.07	2.97
Energy consumption (MMBTUD)			
Pipeline and compression	91.81	33.53	50.30
Gas sweetening	90.86	0.00	91.20
Fractionation	51.72	29.74	0.00
ANG filling station	11.72	3.16	7.08
Total	**246.12**	**66.43**	**148.58**

Table 3. Estimated capital cost.

Capital cost component (US$)	X	Y	Z
Direct cost	16,185,723	26,386,124	27,008,895
Main equipment	12,999,064	19,751,993	20,658,988
Compressor facility	893,750	536,250	715,000
Flowline and gathering	285,000	14,915,000	11,275,000
Gas sweetening facility	4,510,289	-	3,854,956
Fractionation facility	1,932,981	1,264,974	-
ANG filling facility	1,430,925	905,761	1,280,775
Transportation facility	3,946,120	2,130,008	3,533,257
Installation	1,677,189	3,491,648	3,342,056
Piping and instrumentation	1,509,470	3,142,483	3,007,850
Indirect cost	4,279,574	8,330,393	8,095,800
Engineering	1,975,188	3,844,797	3,736,523
Building and construction	1,316,792	2,563,198	2,491,015
Contingency	987,594	1,922,398	1,868,262
Total capital cost	20,465,296	34,716,517	35,104,695

Table 4. Estimated operating cost.

Operating cost component (US$/year)	X	Y	Z
Plant fixed cost	799,557	247,654	534,097
Labor cost	137,130	137,130	137,130
Plant maintenance cost	511,632	104,150	175,523
Miscellaneous cost	150,795	6,374	221,443
Energy cost (variable cost)	239,230	64,573	144,416
Pipeline and compression	89,244	32,592	48,892
Gas sweetening facilities	88,320	-	88,647
Fractionation facilities	50,275	28,906	
ANG filling facilities	11,392	3,075	6,877
Transportation facilities	1,098,587	592,988	983,648
Fuel cost	260,059	140,373	232,850
Vehicle maintenance	557,990	301,188	499,611
Driver cost	258,958	139,779	231,865
License cost	21,580	11,648	19,322
Total operating cost	**2,137,375**	**905,214**	**1,662,161**

Table 5. Economic simulation result.

Result	X	Y	Z
Total cashflow, US$	3,239,789	8,791,045	(10,172,147)
POT (PBP), year	10.64	9.73	0.00
IRR	1.77%	3.75%	-4.37%
NPV, US$	(9,111,223)	(9,093,169)	(20,920,597)
PI (BCR)	0.63	0.74	0.43

Table 6. Field X optimization result.

Scenario	Government capital share	ANG price (US$/MMBtu)	Flare gas price (US$/MMBtu)	Private cashflow		Government cashflow	
				IRR	NPV (MUS$)	IRR	NPV (MUS$)
7	0%	10	2.7	10.81%	957	-	40,327
22	0%	7	1.5	10.42%	445	-	24,419
33	0%	9	1	20.37%	11,139	-	23,946
344	70%	7	2.7	12.44%	1,511	30.61%	21,002
347	70%	8.5	2.7	20.30%	6,546	33.62%	23,843
358	70%	7	2	22.45%	7,086	24.08%	15,037

Table 7. Field Y optimization result.

Scenario	Government capital Share	ANG price (US$/MMBtu)	Flare gas Price (US$/MMBtu)	Private cashflow		Government cashflow	
				IRR	NPV (MUS$)	IRR	NPV (MUS$)
14	0%	10	2.5	10.46%	699	#NUM!	24,454
197	40%	7	2.7	11.20%	1,204	14.92%	7,058
210	40%	10	2.5	20.40%	10,996	20.12%	11,229

Table 8. Field Z optimization result.

Scenario	Government capital share	ANG price (US$/MMBtu)	Flare gas price (US$/MMBtu)	Private cashflow		Government cashflow	
				IRR	NPV (MUS$)	IRR	NPV (MUS$)
34	0%	9.5	1	11.36%	2,113	#-!	18,830
345	70%	7.5	2.7	11.64%	1,169	7.57%	3,006
348	70%	9	2.7	20.35%	7,660	9.71%	5,645

4. Conclusion

The ANG technology would be optimal for fields having flare gas with high methane content and low impurities. The economics of flare gas utilization using ANG technology will be feasible if the financing pattern being the cooperation between privates and the Government where the part of Government cost capital is more than 40%.

5. Acknowledgements

We express our gratitude to the University of Indonesia, which funded this research through the scheme of Hibah Publikasi Internasional Terindeks untuk Tugas Akhir Mahasiswa No 864/UN2.R3.1/HKP.05.00/2017.

6. References

[1] KESDM 2016 *Handbook Economic & Energy Statistic of Indonesia* Jakarta
[2] Zheng Q R, Feng Y L, Zhu Z W and Wang X H 2016 *Appl. Therm. Eng.* **98** 778
[3] Vasiliev L L, Kanonchik L E, Mishkinis D A and Rabetsky M I 2000 Int. J. Therm. Sci. **39** 1047
[4] Nie Z, Lin Y and Jin X 2016 *Front. Mech. Eng.* **11**(3) 1
[5] Peng Y, Krungleviciute V, Eryazici I, Hupp J T, Farha O K, Yildirim T 2013 *J. Am. Chem. Soc.* **135** 11887

IOP Conf. Series: Earth and Environmental Science **105** (2017) 012023 doi:10.1088/1755-1315/105/1/012023

Use of carbon pyrolyzed from rice husk in $LiFePO_4$/V/C composite and its performance for lithium ion battery cathode

N Sofyan[1,2], P Sekaringtyas[1], A Zulfia[1,2] and A. Subhan[3]

[1] Department of Metallurgical and Materials Engineering, Faculty of Engineering, Universitas Indonesia, Depok 16424, Indonesia
[2] Tropical renewable Energy Center, Faculty of Engineering, Universitas Indonesia, Depok 16424, Indonesia
[3] Research Center for Physics, Indonesian Institute of Science (LIPI), Puspiptek Serpong, Cisauk-Banten, Indonesia 15314

E-mail: nofrijon.sofyan@ui.ac.id

Abstract. The characteristics of activated carbon pyrolyzed from rice husk used in the synthesis of $LiFePO_4$/V/C for the development of lithium ion battery cathode has been examined. The synthesis was begun by synthesizing $LiFePO_4$ (LFP) via hydrothermal route using the precursors in stoichiometric amounts of LiOH, $NH_4H_2PO_4$, and $FeSO_4.7H_2O$. The as-synthesized LFP was then added with variation of vanadium concentrations and a fix concentration of the carbon pyrolyzed from rice husk to form a composite of $LiFePO_4$/V/C. The composites were characterized using X-ray diffraction (XRD), scanning electron microscope (SEM), and electrochemical impedance spectroscopy (EIS). The XRD results showed that the $LiFePO_4$/V/C has been successfully formed whereas SEM results showed a difference in morphology of vanadium and activated carbon addition. The EIS results showed that the conductivity of LiFePO4/C-0 wt.% V is 1.0196×10^{-2} S/cm, LiFePO4/C-3 wt.% V is 1.0302×10^{-2} S/cm, LiFePO4/C-5 wt.% V is 6.1282×10^{-3} S/cm, and LiFePO4/C-7 wt.% V is 8.3843×10^{-3} S/cm. The best performance for lithium ion battery cathode was given by LiFePO4/V/C at 3 wt.% vanadium. This result indicated that rice husk can be used as a cheap resource for activated carbon in the development of lithium ion battery cathode.

1. Introduction

Indonesia is the third largest rice producer in the world, after China and India [1]. In addition to producing rice, grain harvest will also produce rice husks. Assuming the grain-husk ratio is 1: 0.24, every ton of grain will produce 0.24 tons of rice husks. Hence, disposal of this vast amount of rice husk has been a major problem being faced by rice milling industry. The traditional method of disposing this rice husk is by burning it, however, other problem arises from this burning due to environmental issue.

Rice husk mostly contains cellulose and silica of around 20-25%, which when pyrolyzed under an inert atmosphere will produce carbon [2]. Many attempts have been performed by many investigators to make use of this rice husk such as precursors to produce zeolite [3], silica [4], concrete [5], activated carbon [6], and carbon for lithium ion battery anode [7].

Driven by the increasing need of energy demand in the midst of decreasing availability of the fossil fuel and the resulting emission of the fossil fuel that is not environmentally friendly, there is an intensive research of material for energy storage. For this purpose, lithium ion battery has drawn much attention due to its good performance and relatively high capacity. The use of $LiFePO_4$ as a cathode has the advantage of low manufacturing cost, non-toxic, and quite high theoretical capacity, high charge-discharge cycles and environmentally friendly [8, 9]. The major problem is that it has slow lithium ions diffusion between the interface that results in low electrical conductivity (~10^{-9} S/cm) and thus low rate capability [10]. Many efforts have been carried out by many investigators to improve the conductivity of this $LiFePO_4$ such as metal doping [11, 12] and carbon coating [13].

Carbon is usually used to improve the conductivity of $LiFePO_4$, and the most use material is carbon black. Carbon black is a material produced by the incomplete combustion of heavy petroleum products that has a high surface-area-to-volume ratio. However, the use of carbon black has been limited due to its carcinogenic effect [14]. Many efforts have been made to replace this carbon black with the one from natural resources such as table sugar and bamboo. In this work, rice husk was used as a precursor to produce activated carbon and the result was applied to increase the conductivity of vanadium doped $LiFePO_4$ in the form of $LiFePO_4$/V/C composite for lithium ion battery cathode. The characteristic of the material was examined and the results are presented and discussed in detail.

2. Experimental Setup

2.1. Carbon pyrolysis

The raw material of rice husk was collected from local rice mill in Wonogiri, Central Java. The rice husk was thoroughly washed with tap water to remove any adhering dirt and dried under room temperature condition. The dried rice husk was heated at 700 °C and the result in the form of rice husk charcoal was crushed into fine carbon fragments. The as-prepared charcoal was further blanched in NaOH (Merck) solution with a weight ratio of 1: 5 and stirred using a magnetic stirrer at 100 °C for 2 hours. After the stirring, the carbon was allowed to stand for 22 hours before filtering it with filter paper and the carbon deposit was transferred to the crucible and was further heated at 450 °C for 1 hour and cooled down to room temperature. After cooling down, the carbon was washed using distilled water to obtain a pH 7 and subsequently dried at 120 °C for 12 hours before characterization using X-ray diffraction (XRD, ARL OPTX-2050) with Cu Kα radiation (λ = 1.5406 Å).

2.2. Synthesis of $LiFePO_4$

In this work, $LiFePO_4$ was prepared from the precursors of LiOH, $NH_4H_2PO_4$, and $FeSO_4.7H_2O$. All of these precursors were analytical grade reagents from E-Merck. The preparation was in accordance to our previous work published elsewhere [15] and is explained as following. The stoichiometric amounts of the precursors were diluted with 30 ml distilled water in a separated glass beaker. After dilution, the three solutions of LiOH, $NH_4H_2PO_4$, and $FeSO_4.7H_2O$ were mixed into a glass beaker and stirred until homogenized under a magnetic stirrer for 30 minutes. The mixed solution was then transferred into a Teflon autoclave and hydrothermally heated at 180°C for 20 hours.

After the hydrothermal process, the solution was transferred into a beaker and left to form precipitates. The precipitates were then separated from the solution and washed with deionized water before being dried in an oven at 80°C for 18 hours. The characteristics of crystal structure of the obtained materials were examined by using X-ray diffraction (XRD, ARL OPTX-2050) with Cu Kα radiation (λ = 1.5406 Å) and a scanning electron microscope (SEM, FEI Quanta-650).

2.3. Synthesis $LiFe_{1-x}V_xPO_4/C$

The synthesis was begun by preparing the as-synthesized $LiFePO_4$ powders from the previous step with variations of 0, 3, 5, and 7 wt.% H_4NO_3V and 3 wt.% of activated carbon from the rice husk. Detail of the process has been published elsewhere [15] and is explained as following. A stochiometric amount $LiFePO_4$ at various concentrations of H_4NO_3V powders were ball-milled until homogenized. An amount of 3 wt.% carbon was then added into the mixture and the ball-milling process continued until a homogenized mixture was obtained.

After the milling process, a small amount of specimen was tested with a thermal analyser (PerkinElmer STA 6000) to examine the crystallization process. The results from the thermal analysis showed that the sintering process needs to be performed at 750°C. Therefore, all the samples were then sintered at 750°C for four hours under argon environment. The characteristics of the resulting cathode active materials were examined by using X-ray diffraction (XRD, Philips PW3020) with Cu Kα radiation (λ = 1.5406 Å) and a scanning electron microscope (SEM, FEI Quanta-650) equipped with energy dispersive X-ray spectroscopy (EDX).

2.4. Conductivity measurement

The cathode was prepared by mixing the material with poly-vinylidene fluoride (PVDF, MTI) in N-methyl pyrrolidone (NMP, MTI) solvent inside a vacuum mixer. The mixture in the form of paste was then applied onto a current collector of aluminium foil (MTI), dried, hot-rolled and heated in a vacuum oven at 80°C. The conductivity was measured using an electrochemical impedance spectroscopy (EIS, Hioki LCR 3532-50).

3. Results and Discussion

X-ray diffraction characterization was performed to confirm the crystal structure of the samples both for the activated carbon from the rice husk and $LiFePO_4$. The diffraction pattern for the activated carbon is shown in figure 1, whereas the diffraction patterns of the reference $LiFePO_4$, the commercial $LiFePO_4$, and the synthesized $LiFePO_4/C$ with various vanadium concentrations are given in figure 2. For the activated carbon, as seen in figure 1, the diffraction pattern shows a crystalline structure indicated by the presence of several peaks at 2θ = 26.50, 44.06, 64.41, 77.53, and 98.32. The pattern is indexed to a rhombohedral R3 space group (JCPDS 026-1079).

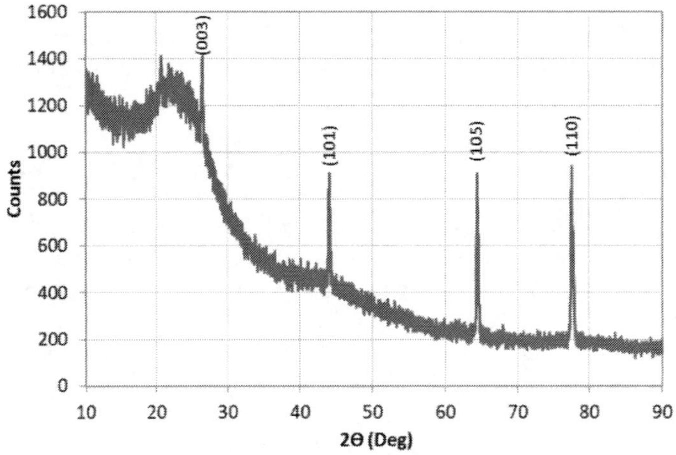

Figure 1. X-ray diffraction pattern of the activated carbon from the rice husk.

For the synthesized $LiFePO_4$ samples with carbon addition are still in agreement with that of $LiFePO_4$ reference with a structure crystal indexed to the Pnma orthorhombic (JCPDS No.083-2092). The same result has been obtained by other investigators [16, 17]. For the $LiFePO_4$ with carbon addition and various concentrations of up to 7 wt.% vanadium, there is no change in the structure as indicated by the diffraction peaks. Some differences in the diffraction intensities exist, however, due to the carbon and or vanadium addition as can be seen in figure 2(e) and (f). In all of vanadium-doped $LiFePO_4/C$, high crystallinity has been obtained with no typical diffraction patterns of carbon nor vanadium. The non-existence of carbon and vanadium diffraction patterns within the materials is an indication that carbon is in an amorphous form whereas vanadium has substituted and occupied Fe sites in $LiFePO_4$.

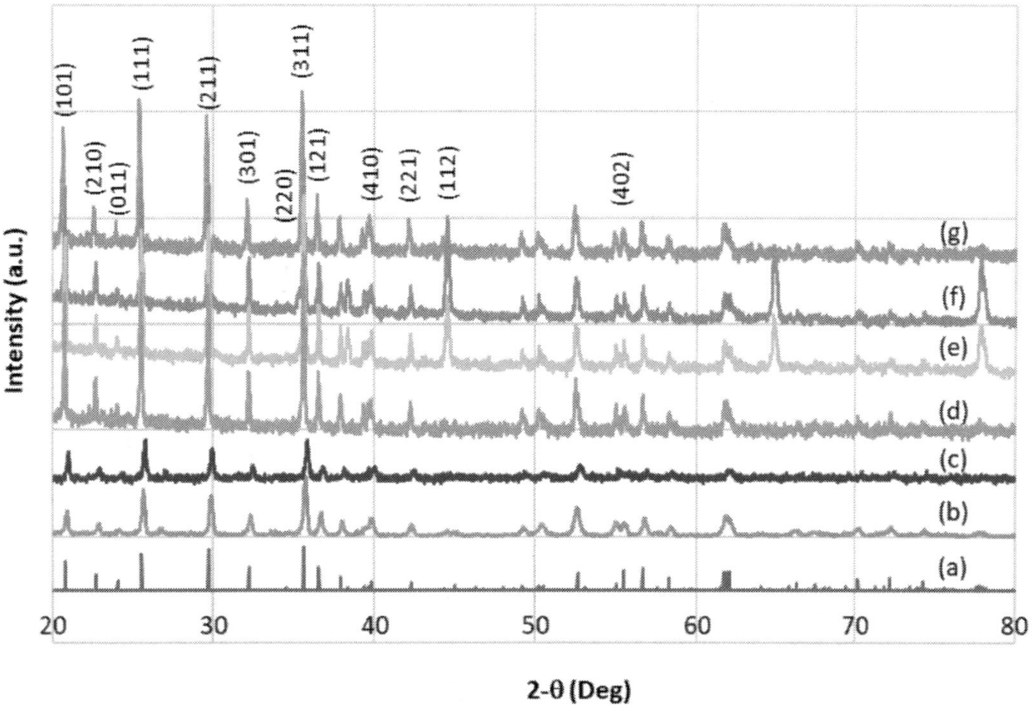

Figure 2. Diffraction patterns of (a) reference $LiFePO_4$ (JCPDS No.083-2092), (b) commercial $LiFePO_4$, (c) as-synthesized $LiFePO_4$, (d) 0 wt.% V-doped $LiFePO_4/C$, (e) 3 wt.% V-doped $LiFePO_4/C$, (f) 5 wt.% V-doped $LiFePO_4/C$, and (d) 5 wt.% V-doped $LiFePO_4/C$.

In order to support the results from the X-ray diffraction investigation, scanning electron microscope observation was performed to reveal the surface morphology of the samples. The results from the secondary electron image for $LiFePO_4/V/C$ composites are given in figure 3. As seen in the micrograph, figure 3(a) is at 0 wt.% V-doped $LiFePO_4/C$, figure 3(b) is at 3 wt.% V-doped $LiFePO_4/C$, figure 3(c) is at 5 wt.% V-doped $LiFePO_4/C$, and figure 3(d) is at 7 wt.% V-doped $LiFePO_4/C$. In all of the micrographs, most of the particles show the morphology of flake shape and the particles distribute homogeneously.

Electrical conductivity was measured using electrical impedance spectroscopy to reveal the change in conductivity through a plot diagram depicting the material's capacitance. The results in the forms of Cole plots are shown in figure 4. Analysis of this diagram shows that the resulting conductivity of $LiFePO_4/C$-0 wt.% V is $1.0196x10^{-2}$ S/cm, $LiFePO_4/C$-3 wt.% V is $1.0302x10^{-2}$ S/cm, $LiFePO_4/C$-5 wt.% V is $6.1282x10^{-3}$ S/cm, and $LiFePO_4/C$-7 wt.% V is $8.3843x10^{-3}$ S/cm. These results indicate that the addition of activated carbon alone could increases the conductivity as compared to the pure $LiFePO_4$. In the case of vanadium, at 3 wt.% V, the conductivity increases; however, at 5 wt.% V and 7 wt.% V results in decreasing the conductivity. It seems that when vanadium concentration is higher than 3 wt.%, it is suspected that, the particles tend to agglomerate results in more pores formation and thus decreases the conductivity. See the surface morphology of the composite from the secondary electron image in figure 3 for more detail.

The cyclic voltammetry and charge/discharge testing were performed for 3 wt.% V and 7 wt.% V only and the curves are given in figure 5. As seen in figure 5, cyclic voltammetry curves shown at the top have reduction-oxidation peaks at 3.59 and 3.22 volts for 3 wt.% V (left) and at 3.59 and 3.20 volts for 7 wt.% V (right), respectively.

Figure 3. Secondary electron images of (a) 0 wt.%, (b) 3 wt.%, (c) 5 wt.%, and (c) 7 wt.% vanadium-doped $LiFePO_4/C$.

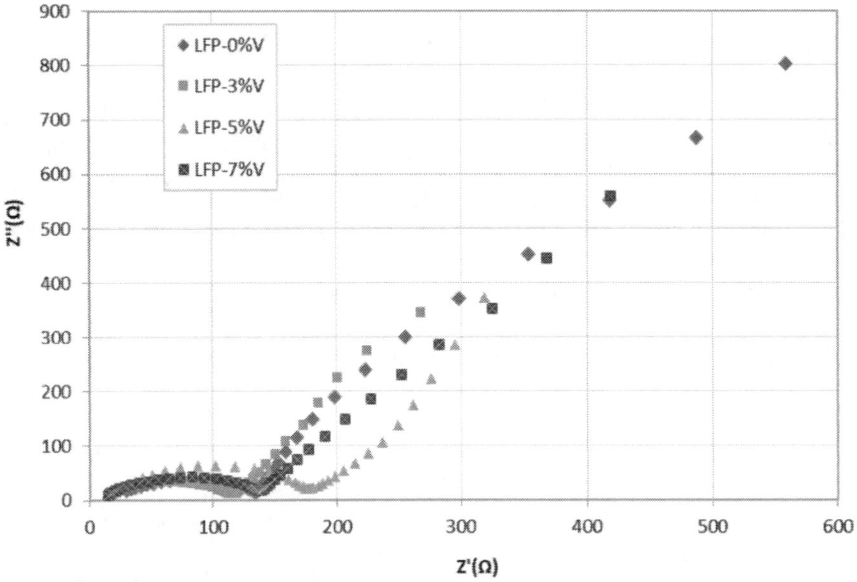

Figure 4. Nyquist plots for the LiFePO4/C (blue diamond), 3 wt.% V (red square), 5 wt.% V (green triangle), and 7 wt.% V LiFePO4/C (blue square).

The charge/discharge testing was performed at the potential range of 2.5–4.2 V at two capacities, i.e. 0.5C and 4C, and the curves are shown at the bottom in figure 5 for the LiFePO4/V/C at 3 wt.% V (left) and 7 wt.% V (right). The initial charge and discharge capacities of both 3 wt.% V (left) and 7 wt.% V (right) at 0.5C are almost the same (8 mAh g^{-1}), however, the charge and discharge capacity of

7 wt.% V drops below 3 mAh g^{-1} at 4C. It seems that, as has been mentioned previously, when the vanadium concentration is higher than 3 wt.%, the particles tend to agglomerate. This agglomeration results in more pores formation that decreases the conductivity and thus the performance.

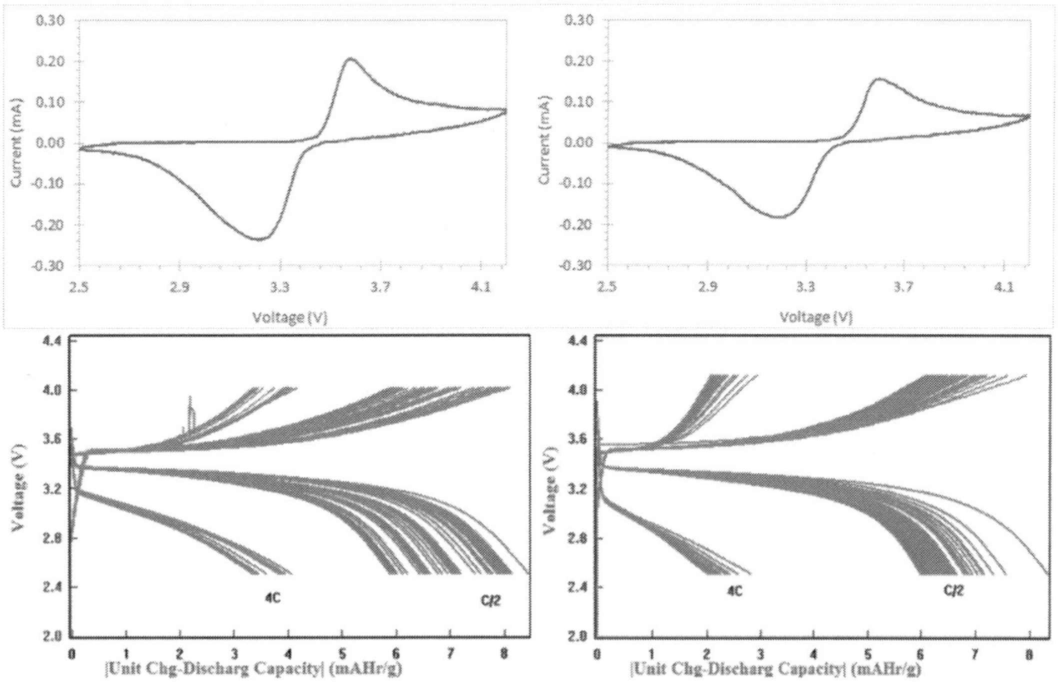

Figure 5. Cyclic voltammetry (top) and charge and discharge (bottom) testing for the LiFePO4/V/C at 3 wt.% V (left) and 7 wt.% V (right).

4. Conclusion
In this work, a relatively high purity of $LiFePO_4$ has been obtained from the hydrothermally synthesis. The addition of activated carbon from the rice husk did not affect the purity of the $LiFePO_4$ particles. The electrochemical impedance spectroscopy test results showed that the conductivity of $LiFePO_4$/C-0 wt.% V is 1.0196×10^{-2} S/cm, $LiFePO_4$/C-3 wt.% V is 1.0302×10^{-2} S/cm, $LiFePO_4$/C-5 wt.% V is 6.1282×10^{-3} S/cm, and $LiFePO_4$/C-7 wt.% V is 8.3843×10^{-3} S/cm., respectively. These results indicate that the addition of carbon increases the conductivity whereas the vanadium addition also increase the conductivity but decreases at vanadium concentration >3 wt.%. The best performance was given by LiFePO4/V/C at 3 wt.% vanadium.

5. Acknowledgements
The authors would like to express their gratitude and appreciation for the funding from the Directorate of Research and Community Services (DRPM), Universitas Indonesia, through Hibah PITTA No. 822/UN2.R3.1/HKP.05.00/2017.

6. References
[1] United Nations: Food and Agriculture Organization *Rice Market Monitor* **20** 2017
[2] Real C, Alcala M D and Criado J M 1996 *J. Am. Ceram. Soc.* **79** 2012
[3] Panpa W and Jinawath S 2009 *Appl. Catal.* **B90** 389
[4] An D, Guo Y, Zou B, Zhu Y, Wang Z 2011 *Biomass Bioenergy* **35** 1227
[5] Gorhan G and Simsek O 2013 *Constr. Build. Mater.* **40** 390
[6] Liou T and Wu S 2009 *J. Hazard. Mater.* **171** 693

IOP Conf. Series: Earth and Environmental Science **105** (2017) 012023 doi:10.1088/1755-1315/105/1/012023

[7] Fey G T-K and Chen C-L 2001 *J. Power Sources* **97-98** 47
[8] Doeff M M 2012 *Battery Cathodes* in Meyers R A (Ed) The Encyclopedia of Sustainability Science and Technology (New York: Springer-Verlag)
[9] Julien C M, Mauger A, Zaghib K and Groult H 2014 *Inorganics* **2** 132
[10] Chung S Y, Bloking J T and Chiang Y M 2002 *Nat. Mater.* **1** 123
[11] Arumugam D, Kalaignan G P and Manisankar P 2008 *J. Solid State Electrochem.* **13** 301
[12] Ou X, Liang G, Wang L, Xu S and Zhao X 2008 *J. Power Sources* **184** 543
[13] Väyrynen A and Salminen J 2012 *J. Chem. Thermo.* **46** 80
[14] Ramanakumar A V, Parent M-É, Latreille B and Siemiatycki J 2008 *Int. J. Cancer* **122** 183
[15] Sofyan N, Putro D Y and Zulfia A 2016 *Int. J. Technol.* **7** 1307
[16] Teng T H, Yang M R, Wu S H 2007 *Solid State Commun.* **142** 389
[17] Yang M R and Ke W H, 2008 *J. Electrochem. Soc.* **155** A729

IOP Conf. Series: Earth and Environmental Science **105** (2017) 012024 doi:10.1088/1755-1315/105/1/012024

Impregnation of activated carbon-TiO_2 composite and its application in photodegradation of procion red synthetic dye in aqueous medium

M O Ruliza[1], T E Agustina[1] and R Mohadi[2]

[1] Chemical Engineering Magister Program, Faculty of Engineering, Universitas Sriwijaya, Palembang 30139, Indonesia
[2] Chemical Program, Faculty of Math and Natural Science, Universitas Sriwijaya, Indralaya 30662, Indonesia

E-mail: tuty_agustina@unsri.ac.id

Abstract. Synthetic dyes are often used for textile dyeing process. The dye wastewater containing hazardous materials, toxic, and also harmful to the environment. Among the existing technologies,photodegradation using semiconductor catalyst is a promising alternative method. However, the weak adsorption capacity of photocatalyst is an issue for photocatalysis process. To overcome this lack, photocatalyst material needs to combine with an adsorbent. The composite was synthesized by combining the function of activated carbon as adsorbents and TiO_2 as a catalyst. The objective of the study was to synthesize a composite of activated carbon and TiO_2 through impregnation method. The activated carbon-TiO_2 composite produced was characterized to determine the surface morphology and the elements contained in the composite by using SEM/EDX. The composite activity to photodegrade various concentrations of Procion red, as a model of synthetic dye, was also investigated. To compare the activity of activated carbon-TiO_2 composite in degrading Procion red, different conditions were applied by using UV lamps, sunlight irradiation, and without irradiation (darkroom). The activated carbon-TiO_2 composite was able to photodegrade of Procion red by UV light irradiated at optimum wavelength 530 nm. The highest photodegradation percentage of 58.9% and 71.7 % were obtained when using the activated carbon-TiO_2 composite ratio of 3:7 and 1:1, respectively.

1. Introduction

Dye is an organic compound containing a conjugated chromophore group. Reactive dyes are a dye used for coloring textiles, for example Remazol brilliant orange 3R, Remazol golden yellow, procionred, and methylene blue. The dyes were often used for batik dyeing process both in large-scale industry and home industry.

The dye wastewater containing hazardous materials and toxic presence in water can block the sunlight to penetrate the aquatic environment, so that disrupting the biological processes that occur in the water body. In addition, it is also disturbing the aesthetics of the water due to the appearance of a foul odor, and also harmful to the environment [1].

It is necessary to set up wastewater treatment technology, so that the effects of pollution can be prevented and controlled. Some wastewater treatment technology with relatively low cost have been investigated to overcome the problems of waste water, synthetic dyes such as adsorption using

IOP Conf. Series: Earth and Environmental Science **105** (2017) 012024 doi:10.1088/1755-1315/105/1/012024

activated carbon from coconut shell [2], or a combination of filtration and adsorption [3]. Some other methods can be used to treat wastewater, such as by coagulation/adsorption [4], chemical and biological oxidation [5], and photodegradation by using semiconductor photocatalyst [6].

Among the existing methods, photodegradation is an alternative method that is relatively cheap and easy to apply[7]. This method requires a semiconductor material include TiO_2, ZnO, or Fe_2O_3 and ultraviolet radiation (UV). Among semiconductor materials exist, TiO_2 is a semiconductor material availability in the market and classified the best of quality [8]. The TiO_2 is often used as a photocatalyst to decompose organic compounds such as pentaklorofenol [9], cibracron yellow, orange metal, and procion red.

Weak adsorption capacity is an issue for photocatalytic process. To cover the shortfall, photocatalyst material needsto combine with an adsorbent [10]. Adsorbent which can be used include activated carbon, which is synthesized from coconut shell.Carbon which is the enhanced energy through activation of so-called activated carbon, has a sizeable surface and a good adsorption properties, so that it can be used as an adsorbent, active carbon properties can help the process of adsorption-catalytic [11].

The objective of thisresearch was to synthesize a composite of activated carbon and TiO_2 through impregnation method and to study the application ofactivated carbon-TiO_2 composite in photodegradation of procion red synthetic dye.In addition, the activated carbon-TiO_2 compositeproduced is characterized to determine the morphology by SEM/EDX, then tested for activity to degrade the procion red dye. Activated carbon-TiO_2 composite is expected to be an effective alternative for dealing with industrial wastewater containing the dying stuff.

2. Materials and method

2.1. Materials
The activated carbon of 200 mesh, TiO_2, aquadest, distilled water, and procion red dye were used as materials.

2.2. Procedure
Activated carbon-TiO_2 composite was varied with the composition of TiO_2 50% and 70% by weight.The composite was synthesized by dissolving the amount of TiO_2 into 100 mL of distilled water. Then the amount of activated carbon of 200 mesh was added to the TiO_2, then stirred by using a magnetic stirrer for 5 hours, and ultrasonic for 30 minutes. Activated carbon-TiO_2formed was dried in an oven for 5 hours with a temperature of 120°C until dry. Activated carbon that has been coated with a photocatalyst is then calcined at a temperature of 400°C for 2 hours. Composite of activated carbon-TiO_2 characterized by using SEM-EDX to determine the morphology of the composite.

Procion red solution with a concentration of 150, 200, 250, 300 mg/L was added with activated carbon–TiO_2 composite at a contact time of 2 hours, 4 hours, 6 hours and 8 hours with 3 different conditions: with UV light, sunlight and darkroom at various time intervals of 2 hours between 10 am until 2 pm for 2 days. The mixture was filtered to separate activated carbon-TiO_2 composite from procion red wastewater.

The absorbance value of procion red that has been photodegradated or adsorbed by activated carbon–TiO_2 composite was measured by using a UV-Vis spectrophotometer at the maximum wavelength. From these absorbance values, the concentration of red procioncan be calculated, as well as the percentage of photodegradation and adsorption of procion red.

2.3. Characterization and analysis
The intensity of sunlight was measured by using a Light meter (Luxtron Lx-103, Lux, Taiwan) at various time intervals of 2 hours between 10 am until 2 pm. The intensity of sunlight was 123.9 W/ m^2. The intensity of UV light was 156 W/ m^2. And the intensity of darkroom was 0-5 W/ m^2. Composite of activated carbon-TiO_2 characterized using Scanning Electron Microscope-Energy

Dispersive X-ray(SEM-EDX) to determine the morphology of the composite and to identify the elemental composition of materials.

3. Results and discussion

3.1. Characterization of activated carbon-TiO_2 composite by SEM/EDX

Activated carbon-TiO_2(AC-TiO_2) composite were characterized by SEM/EDX to see the surface morphology and the elements contained in the composite. Figure 1 and 2 shows the results of SEM.

Figure 1. SEM images of AC-TiO_2 (3:7).

Figure 2. SEM images of AC-TiO_2 (1:1).

From the figure, it can be seen that TiO_2 with the characteristics of white powder are dispersed evenly cover the surface of activated carbon. The results show the texture of activated carbon-TiO_2 composite having an aggregate structure, but the morphology of the surface looks uniform. The SEM results of activated carbon-TiO_2 composites are then compared with the results of another researcher. The research of Tang, *et al* (2012) shows similarities to the surface morphology of activated carbon-TiO_2 composite in this research [12]. This means that the activated carbon-TiO_2 composites have been successfully synthesized.

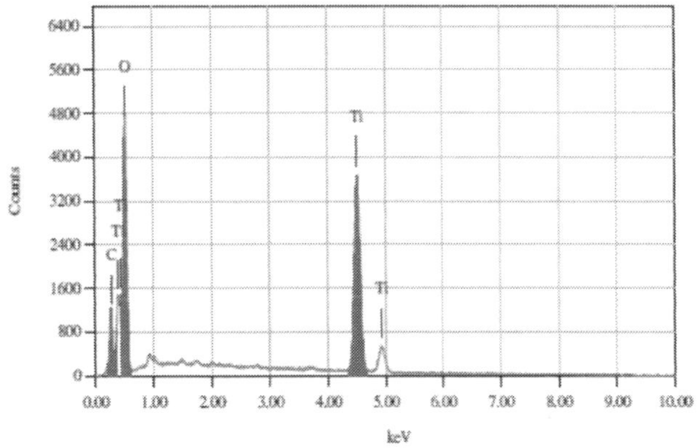

Figure 3. EDX images of AC-TiO_2 (3:7).

IOP Conf. Series: Earth and Environmental Science **105** (2017) 012024 doi:10.1088/1755-1315/105/1/012024

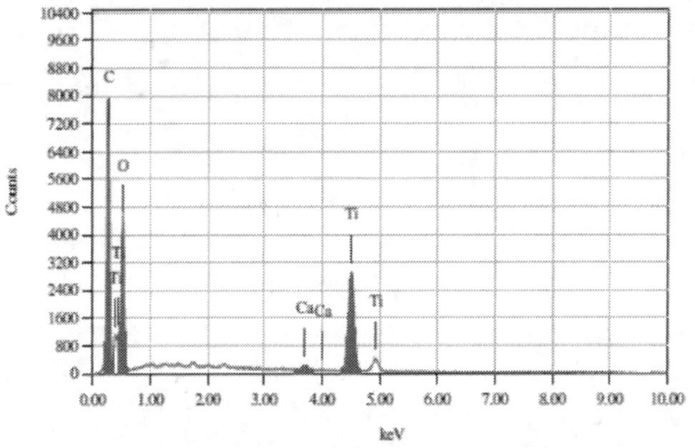

Figure 4. EDX images of AC-TiO_2 (1:1).

Characterization of activated carbon-TiO_2 byusing EDX was shown in figure 3 and 4. The EDX analysis was used to determine which chemical elements are present in the composite.The EDX Spectra showed the peaks of Ti and O.It means the composite comprise of TiO_2photocatalyst.

3.2. Photodegradation of procion red by activated carbon-TiO_2 composite

Photodegradationprocess conducted at a contact time of 8 hours, and by varying the concentration of procion red namely 150, 200, 250, 300 mg/L with 100 mg weight of composite irradiated by UV Light 254 nm. The intensity of UV light was 156 W/m^2. Activities photodegradation of procion red dye by activated carbon-TiO_2 composite irradiated by UV Light can be seen from the percentage value of photodegradation shown in figure 5.

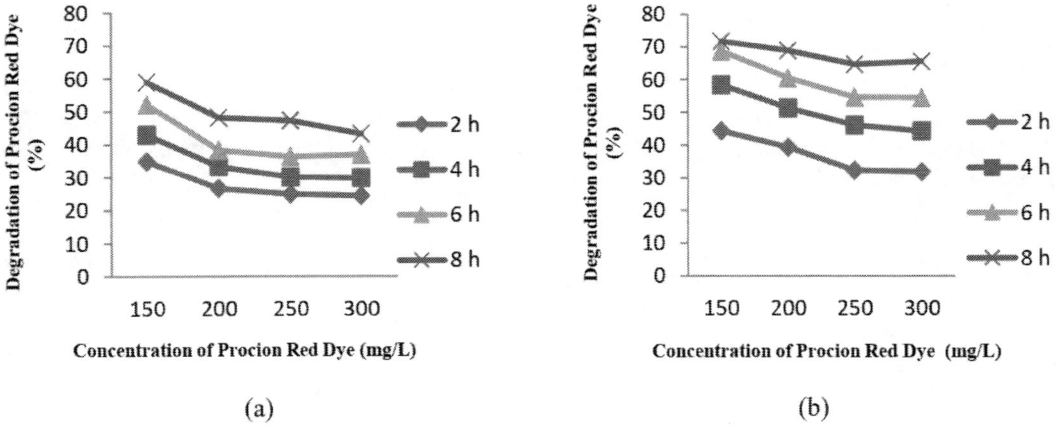

Figure 5. The relationship between concentration and percentage of procion red photodegradation by UV light by a composite of: (a) AC-TiO_2 (3:7); (b) AC-TiO_2 (1:1).

Procion Red were used as controls irradiated with UV light has a photodegradation percentage of 4.6%; 3.3%; 2.7% and 1.9% for each concentration procion red 150, 200, 250, and 300 mg/L, respectively. Photodegradation of procion red by activated carbon-TiO_2 composite ratio of (3:7) and (1:1) irradiated by UV light produce photodegradation percentage respectively 58.9% and 71.7%.

In the photocatalytic process, when TiO_2 absorbs light energy equal to or greater than the band gap energy there will be a charge separation or photo excitation in the semiconductor. Electrons excited into the conduction band leaving a positive hole (H^+) in the valence band. The positive hole has a high affinity for oxygen in H_2O molecules adsorbed on the surface of TiO_2, so it will react into OH^- and H^+. Hydroxyl radicals are highly reactive species that attack organic molecules and degrade into CO_2 and

H_2O, and halide ions if organic molecules containing halogen atoms. The oxidation of organic molecules are not selective. The reaction is written in the equation below :

$$TiO_2 + hv \rightarrow TiO_2(e^- + h^+) \quad (1)$$

$$h^+ + H_2O \rightarrow H^+ + OH\bullet \quad (2)$$

$$h^+ + OH^- \rightarrow OH\bullet \quad (3)$$

$$e^- + O_2 \rightarrow \bullet O_2^- \quad (4)$$

$$OH\bullet + dye \rightarrow oxidation\ of\ the\ dyecompound \quad (5)$$

$$\bullet O_2^- + dye \rightarrow reduction\ of\ the\ dyecompound \quad (6)$$

Photodegradationprocion red was done by sunlight for 8 hours. The intensity of sunlight was measured using Light meter (Luxtron Lx-103, Lux, Taiwan) at various time intervals of 2 hours between 10 am and 2 pm for 2 days. The first day was done 0-4 hours and the second day 4-8 hours. The mean intensity of sunlight irradiation was 123.9 W/m^2. Figure 6 shows the relationship between concentration and percentage of photodegradation procion red irradiated by sunlight by activated carbon- TiO_2 composite (3: 7) and activated carbon- TiO_2 (1: 1).

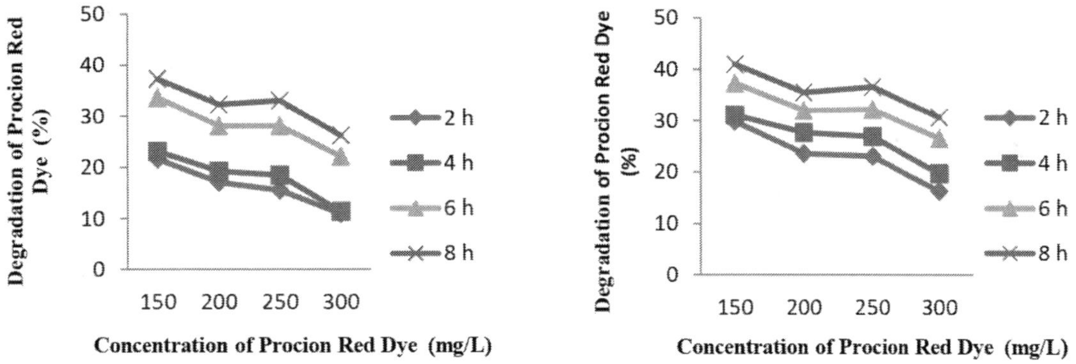

(a) (b)

Figure 6. The relationship betweenconcentration and percentage of photodegradation procion red by sunlight by composite of : (a) AC-TiO_2 (3: 7), (b) AC-TiO_2 (1: 1).

Photodegradation of procion red by activated carbon-TiO_2 composite ratio (3:7) and (1:1) irradiated sunlight produces photodegradation percentage respectively 37.3% and 41%.

To compare the activity of activated carbon-TiO_2composite in degrading procion red, the condition without irradiation (darkroom) was applied. In the activated carbon-TiO_2 composite that is not exposed to UV light and sunlight, the photodegradation processdoes not occur, but only the adsorption process of composite activated carbon-TiO_2 as adsorbent. Adsorption of Procion red by activated carbon- TiO_2 composite ratio (3:7) and (1:1) in a darkroom reaches percentage 23.4% and 24.3%.

Activated carbon-TiO_2 composite having a good photodegradation activity to procion red, by UV light and sunlight. Comparison of the activity of photodegradation and adsorption on activated carbon-TiO_2 composite can be seen in figure 7 below.

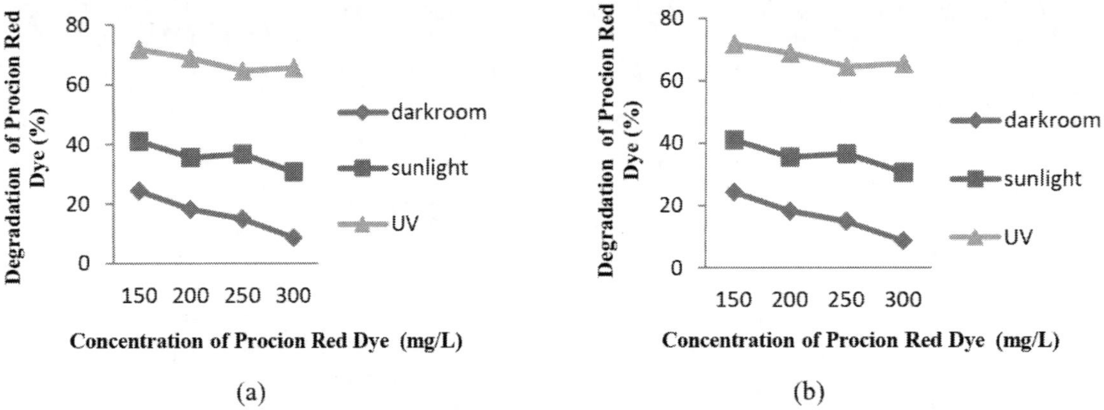

Figure 7. The relationship between procion red concentration and percentage of photodegradation and adsorption procion red by composite of: (a) AC-TiO_2 (3:7); (b) AC-TiO_2 (1:1).

From figure 7 (a), adsorption percentage of activated carbon-TiO_2 composite (3:7) to procion red reached 23.4 %, while photodegradation percentage by UV light and sunlight reached 58.9 % and 37.3% for 8 hours irradiation,respectively. And from figure 7 (b), adsorption percentage activated carbon- TiO_2 composite of (1:1) to procion red reached 24.3 %, while photodegradation percentage by UV light and sunlight reached 71.7 % and 41%, respectively, for 8 hours irradiation. Activated carbon-TiO_2 (1:1) has a higher value than activated carbon-TiO_2 composite (3:7) because of the difference in the composition. The amount of activated carbon to the activated carbon-TiO_2 composite (1:1) more so as to assist the adsorption capacity of the weak which is a problem for the photocatalytic process.

4. Conclusion

Activated carbon-TiO_2 composite has been succeeded in impregnating 86.5% for activated carbon-TiO_2 (3:7) and 84% for activated carbon-TiO_2 (1:1) with a gray powder texture.Photodegradation of procion red by activated carbon-TiO_2 composite ratio of (3:7) and (1:1) irradiated by UV light produce photodegradation percentage of 58.9% and 71.7%, respectively. Whereas photodegradation of procion red by activated carbon-TiO_2 composite ratio of (3:7) and (1:1) irradiated by sunlight produces photodegradation percentage of 37.3% and 41%, respectively. Adsorption of procion red by activated carbon-TiO_2 composite ratio of (3:7) and (1:1) in a darkroom, reaches the photodegradationpercentage of 23.4% and 24.3%, respectively.

5. Acknowledgment

The authors would like to thankful the Waste Treatment Technology Laboratory, Chemical Engineering Department, Faculty of Engineering, UniversitasSriwijaya, for laboratory support.

6. References

[1] Moura I M A 2004 *Adsorption of Yellow Lanasol 4G Reactive Dye in a Simulated Textile Effluent on Gallinaceous Feathers* (Official Publication of The European Water Association : European Water Management Online)

[2] Agustina T E 2011 Pengolahan air limbah pewarna sintetis dengan metode adsorpsi menggunakan karbon aktif *J. Rekayasa Sriwijaya* **20** 36-42

[3] Agustina T Eand Komala R 2015 Application of TiO_2 nanoparticles photocatalyst to degrade synthetic dye wastewater under solar irradiation *Contemporary Eng. Sci.* **8** 1625-36

[4] Papic S, Koprivanac N, Bozic A L and Metes A 2004 Removal of some reactive dyes from synthetic wastewater by combined Al (III) coagulation/carbon adsorption process *Dyes and Pigments* **62** 291
[5] Ghoreishi M and Haghighi R 2003 Chemical catalytic reaction and biological oxidation for treatment of non-biodegradable textile effluent *J. Chem. Eng.* **95** 163-169
[6] Kansal S K, Singh M and Sud D 2007 Studies on photodegradation of two commercial dyes in aqueous phase using different photocatalysts *J. Hazardous Mater.* **141** 581-590
[7] Das R 2014 Application photocatalysis for treatment of industrial waste water—a short review *Open Access Library Journal* **1** e713
[8] Fatimah I and Wijaya K 2005 Sintesis TiO_2/Zeolit sebagai fotokatalisis pada pengolahan limbah cair industri tapioka secara adsorbsi-fotodegradasi *TEKNOIN* **10**(4) 257-267
[9] Krisdaningrum N 2003 *Pemanfaatan Suspensi TiO_2 untuk Proses Degradasi Fotokatalisis Senyawa Pentaklorofenol,* Skripsi, Jurusan Kimia, FMIPA UNAIR.
[10] El-Maazawi M, Finken A N, Nair A B and Grassian A V 2000 Adsorption and photocatalytic oxidation of acetone on TiO2 : An in situ transmission FT-IR Study *J. Catal.* **191** 138-146
[11] Andayani W and Agustin S 2007. Karakterisasi katalis TiO_2 dan TiO_2/karbon aktif yang diimobilisasikan pada pelat Titanium. Indo. *J. Chem.* **7**(3) 238-242
[12] Tang S K, Teng T T, Alkharki, Abbas F M, Li Z 2012 *Sonocatalytic Degradation Of Rhodamine B In Aqueous Solution In The Presence of TiO_2 Coated Activated Carbon* Hong Kong ICESD 5-7

IOP Conf. Series: Earth and Environmental Science **105** (2017) 012025 doi:10.1088/1755-1315/105/1/012025

Adsorption capacity study of ethanol-water mixture for zeolite, activated carbon, and polyvinyl alcohol

J A Laksmono[1,2], U A Pangesti[1], M Sudibandriyo[1], A Haryono[2] and A H Saputra[1]

[1] Department of Chemical Engineering, Faculty of Engineering – Universitas Indonesia, Kampus UI, Depok, West Java, Indonesia, 16424
[2] Research Center for Chemistry – LIPI, Kawasan PUSPIPTEK Serpong, Tangerang Selatan, Banten, Indonesia, 15314

Email : msudib@che.ui.ac.id

Abstract. Global energy issue is no longer a new topic. The expansion of energy production proven to show significant influence is the fossil fuel modification by blending it with liquid renewable fuel, such as bioethanol. Bioethanol must achieve fuel-grade standard to qualify as gasoline, one of the specification is to have moisture content of 1.0% v/v or less, as regulated by ASTM D4806. This parameter is a challenging one to achieve, because water-ethanol mixture will encounter the azeotrope phenomenon when the mixture undergoes a common distillation process and reach 95.6% v/v of ethanol. One of the dehydration method that use less energy is adsorption. One of the efficiency consideration of bioethanol dehydration with adsorption is its adsorption capacity. Adsorption capacity is influenced by the material of adsorbent, operational temperature and time. The material being tested in this research is poly vinyl alcohol (PVA), zeolite, and activated carbon. This research will analyze the dependency and influence of mixture's initial concentration and operational temperature condition towards the final concentration of ethanol and adsorption capacity utilizing a Langmuir model. The result of this study showed that the activated carbon has the highest parameter capacity, which is twice as much than zeolite and three times larger than PVA. Whereas the result of selectivity study between the three prove that zeolite has better selectivity.

Nomenclature

V_S	initial stock volume of ethanol
C_S	initial stock concentration of ethanol
V_B	blended volume of ethanol
C_B	concentration of ethanol after blending
C_0	initial concentration
V_0	initial sample volume
t	time variables in minutes
m	mass of adsorbent
V_t	volume of sample after a particular time
C_t	concentrationafter a particular time
T	temperature
q_m	maximum adsorption capacity
K_L	Langmuir constant
q_e	quantity of adsorbate in adsorbent when reaching an equlibria state
C_e	concentrationwhen reaching an equlibria state
D_{water}	dipole moment or polarity of water
D_{EtOH}	dipole moment or polarity of ethanol
ΔV_t	adsorbed volume at a given time
V_{TP}	total pore volume in sample
V_{pore}	pore volume in every gram of adsorbent
ε_{water}	relative permittivity of water
ε_{EtOH}	relative permittivity of ethanol
S_{BET}	BET surface area
AD	adsorption degree

IOP Conf. Series: Earth and Environmental Science **105** (2017) 012025 doi:10.1088/1755-1315/105/1/012025

1. Introductions

The use of bioethanol (CH_3CH_2OH) as a substitute for fossil energy is quite promising [1]. The natural and renewable sources of it are relatively abundant yet the production process is rather simple. The local prospect of its market in Indonesia is especially supported by the blueprint of national energy policy (Presidential regulation No 5 of 2006) in which stated that the percentage of biofuel role in national energy mix must reach 5% by 2025 [2].

The ethanol that may alternate the use of fossil fuel must fulfill a standard set in ASTM D 4806-16a. Based on the standard, fuel grade ethanol is ethanol with minimum content of water, 1.0% v/v or 1.26 m/m (measured at the temperature of 60°F) at most [3]. To fulfill this requirement, the distillation procedure is not sufficient to separate the water from ethanol because when the water-ethanol mixture reach 3.37:95.63 (in % b/b, or 2.80:97.20 in % v/v) an azeotrope phenomenon occurred. Thus, the vapor of this composition has the same proportions of constituents as the unboiled mixture, and hence the composition cannot be altered by simple distillation anymore [4]. Therefore, the bioethanol produced by fermentation must undergo extra steps after being distilled from the beer product in order to dehydrate the water content.

The dehydration of water-ethanol azeotrope mixture can be done by several options, such as pervaporation, extraction, distillation, adsorption, or the combination of the available options, but the high-energy cost associated to dehydration is known as a major challenge [1].

A review by Frolkova and Raeva (2009) stated that adsorption by molecular sieves are most efficient for water removal from solutions with low water content, and if the process is carried in vapor phase the adsorption would excludes the wetting of molecular sieves and reduce heat and energy consumption required in desorption drying of the sieves. Despite its advantages, this method is not usually employed in large-scale ethanol dehydration [1]. The present day commercial-scale process utilized for water dehydration by adsorption is aadsorber-desorber complex of Pressure Swing Adsorption (PSA). The pressure swing adsorption (PSA) process is attractive for the final separation since it requires little energy input and is capable of producing a very pure product [5]. The adsorbent employed in PSA is currently being researched to determine which adsorbent possess the largest capacity and best kinetic properties to support the dehydration of water from water-ethanol solutions.

Various kind of zeolite and activated carbon are investigated. The recent research objectives vary from the raw material, activation factor, throughout the kinetics of various zeolite and activated carbon. Zeolite 3A is proven to be one of the best zeolite to be applied in water adsorption process, because the size of its sieve (3 nm) is only slightly bigger than average water molecule (0.28 nm), and smaller than the size of ethanol molecule (0.4 nm) [6].

Carbon is another efficient water adsorbent. It could dehydrate water–ethanol vapor mixtures containing 1.6–50.9% water. Carrying out the process in an adiabatic carbon adsorber demonstrated that the adsorbent is stable over 85 adsorption–desorption cycles using hot air or nitrogen (80–120°C) [1].

PVA, a hydrophilic polymer, is known to have irregular hydroxyl groups which may easily bonded with another functional group. It is non-toxic, non-carcinogenic, and known to be able to adsorb water in high quantity (up to 12 times its own mass). PVA is an uncommon water adsorbent for water removal from water-ethanol system, but its potential is promising enough to be inspected further due to the high number of hydroxyl groups content [7].

There has been a growing interest for utilizing adsorption in the means of ethanol dehydration, and the need to compare the available and/or new adsorbent arise. The effects of adsorption capacity of adsorbent towards the adsorption process is important, because its influence in the efficiency of the process is clear: the smallest required amount for an adsorption process with the highest capacity will minimize the energy requirement thus increase the economic and time efficiency. The selectivity of adsorbent must be considered too due to the adsorption process may not only take a target adsorbate from the mixture, and thus in the case of water-ethanol mixture dehydration, adsorbent must have adsorbed water rather than ethanol. This research aims to compare the capacity of PVA, activated carbon, and zeolite as water adsorbent for dehydration from ethanol-water system by the adsorption capacity ($q_m$and K_L) of the Langmuir adsorption isotherm, and then analyze the selectivity of each adsorbent.

2. Materials and method

2.1. Materials

The ethanol (>>95% v/v) was obtained from CV Harum Kimia, while aquadest water was attained from local water which processed by a reverse osmosis (NESCA™ YDO 6800). The stock ethanol was then diluted in the aquadest to provide ethanol with four variations of concentration in addition to the original stock; 88%, 90%, 92%, and 94%. The dilution was done by blending stock ethanol and aquadest together.

There were three kind of adsorbent being tested in this research:80 mesh pellet activated carbon bought from Ajax Chemicals, 40 mesh natural zeolite from Yogyakarta, and 14 mesh fully hydrolyzed polyvinyl alcohol (EMD Millipore Corporation, MW 145,000). Pellet carbon and zeolite was activated using Thermolyne 46200 high temperature furnace in gradually increasing room temperature up to 150°C and then held for 2 hours. Afterwards, the temperature increased by 10°C/minutes up to 800°C and then held for 4 hours. Subsequently, the cooling down process was done in 10°C/minutes. Both pellet activated carbon and zeolite kept in a desiccator at all times outside experiment usage.

After activation was done to pellet carbon and zeolite, each adsorbent was characterized using Fourier Transform InfraRed (FTIR), Brunauer, Emmett and Teller Surface Area Analysis (BET), Scanning Electron Microscopy with Energy Dispersive X-Ray Analysis (SEM EDX), and Hydroxyl Test (ASTM E222, Test Method B) [8]. Iodine Number Test (ASTM D4607) was also done as a standard measure for liquid phase applications characterization for activated carbon.

FTIR characterizations was recorded by Shimadzu FTIR Spectrophotometers in Research Center for Chemistry LIPI, PUSPIPTEK, Serpong, to determine the functional groups existing in each adsorbent. The BET analysis was done utilizing Micromeritics' TriStar II 3020 in Research Center for Chemistry LIPI, PUSPIPTEK, Serpong, using N_2 as analysis adsorptive in -195.850°C to characterize the surface and pore structure of adsorbent. SEM EDX 6510 (LA) instrument in Engineering Faculty of Universitas Negeri Jakarta with 20kV accelerating voltage was used to characterize the morphology of the adsorbents.

2.2. Method

The adsorption experiment was done in batch system, utilizing New Brunswick Scientific controlled environment incubator shaker set in 100 rpm. Each of the adsorbent was tested at the temperature of 30°C.The initial concentration of ethanol (C_0) was varied to the prepared five; 88%, 90%, 92%, 94%, and 95%.

One experiment run is defined for the adsorption process of ten 100ml Schott's Duran bottles inside an incubator with one temperature variation. Each bottle contained an amount of adsorbent, and a volume of one initial concentration variation poured right before the bottles was settled inside the incubator. Every two minutes, one bottle was taken out of the incubator as sample. The volume after particular adsorption contact time is measured. The concentration of ethanol in sample after adsorption contact time is measured Kyoto KEM Densito/Specific Gravity Meter DA-640.

The mass of PVA for each sample is 3.75gr and 25ml of ethanol with a particular C_0. Meanwhile, AC and zeolite experiment prepare 7.5gr of the adsorbent, and 50ml of ethanol with a particular C_0.

2.3. Experimental adsorption capacity calculation

The calculation of adsorption capacity was evaluated with experimental data based on the process mass balance. Applying the assumption that $q_0 = 0$, obtained:

$$q = \frac{(V_0 C_0 - V_t C_t)}{m} \quad (1)$$

From the experimental data, the $q_e$was also calculated by equation (1), determined for each experiment as the highest q achieved. $q_e$is defined as the highest amount of adsorbatein the surface an adsorbent for an experiment run, typically achieved when the adsorption process reach the equilibrium state. The concentration of water measured and the q_e was then used in the Langmuir adsorption isotherm to determine the q_m and K_L.The Langmuir isotherm is an adsorption isotherm with a theoretical or rational basis, given by:

$$q_{\text{mod}} = \frac{q_m K_L C_e}{(1 + K_L C_e)} \quad (2)$$

Where q_m and K_L are empirical constant, and considered as Langmuir capacity parameters. After adsorption experiments were carried out to determine which adsorbent gave the highest q_e in solution, the adsorbent with highest selectivity of water-to-ethanol adsorption was determined.

The adsorption selectivity of adsorbents was determined by adsorbed percentage comparison of water to the adsorbed percentage of ethanol, by considering the moment dipole and relative permittivity of each. The adsorbed volume percentage must include a correction factor of adsorbate dielectric constant or relative permittivity. Permittivity is a property of material corresponds to the coulomb force between two point charges in the material. The relative permittivity is the permittivity factor itself which decreased relative to vacuum. ε_{water} is 78.4 and ε_{EtOH} is 26.6[10]. By considering relative permittivity, the selectivity comparison in percentage was transformed into a comparison of molecule adsorbed degree (AD), determined by the equation below:

$$AD_{water} = \frac{\Delta V_{t\,water}}{C_{0\,of\,water} V_0} D_{water} \varepsilon_{water} \quad (3)$$

$$AD_{EtOH} = \frac{\Delta V_{t\,EtOH}}{C_{0\,of\,EtOH} V_0} D_{EtOH} \varepsilon_{EtOH} \quad (4)$$

3. Results and discussion

3.1. Characteristics of the adsorbents

3.1.1. Surface morphology and element composition of adsorbent. The summary of SEM EDX element characterization results for each adsorbent is shown below in Table 1, while the SEM EDX visual result can be seen in Figure 1. Figure 1 shows distinctive surface morphology of each adsorbent. The micrographs of zeolite and activated carbon shows visible pores, while the PVA micrographs invisible. SEM image of PVA shows a fairly smooth surface topology, with few skin epidermis-like layers, thus differentiate it from the rough looking surface of zeolite and activated carbon. However, less impurities are shown in PVA image rather than activated carbon and zeolite. The image is consistent to the element reading of SEM EDX in, which confirms the existence of various impurities in activated carbon and zeolite. The element reading also shows that the ratio of Si:Al in zeolite is 6.4, thus considered hydrophilic.

IOP Conf. Series: Earth and Environmental Science **105** (2017) 012025 doi:10.1088/1755-1315/105/1/012025

Table 1. Element reading by SEM EDX result.

AC		Zeolite		PVA	
Element	Mass %	Element	Mass %	Element	Mass %
C,	97.87%	Si,	39.44%	C,	85.78%
Cu,	0.82%	O,	30.91%	O,	10.77%
Ca,	0.40%	Ca,	14.73%	Au,	3.19%
Si,	0.32%	Al,	6.14%	Na,	0.26%
Mg,	0.30%	Au,	3.78%		
S,	0.29%	Cl,	2.78%		
		K,	1.17%		
		Na,	1.05%		

Figure 1. SEM micrographs of (a) PVA (5,000x); (b) Zeolite (5,000x); and (c) Activated carbon (1,000x).

3.1.2. Functional groups of the adsorbents. The FTIR results in table 2 shows that the sample examined in this study contains O-H groups in various forms. The result of this characterization corresponds with what bond the adsorbent might form with the adsorbate. The existence of OH in every adsorbent indicates that all three adsorbents is able to adsorb water, because the oxygen in OH will form a bond with both water and ethanol molecule to form a hydrogen bond. Water molecule interact by means of hydrogen bonding, because their own hydrogen atom is attached to a highly electronegative element and this hydrogen atom able to bound with the electronegative atom of a neighboring molecule, forming a hydrogen bridge. The FTIR characterization results also confirms the essentials functional groups of each adsorbent, such as Si-O and Al-O in zeolite.

The existence of OH groups is then affirmed by the hydroxyl value test, along with the determination of the unbounded hydroxyl group. Hydroxyl value (HV) in analytical chemistry is defined as the number of milligrams of potassium hydroxide (KOH) necessary to neutralize the acetic acid taken up on acetylation of one gram of sample that contains free hydroxyl groups.[4] It measures the content of free hydroxyl groups in the sample in units of the mass of KOH in milligrams equivalent to the hydroxyl content of one gram of the sample.

The HV value measurement of PVA result is 893.23 mg KOH/gr PVA, the highest compared to zeolite (100 mg KOH/gr zeolite) and activated carbon (156.06 mg KOH/gr activated carbon). The comparison between each adsorbent's HV shows that PVA has almost 9 times more free or unsubstituted hydroxyl groups than zeolite, and almost 6 times more than activated carbon. This may indicate that PVA have a better hydrophilic property than those of activated carbon or zeolite.

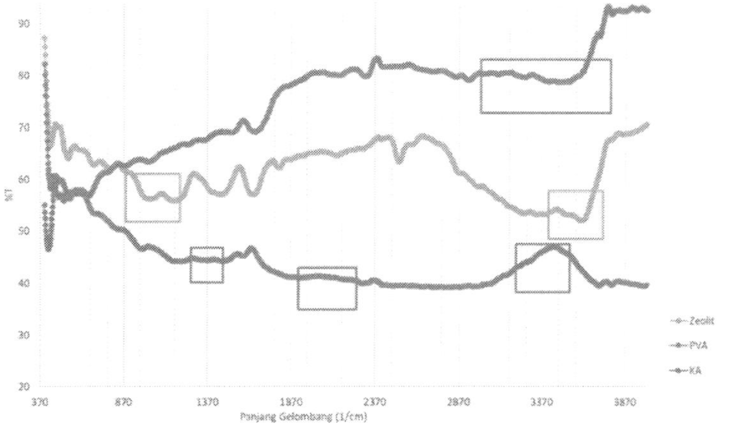

Figure 2. FTIR characterization results summary.

3.1.3. Surface area and pore volume. The BET characterization results in S_{BET} and pore volume of each adsorbent, shown in Table 2. The S_{BET} and pore volume may correspond to the amount of adsorbate that the adsorbent can adsorbs, because the nature of adsorption as an interface reaction in which adsorbate attached to the surface of adsorbent. The BET test show that activated carbon has the largest surface area and pore volume, followed by zeolite and then PVA. This result agrees to the particle diameter of each adsorbent, which is 80 mesh for activated carbon, 40 mesh for zeolite, and 14 mesh for PVA.

Table 2. Characterization result of BET.

Measurement	AC	Zeolite	PVA
S_{BET} (m^2/g)	1149.644	12.9068	0.1209
Pore Volume (cm^3/gr)	0.33028	0.003708	0.00002

Iodine number is a parameter used to characterize activated carbon performance, which measures activity level, where higher number indicates higher degree of activation. This characterization measures the microspore content of the activated carbon (0 to 20 Å, or up to 2 nm) by adsorption of iodine from solution, and is equivalent to surface area of carbon between 900 m^2/g and 1100 m^2/g. Iodine number of the carbon utilized in this research is reported as 2209.67 mg/g, higher than the typical range of 600–1450 mg/g.[5] Corresponding to the result of BET analysis, the surface area of the activated carbon results at 1149.64 m^2/g is also higher than the typical 1100 m^2/g.

3.2. Equilibrium adsorption capacity

The equilibrium adsorption capacity of each experiment is determined in each experiment run by calculating the q_e using equation (1), and then the result is presented below in Table 3.

Table 3. Results of q_e determination for each experiment run.

Temperature (°C)	C_0 of water (% v/v)	q_e (mlof water/gr of adsorbent)		
		AC	Ze	PVA
30°C	12.00	0.1807	0.1519	0.0977
	10.00	0.1799	0.1256	0.0800
	8.00	0.1372	0.0943	0.0651
	6.00	0.0745	0.0743	0.0514
	5.00	0.0669	0.0561	0.0415

The q_e obtained by each adsorbent ranged differently; q_e for activated carbon is ranged 0.0669-0.1807ml/gr, 0.0561-0.1519ml/gr for zeolite, and 0.0415-0.0977ml/gr for PVA. The highest $q_e$obtained by activated carbon with initial concentration of 12.0%, while the lowest is shown by PVA with initial concentration of 5.0%. A decreasing trend of q_e as the C_0 decreases is shown in all adsorbent variant. The decreasing trend of q_e in all adsorbent is consistent to the theory of mass transfer. When the initial concentration increased, the mass transfer driving force would become larger, hence resulting in higher adsorption capacity.

The adsorption capacity is typically increase as the initial water concentration increase, until it reaches an optimum concentration. If the initial water concentration is above the optimum, the adsorption capacity will either be the same as the adsorption capacity produced by the optimum initial water concentration or lower. The data in table 4 shows that the largest adsorption capacity produced by 12.0%, thus the C_0 12.0% is the more optimum initial water concentration for zeolite, activated carbon, and PVA compared to the lower C_0. An extended research to observe higher concentration than 12.0% may be conducted to see whether if the higher concentration is more optimum.

The value of adsorption capacity in activated carbon is larger than zeolite or PVA. This is a valid prove that activated carbon has the largest adsorption capacity compared to PVA and zeolite. This prove is supported by the pore volume comparison in which activated carbon is indeed have the largest pore volume and surface area, as shown in BET test. Zeolite also have larger pore volume and surface area compared to PVA, while PVA compared to the other two is only better in the hydroxyl value comparison. The surface area and pore volume is correlated to the physisorption adsorption mechanism while the number of free hydroxyl groups is correlated to chemisorption mechanism. This means that the water adsorption is more influenced by physisorption rather than chemisorption.

3.3. Isotherm adsorption model

The adsorption data can be represented well by Langmuir model, with average absolute deviation percentage of 7.72%. A comparison of adsorption data and the Langmuir model is presented in figure 3 for all adsorbent. The Langmuir model parameters are obtained, as presented in Table 4.

The model curve comparison shows that as seen in figure 3and table 5, activated carbon has the highest q_m value, followed by zeolite, and then PVA as the smallest q_m between the three adsorbents. This comparison is identical with the adsorption capacity resulted from the experiment.

Table 4. Adsorption capacity constant obtained from Langmuir model.

Adsorbent	q_m	K_L	%AAD
PVA	0.0093	9.6878	4.63
Zeolite	0.0138	10.0107	2.23
AC	0.0158	6.5553	16.30
Average			7.72

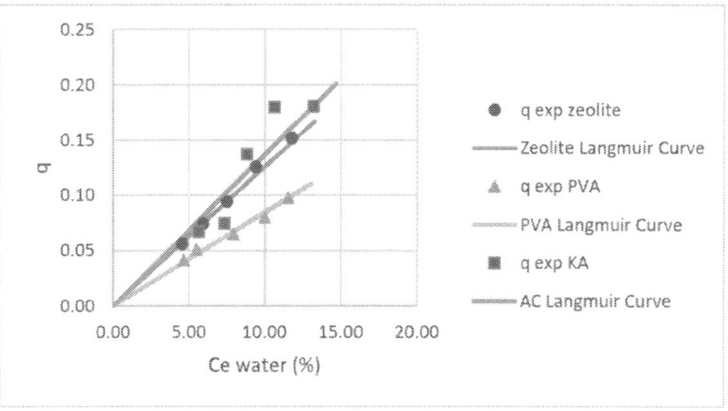

Figure 3. Comparison of adsorption capacity experimental data and the Langmuir model.

3.4. Adsorbent selectivity

A selectivity evaluation is done to the data obtained in the experiment with operational temperature of 30°C. The comparison between adsorbed water and ethanol is shown in figure 4 (a) for activated carbon, (b) for zeolite, and (c) for PVA.

The graph of adsorbed percentage for adsorbent with good selectivity for water adsorption ideally shows that the water adsorbed percentage is higher than the ethanol adsorbed percentage. All of the adsorbent researched in this study shows that the water adsorbed degree is higher than the ethanol adsorbed percentage at all times. The good comparison amount of adsorbed water and ethanol is clearly better in zeolite and PVA than activated carbons. Zeolite shows an average of 20 degree higher water adsorption than ethanol and PVA shows 13 degree average difference, while activated carbon only show only 8 degree.

The selectivity of zeolite may be supported by its molecular shieving structure, which is more selective to the larger sized water molecules rather than the smaller ethanol. The pore structure in activated carbon is rather irregular and consist of smaller sized pores. These pore may have bind ethanol molecule first, and leaving less space for water in its surface. Meanwhile, PVA is more selective because the tendency of hydrogen bridge formation from the free hydroxyls found in PVA with water.

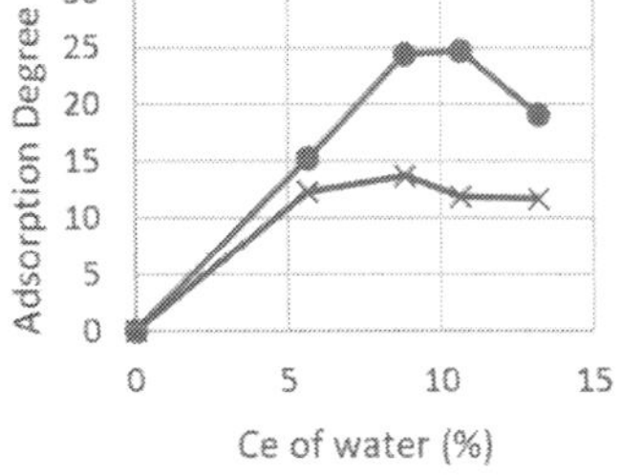

(a)

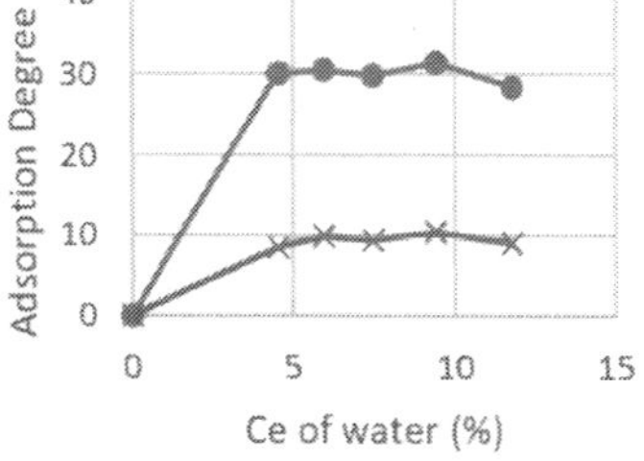

(b)

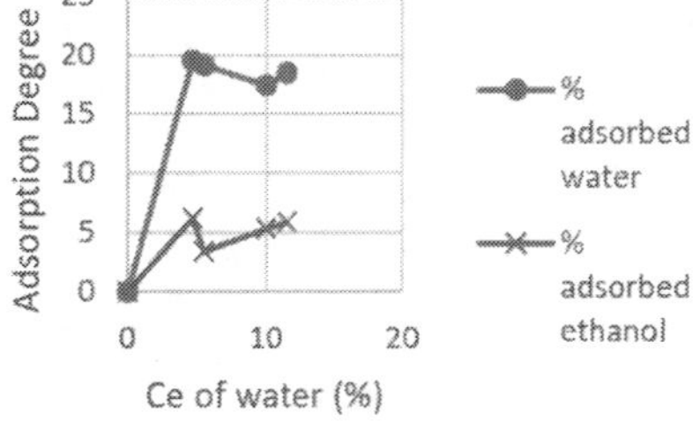

(c)

Figure 4. Adsorben selectivity graph of (a) activated carbon; (b) zeolite; and (c) PVA.

4. Conclusions

The data of adsorption can be represented by Langmuir model. This research has shown that activated carbon has a higher adsorption capacity compared to PVA and zeolite, which is 20% more than zeolite and 89.21% larger than PVA. Even though activated carbon has the largest capacity, the selectivity of zeolite is better than activated carbon or PVA. PVA potential as adsorbent that mainly apply chemisorption mechanism is proven, but did not cause its adsorption capacity to be larger than activated carbon nor ethanol. The initial concentrations of ethanol also affect the adsorption capacity of adsorbent, as the initial concentration increase, the adsorption capacity will decrease.

5. References

[1] Frolkova K A and Raeva V M 2010 Bioethanol dehydration: State of the art *Theoretical Foundations of Chem. Eng.* **44** no. 4 pp 545-556

[2] Presidential regulation Republic of Indonesia No 5 of 2006

[3] Anton A Kiss, David J, Suszwalak P C 2012 Enhanced bioethanol dehydration by extractive and azeotropic distillation in dividing wall column *Separation and Purification Tech.* **86** 70-78

[4] Huang Hua-Jiang, Ramaswamy Shri, Tschirner U W and Ramarao B V 2008 A review of separation technologies in current and future biorefineries *Separation and Purification Technol.* **62** 1-21

[5] Simo Marian, Sivashanmugam Siddarth, Brown Christopher J and Hlavacek Vladimir 2009 Adsorption/desorption of water and ethanol on 3A zeolite in near-adiabatic fixed bed *Ind. Eng. Chem. Res.* **48** 9247-60

[6] Takuji Yamamoto, Young Han Kim, Byoung Chul Kim, Akira Endo, Napawon Thongprachan, and Takao Ohmori 2012 Adsorption characteristics of zeolites for dehydration of ethanol: evaluation of diffusivity of water in porous structure *J. Chem. Eng.* **181-182** 443-448

[7] Kadajji V and Betagari G 2011 Water soluble polymer for pharmaceutical application *Polymer* **3** 1972-2009.

[8] Laksmono J A, Sudibandriyo M, Saputra A H and Haryono A 2017 *Development Of Porous Structured Polyvinyl Alcohol/Zeolite/Carbon Composite As Adsorbent* Int. Conf. on Key Engineering Materials (Penang, Malaysia: In Pres)

[9] Jan Rydberg, Claude Musikas, and Gregory R. Chopin 1992 *Principles and Practices of Solvent Extraction* (New York: Marcel Dekker, Inc)

[10] Anozie A N, Okuhon E E, Osuolale F N and Adewole J K 2010 Dehydration of ethanol-water mixture using activated carbons from sawdust and palm kernel shells *Separation Sci. and Technol.* **45:10** 1482-89

Utilization of steel wool as removal media of hydrogen sulfide in biogas

U Riyadi[1], G A Kristanto[1] and C R Priadi[1,2]
[1] Environmental Engineering Program, Faculty of Engineering, University of Indonesia, Depok, 16424, Indonesia
[2] Tropical Renewable Energi Center, Faculty of Engineering, University of Indonesia, Depok, 16424, Indonesia

E-mail: urip.riyadi@ui.ac.id

Abstract. In order to utilize biogas, hydrogen sulfide/H_2S impurities reduction is needed to maximize methane/CH_4 content. Although H_2S amount is relatively non-dominant, its presence can trigger corrosion, and is harmful to health and environment. Steel wool can be used as adsorption media to reduce H_2S content. This study was conducted to identify the characteristics of steel wool and H_2S concentration reduction efficiency. Biogas was flown to a PVC column (2 inches diameter) containing steel wool. The results showed that steel wool media contain active elements of Fe and Zn which are spread evenly on the media surface with a total amount of 97.5% mass. The concentration of H_2S at inflow ranged from 68 to 111 ppm with the outflow of 21.2-0 ppm, and the temperature in the system varied between 29-33 °C. Optimal H_2S removal efficiency reaches 97% in average, obtained at 100 cm column height and flow rate of 0.1 L/min. It can be concluded that the steel wool media has high content of active element and can reduce H_2S content in biogas at ambient temperature condition.

1. Introduction

As a form of coping with energy and solid waste problems in Indonesia, Anaerobic Digestion (AD) has been developed and implemented widely. When the AD process takes place, the formation of energy in the form of biogas occurs through three stages, namely hydrolysis, acidogenesis, acetogenesis, and methanogenesis [1]. The biogas content is not only CH_4 but also contains CO_2, H_2O, and H_2S. Although the composition of hydrogen sulfide in biogas is relatively non-dominant, its presence can trigger corrosion. Burning biogas containing hydrogen sulfide will also emit sulfuric acid during its combustion process, and most importantly hydrogen sulfide is highly toxic and can pose serious health risks [2]. Thus, it is necessary to process the purification of biogas produced so that the corrosion and health risks decreased, so it can be utilized better.

There are various purification processes that can be done to remove impurities, such as water scrubbing system, chemical absorption, pressure swing adsorption (PSA), membrane, biofilter, and cryogenic separation [3]. In particular, the process of removing H_2S gas from biogas is very important before the installation of upgrading such as PSA. Various methods of removing H_2S include adsorption, absorption, membrane separation, biological processes, and claus process [2] [4] [5] [6]. One of the various methods available which has long been used is in the form of chemical adsorption using iron oxide [6] [7] [8].

The chemical reactions occurring in the process of iron oxide reducing hydrogen sulfide content are expressed in Equations (1) and (2) below [6].

$$Fe_2O_3 + 3\ H_2S \ \ \text{->} \ \ Fe_2S_3 + 3\ H_2O \tag{1}$$

$$Fe_2S_3 + 3/2\ O_2 \ \ \text{->} \ \ Fe_2O_3 + 3\ S \tag{2}$$

Therefore, this research uses iron oxide media in the column for the removal of H_2S from biogas. Iron oxide comes in various shapes/types and in this study used steel wool or iron sponge (iron fiber) that exist in everyday life as a form of iron oxide. This study identifies the characteristics of the steel wool that never been conducted before and its efficiency to reduce H_2S concentrations.

2. Methodology

The dependent variable of the research is the hydrogen sulfide concentration measured before and after biogas through the adsorption column which will be obtained the efficiency value of the removal. In this study, the biogas flows through the columns at ambient/room temperature. The first independent variable that the authors used in the study were the adsorption column height (50 cm, 75 cm, and 100 cm) with medium density to the adsorption column and the controlled flow rate was approximately 0.0943 g/cm^3 and 0.5 L/minute, respectively. Furthermore, after the optimum column height is obtained, the experiment is continued to optimize flow rate (1.5 L/min, 1.0 L/min, 0.5 L/min and 0.1 L/min) to be tested at density and columns height that are controlled equally. The method used in the measurement of H_2S gas in this study is by using SNI Method 19-7117.7-2005. In addition, this study also conducted characterization tests on adsorption media before and after use (media sample from optimum condition). Characterization test using SEM-EDS test methods. The series of research equipment is illustrated in figure 1.

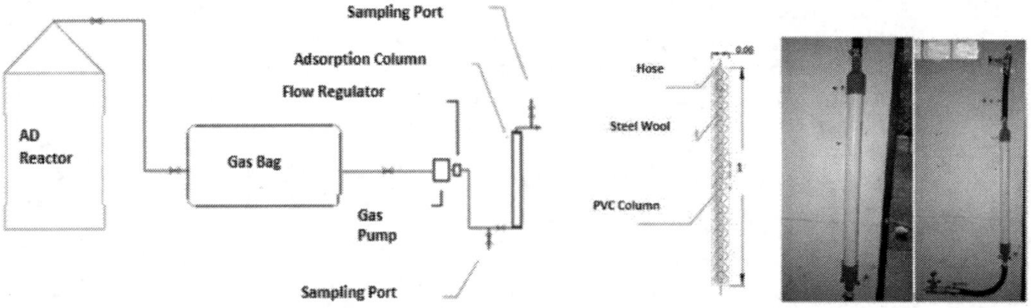

Figure 1. Series of adsorption columns.

3. Result and Discussion

The result of the experiment of column height variation can be seen in figure 2.

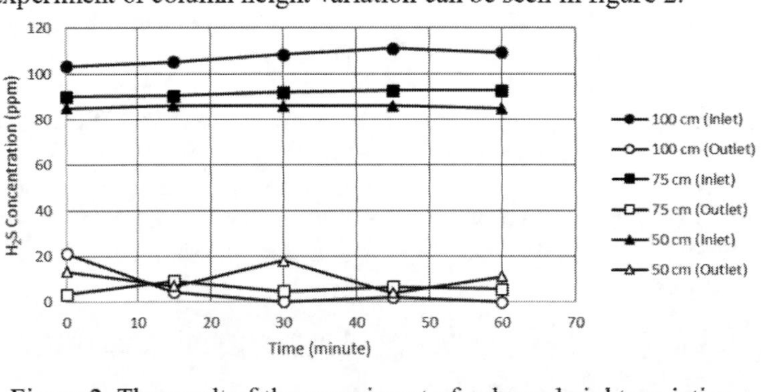

Figure 2. The result of the experiment of column height variation.

IOP Conf. Series: Earth and Environmental Science **105** (2017) 012026 doi:10.1088/1755-1315/105/1/012026

Based on the data obtained that as the height increases, efficiency of media/columns in reducing the concentration of H_2S gas is relatively bigger. As shown in figure 2, the average removal efficiency of H_2S at 50 cm column height reaches 88±6%, then reaches 94±2% at a height of 75 cm and becomes 95±9% at height of 100 cm.

However, ANOVA test results show the significance of 0.202 whose value is greater than 0.05indicating that the column height variations applied in the study of H_2S removal efficiency have no significant impact on H2S adsorption. Nevertheless, the height of 100 cm is applied in the next experiment of flow rate variation because 100% removal efficiency was observed. In addition, this is also based on the value of H_2S gas concentrations at outlets that have been below 10 ppm so as to be designated as stove fuel, even at this 100 cm height can reach natural gas standards under 4 ppm [5].

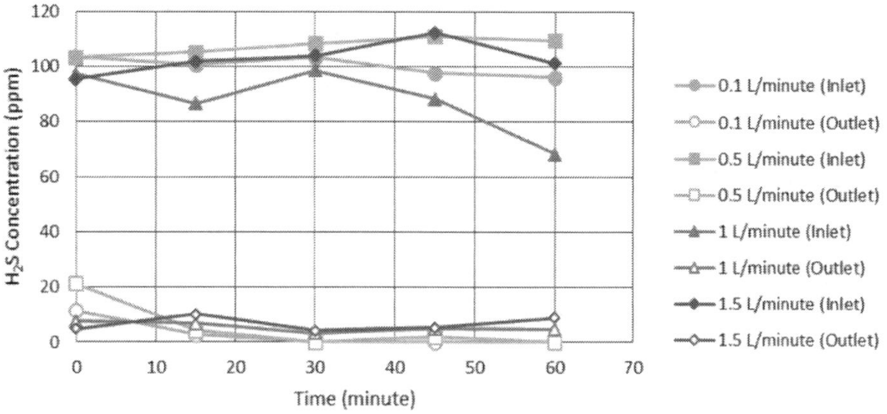

Figure 3. Relation of H_2S Concentration (inlet and outlet) to Time at Flow Rate 0.1 L / min; 0.5 L / min; 1 L / min; and 1.5 L / min.

The second set of experiment demonstrates that as the flow rate increases, the efficiency of media/columns in reducing the concentration of H_2S gas are relatively inversely proportional to the decrease in H_2S decreases with increasing flow rate. The average removal efficiency of H_2S at a flow rate of 0.1 L/min reached 97±5%, then reached 95±9% at a flow rate of 0.5 L/min, 94±2% at a flow rate of 1 L/min and became of 93±3% at a flow rate of 1.5 L/min.

ANOVA test of 0.662 (greater than 0.05) demonstrated that the variation of the flow rate applied to the study on H_2S removal efficiency has no significant difference. This is different compared to [9], that showed that the smallest flow rate variation resulted in the smallest impurity content (in research in the form of CO_2) in biogas due to biogas having longer contact time with the adsorption media. Although compared with [10], optimum removal of H_2S at 3.5 L/min flow rate (variations used 2.5, 3.5 and 5 L/min). Then when compared with [11], there is also a correspondence that the lower the flow rate the greater the H_2S is absorbed although only seen the effect of very small results.

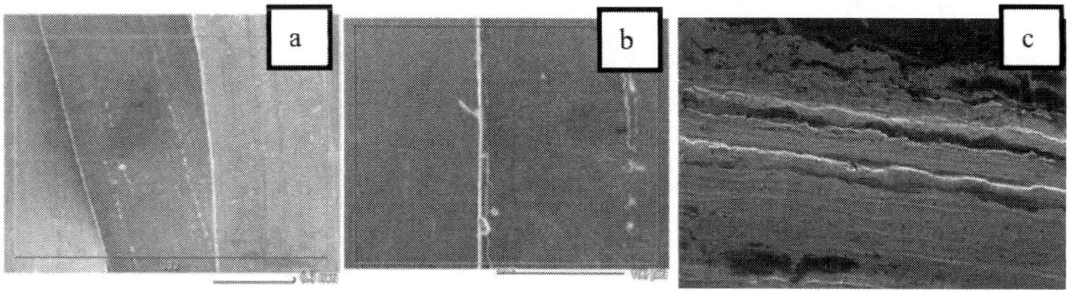

Figure 4. SEM Test Result of Media Before Used with Magnification (a) 100x; (b) 500x; and (c) 1000x.

Figure 4 shows the active area in the adsorption process located on the surface of the medium with a size of about 0.3 mm along the fiber of the adsorption medium. The surface of the media also looks relatively rough, uneven, and has a hollow pore as seen at 1000x magnification, with the illumination of a bright pattern indicating a higher elevation position than the dark one.

EDS media test results showed a large amount of iron (Fe) and Zinc (Zn) elements, respectively reaching 75.49% mass and 22.04% mass. These Fe and Zn elements are the principal active metal elements which can react with H_2S in biogas so that the amount is reduced. The distribution of these elements is even on the surface of the media as shown in figure 5. This illustrates that almost the entire surface of the media is comprised of active element.

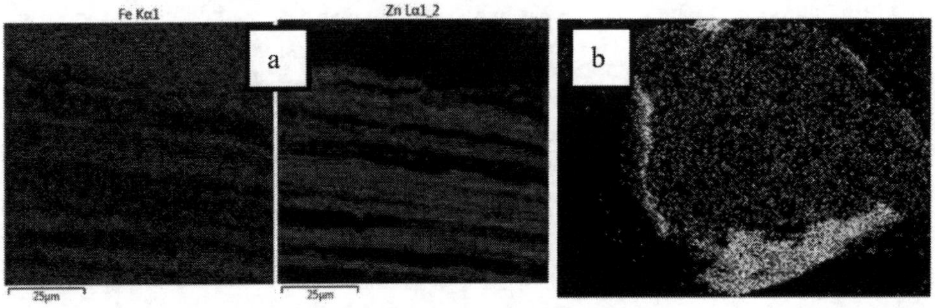

Figure 5. Active elements distribution on surface media : (a) *Steel Wool*; (b) Sulfatreat 410-HP®.

The relatively uniform distribution makes the steel wool media a potentially no lesser medium in the H_2S gas adsorption process when compared to existing commercial media such as Sulfatreat 410-HP® with the distribution of Fe elements as illustrated. With a thorough dissemination on the surface of the media, it illustrates that almost the entire surface of the media is enveloped by an active element.

From the test result, the amount of iron oxide reach 56%, theoretically (stoichiometry), of steel wool media can absorb H_2S with the amount equal to 64% of Fe_2O_3 available. However, there is 0.02% Pb as activator. Activators are elemental ingredients such as Platine, Gold, Silver, Copper, Cadmium, Nickel, Tanker, Mercury, Zinc and Cobalt with amounts of 0.125%-5% on commercial media such as Sulfatreat 410-HP® [4]. So, steel wool media contain fewer activators than the commercial media.

After the process of adsorption, the media used for the process is also tested analysis of SEM-EDS characteristics. Figure 6 is the result of SEM-EDS analysis of the media after the adsorption process. Based on the result of figure 6 there is a defect on the surface of the steel wool media after being used as an adsorption medium. The bright gradient of the image shows the elevation position of the higher surface than the dark portion, indicating the adsorbed material when compared to the medium before it is used with relatively homogeneous or uniform morphological adsorption.

Based on the test sample it is clear that there is a decrease of iron oxide compound which shows that there has been a reaction between H_2S and the compound so that its content in biogas decreases. This suggests that the Fe_2O_3 content decreases from 56.01% to 23.03%, obtained after flowing about 6 L of biogas through the column. This indicates that about 33% of Fe_2O_3 reacts with H_2S gas at the certain point of sample. The remaining Fe_2O_3 content in the media illustrates that the media still have the capacity to reduce H_2S gas further.

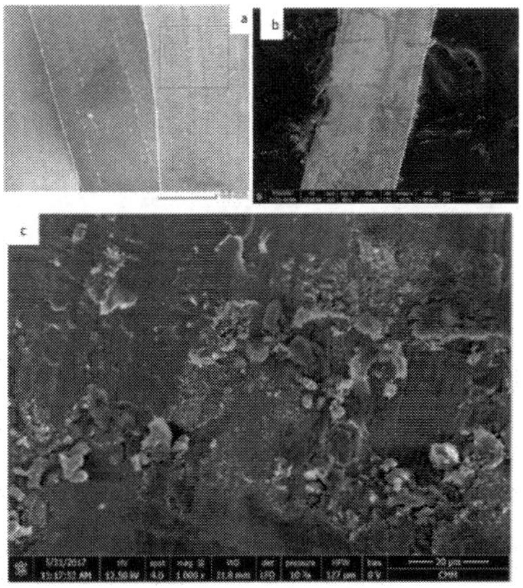

Figure 6. Morphology Test Results of Media : (a) Before Adsorption and (b&c) After Adsorption Process Using SEM (b) magnification 85x (c) 1000x magnification.

4. Conclusions

The active content in this steel wool (Fe and Zn) reaches a total of 97.5% mass with a uniform Fe distribution across the surface and a uniform Zn spread but there are some surfaces that do not contain Zn. The maximum removal efficiency of H_2S gas in Biogas reaches 100% (95% average) obtained at a height of 100 cm. Then 100% H_2S removal efficiency was also obtained at flow rate of 0.1 L/min with 100 cm column height (mean removal efficiency of H_2S was 97%).

5. Acknowledgment

This study is funded by the grant of Green FT Universitas Indonesia and *Publikasi Internasional Terindeks untuk Tugas Akhir Mahasiswa* (PITTA) 2017 with number of grant contract 756/UN2.R3.1/HKP.05.00/2017. Biogas was supplied from Septiana Kurnianingsih and Dwita Fitriani Wijayanti research and experiment location was provided by Faculty of Engineering, University of Indonesia.

6. References

[1] Haryati T 2006 Biogas : Limbah peternakan yang menjadi energi alternatif *WARTAZOA* **16**(3) 160-169

[2] Petersson A 2013 Biogas cleaning. Book Chapter in *The Biogas Handbook- Science, Production and Applications* (Chambridge: Woodhead Publishing Ltd) pp 329-341

[3] Zhao Q, Leonhardt E, MacConnell C, Frear C and Chen S 2010 *Purification Technologies for Biogas Generated by Anaerobic Digestion* Climate friendly farming CSANR Research Report 2010-001

[4] Abatzoglou N and Boivin S 2009 A review of biogas purification processes. *Biofuels, Bioproducts And Biorefining* **3**(1) 42-71

[5] Allegue L and Hinge J 2014 *Biogas Upgrading Evaluation of Methods for H_2S Removal* (Danish Technological Institute)

[6] Siefers A 2010 *A Novel And Cost Effective Hydrogen Sulfide Removal Technology Using Tire Derived Rubber Particles* (Graduate) Iowa State University

[7] Magomnang A and Villanueva E 2014 *Removal Of Hydrogen Sulfide from Biogas Using Dry Desulfurization Systems* Int. Conf. On Agricultural, Environmental And Biological Sci. (AEBS-2014) pp 65-68

[8] Magomnang A and Villanueva E 2015 Utilization of the uncoated steel wool for the removal of hydrogen sulfide from biogas *Int. J. Of Mining, Metal. And Mech. Eng.* **3(3)** 108-111.

[9] Shella 2016 *Proses Simultan Absorpsi-Adsorpsi dalam Purifikasi Biogas Berbasis Limbah Cair Kelapa Sawit* Master Thesis (Universitas Indonesia)

[10] Harihastuti N 2016 *Pemurnian Biogas untuk Mencapai Pipeline Quality Gas sebagai Sumber Energi Terbarukan yang Ramah Lingkungan* (Doktoral) Universitas Diponegoro.

[11] Ho N 2012 Modeling Hydrogen Sulfide Adsorption by Activated Carbon made from Anaerobic Digestion By-Product (Graduate) *University of Toronto*

Experimental study of slope angle and low E glazing effects on photovoltaic module

Arrad Ghani Safitra[1,2], Fifi Hesty Sholihah[1] and Inas Nabilah Fauziyyah[1]

[1] Electronic Engineering, Polytechnic Institute of Surabaya, Indonesia

E-mail: arradgs@pens.ac.id

Abstract. The intensity of solar radiation received by the PV Module can be maximized by installing a PV module with a slope angle. Besides, when the intensity of solar radiation increases, the surface temperature of PV also tends to increase, which decrease PV power output. Temperature Surface can also be reduced by the additional glazing with low emissivity. By knowing the appropriate angle of inclination and the glazing addition, it will able to maximize light energy (photon) and minimize heat energy received by PV surface. So the system can obtained maximum output power and minimum surface temperature. The heat transfer can be determined by analyzing the thermal resistance occurring from the sun down to the PV surface and from the PV surface to the surroundings. The test was performed using a commercial PV module with 180 Watt Peak power, where the test results were discussed and presented. The results shown that PV module facing North with slope angle 30° has the highest output power. The heat release was large enough compared to the other slope angle variation. Besides, the addition of glazing system decreases the output power due to its medium transmittance about 70% and the temperature of PV module also decrease due to low emissivity about 0.26.

1. Introduction

Solar panel or photovoltaic module (PV) is the example of solar energy application. Photovoltaic has been widely used and developed as one of the power plant in many countries as a solution to suppress the use of fossil resources. But the output power of PV module is not constant, depend on the intensity of solar radiation. The intensity of solar radiation that reach the PV module is certainly not constant.

This is because of the pseudo motion of the sun (daily and monthly), and weather conditions (sunny, cloudy and rainy), as well as the geographic location of PV itself. So the positioning of the PV surface greatly affect the amount of radiation received by the solar cells. The intensity of solar radiation received on the PV Module can be maximized by installing a PV module with the correct slope angle (β) and azimuth angle (γ) [1].

Kandil [2] concluded that the increasing solar radiation received on PV modules, increases almost all parameters affecting the performance of module PV, particularly current, maximum power and module efficiency. So it is very important to conduct further research on PV module performance at different angle and orientation. Gopinathan [3] conducted research on several surface orientations with three inclinations for six different azimuth angles. The research found different angle and optimum orientation for summer, cold and yearly. Gunerhan [4] reported the optimal orientation for solar

[2] Author to whom any correspondence should be addressed

collectors in Izmir is south facing. To improve the efficiency of solar collectors, it was recommended, if possible, that solar collectors should be positioned at the average angle of inclination and the slope was adjusted once a month. Ekadewi [5] concluded that for the installation of solar collectors in Surabaya, the maximum inclination angle for March12th– September 30th was varied between the angle 0° - 40° (north facing) and for October1st – March 11th installed between 0° - 30° angles (south facing).

Wasted energy such as heat causes an increase in PV module temperature and there is a correlation between efficiency and temperature [6]. So it is necessary to reduce the PV module temperature. Tripanagnostopoulos [7] concluded that cooling PV modules can improve the electrical efficiency of PV modules. The addition of glass is to raise the thermal output, diffuse reflector can increase the electrical and thermal output. In solar cells there are also reflection losses that occurring above the cell surface which can be reduced by the addition of antireflective coating glass [8].

The solar radiation exposed to PV modules can increase the module temperature because the excess photon absorbed by the PV module is converted to heat. This phenomenon can reduce the performance of PV module. Many research analyzed heat transfer on PV/T module or solar collector. Thus, this experiment analyze the heat transfer happened between PV module and surrounding air, Beside knowing the performance of PV module with slope angle variation. Special glass with low emissivity added to the PV module, expecting that we can know its effect to the heat transfer and the performance. This experiment aims to determine the effect of slope angle and the addition of glass to heat transfer in PV module.

2. Experimental method

PV module used for this research was monocrystalline type with maximum power 180 Watt. This research was take place in Electronic Engineering Polytechnic Institute of Surabaya on July 3rd until July 5th 2017and July 21st2017 (test with glazing addition)from 09.00 to 15.00 WIB. The PV module was directed towards the north because the position of Surabaya is south of the equator line. Then the module was tilted according to the angle variation that is 0°, 10°, 20°, 30°, and 40°. The test system is shown in figure 1.

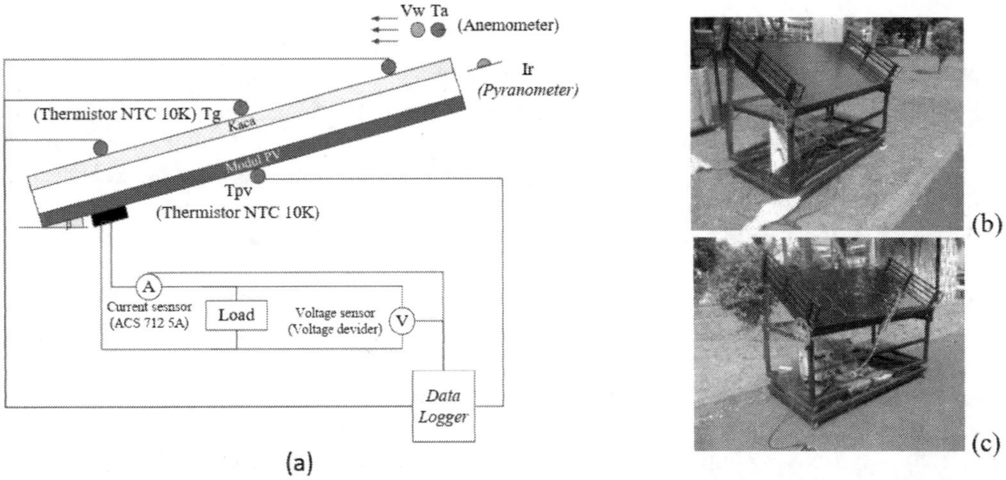

Figure 1.(a) Experimental models, PV system (b) without glass and (c) with glass.

Pyranometer was used to measure solar radiation received by the PV surface. The measurement of output current using ACS712 5A sensor and output voltage was measured by using an integrated voltage divider with data logger to record all data from the sensors.

To calculate the heat release, several parameters required, such as PV and glass temperature, wind velocity, and ambient temperature. The module temperature was measured using a thermistor NTC 10k mounted below the PV module that was also integrated with the data logger and it was recorded by data logger every three seconds. Wind velocity and ambient temperature were measured using an anemometer. The data was recorded by data logger every three seconds. The data from pyranometer, wind velocity and ambient temperature was written manually every one hour.

Calculation of heat release occurred in the PV module was done in steady state operation and analyzed in one dimension at y axis. There is a multistage heat transfer process on the cover glass, including natural convection heat transfer and radiation to the surrounding air. Due to perfect heat transfer conduction happened in the PV module and the glass cover, the conduction was not taken into account. This the following equations can be used to calculate the heat transfer that occurs in the PV module [8].

$$h_{c,pv-a} = h_{c,g-a} = h_w = 8.55 + 2.56V_w \quad (1)$$

$$h_{r,pv-a} = h_{r,g-a} = \varepsilon_{pv}\sigma\,(T_{pv} + T_a)(T_{pv}{}^2 + T_a{}^2) \quad (2)$$

$$h_{r,pv-g} = \frac{\sigma\,(T_{p-pv}+T_g)(T_{p-pv}{}^2+T_g{}^2)}{{}^1/_{\varepsilon_{pv}}+{}^1/_{\varepsilon_g}-1} \quad (3)$$

Where $h_{c,pv-a}$ [9]and $h_{r,pv-a}$ are convection and radiation heat transfer coefficients between PV module to environment, $h_{c,g-a}$ and $h_{r,g-a}$ are convection and radiation heat transfer coefficients between glass cover to environment, $h_{c,pv-g}$ and $h_{r,pv-g}$ are convection and radiation between PV module and glass cover. v is environmental wind velocity; T_{pv}, T_a, T_g are temperatures of PV module, ambient air, and glass cover; ε_{pv} and ε_g are the emissivity of PV module and glass cover;$\sigma$is the Stefan-boltzmann constant.

$$h_{c,pv-g} = \frac{Nu\ .\ k}{L} \quad (4)$$

In the enclosed air gap or parallel flat plates for natural convection at various slopes from 0° to 75° ,Nusselt number from duffie [8] can be applied :

$$Ra = \frac{g\,\beta'\Delta T\,L^3}{v\alpha} \quad (5)$$

Ra = *Rayleigh number*; g = gravitational constant (m/s^2); β'= volumetric expansion coefficient (for an ideal gas), $\beta' = 1/T$), (1/K); ΔT = temperature difference between plate (K); v = kinematic viscousity (m^2/s); α = thermal diffusivity (m^2/s)

$$Nu = 1 + 1.44\,1 - \left[\frac{1708(\sin 1.8\beta)^{1.6}}{Ra\cos\beta}\right]\left[1 - \frac{1708}{Ra\cos\beta}\right]^{+} + \left[\left(\frac{Ra\cos\beta}{5830}\right)^{1/3} - 1\right]^{+} \quad (6)$$

Then, the overall heat transfer coefficient of PV module with (U_{T0}) and without glass cover (U_{Tg}) can be expressed as :

$$U_{T0} = \left(\frac{1}{h_w+h_{r,pv-a}}\right)^{-1} + \left(\frac{1}{h_w+h_{r,pv-a}}\right)^{-1} \quad (7)$$

$$U_{Tg} = \left(\frac{1}{h_w+h_{r,g-a}} + \frac{1}{h_{c,pv-g}+h_{r,pv-g}}^{-1}\right) + \left(\frac{1}{h_w+h_{r,pv-a}}\right)^{-1} \quad (8)$$

So, the heat transfer (Q) in PV module :

$$Q = Ac\ U_T\left(T_{pv} - T_a\right) \tag{9}$$

3. Result and discussion

For the first experiment, PV module without glass at various slopes, the result is shown in figure 2 and figure 3. Figure 2 shows the poweroutput PV module at various slopes and figure 3 showsthe electrical efficiencyof PV module at various slope. The highest power output was achieved at 30° slope angle with total power 700 W and the highest efficiency was 9.20 % at the same slope. The lowest power output was at 0° slope angle with total power 628.5 W, but the efficiency of 0° was not the lowest. The lowest efficiency was about 8.69% at 40° slope. Both figure show that by increasing the slope angle facing to north, the power output and electrical efficiency increase just until 30° slope. But then, if the slope angle is bigger than 30° slope, the power ouput and the efficiency decreasing gradually. At slope angle 40°, the power output decrease into 640.36 W. The decreasing caused by the position of the PV module that nearly vertical so the PV module could not get the maximal irradiance.

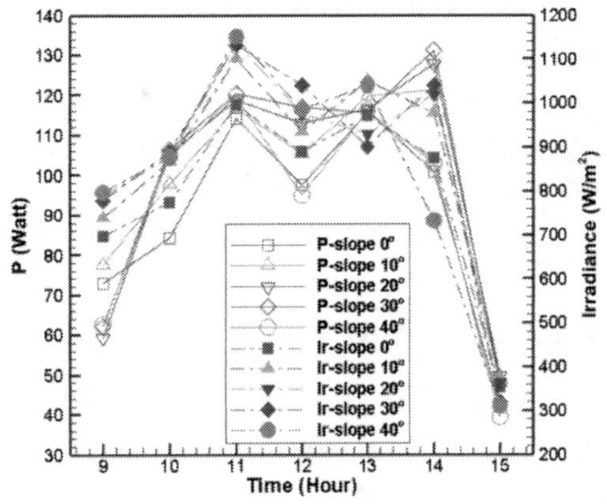

Figure 2. Power output at various slope angle.

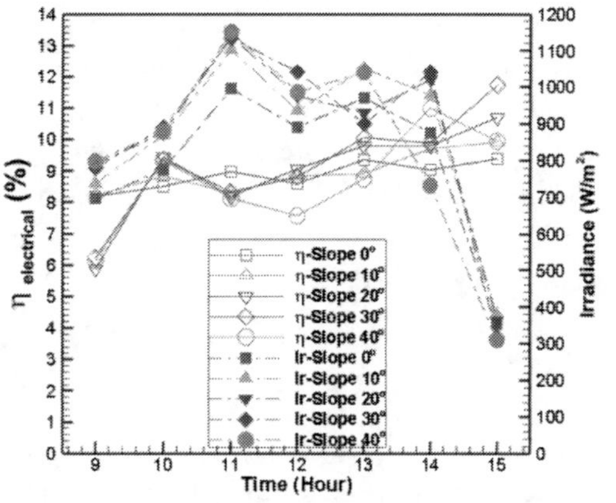

Figure 3. Electrical efficiency at various slope angle.

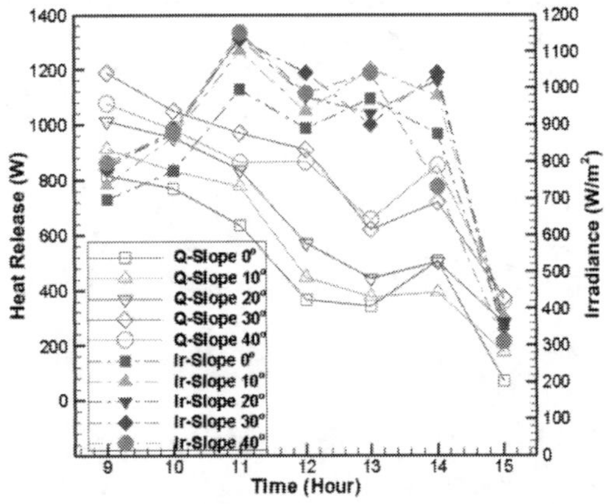

Figure 4. Heat release at various slope angle.

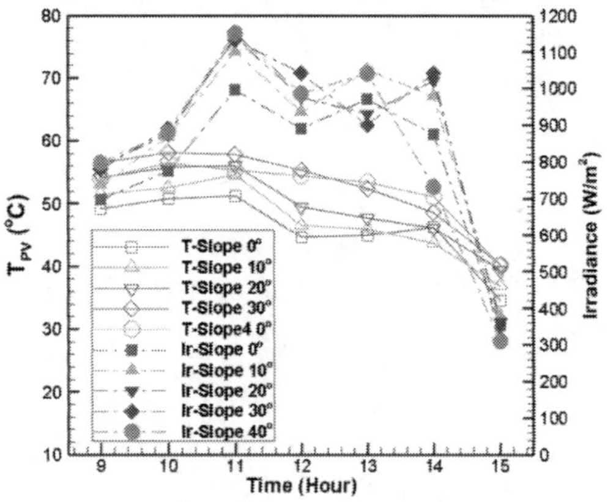

Figure 5. Temperature PV moduleat various slope angle.

In figure 4, The heat release also shows thatat slope angle 30° has the highest heat release value, about 832.91 W. An increase in irradiation that also carries heat energy leads to an increase in PV module temperature as shown in figure 5. Heat transfer is influenced by wind speed and ambient temperature. Higher PV module temperatures cause larger temperature differences with ambient temperatures. The temperature difference is proportional to the value of heat transfer, so the value of heat transfer is also greater. Increasing the temperature gives an impact by reducing the voltage, followed by a decrease in output power. In PV modules, temperature increases can lead to some failure and decrease in function, as temperature increases increase the pressure associated with thermal expansion and also increase the functional degradation value by a factor of about 2 for every 10°C temperature rise [10].

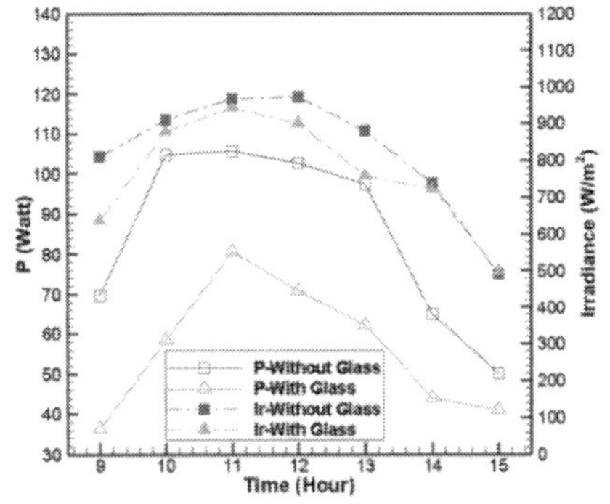

Figure 6. Output power on PV module with and without glass.

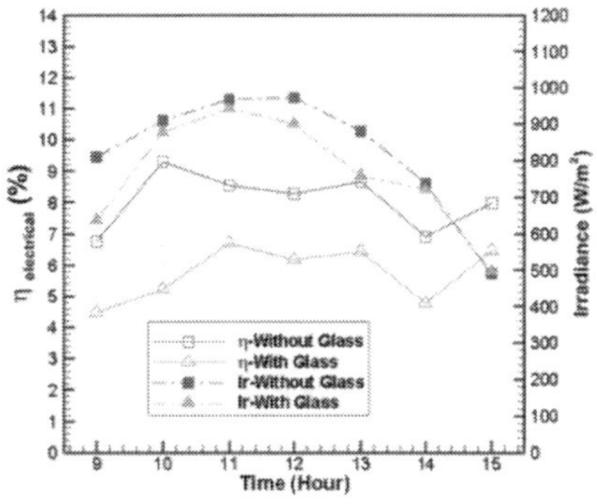

Figure 7. Electrical efficiency on PV module with and without glass.

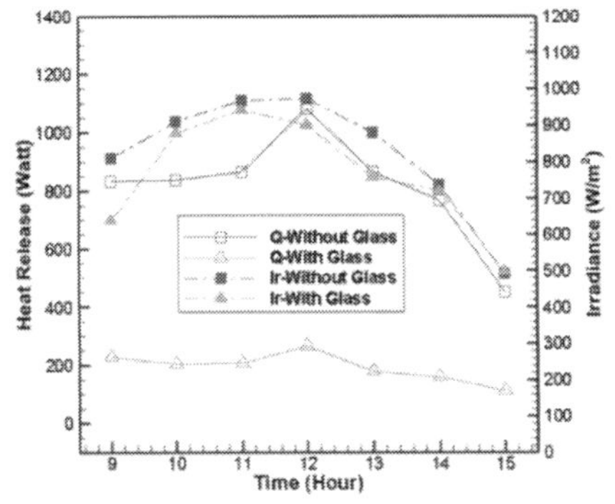

Figure 8. Heat Release on PV module with and without glass

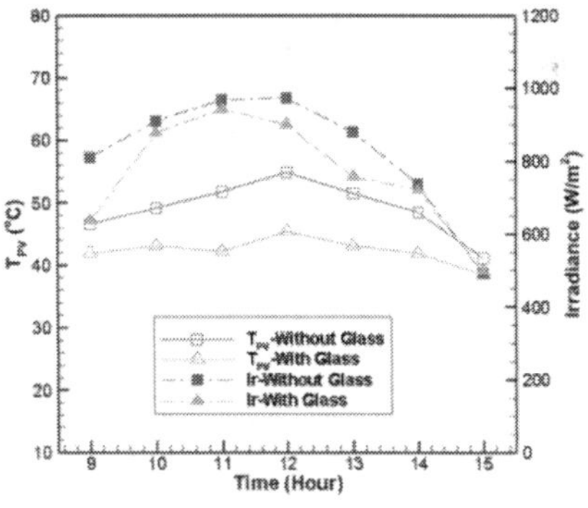

Figure 9. Module PV Temperature with and without glass.

To improve the performance of PV module, the experiment with glazing system with 5 mm gap performed on 30° slope angle which has the best performance. The result shown in figure 6 and figure 7. In figure 6 shows the glazing effect on power output and figure 7 shows the glazing effect on electrical efficiency. Both figure show that PV module with glazing decrease the power output and electrical efficiency. The power output was decreasing from 595 W to 393.4 W and the electrical efficiency from 8.1% to 5.74%.

Output Power decreasing that happened in the PV module with glazing system caused by the glass characteristic which has transmisivity about 70%. So the photon that reach PV surface could not be maximized. But, PV module needs higher light intensity to produce maximum output power, although temperature is also a significant factor in PV module performance.

But the different result on PV temperature. It can be seen in figure 9 that the PV module with glass cover decrease the module temperature up to 12°C. The average of Tpv are 49.1°C without glass cover and 42.27 °C with glass cover. This is caused by the radiation heat transfer that is affected by the emissivity value of an object. Characteristics of dark-colored PV modules that have a 0.8 emissivity cause the surface of the PV module to have a large heat-absorbing ability.

Reduced surface temperature of PV modules causes the heat transfer also to decrease. It can be seen in figure 8, The value of heat treatment without using glass is much higher than the PV module given the addition of glass. Without the addition of glass, the average heat release value is 595 W, whereas with the addition of glass the average value of heat release to 194 W. In glass PV module has a low heat release value due to the characteristics of glass that has a low emissivity of 0.26. So the ability of the glass to absorb heat is lower and make the temperature of the cover glass does not increase significantly. So the module heat transfer is also low and causes low PV module temperature.Although temperature also affects the performance of PV modules but still has to be considered the characteristics of glass that have high transmissivity and low emissivity. Because both light and temperature factors are important to improve the performance of PV modules.

Based on the above discussion can be seen that the addition of glass can reduce the temperature. So the heat transfer to the surrounding air is also lower. In addition, the addition of cover glass also lowers the output power and efficiency. This is in accordance with Wu's research [11] which states that the efficiency of PV-TE without glass cover is always greater than that using cover glass. However, the glass closure system can compete or be superior to that without glass if at the condition of a figure of merit Z, concentration ratio C and high glass transmissivity.

4. Conclusions

The results showed that in the PV module facing North with slope angle 30° has the highest output power and large heat release compared to the other slope angle. In addition, the addition of glazing decrease the output power due to medium transmittance 70% and the temperature of PV module decrease due to low emissivity 0.26

5. Acknowledgements

The author thanks to Electronic Engineering Polytechnic Institute of Surabaya (EEPIS) who has provided funding to local research scheme fiscal year 2017. The authors hope this research can be an additional reference about the utilization of solar energy in the field of energy conversion.

6. References

[1] Pangestuningtyas D L and Karmoro H *Analisis Pengaruh Sudut Kemiringan Panel Surya Terhadap Radiasi Matahari yang Diterima Oleh Panel Surya Tipe Array Tetap* (Semarang: Universitas Diponegoro)

[2] Kandil S A, Mohammad M A and Ahmad G E 1992 *Effect Of Local Meteorological Conditions On The PV Module Performance* In: 2nd World Renewable Energy Conference. Reading, Uk, pp 467-472

[3] Gophinatan K K 1991 Solar radiation on variously oriented sloping surfaces *Solar Energy* **47** 173-179

[4] Gunerhan H and Hepbasli A 2007 Determination of the optimum tilt angle of solar collectors for building applications *Building and Environment* **47** 779-783

[5] Ekadewi A H, Djatmiko I and Prabowo 2013 The optimal tilt angle of a solar collector *Energy Procedia* **32** 166-175

[6] Royne A, Dey C J, and Mills D R 2005 Cooling of photovoltaic cells under concentrated illumination : A critical review *Sol Energy Mater Sol Cells* **86** (4) 451-483

[7] Tripanagnostopoulos Y, Nousia T H, et al 2002 Hybrid photovoltaic/thermal solar system *Solar Energy* **72**(3) 217-234

[8] Duffie J A and Beckman W A 2013 *Solar Engineering Of Thermal Processes. Fourth edition.* (New Jersey: John Wiley & sons, inc)

[9] Test F L, et al. 1981 Heat transfer during wind flow over rectangular bodies in natural environment, Transactions of the ASME *J. of Heat Transfer* **103** 262–267

[10] PV Education, PV Module Temperature, http://www.pveducation.org/pvcdrom/modules/pv-module-temperature

[11] Wu, Ying Ying, et al. 2015 Performance analysis of photovoltaic-thermoelectric hybrid system with and without cover *Energy and Management* **93** 151-159

IOP Conf. Series: Earth and Environmental Science **105** (2017) 012028 doi:10.1088/1755-1315/105/1/012028

Thermal properties of paraffin based nano-phase change material as thermal energy storage

Muhammad Amin[1], Fitri Afriyanti[1] and Nandy Putra[1,2]

[1]Applied Heat Transfer Research Group, Department of Mechanical, Faculty of Engineering, Universitas Indonesia, Depok,16424, Indonesia

E-mail: nandyputra@eng.ui.ac.id

Abstract. One way to save electrical energy is by directly reducing the energy consumption and using materials that able to absorb heat. The best material in absorbing heat is paraffin. Paraffin is a group of organic Phase Change Material (PCM) which has high latent heat. Adding nanoparticles to the paraffin is expected to increase the latent heat of nano-PCM. The research aims to find out the thermal properties of nano PCM based paraffin and engineered to improve its latent heat. In this research, PCM material used is paraffin with Fe_3O_4, CuO, TiO_2, and ZnO nanoparticles are added. Nano – PCM is synthesized using sonification methods with variations of 5, 10, and 15 wt%. Latent heat of thermal properties and a melting point of paraffin nano-PCM are measured using Differential Scanning Calorimetry (DSC). The results show the latent heat of paraffin nano-PCM has increased by 20.67%, 78.89%, 7.5%, and 20.17% for the addition of Fe_3O_4 (5 wt%), CuO (10 wt%), TiO_2 (15 wt%), and ZnO (5 wt%) respectively. The better nano PCM in storing latent heat is paraffin-CuO at a mass fraction of 10 wt%. Meanwhile, the addition of nanoparticles has no significant effect on the melting point. These results showed that paraffin based nano-PCM is an excellent thermal energy storage.

1. Introduction

Electricity utilization for building air system application has been very improvident, especially when it comes to peak demand. However, this electricity consumption is also inevitable for instance, the energy consumption for air conditioner (AC). Thus, saving the energy by minimizing its consumption can be a potential solution. One of the tricks to minimize the energy consumption is using building materials which are able to absorb heat, so that the burden of air conditioner system becomes less and its performance will be more stable, resulting lower cost.

The most recommended building material is Phase Change Material (PCM) [1], which is able to store heat when an amount of heat energy, which exceeds its melting point, is applied to it [2]. Room temperature is maintained well since the excessive heat will be absorbed by PCM, which leads to a lower burden for the AC system [3].

PCM plays an important role as a thermal energy [4]. When a phase change occurs, a great amount of latent heat energy will be absorbed. The amount of latent heat depends on the PCM material type. Generally, PCM is categorized into 3 groups, specifically organic, inorganic, and eutectic [5]. Among

[2] To whom any correspondence should be addressed.

the three groups, organic PCM is the most widely used. Paraffin is one of the organic PCM [6]. PCM plays an important role as a thermal energy [4]. When a phase change occurs, a great amount of latent heat energy will be absorbed. The amount of latent heat depends on the PCM material type.

Besides that, nanoparticles can be added to PCM to enhance its latent heat. Nanoparticles in thermal energy storage research using PCM are usually known as nano-PCM. Several researches explain that the addition of nanoparticles to PCM can decrease latent heat. But some other researches succeed to enhance latent heat of PCM by adding nanoparticles, for example, Eanest J.B [8] who added exfoliated graphite nanoplatelets (xGnPs) particles to PCM in fatty acid group was successful to enhance nano-PCM latent heat as much as 0.83% at 10 wt% mass fractions. Sahan N et al. [9] who added Fe_3O_4 to paraffin, was successful to enhance latent heat as much as 8.8% at 20 wt% mass fractions. Amin M et al. [10] who added graphene nanoplatelets to beeswax was successful to enhance beeswax latent heat as much as 22.32% at 0.3 wt% mass fraction. Liu Y and Yang Y [11] who added TiO_2 to PCM salt hydrate was successful to enhance nano-PCM latent heat as much as 6.4% at 0.3 wt% mass fraction. Kim S. and Drzal L.T [12] who added xGnPs to paraffin was successful to enhance paraffin/xGnPs latent heat as much as 4% at 1 wt% mass fraction.

Therefore, other potential nanoparticles, which can be added to paraffin should be developed to obtain more nano-PCM thermal property data. The objective of this research is to observe thermal properties of nano-PCM based on paraffin for reference enrichment, which is expected to be useful for further research reference especially in the field of energy research.

2. Methodology

2.1. Materials

The PCM material used in this research was paraffin with 100% of purity. While Fe_3O_4, CuO, TiO_2, ZnO were used as nano-PCM with size 19, 52, 14 and 39 nm respectively. Fe_3O_4 and CuO are black while TiO_2, ZnO are white.

2.2. Preparation of the nano-PCM

Nano-PCM was synthesized by mixing the nano-PCM into paraffin at various mass fractions: 5, 10, dan 15 wt% respectively. Sonication was used in the synthesized process to avoid agglomeration. The same technique also was used by Putra N et al. [13].

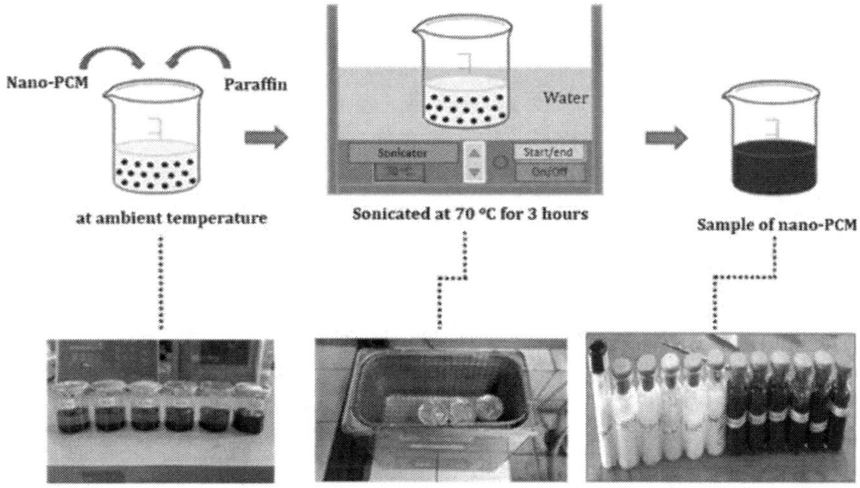

Figure 1. Synthesis of nano-PCM.

Paraffin was placed in glass beaker then mixed with nano-Fe_3O_4. Then this nano-PCM was stirred well and placed in a sonicator. Sonicator was set at temperature 70 °C to maintain nano-PCM at liquid stage. Sonication was done for 3 hours to have Fe_3O_4 dispersed well into paraffin. The same procedures were done to other nano-PCM, specifically paraffin-CuO, paraffin-TiO_2, and paraffin-ZnO. When all the steps were done, the colours of nano-PCM changed to white and black. Nano-PCM synthesis process is described in figure 1.

2.3. Characterization of nano-PCM

In this research, thermal properties of nano-PCM, specifically melting point and latent heat were measured by Differential Scanning Calorimetry (DSC, Perkin Elmer). Each sample as much as 9.5 mg was prepared to be measured by the DSC instrument with heating rate of 10 °C/minute at temperature range 28 °C to 140 °C. Melting point and latent heat were obtained using numerical integration in DSC software.

3. Results and discussions

3.1. Thermal properties of paraffin based nano-PCM

Figure 2 shows a graph of DSC analysis result of PCM material that consists of pure paraffin. The graph shows that the paraffin started to melt at onset temperature 41.92 °C. Paraffin kept on melting until the temperature reached its peak at 51.13 °C. Paraffin was completely melted at endset temperature of 54.29 °C. From onset, peak, and endset temperature data, the area can be calculated. Latent heat of paraffin was calculated by dividing the area with the amount of sample used. It revealed that the value of paraffin latent heat was 207.22 J/g or 207.22 kJ/kg. This value is matched with numerical integration calculation by DSC software.

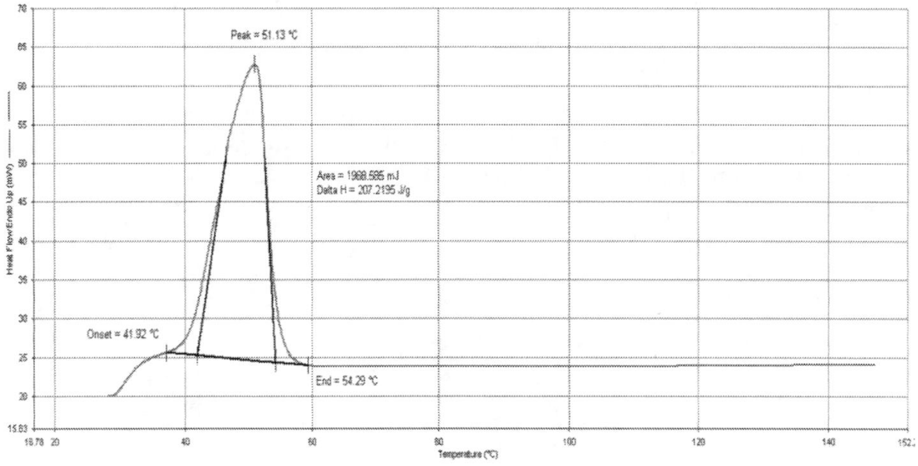

Figure 2. DSC graph of pure paraffin.

Figure 3 to 6 reveal DSC graph for each nano-PCM specifically paraffin-Fe_3O_4, paraffin-CuO, paraffin-TiO_2, and paraffin ZnO. Each DSC graph describes various mass fractions of pure paraffin and nano-PCM with 5, 10, and 15 wt% mass fractions respectively. Nano-PCM affects its increasing latent heat. The increasing amount of latent heat depends on the number of nano-PCM mass fractions. Based on calculation results of numerical integration of DSC software, latent heat of paraffin-Fe_3O_4 nano-PCM was 250.06 kJ/kg at 5 wt% mass fractions. Due to addition of Fe_3O_4 5 wt% to paraffin, latent heat of paraffin- Fe_3O_4 nano-PCM increased as much as 20.67% compared to pure paraffin. This phenomenon also occurred to the latent heat of paraffin-CuO nano-PCM with latent heat as much as 372.77 kJ/kg at 10 wt% mass fractions. It revealed the increment of paraffin-CuO nano-PCM latent

IOP Conf. Series: Earth and Environmental Science **105** (2017) 012028 doi:10.1088/1755-1315/105/1/012028

heat was as much as 78.89%. While for paraffin-TiO_2 nano-PCM, the latent heat was 222.87 kJ/kg at 15 wt% mass fractions. This value revealed the increment of paraffin-TiO_2 nano-PCM latent heat was as much as 7.5%. The last was paraffin-ZnO nano-PCM with latent heat 249.01 kJ/kg, indicating latent heat increment as much as 20.17%. It can be explained that the increased latent heat of nano-PCM is because of the Brownian motion [14]. The nano-PCM random movement can increase the probability of agglomeration in the paraffin base fluid. In addition, Van der Waals forces between nano-PCM will attract each other and forms particles grouping [14]. However, low concentration of each nanoparticles enables thermal storage to be more operational per volume unit.

According to the data, the greatest increase in latent heat is paraffin-CuO nano-PCM with 5 wt% mass fractions.

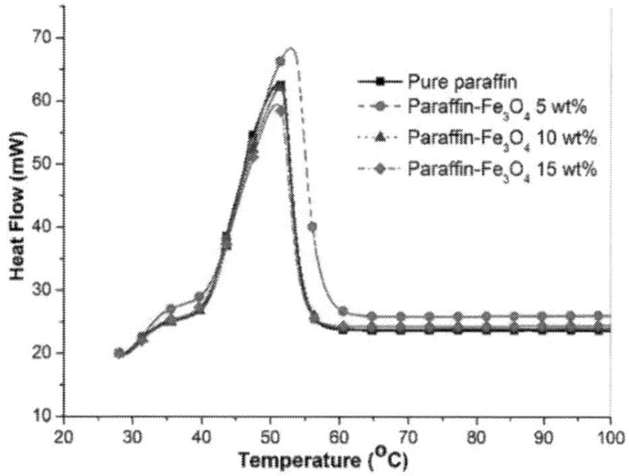

Figure 3. DSC graph of paraffin-Fe_3O_4.

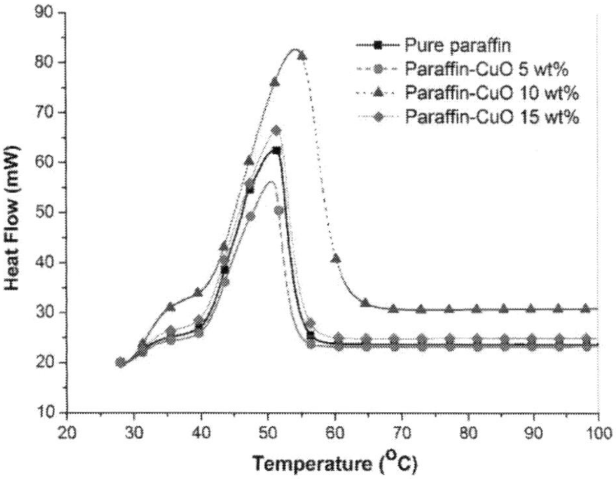

Figure 4. DSC graph of paraffin-CuO.

On the other hand, nanoparticles addition to paraffin did not affect its melting point significantly. Melting point of each nano-PCM was around 51-54.3 °C. Thermal properties of nano-PCM-based paraffin specifically latent heat and melting point are summarized in table 1.

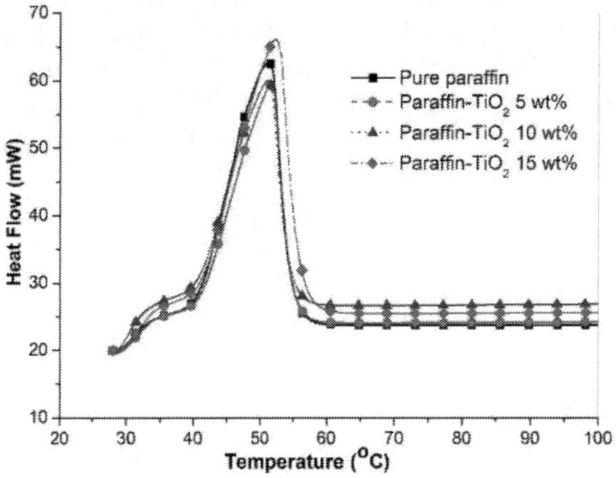

Figure 5. DSC graph of paraffin-TiO_2.

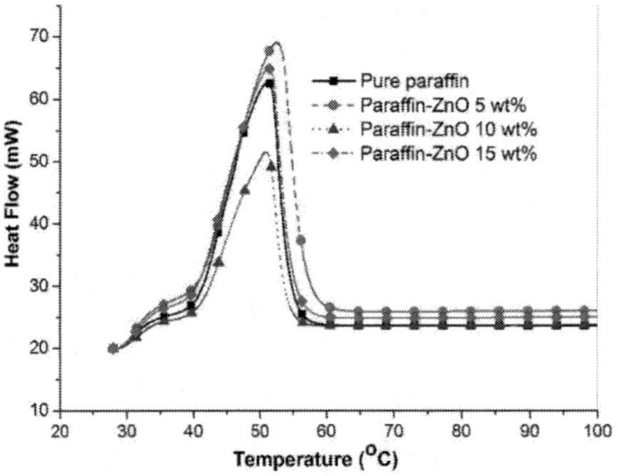

Figure 6. DSC graph of paraffin-ZnO.

Tabel 1. Thermal properties of paraffin based on nano-PCM

Mass fraction [wt%]	Paraffin-Fe_3O_4		Paraffin-CuO		Paraffin-TiO_2		Paraffin-ZnO	
	Melting point [°C]	Latent heat [kJ/kg]	Melting point [°C]	Latent heat [kJ/kg]	Melting point [°C]	Latent heat [kJ/kg]	Melting point [°C]	Latent heat [kJ/kg]
0	51.13	207.22	51.13	207.22	51.13	207.22	51.13	207.22
5	52.85	250.06	50.77	158.58	51.53	175.09	52.51	249.01
10	51.47	198.51	54.39	372.77	51.01	173.81	50.86	134.80
15	51.03	176.53	51.39	225.03	52.39	222.87	51.27	216.37

4. Conclusions

Based on experimental results, it is known that nano-particle addition into paraffin could enhance latent heat of nano-PCM. From all nano-PCM, the greatest increase of latent heat is paraffin-CuO nano PCM at 5 wt% with 78.89% increase, followed by paraffin- Fe_3O_4 nano-PCM at 5 wt% increases

20.67%, paraffin-ZnO nano-PCM at 5 wt% increases 20.17%, and paraffin- TiO_2 nano-PCM at 15 wt% increases 7.5%.

5. Acknowledgments
The authors express gratitude to DRPM Universitas Indonesia for funding this research through the "PITTA 2017" scheme.

6. References
[1] Wahid M A, Hosseini S E, Hussen H M, Akeiber H J, Saud S N and Mohammad A T 2017 An overview of phase change materials for construction architecture thermal management in hot and dry climate region *Applied Thermal Eng.* **112** 1240-59

[2] Yamaha M and Misaki S 2006 The evaluation of peak shaving by a thermal storage system using phase-change materials in air distribution systems *Hvac&R Research* **12** 861-869

[3] Yang L, Lam J C and Tsang C L 2008 Energy performance of building envelopes in different climate zones in China *Applied Energy* **85** 800-817

[4] Zalba B, Marín J M, Cabeza L F and Mehling H 2003 Review on thermal energy storage with phase change: materials, heat transfer analysis and applications *Applied Thermal Eng.* **23** 251-283

[5] Sharma A, Tyagi V, Chen C and Buddhi D 2009 Review on thermal energy storage with phase change materials and applications *Renewable and Sustainable Energy Reviews* **13** 318-345

[6] Khan Z and Ghafoor A 2016 A review of performance enhancement of PCM based latent heat storage system within the context of materials, thermal stability and compatibility *Energy Conversion and Management* **115** 132-158

[7] Tatsidjodoung P, Le Pierrès N and Luo L 2013 A review of potential materials for thermal energy storage in building applications *Renewable and Sustainable Energy Reviews* **18** 327-349

[8] Eanest Jebasingh B 2016 Preparation of organic based ternary eutectic fatty acid mixture as phase change material (PCM), optimizing their thermal properties by enriched solar treated exfoliated graphite for energy storage *Materials Today: Proceedings* **3** 1592-98

[9] Şahan N, Fois M and Paksoy H 2015 Improving thermal conductivity phase change materials—A study of paraffin nanomagnetite composites *Solar Energy Materials and Solar Cells* **137** 61-67

[10] Amin M, Putra N, Kosasih E A, Prawiro E, Luanto R A and Mahlia T M I 2017 Thermal properties of beeswax/graphene phase change material as energy storage for building applications *Applied Thermal Eng.* **112** 273-280

[11] Liu Y and Yang Y 2017 Investigation of specific heat and latent heat enhancement in hydrate salt based TiO2 nanofluid phase change material *Applied Thermal Eng.* **124** 533-538

[12] Kim S and Drzal L T 2009 High latent heat storage and high thermal conductive phase change materials using exfoliated graphite nanoplatelets *Solar Energy Materials and Solar Cells* **93** 136-142

[13] Putra N, Amin M, Kosasih E A, Luanto R A and Abdullah N A 2017 Characterization of the thermal stability of RT 22 HC/graphene using a thermal cycle method based on thermoelectric methods *Applied Thermal Eng.* **124** 62-70

[14] Zabalegui A, Lokapur D and Lee H 2014 Nanofluid PCMs for thermal energy storage: Latent heat reduction mechanisms and a numerical study of effective thermal storage performance *Int. J. of Heat and Mass Transfer* **78** 1145-54

IOP Conf. Series: Earth and Environmental Science **105** (2017) 012029 doi:10.1088/1755-1315/105/1/012029

Signal detection for identification energy and behaviour of male dolphin bottle nose (*Tursiopsaduncus*) using NTD Model

H M Taki[1,2] and M Z Lubis[3,4]

[1] King Abdulaziz University, Department of Urban and Regional Planning, Jedah, Saudi Arabia
[2] University of Indonesia, Department of Geography, Depok, Indonesia
[3] Geomatics Engineering, Politeknik Negeri Batam, Batam 29461, Indonesia

Email: zainuddinlubis@polibatam.ac.id

Abstract. Noisy Time Domain (NTD) model and behaviour of male dolphin using the underwater camera, and also comparing between sound and time from the spectrum.Collecting data was taken at the Safari Park of Cisarua Bogor in Indonesia, in show pool. Data at the show with replications 1,2, and 3 has a salinity of 29 ‰. Sound whistle 3 is 28.03 dB with the frequency interval of 14.64 kHz-16 kHz. The results showed that salinity in the show pool with replication 1,2, and 3 has salinity value that is equal to 30 ‰. F-test at the show pool has heterogeneous value. Treat at the show pool has a value of $P < 0.001$ and $P < 0.001$. Noisy Time Domain have differences with each other and have a lot of different sound patterns and detection of energy from power spectral density (PSD) signal.

1. Introduction

The application of science in the detection of acoustic signals generated in the present era which performs experimental control systems and methods is very limited or does not have the activity or result to test the sound results of echolocation derived from the object source the male dolphin bottle *Tursiopsaduncus* [1-2]. Conditions in training on the Laboratory scale must produce data that has a high level of accuracy to see whether the object is taken for the treatment of feeding and without feeding using sonar system. This is very a challenge in the field of energy that takes the source of sound from marine mammals or marine animals. Bioacoustics is a science that answers the weakness of it, by doing voice detection using passive acoustic instruments commonly called hydrophones [3-6].

From the result of data obtained by using passive sonar with the object that is dolphin males, on vocalization that is click sound which produces energy with intensity range that is 171 dB with the reference that is 1 lPa at the peak to peak, with frequency interval 36-61 kHz [8]. Measuring the sound energy generated from the object of marine animals with the location or container in the captive pond is a very good thing. This is demonstrated by the support of the object of having a biosonar system in the body, capable of making a reliable experiment capable of providing location and energy information from biosonar reflections [9]. In this research object that is sea mammal animal using dolphin males as energy producer at vocalization treatment that is whistle sound with maximal intensity value (dB) that is 229 dB return 1 lPa and possesses the highest frequency that is <100 kHz that is when treatment at echolocation in a pond/captivity.

[4] To whom any correspondence should be addressed.

The process of data retrieval/data recording is done that is when the object is observed and viewed at the bottom and the surface of the pool (on the surface of the pool is 4.5m kea rah array). This is done until the object passes the position of the recording platform using hydrophone, as well as to stop data recording every 5 minutes in the process of data retrieval. The noise value generated in this study is using a tool specification with 184 dB returning 1 lPa / V, 35 dB re 1lPa with intervals of 200 Hz-2.3 kHz. The resulting ambient sound is a long spectral [10-11].

2. Materials and Methods

2.1. Power Spectral Density (PSD)

PSD is useful to create similarities in the number of rows and columns in the matrix data in the m-file data in the result of the voting process. The value of spectral power density is a useful concept to show the band value at the optimum frequency of the existing signal transmission system. The useful energy power for the frequency spectrum which is a calculated density can be processed using the FFT method[12]. Power Spectral value can be obtained equation (1,2 and 3):

$$\lambda = \frac{v}{f} \tag{1}$$

$$\lambda = \frac{C}{f} \tag{2}$$

$$PSD = \frac{|X_n|^2}{f} \quad \left(\frac{(Amplitudo)^2}{Hz} \right) \tag{3}$$

2.2. Recording Whistle

The results of the research on the energy derived from the sound of male dolphins obtained using passive acoustic method and taken underwater video data that is in July - August 2015, using GoPro Hero 3+ and hydrophone SQR 3, which has a range of hydrophone frequency is the maximum 22,000 Hz with a reference of -192 dB re 1 μPa. Energy derived from whistle sounds affirmed to a particular individual when a dolphin is around the camera/hydrophone, one-way proximity (<1.3 m) by simultaneously showing bubble emission and having a correlation with whistle sound results.

The sound produced from the male dolphin with a distance of more than 4.5 m from the recording equipment produces energy and whistle sound with a comparison of the energy obtained from the male dolphins at a distance of 1.5 m. In contrast to a group of dolphins in the dating, it will produce a whistle sound intensity value derived from the whistling individual sound energy that is leading to the center of the hydrophone (<1.5 m), this is estimation of the position conditions of a group of dolphins the male dolphin away (≥ 5 m). In the previous study using sound and seeing the energy produced from male dolphins, only using the identification of sound bubbles removed from the object alone to identify the sound origin of each individual male dolphins [13-15]. In this study not only do that but do the recording and see the behavior of the dolphins are centered on the acquisition of energy generated in the park Safari, Cisarua Bogor in Indonesia. The sound recording of the object is using Raven Pro 1.5 software with a sampling frequency of 44.1 kHz (1-4 whistle sound). Analysis of data from the sound energy data is using F Test conducted in accordance with statistical methods with the software SPSS 14.0, and NTD model that is processed with Matlab R 2010.

3. Results and discussion

The result of the PSD in the data result with M-file in the form of the matrix of the sound recording of the object has the energy of the spectrum frequency in the form of a density value which will be estimated by using FFT data processing. The Power Spectral Density (PSD) method is a useful method for seeing the amount of energy produced that is very widely used in the present era [16-17].There are 4 whistle sounds that produce different energy from each other, with the highest PSD value being

IOP Conf. Series: Earth and Environmental Science **105** (2017) 012029 doi:10.1088/1755-1315/105/1/012029

found with the highest 28.03 dB seen with the green line (figure 1). While the intensity value (dB) or calculated energy has a frequency interval of 14.64 kHz-16 kHz, while the intensity value (dB) of energy produced is at a frequency of 9300 Hz with a value of 21.97 dB PSD. In the voice data recorded on the sound of the whistle 3 produces a sound frequency with intervals of 14.64-16 kHz which is the highest sound compared with the intensity (dB) results at sounds 1.2, and 4 with the same frequency interval.

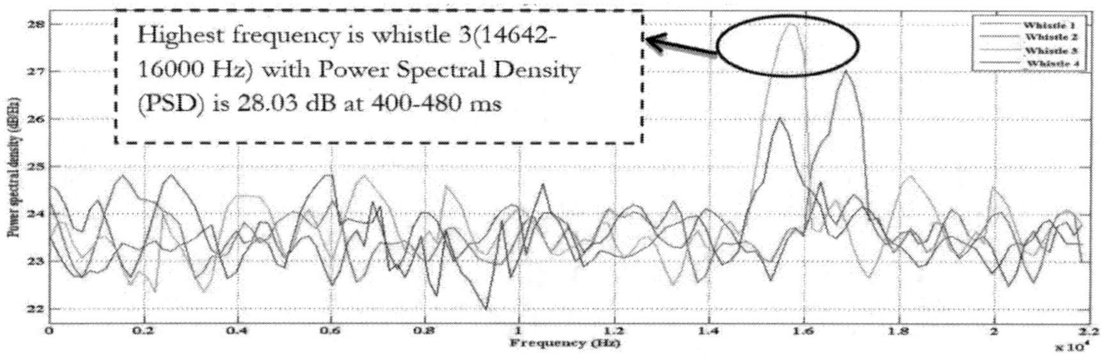

Figure 1. Whistle sound (PSD 1- 4).

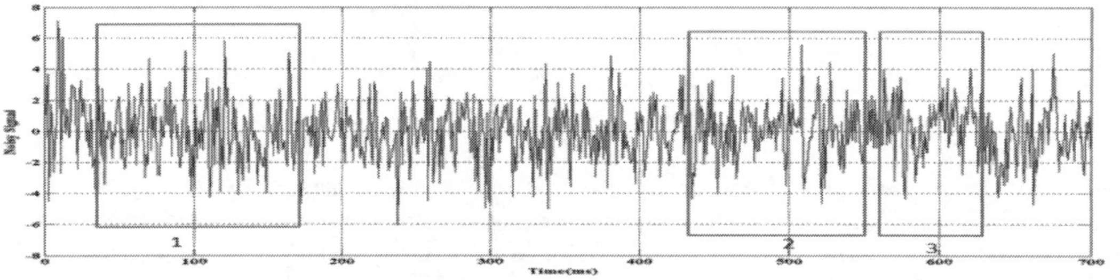

Figure 2. NTD whistle sound 1.

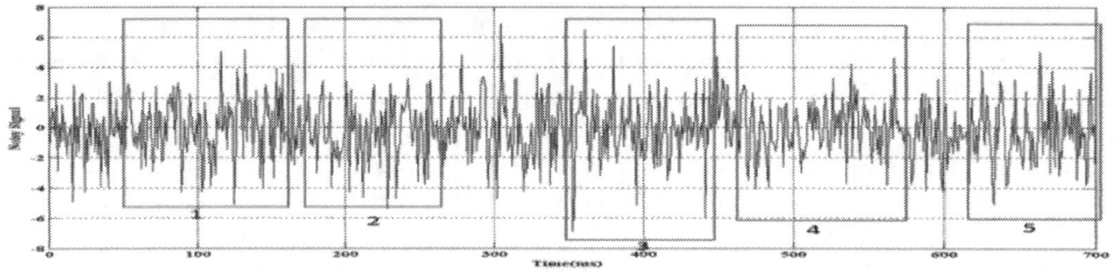

Figure 3. NTD whistle sound 2.

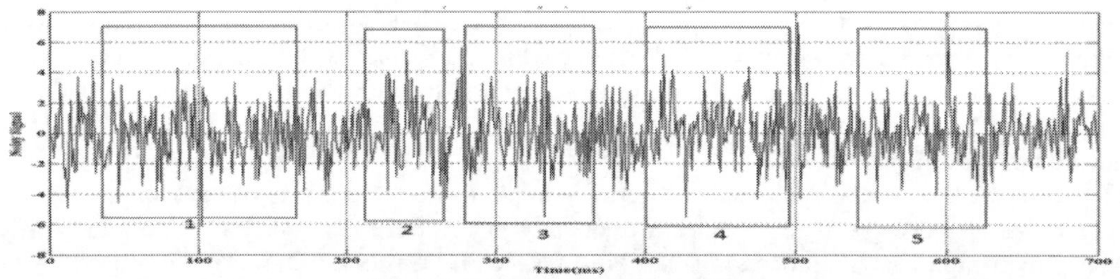

Figure 4. NTD whistle sound 3.

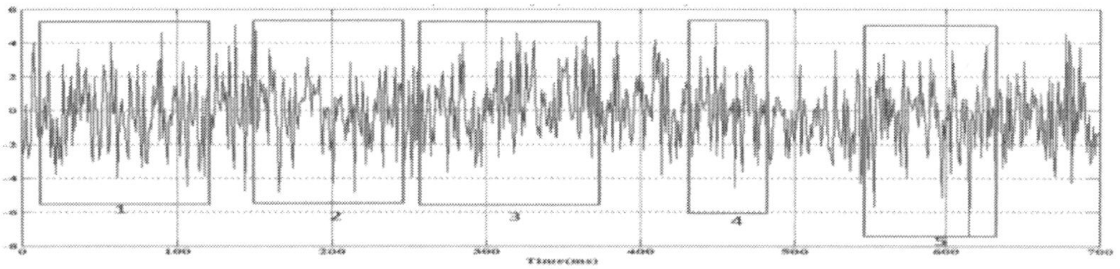

Figure 5. NTD whistle sound 4.

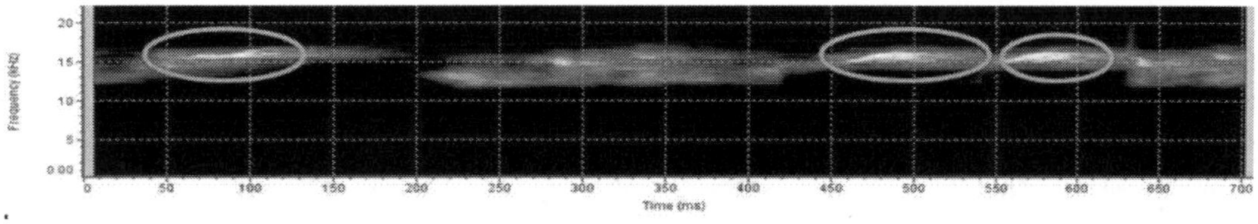

Figure 6. Whistle sound Spectogram 1.

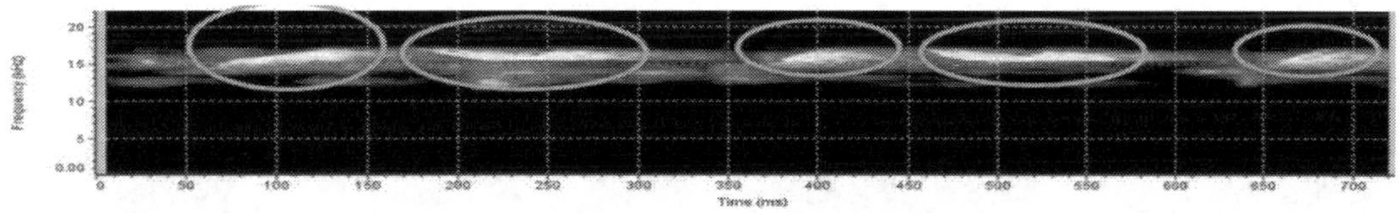

Figure 7. Whistle sound Spectogram 2.

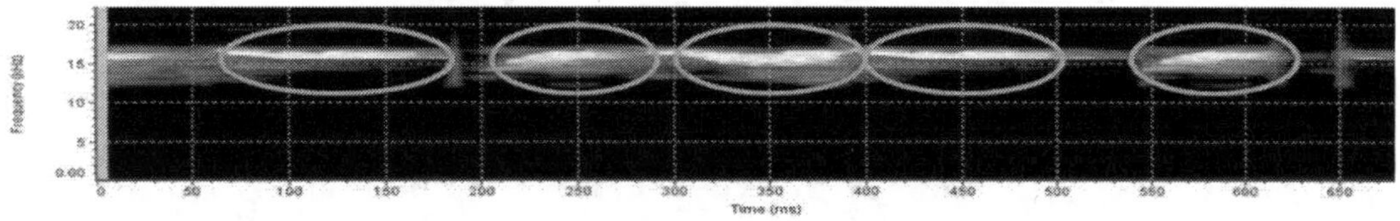

Figure 8. Whistle sound Spectogram 3.

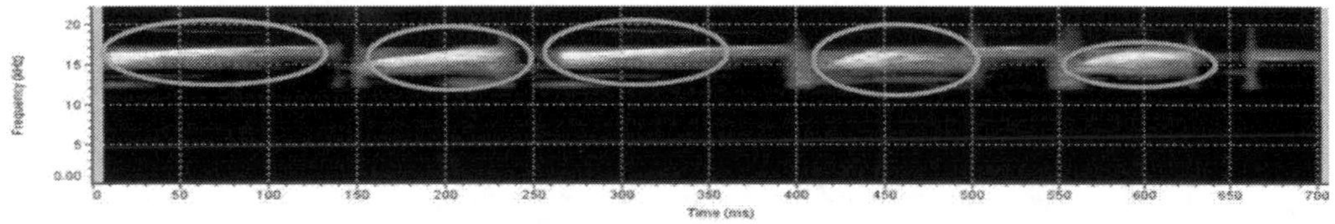

Figure 9. Whistle sound Spectogram 4.

It can be seen that the whole results have 6 sound energy patterns generated at intervals of 71-141 ms, ms 1191-261, 3901-441 ms, 461-561 ms, and 6601-721 ms (figure 7). The results obtained from figure 8 obtained sound energy by the number of patterns ie 5 whistle sound patterns with intervals of 71-161 ms, 231-261 ms, 281-301 ms, 401-481 ms and 561-621 ms. In the sound energy output also produces 5 sound energy patterns whistle with intervals of 21-101 ms, 151-271 ms, 271-351 ms, 441-471 ms and 561-631 ms. The pattern has a sign with circled on the spectrogram result using a green circle, with a maximum of the original male dolphin sound results at 1, 2, 3, 4, which is 701 ms. The result of frequency value at sound energy has a range of 15.1-16.1 kHz.

Table 1. F Test result

No	Time interval (ms)	Energy of whistle sound			
		1	2	3	4
1	101 with 201	*	-	-	-
2	201 with 301	-	-	*	-
3	301 with 401	*	*	-	-
4	401 with 501	-	-	*	*
5	501 with 601	-	-	*	-
6	601 with 701	*	*	-	*

-) Nothing significant effect
*) Have significant effect

The test results using the statistical methods in F Test can be seen in table 1 on the sound energy of whistle 1 at the time interval of 101-201, 201-301 ms, 301-401 ms, 401-501, 501-601 ms, and 601-701 ms, which has a value of the test result is P> 0.001. Sound energy whistle at 101-201, 201-301 ms, 301-401 ms, 401-501, 501-601 ms, and 601-701 ms, which has a value of the test result is P> 0.001. with the result of the sound energy analysis 1-4 seen have a very significant effect on the value of the sound energy produced with the resulting time interval, while the behavior of the male dolphins during data recording tends to be above the surface of the pond. Results of photos of the dolphin behavior can be seen in figure 10.

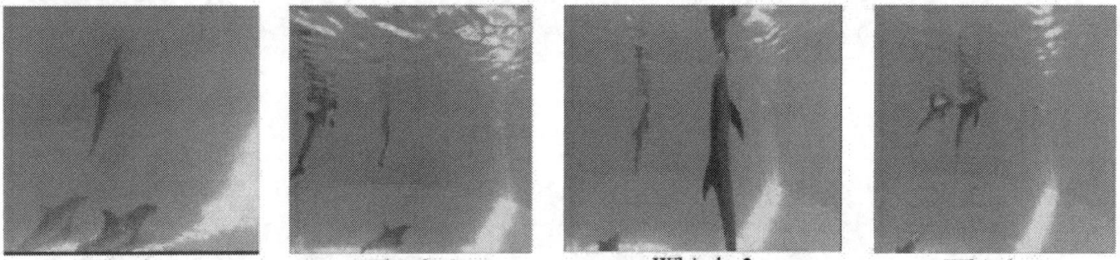

Whistle 1 Whistle 2 Whistle 3 Whistle 4

Figure 10. Behaviour of male dolphin bottlenose.

4. Conclusion

The results of the processing and identification of the whistle sound energy produced from the male dolphin nose bottle (*Tursiopaduncus*) carried out at the Safari Park, Bogor resulted in the PSD value, and the value of the NTD model which has a difference in the magnitude of the load (dB) other (1-4). The result of the obtained energy signal has a significant difference by using the statistical method (Test F) with the time interval of the sound energy of the whistle 1-4. This study can prove that the dolphins are used as objects have energy that is not the same as seeing the sound of the whistle 1-4.

5. Acknowledgments

The study was financed by Safari Park, Cisarua Bogor, Indonesia, and we would like to thank to Politeknik Negeri Batam, for supporting all activities.

6. References

[1] Lubis M Z 2014 *Bioakustik Stridulatory Gerak Ikan Guppy (Poecilia Reticulata) Saat Proses Aklimatisasi Kadar Garam* Skripsi (Bogor: Institut Pertanian Bogor)

[2] Lubis M Z, Wulandari P D, Pujiyati S, Hestirianoto T, Moron J R and Mahdi D P I 2016 Spectral analysis using haar wavelet (original signal, denoised signal, residual signal) and source level for whistle sound of dolphin (Tursiops aduncus) *J. of Fisheries Sciences* **10**(3) 9

[3] Lubis M Z and Pujiyati S 2015 Influence of addition of salt levels against study of bio-acoustic sound stridulatory movement fish guppy (Poecilia reticulata) *The 1st Int. Conf. on Maritime Development Proceeding* p 1-7

[4] Lubis M Z, Wulandari P D, and Hestrianoto T 2017 Karakteristik bioakustik dan tingkah laku lumba-lumba jantan hidung botol (Tursiops aduncus) *J. Teknologi Perikanan dan Kelautan* **7**(2) 179-190

[5] Lubis and Zainuddin M 2016 Produksi suara dan tingkah laku lumba-lumba jantan hidung botol (Tursiops aduncus) dengan metode bioakustik *J. Integrasi* **8**(1) 42-49

[6] Lubis M Z 2016 *Identifikasi Karakteristik Whistle Dan Tingkah Laku Lumba-Lumba (Tursiops Aduncus) Di Taman Safari Indonesia, Cisarua Bogor* Tesis Program Pascasarjana (Bogor: Institut Pertanian Bogor)

[7] Caldwell M C 1965 Individualized whistle contours in bottle-nosed dolphins (Tursiops truncatus) *Nature* **207** 434-435

[8] Norris K S, Prescott J H, Asa-Dorian P V and Perkins P 1961 An experimental demonstration of echolocation behaviour in the porpoirs, Tursiops truncatus (Montagu) *Biol. Bull.* **120** 163–176

[9] Madsen P T, Johnson M, De Soto N A, Zimmer W M X and Tyack P 2005 Biosonar performance of foraging beaked whales (Mesoplodon densirostris) *J. of Experimental Biology* **208**(2) 181-194

[10] Jensen F H, Bejder L, Wahlberg M and Madsen P T 2009 Biosonar adjustments to target range of echolocating bottlenose dolphins (Tursiops sp.) in the wild *J. of Experimental Biology* **212**(8) 1078-1086

[11] Wisniewska D M, Johnson M, Beedholm K, Wahlberg M and Madsen P T 2012 Acoustic gaze adjustments during active target selection in echolocating porpoises *J. of Experimental Biology* **215**(24) 4358-4373

[12] Kostenko P Y and Vasylyshyn V I 2014 Signal processing correction in spectral analysis using the surrogate autocovariance observation functions obtained by the ATS-algorithm *Radioelectronics and Communications Systems* **57**(6) 235-243

[13] Mc Cowan B and Reiss D 2001 The fallacy of 'signature whistles' in bottlenose dolphins: a comparative perspective of 'signature information'in animal vocalizations *Animal behaviour* **62**(6) 1151-1162

[14] Sayigh L S, Esch H C, Wells R S and Janik V M 2007 Facts about signature whistles of bottlenose dolphins, Tursiops truncates *Animal Behaviour* **74**(6) 1631-1642

[15] Janik V M and Sayigh L S 2013 Communication in bottlenose dolphins: 50 years of signature whistle research *J. of Comparative Physiology A* **199**(6) 479-489

[16] López B D 2011 Whistle characteristics in free-ranging bottlenose dolphins (Tursiops truncatus) in the Mediterranean Sea: Influence of behaviour *Mammalian Biology-Zeitschrift für Säugetierkunde* **76**(2) 180-189

[17] Killebrew D, Mercado E, Herman L and Pack A 2001 Sound production of a neonate bottlenose dolphin *Aquatic Mammals* **27**(1) 34-44

[18] May-Collado L J and Wartzok D 2008 A comparison of bottlenose dolphin whistles in the Atlantic Ocean: factors promoting whistle variation *J. of Mammalogy* **89**(5) 1229-1240

[19] Dudzinski K M, Sakai M, Masaki K, Kogi K, Hishii T and Kurimoto M 2003 Behavioural observations of bottlenose dolphins towards two dead conspecifics *Aquatic Mammals* **29**(1) 108-116

[20] Swift K N and Marzluff J M 2015 Wild American crows gather around their dead to learn about danger *Animal Behaviour* **109** 187-197

[21] Warren-Smith Á B and Dunn W L 2006 Epimeletic behaviour toward a seriously injured juvenile bottlenose dolphin (Tursiops sp.) in Port Phillip, Victoria, Australia *Aquatic Mammals* **32**(3) 357

[22] Miksis J L, Tyack P L and Buck J R 2002 Captive dolphins, Tursiops truncatus, develop signature whistles that match acoustic features of human-made model sounds *The J. of the Acoustical Society of America* **112**(2) 728-739

IOP Conf. Series: Earth and Environmental Science **105** (2017) 012030 doi:10.1088/1755-1315/105/1/012030

Effect of graphenenano-fluid on heat pipe thermal performance for passive heat removal in nuclear spent fuel storage pool

Ainur Rosidi[1,2], **Nandy Putra**[1] **and Mukhsinun Hadi Kusuma**[1,2]

[1] Heat Transfer Laboratory, Department of Mechanical Engineering, Universitas Indonesia, Depok 16424, Indonesia
[2] Centre for Nuclear Reactor Safety and Technology, National Nuclear Energy Agency of Indonesia (BATAN), Kawasan Puspiptek Serpong, Tangerang Selatan, 15314, Indonesia

E-mail: nandyputra@eng.ui.ac.id

Abstract. After Fukushima Dai-ichi nuclear power reactor accident, spent fuel in spent fuel storage pool (SFSP) became an important thing to pay attention to, due to its decay heat release. If station blackout occurred, the active system for SFSP cooling system will experience malfunction to remove the product of decay heat. To keep spent fuel safe, heat pipe as passive cooling system device can be used to remove spent fuel decay heat even if the active cooling system failed. Heat pipe with 6 m on length will be proposed as apassive cooling system in SFSP. For that, it is necessary to analyze the effect of Graphene nano-fluid as heat pipe working fluid. The objective of this research is to know the effect on Graphene nano-fluid to enhance the heat pipe thermal performance and to know the heat transfer phenomena inside the heat pipe based on Graphene nano-fluid. Graphene nano-fluid with 1% weight concentration was used as working fluid with filling ratio of 80%. The experimental investigation is conducted with varying the evaporator heat load of 1000, 1500, 2000, and 2500 W. Water as coolant flows in thecondenser with aconstant volumetric flow rate of 8 L/min. The experiment results show that there was anovershoot, zigzag and stable phenomena inside the heat pipe. The thermal resistance of heat pipe is obtained at 0.015 °C/W. The use of Graphene nano-fluid as working fluid can enhance the heat pipe thermal performance significantly and can be used as an alternative working fluid in heat pipe for thepassive cooling system in SFSP of nuclear power plant.

1. Introduction

The Fukushima Dai-chi nuclear reactor accident becomes a valuable lesson for the safety design of nuclear power plants in the world. The safety concept that is fully dependent on the active cooling system is being abandoned so that when the similar accident occurred,it can be addressed with a passive safety system [1].The passive cooling system as and additional safety system is utilized when the accident began to be noticed in the nuclear field.Passive cooling systems are expected to operate when all reactor systems undergo stations blackout. One of the passive cooling systems that can be applied as an additional safety system in spent fuel storage pool is the use of heat pipe technology.

The heat pipe is a closed-loop heat exchanger that works by utilizing the natural circulation of the phase change of the working fluid. The heat pipe is a technology that naturally absorbs heat in the evaporator part and discharges heat in the condenser. The steam generated in the evaporator section will rise towards the condenser section. The heat from the steam in the condenser will be absorbed by the cooling system in the condenser until it becomes condensate. The condensate will fall back to the evaporator by utilizing the capillary force or the gravitational force [2]. The heat pipe is a highly efficient thermal conductor for use in applications requiring high heat flux discharges, in very limited airflow environments, non-uniform heat generation, and weight or space limitation[3].

Research on the use of heat pipe as a passive cooling system in the nuclear field has been largely done by previous researchers. Sabharwall et al. conducted research on the performance of heat transfer in a two phase thermosyphon which will be proposed as a residual heat result in advanced nuclear reactors. The thermosyphon used are made of alkali metals. The results of the analysis show that thermosyphon has a good heat transfer capability and can be recommended as a heat exchanger that works passively in nuclear reactors[4]. Jeong et al. perform simulations using CFDs in order to determine the thermal performance of heat pipes as a passive cooling system to remove residual heat on the nuclear reactor core during an accident. The simulated results show that the heat pipe can be used to reduce the temperature at the reactor core from 290°C to 195.1°C [5].Mochizuki et al. have studied the cooling system using a heat pipe loop for a decay heat dissipation system on a BWR reactor core. The evaporator part of the heat pipe loop is located near the reactor core, and the condenser section with fin is located outside the containment. The released heat in the condenser is then cooled by the outside air. The results show that the heat pipe has a small total thermal resistance and is able to heat the resulting decay to keep the reactor core safe [6].

We have also conducted experiments and simulations using the RELAP5/MOD3.2 code on the use of wickless-heat pipes as apassive cooling system in the spent fuel storage pool. The heat pipe used is made of copper using demineralized water as the working fluid. Our results show that a straight wickless-heat pipe has excellent thermal performance and can be used as an alternative passive cooling technology to remove residual decay heat generated in the spent fuel storage pool when the blackout station occurs[7-9].

The selection of the working fluid is an important aspect of the heat pipe design. One of the alternatives for a heat pipe's working fluid is Graphenenano-fluid. Graphene is a flat monolayer of carbon atoms tightly packed into a two-dimensional honeycomb lattice. The particular 2D structure leads to high electric and thermal conductivity providing easymovement of the charge carriers[10].Graphenenano-fluid has been studied in the form of nano-fluid to use the excellent thermal conductivity of grapheme while maintaining the stability of the nanoparticles in the suspensions. The use of Graphenenano-particles in pool boiling could enhance boiling heat transfer and critical heat flux[11]. Researches on the use of nano-fluids as working fluids in heat pipe have been done by many researchers. Nandy Putra et al. has conducted experiments using Al_2O_3nano-fluid as a working fluid in a heat pipe with a capillary axis of screen mesh. The experimental results show the use of nano fluid as a working fluid improves the thermal performance of heat pipe[12, 13].Asirvatham et al. have conducted experiments using graphene-acetone nano-fluid as a working fluid on a thermosyphon to determine the performance of heat transfer. Their results showed that the overall thermal resistance of the heat pipe was reduced by 70.3% in comparison to the acetone under the same condition. At the same time, the heat transfer coefficient in the evaporator section was increased by 61.25%[14].

From several previous studies, it can be seen that the selection of working fluid is important to improve the thermal performance of the heat pipe. Graphenenano-fluid is expected to improve thermal performance for several reasons including high stability for a long time span and high thermal conductivity. The objective of this research is to know the effect on Graphenenano-fluid to enhance the heat pipe thermal performance and to know the heat transfer phenomena inside heat pipe based on Graphenenano-fluid. Graphenenano-fluid with 1% weight concentration was used as working fluid with filling ratio of 80%. The experimental investigation is conducted with varying the evaporator heat load of 1000, 1500, 2000, and 2500 W. Water as coolant flows in the condenser with aconstant volumetric flow rate of 8 L/min.

2. Methodology

2.1. Experimental setup

In figure 1, it is shown the experimental setup used in this study. The heat pipe is made of copper with a length of 6 m, the inner diameter of 0.1016 m and outside diameter of 0.1031 m. The heat pipe consists of 3 sections with an aspect ratio of 2 m: evaporator, adiabatic, and the condensersection. In the evaporator section, 8 heater belts as a heat load source are installed with a maximum heat load capacity of 2800 W. The condenser section is installed in a water jacket with the coolant flow rate of 8 L/min. It is circulated from the thermostatic bath to the condenser. The flow rate is controlled by 2 flowmeters with an accuracy of ± 4%.Heat loads with variations of 1000, 1500, 2000 and 2500 W are supplied to the heater in the evaporator section that is controlled by the voltage regulator. The heat pipe is filled with graphenenano-fluid with a weight concentration of 1% as the working fluid. Before the working fluid is inserted, the heat pipe was made vacuum first by using the vacuum pump. Once the pressure inside the heat pipe reaches a pressure of -74 cm Hg, the working fluid is inserted with filling ratio of 80%. Filling ratio is the volume ratio of the working fluid that is charged to the evaporator volume. To keep the safety of heat pipe system, a pressure relief valve was installed at the top of the condenser section to control the operating pressure of the steam. Temperature measurements,which are used in this experiment,arethe National Instruments data acquisition system of cDAQ 9174. Fourteen channels of K type thermocouples are installed on the system:3 onthe condenser'soutside wall, 3 onthe adiabatic'soutside wall, 3 onthe evaporator'soutside wall, 1 on the cooling inlet, 1 on the cooling outlet, 2 onthe insulation wall and 1 channel for the ambient temperature. To reduce heat loss during the experiment, the heat pipe is insulated with glass wooland then covered with aluminum sheet. The position of the thermocouple, which is used to measure the wall temperature of the heat pipe is shown in figure 2.

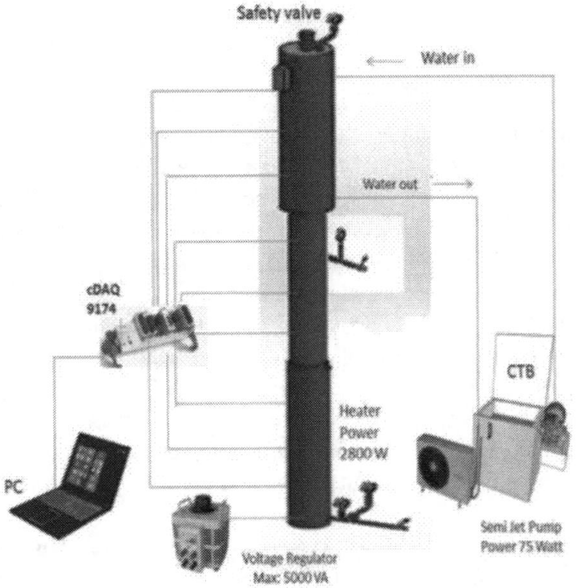

Figure 1. Experimental setup.

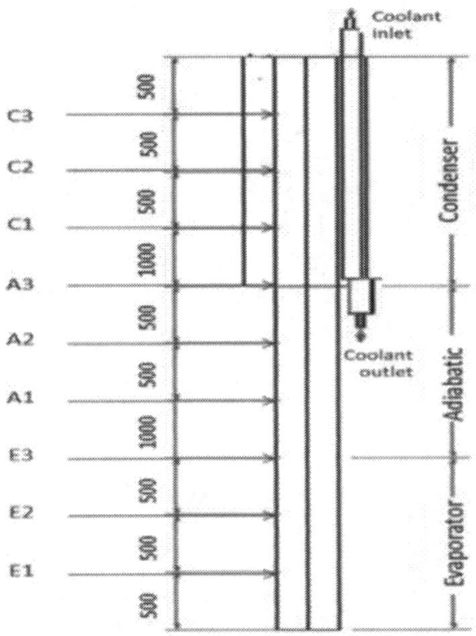

Figure 2. Placement of thermocouples.

2.2. Preparation of Graphenenanofluid

The preparation method of the Graphenenano-fluid working fluid isshown in figure 3. To make a working fluid with a concentration of 1% weight Graphene nanoparticles weighed by using a digital scale of 10 gr, the basic fluid used for mixing is demineralized water with a volume of 990ml. TheGraphenenano particles and demineralized water weremixed in the ultrasonic processor for 4 hours of sonification process.

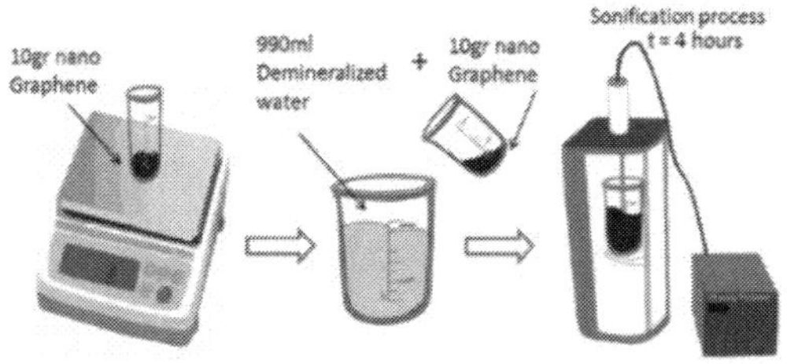

Figure 3. Preparation process of Graphenenano-fluid.

3. Result and discussion

3.1. Transient temperature distribution

The transient temperature as the resultof an experiment with filling ratio of 80%,volumetric cooling rate of 8 L/min with varying the evaporator heat load of 1000, 1500, 2000, and 2500 W can be seen in figure 4.When heat load was given to the evaporator at 1000 W, the evaporator temperature 3 (Te-3) was increased drastically, while at evaporator temperature 1 (Te-1) and evaporator temperature 2 (Te-2) was also increased but not as high as in Te-3. This happened because the evaporator wall 3 received the heat coming from evaporator wall 1 and 2 so that Te-3 has a higher temperature. In this condition, the vapor from the Graphenenano-fluid remains largely in the evaporator and has not risen entirely to the condenser. This situation also occured in the adiabatic and condenser section. The phenomenon in this state is called the overshoot phenomenon.

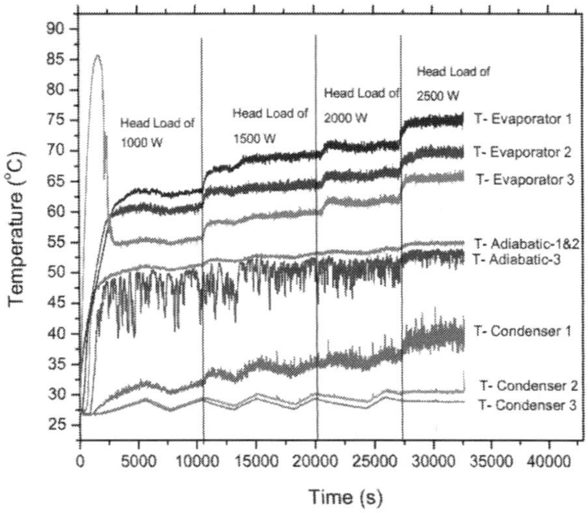

Figure 4. Transient temperature at filling ratio 80% and cooling flow rate 8 Liter/min.

After condensation takes place in the condenser, condensate having a greater density than the steam will descend through the gravitational wall of the heat pipe. The condensate passes through the adiabatic wall to the evaporator. From the figure, it can be seen that the temperature of evaporator wall

3 has decreased previously because condensate was passed foremost compared to the evaporator wall 2 and 1. In this situation zig-zag phenomenon began, where the steam and condensate circulation still has not happened smoothly in the heat pipe.

Zig-zag state continued until there was a more smooth circulation state. This state is called stable phenomenon. The natural circulation is already happening well in the stable phenomenon. The vapor in the evaporator rised to the condenser and the condensate that went down to the evaporator flows in a balanced and continuous way.

These phenomena occur similarly in the heat load given of 1500, 2000, and 2500 W. The experimental results showed that the higher heat load given will increase the evaporator temperature. This happened because the greater the heat load is given to the evaporator will increase the boiling and evaporation process in the working fluid. The boiling and evaporating process on the greater heat load will increase the temperature of each section of the heat pipe to be higher than the smaller heat loads.

3.2. Steady state temperature distribution

Figure 5 shows the steady state temperature distribution on the heat pipe wall at evaporator heat load of 1000, 1500, 2000, and 2500 W.

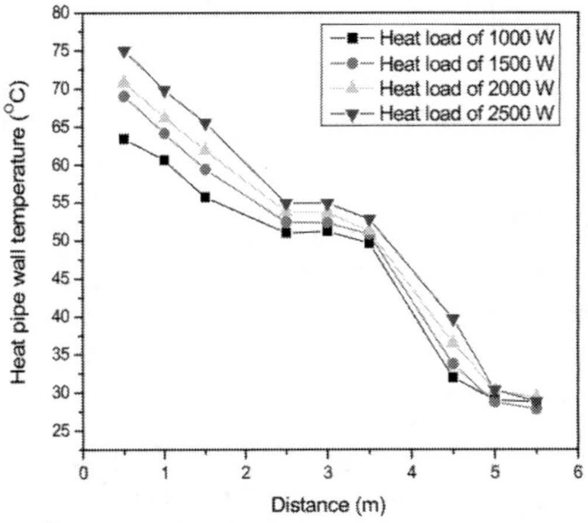

Figure 5. Steadystate temperature distribution at evaporator heat load of 1000, 1500, 2000, and 2500 W.

It can be seen from figure 5 that the steady state temperature of the heat pipe wall increases with the increase of evaporator heat load. The increasing heat load will increase the temperature of the evaporator wall, resulting in more heat transfer from the evaporator wall to the graphenenano-fluid in the evaporator pool, causing faster boiling and vaporizing graphenenano-fluid, increasing the working fluid saturation temperature, generating more steam in the evaporator pool which will be brought to the condenser, and raising the latent heat transferred by the condenser to the cooling system.

3.3. Thermal performance

Thermal resistance is one of theparameters for evaluating heat pipe thermal performance. This thermal resistance is affected by the temperature difference of the evaporator and condenser sections, and the amount of heat load given to the evaporator section. It is generally known that thermal resistance tends to decrease with every increase in the heat load given in the evaporator section. The lower of the thermal resistance obtained, the better the heat pipe thermal performance. The thermal resistance obtained in the experiment is shown in figure 6.

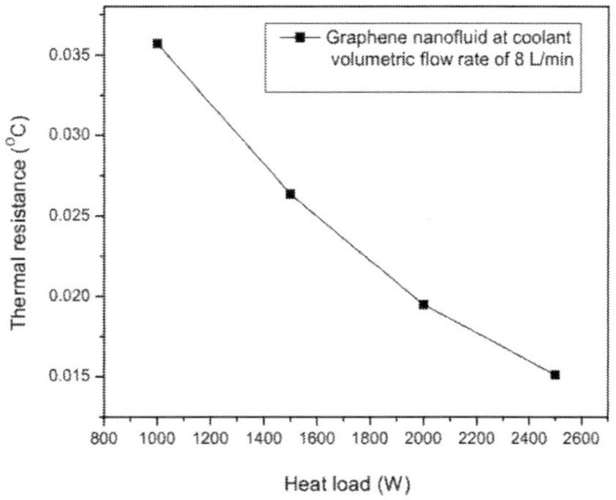

Figure 6. Thermal resistance of the wickless-heat pipe.

Figure 6 shows that the lowest thermal resistance obtained is 0.015°C/W. In the previous investigation of wickless-heat pipe using demineralized water as working fluid, it is obtained that the thermal resistance of wickless-heat pipe is 0.0163°C/W for large geometry and 0.22°C/W for small geometry.

From the experiment results, it can be seen that the use of Graphenenano-fluid as a working fluid can improve the thermal performance of heat pipe. This is because the addition of Graphenenano leads to acollision with large bubbles and it creates a number of smaller bubbles, which means Graphenenano-fluid increases the thermal conductivity and convective heat transfer coefficient of working fluid in the wickless-heat pipe. The collision of nanoparticles with large vapors results in more number of smaller vapor bubbles. Similarly, the Brownian motion of nanoparticles increases the convection effects at the evaporator and condenser. This is the reason for lower thermal resistance with nano-fluid[11].

4. Conclusion

The experimental investigation on the effect of Graphenenano-fluid on heat pipe thermal performance for passive heat removal in nuclear spent fuel storage pool has been done. The results obtained are as follows:

1. The use of working fluid is very influential on thermal performance in heat pipes. It can be concluded that Graphenenano-fluid enhanced the heat pipe thermal performance, with the thermal resistance values obtained at 0.015 °C/W. This value is smaller when compared with previous research results that used demineralized water as a working fluid
2. The temperature distribution phenomena shows the phenomena of overshoot, zigzag, and stable, as also found in our previous investigation.

5. Acknowledgements

The authors express gratitude to DRPM Universitas Indonesia for funding this research through the "PITTA 2017" scheme.

6. References

[1] Darnowski P, Potapczyk K and Świrski K 2017 Investigation of the recriticality potential during reflooding phase of Fukushima Daiichi Unit-3 accident *Annals of Nuclear Energy* **99** 495-509
[2] Vasiliev L L 2005 Heat pipes in modern heat exchangers *Applied Thermal Eng.* **25** 1-19

[3] Wang C, Zhang D, Qiu S, Tian S, Wu Y and Su G 2013 Study on the characteristics of the sodium heat pipe in passive residual heat removal system of molten salt reactor *Nuclear Eng. and Design* **265** 691-700
[4] Sabharwall P, Patterson M, Utgikar V and Gunnerson F 2010 Phase change heat transfer device for process heat applications *Nuclear Eng. and Design* **240** 2409-14
[5] Jeong Y S, Kim K M,Kim I G, and Bang I C 2015 Hybrid heat pipe based passive in-core cooling system for advanced nuclear power plant *Applied Thermal Eng.* **90** 609-618
[6] Mochizuki M, Singh R, Nguyen T and Nguyen T, 2014 Heat pipe based passive emergency core cooling system for safe shutdown of nuclear power reactor *Applied Thermal Eng.* **73** 699-706
[7] Kusuma M H, Putra N,Antariksawan A R, Susyadi and Imawan F A Investigation of the thermal performance of a vertical two-phase closed thermosyphon as a passive cooling system for a nuclear reactor spent fuel storage pool 2017 *Nuclear Eng. and Tech.* **49** 476-483
[8] Kusuma M H, Putra N, Widodo S and Antariksawan A R 2016 Simulation of heat flux effect in straight heat pipe as passive residual heat removal system in light water reactor using RELAP5 Mod 3.2 *Applied Mechanics and Materials* 122-126
[9] Kusuma M H, Putra N, Ismarwanti S and Widodo S 2017 Simulation of wickless-heat pipe as passive cooling system in nuclear spent fuel pool using RELAP5/MOD3. 2 *Int. J. on Advanced Sci. Eng. and Information Tech.* **7** 836-842
[10] Mehrali M, Sadeghinezhad E, Rosen M A, Tahan Latibari S, Mehrali M, Metselaar H S C, *et al.* 2015 Effect of specific surface area on convective heat transfer of graphene nanoplatelet aqueous nanofluids *Experimental Thermal and Fluid Sci.* **68** 100-108
[11] Kim K M and Bang I C 2016 Effects of graphene oxide nanofluids on heat pipe performance and capillary limits *Int. J. of Thermal Sci.* **100** 346-356
[12] PutraN, Saleh R, Septiadi W N, Okta A and Hamid Z 2014 Thermal performance of biomaterial wick loop heat pipes with water-base Al2O3 nanofluids *Int. J. of Thermal Sci.***76** 128-136
[13] Putra N, Septiadi W N, Rahman H and Irwansyah R 2012 Thermal performance of screen mesh wick heat pipes with nanofluids *Experimental Thermal and Fluid Sci.* **40** 10-17
[14] Asirvatham L G, Wongwises S and Babu J 2015 Heat transfer performance of a glass thermosyphon using graphene–acetone nanofluid *J. of Heat Transfer* **137** 111502

The effect of bucketnumber on breastshot waterwheel performance

Warjito[1], Dendy Adanta[1,2], Budiarso[1] and Aji P Prakoso[1]

[1] Department of Mechanical Engineering, Faculty of Engineering, Universitas Indonesia, Kampus Baru UI Depok 16424, Indonesia

E-mail: dendyadanta@ymail.com

Abstract. More than 17% of people in remote areas of Indonesia still do not have electricity. With the abundance of water potential in Indonesia, hydro pico is consideredas the right solution because it has cheap investment and operational cost. The breastshot waterwheel is a suitable turbine that can be applied because it does not affect the garbage in the water. Characteristic of the breastshot waterwheel especially effect of bucket numberon performance have been study in this paper. CFD method isused to evaluate the flow field and explain the effect of the number of the buckets on the waterwheel performance qualitatively. The turbulent model used is k-ε. Simulations were performed with the number of buckets 11, 12, 13, 14 and 15. Analysis using ANOVA block design explains that there is a relationship between the number of buckets to the generated power ($F_0 > F_{0.01,2,8}$). From the analysis, the wheel with the number of bucket 13 produces more stable power than other buckets. This is because the shape of the radial blade is not so steeply upward that the weight energy of water and kinetic energy can be utilized maximally.

1. Introduction

Water turbines with hydro-pico scale (<5 kiloWatt) are suitable for remote areas in Indonesia. Pico hydro is considered suitable because it has the cheapest investment and operational cost when compared with wind turbine and solar PV[1]. Therefore, some developing countries use hydro pico for electrification in remote areas such as Cameroon, Nepal, Laos, Rwanda, Honduras, Bolivia and Peru[1–6].

Indonesia has abundant water energy potential. However, there are major constraints for the utilization of hydro pico in Indonesia, namely garbage. The previous work shows that the garbage caused the turbine blade stuck so that the turbine can not operate. Waterwheel overshot, breastshot and undershot are seen as one turbine that has little effect on the garbage contained in water. This is because the shape of the bucket or blade is larger than the existing garbage.

The difference between the three types of waterwheels lies in the type of energy transferred to the wheel through the blade or bucket. The overshot waterwheel is a hydroelectric power plant that utilizes the gravity of water to rotate the shaft[7]. The undershot waterwheel is a hydroelectric power station that utilizes the compressive energy and boost of water that is transferred through the blade to

[2] Author to whom any correspondence should be addressed.

rotate the shaft[8]. While breastshot waterwheel is a hydroelectric power station that utilizes kinetic energy and weight water to rotate the wheel through the bucket[9].

Due to the available head of 1.5 meters and with a flow rate of 41 l / s, the breastshot waterwheel is considered to be very suitable for application. Many studies have been done to improve breastshot waterwheel performance for maximizing kinetic energy and weight water.Muller and Wolter (2004) observed water velocity on the upstream and downstream sides to find out the loses that occur on the breastshot waterwheel [10]. TevataandImprasit (2011) designed breastshot waterwheel using non-dimensional parameters, Paddle number and immersed radius ratio [11]. Paddle number is a comparison of water velocity and wheel speed. Paudel etal. (2012) to modify the bucket by making the bucket flexible so as not to kill the biota that lives in the water [12]. Yelguntwar et al. (2014) determines the relationship of water velocity and wheel to the resulting power[13]. Emanuele Quaranta&Revelli (2016) determine the type of energy transferred to the bucket breastshot to easier improve performance[14], by 2017, engineering is done by flowing water directly to the bucket for reducing loses[15].

There are manyparameters that influence the performance of breastshot waterwheel including are the number of bucket, wheel width, bucket shape and outer and inner diameter ratio. Bucket shape is very influential in the energy transfer process that occurs in the breastshot because the bucket is the place of the energy transfer process. On the other hand, the number of buckets is a function of the bucket form. To know the effect of bucket quantity on the breastshot waterwheel,an analysis was done by using CFD.

This study aims to evaluate the amount of bucket wheel to be used, to study the behavior of unexpected conditions and to know the relation of the number of buckets to the bucket shape. The CFD method is chosen because it is cheaper and can clearly describe the flow field which can not be done by another method[16].

2. Configure the dimension of breastshot waterwheel

Breastshot waterwheel design begins with determining wheel speed rotation after known rotation speed will be known size parameters of the wheel design to be used[17]. Bucket breastshot waterwheel schematic drawing can be seen in figure 1:

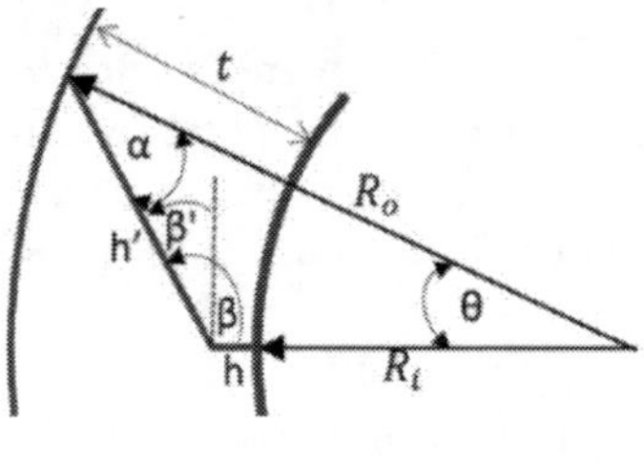

Figure 1. Bucket breastshot waterwheel scheme[17].

Wheel rotation speed is the main parameter that must be set. The rotational speed (n) has an inverse relationship with the width of the mill (w). So to determine the value of w is done iteration with the value of n between 2-20 RPM. Equation 1 is used to determine the value of w and the value of n[17]:

$$w = \frac{4Q60}{\pi({D_o}^2 - {D_i}^2)n} \tag{1}$$

The next step, determine the size of the outer diameter (D) of the wheel[17]. To determine the value of D used the equation 2:

$$D = 2H \tag{2}$$

Where H is the head. The basic difference between breastshot, undershot and overshot is that the diameter of the breastshot turbine wheel should be twice from the available H, this is because the waterwheel breastshot absorbs the water energy through the center of the wheel[17].

To determine the magnitude of the annulus width (t), the best efficiency of waterwheel has a ratio value of t/R_0=0.23 [18]. Determining the angular length of the road (s), the length of the radial side blade (h) and the width or length of line B to C (h') using equation 3[17]:

$$\frac{R_0}{Sin\,\beta} = \frac{(h+R_i)}{Sin\,\alpha} = \frac{h'}{Sin\,\theta} \tag{3}$$

$$Sin\,\alpha = \frac{Sin\,\beta\,(h+R_i)}{R_0} \tag{4}$$

$$\theta = 180^0 - \beta - \alpha \tag{5}$$

for (s) is:

$$s = h' + h \tag{6}$$

Number of bucket using equation 7 [19]:

$$z = \frac{\pi\,D_0\,Sin\,\beta'}{(a.s)+t'} \tag{7}$$

$\beta' = 180 - \beta$, $t'$is the thickness of the iron plate to be used planned (3 mm) and a is the obesity channel between the blades (1/2.5)[17].

3. Method

To design the wheel, the boundary conditions used is 1.5 meters, flow rate of 41 l/s (derived from laboratory conditions), wheel rotational speed of 7.5 RPM, gravity 9.81 m/s^{2}and density of 1000 kg/m^3. The results of the calculations performed by using the previously described equations obtain the value of the width of the (w) 0.36 - 0.03 meters wheel with a wheel speed of between (n) 2 - 22 RPM. The wider the wheel the lower the speed of the turbine, and vice versa. So it takes the ideal width to be used. Wheel width determination is done by averaging the results obtained and then the average value is used to determine the value of rotation speed to be used [20]. Furthermore, the outer diameter of 3 meters is measured, while for inside diameter is 2.15 meters, the width of the wheel is 10 cm, the length of the radial road blade is 0.11 meters, the outer side of the outer blade is between 0.61 - 0.55 meters and the number of buckets used 11-15.

Computational method in this study is using CFD software. Computational method is done to see waterwheel performance. The boundary conditions given in the computational method are inlet velocity is 1.25 m /s, acceleration gravity is -9.81 m/s^2, rotational speed 5, 7.5 and 10 RPM. The flow is assumed to be turbulent. The turbulent model used is k-ε. k-ε k-e can be used to predict the turbulent flow that occurs in the breastshot waterwheel[16]. In the computation method also consider the mesh independency to optimize the number of mesh used. Mesh independency is performed on one of the buckets, which in turn, the optimized number of mesh is applied to the number of other buckets. Furthermore, to save computational power, 2D assumptions were used in this study. The flow chart of the method can be seen in figure 2.

There are two data of simulation result which will be analyzed that is torque (power) and contour velocity. Torque (power) is used to determine the relationship of bucket number with power using ANAVO block design. Prior to analysis using ANOVA block design, Chauvenet's criteria were used as a reference for data filtering (power). While the speed contour is used to see whether the kinetic energy and water weight is absorbed by the bucket. In addition, visual contours are also used to determine the number of buckets recommended for use.

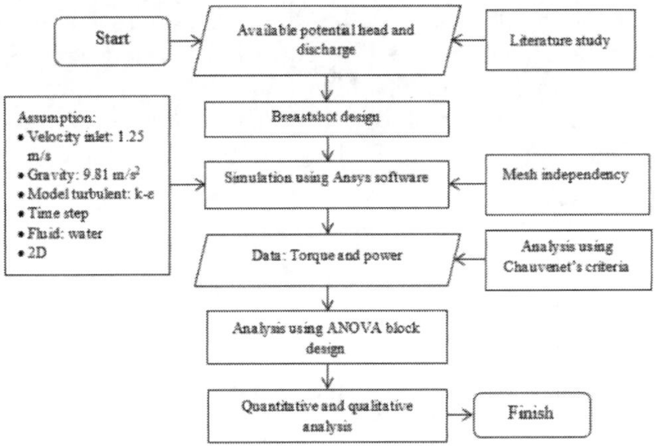

Figure 2. Flow chart simulasion.

4. Result and discussion

ANOVA blocking design is used to determine the relation of the number of buckets with power. Furthermore, Stability of power generated is used as base parameter to determine the ideal number of bucket, so that when the wheel is implemented the turbine can still operate even though the discharge decreases suddenly.

4.1. Statistic analysis

The grid independence process is performed to verify the number of mesh. This process is done by continuously reducing the mesh size, so the level of accuracy of the solving process in the simulation will increase. The grid independency process starts with medium mesh size, then fine mesh, followed by ½ size of fine mesh size. To identify the grid independence result, tested at a point, the test is done by using pressure variable at the point (2.7,0,0) or after the water passes through the wheel. for more details please note figure 3 and table 1.

From the grid independency results, all simulations are performed on mesh number ½ times the number of mesh fine. This number is chosen because when compared with the higher number of mesh (fine) has a smaller difference of 1.3% when compared with the lower mesh (medium) is 4%.

Each wheel with different RPM values is simulated per time step of 0.005 seconds for 4 seconds and with ten simulations. It is necessary to obtain the distribution of data, whether there are samples that are not in the normal distribution. Filtering the data obtained from the simulation results is done using Chauvenet's criterion. Filtering data is used to know the data obtained, what if a data has local standard deviation value ($|(x_i - \bar{x})/stdev|$) is greater than the value of Chauvenet's criteria (t) with the sample value of ten ($N= 10$) is 1.96, then the data obtained must be eliminated because it exceeds the threshold of Chauvenet's criteria [21].

Table 1. Test result mesh number.

Description	Mesh size		
	Medium	½ Fine	Fine
No. of nodes	2540	3650	8496
No. of elements	2172	3185	7748
Pressure (Pa)	358.77	361.3	373.83

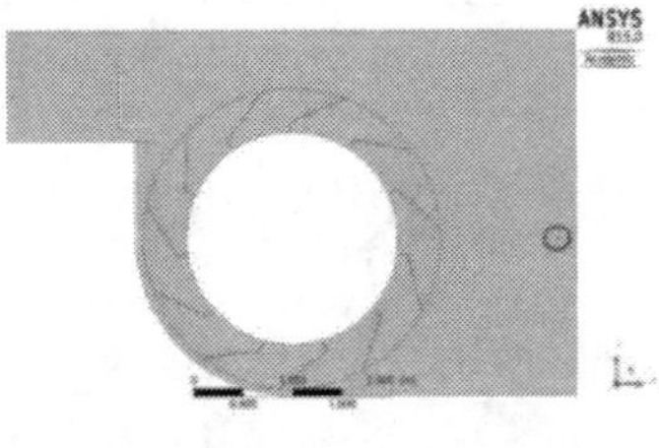

Figure 3. The point of testing.

To determine whether there is a correlation to the number of buckets on the generated power, the analysis is done using ANOVA Blocking Analysis. ANOVA Blocking is chosen because the number of buckets and rotational speed can be controlled[22]. Table 2 is a summary of the results of the analysis using ANOVA Block Design:

Table 2. Resume analisis ANOVA Block design.

Source of variation	Sum of square	df	Mean square	F_0
SSTreatment	73147.72	4.00	18286.93	22.44
SSblock	12074.34	2.00	6037.17	
SSe	6518.56	8.00	814.82	
SSTotal	91740.62	14.00		

$F_{0.01,4,8} = 7.59$. This explains that the value $F_0 > F_{0.01,4,8}$which means that H_0 rejected. This means that statistically proven that there is influence of the number of bucket breastshot waterwheel to the power to be generated by the turbine.

4.2. Computational result

The results of the data obtained from the simulation found that the highest power value obtained by the wheel with the number of bucket 11. This is because the volume of water accommodated by the bucket is much more than the other bucket. However, the percentage decrease or increase in power when the rotational speed decreases or increases, the least changed is the wheel with the number of bucket 13. This is presumably because the distance between buckets is relatively stable so that the water entering into the bucket is more stable. In the wheel with the number of bucket 15, the absorbed power is lower than the other bucket due to the small angle of the blade (β) so that the kinetic energy of water is not maximally utilized (see figure 5). The data of simulation result can be seen in figure 4.

Figure 4 and figure 5 describes both quantitatively and qualitatively the wheel with the number of bucket 11 is the bucket that has the highest efficiency and power. Technically the bucket with the number of bucket 11 is the best bucket but when viewed more clearly, bucket 13 is a bucket that is slightly affected when turbine rotation decreases and increases as result is stable generated torque. This is very advantageous when the wheel is implemented, this condition will extend the life of the generator to be used because the rotation and torque are relatively stable. Such as, when there is rain that causes the discharge increase significantly, this reduces the risk of over-round generator and also when the generator's rotation and torque decreases suddenly, the generator can still generate power because the required rotation of the generator remains met.

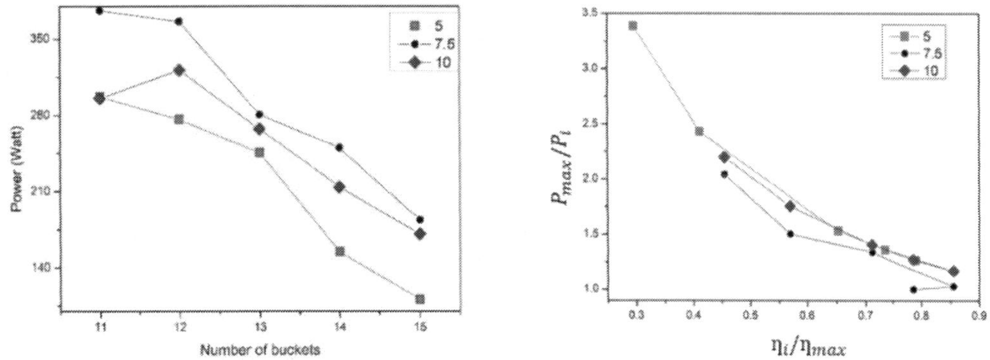

a b

Figure 4. a. Graph of the bucket number relation to power, b. Graph of non-dimensional efficiency and power.

Figure 5. Water volume contour accomodatedin wheel with 13 buckets.

4.3. Discussion

The breastshot waterwheel performance is influenced by the number of buckets because the waterwheel breastshot utilizes weight energy and kinetic energy. To maximize energy absorption, the bucket breastshot design of a minimum value of β is 112.5^0. This is because when the β angle is too high it will block the access of water entering into the bucket thus causing the kinetic energy not to be maximally utilized.

Simulation results obtained similarly to this research was conducted by Quaranta and Revelli (2015). Quaranta and Revelli (2015) discussed the effect of the floodgates on wheel performance. Quaranta and Revelli explained that the smaller the water gate the higher the water velocity resulting in increased kinetic energy, increased kinetic energy impact on minimizing the water filling time in the bucket thereby increasing the torque and getting a positive effect on the performance of the wheel[23].

Research conducted by Helmizar (2016) indicated different result.Helmizar (2016) explains that the number of buckets does not contribute significantly to the performance of the turbine, in the case of 12 buckets obtaining torque of 0.97 Nm, while with buckets of 6 obtaining a torque of 0.91 Nm[9]. This is alleged because of the different forms of bucket used, Helmizar uses a curvature-shaped bucket that leads upwards, causing the kinetic energy of water not maximally utilized yet.

5. Conclusion

The results of the investigation using ANOVA block design explains that there is a relationship between the number of buckets used with the power to be generated ($F_0 > F_{0.01,4:8}$). This is because the angle of the bucket arm greatly affects the absorption of kinetic energy contained in water. The bucket breastshot design is recommended not to create a bucket shape with a value of β below $112{,}5^0$due to the bucket sleeve shape (h') pointing upward, the outer side shape of the upwardly highway blade that will cause the kinetic energy of the water can not be absorbed maximally because it is blocked by bucket arm.

Bucket number 13 is a bucket that is slightly affected when turbine rotation decreases and increases as result is stable generated torque. Thus it can be concluded that for head of 2.71 meters and flow rate of 41 l / s the number of bucket 13 is recommended to use.

6. Referensi

[1] Ho-Yan B 2012 *Design Of A Low Head Pico Hydro Turbine For Rural Electrification In Cameroon* Unpublished Masters Thesis presented to The University of Guelph, Ontario, Canada

[2] Adhikari P, Budhathoki U, Timilsina S R, Manandhar S and Bajracharya T R 2014 A Study on developing pico propeller turbine for low head micro hydropower plants in Nepal 2014 *J. Inst. Eng.* **9** no.1 36–53

[3] Thomas B 2011 Pico-Hydropower franchising in rural Honduras 2011 *Int. J. Serv. Learn. Eng. Humanit. Eng. Soc. Entrep.* **6** no.1 46–63

[4] Williams A A and Simpson R 2009 Pico hydro–reducing technical risks for rural electrification 2009 *Renew. Energy* **34** no.8 1986–91

[5] Pigaht M and Van der Plas R J 2009 Innovative private micro-hydro power development in Rwanda 2009 *Energy Policy* **37** no.11 4753–60

[6] Vicente S and Bludszuweit H 2012 Flexible design of a pico-hydropower system for Laos communities 2012 *Renew. Energy* **44** 406–413

[7] Saputra P A 2017 *Perancangan Volume Bucket Kincir Air Piko Hidro Overshot dengan Metode Analitik dan Numerik* (Universitas Indonesia)

[8] Pradito R 2017 *Perancangan Kincir Air Piko Hidro Undershot dengan Metode Analitik dan Numerik* (Universitas Indonesia)

[9] Helmizar H 2016 *Turbine Wheel-A Hydropower Converter For Head Differences Between 2.5 And 5 M* (University of Southampton)

[10] Muller G and Wolte C 2004 The breastshot waterwheel: design and model tests *ICE Proceedings-Engineering Sustain.* **157(4)** pp 203–211

[11] Tevata A and Inprasit C 2011 The effect of paddle number and immersed radius ratio on water wheel performance *Energy Procedia* **9** 359–365

[12] Paudel S, Linton N, Zanke U C E and Saenger N 2013 Experimental investigation on the effect of channel width on flexible rubber blade water wheel performance *Renew. Energy* **52** 1–7

[13] Praful Yelguntwar A B, Bhange O and Lilhare Y 2014 Design, fabrication & testing of a waterwheel for power generation in an open channel flow *Int. J. Res. Eng. Adv. Technol* **2** no.1

[14] Quaranta E and Revelli R 2016 Optimization of breastshot water wheels performance using different inflow configurations *Renew. Energy* **97** 243–251

[15] Quaranta E and Revelli R 2017 CFD simulations to optimize the blade design of water wheels *Drink. Water Eng. Sci.* **10** no.1 p 27

[16] Adanta D, Budiarso, Warjito and Siswantara A I 2017 Study Of Turbulence Modelling On Pico-Hydro Turbine *Quality in Research*

[17] Biantoro K 1989 *Kincir Air Langkah Atas dengan Daya Keluaran 1.4 kW* (Depok)

[18] Müller G and Kauppert K 2004 Performance characteristics of water wheels *J. Hydraul. Res.* **42** no.5 pp 451–460

[19] Popov E 1984 *Mekanika Teknik* (Jakarta: Erlangga)

[20] Senior J, Wieman P and Muller G 2008 The Rotary Hydraulic Pressure Machine for Very Low Head Hydropower sites *Hidroenergia Conf.* pp 1–8

[21] Hugh W G S and Coleman W 1999 *Experimentation and Uncertainty Analysis for Engineering* (Canada: Wiley Inter Science)

[22] Sujandari I *Design of Experiments: Blocking in One-Factor ANOVA* (Depok)

[23] Quaranta E and Revelli R 2015 Performance characteristics, power losses and mechanical power estimation for a breastshot water wheel *Energy* **87** 315–325

IOP Conf. Series: Earth and Environmental Science **105** (2017) 012032 doi:10.1088/1755-1315/105/1/012032

Study and model development of renewable energy investment feasibility under willingness and ability to pay approach

A Murnandityo[1], E A Setiawan[1,3] and Fajardhani[2]

[1] Tropical Renewable Energy Center, Universitas Indonesia, Depok City, West Java 16424, Indonesia
[2] Lecturer, Electrical Engineering Department, Universitas Indonesia, Depok City, West Java 16424, Indonesia (period 2008-2017)

E-mail : ekoas@eng.ui.ac.id

Abstract. Solar PV power plant investment in Indonesia is one amongst the alternative for electricity supply to maintain electrical reliability supply and to maximize local energy potential for remote areas. An objective approach in terms of ability and willingness to pay of the community at study location becomes the basis for revenue potential calculation of on this study. Assessment of investment feasibility including economic incentive calculation becomes more important if incentive is required in order to improve return of investment and at the time to provide good quality of energy supply at the lowest possible cost. Community characteristics data on study location is processed using statistical methods and micropower optimization simulation to generate PV system specification and all component needed to invest. To increase the return of investment the economic incentive scenarios are generated to provide alternatives to optimize the economic feasibility while maintaining the technical feasibility based on the ability and willingness to pay of the community, and not inflicting financial loss to the incentive giver. The results of investment feasibility calculation, with various scenarios built for investment alternatives, PV investment is economically feasible based on common investment analysis criteria such as expected NPV (net present value), IRR (internal rate of return), and PBP (payback period).

1. Introduction

This paper brings together two stakeholders at the same time, both Government and investors which has different interests, in the development of renewable energy. Renewable energy is unlikely to develop without involving both party. The Government needs funds from investors for renewable energy project's financing while investors will only finance projects if the rate of return is considered feasible.

In several cases when development projects or programs have to be implemented, to achieve a feasible rate of return, the Government needs to provide incentives in a certain amount and over a certain period. The problem is how much and for how long reasonable incentives need to be provided by the Government to achieve a reasonable investment feasibility or to help improve its feasibility.

[3] To whom any correspondence should be addressed.

In principle, the development of engineering projects always considers two things: technical feasibility and economic viability, generally known as investment feasibility assessment or techno-economic assessment or engineering economics[1]. This method has been conducted into various projects and researches and give various results [2] [3]. The model built on this research brings together technical feasibility and economic viability for the Government that provides incentives and investors. As for the state-of-the-art, the model in this research can deliver the best alternative options for providing incentives and investment value to obtain optimum returns for the Government and investors while still meeting the technical feasibility and economic viability principles.

2. Theoretical background

2.1. Electric utility organization

Electric utility organization is part of utility organization operates to provide high-quality electric service for is designated service area. In doing its operation, it invests initial capital, consumes labor and materials with cost of electricity accumulated to be total cost including return to investors. In these operations its objective is to produce electricity at the lowest possible cost[4]. Based on ownership, there are two types of electric utility organization: government owned and private owned.Regardless of the differences of utility ownership type, there are no effect on the society since their role is on the power cost side as shown in figure 1. Therefore, society's willingness and ability to pay the power cost, must be considered when assessing the feasibility of electric utility investment.

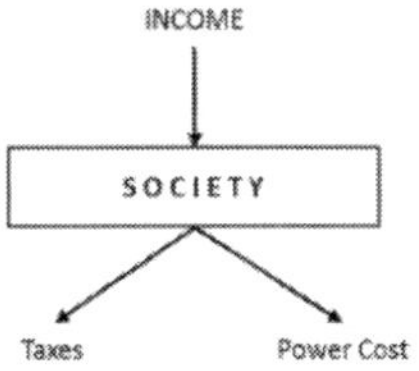

Figure 1. Cost to Society depicted Society's role in electric utility revenue potency. The power cost paid by society becomes revenue for electric utility organization.

2.2. Willingness and ability to pay

When objectively assessing revenue potency from community side, it is necessary to consider willingness to pay (WTP) and ability to pay (ATP). As the name implies, ATP describe how much community can pay the price given to them, therefore the price applied for a product shall not exceed the value of ATP. On the other side, WTP contained user's perception of product related to the level of product quality, availability to meet the needs of society, and usage level of the community which also serve as an important variable. Various research regarding WTP and ATP is presented in [5]–[7]. There are three conditions related with WTP and ATP relationship r:ATP bigger than WTP, ATP smaller than WTP, and ATP equals WTP[8]. Ideal condition is when ATP equals with WTP.

2.3. Engineering economics

In economic activity, cash in and cash out is depicted using cash flow diagram[1]. Using the cashflow diagram, the investment feasibility assessment is performed using the economic indicators such as Net Present Value (NPV), Internal Rate of Return (IRR), and Discounted Payback Period (DPBP).

2.3.1. Net Present Value (NPV). NPV is a simple calculation of difference between the present value of cash inflows and outflows. NPV compares the value of an investment today to the future value of the money based on inflation and returns. It is one of the most effective value to assess the profitability of a long-term project. A positive value of NPV indicates a favourable investment.

$$NPV = \sum_{k=0}^{N} R_k\left(\frac{P}{F}, i'\%, k\right) - \sum_{k=0}^{N} E_k\left(\frac{P}{F}, i'\%, k\right) \tag{1}$$

Among them, R_k is *net revenue potency* on year k, E_k is *net expenditure*, include investment done on year k, i'% is annual interest rate/expected internal rate of return, and *N* isstudy period.

2.3.2. Internal Rate of Return (IRR). Internal rate of return (IRR) is an economic indicator that analyses the profitability of a project by comparing to the discount rate. Investment is more favourable as the difference between IRR and discount rate becomes larger.

$$\sum_{k=0}^{N} R_k (P/F, i'\%, k) = \sum_{k=0}^{N} E_k (P/F, i'\%, k) \tag{2}$$

2.3.3. Discounted Payback Period (DPBP). Payback period (PBP) describes the time needed to return the initial investment. This indicates the risk and liquidity instead of standing as a base of investment decision making. When the PBP calculation consider the time value of money, terms discounted payback period is used.

$$DPBP = N \quad \text{when} \quad \sum (R_k - E_k) - Investment \geq 0 \tag{3}$$

Among them, I_0 is initial investment cost, and NCF is net cashflow for each year.

2.4. Multiple linear regression
Multiple linear regression is a regression analysis that explains the relationship between response variables (dependent variable) with more than one predictor (independent variable) to measure the intensity of the relationship between two or more variables and make predictions of the estimate of the value of the dependent variable based on the independent variable.
The general additive multiple regression model equation is

$$Y| = \beta_0 + \beta_1 x_1 + \beta_2 x_2 + \cdots + \beta_k x_k + \epsilon \tag{4}$$

Where $E(\epsilon) = 0$ and $V(\epsilon) = \sigma^2$[9]. It is assumed that ϵ is normally distributed.

3. Methodology

The investment feasibility assessment in this study begins with investment and O/M cost calculation using HOMER 2.68 program. In parallel, The ATP model for revenue potency is built using descriptive statistics and multiple linear regression equations. The tool used to build the ATP model is the IBM SPSS Statistics 22 software. Then the model is validated using the classic assumption test and model significance test. Various alternative is developed using techno-economics assessment to give the best alternative options for providing incentives and investment value to obtain optimum returns. overall work flow described above is shown in figure 2.

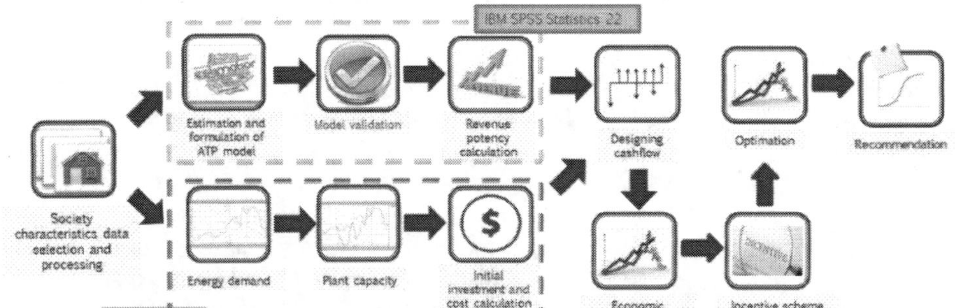

Figure 2. Workflow of investment feasibility study.

In this study, several assumptions to represent real investment condition were made for the development model as follows:
(1) Before tax calculation is assumed for the economic feasibility assessment
(2) Data on community's ability to pay is obtained from fishing and plantation villages
(3) Willingness to pay of community in study location equal with their ability to pay
(4) Investment is totally capitalized by investor internal funding
(5) This study assumes the build-operate-transfer (BOT) investment scheme, the concession period of the BOT project is ten years exclude construction period, and that the salvage value of the fixed asset of the BOT project is not considered. After the concession period expires, the facilities of the BOT project should be returned to the Government unconditionally.

4. Model validation

There are seven principles as foundation that provide a comprehensive doctrine for developing the investment feasibility assessment: (i) develop the alternatives; (ii) focus on the differences; (iii) use a consistent viewpoint; (iv) use a common unit of measure; (v) consider all relevant criteria; (vi) make risk and uncertainty explicit; (vii) revisit your decisions [1]. Various alternatives developed in this model satisfytechno-economic feasibility principles above.

5. Result and analysis

5.1. Initial investment and cost estimation

Initial investment and cost for PV system is shown in table 1. Investment cost consists of PV array, batteries, bi-directional converter include all accessories, balance of plant, and other investment fixed and variable cost. O/M cost consists of fix and variable cost includes land leasing, operator salary, etc.

Table 1. Simulation result for investment and cost estimation

Component	Remarks	Capital ($)	O&M ($/year)
PV	270 kW	901,796	32,150
Battery	264 cells	410,256	67,241
Converter	72 kW	72,388	2,712
BOS and Others		37,216	21,510
Total Investment		1,421,656	123,612
Energy sold	169,515 kWh/yr.		

5.2. Revenue potency

Optimized linier regression model for ATP calculated using IBM SPSS Statistics 22 is shown in equation (5).

$$ATP_i = 274.617{,}106 + 4.500{,}237C \quad (5)$$

Among them, ATP_i isrespondent ATP (IDR/Month/household) and C is household energy consumption per month (kWh/month).

Considering household energy consumption and household population in study location, it is found that revenue potency in USD is $283,716.46/year.

5.3. Incentive and investment scheme alternatives

Using techno-economic principle, a model interface which calculate cashflow component found on section 4.1 and 4.2 is built.When investment feasibility analysis shows that the investment is economically not feasible, seven scenarios of reasonable incentive scheme as shown in table 2 were delivered to make investment economically feasible, objective according to the ability and willingness to pay of the community, and not inflicting financial loss to the incentive giver.

Table 2. Incentive scenarios built for not feasible investment.

Type	Description
Scenario 1	high flat incentive is given during whole study period or until expected PBP achieved
Scenario 2	higher flat incentive is given for the first five years continued with lower flat incentive for the rest five years or until expected PBP achieved
Scenario 3	positive gradient incentive during whole study period or until expected PBP achieve
Scenario 4	negative gradient incentive during whole study period or until expected PBP achieved
Scenario 5	negative gradient incentive is given for the first five years continued with lower flat incentive for the rest five years or until expected PBP achieved
Scenario 6	high flat incentive is given only for the first five years
Scenario 7	negative gradient incentive is given only for the first five years

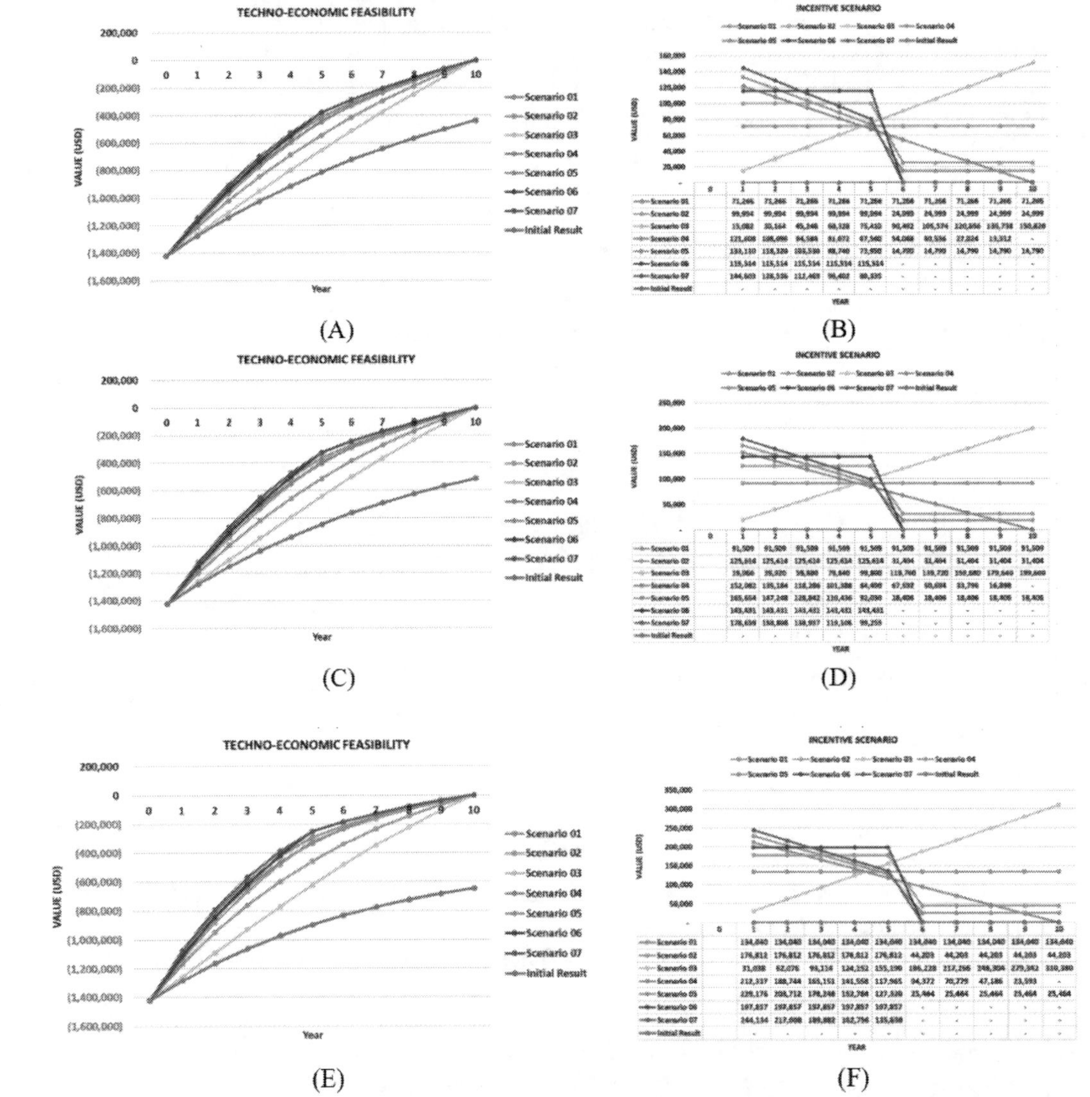

Figure 3. Result of investment feasibility assessment.

Figure 3 (left-hand side) show techno-economic feasibility assessment initial result for investment without incentive and when investment is optimized with seven incentive scenarios to achieve reasonable feasibility at 10%, 12%, and 16% IRR (A – C – E) respectively.It is shown that when initial result investment which is not feasible is optimized by seven incentive scenarios, the investment becomes economically feasible.

Seven incentive scenarios in the right-hand side gives alternatives for investment scheme at 10%, 12%, and 16% IRR (B – D – F) respectively. Result show that with increasing of the expected IRR, the incentive value needed to optimize the initial result investment significantly increases.

6. Conclusion

Taken the ATP of community and proposed incentive scenarios, renewable energy-based power plant investment feasibility model is built and fulfill economic feasibility/techno-economic criteria, objective according to the community condition, and not inflicting financial loss to the incentive giver. ATP linier regression model on study location is a function of household monthly energy consumption with positive correlation. With ATP model, revenue potency calculation is objective according to community condition. The economic feasibility model built in this study is flexible enough to calculate many investment alternativesusing incentive scenario so the optimized assessment results is economically feasible and not inflicting financial loss to the incentive giver.

7. Acknowledgement

This research was supported by Hibah Kompetitif Nasional Kemenristekdikti PUPT 2017. We thank our colleagues from Tropical Renewable Energy Center, Faculty of Engineering Universitas Indonesia that provided insight and expertise that greatly assisted the research.

8. References

[1] Sullivan W G, Wicks E M and Koelling CP 2014 *Engineering Economy*, 16th ed. Upper Saddle River (New Jersey: Pearson Higher Education Inc.)

[2] Hiendro A, Kurnianto R, Rajagukguk M, Simanjuntak Y M and Junaidi 2013 Techno-economic analysis of photovoltaic/wind hybrid system for onshore/remote area in Indonesia *Energy* **59** 652–657

[3] Mat Isa N, Wei Tan C and Yatim A 2017 A techno-economic assessment of grid connected photovoltaic system for hospital building in Malaysia *IOP Conf. Ser. Mater. Sci. Eng.* **217** 12016

[4] Marsh W D 1980 *Economics Of Electric Utility Power Generation*, First Ed. (Oxford: Oxford University Press)

[5] Nababan TS and J Simanjuntak 2008 Aplikasi willingness to pay sebagai proksi variabel harga : suatu model empirik dalam estimasi permintaan energi listrik rumah tangga *VISI* **16** 443–457

[6] Nomura N and Akai M 2004 Willingness to pay for green electricity in Japan as estimated through contingent valuation method *Appl. Energy* **78** no. 4 pp 453–463

[7] Guo X, Liu H, Mao X, J Jin, Chen D and Cheng S 2014 Willingness to pay for renewable electricity: A contingent valuation study in Beijing, China *Energy Policy* **68** 340–347

[8] Elfitra R A 2014 Ability to Pay (ATP) dan Willingness to Pay (WTP) Buruh terhadap Sewa Rusunawa: *Studi Kasus Rencana Pembangunan Rusunawa di Lokasi PT. Kawasan Berikat Nusantara (Persero) Jakarta Utara*, Universitas Indonesia

[9] Devore J L 2012 *Probability and Statistics for Eng. and the Sci* 8th ed. (Boston: Brooks)

IOP Conf. Series: Earth and Environmental Science **105** (2017) 012033 doi:10.1088/1755-1315/105/1/012033

Influence of temperature on conversion of plastics waste (polystyrene) to liquid oil using pyrolysis process

Nasruddin A Abdullah[1], Amaranggana Novianti[1], Imansyah Ibnu Hakim[1], Nandy Putra[1,2] and Raldi A Koestoer[1]

[1] Heat Transfer Laboratory, Department of Mechanical Engineering, Universitas Indonesia, Kampus UI Depok 16424, Indonesia

E-mail: nandyputra@eng.ui.ac.id

Abstract. Plastic waste has been a problem in waste treatment.The growth of plastic waste is increasing especially in Jakarta and needs more serious attention in waste processing. To convert polystyrene to be liquid much heat energy is needed. In this research, pyrolysis is used as the method to process polystyrene to be aliquid product. The objective of this research is to obtain the characteristics of the heating process, and the properties of the liquid product. This liquid can be used as a fuel. Fixed-bed reactor with SUS 316L as the base material was constructed to decompose the polystyrene using an electric heater, which was controlled using Digital PID controller. Power sensor was mounted in the electrical circuit to monitor the power that entered to the heater and recorded using data acquisition. The reaction temperature was varied from 350 °C – 550 °C. No sweep gas injected into the system. The vapor flows naturally based on their partial pressure. The temperature of cooling water was varied into two conditions, water ambient temperature, and cold water. To condense pyrolysis vapor to be liquid oil, the double-pipe condenser was constructed. The thermocouples were installed at many points of the system to monitor temperature change in the system. The maximum liquid yield was obtained at reaction temperature 500 °C with cooling water temperature at 16.59°C.The operating temperature Below 500 °C will produce more wax, and above 500 °C will produce much gas. The liquid can be applied as fuel with heating value 43.83 mJ/kg, density 0.89 g/ml and0.78 cSt of kinematic viscosity.

1. Introduction

DKI Jakarta as the capital city of Indonesia becomes the center of activities and population in Indonesia. So, some people from other provinces move to Jakarta. Together with the increase of population in Jakarta, the consumption, and production of plastics increases. Waste volume in Jakarta is 7000 ton/day and predicted to keep growing each year [1].

Generally, there are three ways to manage plastic waste which is landfilling, incineration with or without energy recovery and recycling[2]. The incineration will increase the emission of harmful gases, e.g.NO_x, SO_x, dan CO_x[3]. The most commonly used method for energy recovery is thermal cracking (Pyrolysis). Pyrolysis process is one of the finest techniques for the conversion of mass to liquid and gaseous products with high energy values [4]. Previous researches have been carried out to convert plastic wastes into oil by means of the pyrolysis process. Temperature has a strong effect on

[2] Author to whom any correspondence should be addressed

the characteristics of pyrolysis liquid, such as viscosity and hydrocarbon chain [5].The pyrolysis process is carried out at the temperature between 300°C to 600°C [6]. Pyrolysis processes are influenced by the reaction temperature, residence time, heat rate and the composition of raw material [7]. Polystyrene has been a good raw material to convert into fuel [8].The reaction temperature needed to degrade this material into vapor and liquidbetween 350 °C-480 °C[9].

M. Artetxeet al, was used spouted reactor to enhance styrene monomer recovery, the liquid yield was influenced by nitrogen flow rate and temperature [10]. The effect of the reaction time and temperature on the liquid yield and quality were investigated. Nitrogen was flowed into the reactor and the vapor was condensed in the condenser using chiller water with low temperature[11]. The aims of this research areto obtain the characteristics of the heating process experimentally in the reactor with natural flow, and the properties of the liquid product with polystyrene as raw material.

2. Materials and methods

2.1. Sample preparation

Polystyrene (PS) plastic was used as feedstock in the thermal pyrolysis process to produce liquid oil, non-condensable gas, and wax. 30 g of PS samples were used for each experiment and it was weighed using digital scales with 0.001 g of accuracy. The raw material (PS plastic waste) was prepared by cutting it into small pieces around 1-2 mm^2, thickness 1-2 mm, and moister content below 10%.

2.2. Experimental setup

A fixed bed pyrolysis reactor was commissioned and used for the conversion of plastic waste into pyrolysis vapor and wax. The reactor was made of stainless steel and covered with an electric heater capacity 20 A and 220 V that allows achieving a maximum temperature of 650 °C. The temperature of the electric heater was controlled by Proportional-Integral-Derivative (PID) digital controller to manage the constant temperature at the outer diameter of the reactor as seen in figure 1. The height of the reactor was 250 mm with 73.6 mm inner diameter and 88.9 mm outer diameter.

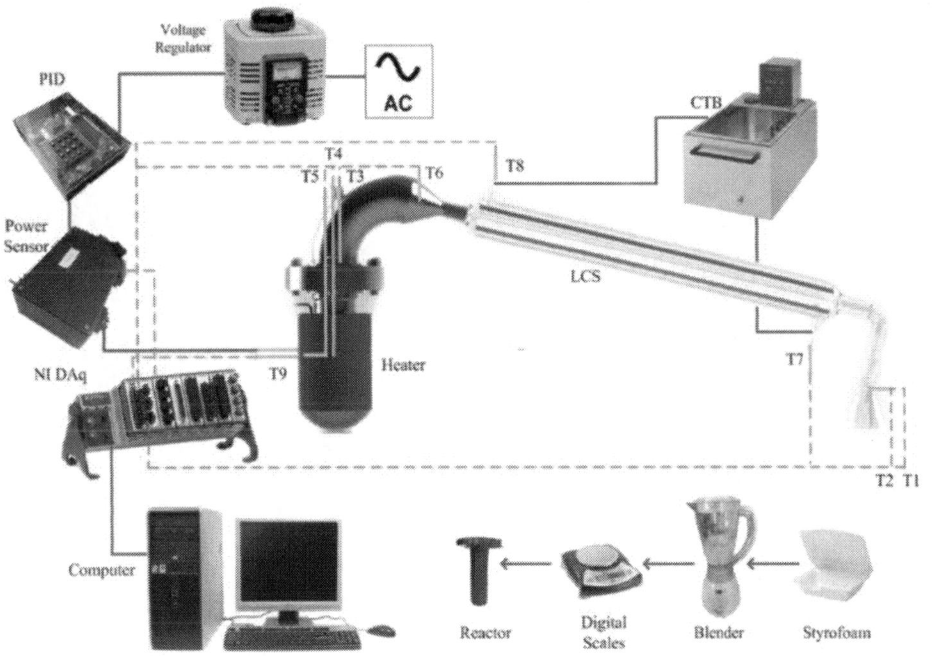

Figure 1. Experimental setup.

A double pipe type condenser as heat exchanger was made from Pyrex material named as liquid collection system (LCS) with water as the cooling fluid to condense the pyrolysis vapor into liquid. The use of Heat exchanger will easier to control the temperature and to separate the heavy and light fraction in liquid collection system [12]. The feedstock was converted into vapors, which were condensed into liquid oil after passing through the condenser and collected in the collection tank at the bottom. The vapor was condensed by using varied temperature which is cold water around 15-17°C and ambient temperature of the water around 28-30 °C. K-type thermocouple was used to measure the temperature of the system. The thermocouple accuracy was 0.1 °C. Each sample of PS was heated in the pyrolysis reactor from room temperature to a varied wall temperature of 350°C, 400°C, 450°C, 500°C, and 550°C respectively. The fractions of liquid oil, non-condensable gas (NCG), and wax were estimated on their weight basis. The yield of liquid oil was further characterized to study the characteristics and its compositions. The compositions were analyzed by GC.MS Agilent 7890A, column DB-5MS. The GC oven temperature is maintained at 40^{0}C for 4 minutes and is programmed to rise to 300^{0}C with the rate of 10^{0}C/min kept for 8 minutes. The carrier gas used was Helium.

3. Result and discussion

3.1. Effect of pyrolysis, vapor and cooling water temperature to yield percentages of product liquid

Yield percentage of product liquid due to pyrolysis temperature and cooling fluid temperature are shown in Table 1. The maximum liquid yield 84.4% was obtained on pyrolysis wall temperature at 500 °C with cooling fluid using cold water at temperature 16.59 °C and heating rate 28.76(°C/min).

Table 1. Comparison of different pyrolysis temperature for product yield.

Mark	Pyrolysis temperature(°C)	Water temperature of LCS (°C)	Power (kWh)	Heating-rate(°C/min)	Product Yield (wt%)		
					Liquid	Wax	NCG
(a)	350	15.40	1.04	29.13	7.11	70.69	22.20
(b)	400	16.96	0.85	30.85	24.99	52.71	22.30
(c)	450	16.15	0.86	31.14	63.82	2.50	33.67
(d)	500	16.59	1.35	28.76	84.80	7.63	7.57
(e)	550	16.93	1.27	27.83	56.59	5.67	37.73
(f)	350	31.18	0.86	31.34	5.88	72.72	21.41
(g)	400	25.48	0.79	29.20	24.86	42.95	32.19
(h)	450	26.28	0.84	30.06	60.77	8.36	30.87
(i)	500	26.58	1.04	28.13	79.17	4.97	15.86
(j)	550	29.37	1.23	29.29	52.04	2.39	45.56

Meanwhile, the temperature distribution of pyrolysis process using PS as raw material shown in figure 2. The processes were done by various heating temperatures using cold water and ambient temperature of water. In the pyrolysis process, PS as the raw material is decomposed by using heat to produce vapor and convert it into liquid oil, non-condensable gas, and wax. We focused on the optimum liquid oil production from several heatingtemperatures using cold water and ambient temperature of water. In figure 2 also shows the chart that shows distribution temperature of the noncondensable gas, liquid product, vapor in the reaction zone, raw material, and vapor in the condenser zone, input and output of cooling water and outside wall (heater). Figure 2 (a) shows the distribution temperature on pyrolysis process at reaction temperature 350 °C. Outside wall temperature was maintained to remain constant at this temperature using PID controller. In this process, the yield liquid product was 7.11%, polystyrene did not decompose completely at this temperature [13, 14]. The yield dominant product was wax 70.69%, and the vapor temperature was below 100 °C. Meanwhile, the Figure 2(b), 2(c), 2(d) and 2 (e) shows distribution temperature of pyrolysis process at

IOP Conf. Series: Earth and Environmental Science **105** (2017) 012033 doi:10.1088/1755-1315/105/1/012033

400 °C, 450 °C, 500 °C and 550 °C respectively using cold water as cooling fluid. Starting from reaction temperature at 400 °C, vapor temperature increased and indicated to produce much vapor. Reaction temperature at 500 °C produced maximum liquid yield. The reaction temperature above 500 °C produced more gases. Some of the vapor, which has low boiling point temperature was condensed at the reaction zone. This phenomenon was indicated with a sudden drop of vapor temperature. The maximum liquid yield using ambient temperature of cooling water was 79.17%. There is a liquid yield reduction around 5.63% when using ambient temperature of water compared to cold water as cooling fluid.

Figure 3 shows the stacked column as the comparison between liquid, wax and NCG product yield due to outside wall temperature. The maximum liquid was obtained at the temperature of 500 °C, either using cold water or ambient temperature of the water as cooling fluid. The quantitative data in the stuck column comes from t table 1.

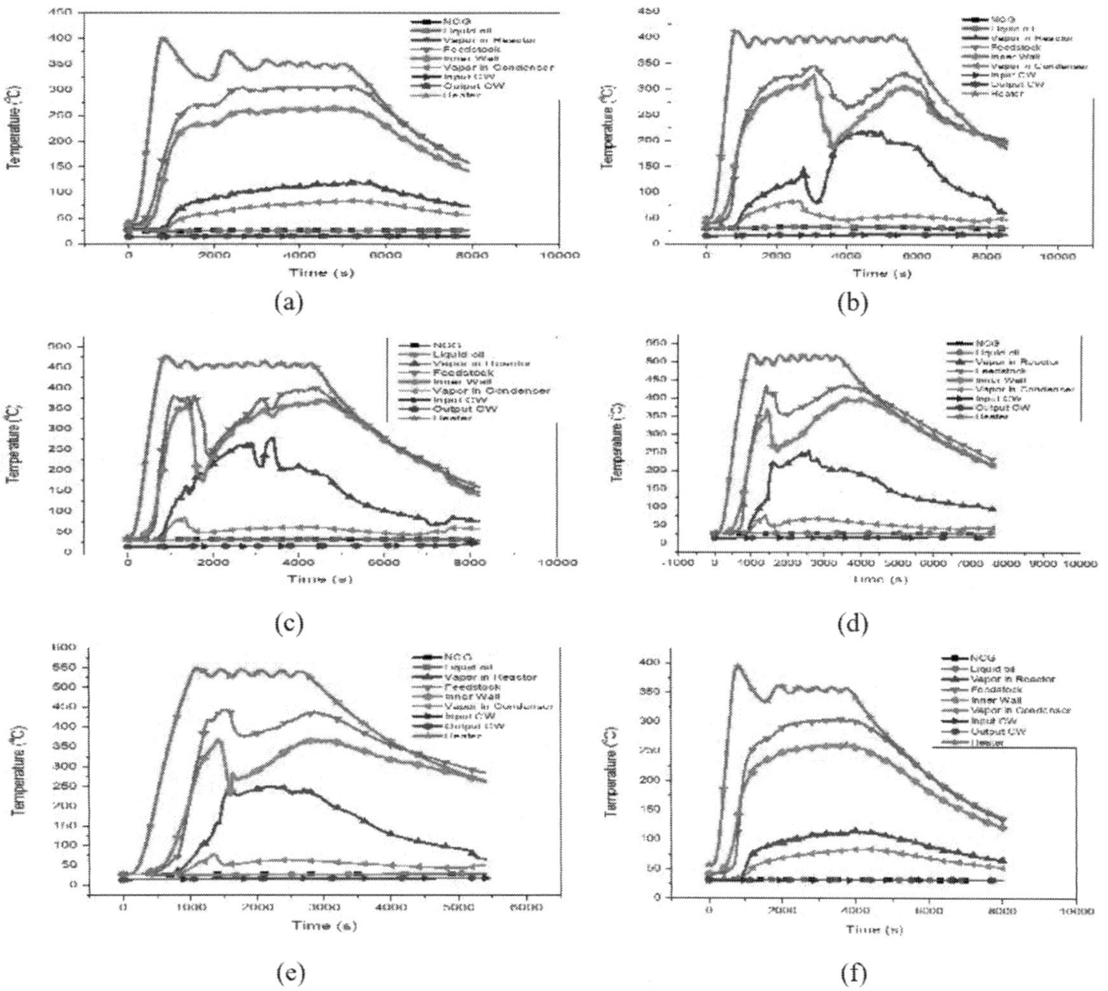

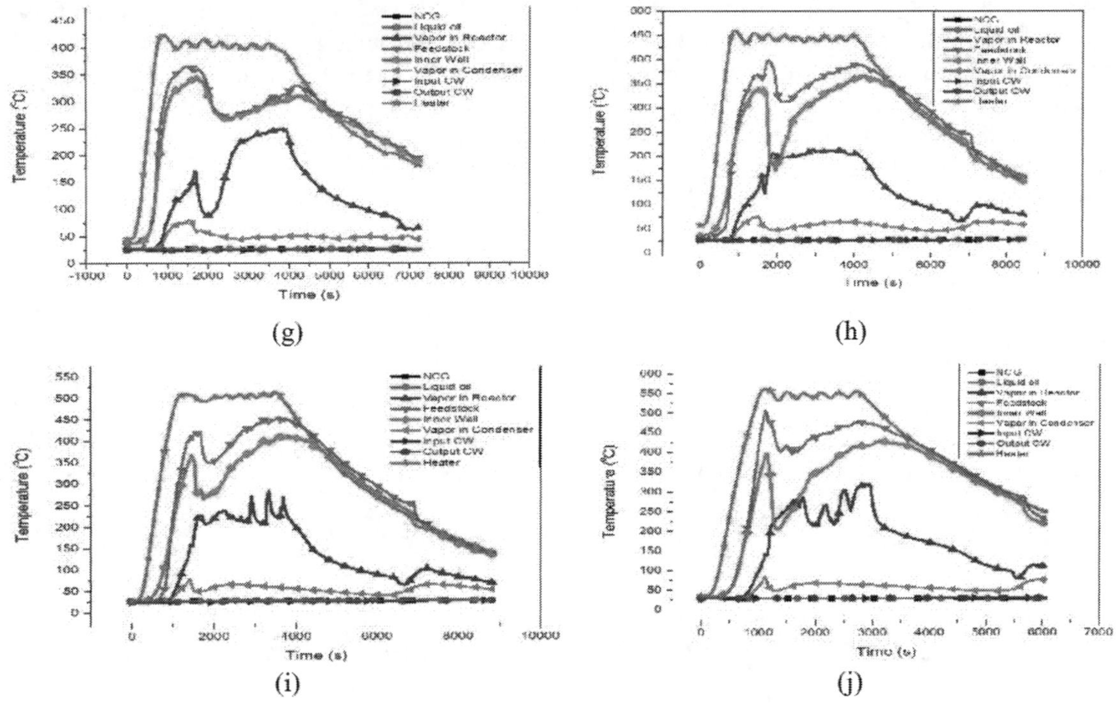

Figure 2.Temperature distribution in pyrolysis process using cold water as cooling fluid and wall temperature at (a)350 °C, (b) 400 °C, (c) 450°C, (d) 500 °C, and (e) 550 °Cand using ambient temperature at (f) 350 °C, (g) 400 °C, (h) 450°C, (i) 500 °C, and (j) 550 °C.

There is no significant difference in liquid yield on the use of cold water or regular water, but reaction temperature becomes the main factor in determining the liquid yield.

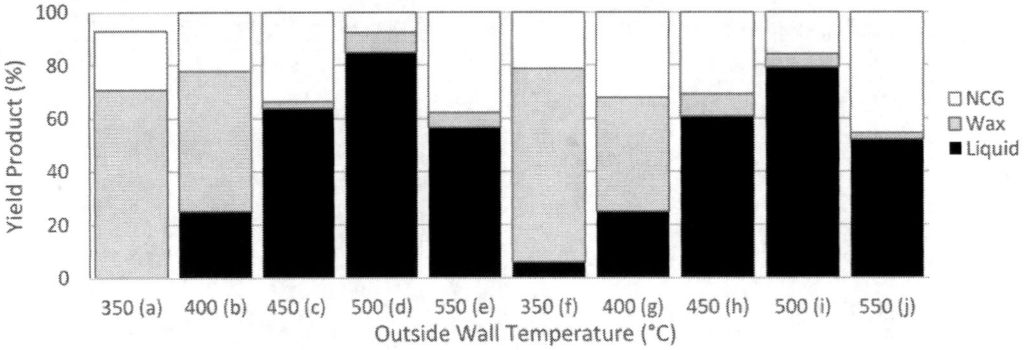

Figure 3.Yield product percentage for each outside wall temperature.

The characterization of liquid product has been conducted to obtain physical properties, such as density, viscosity and heating value. The sample was taken for the maximum liquid yield. The density is the most important parameter for crude oil product [14], low density will consume more fuel and will damage the engine[15]. Heating value is a significant parameter and must be measured in the liquid oil. This value determines the amount of energy released when a fuel is completely burned [15]. Characterization of liquid oil can be seen in table 2. This characteristic of heating value is identical to the character of diesel fuel, that is 43.0 mJ/kg[16]. The chemical compound and content percentage in liquid oil were measured using GC-MS as shown in table 3.

Table 2. Characterization of liquid oil.

No	Characteristic	Result	Unit
1	Density	0.89	g/ml
2	Viscosity at 40°C	0.78	cSt
3	Heating Value	43.83	mJ/kg

Table 3. The chemical compound of liquid oil measured using GC-MS.

No	Retention Time	Compound	% Area	Molecular Formula
1	4.44	Toluene	9.53	C_7H_8
2	6.788	Ethylbenzene	3.96	C_8H_{10}
3	7.804	Benzocyclobutene	60.33	C_8H_8
4	8.264	Isopropyl benzene	0.34	C_9H_{12}
5	9.04	Benzaldehyde	0.19	C_7H_6O
6	9.51	alpha.-Methyl styrene	14.33	C_9H_{10}
7	10.297	cis-beta-methyl styrene	0.15	C_9H_{10}
8	17.667	1,2-Diphenylethane	0.17	$C_{14}H_{14}$
9	19.266	1,3-Diphenylpropane	2.54	$C_{15}H_{16}$
10	19.555	p-Xylene	1.16	C_8H_{10}
11	20.031	3-Butynylbenzene	6.17	$C_{10}H_{10}$
12	20.186	1,2-	1.13	$C_{15}H_{14}$

4. Conclusion

Polystyrene has been used as raw material to produce liquid oil. Pyrolysis process was applied to convert polystyrene to be liquid,gas, and wax. The influence of temperature on conversion of plastics waste (polystyrene) to liquid oil using pyrolysis process and no sweep gas injected into the system has been investigated. . The maximum liquid was obtained at reaction temperature 500°Cand heating rate 28.76(°C/min). In this experiment The operating temperature Below 500 °C will produce more wax, and above 500 °Cwill produce much gas.The rendement at reaction temperature 500 °Cwas 84.80% base on their weight with cooling water temperature of 16.6 °C, and 79.17 wt% with ambient temperature of cooling fluid. The heating value of this liquid is 43.83 mJ/kg, density 0.89 g/ml and 0.78 cSt of kinematic viscosity.

5. Acknowledgments

The author wanted to express gratitude to DRPM Universitas Indonesia for funding this research through "Hibah PUPT"scheme.

6. References

[1] B P D Jakarta 2015 *Statistik Daerah Provinsi DKI Jakarta 2015* (Jakarta: BPS Provinsi DKI Jakarta)

[2] Toraman H E, Dijkmans T, Djokic M R, Van Geem K M and Marin G B 2014 Detailed compositional characterization of plastic waste pyrolysis oil by comprehensive two-dimensional gas-chromatography coupled to multiple detectors *J. of Chromatography A* **1359** 237-246

[3] Kyaw K T and Hmwe C S S 2015 Effect of various catalysts on fuel oil pyrolysis process of mixed plastic wastes *Int. J. of Advances in Eng. & Technol* **8** 794

[4] Buah W, Cunliffe A and Williams P 2007 Characterization of products from the pyrolysis of municipal solid waste *Process Safety and Environmental Protection* **85** 450-457

[5] López A, Marco I D, Caballero B, Laresgoiti M and Adrados A 2011 Influence of time and temperature on pyrolysis of plastic wastes in a semi-batch reactor *Chem. Eng. J.* **173** 62-71

[6] Van Swaaij W P, Kersten S R and Palz W 2015 *Biomass Power for the World* (US: CRC Press)
[7] Miandad R, Barakat M, Aburiazaiza A S, Rehan M, Ismail I and Nizami A 2017 Effect of plastic waste types on pyrolysis liquid oil *International Biodeterioration & Biodegradation* **119** 239-252
[8] Miandad R, Barakat M, Aburiazaiza A S, Rehan M and Nizami A 2016 Catalytic pyrolysis of plastic waste: a review *Process Safety and Environmental Protec.* **102** 822-838
[9] Lee K H and Shin D H 2007 Characteristics of liquid product from the pyrolysis of waste plastic mixture at low and high temperatures: Influence of lapse time of reaction *Waste Management* **27** 168-176
[10] Artetxe M, Lopez G, Amutio M, Barbarias I, Arregi A, Aguado R, *et al.* 2015 Styrene recovery from polystyrene by flash pyrolysis in a conical spouted bed reactor *Waste Management* **45** 126-133
[11] Miandad R, Nizami A S, Rehan M, Barakat M A, Khan M I, Mustafa A, *et al.* 2016 Influence of temperature and reaction time on the conversion of polystyrene waste to pyrolysis liquid oil *Waste Management* **58** 250-259
[12] Abdullah N A, Putra N, Hakim I I and Koestoer R A 2017 *A Review On Liquids Collection System Improvement Process For Producing Liquid Smoke*, ed. 15th QIR 2017
[13] Onwudili J A, Insura N and Williams P T 2009 Composition of products from the pyrolysis of polyethylene and polystyrene in a closed batch reactor: Effects of temperature and residence time *J. of Analytical and Applied Pyrolysis* **86** 293-303
[14] Miandad R, Nizami A, Rehan M, Barakat M, Khan M, Mustafa A, *et al.*2016 Influence of temperature and reaction time on the conversion of polystyrene waste to pyrolysis liquid oil *Waste Management* **58** 250-259
[15] Khan M, Sultana M, Al-Mamun M and Hasan M 2016 Pyrolytic Waste Plastic Oil and Its Diesel Blend: Fuel Characterization *J. of Environmental and Public Health* vol. **2016**
[16] Sharuddin S D A, Abnisa F, Daud W M A W and Aroua M K 2016 A review on pyrolysis of plastic wastes *Energy Conversion and Management* **115** 308-326

IOP Conf. Series: Earth and Environmental Science **105** (2017) 012034 doi:10.1088/1755-1315/105/1/012034

Pyrolysis characteristic of rice husk with plastic bag as fuel for power generation by using a thermogravimetric analysis

Muhammad Anshar[1,4], Dahlang Tahir[2], Makhrani[2] and Farid Nasir Ani[3]

[1] Department of Mechanical Engineering, State Polytechnic of Ujung Pandang, Makassar, Indonesia.
[2] Department of Physics, Hasanuddin University, Makassar, Indonesia.
[3] Department of Thermodynamics and Fluid Mechanics, Faculty of Mechanical Engineering, Universiti Teknologi Malaysia, 81310 Skudai, Johor, Malaysia.

E-mail: M.Anshar60yahoo.com

Abstract. Determination of the characteristics of rice husk (RH) with plastic bag (PB) mixture as a source of energy is carried out in pyrolysis study. Test characteristics of RH with PB mixture is performed by using a thermogravimetric analysis at temperatures from 30 - 800°C, heating rate of 10°C/min, and nitrogen flow rate of 50 ml/min. The dehydration process occurs at temperatures from 40 - 100°C, while for thermal degradation at temperatures from 200 - 340°C, 400-500°C, and 580 - 670°C. Activation energy and calorific value of RH with PB mixture increase significantly with the addition of 10%, 30%, and 50% PB in RH. It is concluded that pyrolysis characteristic of RH with PB mixture is better as fuel for power generation if compared with its original state.

1. Introduction

Fossil fuel consumption in Indonesia recorded a high value of approximately 95% in 2011. The use of oil is the largest portion of the total energy i.e. reaching up to 49.5%, where as coal and gas were in the proportion of 26% and 20.4%, respectively. Continuous utilization of fossil energy sources particularly oil, natural gas and coal will result in an expected deflection of these resources approximately 23, 55, and 83 years from the year 2011, respectively. The same time, electrification ratio in Indonesia was recorded at 72.95%, which meant that there were still 27.05% of households that did not receive electricity, especially in remote areas [1]. These conditions prompt optimization on renewable energy consumption in order to meet the needs for electrical energy, which also implies that the availability of electrical energy in Indonesia is still lacking. Base on previous studies result [2-3] show Indonesia has a great potential of rice husk (RH) is about 13.81 $\times$ 10^6 tons in 2013 with an energy potential of approximately 53,702 GWh. the CV of RH about 13.44 MJ/kg is within the range of 13 - 16 MJ/kg [4]. Meanwhile, the potential of plastic solid waste (PSW) of about 693 tons with energy potential of 28,752,528 MJ, equivalent to 1,065 tons of coal or 593 ton of oil. PSW, specially plastic bags (PB) has great energy with a calorific value (CV) of about 41.2 MJ/kg. This value is within the range of 39.03 – 41.5 MJ/kg [5]. Pyrolysis methods have received much

[4] To whom any correspondence should be addressed.

attention because these methods have the potential in reducing the volume of waste, recovery of various chemicals and as replacement for fossil fuels [6].

2. Materials and methods

2.1. Materials

The sample of PB was dried in an oven at a temperature of 105°C for 1 hour and 7 hours for RH, as from previous studies [7-9]. The PB and RH mixture was sieved to obtain homogeneous sizes within the range of 0.125 - 0.3 mm to allow for an even mix, as in previous studies [10]. The sample was mixed with the following mass ratios of RH to PB: 90% RH: 10% PB (R_9P_1), 70% RH: 30% PB (R_7P_3) and 50% RH: 50% PB (R_5P_5).

2.2. Proximate and ultimate analysis

Testing characteristics of the sample is done with proximate and ultimate analysis bese on previous studies [9]. The CV was obtained by using a bomb calorimeter, model IKA C 2000, according to DIN 51900 methods. The proximate analysis was done according to standard methods i.e. ASTM D3175, ASTM D 3172, ASTM D3173 and ASTM D3174. The ultimate analysis of the samples was done according to standard methods, i.e. ASTM D 4239, ASTM D 3176, and ASTM D 5373 as in previous studies [11-12].

2.3. Thermogravimetric analysis

Thermogravimetric analysis was conducted using the TGA/SDTA851, Mettler Toledo to analyze pyrolysis studies. The heating rates were controlled at 10°C/min from 30 - 800°C. The pyrolysis process used nitrogen flow rate of 50 ml/min as previous studies [9, 13-15]. The dehydration process and thermal degradation on pyrolysis process were analyzed using TGA and DTG curves with the *Sma4Win software.*

2.4. Kinetics study

Kinetic study on pyrolysis process was done to investigate firstly the dehydration process (DP) from 40 - 100°C, thermal degradation (TD) of hemicelluloses from 200 - 260°C and followed by cellulose from 240 - 340°C and finally lignin from 280 - 500°C. The pyrolysis reaction is complete as the temperature reaches the range of 500 - 600°C [16], as in equation (1):

$$\underset{Biomass}{(C_6H_{12}O_6)_m} \xrightarrow{Heat\ (500\text{ - }600^oC)} \underset{Gas}{(H_2+CO+CH_4+\ldots+C_5H_{12})} + \underset{Liquid}{(H_20+\ldots+CH_3OH+CH_3COOH+\ldots)} + \underset{Char}{C} \quad (1)$$

The degradation of hemicelluloses, cellulose and lignin in RH were approximately 30 - 35%, 21 - 31%, and 4 - 19%, respectively [16-17]. The kinetics of the degradation reactions are largely described by the first-order Arrhenius law. Activation energy (Ea) that occurs at every TD process can be obtained from the linear regression equation of the Arrhenius law [9-10, 14-15] as shown in the following equation:

$$k = Ae^{(-Ea/RT)} \quad (2)$$

where; Ea is the activation energy for the reaction (J/mol), A is the pre-exponential factor (min^{-1}), R is the gas constant (8.314 J/K mol), T is the absolute temperature (K). The equation can be written in logarithmic form:

$$\ln k = \ln A - (\frac{Ea}{RT}) \ln e, \textit{ where } \ln e = 1 \quad (3)$$

$$\ln k = -\frac{Ea}{R} \times \frac{1}{T} + \ln A \tag{4}$$

Based on previous studies [9, 13] the following equations are obtained:
$\ln k = -\ln[\frac{\ln(1-\alpha)}{T^2}]$ and $\ln A = \ln \frac{AR}{\beta E}$. Thus, equation (5) can be written as:

$$-\ln[\frac{\ln(1-\alpha)}{T^2}] = \frac{E}{R} \times \frac{1}{T} + \ln \frac{AR}{\beta E} \tag{5}$$

where; β is the heating rate (K/min), the mass loss fraction α for every stage of the thermal degradation is defined as:

$$\alpha = \frac{Mi - Ma}{Mi - Mf} \tag{6}$$

where; α is mass loss fraction (mg), M_i is the initial mass of the sample at the beginning of the reaction (mg), M_a is the actual mass of the sample (mg), M_f is the mass after pyrolysis of the sample (mg). The Ea may then be extracted from a plot of ln k vs. 1/T, which should be linear. This plot is known as the "Arrhenius plot". It is often symbolized by the linear regression equation: y = mX + b. Thus, the $-\ln[\frac{\ln(1-\alpha)}{T^2}]$ as y-axis, $\frac{1}{T}$ as x-axis, m = $\frac{E}{R}$ as slope, and b = $\ln \frac{AR}{\beta E}$ as the intercept of the line with y-axis.

3. Results and discussion

3.1. Characteristic of fuels

The results of proximate and ultimate analysis are presented in Table 1, which shows the characteristic parameters of the fuel. The CV of RH is 13.44 MJ/kg; the addition of PB 10%, 30% and 50% into RH increases the CV of R_9P_1, R_7P_3, and R_5P_5 to approximately17.85 MJ/kg, 23.97 MJ/kg and 28.93 MJ/kg, respectively. Meanwhile, VM, C and H increases with the amount of percentage of PB into RH. In contrast, the percentage of O, S, N, FC, MC, and ash decreased with the increase of PB percentage. The addition of PB into RH can improve the quality of fuel as such CV and VM increase whereas MC, S, N and ash decrease. Small amounts of N will reduce NO_x emissions into the air while small amounts of S can reduce or minimize pollution and corrosion.

Table 1. Test result on calorific value, proximate and ultimate analysis of RH and PB.

Parameters	Unit	RH	R_9P_1	R_7P_3	R_5P_5	PB
Calorific value (CV)	(MJ/kg)	13.44	17.85	23.97	28.93	41.21
Proximate Analysis						
a. Fixed carbon (FC)	(%)	14.81	13.62	10.89	8.00	0.72
b. Volatile matter (VM)	(%)	55.62	58.65	65.21	72.39	89.04
c. Moisture content (MC)	(%)	10.46	9.68	7.53	5.26	0.22
d. Ash	(%)	19.11	18.05	16.37	14.35	10.02
Ultimate Analysis						
a. Carbon (C)	(%)	37.48	40.21	47.00	53.60	67.49
b. Hydrogen (H)	(%)	5.08	5.82	7.10	8.61	10.25
c. Oxygen (O)	(%)	37.81	35.45	29.13	23.08	11.92
d. Nitrogen (N)	(%)	0.43	0.38	0.31	0.28	0.26
e. Sulfur (S)	(%)	0.09	0.09	0.09	0.08	0.06
f. Ash	(%)	19.11	18.05	16.37	14.35	10.02

3.2. Thermal degradation of pyrolysis

The pyrolysis result of RH, PB, R_9P_1, R_7P_3, and R_5P_5 can be seen the TGA curve, which was obtained using the *Sma4Win software* as shown in Fig. 1. The pyrolisis process for RH occurs at temperatures from 40°C - 100°C for dehydration process (DP) and 250°C - 390°C for (TD process), whereas PB occurs at a temperature range from 380°C - 500°C and 620°C - 680°C without the DP because PB is a non-polar material. Meanwhile, for the pyrolysis process of R_9P_1, R_7P_3, and R_5P_5, the TD occurs at temperatures from 40°C - 100°C, 250°C - 390°C, 400°C - 510°C, and 580°C - 680°C.

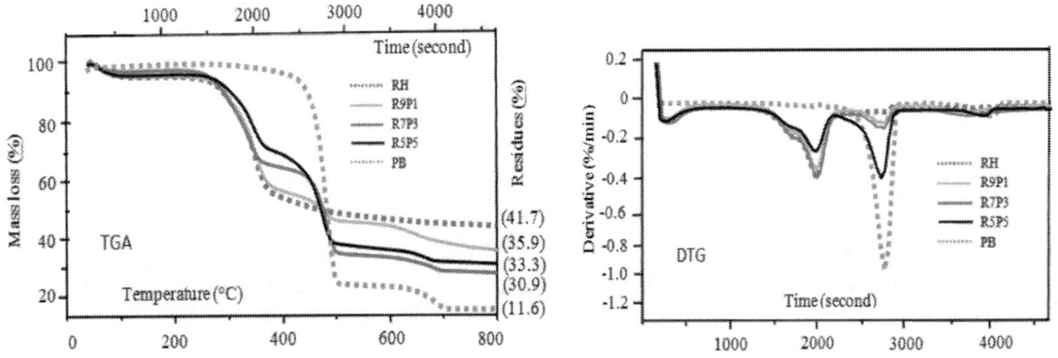

Figure 1. TGA and DTG curve of RH, R_9P_1, R_7P_3, R_5P_5, and PB for pyrolysis at heating rate of 10°C/min, temperatures range of 30°C - 800°C, and nitrogen flow rate of 50 ml/min.

Table 2 shows the pyrolysis process of RH in DP occurs in the temperature range of 40°C - 100°C with a mass loss (ML) of about 5.5%. The TD process zone 1 (TDZ 1) occurs in the temperature range of 250°C - 390°C with ML of about 52.8%, which is volatile to produce energy. Thus, the remaining mass of the char residue is approximately 41.7%. The pyrolysis process of R_9P_1, R_7P_3, and R_5P_5 in zone 1 showed the TD occurs in the temperature range of 250°C - 390°C with a ML of about 42.5%, 35.8% and 27.6% (respectively), which is volatile to produce energy. This is in contrast to the TDZ 2, where mass loss (energy released) of R_9P_1, R_7P_3, and R_5P_5 becomes greater with the increase in percentage of PB in to RH i.e. about 11.4%, 22% and 31.4%, respectively and occurs in the temperature range from 420°C - 510°C.

Table 2. The parameter results of kinetic study for pyrolysis.

Fuels	SM (mg)	DP		TDZ 1		TDZ 2		TDZ 3		Char residues (%)
		ML (%)	T (°C)	ML (%)	T (°C)	ML (%)	T (°C)	ML (%)	T (°C)	
RH	5.6	5.5	40 – 100	52.8	250 – 380	-	-	-	-	41.7
R_9P_1	5.4	4.8	40 – 100	42.5	250 – 390	11.4	420 - 490	5.2	580 – 680	35.9
R_7P_3	5.5	4.5	40 – 100	35.8	260 – 390	22.0	420 - 510	4.4	600 – 660	33.3
R_5P_5	5.4	4.4	40 – 100	27.6	260 – 370	31.4	400 - 500	5.7	600 – 670	32.9
PB	5.6	-	-	79.7	380 – 500	8.8	620 - 680	-	-	11.6

The TDZ 3 occurs in the temperature from 580°C - 680°C with mass loss of about 5.2%, 4.4%, and 5.7%, respectively. Similarly, the TDZ 3 occurs for R_9P_1, R_7P_3, and R_5P_5 in the temperature range of 580°C - 680°C, 600°C - 660°C, 600°C - 670°C, with mass loss of 5.2%, 4.4%, and 5.7%, respectively. In total, the pyrolysis of R_9P_1, R_7P_3, and R_5P_5 where ML occurs about 64.1%, 66.7%, and 67.1%; therefore, the char residues generated is about 35.9%, 33.3%, 32.9%, respectively. This shows smaller ash content with the increasing percentage of PB in the fuel. These characteristics indicate that the use of R_9P_1, R_7P_3, and R_5P_5 as fuels requires minimum temperatures of 680°C.

3.3. Kinetic parameters analysis

The linear regression of the Arrhenius equation of pyrolysis for RH, R_9P_1, R_7P_3, R_5P_5, PB at every stage of TD was obtained using Microsoft Excel. Based on the TD occurred at temperatures from 250°C - 390°C was obtained Ea_1 = 57.01 kJ/mol based on the linear regression equation: $y_1 = 6857x + 2.144$, $R^2 = 0.99$. The linear regression curve of R_9P_1 is presented in figure 2 (a). By the same method, the linear regression equation and Ea according to Arrhenius can be obtained for R_7P_3 and R_5P_5 as shown in Figs. 2B and 2C. Similarly, the linear regression curve for RH and PB can be obtained by the same method.

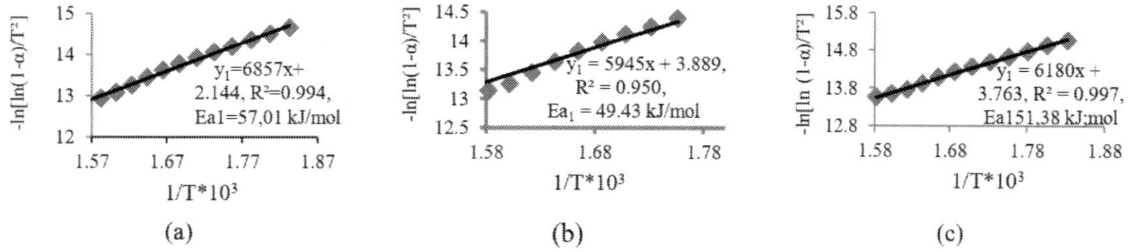

Figure 2. The linier regression curve of (a) R_9P_1, (b) R_7P_3, and (c) R_5P_5 for pyrolysis.

Based on the analysis of kinetic studies, the determination of Ea for each TD was conducted based on the linear regression equation. The value of R^2 shows the relationship between variables x and y. The linear regression equation can be used to determine the value of Ea when $R^2 > 0.9$. The results of PB pyrolysis gave Ea of about 255.5 kJ/mol, which is relatively large if compared with the Ea of RH, which is only 52 kJ/mol. The PB addition of approximately 10%, 30% and 50% into RH improve the Ea of R_9P_1, R_7P_3, and R_5P_5. The linear regression equation, Ea value in pyrolysis process for RH, R_9P_1, R_7P_3, and R_5P_5, PB is presented in table 3.

Table 3. Linear regression of Arrhenius function and activation energy for pyrolysis.

Fuels	TD (°C)	A	Linier regression	(R^2)	Ea (kJ/mol)			
					Ea_1	Ea_2	Ea_3	Total
RH	250 – 380	0.11-0.79	$y_1 = 6934x + 1.763$	0.99	57.7	-	-	57.7
R_9P_1	250 – 390	0.12-0.65	$y_1 = 6857x + 2.144$	0.99	57.1	-	-	-
	420 – 490	0.72-0.85	$y_2 = 1889x + 10.22$	0.91	-	15.7	-	-
	580 - 680	0.80-0.97	$y_3 = 1908x + 10.62$	0.96	-	-	15.9	88.6
R_7P_3	260 – 390	0.14-0.60	$y_1 = 5945x + 3.889$	0.95	49.4	-	-	-
	420 – 510	0.64-0.92	$y_2 = 5194x + 5.728$	0.92	-	43.2	-	-
	600 - 660	0.95-0.98	$y_3 = 1870x + 10.35$	0.93	-	-	15.6	108.2
R_5P_5	260 – 370	0.08-0.4	$y_1 = 6180x + 3.763$	0.99	51.4	-	-	-
	400 – 500	0.50-0.83	$y_2 = 7127x + 3.293$	0.91	-	59.3	-	-
	600 – 670	0.93-0.98	$y_3 = 4695x + 7.394$	0.93	-	-	39.0	149.7
PB	380 – 500	0.01-0.87	$y_1 = 25430x - 20.44$	0.99	211.4	-	-	-
	620 – 680	0.92-0.97	$y_2 = 5296x + 6.935$	0.97	-	44.1	-	255.5

Based on data analysis, the pyrolysis process showed greater content of plastic in the sample, produces a large and increasing the Ea and the residues were getting smaller. This is due to plastic has a CV and VM are large with a small ash residue.

4. Conclusion

Pyrolysis characteristics of RH with PB shows an increase in activation energy, calorific value, volatile matter, and carbon content. On the contrary, ash, nitrogen, sulfur, and moisture show a decrease. The addition of PB into RH increases the activation energy and calorific value of R_9P_1, R_7P_3, and R_5P_5, and reduced char residues, significantly. Pyrolysis results of RH and PB showed improve the thermal characteristics of the RH for improving the performance, resulting in large energy release and lowers the ash residue. Pyrolysis results of RH and PB show feasibility in the implementation as an alternative energy source as fuel in power generation.

5. Acknowledgement
The autors would like also to thank the Ministry of Higher Education of Republic of Indonesia, which has provided funding for research and development in Indonesia.

6. References

[1] Directorate of Energy and Minerals Resources 2012 Alignment of National Energy Policy (KEN) with the General Plan National Energy (RUEN) *Policy paper*

[2] Anshar M, Kader A S and Ani F N 2014 The utilization potential of rice husk as an alternative energy source for power plants in Indonesia *Advanced Materials Research, Trans Tech Publications, Switzerland* **845** 494-498

[3] Anshar M, Ani F N and Kader A S 2017 Co-Combustion Modeling Of Rice Husk And Plastic Bag As Energy Source In Indonesia *Asian Research Publishing Network (ARPN) J. of Eng. and Applied Sci.* **12**(10) 3105-11

[4] Rozainee M, et al. 2008 Fluidized bed combustion of rice husk to produce amorphous siliceous ash *Energy for Sustainable Development* **12**(1) 33-42

[5] Anshar M, Ani F N and Kader A S 2015 The potential energy of plastic solid waste as alternative fuel for power plants in Indonesia *Applied Mech. and Materials, Trans Tech Publications, Switzerland* **699** 595-600

[6] Chattopadhyay J, et al. 2008 Thermogravimetric characteristics and kinetic study of biomass co-pyrolysis with plastics *Korean J. Chem. Eng.* **25**(5) 1047-53

[7] Yuan S, et al. 2012 Rapid co-pyrolysis of rice straw and a bituminous coal in a high-frequency furnace and gasification of the residual char *Bioresource Technol.* **109** 188–197

[8] Xie Z and Ma X 2013 The thermal behaviour of the co-combustion between paper sludge and rice straw *Bioresource Technol.* **146** 611–618

[9] Anshar M, Ani F N and Kader A S 2017 Co-pyrolysis of rice straw with high density polyethylene bag *Chiang Mai J. Sci.* **44**(3) 9877-79

[10] Chen D, Zheng Y and Zhu X 2013 In-depth investigation on the pyrolysis kinetics of raw biomass. Part I: Kinetic analysis for the drying and devolatilization stages *Bioresource Technol.* **131** 40–46

[11] Maiti S, et al. 2006 Physical and thermochemical characterization of rice husk char as a potential biomass energy source *Bioresource Technol.* **97** 2065–70

[12] Sait H H, et al. 2012 Pyrolysis and combustion kinetics of date palm biomass using thermogravimetric analysis *Bioresource Technol.* **118** 382–389

[13] Said M M, John G R and Mhilu C F 2014 Thermal characteristics and kinetics of rice husk for pyrolysis process *Int. J. of Renewable Energy Research* **4**(2)

[14] Kumar S and Singh R K 2014 Pyrolysis kinetics of waste high-density polyethylene using thermogravimetric analysis *Int. J. of Chem Tech Research* **6(1)** 131-137

[15] Chattopadhyay J, et al. 2008 Thermogravimetric characteristics and kinetic study of biomass co-pyrolysis with plastics *Korean J. Chem. Eng* **25**(5) 1047-53

[16] Yokoyama S 2008 The Asian Biomass Handbook, *A Guide for Biomass Production and Utilization* (Japan: The Japan Institute of Energy)

[17] Zhou H, et al. 2013 The pyrolysis simulation of five biomass species by hemi-cellulose, cellulose and lignin based on thermogravimetric curves *Thermochimica Acta* **566** 36-43

IOP Conf. Series: Earth and Environmental Science **105** (2017) 012035 doi:10.1088/1755-1315/105/1/012035

The potential utilization of coconut oil and palm oil as raw material of alkanolamide under alkaline conditions

F Sari[1], B H Susanto[1] and S Bismo[1,2]

[1] Chemical Engineering Department, Universitas Indonesia, Kampus Baru UI Depok, 16426, Indonesia

E-mail: setijo.bismo@ui.ac.id

Abstract. Indonesia has a high production of vegetable oils. Coconut oil and palm oil are vegetable oils that can be source of raw material in Industry. The purpose of this research is potential use of coconut oil and palm oil as raw materials of alkanolamide in alkaline condition. Alkanolamide is produced from the reaction between triglycerides and diethanolamine. Phosporic acid is used as a solvent for the purification of alkanolamide from its mixture with glycerin. So it is expected that alkanolamide produced has a high purity and the resulted alkanolamides have dispersant and detergency properties that are widely used in various industries. The amidation reaction between triglyceride and diethanolamine was mixing in a glass batch reactor for 2 hours using potassium hydroxide catalyst (KOH) 1% (w/w). The temperature reaction was 125 °C. The second step is purification of alkanolamide. The operation condition is preferably performed at 82 °C. The synthesis of palm oil in this study is constrained because it is easily oxidized and the viscosity was higher than coconut oil. So this research is focused on using coconut oil as raw material. From the purification analysis, resulted 85% of alkanolamide (top layer) and 15% glycerine (bottom layer). From FTIR analysis, The amide group (C=O) is formed at a wavenumber of 1618,34 cm^{-1} for alkanolamide from coconut oil and 1617,45 cm^{-1} for alkanolamide from palm oil. GC-MS analysis resulted the compound of diethanolamide laurate with the greatest quality percentage of 93% and the largest percentage of 31, 20%.

1. Introduction

Coconut oil and palm oil are potential raw material to be developed in the industry. Coconut oil and palm oil are easily available and the process is easy and fast. Coconut oil and palm oil can be used to replace oil because they are environmentally friendly and biodegradable. Coconut oil and palm oil can be used as raw materials for alkanolamide. Alkanolamides can be made using a variety of reactants. Alkanolamide is produced from the reaction between the reactants is that one reactant is a fatty acid, methyl ester or triglyceride and the other is an alkanolamine, most commonly diethanolamine [1]. The commonly used raw material is methyl ester. Methyl ester will become a serious consideration if it is used because methyl ester produces abundant foam [2]. In the previous research, acid catalyst was used. Acid catalyst has shortcoming because the rate of reaction is low and it needs high temperature reaction and it is corrosive. In this research, the catalyst used is alkaline catalyst potassium hydroxide

[2] To whom any correspondence should be addressed

(KOH). It was because the alkaline catalyst has the advantage of being cheap and not corrosive. However, the weakness of this catalyst is that it needs relatively difficult separation process from the product. In the purification process, phosporic acid is used as a solvent to separate alkanolamide from glycerin so the production of alkanolamide will have high purity. The resulted alkanol amides have dispersant and detergency properties that are widely used in various industries. In its use as a fuel additive, it attracts impurities derived from deposits that form on the intake system or combustion chamber to reduce friction on the engine and can save fuel consumption [3]. Dispersant and detergents in fuel additives are classified as deposit control agent (DCA). Alkanolamide is produced from the amidation reaction. Based on the reaction alkanolamide is produced from the reaction of alkanolamine and the source of fat/oil [1]. The alkanolamide reaction is as follows:

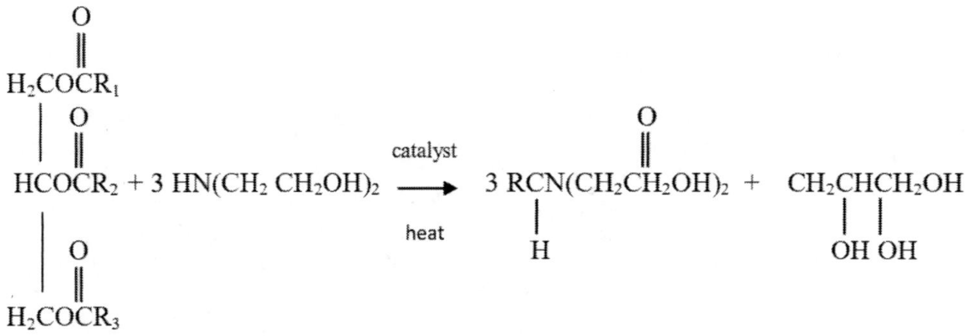

Figure 1. Reaction of alkanolamide.

2. Methodology/experimental

2.1. Material and equipments

The following chemicals were commercially available and used as received Coconut oil, palm oil as source of triglyceride, Diethanolamine as source of amine, KOH (potassium hydroxide) as agent catalyst, Phosphoric acid as purification of alkanolamide. While the equipment are a glass batch reactor, a stirring hot plate, and thermometer. All chemicals used are pro analysis grade.

2.2. Procedure

The method of making alkanolamide has two steps, the first is the reaction of diethanolamine and triglyceride and second is the purification of the alkanolamide from its mixture of with glycerin.

2.2.1 Synthesis of alkanolamide. The amidation, the reaction between triglyceride and diethanolamine, was held in a beaker glass reactor on the top of a stirring hot plate for 2 hours with mole ratio of 1: 1.5 using alkaline catalyst (KOH) 1 % (w/w). The temperature reaction at 125 °C. Under these conditions, the mixture becomes transparent (homogeneous) as a liquid phase. Temperature during mixing is maintained. The reaction is continued until the total reaction time corresponds to the desired variation until the solution is dark yellow. After the reaction time is completed, one-phase solution will be formed, and temperature decreased until 82 °C (should be below 100 °C to prevent condensation).

2.2.2 Purification of alkanolamide. One of the disadvantage of using an alkaline catalyst is the difficulty of separating the product from the reaction. A method for alkanolamide purification is required. H_3PO_4 is used as a solvent for the purification of alkanolamide from its mixture of with glycerin. Phosphoric acid is used to remove the free amine from glycerin. Phosphoric acid used 6-9

IOP Conf. Series: Earth and Environmental Science **105** (2017) 012035 doi:10.1088/1755-1315/105/1/012035

parts by weight of phosphoric acid to 100 parts by weight of crude amide. The operation is preferably performed at 82 °C. The layers to be formed consist of the top layer (organic) which is the amide compound and the bottom layer (inorganic) which is glycerin compound. There should be no fatty compound in this lower (inorganic) layer. Analysis of top layer results (amide compounds) is showed below.

2.2.3 Analysis of samples. The analysis include Fourier Transfer Infra-Red (FTIR) (Thermo scientific Nicolet iS5 with iD5 ATR) to know the structure of alkanolamide, Gas Chromatograpy-Mass Spectrometry (GCMS) (GC 6890 MS 5973 Agilent) that use helium carier gas to determine the moleculer weight of alkanolamide.

3. Results and disscussion

3.1. Analysis of purification
In alkanolamide synthesis, the reaction between triglyceride and diethanolamine by using an alkaline catalyst will form one phase, it is necessary to purify to separate the alkanolamide from the byproducts of glycerin. In this case the amide purification is carried out by a simple method by adding a solvent to the amide solution formed. Then used compounds containing phosporic acid to bind alkanolamides formed. Then the compound used is phosphoric acid, its group contains a proton that can dissociate, leaving a residual molecule with a negative charge. In this case the transfer of grup occurs is phosphorylation. Alkanolamides formed in one phase when added phosphoric acid compounds will form two layers of phosphorylated glycerin and alkanolamine phosphate.Phosphoric acid used 6-9 parts by weight of phosphoric acid to 100 parts by weight of crude amide. The operation condition is preferably performed at 82 °C [2]. The layers to be formed consist of the top layer (organic) which is the amide compound and the bottom layer (inorganic) which is glycerin compound. The phosphate will bind *mono, di,* triglycerides and will be bonded to the bottom layer, or precipitation forming phosphoglycerides. The resulting end product consists of substantially 85% of alkanolamide (top layer) and 15% glycerine (bottom layer) . The results of this top layer will be analyzed using FTIR and GC-MS.

3.2. FTIR Analysis
The characteristic of alkanolamide synthesis with FTIR was used to identify the groups contained from the result of alkanolamide synthesis. Alkanolamide formed can be proven by comparing the results of analysis with raw materials. The FTIR results can be seen in figure 2.

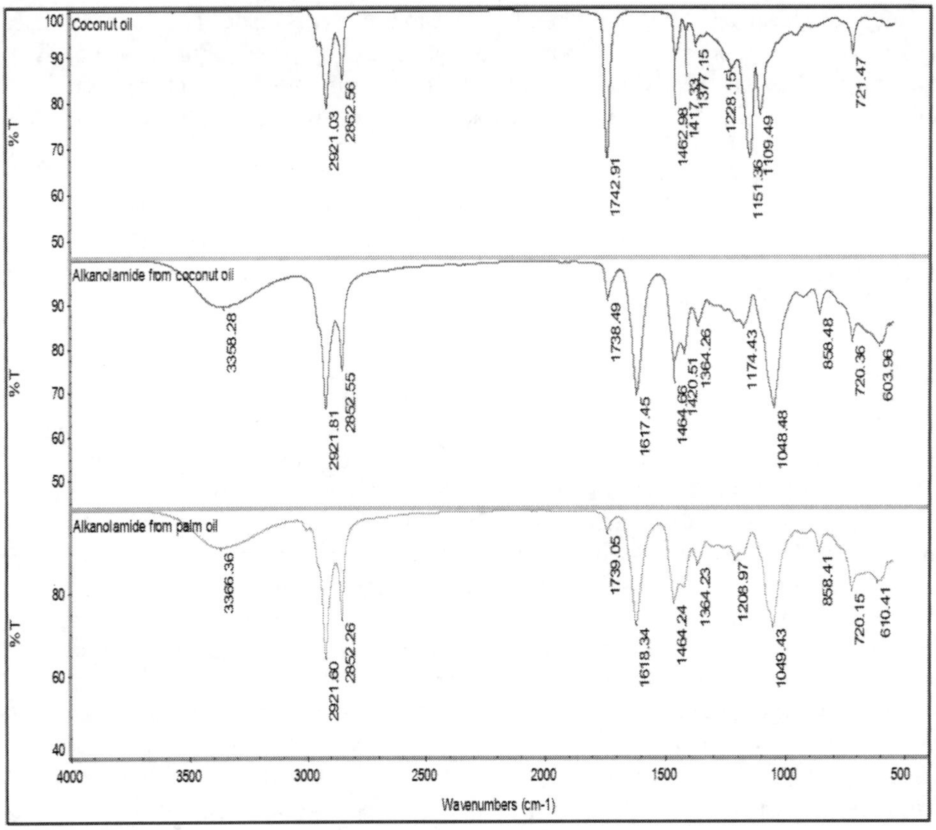

Figure 2. Comparison of IR spectra of alkanolamide and coconut oil.

In this case the results of FTIR alkanolamide analysis will be compared with coconut oil. Figure 2 shows that the alkanolamide has been formed. It can be seen that there is a difference in spectrum results among coconut oil, coconut oil alkanolamide and palm oil. From literatures, alkanolamides were formed by the existence of strong absorption at C=O at 1630–1660 cm^{-1}. If there is hydrogen bonded with nitrogen, there will be formed N-H absorptions at around 3300 cm^{-1} [4]. The difference in spectrum results among coconut oil, alkanolamide from coconut oil and alkanolamide from palm oil spectrum. The differences in the results of the spectrum between coconut oil and alkanol amide with coconut oil and palm oil are characterized by the formation of wave numbers of 3358.28 cm^{-1} for alkanolamide from coconut oil and 3366.36 cm-^{1} for alkanolamide from palm oil for which the absorption of O-H groups are widened and strong, and the N-H group also appears in this area and near with the O-H group so that it is not apparent on the spectrum. The C-H aliphatic group may form at wavenumbers of 2921.81 cm^{-1} for alkanolamide from coconut oil and 2921.60 cm^{-1} for alkanolamide from palm oil. The C-H aliphatic group supports that fatty acids in palm oil and coconut oil are long-chain carboxylic acids. The amide group C=O is formed at a wavenumber of 1618.34 cm^{-1} for alkanolamide from coconut oil and 1617.45 for alkanolamide from palm oil. While, from the FT-IR spectrum of coconut oil triglycerides, it can be seen that the wavenumber of 2921.03 cm^{-1} shows the stereched C-H bond. The typical absorption from triglycerides also appears at wavenumber of 1742.91 cm^{-1} which is an absorption of C=O ester. From two samples of alkanolamide can be seen results that not much different, the difference of intensity only. This is because the constituent groups of each of the same alkanolamides are the OH groups, the NH groups,the C-H aliphatic groups and the C=O amide groups. In this case it can be seen in table 1.

Table 1. Wavenumber of IR spectra of alkanolamide and coconut oil

Sample	Wavenumber (cm^{-1})		
	N-H	C-H	C=O
Coconut oil	-	2921.03	1742.91
Alkanolamide from coconut oil	3358.28	2921.81	1618.34
Alkanolamide from palm oil	3366.36	2921.60	1617.45

3.3. GC-MS Analysis

The GC-MS analysis was performed to determine the alkanolamide composition resulting from the amidation reaction between oil and diethanolamine. The result of alkanolamide synthesis from palm oil was not analyzed using GC-MS because the viscosity of alkanolamide from palm oil was high, which performed from the fatty acid structure of the palm oil composite that having a fairly high carbon chain of palmitic acid with the C_{16} group. The results of GC-MS alkanolamide analysis from coconut oil can be seen in figure 3 and table 2.

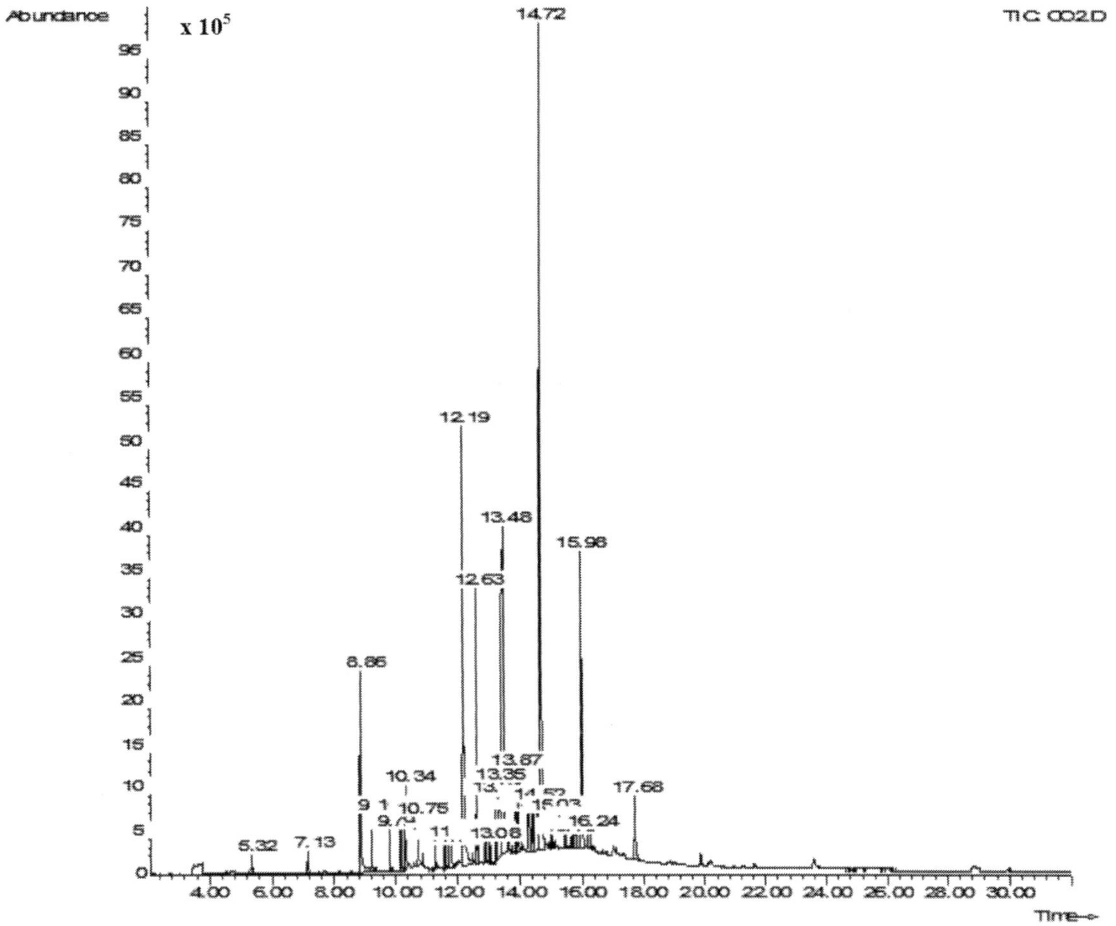

Figure 3. GC-MS spectra analysis of alkanolamide synthesis from coconut oil.

Table 2. GC-MS analysis alkanolamide from coconut oil.

Time (min)	Area (%)	Library	Qual/Similarity(%)
8.86	6.80	Dedocanoic acid	99
10.34	1.56	Tetradecanoic acid	99
14.72	31.20	Dedocanamide	93

The most commonly peak analysis result was the dedocanamide or diethanolamide laurate compound with the greatest quality percentage of 93% and the largest percentage of 31.20% which having molecular weight of 287.44 gr/mol with molecular formula $C_{16}H_{33}NO_3$. Diethanolamide laurate formed from derived fatty acid of coconut oil, lauric acid (C_{12}). Diethanolamide laurate is a detergent in bioadditive gasoline because the amine content contained in alkanolamide will attract impurities derived from the deposit formed in the intake system or combustion chamber. Dispersant and detergent in fuel additives are classified into a deposit control agent (DCA). The amide group is used as deposit control agent (DCA), based on the polar head formed from the alkanolamide compound. Where the hydrocarbons formed are hydrophobic and the amide compounds are polar and hydrophilic.The action of the DCA is that the polar part (amine) attracts to the solid particles (such as soot in combustion, wear and dust products) and forms a protective layer on the surface of the deposit precursor thus preventing the deposit from being difficult to agglomerate and difficult to accumulate on the intake valve, injector and combustion chamber. It also acts as a nonionic surfactant as a foam stabilizer, emulsifying, coagulant material [5]. The lauric acid content of the diethanolamide has a function as a lubricant, in which the amount of carbon 8 to 24 is antiwear, this lubricant composition provides strong adhesion and oil wetting on the metal part, thereby reducing wear and tear, improving lubrication and fast pit drilling [6].

4. Conclusion

As mention before about the characteristic of coconut oil and palm oil, this research is focused on using coconut oil as raw material. From the analysis purification, resulted 85% of alkanolamide (top layer) and 15% glycerine (bottom layer). The amide group C=O is formed at a wavenumber of 1618.34 cm^{-1} for alkanolamide from coconut oil and 1617.45 cm^{-1} for alkanolamide from palm oil, GC-MS analysis is the compound diethanolamide laurate with percent area of 32.10%. Diethanolamide laurate fuction as detergent in bioadditive gasoline, non-ionic surfactant surfactant as a foam stabilizer, emulsifying, coagulant material. The lauric acid content of the diethanolamide has a function as a lubricant.

5. Acknowledgments

The authors acknowledge Universitas Indonesia and PITTA UI-2017 program for the financial support. We are also thankful to Laboratorium Intensifikasi Proses, Departemen Teknik Kimia, Fakultas Teknik Universitas Indonesia; Pusat Laboratorium Forensik (Puslabfor), Jakarta for the instruments to gas chromatography analysis

6. References

[1] Anthony J. O'Lenick, Jr 2014 *Surfactants: Strategic Personal Care Ingredients* (USA: Allured Publishing Corporation)
[2] Lohr, J W, Cincinnati, Ohio 1968 *US Patent* **3,040,075** 1-4
[3] Colucci, William J, Kadkhodayan, Abbas 2013 *US Patent* **8,444,720 B2** 1-2
[4] Wade L G Jr 2010 *Organic Chemistry Seventh Edition* (Boston : Prentice hall)
[5] Manurung, Renita, Sinaga, Rahmat Akbar, SimaTupang and Rahmat Taufik. 2013 *Int. J. of Innovative Research in Sci. Eng. and Technol* **2** 4205-10
[6] Gatlin and Larry W 2001 *US Patent* **6,194,361 B1** 1-3

Utilization of waste cooking oil as raw material for synthesis of Methyl Ester Sulfonates (MES) surfactant

Resi Levi Permadani[1], Muhammad Ibadurrohman[1] and Slamet[1,2]

[1] Department of Chemical Engineering, Faculty of Engineering, Universitas Indonesia, Depok 16424, Indonesia

E-mail: slamet@che.ui.ac.id; slamet.dtk.ui@gmail.com

Abstract. In this study, MES surfactant was synthesized from waste cooking oil (WCO) which can lower the synthesis cost and represent efficient utilization of waste.MESis an eco-friendly anionic surfactant, which can be used for our detergent application. The purpose of this study was to obtain optimum MES surfactant from purification of WCO. WCO is purified first to reduce high impurities and free fatty acid (FFA). Purification steps of WCO consist of filtration to separate food residues, neutralization with various concentrations of NaOH solution13%;14%;15%;16% and bleaching with activated carbon 7.5% (wt. % of oil). After purification followed by trans-esterification process with variation of mole ratio oil and methanol 1:8; 1:9; 1:10 then synthesized MES surfactant with sodium bisulfite ($NaHSO_3$) as sulfonating agent. Neutralization results showed high reduction percentage of FFA was 31.31% for 15% NaOH solution; bleaching results showed reduction percentage of FFA was 20.27% and WCO color was originally dark brown to be light yellow. The results of trans-esterification showed highest yields 94.15% for mole ratio of oil and methanol 1:9 and sulfonation results will be described further in this paper.

1. Introduction

Waste cooking oils (WCO) are generated from cooking and frying food with edible vegetable oils, which is usually disposed into the environment [1, 2]. In Indonesia, most of waste cooking oil (WCO) from households is being disposed to drainage and soil causing environmental pollution [3]. WCO can not be consumed again due to presence of toxic components which can damage cytochrome enzymes and other enzymes in human body and become carcinogenic caused by aflatoxin, due to high oil oxidation [1]. The tremendous growth in human population and increase in food consumption contributed to the production of huge amounts WCO[2].Nowadays, WCO have been widely used as raw material for methyl ester (biodiesel) production [4-6], which represent an economical and environmentally friendly disposal method. The cost of WCO is two to three times cheaper than vegetable oils, therefore can reduce the production cost [2, 7]. Another alternative of WCO utilization is used for MES (methyl ester sulfonate) production.MES is an eco-friendly oleo chemical based anionic surfactant, can be easily synthesized using bio-oil feedstock and can be used for detergent application [8].The advantages of MES are higher detergency at lower doses, stable in hard water, good biodegradability, harmless oral toxicity (2.2-3.8 g/kg weight), low toxicity on animal (low toxicity range) and good skin compatibility [8-10].

[2] To whom any correspondence should be addressed.

Raw material such as palm oil still widely used to MES synthesis, but nowadays the use of vegetable oils have to considered seriously because they compete with food production [11]. For sulfonating agent such as sulfur trioxide (SO_3) still widely used in industries to MES production [12, 13]. The use of SO_3 produced MES with dark color and require special equipment and instrumentation that allows tight control and rapid removal of the heat of reaction [14]. WCO can be consideration for developing a new low cost oil source in MES production and sodium bisulfite ($NaHSO_3$) can be alternative to replace SO_3 as sulfonating agent because it produced MES with light color so blanching process is not necessary. In few studies, WCO and $NaHSO_3$ have been used for synthesis of MES [15, 16], but the majority of these works are focused on conditions of sulfonation process such as temperature, times, mole ratios and speed of agitation with different sulfonating agents. On the other hand, pH optimization on the neutralization process has not be done, even though pH will affect the disalt content that will impact on MES quality. In the MES product, there is two kinds of active matters: MES ($RCH(CO_2Me)SO_3Na$) and disalt ($RCH(CO_2Na)SO_3Na$) which is obtained from neutralization of fatty acid methyl ester sulfonic acid (MESA) with NaOH[17]. Disalt content can lower surface activity of MES and lower resistance to hard water [8], to avoid hydrolysis of MES to disalt, it is important to avoid extremes pH in neutralization [12].

In the present study, variation of pH in neutralization step of MES was studied to optimize the quality ofMES. WCO and sodium bisulfite ($NaHSO_3$) used for synthesis of MES.WCO was purified first to remove impurities from food residue, reduce FFA content and get oil with lighter color. Therefore synthesis of methyl ester is not required esterification process. Then followed by trans-esterification and sulfonation process. After synthesis process, the chemical composition of MES and the chemical properties were determined.

2. Experimental section

2.1. Materials

The type of WCO was palm oil collected from food stalls in Kukusan, Depok, West Java, Indonesia. Sodium hydroxide (Merck, P.A), activated carbon were provided by local supplier (East Jakarta, Indonesia) with material base from coconut shell, particle size 200 mesh and iodine number of 802 mg/g, potassium hydroxide (Merck, P.A), methanol (Merck, P.A), ethanol (Merck, 99.9%), phenolphtalein (Merck, pure), hydrochloric acid (Merck, 37%), sodium bisulfite (Smartlab, P.A) and aquadest (Wiloso) was used for all experiments.

2.2. Purification of WCO

WCO was purified by filtration, neutralization and adsorption (bleaching) process. Filtration used to remove the food residues, neutralization with NaOH solution 13%, 14%, 15% and 16% at temperature 70°C for 15 min. Adsorption was done by activated carbon 7.5% (wt. % of oil) at 80-110°C for 1 h, then filtrated to separate activated carbon from oil.

2.3. Synthesis of MES

2.3.1. Preparation of Methyl Ester. If FFA (free fatty acid) in oils below 2.5%, it can be done immediately with trans-esterification [4]. Trans-esterification process was carried out in 500 ml two neck flask which equipped with reflux condenser, magnetic stirrer and hot plate. Mole ratio of WCO and methanol was 1:8, 1:9 and 1:10 with catalyst KOH 1% (wt. % of oil). The flask was heated at temperature reaction 60°C for 1 h and stirred at constant speed. After reaction, the mixture was put into separating funnel to separate glycerol from methyl ester. Then methyl ester washed with hot water to separate glycerol and soap residue, to remove water was done by evaporation.

2.3.2. Sulfonation of Methyl Ester. Sulfonation was carried out at temperature 100°C for 4.5 h with mole ratio of methyl ester and $NaHSO_3$ was 1:1.5. After reaction, the mixture was centrifuged at 1500 rpm for 30 min to separate residual $NaHSO_3$. Then, followed by purification with methanol 30% (v/v MESA) at 50°C for 1.5 h, to remove methanol excess was done by evaporation. Neutralization process with NaOH solution 20% at 55°C for 30 min, with variation of pH 5, 7 and 8.

2.4. Analysis and characterization
The composition of WCO and methyl ester was analysed by a GC-MSequipped with 5975C mass selective detector and capillary column(HP Innowax, 30 m x 0.25 mm (I.D) x 0.25 μm). Helium was used as carrier gas and about 1 μL of the solution was injected in the column. The injection temperature was 260°C and the oven was started at a temperature of 40°C, which was increased to 195°C at rate 8°C/min held for 0 min, finally raised to 225°C at rate of 1°C/minand held for 22 min.

The chemical structure of MES was checked by FTIR and LC-MS. The FTIR analysis was performed in the range 4000-400 cm^{-1} using KBr pellets. LC-MS Waters ACQUITY equipped with an electro spray interface (ESI) and a column (UPLC BEH C18, 2.1 x 50 mm x 1.7 **μm**). Mass Spectrometry was controlled by Mass-Lynx software. FFA content of WCO was analysed [18] and chemical properties of MES was determined by surface tension using automatic surface tensiometerand solubility in water. Solubility of MES in water was determined by separation time between phases of MES and water. MES and water mixed with volume 1 mL and 6 mL respectively, the mixture then stirred at rate of 250 rpm for 5 min. Separation time between phases measured using stopwatch.

3. Results and discussion

3.1. Purification of WCO
Purification of WCO through three-step process i.e. filtration, neutralization and adsorption (bleaching). Suspended solids, phospholipids and other impurities in WCO can be removed by paper filtration [2], the result showed in figure 1.Originally the color of WCO was dark brown because it still contained lot of solids from food residues, after filtration the color to be light brown (see figure 1).

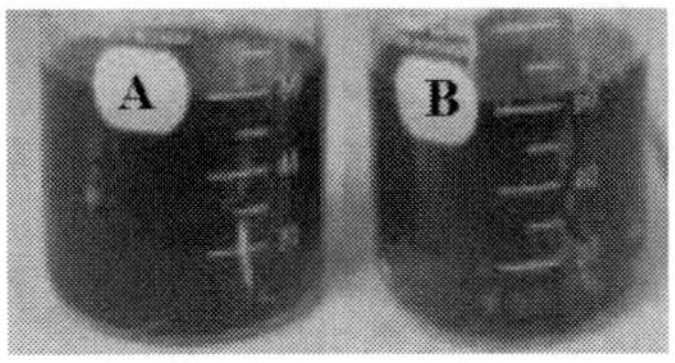

Figure 1. The color of WCO (A) dark brown before filtration and (B) light brown after filtration.

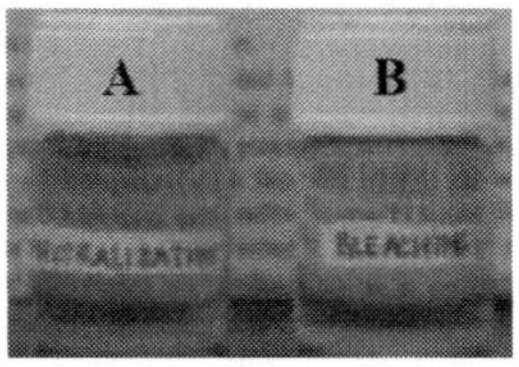

Figure 2. WCO color (A) after neutralization (orange) and (B) after purification (light yellow).

The second step was neutralization, to reduce high FFA content in oil. FFA reacted with alkalis forming soaps, then soaps separated by filtration. In this study, variation of NaOH concentrations were done to get high percentage of FFA reduction. The result in table 1 showed that NaOH 15% give the high percentage reduction 31.31%, therefore it used for the neutralization process.

Table 1. Percentage of FFA reduction at various concentrations of NaOH for neutralization.

NaOH(%)	FFA reduction (%)
13	16.22
14	16.99
15	31.31
16	14.88

Table 2. The dominant compositions of WCO.

Fatty acid	Composition (%)
Palmitic acid	43.34
Oleic acid	40.30
Linoleic acid	9.23
Stearic acid	4.69

The last process was adsorption with activated carbon to bleaching the color of WCO. Other than that, this process also can reduce FFA but percentage reduction lower than neutralization. The result showed WCO color was originally dark brown to be light yellow (see figure 2) and percentage of FFA reduction was 20.27%. After purification, FFA content in WCO was originally 2.94% reduced to be 1.58%. It showed that FFA in WCO below 2.5% so it can be directly followed by trans-esterification process. The composition of WCO were analyzed by GC-MS and reported in table 2. The result show that the main components from WCO are palmitic acid, oleic acid and linoleic acid with the highest percentage is palmitic acid 43.34%. Therefore WCO contain mostly unsaturated fatty acid.

3.2. Effect of mole ratio on Methyl Ester Yield

Methyl esters were prepared by trans-esterification process of WCO with methanol and KOH catalyst. Some of parameters which significantly affect on final conversion and yield of methyl ester are temperature, FFA in oil, reaction time and mole ratio of oil to alcohol [7].In this study, variation of mole ratios were done to get high yield. The result in table 3 showed that the high yield 94.15% obtained at mole ratio of oil and methanol 1:9.

Table 3. The methyl ester yields at various mole ratios of oil and methanol.

Mole ratio	Yield (%)
1:8	91.49
1:9	94.15
1:10	88.12

Table 4. The dominant compositions of methyl ester wereanalyzed by GC-MS at mole ratio of 1:9.

Compound	Composition (%)
Methyl-7-octadecenoate	46.70
Methyl palmitate	30.34
Methyl myristate	1.36

3.3. Synthesis of Methyl Ester and MES

Mole ratio of oil and methanol 1:9 was used for methyl ester synthesis because it gave the highest yield. The composition of methyl ester were analyzed by GC-MS and reported in table 4. The highest percentage of components in MES was methyl-7-octadecanoat with percentage of 46.70%. Methyl-7-octadecanoat was used as the basis for calculating mole of methyl ester due to the highest percentage.

The content of methyl esters obtained was dominated by atom of C_{18} so it is suitable for MES synthesis. MES C_{16}-C_{18}shows good surface activity and best detergency power. MES from oil with C_{16}-C_{18} atom commonly used for powder detergent and liquid detergent (liquid laundry detergent)[19, 20]. After trans-esterification, followed by sulfonation of methyl ester with $NaHSO_3$. Yield of methyl ester sulfonic acid (MESA) after sulfonation was obtained 77.20%. Then methanol was added to reduce substitution of methyl groups on the structure of MES [21, 22]. The disalt content would be too high, if methanol is not added [12]. Then MES was neutralized with NaOH solution until pH of MES was 7.

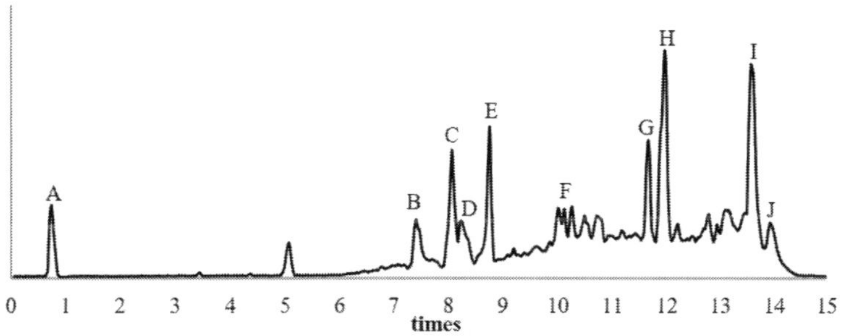

Figure 3. LC-MS chromatogram of MES at pH 7.

The LC-MS/MS of MES was analysed by ESI diluted in methanol revealed the presence of several difference peaks from methanol blank (see figure 3). Some of those peaks were performed by spectrum analysis of m/z, the spectrum revealed the fragments correspond to MES surfactant compounds at retention time 0.706 s (A), 7.44 s (B), 8.058 s (C), 8.172 s (D), 10.253 s (F), 11.945 s (H) and 13.546 s (I). Respectively, the m/z value at those retention time are 283.0587 m/z correspond to $C_{11}H_{15}O_5NaS$, 429.1631 m/z correspond to $C_{17}H_{32}O_8S_2$, 429.1639 m/z correspond to $C_{16}H_{28}O_{13}$, 441.2822 m/z correspond to $C_{22}H_{41}O_7Na$, 423.2708 m/z correspond to $C_{26}H_{39}ONaS$, 293.2118 m/z correspond to $C_{18}H_{28}O_3$ and 319.1947 m/z correspond to $C_{19}H_{26}O_4$. This indicates that MES has been successfully synthesized.

3.4. Effect of pH on MES Synthesis

In this study variation of pH at neutralization stepwere done to get MES with good quality. The infrared spectrum of MES in various pH showed in Figure 4. The strong band at 1741 cm^{-1} is typical for carbonyl group C=O. The peaks appearing at 2922 cm^{-1} and 2852 cm^{-1} are attributed to the symmetric and asymmetric $-CH_2$ stretching. The presence of sulfonate group in surfactant structure at 1195-1117 cm^{-1}. The peaks at 721-848 cm^{-1} is typical for S–O group. There is slightly difference of peak at pH 8 was located in the area 1589 cm^{-1}, it showed the presence of carboxylic ions in form of carboxylic salts, which supposedly derived from the disalts in MES with pH 8. The absorption bands of carboxylic ion are in area 1600-1560 cm^{-1} [23].FTIR spectra showed that the disalt has not much formed in MES at various pH.

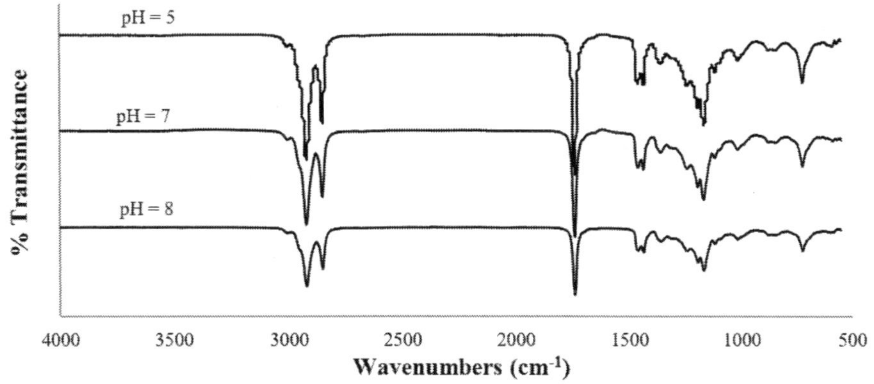

Figure 4. FTIR spectra of MES.

Capability of MES to decrease surface tension were measured by surface tensiometer, the result showed in table 5. The lowest surface tension of water obtained at addition MES with pH 7 was 32.4 mN/m. it showed that MES can lower surface tension so it can be used for detergent formulation. This surface tension value almost same with MES prepared from WCO with chlorosulfonic acid as sulfonating agent by Jin, et. al. which is 32.3 mN/m [8]. The surface tension of water will be reduced by adding MES to 30-40 dynes/cm[24].

Table 5. Surface tension of water by adding MES at various pH.

pH	Surface tension (mN/m)
5	35.8
7	32.4
8	34.6

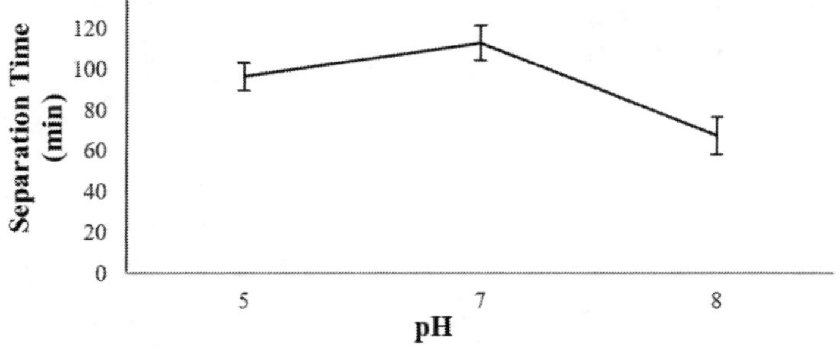

Figure 5. Separation time between MES and water at various pH.

The pH conditions of MES are very important to know because extreme pH values (highly alkaline) can lead to hydrolysis of MES which can form disalt and methanol. Disalt can reduce solubility of MES in water [25]. Therefore solubility of MES in water at various pH was studied and reported in figure 5. Solubility in water was expressed in terms of separation time between MES and water. The result showed that MES at pH 7 took longer time to separate from water. From both of those properties at pH 7 give the best result.

4. Conclusion

Methyl Ester Sulfonate (MES) has been successfully synthesized from WCO through purification, trans-esterification and sulfonation process followed by purification with methanol and neutralization step. Purification of WCO produced WCO with low FFA content and better colors, so esterification process is not necessary. Trans-esterification produced methyl ester with C_{16}-C_{18}which is suitable for MES synthesis. MES as final product are obtained with clear color without bleaching process. From the results obtained, it appears that the surface tension value is close to those in literature data and MES at pH 7 give the best result. WCO as raw material for MES production can be considered as an alternative method for waste oil utilization.

5. Acknowledgments

This study was financially supported by "Hibah Publikasi Internasional Terindeks untuk Tugas Akhir (PITTA)" 2017 from Universitas Indonesia.

6. References

[1] Vinyes E, Gasol C M, Oliver-Solà J, Ugaya C and Rieradevall J 2011 *Proceedings of the LCM 2011 Conference Berlin* p. 29-31
[2] Yaakob Z, Mohammad M, Alherbawi M, Alam Z and Sopian K 2013 *Renewable and Sustainable Energy Reviews* **18** 184-93
[3] Fujita H, Iijima W, Nakamo K, Prayitno J, Tsubaki H and Kitagawa G 2014 *Proceedings of the 3rd International Conf. on Environmental Science and Development*
[4] Ehsan M and Chowdhury M T H 2015 *Procedia Engineering* **105** 638-45
[5] Demirbas A 2009 *Energy Conversion and Management* **50** 923-7
[6] Raqeeb M A and Bhargavi R 2015 *J. of Chemical and Pharmaceutical Research* **7** 681
[7] Talebian-Kiakalaieh A, Amin N A S and Mazaheri H 2013 *Applied Energy* **104** 683-710
[8] Jin Y, Tian S, Guo J, Ren X, Li X and Gao S 2016 *J. of Surfactants and Detergents* **19** 467-75
[9] Martinez D, Orozco G, Rincon S and Gil I 2010 *Bioresource technology* **101** 8762-71
[10] Siwayanan P, Aziz R, Bakar N A, Ya H, Jokiman R, Chelliapan S 2014 *J. of Surfactants and Detergents* **17** 871-80
[11] Gashaw A and Teshita A 2014 *International J. of Renewable and Sustainable Energy* **3** 92-8

[12] Roberts D W, Giusti L and Forcella A 2008 *Biorenewable Resources* **5** 8
[13] Sheats W B and MacArthur B W 2005 *The Chemithon Corporation* **19**
[14] Foster N C 1996 *Sulfonation And Sulfation Processes Soap and Detergents: A Theoretical and Practical Review* Spitz, L. ed (Champaign, Illinois: AOCS Press)
[15] Sari A I, Widodo W G and Kusmartono B 2016 *J. Inovasi Proses* **1**
[16] Hidayati S, Gultom N and Eni H 2012 *REAKTOR* **14** 165-72
[17] Xie T, Zeng C, Wang C and Zhang L 2013 *Industrial & Engineering Chemistry Research*
[18] IUPAC C P 1979 *Standard Methods For The Analysis Of Oils, Fats And Derivatives* (Oxford Pergamon Press)
[19] Hui Y 1996 *Bailey's Industrial Oil and Fat Products. V. 2 Edible Oil and Fat Products Oils and Oil Seeds* (USA: John Wiley and Sons)
[20] Watkins C 2001 *INFORM-CHAMPAIGN* **12** 1152-60
[21] Hovda K 1997 *The Challenge of Methylester Sulfonation* (Seattle, WA: Chemithon)
[22] Hariani P L, Riyanti F and Fadilah A 2016 *Indonesian J. of Fundamental and Applied Chemistry* **1** 14
[23] Silverstein R M, Webster F X, Kiemle D J and Bryce D L 2005 *Spectrometric identification of organic compounds* Seventh ed (USA: John Wiley & Sons)
[24] Shaw D J 1992 *Introduction to colloid and surface chemistry* Fourth ed (Great Britain: Butterworth-Heinemann)
[25] Widyaningsih W, Budiasih I B I, Kurniawan W K W, Chasani M C M and Nursalim V N V 2014 *Molekul* **9** 63-72.

IOP Conf. Series: Earth and Environmental Science **105** (2017) 012037 doi:10.1088/1755-1315/105/1/012037

Effects of reduction time on carbothermic reduction of lateritic nickel ore using palm kernel shell as green reducing agent

Dwipuji Rahayu[1], **Ahmad Maksum**[1, 2] **and Johny Wahyuadi Soedarsono**[1,3]

[1] Center of Minerals Processing and Corrosion Research, Department of Metallurgical and Materials Engineering, Universitas Indonesia, Depok 16424, Indonesia
[2] Department of Mechanical Engineering, Politeknik Negeri Jakarta, Depok 16425, Indonesia

Email : jwsono@metal.ui.ac.id

Abstract. Lateritic nickel ore is one of the biggest mineral resources in Indonesia. This ore has a huge potential, i.e., economic value for processing and purification, however requires high energy in the separating of mineral and gangue minerals, in addition the production cost is also high. Therefore, to resolve the problems, a pre-reduction stage called carbothermic reduction process is carried out. Carbothermic reduction process is usually used for saprolite-type lateritic nickel ore which needs a reductor for the reduction reaction to produce pure nickel. Common reductors used are coal and cokes. In this study, a development on carbothermic reduction of saprolite, type of lateritic nickel ore using biomass reductor palm kernel shell, is proposed. The lateritic nickel ore used is obtained from East Halmahera and the palm kernel shells are obtained from the waste of palm oil plantation at Palangkaraya, Central Kalimantan. The purpose of this study is to find out the effect of reduction time variation on carbothermic reduction result of lateritic nickel ore with constant temperature and mass ratio value. Reduction time variation used in this study was 1, 2, 3, and 4 hours. All samples were tested at 800°C with the mass ratio of 1:4 (lateritic nickel ore : palm kernel shell) which were put into a crucible and then the carbothermic reduction process was done in a melting furnace. Peak formed on the XRD pattern showed that the process was able to reduce hematite or magnetite to wustite within one hour. XRF and recovery calculation results showed that the reduction time of one hour is the optimum time because nickel oxide (NiO) content was the highest i.e. 2.68% compared to other time variation.

1. Introduction

Nickel lateritic ore deposits are commonly found in tropical regions such as Indonesia, particularly in Pomalaa (Southeast Sulawesi) and Halmahera (North Maluku). Lateritic nickel is divided into two types of ore i.e., saprolite and limonite. The major mineral constituents of the limonite layer are geothite [FeO(OH)] and hematite (Fe_2O_3), while the saprolite layer is composed by serpentine minerals [(Ni,Mg)SiO_3.nH_2O] [1]. Most of the lateritic nickel ore deposits in Indonesia have not been optimally processed due to several difficulties in the processing of lateritic nickel ore. One of the problems is the high energy usage in the mineral separation process or gangue minerals at recovery

[3] To whom any correspondence should be addressed.

process [2]. Nowadays, pyrometallurgy process is still used for saprolite nickel ore because saprolite is generally more heterogeneous in terms of its chemical composition and mineralogy. However, this process requires high amount energy and produce air pollution due to the use of coal and cokes as reducing agent.

Considering the environmental problems generated from the pyrometallurgical process, a new technology innovation has been developed to reduce nickel ore by using biomass energy or waste from natural resources. Utilization of waste has begun developing further as a major alternative to replace coal as the reductor of nickel ore [3–5]. Biomass can be converted to fuel as reductor through pyrolysis and gasification processes. The process of pyrolysis and gasification will produce reducing gases such as carbon monoxide (CO), hydrogen (H_2), carbon dioxide (CO_2), methane (CH_4), and other hydrocarbons gas. Gas produced in the form of CO and H_2 served as the reducing agents [6]. Palm kernel shell is studied to be a reductor of low grade iron ore because its C, H, O element contents; low ash content; high calorific value and high amount of fixed carbon [7]. In addition, this research will study the effect of reduction time variation on carbothermic reduction of saprolite lateritic nickel ore using palm kernel shell as a reductor.

2. Materials and method

2.1. Materials

The lateritic nickel ore used in this study is initially characterized by XRD testing. Figure 1 presents the results of XRD testing of lateritic nickel ore.

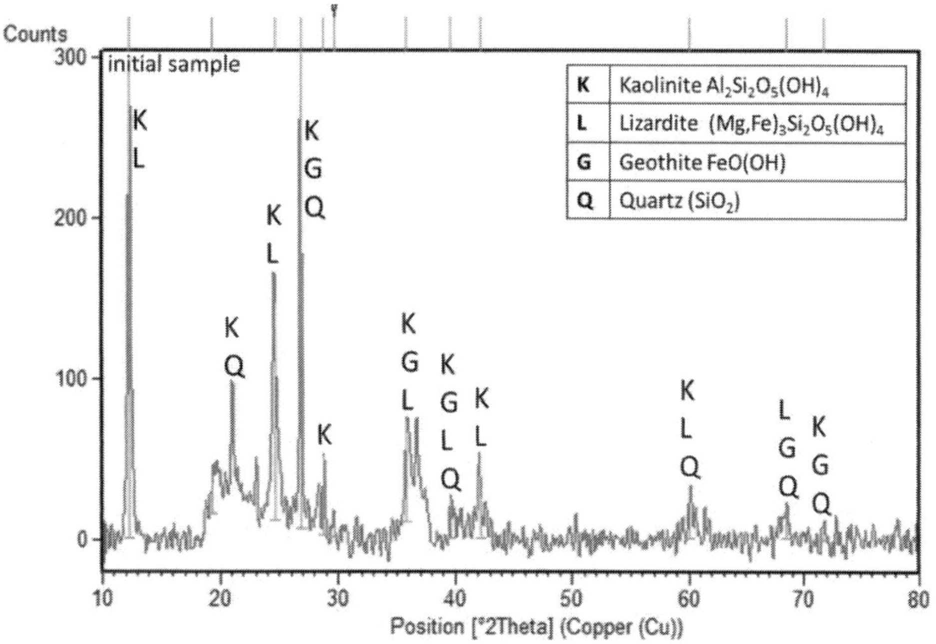

Figure 1. XRD pattern of initial lateritic nickel ore.

Proximate and ultimate analysis tests, as shown in table 1, have been done on the palm kernel shell is used as the reductor in this research.

2.2. Experiment

The lateritic nickel ore sample which has been prepared to form fine powder (#270) and mixed with palm kernel shell was inserted into crucible in a melting furnace. The process of lateritic nickel ore

IOP Conf. Series: Earth and Environmental Science **105** (2017) 012037 doi:10.1088/1755-1315/105/1/012037

reduction was done in variation of increasing reduction time, respectively 1, 2, 3, and 4 hours. When the carbothermic reduction occured, only the reduction time variables were changed, while the temperature and mass ratio is set at a fixed value; 800°C with a mass ratio of 1 : 4 (lateritic nickel ore : palm kernel shell). Carbothermic reduction process took place in the melting furnace.

Table 1. Proximate and Ultimate Analysis of palm kernel shells.

Parameter	Content (% adb)
Proximate Analysis	
Moisture in air dried sample	3.70
Ash	2.09
Volatile Matter	74.04
Fixed Carbon	20.17
Ultimate Analysis	
Carbon	49.90
Hydrogen	6.00
Nitrogen	0.28
Oxygen	41.66

After the carbothermic reduction process, XRD and XRF tests are performed to analyze possible formed phase and to find out the amount of compounds like oxides or other elements contained in the reduced lateritic nickel ore.

3. Result and discussion

Figure 2 shows the XRD test result comparison of samples after carbothermic reduction at all reduction time variation. It could be seen that during the process of carbothermic reduction with reduction time variation, there was no significant change in the phase formed; the only difference was the amount of the peak number of certain phase that increased or decreased.

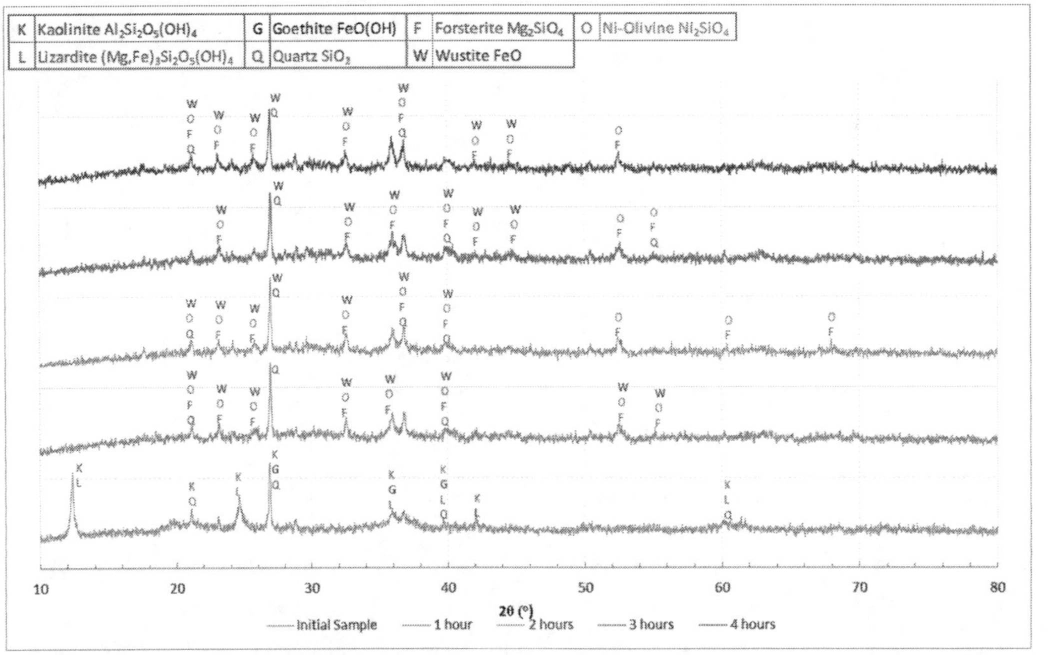

Figure 2. XRD patterns of carbothermic reduction from lateritic nickel ore at all reduction time variations.

In the initial sample of lateritic nickel ore shown in figure 1, there are peaks that identified as lizardite, kaolinite, geothite and quartz phases. After one hour of reduction, the result showed the formation of quartz (SiO_2), wustite (FeO), forsterite (Mg_2SiO_4), and Ni-olivine (Ni_2SiO_4). Peak phases present in the initial sample such as lizardite, geothite and kaolinite phases is no longer formed because it has been dehydroxylated to form new phases described in equation (3.1) to (3.8). The dehydroxylated goethite phase formed the hematite phase (Fe_2O_3), then the hematite was reduced entirely as it reacted with the reductor to form magnetite (Fe_3O_4) which would be reduced to wustite (FeO) [8].

In equation (3.4), the dehydroxylated lizardite phase formed ferrous forsterite phase [$(Mg,Fe)_2SiO_4$] at the temperature of 700°C and clinoenstatite phase ($MgSiO_3$). As the reduction temperature increased to 800°C, the ferrous forsterite turned into forsterite [9]. The absence of clinoenstatite phase possibly because it has reacted with MgO to form forsterite phase, increasing the forsterite phase amount after one hour reduction time.

Additionally, there was also an identified Ni-olivine phase shown in equation (3.8). The phase was formed by dehydroxylation of serpentine $(Ni,Fe,Mg)_3Si_2O_5(OH)_5$ during the reduction process which formed the olivine phase $(Mg,Fe,Ni)_2SiO_4$ [10]. The nickel cations might be substituted into magnesium cations (because of similar atomic radius; 0.068 and 0.065 nm) and replaced the magnesium sites presented in the olivine crystals, causing the formation of Ni-olivine. However, Ni-olivine was a phase that was difficult to reduce because the elements present in olivine such as Ni, Mg and Fe were covered by other non-reactive minerals that prevents further reduction of both nickel and iron oxide.

$$2\alpha - FeO(OH) \rightarrow \alpha Fe_2O_3 + H_2O \quad (3.1)$$

$$3Fe_2O_3 + CO \rightarrow 2Fe_3O_4 + CO_2 \quad (3.2)$$

$$Fe_3O_4 + CO \rightarrow 3FeO + CO_2 \quad (3.3)$$

$$[(Mg,Fe)]_3Si_2O_5(OH)_4 \rightarrow [(Mg,Fe)]_2SiO_4 + MgSiO_3 + H_2O \quad (3.4)$$

$$[(Mg,Fe)]_2SiO_4 \rightarrow Mg_2SiO_4 \quad (3.5)$$

$$MgSiO_3 + MgO \rightarrow Mg_2SiO_4 \quad (3.6)$$

$$Al_2Si_2O_5(OH)_4 \rightarrow Al_2Si_2O_7 + H_2O \quad (3.7)$$

$$(Ni,Fe,Mg)_3Si_2O_5(OH)_5 \rightarrow 2(Ni,Fe,Mg)_2SiO_4 + 4H_2O + SiO_2 \quad (3.8)$$

Meanwhile, the quartz phase remained after one hour of reduction process. Quartz phase was not able to be reduced at the temperature of 800°C because quartz required temperature above 1000°C to change phase.

The phases formed in the reduction time variation of 2 to 4 hour were similar to the one-hour reduction time such as quartz, wustite, forsterite, and Ni-olivine. At the reduction time of 4 hours, the number of wustite peak, forsterite and Ni-olivine phase peak were the highest compared to other reduction times. Forsterite formation caused Ni to be stable or inert at the reduction of saprolite nickel ore and the role of sulfur was very important to limit the formation of the forsterite phase here [11]. In addition, the formation of forsterite allowed the closure of the oxide to grain and to inhibit the development of further reduction of nickel and iron. Due to the increasing number of forsterite phases, wustite became difficult to reduce to Fe metallic on the variation of reduction time. This was probably because of the longer reduction time used, the amount of carbon used to reduce was decreasing.

Based on the XRF test results, as carbothermic reduction time increased, the element or compound content in the sample did not change significantly or remains constant. After 1 hour of carbothermic reduction process, the NiO content increased from 2.662% to 2.68%.

The result showed that decreasing percentage of NiO after 2, 3 and 4 hours during experiment. This condition was reported by Zhu et al.[18] in his research which states that as the reduction time increases, the nickel content obtained will increase until it reaches the optimum point. If the reduction time exceeds the optimum point, it will result in a decreasing of nickel content which corresponds to the increasing Fe content in the lateritic nickel sample. In addition, a decrease in nickel content might occur because the reductor used was decreasing due to an increase in CO_2 gas concentration which caused the gas-solid reaction limit to be filled by CO_2 gas and might inhibit the progress of the reduction reaction [12].

The carbothermic reduction process using the palm kernel shell as reductor results then studied further to find out how much % recovery of Ni in concentrate at each time variation, as shown in figure 3.

Table 2. XRF test results of lateritic nickel ore after carbothermic reduction at all reduction.

Sample	NiO (%)	Fe_2O_3 (%)
Initial Sample	2.662	15.82
1 hour Roasting	2.680	15.90
2 hours Roasting	2.392	15.51
3 hours Roasting	2.417	15.63
4 hours Roasting	2.518	16.33

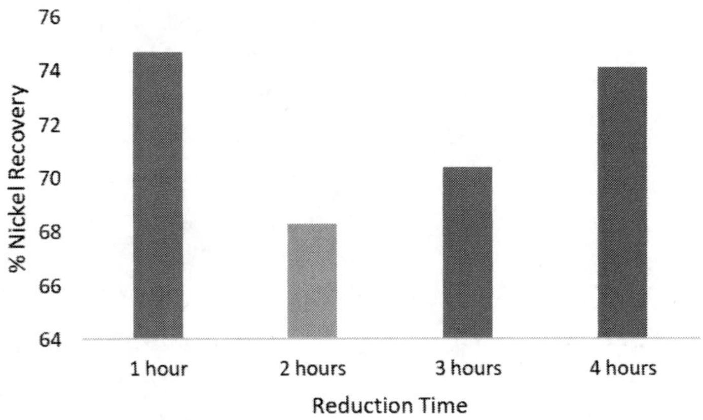

Figure 3. Nickel recovery.

Figure 3 showed that the highest recovery percentage of nickel is obtained after 1 hour reduction time. Therefore, 1 hour of reduction with 74.7% recovery, which is the highest % recovery of Ni among other reduction time variations, was already able to reduce the nickel. So, it did not require a longer reduction time than 1 hour to reduce the lateritic nickel ore used in this study.

In this study, the palm kernel shell has been able to reduce hematite or magnetite to wustite. However, wustite required more CO gas to be reduced to Fe metallic. This was possible because the reducing gas has not reached the maximum value to be able to reduce the wustite. Therefore, the resulting reduction result in this study was limited to form wustite. In addition, there was no identification of the NiO peak phase in the sample characterization results using XRD. This was because NiO in lateritic nickel ore is a type of mineral complex (silicate-bearing) that has a tendency to bind to form Ni-olivine making it more difficult to reduce.

4. Conclusion

The study on effects of reduction time on carbothermic reduction of lateritic nickel ore using palm kernel shell as green reducing agent yields results that can be concluded as follows:

a. Palm kernel shell can be used as reductor as an alternative replacements of coal or coke with the fixed carbon amount of 20.17%.
b. The XRD test results revealed that in the reduced lateritic nickel ore at reduction time variation of 1 hour, 2 hours, 3 hours, and 4 hours, NiO reduction process to Ni didn't occur because the ore classified as complex minerals type (silicate-bearing), so Ni is difficult to reduce. However, 1 hour reduction time can reduce hematite or magnetite to wustite.
c. XRF test results revealed that in the variation reduction time of 1 hour there was an increasing of nickel oxide composition (NiO) from 2.662% to 2.68%.
d. The highest % recovery of Ni in lateritic nickel ore was 74.7%, obtained at 1 hour reduction time. Thus, 1 hour reduction time is enough to reduce laterite nickel ore with palm kernel shell as reductor without requiring longer reduction time.

5. Acknowledgements

Special thanks are due to Mr. Agus Budi Prasetyo (Research Centre for Metallurgy, Indonesian Institute of Sciences - LIPI, Indonesia) for helpful supports. The authors also wish to thank the Ministry of Research Technology and Higher Education and Directorate of Research and Community Engagement, Universitas Indonesia, under PUPT Research Grants Contract No. 2747/UN2.R3.1/ HKP05.00/2017 for the financial support of this research.

6. References

[1] Soedarsono J W, Kawigraha A, Sulamet-Ariobimo R D, Johansyah D, Kusuma G D, Suprayogi S, Yosi A, Saputro N L, Sidiq A T, Erwin E and Natanael D 2013 Potential indonesia ores as raw material for producing iron nugget *3rd Int. Conf. on Advances in Materials and Manufacturing Processes, ICAMMP 2012* **652–654** pp 2529–33

[2] Rao M, Li G, Jiang T A O, Luo J U N, Zhang Y and Fan X 2013 Carbothermic Reduction of Nickeliferous Laterite Ores for Nickel Pig Iron Production in China : A Review **65**

[3] Guo D, Hu M, Pu C, Xiao B, Hu Z, Liu S, Wang X and Zhu X 2015 Kinetics and mechanisms of direct reduction of iron ore-biomass composite pellets with hydrogen gas *Int. J. Hydrogen Energy* **40** 4733–40

[4] Guo D, Zhu L, Guo S, Cui B, Luo S, Laghari M, Chen Z, Ma C, Zhou Y, Chen J, Xiao B, Hu M and Luo S 2016 Direct reduction of oxidized iron ore pellets using biomass syngas as the reducer *Fuel Process. Technol.* **148** 276–81

[5] Srivastava U, Kawatra S K and Eisele T C 2013 Production of pig iron by utilizing biomass as a reducing agent *Int. J. Miner. Process.* **119** 51–7

[6] Dahlquist E 2013 *Technologies for Converting Biomass to Useful Energy-Combustion, Gasification, Pyrolysis, Torrefaction and Fermentation* ed E Dahlquist (Leiden: CRC Press)

[7] Abd Rashid R Z, Mohd. Salleh H, Ani M H, Yunus N A, Akiyama T and Purwanto H 2014 Reduction of low grade iron ore pellet using palm kernel shell *Renew. Energy* **63** 617–23

[8] Lu J, Liu S, Shangguan J, Du W, Pan F and Yang S 2013 The effect of sodium sulphate on the hydrogen reduction process of nickel laterite ore *Miner. Eng.* **49** 154–64

[9] Li G, Shi T, Rao M, Jiang T and Zhang Y 2012 Beneficiation of nickeliferous laterite by reduction roasting in the presence of sodium sulfate *Miner. Eng.* **32** 19–26

[10] Rao M, Li G, Zhang X, Luo J, Peng Z and Jiang T 2016 Reductive roasting of nickel laterite ore with sodium sulphate for Fe-Ni production . Part II : Phase transformation and grain growth *Sep. Sci. Technol.* **6395**

[11] Elliott R, Pickles C A and Forster J 2016 Thermodynamics of the Reduction Roasting of Nickeliferous Laterite Ores *J. Miner. Mater. Charact. Eng.* **4** 320–46

[12] Bunjaku A, Kekkonen M, Pietilä K and Taskinen P 2012 Effect of mineralogy and reducing agent on reduction of saprolitic nickel ores *Miner. Process. Extr. Metall.* **121** 156–65

Effect of electrolyte type on the morphology and crystallinity of TiO_2 nanotubes from Ti-6Al-4V anodization

Khalil Gibran[1], Muhammad Ibadurrahman[1] and Slamet[1,2]

[1] Chemical Engineering Department, Faculty of Engineering, Universitas Indonesia Depok 16424, Indonesia

Email: slamet@che.ui.ac.id ; slamet.dtk.ui@gmail.com

Abstract. A good biocompatibility material such as Ti-6Al-4V is neededfor a dental application.However,Ti-6Al-4V has no antibacterial properties. Therefore Ti-6Al-4V is modified using anodizing methods to improve its anti-bacterial properties. One of the most influential factors in the anodizing process is the type of electrolyte solution. The influence of electrolyte solution based on glycerol and ethylene glycol to morphology and crystallinity of TiO_2 nanotubes is studied. Smoothing and chemical polishing was performed against the Ti-4Al-6V plate. Then, the anodizing process for each plate in glycerol and ethylene glycol solution was done for 2 hours and 50 volts. Calcination was carried out for 3 hours at a temperature 500°C. FESEM-EDX and XRD were employed to analyze the surface morphology and crystallinity of TiO_2 nanotubes. Viscosity of the electrolyte solution influenced surfacemorphology. Calcination process influenced crystallinity of TiO_2 nanotubes.The average tube diameter and thickness from glycerolelectrolyte solution was 135.45 nm and 26.27 nm, while ethylene glycol was 68.52 nm and 35.62 nm respectively. The size of anatase crystal was 34.95 nmfor glycerol and 44.77 nm for ethylene glycol.

1. Introduction

Many problems about teeth have been widely prevalent in society. To overcome this problem, thedental implantation technology is developed with the aim to replace broken or missing teeth [1]. In fact, the technology is quite high, but sometimes, it can cause some problems that occur after the installation process. One of which is *Peri-implantitis* which cause bone erosion around the implant, sothat it can cause implant failure [2]. *Peri-implantitis*is caused by the formation of bacterial biofilms on the surface of implants withoutantibacterial activity. The Ti-6Al-4V alloy is a titanium-based dental implant material that is widely used in orthopedics and dentistry because of its high biocompatibility, low density, high mechanical resistance, high chemical inertness and toxicity and high corrosion resistance[3-7].

Surface modification methods are so important to improve the biocompatibility and corrosion resistance to get ideal surface properties for implant application [8]. Modifications of Ti-6Al-4V can be performed to enhance antibacterial properties to prevent *Peri-implantitis* formation.The modification is done by anodizing method aimed at growing TiO_2nanotubes array so that photocatalytic activity to disinfect bacteria can occur effectively. One of the important factors

[2] To whom any correspondence should be addressed.

affecting the anodized form of nanotubes is the electrolyte solution used during the nanotubes synthesis.

In this study,nanotubes synthesis was performed using electrolyte solution of ethylene glycol and glycerol to compare its surface morphology and crystallinity.The previous study investigated that the length and diameter of the tube formed were influenced by the concentration of fluoride in the electrolyte solution[9]. Ethylene glycol is an organic solution with high viscosity has a great possibility to get TiO_2nanotubes with smooth and homogeneous walls [10, 11]. The use of ethylene glycol electrolyte solution yielded a more uniform diameter of nanotubes, finer nanotubes wall, and better photocatalytic degradation in Methyl Orange (MO) compared to glycerol [12]. However, this study has not yet led to the application of dental implants for anti-bacterial properties. The use of this type of organic electrolyte solution can produce different forms of nanotubes, so further research is necessary to determine the effect of surface morphology produced against anti-bacterial properties by using Ti-6Al-4V. The obtained result is expected to be different from the previous research and can be used for dental implant application.

2. Experimental

2.1. Preparation of TiO_2 nanotubes

The Ti-6Al-4V plate was cut to the size of 4 cm x 1 cm. Then, smoothed using 1000cc abrasive paper. The plate was washed thoroughly with soap and then the chemical polishing method was performed using the ratio of HNO_3: HF: H_2O = 3: 1: 6. The anodizing process was carried out with a variation of an electrolyte solution of ethylene glycol with a composition of 0.5% NH_4F and 3% water content and glycerol with a composition of 0.5% NH4F and 25% water content. The electrolyte solution was mixed using a magnetic stirrer. The electrodes used during the anodizing process were Pt electrodes as cathodes and Ti-6Al-4V materials as cathodes. The anodizing process was carried out on a beaker glass with an electrolyte amount of 60 ml. The constant process voltagewas kept at 50 V for 2 hours at room temperature.This processwas also donewhile stirred using a magnetic stirrer. After these processes had been completed, the material was rinsed with aquadestand then dried with a hair dryer. Then the calcination processwas performed at 500^0 C for 3 hours. All procedures were the same for both variations of the electrolyte solution.

2.2. Characterization of TiO_2nanotubes

The surface morphological structure and elemental of the synthesized TiO_2 nanotubes can be seen with the characterization using Field Emission Scanning Electron Microscopy/Energy Dispersive X-ray (FESEM-EDX JEOL Type JIB 4610F) at 15 kV with magnification up to 100000 times. The length of the tube formed can also be seen through the measurement of its cross section., X-Ray Diffraction (XRD) was used to determine the crystal structure and composition of the material. The XRD patterns were obtained using PANanalytical-Empyreanwith Cu anode tube (λ = 0.15406 nm) runs at voltage 40 kV and current 30 mA, with the range of 2Θ angle values 10° to 80°. The crystallite size of the specimen was estimated from FWHM (full-width at half-maximum) of XRD by Scherrer equation.

3. Result and discussion

It is necessary to know the morphology and crystal structure of TiO_2 nanotubes because it will certainly affect the photocatalytic performance and the stability of the materials when it used or medical application, such as dental implants. The result of FESEM analysis for TiO_2 anodized by electrolyte solutionglycerol and ethylene glycol with the same calcination process conditions are shown in figure 1.

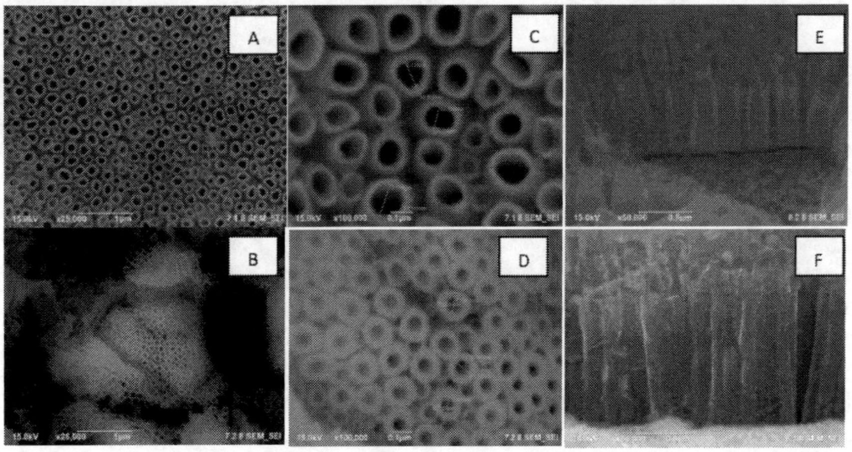

Figure 1. FESEM images of TiO_2nanotubes from electrolyte solution glycerol (left) and ethylene glycol (right) with different magnification rates: (A and B) 25000x, (C and D) 100000x, (E and F) cross-sectional.

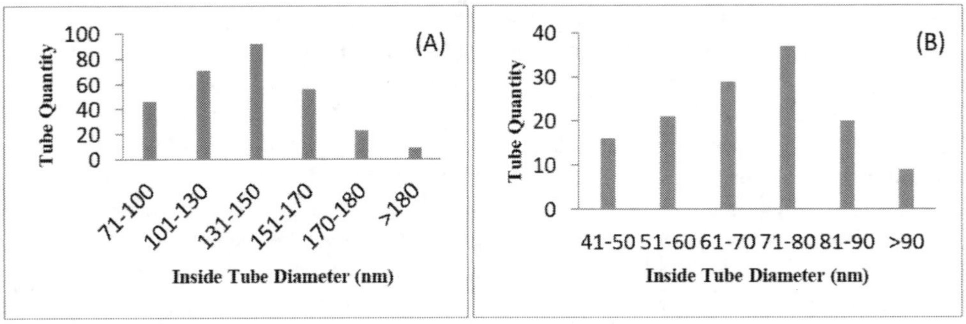

Figure 2. Graphic of diameter size from (A) Gycerol and (B) Ethylene Gycol.

Tabel 1. Diameter and thickness sizeof TiO_2 nanotubes.

Electrolyte solution	Inside diameter tube average	Thickness wall average
Glycerol	135.45 nm	26.27 nm
Ethylene Glycol	68.52 nm	35.62 nm

FESEM with 25000x magnification was used for estimating the tube diameter and the thickness of TiO_2. It shows various nanotubes dimensions.Thus, the estimated tube average diameter and thickness is more representative. In glycerol, the structure of TiO_2nanotubes formed has a large diameter in tube size. While on ethylene glycol, theTiO_2 structure has a tube diameter size smaller than glycerol. In table 1, the average diameter of Ti-6Al-4V nanotubes using glycerol electrolyte was 135.45 nm. While using ethylene glycol, the average diameter of Ti-6Al-4V nanotubes was 68.52 nm. But when viewed from the thickness of the formed tube, the synthesized TiO_2nanotubes using ethylene glycol electrolyte solution have greater thickness compared to using glycerol. Its value from electrolyte glycerol and ethylene glycol were 26.27 nm and 35.62 respectively. The thickness is affected by the water composition in the electrolyte solution where a small amount of water will increase TiO_2nanotubes thickness [13]. The photocatalyticperformance will be better if the size of the photocatalyst is smaller because it will increase the active surface area to react.

The formed tube shape from TiO_2nanotubes can be seen from the cross-sectional FESEM analysis. In TiO_2nanotubes using ethylene glycol, tube forms look longer compared to those using glycerol. The structure also looks ruder than the more refined tube shape created using glycerol. The difference in length between the two caused by the viscosity of the two solutions. In the viscosity measurement results, the ethylene glycol electrolyte solution has a lower viscosity value than glycerol, 15.1 and 35.2 CSt, respectively.The lower viscosity electrolyte solution permits the dissolution rate of F^- ions during the anodizing process. It can grind down to the lower part of TiO_2 so that the tube length produced on TiO_2 using ethylene glycol is longer.

Tabel 2. Elemental composition of modified Ti-6Al-4V.

Elements	Elemental Composition (%wt)	
	Glycerol	Ethylene Glycol
Ti	60.9	51.9
O	31.9	41.6
Al	3.8	4.3
V	3.4	2.2

Characterization using Energy Dispersive X-Ray produces data as in table 2. It shows that Ti on Ti-6Al-4V successfully oxidized so that it can be said that TiO_2nanotubes successfully formed. But in each electrolyte solution, there is a difference of %wt from each element. It is influenced by the height, diameter and thickness of the tubes produced using glycerol and ethylene glycol electrolyte solutions. The result of previous FESEM analysis seen from the cross-section shows that tube height formed from glycerol electrolyte solution is shorter than ethylene glycol. It causes the EDX current into the inside of the tube to the alloy portion of Ti-6Al-4V. Thus, the % wt of the alloy of glycerol electrolyte solution is greater than using ethylene glycol.

The EDX analysis also shows that Al and V are also oxidized to form thin surfaces of Al_2O_3 and V_2O_5. The presence of these two oxide compounds improves the stability of theTi-6Al-4V material. When viewed from the results of EDX analysis, TiO_2 nanotubes formed either by using electrolyte solution of ethylene glycol or glycerol can be utilized for dental implant applications [14].

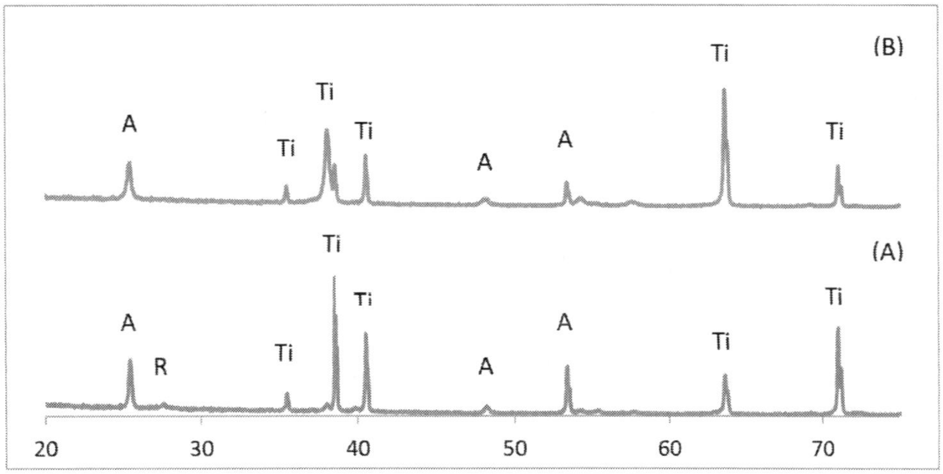

Figure 3. The XRD patterns of TiO_2nanotubes prepared from electrolyte solution (A) ethylene glycol, (B) glycerol.

Figure 3 shows that the TiO_2 crystal structure formed through reaction with glycerol electrolyte solution is entirely anatase. The anatase peaks are seen at an angle of $2\theta = 25.3°$, $48.0°$ and $55.1°$[15].

No crystal rutile structures are formed. The new rutile crystalline structure is formed at a calcination temperature of 600°C with a calcination time of 3 hours when using a glycerol electrolyte solution at the anodizing process [16], using an ethylene glycol electrolyte solution wherein the same calcination time as glycerol is 500° C for 3 hours, the rutile crystals begin to form even slightly. The rutile peak is at an angle of $2\theta = 27.45°$ [17].

Tabel 3. Crystallite sizes of modified Ti-6Al-4V.

Crystal	Crystallite Size (nm)	
	Glycerol	Ethylene glycol
Anatase	34.95	44.77
Rutile	-	152.19

The crystallite size can be calculated by Scherrer equation[18]:

$$t = \frac{0{,}9\lambda}{B \cos \theta} \qquad (1)$$

Where t is the average crystal size (nm), λ is the X-ray's wavelength, B is the width of thehalf-maximum peak (FWHM in radian unit), and Θ is diffraction angle in degree (°). XRD data calculation can be seen in table 3. Anatase crystals formed on TiO_2 nanotubes use glycerol electrolyte measuring 34.95 nm while with ethylene glycol of 44.77 nm and rutile of 152.19 nm. Although by using ethylene glycol the formed anatase crystalline structure is larger, The difference between both is not too significant.

The difference in the formation of this rutile crystalline structure using glycerol and ethylene glycol are present in the crystallization process of TiO_2nanotubes at the time of calcination. The anatase crystals of TiO_2isformed from amorphous crystals while the rutile crystalsare formed from anatase. Rutile crystals begin to form at high calcination temperatures where the amorphous crystals are no longer present. In the XRD patterns, it can be seen that by using glycerol electrolyte solution, the anatase crystals intensity is less than using ethylene glycol. It occurs because the amorphous crystals on TiO_2nanotubes using glycerol still exist in them. As with the use of ethylene glycol electrolyte solution, it is seen that the intensity of more and more crystalline anatase crystals and rutile crystals has been formed even with little intensity. It is because the amorphous crystals of TiO_2 nanotubes are depleted to form a rutile crystal structure.

With the same calcination process, the rutile crystalline structure is formed more rapidly in TiO_2using an electrolyte solution ethylene glycol than glycerol. It is one of them influenced by the boiling point of both types of the electrolyte solution. Glycerol has a higher boiling point than ethylene glycol, 290° Cand 197.3° C, respectively[19]. When the anodizing process is completed, the electrolyte solution of glycerol and ethylene glycol on each plate is indicated to be still attached to its TiO_2 surface. At the time of the calcination process, the heating first evaporates the electrolyte solution still attached to TiO_2. Because of the higher boiling point of glycerol, the heating will first vaporize the electrolyte that is still attached to TiO_2 so that the crystallization process becomes longer. It is why the rutile crystal structure of the ethylene glycol electrolyte is formed first with the same temperature and the length of the calcination process.

Previous research has been reported that the rutile crystal structure begins to form at 580° C for 2 hours [20]. But in other studies, it was found that the rutile crystal structure began to form at a temperature of calcination of 600° C for 2 hours [21]. Of the two studies mentioned, Rutile crystals began to form at a temperature of calcination above 500° C for 2 hours. At the calcination temperature of 450° C for 3 hours, the rutile crystal structure was not formed [22]. In this study using a 500° C calcination temperature for 3 hours where the calcination time is longer than the previous study has been obtained rutile crystal structure of TiO_2 nanotubes.

4. Conclusion

TiO_2nanotubes were synthesized by anodization method form twotype electrolyte solutions which produced significant differences in its morphological structure. The morphological differences are caused by the difference in the viscosity value of the two solutions. The rate of F^- ion dissolution during the anodizing process affects tube size, thickness, and length of TiO_2 nanotubes. The average tube diameter and thickness thatare resulting from electrolyte solution glycerol was 135.45 nm and 26.27 nm, while ethylene glycol was 68.52 nm and 35.62 respectively. During the calcination process, the combination of temperature, time, and the type electrolyte solution will affect the form of the produced TiO_2 crystals. Rutile is not formed when using glycerol electrolyte. The size of the anatase crystals formed by using glycerol and ethylene glycol electrolytes was 34.95 nm and 44.77 nm. While the size of the rutile crystals formed using electrolyte ethylene glycol amounted to 152.19 nm

5. Acknowledgement

The authors would like to sincerely acknowledge to *Publikasi Internasional Terindeks Untuk Tugas Akhir Mahasiswa Universitas Indonesia* (PITTA UI 2017) and thanks to Ahmad Husein Alkaff as the English editor of this paper.

6. References

[1] Herda S E 2008 *Dental J.* **13** pp 28-32
[2] Dhir S 2013 *J. of Indian Society of Periodontology* **17**(1) pp 5–11
[3] Macak J M, Tsuchiya H, Ghicov A, Yasuda K, Hahn R, Bauer S, *et al.*2007 *Current Opinion in Solid State and Materials Sci.* **11** 3-18
[4] Liu X, Chu P K and Ding C 2004 *Materials Sci. and Eng. R: Reports* **47** 49-121
[5] Kokubo T 1998 *Acta Materialia* **46** 2519-27
[6] Sul Y T, Johansson C B, Petronis S, Krozer A, Jeong Y, Wennerberg A, *et al.* 2002 *Biomaterials* **23** 491-501
[7] Mohan L, Durgalakshmi D, Geetha M, Narayanan T S and Asokamani R 2012 *Ceramics Int.* **38** 3435-43
[8] Ahmed R A, Fekry A M and Farghali R A 2013 *Applied Surface Sci.* **285** 309-316
[9] Wang Z L J 2009 *J. Phys. Chem.* **C 113** 4026–30
[10] Albu A G S P, Macak J M and Schmuki P 2007 *Phys. Status. Solidi R* **1** pp R65-R67
[11] Macak B G G J M, Hueppe M and Schmuki P 2007 *Adv. Mater.* **19** 3027–31
[12] Sreekantan S, Saharudin K A, Lockman Z and Tzu T W 2010 *Nanotechnology* **21** p.365603
[13] Hilario F, Roche V, Nogueira R P and Junior A M J 2017 *Electrochimica Acta* **245** 337-349
[14] Slamet B M B, Wulan P P D K, Setiadi and Aprianto B 2016 *Word Scientific Publishing Company* **X** 1-10
[15] Xia X H, Jia Z J, Yu Y, Liang Y, Wang Z and Ma L L 2007 *Carbon* **45** 717-721
[16] Das S, Zazpe R, Prikryl J, Knotek P, Krbal M, Sopha H, *et al.*2016 *Electrochimica Acta* **213** 452-459
[17] Li G, Liu Z Q, Lu J, Wang L and Zhang Z 2009 *Applied Surface Sci.* **255** 7323-28
[18] Tristantini D and Ibadurrohman M 2013 *Int. J. of Energy Research* **37** 1372-81
[19] Riddick W B B J A 1970 *Organic Solvents, 3rd Edition.* (New York: Wiley-Interscience Publication)
[20] Li Y, Yu H, Zhang C, Song W, Li G, Shao Z, *et al.* 2013 *Electrochimica Acta* **107** 313-319
[21] Roy P, Berger S and Schmuki P 2011 *Angewandte Chemie International Edition* **50** 2904-39
[22] Song H, Cheng K, Guo H, Wang F, Wang J, Zhu N, *et al.* 2017 *Catalysis Communications* **97** 23-26

IOP Conf. Series: Earth and Environmental Science **105** (2017) 012039 doi:10.1088/1755-1315/105/1/012039

Catalytic oxidation of benzene using nano-CuO/ γ-Al_2O_3 and commercial catalysts

Ryan A. Rahardi[1], Retno Maharsi[2], Ferry Iskandar[2], Hary Devianto[1] and Yogi Wibisono Budhi[1,3]

[1] Chemical Engineering Study Program, Institut Teknologi Bandung, Jl. Ganesha 10, Bandung, Indonesia

[2] Physic Study Program, Institut Teknologi Bandung, Jl. Ganesha 10, Bandung, Indonesia

E-mail: Y.Wibisono@che.itb.ac.id

Abstract. Volatile organic compounds (VOC) such as benzene are among the most dangerous air pollutants emitted by chemical industry stacks, as they may contribute to environment and health issues. Lean catalytic oxidation of benzene has been considered as most proper method to abate it from the flue gas. This work developed nano-based copper oxide catalysts for lean oxidation of benzene. The aim of this study was to evaluate the activity performance of the nano-based copper oxide catalyst and compare to commercial catalyst. On the basis of the commercial catalyst, this study was also aimed to determine the reaction rate and its kinetic parameter. The oxidation of benzene was conducted in a fixed bed reactor at 300°C, 1 atm, and GHSV of 15,000 h^{-1}. The concentration of benzene in the feed and product were measured using online gas detector (Cosmos Gas Detector). The catalyst activity of nano-based copper oxide catalysts showed 20-30% conversion of benzene, while for commercial catalyst showed 86%. The reaction rate determination for first order reaction of benzene indicated that the activation energy was 48 kJ/mol with Arrhenius constant of $3x10^4$ s^{-1}.

1. Introduction

Volatile Organic Compounds (VOCs) have been widely recognized as one of the most dangerous air pollutants that may contribute to environment and health issues. VOCs may cause the photochemical smogs formation, ozone layer depletion, and global warming as greenhouse gases [1]. VOCs are frequently emitted by various sources such as vehicle emissions, industrial emissions, and household emissions [2]. Benzene is one of VOCs that has carcinogenic properties and very harmful to human health. Benzene is generally emitted from chemical industry such as petrochemical and oil refinery [3]. There are many methods to convert VOCs into unharmful substance, such as adsorption, absorption, thermal oxidation, and catalytic oxidation [4]. The catalytic oxidation is well recognized as one of the most appropriate methods for treating VOCs due to its catalytic performance, large capacity, and moderate operating cost [4,5]. In addition, the catalytic oxidation can efficiently treat VOCs at very dilute concentration (below 1%) with operating temperature much lower than thermal oxidation [6].

[3] Author to whom any correspondence should be addressed.

In general, there are two types of catalyst used in VOCs catalytic oxidation, i.e. noble metal catalyst and metal oxide catalyst. Noble metal catalysts, such as platinum and palladium, have high catalyst activity but they are quite expensive and can be easily poisoned. Metal oxide catalysts, such as CuO, FeO, NiO, and MnO, have high thermal resistance, not expensive, and not easily poisoned [7,8,9,10], although the catalyst activity is much lower than noble metal catalyst [11]. In recent years, many researchers have shown that metal oxide catalyst with the highest catalyst activity are copper-based metal oxide catalyst (CuO) [6,10,11,12]. Addition of ceria and mangan oxide as catalyst support also increases catalyst activity [11].

In recent years, nanomaterial has been used as catalyst and called as nanocatalyst. Nanocatalysts have very active site and high surface area, thus the activity of nanocatalysts are much larger than the conventional catalysts [13]. In addition to high surface area, nanocatalysts also have high selectivity and good stability [14]. Copper based nanocatalyst has been developed by many researches and has been shown a high catalyst activity property [13]. Nano-CuO catalyst has been developed in previous research [15] by impregnating copper into alumina mesoporous which is made by urea combustion method. The aim of this work was to investigate the activity of nanocatalyst that has been developed in previous research, compare the result with commercial CuO catalyst, and determined kinetic parameter for nanocatalyst.

2. Materials and methods

2.1. Catalyst preparation

The details of the preparation of nano-CuO /γAl_2O_3 catalyst using urea nitrate combustion and Cu impregnation have been described in the previous work [15]. Gamma alumina was prepared by mixing specific ammounts of alumunium nitrate nonahydrate [$Al(NO_3)_3.9H_2O$] and urea at spesific rasio, in water under vigorous stirring for 1 hour. The mixture was then dried in an oven at 120°C for 5 h and at 250°C for 3 h. The sample was calcined in air atmospheric at 500°C for 6 h. The resulted powder was then added into copper nitrate solution at spesific concentration while vigorously stirred for 45 minute. The sample was subsequently dried in oven at 120°C until water content was removed. Then the sample was calcined at 700°C or 800°C for 6 h.

2.2. Experimental setup

The experimental runs for catalysts activity test were performed in a quarts tubular fixed bed reactor (∅-10 mm) operated at atmospheric pressure. The reaction mixture was composed of nitrogen, oxygen (±21%), and 5,000 ppm of benzene. For catalyst activity test, 200 mg of catalyst was diluted with glass bead to give porosity around 0.4 and loaded into the fixed-bed reactor. The reaction temperature and GHSV were kept constant at 300°C and 15,000 h^{-1}. The benzene concentration in gas stream was measured by an online gas detector (Cosmos XP3160 type). The schematic view of the experimental set up is shown in figure 1. The activity test was conducted for 9 variations of nanocatalyst, 1 commercial catalyst, 2 types of alumina mesoporous suport without impregnated Cu, and 1 commercial alumina, which are shown in table 1.

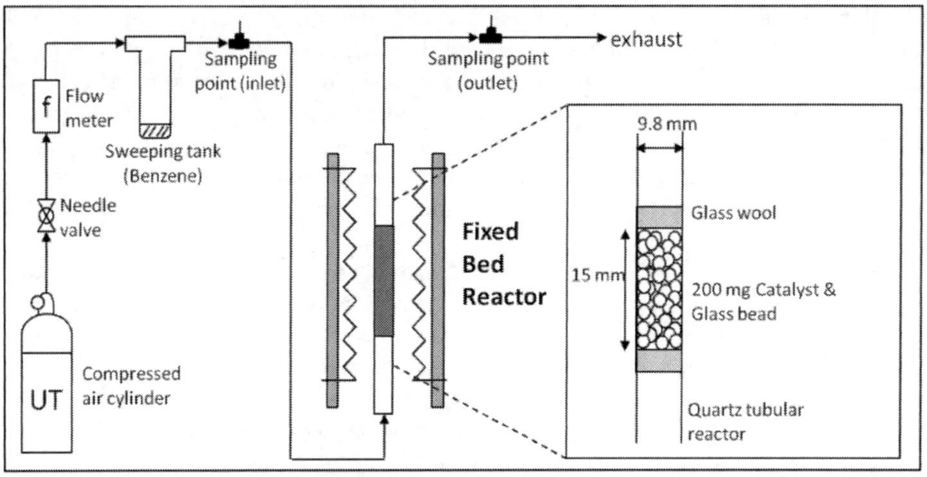

Figure 1. Experimental Setup.

Table 1. Variations of catalyst in activity test.

Catalyst no.	Variation		
	Al : Urea	Calcination Temperature (°C)	Cu loading
1	1:2	800	30%-w
2	1:2	800	10%-w
3	1:2	800	5%-w
4	1:2	700	30%-w
5	1:2	700	5%-w
6	1:3.5	700	5%-w
7	1:3.5	800	30%-w
8	Alumina Commercial	800	10%-w
9	1:2	800	30%-w Ni
10	Commercial CuO Catalyst		
11	Alumina mesoporous with Al : Urea = 1:2		
12	Alumina mesoporous with Al : Urea = 1:3.5		
13	Alumina commercial		

3. Results and discussion

3.1. Activity test

The results of activity test for 13 types of catalysts are shown in figure 2. The activity of catalyst is measured as benzene conversion for benzene oxidation reaction. From figure 2, the highest benzene conversion is 86% obtained from the commercial catalyst. Nanocatalysts resulted in benzene conversion much lower, in the range of 16%-27%. The alumina mesoporous gave benzene conversion in the range of 20%-40% with the highest conversion coming from alumina mesoporous with Al:urea 1:2, and the lowest was from commercial alumina. The commercial catalyst indicated higher activity than nanocatalyst because of the existence of other components in the commercial catalyst instead of CuO which acted as promoter.

Table 2 shows the surface area for alumina mesoporous before and after impregnation. Before impregnation, the alumina mesoporous had surface area around 150 – 200 m^2/g. After impregnation, the alumina mesoporous had surface area around 40 m^2/g. The decrease in the surface area indicated that there was a failure in the method of making nanocatalysts. One possible cause of failure was the higher calcination temperature, which was in the range of 700 – 800°C. The high calcination temperature might cause several failures in making nanocatalyst. There might be sintering of the copper that caused blockage or agglomeration in catalyst pore which reduced surface area greatly.

There might be a change of alumina phase from gamma alumina, which had high surface area to alpha alumina, which had lower surface area. It could also cause a chemical reaction between alumina solids with copper to produce inactive component, which lowered catalyst activity.

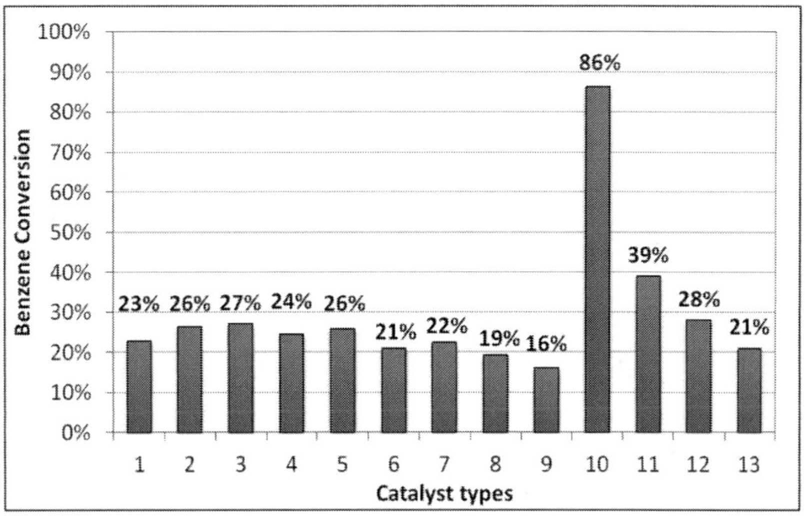

Figure 2. Benzene Conversion for each type of catalysts.

Table 2. Surface area of Al_2O_3 before and after impregnation with 30%-Cu and calcined at 800°C.

Alumina type	Surface area	
	Before impregnation	After impregnation
Alumina mesoporous with Al:urea = 1:2	213	44
Alumina mesoporous with Al:urea = 1:3.5	126	44
Alumina commercial	150	44

From XRD data at figure 3, there were other components beside CuO in nanocatalyst. The copper oxide (CuO) components acted as active phase in nanocatalysts [6,10,11,12]. The other component formed was $CuAl_2O_4$ or copper aluminat. The copper aluminat was formed at high calcination temperature at 700 and 800°C. This also indicated that at calcination temperature higher than 700°C, copper would react with alumina to produce copper aluminat, which was inactive component.

Figure 2 also showed that alumina mesoporous before impregnation (catalyst number 11 and 12) has higher activity than after impregnation. This indicates that the calcination temperature was too high. Several researches usually calcined the catalyst at 500 – 600°C [6,10,11]. The effect of copper loading in catalyst also had bigger role in decreasing catalyst activity. Figure 2 shows that the higher the copper loading in catalyst, the lower the benzene conversion. This indicated that at high amount of copper loading, the agglomeration in surface area became more significant thus lowered the activity. Other research [6] indicated that copper loading above 7%-weight would lower the activity.

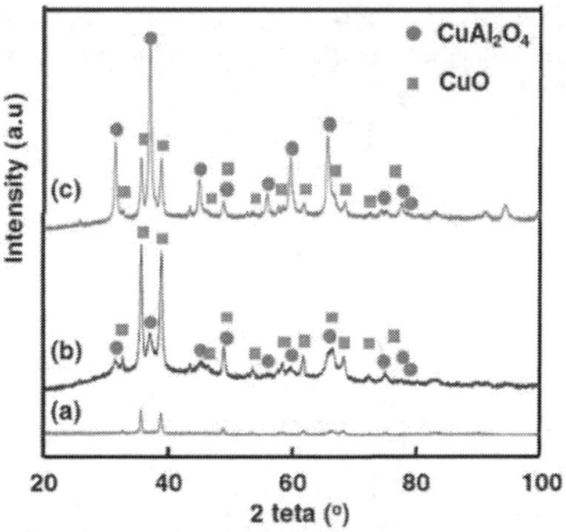

Figure 3. Diffraction patterns of Al_2O_3 mesoporous with Al:urea = 1:2 after Cu impregnation and calcined at (a) 600°C, (b) 700°C, (c) 800°C. The insert shows a magnification of the diffraction pattern of the sample calcined at 600°C.

3.2. Determination of kinetics rate equation

Kinetics rate determination was performed for catalyst 3, which had highest activity among nanocatalyst. The power law model was used as kinetic equation for benzene oxidation. The catalytic oxidation of benzene operated in the excess of oxygen can be described by a first-order kinetic equation (1):

$$r_{bz} = k.C_{bz} = A.\exp\left(-\frac{Ea}{RT}\right).C_{bz} \tag{1}$$

where r_{bz} is kinetic rate, A is Arrhenius constant, E_a is activation energy, and C_{bz} is benzene concentration. The reactor was assumed as plug flow reactor, which has reactor model as written in equation (2) with τ is gas retention time.

$$\frac{dC_{bz}}{d\tau} = -r_{bz} \tag{2}$$

At GHSV 15,000 h^{-1}, experiment was conducted by varying reaction temperature from 300 to 400°C. The data treatment resulted in conversion and reaction rate constant (k) as presented in Table 3. The reaction rate constant was obtained by combining equation (1) and (2). The activation energy and Arrhenius constant were calculated from the plots of ln k versus $1/T$, where E_a was obtained from the slope of the resulting linear plots, and A was obtained from the intercept. From the calculation, E_a for benzene oxidation with nanocatalyst was 48 kJ/mol and A was 3 x 10^4 s^{-1}. From this kinetic parameter, the conversion of benzene at several temperatures were calculated and shown in figure 4.

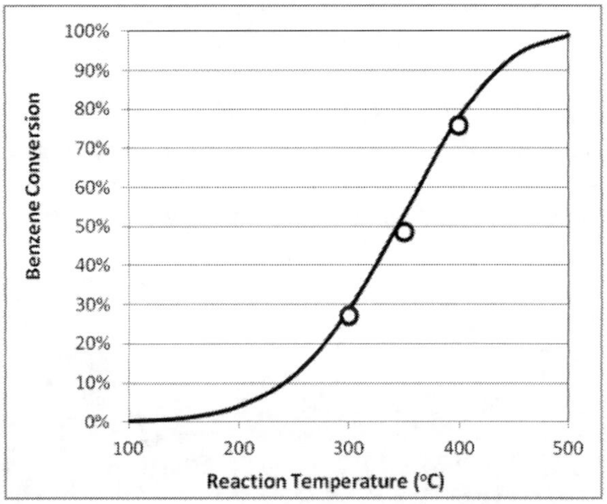

Figure 4. Benzene conversions at GHSV 15,000 h^{-1} and varied reaction temperatur using nano-catalyst. Line (–) shows data from simulation and Dots (**o**) shows data from experiment.

Table 3. Benzene conversion of varied reaction temperature at GHSV 15,000 h^{-1}

Temperature (°C)	Benzene Conversion	k (s^{-1})
300	27.0%	1.31
350	48.6%	2.77
400	75.8%	5.91

4. Conclusion
The activity of nano-CuO/γ-Al_2O_3 and commercial catalyst has been tested in the fixed bed reactor. The result shows that nanocatalyst has much lower catalyst activity than commercial catalyst. There has also been done for determining the kinetic rate parameter of the nanocatalyst.

5. Acknowledgments
The financial support provided by Ministry of Research, Technology and Higher Education, Republic of Indonesia, through National Competitive Research Grant "National Strategic Research" no. 009/SP2H/LT/DRPM/IV/2017, is gratefully acknowledged.

6. References
[1] Kamal M S, Razzak S A and Hossain M M 2016 *Atmospheric Environment* **140** 117-134
[2] Liotta L F 2010 *Applied Catalysis B: Environmental* **100** 403-412
[3] Özçelik Z, Soylu G S P and Boz I 2009 *Chem. Eng. J.* **155** 94-100
[4] Santos V P 2010 Catalytic Oxidation of Volatile Organic Compounds *Phd disertation from University of Porto* 9 – 12
[5] Kim H S, Kim T W, Koh H L, Lee S H and Min B R 2005 *Applied Catalysis A: General* **280** 125-131
[6] Yang J S, Jung W Y, Lee G D, Park S S, Jeong E U, Kim H G and Hong S S 2008 *J. of Industrial and Eng. Chemistry* **14** 779-784
[7] Wu H, Wang L, Zhang J and Shen Z 2011 *Catalysis Communications* **12** 859-865
[8] Zeng J, Liu X, Wang J, Lv H and Zhu T 2015 *J. of Molecular Catalysis* **408** 221-227
[9] Mu Z, Li J J, Duan M H, Hao Z P and Qiao S Z 2008 *Catalysis Communications* **9** 1874-1877
[10] Li S, Wang H, Li W, Wu X, Tang W and Chen Y 2015 *Applied Catalysis B: Environmental* **166-167** 260-269
[11] Li T Y, Chiang S J, Liaw B J and Chen Y Z 2011 *Applied Catalysis B: Environmental* **103** 143-148
[12] Tang X, Xu Y and Shen W 2008 *Chemical Engineering J.* **144** 175-180
[13] Hu C, Zhu Q and Jiang Z 2008 *Powder Technol.* **194** 109-114
[14] Singh B S and Tandon P K 2014 *J. of Energy and Chem. Eng.* **2** 106-115
[15] Maharsi R, Septianto R D, Rohman F, Iskandar F, Devianto H and Budhi Y W 2017 *Material Research Express* **4** 044002
[16] Xing T, Wan H, Shao Y, Han Y, Xu Z and Zheng S 2013 *Applied Catalysis A: General* **468** 269-275

Reading sustainability post annexations of Rumah Gadang architecture

D Trisnawan[1,2] **and A Rifqih**[1]

[1] Department of Architecture, Faculty of Engineering, Universitas Indonesia, UI Depok Campus, Depok 16424, Indonesia

E-mail: ditadesign1@gmail.com

Abstract. Many of Rumah Gadang of Minangkabau in West Sumatra has strived through many years of extreme tropical climate. Through times this type of vernacular house has remarkably endure and stay, hence the increasing numbers of maintenance cost. One family generation to its next, as many other normal family houses may have experienced, the annexation phase (adding spaces) of living is inevitable. As Rumah Gadang design has been considered as one model of eco-tropical architecture fine example, its annexation could make the green compactness of its majectic design changed. This research analyzes major typical modifications and changes that could happen to Rumah Gadang's original design. Field research was conducted direct to its site's origin in West Sumatra. Its actual and recent plans (after annex) are documented to study the basics and its upgrading modification tendencies. This information then could become the base for further research on exploring the local wisdom, its traditional building values, material development and changes, and also its future endurance.

1. Introduction

The effort to make Minangkabau traditional house (Rumah Gadang) sustainable has been done by the people of Minang, following the availability of resources, surrounding nature, and settling characteristic of the society. People of Minangkabau have been trying to achieve a sustainable dwelling design even before the development of technologies that support it. The design development of Rumah Gadang continues to evolve following the needs of its more modern society. Natural resource exploitation has caused negative impacts to the environment following the fast growth of the population which requires more provided land for settlements. The settling characteristic of the society has been evolving which then caused changes of the settlement spatial program. The objectives of the writing are to deepen understandings on Rumah Gadang architecture, its sustainability concepts and the applications after its physical changes, reflected by its spatial and functional changes.

[2] To whom any correspondence should be addressed.

Figure 1. Rumah Gadang of Minangkabau.
http://putriaulina.blogspot.co.id/2013/03/rumah-gadang-minangkabau.html

2. Theory of sustainability

2.1. Sustainability
The sustainable architecture approach is used in general to explain a building design which its technology, material, ecology, and environment are all in balance (Attmann, 2010). In context, this approach is divided into three main components: sustainable elements (material and technology), sustainable resource, and sustainable environment.

According to Attmann, sustainable elements consist of material and technology in which fulfil the following aspects: durability, economical, low maintenance, and recyclability. These aspects must exist as parameters in considering the architecture's level of sustainability. On sustainable resource, it refers to the supports how the building works, such as material resource, site potential, accessibility, and the nature response within the environment. This sustainable factor can be achieved whenever these resources are applied effectively to support the physics and the actual functions of the architecture. Sustainable environment is achieved when the surroundings are proven healthy, habitable, responsive to social capacity, and protected physically from faults or mistakes due to and during the development process, and/or private security threats.

2.2. Cultural base of Minangkabau society
Minangkabau people sees nature as a knowledge resource. *Alam takambang jadi guru* (nature developed into teach(ing) – as a teacher), this is a wisdom teaching of the society on how they value their nature (Navis, 1984). This philosophical inheritance goes down to generations within the society. Furthermore, a local people named Taufik explained that there are some rules which cannot be disturb or change as it already stated (by the society), and rule which can be changed base on agreement within the society, as long as it does not in contrary to the firmed and already stated one (interview, Nov 22, 2016).

In connection to Rumah Gadang, anything which rule about its architectural function and elements are stated as law that can be changed (with agreement). Building a traditional house Rumah Gadang nowadays can be modified as long as it does not change its identity. The most important matter is how the house meet the needs of the residents and its function for traditional local custom/practice (*adat*).

3. Post annexation of the original Rumah Gadang
At the beginning, variant of Rumah Gadang are differed from its harmonious style of every society group. According to Navis (1984), it really depends on the leaderships of either *Datuk Katumangguangan* and *Datuk Parpatiah nan Sabatang.* This difference has evolved and make unique of the Rumah Gadang within the territory of Minangkabau. Couto and Darwis (1999) explained the schematic development and changes of Rumah Gadang on their research. This diagram shows the

morphology of the traditional house forms. We assume that based on the society's philosophy of *Alam Takambang Jadi Guru,* that there are applications of sustainable architecture on the very house. Ecological environment has come to their intention relatively in balance with material and technological aspect of sustainability.

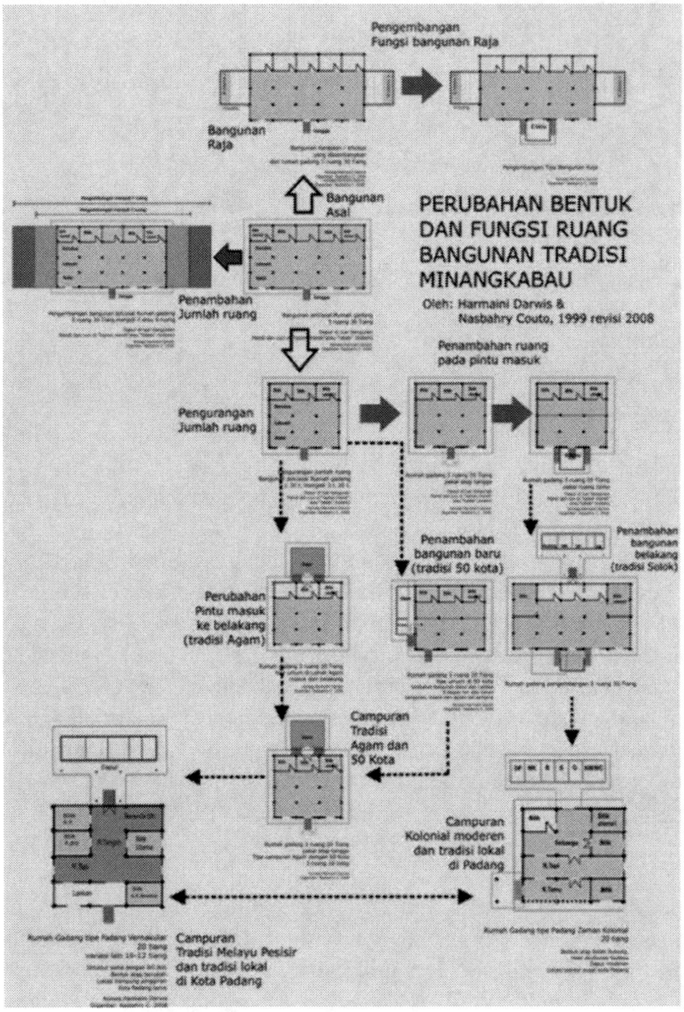

Figure 2. Morphological and functional changes of Rumah Gadang according to Couto and Darwis (in Indonesian).

4. Research method

We tried to find the conceptual practice of sustainability applied to Rumah Gadang, especially the ones that have spatially changed. Changes refer to Couto and Darwis's research outcome as research scope limitation. The actual parameters to reach for sustainability are based on Attmann theory. Case studies research were done to collect data needed for analysis. Two classified houses were chosen based on the changes classification made by Couto and Darwis. The two houses are: *Kaum Datuk Amat Dirajo* House at Batusangkar, and *Kaum Bawah Balai* House, also at Batusangkar. Same locus is arranged for the reason on comparing how the two house response to the similar natural environment around. The two houses were observed to understand the logic of it sustainability shifted changes.

5. Research results

IOP Conf. Series: Earth and Environmental Science **105** (2017) 012040 doi:10.1088/1755-1315/105/1/012040

Analysis based on observation and diagrammatic data were done to define the sustainable aspects of Rumah Gadang. The annexations were built according to the current needs, expanding and shifting spatial changes. To some extend the results are still attached to the main house, connected by roofed staircases.

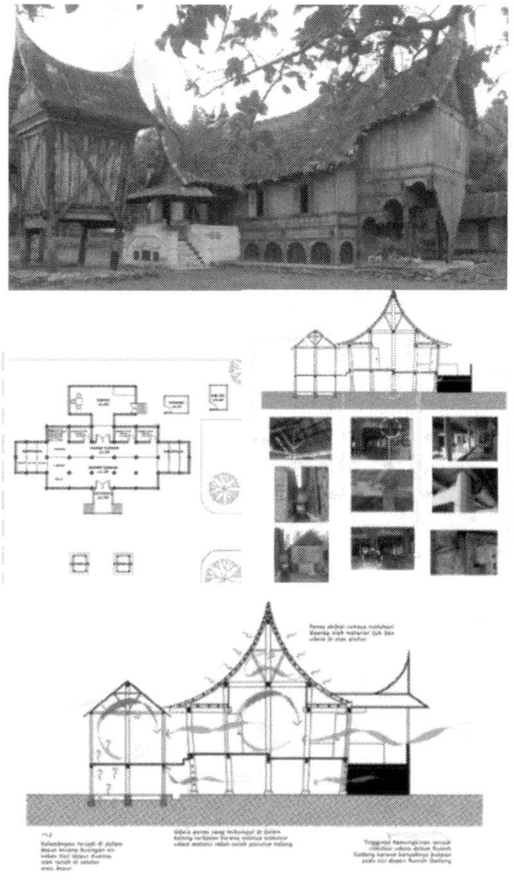

Figure 3. Rumah Gadang Kaum Datuk Amat Dirajo + Annexation. Plan and sustainability aspect approaches analysis.

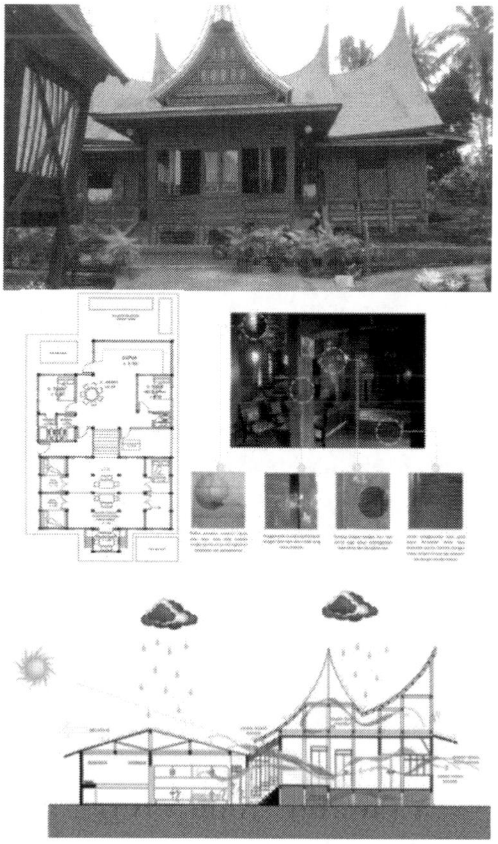

Figure 4. Rumah Gadang Kaum Bawah Balai + Annexation. Plan and sustainability aspect approaches analysis.

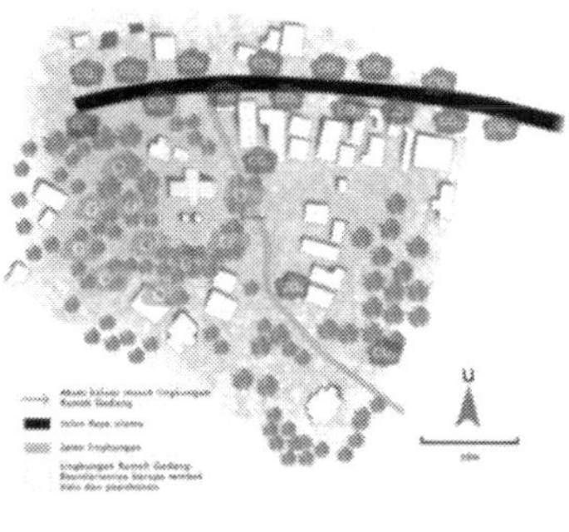

Figure 5. Rumah Gadang Kaum Datuk Amat Dirajo + Annexation. Sustainability aspect: Environment, efficiency and its potential site.

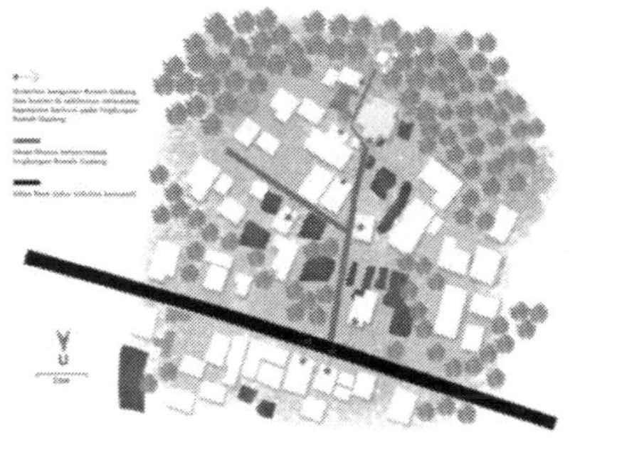

Figure 6. Rumah Gadang Kaum Bawah Balai + Annexation. Sustainability aspect: Environment, efficiency and its potential site.

The natural resources sustainability aspects are mapped diagrammatically to illustrate how the building works. The sun path is drawn to rationalize the logic of natural lighting, as same as the wind directions towards the house. The materials and structures are explained as rare wooden material with ornamental carvings, with finishing. The sustainability aspect of the material would be weakened, depending on the durability of the woods. Replacements are most likely would be differed from the originals. Among those changes, these case houses are the ones that match the classifications from Couto and Darwis, where annexations were taken.

6. Conclusion

The two houses show sustainable architecture aspects inherent to Attmann theory. On Rumah Gadang Kaum Bawah Balai, the requirements for durability, efficiency, economical on maintenance are intacted. The annex does not use natural material, which freed the house owner from dependency. In terms of recyclability, the house cannot fully meet the requirement, since the existing original material cannot be replaced by other material resources of the same kind.

The two houses responses to the natural resources and site potential as supporting energy for the inside space of Rumah Gadang. Healthiness and habitable aspects are supported by the durability of the elements and the use of surrounding resources. The houses are nicely responsed activities of the residents, which include networking of them within the society. Local customs held using the houses are still meeting the needs occasionally. The only aspect that short is the poor waste management system, in which at the end will affect the cleanliness factor of the environment. After further analysis, the two houses demonstrate the shifted archetypes of Rumah Gadang after annexations, where changes found without jeopardizing the existence of the main house. The philosophical value of the house is maintained with changes on its façade appearances. Its sustainability aspects are in place.

Further empiric research is suggested to calculate the shifted parameters used for its sustainability. This research as a foundation has shown the mapping of sustainable aspects of Rumah Gadang while at the same time put the recent dynamics facing the future of these types after annexation. Models are described in details according to its related sustainability theory. By reading it, we can further read other vernacular houses sustainability, from different ethnic groups in Indonesia or any other countries.

7. References

[1] Attmann O 2010 *Green Architecture: Advanced Technologies and Materials* (USA: The McGraw-Hill Companies, Inc)
[2] Bartlett A A 2012 *The Meaning of Sustainability* **31** No 1 pp 1
[3] Keeler M and Burke B 2009 *Fundamentals of Integrated Design for Sustainable Building* (Chichester: John Wiley and Sons Ltd)
[4] Kim J 1998 *Sustainable Architecture Module: Introduction to Sustainable Design* (Michigan: National Pollution Prevention Center for Higher Education)
[5] Nasbahry C 2010 *Morfologi Bentuk Bangunan Tradisi Minangkabau Sebagai Refleksi Budaya*. 26 Juli 2010. Accessed on November 6th, 2016, at 2 PM. http://visualheritageblog.blogspot.com/2010/12/morfologi-bentuk-bangunan-tradisi.html
[6] Navis A A 1984 *Alam Terkembang Jadi Guru*. (Indonesia: Grafiti Pers)
[7] Syafwandi 1993 *Arsitektur Tradisional Sumatera Barat* (Jakarta: Departemen Pendidikan dan Kebudayaan)
[8] Thaib Taufik 2016, November 22. Personal Interview

Power consumption analysis on large-sized electric bus

G Heryana[1], S Prasetya[1], M Adhitya[1,2] and D A Sumarsono[1]

[1] Department of Mechanical Engineering, Faculty of Engineering, Universitas Indonesia, Kampus Baru UI Depok 16424, Indonesia

E-mail: madhitya@rocketmail.com

Abstract. The increase of greenhouse gasses effect is one of the causes of climate change. The use of vehicles with fossil fuels is one of the contributors to pollution and global warming. Another reason why dependence on it should be reduced is the lack of Indonesia's petroleum reserves compared to other OPEC countries. Researchers are trying to anticipate this by developing electric vehicles capable of operating without pollution. Implementation of electric vehicles has begun with the electric train, tram, electric bus, and others. Vehicles with rails get electricity supplies from the grid along the tracks, but the type of freely moving vehicle without a rail requires another way to get electricity supplies. Electric vehicles with power storage (battery) have an advantage in the roaming area if supported by proper recharging techniques. In electric vehicles, the battery is recharged by the direct charging process or swapped with other one (swapping). If the vehicle has high mobility, such as public transport (bus) then the shortest charging time may be very important.

1. Introduction

Global warming, air pollution, and reduced fossil fuel sources are global issues that need to be sought immediately. USA became the largest contributor of carbon dioxide in the world by 28%. The biggest source of pollution caused the greenhouse effect is CO_2 exhaust from conventional motor vehicles with fossil fuels. Not using vehicle based on hydrocarbon fuel (HC) is one way to reduce CO_2[1].
Another thing that is not less important is the decline in oil reserves Indonesia, whereas as a member of OPEC Indonesia's oil reserves are among the smallest [2].

There is no 100% perfect solution, this is because the vehicle needs of each person or area will be different. Technical studies cannot be separated with socio-cultural studies. The results of a comprehensive study stated that the development of electric vehicles in the form of buses is the most strategic in Indonesia. The bus was developed to be a comfortable public vehicle so that the use of buses and private vehicles with fossil fuels is significantly reduced, especially in urban areas.

2. Research planning

The objective of this research is to design prototype of electric bus and to test its performance to fit the needs of load, route, and bus driving character. Thus, the most appropriate propulsion system, battery type and charging mode can be determined. The speed of the vehicle when testing is set maximum 40 kph. This speed can be a profile for electric buses with frequent stop and shorter shelter distance.

[2] To whom any correspondence should be addressed

Frontal cross-sectional area (appx.) 8,16 m^2, aerodynamic resistance (C_d) 0,5328 [3], and rolling resistance 0,002. In addition, there is the assumption of the fluid type mass, and the elevation angle.

Bus is tested without passenger load and without operating cabin air conditioner (AC). Thus the devices that must be functioned are the main motor, compressed air motor, and power steering motor. This is a minimal condition for running a bus. The distance traveled about 4.4 km. This route has short distance between stops and short stop time.

3. Theory

The needs of the main electric motor (bus drive) is calculated by the traction power analysis of the vehicle. The traction force required to drive the vehicle is influenced by several resistances that act while the vehicle is moving as illustrated by the following equation [4] (figure 1).

$$F = m\,a + R_a + R_{rl} + R_g \tag{3.1}$$

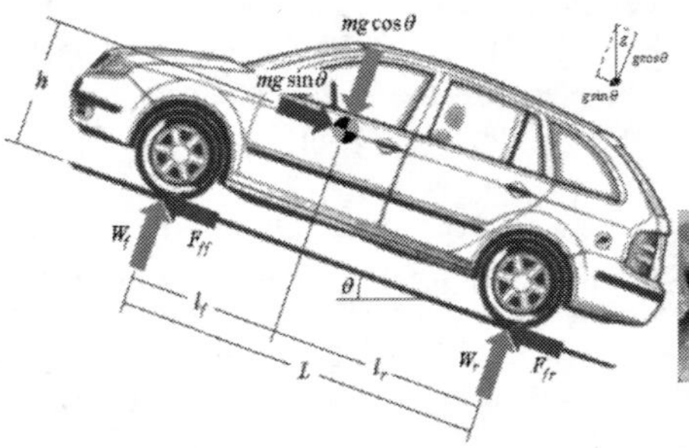

Figure 1. Forces for traction calculation .

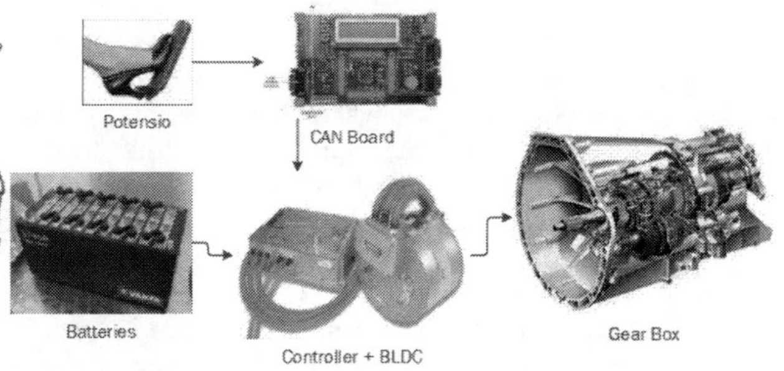

Figure 2. Main motor integration system.

The value of vehicle traction with a certain speed is obtained by the formula [5]:

$$Po = \left(m\,a + \frac{1}{2}\rho\, C_d A_f v^2 + f_{rl}\, m\, g + m\, g \sin\alpha\right).v \tag{3.2}$$

The main voltage of the battery is supplied to the controller then to the motor. The bus uses a multi-motor system in which each mechanism is driven by one motor.

The BLDC 200 KW 380V is used as prime mover. Maximum torque that can be achieved around 700 Nm. Motor works on a voltage between 230 ~ 400 VDC. The motor rotation is then coupled and forwarded to the gear box. How fast is the motor rotates depending on the number of resistance froman acceleration pedal.This resistance value is still analog, then translated by CAN board into digital signal, rocessed and given priority to send. The signal value is forwarded to the BLDC Controller with the CANbus protocol (figure 2).

Power steering using hydraulic system (figure 3). The hydraulic pump is rotated by a 3 phase 380 VAC induction motor with a capacity of 7.5 KW. In the design of power steering that has been installed, although already using an electric motor as a driver but has not been controlled by sensors. The electric motor keeps rotating even when the bus is idle. Motor Power Steering is enabled with a button on the bus dashboard. This button (switch type) provides an enable or a disable signal to the motor inverter.

Compressed Air is filled into the reservoir with reciprocating pump (figure 4). The pump is rotated by an electric motor with a power capacity specification of 4.4 KW 380 VAC 3 phase.For power steering currents limited to a maximum of 22 Ampere, while air compressor limited to a maximum of 10 A.

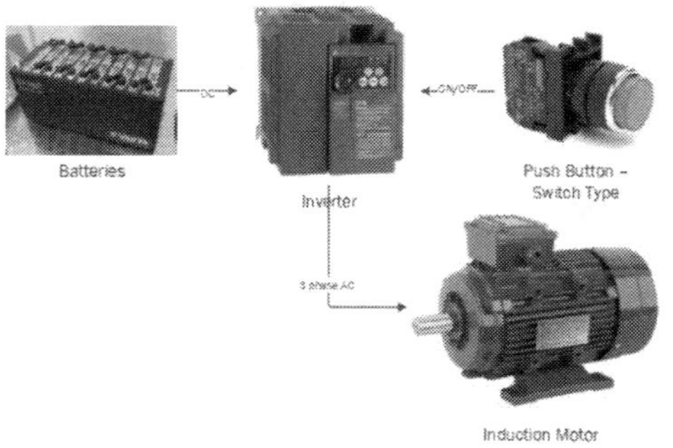

Figure 3. Power Steering System.

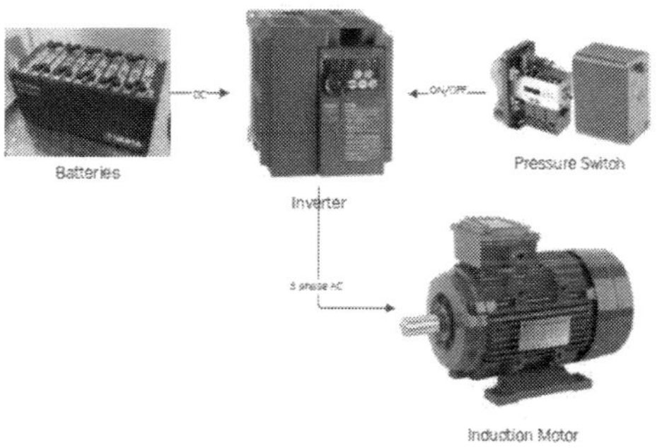

Figure 4. Compressed Air System.

The value can be calculated with guidance of specification data of each motor. The main values to be input to the inverter are the working voltage of the motor, the maximum current, the working frequency of the motor, the inverter activation method, the rotation direction, the acceleration and the ramp up & ramp down time. Address parameters can be seen in the inverter book manual [6].

The motor for compressed air operates under pressure sensor trigger. The motor works when the minimum pressure is reached and will stop when the maximum pressure is reached (intermittent rotating). While the power steering rotates continuously (non-stop).

4. Result and discussion

The bus is tested at 334 Volt (BLDC) / 339 Volt (PQA) battery voltage position. The result of the test of the electric bus that runs for about 4.4 km obtained the data of 108 volts, voltage drop based on data from BLDC Controller logger, and 109 volts based on Power Quality Analyzer (PQA) data. Running duration is 908 seconds or 15 minutes and 8 seconds (figure 5).

Recording results of both devices show similar results. At a 230 volts, voltage drop is almost no more. During the driving, power steering is continuously active while the compressed water is actively start and stop in accordance with the reservoir volume condition.

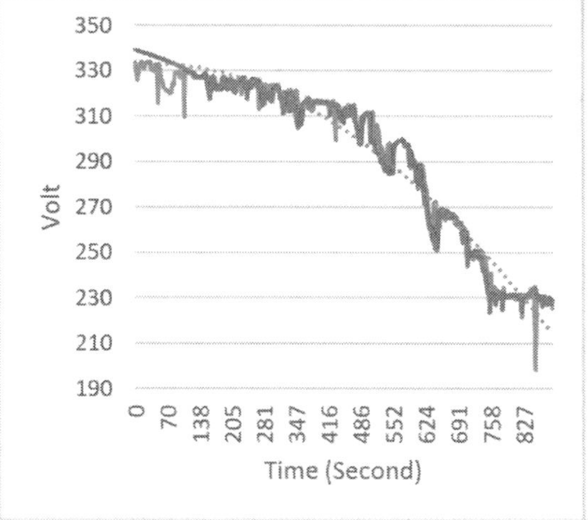

Figure 5. Battery Voltage Drop - Main Motor.

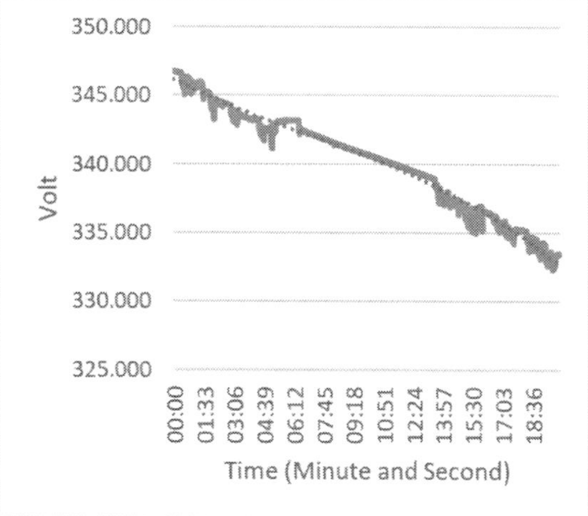

Figure 6. Voltage Drop – P/S and Compressed Air.

Power steering and compressed air loads make the impact of linear voltage drops tend to be linear. This shows that the load of the two devices is relatively constant throughout the test (figure 6).

IOP Conf. Series: Earth and Environmental Science **105** (2017) 012041 doi:10.1088/1755-1315/105/1/012041

Ripple or spike on the chart is a condition in which the steering wheel is rotated left and right actively as if it were a winding path. Voltage drops with "interference" because the battery load changes. Moments later the system was tested with the condition as if it were a straight path, figured on the graph that the relative voltage drop is subtle.

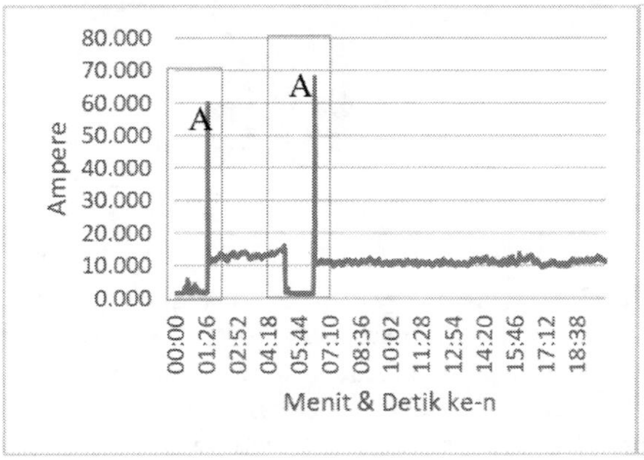

Figure 7. Comp. Air Current Profile VS Time.

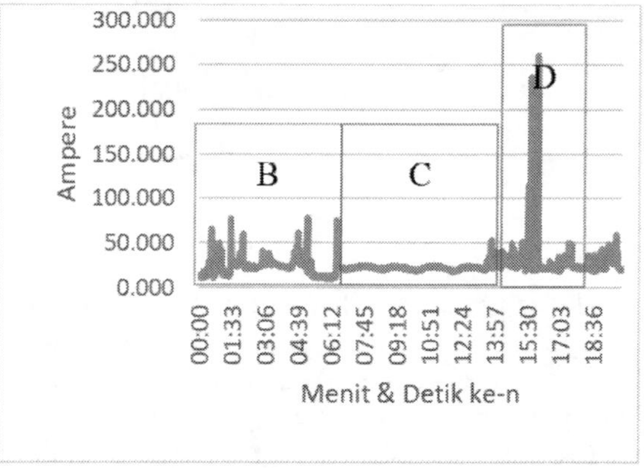

Figure 8. P/S Current Profile VS Time.

During the test, the motor compressor is only 2 times off. The rest of the motor continues to rotate. The load during the active motor is relatively constant, except when the reservoir is almost full the graph indicates a slight increase in current. Another phenomenon is the existence of spike when the motor is active (figure 7 - A).

The minimum pressure on the pressure switch is set to 6 bar. At this pressure the motor will be active. The motor will stop filling the reservoir when the pressure reaches 8.5 bar. The test results do not show any lack of air supply due to the frequency of braking, clutching, or leakage.

Power steering (P/S) motor currents show almost uniform fluctuations in almost equal ranges (figure 8). Spike in range B is a test condition with a sharp twisty road. Range C when road conditions are relatively straight with no sharp turns or just a lane move. The spike occurs because of more load when sharp turn. The rest of the current consumption is relatively stable.

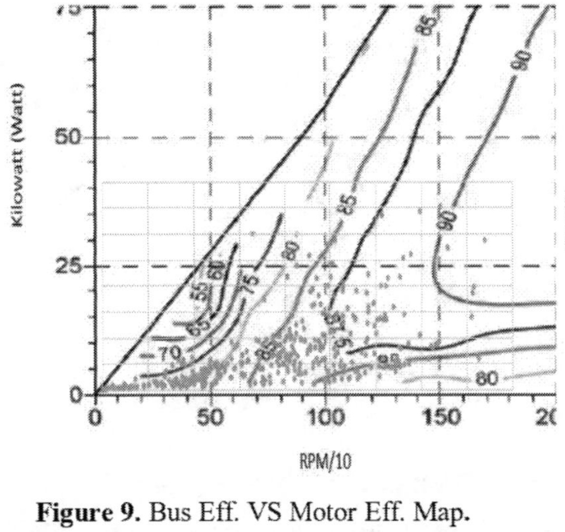

Figure 9. Bus Eff. VS Motor Eff. Map.

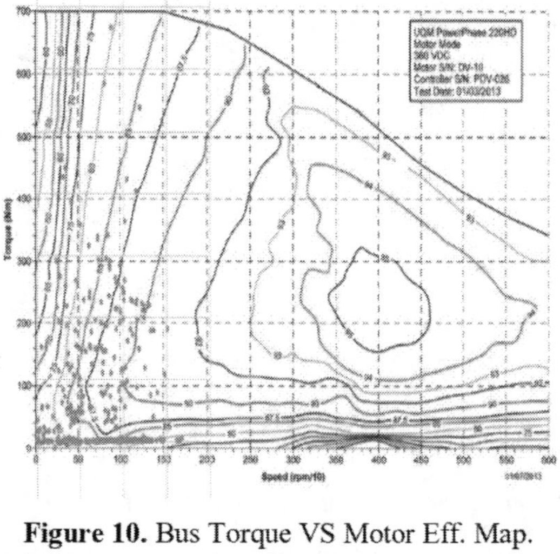

Figure 10. Bus Torque VS Motor Eff. Map.

When turning the steering wheel for parking maneuvers (figure 8 - range D), bus speeds are often almost zero, so the front wheel load proportion is still relatively large. In this condition the required torque from hydraulic power steering pump jumps high, in other words the electric current of the pump motor must be larger. With improper driving techniques the motor will often over load and die. The calculation or redesign of the driving force is necessary.

The BLDC motor character attached to the UI electric bus has the best power efficiency (95%) in the range of about 3800 to 4800 RPM. At that rotation the power output is about 80 to 130 kW.

In the UI electric bus test, the required power is recorded on the logger. If the power value along the way is spread on the Power Efficiency Map [7], it can be seen that the efficiency value is in the range of 70% to 87.5%. Only a small percentage is in the 90% area (figure 9). This is in accordance with the conditions at the time of the test in which the internal bus route UI only requires low speed with engine speed around 1500 RPM.

In terms of torque, the BLDC motor character attached to the UI electric bus has the best efficiency (95%) in the range of about 3800 to 4800 RPM. At that rotation the torque is about 150 to 320 Nm.

If the value of torque along the way spread on the Torque Efficiency Map can be seen that the efficiency value is in the range of 70% to 87.5%. The required torque is spread below 300 Nm (figure 10). The UI bus test route and the relative load when test does not require large torque and the bus does not require high rotation due to the character of DC motor which has sufficient torque at low rotation.

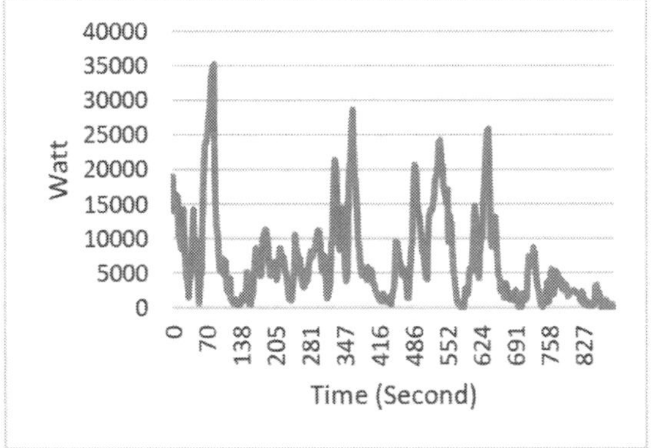

Figure 11. Power from BLDC Logger.

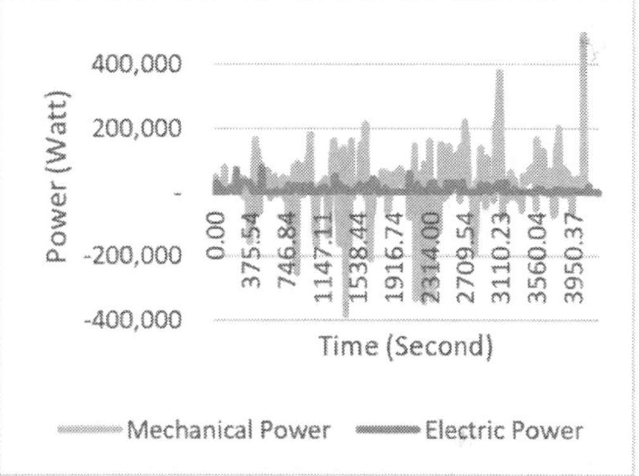

Figure 12. Mech. Power vs Elec. Power.

The power during testing indicates irregular fluctuations (figure 11). Among the test time ranges are 4 to 5 times with the highest spikes. In the last minute the fluctuations were not too sharp. This completes other test data that indicates an almost ending battery voltage. If this data is combined with road condition data then it will get a more sharp conclusion.

Mechanical Power is theoretically power calculation while Electrical Power is the actual power taken from the BLDC motor data. In Mechanical Power chart shown that when the bus requires thrust or uphill then the value will go positive, while when the road decreased power also decreased even to the negative (figure 12). While on the Electric Power chart shown that the motor provides power in accordance with the needs with value from zero to the positive direction, whereas if the road decreases the power of the bus motor cannot be changed as regenerative power. That's why the lowest value on the Electric Power chart is zero.

If observed, the output power of the calculation result (theoretical) in general has a higher average than the actual power (input) logged by the motor. Whereas in theory the output power should be smaller than the input power. This phenomenon will be investigated and analyzed further. The hypothesis is that the theoretical formula does not consider that in actual conditions the bus is only need power impulsively. Power is delivered until the desired speed is reached. After the acceleration pedal is stepped on and released, there is actually some power that causes the bus to stay up for a certain time.

5. Conclusion

1. The voltage drop is quite significant if the battery voltage is not in the maximum position
2. In pressurized air filling, an increase in current occurs when the charge is almost full due to an increase in pressure as a load
3. Power steering loads are relatively constant and small compared to main motor loads
4. Power efficiency is in the range of 65% to 80%
5. Torque efficiency is in the range of 75% to 82.5%
6. Further analysis is required to compare theoretical test results

6. References

[1] Kim J, Song I and Choi W 2015 An electric bus with a battery exchange system *Energies* **8** 6806-19
[2] http://katadata.co.id/berita/2015/11/16/cadangan-minyak-turun-gas-meningkat
[3] Ridho M 2015 *Studi Efek Hambatan Aerodinamik Pada Performa Bus Listrik Universitas Indonesia* (Skripsi, Fakultas Teknik Departemen Teknik Mesin, Universitas Indonesia)
[4] Sutantra, I Nyoman and Sampurno B 2010 *Teknologi Otomotif Edisi Kedua* (Surabaya: Guna Widya) p 56
[5] Helmi A S 2016 *Perancangan dan Uji Performa Sistem Powertrain Konversi Listrik Pada Mobil Listrik Nasional UI* (Skripsi, Fakultas Teknik Departemen Teknik Mesin, Universitas Indonesia) p 21
[6] http://www.MitsubishiElectric.co.jp/melfansweb
[7] UQM Datasheet 2013 *UQM Power Phase® HD 220 System, Revision: X6* (UQM Technologies, Inc., Longmont, CO)

Characterization of shape-stabilized phase change material using beeswax and functionalized multi-walled carbon nanotubes

S Rawi[1], M Amin[1], E Kusrini[2] and N Putra[1]

[1] Applied Heat Transfer Research Group, Department of Mechanical Engineering, Faculty of Engineering, Universitas Indonesia, Kampus Baru-Depok, Jawa Barat, Indonesia
[2] Department of Chemical Engineering, Faculty of Engineering, Universitas Indonesia, Kampus Baru-Depok, Jawa Barat, Indonesia

E-mail: nandyputra@eng.ui.ac.id

Abstract. One of the promising solutions for energy management is thermal energy storage. Beeswax is type of wax from beehive which can also be performed as thermal storage with high latent heat but it has low thermal conductivity. On the other hand, multi-walled carbon nanotubes (MWCNTs) are known as an advanced nanoparticle which has high thermal conductivity. In this research, modification of Beeswax/Acid-treated CNTs (A-CNTs) composite was conducted in order to determine shifting on its structure and thermal performance. MWCNTs were functionalized with strong acidic mixture by ratio 3:1 of H_2SO_4 and HNO_3 becoming A-CNTs. Impregnation method was being used to shape the composite with mass ratio of 5% A-CNTs. Several tests were conducted, functional group formation was determined by Fourier Transform Infrared (FTIR) test and thermal performance was observed by Differential Scanning Calorimetry (DSC) test, and thermal conductivity test. Test results show that modification of stable Beeswax/A-CNT composite was successfully formed and indicate that with addition of 5 wt.% A-CNTs, the latent heat of Beeswax decreases 25% and significantly escalate the Beeswax's thermal conductivity by up to 84%. It supports that Beeswax is a promising solution for energy management system near in the future that is available abundantly and affordably.

1. Introduction

Today we are facing two major challenges in this world which are reducing gas emissions from fossil fuels and finding alternative energy source to fulfil our needs. This condition forces people to utilize the energy usage and it also drives researcher to develop a new solution. One of the most significant advancement for energy efficiency is Thermal Energy Storage (TES). A large number of Phase Change Materials (PCM) such as organic, inorganic compound, paraffin, and non-paraffin organic acids have been considered to be applied for latent heat thermal energy storage (LHTES) applications [1-4]. It has been recently studied that the PCMs organic, inorganic and their mixtures are prospective LHTES materials that can be used for

IOP Conf. Series: Earth and Environmental Science **105** (2017) 012042 doi:10.1088/1755-1315/105/1/012042

building materials impregnation [3]. Organic PCMs have been attracted more than the others due to its good characteristics [5, 6].

In Indonesia, one of the most popular organic PCMs is Beeswax. Beeswax is comprised of esters of fatty acids and long chain alcohols. Beeswax belongs to category of organic non-paraffin PCMs. Its development as PCM described by Sinaringati, et al. 2016 for infant incubator heat source [5]. Amin, et al. 2017 tested the modification of nano-PCM using Beeswax and graphene with variative mass ratio to determine its thermal characteristics and the result is thermal conductivity of this nano-PCM increased by 12% [6].

Nevertheless, Beeswax as organic PCM possess some drawbacks that it has low thermal conductivity, and possesses leakage during phase transition. In order to eliminate these disadvantages, there are several methods to advance the process, for instance, encapsulation and formation of Shape Stabilized Phase Change Material (SSPCM) [7]. Type of material that is frequently prepared and characterized as SSPCM are porous and carbon-based materials. This method is an effective way to resolve the leaking and low thermal conductivity problems of PCMs because carbon has high thermal conductivity of 2000–6000 W/m K [4].

The objective of this paper is to prepare a novel form-stable PCM utilizing Beeswax as PCM and Multi-walled Carbon Nanotubes (MWCNTs) as supporting material. Characterization of the composite was conducted with FTIR test, and DSC test to capture the thermal properties and thermal reliability of the SSPCM. It was discovered that the thermal conductivity of SSPCM is increased by the addition of MWCNTs.

2. Methodology

2.1. Materials

The pristine MWCNTs were procured from Chengdu Organic Chemical Co., Ltd in diameter of 10-30 nm and length of 10-30 μm with surface area of 60 m^2/g and purity: 95 wt.%. Beeswax as PCM has melting temperature of 60–62 °C [6], and the chemical compounds of nitric acid (HNO_3, 65 wt.%), sulfuric acid (H_2SO_4, 98 wt.%), and absolute ethanol were purchased from Merck Millipore Co.Ltd.

2.2. Preparation of MWCNTs

To begin the acid oxidation, two grams of Pristine CNTs and 400 ml of acid mixture which consisted of nitric (6M) and acids (6M) with volume ratio of 1:3 were stirred altogether in a beaker for 1 hour at temperature of 50 °C. Then the sample in beaker was diluted with aquadest for several times until the solution show no acidity (pH = 6.5), next it was filtrated by using a polycarbonate membrane filter with pore diameter of 45 nm. The neutral MWCNTs were put in a stainless steel container covered by alumunium foil and heated in an oven to remove the attached water at 100 °C for 24 h. The MWCNTs that were treated by the aforestated method of acid oxidation as shown in figure 1 named as A-CNTs.

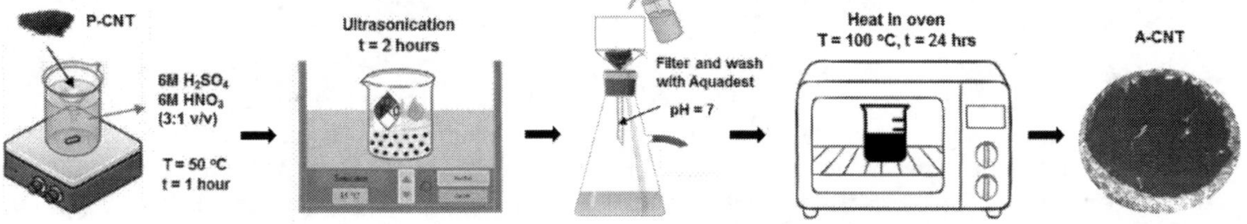

Figure 1. Oxidation of carbon nanotubes using acidic mixture.

IOP Conf. Series: Earth and Environmental Science **105** (2017) 012042 doi:10.1088/1755-1315/105/1/012042

2.3. Preparation of Beeswax/MWCNTs composite shape-stabilized PCMs

As shown in figure 2, the preparation of Beeswax/MWCNTs composite implemented by using impregnation method [8]. A fixed proportion of A-CNTs/ethanol (1 gr/150 mL) suspensions was blended with Beeswax/ethanol solution with weight ratio of Beeswax/A-CNTs is 5%. Then the mixture was treated in an ultrasonic bath for 30 minutes and stirred for 4 h at 70°C by using magnetic stirrer [9]. Sonication is important in order to make a well dispersed and homogeneous Beeswax/A-CNTs composite. In order to evaporate the attached ethanol, the mixture was heated in an oven at 80 °C for 24 h and finally the sample of Beeswax/A-CNT 5 wt.% was successfully prepared for properties identification.

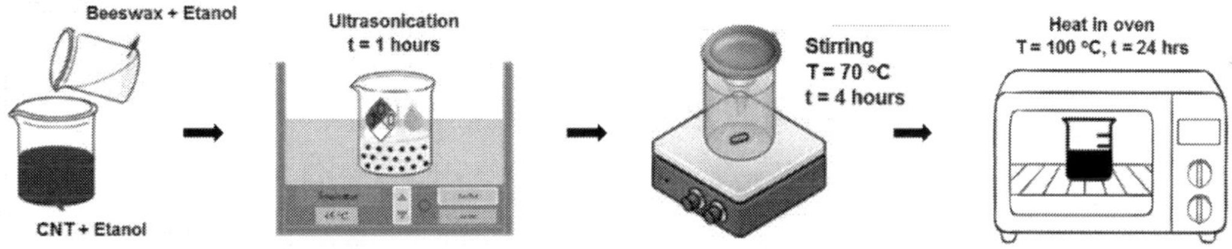

Figure 2. Composite preparation by impregnation method.

2.4. Characterization

Fourier Transform Infrared Spectroscopy (FTIR) was occupied to identify the shifting and formation of chemical groups in MWCNTs after treated by axid oxidation. Differential Scanning Calorimetry (DSC) ASTM D3418:15 was occupied to measure the thermal properties of SSPCM including its melting temperature, freezing temperature and latent heat capacity. The DSC assessments was completed at a 5 °C/min heating and cooling rate and temperature ranges of 0 to 100 °C and 100 to 0 °C. Physical phase stabilization of SSPCMs was potrayed by using a hot stage-digital camera at temperature of 100 °C [8]. Finally, thermal conductivity of SSPCMs were measured by comparative cut-bar method [10].

3. Results and discussion

3.1. Physical, Chemical and Morphological properties

Based on the observation of each sample that had been heated at 80°C for 5 different time variations: 0 minute and 15 minutes as seen in table 1, it was shown that Beeswax/A-CNTs composite is able to maintain its shape without any lamination better than pure Beeswax. Composite using A-CNT 5 wt% was able to maintain its shape until the tenth minute without leakage. Oxidized carbon nanotubes using strong-acid resulted to the most optimal CNT that could be used for SSPCM formation. This type of CNT traps beeswax inside the A-CNT cavity, therefore the beeswax did not leak out during the heating process.

Table 1. Phase changing process for period of 15 minutes.

	1 minute	5 minutes	10 minutes	15 minutes
Pure Beeswax	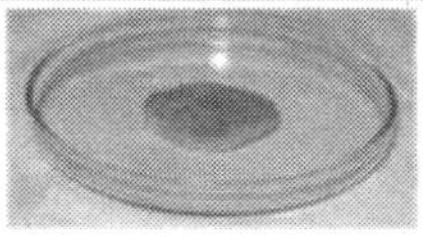			
Beeswax/ A-CNTs 5 wt.%		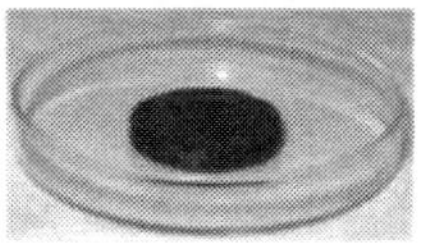		

Based on the physical examination, Beeswax/A-CNTs composite tends to maintain its form due to its chemical structure changes. Oxidizing CNT using strong acid allows the CNT to form certain functional groups. Therefore, the success of CNT oxidation could be observed using Fourier Transform Infrared (FTIR) Spectrophotometer, Shimadzu IR Prestige 21. The FTIR test graph shows waves' crests that can be used to identify the functional groups of samples. Figure 3 shows the FTIR test results of CNT and A-CNT that were used as the base materials for SSPCM. The FTIR spectrum of pure CNT does not show peak on the characteristic band, meanwhile A-CNT result shows several peaks.

Carboxylic acid C=O is shown from 1535cm^{-1} wave, which is a specific carbonyl of CNTs. If there is >3200 cm^{-1} wave, there is a high chance that the CNT contains carboxyl functional group (-COOH), as 3442 cm^{-1} indicates the existence of O-H bond. Between 1750 – 1550 cm^{-1} wave, there are C=O functional group which has alkanoic acid and C=C bond. Wavenumber shown between 3300 – 2700 cm^{-1} introduces C-H bond as the sample has at 2915 cm^{-1}. Lastly, 1300 – 950 cm^{-1} wave proves C-O bond does exist [11]. The peak formations on the FTIR graph indicate that A-CNTs, which was used as supporting material for composite, has new functional groups after being oxidized with strong acid. The desired result is for the bond between A-CNTs and Beeswax to become stronger and still be able to maintain its form although has reached its melting point to meet the desired characteristics of SSCPM.

3.2. *Thermal properties*

The thermal properties of pure Beeswax and Beeswax/MWCNTs composite were identified for the further analysis by using Differential Scanning Calorimetry (DSC). These samples' freezing points, melting points, characteristics temperatures and phase change enthalpies were examined and compared in order to distinguish the thermal performance of Beeswax with and without the addition of MWCNTs.

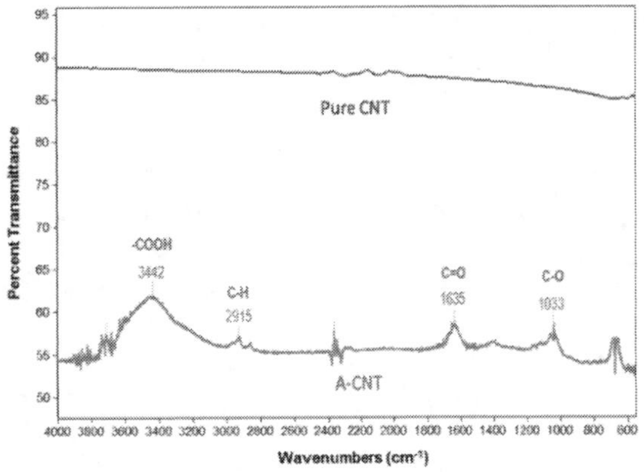

Figure 3. FTIR spectra of Pure CNT and A-CNT.

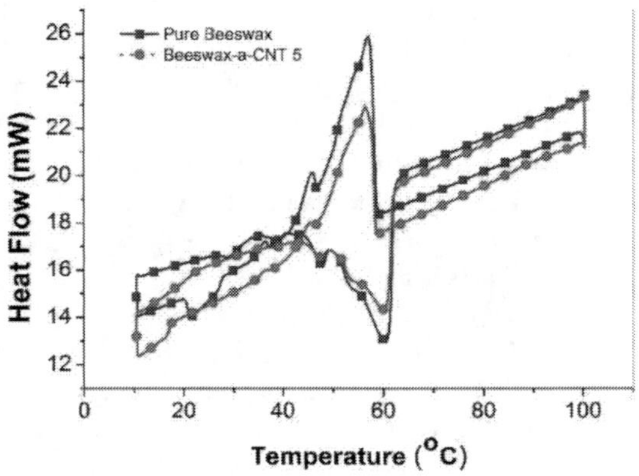

Figure 4. DSC curves of Beeswax/ A-CNTs 5 wt.%.

The cooling and heating rate of DSC was set at a constant rate of 5 °C per minute and at the beginning of DSC analysis, both of pure Beeswax and Beeswax/A-CNTs composite were at an initial temperature of 10 °C. These samples experience the phase transition processes through melting – freezing cycles within temperature range of 10-100 °C. Occupied calorimeter has accuracy within 0.2% and temperature error measurement within 0.05 °C. The working principle of DSC shows that the heat flow signal of a DSC curve could demonstrate characteristic temperatures of the inserted sample which are melting point, freezing point, onset heating temperature and onset cooling temperature. Figure 4 shows the comparison of DSC analysis for pure Beeswax and Beeswax/A-CNTs 5 wt.% during the melting and cooling process, and the value of characteristic temperatures are shown in table 2.

Table 2. Phase change temperature and latent heat of the Beeswax and the Beeswax/CNTs composite.

Sample	Heating			Cooling		
	Tm (°C)	Onset (°C)	ΔH (J/g)	Tc (°C)	Onset (°C)	ΔH (J/g)
Beeswax (Bwx)	60.59	51.36	153.21	57.00	58.71	-154.01
Beeswax / A-CNT 5 wt.%	60.18	49.38	115.54	56.39	58.46	-126.20

From the resut, there is no substantial temperature change between sample with and without addition of A-CNTs. The melting and freezing temperatures are determined as 60.59 °C and 57 °C for the pure Beeswax whereas Beeswax/MWCNTs composite exhibited modest changes in the phase transition temperatures because there is a confinement of the Beeswax molecules into the pores of MWCNTs. Form stable composite with impregnation of A-CNTs have melting point and cooling point of 60.18 °C and 56.39 °C for Beeswax/A-CNTs composite. The melting and freezing temperatures of the Beeswax/A-CNTs composite are suitable for TES applications in buildings [12] . However, the slight variation during this phase change process due to MWCNTs addition could be affected by the heat transfer enhancement and crystallization effect [13].

Moreover, DSC results show the latent heat for melting and freezing are found to be 153.21 J/g and 154.01 J/g for the pure Beeswax, 115.54 J/g and 126.2 J/g for the Beeswax/A-CNTs 5 wt.% composite. Latent heat is decreasing by addition of MWCNTs. This thermal performance is appropriate to the initial hypothesis that in order to enhance the thermal conductivity of PCM, addition of nanoparticle MWCNTs is required.

One of important parameters that must be considered in the selection of a composite PCM is thermal conductivity, especially for TES system application. Therefore, improvement of this parameter is efficient on enhancing the storage and release performance and designing of TES system [1]. Knowing this fact, this research is also considering on increasing the thermal conductivity of the fabricated Beeswax/A-CNTs composite. The thermal conductivity performance of SSPCM was measured by thermal conductivity meter Ogawa Seiki OSK 4565-A.

Thermal conductivity of samples are shown in table 3 which the thermal conductivity of Beeswax is approximately 0.25 W/m K at temperature 40 °C as reported from literatures [6]. Increasement of MWCNTs into Beeswax improve the thermal conductivity of the composite. Based on test result, thermal conductivity of form-stable Beeswax/A-CNTs 5 % wt is 0.46 W/m K, meaning that the thermal conductivity of composite is increased by 84% compared to the pure Beeswax . This shows that the MWCNTs affects the enhancement of the thermal conductivity of Beeswax. Through the thermal conductivity analysis, it is confirmed that the modified composite in this research is useful for TES applications.

Table 3. Thermal conductivity of samples at T = 40 °C.

Material	K (W/mK)
Beeswax [6]	0.25
5 wt.% A-CNT	0.46

4. Conclusion

The preparation and characterization of physical/chemical/morphological properties of Beeswax/A-CNTs as SSPCM with enhanced thermal conductivity were presented in this paper. Two samples were utilized which are pure Beeswax and Beeswax/MWCNTs composite. MWCNTs were modified, tested, and added in order to form a shape-stable composite. From experiments conducted, it is known that the SSPCM was successfully fabricated by the impregnation of A-CNTs into Beeswax. Thermal conductivity of composite is increased by the addition of 5 wt.% A-CNTs. Based on FTIR test, it is proven that A-CNT has a good physicochemical compatibility as supporting material for composite. The DSC results confirm that the prepared Beeswax/A-CNTs composite has melting temperatures of 59.79 °C, and latent heat value of 91.64

J/g. The composite posess a good realibility as well as shape stability for TES material. Moreover, thermal conductivity of Beeswax is increased considerably as 84% by the addition of 5 wt.% A-CNTs. In conclusion, Beeswax/A-CNTs 5 wt.% composite is a promising TES material that could reduce energy usage in buildings that unintentionally could lessen the climatic catastrophe. However, the large-scale TES performance of this Beeswax/A-CNTs composite is necessary for further consideration.

5. Acknowledgement

The authors express gratitude to *Direktorat Riset dan Pengabdian Masyarakat* (DRPM) Universitas Indonesia for funding this research through the "PITTA 2017" scheme.

6. References

[1] Zalba B, Marm J M , Cabeza L F and Mehling H 2003 Review on thermal energy storage with phase change: materials, heat transfer analysis and applications *Appl. Therm. Eng.* **23** 251-83
[2] Pasupathy A, Velraj R and Seeniraj R V 2008 Phase change material-based building architecture foe thermal management in residential and commercial establishments *Renew. Sustainable Energy Rev.* **12** 39-64
[3] Sharma A, Tyagi V V, Chen C R and Buddhi D 2009 Review on thermal energy storage with phase change materials and applications *Renew. Sustainable Energy Rev.* **13** 318-45
[4] Pielichowska K and Pielichowski K 2014 Phase change materials for thermal energy storage *Prog. Mater Sci.***65** 67-123
[5] Sinaringati S, Putra N, Amin M, and Afriyanti F 2016 Utilization of paraffin and beeswax as heat energy storage in infat incubator *ARPN-JEAS* **11** 800-4
[6] Amin M, Putra N, Kosasih E A, Prawiro E, Luanto R A and Mahlia T M I 2017 Thermal properties of beeswax/graphene phase change material as energy storage for building applications *Appl. Therm. Eng* **112** 273-80.
[7] Fang G, Tang F and Cao L 2014 Preparation, thermal properties and applications of shape-stabilized thermal energy storage materials *Renew. Sustainable Energy Rev.* **40** 237-59
[8] Li B, Liu T, Hu L, Wang Y and Nie S 2013 Facile preparation and adjustable thermal property of stearic acid–graphene oxide composite as shape-stabilized phase change material *Energy Convers. Manag.* **215** 819-26
[9] Li B, Nie S, Hao Y, Liu T, Zhu J and Yan S 2015 Stearic-acid/carbon-nanotube composites with tailored shape-stabilized phase transitions and light–heat conversion for thermal energy storage *Energy Convers. Manag.* **98** 314-21
[10] Zhao D, Qian X, Gu X, Jajja S A and Yang R 2016 Measurement techniques for thermal conductivity and interfacial thermal conductance of bulk and thin film materials *J. Electron Packag.* **138** 1-19
[11] Stobinski L, Lesiaka B, Kover L, Toth J, Biniak S and Trykowski G 2010 Multiwall carbon nanotubes purification and oxidation by nitric acid studied by the FTIR and electron spectroscopy methods *J. Alloys Compd.* **501** 77-84
[12] Cabeza L F, Castell A, Barreneche C, Gracia A D and Fernandez A I 2011 Materials used as PCM in thermal energy storage in buildings: a review *Renew. Sustainable Energy Rev.* **15** 1675–95
[13] Li T, Lee J H, Wang R and Kang Y T 2013 Enhancement of heat transfer for thermal energy storage application using stearic acid nanocomposite with multi-walled carbon nanotubes *Energy* **55** 752-61

IOP Conf. Series: Earth and Environmental Science **105** (2017) 012043 doi:10.1088/1755-1315/105/1/012043

Pentamaran configuration design with modeling hull form for resistance minimization

Yanuar[1] and Wiwin S[2]
[1]Department of Mechanical Engineering, Universitas Indonesia, Depok 16424, Indonesia
[2]Doctoral Student of Department of Mechanical Engineering, Universitas Indonesia, Depok 16424, Indonesia

E-mail: yanuar@eng.ui.ac.id

Abstract. Multi-hull ship had characteristics with complex geometry configurations. In the early stage design, multi-hull ship with proper configurations can be producing both minimum resistance and power consumption. Increasing number of hull as in pentamaran will increase frictional resistance, but total resistance can be reducing by decreasing wave making resistance with a slender hull form. With CFD tool used to examine minimum resistance with modified modelling of chine hull form on variation i.e. deadrise angles, waterline section of entrance angle and longitudinal section of stem angle. All simulations were conducted with fix separation on S/L 3/16 and Froude number up to 0.7. Comparison with the original hull form, chine hull form C35 II in deadrise 35^0 and angle of entrance 15^0 more advantageous with maximum resistance drag reduction 20.38%. The results of the study had been displayed in table and charts of chine pentamaran which having the smallest total resistance expressed in coefficient of friction and wave resistance.

1. Introduction

Experimental and numerical modeling techniques were essential to obtaining designs with good hydrodynamics, both performance and safety. Experimental for multi hull ships was first performed by [1] on a catamaran to obtain optimal hull form with minimum resistance. [2] had tested outrigger configurations of multihull to investigated interference resistance for catamaran, trimaran and tetramaran. Experimentally studied about pentamaran configurations had been studied by: [3], [4], [5], [6] and [7]. Numerical approaches to investigation of wave interference effects, seakeeping performances and the optimal separation distance between the hulls were founded in [8], [9], [10].

Due to a hull form is very influential on the ship's resistance, it is necessary to getting design with proper configuration on the multi hull ship. Researches on multihull generally used Wigley hull form, while chine hull form in some studies shown advantages such as: reducing resistance, easier and faster in the building process. [8] had compared chine and Wigley model on catamarans, which Wigley had a higher value of Cw (wave coefficient) at Fr <0.6, while chine hull form Cw tends to decrease at Fr > 0.6 and lift coefficient increases with increasing angles of deadrise. Agree with [8], [11] had used multihull with chine on variations deadrise angle 0°-20°, which the lift coefficient improved with increasing deadrise angle and Fr. This research is investigating resistance components and drag reduction to the position effect of the side-hulls from both Wigley and chine hull form with modelling modified at variation of deadrise angles on 25^0, 30^0, 35^0; angle entrance: 5^0,10^0,15^0 and stem angle: 0^0.

IOP Conf. Series: Earth and Environmental Science **105** (2017) 012043 doi:10.1088/1755-1315/105/1/012043

The experimental data were using the results of [6], which the original Wigley hull form changed to variations chine hull form with outriggers position configuration. From the experimental results will be compared with the results of numerical analysis by using CFD tool.

2. Hull form

This research is a comparative study between Wigley and chine with modification of hull form and outriggers configuration. The modification of hull form conducted to chine with dead rise angle: 25°, 30°, 35°, stem angle: 0^0 (vertical straight), and entrance angle: 5^0-15^0. Preliminary analysis was performed on Wigley and chine with dead rise angle: 25°, 30°, 35°, while angle of entrance (15^0) and stem angle actually used model of [6] experiment. The hull form of both Wigley and chine characteristics are listed in table 1.

Table 1. Hull form Characteristics

	Wigley			**Chine 25⁰**		
	Main	Inner side	Outer side	Main	Inner side	Outer side
Length	2.00 m	1.00 m	0.75 m	2.40 m	1.02 m	0.93 m
Beam	0.20 m	0.10 m	0.07 m	0.20 m	0.10 m	0.07 m
Draft	0.07 m	0.07 m	0.06 m	0.07 m	0.07 m	0.06 m
Depth	0.15 m	0.15 m	0.14 m	0.15 m	0.15 m	0.14 m
WSA	0.39 m^2	0.18 m^2	0.11 m^2	0.55 m^2	0.30 m^2	0.21 m^2
B/T	2.86 m	1.43 m	1.17 m	2.86 m	1.43 m	1.17 m
Δtot		24.5 kg			≈24.5 kg	
	Chine 30⁰			**Chine 35⁰**		
	Main	Inner side	Outer side	Main	Inner side	Outer side
Length	2.75	1.08 m	0.93 m	3.30 m	1.12 m	0.93 m
Beam	0.20 m	0.10 m	0.07 m	0.20 m	0.10 m	0.07 m
Draft	0.07 m	0.07 m	0.06 m	0.07 m	0.07 m	0.06 m
Depth	0.15 m	0.15 m	0.40 m	0.15 m	0,15 m	0.14 m
WSA	0.61 m^2	0.31 m^2	0.21 m^2	0.69 m^2	0.32 m^2	0.21 m^2
B/T	2.86 m	1.43 m	1.17 m	2.86 m	1.43 m	1.17 m
Δtot		≈24.5 kg			≈24.5 kg	

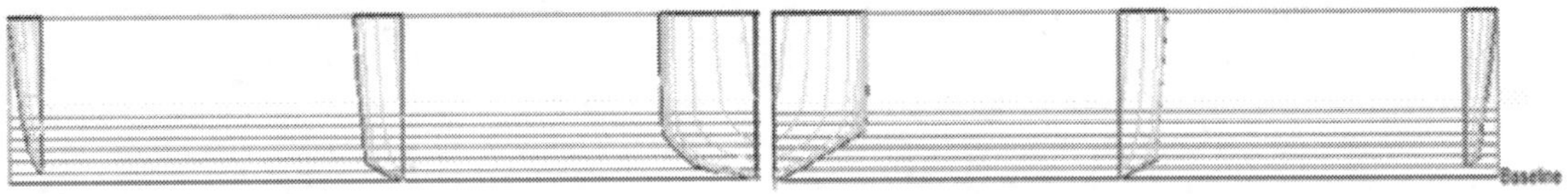

a) Wigley hull form b) Chine hull form

Figure 1. Centre line half view of pentamaran hull form

The experimental data of [6] on the smallest resistance Wigley pentamaran configuration IIa (S/L 3/16 and R/L 1/20) used as a configuration reference for this research. Transformation Wigley to chine hull form with differences L/B of main hull, constant B/T and similar displacement have been produced 6 new configurations. The cross section of both pentamaran hull form of Wigley and chine at midship is shown in figure 1. And the six new configurations position of configuration is defined in table 2.

Table 2. Six configurations position of chine pentamaran

	Chine 25^0 (C25)		Chine 30^0 (C30)		Chine 35^0 (C35)	
	I	II	I	II	I	II
S/L	3/16	3/16	3/16	3/16	3/16	3/16
R/L	1/16	1/33	1/8	1/14	1/5	1/7

3. Numerical Study

3.1. CFD approach

CFD approach was a practical step to predictions of the flow around the hull, flow separation, etc., which solved by the Reynolds averaged Navier-Stokes (RANS) equations. The first analysis was investigation of the accuracy turbulent model which corresponds to the experimental data results. The investigations were on K-ε model, Shear Stress Transport (SST) and Baseline (BSL) Reynolds. The K-ε model used to simulate characteristics of turbulent flow, where K is turbulent kinetic energy determines in the turbulence, ε. The SST model and BSL Reynolds stress models were a combination both of k-ω model and k-ε model depending on the fluid regions. [12] identified SST model was a variant of k-ω model in the inner boundary layer and a transformed of k-ε model in the outer boundary layer and the free stream. [12] showed BSL Reynolds stress allows anisotropic turbulence effects to separation off curved surfaces of model complex flow.

3.2. Correlation with experiment

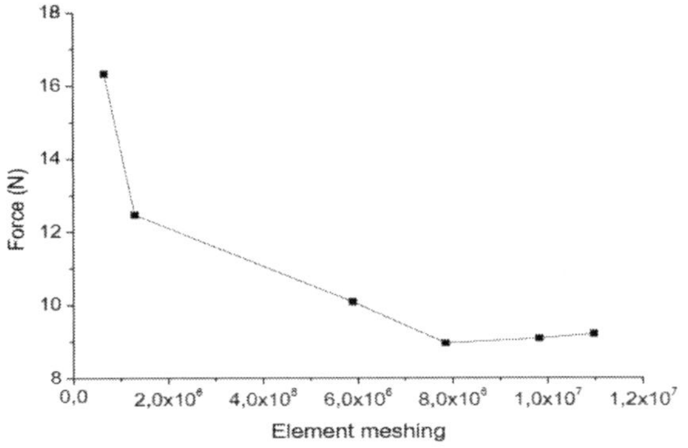

Figure 2. Convergence of Force versus the number of element; Fr 0.7

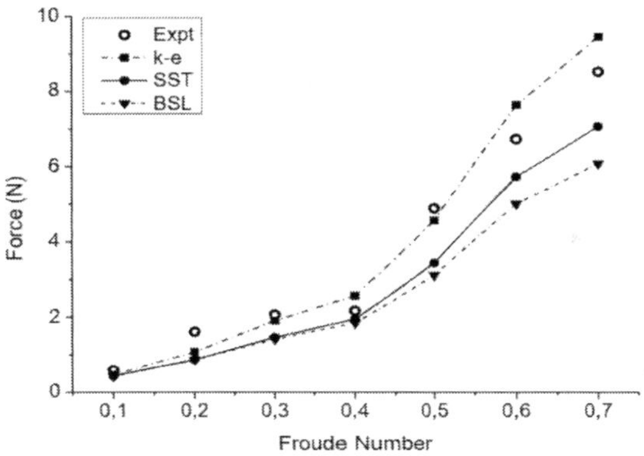

Figure 3. Comparison between simulation and experiment: Force versus Fr

Both modeling and meshing process generated in ICEM well done with repetitive process. For verification, i.e. convergence and the assessment of numerical uncertainty with increasing number of elements of model, which constitute the solution to approach the analytical solution. Processing mesh was repeated 6 times with the number of elements obtained ranging from 0.6 M, 1 M, 6 M, 8 M, 10 M and 11 M, respectively. It has been carried out to the total resistances (force) which element size both boundary and ship hull i.e.: 0.1; 0.005 to 0.003. Evaluation result of meshing convergence is shown in figure 2. The result for 3 turbulent models at variation of Froude number 0.1 – 0.7 is shown in figure 3, which the K-ε model provided a closer match with the experimental data. The highest deviation of total resistance (force) value occurs at Fr 0.7 with error for SST and BSL to over 10%. Then, based on the comparison result of turbulence models with the experimental data of [6], further analysis used ICEM settings with number elements of 10 M and K-ε model.

3.3. Multi Hull Resistance Components

Following the diagram of [13] explains the total resistance consists of two components i.e.: viscous resistance (R_V) as energy lost in wake and wave resistance (R_W) as energy in the form of water waves.

$$R_T = R_V + R_W \tag{1}$$

With ITTC-57, viscous resistance (R_V) can be estimated by equation:

$$R_V = (1 + k)R_F \tag{2}$$

where R_F = frictional resistance; $(1+k)$ = form factor of hull.

Numerical computation in CFX- Post, frictional resistance (R_F) could be calculated by performing an area integral of the wall shear in the x-direction. While determine the magnitude of form factors (1+k) of six chine models was using [14] plot, a regression method from slow speed data. The form factor (1+k) of six chine models of pentamaran predicted from Prohaska method that are: C25_I=1.11; C25_II=1.14; C30_I=1.24; C30_II=1.2; C35_I=1.44; C35_II=1.37, respectively. In figure 6 to 8, the measurement nondimensionalized of total resistance coefficient C_T, frictional coefficient C_F and wave coefficient C_W as follows,

$$C_T = \frac{R_T}{\frac{1}{2}\rho V^2 S}\ ;\ C_F = \frac{R_F}{\frac{1}{2}\rho V^2 S}\ ;\ C_W = \frac{R_W}{\frac{1}{2}\rho V^2 S} \tag{3}$$

where V = velocity; S = wetted area of hull. Furthermore, figure 7 expressed drag reduction of Wigley pentamaran to chine hull form was calculated by the equation,

$$D_R(\%) = \left|\frac{C_{T_{wigley}} - C_{T_{chine}}}{C_{T_{wigley}}}\right| \tag{4}$$

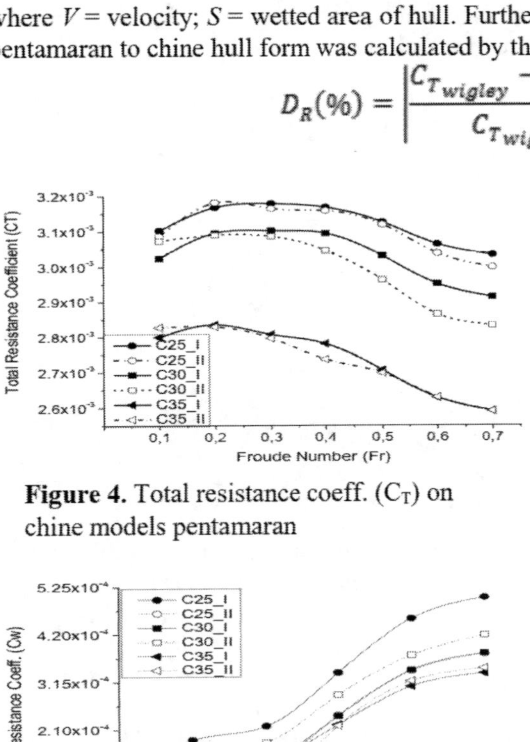

Figure 4. Total resistance coeff. (C_T) on chine models pentamaran

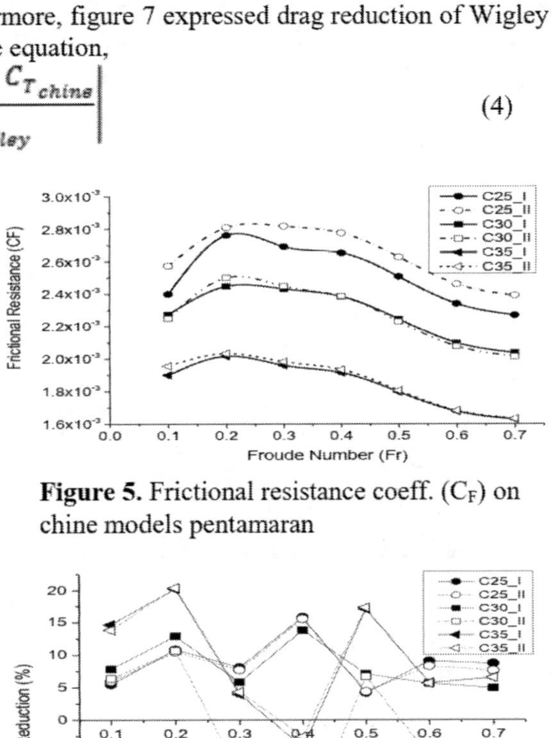

Figure 5. Frictional resistance coeff. (C_F) on chine models pentamaran

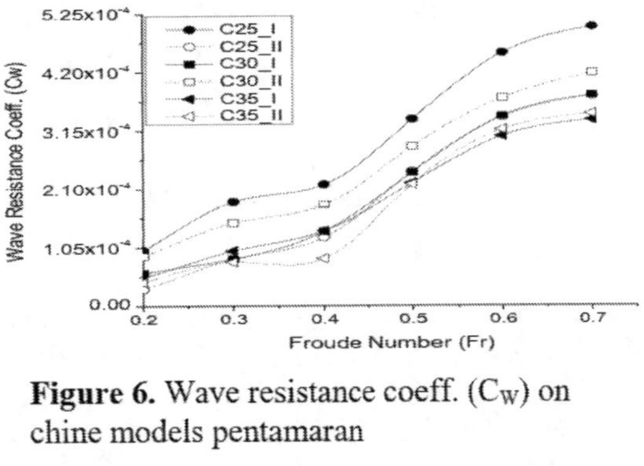

Figure 6. Wave resistance coeff. (C_W) on chine models pentamaran

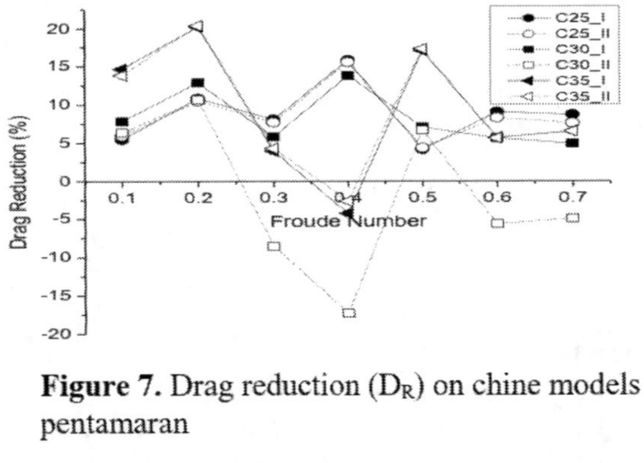

Figure 7. Drag reduction (D_R) on chine models pentamaran

Based on the comparison of six configurations in figure 4 to 6 showed C35_II had the smallest C_T and C_F, also generates smaller C_W than the others. All of six configurations appeared hump and hollow phenomenon on percentage of drag reduction. A negative percentage drag reduction indicates a detrimental hull form of chine model than Wigley. The drag reduction of C35_II had the highest percentage values of 20,38% at Fr 0.2 though not as a whole at increasing Fr. Even the model C25_I at the highest Fr generated the biggest drag reduction: 8,67% and smaller C_W than C35_II, but had the largest C_T and C_F values. Overall the economics of a ship was seen from the smallest value of total resistance. The six models have been analyzed and the results model C35_II more advantageous compared with Wigley and other chine models.

4. Modeling Hull Form

From preliminary analysis of chine pentamaran, the smallest total resistance coefficient has been generated by model C35_II. For further analysis, this model is selected as initial model then changed its shape with variations of entrance angle. Modeling variations of entrance angle are 5^0-15^0 with stem vertical straight (stem angle 0^0) in similar displacement and configurations as C35_II. From these combinations generate 3 new hull forms with shape characteristics are shown in the table 3.

Table 3. Chine models with changed stem and variations of entrance angle

	Deadrise (degree)	Stem (degree)	Entrance (degree)
C35_II (as initial)	35	47	15
C35_IIA	35	0	5
C35_IIB	35	0	10
C35_IIC	35	0	15

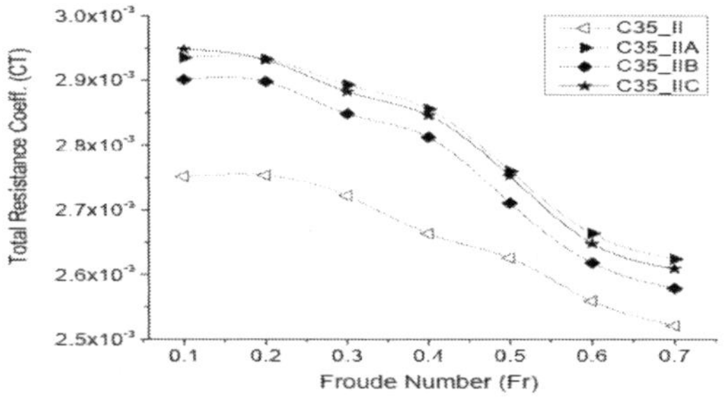

Figure 8. Total resistance coeff. (C_T) of chine with vertical straight stem

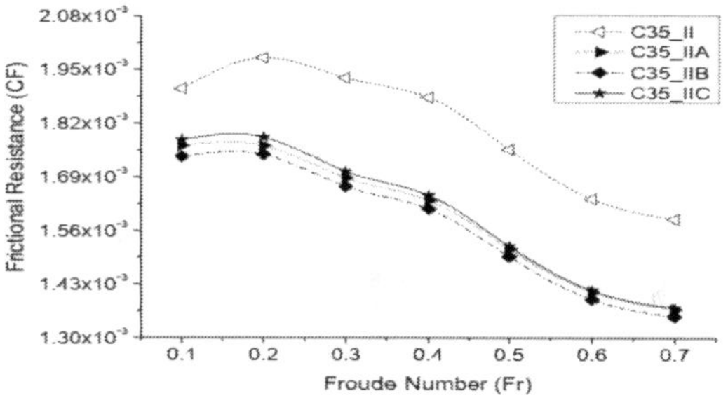

Figure 9. Frictional resistance coeff. (C_F) of chine with vertical straight stem

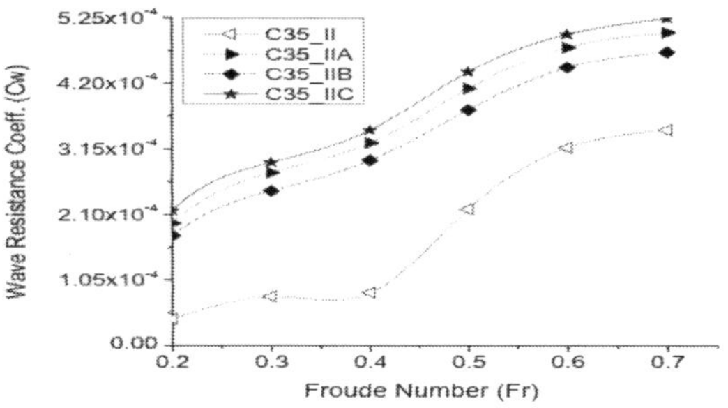

Figure 10. Wave resistance coeff. (C_W) of chine with vertical straight stem

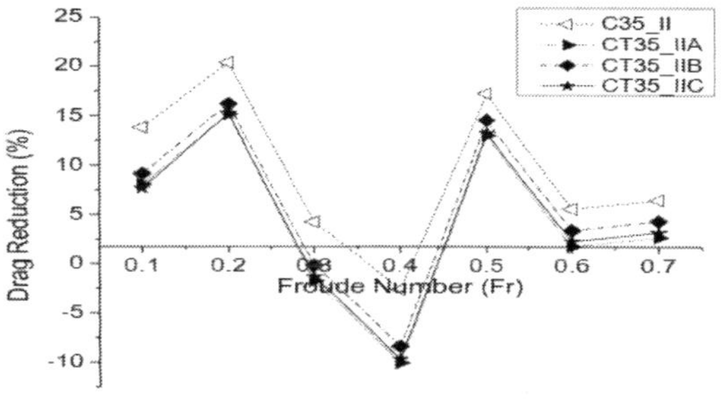

Figure 11. Drag reduction (D_R) of chine with vertical straight stem

Stem angle with vertical straight rise effect to larger form factors, which the form factor (1+k) of new chine hull form that are: C35_IIA=1.55; C35_IIB=1.56; C35_IIIC=1.52, respectively. With the change of stem angle and variation of the entrance angle on initial model C35_II does not indicate the reduction of components resistance. The smaller stem angle raises increasing wave resistance, which means increasing total resistance. Figure 8 to 10 shows model C35_II still provides the smallest both

C_T and C_W but the highest C_F. In figure 11, the drag reduction graph of all models have similar trend with hump and hollow phenomenon, where drag reduction (D_R) of model C35_II bigger than other model and the biggest of 20,38% at Fr 0.2.

5. Conclusion

Total resistance tends to be smaller on higher L/B with constant B/T on similar displacement. Outriggers on longitudinal position give a strongly effect to the components resistance especially on wave resistance. Pentamaran with chine hull form at high speeds, Fr 0.4 to 0.7, there are in general two favorable with increasing deadrise: decreasing total resistance and increasing drag reduction. However, change of entrance with smaller angle and stem on vertical straight give detrimental effect to the components resistance, both wave and drag reduction. All the results comparison of chine models shows model C35_II better than the others with drag reduction of 20.38%. Further it is planned to study pentamaran with chine hull form on position optimization of outriggers and optimum angle of: deadrise; stem; and entrance.

6. References

[1] Hsiung C C and Xu H 1988 Determining Optimal Forms of a Catamaran for Minimum Resistance by the Mathematical Programming *Meth. Schiffstech*. Vol **35**
[2] Tuck, Ernest O, and Leo Lazauskas 1998 Optimum hull spacing of a family of multihulls
[3] Dudson, E, Gee, N 2001 Optimisation of the Seakeeping and Performance of a 40 Knot Pentamaran Container Vessel *6th Inter. Conf. on Fast Sea Trans.* Southampton
[4] Begovic, E, et al 2004 Pentamaran hull for medium size fast ferries *Hydrodynamics VI* (Eds L Cheng and K Yeow):23-28
[5] Ikeda, Yoshiho, Emiko Nakabayashi, and Ai Ito 2005 Concept design of a pentamaran type fast RoRo ship *Ocean Eng* **1**:35-42
[6] Yanuar, Gunawan, Kurniawan T, and A Jamaluddin 2015 Experimental Study Resistances of Asymmetrical Pentamaran Model with Separation and Staggered Hull Variation of Inner Side-Hulls *Inter. J. of Fluid Mech. Research.* Vol **42** No 1
[7] Yanuar, Ibadurrahman, Kurniawan T Waskito, S Karim, and M Ichsan 2017 Interference Resistance of Pentamaran Ship Model with Asymmetric Outrigger Configuration *JMSA.* **16**:42-47
[8] Moraes, HB, JM Vasconcellos, and RG Latorre 2004 Wave resistance for high-speed catamarans *Ocean Eng.* **31**(17):2253-2282
[9] Tarafder, Md Shahjada, and Kazuo Suzuki 2008 Numerical calculation of free-surface potential flow around a ship using the modified Rankine source panel method *Ocean Eng.* **35**(5):536-544
[10] Tarafder, Md Shahjada, Mir Tareque Ali, and Md Shahriar Nizam 2013 Numerical prediction of wave-making resistance of pentamaran in unbounded water using a surface panel method *Proc. Eng.* **56**:287-296
[11] Bari, Ghazi S, and Konstantin I Matveev 2016 Hydrodynamic modeling of planing catamarans with symmetric hulls *Ocean Eng.* **115**:60-66
[12] Ferziger JH and Peric M 2002 Computational Methods for Fluid Dynamics 3rd ed. Springer Verlag pp 157–206
[13] Molland, A F, Wilson, P A, Turnock, S R, Taunton, D J and Chandraprabha, S, 2001 The prediction of the characteristics of ship generated near-field wash waves *Proc. of 6th Inter. Conf. on Fast Sea Trans.* Southampton pp 149–164
[14] Prohaska, C W 1996 A Simple Method for Evaluation of Form Factor and the Low Speed Wave Resistance *Proc. 11th ITTC* p 65-66

Acquisition of Co metal from spent lithium-ion battery with emulsion liquid membrane technology using cyanex 272 as extractant

Yuliusman[1], P T Wulandari[1], R A Amiliana[1], M Huda[1], Silvia[1] and F A Kusumadewi[1]

[1] Chemical Engineering Department, University of Indonesia, Depok 16424, Indonesia

E-mail: usman@che.ui.ac.id

Abstract. Lithium-ion batteries are the most common type to be used as energy source in mobile phone. The amount of lithium-ion battery wastes is approximated by 200 – 500 ton/year. In one lithium-ion battery, there are 5 – 20% of cobalt metal, depend on the manufacturer. One of the way to recover a valuable metal from waste is leaching process then continued with extraction. Spent lithium-ion batteries will be characterized with EDX and AAS, the result will show the amount of cobalt metal with form of $LiCoO_2$ in the cathode. Hydrochloric acid 4 M, temperature 80°C, and reaction time 1 hour condition give out the best leaching efficiency for both Co and Mn metals, 88.54% and 89.28% respectively. For extraction, the best extraction efficiency achieved when feed phase pH is 5 for Co and 6 for Mn. This study will discuss the leaching and extraction process and compare the efficiency for both processes.

1. Introduction

Every year, 2 billion cells of lithium ion battery are produced. Lithium ion is used because it has a lot of advantages compared to other material such as Ni Cd (Nickel Cadmium). It has bigger capacity, no memory effect, does not contain toxic substances compared to others so it is more environmentally friendly [1]. Lithium ion batteries speed of charging is considered fast if compared to other type of battery, durable, and has a high power compact in a lighter packaging. However, behind all of its advantages, there is a potential hazard in lithium ion battery and can harm the user because it contains flammable substances and high pressure so it can't be thrown away into the environment without any proper treatment beforehand.

Lithium ion batteries consist of heavy metals, organic chemicals and plastics in the proportion of 5-20% cobalt, 5-10% nickel, 5-7% lithium, 15% organic chemicals and 7% plastics, the composition varying slightly with different manufacturers. Some methods to recover cobalt are liquid-liquid extraction, electrowinning, emulsion liquid membrane, vaporization, and hydroxide precipitation. In this study, leaching is used to recover cobalt, to dissolve lithium ion battery wastes with leaching agent and continued with emulsion liquid membrane to extract cobalt metal from leachant.

Leaching is a solid-liquid extraction method where the separation of solid cobalt mixed with leaching agent so the metal inside can be dissolved into the leaching agent. Strong acids like sulphuric acid and hydrochloric acid were often used as leaching agent. There is a previous study about recovering cobalt

[1] To whom any correspondence should be addressed.

metal from lithium ion battery, comparing three reagents as leaching agents which are sulphuric acid, hydroxylamine hydrochloride, and hydrochloric acid [2]. Between three reagents stated, hydrochloric acid is the most suitable and effective with more than 99% cobalt metal leached in a hydrochloric acid 4 M in 80°C temperature, and stirring time of 1 hour.

Emulsion liquid membrane have some attractive features, for example, simple operation, high efficiency, extraction and stripping in one stage, larger interfacial area, scope of continuous operation. By extracting and stripping in one stage, it will reduce reaction time and reagents needed unlike liquid-liquid extraction method [3]. Emulsion liquid membrane is the right choice because it has the highest degree of separation and can be applied in a large scale [4].

Extractants have been used in prior studies to recover cobalt metal with emulsion liquid membrane system are Cyanex 272, Cyanex 921, tributyl phosphate (TBP), and PC-88A. From three extractants stated, Cyanex 272 is the most selective toward cobalt existence. Each extractant has different efficiency depend on leaching agent used.the reason of using Cyanex 272 is to investigate cobalt metal extraction efficiency with hydrochloric acid as leaching agent and using kerosene as diluent in emulsion liquid membrane system. In this study, hydrochloric acid is used as leaching agent, Cyanex 272 as extractant, kerosene as diluent, SPAN 80 as surfactant, and H_2SO_4 as stripping agent.

2. Experimental

2.1. Materials

Lithium ion batteries that are used are from cellphone batteries various brand sold at the market (Samsung, iPhone, Xiaomi, and LG). Hydrochloric acid used as leaching agent, Cyanex 272 as extractant. Commercial kerosene with density of 830 kg/m^3 used as diluent. SPAN 80 used as surfactant, and H_2SO_4 as stripping agent. All the materials were immediately used without reprocessing to purify the materials.

2.2. Experimental procedure

Initial step is preparation of materials. One of the preparation that needed to be done is soaking lithium ion batteries into sodium chloride solution to emptied the remaining electrical charge inside the batteries which may still be inside. Next step is leaching process with varying leaching agent which has been specified before. Solution after leaching (called leachant) then filtered to get rid of solid components which can't dissolve. After filtered, leachant then analysed of cobalt and manganese metal content with AAS (Atomic Absorption Spectroscopy). Next step is to extract cobalt and manganese metals using emulsion liquid membrane technology. The extraction process done with mixing feed phase, membrane phase, and receiving phase with pH of feed phase being varied. After extraction process is done, solution from feed phase will be analysed with AAS to determine cobalt and manganese metals extraction percentage.

3. Results and discussion

3.1. Characterization lithium-ion battery waste

3.1.1. Soaking in sodium chloride solution. Before opening the battery's casing, the battery need to be soaked in sodium chloride solution 1%wt first with the intention of emptying the remaining electrical charge which may still be inside the battery for 24 hours. There is a chance that there is a remaining electrical charge, even though the batteries used are damaged. The remaining electrical charge need to be emptied so when the battery casing is opened, it won't cause any harm such as the ignited fire. Using sodium chloride solution with too high concentration will decrease battery's voltage faster but a leakage may happen, which is the sodium chloride solution entered the casing. This leakage affecting electrodes inside the battery contaminated with salt solution. Besides, $LiPF_6$ in the electrolyte may react with water and forming HF and will bring negative effect for the next process. To reduce reactivity during soaking process, sodium chloride 1%wt solution is used because it won't cause a leakage. The reactions are electrolysis with sodium chloride solution as shown below:

$$2NaCl \longrightarrow 2Na + Cl_2 \quad (1)$$
$$2H_2O + 2Na \rightleftharpoons 2NaOH + H_2 \quad (2)$$
$$2H_2O \longrightarrow 4H_2 + O_2 \quad (3)$$
$$Cl_2 + H_2O \rightleftharpoons HCl + HClO \quad (4)$$
$$2Al + 2NaOH + 2H_2O \rightleftharpoons 2NaAlO_2 + 3H_2 \quad (5)$$
$$NaAlO_2 + HCl + H_2O \rightleftharpoons Al(OH)_3 + NaCl \quad (6)$$
$$NaOH + HCl \rightleftharpoons NaCl + H_2O \quad (7)$$

Reaction (1) and (3) are electrolysis reactions and primer reactions of soaking lithium ion battery into sodium chloride solution. During soaking process, there are bubbles formed. Based on reaction (1) until (7) there are some gases formed, Cl_2, H_2, and O_2. Those reactions also formed white foam which is $Al(OH)_3$ around the battery and the cause of leakage is reaction (3) [5].

3.1.2. Characterization of battery cathode powder. To characterize lithium ion battery waste with EDX and AAS analysis, cathode sheet scraped and crushed, after that filtered with 100 mesh filter so the result is cathode powder with size as small as possible so leaching process will be more effective, to make leaching efficiency high, it needs a particle with the size 100 mesh [1]. Table 1 shows AAS analysis result with the amount of cobalt metal in cathode battery powder is 164.38 mg/g and manganese metal is 77.79 mg/g. This goes with previous study by Shin, et al. (2005), in general range of cobalt metal content in lithium ion battery is 5 – 20% weight, depends on the battery's manufacturer.

Table 1. AAS analysis result

Substances	Analysis Result (ppm)	mg/g Sample
Co (*cobalt*)	8218.75	164.38
Mn (*manganese*)	3889	77.79

3.2. Leaching with hydrochloric acid

Leaching process is perform to dissolve solid cobalt metal contained in the battery cathode powder by using hydrochloric acid as a leaching agent to dissolve the cobalt metal. The usage of hydrochloric acid is varied in concentration, while other operating conditions such as stirring speed, temperature, solid/liquid ratio, and stirring time are maintained according to the optimum results of the prior study. It is aimed to obtain optimum hydrochloric acid concentration with other parameters are kept constant so that the results obtained will be accurate.

Leaching, or sometimes called solid-liquid extraction, is one of the method to separate the dissolved fraction (solute or leachant) from a solid material by a liquid solvent. Solute (in this study is cobalt metal) diffuses from within the solid to the solvent environment (in this study is hydrochloric acid) so that in the solvent, it will contain cobalt metals that are successfully leached [6]. The chemical reaction that occurs in the leaching process using hydrochloric acid is shown in equation (8) [7].

$$4\,LiCoO_2 + 12\,HCl \longrightarrow 4\,LiCl + 4\,Co(Cl)_2 + 6\,H_2O + O_2 \quad (8)$$

In this study, the ratio of solid : liquid is 1 g : 20 ml due to research of Zhang et al. (1998), the optimum result of leaching using hydrochloric acid is when the ratio of solid : liquid is 1 g : 20 ml. figure 1 shows effect of hydrochloric acid concentration to leaching cobalt and manganese metals efficiency.

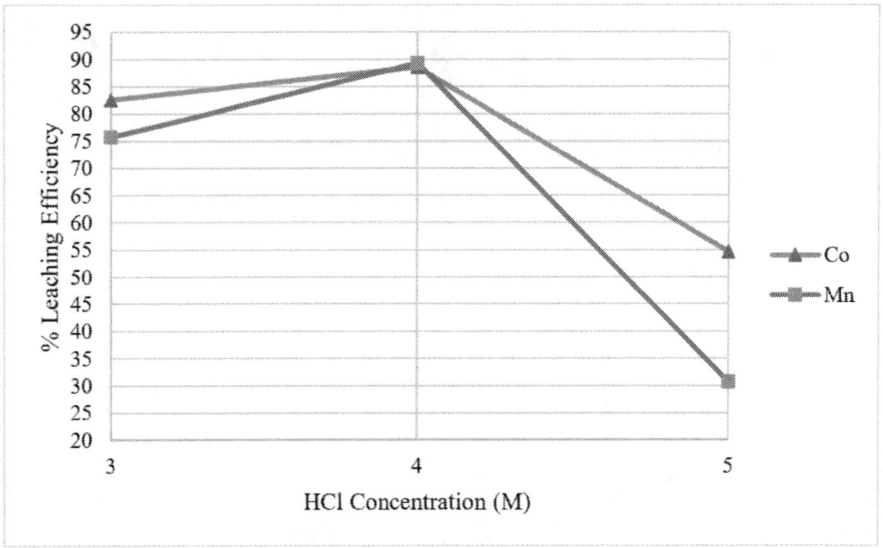

Figure 1. Effect of hydrochloric acid concentration to leaching cobalt and manganese metals efficiency (stirring speed 500 rpm, S/L = 1g/20ml, temperature 80°C, stirring time 60 mins)

The result of efficiency calculation shows that hydrochloric acid with concentration 4 M is the most efficient concentration to dissolve cobalt metal that is up to 88.54%. This efficiency is not the same with previous study conducted by Zhang et al. (1998) when leaching using hydrochloric acid with operating conditions of 80°C, 500 rpm stirring rate, stirring time for 1 hour, that can be produced 99% efficiency to cobalt. This can occur due to the possibility of a temperature deviation during leaching process. However, when compared to leaching efficiency of other hydrochloric acid concentrations, the results of this study are still appropriate because the highest efficiency is when using 4 M hydrochloric acid.

The solution of leaching of cobalt metal using hydrochloric acid is colored pink. Pink color indicates the presence of cobalt metal ions in aqueous solutions of hydrochloric acid and water, giving rise to a pink color. The darker the color of the leaching result, the more the cobalt metal ion content in the solution [8].

While leaching cathode powder, not only cobalt metal get leached but also manganese. As shown from figure 1, the highest leaching efficiency for manganese reached 89.28%. This can happen because there is manganese metal in the cathode battery so when the solution is mixed and heated, manganese metal got leached. For manganese metal leaching, the reaction shows in equation (9) [9].

$$4\ LiMnO_2 + 12\ HCl \longrightarrow 4\ MnCl_2 + 4\ LiCl + 6\ H_2O + O_2 \quad (9)$$

If the percentage of leaching using 3 M and 4 M hydrochloric acid is compared, the 3 M concentration has a lower efficiency. This can happen because when the acid concentration is increased, the number of protons will increase as well and replace more Co^{2+} ions. On the other hand, when hydrochloric acid is present between the solid and liquid phases, the gradient of H^+ concentration between the solid and liquid phases with the solution increases significantly and causes increased diffusion velocity. Both of these can speed up the leaching process and provide higher leaching efficiency values. When the leaching reaction has reached equilibrium, the efficiency of the Co and Mn metal leaching reaches its maximum point. This occurs because the speed of diffusion will not change again even when the concentration of hydrochloric acid used is increased, and the Co^{2+} and Mn^{2+} ion has been replaced [10]. It appears that in this study, the equilibrium point of leaching using hydrochloric acid was at a concentration of 4 M because when using a concentration of 5 M hydrochloric acid, leaching efficiency decreased.

3.3. Extraction with emulsion liquid membrane technology

Extraction process begins by making an optimum emulsion phase (10% w/v SPAN 80 and 0.7 M Cyanex 272). This extraction process goes in one stage with the stripping process because the extracting technology used is the emulsion liquid membrane. Then the emulsion phases that have been formed are mixed with the optimum leachant that is the result of leaching using 4 M hydrochloric acid which varied the pH of the solution, ie 3, 4, 5, and 6. The pH variation is done because Cyanex 272 is a slightly acidic extractant so the pH of the stripping solution should be less than the pH of leachant metal during the extraction process with the emulsion liquid membrane. NaOH solids added gradually to change feed phase pH until the pH reaches 3, 4, 5, and 6. The ratio of the mixing volume between the emulsion phase and the feed phase is 1 : 2. After that, the solution is stirred at 250 rpm for 15 minutes. After stirring, the mixture is both transferred to the separatory funnel for extraction and held for 15 - 30 minutes and there is a separation between the feed phase and the emulsion phase. The calculation of the efficiency of the cobalt extraction process is shown in figure 2.

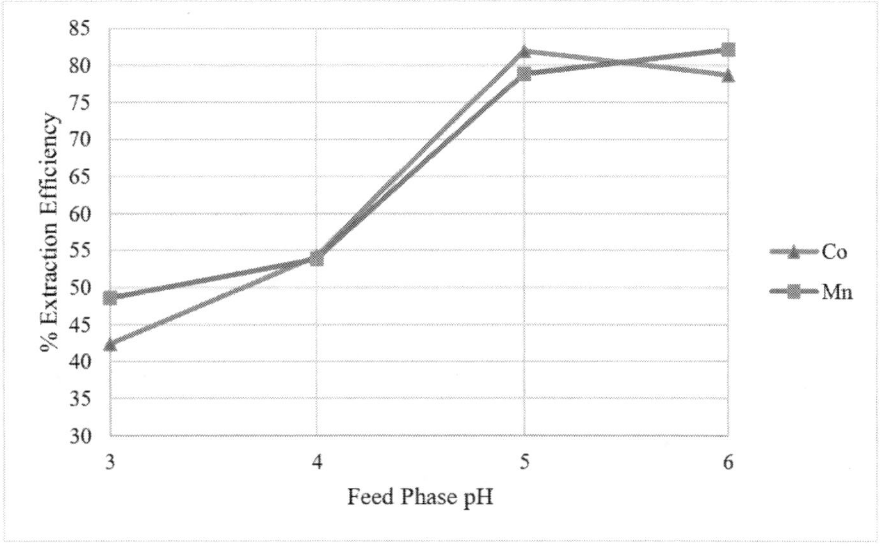

Figure 2. Extraction efficiency with pH 3, 4, 5, and 6 (stripping speed 250 rpm, stirring time 15 mins, room temperature)

Based on figure 2 shows that extraction efficiency for cobalt extraction for cobalt metal increasing from pH 3 to 5 but decrease at pH 6. Feed phase pH 5 is the highest between other variation which is 81.94%. Meanwhile for the highest manganese extraction efficiency is reached when feed phase pH is 6 and give out efficiency of 82.17%. A low extraction efficiency may occur due to emulsion instability. Because of when the pH is high, emulsion stability changes with the emulsion, whereas at too low pH can be explained by competing H^+ ions with the solute due to the release of H^+ ions from extractant to leachant [11]. Extraction rate will decrease when the pH is greater than 5 due to emulsion instability because H^+ ions transfer phenomenon from acidic feed solution to membrane phase so that the membrane can swell and decompose. pH value between stripping solution and feed solution should be different to make a difference of chemical potential for H^+ ions. Physical properties of the extractant will cause a reduction of the ion activity difference on both phases. Cobalt extraction process increases with increasing amount of proton concentration up to a certain concentration [12]. According figure 2, at pH 6 for cobalt extraction, the emulsion has swelled so the extraction efficiency is lower than that of pH 5.

Type and concentration selection of stripping solution appropriately is one of the determinants of extraction process efficiency with emulsion liquid membrane technology. Reaction by the stripping solution plays an important role in the extraction of solute from feed solution to stripping solution in emulsion liquid membrane process. Based on prior study by Kumbasar (2012), performance of Na_2CO_3, H_2SO_4, and HCl as stripping solution was compared. In the first ten minutes, the best stripping solution sequence is $Na_2CO_3 > H_2SO_4 > HCl$. However, after the first ten minutes, emulsion stability while using Na_2CO_3 is changed so that its extraction efficiency is reduced so that the sequence of extraction efficiency at the end of process from highest to lowest is $H_2SO_4 > Na_2CO_3 > HCl$.

4. Conclusion

- Concentration of hydrochloric acid as leaching agent to produce the best leaching efficiency both for Co and Mn is 4 M with operating conditions 500 rpm stirring speed, stirring time 1 hour, temperature 80°C, and S/L = 1 g/20 ml.
- Highest extraction efficiency was obtained when extractant concentration Cyanex 272 was 0.7 M, SPAN 80% w/v surfactant, stripping agent concentration H_2SO_4 0.1 M, and pH of the feed phase for cobalt is 5 and for manganese is 6.

5. References

[1] Shin S M, Kim N H, Sohn J S, Yang D H and Kim Y H 2005 *Hydrom.* **79** 172-81
[2] Zhang P, Yokoyama T, Itabashi O, Suzuki T and Inoue K 1998 *Hydrom.* **47** 259-71
[3] Harimu L, Matsjeh S, Siswanta D and Santosa SJ 2010 *Indo. J. Chem.* **10**(1) 69-74
[4] Kislik V S 2010 *Liquid Membranes: Principles & Applications in Chemical Separations & Wastewater Treatment* (Amsterdam: Elsevier B.V.)
[5] Mi L, Houan Z, Bingchen W, Xiadong Z and Changsong D 2013 *Int'l J. Elec. Chem. Sci.* **8** 8201 09
[6] Treybal R 1981 *Mass-transfer Operations* (Singapore: McGraw-Hill)
[7] Li L, Jing G, Renjie C, Feng W, Shi C and Xiaoxiao Z 2010 *Waste Manag.* **176** 288-93
[8] Jim C 2015 *Chem Guide* Online http://ihome.ust.hk/~kexhu/ceng3210/221-07.pdf
[9] Nayl A A, Elkashab R A, Badawy S M and El-Khateeb M A 2017 *Arab. J. Chem.* **10** S3632-39
[10] Yang G, Feng L, Haochen Z, Guangming L, Juwen H and Wenzhi H 2016 *Renew. Sust. Ener. Rev.* **64** 106-28
[11] Kumbasar R A 2009 *Separ. Purif. Tech.* **64** 273-79
[12] Volkan E and Recep K A 2015 *J. Ind. Eng. Chem.* **21** 303-10

IOP Conf. Series: Earth and Environmental Science **105** (2017) 012045 doi:10.1088/1755-1315/105/1/012045

Battery thermal management system using loop heat pipe with LTP copper capillary wick

Bambang Ariantara[1,2], **Nandy Putra**[2] **and Sugeng Supriadi**[3]

[1]Mechanical Engineering Department, Universitas Pasundan, Jl. Dr. Setiabudi 193 Bandung 40153, Indonesia

[2]Applied Heat Transfer Research Group, Mechanical Engineering Department, Universitas Indonesia, Depok 16424, Indonesia

[3]Manufacturing Laboratory, Mechanical Engineering Department, Universitas Indonesia, Depok 16424, Indonesia

E-mail: nandyputra@eng.ui.ac.id

Abstract. Loop heat pipes (LHPs) with Lotus-Type Porous Copper (LTP Copper) capillary wick are expected to be applied to battery thermal management systems for safe operation at high performance with a long service life. Sintered LTP Copper is a high permeability porous metal with an excellent capillary pumping characteristic. The objective of this work is to determine the performance of the battery thermal management system using LHP with sintered LTP Copper capillary wick experimentally. The experiment used two battery simulators made of aluminum. The heat generation of the battery was simulated using cartridge heaters. The LHP was made of 10 m OD copper tube, and the sintered LTP Copper capillary wick was placed in the liquid line. Water was used as working fluid with filling ratio of 50%. The evaporator section of the LHP was inserted between the battery simulators surfaces. A thermostatic bath was used to regulate the condenser cooling fluid temperature. K-type 0.3 mm thermocouples were used for temperature measurement, and a digital power meter was used to measure the electric power. Experiments were conducted with various heating power with the condenser cooling fluid temperature was kept at 28°C. At a heat generation of 20 W, the LHP was capable of maintaining the battery surface temperature below 50°C. At a heat generation of 40 W, the utilization of LHP with LTP Copper can reduce the average battery simulator surface temperature from 93°C to 65°C.

1. Introduction

The development of battery technology for electric vehicles has produced lithium-ion batteries with high energy density. However, this progress is accompanied by the risk of thermal runaway which can cause serious accidents such as that experienced by a commercial aircraft Boeing 787 Dreamliner on January 16, 2013, in Japan [1]. Chemical reactions that occur in a lithium-ion battery, either at the time of charging or discharging, are always accompanied by the generation of heat that causes the rise in battery temperature. Thermal runaway begins when the high battery operating temperature triggers a damage to one cell and causes a sharp rise in the temperature of the cell. In the absence of an effective heat release mechanism to the surrounding, the increase in temperature of the cell may propagate to the neighboring cells, causing an uncontrollable chain reaction and may cause the battery to burn and explode. Figure 1 shows the Boeing 787 Dreamliner of All Nippon Airways battery, which burned and exploded caused by thermal runaway [1]. Also, overly high operating temperatures can lead to lower initial battery capacity and shorten battery life [2].

Conventional thermal management systems such as vapor compression cycles or heat exchangers that use circulated cooling fluid will increase battery power consumption to reduce the cruising range of electric vehicles. The battery operation at a safe temperature range which delivers high performance with long service life requires a reliable thermal management system with compact size, light weight, and energy-saving.

Figure 1. The burned and exploded Boeing 787 Dreamliner battery [1].

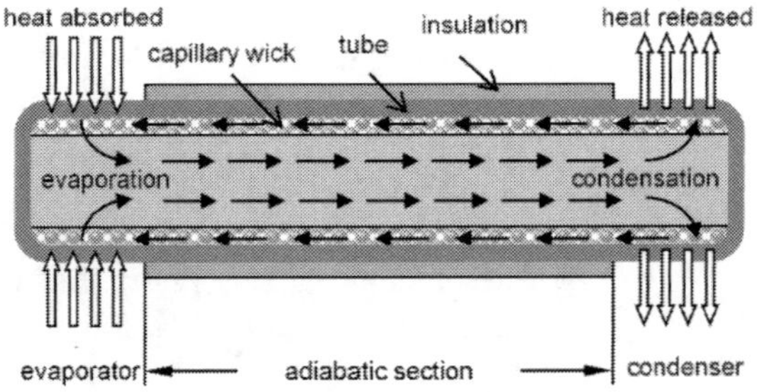

Figure 2. Straight heat pipe [3]

A heat pipe is a thermal device made of a closed tube comprising the evaporator, adiabatic and condenser section. The heat absorbed at the evaporator causes the working fluid to boil, and the vapor then flows through the adiabatic section toward the condenser section. In the condenser section, there is a release of heat to the surroundings and causes the working fluid to condense. The condensate fluid is then pumped back to the evaporator section via a capillary wick located on the inner surface of the tube as shown in figure 2 [3]. The heat pipe has a high heat transfer per unit area capacity because it involves the phase change of the working fluid. A heat pipe has a compact size and lightweight because it requires only a small amount of working fluid and does not require a pump or compressor to circulate it. Moreover, the heat pipe is a passive thermal device because it does not require an external power supply to circulate its working fluid.

The performance of a heat pipe depends on the design of the structure, the working fluid, and the capillary wick structure. LHPs have a higher cooling capacity compared to straight pipes [4]. The addition of nanoparticles to the working fluid can improve the performance of a heat pipe [5-7]. Putra et al. have found that a capillary wick made of corals combined with a nanofluid can significantly improve an LHP performance [7]. The smooth, uniform and continuous pore structure of the coral provides high permeability and capillarity to provide high capillary pressure with high mass flow rates [8]. However, the environmental consideration limits the use of coral. Supriadi et al. have succeeded in making Lotus Type Porous (LTP) Copper through slip casting and sintering techniques to mimic the coral pore structure [9]. In this study, LTP Copper is applied as a capillary wick of an LHP that is used in a battery thermal management system. The objective of this study was to determine the performance of battery thermal management system using LHP with LTP Copper capillary wick experimentally.

2. Methods

The manufacture of Lotus-type porous copper (LTP Copper) using Gasar method is quite complicated and requires expensive equipment [10, 11]. To simplify the manufacturing process, LTP Copper was made through slip casting and sintering process according to the method developed by Supriadi et al. [9]. This method used pore former such as implemented in bi-porous capillary wick manufacturing techniques performed by Li et al. [12]. However, instead of using Natrium Carbonate, Supriadi et al. use nylon threads as a pore former to produce directional pore structure. In this study, the LTP copper wick was fabricated using 200 mm copper powder and 120 mm nylon threads pore former. The copper slurry was poured into a cylindrical copper tube of 10 mm in diameter and 60 mm in length. After drying, the green product was sintered at 850°C for 60 minutes. The final product of LTP Copper wick is then installed as part of the liquid line section of an LHP as shown in figure 3.

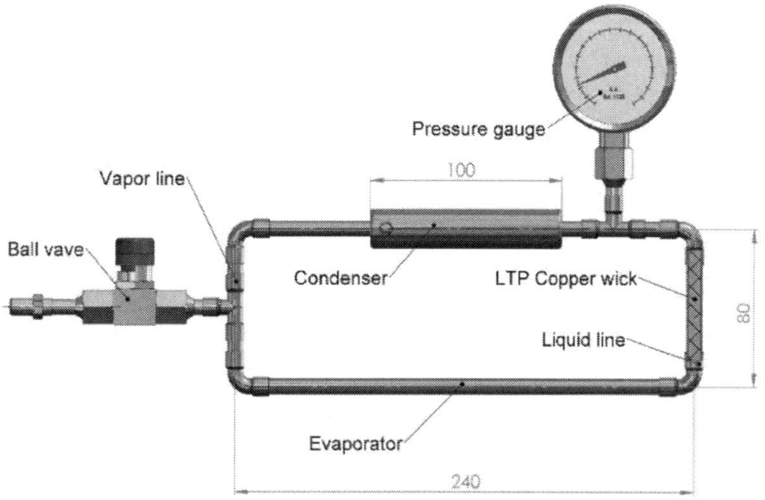

Figure 3. Loop heat pipe with LTP Copper wick

The LHP was made of 10 mm OD copper tubes with a length of 240 mm and height of 80 mm. The condenser section had a length of 100 mm and was equipped with a 25 mm OD cooling water jacket. The evaporator section was clamped between two grooved copper plates. Thermal paste was filled into gaps between the grooves and the evaporator surface to reduce the thermal contact resistance. The LHP with the holder plates was inserted between two battery simulators made of aluminum alloy measuring 48 mm x 82 mm x 138 mm. The battery simulator dimension was determined based on the dimension of a 12 V – 20 Ah Lithium-Ion battery. Each battery simulator was equipped with a 200 W cartridge heater for simulating the heat generation inside the battery. An AC voltage regulator was used to vary the heat generation. K-type thermocouples with 0.3 mm in diameter are connected to a data acquisition system to measure the surface temperature of the battery. A thermocouple is mounted on each surface of the upright wall of the battery simulator, including the inner surface, i.e., surfaces in contact with the heat pipe holder plate. A thermostatic bath is used to supply the condenser cooling water with a constant temperature of 28°C. For electric power measurement, a digital power meter which was connected to the computer was applied.

Figure 4 shows the experimental setup. The experiments were carried out in a room where the air was conditioned at 28°C. The battery simulator was placed on a flat surface so that its base surface is not exposed to the surrounding air. The heat transfer to the surrounding air occurred from the side and top planes by natural convection. All parts of the LHP exposed by the surrounding air are isolated using Rockwool, and the outer surface is covered with aluminum foil as shown in figure 5. Thus the heat absorbed by the evaporator can be approximated as the heat released by the condenser. The first experiment was done without heat pipe, and the second was done using the LHP. In each experiment, the heat generation of 20 W, 30 W, and 40 W were applied. The applied values correspond to the high heat generation of a Lithium-Ion battery. Changes in heat generation values are made after the steady state was reached. For the experiment using heat pipe, the volumetric flow rate of the cooling fluid was kept at 180 ml/minute. The volumetric flow rate was measured by using a measuring glass and a stopwatch.

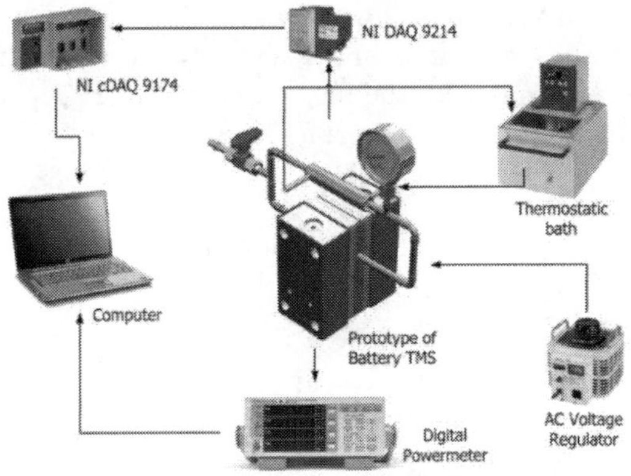

Figure 4. Experimental setup

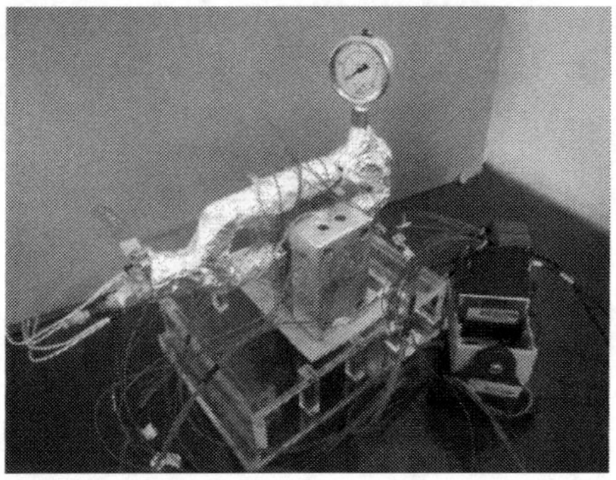

Figure 5. Prototype of battery thermal management system

3. Results and Discussion
Figure 6 shows the results of temperature and heating power measurements of the experiment without heat pipe. It appears that the surface temperature of the battery simulator is almost uniform. The ambient air temperature is slightly fluctuating but can be considered constant. The battery heat generation is simulated by the electric heating power can be kept constant at low voltage but tends to fluctuate at the higher voltage. It is due to the fluctuation of the electric voltage of the PLN, which cannot be overcome by the use of a voltage regulator. Fluctuations in heating power increase sharply with the increase of heating power since it is directly proportional to the square of the electric voltage.

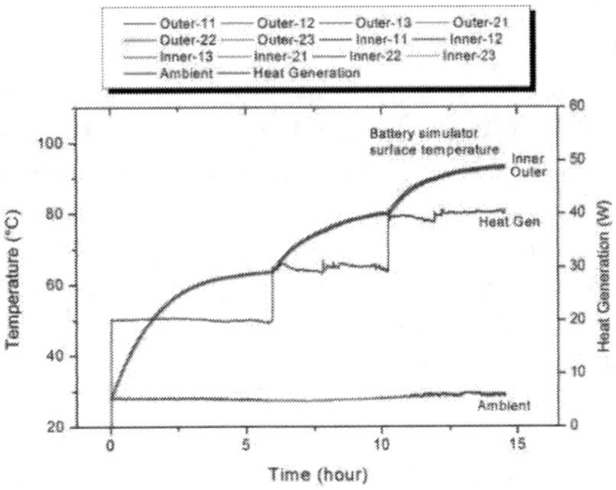

Figure 6. Results of temperature and heat generation measurement without heat pipe

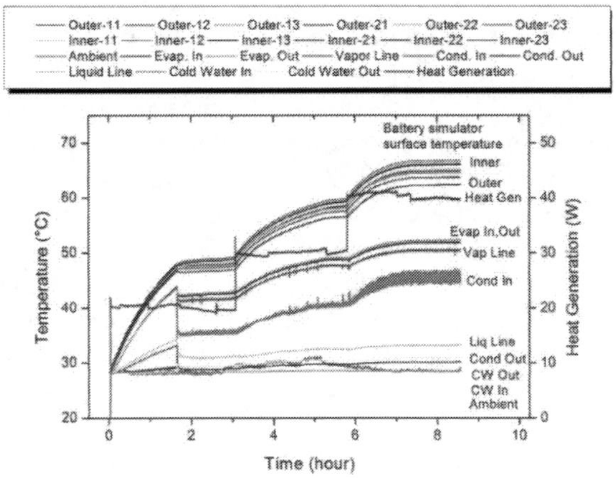

Figure 7. Results of temperature and heat generation measurement with heat pipe

Figure 7 shows the results of temperature and heating power measurements of the experiment with the heat pipe. At 1.65 hour, the temperature of the condenser wall on the inlet side increased sharply indicating the LHP successfully start-up. At the same time, the slope of the surface temperature curve of the battery simulator suddenly turns flatter. The ambient temperature and the heating power is similar to that of the previous experiment. The use of LHP causes some of the generated heat to flow to it. The low thermal resistance of LHP caused the battery simulator temperature at a position close to it to be lower than elsewhere. Thus, the battery temperature of the battery is not as uniform as in the first experiment.

Figure 8 shows the average surface temperatures of the battery simulators, both those without LHP and those using LHP for every heating power. At the heat generation of 20 W, the temperature of the battery simulator can be kept below 50 °C. At the 40 W heat generation, the battery simulator surface temperature can be lowered from 93°C to 65°C, or a decrease by 28°C. It appears that the LHP with LTP Copper wick was capable of providing a significant temperature reduction. These results indicated that the LHP was able to transfer heat effectively from the surfaces between the battery simulator through the evaporator toward the condenser. To know how much heat is transferred through LHP, an energy balance analysis is done as follows.

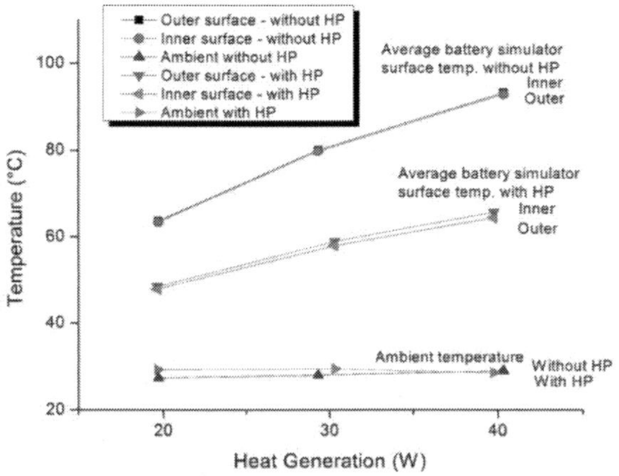

Figure 8. Effect of heat pipe utilization on battery simulator surface temperature

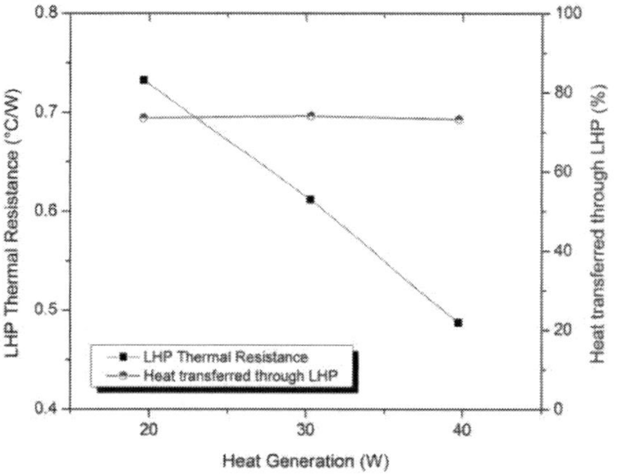

Figure 9. LHP performance

With ignoring the heat loss to the surrounding air, the heat transfer through the LHP was obtained using heat balance at the condenser section according to Eq. 1. The percentage of heat transfer is obtained by dividing the heat transfer through the LHP by the heat generation. Figure 9 shows the calculations results of the heat transfer through the LHP. It appears that for all heat generation, the percentage of heat transferred is always constant about 75%. The LHP thermal resistance can be obtained from equation 2. From figure 9 it appears that the thermal resistance tends to descend further at a higher heat load. It suggests that the LHP is still able to handle any higher heat load.

$$Q = \rho \dot{V} c_p (T_{co} - T_{ci}) \qquad (1)$$

$$R = \frac{(T_e - T_c)}{Q} \qquad (2)$$

Q : Heat transfer through the LHP (W)
R : Thermal resistance of LHP (°C/W)
ρ : Specific mass of cooling fluid (W)
$\dot{V}$: Volumetric flow rate of cooling fluid (m^3/s)
T_{co}, T_{ci} : Outlet and inlet temperature of cooling fluid (°C)
T_e, T_c : Average temperature of evaporator and condenser (°C)

4. Conclusion

The experiments to determine the performance of battery thermal management system using loop heat pipe with LTP Copper capillary wick have been successfully performed. The loop heat pipe with LTP Copper capillary wick was capable of significantly decreasing battery simulator temperature. At the heat generation of 20 W, the temperature of the battery simulator can be kept below 50 °C. At the heat generation of 40 W, the battery simulator temperature can be lowered from 93 °C to 55 °C, or a decrease by 38 °C. The loop heat pipe with LTP Copper capillary wick was able to effectively remove the heat from the battery simulator as much as 75% of the heat generated by the heater.

5. Acknowledgments

The authors would like to thank DRPM - Ministry of Research, Technology and Higher Education of the Republic of Indonesia for funding this research through the Research Grant of Doctoral Dissertation.

6. References

[1] Ross PE. 2013 Boeing's battery blues [News] *Spectrum, IEEE* **50(3)** 11-2

[2] Ramadass P, Haran B, White R and Popov B N 2002 Capacity fade of Sony 18650 cells cycled at elevated temperatures: Part II. Capacity fade analysis *Journal of Power Sources.* **112(2)**:614-20

[3] Chaudhry H N, Hughes B R and Ghani S A 2012 A review of heat pipe systems for heat recovery and renewable energy applications. *Renewable and Sustainable Energy Reviews* **16(4)** 2249-59

[4] Reay D, McGlen R and Kew P 2013 Heat pipes: theory, design and applications: *Butterworth-Heinemann*

[5] Putra N, Septiadi W N, Rahman H and Irwansyah R 2012 Thermal performance of screen mesh wick heat pipes with nanofluids. *Experimental Thermal and Fluid Science* **40** 10-7

[6] Putra N, Yanuar and Iskandar F N 2011 Application of nanofluids to a heat pipe liquid-block and the thermoelectric cooling of electronic equipment. *Experimental Thermal and Fluid Science* **35(7)**1274-81

[7] Putra N, Koestoer R A, Saleh R and Septiadi W N 2013 The Effect of CuO-water Nanofluid and Biomaterial Wick on LHP Performance. *Advanced Material Research*

[8] Putra N, Saleh R, Septiadi W N, Okta A and Hamid Z 2014 Thermal performance of biomaterial wick LHPs with water-base Al2O3 nanofluids *International Journal of Thermal Sciences.* **76(0)**128-36

[9] Supriadi S, Putra N, Ariantara B and Rahmanto D D 2016 Fabrication of Lotus-Type Porous Copper Using Slip Casting and Sintering Techniques for Heat Pipe Applications *Applied Mechanics and Materials Trans Tech Publ*

[10] Nakajima H 2007 Fabrication, properties and application of porous metals with directional pores *Progress in Materials Science* **52(7)** 1091-173

[11] Nakajima H, Hyun S K, Ohashi K, Ota K and Murakami K 2001 Fabrication of porous copper by unidirectional solidification under hydrogen and its properties *Colloids and Surfaces A: Physicochemical and Engineering Aspects* **179(2–3)** 209-14

[12] Li H, Liu Z, Chen B, Liu W, Li C and Yang J 2012 Development of biporous wicks for flat-plate LHP *Experimental Thermal and Fluid Science* **37** 91-7

Design of a combination package of heat exchanger and heater for organic rankine cycle power plant

M Muslim[1], M Idrus Alhamid, Nasruddin, Dieter R, M Zaky S, Edi Marzuki, and Nyayu Aisyah
[1]Faculty of Engineering, Universitas Indonesia, Kampus Baru UI Depok, Depok 16424, Indonesia

E-mail: muswar.muslim@ui.ac.id

Abstract. This paper presents the capacity of a heat exchanger with a plate heat exchanger model as evaporator and condenser to liquefy and evaporate R-134a refrigerant as the working fluid in a power plant system by operating an Organic Rankine Cycle (ORC) system using a scroll type expander. The evaporating and liquefying refrigerant as the working fluid uses hot water at a temperature between 40 ^{0}C - 80 ^{0}C and an evaporator inlet temperature at between 24.1 ^{0}C – 28.5 ^{0}C. The model design of the hot water production system is a combination of heat exchanger and heater with a shell and tube construction, where the heater is immersed in the heat exchanger. Experimental results show that the average revolutions of the turbine expander scroll is about 348.2 rpm and thermal efficiency between the evaporator and condenser as the heat exchanger ranges from 2 % to 8.5 %.

1. Introduction

Nowadays there is a rapid development of power generation technology that uses the basic theory of Organic Rankine Cycle system (ORC), especially in small scale power plants that range from 1 - 5 kW. Research on ORC by Sylvain Quoilin et al., used a scroll compressor as an expander to circulate working fluid HCFC-123 with the purpose of comparing predicted results with real results [1]. Researcher Mingshan Wei et al., conducted small scale experiments on waste heat recovery using strict Organic Rankine Cycle system rules and showed that flow loss has a great effect on the suction scroll expander [2]. Another researcher, Janette Hogerwaard et al. [3] conducted an experiment on heat engines to control heat at low temperatures aiming to measure power and effective heat production for small-scale applications. The system processes used renewable energy such as biomass / biofuel / biogas combustion, geothermal heat, solar and industrial waste heat with the R134a refrigerant with a low temperature heat source below 150 °C. Researcher Noboru Yamada et al. [4] conducted experiments using the Pumpless Rankine type cycle tool called (PRC) and also used the expander for the production of electricity as an organic rankine cycle system. His research used the pumpless model scroll compressor type expander to produce 20 watts of electricity. Jen-Chieh Chang et al. [5] focused on compressor scroll expander experiments that can be applied to low temperature organic cycle systems. This research used refrigerant R245fa as the working fluid and two compressors to determine the difference of volume ratio. The maximum power output from the expander is 1.77 kW while the amount of electricity delivered to the generator is 1,375 kW.

2. Research Methodology

This study focuses on the design of a package of heater and heat exchanger with a diameter of 38 cm and height of 50 cm. The 1200 watt electricity heater reaches an evaporator temperature of 80°C. The heater serves to heat the water which channeled to the plate heat exchanger (PHE) to heat the refrigerant that will evaporate at temperature and pressure conditions specified in the ORC system. The evaporator, which is one of the main components in the ORC system, uses low temperatures. The detailed diagram of the system and the flowchart of the experiment as follows.

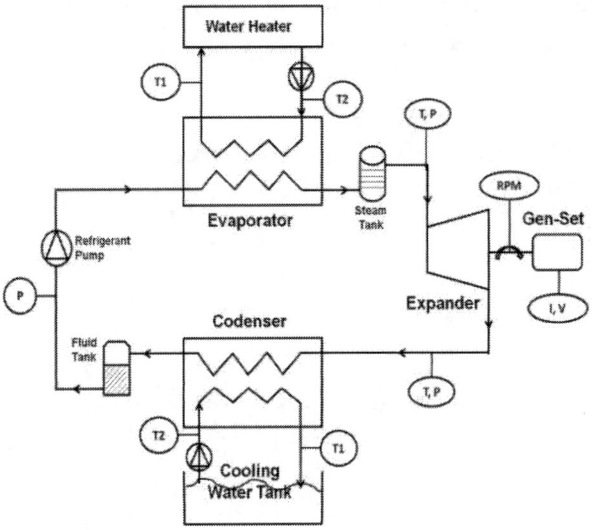

Figure 1. Diagram Cycle for Organic Rankine Cycle

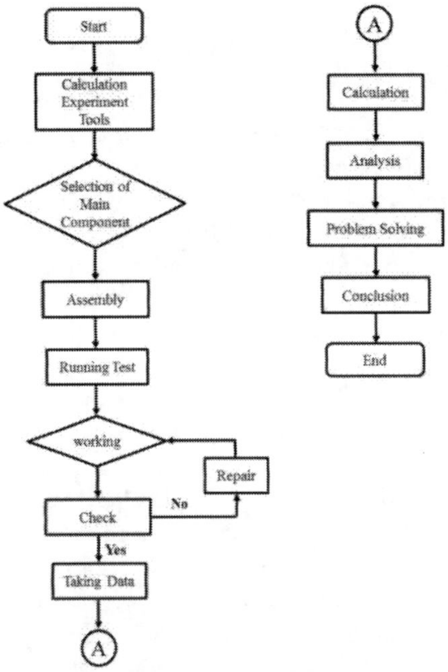

Figure 2. Flowchart of ORC Experiment

For the purpose of simplifying analysis, complicated and complex heat transfer can be avoided between the generating component and its surroundings. Changes in kinetic and potential energy are also ignored. Each component is considered to be stable. By using the principle of mass conservation and energy conservation along with the idealization, equations for energy transfer are designed for each of the generating components. The main components used in this research are the Sanden type TRSE09 expander, the Kaori type K70 condenser and evaporator and Motor pump 3 phase with 1600 rpm as the refrigerant pump working fluid. Here follows an explanation of the basic Organic Rankine Cycle work process theory using T-s diagram.

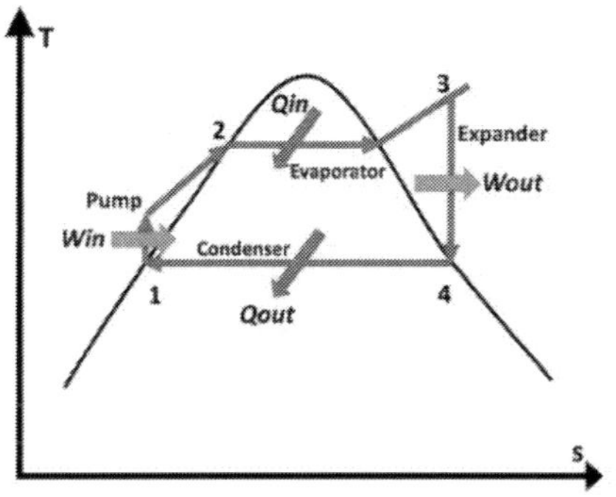

Figure 3. Diagram of ORC cycle

2.1. Working fluid pump

At point 1 the refrigerant pump applies suction and pressure to refrigerant R-134a which increases the pressure so that it moves from the condenser towards the evaporator. It is assumed that no heat transfer occurs around the pump and there is an equilibrium equation of mass and energy rate as follows.

$$Q - w = \left((h_3 - h_2) + \frac{(v_1^2 - v_2^2)}{2} \right) + g\,(z_4 - z_3) \tag{1}$$

or

$$w_p = h_2 - h_1 \tag{2}$$

Where w_p is the incoming power per unit of mass through the pump

2.2. Heater as producer of hot water for evaporator

The evaporation process of the refrigerant occurs in the heat exchanger which combines with a heater to produce hot water. Hot water releases heat to the refrigerant pipe in the evaporator at points 2 and 3 enabling the refrigerant to change state from liquid to gas. The thermal change process is the equilibrium of mass and energy rate as given in the following equation,

$$Q_{in} = h_3 - h_2 \tag{3}$$

where Q_{in} is the rate of heat transfer from the energy source into the working fluid per unit of mass through the evaporator, and then the refrigerant vapor flows to the expander.

2.3. Expander scroll as turbine

The refrigerant vapor 134a that exits the evaporator at point 3 has high temperature and high pressure which will expand to rotate the expander to create work before discharging it to the condenser at point 4 with relatively low vapor working fluid pressure. Heat transfer around the expander is not calculated, so the equilibrium mass energy rate around the expander is given in the equation,

$$w_t = h_3 - h_4 \tag{4}$$

where w_t denotes the rate of work produced per unit of vapor mass through the expander.

2.4. Condenser

At point 4 - 1 the condenser takes heat from the refrigerant flow pipe through a cooling water pipe at a specific temperature and amount of heat. The energy equation for the condenser is given as follows.

$$q_{out} = h_4 - h_1 \tag{5}$$

where, q_{out} is the rate of energy transfer from the working fluid to the cooling water per unit of fluid mass working through the condenser. Thermal efficiency is used to measure how much energy enters the refrigerant by the following equation.

$$\eta_{th} = \frac{(w_t - w_p)}{Q_{in}} = \frac{(h_3-h_4)-(h_4-h_3)}{(h_1-h_4)} \tag{6}$$

There is heat loss into the environment so heat *(Q_{in})* is needed to increase the heat transfer process while thermal efficiency is reduced. This condition occurs due to the irreversibility of the pump and expander so that as greater work is required *(Win)* the expander produces more work *(W_{out})* .

Table 1. Properties of R134a

Characteristics	Conditions
Boiling Point	-14.9 °F or -26.1 °C
Auto-Ignition Temperature	1418 °F or 770 °C
Ozone Depletion Level	0
Solubility In Water	0.11 % by weight at 77 °F or 25 °C
Critical Temperature	252 °F or 122 °C
Cylinder Color Code	Light Blue
Global Warming Potential (GWP)	1200

3. Preparation for ORC experiment

Equipment & devices created for the ORC experiments use several components that relate to electricity generation and refrigeration work processes. The components used in the ORC consist of working fluid pump, plate heat exchanger as an evaporator and condenser, a heater and heat exchanger drum, cooling water drum liquid receiver, oil water separator and a control panel. Installation of these components is performed in several steps:

- Step 1: calculate piping and connections required such as napples and nuts to link the main components mentioned above before installation.
- Step 2: install and connect pipes that serve as a liaison between the main components with valves and nuts and perform any welding required.
- Step 3: after all components are installed and connected with the pipe, conduct test runs and collect data.

The heater and heat exchanger drum seen in figure 5 is 38 cm in diameter and 50 cm in height. This heater produces hot water which is used to vaporise the working fluid in the evaporator to the desired amount of volume. Hot water re-enters the heat exchanger drum and circulates constantly.

Figure 4. ORC experiment equipment

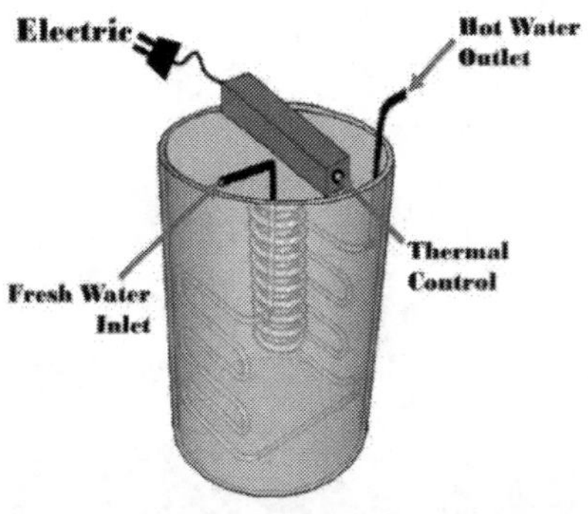

Figure 5. Combination packet of heater & heat exchanger

4. Result and Discussion

The design of combining a heater and heat exchanger device in the form of a drum begins with setting the heater temperature at between 40 °C – 80°C, then determining data for the inlet temperature into the evaporator, the amount of pressure in the expander and the number of expander rotations generated by this ORC system. The first step is to calculate the energy values from the evaporator, condenser, and expander. Then, data from the measuring equipment and from data acquisition are collected and processed.

Finally, the flow of working fluid is assumed to equal the airflow measurement at the expander and a wind compressor measures pressure and fluid flow in the superheat phase of 1.38 Bar and 0.006 kg / s, while the working fluid mass at liquid state is 1 kg / S. Results from data processing is used to record evaporator inlet (°C) temperatures every second up till 15 seconds as shown in figure 6. Next, the data on vapor pressure of working fluid (psi) in the expander is shown in figure 7 using the same time parameters. Data for measuring temperatures are obtained by using a thermocouple which is connected to a data logger to record all measurements. And figures 6 and 7 show the working fluid temperature variation between 24 ^{0}C to 28.5 ^{0}C on the variations of temperature control settings between 40 ^{0}C - 80 ^{0}C and the working fluid pressure entering the expander between 120 psi to 175 psi.

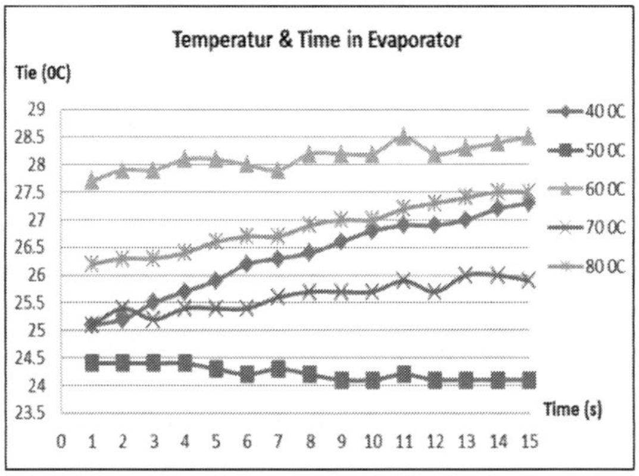

Figure 6. Temperature & time in evaporator

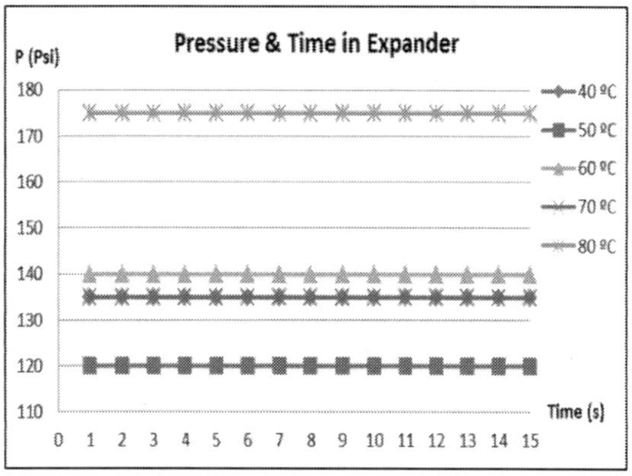

Figure 7. Pressure & time in expander

Figure 8 shows how the rotating expander produces various pressure variations that occur between 316.7 rpm to 377.9 rpm with heater temperature settings between 40 ^{0}C to 80 ^{0}C.

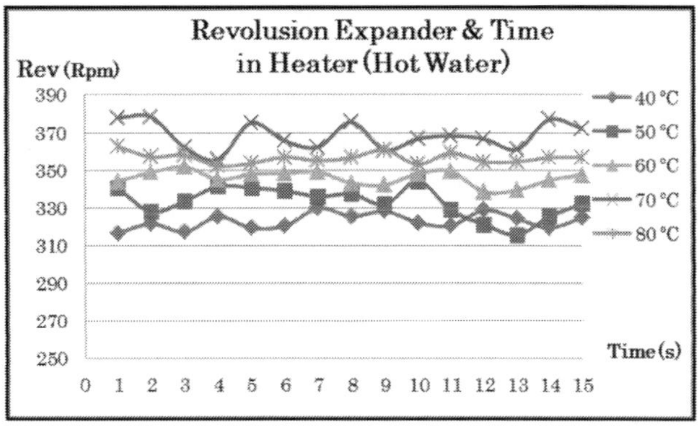

Figure 8. Variations of expander revolutions and time in the heater (hot water)

The calculated thermal efficiency of heater settings between 40 ^{0}C - 80 ^{0}C that produce hot water to create energy and work can be shown in table 2.

Table 2. Values of energy, power, and thermal efficiency in each temperature variation

No.	T_{heater}	Q_E	W_P	W_E	Q_C	ηth
1.	40	185.87	1.4880	0.03	182.10	2.0%
2.	50	200.89	1.4893	0.05	194.66	3.1%
3.	60	203.02	1.5220	0.07	193.13	4.9%
4.	70	218.53	8.6547	0.10	206.00	3.7%
5.	80	222.79	3.1893	0.13	203.91	8.5%

5. Conclusion

The Organic Rankine Cycle system experiment has enough data to determine the performance of the system. The experimental results show that based on the heater temperature set between 40 ^{0}C and 80 ^{0}C the inlet temperature of the evaporator is between 24.1 ^{0}C and 28.5 ^{0}C, while the vapor pressure of the working fluid entering the expander is between 120 psi to 175 psi. Some parameter results show the average expander rotation is 348.2 rpm and thermal efficiency is between 2 % - 8.5 %.

6. Acknowledgement

Acknowledgments are addressed to PITTA Program Universitas Indonesia Fiscal Year 2017 based on Letter of Agreement Number 810/UN2.R3.1/HKP.05.00/2017 which has funded this research program.

7. References

[1] Quoilin S, Lemort V and Lebrun J 2010 Experimental study and modeling of an Organic Rankine Cycle using scroll expander *Applied Energy* **87** 1260.

[2] Wei M, Song P, Zhao B, Shi L, Wang Z and Ma C 2015 Unsteady flow in the suction process of a scroll expander for an ORC waste heat recovery system *Applied Thermal Engineering* **78** 460 .

[3] Hogerwaard J, Dincer I and Zamfirescu C 2013 Analysis and assessment of a new organic Rankine based heat engine system with/without cogeneration *Energy* **62** 300.

[4] Yamada N, Watanabe M and Hoshi A 2013 Experiment on pumpless Rankine-type cycle with scroll expander *Energy* **49** 137.

[5] Chang J C, Chang C W, Hung T C, Lin J R, Huang K C 2014 Experimental study and CFD approach for scroll type expander used in low temperature organic Rankine cycle *Applied Thermal Engineering* **73** 1444.

[6] Teng H, Regner G and Cowland C 2007 Waste Heat Recovery of Heavy-duty Diesel Engines by Organic Rankine Cycle Part I: Hybrid Energy System of Diesel and Rankine Engines *SAE Tech. Pap.*, **01** 0537.

[7] Wei M, Fang J, Ma C and Danish S 2011 Waste heat recovery from heavy-duty diesel engine exhaust gases by medium temperature ORC system, Sci *China Technol. Sci.* **54** 2746.

[8] Tchanche B F, Papadakis G, Lambrinos G and Frangoudakis A 2009 Fluid selection for a low temperature solar organic Rankine cycle *Appl. Therm. Eng.* **29** 2476.

[9] ASHRAE Handbook. 2006. Refrigeration System and Applications (SI). *American Society of Heating*, Refrigerating and Air Conditioning Engineer. Atlanta. Georgia.

[10] Musthafah B, Mohd.Tahir, Noboru Yamada and Tetsuya Hoshino. 2010 Efficiency of Compact Organic Rankine Cycle System with Rotary-Vane-Type Expander for Low-Temperature Waste Heat Recovery *International Journal of Mechanical, Aerospace, Industrial, Mechatronic and Manufacturing Engineering* **4** No:1

[11] Incropera F P, DeWitt D P, Bergman T L and Lavine A S 2007 *Fundamentals of Heat and Mass Transfer (6th ed.)*. United States of America: John Wiley & Sons.

The properties of vegetable cooking oil as a fuel and its utilization in a modified pressurized cooking stove

Suhartono[1], Suharto[2] and Andisa Eka Ahyati[1]

[1]Department of Chemical Engineering, Faculty of Engineering, Universitas Jenderal Achmad Yani, Cimahi, 40533, Indonesia

E-mail: suhartono@lecture.unjani.ac.id, andisaekaahyati@ymail.com

[2]Research and Development Division for Mineral Technology, Indonesia Institute of Sciences, Jln. Ir. Sutami KM. 15, South Lampung 35361, Indonesia

E-mail: harto_berg@yahoo.com

Abstract. Vegetable cooking oils (VCOs) and used vegetable cooking oils (UCOs) can be considered as alternative fuels which will provide the household with low price fuel and may solve the problem of getting rid of waste VCOs. Their utilization as cooking fuel can bring numerous benefits not only for urban but also for rural communities in Indonesia. The paper focuses on characterizing VCOs and UCOs as fuels for the household cooking application using a modified pressurized cooking stove. Physical properties such as auto-ignition point, auto-ignition time, flash point, density and viscosity of VCOs play the vital role in the combustion. Some properties of these oils were measured and characterized according to ASTM standards. The oil was used directly as a fuel using in a design modified pressurized cooking stove. Adjusting the temperature of vegetable cooking oils used as fuel, it is possible to improve their combustion performance, thus reducing ignition time and incomplete combustion. The main target of the research is to determine the quality and performances of these oils combustion. The auto-ignition point for the several oils was determined to be as follows: UCO: 460 °C, crude VCO: 406 °C, fresh VCO: 405 °C and the peanut cooking oil did not auto-ignite. Crude VCO gave the shorter auto-ignition time than other oils within 30 s. The efficiency of the pressurized stove using UCO, crude VCO and fresh VCO as fuel were observed 23.65%, 25.99%, and 31.57%, respectively. The highest flame temperature of 942°C in these experiments was achieved by burning fresh VCO as fuel in this modified pressurized cooking stove. UCOs tended to produce luminous flames compared other oils.

1. Introduction

Indonesia is an agro based country and it has all the potential to produce vegetable cooking oil in huge quantities. Fuels of bio-origin such as vegetable cooking oils (VCOs) are a new alternative cooking fuel resource securing a sustainable and independent cooking energy supply. These alternative fuels can be produced from renewable energy sources such as soybean oil, sunflower oil, groundnut oil, palm kernels oil etc. Vegetable cooking oil (VCO) has the added safety advantage of having a much higher flash point than any other. Such fuels can be used directly as fuel for the household cooking application. The using of VCO has several advantages as combustion fuel due to relatively simple and low-cost technology for expelling and filtering, the VCO can be processed on the farm itself, thus saving the transport cost, time and energy [1].

Physical properties such as ignition point, of VCO, play the vital role in the combustion. It would be very beneficial to study the various factors surrounding the ignition of VCO. The properties of the oil would determine the auto-ignition temperature of the VCO due to different chemical compositions. Therefore different types of VCO would have different auto-ignition temperatures, resulting in different ignition times [2], [3].

1.1. Physical and thermal properties of fuel

Kerosene is commonly used as a liquid fuel where VCOs can be used as substitutes. Kerosene composed of hydrocarbon molecules. The kerosene fraction belongs to the group of hydrocarbon called paraffin, which has lower specific gravity than aromatic hydrocarbon of the same boiling point. The main components of kerosene are paraffin, cycloalkanes (naphtha) and aromatic compounds, where paraffin is the highest composition. VCO are tri-glycerol of fatty acids, with distinct chemical and physical properties, and different combustion characteristics than those of kerosene. VCO have a chemical composition that corresponds in most cases to a mixture of 95% triglycerides and 5% free fatty acids, sterols, waxes and various impurities [4]. The viscosity of VCO can be up to about 30 to 50 times higher than that of kerosene. The flash point of VCOs ranges from 180 to 300°C, compared to 80°C for kerosene, as shown in Table 1. [2], [3], [6-8]. This means the operating risks of kerosene are much higher due to its easy inflammation [5]. All properties of VCO are close to kerosene except the viscosity. High-value viscosity of the VCOs is considered to be the major constraint as combustion fuel, such as causing incomplete combustion and the formation of deposits on the burner nozzle [5].

Table 1. The properties of VCOs and kerosene

Fuel	Fuel ignition (°C)	KinematicViscosity (cSt)	Gross calorific value (MJ/kg)
Kerosene	50-55	2.2	43.50
Palm oil	314	88.6	39.54
Coconut oil	270-300	51,9	37,54
Sunflower oil	316	65,8	39,81
Peanut oil	340	57,4	39,65
UCO	164	5.16	36,59

High ignition points of VCO in connection with extremely high viscosity require special adaptation mechanism of the cooking stoves. Many techniques have been developed to reduce the density and viscosity of VOC. Preheating of the VCO reduced the viscosity which resulted in improved atomization property [6]. Therefore these must be treated prior to the use of VCOs in the cooking stove. By adequately heating the VOC in the horizontal spiral coil before burning on burner spoiler, its physical parameters can reach values very close to that of kerosene. Consequently, by properly adjusting the temperature of VOC used as fuel, it is possible to improve their combustion performance, thus reducing ignition time and incomplete combustion.

1.2. Pressurized cooking stove

The cooking stove, in general, is classified in two main categories; vapor jet burner (pressurized) and wick burners. The thermal efficiency of kerosene stove is between 20–40 % depending on stove and cooking equipment design [8]. Since the viscosity of VCOs is higher than the viscosity of kerosene, directly usage of kerosene wick-type stoves is not suitable for the use with VCOs. Due to its high viscosity, VOC has difficulty to be used as direct fuel for cooking. Also, the presence of other component forming coke and higher ignition temperature of VCOs compared to kerosene make it difficult to ignite the fuel. Therefore, the straight use of VCO as fuel in pressurized cooking stove entails adjusting several physical properties of density and viscosity. Designs of stoves using were based on the methods to vaporize and spray under pressure into a specially designed stove [5]. In this experiment, VCO was used directly as a fuel using in a design of modified pressurized cooking stove equipped with the spiral coil as a preheater. Heating the VCO before injecting them into the combustion stove is one of the methods to reduce the viscosity number [6] and reducing the ignition time. VCO was auto preheated before being burned to decrease its viscosity near to kerosene. Due to different chemical compositions, therefore different types of VCOs would have different auto-ignition temperatures, resulting in different ignition times. The results could give insight into safer VCOs to use that have higher auto-ignition temperatures. In this work, auto-ignition temperatures of these VCOs were measured and characterized according to ASTM standards.

The main objective of present study was to analyze the VCOs and UCO as fuel and to reduce the viscosity of oils by auto preheating in a modified pressurized stove.

2. Materials and methods

This work emphasizes the importance of the need to understand the vegetable cooking oil fires, especially the ignition characteristics of common VOCs. These experiments were researched to give insight into the temperatures ranges at which cooking oil will auto-ignite. The purpose of this research is to experimentally determine the combustion characteristics of three vegetable oils, namely VOC, UCO, crude VOC. To determine its usefulness as an alternative fuel, the combustion characteristics of vegetable oils are compared with kerosene.

The physical properties of VCOs and UCO were tested according to ASTM methods. An experiment was designed and several tests were conducted to determine the density, viscosity, auto-ignition temperatures, ignition time and smoke point of vegetable oils. To determine auto-ignition temperature, a small amount (5 mL) of cooking oil was heated in a small pan on a hot plate until the oil auto-ignited. The temperature of the oil in the pan was measured by using digital thermocouple Krisbow-KW06-283. The temperature of a smoke point was observed if enough volatile compounds emerge when a bluish smoke becomes clearly visible from the oil. The auto-ignition temperature and ignition time were measured when the minimum temperature at which an oil will start to burn without additional application of external heat (figure 1).

Figure 1. Test set up of ignition values

Three fairly common household cooking oils of UCO, crude VCO, fresh VCO and the peanut cooking oil were tested. The density measurements were performed gravimetrically at 30°C using volumetric glassware. Three test runs were taken to ensure the repeatability and accuracy of the data according to ASTM D1298 methods. The viscosity of household cooking oils was measured by using a Cannon-Fenske capillary viscometer immersed in a thermostatic bath. The measurements of viscosity were carried out at temperatures of 30°C, for a total of 3 measurements, each of which was obtained by averaging the values of three repetitions. The Cannon-Fenske type 200 viscometer with the viscosity range of 20-100 mm^2/sec. (centi-Stokes or cSt) was utilized to measure the viscosity of the oil using method suggested by ASTM D445. They allowed measuring the viscosity of all VCOs with flow times within the recommended range set by the standard ISO 3105:1994. The viscosity is obtained from the following equation (1) [9].

$$\upsilon = tc_0(T) \tag{1}$$

ν is the kinematic viscosity in centistokes, t the flow time expressed in seconds and $c_0(T)$ the characteristic calibration constant of each viscometer measured at the working temperature.

The relationship between the chemical structure and composition of VCOs and heating value can be expressed by a mathematical equation [10]. In this case, heating values of various VCOs were estimated based on kinematic viscosity, v instead of chemical structure and composition effected by using equation (2) [11], [12].

$$HV = 37.946 + 0.0491\upsilon \tag{2}$$

HV being heating value in MJ/kg.

The main obstacle of using VCOs in cooking stoves is its high viscosity, which often leads to clogging of the fuel pipe or burner nozzle. The stove has been adapted to or modified for utilizing of VCOs. The design of stove using the VCOs was based on the method to utilize the VCOs, vaporized, auto preheated and sprayed under pressure. Several stoves which have been adapted to or specifically designed for vegetable oils, but it is not reported yet, the stove design with spiral pipe for auto preheating to produce good atomization and combustion [1], [2], [5-8], [13-15].

The modified stove consists of a fuel pressure vessel, a hand-pressure pump that develops a pressure of around 3 bar by keeping the fuel pressure vessel at higher levels for supplying the fuel, a spiral-shaped pipe that serves as oil preheating (vaporizer) and nozzle sprayer as burner tip with three different sizes of diameter. The diameter and height of modified pressurized stove were 20 cm and 30 cm, respectively. The dimensions of vegetable oil pressure vessel, which is 15 cm in diameter and 35 cm in height, while sprayer nozzle of 0.5 mm; 0.7 mm and 0.8 mm (figure 2). Due to the pressure in the tank, the fuel flows through the horizontal spiral oil pipeline. The flow of the fuel was regulated with a valve provided in the oil line.

The ignition was conducted by burning alcohol with variations in the amount of 10 mL, 15mL, and 20 mL, while the injecting VOC on horizontal spiral pipeline start to auto-preheat to vaporize and raise the temperature above the fire (ignition) point. The hot fuel mixes with the ambient air and gave out a yellow-blue flame. The ignition point, ignition time, flame temperature, thermal efficiency and fuel consumption rate were measured in these works based on ASTM 90-92D method. The ignition temperature and ignition time were measured when the oil started to burn. The flame temperature was measured as long as the burning oil.

During the experimentation, some observations of the amount of water evaporated and mass of fuel burned have been taken to calculate stove performance indicators as thermal efficiency. The thermal efficiency of the fuel/stove/pot combination were determined through calculations based on the modified version of the well-known water-boiling test (WBT) revised calculation procedure [13-17]. The test calculates this as the ratio of enthalpy change of the water in the pot to the maximum energy theoretically available from combustion. Thermal efficiency is calculated using the formula given below:

$$\eta_T = \frac{m_w \times 4.186(T_2 - T_1) + m_{cv} \times 2.257 \times 500(T_2 - T_1)}{m_f \times NHV} \tag{3}$$

where m_w is the initial mass of water in the vessel, (4.186 kJ/kg.°C) is the specific heat of water, (T_2-T_1) is the change in water temperature, m_{cv} is the mass of water evaporated from vessel, (2257 kJ/kg) is the latent heat of evaporation of water, m_v is the mass of vessel, (500 kJ/kg. °C) is the thermal conductivity of vessel, m_f is the mass of the fuel used, and LHV is the lower heating value of fuel.

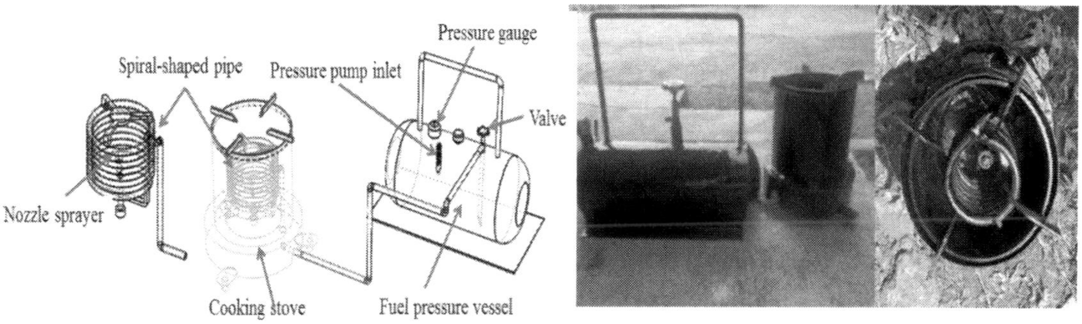

Figure 2. Sketch and photo of pressurized cooking stove with spiral pipe line as preaheater

3. Results and discussions

3.1. Density and viscosity

Among the VCOs properties as fuel are kinematics viscosity, density and heating value are the most important parameters that affect the combustion performance. The density of the VCOs as fuel is one of the important parameter since other performance parameters of combustion such as heating value have been correlated with it. The density values can also be used to measure the amount fuel in pressured vessel system by volumetric method. The variation of the density affects the power and the fuel spray characteristics during combustion in the pressurized cooking stove.

Table 2. Density and viscosity of VCOs and kerosene

Fuel	Auto-Ignition time (sec.)	Smoke Point (°C)	Auto-Ignition temperature (°C)
UCO	30	238	460
Crude VCO	45	215	412
Fresh VCO	54	200	405
Peanut cooking oil	-	182	-

The variation in density of vegetable oils is shown in table 2. Density is found to be more or less same for all vegetable oils, which is 50 % above that of kerosene (0.79 mg/mL). Since the density varies with composition and temperature and has a great influence on the atomization process. Thus the knowledge of the temperature dependence of this parameter is a highly important component of the process of the study. The fuel density required to be reduced more, in order to make it suitable for the pressurized cooking stove and makes it spray characteristics more like those of kerosene. From the different kinds of literature, it has been concluded that heating of the fuel which is the direct reduction of density [14,18].

Kinematic viscosity is also an important physical property affecting pressure drop in the fuel line of the pressurized cooking stove as well as the fuel atomization in nozzle sprayer. High VCOs viscosity can result in excessive pressure drop and produce spray with large droplets, which deteriorates the combustion performance. High VCOs viscosity also results in poor fuel atomization, incomplete combustion, deposits in the combustion vessel. In table 2. the kinematic viscosity of VCOs from this experiment is about 20-25 times higher than that of kerosene at 30 °C. At this temperature kinematic viscosity of vegetable oils varies between 50-60 centistokes. Higher ranges of kinematic viscosity can clog fuel lines, affects the start of injection and adversely affect fuel atomization. The higher viscosity of VCOs than kerosene is directly related to the level of unsaturation and the length of the fatty acid chains. The viscosity tends to decrease when there is an increased presence of double bonds and grows with an increase in the length of the hydrocarbon chain and according to the level of polymerisation in the oil. The kinematic viscosity of these VCOs required to be reduced more, in order to make it suitable as kerosene in a pressurized cooking stove. From the different kinds of literature, it has been concluded that heating of the fuel makes it spray characteristics more like those of kerosene, which is the direct reduction of viscosity [18]. It was also reported that heating the vegetable oils to 140 °C would reduce the viscosity to near that of kerosene at 20 °C and improve the spray characteristics [18]. There are several ways to reduce VCOs kinematic viscosity, and heating oil is one of them. Therefore an effort has been tried to reduce the kinematic viscosity by auto preheating the VCOs in a design pressurized cooking stove using horizontal spiral pipeline as a preheater.

Predicted heating values of VOCs using equation (3), as presented in table 2. shows a variation of the calorific value of kerosene with different VCOs. This may be due to the difference in the chemical composition or presence of oxygen molecule in the molecular structure of the oil. It can be explained that the viscosity increases when the presence of double bonds decreases and when the length of the hydrocarbon chain increases. The energy content (heating value) increases in accordance with the increase in the length of the chain, and therefore with the increase in carbon atoms, but decreases when the percentage of carbon decreases with respect to the oxygen. Moreover, the calorific content of the fuel decreases in accordance with the reduction in the hydrogen content, and therefore as the number of double bonds increases [18].

3.2. Fuel properties of VCOs

The liquid VCOs were tested to establish the auto ignition temperature. The VCO was heated, the temperature increase slows down before it reaches its boiling point, oil will start to emit smoke and pyrolysis. This is called the smoke point, at which the temperature of an oil under defined conditions, enough volatile compounds emerge from the oil that a bluish smoke becomes clearly visible. Then, once boiling begins, the temperature increases more rapidly as the boiling and smoke emission becomes more intense. When this is observed, the oil is becoming close to its auto-ignition temperature (without fire source).

Table 3. Specification of VCOs charaterization

Fuel	Auto-Ignition time (sec.)	Smoke Point (°C)	Auto-Ignition temperature (°C)
UCO	30	238	460
Crude VCO	45	215	412
Fresh VCO	54	200	405
Peanut cooking oil	-	182	-

Table 3 summarizes characteristics of some common vegetable cooking oil. The auto-ignition temperature of UCO, 460 °C is highest than other oils. It can be understood because during frying, vegetable cooking undergoes various physical and chemical changes, and many undesirable compounds are formed. These include free fatty acid and some polymerized triglycerides which increase the molecular mass and reduce the volatility of the oil. Therefore, fatty acid esters obtained from frying oil influences the fuel characteristics, such as the viscosity and burning characteristics [19], [20]. UCO is also full of bits of food and carbon that will reduce ignition time, the 30s shorter than three other oils. The lower auto-ignition temperature of crude VCO and fresh VCO were observed 412 °C and 405 °C, respectively, related to the level of double bonds and the length of the hydrocarbon chain on viscosity. The viscosity increases when the presence of double bonds decreases and when the length of the hydrocarbon chain increases. The increase in the viscosity of a vegetable oil subjected to rapid heating is determined by the breaking of the double bonds and a consequent reduction in the level of unsaturation [18]. The peanut cooking oil did not auto-ignite in this test. It was observed that peanut oil would vapor and dry without igniting due to unbroken of double bonds of fatty acid [21]. Results of this research could give an assumption that correlates to higher auto-ignition temperatures due to different chemical compositions. Therefore different types of oil would have different auto-ignition temperatures, resulting in different ignition times. The results could give insight into safer oils to use that have higher auto-ignition temperatures.

3.3 Performance of modified pressurized cooking stove

The main pressurized cooking stove design was manipulating the viscosity of the oil to make it behave like kerosene. This was done by auto preheating the oil on the horozontal spiral pipeline. At above room temperature (30 °C) VCOs have a theoretical viscosity of around 16-20 cSt, the same as kerosene. The viscosity of VCOs is closely dependent on temperature. An increase in temperature corresponds to a non-linear reduction in viscosity.

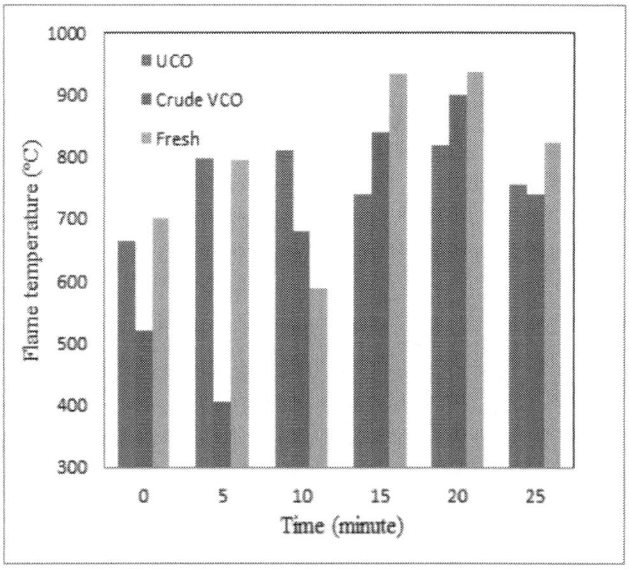

Figure 3. Flame temperature profile of VCOs burning

The water-boiling tests (WBT) were conducted to demonstrate the VCOs as potential cooking liquid fuels. The results show that the VCOs burnt transparent a yellow-blue flame. UCOs tended to produce luminous flames compared other oils. In these tests, the minimum temperature at which the vapors of a fuel catch fire if in contact with a flame was measured as ignition temperature. Figure 3 is the samples of flame temperature profile ignite by 20 mL of alcohol. The surface flame temperature different versus time was because of a little bit of pressure vessel changes. The highest surface flame temperature of 942°C in these experiments was achieved by burning fresh VCO as fuel in this modified pressurized cooking stove. It is understood because of less fatty acid contained in fresh VCO rather than crude VCO and UCO. So that, the breaking of the chain in fatty acid become easier and faster [21]. The maximum temperature generated from combustion of VCOs an UCO using this pressurized cooking stove with horizontal pipe line preaheter is lower then than that of using a pressurized stove with vertical pipe line preaheter. The maximum surface flame temperatures of UCO, crude VCO and fresh UCO as fuel using the pressurized stove with vertical pipe line preaheter of 1100 °C, 1042 °C and 920 °C, respectively [22]. It is possible that the vertical spiral pipe provides greater heat transfer than the horizontal spiral pipe, since the surface of the heat transfer is larger.

A fuel/pressurized stove combination was characterized by a thermal efficiency. The following table 4 provides characteristics of VCOs. Auto preheating the oil in spiral coil pipeline will decrease viscosity and increases the mixing rate of air and vegetable oil due to increase the fuel mobility, assisting completion of combustion to produce higher flame temperature and increased the ignition temperature. The higher ignition of VCOs than kerosene is a beneficial safety feature, as the fuel can be safely stored and transported at the room temperature.

Table 4. Pressurizes of modified stove characterization

Fuel	Ignition temperature (°C)	Fuel consumption (L/h)	Thermal efficiency (%)
UCO	478	4.10	23.65
Crude VCO	429	1.57	25.99
Fresh VCO	415	1.19	31.57
Kerosene	55	1.30	45.84

The specific fuel consumption seemed to be affected by the combined effects of fuel heating values and the viscosities. Low viscosity combined with the low-energy content of fuel resulted in much fuel consumption and time to boil. The thermal efficiency of all VCOs was found to be in the same range (23-32%) and about 1.3 times lower than thermal efficiency of kerosene.

Results of this research show that the VCOs require preheating to match the physical properties of kerosene. Heating of the VCOs to attain a lower viscosity has helped with the spraying (atomizing) of the oil on a pressurized stove. The reduced viscosity is not only to improve the combustion performance but is necessary for an even flow of the fuel through the fuel pipelines and nozzle spoiler.

4. Conclusion

It was found that VCOs and UCO as fuel had much higher ignition temperature (405 °C-478 °C) compare to kerosene whose ignition point only 55 °C. The VCOs and UCO utilization as fuel have several advantages as combustion fuel due to much safer to use than kerosene, little risk of fire hazards in the household application and other numerous ecological and sociological benefits.

The much higher kinematic viscosity of VCOs and UCO (50.14 cst-59.67cst) compare to kerosene (2.4 cSt) that affecting pressure drop in the fuel line of the pressurized cooking stove and the fuel atomization in nozzle sprayer could be manipulated to make it behave like kerosene by spiral pipe line as auto-preheater in the pressurized cooking stove.

It was observed that auto preheating the oil in a pressurized cooking stove and only a small amount of ethanol is needed for start-up, decreased the viscosity of VCOs and UCO by an indicator of higher flame temperature (about 900°C) and increased the ignition temperature (415 °C-478 °C). The VCOs combustion produce transparent a yellow-blue flame. UCOs tended to produce luminous flames.

Generally, all VCOs fuel and kerosene showed average fuel consumption of ± 1.3 L/h, except UCO fuel consumption of 4.10 L/h. UCO is full of bits of food and carbon that will increase fuel consumption due to decreasing of heating value and density and viscosity.

The fuels tested using the standard water-boiling tests show that VCOs fuels have an average of thermal efficiency of around 23.65%-31.57%, which is ±30% lower than that of kerosene. Utilize UCO as fuel using this modified pressurized cooking stove with spiral pipe line auto-preaheater gives the thermal efficiency of 31.57% near to kerosene (45.84%).

Based on these findings, VCOs are a sustainable energy source as fuels have great potential to provide an alternative source of cooking energy supply not only for urban but also for rural communities in Indonesia. The introduction of a new design of prototype pressurized cooking stove with spiral pipe line auto-preaheater can be readily acceptable since its operation is similar to the well-known pressurized kerosene stoves.

5. Acknowledgment

We wish to acknowledge the financial support from DRPM-RISTEKDIKTI sponsorship through the PUPT-program for the year 2016-2018. We also wish to thank Universitas Jenderal Achmad Yani for supporting this work through the provision of laboratory space and equipment. We also wish to thank all those who supported this work, directly and indirectly, are greatly acknowledged.

6. References

[1] Singh R N 2011 Straight Vegetable Oil: An Alternative Fuel for Cooking, Lighting and Irrigation Pump *IIOAB Journal* vol. **2** no. 7 pp. 44–49

[2] Ramkumar S and Kirubakaran V 2016 Review on Admission of Preheated Vegetable Oil in C.I. Engine *Indian J. Sci. Technol.* **9** (2) 1-11.

[3] Giuseppe Toscano and Eleonora Maldini 2007 Analysis of the Physical and Chemical Characteristics of Vegetable Oils as Fuel *J. of Ag. Eng. - Riv. di Ing. Agr.* 3 pp. 39-47

[4] Joel Blin, Christel Brunschwig, Arnaud Chapuis, Odilon Changotade, Sayon Sidibe, Eric Noumi and Philippe Girard 2013 Characteristic of Vegetable Oils for Use as Fuel in Stationary Diesel Engines-Towards Specifications for a Standard in West Africa *Renewable and Sustainable Energy Reviews* 22 pp. 580-597

[5] Ashok Yadav and Prakash Chandra Jha 2013 A Case Study On Biofuel Stove Technology: Jatropha As A Biofuel *Int. J. Tech. Enhancements and Emerging Engineering Research* vol.**1** no. 2 pp. 14-18

[6] Shahid E M, Jamal Y, Shah A N, Rumzan N and Munsha M 2012 Effect of Used Cooking Oil Methil Ester on Compression Ignition Engine *Journal of Quality and Technology Management* **8** (2) 91–100

[7] Sjaffriadi, B. Nurachman, N. A. Sasongko, and Masfuri Imron 2012 Waste Cooking Oil Pressurized Multifuel Stove *Proc. of National Seminar of chemical Engineering of Parahyangan Chatoclic University*

[8] Ejilah I R, Olorunnishola A A G and Enyejo L A 2013 A Comparative Analysis of the Combustion Behavior of Adulterated Kerosene Fuel Samples in a Pressurized Cooking Stove *Global Journal of Researches in Engineering Mechanical and Mechanics Engineering* vol.**13** no.6 pp. 35-43

[9] Bernat Esteban, Jordi-RogerRiba, Grau Baquero Antoni Rius and Rita Puig 2012 Temperature Dependence of Density and Viscosity of Vegetable Oils *Biomass, and Bioenergy* **4** pp.164-171

[10] DIN-V-51605 2006 Fuels for Vegetable Oil Compatible Combustion Engines–Fuel from Rapeseed Oil–Requirements and Test Methods 07

[11] A Demirbas 2000 A Direct Route to the Calculation of Heating Values of Liquid Fuels by Using their Density and Viscosity Measurements. Energy Convers Manage vol. **41**, pp.1609–14

[12] M. Kratzeisen, and J. Muller 2010 Influence of free Fatty Acid Content of Coconut Oil on Deposit and Performance of Plant oil Pressure Stoves *Fuel.* vol. **89** pp.1583–1589

[13] R. Natarajan, N.S.Karthikeyan, A. Agarwaal and K. Sathiyanarayanan 2008 Use of Vegetable Oil as Fuel to Improve the Efficiency of Cooking Stove *Renewable Energy* **33** pp. 2423–2427

[14] Wagutu, Agatha W, Thoruwa Thomas F N, Mahunnah R L A, Chhabra Sumesh, C Caroline, Thoruwa Lang'at and Mahunnah R L A 2010 Performance of a Domestic Cooking Wick Stove Using Fatty Acid Methyl Esters (FAME) from Oil Plants in Kenya *Biomass and Bioenergy*, pp.1250-1256

[15] Tafadzwa Makonese, Crispin Pemberton-Pigott, James Robinson, David Kimemia, and Harold Annegarn 2012 Performance Evaluation and emission Characterisation of Three Kerosene Utoves using a Heterogeneous Stove Testing Protocol (HTP) *Energy for Sustainable Development* vol.16 no. 3 pp. 344–351

[16] Gaurav Jambhulkar,Vibhor Nitnaware, Manisha Pal, Neha Fuke, Purva Khandelwal, Pallavi Sonule, Sneha Narnawre and V. P. Katekar 2015 Performance Evaluation of Cooking Stove Working on Spent Cooking Oil *Int. J. Emerging Sci. Eng.* **3** no. 4 pp. 26-31

[17] Suhartono, Bayu Dwi Prasetyo and Ikrimah Nur Azizah, 2016 Synthetic Gas (syngas) production in downdraft corncob gasifier and its application as fuel using conventional (LPG) domestic stove *J. Eng. Appl. Sci.* **11** no. 8 pp.5238-5243

[18] Giuseppe Toscano and Eleonora Maldini 2007 Analysis of the Physical and Chemical Characteristic of Vegetable Oils as Fuel *J. of Ag. Eng.-Riv. di Ing. Agr.* vol. **3**, pp. 39-47

[19] S. K. Acharya and M. K. Mohanty R. K. Swain 2009 Kusum Oil as a Fuel for Small Horse Power Diesel Engine *Int. J. Eng. Technol.* **1** 3 pp. 1793-8236

IOP Conf. Series: Earth and Environmental Science **105** (2017) 012047 doi:10.1088/1755-1315/105/1/012047

[20] K. Naima and A. Liazid 2013 Waste Oils as Alternative Fuel for Diesel engine: A review, *J. of Petroleum Technol. and Alternative Fuels* **4** no.3 pp. 30-43
[21] Krystyna Buda-Ortins. B.S. 2010 Auto-Ignition of Cooking Oil. Dept. of Fire Protection Engineering, University of Maryland
[22] Suhartono, Tiara Arini Putri, Lutfi Fauziah 2017 Performance Evaluation of a Pressurized Cooking Stove Using Vegetable Cooking Oils as Fuel *Int. J. on Advanced Science, Engineering and Information Technology* **7** no 4 pp. 1234-1241

Manufacture of solid soap based on crude papain enzyme and antioxidant from papaya

Setiadi[1], Putri and Famila Anindia[2]
[1]Chemical Engineering, Universitas Indonesia, Kampus Baru UI, Depok, Indonesia
[2]Bioprocess Engineering, Universitas Indonesia, Kampus Baru UI, Depok, Indonesia

Abstract. The uncontrolled formation of free radical formation released to the air especially in the high urban area lead to resulting negative effect on the air quality. This condition is consequently made the human skin continuously contacts to those radicals and may cause a skin diseases. Therefore, one way to overcome this problem is to manufacture a soap formulation which provide a new function of its ingredient as radical scavenger or to keep the skin in health. Papain is a plant proteolytic enzyme for the cysteine protease enzyme which found naturally in papaya (Carica papaya) manufactured from the latex of raw papaya fruits. Proteolytic enzyme has a function which is able to break down organic molecules made of amino acids, known as polypeptides. In this research crude papain enzyme is developed into active ingredient of solid soap formulation with the antioxidant originated from fresh papaya to enhance the useful soap for human skin health. This study aims to produce a solid soap formula that is safe for the skin, meets the standard of SNI 1996 and tested the benefits in the addition of crude papain enzymes. The result showed that Formula IV is a formula soap that meets standard SNI 1996 and have highest antioxidant activity among other formulas indicated by the value of IC50 is 13,657 ppm. Compared soap with positive control (without enzymes) than negative control (with enzyme) has value of percentage dirt removal higher than positive control, it is 19% with measurement absorbance of wash water substrate and 32% measured by mass substrate after washed.

1. Introduction

As protection of the body from the outside, the skin has various problems such as dry skin, premature aging, chronic diseases like skin cancer. This can be caused by air pollution resulting in many free radicals, exposure ultraviolet, and also less care for cleanliness. The main treatment for maintaining skin health is to use soap regularly. Soaps are sodium or potassium salts of fatty acids derived from vegetable oils or animal fats [1]. The use of chemicals will provide side effects on sensitive skin, while soaps with natural ingredients will provide the maximum nutrition on the skin.

Papain enzymes found in papaya's soap have high activity in the breakdown of proteins into peptides and amino acids. This makes papain can be used as an active ingredient in the manufacture of skin cleanser because it can dissolve dead cells that are attached to the skin and difficult to get rid of in a physical way. Antioxidants are substances that the body needs including the skin to neutralize free radicals and prevent damage caused by free radicals. In this study we make solid soap containing antioxidant as crude from papaya fruit and crude enzyme papain from latex of papaya.

2. Material and Methods

2.1. Manufacture Solid Soap

The raw material for solid soap were coconut oil, NaOH, water, citric acid, glycerin, papaya fruit, crude papain enzyme. The oil mixed gently with lye solution and kept warm at ambient temperature then placed in the stainless steel pot soap, stirred frequently for 5-10 min. while stirring the mixture added the other ingredients and next pour the solution into the moulds in the ambient temperature for 24 hours until solution change into solid. Formulation of soap presented in Table 1 below

Table 1 Solid soap formulation

Ingredients	Formula (%w/v)					
	I	II	III	IV	V	VI
Coconut Oil	55	48	42	42	55	48
NaOH	7	6	6	6	7	6
Water	21	19	17	17	21	19
Stearic Acid	0.8	0.7	0.6	0.6	0.8	0.7
Glycerin	1.5	1.2	1.1	1	1.5	1.2
Papaya	15	25	30	30	15	25
Perfume	0.5	0.5	0.5	0.5	0.5	0.5
Crude enzyme papain	0.5	1	1.5	1.5	1	0.5

2.2. pH Determination

The pH was determined using pH meter and prepare 10% soap solution with distilled water. The electrode of the pH meter was inserted into solution after calibration with buffer solution. The pH reading was recorded.

2.3. Foam Ability test

Around 1 gram of the sample soap was added to 20 mL measuring cylinder containing 10 mL of distilled water. The mixture was shaken using vortex vigorously so as to generate foams. After shaking for about 30 sec, the measuring cylinder was allowed to stand for 5 min (t_5). The height of the foam in the solution was then measured and recorded.

2.4. Fatty Acid Determination

10 gram of sample soap was added into 100 mL distilled water in the 250 mL erlenmeyer flask and heated in water bath. Subsequently the sample soap titrated with H_2SO_4 20% using methyl orange indicator until the solution become red. Stir solution to homogenized and heated until two phased was formed. Next, get 10 gram micro paraffin (beeswax) into that solution and heated for an hour until the solution was formed in one phased and then cooling the erlenmeyer flask in the water bath quickly. Mixture of beeswax (wax cake) and fatty acid was change into solid state in the flask, and measured the weight of sample (1). Fatty acid can be calculated with the formula :

$$\textbf{Fatty acid} = \frac{m_w - m_b}{m_s} \times 100\% \qquad (1)$$

Where m_w is weight of wax cake, m_b is weight of micro paraffin beginning, m_s is weight of sample soap.

2.5. Alkali Free Determination

First step to determination alkali free was to prepare 100 mL alcohol 96% into flask and then added indicator phenolphthalein and subsequently titrate with KOH 0,1 N until appear pink colour in solution. 5 gram of sample soap was added into alcohol neutral solution and also boiling stone or flakes of porcelain it use to avoid an explosion. Second step boiled until soluble solution dissolves for 30 minutes. If it doesn't contain alkali (colour of solution not pink), cooled it to 70°C and titrates with

0.1N KOH until pink colour stand for 15 seconds. Next step titrated the solution with 0.1N HCL in the alcohol with micro burette until the pink colour disappears. Alkali free can be calculated in formula

$$\textit{Alkali free} = \frac{V \times N \times 0.04}{W} x\ 100\% \quad (2)$$

Where V is volume of HCL, N is Normalitas of HCL, W is mass of sample soap, 0.04 is mass of relative NaOH

2.6. Specific Density Determination(25°C)

1 mL of the sample soap was measured in a clean micro tube and stand for a while into incubation until get 25°C (room temperature). The weight of both sample soap and the micro tube was also measured then the weight of the sample soap was obtained by subtracting the weight of micro tube from the weight of both the sample soap and measuring micro tube. The specific density of the sample soap is then obtained using: stand for 5 min (t_5) and 30 min (t_{30}). The height of the foam in the solution was then measured and recorded

$$\textit{Density}\ (25°C) = \frac{W_1 - W_0}{W_0} x\ 100\% \quad (3)$$

Where W_1 is weight of measuring micro tube and the sample, W_0 is weight of measuring of micro tube, and V_0 is volume of the sample used. stand for 5 min (t_5) and 30 min (t_{30}). The height of the foam in the solution was then measured and recorded.

2.7. 2,2-Diphenyl-1-picrylhydrazyl (DPPH) free radical scavenging assay

Various amounts of samples dissolved in methanol and have concentration 500, 1000, 1500, 2000, and 2500 ppm each sample were added to DPPH solution 100 ppm. After 30 min of incubation at room temperature, the absorbance at 516 nm was measured using spectrophotometer UV-VIS. Percent inhibition of free radical DPPH was calculated as follows:

$$\textit{Antioxidant activity}\ (\%) = \frac{A_{control} - A_{Sample}}{A_{control}} x\ 100\% \quad (4)$$

The quality of radical scavenging property of the soap was determined by calculating IC_5. The IC_{50} value is the concentration of each sample of soap required to scavenge the DPPH radical to 50% of the control. The concentration of sample soap necessary to decrease the initial DPPH calibration curve. Radical scavenging activity of sample soap was calculated as IC_{50}. The lower IC_{50} value indicates the higher DPPH radical scavenging activity.

2.8. Effectiveness in Cleaning

First step is prepare clean white fabric as a substrate 5cm x 5cm. The dirt is formed by protein from egg and milk mixture. Paste the dirt on the substrate by dipping for a minute and drying it. Wash the substrate with control positive soap (without crude enzyme) and control negative soap (with crude enzyme) for 2,4,6,8,10 minutes using stirrer. Performance test which aims for effectiveness in cleaning consist of 2 methods there are removal dirt test which aims to know the capability of the detergent to remove dirt and by measuring the absorbance of wash water substrate using spectrophotometer UV-VIS at wavelength 270 nm. For removal dirt test

$$\textit{removal dirt}\ (\%) = \frac{m_d - m_c}{m_d} x\ 100\% \quad (5)$$

Where m_d is mass of dirt substrate and m_c is mass of clean substrate after washing with soap sample.

3. Result and discussion

3.1. Soap Analysis

Table 2 present the quality criteria of soap based on BSN (*Badan Standarisasi Nasional*) which is safe for human health skin.

Table 2 Soap quality terms

No	Parameters	
1	Fatty acid (%)	>70
2	Alkali free (%)	
	as KOH (liquid soap)	Max 0,22
	as NaOH (solid soap)	Max 0,1
3	Specific density (25°C)	1,01-1,1
4	pH	8-11

Sources (1)

Saponification occurs when, first, three molecule of sodium hydroxide (NaOH) are dissolved in water (H_2O) and are splits apart, which results in three sodium ions (Na) and three hydroxyl groups (OH). Second, a triglyceride (fat) molecule ($C_3H_5(COOR)_3$) is hydrolyzed reaction, which results three fatty acid tails (COOR). Third, the hydroxyl groups (OH) all bond to the free glycerol (C_3H_5) to form a molecule of glycerin ($C_3H_5OH_3$). Fourth, the three fatty acids (COOR) each bond with one of the three sodium ions (Na) to form three molecules of soap (3NaCOOR) (2). The result saponification was measured in table 3 all sample (F) and physiochemical properties some parameter.

Table 3 Result of physiochemical properties of solid soap for all formula

Sample	pH	Fatty acid (%)	Alkali free (%)	Specific density (gr/mL)	Foam ability (t_5) (%)
F I	10.3	87	0.110	1.0873	82
F II	10.4	78	0.219	0.9120	83
F III	10.3	86	0.217	1.0619	86
F IV	10.35	79	0.108	1.0595	78
F V	9.95	78	0.109	1.0278	95
F VI	10.2	82	0.108	1.0271	67

As measured in table 3 the high pH content indicating high percentage amount of unspecified and unsaponificable matter due to incomplete alkaline hydrolysis. The pH value measurement of the whole formula has an alkaline formed because the amount of alkali present in the soap preparation affects the pH value. The basic ingredients used are NaOH, one of the strongest bases, which play an important role in producing saponification reaction with fatty compound. Based on criteria from (1) all formula enter the threshold.

Fatty acids composition of oil used in soap formation can also determine the nature and stability of leather produced by different soaps, it was reported that, lauric acid, myrisic acid, which are all saturated fatty acids produce soap with fatty leather, higher cleansing power but of low stability (3). The principle of measuring the amount of fatty acids containing in the soap is by breaking the bonds between fatty acids with sodium in soap using strong acid, then extract it with micro paraffin wax to form a cake containing mixture micro paraffin wax, the amount of fatty acids, and mineral oil may exist (1). SNI establishes a standard amount of fatty acid of at least 70% which means that the ingredients added as a filler in the soap should be less than 30%. With the aim to be able to accelerated the dirt removal form of oil or fat when used as ingredients soap.

Alkali free is alkali that is not bound as a compound at the time of making soap. Table 3 shown that there are two formula that doesn't meet the range of requirement that determined by SN, it is cause due to the excessive addition of alkali in the preparation and the lack of the amount of fatty acids in the soap. Free alkali that exceeds the standard will cause skin irritation and it will also cause dry skin (4). Oil with low density is an indication that it contains low molecular weight fatty acid, likewise it will have high saponification value which makes it suitable for soap production. The density for formula II doesn't meet range standard BSN, which has a papaya content of 15% and a crude papain enzyme 0.5%. The result can be due to the soap composition of the unbalanced formula so it is not included in the SNI standard range. Most ingredients such as sugars and salts may cause increased density, but density may also decrease if there is fat or ethanol in solution.

Most consumer like cleaning products that have abundant foams, but the amount of foam in the cleaning products due to surfactant is sodium lauryl sulfate (SLS) which can cause skin irritation (6). A good foam stability criterion is when subjected to a foam stability range of about 60-70% in the first 5 minutes. In each soap formula containing stearic acid so that the resulting foam stability is good enough to be proved in the foam stability data within the first 5 minutes.

3.2. Antioxidant Activity

Based on the analysis of variance, the addition papaya on solid soap, was no significantly affected to antioxidant activity. Antioxidant properties of sample soap caused by active substance from papaya fruit such as α-tocopherol, ascorbic acid, beta carotene, flavonoids, vitamin B1, and niacin (7). Polyphenols as antioxidants work through four mechanisms, such as damaging free radicals, preventing the formation of free radicals through hydrogen bonding, deactivate singlet oxygen which acts as a free radical in the body, and to be bonded with metals. We're using solvent methanol because in flavonoid raw materials and phenolic acids will dissolve in methanol and is recognized as the main phenolic fraction with antiradical activity (8). The The average of antioxidant activity and ability of sample soap to scavenge the DPPH radical measured as IC_{50} will shown in table 4.

Table 4 Antioxidant activity on solid soap added papaya from different concentration

Formula	Antioxidant activity (%)					IC_{50} (ppm)
	500	1000	1500	2000	2500	
F I	13.41	16.01	15.88	17.18	17.44	20,434
F II	14.19	14.97	15.36	18.09	20.69	11,929
F III	15.62	18.22	18.09	20.17	26.15	8,098
F IV	15.36	15.75	16.27	17.05	21.47	13,657
F V	17.05	16.66	18.35	19.39	18.87	26,034
F VI	13.28	13.41	16.53	20.17	20.3	19,405

This suggests that the antioxidant activity in soap samples is very small and has significantly decreased antioxidant activity after formulated into soap. Based on research from the Journal International Food Research obtained IC_{50} value of mature papaya fruit about 4300 ppm with 90.67% antioxidant activity (9).

3.3. Effectiveness in Cleaning

Dirt removal test by comparison of control positive soap (without enzymes) with control negative (with enzyme). The process of lifting the dirt is removal of impurities from the fabric fibers (substrate) which includes the interaction between surfactant, dirt, and fabric surface. Wavelength 270 nm was chosen based on the maximum absorbance result of the washing solution and according to the Journal International of Research in Pharmacy and Science Department of Quality Assurance, B.N College of Pharmacy papain solution has a maximum absorption at wavelengths of 270 - 280 nm (11).

Table 5 Differences of control soap negative and positive, measurement removal dirt with absorbance of wash water substrate

Negative control	Absorbance	Removal dirt (%)	Positive control	Absorbance	Removal dirt (%)
E_2	0.1963	7	B_2	0.1537	5
E_4	0.2994	11	B_4	0.194	7
E_6	0.3108	11	B_6	0.2061	7
E_8	0.3509	13	B_8	0.2252	8
E_{10}	0.526	19	B_{10}	0.2799	10
Absorbance dirt				2.8021	

Table 5 and 6 shown that negative soap control ability has higher dirt removal ability compared with positive control soap. So the rough papain enzyme contained in negative control soap has a role in removing more impurities in the washing process by the same treatment. But differences in soap removal ability of positive and negative controls are not significant. Because the use of the active substance rough enzyme papain is not a pure papain enzyme that is by tapping the sap from young papaya fruit. In crude papain enzyme there are still components of contaminants that interfere with the work of papain enzyme. According to the journal Biochem J. University of London papain is heavily contaminated by chymopapains A and B and there are other impurities that solution in crude papain enzyme during tapping of papaya fruit.

Table 6 Differences of control soap negative and positive, measurement removal dirt with mass of substrate

Negative control	Mass (gr)	Removal dirt (%)	Positive control	Mass (gr)	Removal dirt (%)
E2	1.2041	19	B2	1.3038	13
E4	1.1316	24	B4	1.2117	19
E6	1.0661	28	B6	1.0705	28
E8	1.0361	30	B8	1.0658	28
E10	1.0142	32	B10	1.0504	29
Mass of dirt substrate (gr)				1.4968	

4. Conclusion

Based the research it can be concluded that the addition of crude enzyme papain on the solid soap have affected to removal dirt higher than soap without enzyme, and did not significantly affected to the fatty acids, alkali free, pH, specific density and foam stability. The best formula based on the activity antioxidant which safe for skin and full meet criteria SNI 1996 was formula IV. Formula IV (addition 30% papaya fruit as antioxidant and 1.5% crude papain enzyme) had pH 10.35; value of fatty acid 79%; alkali free 0.108%; specific density 1.0595 gr/mL; and foam ability 78% after stand for 5 minutes.

5. Reference

[1] BSN 1996 *Standart Mutu Sabun Mandi SNI 06-4085-1996.* Jakarta : Dewan Standarisasi Nasional

[2] Warra A A, et al 2010 Cold- Process Synthesis and Properties of Soaps Prepared from Different Triacylglycerol Sources *Nigerian J. of Basic and Appl. Sci.* Vol. **18**(2), pp. 315-321.

[3] Atiku F A 2014 Production of Soap Using Locally Available Alkaline Extract from Millet Stalk: A Study on Physical and Chemical Properties of Soap *Int. J. of Advanced Research in Chemical Sci. (IJARCS)* Vol. **I**, pp. 1-7.

IOP Conf. Series: Earth and Environmental Science **105** (2017) 012048 doi:10.1088/1755-1315/105/1/012048

[4] Hernani, Tatit K, Bunasor and Fitriati 2010 Formula Sabun Transparan Antijamur dengan Bahan Aktif Ekstrak Lengkuas (Alpinia galanga L.Swarts) *BUl. Littro* Vol. **21** No. 2, pp. 192-205.

[5] Gaman, P. M and Sherrington 1990 K. B. *The Science of Food.* 3nd ed. Oxford : Pergamon Press

[6] Schwitulla J, et al. 2014 Skin Irritability to Sodium Lauryl Sulfate os Associated With Increased Positive Patch Test Reactions *British Journal of Dermatology* pp. 115-123.

[7] Ozkan, Aysun, et al. 2011 Antioxidant Capacity of Juice from Different Papaya (Carica papaya L.) Cultivars Grown Under Greenhouse Condititions in Turkey *Turk J Biol* pp. 619-625.

[8] Shalaby, Emad A and Shanam, Sanaa M M 2013 Comparison of DPPH and ABTS assays for determingin antioxidant potential water and methanol extract of Spirulina platensi *Indian Journal of Geo-Marine Sciences* Vol. **42**(5), pp. 556-564.

[9] Maisarah, A.M, et al 2013 Antioxidant analysis of different parts of Carica papaya *International Food Research Journal* Vol. **20**(3), pp. 1043-1048.

[10] Holmberg, K, et al. 2003 *Polymers in Aqueous.* 2n ed. Chichester : J Willey

[11] Trivedi, Vishal, et al. 2013 Pepsin, Papain and Hyaluronidase Enzyme Analysis: A Review *Int. J. of Research in Pharmacy and Sci.* Vol. **3**(1), pp. 01-18.

Effect of feed composition of co-pyrolysis of corncobs–polypropylene plastic on mass interaction between biomass particles and plastics

D Supramono[1], M A Nabil[1], Setiadi[1] and M Nasikin[1]
[1]Department of Chemical Engineering, Universitas Indonesia, Kampus Baru UI Depok 16424, Indonesia

Email: setiadi@che.ui.ac.id

Abstract. Polypropylene plastic (PP), which has higher hydrogen content compared to that in biomass, is potential to be used as a cheap hydrogen source for pyrolysis of biomass, such as corncobs, to remove part of oxygen content in the biomass pyrolysis oil. By charging these two materials, synergistic effects will appear that improve quality and quantity of bio-oil produced. The aim of the present work is to investigate possible mass interaction between biomass and plastic materials leading to the synergistic effects using slow co-pyrolysis. The experiment was performed in a displacement reactor. Feed composition was varied at 12.5%, 25%, 37.5%, 50%, 62.5%, 75%, and 87.5% weight of PP. This phenomenon was observed through expansion-contraction of the reactor prior to large mass decomposition of the reactor feed. In experiment involving biomass-dominated feeds, i.e. PP contents $< 50\%$ (regime 1), there was reactor bed contraction due to slight biomass decomposition followed by no-displacement of reactor piston. During the contraction, it is estimated, there was low velocity ejection of biomass volatile matter, while during no-displacement stage, there was a phase change of plastic granules to plastic melt. In experiment with plastic-dominated feeds, i.e. PP $\geq 50\%$ (regime 2), the reactor bed underwent contraction due to phase change of plastic granules which reduced the bed volume, followed by bed expansion due to the swelling of biomass particles estimated in the presence of hydrogen bonding in the phenolic environment in biomass material. The last stage of co-pyrolysis in regime 2 was the bed contraction and it is predicted that there was high velocity gas ejection through the small pores which intensified interaction of biomass-plastic radicals in hydrogen-rich environment favourable for exerting synergistic effect.

1. Introduction

Energy is an important supporting aspect of human life. In fact, widely-used fossil fuels are derived from non-renewable energy sources, and this disadvantage encourages various research and development efforts of finding alternative energy sources. One way of these efforts is by conducting co-pyrolysis of biomass and plastics especially those with high hydrogen content, such as polypropylene (PP). The addition of plastics to the co-pyrolysis at sufficient amount will decrease the oxygenated compounds in the bio-oil product and improve the bio-oil yield [1]. The quality improvement of bio-oil is induced by hydrogen donation from free radicals of plastic pyrolysis to those of biomass to remove oxygen from pyrolysis volatile matter and form more non-oxygenate compounds in bio-oil [2]. The quantity improvement of bio-oil occurs as a result of bonding of small radicals into larger molecules which increases bio-oil yield [1]. Synergistic effect occurs if the yield of bio-oil or non-polar phase of bio-oil as a result of oxygen removal produced by co-pyrolysis of 2 materials is higher than that obtained by summing yields of pyrolysis of pure materials. Therefore, synergistic effect improves both quantity and quality of bio-oil.

There are 2 different feeds which affect yields of bio-oil and non-polar phase, i.e. biomass-dominated feeds (regime 1) where the PP plastic provides no sufficient hydrogen to remove any oxygen radicals from bio-oil and plastic-dominated feeds (regime 2) where PP plastic provides enough hydrogen to remove part of oxygen radicals of biomass pyrolysis. In regime 1 no synergistic effects have been observed both on bio-oil and non-polar phase yields in co-pyrolysis, while in regime 2 they occur [3]. It is suspected that there is mass interaction between biomass and plastic materials which governs the bio-oil and non-polar phase yields. However, there has been no research in co-pyrolysis of biomass-plastic investigating what possible mass interaction occurs affecting the yields. It seems that Fei et al (2012) investigated for the first time relationship between swelling-shrinking phenomena of bituminous coal and synergistic effect in co-pyrolysis of lignite and bituminous coal [4]. They observed combined effect of swelling of bituminous coal particles and calcium content in lignite particles. They explained that the swelling assists coal fluidity and improves contact area between lignite and coal particles, while calcium could provide catalytic effect on tar yield of bituminous coal so that the bituminous coal particles shrink. Therefore, calcium in lignite particles are favourable for synergistic effect on bio-oil yield. This synergistic effect mechanism may have been different for the case of co-pyrolysis of biomass-plastic in the present work. Biomass may undergo swelling due to the presence of hydrogen bonding [5]. This physical character of biomass particles may influence synergistic effects on bio-oil and non-polar phase yields in co-pyrolysis of biomass-plastic.

The aim of the present work is to investigate possible mass interaction between biomass and plastic materials leading to the synergistic effects. The experiment was carried out in a displacement reactor to perform slow co-pyrolysis. To investigate the swelling-shrinking of biomass particles in the environment of plastic melt attributed to the mass interaction, the present work has used in a small tubular reactor of 24mm diameter and 60mm height filled with a feed blend of corn cob particles and PP plastic granules to simulate the swelling-shrinking of biomass particles leading to synergistic effects in a stirred tank reactor for larger yield of bio-oil. The independent variable used is feed compositions of corn cob and polypropylene plastic in the feed. The swelling-shrinking of the biomass particles was investigated in relationship with the synergistic effects.

2. Experimental

Figure 1 shows schematic design of the displacement reactor. The movement of the piston reactor was observed to measure the expansion-contraction of the reactor piston at the top of the reactor using a ruler in which swelling-shrinking of biomass feed may have occurred in the reactor. The cylindrical reactor and its piston are made of stainless steel (SS 316). When the piston was placed into the reactor, there was a very narrow gap between the rim of the piston and the inner wall of the reactor to allow the release of volatile matter and the free up and down movement of the piston when the biomass particles swelled and shrank. On the side of the reactor, there is a ruler to measure of the piston displacement during co-pyrolysis.

The reactor feed composition was varied at 0%, 12.5%, 25%, 37.5%, 50%, 62.5%, 75%, and 87.5% weight of PP and the feeds were charged into a displacement reactor at different batches. In each batch, the co-pyrolysis process was carried out at a slow heating rate (slow co-pyrolysis) of 5°C/min from room temperature to temperature just before the mass decomposition occurred estimated by TGA curves [6]. When the co-pyrolysis process took place, the magnitude of reactor piston displacement was measured every 1 minute. The initial total feed weight and final temperature of heating for each composition are given in Table 1.

To investigate possible relationship between the swelling-contraction of the biomass particles and synergistic effect, SEM image of char surface of each co-pyrolysis regime has been taken and compared to that taken from pyrolysis of pure biomass particles. All batches were conducted prior to large mass decomposition of the co-pyrolysis to observe how synergistic effects were prepared.

Table 1. Initial total feed weight and final temperature of heating

Composition (Corncob : PP Plastic)	Total Feed Weight (gr)	Final Temperature of Heating (°C)
87.5%:12.5%	12.00	215
75%:25%	12.00	220
62.5%:37.5%	12.00	225
50%:50%	10.83	300
37.5%:62.5%	8.96	310
25%:75%	8.05	315
12.5%:87.5%	7.44	320

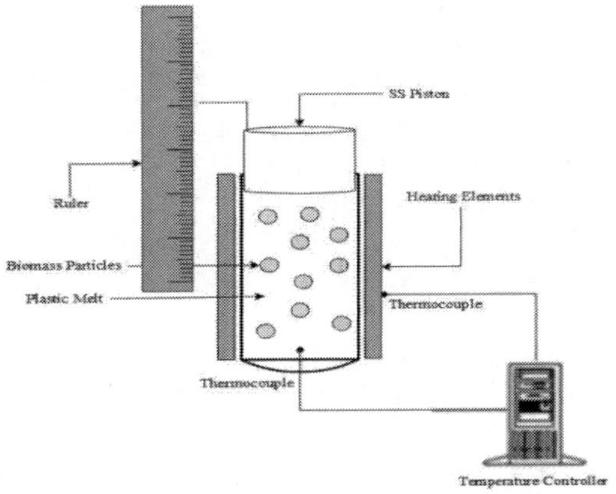

Figure 1. Schematic displacement reactor

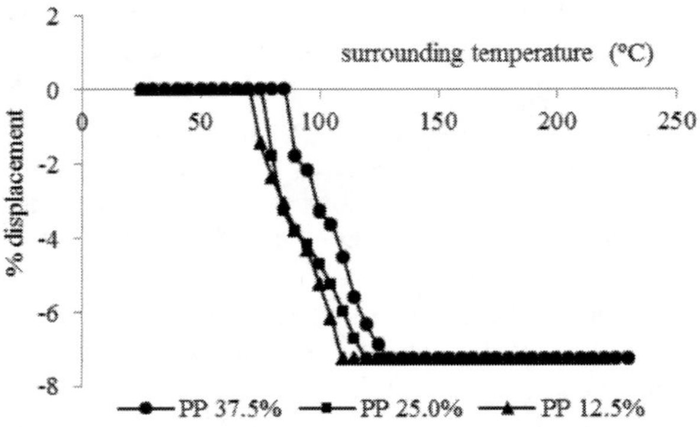

Figure 2. Biomass temperature vs % displacement in regime 1

3. Result and discussion

For biomass-rich co-pyrolysis feeds (regime 1), i.e. those with PP < 50% weight, the piston displacements during co-pyrolysis are shown in figure 2. The figure shows negative displacements (contraction) in the reactor followed by no displacement. These indicate that there was only shrinking of biomass particles. It was observed that during the contraction, the release of vapour was seen from the reactor. Figure 2 shows that the contraction ends at maximum temperature of 125°C. The measurement of melting point of PP granular in the present work shows that it started at 130°C. Therefore, contraction of all samples occurred before the melting of PP granules occurred and the heating of biomass particles were still in gas environment. In such a condition, heat radiation was predominant during biomass particle shrinkage and the feed with higher biomass composition received higher heat flux of radiation from the reactor wall. Consequently, the feed with higher biomass composition or lower PP content started shrinking at lower surrounding temperature as shown in figure 2 due to the less required temperature driving force. The contraction occurred mostly due to the decomposition of cellulose and hemicellulose to form vapour phase starting at low temperature [7]. All the contraction curves end before the plastic melted.

During the contraction stage, where the heat radiation was predominant, the heat sensed by the biomass material depended on the depth of heat penetration. Because the density of biomass is low, the radiative heat can penetrate into the inner part of the biomass [8]. Therefore, cellulose and hemicellulose, which were more susceptible to crack by heating than lignin [9], decomposed into smaller molecules, while lignin may have been still intact because delignification could occur in the presence of small molecule carboxylic acid such as acetic acid and formic acid as pyrolysis products of hemicellulose starting at 200°C [10]. To support the explanation of the above displacement mechanism, it is necessary to observe the surface morphology of the biomass particles by taking a char morphology image through a Scanning Electrone Microscope (SEM) test on the char. As representation of regime 1, the char tested was of co-pyrolysis using 25% PP feed. In addition, for comparison, char morphology by SEM originating from pyrolysis of pure biomass particle surfaces was also observed. These two char samples were taken from pyrolysis runs at temperatures prior to large mass decomposition according to the TGA curves measured by Miranda (2017) [6], which temperature data are presented in Table 1. Pure biomass char of pyrolysis results was taken from a pyrolysis run ending at 205°C, while char derived from 25% PP feed co-pyrolysis was taken from a pyrolysis run ending at 220°C. SEM images of both char samples are shown in figures 3a and 3b. SEM images show that the morphology of both char samples are similar wherein they exhibit sufficiently large char pores allowing vapor phase of pyrolysis to come out easily. The dark parts of the SEM images show the pore locations, while the light parts the char surface. It is expected that vapour phase ejection from large-sized pores occurred at low velocities.

Figure 3. SEM images of char (a) from pure biomass pyrolysis, (b) from co-pyrolysis with 25% PP feed (regime 1), (c) from co-pyrolysis with 75% PP feed (regime 2)

As the contraction curves end, PP granules were expected to melt because they end about the melting point of PP. Figure 2 shows no displacement stage after contraction stage describing the phase change of PP granules from solid to melt. During no displacement stage, it is expected that the heat radiation from the reactor wall to biomass material was obstructed by plastic melt surrounding the biomass material. As a consequence, the production of volatile matter from biomass decomposition was considered to be non-appreciable to the contraction of the reactor.

Figure 4 describes contraction-expansion of the displacement reactor. Contraction stage may have started from the melting point of plastic considering that there was temperature difference between surrounding gas temperature and plastic melt temperature. Exception occurred for the feed containing 50%PP which started at a temperature less than PP melting temperature because part of biomass material was not obstructed against heat radiation by PP granules. The difference in temperature ranges of the contraction stage between regimes 1 and 2 because of the difference in heat transfer mechanism. In regime 1 where the feed was rich in biomass, radiative heat transfer predominated due to its high emissivity [11], whereas in regime 2 where the feed was rich in plastic mass, the heat flux from the reactor wall to biomass was obstructed by plastic granules and heat passed through plastic melt before reaching the biomass particles. Large contraction in regime 2 may have been caused by melting of PP from PP solid granules which reduced the bed porosity inside the reactor thus contracting the feed bed volume. In this case, decomposition of biomass may have been too small to contribute to the contraction of the reactor.

Co-pyrolysis with feed comprising 50%PP is a transition one between regime 1 and regime 2. The biomass particles were partly surrounded by PP granules against heat radiation from the reactor wall. However, because mass decomposition occurred at temperatures larger than those at regime 1, the bed contraction partly contributed by biomass mass decomposition was larger than that in regime 1. The remaining contraction was contributed by phase change of PP granules to molten phases.

The contraction stage reached a point thereafter the reactor underwent expansion stage. This stage may have occurred due to the swelling of biomass induced by the presence of hydrogen bonding [12] between cellulose and lignin and between cellulose and hemicellulose [13]. Phenolic compounds as monomers produced by the first stage of lignin thermal cracking [14] contribute to strong hydrogen bonding [5]. In terms of physical structure, lignin is located in the outer cell of biomass, cellulose is located within lignin shell and hemicellulose with amorphous structure is located within the cellulose and between cellulose and lignin [15]. This structure allows the conductive heat from plastic melt to transport heat flux in the first instance into lignin, which decomposes into phenolic compounds.

The swelling resulting from the presence of phenols inhibited the evolution of volatile matter to the surrounding plastic melt. As a consequence, with time and increased temperature, more and more cellulose and hemicellulose decompose until at a temperature that the swelling biomass could not resist the ejection of the volatile matter. When the pressure inside the biomass particle was high, the vapour would be pushed out of the swelling biomass at high velocity. The discharge of this vapour in figure 4 is illustrated by declining curves marking the second contraction.

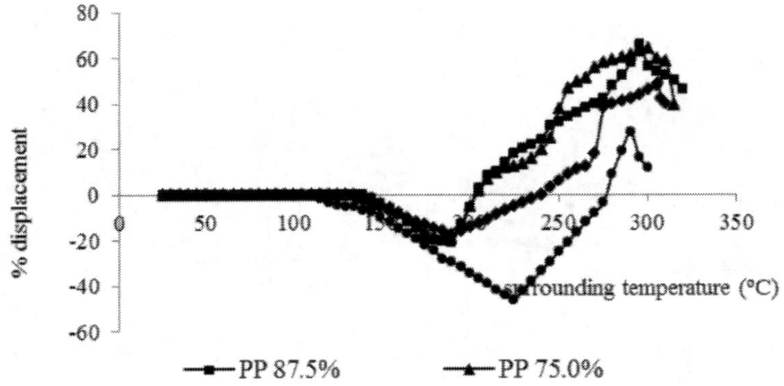

Figure 4. Temperature vs % displacement in regime 2

To support the explanation of the presence of high velocity in regime 2, char sampled from co-pyrolysis in a stirred tank reactor involving PP 75% feed was taken and observed by SEM to image the char surface morphology. The char was taken at a temperature before the large mass decomposition based the TGA curve [6], i.e. before reaching 315°C. This high velocity ejection is indicated by the existence of small pores on the char surface (see figure 3c), which are smaller than the pores found in regime 1 (see figure 3b). This high velocity vapour phase then penetrated strongly inside the plastic melt surrounding the biomass particles. In other words, there is a strong and intensive interaction between the vapour phase of biomass pyrolysis and the plastic melt. As far as fluid dynamic point of view, this ejection would generate interface between turbulent side generated by the ejection and the laminar side exerted by plastic melt which entrains plastic melt and mixes with biomass volatile matter [16]. This entrainment through the turbulent-laminar interface enhances reactions between biomass and plastic radicals during large mass decomposition of co-pyrolysis [17]. Therefore, high velocity ejection in regime 2 where the feed blend is rich in hydrogen allows hydrogen transfer to generate synergistic effects in the co-pyrolysis of biomass-plastic [1].

4. Conclusions

The present work on co-pyrolysis of corn cob-PP plastic prior large mass decomposition has resulted in some conclusions. For reactor biomass dominated feeds, i.e. those with PP <50% (regime 1), co-pyrolysis feed bed undergoes contraction due to slight biomass decomposition prior to plastic melting, followed by no-displacement stage. For reactor plastic-dominated feeds, i.e. those with PP $\geq$ 50% (regime 2), co-pyrolysis bed undergoes contraction due to phase change of plastic granules to plastic melt, followed by bed expansion estimated due to the presence of hydrogen bonding in the phenolic environment in biomass material. The last stage of co-pyrolysis bed in regime 2 is the reactor bed contraction attributed to the existence of small pores on biomass sample surface and it is predicted that there is high velocity gas ejection through the small pores which intensifies interaction of biomass-plastic radicals in hydrogen-rich environment favourable for synergistic effect.

5. Acknowledgement
Authors wish to express their gratitude to PITTA research grant 2017 from Universitas Indonesia for financing the present work.

6. References
[1] Abnisa F and Wan Daud W M A 2014 A review on co-pyrolysis of biomass: an optional technique to obtain a high-grade pyrolysis oil *Energy Conversion and Management* **87** pp 71-85.

[2] Sharypov V I, Marin N, Beregovtsova N G, Baryshnikov S V, Kuznetsov B N, Cebolla V L and Weber J V 2002 Co-pyrolysis of wood biomass and synthetic polymer mixtures. Part 1: Influence of experimental conditions on the evolution of solids, liquids and gases *J. Anal Appl. Pyrolysis* **64** pp 15–28.

[3] Supramono D, Julianto, Haqqyana, Setiadi, Nasikin M 2017 Phase separation of bio-oil produced by co-pyrolysis of corn cobs and polypropylene in *The 2nd International Conference on New Energy and Future Energy System*, Kunming, Yunnan, China.

[4] Fei J, Zhang J, Wang F and Wang J 2012 Synergistic effects on co-pyrolysis of lgnite and high-sulfur swelling coal *Journal of Analytical and Applied Pyrolysis* **95** pp 61-67.

[5] Gardner D J and Tajvidi M 2016 Hydrogen bonding in wood-based materials: an update *Wood and Fiber Science* **48** 4 pp 1-11.

[6] Miranda Y 2016 Efek sinergetik pada slow co-pyrolysis bonggol jagung dan plastik prolipropilena murni untuk peningkatan kualitas dan yield bio-oil in *Skripsi* Dept. of Chemical Engineering, Universitas Indonesia, Depok.

[7] Basu P 2013 Biomass gasification, pyrolysis and torrefaction, Amsterdam: Academic Press.

[8] Kommareddy A and Anderson G 2004 Study of light requirements of a photobioreactor in *North Central ASAE/CSAE Conference.*

[9] Nunn T R, Howard J B, Longwell J P and Peters W A 1985 Product compositions and kinetics in the rapid pyrolysis of sweet gum hardwood *Ind. Eng. Chem. Proc. Des. Dev.* p 24.

[10] White R H and Dietenberger M A 2001 in: Buschow K H J and Chan R W in *The Encyclopedia of Materials: Science and Technology*, Elsevier, Amsterdam pp 9712-9716.

[11] Klason T, Bai X S, Bahador M, Nilsson T K, Sunde B 2008 Investigation of radiative heat transfer in fixed bed biomass furnaces *Fuel* **87** pp 2141–2153.

[12] Mantanis G I, Young R A and Rowell R M 1994 Swelling of wood, Part I: Swelling in water *Wood Sci.Technol.* **28** pp 119-134.

[13] Jin Z, Katsumata K S, Lam T B T and Iiyama K 2006 Covalent linkages between cellulose and lignin in cell walls of coniferous and nonconiferous woods *Biopolymers* **83** pp 103−110.

[14] Patwardhan P R, Brown R C, Shanks B H 2011 Understanding the fast pyrolysis of lignin *Chemsuschem* **4** pp 1629–1636.

[15] Zhang J, Choi Y S, Yoo C G, Kim T H, Brown R C and Shanks B H 2015 Cellulose−hemicellulose and cellulose−lignin interactions during fast pyrolysis *ACS Sustainable Chem. Eng* **3** pp 293-301.

[16] Philip J and Marusic I 2012 Large-scale eddies and their role in entrainment in turbulent jets and wakes *Phys. Fluids* **24** 055108.

[17] Watanabe T, Sakai Y, Nagata K and Terashima O 2013 Visualization of turbulent reactive jet by using direct numerical simulation *International Journal of Modeling, Simulation, and Scientific Computing* **4** Suppl. 1pp 1-16.

Effect of coconut fiber suspensions on drag reduction in circular pipe

W Wulandari[1], K T Waskito[2], S Mau[1], Yanuar[2] and Marcus A Talahatu[2]

[1] Graduate Student Department of Mechanical Engineering, Faculty of Engineering, Universitas Indonesia, Kampus Baru UI Depok, Depok, 16424, Indonesia
[2] Department of Mechanical Engineering, Faculty of Engineering, Universitas Indonesia, Kampus Baru UI Depok, Depok, 16424, Indonesia

E-mail: yanuar@eng.ui.ac.id

Abstract. Pipeline system as fluid transport is very commonly used. Thus, the drag reduction becomes very interesting to be studied because it is related to energy efficiency. One method of drag reduction is the active method by adding drag reduction additive agents such as surfactants, polymers, nanofluids, and fibers. The purpose of this study is to analyze the effect of coconut fiber on drag reduction by adding coconut fiber with varied concentration. The experimental was experimented by using circular pipe ID 38 mm with 1200 mm length. The test fluid was coconut fiber suspension with concentration 300, 500, and 1000 ppm. This study was conducted from low Reynolds Number until Reynolds Number about 25,000. In this research condition, the results showed that the drag reduction on circular pipe ID 38 was about 7.6% in the Reynolds Number about 25,000. The maximum drag reduction was for coconut fiber suspension with concentration 1000 ppm. The drag reduction increases with the increase of coconut fiber suspension concentration. Based on this research, it can be concluded that coconut fiber can be used as a drag reducing agent which save natural resources and environmentally friendly.

1. Introduction

Drag reduction in pipes becomes very attractive to investigate because pipeline is very commonly used for fluid transport. By reducing drag in pipes, it can reduce energy usage and it means energy efficiency. There are two methods used in the drag reduction, namely the passive method and the active method. The passive method is by utilizing geometry in a solid object flowed by fluid, such as utilizing the geometry of the spiral pipe [1], coating the pipe with the anti-wet agent [2], injecting the micro bubble as in the marine application [3], and using the riblets such as shark skin [4]. The active method is by adding drag reduction additive agents such as surfactants, polymers, nanofluids, and fibers. The effect of using surfactant, cetyltrimethyl ammonium chloride (CTAC), in fluid can reduce the drag up to 51% that occurs at concentration 75 ppm [5]. Polymethyl methacrylate suspension on monochlorobenzene have been added to the fluid can even reduce the reduction of resistance up to 80% [6].

In addition to surfactants and polymers, nanoparticles have also been experimented. The results showed that the drag reduction by adding Nano silica- particles (nano-SiO2) reaches 24% occurred at 1 Wt% nanofluids concentration [7]. Fibers have been studied such as asbestos, the result showed that

IOP Conf. Series: Earth and Environmental Science **105** (2017) 012050 doi:10.1088/1755-1315/105/1/012050

asbestos reduced the drag up to 40% that occurred at about concentration 100 wppm on Reynolds number around 20,000 [8]. Of the three common types of drag reduction additive agents above, it can be concluded that the active method by using additives is effective in reducing the drag of fluids. However, the lack of these additives is that those additive agents can pollute the environment and are unsafe for the environment, due to those additives are chemical. Therefore, it is necessary to find some kind of additive reducing agent which is certainly effective in reducing the drag, yet safe for the environment.

The active method using natural fibers has already been done by adding a wood pulp fiber which showed a drag reduction up to 40% [9]. The bamboo fiber also used to reduce the drag up to 20% [10]. In addition, paddy husk fiber which also reduced the drag in pipes up to 32% [11]. Hence in this study, coconut fiber is chosen for the experiment to determine the effect in reducing the drag in circular pipe. Coconut fiber is a natural fiber that is widely available in nature, especially in Indonesia, thus it is very easy to get and relatively inexpensive. The important thing is the coconut fiber is environment friendly, due to its characteristics are easy to decompose. Hence, the purpose of this study is to analyze the effect of coconut fiber solutions on drag reduction with concentration of 300, 500, and 1000 ppm in circular pipe ID 38 mm with length 1200 mm. This study was conducted from low to high Reynolds number about 25,000. The results explain that there is an effect of drag reduction in the circular pipe by using coconut fiber solution with varied concentrations.

2. Methodology

Figure 1 is coconut fiber that has been processed into a very fine fiber. The process for obtaining coconut fiber for the experimental is done by several processes. The outer shell of coconut husk that is old and brown is separated from the inner coconut husk manually. The coconut husk is then cut into smaller pieces to make it easier for the next process. After that, the cut coconut husk is fed into a crusher to get a finer fiber about 1 mm length. In this experimental, the coconut fiber is mixed with tap water to get concentration 300, 500, and 1000 ppm. The temperature of coconut fiber suspension is constantly maintained at 27 °C.

Figure 1. Coconut fiber.

Figure 2 is a schematic diagram of an experimental set-up with a closed circulation system consisting of storage tank, pump, voltage regulator, horizontal pipe ID 38 mm with 1200 mm length, valve, pressure transducer, Data Acquisition (DAQ), computer, and measuring glass. The coconut fiber suspension in the storage tank is pumped by centrifugal pump to the test pipe. Two pressure transducers are mounted on the inlet and outlet of the test pipe to measure the pressure. Pressure transducer is connected to DAQ which is then recorded by the computer. The fluid through the outlet of the test pipe was measured using a measuring glass. The time required to fill the measuring glass with a certain volume can be determined by using a stopwatch. The fluid weighed using a digital scale to determine the mass of the fluid. The test is carried out from Reynolds numbers about 6,000 to Reynolds about 25,000 by using adjustable flow rate variations using a voltage regulator arranged from 100 volts to 200 volts.

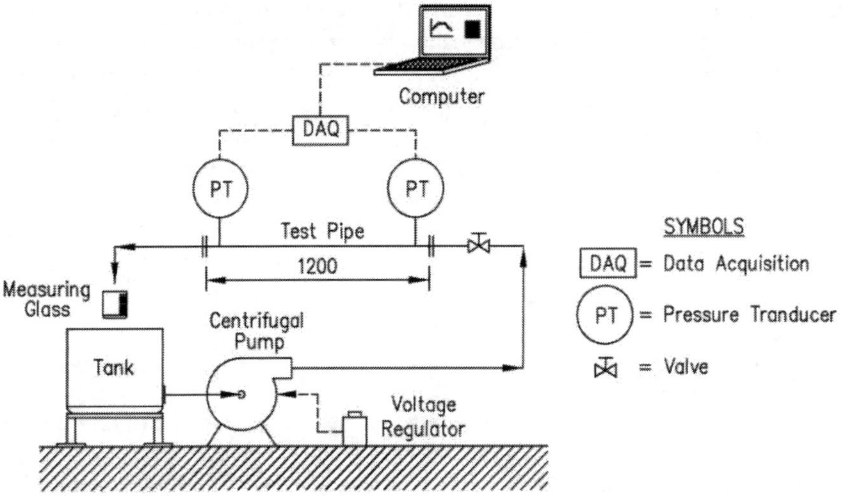

Figure 2. Schematic diagram.

3. Rheological model

Experimental data is plotted in the moody chart. The laminar region refers to Hagen-pouiseuille line and the turbulent region refers to Blasius line. Drag reduction is defined by decreasing of the coefficient of friction on the flow. The friction factor is caused by shear stress that occurs between each layer of the velocity. The relationship between shear stress (τ) in fluid is proportional to the velocity gradient or shear rate (γ) and can be described by Newtonian model (equation 1) as stated bellow:

$$\tau = \mu \frac{du}{dy} \tag{1}$$

where the constant of proportionality μ is known as the viscosity. The Newtonian viscosity only depends on temperature and pressure. It is not affected by the shear rate. Viscosity is described as the ratio between the shear stress and the shear rate. Other than that, some fluids show a non-linear relationship between shear stress and shear rate. It shall follow Power Law model or non-Newtonian fluid. Non-Newtonian fluid is divided into Bingham, Pseudoplastics, and Dilatant. It does not depend on temperature and pressure, but affected by shear rate. The relationship between shear stress and shear rate can be described by measuring the pressure drop and the flow rate by using equation 2 stated below:

$$\frac{D\Delta P}{4L} = \mu \frac{8u}{D} \tag{2}$$

where D is inner diameter, ΔP is pressure drop, L is length of the experimental pipe, u is average velocity. Whereas, coefficient of friction (f) can be expressed by Darcy equation (equation 3) as stated below:

$$f = \frac{2Dg\Delta h}{Lu^2} \tag{3}$$

where f is friction factor, g is gravity, and Δh is different height of head between two measuring point.

Drag reduction can be obtained by using equation 4 as stated below:

$$DR = \frac{f_{water} - f_{fiber}}{f_{water}} \times 100\% \tag{4}$$

IOP Conf. Series: Earth and Environmental Science **105** (2017) 012050 doi:10.1088/1755-1315/105/1/012050

4. Result and discussion

4.1. Characteristic of coconut fiber suspensions

In order to know the type of fluid suspension of coconut fiber, it can be done by comparing with water. From figure 3 it can be seen that the relationship between shear stress and shear rate shows a linear relationship for the coconut fiber suspension with concentrations 300, 500, and 1000 ppm. The diagram is formed to resemble a straight line and close to the water data. The straight line is a shear stress and shear rate line for Newtonian fluid, where water data indicates conformity with the straight line. Thus it can be seen that the coconut fiber suspension is a type of Newtonian fluid.

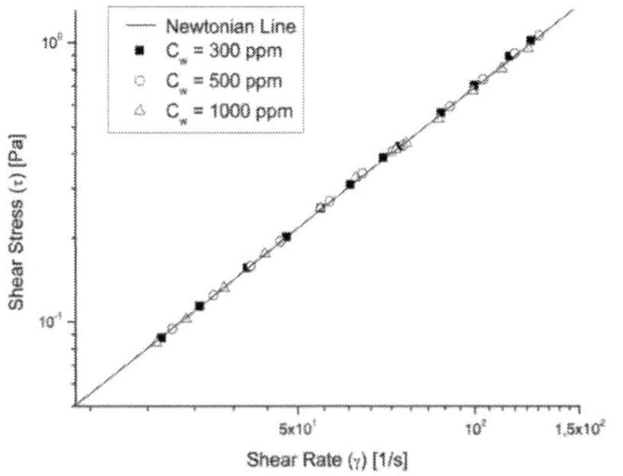

Figure 3. Flow curves of Coconut fiber suspensions.

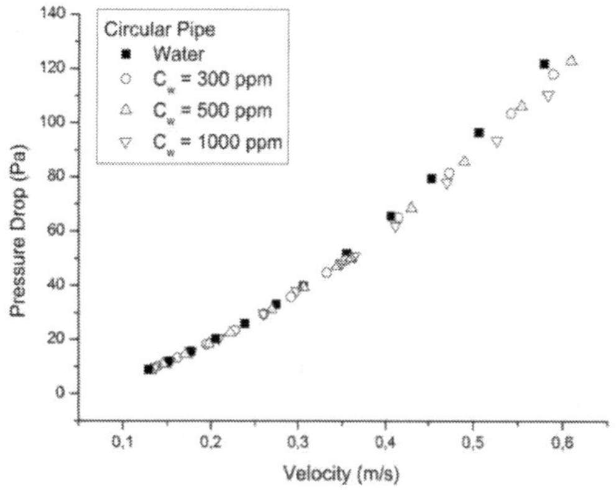

Figure 4. Pressure drop curves of water fluid and coconut fiber suspension.

4.2. Pressure drop

The pressure drop in the flowing fluid is influenced by several factors such as the pipe length, the medium that the fluid flows and the flow velocity. Pressure drop in the fluid flow is kept as small as possible to reduce losses in fluid flow. The value of the pressure loss is an important parameter in the fluid flow because it corresponds to the energy required in the fluid distribution. The relationship between velocity and pressure drop of coconut fiber suspension can be seen in figure 4. Figure 4 shows that at low velocities starting at about 0.13 m/s to 0.35 m/s, the pressure drop of the coconut fiber suspension remains relatively similar as the water. However, at the flow rate above 0.35 m/s to 0.6 m/s, the pressure drop of coconut fiber suspensions decrease and lower than the pure water.

Other than that, the higher the concentration value of coconut fiber suspension, the pressure drop is lower than the water. This may occur because coconut fiber suspension has a lower viscosity than water, especially at higher concentrations 1000 ppm. Thus causing the shear stress to occur is also lower than the water at similar velocity. Another influencing factor is the flow velocity. The higher the flow rate, the pressure drop of coconut fiber suspension also becomes lower than the water.

4.3. Friction factor and drag reduction

From figure 5, it can be seen that the water fluid is in the turbulent Blasius equation. The relation between Reynolds number to friction coefficient of coconut fiber shows that friction coefficient of coconut fiber is smaller than water starting from Reynolds number about 15,000 to Reynolds number higher which is about 25,000 in this research. The higher the Reynolds number, the lower the friction coefficient. Low friction coefficient leads to lower energy use, thus saving energy.

4.4. Drag reduction

Based on figure 6, the drag reduction ratio of coconut fiber suspension began to occur in the Reynolds number about 15,000 to the highest Reynolds number about 25,000 in this study. Drag reduction occurs for all concentrations of coconut fiber suspension. The drag reduction ratio increases with increasing Reynolds numbers. There are two factors that influence the drag reduction ratio, which is the Reynolds number and the concentration of coconut fiber suspension. The higher the concentration of coconut fiber suspensions get, hence the drag reduction ratio also gets higher. Then it can be concluded that the highest drag reduction ratio in circular pipe ID 38 mm occurs in coconut fiber suspension with concentration 1000 ppm. The drag reduction ratio is about 7.6% and occurs in the Reynolds number about 25,000.

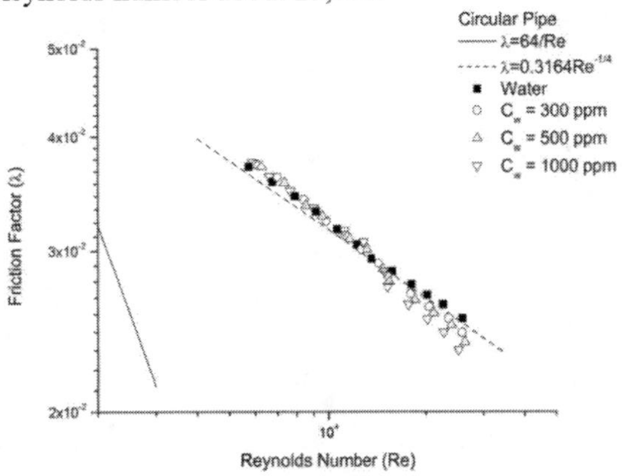

Figure 5. Friction coefficient of coconut fiber suspensions.

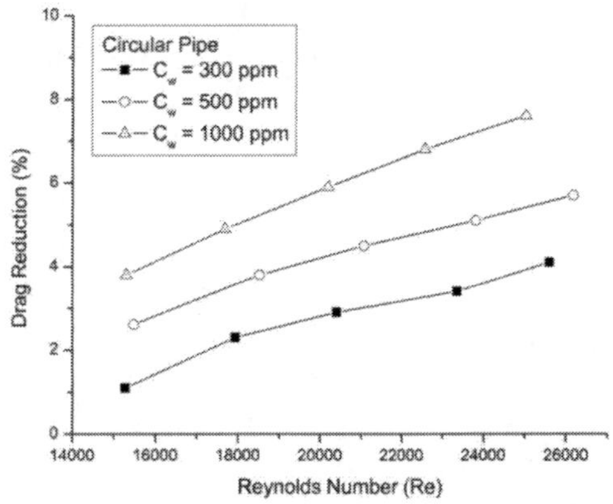

Figure 6. Ratio of drag reduction of coconut fiber suspensions.

4.5. Characteristics of drag reduction of coconut fiber suspensions

The drag reduction is divided into two types: type A and type B [12]. In the drag reduction type A, the additive suspension produces a friction factor segment expanding outward from the "common" point on the Prandlt-Karman line, the curve rising with increasing concentration. In addition, the drag reduction also increases with the increase of $Re.f^{1/2}$. This happens due to the characteristic of random macromolecules. While in the drag reduction type B, the additive suspension produces a segment relatively parallel to the Prandtl-Karman (P-K) line where the drag reduction does not depend on $Re.f^{1/2}$, but increases with increasing of concentration. This behavior is indicated by various types of additives including extended polyelectrolytes, fibers, soaps, and clays. This reduction of resistance occurs without a clear onset effect in which the macromolecules causing the effect are already in an extended state and need not be stretched by a strong flow.

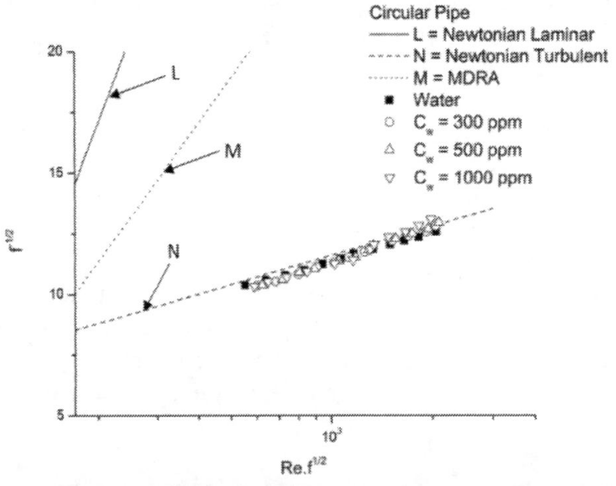

Figure 7. Friction factor of Prandalt-Karman coordinate $f^{1/2}$ vs $Re.f^{1/2}$.

Figure 7 is the experimental results for coconut fiber suspension in circular pipe using the semi-logarithmic coefficient of Prandtl-Karman $f^{1/2}$ vs $Re.f^{1/2}$. Three straight lines characterized by L, N, and M. L are Newtonian laminar flow, N is Newtonian turbulent flow, and M is maximum possible drag reduction asymptote (MDRA) [13]. From figure 7, it can be seen that the coconut fiber suspension results on turbulent flow parallel to the N line where higher concentration is at higher curve. This indicates that coconut fiber suspension is the drag reduction type B that can be seen in fiber suspensions. In general, the drag reduction type B mechanism is associated with suppression of vortices [14]. In the fiber suspension, it is thought that the floc produced in the flow field affects fluid resistance and the fiber suppresses the vortices as they are distributed evenly toward the destination, resulting in drag reduction.

4.6. Drag reduction mechanism of fiber suspensions

The drag reduction mechanism of fiber suspensions is affected by fiber tissue and the influence of tidal flow, both for synthetic fibers and natural fibers [15].

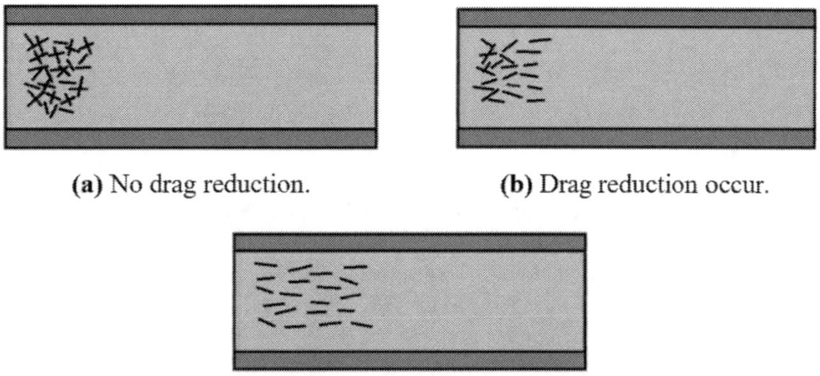

(a) No drag reduction. **(b)** Drag reduction occur.

(c) Maximum drag reduction.

Figure 8. Drag reduction mechanism of fiber suspensions.

At low velocity, plug flow occurs where the fiber suspension friction on the pipe wall and form fiber network on the wall. The friction of the fibers on the pipe wall causes an increase in the friction coefficient as shown in figure 8(a) However in figure 8(b), when the flow velocity is increased, the annulus of water fluid is formed between the fibers and walls. Shear is concentrated in the annulus. As the velocity increases, the size of the annulus also increases with a larger size than the velocity increase, causing shear on the pipe wall to decrease and resulting in a decrease in friction loss. In this phase, drag reduction starts to occur but the ratio is still very low. Furthermore when the velocity is increased higher, the annulus will become turbulence. This annulus will attract and stir the fibers. At this stage, the mixed flow regime begins. By increasing the velocity even higher, the size of the turbulent annulus increases and the plug core decreases. In this mixed flow regime, the friction loss at the suspension becomes lower than the water at the same rate. This phenomenon is called the drag reduction. Fiber tends to spread and flows relatively parallel to the axis of the test pipe as shown in figure 8(c) and the water layer develops on the wall. This causes a decrease in friction on the wall and leads to maximum drag reduction. If the velocity increases even higher, there will be a fully turbulent flow regime reaching the overall diameter of the pipe.

5. Conclusion

From the results of analysis of coconut fiber suspension characteristics in the horizontal circular pipe ID 38 mm, it can be concluded that the coconut fiber suspensions has an effect on the drag reduction about 7.6% in the Reynolds number around 25,000. Drag reduction is influenced by fiber suspension concentrations. The higher the concentration of coconut fiber suspensions, the greater the drag reduction ratio occurs. In this study, the concentration of fiber with the highest drag reduction is coconut fiber with concentrations 1000 ppm.

6. References

[1] Yanuar, Ridwan, Budiarso and Koestoer R A 2009 Hydraulics conveyances of mud slurry by a spiral pipe *J. Mech. Sci. Technol.* (South Korea: Korean Society of Mechanical Engineers) vol 23(7) pp 1835-1839

[2] Yanuar, Gunawan, Waskito, K T 2017 Effect of Agar Jelly Coating in Rectangular Pipe to Flow Drag Reduction *J. Appl. Fluid Mechanics* Vol 10(4) pp 1161-1166

[3] Yanuar, Gunawan, Sunaryo and Jamaluddin A 2012 Micro-bubble drag reduction on a high speed vessel model *J. Marine Sci. Appl.* (China: Haerbin Gongcheng Daxue/Harbin Engineering University) vol II(3) pp 301-304

[4] Fu Y F, Yuan C Q and Bai X Q 2017 Marine drag reduction of shark skin inspired riblet surfaces *Biosurface and Biotribology* (Amsterdam: Elsevier) vol 3(1) pp 11-24

[5] Li F, Kawaguchi Y, Yu B, Wei J and Hishida K 2008 Experimental study of drag-reduction mechanism for a dilute surfactant solution flow *Int. J. Heat and Mass Transfer* (United Kingdom: Elsevier) vol 51(3) pp 835-843

[6] Toms B A 1949 Some observations on the flow of linear polymer solutions through straight tubes at large Reynolds numbers *Proc. of the 1st Int. Congress on Rheology* vol 2 pp 135-141

[7] Pouranfard A R, Mowla D and Esmaeilzadeh F 2014 An experimental study of drag reduction by nanofluids through horizontal pipe turbulent flow of a Newtonian liquid *J. Industrial and Engineering Chemistry* (South Korea: Korean Society of Industrial Engineering Chemistry) vol 20(2) pp 633-637

[8] Sharma R S, Seshadri V and Malhotra R C 1979 Drag reduction in dilute fibre suspensions: Some mechanistic aspects *Chemical Engineering Science* (United Kingdom: Elsevier) vol 34(5) pp 703-713

[9] Kazi M S N, Duffy G G, Chen X D 1999 Heat transfer in the drag reducing regime of wood pulp fibre suspensions *Chemical Engineering Journal* (Netherland: Elsevier) vol 73(3) pp 247-253

[10] Kubo T and Ogata S 2012 Flow properties of bamboo fiber suspensions *in ASME 2012 Int. Mechanical Engineering Congress and Exposition, IMECE* (Houston)

[11] Hayder A A B, Mohd A A and Rosli B M Y 2010 Experimental Study On The Reduction Of Pressure Drop Of Flowing Water In Horizontal Pipes Using Paddy Husk Fibers *Canadian Journal of pure & applied sciences* (British Columbia: SENRA Academic Publishers) vol 4(2) pp 1221-1225

[12] Virk P S 1975 Drag reduction fundamentals *AIChE Journal* (United States: Wiley-Blackwell) vol 21(4) pp 625-656

[13] Virk P S, Mickley H S and Smith K A 1970 Ultimate Asymptote and Mean Flow Structure In Toms' Phenomenon *J. Applied Mechanics, Transactions ASME* (United States: ASME) vol 37(2) pp 488-493

[14] Ogata S, Tetsuya N and Takuya K 2011 Drag reduction of bacterial cellulose suspensions *Advances in Mechanical Engineering* (United States: SAGE Publications) vol 2011

[15] Kazi S N, Duffy N N and Chen X D 2015 The Effect of Varying Fiber Characteristics on the Simultaneous Measurement of Heat and Momentum Transfer to Flowing Fiber Suspensions *J. of Heat Transfer* (United States: ASME) vol **137(1)**

IOP Conf. Series: Earth and Environmental Science **105** (2017) 012051 doi:10.1088/1755-1315/105/1/012051

Synthesis of green diesel through hydrolysis and hydrodeoxygenation reaction from waste cooking oil using NiMo/Al_2O_3 catalyst

R U Putri[1], B H Susanto[1], D F Pratama[1] and M B Prakasa[1]
[1]Chemical Engineering Department, Universitas Indonesia
Kampus Baru Universitas Indonesia, Depok 16424, Indonesia

E-mail: bambanghs@che.ui.ac.id

Abstract. Green diesel is a second generation of biofuel that has a potential to answer the energy needs in Indonesia, also in the world. The reactions used to produce green diesel are hydrolysis and hydrodeoxygenation by using NiMo/Al_2O_3 catalyst. Hydrolysis will change the triglycerides in the raw material, which is waste cooking oil into free fatty acids (FFA) and then converted into green diesel through hydrodeoxygenation. Hydrolysis of waste cooking oil carried out at temperature of 200°C and pressure of 16 bar for 3 hours with water and oil volume ratio of 1:1. FFA product of hydrolysis is as much as 73.89%. For hydrodeoxygenation, variation in operating condition used are 375°C with pressure of 12 bar and 400°C with pressure of 15 bar. Characterizations of green diesel have shown that both of hydrodeoxygenation products have met the specifications of diesel, except the acid number with optimum operating condition at 400°C and pressure of 15 bar. These conditions with NiMo/Al_2O_3 activated led to dominant yield of 68.87% of diesel fraction, selectivity of 69.58%, and conversion of 82.15%.

1. Introduction

Due to depletion of crude reserves, shortage in petroleum supply, and increased demand for clean hydrocarbon fuels, the production of renewable fuels has emerged to solve this problem. As a result, nowadays more consumers are turning to biodiesel instead of petroleum-based diesel for transportation. Unfortunately, biodiesel has low oxidative stability, higher viscosity, higher cloud point and pour point, higher nitrogen oxides (NO_x) emissions, lower energy density, and higher injector or engine wear [1]. To avoid the disadvantageous physical properties of oxygenated fuels, hydrodeoxygenation has been investigated as an alternative pathway of biofuel production.

Hydrodeoxygenation is a series of reactions that convert the molecular structure of triglyceride in oil to linear hydrocarbon and it means removal of oxygen from an oxygen-containing compound under hydrogen pressure at high temperatures and pressure with heterogenous catalyst, which product known as green diesel [2]. The hydrodeoxygenation process consisted of decarboxylation, decarbonylation, and hydrogenation reactions [3]. The main drawback of this reactions seem to be large consumption of hydrogen during the process, which increase the production cost.

Decarboxylation : $C_{17}H_{35}COOH \rightarrow C_{17}H_{36} + CO_{2\,z}$ (1)

Decarbonylation : $C_{17}H_{35}COOH + H_2 \rightarrow C_{17}H_{36} + H_2O + CO$ (2)

Hydrogenation : $C_{17}H_{35}COOH + 3H_2 \rightarrow C_{18}H_{38} + 2H_2O$ (3)

The green diesel that obtained directly by hydrodeoxygenation is mainly composed by C17 and C18 n-paraffins that have a high cetane number. To produce green diesel, the reaction needs catalyst, as the acid sites of the catalyst promote the conversion of larger alkyl chains into smaller chains, so that the $NiMo/Al_2O_3$ is used. Besides, it has been observed that $NiMo/Al_2O_3$ catalyst promote hydrodeoxygenation to have high selectivity toward diesel range products [4]

The feeds of hydrodeoxygenation are free fatty acids, so they need hydrolysis as a pre-treatment to increase the free fatty acids which in the end will increase selectivity, yield, and conversion of green diesel product. When water is maintained at temperatures within the range of its boiling point (100°C) and its critical point (374°C) and kept in the liquid state, this is referred to as the subcritical water condition. For the hydrolysis of the oil, subcritical water presents higher solubility in the oil due to the modification of its properties, such as a reduction in its dielectric constant with increasing temperature. Thus, water in subcritical conditions in hydrolysis acts both as reactant and as a solvent for a reaction. Free fatty acids which are formed in the hydrolysis reaction of vegetable oils, they act as acid catalysts and are able to accelerate their own reaction [5].

$$C_3H_5(COOR)_3 + H_2O \rightarrow C_3H_5(COOR)_2(OH) + RCOOH \quad (4)$$

$$C_3H_5(COOR)_2(OH) + H_2O \rightarrow C_3H_5(COOR)(OH)_2 + RCOOH \quad (5)$$

$$C_3H_5(COOR)(OH)_2 + H_2O \rightarrow C_3H_5(OH)_3 + RCOOH \quad (6)$$

$$C_3H_5(COOR)_3 + C_3H_5(COOR)(OH)_2 \rightarrow 2C_3H_5(COOR)_2(OH) \quad (7)$$

In the reaction above, triglycerides are converted to diglycerides and FFA (RCOOH), then to monoglycerides and FFA, and ultimately into glycerol and FFA.

The aims of this experiment are to know the best operating condition of hydrolysis and hydrodeoxygenation reactions and also to produce green diesel in high conversion, selectivity, and yield from waste cooking oil.

2. Methodology

2.1. Materials

The materials needed in this experiment are: waste cooking oil (household waste), NiMo /Al_2O_3, ethanol (Merck), potassium hydroxide (Merck), phenolphthalein (Merck), Wijs solution (Merck), potassium iodide (Merck), chloroform (Merck), amylum (Merck), sodium thiosulfate (Merck), and aquadest (Merck).

2.2. Pre-treatment of Waste Cooking Oil Through Hydrolysis

Hydrolysis was performed in 330 mL batch stirred autoclave reactor integrated with a condenser and furnace. Extensive stirring of the oil and water in the reactor was achieved using a magnetically driven stirrer with a DC geared motor. Waste cooking oil and water was used as feed with ratio of 1:1 %wt. Temperature and pressure during the reaction was kept constant at 200°C and 16 bars, also the stirrer speed at 500 rpm. Nitrogen was used to give the reactor pressure [4]. Reaction time was varied from one to three hours. The best result of hydrolysis with optimum reaction time would be used as hydrodeoxygenation feed.

2.3. Hydrodeoxygenation of Waste Cooking Oil

The reaction was performed in the same reactor with same equipment as hydrolysis. Hydrolysis product and $NiMo/Al_2O_3$ catalyst was used as feed with ratio of 100:1 %wt. Nitrogen gas purged for 30 minutes. Stirrer speed during the reaction was kept constant at 800 rpm. The reaction has two different operating condition i.e. 375°C at 12 bar and 400°C at 15 bar. The reactions were performed at temperature of the reactor reached the target value and the hydrogen pressure was adjusted to desired value. From this time onwards, the reaction time was occurred. In this step, the double bonds that were present in the triglycerides were saturated with hydrogen gas. The gas products were condensed and collected as condensate. The bottom product was distilled according to the boiling range of petroleum products in three categories i.e. IBP-200°C, 200-300°C, and 300-EP [5].

2.4. Hydrolysis and Hydrodeoxygenation Product Characterization

Hydrolysis and hydrodeoxygenation products were characterized by GC-MS (Agilent) to identify fraction component of sample, FTIR (Termo Scientific) to know the functional groups of the product, and GC-TCD to know the gas consumption and formation. Chemical properties of green diesel were investigated through iod number and acid number to know the amount of double bonds and free fatty acids in the products. Acid numbers were determined by titration with KOH whereas iod number with $Na_2S_2O_3$. Physical properties of green diesel were investigated through density, viscosity, and cetane index to know the quality of the products.

3. Result and Discussion

3.1. Pre-treatment of Waste Cooking Oil Through Hydrolysis

Waste cooking oil has a low free fatty acid content caused by the presence of triglyceride bonds in large quantities. Figure 1 shows that the longer the reaction time, the more triglycerides are changed to free fatty acids, so it can increase the feed for the hydrodeoxygenation process. The optimum reaction time for hydrolysis is at 3 hours.

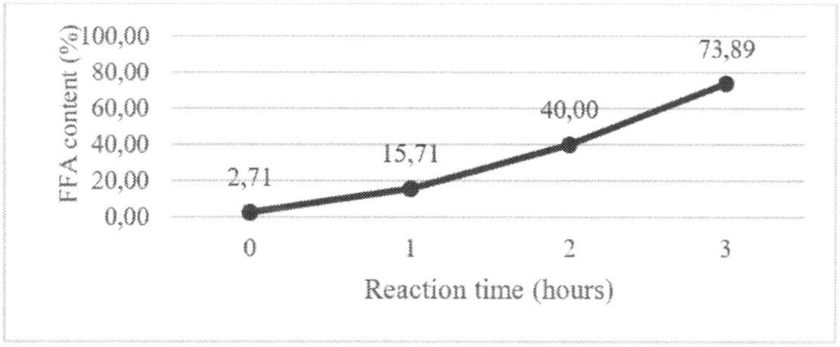

Figure 1. Effect of hydrolysis time on free fatty acid content.

The success of waste cooking oil pre-treatment through hydrolysis is indicated by the result of purer free fatty acids and the loss of glycerols. Table 1 shows an increase in the amount of fatty acids characterized by an increase in the percentage of palmitic acid, also the formation of lauric acid and myristic acid. This indicates the breakdown of triglycerides to free fatty acids and glycerol, but glycerol cannot be seen in GC-MS results because it has dissolved in water during heating, then evaporated with water.

Table 1. GC-MS results of WCO and hydrolysis products

Substances	WCO		Substances	HDR	
	RT	Content (%)		RT	Content (%)
Palmitic Acid	11.77	27.89	Lauric Acid	8.88	33.72
Oleic Acid	12.93	58.83	Myristic Acid	10.41	56.81
Palmitic Acid Chloride	13.82	2.57	Palmitic Acid	11.99	0.12
Oleylaldehyde	14.83	3.81	Oleic Acid	13.17	0.19
16-Dehydropregnolone	14.91	2.46	Petroselenic Acid	13.62	1.57
1-Monopalmitin	15.02	1.89	1-Mono-Olein	14.62	0.18
2,3-Dihydroxypropyl elaidate	16.22	1.37	1-Methyl-4-Phosphorinanone	14.88	0.48
Squalene	17.34	1.18	2-Ethylbutyric Acid, eicosyl ester	14.99	0.75
			1-Monopalmitin	15.12	4.99
			3-butyl-thiophene-1,1-dioxide	16.01	0.38
			1-Mono-Stearin	16.49	0.44
			(acetoxymethyl)methyl [(trimethylsilyl)methyl]silane	16.13	0.36

3.2. Green Diesel Synthesis Through Hydrodeoxygenation

Hydrodeoxygenation was carried out under two different operating conditions, at 375°C with 12 bar pressure (HDO12) and 400°C with 15 bar pressure (HDO15). Table 2 shows that green diesel has met all the specifications of diesel fuel based on Indonesia National Standard (SNI) of biodiesel and diesel fuel, except the acid number. It means that there are still many free fatty acids that have not been converted into alkanes, so the reactions have not run optimally. A decrease in the iod number indicates that the double bond on green diesel has been reduced compared to waste cooking oil feedstock.

The decrease in the density value due to long carbon chain bonds in the waste cooking oil has been broken into short carbon chains so that the molecular weight of green diesel products will decrease. While the decrease of viscosity caused by the breaking of carboxylic bond from the waste cooking oil containing carbonyl (CO) and hydroxyl (-OH). The increase of cetane index on green diesel shows better combustion quality compared to SNI of biodiesel and diesel.

Table 2. Properties comparison of Green Diesel and Other Product

Sample	Iod Number	Acid Number	Density	Viscosity	Cetane Index
HDO12	80.37	3.14	0.87	4.80	51.23
HDO15	66.83	1.35	0.83	4.50	55.68
HDR (hydrolysis product)	-	146.31	0.92	40.26	-
WCO (feedstock)	106.60	4.94	0.93	40.80	-
SNI of Biodiesel	115.00	0.80	0.85-0.89	2.30-6.00	51.00
SNI of Diesel	170.00	0.60	0.82-0.87	2.00-5.00	40.00

Functional groups of the product investigated by FTIR. Figure 2 shows that on the spectra of waste cooking oil (WCO) and hydrolysis product (HDR), there are still a C=O and C-O groups which are carboxylic acid. Whereas in the spectra of HDO12 and HDO15, the C=O group has been decreased and C-O group has disappeared. It means that the carboxylic acid has been well disentangled. There is also a decrease in the $=CH_2$ group of the waste cooking oil, the hydrolysis reaction, to the hydrodeoxygenation reaction, in which the group is identified as an alkene. It can be concluded that the alkene has been able to broke properly through hydrodeoxygenation into alkane. From properties of green diesel and FTIR result, the optimum condition of green diesel production is at 400°C and 15 bar.

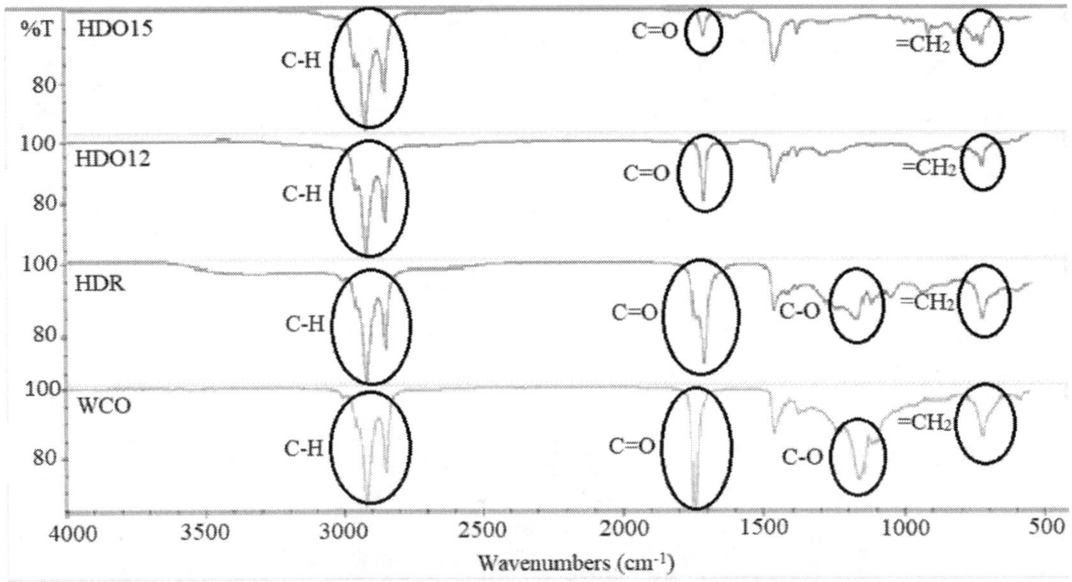

Figure 2. FTIR result of green diesel

Figure 3 and 4 shows that CO gas formation is larger than CO_2 and CH_4 which means $NiMo/Al_2O_3$ catalyst leads to the catalytic decarbonylation reaction compared to decarboxylation and methanation, where the reaction requires hydrogen to produce heptadecane, CO gas, and water. The presence of gas and water generated from the main reaction encourages the formation of water-gas shift side reactions, where the CO gas and water produced from the decarbonylation reaction reach equilibrium conditions and produce CO_2 and H_2 gases. The use of a particular catalyst may cause the reaction to be reversible, so that the CO gas will increase. From the GC-TCD test results, it can be concluded that the use of $NiMo/Al_2O_3$ catalysts has been able to direct to the main reaction, although it still requires a lot of H_2 gas for the decarbonylation reaction compared with decarboxylation.

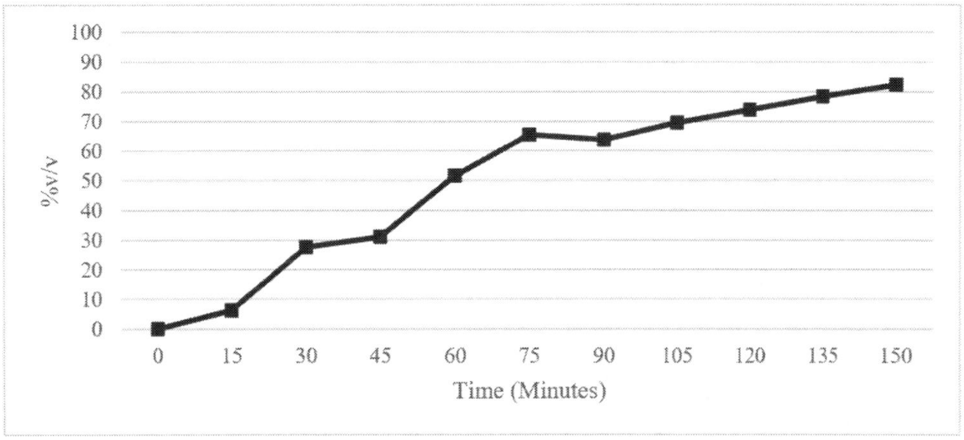

Figure 3. Graph of gas consumption rate

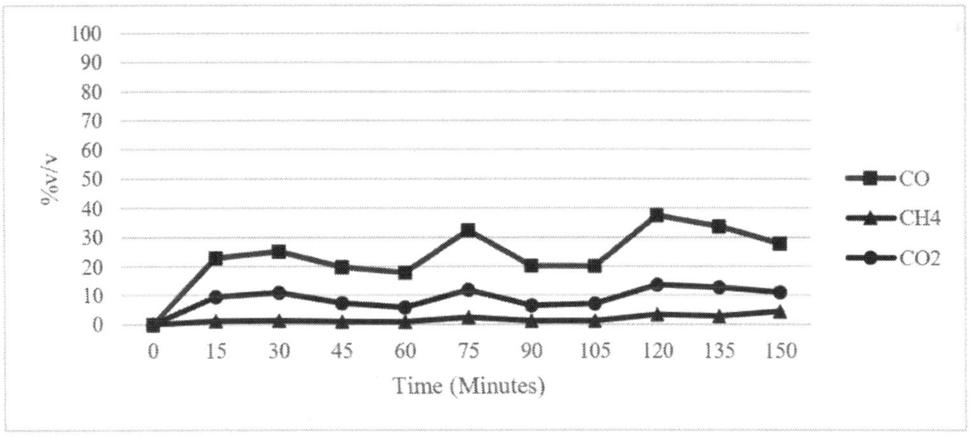

Figure 4. Formation of hydrogenated gas

Conversion, selectivity, and yield are the success parameters of hydrodeoxygenation process in producing green diesel product. The greater the value of the three parameters, the better hydrodeoxygenation process. Table 3 shows that the largest percentage of selectivity, yield, and conversion of green diesel occurred at HDO15. It means that the optimum operating condition of hydrodeoxygenation achieved at 400°C and pressure of 15 bar.

IOP Conf. Series: Earth and Environmental Science **105** (2017) 012051 doi:10.1088/1755-1315/105/1/012051

Table 3. Selectivity, yield, and conversion of distillation product

Product	% Selectivity	%Yield	%Conversion
Biogasoline	8.44	8.36	82.15
Biokerosene	7.17	7.10	
Green diesel	69.58	68.87	
Lubricant	11.93	11.81	

4. Conclusion

In this study, green diesel is produced through hydrolysis and hydrodeoxygenation reaction using $NiMo/Al_2O_3$ catalyst. Characterizations of green diesel have shown that both of hydrodeoxygenation products have met the specifications of diesel, except the acid number with optimum operating condition at 400°C and pressure of 15 bar. These conditions with $NiMo/Al_2O_3$ activated led to dominant yield of 68.87% of diesel fraction, selectivity of 69.58%, and conversion of 82.15%.

5. Acknowledgement

Director of Research and Community Service through International Indexed Publication Grant for Student Thesis (PITTA) Universitas Indonesia 2017 is acknowledged for the financial support to carry out the present study.

6. References

[1] Rogelio S B, Fernando T Z, and Felipe D J H L 2012 *Hydroconversion of Triglycerides into Green Liquid Fuels. InTech* 338

[2] Rogelio S B, Yanyong L and Tomoaki M 2010 *Production of Green Diesel by Hidrocracking of Canola Oil on Ni-Mo/Al₂O₃ and Pt/zeolitic based catalysts* 2-7

[3] Bambang H S, Muhammad B P, and Muhammad H S 2016 *Synthesis of renewable diesel through hydrodeoxygenation reaction from nyamplung oil (Calophyllum Inophyllum oil) using NiMo/Z and NiMo/C catalysts with rapid heating and cooling method. IOP Conf. Ser. Mater. Sci. Eng.* **162** 12009

[4] Luciana P T, Carine T A, Ednildo A T, Heloysa M C A, Fernando L P P and Silvio A B V D M 2015 *Hydrolysis of waste frying oils in subcritical water for biodiesel production by esterification using a heterogeneous catalyst. Chem. Eng. Trans.* **43** 565–570

[5] Bambang H S, Mohammad N and Andri W 2014 *Synthesis of Renewable Diesel through Hydrodeoxygenation Using Pd/zeolite Catalysts. Procedia Chem.* **9** 139–150

IOP Conf. Series: Earth and Environmental Science **105** (2017) 012052 doi:10.1088/1755-1315/105/1/012052

Biodiesel synthesis optimization from wet and dry based microalgae *chlorella vulgaris* by reaction time arrangement

Dianursanti[1], D N Putri[1], P Religia[1] and P Anggraini[1]
[1] Chemical Engineering Department, Universitas Indonesia, Kampus Baru UI Depok 16424, Depok, Indonesia

E-mail: dianursanti@yahoo.com

Abstract. Microalgae have the potential to be used as the raw material for biodiesel synthesis. One of promising microalgae species to be used is *Chlorella vulgaris*. However, in the process of biodiesel synthesis from microalgae needs to be optimized. In this study, optimization of biodiesel synthesis from wet and dry based microalgae *Chlorella vulgaris* has been studied. Optimization done by arranges the time of transesterification reaction. Transesterification reaction has been done using KOH catalyst. The reaction time was varied at 20, 40, and 60 minutes. From the results, the highest yields of biodiesel are obtained on the 40 minutes reaction time with biodiesel yield is 75% for wet based microalgae and 62.8% for dry based microalgae. Biodiesel from dry based microalgae *Chlorella vulgaris* composed by several types of FAME, including saturated FAME such as methyl myristate (4.48%), methyl palmitate (28.3%), and methyl stearate (17.1%), and also composed by unsaturated FAME such as methyl palmitoleate (11.1%) and methyl oleate (39.1%).

1. Introduction

Biodiesel is biofuel composed by monoalkyl esters derived from organic oils from plants or animals through transesterification process [1]. Several feed stocks from plants commonly used to produce biodiesel such as corn, soybean, peanut, canola, rapeseed, castor, karanj (*Pongamia pinnata*), coconut, oil palm, and microalgae. However, microalgae are the highest oil producer compared with other feed stocks. Based on [2], microalgae are able to produce 58.700-136.900 L of oil per hectare area. Moreover, microalgae are also doesn't compete with food needs. Therefore, microalgae will become very potential to be used as the main raw material for biodiesel production. In addition, microalgae are fairly easy to be developed in Indonesia due to its tropical climate. One of microalgae species that exist in Indonesia is *Chlorella vulgaris*. *Chlorella vulgaris* has lipid content about 5-58 wt% [3]. Due to its lipid content, *Chlorella vulgaris* are promising to be used as raw material of biodiesel production.

In the process of biodiesel production from microalgae, initial condition of microalgae is one of the influential factors. All this time, dry based microalgae are usually used to produce biodiesel. However, drying process of microalgae takes a long time and may damage the lipid content of microalgae. Meanwhile, by using wet based microalgae, the water content that remain in the extracted lipid may cause an increase of Free Fatty Acid (FFA) values. This condition will results saponification reaction which able to inhibit transesterification process. However, wet based microalgae can be used for in-situ transesterification methods. By using in-situ transesterification, lipid extraction process can be removed.

Moreover, in conducting biodiesel synthesis from microalgae through transesterification process, there are several important factors that affect, such as reaction temperature, reaction time, type of catalyst, catalyst concentration, and mole ratio between oil and methanol. These factors mostly affect to the yield of biodiesel produced. In order to optimize biodiesel synthesis from microalgae through transesterification process, arrangement of reaction time and temperature should be done to obtain the best yield of biodiesel. Therefore, this present study will analyze the effect of initial condition of microalgae to the character and amount of biodiesel produced. The synthesis will be conducted by arrange the reaction time, in order to optimize the operating condition of transesterification process to obtain the highest yield of biodiesel.

2. Materials and Methods

2.1. Microalgae Preparation

Microalgae *Chlorella vulgaris* were cultivated in a photobioreactor at Bioprocess Engineering Laboratory, Department of Chemical Engineering, Universitas Indonesia using Walne medium. Microalgae were then separated from the medium to obtain wet base biomass of *Chlorella vulgaris*. In order to obtain dried biomass, wet biomass of *Chlorella vulgaris* was then dried by drying process to decrease the water content.

2.2. Lipid Extraction

1 gram of microalgae *Chlorella vulgaris* both wet and dried based was extracted each using 50 mL n-hexane solvent. The extraction was done by ultrasonication method for 60 minutes, at a temperature of 60°C. Lipid extracted was then separated and evaporated to remove solvent content and obtain crude lipid of microalgae.

2.3. Calculation of FFA values

Lipid extracted from microalgae was then calculated by FFA value. FFA value of the extracted lipids can be calculated using the following equation:

$$FFA = \frac{56{,}1\ x\ N_{KOH}\ x\ V_{KOH}}{10\ x\ w}\ x\ 100\% \tag{1}$$

where,
56.1 = FFA calculation constant
NKOH = Normality of KOH standardized
VKOH = volume of KOH used during titration (mL)
w = mass of the test sample (g)

2.4. Transesterification

Lipid from microalgae *Chlorella vulgaris* were reacted with 200 mL methanol to carry out transesterification reaction at 60°C. In this transesterification reaction, 1 gram of KOH was added as the catalyst used. Transesterification was done by varying the reaction time at 20, 40, and 60 minutes. Biodiesel produced were then purified to remove the remaining solvent and catalysts. The biodiesel yield was calculated using the following equation:

$$Y = \frac{m_B}{m_L}\ x\ 100\% \tag{2}$$

where:
Y = yield of biodiesel (%)
mB = mass of biodiesel (g)
mL = mass of lipid extracted (g)

2.5. Product Analysis

Biodiesel from microalgae were analyzed its Fatty Acid Methyl Esters (FAME) content. Analysis was done by using GC-MS. This analysis will show the composition of FAME on biodiesel product from microalgae.

3. Results and Discussions

3.1. Effect of Microalgae Condition to Total Lipid Extracted

The amount of lipid extracted from different microalgae base can be seen in table 1.

Table 1. The Number of Lipid Extracted

Condition	Lipid (%)
Wet based microalgae	0,238
Dry based microalgae	1,57

Based on Table 1, it can be seen that dry based microalgae produce more lipids. This is because on wet based microalgae, there is remained water in microalgae biomass. In the extraction process, n-hexane as a solvent is non-polar whereas water is polar. This difference, causing n-hexane is not able to contact the biomass maximally. Therefore the lipids extracted from the wet based microalgae were also not optimal. Meanwhile, on dry based microalgae, the water content in microalgae biomass has been relatively small. This condition causes the non-polar n-hexane are able to contact with microalgae biomass maximally. Therefore the lipids extracted from dry based microalgae can be optimized.

3.2. Effect of Time Reaction on the Number of Biodiesel Produced

In order to observe the effect of reaction time to the yield of biodiesel, temperature reaction is kept constant at 60°C. This temperature is chosen according to the transesterification of *Chlorella vulgaris* microalgae lipid by [4] that produce the highest yield of biodiesel with the amount is 36.79%.

Effect of reaction time on the number of biodiesel produced can be seen in figure 1. Based on figure 1, it is clear that the reaction time is very influential to the amount of biodiesel produced. On wet based microalgae, biodiesel yield are 60%, 75%, and 48.9% respectively for the reaction time of 20, 40, and 60 minutes. As well as for dry based microalgae, the difference of reaction time also poses a considerable difference in biodiesel produced. Yield of biodiesel obtained are 18.4%, 62.8%, and 51.9% for the reaction time of 20, 40, and 60 minutes.

Based on these results, the highest biodiesel yield obtained at 40 minutes of reaction time, both on wet and dry based microalgae. At 20 minutes, biodiesel yield is lower than at 40 minutes. This is due to the formation of biodiesel is not completed yet. The amount of reactant is more than the amount of product. Therefore, due to transesterification is reversible reaction, the equilibrium is shifted toward the formation of the product. Meanwhile, at 60 minutes, production of biodiesel decreased as compared to the 40 minutes. This is due to the amount of the product is much more than the reactant. Therefore the equilibrium shifts towards to the reactants resulting in a reduced number of products that have been formed. This result is in line with a study on *Nannochloropsis* sp. microalgae direct transesterification by [5]. Yield of biodiesel is increase when the reaction time up to 4 hours. However, the yield is decrease when the reaction time up to 6 hours. The decrease of biodiesel yield is also able to occur because the catalysts used are fewer after long time reaction [6].

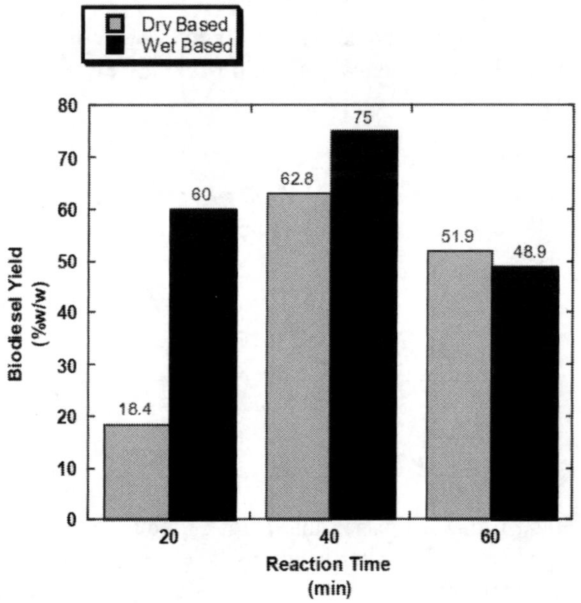

Figure 1. Effect of Time Reaction on the Number of Biodiesel Produced at 60°C

Based on the results, it is also known that the highest biodiesel yield is obtained from wet based microalgae. However, this is due to wet based microalgae done without any drying process. Therefore, lipid quality and quantity of microalgae are avoided from damage caused by drying using heating process. Therefore, biodiesel produced from wet based microalgae are higher than dry based microalgae.

3.3. Effect of FFA Value on the Number of Biodiesel Produced

Free Fatty Acid (FFA) value is a value that plays role in the selection of catalyst on biodiesel synthesis. If the FFA content less than 2.00%, alkaline catalyst can be used in the reaction. If the FFA value more than 2.00%, acid catalyst have to be used before the alkaline catalyst.

figure 2 shows the FFA value from wet base, dry base, and Bligh-Dyer method. Effect of FFA value on the number of biodiesel produced can be seen in Table 2, where the operation condition is 60°C and 40 minutes.

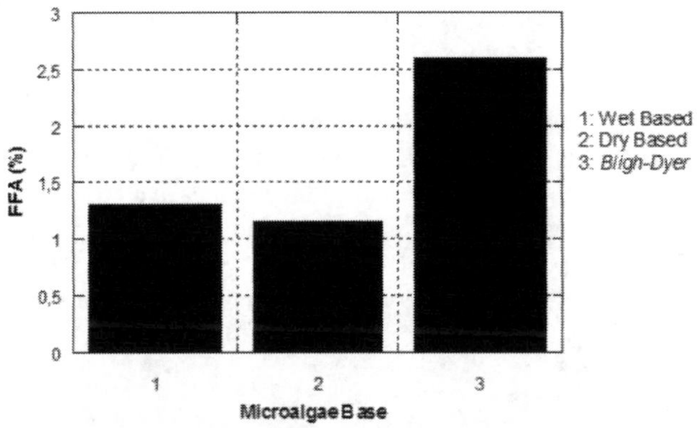

Figure 2. FFA Value (1) Wet Based, (2) Dry Based, (3) Bligh-Dyer

Based on figure 2, it appears that microalgae lipid extracted by the Bligh-Dyer method has higher FFA levels than others. FFA level is influenced by the amount of water from the source. On wet based, the water does not react on microalgae due to the solvent used was non-polar n-hexane. While the Bligh-Dyer method, the solvent used is chloroform which is slightly polar. Slightly polar nature of chloroform causing water levels of lipids from microalgae becomes larger.

Table 2. Effect of FFA Value on the Number of Biodiesel Produced

	FFA (%)	Biodiesel (%)
Wet Base	1.30	75.0
Dry Base	1.15	62.8
Bligh-Dyer Method	2.60	50.0

Based on table 2, the amount of biodiesel produced in the Bligh-Dyer method are smaller because more microalgae lipids react with the catalyst to form soap through a saponification reaction so that the effectiveness of the catalyst decreases due to only a portion of catalyst reacts with the fatty acids. On dry based microalgae, the amount of biodiesel produced is also smaller than the wet based microalgae. This is due to some content of microalgae are damaged as a result of drying process with the use of heat.

3.4. Composition of Fatty Acid Methyl Ester (FAME) from Biodiesel Synthesis Process

FAME compositions of biodiesel were tested using GC-MS. The tested biodiesel were obtained from dry based microalgae lipid which was extracted using Bligh-Dyer method and reacted with methanol at 60°C and 40 minutes. FAME compositions from biodiesel can be seen in the table 3.

From the results, composition of biodiesel produced from *Chlorella vulgaris* microalgae are varied for each method with the fixed component are methyl palmitate and methyl stearate. The most dominant component in both types of FAME composition is methyl palmitate.

Fatty acids in microalgae consisting of saturated fatty acids and unsaturated fatty acids. Based on Table 3, the content of saturated FAME includes methyl myristate, methyl palmitate, and methyl stearate. For biodiesel derived from dry based microalgae, the amount of saturated FAME content was 49.9%. Meanwhile, biodiesel derived from microalgae through Bligh-Dyer extraction process has saturated FAME content of 42.0%.

Table 3. Composition of FAME from Dry Base and Bligh-Dyer Method

Type of FAME	FAME (%)	
	Dry Based	***Bligh-Dyer***
Methyl Myristate$^{(Sat)}$	4,48	-
Methyl Palmitoleate	11,1	-
Methyl Palmitate$^{(Sat)}$	28,3	30,9
Methyl 9-Octadecanoate	-	34,8
Methyl Linolenate	-	23,3
Methyl Oleate	39,1	-
Methyl Stearate$^{(Sat)}$	17,1	11,0

4. Conclusion

The highest yields of biodiesel are obtained on the 40 minutes reaction time with biodiesel yield is 75.0% for wet based microalgae and 62.8% for dry based microalgae. Based on the initial condition of microalgae, wet based microalgae produced higher yield of biodiesel than dry based microalgae. From the results, biodiesel from dry based microalgae *Chlorella vulgaris* composed by several types of FAME, including saturated FAME such as methyl myristate (4.48%), methyl palmitate (28.3%), and methyl stearate (17.1%), and also composed by unsaturated FAME such as methyl palmitoleate (11.1%) and methyl oleate (39.1%).

5. Acknowledgement

The authors would like to thank the financial support provided by Universitas Indonesia through the Publikasi Internasional Terindeks Tugas Akhir (PITTA) 2017 funding scheme under grant no. 762/UN2.R3.1/HKP.05.00/2017 managed by the Directorate for Research and Public Services (DRPM) Universitas Indonesia.

6. References

[1] Demirbas A 2007 Importance of biodiesel as transportation fuel *Energy Policy* **35** 4661-4670

[2] Gouveia L and Oliveira AC 2009 Microalgae as Raw Material for Biofuels Production *J. Ind. Mocrobiol Biotechnol.* **36**:269-274

[3] Mata TMM, António A, Caetano, and Nidia S 2010 Microalgae for biodiesel production and other applications: A review *Renewable and Sustainable Energy Reviews* **14**(1):217-32

[4] Dianursanti, Delaamira M, Bismo S, and Muharam Y 2017 Effect of reaction temperature on biodiesel production from Chlorella vulgaris using CuO/zeolite as heterogeneous catalyst *IOP Conf. Series: Earth and Environmental Science* **55** 012033

[5] Dianursanti, Religia P, and Wijanarko A 2015 Utilization of n-Hexane as co-solvent to increase biodiesel yield on direct transesterification reaction from marine microalgae *Procedia Environmental Sciences* **23**:412-20

[6] Ma X 2012 Biodiesel production from algae through in situ transesterification technology *University of Minnesota*

Biodiesel synthesis from *nannochloropsis oculata* and *chlorella vulgaris* through transesterification process using NaOH/zeolite heterogeneous catalyst

Dianursanti,[1] A G Sistiafi[1] and D N Putri[1]
[1]Department of Chemical Engineering, Faculty of Engineering, Universitas Indonesia, Depok 16424, Indonesia

E-mail: danti@che.ui.ac.id or dianursanti@yahoo.com

Abstract. Microalgae are promising sources of biofuel due to its production capacity of lipid that can be utilized as raw material for biodiesel production, especially *Nannochloropsis oculata* and *Chlorella vulgaris*. The lipid produced can be converted into biodiesel through transesterification reaction using homogenous or heterogeneous catalysts. Heterogeneous catalysts are more advantageous than homogeneous catalysts due to its solid form that eases the separation of catalysts from the products. In this research, NaOH/zeolite heterogeneous catalyst is utilized with varying Na loadings in the zeolite to observe its effect towards the *yield* of biodiesel produced from *N. oculata* and *C. vulgaris*. The best result was obtained with Na loading concentration of 20.5%. The biodiesel yields obtained from the lipids are 83.5% from *N. oculata* and 98% from *C. vulgaris*. The biodiesels contain 47.15% of saturated fatty acid methyl esters from *N. oculata* and 56.41% from *C. vulgaris*.

1. Introduction

Microalgae are currently seen as an ideal feedstock for third generation biofuel mainly due to its high growth rate, ability to thrive in non-arable land or saline water, and high lipid productivity [1]. The lipid produced, especially triglycerides, can be utilized as a raw material for biodiesel production. Some of the most potential microalgae to be utilized as feedstock for biodiesel are *Nannochloropsis oculata* which contains 22.7-29.7 wt% lipid and *Chlorella vulgaris* which contains 5-58 wt% [2]. Biodiesel is commonly produced through transesterification process in which triglyceride and methanol react to produce methyl ester and glycerol. The process requires a catalyst otherwise it would require supercritical conditions that are difficult and expensive to achieve [3]. Homogenous base catalysts are commonly used in an industrial scale [4]. However, they are difficult to separate from the products such that biodiesel production processes suffer from environmental concerns [5]. These problems can be alleviated by using basic heterogeneous catalysts which are solid in form and thus easier to separate from the products. It would also be much more beneficial if the selected material for the catalyst is lower in cost, such as using local natural zeolite from Lampung, Indonesia, instead of synthetic zeolite or resin which is more expensive. Aside from reducing costs, using local natural zeolite will also maximize its use as one of the most abundant and highly potential local natural resources. Previous researches suggest that the amount of heterogeneous catalyst loading that is used can significantly affect the yield of the biodiesel produced [6, 7, 8, 9]. Biodiesel production

[1] To whom any correspondence should be addressed.

from microalgae has been researched for years, but this research aims to provide a basic insight on the potential developments of microalgae in Depok, Indonesia, from which the microalgae were obtained to be cultivated in the Bioprocess Engineering Laboratory, Department of Chemical Engineering, Universitas Indonesia. Depok is a city surrounded by many lakes inhabited by microalgae [10], therefore this research can contribute in the utilization of microalgae in lakes surrounding Depok for a higher economic value. Considering that, this paper intends to find the highest *yield* of biodiesel produced from *N. oculata* and *C. vulgaris* by varying amounts of NaOH/zeolite catalyst loading.

2. Methods

2.1. Microalgae cultivation

Cultivation was conducted at Bioprocess Engineering Laboratory, Department of Chemical Engineering, Universitas Indonesia. *N. oculata* and *C. vulgaris* were cultivated in 6 L photobioreactors using Walne medium. Microalgae growth was measured using BEL UV-M90 Double Beam Spectrophotometer to determine its change in cell density and harvested to be used as feedstock for the transesterification process.

2.2. Lipid extraction

Lipid extraction was conducted at Bioprocess Engineering Laboratory, Department of Chemical Engineering, Universitas Indonesia. Microalgal lipid was extracted using a modified Bligh-Dyer method [11]. The harvested microalgae were centrifuged at 4600 rpm for 15 minutes. The separated biomass was extracted using methanol and chloroform with biomass:methanol:chloroform ratio of 1:4:2 v/v/v. The mixture was then ultrasonicated in Elmasonic S 30 H Ultrasonic Bath for 20 minutes. The mixture was then added with chloroform and water with biomass:chloroform:water ratio of 1:2:2 v/v/v and ultrasonicated again for another 20 minutes. Finally, the mixture was centrifuged again at 4600 rpm for 15 minutes which results in two separate layers. The bottom layer, which was microalgal lipid dissolved in chloroform, was separated from the mixture using a pipette. The chloroform solvent was evaporated in a fume hood such that only the lipid remained.

2.3. Catalyst preparation and characterization

Catalyst preparation was conducted at Bioprocess Engineering Laboratory, Department of Chemical Engineering, Universitas Indonesia. Natural clinoptilolite zeolite from Lampung was immersed in HF 1% solution and stirred for 30 minutes. It was then washed with water and dried in the oven at 120 °C for two hours. The dried zeolite was then impregnated by immersing it to NaOH solutions with concentrations of 1 M, 1.5 M, and 2 M for 24 hours. The zeolite was then separated and dried again in the oven at 120 °C for two hours, resulting in NaOH/zeolite. The catalysts were then characterized using X-ray fluorescence (XRF) at PTPSM-BPPT (Pusat Teknologi Pengembangan Sumberdaya Mineral - Badan Pengkajian dan Penerapan Teknologi) to determine the amount of Na that was successfully impregnated into the zeolite.

2.4. Biodiesel synthesis and characterization

Biodiesel synthesis was conducted at Bioprocess Engineering Laboratory, Department of Chemical Engineering, Universitas Indonesia. The process was done with lipid:methanol ratio of 1:50 v/v with 5 wt% of NaOH/zeolite catalyst. The mixture was heated to 60 °C and stirred at 300 rpm for three hours. The catalyst was then separated by centrifugation at 4600 rpm for 15 minutes. The remaining excess methanol was heated to 65 °C in a fume hood, leaving biodiesel as the main product. The biodiesel will be calculated for its yield using equation 1.

$$Yield\,(\%) = \frac{Mass\ of\ Biodiesel}{Mass\ of\ Lipid}(100\%) \qquad (1)$$

It was then characterized using gas chromatography - mass spectrometry (GC-MS) at PUSLABFOR MABES POLRI (Pusat Laboratorium Forensik Markas Besar Kepolisian Negara Republik Indonesia) to determine its composition.

3. Results and discussion

3.1. Microalgae growth

The changes in cell density of both *N. oculata* and *C. vulgaris* throughout the cultivation are shown in figure 1 and figure 2. For a cultivation period of 318 hours, the biomass productivity shown is 0.104 g/L/day for *N. oculata* and 0.128 g/L/day for *C. vulgaris*. A previous research stated that the range of biomass productivity for *Nannochloropsis* sp. is 0.17-1.43 g/L/day and 0.02-2.5 g/L/day for *Chlorella* sp. [2]. The biomass productivity obtained from this research is approaching closely to the aforementioned range for *N. oculata* and falls within range for *C. vulgaris*. From their sufficient biomass productivity, both microalgae have shown that they are viable feedstock to be utilized for the process of biodiesel synthesis.

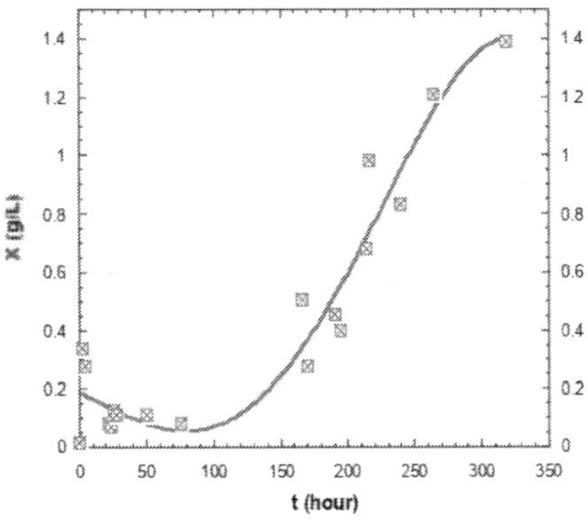

Figure 1. *Nannochloropsis oculata* growth curve in Walne medium.

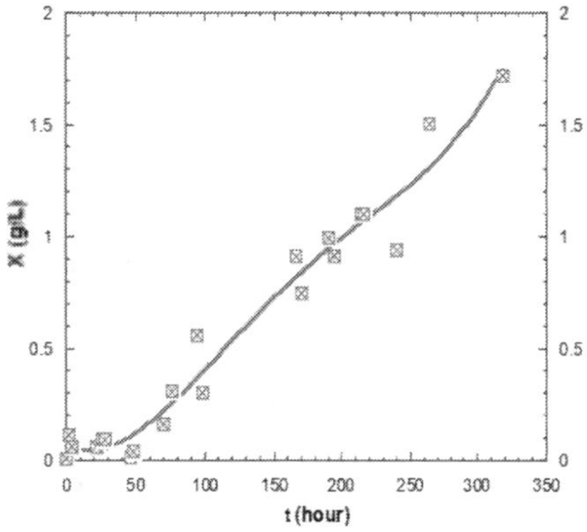

Figure 2. *Chlorella vulgaris* growth curve in Walne medium.

3.2. Catalyst characterization

Catalysts that have been impregnated by NaOH are analyzed using X-ray fluorescence. As shown in table 1, the amount of Na successfully impregnated into the zeolite increases as the concentration of NaOH is increased. A similar trend was also obtained in a previous research that used NaOH concentrations of 1.4, 4, and 10 wt%-eq and resulted in the amount of Na loading of 1.16, 3.09, and 3.64 wt%-eq [12]. The amount of impregnated Na is important since it is the element that reacts with methanol to form an alkoxide as the active species that will catalyze the transesterification process [13, 14].

Table 1. Variations of catalyst loading analyzed at Pusat Teknologi Pengembangan Sumberdaya Mineral - Badan Pengkajian dan Penerapan Teknologi.

Catalyst	**Na loading (%)**
NaOH/Zeolite (1 M)	15.2
NaOH/Zeolite (1.5 M)	16.4
NaOH/Zeolite (2 M)	20.5

3.3. Biodiesel yield

The resulting biodiesel yield calculated using equation 1 is presented in figure 3. In overall, *C. vulgaris* gave a higher yield of biodiesel in comparison to *N. oculata*. For both *N. oculata* and *C. vulgaris*, the yield obtained increased as the amount of catalyst loading is increased. The highest yield is 83.5% from *N. oculata* and 98% from *C. vulgaris* using 20.5% of catalyst loading. This is most likely due to the higher amount of Na available to catalyze the transesterification process which in turn produces a higher amount of biodiesel. The same increasing trends were also obtained in previous researches in which biodiesel was produced from *N. oculata* and *C. vulgaris* and various amount of loadings were used in heterogeneous catalysts [6, 7, 8, 9].

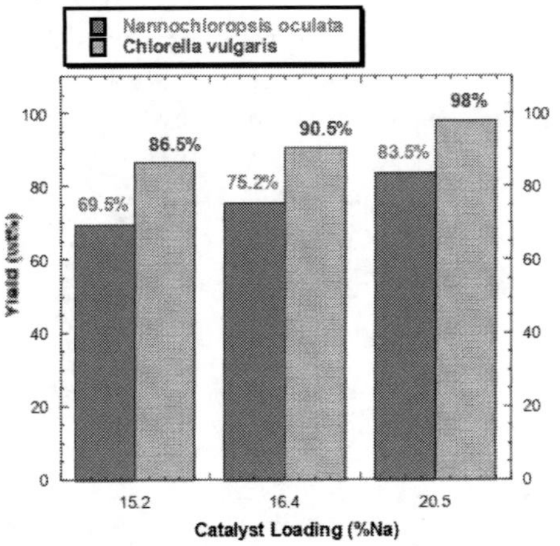

Figure 3. Biodiesel yield from *N. oculata* and *C. vulgaris*.

3.4. Composition of biodiesel

The biodiesel, which is a mixture of various types of fatty acid methyl esters, is analyzed using gas chromatography - mass spectrometry and their composition is presented in table 2. The two types of methyl ester detected in the highest amount from *N. oculata* are methyl linoleate (46.24%) and methyl palmitate (29.45%) while those from *C. vulgaris* are methyl palmitate (56.41%) and methyl linoleate (33.40%). For *N. oculata*, a previous research which analyzed its biodiesel composition stated that the two methyl esters present with the highest amount are methyl oleate (63.8%) and methyl palmitoleate (26.77%) [15]. This difference may be the result of various growth factors such as different nutritional and environmental factors, cultivation conditions, and growth phases which may result in different fatty acid composition produced by microalgae [2] and therefore result in different methyl ester composition. However, for *C. vulgaris*, a previous research which analyzed its biodiesel composition had a more similar result with this research, stating that the two methyl esters present with the highest amount are methyl palmitate (56.34%) and methyl linoleate (25.79%) [16].

The total amount of saturated methyl esters is also calculated which is 47.14% for *N. oculata* and 56.41% for *C. vulgaris*. The amount of saturated methyl ester is a significant aspect since it determines the oxidative stability of biodiesel which is an important property to ensure its acceptability of use especially during a long duration of time in storage [17]. From the results in table 2, biodiesel from *C. vulgaris* has a higher amount of saturated methyl esters compared to *N. oculata*. Unsaturated fatty acid is commonly found to be produced by microalgae [17] which in turn may result in higher amount of unsaturated methyl ester. However, this can be easily reduced by partial catalytic hydrogenation of the fatty acids [17] before biodiesel production. Moreover, this property is not entirely disadvantageous, as a higher amount of unsaturated methyl ester makes it suitable for usage in countries with lower ambient temperature therefore reducing the need for additives [18].

Table 2. Composition of biodiesel (FAME) from *N. oculata* and *C. vulgaris* analyzed at Pusat Laboratorium Forensik Markas Besar Kepolisian Negara Republik Indonesia.

Fatty acid methyl ester (FAME)		**Amount (%)**	
		N. oculata	***C. vulgaris***
Methyl myristate	C14:0	0.32	-
Methyl palmitate	C16:0	29.45	56.41
Methyl palmitoleate	C16:1	-	3.82
Methyl 7,10-hexadecadienoate	C16:2	5.91	-
Fatty acid methyl ester (FAME)		***N. oculata***	***C. vulgaris***
Methyl stearate	C18:0	15.23	-
Methyl oleate	C18:1	-	6.37
Methyl linoleate	C18:2	46.24	33.40
Methyl 6,9,12-octadecatrienoate	C18:3	0.70	-
Methyl arachidate	C20:0	2.15	-
Total saturated FAME		**47.14**	**56.41**
Total unsaturated FAME		**52.86**	**43.59**

4. Conclusions

N. oculata and *C. vulgaris* are shown to be suitable feedstock for biodiesel production. An increasing amount of NaOH/zeolite catalyst loading (15.2%, 16.4%, and 20.5%) can be achieved by using an increasing concentration of NaOH solution (1 M, 1.5 M, and 2 M). The amount of catalyst loading significantly affects the yield of biodiesel produced. The best yield obtained is 83.5% from *N. oculata* and 98% from *C. vulgaris*, both using 20.5% of catalyst loading. The composition of biodiesel produced from both microalgae contains methyl palmitate and methyl linoleate as the two types of methyl esters present with the highest amount. The amount of unsaturated methyl ester detected in the biodiesel is 47.14% from *N. oculata* and 56.41% from *C. vulgaris*.

5. Acknowledgment

The authors would like to thank the financial support provided by Universitas Indonesia through the Publikasi Internasional Terindeks Tugas Akhir (PITTA) 2017 funding scheme under grant no. 762/UN2.R3.1/HKP.05.00/2017 managed by the Directorate for Research and Public Services (DRPM) Universitas Indonesia.

6. References

[1] Alam F, Saleh M and Chowdhury H 2015 Third generation biofuel from algae *Procedia Engineering* **105** 763-8

[2] Mata T M, Martins A A and Caetano NS 2010 Microalgae for biodiesel production and other applications: *A review Renewable and Sustainable Energy Reviews* **14** 217-32

[3] Atadashi I M, Aroua M K, Aziz A R A and Sulaiman N M N 2012 The effects of water on biodiesel production and refining technologies: *A review Renewable and Sustainable Energy Reviews* **16** 3456-70

[4] Sivasamy A, Cheah KY, Fornasiero P, Kemausuor F, Zinoviev S and Miertus S 2009 Catalytic applications in the production of biodiesel from vegetable oils *ChemSusChem* **2** 278-300

[5] Saifuddin N, Smiuddin A and Kumaran P 2015 A review on processing technology for biodiesel production *Trends in applied sciences research* **10** 1-37

[6] Umdu ES, Tuncer M and Seker E 2009 Transesterification of Nannochloropsis oculata microalga's lipid to biodiesel on Al2O3 supported CaO and MgO catalysts *Bioresource Technology* **100** 2828-31

[7] Li Y, Lian S, Tong D, Song R, Yang W, Fan Y, Qing R and Hu C 2011 One-step production of biodiesel from Nannochloropsis sp. on solid base Mg–Zr catalyst *Applied Energy* **88** 3313-7

[8] Carrero A, Vicente G, Rodriguez R, Linares M and Peso GL 2011 Hierarchical zeolites as catalysts for biodiesel production from Nannochloropsis microalga oil *Catalysis Today* **167** 148-53

[9] Ma G, Hu W, Pei H, Jiang L, Ji Y and Mu R 2015 Study of KOH/Al2O3 as heterogeneous catalyst for biodiesel production via in situ transesterification from microalgae *Environmental Technology* **36** 622-7

[10] Prihantini NB, Wardhana W, Hendrayati D, Widyawan A, Ariyani Y and Rianto R 2008 Biodiversitas Cyanobacteria dari beberapa situ/danau di kawasan Jakarta-Depok-Bogor, Indonesia *Makara* **12** 44-54

[11] Li Y, Naghdi FG, Garg S, Adarme-Vega TC, Thurecht KJ, Ghafor WA, Tannock S and Schenck PM 2014 A comparative study: The impact of different lipid extraction methods on current microalgal lipid research *Microbial Cell Factories* 13

[12] Wang YY, Chou HY, Chen BH and Lee DJ 2013 Optimization of sodium loading on zeolite support for catalyzed transesterification of triolein with methanol *Bioresource technology* **145** 248-53

[13] Serio MD, Tesser R, Pengmei L and Santacesaria E 2008 Heterogeneous catalysts for biodiesel production *Energy & Fuels* **22** 207-17

[14] Ningtyas D P, Budhiyanti S A and Sahubawa L 2013 Pengaruh katalis basa (NaOH) pada tahap reaksi transesterifikasi terhadap kualitas biofuel dari minyak tepung ikan sardine *Jurnal Teknosains* **2** 71-158

[15] Dianursanti, Religia P and Wijanarko A 2015 Utilization of n-Hexane as co-solvent to increase biodiesel yield on direct transesterification reaction from marine microalgae *Procedia Environmental Sciences* **23** 412-20

[16] Dianursanti, Anggraini P, Nilawati D and Wijanarko A 2013 Preliminary study of biodiesel synthesis from microalgae lipid of Chlorella vulgaris based Walne medium through esterification-transesterification and transesterification reaction *Proc. of the 13th Int. Conf. on QIR*

[17] Chisti Y 2007 Biodiesel from Microalgae *Biotechnology Advances* **25** 294-306

[18] Dianursanti, Delaamira M, Bismo S and Muharam Y 2016 Effect of reaction temperature on biodiesel production from Chlorella vulgaris using CuO/Zeolite as heterogeneous catalyst *IOP Conf. Series: Earth and Environmental Science* 55

IOP Conf. Series: Earth and Environmental Science **105** (2017) 012054 doi:10.1088/1755-1315/105/1/012054

Influence of anodizing concentration and electric potential on surface morphology and corrosion behavior of anodized magnesium in seawater activated battery

Supriyono[1], R Oktavian[1], L P M I Wulaningfitri[1], J C Pradana[1] and I Feliana[1]
[1]Department of Chemical Engineering, Engineering Faculty, Brawijaya University, Jl. Mayjen Haryono number 167, Malang, postal code 6541, Indonesia

E-mail: rama.oktavian@ub.ac.id

Abstract. Magnesium is one of the newly developed anode materials for seawater activated batteries. Anodizing is usually performed to expand the surface of anode materials which can improve the battery discharge performance. This work used Oxalic acid with concentration varied from 0.1 to 0.3 as an anodizing agent with electric potential varied from 1.5 to 6.0 Volt for an anodizing process. Electric potential used in the anodizing was varied from 1.5 to 6.0 Volt. Microstructure and corrosion behavior of anodized magnesium used in this work as battery anode and graphite as a cathode in 3.5% NaCl electrolyte solution were studied. Anodized magnesium in 0.3 M oxalic acid has the most uniform pores of microstructures than those anodized at other concentrations. Anodized magnesium at electric potential from 1.5 Volts tend to has higher value of potential (Ecorr = -1.754 to -1.772 Volt) and corrosion current (Icorr = 2.790 to 11.690μA/cm2), and lower polarization resistance (Rp = 425.170 to 1024.700 Ω) than non-anodized magnesium (Ecorr = -1.649 Volt; Icorr = 10.984 μA/cm2; Rp = 895.510 Ω). Anodized magnesium at the potential from 3.0 until 6.0 Volt indicates the presence of oxidation on the surface which leads to the less good characteristic pattern of dynamic potential.

1. Introduction

Recently, seawater batteries have been gaining a considerable attention from several countries for addressing the problem increasing energy demand. The principle of this battery is utilizing seawater as an electrolyte solution. It contains electrolyte NaCl when it is dissolved in water (Koontz *et al*, 2002; Petter *et al*, 2012). NaCl in water is ionized into Na^+ and Cl^- ions in which the existence of these ions can cause sea water becomes an electrolyte solution. The abundant amount of seawater becomes the advantage of the use of seawater as a electrolyte. Moreover, seawater battery can be used directly by fishermen in the sea since they can find an electrolyte easily.

Magnesium is considered as anode material for seawater battery. It is because magnesium has several benefits, such as: (1) Its relative negative standardized potential value is -2.367 Volt vs. Standart Hydrogen Electrode (SHE). This enables magnesium to generate relatively large potential electrode potential value; (2) It has high theoretical value of gravimetric charge density of 2.233 mAh/gram allowing magnesium to store large electric current; (3) Its high theoretical value density of volumetric density of 3.881 mAh/cm^3 allowing magnesium to store large currents per small volumes; (4) Its abundant amount on earth. Magnesium alloys have been successfully used as marine battery anodes (Davis, 2006; Kun, *et al* 2012). These alloys were produced by liquefying magnesium, gallium, aluminum, zinc, lead, each having 99% purity at 973 K with atmospheric Ar_2+SF_6 conditions to prevent magnesium burning. An alloy consisting of Mg, 3%wt Ga, 2%wt Hg has larger current density discharge than AZ31 (Mg, 2.94wt% Al, 0.93%wt Zn) and AP65 (Mg, 6%wt Al, 5.08%wt Pb, 0.6%wt Mn). Wang, *et al* (2012) also succeeded in using magnesium alloy as an anode of seawater battery. These alloys were produced by melting the magnesium, aluminum, zinc, lead metals each having 99% purity at 750°C under Ar atmosphere to prevent magnesium burning. Wang, *et al* (2012) concluded that the AP65 microstructure is very influential on battery discharge performance. The more homogeneous the alloy microstructure the better the discharge performance. Cheng, *et al* (2013) successfully synthesized magnesium with the porous morphology which has better discharge performance than non-porous magnesium.

Anodizing is an electrochemical process that is performed with help of the electric current to form pores on the metal surface. The metal is connected to the positive current (anode) from a power source, while an inert electrode is used as a cathode. This anodizing process causes anode metal turns into its oxide so that the surface of the anode will be eroded and will run out at a certain time. Therefore, the pores on the metal surface will emerge as an anodizing impact. Feiyue, *et al* (1998) succeeded in making porous aluminum by anodizing method in weak acid solution. This study aims to develop and evaluate the anodizing method on magnesium for its surface morphology and its discharge performance as an activated seawater battery anode material.

2. Experimental

2.1 Materials and Apparatus Research

The materials research used in this study include: (1) Magnesium foil with purity of 99.50%, with the dimension of 0.15-0.30 mm thickness, and 3 mm wide supplied by Merck was used as anode material; (2) Demineralized water was used as diluent of anodizing solvents and washers; (3) Oxalic acid with purity of 99% supplied by Sigma-Aldrich was used as anodizing electrolyte solution; (4) Sodium chloride aqueous solution with concentration of 3.5% wt. was used as electrolyte solution for battery performance test; and graphite with dimension of (5 x 1 x 0.5 cm) was used as cathode when anodized. The apparatus used in this study include: (1) Electrolysis vessel that was used as anodizing reactor; (2) Power supply with constant discharge current of 1200 mA that was used as electrical energy source in anodizing process; (3) Oven (Redline Brand Binder RI53) was used as dryer.

2.2 Research procedure

Magnesium foil with purity of 99.5% was polished with 1000 grid Si-C emery paper, it was then rinsed with demineralized water to clean the dirt on the surface. Furthermore, magnesium foil was dried in oven at 105°C until constant weight. Anodizing was performed in 0.1 M oxalic acid solution using magnesium foil as anode and graphite as cathode. Two electrodes were immersed in electrolytic solution at depth of 2 cm. The distance of between two electrodes was set at 1 cm. Anodizing potential was varied from 1.5 to 6.0 Volts for 7 minutes. The schematic diagram for experimental set is shown in figure 1.

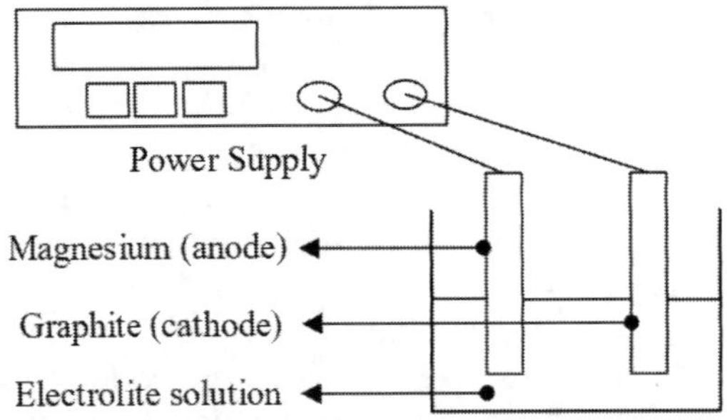

Figure 1. Schematic diagram for experimental setup

After anodization process, magnesium was then cleaned by rinsing with demineralized water, and then it was stirred at the temperature of 105°C to constant weight. Surface morphology was observed using

Scanning Electron Microscopy test (SEM, FEI Inspect-S50). The corrosion behavior was observed using linear polarization test (Metrohm Autolab PGSTAT 302).

3. Results and Discussion

3.1 Microstructure

SEM images of non-anodized and anodized magnesium anodes are shown in figure 2. It can be seen that there is anodized magnesium pore in all oxalic acid concentration variables. The anodized magnesium at 0.3 M oxalic acid concentration has the most pores among the other oxalic acid concentrations of 0.1 M and 0.2 M. The more pores on a battery anode will generate larger surface area. The larger surface area will increase the discharge performance of an anode battery. As the supporting data used data of linear polarization test results.

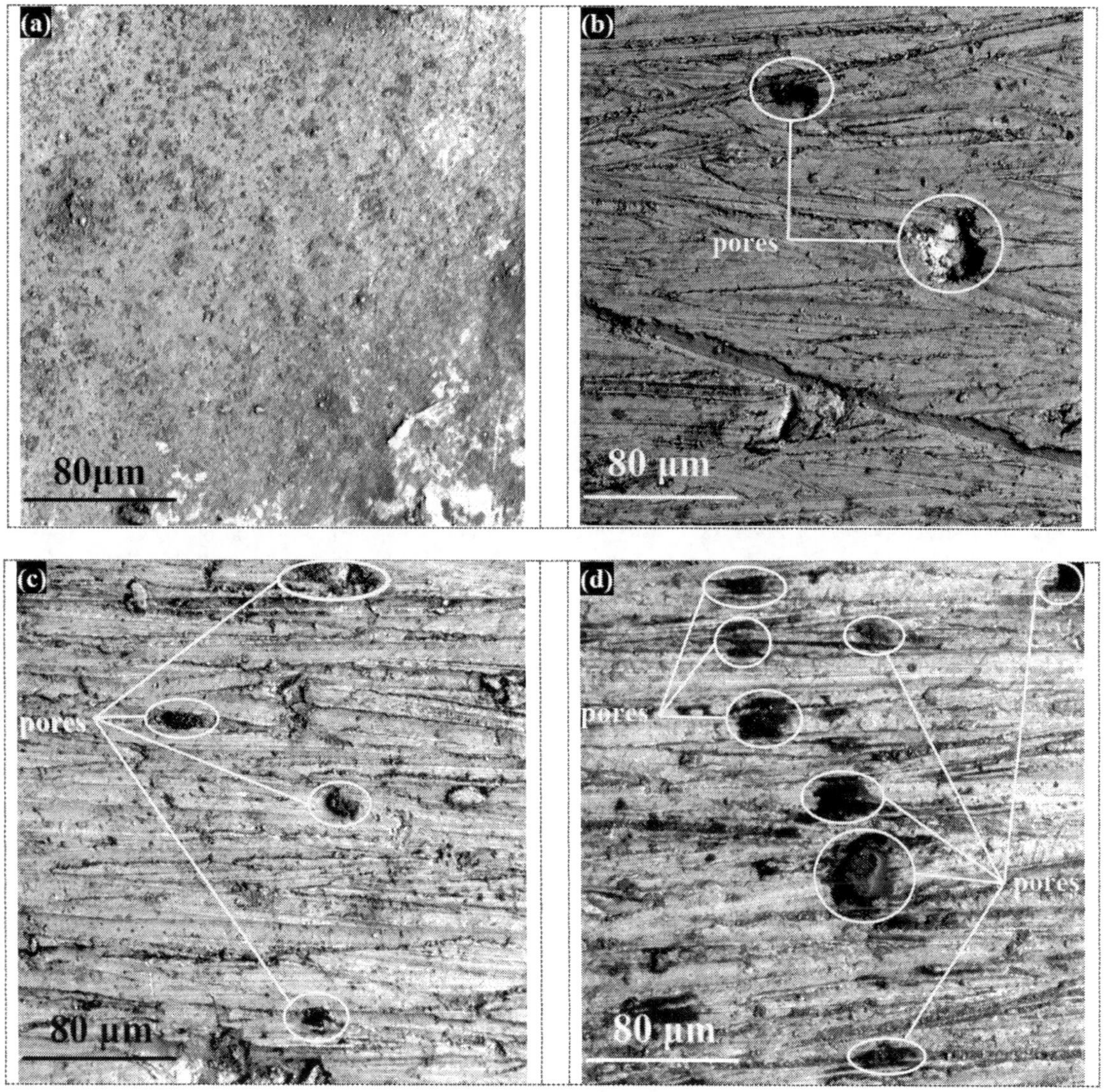

Figure 2. SEM photos of magnesium anode without anodizing (a) and it's anodized at potential of 1.5 Volt for 7 minutes in 0.1 M Oxalic acid (b), 0.2 M (c), and 0.3 M (d)

3.2 Linear Polarization

The linear polarization testing method was used to determine the magnesium anode corrosion behavior in 3.5% NaCl solution. The dynamic potential polarization graph of the magnesium anode at varied oxalic acid concentration and anodizing potential are shown in figure 3. The anodic branch represents the oxidation process of the magnesium anode discharging during the anode polarization. The cathodic branch represents the reduction process of the hydration of the protons. The corrosion current is evaluated from the linear polarization graph using the extrapolation of the cathodic branch returning to the corrosion potential. The measured parameters of the dynamic potential polarization graph are shown in Table 1. These parameters are the corrosion potential (E_{corr}), the corrosion current density (I_{corr}), and the polarization resistance for the magnesium anode in 3.5% NaCl solution. The standard corrosion potential for magnesium is -2.57 Volt vs. Ag/AgCl, while for non-anodized magnesium the value is -1.649 Volt at steady state condition. This difference occurs due to film formation on the magnesium surface for non-anodized one and pore formation for anodized one. Magnesium with porous morphology has better discharge performance than non-porous Magnesium (Cheng et al, 2013). It is shown from this study that anodized magnesium has more negative value of corrosion potential compared to non-anodized magnesium. For anodized magnesium with anodized potential of 1.5 Volts showed more regular dynamic potential test pattern than the one with the potential of 3.0 to 6.0 Volts. At the anodizing potential of 3.0-6.0 Volts, corrosion on the magnesium surface occurs too rapidly so that the release rate of magnesium oxide from the magnesium surface is less than the corrosion rate. This will lead to the thickening of the insulator's corroded layer. This insulator layer is not expected when it is applied as an anode battery since it will degrade the performance of its discharges. At the anodizing potential of 1.5 Volts, anodized magnesium at 0.3 M of oxalic acid concentrations showed the most negative corrosion potential value (E_{corr} = -1.772 Volt vs Ag/AgCl) among all concentrations variable (0.1 M and 0.2 M with E_{corr} = 1.754 Volt and 1.765 Volt, respectively). This anodized magnesium also exhibits greater corrosion current density than both non-anodized magnesium and anodized magnesium at oxalic acid concentration of 0.1 M and 0.2 M. Of the two parameter (potential and corrosion current) shows anodized magnesium at potential of 1.5 Volt is the most activated in 3.5% NaCl solution rather than non-anodised and anodized magnesium from potential of 3.0 Volt to 6.0 Volt.

The corrosion rate is very much related to polarization resistance (Rp). High polarization resistance (Rp) value indicates low corrosion rate and vice versa. From Table 1 it can be seen that the anodized magnesium at anodizing potential of 1.5 Volt tends to have lower polarization resistance value (Rp = 425.170 until 1024.700 Ω) than the non-anodized magnesium (Rp = 895.510 Ω) and anodized magnesium at the anodizing potential of 3.0-6.0 Volt (Rp = 783.630 until 2388.100 Ω). The increasing value of polarization resistance (Rp) is directly proportional to the increasing potential used for anodizing. The value of polarization resistance on anodized magnesium that is relatively smaller than non-anodized magnesium. Low polarization resistance indicates the formation of the film on the surface of magnesium (Yu *et al*, 2011; Yu *et al*, 2012; Yu, *et al*, 2015). It is indicated by thinner film formation on the anodized magnesium anode surface compared to that of non-anodized magnesium anode. The polarization resistance value of anodized magnesium at potential of 1.5 Volts has smaller tendency than the anodized magnesium anode at the potential of 3.0-6.0 Volts. This shows that the film formed on the surface of the anodized magnesium anode at potential of 1.5 Volts is thinner than that of anodized magnesium anode at potential of 3.0-6.0 volts. The charge value of the polarization resistance (Rp) determines the corrosion behavior pattern of the magnesium anode. During corrosion process, the films formed allow to inhibit activation of magnesium in 3.5% NaCl electrolyte solution. The discharge process of the magnesium anode will be inhibited. At the anode magnesium anode at potential 1.5 V volt, the Mg^{2+} ion activation process is easier and faster than in the anodized magnesium anode at 3.0-6.0 Volt and non-anodized magnesium due to the thin film formed which is proved by the lowest polarization resistance (Rp) value.

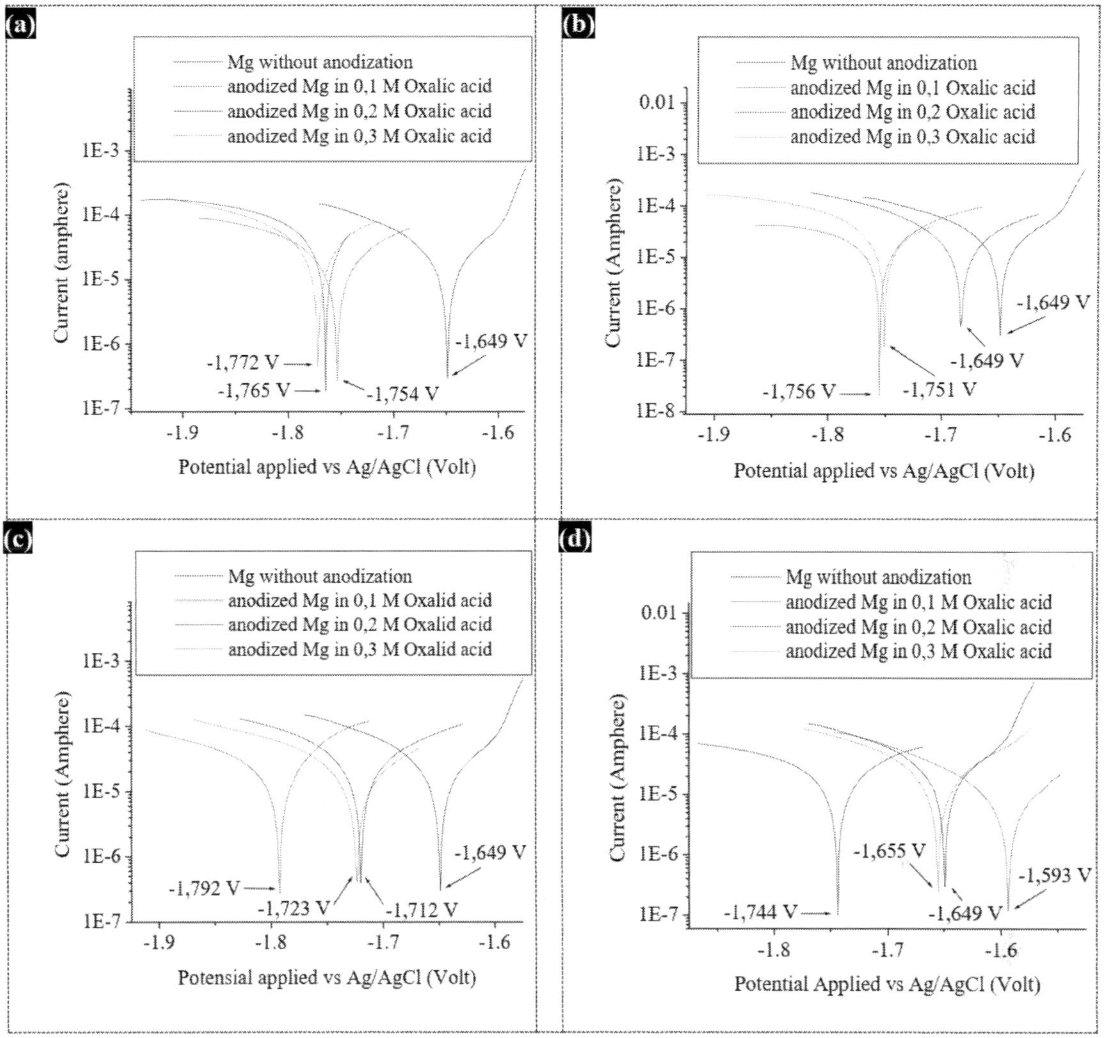

Figure 3. Potensiodynamic polarization behavior of the magnesium anode in NaCl 3.5%wt at different anodizing potential: 1.5 Volt (a), 3.0 Volt (b), 4.5 Volt (c), 6.0 Volt (d)

4 Conclusion

This study shows that the anodized magnesium at anodizing potential of 1.5 Volt and 0.3 M in Oxalic acid concentration generates more pore than the anodized magnesium at concentrations of 0.1 M and 0.2 M. Anodized magnesium at anodizing potential of 1.5 Volts tends to have higher value of corrosion potential (E_{corr} = -1.754 until -1.772 Volt) and current (I_{corr} = 2.790 until 11.690 μA/cm^2), and lower polarization resistance (Rp = 425.170 to 1024.700 Ω) than non-anodized magnesium (Ecorr = -1.649 Volt; I_{corr} = 10.984 μA/cm^2; Rp = 895.510 Ω). Anodized magnesium at the anodizing potential from 3.0 until 6.0 Volt indicates the presence of oxidation process on the surface which leads to the less good characteristic pattern of dynamic potential. Anodized Magnesium at anodizing potential from 1.5 Volt exhibited satisfactory discharge performance due to its microstructure and potential dynamic characteristics.

Table 1. The parameters obtained from the graph of dynamic polarization potential

Materials [$H_2C_2O_4$]; Anodizing potential	Ecorr (Volt)	Icorr (μA/cm2)	Rp (Ω)
Mg without anodizing	-1.649	10.984	895.510
0,1 M; 1,5 Volt	-1.754	2.790	1024.700
0,1 M; 3,0 Volt	-1.755	5.916	1147.200
0,1 M; 4,5 Volt	-1.792	8.340	936.900
0,1 M; 6,0 Volt	-1.593	1.957	2388.100
0,2 M; 1,5 Volt	-1.765	9.586	425.170
0,2 M; 3,0 Volt	-1.683	6.101	970.870
0,2 M; 4,5 Volt	-1.720	23.117	783.630
0,2 M; 6,0 Volt	-1.744	3.908	1116.300
0,3 M; 1,5 Volt	-1.772	11.690	586.330
0,3 M; 3,0 Volt	-1.751	9.599	1037.400
0,3 M; 4,5 Volt	-1.723	2.237	1166.600
0,3 M; 6,0 Volt	-1.655	1.583	1037.400

5 References

[1] Chen B, Wang R, Peng C, Feng Y, Wang N 2013 Influence of Al-Mn master alloys on microstructures and electrochemical properties of Mg-Al-Pb-Mn alloys *Transaction of Nonferrous Metals Society of China* **24** p 423-430

[2] Li F, Zhang L, Metzger R M 1998 On the Growth of Highly Ordered Pores in Anodized Aluminum Oxide *Chem. Mater* **10** p 2470-2480

[3] Davis J R 2006 *Handbook of Materials for Medical Devices* [M] ASM international OH: ASM International p 359−377

[4] Koontz R F, David L R 2002 *Water Activated Batteries* United States of America: McGraw-Hill inc.

[5] Ning G, Haran B, Popov B N 2002 Capacity fade study of lithium-ion batteries cycled at high discharge rates *J. of Power Sources* **117** p 160–169

[6] Petter S A, Fride V 2012 *Alloys as Anode Materials in Magnesium Ion Batteries* Norwegian University of Science and Technology: Department of Materials Science and Engineering

[7] Wang N, Wang R, Peng C, Feng Y, Chen B 2012 Effect of hot rolling and subsequent annealing on electrochemical discharge behavior of AP65 magnesium alloy as anode for seawater activated battery *Corrosion science* **64** p 17-27

[6] Yu K, Tan X, Hu Y, Chen F, Li S 2011 Microstructure effects on the electrochemical corrosion properties of Mg – 4.1% Ga – 2.2% Hg alloy as the anode for seawater-activated batteries *Corrosion Science* **53** p 2035–2040

[9] Yu K, Huang Q, Zhao J, Dai Y 2012 Electrochemical properties of magnesium alloy anodes discharged in seawater *Transaction of Nonferrous Metals Society of China* **22** p 2184-2190

[10] Yu K, Xiong H, Wen L, Dai Y, Yang S, Fan S, Teng F, Qiao X 2015 Discharge behavior and electrochemical properties of Mg-Al-Sn alloy anode for seawater activated battery *Nonferrous Met.* **25** p 1234-1240

[11] Zhao J, Yu K, Hu Y, Li S, Tan X, Chen F, Yu Z 2011 Discharge behavior of Mg–4 wt%Ga–2 wt%Hg alloy as anode for seawater activated battery *Electrochimica Acta* **56** p 8224-8231

IOP Conf. Series: Earth and Environmental Science **105** (2017) 012055 doi:10.1088/1755-1315/105/1/012055

Synthesis of kemiri sunan (*reutealis trisperma (blanco)* airy shaw) H-FAME through partially hydrogenation using Ni/C catalyst to increase oxidation stability

E Ramayeni[1], **B H Susanto**[1] **and D F Pratama**[1]
[1]Chemical Engineering Department, Faculty of Engineering, University of Indonesia, Depok 16424, Indonesia

E-mail: bambanghs@che.ui.ac.id

Abstract. Hydrogenated FAME (H-FAME) synthesis of Kemiri Sunan (*Reutealis Trisperma* (*Blanco*) Airy Shaw) using Nickel/Carbon catalyst is one of the methods to improve the oxidation stability of Kemiri Sunan biodiesel. The partial hydrogenation reaction breaks the unsaturated bond on FAME (Fatty Acid Methyl Ester), which is a key component of the determination of oxidative properties. Changes in FAME composition by partial hydrogenation reaction are predicted to change the oxidation stability, so it does not cause deposits that can damage the diesel engine injection system, pump system, and storage tank. Partial hydrogenation reaction is under conditions of 120 °C and 6 bars with 100:1, 100:5, 100:10 % wt catalyst in the stirred semi-batch autoclave reactor. H-FAME synthesis with 100:5 % wt Ni/C catalyst can decrease the iodine number which is the empirical measure of the number of unsaturated bonds from 166.77 to 155.64 (g-I_2/100 g) with an increase of oxidation stability from 650 to 665 minutes.

1. Introduction

Biodiesel is a biofuel that reduces the level of exhaust emissions. In addition, biodiesel mixes perfectly with diesel oil (Solar) and works well on all types of diesel engines. The main biodiesel advantage is that engine modifications are not required to run diesel engines. To ensure the quality of biodiesel, Indonesia has implemented the National Standard SNI 04-7182-2006 which was first ratified in 2006 and revised in 2015 with changes in the value of several parameters.

In fact, biodiesel that has passed through the production process can be degraded into other compounds that result in changes of biodiesel components during the storage process even though the biodiesel meets the established standards. These changes happen because biodiesel is very easily degraded into compounds such as acids, aldehydes, ketones, esters, peroxides and alcohols [1]. Changes in the characteristics of biodiesel caused by the oxidation process include increasing the viscosity value, peroxide number, acid number and decreasing the induction period [2].

Pertamina currently produces Biosolar with 15% biodiesel and 85% of diesel. Furthermore, the government hopes to increase biodiesel percentage by up to 30%. However, this mixture should be further investigated because the mixture results in a low biodiesel oxidation stability effect in which diesel engines are adversely affected by the emergence of deposits and gums that would damage the injection system, pumping systems and storage tanks.

The oxidation process of biodiesel is influenced by long acid chain fatty acids (C_{16} to C_{24}) which contain unsaturated compounds that have lower oxidation stability value. Therefore, to improve the oxidation stability, it is necessary to modify the biodiesel component (FAME). One way that can be done is by a partial hydrogenation reaction in which the reaction will break the unsaturated bond on FAME which is a key component of the determination of oxidative properties. Changes in the composition of FAME will change the oxidation stability, cetane number, and biodiesel viscosity [3].

On the other hand, the raw materials used by the biodiesel industry are vegetable oils from palm oil. However, this is a problem because vegetable oil from palm oil is not only used in the biodiesel industry but also the food industry. Therefore, the biodiesel industry should look for alternative vegetable oils which availability is assured and the price remains stable. Reliable vegetable oil among other oil is Jarak Pagar oil (*Jatropha Curcas L.*) and Kemiri Sunan oil (*Reutealis Trisperma* (Blanco) Airy Shaw) but Kemiri Sunan oil has the largest production number among other oil that has the potential for biofuels, such as Jatropha, Nyamplung, Kosambi, Bintaro, Kranji, and Kepuh. Kemiri Sunan oil gives a 52.17% percentage of each Kemiri Sunan seedlings, while there are 12,187 liters / ha of Kemiri Sunan oil productions annually [4]. In addition, Kemiri Sunan oil has a low FFA (Free Fatty Acid) of 2.6%, so it does not need pre-treatment to produce biodiesel [4].

Thus, this study was conducted to prove the oxidative stability of the product to be compatible with biodiesel standards through the partial hydrogenation of FAME components with the source of biodiesel feedstock from Kemiri Sunan oil.

2. Methodology

2.1. Materials

This study was conducted in two steps namely the synthesis of H-FAME in a hydrogenation reactor and characterization of H-FAME product. The materials needed in this study were: Kemiri Sunan oil (Darma Prema Bioenergy), Palm Biodiesel (Wilmar), Ni/C (Pratama), Aquadest (CV. Dwinika), Chloroform (Merck), Wijs (Merck), Sodium Thiosulfate (Merck), Potassium Iodide (Merck), Starch (Merck), Potassium Hydroxide (Merck), Phenolphthalein (Merck).

2.2. Partially Hydrogenated FAME

The reaction was performed in a 330 mL batch stirred reactor integrated with a condenser and furnace. Extensive stirring of the biodiesel and catalyst in the reactor was accomplished using a magnetically driven, stirred with a DC geared motor. Ni/C catalyst was purged using nitrogen for 30 minutes, and then the Ni/C catalyst was activated by hydrogen at 10 bars and 400 °C for 1 hour. Kemiri Sunan FAME was used as feed and Ni/C catalyst with variation ratio of 100:1, 100:5, and 100:10 % wt. Before the hydrogenation, the feed and the catalyst were purged using nitrogen to remove O_2 for 30 minutes. Stirring speed and pressure during the reaction were kept constantly at 800 rpm and 6 bars respectively. The reactions were performed at temperature 120°C for 2.5 hours. In this step, the double bonds that were presented in the FAME were saturated with hydrogen. To compare the double bond reduction in the Kemiri Sunan H-FAME, palm FAME (B100) that has complied SNI Biodiesel was used in partial hydrogenation using the same catalyst and operating conditions as Kemiri Sunan FAME. Figure 1 shows a schematic diagram of the apparatus.

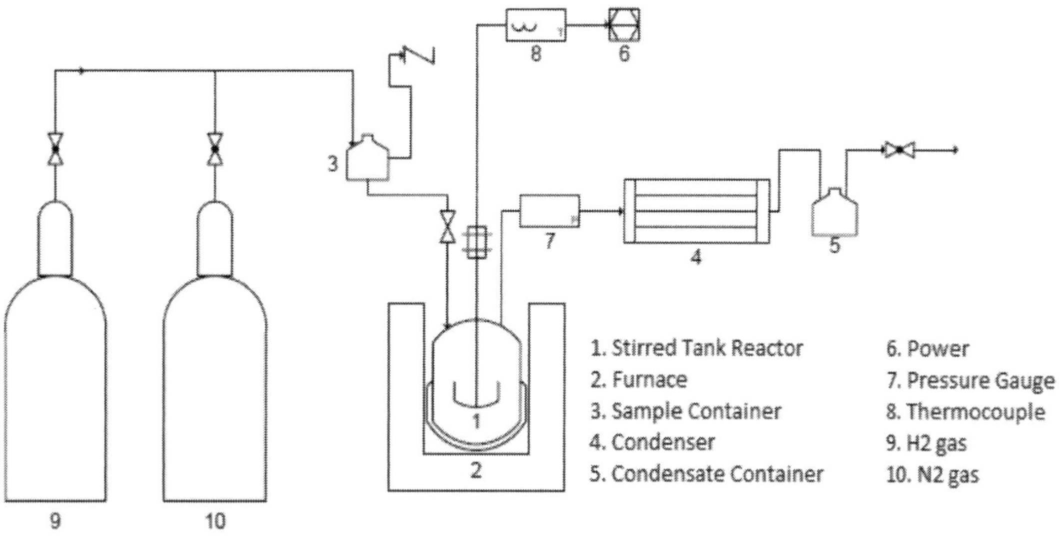

Figure 1. Scheme of Hydrogenation Reaction

The product is characterized by iodine number test to determine the double bond reduction, acid number test to determine the amount of free fatty acid in the oil, FT IR (Thermo Scientific) to know the effect of reaction to the presence of functional group in FAME, GCMS (Agilent Technologies) to identify components of sample, and oxidation stability test to determine the increased oxidative stability. Furthermore, the product is characterized by the density, viscosity, and cetane index to examine quality of combustion of the product.

3. Result and Discussion

The used catalyst was Nickel/Carbon. In test variations, the catalyst was activated first. A physical properties result of the catalyst that had been used in H-FAME synthesis is attached in table 1.

Table 1. Physical Properties of Catalyst Used

Catalyst	Surface Area (m^2/gram)	Pore Size (nm)	Crystal Size (nm)
Ni/C	285.85	2.2	77.78

To facilitate an explanation of the characterization results of the H-FAME samples in the next section, the researcher made a standard naming sample based on the operating conditions presented in table 2.

Table 2. Naming and Sample Operating Conditions

Products of Kemiri Sunan H-FAME	Products of B100	Ni/C (% wt)
H-FAME a	H-FAME d	100:1
H-FAME b	H-FAME e	100:5
H-FAME c	H-FAME f	100:10

3.1. Results of Characterization

Result of chemical characterization of H-FAME product can be seen in table 3.

Table 3. Chemical Properties of H-FAME

Product	Iodine Number ($g-I_2/100g$)	Acidity Number (mg-KOH/g)	Oxidation Stability (minute)
Kemiri Sunan FAME	166.77	0.45	650
B100	91.78	0.33	585
H-FAME a	162.18	0.44	-
H-FAME b	155.64	0.40	665
H-FAME c	152.38	0.40	-
H-FAME d	87.21	0.30	-
H-FAME e	82.38	0.26	602
H-FAME f	80.64	0.25	-
SNI 2015	Max. 115	Max. 0.5	Min. 480

Based on table 3, the iodine number decreased with the increasing of the amount of catalyst used. The highest decrease in iodine number was in the catalyst ratio of 100:5 %wt. The decreasing of iodine number showed a reduction in the number of double bonds in the FAME component.

The oxidation stability characteristic was only performed on H-FAME b, H-FAME e, Kemiri Sunan FAME, and Palm FAME (B100). It was based on the decreasing of the highest iodine number. The oxidation stability of Kemiri Sunan FAME and B100 were examined to know the initial data before the partial hydrogenation process. Based on the table 3, it was proved that the partial hydrogenation reaction could increase the oxidation stability of biodiesel.

Table 4. Identification of Functional Groups on H-FAME Products

Bonds	Functional Groups	Wave Number (cm^{-1})	Kemiri Sunan FAME	B100	H-FAME	
					b	E
$=CH_2$	Alkene	991-723	√	√	√	√
C-O	Alcohol	1169-1243	√	√	√	√
C=O	Carbonyl	1740	√	√	√	√
C-H	Alkane	1360-1459	√	√	√	√
	Alkane	2853-2932	√	√	√	√

Identification of functional groups using FTIR indicated that the partial hydrogenation reaction did not break the functional group bond but only broke the double bond of the FAME component which was a key component of oxidative properties.

Table 5. Component Identification of Sample BKS and H-FAME b and e

Components	Kemiri Sunan FAME	B100	H-FAME	
			b	e
% Methyl Palmitate (18:0)	13.97	33.17	13.36	33.29
% Methyl Stearate (18:0)	8.14	9.78	7.81	9.70
% Methyl Oleate (18:1)	0	38.63	1.44	39.81
% Methyl Linoleate (18:2)	28.17	0.28	26.10	0.22
% Methyl 9,11,13 Octadecatrienoate (18:3)	38.70	0.23	37.82	0.18

According to table 4, the percentage of Methyl Linoleate and Methyl 9,11,13 Octadecatrinoate components were decreased, while the components of Methyl Oleate, Methyl Stearate, and Methyl Palmitate were increased. These results indicated that the partial hydrogenation reaction had broken the saturated bond on the FAME component into an unsaturated bond. These results also confirmed the result of the decrease in iodine number.

Table 6. Physical Properties of H-FAME

Products	Density (kg/m^3)	Viscosity (CSt)	Cetane Index
Kemiri Sunan FAME	878.94	6.5	47.06
B100	856.96	4.5	52.51
H-FAME a	879.56	-	-
H-FAME b	884.78	6.8	49.04
H-FAME c	883.95	-	-
H-FAME d	858.21	-	-
H-FAME e	863.69	4.6	55.59
H-FAME f	864.43	-	-
SNI 2015	850-890	2.3-6.0	Min. 51

Based on the results of physical properties, density of H-FAME product increased, this was because the FAME component bound to hydrogen. While the viscosity of the product increased due to the increasing number of saturated methyl ester components and partial hydrogenation. The breakdown of unsaturated bonds into saturated bonds will increase the oxidation stability of biodiesel but it decreases flow properties, so it is necessary to obtain optimum conditions for both parameters to meet the specification [5]. In addition, the cetane index increased after the partial hydrogenation reaction showed that the more saturated components of the FAME, the more cetane index of the FAME which indicated combustion quality.

Table 7. Conversion. Selectivity and Yield of Product

Products	% Conversion	% Selectivity	% Yield
H-FAME b	92.74	10.87	10.08
H-FAME e	93.26	10.58	9.87

The calculation of selectivity was based on the amount of addition of the methyl oleate component, since methyl oleate is the desired component by the partial hydrogenation reaction in biodiesel. If the FAME component gets saturated, it will cause a high viscosity of biodiesel thereby imparting the flow properties [5].

4. Conclusion

Partial hydrogenation Kemiri Sunan (*Reutealis Trisperma* (Blanco) Airy Shaw) FAME using Ni/C catalyst under conditions of 120 °C and 6 bars with 100:5 % wt catalyst in the stirred semi-batch autoclave reactor decreased the iodine number which was the empirical measure of the number of unsaturated bonds from 166.77 to 155.64 (g-I_2/100 g) with an increase of oxidation stability from 650 to 665 minutes. These conditions with activated Ni/C led to dominant yield of 10.08%, conversion of 92.74%, and selectivity of 10.87% of methyl oleate (18:1). The decrease of iodine number on Palm H-FAME (B100) indicates that the catalyst can react to various types of biodiesel, whether it is compliant with SNI or not yet fulfilled.

5. Acknowledgements

Directorate of Research and Community Service through PITTA UI 2017 is acknowledged for the financial support to carry out the present study.

6. References

[1] Xin, Jiayu, Imahara, Hiroaki, and Saka Shiro 2009 *Kinetics on the oxidation of biodiesel stabilized with antioxidant.* **88(2)** 282-286
[2] Agarwal, Avinash Kumar and Khurana, Deepak 2013 *Long-term storage oxidation stability of Karanja biodiesel with the use of antioxidants. Fuel Processing Technology,* **106** 447-452
[3] Knothe, Gerhard 2007 *Some aspects of biodiesel oxidative stability. Fuel Processing Technology,* **88(7)** 669-677
[4] Ministry of Forestry 2015 Internal Report
[5] Goto S, M Oguma, N Chollacoop 2010 *EAS-ERIA Biodiesel Fuel Trade Handbook* 28-38

Application of modified microwave polyol process method on NiMo/C nanoparticle catalyst preparation for hydrogenated biodiesel production

D F Pratama[1], **B H Susanto**[1] **and E Ramayeni**[1]

[1]Department of Chemical Engineering, University of Indonesia, Depok 16424, Indonesia
Email: bambanghs@che.ui.ac.id

Abstract. The development of renewable feedstock-based diesel fuel is start to come up as the solution of national energy problem. However, the thermal and oxidative stability of biodiesel is not good enough. As a result, biodiesel can only be added to commercial diesel fuel as a mixture with concentration under 20%. To get better thermal and oxidation stability, partial hydrogenation process is applied to biodiesel caused the increase of monounsaturated FAME structure. Activated carbon supported NiMo nanocrystal catalyst was used in partial hydrogenation reaction to get high activity, conversion, and selectivity. In this research, NiMo/C catalyst was prepared by modified microwave polyol process method, which is provided a rapid heating and cooling process. This method can produce nano-sized NiMo/C catalyst with short time and low energy consumption. NiMo/C catalyst produced in this research has 285.85 m^2/gram surface area and 77.79 nm crystal size, resulting 20.41% conversion and 8.87% selectivity of biodiesel product. Further research should be conducted to obtain optimum condition.

1. Introduction

The depletion of petroleum energy reserves and increasing demand for clean and sustainable energy source has triggered the development of renewable energy source, such as biofuels. Biodiesel is one of the biofuels which can be produced by transesterification of triglyceride from various vegetable oil. But, biodiesel still have several problems. Thermal and oxidative stability of biodiesel is not good enough. As a result, biodiesel can only be added to commercial diesel fuel as a mixture with concentration under 20%.

Oxidative stability of biodiesel fuel depends on the degree of saturation of fatty acids in the starting vegetable oil. The presence of polyunsaturated fatty esters is the cause of oxidative stability with biodiesel [1]. One approach to solve this problem is the hydrogenation of polyunsaturated fatty acid methyl esters (FAME) in biodiesel. Since complete hydrogenation of polyunsaturated FAME to saturated FAME dramatically affects the cold flow properties of biodiesel, partial hydrogenation of polyunsaturated FAME to monounsaturated FAME is a promising solution to improve oxidative stability of biodiesel [2]. The product of this reaction is known as hydrogenated biodiesel.

The catalysts used in partial hydrogenation reaction are supported catalysts based on Ni, Cu, and noble metals. In this study, catalyst used is NiMo/C catalyst. NiMo Catalyst preparation process generally is still using incipient-wetness method, which can takes up to 24 hours preparation time and uses high energy consumption [3]. Using microwave as a heating media instead of conductive heating in NiMo catalyst preparation is a simple, fast, and efficient method as a modification of current method [4]. Moreover, rapid time is required to avoid the growth of large metal catalyst.

This study will synthesize carbon supported NiMo catalyst by microwave polyol process method with various microwave power and heating time to obtain optimum results that can provide surface area, crystal size, and activity of partial hydrogenation reaction.

2. Experimental section

2.1. Materials

The materials used in this study are granular activated carbon as catalyst support, precursors $(NH_4)_6Mo_7O_{22}.4H_2O$ and $Ni(NO_3).6H_2O$ (Merck) as the source of the active site, ethylene glycol as polyols, sodium hydroxide (Merck) as the source of alkaline, potassium bromide (Merck) as stabilizer compound, and biodiesel from Kemiri Sunan oil (BKS).

2.2. Procedure

2.2.1. Synthesis of catalyst. NiMo/C catalyst with NiMo content of 5% [5] was synthesized through modified microwave polyol process method in the EG polyols solution. Synthesis of catalyst was done through two steps. The first step is impregnation of Mo into activated carbon. Required amounts of precursor $(NH_4)_6Mo_7O_{22}.4H_2O$ was dissolved in distilled water. EG then added into the solution. Activated carbon then added into the solution and KOH was added dropwise until pH 8-10. KBr was added into the solution as stabilizer compound. The solution was stirred on the magnetic stirrer. Next, the solution was dispersed in the sonication bath. The solution then heated in the microwave with certain power and time. that, the solution was filtered, and the solid product then washed with demineralized water and ethanol before heated in the furnace to produce Mo/C. The second is impregnation of Ni into activated carbon. Required amounts of precursor $Ni(NO_3).6H_2O$ $4H_2O$ was dissolved in distilled water. EG then added into the solution. The Mo/C then added into the solution and repeat the exactly same steps as synthesizing Mo/C to produce NiMo/C.

2.2.2. Catalyst characterization. The resulting catalyst then characterized with BET (Micrometrics ASAP 2020) to determine the surface area, pore size, and pore volume, SEM-EDX (JED-2300 Analysis Station JEOL) to determine the morphology and composition of the catalyst, and XRD (PANalytical Empyrean) to determine the natural and type of crystals.

2.2.3. Partial hydrogenation reaction. Partial hydrogenation reaction was carried out in a 330 ml stainless steel stirred batch reactor, which integrated with condenser and furnace, at temperature and hydrogen partial pressure of 110 °C and 4 bars, respectively. Stirring rate was maintained at 800 rpm. The weight of catalyst used is 1% of feed mass. Biodiesel and NiMo/C catalyst were placed into the reactor. The reaction started by increasing temperature and pressure to the desired point. The reaction occurred in 180 minutes, started when the desired temperature and pressure has achieved.

3. Result and discussion

3.1. BET characterization result

Table 1 shows that the catalyst produced in this study has a small pore diameter and large pore volume. The large pore volume and small pore diameter indicate that the support structure is very porous and has a wide pore space. A catalyst with a large surface area is needed to increase the area of contact of the active nucleus. However, sometimes the overly large surface area is avoided because it is indicated that the active site did not properly impregnated. Small pore diameter also indicated that the active site only attached on the surface, not inside the pore.

Table 1 also shows that with the same heating time, catalyst that is prepared with higher microwave power produced higher surface area. Higher power produces higher microwave intensity, resulting in higher heating rates as well. High heating rates results the ion particles reduced more completely. Meanwhile, with the same microwave power, catalyst that is prepared with less heating time produced higher surface area. Longer heating process will lead to more agglomeration causing the surface area to decrease. The nucleus of the active site is formed at the beginning of the heating process. Furthermore, the active site nanoparticles grow as temperature increases. As the temperature reaches a higher point, the active site nanoparticles grow larger and begins to form clots. The clump will close the pores of the activated carbon so that the catalyst surface area will be reduced [6]. Catalyst that is prepared with rapid cooling procedure also produced higher surface area.

Table 1. BET characterization result

Catalyst	Microwave Power (Watt)	Heating Time (minutes)	Ultrasonic Dispersion	Rapid Cooling	Surface Area (m^2/g)	Pore Volume (cm^3/g)	Pore Diameter (nm)
NiMo/C-1	550	3	-	√	379.95	0.202	2.132
NiMo/C-2	550	4	-	√	354.63	0.191	2.159
NiMo/C-3	800	3	-	√	411.32	0.224	2.179
NiMo/C-4	800	3	√	-	177.51	0.100	2.263
NiMo/C-5	800	3	√	√	285.85	0.157	2.200

3.2. SEM-EDX characterization result

From figure 1 and 2, it can be seen that the morphology of NiMo/C1 and NiMo/C3 tend to be clumping which is indicated by nonuniform particle size. Whereas, figure 3 shows that NiMo/C5, which is dispersed with ultrasonic wave, has better morphology than the other two which is indicated by more uniform particle size. These results indicated that the usage of ultrasonic wave as a dispersion media can produce better morphology of catalyst because of better dispersion method.

Figure 1. The morphology of NiMo/C1 catalyst at 1000x magnification

Figure 2. The morphology of NiMo/C3 catalyst at 1000x magnification

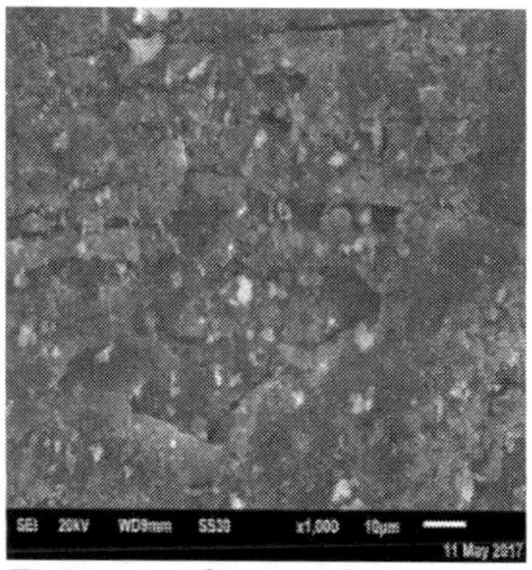

Figure 3. The morphology of NiMo/C5 catalyst at 1000x magnification

Table 2. EDX characterization result

Element	Mass percentage (%)		
	NiMo/C1	NiMo/C3	NiMo/C5
C	90.33	91.53	78.88
O	4.13	4.67	6.24
Al	0.46	0.30	2.11
K	2.42	1.02	5.59
Ni	2.66	2.09	2.72
Mo	-	0.63	4.46
Total	100	100	100

The EDX result in table 3 shows that the molybedenum in NiMo/C1 and NiMo/C3 was not well dispersed because the amount of Mo in NiMo/C1 and NiMo/C3 are scrimp and the loading of NiMo are still far from initial loading. Whereas, NiMo/C5 shows more amount of Mo and the loading of NiMo is approaching the initial loading. This result once again indicated that ultrasonic wave can give better dispersion.

3.3. XRD characterization result

Figure 4 and 5 shows the diffractogram of NiMo/C3 and NiMo/C5. It can be seen from the diffractograms that sharp peaks appear on each catalyst. The intensity of peak represents the crystallinity. The sharper peak obtained, means particle tend to have better crystallinity. High crystallinity also will affect to the catalytic activity and stability at high temperature [5]. Sharper peak also indicated larger crystal size. It can be seen from figure 4 and 5 that the diffraction of NiMoO4 appear at $2\theta = 46$, NiO at $2\theta = 63$ and 78.2, and MoO2 at $2\theta = 37.2$.

Table 3 shows the crystal size of NiMo/C3 and NiMo/C5. The crystal size can be obtained using the Scherrer equation. It can be seen that NiMo/C3 has bigger crystal size than NiMo/C5. This is because NiMo/C5 used ultrasonic wave as dispersion media, which can break up clumps and agglomerates so they have a smaller crystal size. From this result, only NiMo/C5 catalyst formed nanocrystalline.

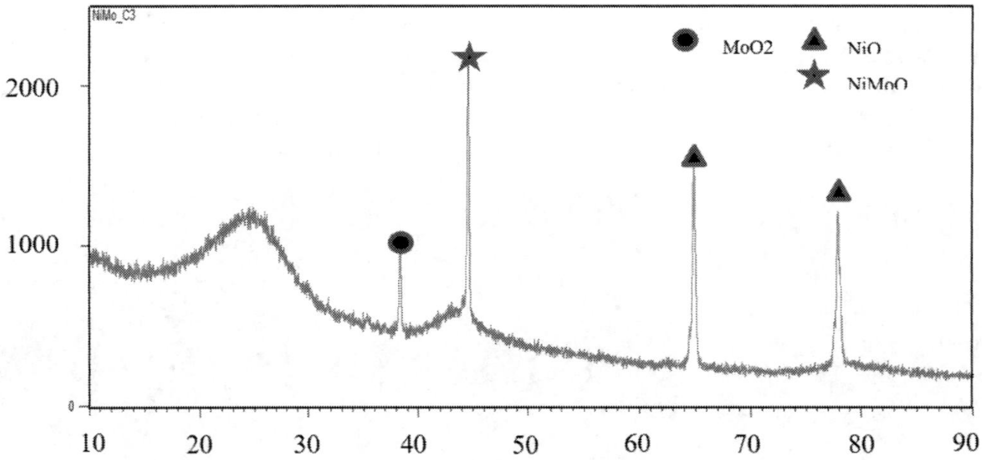

Figure 4. Diffractogram of NiMo/C3

Table 3. Crystal size

Catalyst	Crystal size (nm)			
	NiO	MoO_2	$NiMoO_4$	Average
NiMo/C3	121.866	180.304	64.981	122.857
NiMo/C5	68.105	105.128	69.800	77.785

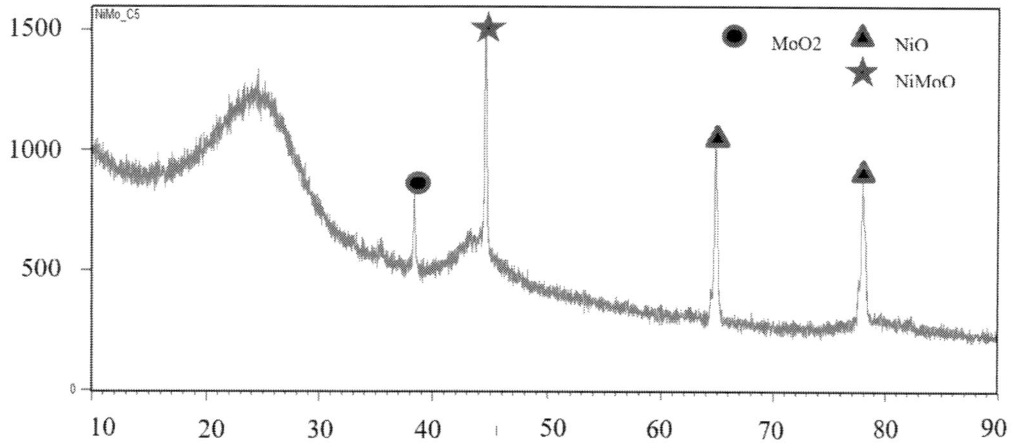

Figure 5. Diffractogram of NiMo/C5

3.4. Partial hydrogenation reaction result

Partial hydrogenation reaction was performed to break the double bond on *Biodiesel Kemiri sunan* (BKS). The activity of NiMo/C5 catalyst in this study was determined by how much double bonds were broke during the partial hydrogenation reaction.

Table 4 shows the data obtained from GC-MS. it can be seen that after the partial hydrogenation process, changes occur when the decreased of Methyl 9,11,13 Octadecatrienoate (18:3) and Methyl Linoleate (18:2) components, as well as the increased of the Methyl Oleate component (18:1) as the desired component of the H-FAME product. This result shows that the NiMo/C5 catalyst used in the partial hydrogenation reaction has been able to break the double bonds C18:3 and C18:2 present in biodiesel. On the other hand, a very small increase of Methyl Oleate (18:1) indicates that although NiMo/C5 catalysts have been able to break the double bonds in a partial hydrogenation reaction, the NiMo/C5 catalyst still cannot aim the reaction into the desired product, which is *Methyl Oleate* (18:1).

Figure 6 shows that partial hydrogenation reaction still results low conversion, yield, and selectivity. This is maybe because the BKS as the raw material contains high amount of C18:3, which causes the difficulties produce C18:1 because the hydrogenation occurred step by step, from C18:3 to C18:2 then from C18:2 to C18:1. Another reason is the amount of catalyst used in this reaction is too small so that the reaction did not run optimally.

Table 4. Component identification for BKS and H-FAME

Component	Mass percentage (%)	
	BKS (%)	H-FAME (%)
Methyl Stearate (18:0)	8.60	8.99
Methyl Oleate (18:1)	5.63	6.81
Methyl Linoleate (18:2)	24.20	20.30
Methyl 9,11,13 Octadecatrienoate (18:3)	42.97	40.74

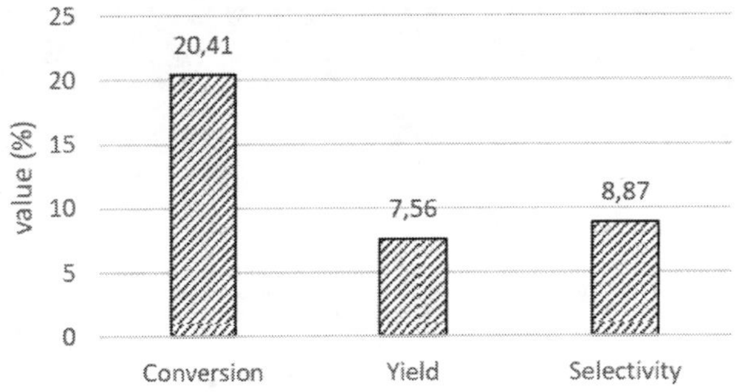

Figure 6. Conversion, yield, and selectivity

4. Conclusion
In this study, NiMo/C catalyst were prepared by modified microwave polyol process. The best catalyst produced in this study was NiMo/C5, which was prepared with 800W microwave power for 3 minutes with ultrasonic dispersion and rapid cooling process. Characterization result have shown that NiMo/C5 has 285.85 m^2/g surface area and 77.79 nm crystal size. The NiMo/C5 catalyst was used in partial hydrogenation reaction of *Biodiesel Kemiri Sunan* at 120 °C and, 4 bar for 180 minutes with 1% catalyst weight, resulting 20.41% conversion and 8.87% selectivity. This result indicated that the catalyst has low activity for partial hydrogenation reaction of *Biodiesel Kemiri Sunan* with those operating condition. Research should be done further using the national standard biodiesel to determine the optimum amount of catalyst used.

5. Acknowledgement
Director of Research and Community Service through International Indexed Publication Grant for Student Thesis (PITTA) Universitas Indonesia 2017 is acknowledged for the financial support to carry out the present study.

6. References
[1] Knothe G 2010 biodiesel and renewable diesel: a comparison *Prog. Energy Combust. Sci.* **36(3)** 364–373
[2] Numwong N, Luengnaruemitchai A, Chollacoop N and Yoshimura Y 2013 effect of metal type on partial hydrogenation of rapeseed oil-derived FAME *JAOCS, J. Am. Oil Chem. Soc.* **90(9)** 1431–38
[3] Sotelo-boyas R, Liu Y and Minowa T 2011 renewable diesel production from the hydrotreating of rapeseed oil with Pt / Zeolite and NiMo / Al 2 O 3 catalysts *Ind. Eng. Chem. Res.* **50** 2791–99
[4] Susanto B H, Nasikin M, Sukirno, Faisal A and Irfan M 2014 preparation and characterization NiMo/Zeolite catalyst using microwave polyol process method for synthesizing renewable diesel from jathropa oil *5th Sriwij. Int. Semin. Energy Environ. Sci. Technol.* 1–7
[5] Susanto B H, Prakasa M B and Shahab M H 2016 synthesis of renewable diesel through hydrodeoxygenation reaction from nyamplung oil (*Calophyllum Inophyllum* oil) using NiMo/Z and NiMo/C catalysts with rapid heating and cooling method *IOP Conf. Ser. Mater. Sci. Eng.* **162(1)** 12009
[6] Li H, Zhang S, Yan S, Lin Y and Ren Y 2013 Pd/C catalysts synthesized by microwave assisted polyol method for methanol electro-oxidation *Int. J. Electrochem. Sci.* **8(2)** 2996–3011

Performance analysis (WHP and torque) on SI engine fueled with low-grade bioethanol and oxygenated fuel additive

Setia Abikusna[2], Bambang Sugiarto[1], Ratna Monasari[1], Raihan Aditya[1] and Dimas Hendrawan[1]
[1]Departement of Mechanical Engineering, University of Indonesia, 16424, Indonesia
[2]Automotive Engineering, Astra Manufacture Polytechnics, Sunter Jakarta 14330, Indonesia
E-mail: ratnamona92@gmail.com

Abstract. Utilization of renewable energy in Indonesia is still relatively low compared to the non-renewable energy source of petroleum, coal, and natural gas. The limited amount of the non-renewable energy encourages the development of alternative fuels. Currently, low-grade bioethanol is one of the alternative fuel being developed. Bioethanol has a higher octane number compared to gasoline. Mixing of bioethanol requires a certain comparison, the mixing is intended to increase the octane number of a fuel mixture. The use of low-grade bioethanol (C2H5OH) as a substitute or mixture of fuels has an impact on engine performance. To get a more optimal effect on the fuel mixture, oxygenated cycloheptanol as an additive can be added to the fuel. This study examined the effect of fuel mixture with the addition of oxygenated cycloheptanol to single cylinder SI engine, 150 cc premix type at 100% throttle position. The tests was carried out on E5, E10, E15 and the addition of 0.5% oxygenated cycloheptanol at each fuel mixture with engine speed variation above 4000 rpm. Performance test will be performed with connecting the machine to a dynamometer. This study aims to obtain efficient performance on the engine with the addition of additives, there is an average increase of 9% in horsepower and 6% in torque.

1. Introduction

Utilization of renewable energy as an alternative fuel continues and becomes an object of interest to be developed. The use of alternative fuels cannot be separated from the two global issues about the availability of crude oil and the issue of exhaust emission are increasingly worrying [1]. Indonesia is a tropical country with the highest potential to develop alternative energy.

Bioethanol (C2H5OH) is a renewable fuel source derived from biomass derivative product from the plant's fermentation containing starch. Bioethanol has a simple molecular structure with easily defined chemical and physical properties. Bioethanol is able to use as fuel either directly or as a mixture of other fuel, such as gasoline. To apply bioethanol as a replacement or mixture fuel for the engine, bioethanol must have a high content at least 85%.

The use of bioethanol as a fuel for Spark Ignition machine has some several advantages compared to gasoline. Bioethanol has better anti-knock characteristic and the improved thermal efficiency of the engine on the compression ratio [2]. A mixture of bioethanol and gasoline at the condition of stoichiometry can reduce knocking effect [3].

Bioethanol burns at lower fire temperatures and luminosity due to a decreasing peak temperature in the cylinder, thereby resulting in wasted heat and decrease NOx emission. Bioethanol has a high latent heat of vaporization. Latent heat cools the air inlet for increased density and volumetric efficiency.

However, the oxygen content in bioethanol reduced caloric value more than gasoline. Therefore, bioethanol can be used as fuel in SI engine [4, 5].

Rodrigo C.Costa et al [6] conducted an experiment on the 1000cc engine, 4 cylinders. Fuel mixture of 78% gasoline and 22% ethanol. And hydrous ethanol (E100). On the variation of 1.500 rpm till 6.500 rpm. The measured ratio compression is 10:1, 11:1, and 12:1. The performance evaluated were torque, brake mean effective pressure, power, specific fuel consumption, thermal efficiency, exhaust gas temperature and volumetric efficiency. The result obtained was on a higher ratio compression that could improve the engine performance to both of fuels in engine rotation.

Gholamhassan Najafi et al [7] tested to the engine in speed variation, besides this study was conducted by using RSM and its tool was DoE to optimize parameter perform and bioethanol – gasoline 5%, 7,5%, 10%, 12,5% and 15%. The obtained result from fuel ethanol mixture was increased compared with the gasoline, brake power, and engine torque, while brake specific fuel consumption declined. In addition, the concentration of CO and NOx decreased, but on CO_2 and NOx increased [8].

Setia abikusna et al [9] conducted the test by using distilled bioethanol low-grade autonomously that utilized the exhaust heat on compact destilator to produce high-grade bioethanol which was ready to use as fuel mixture. From the test was obtained that wheel power and wheel torque generated from a mixture of gasoline and bioethanol had a higher value than pure gasoline as fuel. A mixture of gasoline and bioethanol was able to increase the power by 15%. While the torque value was generated at a mixture of E5, E10, and E 15 respectively amounted to 6.92Nm, 6.64Nm, and 6.92Nm, where the value was higher than pure gasoline by 6.1Nm.

C. Ananda Srinivisan [10] examined the effect of ethanol gasoline which was mixed with oxygen additives on SI engine. The experiment result addressed that there was an increased efficiency thermal brake. On the emission test was found that CO decreased slightly, while HC and O_2 increased enough and CO_2 and NOx decreased. In addition, the analysis of burning was done where the cylinder pressure and heat release rate were analyzed. Combustion analysis in this research was produced with maximum load 40N at 3.000 rpm. If cylinder pressure increased, heat release rate was also increased. For alcohol fuel, if cylinder dropped, heat release rate also declined.

This research was conducted on SI machine with 7 variations of the gasoline-bioethanol fuel mixture and oxygenated cycloheptanol additive. The purpose of this research is for finding out the influence of additive to performance machine based on wheel horsepower and torque, besides the comparison of performance with additive and without additive will be discussed in detail in this paper. Based on the author's knowledge, performance testing on the SI engine using oxygenated cycloheptanol is new research.

2. Experimental Set Up

The engine was tested in this study as a form of SI engine single cylinder 150cc with fuel injection. Fuel system modification was done on fuel tank system with a pre-mixed fuel mixture of gasoline and bioethanol using a mixer on the fuel tank. There were seven variations of mixed fuel used, i.e. E0, E5, E15 without additives and E5, E10, E15 with the addition of additives (Table 1). The test was done by placing the engine on the dynamometer chassis to analyzed the power and torque that were generated on some variation of the fuel mixture. Timing and injection duration were regulated by the ECU. Here is set-up experimental in 150 cc SI engine that was supported to others component.

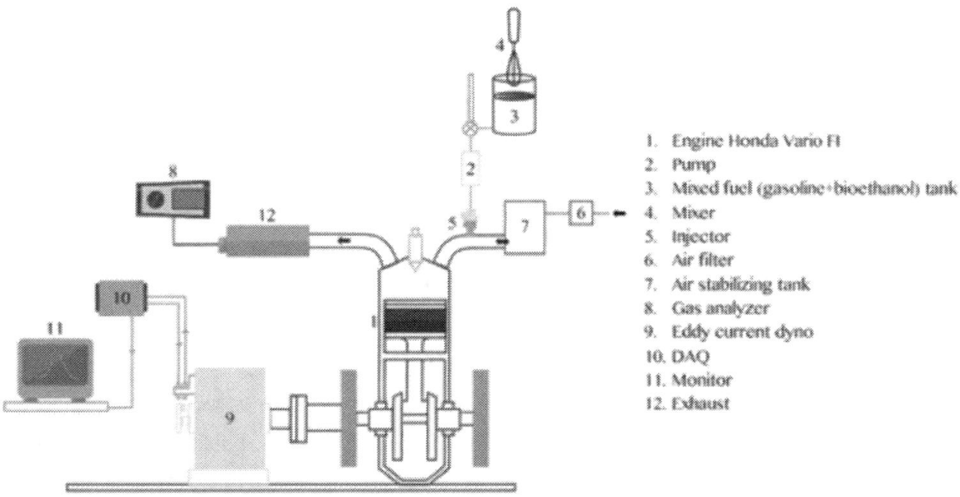

Figure 1. Experimental Set up Engine

Testing was done on the SI 150cc single cylinder engine that was mounted on a dynamometer chassis, where rear wheel of the vehicle served as a driving force that was linked on a roller dynamometer. Dynamometer Calibration process was done by adjusting the rotation of the wheels of the vehicle with the round roller drum to obtain corresponding values between rpm engine and the results of the readings on a dynamometer. Further, testing was done at 100% throttle position variation rounds of 4000 rpm to 10000 rpm. The results of the power and torque testing were issued in the form of a graph and its value on the monitor that were connected to the dynamometer. The test was performed four times for each variation of the fuel mixture. That was intended to get more accurate data.

Table 1. Sample volumetric composition

Sample name	Gasoline	Bioethanol	Oxygenated cycloheptanol
E0	100.00	----	----
E5	95.00	5.00	----
E10	90.00	10.00	----
E15	85.00	15.00	----
E5 +additive	95.00	4.5	0.5
E10 +additive	90.00	9.5	0.5
E15 +additive	85.00	14.50	0.5

3. Fuel Mixing

The method was used in mixing the fuel namely premixture by placing the mixer on the fuel tank which had been modified. The next, gasoline, bioethanol and oxygenated additives cycloheptanol were put into the tank with the appropriate comparison to the types of variables that had been set.

Then the mixer was turned on by pressing the on/off valve on the mixer. By using this method, there was no fuel mixture settles, because, during the process of testing, the fuel continued to stir until well mixed perfect before streamed to the injectors for atomizing towards to the combustion chamber.

IOP Conf. Series: Earth and Environmental Science **105** (2017) 012057 doi:10.1088/1755-1315/105/1/012057

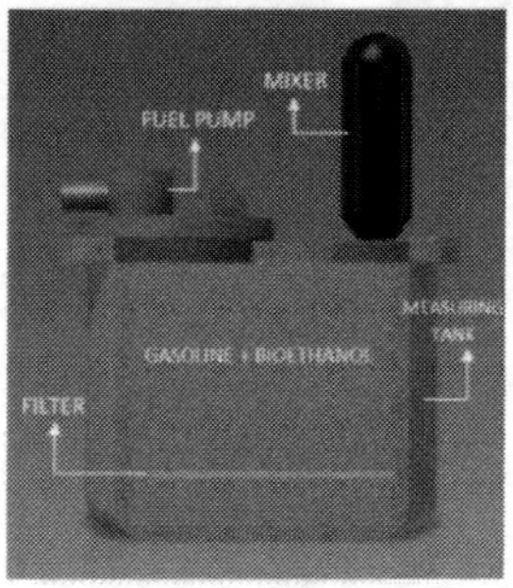

Figure 2. Modification of Fuel Tank

4. Result and Discussion

Overall, the analysis of the engine performance on this test included two things, namely power, and torque. Engine performance testing was conducted by using a chassis dynamometer Dyno Mainline. Based on the study, illustrated that the test results in the form of a comparison between the power and torque of the engines rotation variations for each variation of the fuel that was tested and presented in the form of tables and charts that aimed to make it easier to analyze data.

4.1. The Results of the Testing Power (Wheel Horse Power)

Testing gains value from the round wheel which is measured on the chassis dynamometer. Wheel horsepower is the actual value of the resources obtained through several components such as couplings, transmission and more. While its value is obtained from the measurements at the crankshaft or commonly referred to as Brake Horse Power will produce a comparison of 20%-30% that is larger than the measurements on the wheel (WHP).

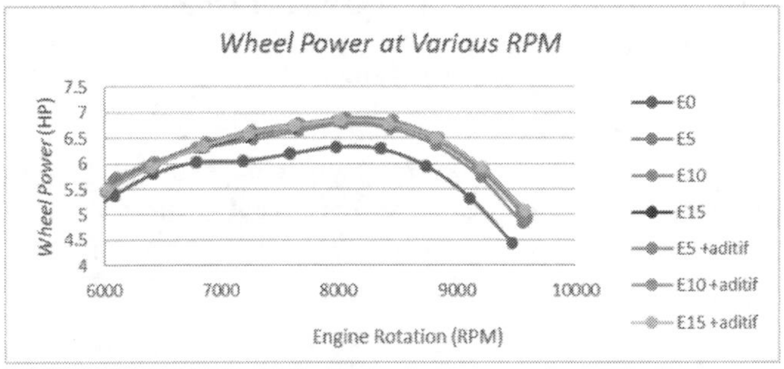

Figure 3. Wheel Power at Various RPM

Further, figure 3 is the result obtained in the test wheel power variation rounds until 10.000 rpm for a wide variety of fuel tested. On the graph to see that power is generated by the gasoline on some variation of the machine rotation lower than that is produced by the fuel mixture. The higher the spin machine, then the higher the delta power which is generated between the E5 to E15 + additive compared to E0. The maximum power is capable to result of E0 equal to 6.33 HP at 8000 rpm with an average difference between E0 – and the variation of fuel mixture of 9%. The difference is caused by the difference in air-fuel ratio (AFR), large or small values of AFR affected the amount of fuel entering the combustion chamber. In addition to the AFR, other things that affected to the power generated by the engine, namely a heat value of fuel, fuel efficiency, volumetrics and conversion of energy. The efficiency of volumetrics on some variation of fuel will be different, besides the round value also affects the efficiency of the machine volumetric.

The addition of additives on mixed gasoline with bioethanol also affects the resulting peak power. The peak power is generated on every variation of fuel at 8.000 rpm. Maximum power is generated on E5(5% bioethanol) equal to 6.78 HP while on E5+ additive (5% bioethanol with additives) is generated power equal to HP 6.78. On E10 (10% bioethanol) produces power equal to 6.90 HP and maximum power on E10+additive equal to 6.93 HP. The variation of fuel E15(15% bioethanol) generates the power equal to 6.83HP and E15+additive equal to 6.8HP. The additive is mixed in fuel variations (E5, E10, and E15) raises the value of the AFR, so the power is produced by adding additives larger than the mixture variation without additives. Figure 4 Shows a power comparison of 7 fuel variations tested at 8.000 rpm.

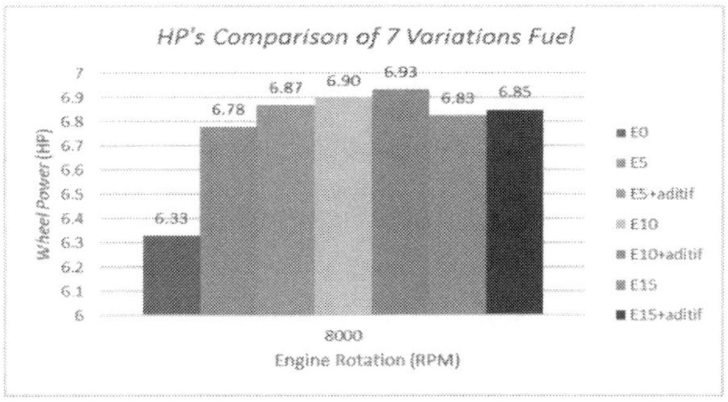

Figure 4. HP's Comparison 0f 7 Fuel Variations

Based on the test results, obtained that the variation of the fuel mixtures was the most effective on E5 (5% bioethanol). The value of the AFR on E5 is richer with the resulting power of 6.78HP. Besides the addition of oxygenated additives, cycloheptanol was conducted on the most optimal variation of fuel mixture E5 which generated a power of 6.78HP. In addition, to produce high-performance value, the detonation effect on the E5+additive variation was also relatively minor so the generated power more optimum.

4.2. The Results of the Testing Torque (Wheel Torque)

This test yielded a value of Wheel Torque (WT) of the wheel rotation that was connected on the chassis dynamometer.

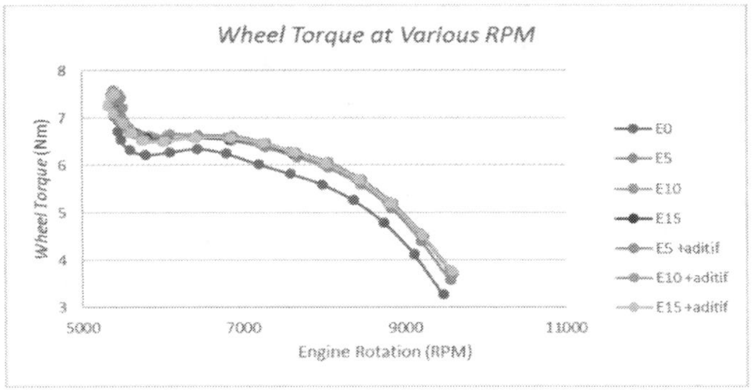

Figure 5. Graph of Torque at Various RPM

In figure 5 shows that torque results on the variation of fuel mixture gasoline with bioethanol has a higher value than gasoline (E0). The maximum torque value occurs at 5.000-6.000 rpm engine. The maximum torque value that can be generated by E0 fuel variation equal to 7.08 Nm at 5.400 rpm. Its value is lower 6% of the mean difference-maximum torque between fuel gasoline and gasoline-bioethanol mixture at rotation about 5.400 rpm. The torque resulted in each fuel variations in sequence on E5, E5+additive, E10, E10+additive, E15, and E15+additive is 7.48 Nm, 7,5Nm, 7.43 Nm, 7,6Nm, 7.45Nm, and 7.53Nm.

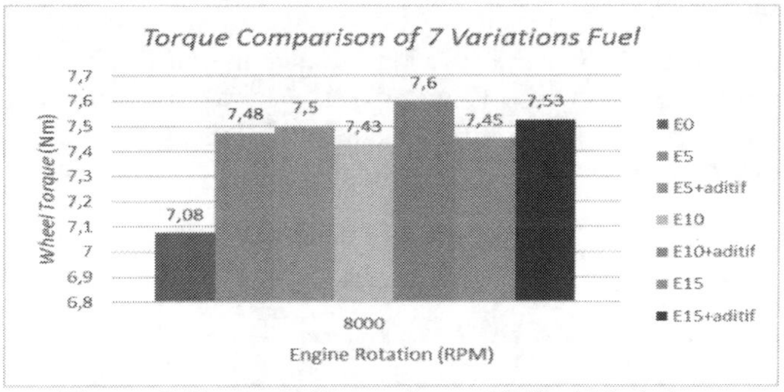

Figure 6. Torque's Comparison of 7 Fuel Variations

Figure 6 shows that fuel variations with additive additions result in greater torque compared to fuel variations with the same amount of ethanol without additives. The largest increase in torque occurs at about 5.400 rpm with a variation of E10and E10 +additives. With the addition of an additive, the torque increased by 2.4%. This is because the AFR at 5.400 rpm round approaches stoichiometry, with the additives the AFR becomes rich and it would decrease engine rotation. The engine rotation affected to torque, low engine rotation would be increasing torque.

Mustafa koc et al [11] tested the performance and emission on the engine with gasoline-ethanol fuel mixture (E50 and E85), maximum torque was obtained on E85 mixture variation at 3500 rpm with a value of 33 Nm higher than E50 and E0. Ethanol mixture in the fuel increased the oxygen content so that the combustion had completed and the torque increased. While based on the test results, the variation of fuel (E5, E10, and E15) with the addition of oxygenated additive cycloheptanol could increase the generated maximum torque of additives on a fuel mixture

5. Conclusions

This research aims to maximize the performance that is obtained in the engine by adding additives to the fuel mixture of gasoline and bioethanol. Through this paper, we demonstrated the results, namely the method of mixing fuel that modified tank and fuel mixer tool rated effective. The mixture of gasoline with bioethanol and oxygenated additives cycloheptanol approached homogenously. Due to the fuel mixture was stirred during the testing process, so the fuel entered the combustion process which had not occurred precipitation. The power value generated on the variation of the fuel mixture is higher than gasoline (E0). The addition of oxygenated additives cycloheptanol is the most optimal variation to E5 fuel variation that generated power equal to 6.78HP, increase equal to 1.3% of the E5 mixture without additives. The torque resulted on the fuel mixture of E5, E10, E15, E5+additive and E15 +additive is greater than gasoline (E0). The maximum torque is obtained at 5.400 rpm engine. The addition of additives enhances the value of the torque n each variation of the fuel (E5, E10, and E15) equal to 0.33%, 2.4%, and 1,1%.

6. Acknowledgement

The authors would like to thank DRPM UI for providing the grant in this research so that it can be completed well.

7. References

[1] Abikusna S, B Sugiarto and A Zulfan 2017 Fuel consumption and emission on fuel mixer low-grade bioethanol fuelled motorcycle. in MATEC Web of Conferences *EDP Sciences*

[2] Keith O and Trevor C 1995 Automotive fuels reference book Society of Automotive Engineers *Inc. Warrendale, PA*: p. 487

[3] Liu H et al. 2015 Dual-Fuel Spark Ignition (DFSI) combustion fuelled with different alcohols and gasoline for fuel efficiency *Fuel* **157**: p. 255-260

[4] Yücesu H S et al. 2007 Comparative study of mathematical and experimental analysis of spark ignition engine performance used ethanol–gasoline blend fuel *Appl. Thermal Engineering,* **27(2):** p. 358-368

[5] Yao Y C, Tsai J H, and I.-T. Wang 2013 Emissions of gaseous pollutant from motorcycle powered by ethanol–gasoline blend *Appl. energy* **102**: p. 93-100

[6] Costa R C and J.R. Sodré 2011 Compression ratio effects on an ethanol/gasoline fuelled engine performance *Applied Thermal Engineering,* **31(2)**: p. 278-283

[7] Gholamhassan Najafi, B.G., Talal Yusaf, Seyed Mohammad Safieddin and R.M. Ardebili 2015 Optimization of performance and exhaust emission parameters of a SI (spark ignition) engine with gasoline-ethanol blended fuels using response surface methodology *Elsevier,* **energy xxx**: p. 15

[8] de Melo, T.C.C., et al. 2012 Hydrous ethanol–gasoline blends–Combustion and emission investigations on a Flex-Fuel engine *Fuel* **97**: p. 796-804

[9] Abikusna, S., et al. 2017 Low grade bioethanol for fuel mixing on gasoline engine using distillation process *AIP Conf. Proc.* AIP Publishing

[10] Srinivasan, C.A. and C. Saravanan 2010 Study of combustion characteristics of an SI engine fuelled with ethanol and oxygenated fuel additives *J. of Sustainable Energy & Environment,* **1(2)**: p. 85-91

[11] Koç, M., et al. 2009 The effects of ethanol–unleaded gasoline blends on engine performance and exhaust emissions in a spark-ignition engine *Renewable energy* **34(10)**: p. 2101-2106

IOP Conf. Series: Earth and Environmental Science **105** (2017) 012058 doi:10.1088/1755-1315/105/1/012058

Analysis of emission gas and fuel consumption on SI engine fueled with low-grade bioethanol and oxygenated cycloheptanol additive

Ratna Monasari[1], Setia Abikusna[2], Bambang Sugiarto[1] and Briano Ajiseno[1]
[1]Departement of Mechanical Engineering, University of Indonesia, 16424, Indonesia
[2]Automotive Engineering, Astra Manufacture Polytechnics, Sunter Jakarta 14330, Indonesia
E-mail: ratnamona92@gmail.com

Abstract. The use of alternative energy source is expected to reduce fossil energy consumption and exhaust emissions. One of the non-fossil energy alternatives introduced for vehicles is bioethanol, it can produce in a simple way and cheap. Low-grade bioethanol has characteristics that depend on exhaust emissions generated. Low-grade bioethanol can be used as a substitute or mixture of fuel. The mixture of bioethanol and gasoline gives an effect to the increase of octane number and reduces the emission of CO2 produces. To get a more optimum effect on fuel consumption and emission, the oxygenated additive can be added to the fuel mixture. In this study, the effect of fuel consumption and emission was carried out by comparison of E5, E10, E15 without additive and with additive oxygenated cycloheptanol. The test will be performed by the method of calculating the amount of fuel consumption against the time to get specific fuel consumption. Emission test will be performed in single cylinder spark-ignition (SI) engine 150cc premix fuel with 100% opening throttle position connected to the gas analyzer in variation speed engine to see the emission gas. This study aims to obtain reduce the emission gas (CO2 and CO) and fuel consumption with the addition of the oxygenated additive.

1. Introduction

The limited resources of fossil energy and the increasing need for fuel causes the necessity to finding another renewable energy source [1]. In addition, the increased fuel consumption that occurs today will cause the amount of exhaust emission generated increased and this will have an adverse impact on the environmental. Because of that problem, the use of regenerative fuels has been developed lately [2, 3]. By utilizing alternative energy that is environmentally friendly and widely available in nature is expected to be a solution that addresses current problems. One of the alternative energy developed today is biofuel. Biofuel can be used as a substitute for fuel or mixture on fuel [4]. Biofuel consists of several types including bioethanol, biodiesel, biobutanol, bio-methanol, biogas, biohydrogen, and vegetable oil. Currently, bioethanol and biodiesel are being developed to be used as fuel for the future.

Bioethanol is one of the renewable fuel sources derived from biomass product. Bioethanol is widely produced from agricultural residue (such as bagasse, rice straw, corn straw, and wheat straw), industrial waste starch processing and lignocellulosic biomass (such as agricultural residues, woody biomass, algae, and industrial solid waste). Indonesia is a country that has abundant raw materials in the manufacture of bioethanol. Looking for the raw material is widely available, bioethanol can be produced on a large scale [5, 6].

Bioethanol is claimed as an environmentally friendly alternative fuel because the emissions produced are clean from the pollutants. Bioethanol/ethanol has the potential to substitute gasoline as the main fuel in the internal combustion engine. The Environmental Protection Agency (EPA) explains that NOx is one of the most harmful pollutants in the respiratory system. The use of ethanol as a fuel can reduce emission from NOx [6]. Ethanol (C2H5OH) is colourless, transparent, neutral, volatile, flammable, oxygenated liquid hydrocarbon, and has a sharp odor [7]. Ethanol is generally more reactive than hydrocarbon fuel. This is because of the content of hydroxyl radical as a polar fraction and carbon chain as a nonpolar fraction, so easily dissolved on nonpolar substances such as gasoline and also on polar substances like water [8].

The use of high-grade ethanol as fuel needs to be modified on an SI machine, whereas low-grade ethanol can be directly used without modifying the machine [9]. The advantages of ethanol compared to gasoline is high octane number, higher latent heat of ethanol that increases volumetric efficiency, contain more oxygen in the combustion process, lower vapor pressure can reduce the evaporative emission, high laminar flame propagation speed which makes combustion process to be finished earlier, ethanol increase thermal efficiency, allows use of high compression ratio without knocking and produce clear emission [10].

The efforts to reduce exhaust emission continue to be developed, one of the efforts is engineered fuel by adding additives to a mixture of the fuel. Methyl tert butyl ether (MTBF) is one of the most popular additives used as a mixture of fuel to increase octane number and reduce carbon monoxide and ozone [11]. However, MTBF has an adverse impact on the environment, there are several complaints related to the impact of MTBF use, including irritation of the eyes and lungs [12].

C. Ananda Srinivasan [13] studied the effect of a mixture of fuel gasoline-ethanol mixed with an oxygenated additive on a spark ignition machine. The experimental results prove that CO2, CO, and NOx emission decrease, while HC and O2 increase.

This study was conducted on SI machine by using another alternative in the form of oxygenated cycloheptanol as an additive used in the gasoline bioethanol fuel mixture. Comparison results emission gas and fuel consumption between additive fuel and without additive fuel can be seen in this paper. Cycloheptanol properties can be found in table 1.

2. Research Methodology

The test was performed on an SI engine 150 cc single cylinder with a fuel injection system using the injector. There are 7 fuel variables with the ratio of gasoline and bioethanol E0, E5, E10, E15 without additive and E5, E10, E15 with add 0.5 ml oxygenated cycloheptanol in every mixture.

2.1. Valve Opening Control

Gasoline engine with fuel injection generates the air-fuel mixture directly in the combustion chamber. Air flow through the intake valve in the open position and the fuel is injected directly into the combustion chamber using high-pressure injector controlled by ECU. Engine achieve the best performance with the right air-fuel ratio. Gasoline has AFR 14.7:1, the leaner mixture tends to degrade performance, while richer the fuel is waste and emission will be increased without substantially increasing performance. The lean setting in carburettors create some problems, the engine in the pre-emission law days was normally set up on the rich side. This made engine quick warm up, easy to start and provided good performance.

Electronic fuel injection systems consist of various sensors to know the exact engine condition. Sensors give an information about engine condition to the computer. After comparing that information to set of known parameters, the computer determines exactly the amount of fuel is required to maximize power while creating the lowest emission, then adjust air-fuel ratio for the combustion process. Electronic fuel injection allows the engine to receive the ideal mixture, neither so lean that is creating problem nor so rich that it exceeds emission standards.

Table 1. Properties of Cycloheptanol

Chemical Name	Cycloheptanol; 502-41-0; Suberol
Molecular Formula	C7H14O
Molecular Weight	114.118 g/mol
Appearance	Clear; colorless liquid
Melting Point	2°C
Boiling Point	185°C
Flash Point	60°C
Density	0.9554 g/cm^3 (20°C)
Heat of Vaporization	48.7 kJ/mol
Heat of Combustion	-4441 kJ/mol
Vapor Pressure	0.24 (25°C)

2.2. Experimental Set Up

Figure 1 is set-up experimental in 150 cc SI engine that was supported to other components. Testing was done on the SI engine single cylinder that was mounted on a dynamometer chassis type MCD400L series by mainline dyno log and connected techno motor gas analyser to exhaust manifold. Dynamometer Calibration process was done by adjusting the rotation of the wheels of the vehicle with the round roller drum to obtain corresponding values between rpm engine and the results of the readings on a dynamometer. The test performs in the various rotation (4.000rpm, 6.000 rpm, and 8.000 rpm) for each variations fuel. The results of the emission gas were issued in the gas analyser monitor. The fuel consumption results issued in the graphics.

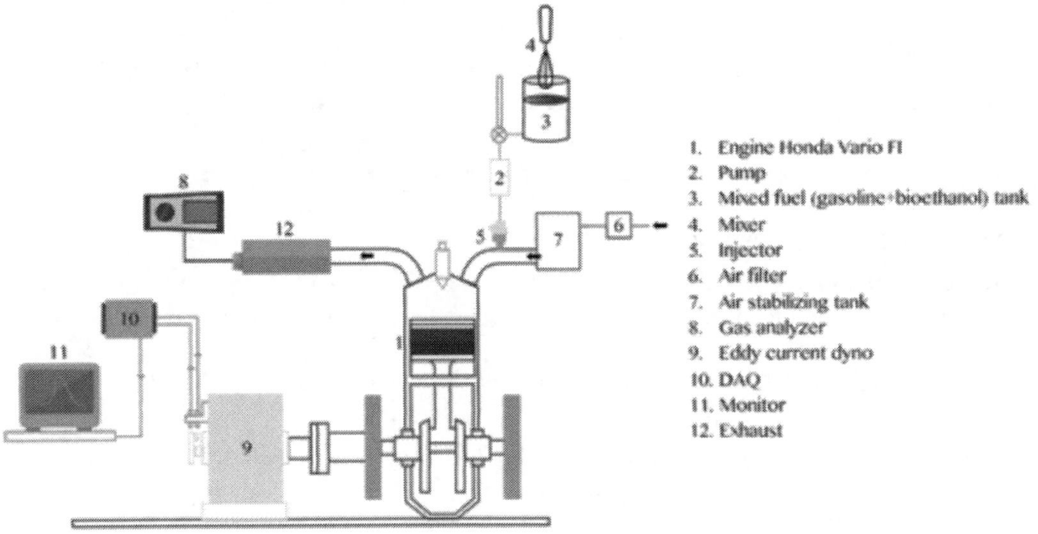

Figure 1. Experimental Set up Engine.

2.3. Exhaust Gas Emission

Techno gas analyser type Gas-810 was used in this study. The first test conducted in the collect exhaust emission data (CO, CO2, and O2). The test starts by connecting the hose from the gas analyser to exhaust manifold and were carried out from E0 to E15 +additives at each rotation of 4.000 rpm, 6.000 rpm, and 8.000 rpm. The results will be issued by the monitor in the gas analyser.

2.4. Fuel Consumption

The fuel consumption test performed by using the rate of fuel consumption unity of the time. Test conducted at 4.000 rpm, 6.000 rpm, and 8.000 rpm with 7 variations of fuel. Here is the procedure for this test:

1. Preparing each fuel variations. Put the fuel into the measuring tank that connected by hose into injectors.
2. Start the engine. Data retrieval starts at 4.000 rpm, the time was counting when the rotation hold in 4.000 rpm and 25ml of fuel used in constant rotation. The time is taken to consume 25ml of fuel will be calculated to get the specific fuel consumption.
3. The test performed on each variation of fuel at the variable rotation.

Specific fuel consumption is a parameter to determine the amount of fuel required to produce a certain amount of power in a certain time interval. The SFC is called the Brake Specific Fuel Consumption (BSFC) if it uses Power brake horse, and if using indicated Power then it is called Indicated Specific Fuel Consumption (ISFC).

3. Result and Discussion

The test was performed on variations of E0, E5, E15, and variations using oxygenated cycloheptanol additives. These tests include exhaust emissions in the form of CO, CO2, and O2. Here is the result of emission test results and fuel consumption of each fuel mixture variation.

3.1. The Result of Carbon Dioxide (CO_2)

The tests were performed at the variations engine rotate. (4.000 rpm, 6.000 rpm, and 8.000 rpm). Here is the graph of carbon dioxide gas emission test results.

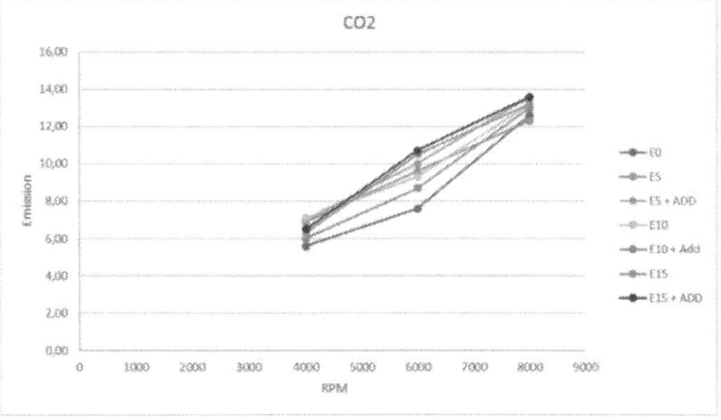

Figure 2. CO_2 vs Engine Speed.

As the engine rotation increases, the CO2 emission increase. This indicates that complete combustion in the combustion chamber. Based on the graph above shows that the addition of bioethanol can increase the CO2 level. In the variation of E5, E10, and E15 obtained the average of each rotation with a percentage respectively amounting to 9.6%, 9.87%, and 10%. For the percentage of gasoline (E0) at each rotation equal to 8.6%. An increase of 11.6% by E5, 14.7% by E10, and 16.2% by E15. As for the variations with additives yielded a percentage of CO2 with an average of 10.2% for E5 +additives, 9.23% for E10 +additives, and 10.27% for E15 +additives variations.

In general, the addition of oxygenated cycloheptanol to fuel variations result in the combustion cycle. The higher percentage of CO2 in combustion process indicates complete combustion in the combustion chamber. This happens because the oxygen molecules (-OH) in bioethanol and the additive reacts with CO and produce more CO2.

3.2. The Result of Carbon Monoxide (CO)

The results of the CO exhaust test in each fuel are averaged to represent the percentage of emissions in each rotation. From the test results, see that mixing of bioethanol and additives in the fuel results lower CO gas emission than E0. In E0 variation obtains the average percentage of each engine speed of 0.74%, it is higher than the average percentage obtained from the variation of E5 – E15 fuel mixture without additive and with an additive. Bioethanol can reduce CO emissions by up to 24%, while with the addition of CO additive has been reduced to 63% on E15 +additives of the CO produced by E0.

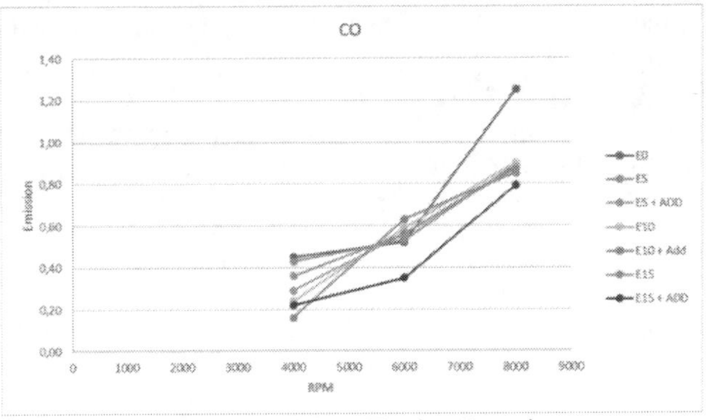

Figure 3. CO vs Engine Speed.

Figure 3 shows the effect of CO on fuel variations without additives and with additives. The addition of the oxygenated cycloheptanol additive to the fuel variations can reduce CO emissions by 0.14% in E15. CO comes from incomplete combustion products because of insufficient oxygen in the air-fuel mixture in the combustion cycle. The air-fuel ratio for ethanol is lower, in the stoichiometric conditions about 8.95:1, while for gasoline is 14.7:1. Therefore, the addition of ethanol and oxygenates into gasoline leads effective leaning on the mixture. The higher bioethanol mixture on the fuel can reduce CO, and by adding the oxygenated cycloheptanol it produced lower CO. This because of the -OH molecules are in bioethanol and additives produces complete combustion. Mixed fuels that contain lots of oxygen will help the process oxidation of CO to CO2. The percentage of CO emissions also depends on the operating conditions of the engine.

3.3. The Result of Oxygen (O_2)

Based on Figure 5, the percentage of oxygen generated from the mixed variation from E0 to E15 +additive while the O2 produced decreased by engine speed. In each engine speed obtained an average percentage of 0.18% for E0, 6.5% for E5, 6.13% for E10, and 6.02% for E15. On a variation fuel with additives yield a percentage of 7.67% for E5 + additives, 7.02% for E10 +additives, and 8.65% for E15 +additives.

The use of fuel variations results is higher O2 although is not too significant. The percentage of oxygen in the emission gas is inversely proportional to the CO2 concentration. Figure 4 can be seen the higher engine speed will reduce the O2 contain in emission gas. At the same time, the fuel variation with additives increases the O2 contain.

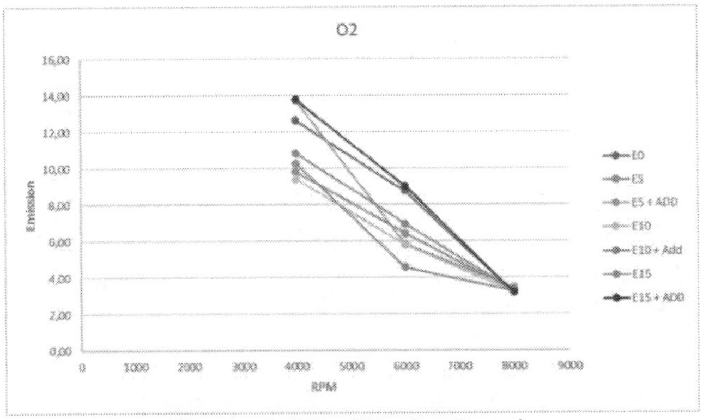

Figure 4. O2 vs Engine Speed.

3.4 The Result of Specific Fuel Consumption

Fuel consumption test was done by observing the amount of fuel consumption at a certain engine speed so that the data of fuel into combustion chamber per unit time was obtained. Figure 5 shows that fuels with bioethanol (E5, E10, E15) produce greater level consumption than gasoline (E0). This explains that the use of bioethanol in the fuel mixture impact on the fuel consumption. The large consumption fuel of bioethanol is due to the AFR stoichiometric of bioethanol is lower than pure gasoline with less air flow into the combustion chamber so that the mass of bioethanol fuel into the combustion chamber is more. Fuel mixture has a large flame speed that allows the fuel burning fast so that less burnt fuel burns and a complete combustion [14]. In the graph, can also be seen that the higher engine rotation the larger fuel consumption. This due to the shortening of combustion time resulting in higher engine rotation requiring greater fuel consumption.

Subsequent test results were carried out by adding an oxygenated cycloheptanol additive to the fuel variations E5, E10, and E15. The addition of additives results in lower fuel consumption, which means fuel consumption saving occurs in every variation of fuel mixed with additives. From the graph, can be seen the least fuel consumption occurs in the variation E5. At 4.000 rpm rotation fuel consumption produces 1L/H for E5, 0.9L/H for E5 +additives. While at 8000rpm rotation produces 2L/H for E5 and 1,3L/H for E5 additive.

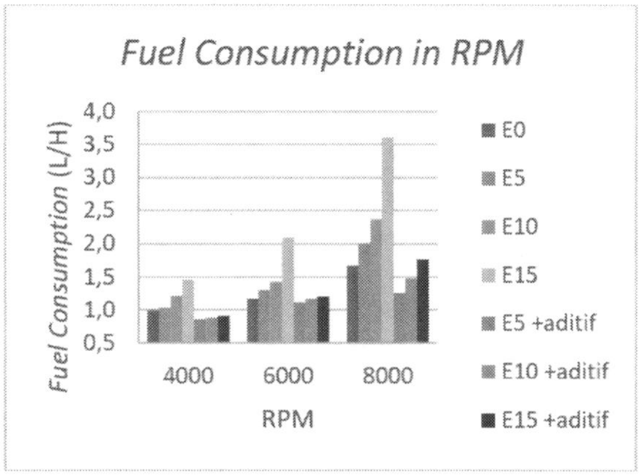

Figure 5. Fuel Consumption in RPM.

The specific fuel consumption (SFC) is defined as the ratio of the fuel mass flow rate to the generated power (output), SFC state how efficiently the fuel is supplied to the engine for output power. A low SFC value indicates an efficient fuel consumption, therefore a low SFC is desirable to achieve fuel efficiency.

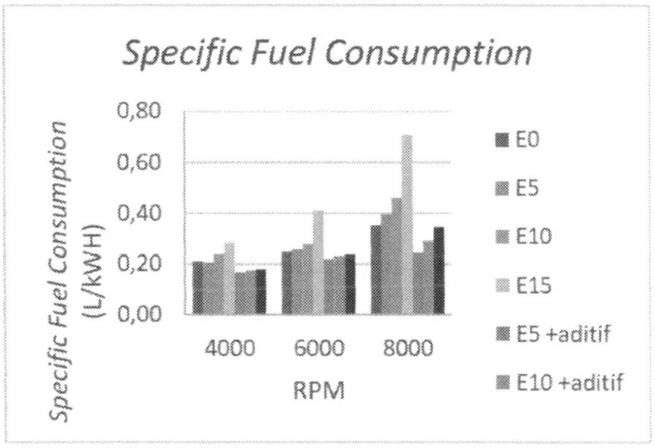

Figure 6. Specific Fuel Consumption.

Figure 6 shows that the smallest SFC is at the E0 variation on each engine speed. The gasoline bioethanol mixture yields a higher SFC value compared to E0. While for variations of fuel with additives produce smaller SFC values. The higher volume of ethanol causes the higher of fuel consumption. This is because the calorific value contained in ethanol (26,000 kJ/kg) is almost half times the gasoline (42,000 kJ/kg) contained because it requires more fuel to reach the equivalent amount of energy.

Fuel consumption is a representation of SFC, where at high engine speed the value of fuel consumption increases due to increased friction force. SFC is calculated by measuring fuel consumption. The addition of bioethanol and additives oxygenated cycloheptanol causes an increase in SFC. This influenced by heating value contained, the lower fuel energy content causing the increase of SFC even without any modification.

4. Conclusions

Based on the results of emission test and fuel consumption of bioethanol hydrate mixing with gasoline through fuel mixer mechanism, and the ratio of emission result and fuel consumption of bioethanol with gasoline, the following conclusions can be drawn as CO2 gas emissions are expected in the combustion cycle. The higher CO2 levels in combustion process indicate complete combustion in combustion chamber. This happens because the oxygen molecule (-OH) in the fuel reacts with CO and produce CO2. The mixture of gasoline and bioethanol can reduce CO emissions greater. The addition of the oxygenated cycloheptanol additive can reduce CO emission better. This occurs because of bioethanol and oxygenated cycloheptanol results in more effective combustion and greater oxygen content. The addition of additives in a fuel variations increase O2 contain in emission gas due to oxygen content in fuel increases. The addition of bioethanol and additives oxygenated cycloheptanol causes an increase in SFC at every engine speed compared to gasoline (E0).

5. Acknowledgement

The authors would like to thank DRPM UI for providing the grant in this research so that it can be completed well.

6. References

[1] Abikusna S, Sugiarto B, and Suntoro D 2017 Low Grade Bioethanol for Fuel Mixing on Gasoline Engine Using Distillation Process in *AIP Conference Proceedings* AIP Publishing.

[2] Demirbas A 2007 Progress and Recent Trends in Biofuels *Progress in energy and combustion science* **33**(1): p. 1-18.

[3] Gao J D Jiang and Z Huang 2007 Spray Properties of Alternative Fuels: A Comparative Analysis of Ethanol–Gasoline Blends and Gasoline *Fuel* **86**(10): p. 1645-1650.

[4] Demirbas A 2011 Competitive Liquid Biofuels from Biomass *Applied Energy* **88**(1): p. 17-28.

[5] Thangavelu S K, A S Ahmed, and F N Ani 2016 Review on Bioethanol as Alternative Fuel for Spark Ignition Engines *Renewable and Sustainable Energy Reviews* **56**: p. 820-835.

[6] Sanchez O J and C A Cardona 2008 Trends in Biotechnological Production of Fuel Ethanol from Different Feedstocks *Bioresource technology* **99**(13): p. 5270-5295.

[7] Ganguly A, P Chatterjee, and A Dey 2012 Studies on Ethanol Production From Water Hyacinth—A Review *Renewable and Sustainable Energy Reviews* **16**(1): p. 966-972.

[8] Costa R C and Sodré J R 2010 Hydrous Ethanol vs. Gasoline-Ethanol Blend: Engine Performance and Emissions *Fuel* **89**(2): p. 287-293.

[9] Balat M and H Balat 2009 Recent Trends in Global Production and Utilization of Bio-Ethanol Fuel *Applied energy* **86**(11): p. 2273-2282.

[10] Masum B, Masjuki H, Kalam, and M Fattah I 2013 Effect of Ethanol–Gasoline Blend on NOx Emission in SI engine *Renewable and Sustainable Energy Reviews* **24**: p. 209-222.

[11] Poulopoulos S and C Philippopoulos 2000 Influence of MTBE Addition Into Gasoline on Automotive Exhaust Emissions *Atmospheric Environment* **34**(28): p. 4781-4786.

[12]. Bodenstein Y and L K Duffy 1998 Expression of Hsp60, a Stress Protein, in Human Nasal Septa Cells After Exposure to MTBE *Environmental toxicology and pharmacology* **5**(1): p. 79-83.

[13] Srinivasan C A and C Saravanan 2010 Study of Combustion Characteristics of an SI Engine Fuelled with Ethanol and Oxygenated Fuel Additives *J. Sustainable Energy & Environment* **1**(2): p. 85-91.

[14] Abikusna, S., B. Sugiarto, and A. Zulfan 2017 Fuel consumption and emission on fuel mixer low-grade bioethanol fuelled motorcycle in *MATEC Web of Conferences* EDP Sciences.

An eco-friendly preparation of cellulose nano crystals from oil palm empty fruit bunches

Arie Wibowo[1], Haroki Madani[2], Hermawan Judawisastra[1], Elvi Restiawaty[3], Chrisella Lazarus[2], Yogi W. Budhi[2]

[1] Material Engineering Study Program, Institut Teknologi Bandung, Jl. Ganesha 10, Bandung, Indonesia

[2] Chemical Engineering Study Program, Institut Teknologi Bandung, Jl. Ganesha 10, Bandung, Indonesia

[3] Bioenergy and Chemurgy Engineering, Institut Teknologi Bandung, Jl. Ganesha 10, Bandung, Indonesia

E-mail: Y.Wibisono@che.itb.ac.id

Abstract. Cellulose Nanocrystals (CNCs) have attracted a lot of attention as one of the promising green nanomaterials because of their interesting properties such as low density, high mechanical properties, high surface area, high biocompatibility and low toxicity. However, sulphuric acids is generally used in conventional method to prepare CNCs which is highly corrosive and harmful to environment. Replacing sulphuric acid with Ammonium Persulfate (APS), which possess lower toxicity, are desirable from environment point of view. In this research, preparation of CNCs from Oil Palm Empty Fruit Bunches (OPEFB) as alternative raw materials and using APS is reported. Three kinds of method were carried out before applying APS solution: (i) without pre-treatment, (ii) with alkali pre-treatment and (iii) with alkali-chloride pre-treatment. White-colored sample with crystallinity reaching 78% (XRD result) and particle size ranging between 31-113 nm (DLS result) was obtained. Those results implied that there were potential existence of CNCs in the samples.

1. Introduction

In the last decades, Cellulose Nanocrystals (CNCs) have attracted a lot of attention as one of the promising green nanomaterials because of their interesting properties such as low density, high mechanical properties, high surface area, high biocompatibility and low toxicity [1]. Those superior properties make CNCs useful for many applications from various fields such as reinforcement of nanocomposites, plasticizer and strength enhancement of cement, films for optical and barrier properties, catalyst support, etc. So, due to its advantages and broad potential of application, many researchers have been studying CNCs extraction from various lignocellulosic source such as wood, cotton, bamboo, hemp, flax, kenaf etc. [2,3,4,5,6,7]. Some CNCs research groups or institutions have reached either a commercial pilot scale (kg/day) or even real industrial scale (ton/day)[8].

Common process used to prepare CNCs uses acid hydrolysis, which results in non-eco-friendly acid waste. Only one of at least five commercial scales of CNCs production does not use the acid hydrolysis method as reported by Reid et al. [8]. Some researchers try to develop new processes, which have less environmental damage risk by reducing dangerous waste in its process. The use of Ammonium Persulfate (APS) as reported by Leung et al. (2011) could be an interesting option to meet better environmental impact [9]. APS is a chemical with high water solubility and low long-term toxicity, which results in free radical SO_4^{2-} and hydrogen peroxide (H_2O_2) in high temperature and acidic condition. Those free radical and hydrogen peroxide can be used to remove the lignin and cut the cellulose to produce CNCs. In their research, Leung et al. succeeded to extract CNCs from flax, hemp, triticale, MCC, wood pulp, and bacterial cellulose [9]. Other researchers have also adopted this method to extract CNCs from other sources as reported by Hu (2014) and Mascheroni (2016) who studied the extraction of CNCs from bamboo and cotton, respectively [11,12]. This method offers simple process, easy to scale-up, and eco-friendly that could also decrease the waste treatment cost.

In this study, the CNCs was synthesized from solid waste containing lignocellulosic component from Oil Palm Empty Fruit Bunch (OPEFB), which is abundant amount of solid waste in palm oil mill. The empty fruit bunch may contain cellulose up to 45%, which is very potential for CNC production. Indonesia is well-known as the second largest of palm plantation area in the world, which has 11 million hectares of palm plantation area producing 33 million ton of palm fruit bunch during 2016 [12]. OPEFB is the largest solid waste of palm oil production in palm oil mill (approximately 23%-wt of palm fruit bunch), which has not been utilized well [13]. Some of researchers have used OPEFB as raw material for CNCs preparation such as Dasan (2015) and Fahma (2010) [14,15]. Herein, the feasibility study of the utilization of ammonium persulfate for producing CNCs from OPEFB is studied. In the present work, the synthesis of CNC from OPEFB as raw material for high-value product through an eco-friendly process is investigated.

2. Materials and Methods

2.1. Materials

Oil Palm Empty Fruit Bunches (OPEFB) used in this research was obtained from PT. Perkebunan Nusantara VIII, Cikasungka, Bogor, West Java, Indonesia. The dried empty fruit bunch was chopped in the form of short fiber. All chemicals used for one-step extraction of CNCs were purchased from Merck KgaA, EMD Millipore Co. with pro analysis grade. The demineralized water was used in all experiments.

2.2. Main Experiment: CNCs Preparation

Three kinds of pre-reacted materials were first prepared consisting of OPEFB without any treatments, OPEFB with alkali pre-treatment, and OPEFB with alkali-chloride pre-treatment. The experiment was carried out by reacting 3 grams of pre-reacted material with 1-2 molar of Ammonium Persulfate (APS) in a refluxed round flask at 70°C for 15 h [16]. The mixture of reaction products obtained after 15 h operation was filtered using vacuum filtration to remove remained APS in the mixture. Remained solid state sample was washed several times until its pH reached 7.

2.3. Alkali Pre-treatment

Three grams of OPEFB was added into 30 mL sodium hydroxide solution (4-wt%) and then vigorously agitated for 2 h at 90°C to purify the cellulose by removing hemicellulose and lignin. The solid was then filtered and washed several times using demineralized water until reached pH of 7.

2.4. Alkali-Chloride Pre-treatment

The alkali-chloride pre-treatment was performed by adding a buffer solution of aqueous chlorite (1.7-wt%), acetic acid, and demineralized water into erlenmeyer. The solution was heated at 90-100°C for 4 h under mechanical stirring. The resulting solution was cooled, filtered, and washed until pH neutral.

2.5. Composition Analysis

Composition analysis was carried out based on SNI 0492:2008, SNI 14-1304-1989, and SNI 0444:2009 for determining the contents of lignin, hemicellulose, and α-cellulose, respectively.

2.6. Crystallinity Measurement

X-Ray Diffraction method was used to determine the Crystallinity Index (CrI) of the sample. X-Ray Diffractometer Bruker D8 Discover with CuKα Radiation (wavelength = 1,54 Å) was used with 2θ range of 5^0-80^0 with 0.010 step size. The Crystallinity Index was determined using empirical method with following equation:

$$CrI = \frac{I_{002} - I_{am}}{I_{002}} \times 100 \qquad (1)$$

where I_{002} and I_{am} are the peak intensities of crystalline and amorphous materials, respectively [16].

2.7. Particle Size and Morphological Analysis
Prediction of particle size and its distribution was carried out by Dynamic Light Scattering (DLS).

3. Results and Discussion

3.1. Visual Observation
It was visually observed that the pre-treatment of OPEFB could reduce the lignin content, which was indicated by discoloration. Figure 1 shows discoloration of OPEFB after typical pre-treatment. The brighter pre-reacted materials, the lower its lignin content. Refer to Figure 1c, OPEFB using alkali-chloride pre-treatment showed that the lignin content could be removed significantly.

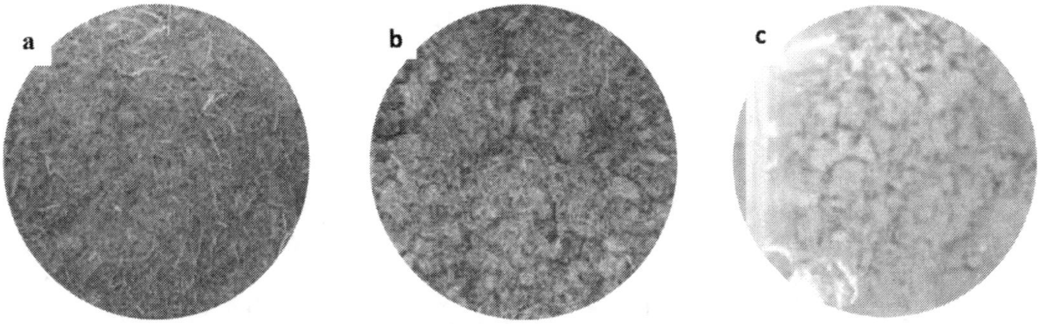

Figure 1. Three kinds of pre-reacted materials: (a) OPEFB without any pre-treatment, (b) OPEFB after alkali pre-treatment, and (c) OPEFB after alkali-chloride pre-treatment.

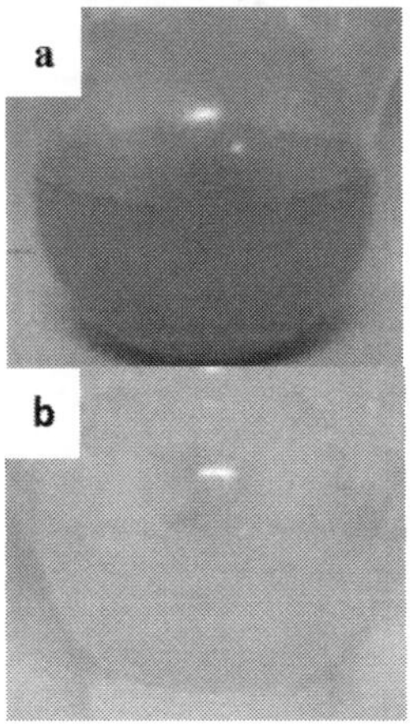

Figure 2. Mixture after reaction with 1 M of APS (a) and 2 M of APS (b) **without any pre-treatment** at 70°C for 15 hours.

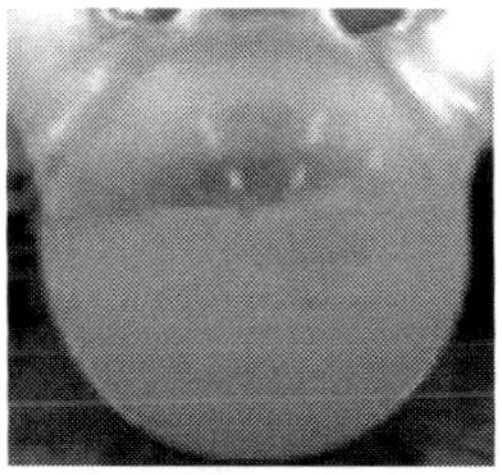

Figure 3. Mixture after reaction with 1 M of APS with **alkali pre-treatment** at 70°C for 15 hours.

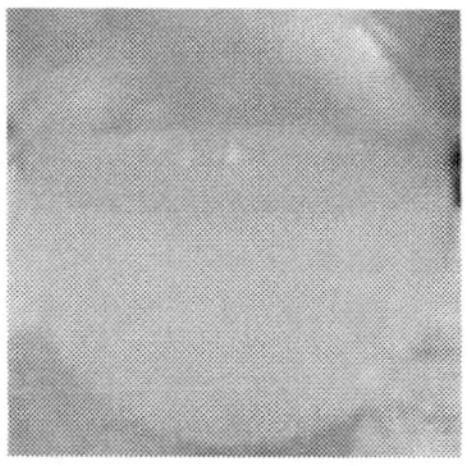

Figure 4. Mixture after reaction with 1 M of APS with **alkali-chloride pre-treatment** at 70°C for 15 hours.

Based on the experiment, dry OPEFB without any pre-treatment has lignin, α-cellulose and hemicellulose content of 17.45%, 36.68%, and 26.26%, respectively. The lignin content of dry OPEFB is high enough compared to lignocellulosic materials used by Leung et al. (2011) [9], who used flax, hemp, triticale, MCC, wood pulp, and bacterial cellulose as raw material for CNCs synthesis and succeeded to yield CNCs from those materials without any pre-treatment through oxidation by 1 M of APS.

In line with Leung et al. (2011) that the oxidation by using APS can produce CNCs from lignocellulosic materials with lignin content up to 20%. Although OPEFB has high lignin content, the experimental results show the existence of CNCs in the product mixture, which was indicated by white-colored mixture as increasing APS concentration and decreasing lignin content by pre-treatment. Figures 2, 3, and 4 shows the mixtures after reaction with APS.

In this work, it was also indicated that reaction between OPEFB without pre-treatment with 1 M of APS did not represent the white-colored mixture (Figure 2a), which was probably induced by high lignin content in the OPEFB. In addition, the extractives content existing in the OPEFB may reduce penetrability of reaction agent [17]. Unlike flax, hemp, triticale, MCC, bacterial cellulose, and water hyacinth, which have small number of extractives, OPEFB has higher extractives content. Therefore, different composition of lignin and extractives content may affect the reaction condition.

Empirically, the white-colored mixture was obtained after increasing APS concentration to 2 M (Figure 2b). Increasing the APS concentration was based due to fact that the OPEFB lignin content is more than twice than other lignocellulosic materials used by Leung et al. (2011). However, increasing APS concentration to 2 M might increase the chemical cost as twice. This implication must be considered for future scale-up processes.

Another alternative to produce CNCs from OPEFB through oxidation by APS was trying by reducing the lignin content, which could be achieved by conducting the pre-treatment for OPEFB. In this research, two kinds of pre-treatment were conducted, consisting of alkali pre-treatment and alkali-chloride pre-treatment. Alkali pre-treatment could loosen the lignocellulosic structure, break lignin-carbohydrate bonds, and increase surface area to ease the penetration of oxidizing agent [18]. When alkali pre-treated OPEFB was used as raw material, the white-colored mixture was obtained which indicate most of lignin have been removed. It may be caused by easier penetration of oxidizing agent due to pre-treatment process. While further delignification by chlorine results sample with least lignin which yield the most promising product when reacted with APS. Thus, decreasing lignin content could give more promising mixture to obtain CNCs.

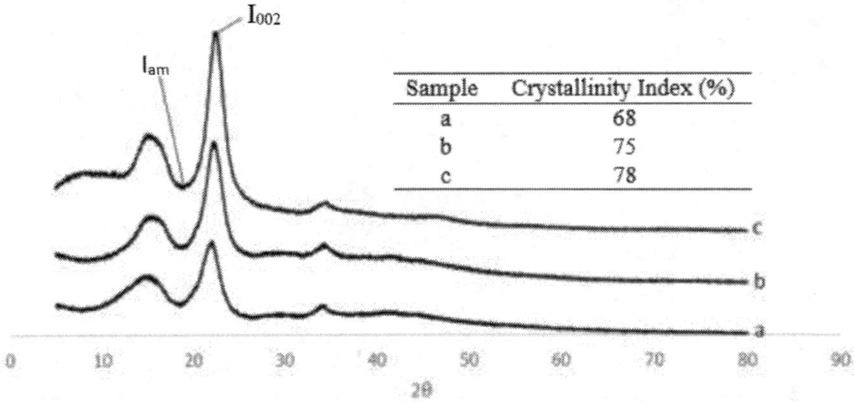

Figure 5. XRD curve of obtained samples following raw materials: (a) OPEFB without any pre-treatment (2M APS), (b) OPEFB after alkali pre-treatment (1M APS), and (c) OPEFB after alkali-chloride pre-treatment (1M APS).

3.2. Crystallinity Measurement
All obtained samples have Crystallinity Index higher than 54%, which is the lower limit of Cellulose Nanocrystals criterion. The XRD Curve and the Crystallinity Index for three samples are showed in Figure 5.

Decreasing lignin content by applying pre-treatment of OPEFB could give higher crystallinity index than increasing the APS concentration. Hence, decreasing the lignin content might be more preferable to yield higher quality of CNCs. In addition, conducting pre-treatment using sodium hydroxide was more cost effective than the oxidative by APS with double amount of concentration, which was more expensive. Thus, decreasing lignin content may be the best choice for next research of CNCs preparation from OPEFB through oxidation by APS by considering the quality of sample and the production cost.

The lignin content can be decreased by conducting alkali and alkali-chloride pre-treatment. The data shows that there was no significant difference of the crystallinity index between sample from alkali and alkali-chloride pre-treated OPEFB. Nevertheless, those two kinds of pre-treatment had different implication in the CNCs preparation process. The alkali pre-treatment was a one-step pre-treatment, which was more practical, time saving, and cost effective than alkali-chloride pre-treatment.

3.3. Particle Size Determination and Morphological Analysis
The result of Dynamic Light Scattering (DLS) characterization shows that obtained samples have particle size ranging between 31 to 113 nm. Table 1 shows particle size distribution and its average size.

Table 1. Particle Size Distribution and Its Average Size

No	Sample	Particle Size (D-90%) (nm)	Average Size (nm)
1	Sample from OPEFB without pre-treatment (2M of APS)	113	90
2	Sample from OPEFB after alkali pre-treatment (1M of APS)	101	85
3	Sample from OPEFB after alkali-chloride pre-treatment (1M of APS)	31	25

The particle size range of all samples was very close to the criterion of CNCs, which had length in the range of 100-250 nm and diameter of 5-70 nm [1]. This was a rough determination of particle size of samples, but those data represented very potential existence of CNCs in the samples.

4. Conclusions
The synthesis of CNCs from Oil Palm Empty Fruit Bunch (OPEFB) has been conducted by using Ammonium Persulfate (APS) as oxidation agent. The produced CNCs by using APS with crystallinity index higher than 54% can be achieved and the size distribution is also very close to criterion of CNCs. It was also found that the lignin content of pre-reacted materials has significant effect in CNCs preparation through oxidation with APS.

5. Acknowledgement
The financial support provided by Badan Pengelola Dana Perkebunan Kelapa Sawit (BPDPKS), Republic of Indonesia, through Research Grant Batch 2 (GRS K16) PRJ-38/DPKS/2016, is gratefully acknowledged.

6. References
[1] Brinchi L, Cotana F, Fortunati E, and Kenny J 2013 *Carbohyd. Polym.* **94** 154-169
[2] Beck-Candanedo S, Roman M, Gray DG 2005 *Biomacromolecules* **6** 1048–54
[3] Chang C P, Wang I C, Hung K J, Perng Y S 2010 *Taiwan J. Fos. Sci.* **25**, 251-264
[4] Yu M, Yang R, Huang L, Cao X, Yang F and Liu D 2012 *BioResources*, **7**

[5] Cao X, Chen Y, Chang P R, Stumborg M, and Huneault M A 2008 *J. Appl. Polym. Sci.* **109** 3804-3810
[6] Cao X, Dong H, and Li C M 2007 *Biomacromolecules*, **8** 899-904
[7] Zaini L H, Jonoobi M, Tahir P M and Karimi S 2013 *J. Biomater. Nanobiotechnol.* **4** 37-44
[8] Reid M, Villalobos M and Cranston, E 2016 *Langmuir* **33** 1583-1598
[9] Leung A C, Hrapovic S, Lam E, Liu Y, Male K B, Mahmoud K A and Luong J H 2010 *Small* **7** 302-305
[10] Hu Y, Tang L, Lu Q, Wang S, Chen X and Huang B 2014 *Cellulose* **21** 1611-1618
[11] Mascheroni E, Rampazzo R, Ortenzi M A, Piva G, Bonetti S and Piergiovanni L 2016 *Cellulose* **23** 779-793
[12] Direktorat Jendral Perkebunan 2015 *Statistik Perkebunan Sawit Indonesia 2014-2016 Kelapa Sawit*
[13] Darnoko Z, Poeloengan and Anas I 1992 *Buletin Penelitian Kelapa Sawit* **2**
[14] Fahma F, Iwamoto S, Hori N, Iwata T and Takemura A 2010 *Cellulose* **17** 977-985
[15] Dasan Y K, Bhat A H and Faiz A 2015 *AIP Conf. Proc.* **1669** 020058
[16] Majdanac L D, Poleti D, Teodorovic M J 1991 *Acta Polym.* **42** 351-357
[17] Rowe J W 2014 *Natural Products of Woody Plants Chemicals Extraneous to the Lignocellulosic Cell Wall* (Berlin : Springer)
[18] Zhao Y, Wang Y, Zhu J, Ragauskas A and Deng Y 2008 *Biotechnol. Bioeng.* **99** 1320-1328

Study of curcuma xanthorrhiza extract as green inhibitor for API 5L X42 steel in 1M HCl solution

Johny Wahyuadi Soedarsono[1], Muhammad Nafies Shihab[1], Muhammad Fikri Azmi[1] and Ahmad Maksum[1, 2]

[1]Center of Minerals Processing and Corrosion Research, Department of Metallurgical and Materials Engineering, Universitas Indonesia, Depok 16424, Indonesia
[2]Department of Mechanical Engineering, Politeknik Negeri Jakarta, Depok 16425, Indonesia

Email: jwsono@metal.ui.ac.id

Abstract. The inhibitory effects of *Curcuma xanthorrhiza* extract on the corrosion of API 5L X42 steel in 1M HCl solution was investigated by using weight loss, potentiodynamic polarization, and electrochemical impedance spectroscopy methods. In this study the concentration of *C. xanthorrhiza* extract used was 100, 250, 500, and 1000 ppm. The results show that *C. xanthorrhiza* inhibit the steel corrosion and acted as mixed type inhibitors. The corrosion rate decreased with the increasing of inhibitors concentrations. At the same time, inhibition efficiency increased with the increase of inhibitors concentrations. The adsorption of the extract on the steel surface was found to obey Langmuir's adsorption isotherm. The free energy value (ΔG_{ads}) indicated that the adsorption of the inhibitor molecules was typical of physisorption. It can be concluded that *C. xanthorrhiza* extract could be used as an alternative and environmental friendly inhibitor for API 5L X42 steel in acidic environment.

1. Introduction

Carbon steels are widely used in petroleum industry, chemical processing, marine applications, chemical processing, mining and metal processing equipment. The type of carbon steel commonly used on the oil and gas industry is API 5L. One reason the use of carbon steel as pipe is because the price that is considered relatively cheaper than the other type of steel. However, the low corrosion resistance of carbon steel will cause corrosion that can lead to damage with more expensive repairs cost [1]. The corrosion inhibitor is one of the best options for corrosion prevention in a closed-loop system. Corrosion inhibitors mostly come from organic and inorganic compounds containing groups which have free electron pairs such as nitrite, chromate, fospat, urea, phenylalanine, imidazoline, and amine compounds. Synthetic chemicals such as phosphate and chromate are not only expensive but also harmful to humans and the environment. Therefore, researchers have conducted research on eco-friendly corrosion inhibitors from plant extracts such as morinda citrifolia [2] and purple sweet potato extracts [3]. One of the plant extract in Indonesia that contains antioxidant substances is *"temulawak"* (*C. Xanthorrhiza*) [4]. So, in the present study, *C. xanthorrhiza* extract was investigated as green inhibitors of API 5L X42 steel corrosion in 1 M HCl using weight loss, potentiodynamic polarization, and electrochemical impedance spectroscopy measurements. The adsorption isotherm was also studied.

2. Materials and method

2.1. Preparation of Steel Specimen

In this research, API 5L X42 steel with a chemical composition (wt%) of 0.178% C, 0.270% Si, 0.519% Mn, 0.014% P, 0.005% Ni, 0.012% Cu and Fe balance was used. The steel specimens were abraded using 80 to 600 grades of abrasive papers. The specimens were well cleaned using distilled water and then again by acetone. The steel specimens with a specific size (30 x 20 x 5 mm) were used for the weight loss method and another specific size (10x10x6 mm) were used for the electrochemical studies.

2.2. Preparation of Test Solution

The test solution was 350 ml 1M HCl, which was prepared by the dilution of technical grade 32% m/v HCl with distilled water. Curcuma xanthorrhiza extracts were added to the acid solution at the varied concentrations 100, 250, 500, and 1000 ppm, and the solution in the absence of inhibitor was taken as blank for comparison purposes.

2.3. Weight Loss Measurement

In the weight loss measurements test, each sample was immersed in 350 ml 1M HCl containing inhibitors with the concentrations of 0 ppm (blank), 100 ppm, 250 ppm, 500 ppm, and 1000 ppm. The sample was immersed for 7 days. Then the corrosion rate was calculated using Eq. 1 [5]:

$$CR_{(mm/yr)} = \frac{87.6\ x\ w}{D\ x\ A\ x\ t} \quad (1)$$

Where, w is calculated from the initial weight minus the final weight (mg), A is the specimen area (cm^2), t is the immersion time (hour) and D is the density of steel in gcm^{-3}.

The Inhibition efficiency (*IE*) was calculated using Eq. 2 [5]:

$$IE\ (\%) = \frac{CR - CR'}{CR} x100 \quad (2)$$

Where, CR' and CR are the corrosion rate of steel with and without inhibitor.

2.4. Potensiodynamic Polarization Measurement

Autolab PGSTAT302N instrument equipped with Nova 1.10 software was used for the electrochemical measurements by recording Tafel polarization curve. All measurements were done using a 150 ml corrosion cell in which a reference electrode (Ag/AgCl electrode), a counter electrode (C electrode), a working electrode (API 5L X42). Potentiodynamic polarization testing was performed using a scanning rate of 0.01 Vs^{-1} and an open circuit potential (OCP) reading was performed for 120 seconds. The IE was calculated using Eq. 3 [5] .

$$IE\ (\%) = \frac{i_{corr} - i'_{corr}}{i_{corr}} x100 \quad (3)$$

Where, i_{corr} *and* i'_{corr} are current densities (A) of corrosion with and without the inhibitors, respectively.

2.5. Electrochemical Impedance Spectroscopy Measurements

The electrochemical experiments were performed using the PGSTAT302N instrument which was installed GPES and FRA software provided by Autolab. EIS testing was performed using an AC signal with the amplitude of 0.01 and with a frequency range from 0.1 kHz to 10 MHz. This test produces a Nyquist diagram that formed a semicircle graph. The inhibition efficiency (IE) was calculated from the electrochemical impedance diagrams, as described by Eq. 4 [5]:

$$IE\ (\%) = 100\left(1 - \frac{R_p}{R_p'}\right) \quad (4)$$

Where, R_p is the polarization resistance (Ω) with inhibitor, and R_p' is the polarization resistance (Ω) without inhibitor.

3. Result and Discussion

3.1. Weight Loss Measurement

The effect of *C. xanthorrhiza* extract addition on the corrosion of API 5L X42 in 1M HCl solution was studied by weight loss measurements after 7 days of immersion period. And the calculation results of inhibition efficiency and corrosion rate are summarized in Table 1.

Table 1. Weight Weight loss measurements of API 5L X42 steels soaked in 1M HCl solution for 7 days, with the addition of different C. Xanthorrhiza extract concentrations.

Inhibitor Concentration (ppm)	W_o (g)	W_i (g)	Weight Loss (g)	Corr Rate (mm/yr)	IE (%)
0	32.7397	31.2029	1.5368	5.86	0
100	33.0140	32.6527	0.3613	1.37	76
250	28.0764	27.8226	0.2538	0.97	83
500	31.8669	31.6500	0.2169	0.83	86
1000	31.4752	31.3373	0.1379	0.50	91

Table 1 shows that the corrosion rate values were decreased with the increasing of inhibitor concentrations. This indicates that increased concentration will increase the surface area of metal covered by C. xanthorrhiza extract, resulting in a decrease in corrosion rate of steel to reach its lowest value, ie 0.50 mm/yr. While the highest IE value reached 91% occurred when the concentration of C. xanthorrhiza extract is 1000 ppm. This phenomenon can occur due to steel surface was covered by the curcumin molecule of C. xanthorrhiza extract, via an adsorption mechanism. Therefore, the redox reactions between the steel surface with the acid will be inhibited and corrosion can be prevented. That is why, it can be said that xanthorrhiza extract can be used to protect the API 5L X42 steel in 1M HCl solution.

3.2. Potentiodynamic Polarization Measurement

Figure 1 show the Tafel polarization curve of API 5L X42 steel in 1 M HCl with different concentrations of *C. xanthorrhiza* extract. The extrapolation of anodic and cathodic lines were used to determine various electrochemical parameters such as anodic and cathodic Tafel slopes (*ba* and *bc*), potential (E_{corr}), and current density of corrosion (i_{corr}) as given in Table 2. Table 2 shows the i_{corr} value decreases with increasing *C. xanthorrhiza* extract concentrations. As shown by Figure 1, the inhibitors affects both the cathodic and anodic parts of the Tafel curves, and decreases both the cathodic and anodic current densities. Therefore, the presence of *C. xanthorrhiza* extract inhibits the hydrogen evolution and the anodic dissolution processes [5]. This result indicates that the *C. xanthorrhiza* extract act as mixed type corrosion inhibitors. In addition, the maximum displacement in E_{corr} was 32.59 mV. This results show that *C. xanthorrhiza* extract act as mixed type inhibitors since the displacement of $E_{corr} < 85$ mV.

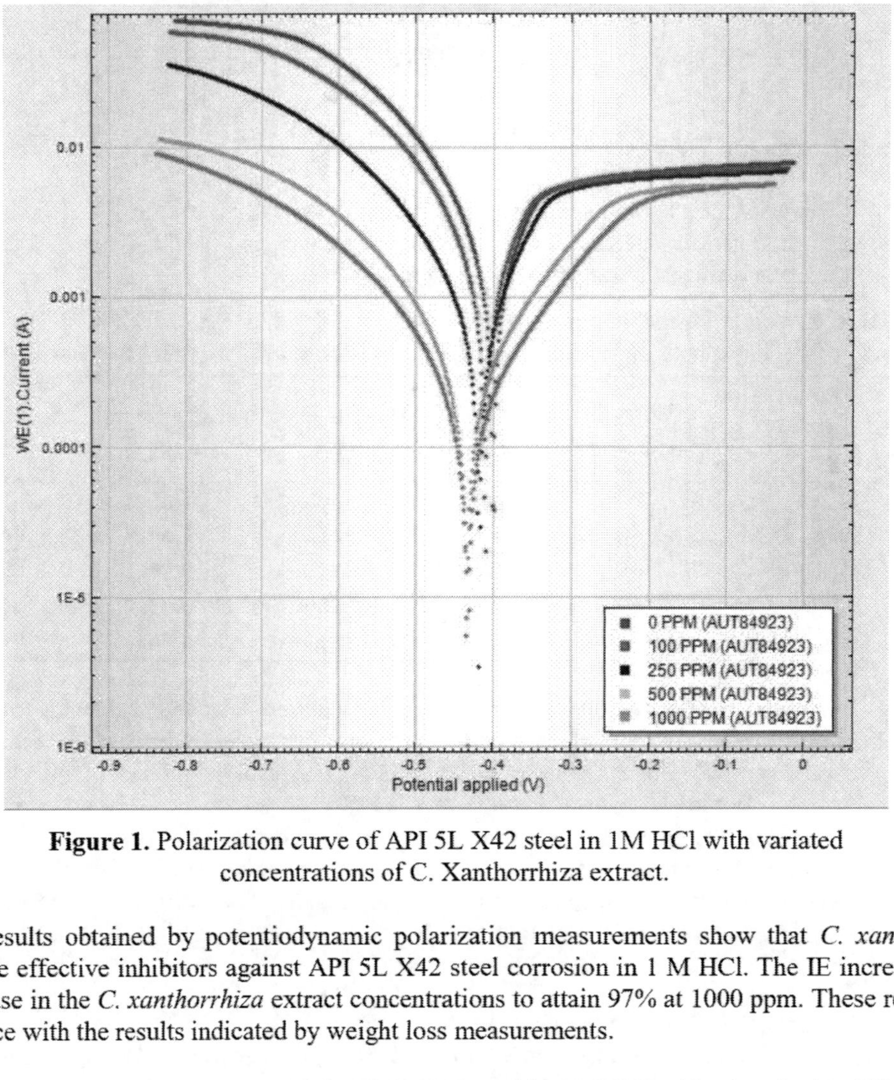

Figure 1. Polarization curve of API 5L X42 steel in 1M HCl with variated concentrations of C. Xanthorrhiza extract.

The results obtained by potentiodynamic polarization measurements show that *C. xanthorrhiza* extract are effective inhibitors against API 5L X42 steel corrosion in 1 M HCl. The IE increases with the increase in the *C. xanthorrhiza* extract concentrations to attain 97% at 1000 ppm. These result is in accordance with the results indicated by weight loss measurements.

Table 2. Polarization parameters of API 5L X42 steel in 1 M HCl without and with different concentrations of C. xanthorrhiza extract.

Inhibitor Concentration (ppm)	***ba* (mV/dec)**	***bc* (mV/dec)**	***Ecorr* (mV)**	***icorr* (μA)**	***Corr Rate* (mm/yr)**	***IE* (%)**
0	35.077	42.802	-401.44	476.23	5.53	0
100	6.2786	14.175	-411.96	106.82	1.24	78
250	12.644	14.949	-418.81	66.795	0.78	86
500	14.932	13.947	-434.03	24.125	0.28	95
1000	12.55	9.8129	-432.91	12.69	0.15	97

3.3. Electrochemical Impedance Spectroscopy Measurement

The Nyquist plot of API 5L X42 steel in 1 M HCl solution with different concentrations of *C. xanthorrhiza* extract are shown in Figure 2. Various parameters obtained from impedance measurements as R_p, R_s, *CPE*, *N* and *IE* are given in Table 3.

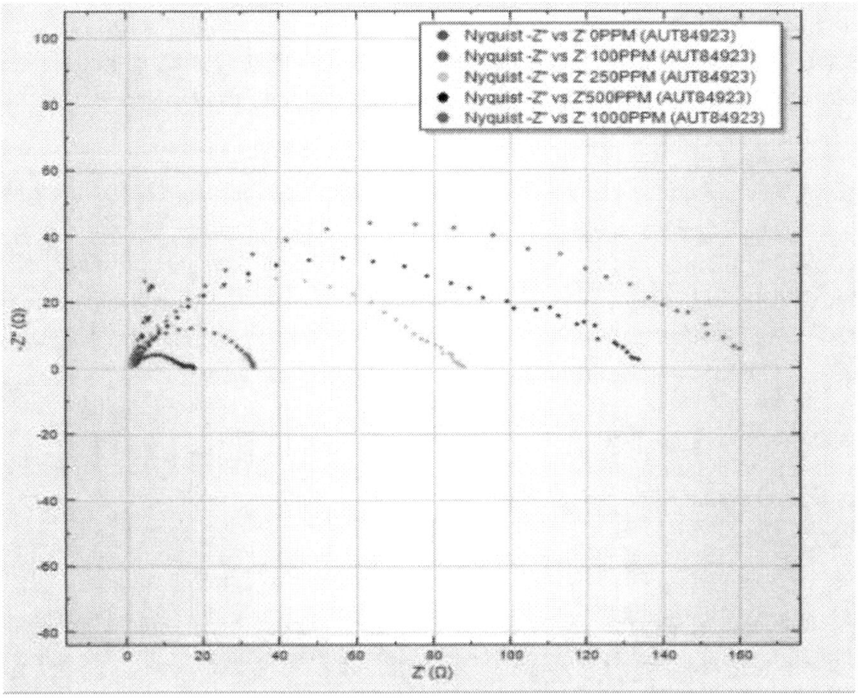

Figure 2. Nyquist plot of API 5L X42 steel in 1M HCl in the absence and presence of different concentrations of C. Xanthorrhiza extract.

The *Rp* values increased with increasing the inhibitor concentration. Increasing of *Rp* values indicated the adsorption of the inhibitor molecules on metal surface form a protective electrochemical double layer. Furthermore, the formation of the protective double-layer may decrease the electron transfer between the metal surface and the corrosive medium. Finally, the corrosion was inhibited.

Table 3. EIS test results in the absence and presence of different concentrations of C. Xanthorrhiza extract.

Inhibitor Concentration (ppm)	Rs (mΩ)	Rp (Ω)	CPE (μMho)	N	*IE* (%)
0	642	14.6	1720	0.611	0
100	519	34.5	791	0.749	58
250	843	80.7	605	0.772	82
500	997	110	473	0.801	87
1000	852	149	229	0.813	90

Table 3 shows that the values of CPE decrease with the increase in *C. xanthorrhiza* extract concentrations, which consequently may decrease the electrical capacity and/or increase in the thickness of the electrical double layer

Table 3 shows a decrease in CPE values along with increased concentrations of C. xanthorrhiza extract, resulting in the decreasing of electrical capacity and / or increasing in the thickness of electrical double layer [6]. This electrical double layer was formed by the C. xanthorrhiza extract adsorption at the metal-solution interface and replaced water molecules gradually [7]

The maximum IE value based on this EIS reaches 90% at 1000 ppm C. xanthorrhiza extract, and the EIS results are also in line with the results indicated by both polarization and weight loss data.

4. Conclusion

Curcuma Xanthorrhiza extract are good inhibitors for API 5L X42 in 1M HCl solution. The C. xanthorrhiza extract acts as a mixed type inhibitor where the reactions on both cathode and anode were inhibited by xanthorrhiza molecules adsorbing on to the metal surface. A maximum efficiency 97% was obtained with the addition of inhibitor 1000 ppm in the potentiodynamic polarization studies. Weight loss method obtained 91% efficiency inhibition of 1000 ppm inhibitor in 7 days immersion time. EIS method obtained 90% efficiency inhibition of 1000 ppm inhbitor.

5. Acknowledgement

The authors wish to thank the Ministry of Research Technology and Higher Education and Directorate of Research and Community Engagement, Universitas Indonesia, under PITTA Research Grants Contract No. 790/UN2.R3.1/ HKP05.00/2017 for the financial support of this research.

6. References

[1] Bastidas-Arteaga E and Stewart M G 2015 Damage risks and economic assessment of climate adaptation strategies for design of new concrete structures subject to chloride-induced corrosion *Struct. Saf.* **52** 40–53

[2] Kusumastuti R, Pramana R I and Soedarsono J W 2017 The use of morinda citrifolia as a green corrosion inhibitor for low carbon steel in 3.5% NaCl solution *AIP Conf. Proc.* **1823** 1–9

[3] Ayende, Rustandi A, Soedarsono J W, Priadi D, Sulistijono, Suprapta D N, Priyotomo G and Bakri R 2014 Effects of Purple Sweet Potato Extract Addition in Ascorbic Acid Inhibitor to Corrosion Rate of API 5L Steel in 3.5%NaCl Environment *Appl. Mech. Mater.* **709** 384–9

[4] Bin Ab Halim M R 2014 *Validation of Gc-Ms Method for Standardization of Curcuma Xanthorrhiza Extracts Using Biochemical Markers , Ar-Curcumene and Xanthorrhizol* (University Sains Malaysia)

[5] Sastri V S 2012 *Green corrosion inhibitors: theory and practice* vol 10(John Wiley & Sons, Inc.)

[6] Kuriakose N, Kakkassery J T, Raphael V P and Shanmughan S K 2014 Electrochemical Impedance Spectroscopy and Potentiodynamic Polarization Analysis on Anticorrosive Activity of Thiophene-2-Carbaldehyde Derivative in Acid Medium *Indian J. Mater. Sci.*

[7] Ibraheem M A, El Sayed Fouda A E A, Rashad M T and Nagy Sabbahy F 2014 Sweet Corrosion Inhibition on API 5L-B Pipeline Steel *J. Dispers. Sci. Technol.* **35**739–752.

IOP Conf. Series: Earth and Environmental Science **105** (2017) 012061 doi:10.1088/1755-1315/105/1/012061

Study of corrosion resistrance of pipeline API 5L X42 using green inhibitor bawang dayak (*Eleutherine americanna Merr.*) in 1M HCl

Muhammad Fikri Azmi[1] **and Johny Wahyuadi Soedarsono**[1]
[1]Department of Metallurgical and Materials Engineering, Faculty of Engineering, Universitas Indonesia, Kampus Baru UI Depok, Depok, 16424, Indonesia

Email : jwsono@metal.ui.ac.id

Abstract. This study aimed to investigate the ability of Eleutherine americanna Merr. extract as an environment friendly inhibitor for API 5L X42 in 1M HCl. Corrosion inhibition ability of this extract was tested using tafel polarization, weight loss and electrochemical impedance spectroscopy methods. FTIR test was used to investigate *flavonoid* and antioxidant compound that play an important role to inhibit the corrosion process. In this study, the concentration of *Eleutherine americanna Merr.* extract used was 0 ppm, 100 ppm, 250 ppm, 500 ppm, and 1000 ppm. The corrosion rate decreases within the increasing of inhibitor's concentration.The highest percentage inhibition efficiency was found to be 84.5% at 1000 ppm. It can be concluded that *Eleutherine americanna Mer.* extract can be used as an alternative and environmental friendly inhibitor for API 5L X42 in 1M HCl.

1.Introduction

In oil and gas industry, corrosion can be seen on the pipe of production oil and gas or the separator machine of oil, gas, and water. The existence of aggressive ion such ion Cl^-can make a pitting corrosion [1].This type of corrosion is the most dangerous attack because we couldn't see clearly the existence of the pit because it's tiny size. On the other hand, the fluid which contained in the pipe could be the cause of erosion corrosion and H_2S corrosion.

Basically, oil and gas industry is using pipeline in which made by steel because its cheap price, whereas, the corrosion resistance of steel is low [1]. Therefore, researchers and engineers are still finding the best way to prevent the corrosion. One of the research found a chemical liquid which can prevent the corrosion called inhibitor and it doesn't cost much. Inhibitor is a chemical liquid which added to the fluid in a small concentration and has an effect to prevent the corrosion [1].The effect of inhibitor for preventing the corrosion was proven in many experiments. However, it has a side effect which can be harmful for the environment. Nowadays, researchers and engineers are developing an inhibitor which called green inhibitor which made by organic compound such as plants and fruits.

It's quite important to do some experiments to know the ability of each kind of green inhibitors to inhibit the corrosion. This experiment is based on the investigation of the ability of green inhibitor Bawang Dayak (*Eleutherine americanna Merr.*) in 1M HCl to inhibit the corrosion using weight loss, potentiodynamic polarization, and electrochemical impedance spectroscopy. The adsorption isotherm was also included.

2. Materials and Method

2.1. Preparation the Sample of Pipeline API 5L X42

The material used in this experiment is pipeline API 5L X42 which made by steel. This type of pipeline is commonly used in oil and gas industry as the pipeline for production of oil and gas. The API 5L X42 is cut into small specimens for weight loss testing, electrochemical testing, and FTIR testing.

2.1.1. Weight Loss Test. For the weight loss samples, the steel is cut using a chainsaw to be 40mm x 30mm x 5mm, then the specimens are drilled with 5mm drill bit which will be used to place the yarn when testing the weight loss. After the cut, the sample is then getting smoothed on the surface using a grinder, followed by sandpaper with grit paper 80, 120, 240, 400, 600, 800. Before the weight loss test, pickling is done in accordance with NACE RP0775-2005 standard. Then calculate the weight of the specimens.

Inhibitory Efficiency is used to know the corrosion rate that occurs with addition of inhibitor and or without addition. The efficiency can be calculated by using this equation [2]:

$$Inhibition\ Efficiency = \frac{Corr.rate(unhibited) - Corr.rate(inhibited)}{Corr.rate(inhibited)} \times 100\% \quad (1)$$

2.1.2. Electrochemical Test. Potentiodynamic polarization is a method to determine the parameters the kinetic of corrosion, such as corrosion potential and current density which are used to find the corrosion rate. The curve between Ecorr and Icorr will be given as the result. Electrodes which used in this method are working electrode, counter electrode, dan reference electrode. The inhibition efficiency is calculated using:

$$IE(\%) = \frac{icorr - i'corr}{icorr} \times 100 \quad (2)$$

where, *icorr and i'corr* are corrosion current densities (A) in the presence and absence of the inhibitors, respectively.

Electrochemical Impedance Spectroscopy (EIS) is a method to analyze the response of a corroded electrode to a potential signal. The results of EIS measurements are expressed in the Nyquist chart. The electrical resistance in the EIS is expressed in impedance (Z). The impedance itself is a measure of the ability of a circuit to resist the flow is a semicircle diagram that flow real impedance to the imaginary impedance. The inhibition efficiency is calculated from the electrochemical impedance diagrams, using:

$$IE\% = 100\left(1 - \frac{Rp}{Rp'}\right) \quad (3)$$

where, *Rp* and *Rp'* are the polarization resistance (Ω) with and without inhibitor.

In addition to weight loss samples, samples are also prepared for electrochemical testing, in example polarization and EIS tests. The sample size for the electrochemical test is 10mm x 10mm x 10mm. After the cut, the sample is getting smoothed on the surface using the sandpaper with grit paper 80 to 1000, after it is connected with copper wire using solder. Then the sample is mounted with resin. Before the test, the sample is sanded back with grit 80 to 600.

2.2. Preparation of Eleutherine americanna Merr. Extract

The green inhibitor which used in the experiments were in powder shaped and will be used in weight loss testing, electrochemical testing, FTIR testing. The extract is first mixing the green inhibitor into HCl 1M.

2.3. Preparation of HCl 1M Solution

The test solution was 350 ml 1M HCl, whichwas prepared by the dilution of technical grade 32% m/v HCl with distilled water. Inhibitors were dissolved in acid solution at required concentrations in mgL^{-1} (ppm) and the solution in the absence of inhibitor was taken as blank for comparison purposes.The concentration range of the *Eleutherine americanna Merr.* extracts varied from blank to 1000 ppm in the electrolyte solution.

3. Result and Discussion

3.1. Weight Loss Test Measurement

The effect of addition of *Eleutherine americanna Merr.* extract tested at various concentrations on the corrosion of API 5L X42 in 1M HCl solution was observed by weight loss measurements after 7 days of immersion. The values of inhibition efficiency and corrosion rate are summarized in Table 1. It is observed that the decreasing corrosion rate is associated with the increase of inhibitor's concentration which indicates, adsorption of inhibitors onto the steel surface or at the solution interface on increasing its concentration and providing wider surface coverage. The highest percentage inhibition efficiency was found to be 83% at 1000 ppm.

Table 1. *Weight Loss*

Concentration of Inhibitor	Initial weight (g)	Final weight (g)	ΔW (g)	Corr. Rate (mm/yr)	Inhibitory Efficiency
0 ppm	32,7397	31,2029	1,5368	5,8692	-
100 ppm	28,4732	27,3538	1,1194	4,2751	27,16%
250 ppm	30,1456	29,2519	0,8937	3,4131	41,84%
500 ppm	32,5805	31,9208	0,6597	2,4091	59%
1000 ppm	31,5104	31,2491	0,2613	0,9979	83%

3.2. Polarization Test Measurement

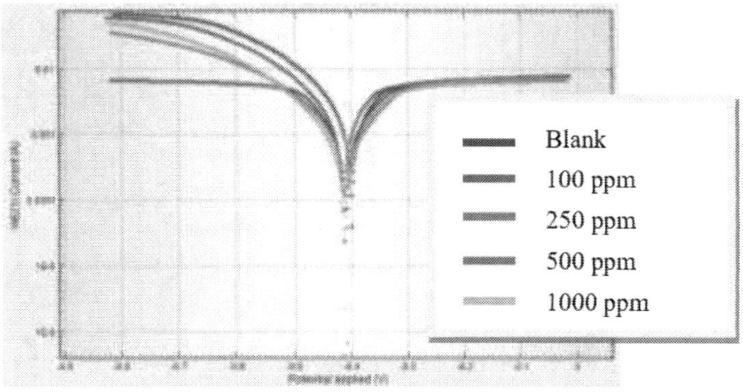

Figure 1. Polarization Tafel Curve

Figure 1 illustrates the cathodic and anodic Tafel polarization curves of sample of pipeline API 5L X42 in 1 M HCl without and with different concentrations of *Eleutherine americanna Merr.*extract. Tafel extrapolations of the anodic and cathodic lines of the polarization curves were used to determine several electrochemical parameters such as anodic slope (*ba*), cathodic slopes (*bc*), potential (E_{corr}), and current density of corrosion (i_{corr}) as given in table 2.

Table 2. Tafel Polazitation

No	Concentration	βa (mV/dec)	βc (mV/dec)	Ecorr (mV)	Icorr (μA)	Corr.rate (mm/yr)	Inhibitory Efficiency
1	Blank	35,0770	42,8020	-401,440	476,230	5,5338	-
2	100 ppm	29,2150	26,8400	-414,390	292,370	3,3973	38.6%
3	250 ppm	31,5230	26,2540	-410,540	210,320	2,4439	55,8%
4	500 ppm	15,9920	16,7930	-408.900	93,697	1,0888	80,3%
5	1000 ppm	10,2030	22,7630	-415,610	81,635	0,94859	84,5%

From table 2, we can see that the value of i_{corr}decreases with increasing inhibitors concentrations. The same condition also happens on both the cathodic and anodic current densities. Therefore, the presence of molecules from inhibitor inhibits the hydrogen evolution and the anodic dissolution processes [6]. This condition indicates that the inhibitors act as mixed type corrosion inhibitors. Moreover, the maximum displacement in E_{corr} was 12.29 mV where was not exceed 85 mV. Thus, this inhibitor is classified as mixed type inhibitors[7].According to the highest percentage inhibition efficiency was 84.5% in 1000 ppm, *Eleutherine americanna Merr*. can be used to inhibit the corrosion process of API 5L X42 in 1M HCl.

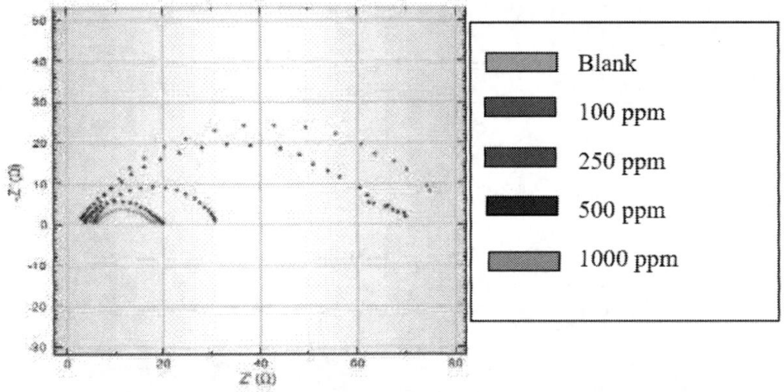

Figure 2. Graph of Electrode Impedance Spectroscopy

3.4. Electrochemical Impedance Spectroscopy (EIS) Test Measurement

In table 3, there are an increasing of Rct value and a decreasing CPE value. Punita [8], increasing of Rct value indicates the presence of a protective layer on the metal interface and solution. CPE is widely used to replace Cdl in analyzing coating phenomenon [9]. Obot [10], decreasing of CPE value indicates an increase in the thickness of the protective layer of the inhibitor molecule on the metal surface. Changes in CPE values indicate a shift between the water molecules on the metal surface. Hazwan [11], a decrease in CPE value is the result of a decrease in surface dielectric constant followed by the addition of thickness of the double layer. The inhibition layer increases with increasing concentration of inhibitor added, where more and more molecules of inhibitor can be adsorbed on the surface of the metal.

Table 3. Result of Fitting and Simulation EIS

Concentration of Inhibitor	Rs(Ω)	Rct(Ω)	CPE (mMho)	Inhibitory Efficiency
0 ppm	5,49	12,6	1,89	-
100 ppm	3,24	16,2	1,09	22,22%
250 ppm	5,25	25,5	1,04	50,58%
500 ppm	1,84	67,3	0,729	81,27%
1000 ppm	3,92	74,8	0,723	83,15%

The following table shows that the values of CPE decrease with the increase in inhibitor concentrations due to the addition of inhibitor that increases the adsorption phenomena, which consequently may decrease the electrical capacity and/or increase in the thickness of the electrical double layer[12]. This double layer was formed by the adsorption of the inhibitor molecules at the metal-solution interface and replaced water molecules gradually [12].

4. Conclusion

Bawang Dayak extract (*Eleutherine americanna Merr.*) is classified as an eco-friendly organic inhibitor of mixed inhibitor type with flavonoid compounds that play a role in corrosion inhibitor process on API 5L X42.The highest inhibitory efficiency occurs when the addition of 1000 ppm inhibitor concentration into HCl 1M solution. Where efficiency of inhibitor from each test result, weight loss equal to 83%, tafel polarization equal to 84,5%, and EIS equal to 83,15%. Addition of inhibitor concentration of 1000 ppm Bawang Dayak is able to decrease the steel corrosion current density of API 5L X42 to 81,635 μA and increase the solution resistance to 74,8 Ω. Effective inhibitory concentrations inhibit the corrosion process in sequence is 100 ppm < 250 ppm < 500 ppm < 1000 ppm. Extract of Bawang Dayak physically inhibits the langmuir isotherm mode, which will predominantly control the charge transfer process on the API 5L X42 steel interface.

5. Acknowledgment

It gives the author great pleasure to offer the sincere thanks to Prof. Dr. Ir. Johny Wahyuadi S., DEA. as a lecturer of metallurgical and material engineering, University of Indonesia.

6. References

[1] Jones D. A. (1992). Principles and Prevention of Corrosion. New York: Macmillan; 568.p.
[2] G. Camila. (2014). F. Alexandre," Corrosion Inhibitors – Principles, Mechanisms and Applications," *InTech*.
[3] Evi Mintowati and Laurentius. (2010). "Structural Development and Bioactive Content of Red Bulb Plant (*Eleutherine Americana*); a Traditional Medicines for local Kalimantan People," *Biodiversitas*, vol. 11, no. 2, pp. 102-106.
[4] Dina Pratiwi. (2013). "The Test of Antioxidant Activity from Bawang Mekah Leaves (*Eleutherine americana Merr.)* Using DPPH (2,2-Diphenyl-1-Picrylhydrazyl) Method," *Trad. Med. J.*, vol. 18, pp. 9-16.

[5] L. Sjahfirdi L. (2015). "Aplikasi *Fourier Transform Infrared* (FTIR) dan Pengamatan Pembengkakan Genital Pada Spesies Primata, Lutung Jawa (*Trachypithecus auratus)* Untuk Mendeteksi Masa Subur," vol. 9, no. 2.

[6] K. Boumhara. (2015). "*Artemisia mesatlantica* Essential Oil as Green Inhibitor for Carbon Steel Corrosion in 1 M HCl Solution: Electrochemical and XPS Investigations," *Journal of Industrial and Eng. Chem.*, vol. 29, pp. 146-155.

[7] Priyanka. (2016). "*Novel qunoline* Derivatives as Green Corrosion Inhibitor for Mild Steel in Acidic Medium: Electrochemical, SEM, AFM, and XPS Studies," *Journal of Molecular Liquids*, vol. 216, pp. 164-173.

[8] M. Punita. (2014). "Corrosion Inhibition of Mild Steel in Acidic Solution by *Taegetes erecta* (Marigold flower) Extract as a Green Inhibitor," *Corr. Sci.*, vol. 85, pp. 352-363. 9

[9] Basak Dogru. (2011). "Experimental and Theoritical Investigation of 3-amino-1,2,4-triazole-5-thiol as a Corrosion Inhibitor for Carbon Steel in HCl Medium," *Corrosion Science*, vol. 53, pp. 4265-4272.

[10] I.B. Obot, "Enhanced Corrosion Inhibition Effect of Tannic Acid in the Presence of Gallic Acid at Mild Steel/HCl Acid Solution Interface," *Journal of Industrial and Engineering Chemistry*, vol. 25, pp. 105-111, 2015.

[11] M. Hazwan. (2016). "The Capability of Ultrafiltrated Alkaline and Organosolv Oil Palm (*Elaeis guneensis*) Fronds Lignin as Green Corrosion Inhibitor for Mild Steel in 0.5 M HCl Solution," *Measurement*, vol. 78, pp. 90-103.

[12] S. Manimegalai. (2015). "Thermodynamic and Adsorption Studies for Corrosion Inhibition of Mild Steel in Aqueous Media by *Sargasam swartzii* (Brown algae)," *J. Mater. Environ. Sci.*, vol. 6, pp. 1629-1637.

Utilization of secang heartwood (caesalpinia sappan l) as a green corrosion inhibitor on carbon steel (API 5L Gr. B) in 3.5% NaCl environment

Erol Efraim Kaban[1], Ahmad Maksum[1, 2], Sulaksana Permana[1] and Johny Wahyuadi Soedarsono[1]
[1]Center of Minerals Processing and Corrosion Research, Department of Metallurgical and Materials Engineering, Universitas Indonesia, Depok 16424, Indonesia
[2]Department of Mechanical Engineering, Politeknik Negeri Jakarta, Depok 16425, Indonesia

Email: jwsono@metal.ui.ac.id

Abstract. This research objective was to investigate secang heartwood extract (caesalpinia sappan l) as a corrosion inhibitor on carbon steel metal (API 5L Gr B) in 3.5% NaCl environment. This research utilized polarization linier for the measurement method. Secang extract was adequate to reduce corrosion rate by 53.18% efficiency (based on polarization measurement). The most effective concentration of inhibitor secang was 2.0 ml/400 ml 3.5% NaCl. Furthermore, inhibitor efficiency of secang has tended to decrease by increasing of temperature. From polarization study, secang was indicated as mixed type inhibitor, with predominant cathodic effectiveness. In addition, the inhibition mechanism of secang was shown as physicsorption mechanism and obeys Langmuir Isotherm model.

1. Introduction

Corrosion inhibitor is one of methods for corrosion control. As material point of view, corrosion inhibitor is divided into two types : organic inhibitor and inorganic inhibitor types [1]. Organic inhibitors have many atoms (heteroatoms) including O (oxygen), N (nitrogen), and S (sulfur). O, N, and S is the center absorption activation process on the metal surface. The research reports show that the organic inhibitor is absorbed on metal surface by the turn of the water molecules and form a barrier compact. The performance of an organic inhibitor related to chemical structure and physicochemical properties of the compound as a functional group, electron density in the atom donor, p-orbital characters and the electron structure of the molecule. Inhibition can be caused by absorption of molecules or ions on anodic or cathodic sites, potential increasing on cathodic and or anodic, a barrier layer / protector forming. Absorption inhibitor mechanism can be through as following ways : physicsorption, caused by electrostatic forces between ions which inhibit and electric charge on metal surface; chemisorption, caused by interaction between pairs free electrons with the metal to form a bond , this takes place during availability of a heteroatom such as P, Se, S, N and O will be pair with electrons itself or aromatic rings in molecule absorption; then the combination of physicsorption and chemistsorption[2].

Secang heartwood (caesalpinia sappan l) can be considered as environmentally friendly corrosion inhibitor on the metal material of carbon steel in sea water containing 3.5% NaCl, because the phenolic compounds contained in the secang heartwood, ie brazilin, chalcone, protossappanin and homoisoflavonoid [3], are expected to inhibit corrosion through physisorption mechanisms.

Figure 1. The organic structure of flavonoids.

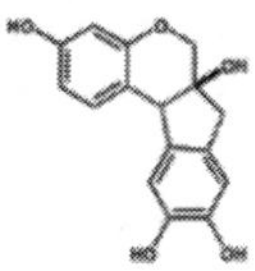

Figure 2. The organic structure of brazilin.

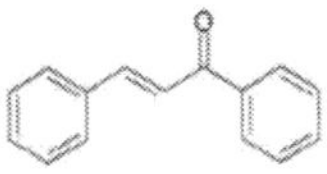

Figure 3. The organic structure of chalchone.

The purpose of this study is to determine the inhibition efficiency of the secang extract as corrosion inhibitor in a 3.5% NaCl solution, find out most effective concentrations of secang extract inhibiting corrosion rate, find out the solution temperature change effects to the inhibitor performance, and to understand the mechanism of secang (caesalpinia sappan l) extract inhibition on metal surface. Evaluation of corrosion rate and inhibition mechanism has been evaluated by linear polarization.methods.

This study uses a variation of extract concentration and temperature as following parameters: extract concentration variation ranging from 0 to 2.5 ml / 400 ml NaCl at temperature 25, 35, and 45°C. Then this study is limited by the following conditions: operating pressure during the experiment is assumed to constant (1 atm), NaCl concentration is 3.5%.

2. Experimental work

2.1. Materials and Solution

Metal samples of carbon steel API 5L Gr. B with composition (%) 0.147 C, 0.199 Si, 0.004 S, 0.013 P, 0.424 Mn, <0.005 Ni, 0:01 Cr, 0.008 Mo, <0.002 Ti, 0.007 Cu, <0.002 Nb, <0002 V, 0.002 Al and free Fe used in this study.

Secang extract was prepared by mixing of secang's slicing with methanol for minimum 3 days, then the initial extract result continued by evaporation to remove methanol and water contents. The evaporation using equipment rotary vacuum evaporator with pressure parameters of 185 bar at room temperature (27-28°C). While NaCl solution was prepared by mixing of salt and distilled water to obtain 3.5% NaCl solution.

2.2. Linear Polarization Measurements

Electrochemical measurements utilized Metrohm Instruments Autolab Potentiostat/Galvanostat PGSTAT302N with NOVA 1.10 software. The cell contained three electrodes; the working, counter and reference electrodes, comprised of mild steel, a carbon bar and AgCl (saturated with KCl) respectively.

2.3. FTIR Measurement

FTIR is used to identify the functional groups contained in the extract of secang's heart-wood. FTIR measurement is also conducted to ascertain whether the functional groups contained in the wooden cup extract can be adsorbed well on the steel surface and form a inhibition layer which would inhibit the corrosion reaction. For this purpose FTIR measurement was done 2 times; the extract of secang heart-wood and a metal surface that has been immersed in a solution of 3.5% NaCl were added to extract the wooden cup at the optimal concentration.

3. Results and Discussions

3.1. Linier Polarization

The polarization measurements result as shown in Figure 4 shows the inhibition effect on the material API 5L Gr. B in 3.5% NaCl environment, while Table 1 is polarization parameters of inhibitor concentration variations. Parameters results on Table 1 show decreasing of current density (Icorr) corresponding addition of inhibitor concentrations up to concentration of 2.0 ml (Icorr = 12.88 μ.Acm-2), then current density rising again at concentration of 2.5 ml with Icorr = 16.94 μ.Acm-2.

The same pattern also applied in corrosion rate (CR) value, the inhibitor concentration addition will reduce the CR of up to 2.0 ml inhibitor concentration (CR= 0.150 MPY) then the CR will increase at concentration of 2.5 ml.

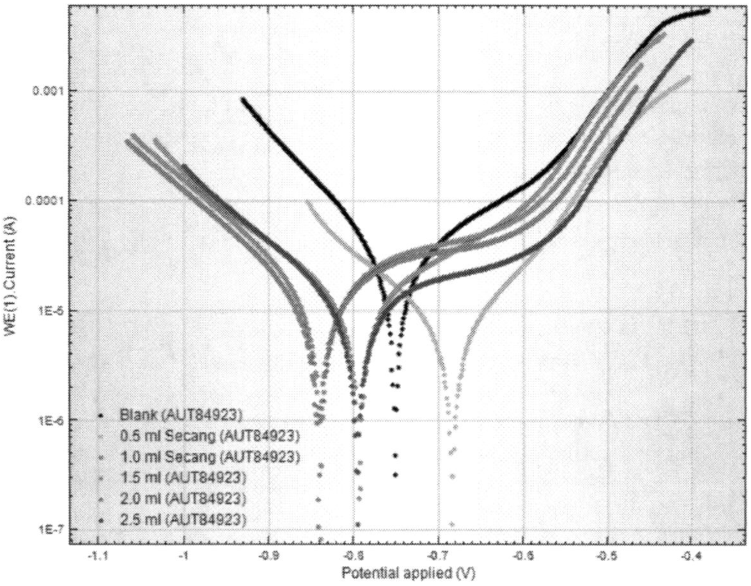

Figure 4. Polarization curve for 3.5% NaCl solution with inhibitor concentration variation.

The inhibition efficiencies for different inhibitor concentrations were calculated from the following equation [6]:

$$\eta_p = \frac{i_{corr(uninh)} - i_{corr(inh)}}{i_{corr(uninh)}} x100 \quad (1)$$

Where, i_{corr} is corrosion current densities.

Table 1. Polarization parameters of inhibitor concentration variations in 400ml NaCl 3.5%.

Secang extract volume (ml)	E_{Corr} (mV)	I_{Corr} ($\mu A.cm^{-2}$)	b_a ($mVdec^{-1}$)	b_c ($mVdec^{-1}$)	C_R (mpy)	EI, η (%)
Blank	-747.51	27.49	121.69	210.19	0.32	0.00
0.5 ml	-683.57	20.05	379.87	333.15	0.23	27.07
1.0 ml	-794.05	17.23	221.39	290.48	0.20	37.33
1.5 ml	-831.16	13.86	152.01	166.89	0.16	49.58
2.0 ml	-837.41	12.88	157.67	196.88	0.15	53.14
2.5 ml	-784.91	15.20	194.79	350.60	0.18	44.71

As the investigation, concentration of 2.0 ml/400 ml NaCl is most effective inhibitor concentration with highest efficiency of 53.14%, the lowest current density (Icorr) and the lowest corrosion rate (CR). In another words, the inhibitor concentration of 2.0 ml /400 ml NaCl solution is the most effective concentration reducing or inhibit corrosion rate on metal sample API 5L Gr B in the 3.5% NaCl environment.

In accordance to Figure 4, polarization curves tend to shift to more negative potential values (cathodic area). Based on Table 1, corrosion potential (E_{Corr}) parameter at concentration of 0.5 ml inhibitor shows shifting to anodic area with value of -683.57 mV, while other inhibitors concentration shifts toward cathodic area ranging from -794.05 mV up to -837.41 mV, maximum E_{Corr} is showed at inhibitor concentration of 2.0 ml. As the investigation, E_{Corr} shifting until 89.9 mV of blank E_{Corr}. According to Ambrish Singh [7], Mc Caferty [8], Bramley [9], secang is cathodic inhibitor type due to E_{Corr} shifting exceeds value of 85 mV.

From Table 1, value of the anodic Tafel constants (b_a) and cathodic Tafel constants (b_c) look change due to inhibitor concentration increasing. Changes value of b_c and b_a indicate that absorption process of inhibitor extract able to change cathodic solubility mechanism [10]. According to Li [11] and Rahmad [12], changes of bc and ba due to inhibitor volume changes also indicate an absorption process on both anodic and cathodic areas.

As above review, secang heartwood is mixed type inhibitor, with predominant cathodic effectiveness.

3.2. Thermodynamic Aspect

Inhibition and absorption mechanism of the inhibitor molecules were investigated using temperature variations measurement. The resulted curve and parameters of linear polarization measurements are shown in Figure 5.

Based on Table 1, Figure 6 and Figure 7 temperature increasing will raise the current density (I_{Corr}), corrosion rate (CR), and reduce inhibitor efficiency. Such tendencies also occur in other plant extracts such as Morinda citrifolia[4], Mentha pulegium [5], Thymus algeriensis [6], and Santolina chamaecyparissus [7]. The increasing of temperature will damage the inhibitor bonding on to surface, then reduce inhibition performance between the inhibitor with metal surface.

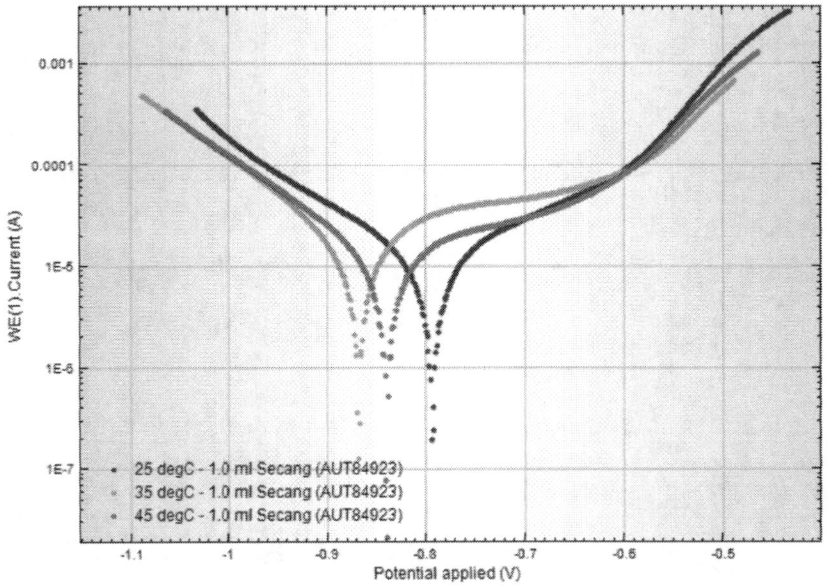

Figure 5. Polarization curve of 3.5% NaCl solution with inhibitor at temperature 25, 35, and 45°C.

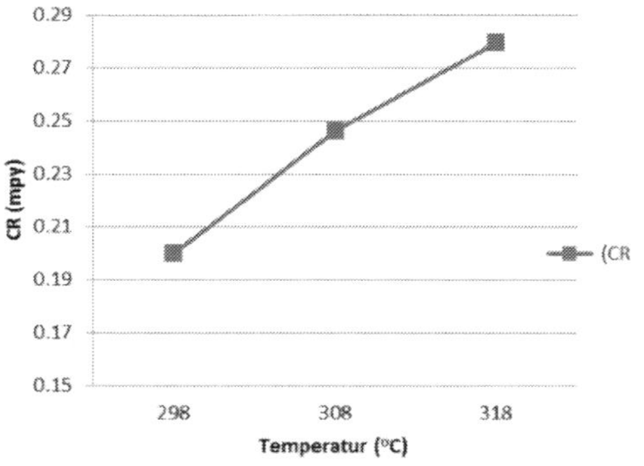

Figure 6. Corrosion Rate (CR) at temperature variations.

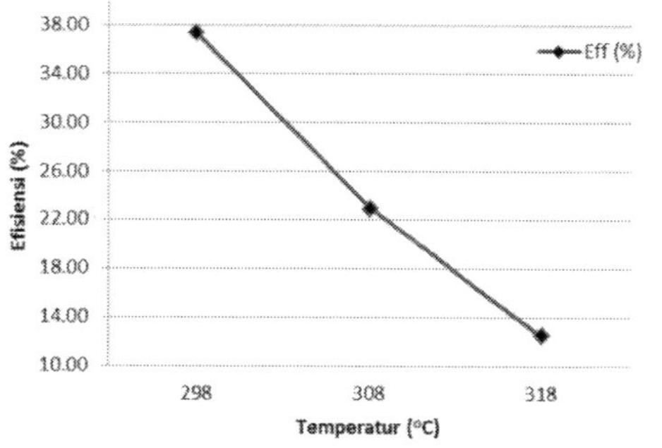

Figure 7. Inhibitor Efficiency at temperature variations.

According to Olivares et al. [8] and Kairi et al. [9], inhibitor efficiency reduction due to temperature increasing proved that absorption mechanism of secang inhibitor molecules acted as physically (physicsorption).

3.3. Absorption Mechanism

Absorption process of the inhibitor molecule will obey the Langmuir modeling based on figures 5 because the correlation coefficient (degree of conformity) of modeling Langmuir has a value > 0.95. As investigation result by Yurt et al. [10], organic molecules to be adsorbed as a monolayer on metal surfaces without any interaction with other absorbed molecules, and according to Behpour investigation result [11], there is only one site (area) where the absorption of molecules. Therefore, molecules of secang inhibitors in this study are strongly indicated run into the formation of a single layer (monolayer) on the metal surface.

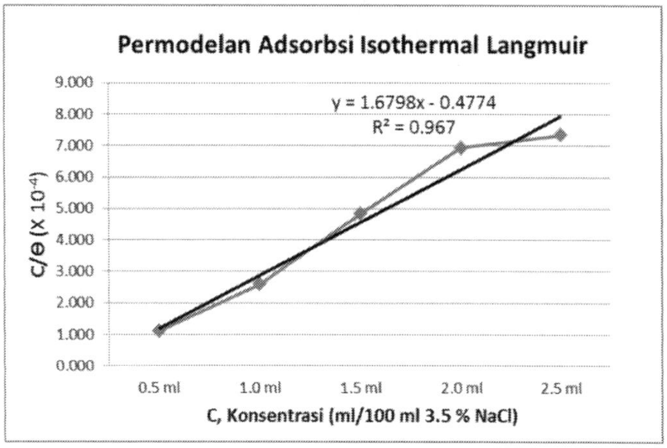

Figure 8. Isotherm Langmuir for secang inhibitor absorption on metal API 5L Gr. B surfaces in 3.5% NaCl solution.

Negative values in the free energy show the corrosion reaction runs spontaneously. According to Ahmed Musa Y. et al [12], when the interaction free value was reached -40kJmol^{-1}, then the interaction of adsorption is chemisorptions where there is charge exchange of organic molecules to the metal surface forming the coordinate bond.

Based on the resulted parameters in above Table 1, absorption mechanism of secang heartwood inhibitor was indicated to react as physicsorption because of free energy value is below -40kJmol^{-1}.

3.4. FTIR Measurement

FTIR Measurement result (spectra) of secang inhibitor absorption layer on the metal surface can be seen in Figure 9.

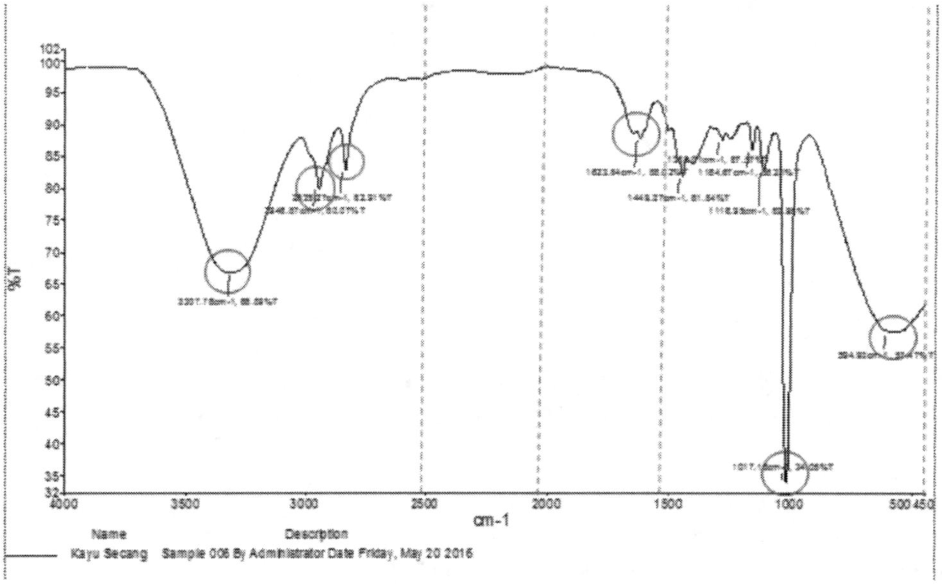

Figure 9. FTIR spectrum of secang absorption on metal API 5L Gr B surface.

As Figure 9 shows absorption widened at wave number 3308 cm^{-1} is the absorption of O-H hydroxyl (single), then looks weak absorptions between area 2947dan 2950 cm^{-1} are typical C-H aldehyde. Furthermore, the weak absorptions double C = O stretch looked at wave number of 1622 cm^{-1}. Then there is strong absorption of the carbonyl group C-O at wave number 1017 cm^{-1}.
Above facts confirm that secang heartwood extract contain braziline compounds [13,14] which has a functional group of O-H and chalchone compounds which has group of C-H and the double bond of C=O act as antioxidant that can be used to inhibit corrosion rate on metal surfaces by absorption mechanism.

4. Conclusion

Secang heartwood extract can be utilized as an alternative of corrosion inhibitor because the inhibitor is proven to inhibit corrosion on metal samples API 5L Gr. B in 3.5% NaCl environment; polarization study indicate secang heartwood is mixed type inhibitor with predominant cathodic effectiveness; inhibition efficiency is influenced by concentration and temperature, the higher inhibitor concentration the higher inhibition efficiency while conversely the higher temperature the lower inhibition efficiency; the inhibitor concentration of 2.0 ml / 400 ml NaCl is the best effective inhibitors concentration reducing corrosion rate; the highest inhibition efficiency is 53.18% based on Polarization evaluation. And, the secang heartwood inhibitor reacts as physicsorption and obeys Langmuir Isotherm model.

5. Acknowledgement

The authors wish to thank the Ministry of Research Technology and Higher Education and Directorate of Research and Community Engagement, Universitas Indonesia, under PITTA Research Grants Contract No. 790/UN2.R3.1/ HKP05.00/2017 for the financial support of this research.

6. References

[1] Sastri V S 2012 *Green corrosion inhibitors: theory and practice* vol 10(John Wiley & Sons, Inc.)

[2] Perez T E 2013 Corrosion in the oil and gas industry: An increasing challenge for materials *Jom* **65** 1033–42

[3] Fu L C, Huang X A, Lai Z Y, Hu Y J, Liu H J and Cai X L 2008 A New 3-Benzylchroman Derivative from Sappan Lignum (Caesalpinia sappan) *Molecules* **13** 1923–30

[4] Kusumastuti R, Pramana R I and Soedarsono J W 2017 The use of morinda citrifolia as a green corrosion inhibitor for low carbon steel in 3.5% NaCl solution *AIP Conf. Proc.* **1823** 1–9

[5] Khadraoui A, Khelifa A, Boutoumi H and Hammouti B 2014 Mentha pulegium extract as a natural product for the inhibition of corrosion. Part I: electrochemical studies *Nat. Prod. Res.* **28** 1206–9

[6] Khadraoui A, Khelifa A, Hachama K and Mehdaoui R 2016 Thymus algeriensis extract as a new eco-friendly corrosion inhibitor for 2024 aluminium alloy in 1 M HCl medium *J. Mol. Liq.* **214** 293–7

[7] Shabani-Nooshabadi M and Ghandchi M S 2015 Santolina chamaecyparissus extract as a natural source inhibitor for 304 stainless steel corrosion in 3.5% NaCl *J. Ind. Eng. Chem.* **31** 231–7

[8] Olivares-Xometl O, López-Aguilar C, Herrastí-González P, Likhanova N V., Lijanova I, Martínez-Palou R and Rivera-Márquez J A 2014 Adsorption and corrosion inhibition performance by three new ionic liquids on API 5L X52 steel surface in acid media *Ind. Eng. Chem. Res.* **53** 9534–43

[9] Kairi N I and Kassim J 2013 The Effect of Temperature on the Corrosion Inhibition of Mild Steel in 1 M HCl Solution by Curcuma Longa Extract *Int. J. Electrochem. Sci.* **8** 7138–55

[10] Yurt A, Duran B and Dal H 2014 An experimental and theoretical investigation on adsorption properties of some diphenolic Schiff bases as corrosion inhibitors at acidic solution/mild steel interface *Arab. J. Chem.* **7** 732–40

[11] Behpour M, Ghoreishi S M, Khayatkashani M and Soltani N 2012 Green approach to corrosion inhibition of mild steel in two acidic solutions by the extract of Punica granatum peel and main constituents *Mater. Chem. Phys.* **131** 621–33

[12] Musa A Y, Kadhum A A H, Mohamad A B, Daud A R, Takriff M S, Kamarudin S K and Muhamad N 2009 Stability of layer forming for corrosion inhibitor on mild steel surface under hydrodynamic conditions *Int. J. Electrochem. Sci.* **4** 707–16

[13] Chu M J, Wang Y Z, Itagaki K, Ma H X, Xin P, Zhou X G, Chen G Y, Li S and Sun S Q 2013 Identification of active compounds from Caesalpinia sappan L. extracts suppressing IL-6 production in RAW 264.7 cells by PLS *J. Ethnopharmacol.* **148** 37–44

[14] Ye M, Xie W D, Lei F, Meng Z, Zhao Y N, Su H and Du L J 2006 Brazilein, an important immunosuppressive component from Caesalpinia sappan L. *Int. Immunopharmacol.* **6** 426–32

IOP Conf. Series: Earth and Environmental Science **105** (2017) 012063 doi:10.1088/1755-1315/105/1/012063

Preparation of cellulose nanocrystals from empty fruit bunch of palm oil by using phosphotungstic acid

Y W Budhi[1], M Fakhrudin[1], N T U Culsum[1], V Suendo[2] and F Iskandar[3]
[1] Chemical Engineering Study Program, Faculty of Industrial Technology, Bandung Institute of Technology, Bandung, 40132, Indonesia
[2] Chemistry Study Program, Faculty of Mathematic and Science, Bandung Institute of Technology, Bandung, 40132, Indonesia
[3] Physics Study Program, Faculty of Mathematic and Science, Bandung Institute of Technology, Bandung, 40132, Indonesia

E-mail : y.wibisono@che.itb.ac.id

Abstract. Empty fruit bunch (EFB) of palm oil, an abundant agro-waste in Indonesia, was used as raw material of Cellulose Nanocrystals (CNCs) preparation. Instead of conventional acid mineral, phosphotungstic acid ($H_3PW_{12}O_{40}$, HPW) was used to hydrolyze cellulose due to recycling ability and easy handling. Before hydrolysis process, dried EFB was treated by 3% NaOH solution at 90°C for 2 hours and then bleached using 2% $NaClO_2$ solution at 80°C for 3 hours to remove hemicellulose and lignin. Hydrolysis reaction parameters such as temperature, acid concentration, and reaction time were optimized with fixed solid-liquid ratio of 1:40. Response surface method was used for experimental design to determine the optimum condition of each parameter by using software Minitab. In this study, pulp from dried EFB produced 44.8% yield of CNCs. Dynamic Light Scattering (DLS) analysis showed that most of CNCs equivalent diameter was 140 nm. Crystallinity index was observed at 73.3% using X-ray Diffraction (XRD) analysis. Thus, a green established process for the preparation of CNCs was achieved.

1. Introduction

Palm Oil is a commodity mainstay of Indonesia which its development is quite rapid. This plantation plays an important role in the Indonesia economic sector with production reaching 33.4 million tons and accounted for 11% of Indonesia's export revenue. According to Directorate General of Estate Crops (2015), oil palm plantations have spread in 25 provinces, and the area reached 10.8 million ha in 2014 [1]. The increasing of plantation area will be followed by the escalating of oil palm production and volume of waste. Empty fruit bunch of palm oil (EFBPO) as the most abundant waste of oil palm production has not yet been widely utilized. This waste contains high cellulose content, which is 40-60% [1,2], therefore, it is potential to be used as raw material for the manufacture of Cellulose Nano Crystals (CNCs).

CNCs is one of advanced material that is currently studied because of their amazing material properties, such as low density, high mechanical strength, high surface area, low toxicity, and other

[1] To whom any correspondence should be addressed.

intriguing optical properties [3,4]. Hence, CNCs attracts researchers and industries to be applied in various fields, such as optical and electronic devices, catalyst support, wound dressing, paint and coating, food and cosmetic, enzyme immobility and controlled drug delivery [4-8].

The common method used for synthesizing CNCs is hydrolysis of cellulose by mineral acids, including sulfuric acid, hydrochloric acid, phosphoric acid, and their mixture [9-11]. Although this method is simple, CNCs produced through sulfuric acid hydrolysis are usually sensitive to temperature because of the presence of acidic moieties at their surface [12]. This condition will limit their applications in some areas. In addition, some issues still need to be addressed, such as large water consumption, serious equipment corrosion and great amount of waste produced [13,14].

Recently, many studies have been done to optimize the hydrolysis parameters, avoid the corrosion, and reduce the waste. In this study, solid acid is investigated to substitute the usage of strong liquid acids to overcome the limitation of CNCs produced by liquid acid. The main advantages of solid acid are the easy recovery of solid acid, less corrosion, high thermal stability, and relatively safe working environment [13,14]. However, the limited contact between solid acid and cellulose significantly increases the hydrolysis time. Previous studies showed that phosphotungstic acid ($H_3PW_{12}O_{40}$, HPW) has many Bronsted acid sites, so it can be used as an alternate catalyst of liquid acids for the hydrolysis of cellulose [15]. Hence, its usage is effective for the hydrolysis reaction. Therefore, it is expected that CNCs can be synthesized through hydrolysis using HPW by controlling the hydrolysis parameters.

2. Materials and Methods

2.1. Materials

EFBPO used as raw material was obtained from PTPN VIII located at Bogor, West Java, Indonesia. It was dried, ground and chipped into 40-80 mesh size.

2.2. Methods

2.2.1. Alkali pre-treatment. The alkali pre-treatment was completed by heating the mixture of EFB and NaOH solution (3-wt%) under mechanical stirring for 2 hours. The solid product was washed several times using distilled water. Neutral solid was dried in oven at 105°C for 8h.

2.2.2. Bleaching. The resulting solid from previous treatment was mixed with solution consisting of equal parts (v:v) of acetic acid buffer (27 g NaOH and 75 mL glacial acetic acid, diluted to 1 L of distilled water) and 2% aqueous sodium chlorite (pH of mixed solution is 3.6-4) [16]. The mixture was heated at temperature of 80°C for 3 h under continuous stirring. Bleached solid (EFB pulp) was filtered and washed using deionized water until neutral pH was reached. Drying of pulp was done in oven at 105°C for 8 h.

2.2.3. Acid Hydrolysis. One gram of pulp was placed in an erlenmeyer and HPW was added with concentration varied from 65-85%-wt. The mixture was heated at temperature 80°C-100°C under mechanical stirring for time range of 25-35 h. The reaction was stopped by rapid cooling in chilled water bath until reached room temperature. The resulting solution was then extracted with excess diethyl ether in separating funnel. Three layers were formed. The lowest layer was the mixture of HPW and diethyl ether, the middle layer was consisted of CNCs, water, and degraded sugar and the upper layer was diethyl ether [17]. The lowest layer was distilled at 45°C to recover HPW and diethyl ether. The middle layer was centrifuged at 4000 rpm for 15 min at room temperature to and the supernatant was decanted. The decanted solid was mixed with deionized water and centrifuged three times. Near neutral pH solid was then mixed with ethanol and centrifuged to remove unreacted cellulose. CNCs was collected by further centrifugation with deionized water. Wet CNCs was dried in oven at 105°C for 8 h to obtain dried CNCs. Yield of CNCS was calculates by equation (1).

$$Yield = \frac{mass\ of\ dried\ CNC}{mass\ of\ pulp} \times 100\% \quad (1)$$

2.2.4. Characterization. The CNCs produced was characterized using X-ray Diffraction (XRD), Dynamic Light Scattering (DLS), and Transmission Electron Microscopy (TEM). XRD pattern for CNCs was perform on Bruker D8 Discover Diffractometer (Bruker Co., USA) using Cu Kα radiation (λ=1.54 Å) at 40 kV and 30 mA. Diffactogram was collected in the range 2θ of 5-60°. Crystallinity Index (CI) was determined by peak deconvolution method. This method extracts individual crystalline peak by a curve-fitting process from diffraction intensity profiles [18]. PeakFit v4.12 trial version (SeaSolve Software, Inc. USA) was used to separate amorphous and crystalline contributions to the diffraction spectrum. CI is calculated from ration of the area of all crystalline peaks to the total area. DLS was used to analyze particle size distribution of CNCs. This analysis used common Delsa TM Nano Submicron particle analyzer (Beckman Coulter, Inc., USA) at fixed angle 165°. Morphology and size of CNCs were observed by TEM with a Hitachi HT-7700 transmission electron microscope (Hitachi High-Technologies Co., Japan) using an accelerated voltage of 80 kV. Diluted suspension of CNCs was deposited on an 20 nm carbon-coated grid.

2.2.5. Experimental Design. Response Surface Method (RSM) and Box-Behnken Design were used for experimental design to optimized CNCs yield (Y, %) with software MiniTab 16.2.4.0 (Minitab Inc., US/Canada/Mexico). The independent variables were HPW concentration (X_1, g/mL), reaction time (X_2, h), and reaction temperature (X_3, °C). Yield data obtained from the experiment were analyzed using Multiple Regression Analysis and fitted to second-order polynomial model expressed by equation (2).

$$Y = X_0 + \sum_{i=1}^{3} \beta_i X_i + \sum_{i=1}^{3} \beta_{ii} X_i^2 + \sum_{i=1}^{2} \sum_{j=i+1}^{3} \beta_{ij} X_i X_j + \varepsilon \quad (2)$$

where, β_0, β_i, β_{ii}, and β_{ij} are regression coefficients, Xi and Xj are independent variables, and ε corresponds to the residue of the model. Test of statistical significance was applied to model and and lack of significance using F and P-test from analysis of variance (ANOVA) at significant level (α) of 0.05.

3. Results and Discussions

3.1. Model Fitting and Statistical Analysis

Design of experiment and yield of CNCs as response was shown in table 1. Second-order polynomial model that relates each parameters and the experimental yield was shown in equation 3. This regression model was then evaluated using analysis of variance (ANOVA). The result of ANOVA was shown in table 2.

IOP Conf. Series: Earth and Environmental Science **105** (2017) 012063 doi:10.1088/1755-1315/105/1/012063

Table 1. Box-Behnken Design and CNCs Yield as Response

Coded Variable			Yield
X_1(%)	X_2(h)	X_3 (°C)	(%)
-1 (65)	-1 (25)	0 (90)	64.47
1 (85)	-1 (25)	0 (90)	54.55
-1 (65)	1 (35)	0 (90)	50.71
1 (85)	1 (35)	0 (90)	70.42
-1 (65)	0 (30)	-1 (80)	53.34
1 (85)	0 (30)	-1 (80)	70.20
-1 (65)	0 (30)	1 (100)	61.92
1 (85)	0 (30)	1 (100)	57.80
0 (75)	-1 (25)	-1 (80)	50.90
0 (75)	1 (35)	-1 (80)	64.69
0 (75)	-1 (25)	1 (100)	55.57
0 (75)	1 (35)	1 (100)	52.48
0 (75)	0 (30)	0 (90)	44.05
0 (75)	0 (30)	0 (90)	47.15
0 (75)	0 (30)	0 (90)	45.04

Table 2. ANOVA for Response Surface Second-Order Polynomial Model

Source	DF	Seq SS	Adj MS	F	P
Model	9	994.34	110.482	32.39	0.001
Linear	3	100.09	33.364	9.78	0.016
X_1	1	63.45	63.45	18.6	0.008
X_2	1	20.51	20.512	6.01	0.058
X_3	1	16.13	16.131	4.73	0.082
Square	3	493.49	164.495	48.22	0
X_1^2	1	303.4	352.051	103.21	0
X_2^2	1	72.76	87.196	25.56	0.004
X_3^2	1	117.33	117.329	34.4	0.002
Interaction	3	400.76	133.586	39.16	0.001
$X_1.X_2$	1	219.48	219.484	64.35	0
$X_1.X_3$	1	110.04	110.04	32.26	0.002
$X_2.X_3$	1	71.23	71.234	20.88	0.006
Residual	5	17.06	3.411		
Lack of fit	3	12.04	4.014	1.6	0.407
Pure error	2	5.01	2.507		
Total	14	1011.39			

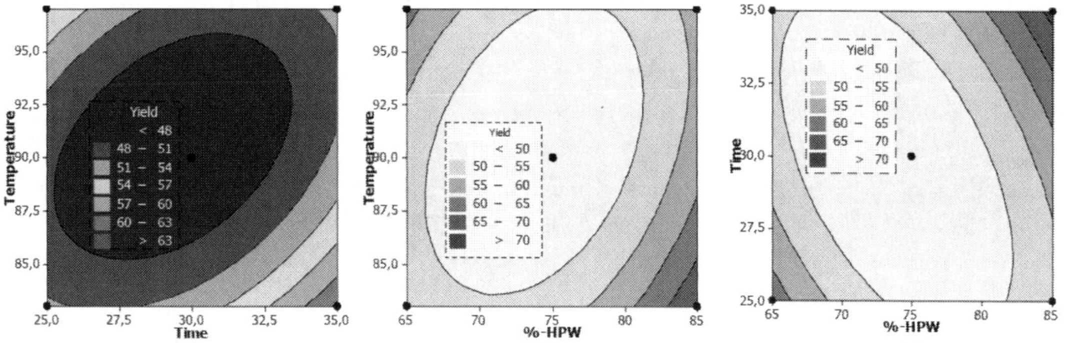

(a) (b) (c)

Figure 1. Contour plots as a function of (a) time and temperature at 75%-HPW, (b) HPW concentration and temperature at 30 h, and (c) HPW concentration and time at 90°C

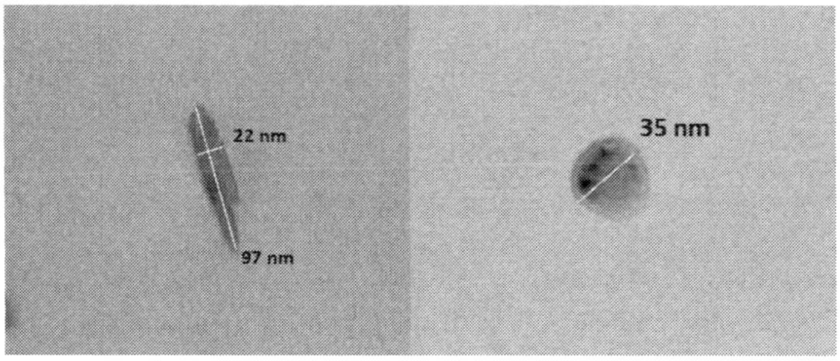

(a) (b)

Figure 2. TEM image of CNCs sample, (a) rod-like and (b) spherical shape

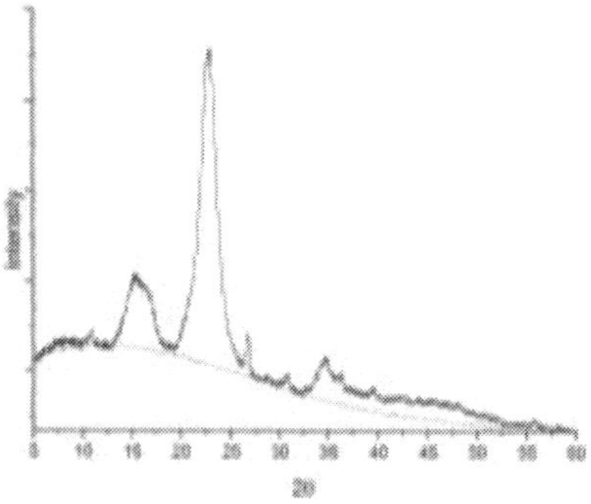

Figure 3. XRD spectra of CNCs sample for parameters 75% HPW, 30h, and 90°C

$$Y = 1191.02 - 12.07X_1 - 11.6026X_2 - 11.6737X_3 + 0.0976X_1^2 + 0.1943X_2^2 + 0.115X_3^2 + 0.1482X_1X_2 - 0.075X_1X_3 - 0.121X_2X_3 \quad (3)$$

Determination coefficient of the model was close to 1 (R^2= 98.31% and adjusted-R^2= 95.28%). Calculated F-value of model was higher than tabulated F-value ($F_{0.05}$=4.77) and P-value was less than 0.05. Lack of fit was insignificant since the calculated F-value is less than tabulated F-value ($F_{0.05}$=19.6) and P-value is higher than 0.05. These results were indicated that the model equation was well fitted to the experimental data [19].

3.2. Optimization of Experimental Parameter

Contour plot (Figure 1) was constructed based on equation (3). These plots correlated the three independent variables and the response keeping one of the independent variables in its central point. According to equation (3), the higher importances on the response were three square terms and term of interaction between HPW concentration and reaction time because of its positive signal. The positive values in the model indicate that an increase of factor values tends to increase the response [20].

Three of the contour plots indicates that initially, yield decreased with increase of each factor. After yield reached its minimum point, yield increased with increase of factor. The decrease of yield was due to further hydrolysis of cellulose on crystalline region. This excessive hydrolysis happened may because of the too high level of reaction parameters. The increase of yield after minimum point was resulted by the formation of amorphous nanoflocks. A tendency of nanoflocks formation is higher when the reaction parameters was too high [13]. These amorphous nanoparticle was about the same size as CNCs that make them difficult to separate by centrifuging. Optimum yield based on equation (3) was 88,38% under %-HPW, time, and temperature value of 85%, 35h, and 80°C, respectively. The possibility of nanoflocks formation made this yield value not convincing. CNCs produced under high level reaction parameter had to be analized further by XRD analysis to make sure the present of nanoflocks in sample.

3.3. Characteristics of CNCs

Diffractogram of CNCs sample produce under middle level parameter value was shown in figure 3. Crystallinity Index of sample was 84,0% and indicated that most of cellulose amorphous regions were hydrolized. Particle size distribution was observed in the range of 23-770 nm and mean value of 140 nm as equivalent diameter by DLS analysis. TEM image of CNCs sample was shown in figure 2. Two types of CNCs morphology were observed, rod-like and spherical. Most CNCs morphology was rod-like with length under 200 nm.

4. Conclusions

In this study, a green established process for the preparation of CNCs from was achieved using phosphotungstic acid hydrolysis with minimum 44.04% yield. Characterization of sample show that the crystallinity and size of CNCs sample was meet the CNCs characteristic. Moreover, the recycling ability and less corrosivity of HPW made this the best among other acid hydrolysis method.

5. References

[1] Abdullah N, Sulaiman F, and Gerhauser H 2011 Characterization of palm oil empty fruit bunches for fuel application *J. Phys. Sci.* **22** 1–24

[2] Richana N, Winarti C, Hidayat T, and Prastowo B, 2015 Hydrolysis of empty fruit bunches of palm oil (*Elaeis Guineensis Jacq.*) by chemical, physical, and enzymatic methods for bioethanol production *Int. J. Chem.Eng.App.* **6** 422-6

[3] Lam E, Male K B, Chong J H, Leung A C W, and Luong J H T 2012 Applications of functionalized and nanoparticle-modified nanocrystalline cellulose *Trends Biotechnol.* **30** 5

[4] Li B, Xu W, Kronlund W, Maattanen A, Smatt J H, Peltonen J, Willfor S, Mu X, and Xu C 2015 Cellulose nanocrystals prepared via formic acid hydrolysis followed by TEMPO-mediated oxidation *Carbohyd. Polym.* **133** 605–12

[5] Roman M, Dong S, Hirani A, and Lee Y 2009 Cellulose nanocrystals for drug delivery ACS Sym. Ser. **1017** 81-91

[6] Kaushik M, Friedman H M, Bateman M, and Moores A 2015 Cellulose nanocrystals as non-innocent supports for the synthesis of ruthenium nanoparticles and their application to arene hydrogenation *RSC Adv.* **5** 53207–10

[7] Ooi S Y, Ahmad S, and Amin M C I M 2016 Cellulose nanocrystals extracted from rice husks as a reinforcing material in gelatin hydrogels for use in controlled drug delivery systems *Ind. Crops and Prod.* **93** 227–34

[8] Singla R, Son, S, Kulurkar M, Kumari A, Mahesh S, Patial V, Padwad Y S, and Yadav S K 2007 In situ functionalized nanobiocomposites dressings of bamboo cellulose nanocrystals and silver nanoparticles for accelerated wound healing *Carbohyd. Polym.* **155** 152–62

[9] Tang Y, Yang S, Zhang N, and Zhang J 2014 Preparation and characterization of nanocrystalline cellulose via low-intensity ultrasonic-assisted sulfuric acid hydrolysis *Cellulose* **21** 335–46

[10] Yu H Y, Qin Z Y, Liang B L, Liu N, Zhou Z, and Chen L 2013 Facile extraction of thermally stable cellulose nanocrystals with a high yield of 93% through hydrochloric acid hydrolysis under hydrothermal conditions *J. Mater. Chem A* **1** 3938–44

[11] Espinosa C S, Kuhnt T, Foster E, and Weder C 2013 Isolation of thermally stable cellulose nanocrystals by phosphoric acid hydrolysis *Biomacromolecules* **14** 1223-30

[12] Lin N and Dufresne A 2013 Supramolecular hydrogels from in situ host−guest inclusion between chemically modified cellulose nanocrystals and cyclodextrin *Biomacromolecules* **3** 871−80

[13] Liu Y, Wanga H, Yua G, Yua Q, Li B, and Mua X 2014 A novel approach for the preparation of nanocrystalline cellulose by using phosphotungstic acid *Carbohyd. Polym.* **110** 415–22

[14] Lu Q, Cai Z, Lin F, Tang L, and Huang B 2016 Extraction of cellulose nanocrystals with a high yield of 88% by silmutaneous mechanochemical activation and phospotungstic acid hydrolysis *ACS Sustain. Chem. Eng.* **4** 2165-72

[15] Shimizu K, Furukawa H, Kobayashi N, Itaya Y, and Satsuma A 2009. Effects of bronsted and lewis acidities on activity and selectivity of heteropolyacid-based catalyst for hydrolysis of cellubiose and cellulose *Green Chem* **11** 1627-32

[16] Neto F W Silvério H Dantas N, and Pasquini D 2013 Extraction and characterization of cellulose nanocrystals from agro-industrial residue – Soy hulls *Ind. Crops Prod.* **42** 480-8

[17] Liu Y, Wanga H, Yua G, Yua Q, Li B, and Mua X 2014 A novel approach for the preparation of nanocrystalline cellulose by using phosphotungstic acid *Carbohyd. Polym.* **110** 415–22

[18] Park S, Baker J, Himmel M, Parilla P, and Johnson D 2010 Cellulose crystallinity index: measurement techniques and their impact on interpreting cellulase performance *Biotechnol. Biofuels* **3** 10

[19] Barizão É, Boeing J, Martins A, Visentainer J, andAlmeida V 2015 Application of response surface methodology for the optimization of ultrasound-assisted extraction of pomegranate (Punica granatum L.) seed oil *Food Anal Methods* **8** 2392-400

[20] Montgomery D C 2001 *Design and analysis of experiments* (New York: Wiley)

IOP Conf. Series: Earth and Environmental Science **105** (2017) 012064 doi:10.1088/1755-1315/105/1/012064

Exergetic-Economic analysis and optimization of solar assisted heat pump using Multi-objective Genetic Algorithm

Nasruddin, M I Alhamid and N Aisyah

Faculty of Engineering, Universitas Indonesia, Kampus Baru UI Depok, Depok 16424, Indonesia

Email: nasruddin@eng.ui.ac.id

Abstract. This study proposes the use of two-stage heat pump systems (SAHPs) for high temperature applications, 105°C. This system integrates solar thermal collectors and heat pumps into a hybrid system to meet the 400 kW heating load. The aim of this research is providing a method to deliver heat with sustainable energy resource, than to improve the performance of the system which is indicated by low exergy destruction. The model creation, performance evaluation and the optimization of solar assisted heat pump system are discussed in this paper. This system used R1234ze (E) as working fluid. A genetic algorithm is employed to optimize operation condition of the system. To ensure that the optimal solution obtained from the proposed method is an optimum condition, three constraints are selected, including evaporation temperatures, condensing temperatures and compressor temperatures while exergy destruction and total cost as the objective functions. The result showed that the system has an optimum condition at evaporating temperature of 317 K, Flash Tank temperature of 353.6 K and condensing temperature of 380.4 K with exergy destruction of 70.21 kW and total cost of 63,441 US$.

1. Introduction

Heat pumps are device that can upgrade heat from low to high temperature level by integrated it with external energy source such as solar energy [1,2]. Heat pumps can greatly reduce the use of energy in the system while producing adequate heating performance [2,3]. Nowadays, heat pumps are available for low temperature applications (below 60° C) in the residential sector, but have yet in industry applications (above 90° C) [4]. To make heat pumps more attractive for industrial applications, it is important to demonstrate their potential.

Some studies have been conducted on analysis of high temperature heat pump. Chaturvedi et al., investigated the capability of two stage heat pump system of meeting required heating load. They used R134a for the refrigerant and one-cover solar collector. The result showed that thermal performance of the system had an improvement by using two stage systems with high condensing temperature [5]. Another research has been carried out by Wu XK [6]. They evaluated the systems for heating process in industry with 80-90° C target temperature. The result showed the COP of the system reached 5.0. It indicates that heat pump is very promising in high temperature application.

In addition, optimum condition of SAHP systems such as good performance and total cost efficient is important to assure. Studies on the optimal condition for SAHP systems are helpful in increasing the performance of the systems. Khorasaninejad performed solar assisted heat pump optimization using particle swamp [7]. Daley and Redlund found possibilities for improving the heat pump system performance by modeling and doing an optimization using MATLAB [8].

Therefore, this paper presents an optimization method by using multi objective Genetic Algorithm (GA). GA is employed to optimize the system in terms of exergy destruction and total cost. This study has been set three constraint, including temperature in evaporator, compressor and condenser to ensure that the optimal condition can be obtained.

2. Methodology

2.1. Model Description

A schematic diagram of solar assisted heat pump cycle which is discussed in this paper is demonstrated in figure 1.

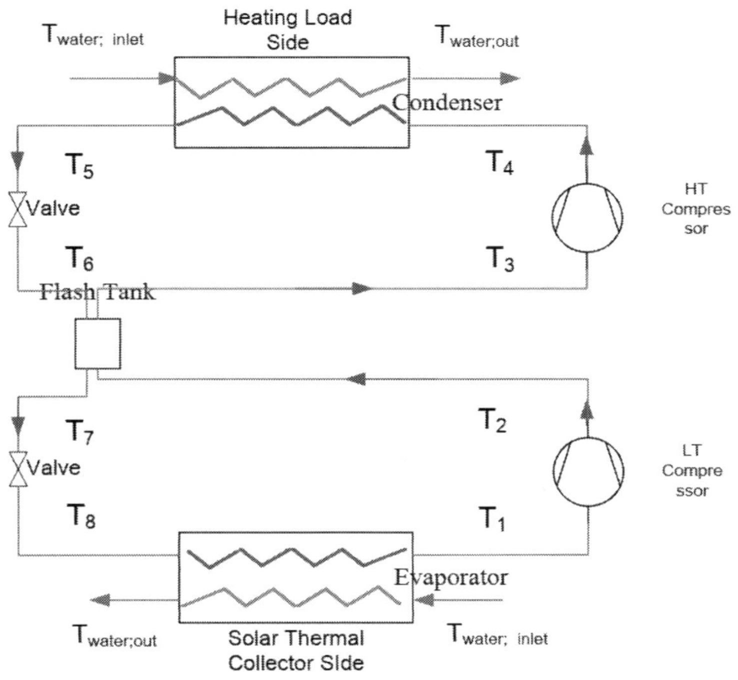

Figure 1. Schematic diagram of Solar Assisted Heat Pump.

The system consists of solar thermal collector and heat pump system. Evacuated tubular solar collector help the heat pump system to absorb energy from the sun in order to raise the temperature of refrigerant in evaporator. The heat pump system consists of evaporator, two compressors, high temperature compressor and low temperature compressor, condenser and two expansion valves. The evaporator absorbs heat from solar thermal collector in the evaporating temperature T_E. The condenser delivers heat to raise the temperature of water up to 105° C at condensing temperature T_C. The heat rejected from the condenser is equal to the sum of the heat absorbed by evaporator and the work of two compressors.

R1234ze(E) working fluid is used as a working fluid in this system. The reason for selecting R1234ze(E) as a refrigerant for this system is because of its thermodynamic properties and their effect for environment. The critical temperature of R1234ze is suitable for a high temperature heat pump, up to 110°C and R1234ze (E) has been nominated as an alternative for R134a, because it has a low GWP of around 6 [9].

2.2. Thermodynamic and exergetic analysis

Thermal modeling includes mass, energy and exergy balance are very important to improve system efficiency [10-12]. In this paper, components modeling are using some assumptions as follows:

1. The system operates under steady state condition.
2. The pressure and heat loss in pipe lines of the system are neglected.
3. Saturated refrigerant occurs at the exit of evaporator and condenser.
4. The kinetic and potential energies are not considered for the exergy analysis.

2.2.1. Energetic and Exergetic Analysis. The energy and exergy balance in the system components are list in table 1.

Table 1. Calculation of thermodynamic analysis.

Component	Mass	Energy	Exergy destruction
Evaporator	$m_1 = m_8$	$Q_{evap} = m_{ref;LT}\,(h_1 - h_8)$	$Ex_{D;evap} = [m_{ref;LT}\,(Ex_8\text{-}Ex_1)]+[m_w(Ex_{win}\text{-}Ex_{w;out})]$
Compressor			
LT	$m_2 = m_1$	$W_{LT;c} = m_{ref;LT}\,(h_2 - h_1)$	$Ex_{D;LTc} = [m_{ref;LT}\,(Ex_1\text{-}Ex_2)]+[\,W_{LT;c})]$
HT	$m_3 = m_4$	$W_{HT;c} = m_{ref;HT}\,(h_4 - h_3)$	$Ex_{D;HTc} = [m_{ref;HT}\,(Ex_3\text{-}Ex_4)]+[\,W_{HT;c}]$
Condenser	$m_4 = m_5$	$Q_{cond} = m_{ref;HT}\,(h_4 - h_5)$	$Ex_{D;cond} = [m_{ref;HT}\,(Ex_4\text{-}Ex_5)]+[1\text{-}\,T_w/T_e]$
Valve			
LT	$m_5 = m_6$	$h_5 = h_6$	$Ex_{D;LTv} = [m_{ref;LT}\,(Ex_5\text{-}Ex_6)]$
HT	$m_7 = m_8$	$h_7 = h_8$	$Ex_{D;HTv} = [m_{ref;HT}\,(Ex_7\text{-}Ex_8)]$

The coefficient of performance and the exergy efficiency [13] are calculated as follows :

$$COP = \frac{Q_{cond}}{W_{comp;LT} + W_{comp;HT}} \tag{1}$$

$$Ex_{in} = W_{HT;comp} + W_{LT;comp} + Q_{solarcollector} \tag{2}$$

$$Ex_{D;tot} = Ex_{D;evap} + Ex_{D;compLT} + Ex_{D;compHT} + Ex_{D;valveLT} + Ex_{D;valveHT} \tag{3}$$

$$Ex_{eff} = 1 - \frac{Ex_{D;tot}}{Ex_{in}} \tag{4}$$

2.2.2. Economic Analysis. The influence of the price of each component of the heat pump system greatly affect the election in consideration of solar assisted heat pump system as the best option for process heating. Total Annual Cost (TAC) will represent the economic value of the system, while TAC consists of investment costs, operating electrical appliances. Before calculating the cost components, it is necessary to do the following assumptions:

1. Age-made systems (lifetime, n) for 20 years.
2. Total operating time (H) for 7000 hours per year.
3. The cost of electricity (Cel) amounted to 0.08 US$ per kWh.
4. Interest rates (interest rate, i) 12%.

The equation for calculating the components of the overall cost of the system can be described as follows [14]:

$$C_{total} = C_{investation} + C_{operational} \tag{5}$$

$$C_{investation} = \left[\left[C_{inv,cond} + C_{inv,evap} + C_{inv,comp} + C_{inv,coll}\right] x\ CRF\right] \tag{6}$$

$$C_{operational} = \left[C_{elec} \cdot \square \cdot \left[C_{comp} + W_p\right]\right] \tag{7}$$

While Capital Recovery Factor (CRF) and cost for other components can be calculated as [14]:

$$CRF = \frac{i*(1+i)^n}{(1+i)^n - 1} \tag{8}$$

With the assumption of Cel is about 0.104 US$/kWh, operation hours 6570, lifetime of operation 20 years and interest rate 0.12 per year.

2.3. Multi-objective Optimization

To determine the optimal operation conditions of heat pump system, the optimization method is developed. This study uses Genetic Algorithm (GA) to optimize the system, where constraints with value: $313 \leq T_{evap} \leq 318$, $350 \leq T_{compHT} \leq 355$ and $379 \leq T_{cond} \leq 382$. The GA parameters used for the optimization procedure are: number of generations = 600, population size = 50, crossover probability = 0.7 and mutation probability = 0.2.

3. Result and Discussion

In this study, low GWP and ODP working fluids, R1234ze(E) is selected. The proposed optimization method was applied for modeling and optimizing the SAHP system. Exergy destruction and total cost are two objective functions in this optimization procedure. Low exergy destruction indicates that the system has a good performance, while as exergy destruction decrease, total cost of the system increases. To minimize exergy destruction and total cost, multi objective genetic algorithm (MOGA) was applied. Figure 2 shows pareto front as multi-objective optimization result.

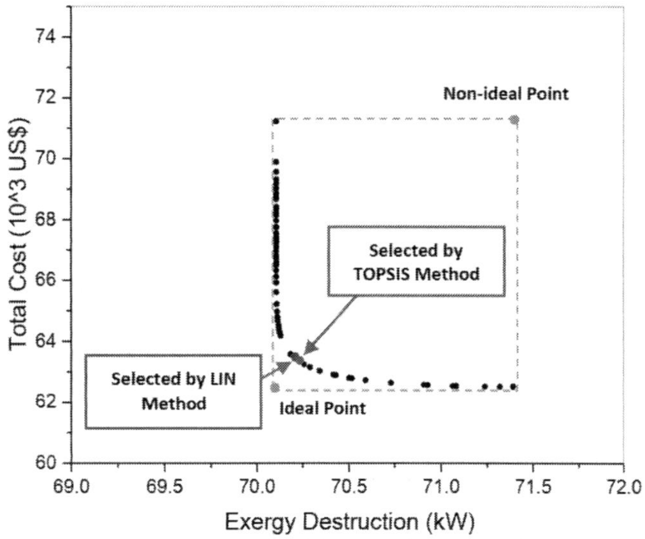

Figure 2. Decision making method.

The result showed that any change in decision variables that decreases the exergy destruction, leads to an increase in total cost and vice versa. Figure 2 gives an optimum condition of the system based on thermodynamic and economic. Due to so many optimum points offer by Pareto front as illustrated in figure 2 and the difficulties to choose the optimum if both thermodynamic and total cost are consider, a selection of a single optimum point is needed. Two common types of decision-making method, TOPSIS (Technique for Order Preference by Similarity to an Ideal Solution) and LINMAP (Linear Programming Technique for Multi-dimensional Analysis of Preference) are applied for selecting the optimum point. TOPSIS method was considered as one of the best method for selecting the optimum value of a number of optimum solutions. The optimum point is determined by selecting a point farthest to the non-ideal point and the point that is closest to the ideal point. The point selected by using the TOPSIS method was chosen as the point with optimum operating conditions. While LINMAP method chooses the best point by considering the distance from ideal point. The point that has a minimum distance from ideal point is selected as optimum point. Figure 4 and table 3 gives information about the optimum point selected by using both TOPSIS and LINMAP method.

IOP Conf. Series: Earth and Environmental Science **105** (2017) 012064 doi:10.1088/1755-1315/105/1/012064

Table 2. Optimum point from Pareto Front.

Parameter	TOPSIS Method	LINMAP Method
Evaporating Temperature (K)	317.95	317.90
HT Compressore inlet (K)	353.64	353.60
Condensing Temperature (K)	380.42	380.38
COP	3.58	3.56
Exergy Destruction (kW)	70.207	70.183
Total Cost (US$)	63,444	63,577

From table 2 it can be seen that both decision-making methods lead to the same values for exergy destruction but 100 US$ difference in total cost. Hence, the selected point by LINMAP method was considered as the final optimal design point. Table 4 shows the result of system optimization with single-objective function I (thermodynamic), single-objective function II (economic) and multi-objective (both thermodynamic and total cost).

Table 3. Single-objective vs Multi-objective.

Parameter	Single Objective I (Thermodynamic)	Single Objective II (Economic)	Multi-Objective
Evaporating Temperature (K)	318	317.4	317.6
HT Compressore inlet (K)	353.66	353.14	353.64
Condensing Temperature (K)	380.26	380.67	380.42
COP	3.58	3.54	3.56
Ex,des evaporator (kW)	0.925	1.11	1.03
Ex,des condenser (kW)	48.08	49.39	48.50
Ex,des LT comp (kW)	7.03	7.03	7.11
Ex,des HT comp (kW)	5.34	4.44	5.48
Ex,des LT valve (kW)	4.96	4.92	5.03
Ex,des HT valve (kW)	6.82	7.27	6.94
LT Compressor (US$)	46,774	46,687	46,971
HT Compressor (US$)	44,807	45,771	45,129
Condenser (US$)	18,288	17,375	17,911
Evaporator (US$)	25,152	17,028	18,916

The results show that the highest amount of exergy destruction occurs in condenser (48.08 kW, 49.39 kW and 48.50 kW for minimizing objective function I, objective function II, and optimization with multi-objective functions I and II respectively). Compressors and valves had the next highest amount of exergy destruction. Then, for economic point of view, LT compressor has the highest cost (46,774 US$, 46,687 US$ and 46,971 US$ for minimizing objective function I, objective function II, and optimization with multi-objective functions I and II respectively) while the condenser has the lowest cost. As is shown, the total cost of the system obtained from objective function I had the highest amount, while the total cost obtained from objective function II had the lowest amount, and the total cost obtained from both objective functions was in between the above two limits. Therefore, in multi-objective optimization technique, simultaneous analysis of economic and thermodynamic aspects was performed. Thus, this approach gives a reasonable trade-off between the first and the second objectives.

4. Conclusion

The optimization of solar assisted heat pump for high temperature application has been investigated in this paper. The aim of the research is to determine operating condition to make an optimum system both in thermodynamic and economic point of view. In optimization procedure there are three decision variables including evaporating temperature, flash tank temperature and condensing temperature. The result showed that Genetic Algorithm is a good solver to obtain the optimum condition of the system. The system has an optimum condition at evaporating temperature of 322 K, outlet compressor LT temperature of 348 K and condensing temperature of 379 K with 40% exergy efficiency, exergy destruction of 56.74 kW and total cost of 47,716 US$.

5. Acknowledgement

The authors would like to thanks to Ministry of Education of Indonesia (RISTEK DIKTI) for PMDSU Project.

6. References

[1] Zhang J, Zhang HH, He YL and Tao WQ 2016 A comprehensive review on advances and applications of industrial heat pumps based on the practices in China *Applied Energy* **178** 800-825
[2] Borgas A 2014 *Development of Hybrid Absorption Heat Pump Process at High Temperature Operation* (Norway: Institutt for energi-og prosessteknikk)
[3] Degrove JM 2015 *The Integration of Heat Resources in a Solar Thermal-Heat Pump Hydronic System* (USA: Purdue University)
[4] Lund RH 2016 *Optimized and Cost Efficient Industrial Heat Pumps* (Norway: NTNU)
[5] Chaturvedi SK, Abdel TM, Sreedharan SS and Gorozabel FB 2009 Two-stage direct expansion solar-assisted heat pump for high temperature application *Applied Thermal Engineering* **29** 2093-2099
[6] Wu X, Xing Z, He Z, Wang X and Chen W 2016 Performance evaluation of a capacity-regulated high temperature heat pump for waste heat recovery in dyeing industry *Applied Thermal Engineering* **93** 1193-1201.
[7] Khorasaninejad E and Hassan H 2014 Thermo-economic and environmental optimization of solar assisted heat pump by using multi-objective particle swam algorithm *Energy* **72** 680-690
[8] Daley B and Redlund O 2012 Mathematical modelling and thermodynamic optimization of CO2 heat pump cycle. (Sweden: Chalmers University)
[9] Fukuda S, Kondou C, Takata N and Koyama S 2013 Low GWP refrigerants R1234ze (E) and R1234ze (Z) for high temperature *International Journal of Refrigeration* **82** 160-173
[10] Bindra H, Bueno P and Morris JF 2014 Sliding flow method for exergetically efficient packed bed thermal storage *Appl. Therm. Eng* **64** 201–208.
[11] Feng H, Chen L, Liu X, Xie Z and Sun F 2016 Constructal optimization of a sinter cooling process based on exergy output maximization *Appl. Therm. Eng* **96** 161–166.
[12] Shojaeizadeh E and Veysi F 2016 Development of a correlation for parameter controlling using exergy efficiency optimization of an Al2O3/water nanofluid based flat-plate solar collector *Appl. Therm. Eng* **98** 1116–1129.
[13] Rezayan O and Behbahaninia A 2011 Thermoeconomic optimization and exergy analysis of CO2/NH3 cascade refrigeration systems *Energy* **36** 888–895.
[14] Bejan A and Tsatsaronis G 1996 *Thermal Design and Optimization* (New York: John Wiley & Sons)

Kinetic study of free fatty acid in Palm Fatty Acid Distillate (PFAD) over sugarcane bagasse catalyst

V A S Augustia[1], R A Djalal[2], B Sutrisno[1] and A Hidayat[1]
[1] Chemical Engineering Department, Universitas Islam Indonesia Jalan Kaliurang km 14,5 Yogyakarta 55584, Indonesia
[2] Industrial Engineering Department, Universitas Islam Indonesia Jalan Kaliurang km 14,5 Yogyakarta 55584, Indonesia

E-mail : arif.hidayat@uii.ac.id

Abstract. Biodiesel production is made by using edible or vegetable oils; however, their high prices and use as food resource are great limiting factors to their application. During the refining of palm oil, a lower-value by-product known as palm fatty acid distillate (PFAD) is generated in the fatty acid stripping and deodorization stages. PFAD is potentially a valuable, low-cost feedstock for the production of biodiesel. In this work, the esterification reactions of free fatty acid (FFA) in PFAD over sugarcane bagasse catalysts were studied. The effects of catalyst concentration (1 to 10 wt%), reaction temperature (30 to 60°C), and molar ratio of oil/methanol (1:6 to 1:12) on the conversion of FFA was studied to optimize the conditions for maximum conversion of FFA in PFAD and to observe the appropriate kinetic model for the experimental data. A kinetic model was developed on the basis of the Eley–Rideal mechanism according to the experimental data. The esterification reaction occurred between methanol adsorbed on solid base active sites and FFA from the liquid phase. The surface reaction of FFA with adsorbed methanol was assumed to be rate-determining. The collision factor value and activation energy for P-H model were found to be 1217.7985 L/mol.min and 24.2950 kJ/mol, then for E-R model were 1492.6375 L/mol.min and 19.5391 kJ/mol for E-R model, respectively.

1. Introduction

Increased demand for energy, global warming due to emission of greenhouse gases, environmental pollution, and fast diminishing supply of fossil fuels are the major key factors leading to search for alternative sources of energy [1]. Currently, biodiesel production is made by using edible or vegetable oils (e.g. palm oil, rapeseed, soybean and sunflower oil); however, their high prices and use as food resource are great limiting factors to their application. Therefore, nowadays lots of research done on the production of biodiesel with this material. Biodiesel has attracted lots of attention recently due to its benefits and alternative renewable energy.

Biodiesel is a product of esterification with triglyceride (vegetable oil or animal fat) and alcohol as raw materials. The products are methyl and ethyl esters [1]. During the refining of palm oil, a lower-value by-product known as palm fatty acid distillate (PFAD) is generated in the fatty acid stripping, deodorization stages, and low-cost feedstock for the production of biodiesel [18].

Esterification produces fatty acid methyl or ethyl esters by using some type of catalysts, those are homogeneous acid-catalyzed process, enzymatic process, supercritical process, and heterogeneous catalyst process [2 – 13, 20]. Heterogeneously catalysed esterification is very important and used in many applications due to its ease of purification products, reduction of water waste, and economic factor [6, 17].

Kusumaningtyas et al. (2013) in their paper examined the kinetics study of fatty acid esterification with sulphated zirconia/zeolite catalyst for producing biodiesel [12]. In this work, the esterification of PFAD was investigated, and the effects of catalyst concentration, reaction temperature, and molar ratio of oil/methanol on the conversion of FFA were studied to optimize the conditions for maximum conversion of FFA in PFAD and to observe the appropriate kinetic model for the experimental data. Moreover, a kinetic model was proposed on the basis of the Eley-Rideal mechanism and the kinetic parameters were determined by fitting model with the experimental data.

2. Experimental

2.1. Catalyst Synthesis

Sugarcane Bagasse Biochar (SCB) catalyst samples were prepared via concentrated sulfuric acid. The SCB were sulfonated according to the method of Toda et al. [9]. Two hundred mL of concentrated sulfuric acid was added to 20 g of biochar in a 500 mL round bottom flask and heated to 150°C. The SCB catalyst was washed with 80°C distilled water until the wash water was neutral and free from sulfate ions and dried in an oven at 70°C for 1h.

2.2. Catalytic activity

The reaction procedure is mixture of catalyst and methanol was heated up to 60˚C under continuous stirring at 500 rpm. Ninety ml of PFAD was heated separately to reach the reaction temperature. Esterification experiment was carried out at reaction time of 2 to 4 h. Biodiesel product was then repeatedly washed using warm distilled water (60°C). The variables affecting the acid-catalyzed esterification such as methanol-to-oil molar ratio (6:1–12:1), catalyst amount (1–10 wt. % of oil), reaction time (15–120 min) and temperature (30–60°C) were studied.
The acid value was calculated using the below equation:

$$AV = \frac{56.1 \times C \times V}{m} \tag{1}$$

where AV is the acid value of the sample (mg KOH.g-1), V is volume of KOH for titration (mL), C is concentration of the KOH for titration (mol/L), m is the weight of the sample (g), and 56.1 is the molecular weight of KOH. The conversion of free fatty acid was calculated using equation (2):

$$X_F = \left(1 - \frac{AV_2}{AV_1}\right) \times 100 \tag{2}$$

where XF is the FFA conversion, AV_1 is the initial acid value of the mixture and AV_2 is the acid value of mixture after reaction

2.3. Reaction kinetics model

The esterification of PFAD and methanol using SCB as a catalyst is represented as follows:

$$\underset{(F)}{RCOOH} + \underset{(A)}{CH_3OH} \leftrightarrow \underset{(E)}{RCOOCH_3} + \underset{(W)}{H_2O}$$

Reaction mechanisms of the esterification processes, the following reaction steps were proposed:

2.3.1. Pseudo-Homogeneous model. The P-H model was developed based on the following assumptions: (1) The rate of the esterification reaction under the operating conditions is kinetically controlled; and (2) The entire reaction system is considered to be an ideal solution in which internal and external mass transfer resistance does not exist. Therefore, the reaction mechanism can be given as follows

$$-\frac{dC_F}{dt} = k_1.C_F.C_A \tag{3}$$

the concentration of FFAs can be expressed as: $C_F = C_{F0}.(1 - X_F)$; while the amount of alcohol is excessive, thus the concentration of alcohol is constant. The X_F and C_{F0} refer to FFAs conversion and the initial concentration of FFAs. Where C_{A0} are the initial concentration of alcohol, respectively.

2.3.2. Eley-Rideal model. Heterogeneous esterification reaction can be described using Eley-Rideal (E-R) model, where, FFA (F) is not adsorbed on the catalyst surface and remains in the liquid medium and methanol (A) is adsorbed on the catalyst surface because of its polarity. Eley-Rideal model can be developed in the following equation below:

$$r_E = \frac{k.K_E.C_F.C_E}{1 + K_F.C_F + K_E.C_E} \tag{4}$$

In this case the criterion is the minimization of the Sum Squares of Errors (SSE) between the experimental and calculated reaction rates using the equation:

$$SSE = \sum (X_{F\,exp} - X_{F\,calc})^2 \tag{5}$$

3. Result and Discussion

3.1. Effect of methanol amount on the esterification reaction

The esterification is a type of reversible reaction. Therefore, to improve the conversion of this reaction was by adding amount of reactants so that the reaction becomes excessive. The effect of methanol amount added on the esterification reaction was shown in Figure 1. That figure showed that FFA conversion was enhanced while increasing the amount of methanol. Excess methanol (alcohol) added to this reaction made a greater collision among the reactants. Ratio 1:12 of PFAD to methanol gave the highest conversion of FFA, that was 91%. However, there was no significant increase in conversions that occur in this reaction using ratio PFAD to methanol 1:10 and 1:12. These results might be attributed to the maximum conversion and to the saturation of the catalytic surface with methanol or prevention of nucleophilic attack by shielding protonated methanol. That was confirmed by Eley-Rideal mechanism with chemisorption of the methanol [2].

3.2. Effect of catalyst loading on the esterification reaction

The amount of catalyst loading in this reaction was varied as 1%; 2.5%; 5%; and 10%. Figure 2 was the research results. It showed that the conversion of FFA increased with increasing catalyst amounts and the reaction for 10% catalyst loading had the highest conversion. This result might be caused by increasing the number of active sites. Low amount of catalyst loaded in the reaction produced a small number of active sites. Therefore, conversion of FFA became small due to a small number of catalytically active sites occupied by the reactants. Something similar to experiment conducted by Liu et al. (2014). By the experimental data, this reaction was dependent on the catalyst loading on the process and it was found that the conversion of FFA increased with the increasing amount of catalyst in range of 1-10%.

3.3. Effect of temperature on the esterification reaction

The other important variable affecting a conversion reaction is temperature. The effect of different temperature was observed in the range of 30-60°C. The results of conversion of FFA in this research was shown in figure 3. As shown in the figure, increasing esterification temperature caused an increase in conversion of FFA. This was because of higher molecule motion speed and mass transfer rate happened during esterification reaction [1]. By using high temperature might reduce the as-formed water molecules on the catalyst surface bounded to the active sites [1].

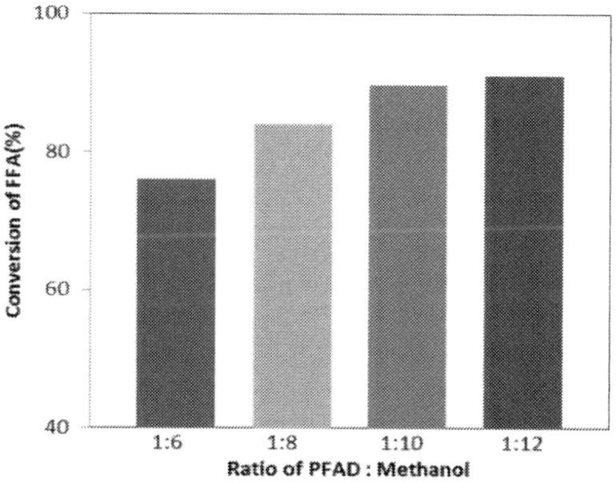

Figure 1. Effect ratio of PFAD to methanol (1:6; 1:8; 1:10; and 1:12).

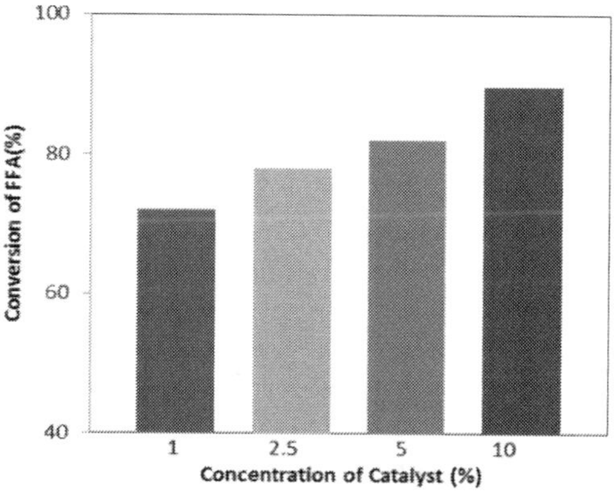

Figure 2. Effect concentration of catalyst on the esterification reaction (1%; 2.5%; 5%; and 10%).

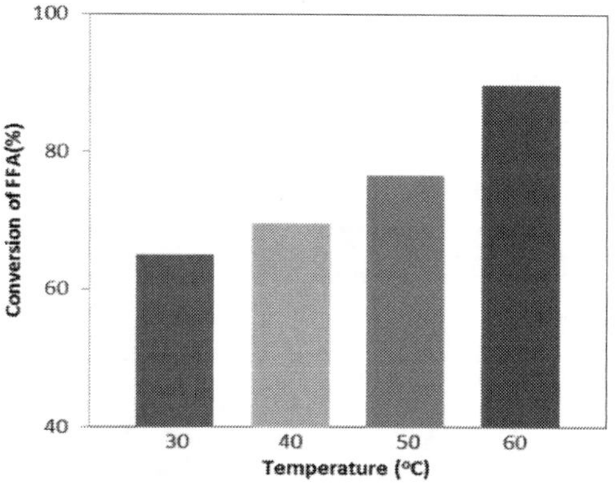

Figure 3. Effect of temperature on the esterification reaction (30°C; 40°C; 50°C; and 60°C).

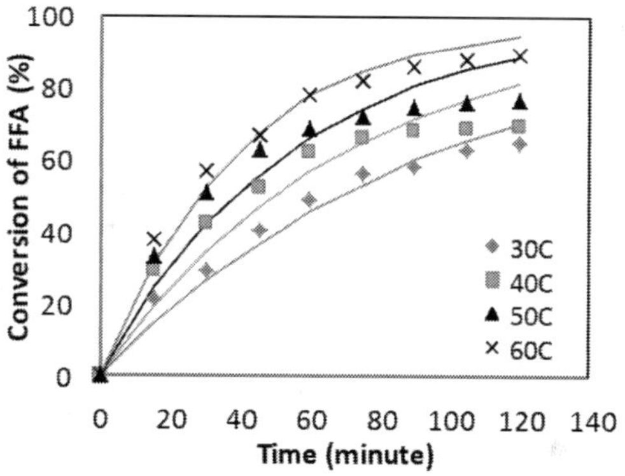

Figure 4. Comparison of simulation and experimental data (P-H Model).

3.4. Kinetics model

Esterification reaction using heterogeneous catalyst consists of some processes; those are diffusion, adsorption/desorption, and surface reaction. By those processes, various models can be used to approach kinetic of reaction. In this work, two models used to express the models are pseudo-homogeneous model and Eley-Rideal (E-R) model. The differences of these models are in the adsorption and desorption processes. Pseudo-homogeneous model neglect the processes and E-R model take an assumption that one of the adsorbed reactants react with the others [12].

Figure 4 and 5 are the comparison between the simulation and experimental data based on P-H model and E-R model. Furthermore, from those figures, the E-R mechanism shows that this model can lead to the esterification reaction which is better than P-H model. This was conclude that the heterogeneous kinetic model (E-R model) was suitable to correctly describe the reaction kinetics.

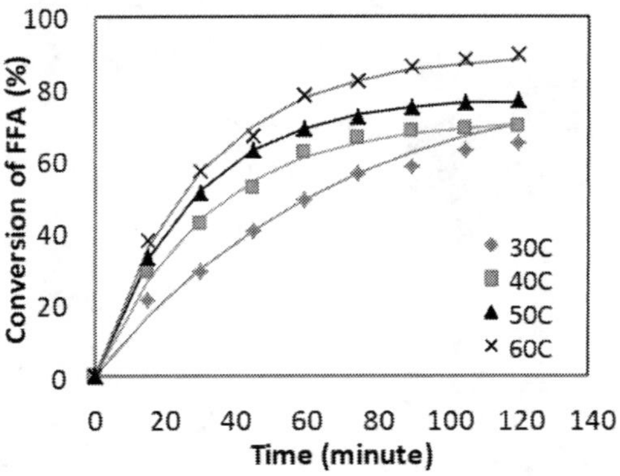

Figure 5. Comparison of simulation and experimental data (E-R Model).

From table 1, the data of reaction rate constant (*k*) shows that increasing temperature will increase the value of *k*. The value of *k* will be used to calculate the value of collision factor (Ar) and activation energy (Ea) by using the Arhenius equation. From that equation, the value of Ar and Ea obtained were 1217.7985 L/mol.min and 24.2950 kJ/mol for P-H model, then 1492.6375 L/mol.min and 19.5391 kJ/mol foe E-R model.

Table 1. Value of *k* in different temperature.

Temperature (°C)	Pseudo-Homogeneous Model		Eley-Rideal Model	
	k (L/mol/min)	*k* Arhenius (L/mol/min)	*k* (L/mol/min)	*k* Arhenius (L/mol/min)
30	0.0792	0.0790	0.5558	0.6392
40	0.1085	0.1075	0.9339	0.8190
50	0.1392	0.1435	1.2138	1.0333
60	0.1915	0.1883	1.0934	1.256

4. Conclusion

The kinetic of esterification reaction was carried out using sugarcane bagasse biochar as catalyst at temperatures from 30 to 60°C. The experimental data from this reaction were fitted to P-H model and E-R model. Increasing the mole ratio of PFAD to methanol, concentration of catalyst, and temperature successfully increases the conversion of FFA. From the simulation results, the value of *k* increase with increasing reaction temperature. The collision factor value and activation energy for P-H model were found to be 1217.7985 L/mol.min and 24.2950 kJ/mol, then for E-R model were 1492.6375 L/mol.min and 19.5391 kJ/mol for E-R model, respectively.

5. References

[1] I. Fatimah, A. Hidayat, and K. H. Setiawan *Asian Journal of Chemistry* **22** (5), 3793-3801 (2010).
[2] Liu W, Zhang J, Tang Q and Qu R 2014 *Energy Conversion and Management* **82** 83 – 91
[3] Edgar L, Liu Y, Lopez D E, Kaewta S, Bruce D A and Goodwin J G 2005 *Ind Eng Chem Res* **44** 5353 – 5363
[4] Saka S and Kusdiana D 2001 *Fuel* **80** 225 – 231
[5] Madras G, Klluru C and Kumar R 2004 *Fuel* **83** 2029 – 2033
[6] Kamini N R and Lefuji H 2007 *Process Biochem.* **37** 405 – 410
[7] Kawashima A, Matsubara K and Honda K 2009 *Bioresour. Technol.* **100** 696 – 700
[8] Jitputti J, Kitiyanana B, Rangsunvigita P, Bunyakiata K, Attanathob L and Jenvanitpanjakulb P 2006 *Chem. Eng. J.* **116(1)** 61 – 66.

[9] Hidayat A and Sutrisno B 2017 *Esterification free fatty acid in sludge palm oil using* ZrO_2/SO_4^{2-} *- Rice husk ash catalyst. 1840*, Article number 050001.
[10] Olivera C F, Dezaneti L M, Garcia F A C, Macedo J L, Dias J A, Dias S C L, et al. 2010 *Appl. Catal.* **372** 153 – 161
[11] Serio M D, Cozzolino M, Tesser R, Patrono P, Pinzari F, Bonelli B, et al. 2007 *Appl. Catal.* **320** 1 – 7
[12] Qu Y, Peng S, Wang S, Zhang Z and Wang J 2009 *Chin. J. Chem. Eng.* **17(5)** 773 – 780.
[13] Hidayat A, Rochmadi, Wijaya K, Nurdiawati A, Kurniawan W, Hinode H, Yoshikawa K and Budiman A 2015 *Energy Procedia* **75** 969-974
[14] Felizardo P, Machado J, Vergueiro D, Correia M J N, Gomes J P and Bordado J M 2011 *Fuel Process Technol.* **92** 1225 – 1229
[15] Kusumaningtyas R D, Masduki, Hidayat A, Handayani P A, Rochmadi, Purwono S and Budiman A 2013 1 – 16.
[16] Toda M, Takagaki A, Okamura M, Kondo J N, Hayashi S and Domen K 2005 *Nature* **438** 178
[17] Hidayat A, Rochmadi, Wijaya K, and Budiman A 2015 *AIP Conference Proceedings* 1699, 050005
[18] Hidayat A, Rochmadi, Wijaya K, and Budiman A 2016 *IOP Conference Series: Materials Science and Engineering* **105** (1), 012026.

IOP Conf. Series: Earth and Environmental Science **105** (2017) 012066 doi:10.1088/1755-1315/105/1/012066

Effects of temperature on the direct reduction of Southeast Sulawesi's limonite ore

Yudianto[1], **Ahmad Maksum**[1, 2], **Delfiendra**[3] **and Johny Wahyuadi Soedarsono**[1]

[1]Center of Minerals Processing and Corrosion Research, Department of Metallurgical and Materials Engineering, Universitas Indonesia, Depok 16424, Indonesia

[2]Department of Mechanical Engineering, Politeknik Negeri Jakarta, Depok 16425, Indonesia

[3]Center for Testing and Identification of Goods, Directorate General of Customs and Excise, Jakarta, Indonesia

Email: jwsono@metal.ui.ac.id

Abstract. Mining industry is one of the most important industries for the improvement in people's value in daily lives. Indonesia has abundant mineral resources. Nevertheless, these minerals are still in the depths of the earth and need to be explored so that the minerals are on the surface of the earth. In addition, it is necessary to also extraction process to be more efficient in the next process. Therefore, the method of mineral extraction needs to be further developed to obtain optimal results with the best use of energy but at the most competitive cost. In this study, the objective is to know the effect of reduction temperature on the increase of nickel contents in nickel laterite process using a pyrometallurgy extraction which will be performed in the varying temperature of 700°C, 800°C, 900°C, and 1000°C. The mixture contains coal reducer which will reduce iron metal and supposedly increase the nickel contents in nickel laterite ore. Moreover, Na_2SO_4 will be added as a constant variable where the sulfur in Na_2SO_4 is known to help extracting the nickel laterite. The tests include XRD, AAS, and proximate and ultimate test from coal. The result of this study shows that 700°C is the optimum temperature in doing the nickel laterite reduction. Apart from some factors that affect the final result, it is only at the temperature of 700°C that the nickel contents facing an increase from 1.16% to 1.18% after being reduced in a muffle furnace for an hour.

1. Introduction

Metallic minerals inside the soil can be considered raw and needs further processes, so that it can be utilized properly. One of the processes is extraction which is a method of purifying the metal by separating it between the valuable minerals and the impurities. The process can also increase the content of valuable metal inside the minerals or unrefined ore. This study will discuss nickel, which its demands keep in rising around the world, and its mining process which cost less energy because it can be found easily on the soil surface.

The majority of the world's nickel demands are supported by nickel sulfide deposit [1]. It also needs to be known that the nickel sulfide deposit is only 28% of the total world's deposits, and the rest of 72% is nickel laterite. One of laterite types is limonite ore which has low-grade of nickel content. Indonesia's nickel deposit in the form of laterite oxide is estimated to reach 12% of world's nickel laterite deposit. In its function, nickel is widely used as an element that contributes to the manufacture of stainless steels, catalysts, and special steels, such as tool steel and protective steel [2].

Until this day, a low-grade nickel ore, like limonite, was still couldn't be processed effectively. Whereas, the volume of nickel limonite usually was 2-3 times bigger than the volume of saprolite. With this condition, the usage of limonite as a nickel sources through the extraction process to increase the grade of nickel inside, would be more possible Limonite ore contains some of metallic elements like nickel, cobalt, and also ferrous with almost same grade like ferrous ores in common, which is around 60% of ferrous inside a limonite. In everyday life, we could see the application of nickel e.g. for stainless steel making, catalyst, and also in making of specific steel (like tool steel and shield steel) [3].

There are some factors that affect the effectiveness in increasing nickel content in laterite ore i.e. the pellets formation process where the weight and the diameter will affect the percentage of adding Na_2SO_4 as a chemical element for leaching process, and the temperature where the oxidative roasting and the reduction of nickel laterite is done. From those factors, an experiment has been done in varying situations and previous measures. In this study, the author will try to do nickel laterite extraction in order to increase the nickel content using some variables that have been mentioned above. The reduction temperature and condition when reducing the laterite will be the prime variable in this study.

2. Materials and method

2.1. Materials

Limonite ore samples used in the study carry out initial characterization by XRD test. Figure 1 provides the XRD test result of nickel ore. From the data in figure 1 it can be seen that the Fe contents in the nickel limonite sample were considered as high. As for the nickel contents, about 1.16% in nickel limonite which is a low-grade ore, have fitted the required range of 0.8% - 1.5%. Based on the composition phase in nickel limonite sample that was used, some phases could be seen, such as goethite ($FeO(OH)$), kaolinite ($Al_2Si_2O_5(OH)_4$), lizardite $(Mg,Fe)_3Si_2O_5(OH)_4$, and quartz (SiO_2). This study was in accordance with Connor and his colleagues' research [4] that inside the nickel laterite, which has not been reduced yet, contains many goethite, kaolinite, and lizardite. Moreover, in the solution, Fe was oxidized and settled as a ferric-hydroxide and eventually formed some minerals like goethite, hematite, magnetite, and limonite near the surface.

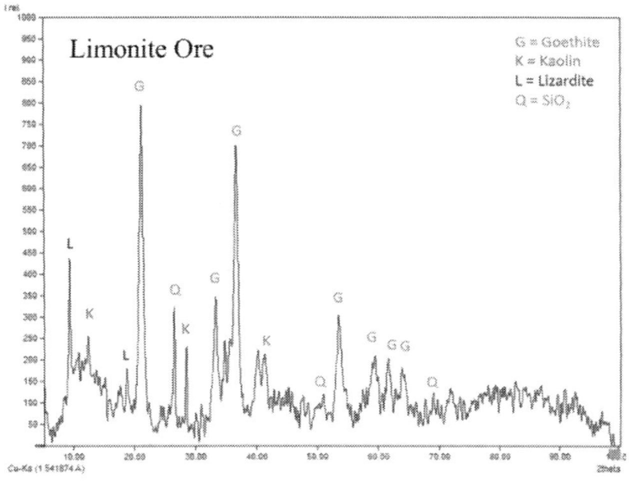

Figure 1. XRD pattern of initial nickel lateritic ore.

While for the quantity test, the AAS test is performed to see the Fe and Ni contents in the limonite ore. The AAS test result can be seen in Table 1.

Beside the limonite, coal as a reductor was also tested with proximate and ultimate test to see the composition of the coal. The test results can be seen in Table 2.

Table 1. Atomic absorption spectroscopy analysis of iron, nickel, and cobalt content from limonite ore.

Element	Content (%)
Fe	41.46
Ni	1.16
Co	0.08

Table 2. Proximate and Ultimate Analysis of Coal.

Parameter	Content (% adb)	Standard Method
Proximate Analysis		
Moisture in air dried sample	7.26	ASTM D.3173
Ash	15.87	ASTM D.3174
Volatile Matter	16.44	ASTM D.3175
Fixed Carbon	60.43	ASTM D.3172
Ultimate Analysis		
Carbon	64.02	ASTM D.5373
Hydrogen	3.11	ASTM D.5373
Nitrogen	0.60	ASTM D.5373
Total Sulphur	0.08	ASTM D.4239
Oxygen	16.32	ASTM D.3176

2.2. Experiment

The process included the sample preparation (initial characterization and ball milling), the reduction, and the final characterization. The sample preparation included preparing nickel laterite ore type limonite that was going to be used. After the laterite ore has been prepared and already measured to the required size and quantity, the sample then was mixed with the coal reducer using ball milling method. After that, the sample was ready for the extraction process to increase the nickel content in laterite ore using muffle furnace. The carbothermic process or what is usually called the reduction was done with the approximate temperature of 700 to 1000°C. Then, it was followed by the sample characterization using XRD and AAS tests and analysis.

3. Result and discussion

In the making of pellets, nickel limonite was mixed with coal reducer using a ball mill. After the nickel limonite and coal reducer have been mixed properly, sodium sulfate (Na2SO4) and bentonite, as the binders, were added. Then, pellets were made with the weight of 10 grams each. In this study, the reduction temperatures were 700°C, 800°C, 900°C, and 1000°C. Each reduction process in each temperature was held for an hour. Then, the obtained product was analyzed using XRD and AAS, as shown in table 3 and figure 2.

In the nickel laterite extraction research performed by M. Valix and colleagues [5], it was mentioned that starting at temperature above 600°C is the most optimum temperature in the nickel limonite reduction and held for about 30-60 minutes. The recovery calculation showed that the temperature of 700°C was the optimum temperature in increasing the nickel contents in the reduction result sample.

Table 3. Comparison of nickel and iron content of limonite between ore and after reduction samples.

Element	Ore	After reduction				Unit
		700°C	800°C	900°C	1000°C	
Fe	41.46	30.57	16.67	7.61	13.49	% w/w
Ni	1.16	1.18	0.82	0.56	1.00	% w/w
Co	0.08	0.07	0.06	0.05	0.06	% w/w

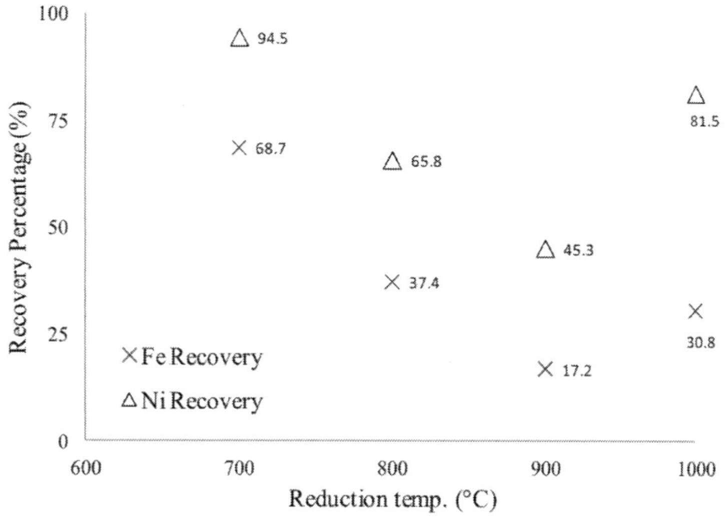

Figure 2. Result of Nickel Recovery in a graphic (Reduction Temp. vs Recovery Percentage).

However, there was no significant increase in nickel contents at all four temperatures. Only at the temperature of 700°C that the nickel contents increased by 0,02% and the recovery of 94,5% occurred because of these following possible reasons. In the nickel laterite extraction research performed by Xueyi Guo and colleagues [6], it is mentioned that starting at temperature above 600°C is the most optimum temperature in nickel limonite reduction and being held for about 30-60 minutes, but unfortunately, it took too long holding time on this research. First, the reduction process took place in the muffle furnace was too long. In each temperature of 700°C, 800°C, 900°C, and 1000°C, the nickel spent approximately 5 hours of reduction in the muffle furnace. This certainly made the reducer, which should reduce the Fe and was expected to increase the nickel content, drained out before it was properly reduced. The heating Fe was also too long that cause the metals to re-oxidize, which probably happened when the temperature was declined after being held for an hour. Consequently, those factors made the coal reducer drain out, the Fe re-oxidize inside the sample, and the nickel metallization obstruction. Besides, the magnetic elements inside the nickel limonite sample that were not separated made it possible for impurities to exist in the sample and eventually decrease the valuable minerals contents that were expected to be obtained.

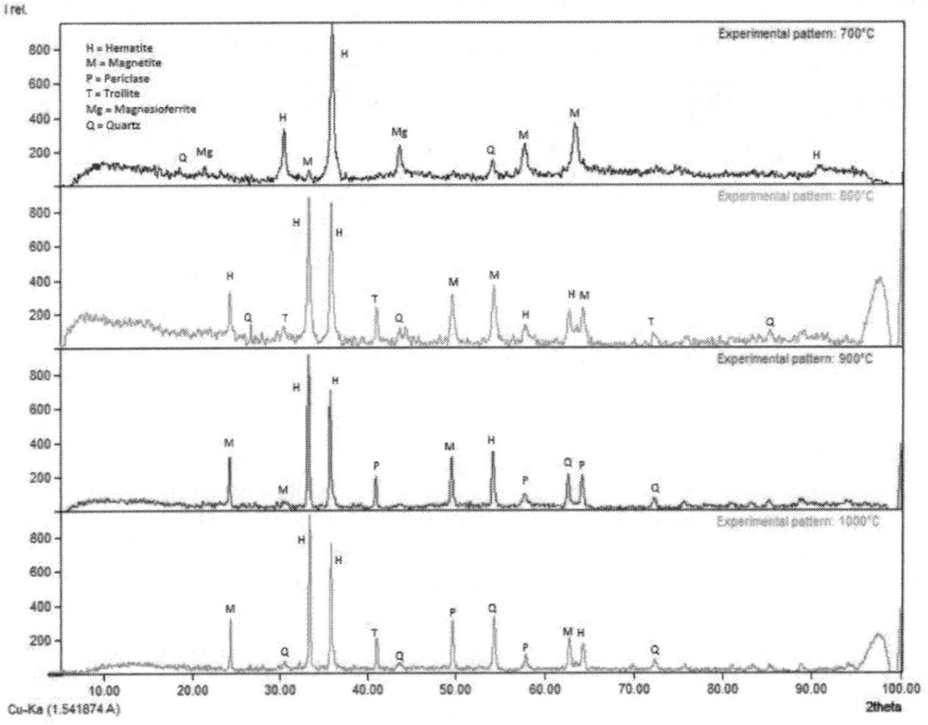

Figure 3. XRD pattern of nickel lateritic after the Reduction at 700°C-1000°C.

The reduction temperature of 700°C was the only temperature that was capable of increasing the nickel contents. It can be seen in the XRD test result in Figure 3. It shows that hematite phase (Fe_2O_3), was reduced quite a lot and it changed most of the hematite into magnetite (Fe_3O_4). According to a research by J. Lu and colleagues [7], goethite phase that existed in nickel laterite since the beginning was going to undergo some dehydroxylation at the temperature around 300°C and being reduced to become a hematite. Afterwards, hematite began to reduce to become magnetite at the temperature of 500°C. As for magnesioferrite (Fe_2MgO_4), it was formed by the reduction of MgO and reacted to Fe. Even so, the quartz phase was not yet being reduced because the minimum temperature of silica reduction was 1200°C which was mentioned by Samadhi [8].

At the temperature of 800°C, hematite phase was still the dominant one. Then, it was followed by troilite, magnetite, and quartz phases. This means that hematite phase has not been reduced properly into magnetite. The phase formed after the reduction from 700°C and above was magnetite which was the result of hematite reduction. For troilite (FeS) phase, after being added by Na_2SO_4 which contains sulfur element combine with a long duration of heat, made the Fe face a breakdown or an expansion with Na_2SO_4 that made the sulfur easily react with other metals inside the nickel laterite ore. Nevertheless, quartz phase was only reduced under 1200°C, so that the quartz phase would still be formed.

The XRD test result at the temperature of 900°C decreased the magnetite contents from about 13%, from temperature of 800°C, to ±6%. Nonetheless, in periclase (MgO) phase, the contents which were 7.2%, was increasing compared to the initial sample. It might happen because the reduction was done imperfectly. The periclase has not been fully reduced and it reacted with other elements in nickel laterite. As what has been seen before at temperature of 700°C, the MgO phase has reacted and reduced better, so that it formed magnesioferite (Fe_2MgO_4) phase. In nickel laterite ore that contained high periclase (MgO) and low magnetite (Fe) reduction samples like the one in XRD test at the temperature of 900°C, the nickel reduction was very low. That was due to the tendency of the nickel to replace the magnesium in magnesium silicate which can be seen in Johny et al. research [9]. Also, the quartz (SiO_2) phase was still formed although it continued to decrease compared to the XRD result at 700°C and 800°C.

In the temperature of 1000°C the nickel and iron contents which were previously decreasing from 700°C to 800°C and 900°C, showed an increase at 1000°C. However it was still lower than the initial sample and the reduction sample at 700°C. The same result appeared in the XRD test also where the content of magnetite and quartz phases were increasing compared to the previous sample at 900°C. Meanwhile, the formation of plenty MgO decreased the Ni reduction, so that Ni reacted with magnesium silicate. Moreover, troilite (FeS) reappeared at the temperature of 1000°C eventhough only a small amount of it. For quartz phase, the contents were quite many and kept on increasing as the theory said that this phase will be reduced when the temperature hits 1200°C.

4. Conclusion

Nickel laterite ore was not properly reduced because the test result shows that the nickel contents do not increase significantly. Even though the reduction was not performed well enough from the four different temperatures, the temperature of 700°C was the most significant one since Ni content has increased from 1.16% to 1.18% and gained the highest nickel recovery of 94.8%. The XRD test result showed that reduction at 700°C was able to reduce hematite (Fe_2O_3) into magnetite (Fe_3O_4) quite well in terms of the magnetite content that was formed.

5. Acknowledgement

Special thanks are due to Mr. Agus Budi Prasetyo (Research Centre for Metallurgy, Indonesian Institute of Sciences - LIPI, Indonesia) for helpful supports. The authors also wish to thank the Ministry of Research Technology and Higher Education and Directorate of Research and Community Engagement, Universitas Indonesia, under PITTA Research Grants Contract No. 791/UN2.R3.1/HKP05.00/2017 for the financial support of this research.

6. References

[1] Anderson C G, Dunne R C, and Uhrie J L 2014 *Mineral Processing and Extractive Metallurgy:100 years of innovation.*
[2] Butt C R M, and Cluzel D 2013 Nickel Laterite Ore Deposits: Weathered Serpentinites Elements *gselements* **9** 123-128
[3] Maliy V M, and Bogdanova I P 1992 High Intensity Magnetic Separation of Limonite Iron Ores **4** 47-59
[4] Connor F O, Cheung W H, and Valix M 2006 Reduction Roasting of Limonite Ores : Effect of Dehydroxylation *International Journal of Mineral Processing* **80** 88-99
[5] Valix M, and Cheung W H 2002 Study of Phase Transformation of Laterite Ores at High Temperature *Miner. Eng* **15** 607-612
[6] Xueyi G, Dong L, Kyung-Ho P, Qinghua T, and Zhan Wu 2009 Leaching Behavior of Metals From A Limonitic Nickel Laterite Using A Sulfation–Roasting–Leaching Process *Hydrometallurgy* **99** 144-150
[7] Lu J, Liu S, Shangguan J, Du W, Pan F, and Yang S 2013 The Effect of Sodium Sulphate On The Hydrogen Reduction Process of Nickel Laterite Ore *Miner. Eng* **49** 154-164
[8] Samadhi T W 2017 Thermochemical Analysis Of Laterite Ore Alkali Roasting: Comparison Of Sodium Carbonate, Sodium Sulfate, And Sodium Hydroxide *AIP Conf. Proc.* **1805** 1-5
[9] Soedarsono J W, Kawigraha A, Sulametariobimo R D, Asy M A, Yosi A, and Putra E M 2013 The Influence of Coal and Reduction Process Parameters in Producing Iron Nugget *Adv. Mater. Res.* **789** 517-521

IOP Conf. Series: Earth and Environmental Science **105** (2017) 012067 doi:10.1088/1755-1315/105/1/012067

Effect of pellet size addition on the selective reduction of limonite ore from Southeast Sulawesi

Muhammad Azif[1], Ahmad Maksum[1, 2], Delfiendra[3] and Johny Wahyuadi Soedarsono[1]

[1]Center of Minerals Processing and Corrosion Research, Department of Metallurgical and Materials Engineering, Universitas Indonesia, Depok 16424, Indonesia
[2]Department of Mechanical Engineering, Politeknik Negeri Jakarta, Depok 16425, Indonesia
[3]Center for Testing and Identification of Goods, Directorate General of Customs and Excise, Jakarta, Indonesia

Email : jwsono@metal.ui.ac.id

Abstract. With the increasing demands of nickel in the world and the decreasing of nickel sulfide deposits, the processing of nickel laterite is the new challenge with selective reduction by using coal as reducing agent and Na2SO4 additives. This research aims to determine the preparation parameters of the pelletizing method before the reduction process in order to increase the content of the nickel in the reduction product. The samples were prepared by mixing limonite ore, coal, bentonite, and Na2SO4 additives, then continued by making the pellets with the variability of mass measurements of 5 grams, 10 grams and 15 grams. After the selective reduction, they were characterized by XRD to determine the formed phase, and an AAS test to determine the content after selective reduction. The test result of XRD showed the forming of magnetite, which were found in the pellets with mass size 5 grams. After all, fayalite phase was found in each size of pellet samples. The largest iron and nickel content was shown in the 5 gram pellets, 25% and 0.85%, respectively

1. Introduction

The processing of nickel laterite ore is about 40% of world total nickel production, therefore it will contribute in the scarcity of nickel sulfide in the future if it is continuously mined and exploited [1,2]. With the increasing demands of nickel in the world and the decreasing of nickel sulfide deposits, the alternative is the exploitation of nickel oxide with various methods and innovation so that the needs and demand for nickel can be maintained. One of the methods to increase the content of nickel laterite is the selective reduction method, which is the process of ore reduction with carbon reductant and addition of the additives which can increase mineral selectivity to be obtained at the reduction stage. One element added as an additive is sulfur because sulfur from some studies is considered capable of increasing the effectiveness of reduction and increasing the acquisition of nickel with other metals [3,4]. In laterite ore, the presence of sulfur accelerates the formation of FeS phase and makes ferronickel phase becomes larger so the value can be increased, as well as suppressing the growth rate of fayalite and forsterite in nickel laterite reduction process [5]. From various studies that have been done, the variables that are often used are the time of reduction, temperature, and concentration of the Na_2SO_4. In the preparation process of specimens conducted such as pellet sample preparation. The purpose of the use of pellets is that the interaction between the laterite nickel ore with the coal reducing agent and the additive progresses better so that the reduction process of nickel oxide and the reaction of iron sulfide formation takes place more effectively, thus the selective nickel reduction process laterit can be better [6]. Further research has not been done and that may affect the course of the reduction process on laterite nickel. This research aims to find the effect of pellet's variable with its physical characteristics possessed against the value of acquisition by the selective reduction method.

2. Materials and method

2.1. Materials

The material used is limonite nickel from South Sulawesi, Indonesia. initial sample of limonite used in this research was initially characterized with the test of XRD, XRF, and AAS. According to Zevgolis et al. [7] there are three types of nickel laterite deposits that can economically be processed and utilized, i.e. limonite, intermediate and saprolite (Garnierite). The three types of nickel mineral deposits have different mineral constituent phases. These three types of nickel mineral deposit have a different phase mineral constituent, as in the type of saprolitic deposit containing Garnieritic $(Ni, Mg)_6Si_4O_{10}(OH)_8$ with high magnesia content but low iron content. Limonite deposits contain iron oxide as a major constituent of such mineral bonds, such elements as goethite (FeO(OH), hematite (Fe_2O_3) or magnetite (Fe_3O_4).

In the figure 1 below, showing the available phases of the laterite nickel XRD data of the sample before the selective reduction process was carried out containing goethite, quartz (SiO_2), kaolinite ($Al_2Si_2O_5(OH)_4$), and magnetite, so it can be said that the mineral used in this research is limonite.

In table 1 the following is the result of the initial XRF test of limonite. Beside the limonite, coal as a reductor was also tested with proximate and the ultimate test, the proximate and ultimate analysis aims to determine the physical parameters and chemical content in the coal [8]. The test results can be seen in table 2.

Table 1. XRF analysis of iron and nickel oxide content of limonite ore.

Oxide	Content (%)
Fe_2O_3	75.3
NiO	1.69

Table 2. Proximate and Ultimate Analysis of Coal.

Parameter	Content (% adb)
Proximate Analysis	
Moisture in air dried sample	7.26
Ash	15.87
Volatile Matter	16.44
Fixed Carbon	60.43
Ultimate Analysis	
Carbon	64.02
Hydrogen	3.11
Nitrogen	0.60
Oxygen	16.32

2.2. Experiment

The research method is done through several stages, starting with the initial characterization of limonite and coal such as XRD test, XRF test, and AAS test. Furthermore, the preparation of lateritic nickel ore, Na_2SO_4, and coal sample will be done. The ratio of the pellet mixture between limonite and coal is 100:6, then mixing with Na_2SO_4 additive is about 10% of the total mixing weight between the limonite and coal which will then be processed into pellet form manually with the addition of water gradually. To increase the binding power between the pellet components, 3% of bentonite was added from the total mixture of three previous materials. The pellets have been varied by 5 grams, 10 grams, and 15 grams in size. After were dried for 2 hours in the oven at a temperature of 100° C, they were proceeded with the reduction process at a temperature of 800°C for 1 hour in the furnace with a gradual rise in temperature. The first stage was at 480 ° C at a rate of 10°C per minute, after reaching the temperature and held for 10 minutes, then the furnace temperature will rise to the target temperature which was then held for 1 hour. After the reduction process is completed, then the cooling process was done in the furnace. The reduction pellet is weighed after the reduction process, then crushed to prepare for the AAS characterization, XRD.

3. Result and discussion

In Figure 1 below explains the XRD graph of pellets before reduction to pellet after reduction with various mass sizes. The graph before the selective reduction process was carried out containing goethite, quartz (SiO_2), kaolinite ($Al_2Si_2O_5(OH)_4$), and magnetite, so it can be said that the mineral used in this research is limonite. After the selective reduction process in the temperature of 800 °C for 1 hour, a phase change in each pellet causes a distinct reducibility of each pellet by various test weight sizes. In pellets weighing 5 grams, the phases formed are fayalite ($Fe_2Si_2O_4$), magnetite, hematite and quartz. In a selective reduction of 5 grams of pellets, the goethite and kaolinite phases are dehydroxylated. Meanwhile, the quartz and magnetite phases do not appear to be noticeable changes in the XRD pattern. When goethite decomposition was occurred, nickel which is in the different solid solution with goethite structure [(FeNi)O.OH-n.H_2O] will be released from its bonds, and become reduced. This mechanism is the reinforced by the results of AAS analysis in table 3, wherein the pellet reduction of 5gr higher amount of Fe and Ni than other pellets size that has been detected. This further strengthens the notion that in 5g pellet size there are more goethite phase changes being hematite at the time of the reduction process. An increase in Ni content of 5 gram reduction results indicates that in the pellet condition there is release of the Nickel bond from oxide to metallic Nickel. There was a decrease content of Fe for 5 grams compared to iron contents in the initial sample, it is because it has not completely decomposed into Fe, it is a desirable condition because it can increase the NI selectivity [9]. This result is similar to the research done by Johny et al [10] where the hematite phase is transformed into a magnetite phase in saprolite reduction using coal as a reducing agent.

Dehydroxilation phase of kaolinite from hematite occurred in the temperature of 700 °C so that it formed the phase of meta kaolinite, however this form of phase is not stable so it could not be seen in the peak XRD of 5 grams pellets reduction results.

At the reduced temperature, the fayalite phase formation occurred, whose intensity is not very significant in the 5 gram pellets. That phase should be avoided as it covers the iron oxide grains. In the presence of a Fayalite phase on a 5-grams pellet it could kinetically inhibit the pellet reduction process. Fayalite phase formation occurred because of several minerals belonging to the serpentine minerals species such as chrysolite, lizardite, and antigorite.

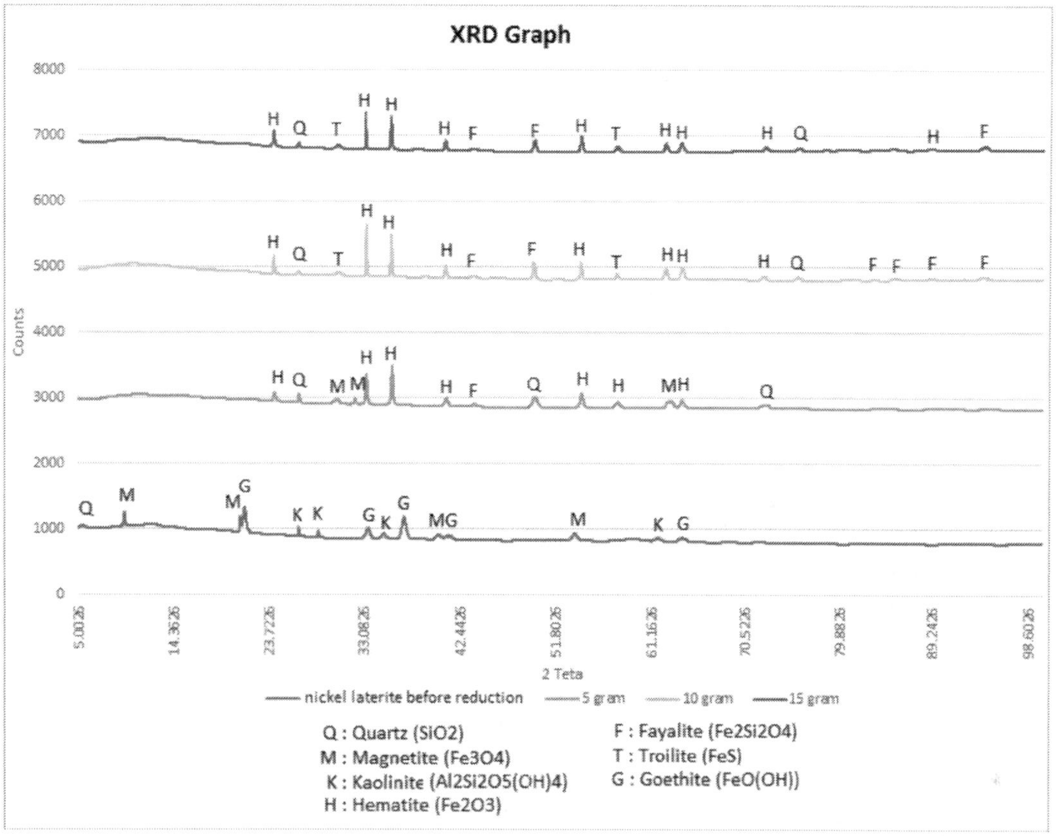

Figure 1. XRD Pattern of initial limonite and limonite after selective reduction with pellet size variations

At a temperature of 700 - 800°C the mineral will decompose and crystallize into olivine, fayalite and enstatite. The phase formation occurs when the reduction rate runs longer in the furnace, although in this study the reduction process has been added with the Na_2SO_4 additive with a 10% additiion of the fayalite phase remaining, the conclusion could mean that the amount of the content has not been able to break the silica bond with the oxide at the pellet and the difference of heat transfer due to the difference of temperature gradient present in each pellet tested.

At the same time the quartz phase is still available in the pellets with 5 grams weight, because decomposition of the quartz phase occured in the temperature of 1000°C. However, in XRD graph of 5 grams pellets occured the growth of the quartz phase, it is possible because the dehydroxylation of serpentine which produce fayalite and quartz. The forming process of that phase will increase along with the increase of temperature, which used to reduce nickel laterite.

In the 10 grams and 15 gram pellets there was a phase transformation into fayalite, hematite and quartz. The formation of the fayalite and hematite phases in both pellets indicated a longer reduction rate than the 5 grams pellet due to the magnitude of the temperature gradient to convert the phase into a metallic phase. In the phase transformation of both pellets, troilite phase was occurred [2]. The Troilite phase acted as an activating agent to accelerate the formation of phase shake and increase the rate of particle aggregation. The existence of such phases will facilitate the separation magnetically. Based on the

IOP Conf. Series: Earth and Environmental Science **105** (2017) 012067 doi:10.1088/1755-1315/105/1/012067

table below, indicates a decrease in the iron content of each pellet along with the increase in the size of the pellet mass, but at the nickel content, there is an increase in the content of pellet's weighing 5 grams, and decreasing content in pellet's size 10 grams and 15 grams.

Table 3. Comparison of nickel and iron content measurement of limonite between before and after reduction by Atomic Absorption Spectroscopy.

Element	Before reduction	After reduction			Unit
		Pellet size 5 gr	Pellet size 10 gr	Pellet size 15 gr	
Fe	41.46	30.57	16.67	7.61	% w/w
Ni	1.16	1.18	0.82	0.56	% w/w

4. Conclusion

Based on the research which has been done to the limonite with various variations of pellets mass with 10 wt% Na_2SO_4 as additives and coal reductor in the temperature of 800 °C, it could be concluded that,

1. Pellets with the weight of 5 grams have Magnetite phase more than other size pellets.
2. There was a decrease content of Fe for 5 grams compared to iron contents in the initial sample it is because it has not completely decomposed into Fe.
3. Fayalite phase in the Pellets with the weight of 5 grams formed not too significantly, it was proven in the intensity of peaks in the graphic of XRD results.
4. Pellets with the weight of 5 grams have gained the higher content than other size pellets, even though the reduction degree of each pellets with various sizes was lower than the reduction degree of initial limonite.

5. Acknowledgement

Special thanks are due to Mr. Agus Budi Prasetyo (Research Centre for Metallurgy, Indonesian Institute of Sciences - LIPI, Indonesia) for helpful supports. The authors also wish to thank the Ministry of Research Technology and Higher Education and Directorate of Research and Community Engagement, Universitas Indonesia, under PITTA Research Grants Contract No. 791/UN2.R3.1/ HKP05.00/2017 for the financial support of this research.

6. References

[1] Butt C R M and Cluzel D 2013 Nickel laterite ore deposits: Weathered serpentinites *Elements* **9** 123–8
[2] Mudd G M 2010 Global trends and environmental issues in nickel mining : Sulfides versus laterites *Ore Geol. Rev.* **38** 9–26
[3] Jiang M, Sun T, Liu Z, Kou J, Liu N and Zhang S 2013 Mechanism of sodium sulfate in promoting selective reduction of nickel laterite ore during reduction roasting process *Int. J. Miner. Process.* **123** 32–8
[4] Li G, Shi T, Rao M, Jiang T and Zhang Y 2012 Beneficiation of nickeliferous laterite by reduction roasting in the presence of sodium sulfate *Miner. Eng.* **32** 19–26
[5] Valix M, Cheung WH 2002 Effect of sulfur on the mineral phases of laterite ores at high temperature reduction *Miner Eng.***15** 523–30.

[6] Subagja R, Budi Prasetyo A, and Mayang Sari W 2016 Peningkatan Kadar Nikel Dalam Laterit Jenis Limonit Dengan Cara Peletasi, Pemanggangan Reduksi dan Pemisahan Magnet Campuran Bijih, Batu Bara, Dan NA2SO4. *Metalurgi.* **2** 103–115

[7] Zevgolis E N, Zografidis C, Perraki T and Devlin E 2010 Phase transformations of nickeliferous laterites during preheating and reduction with carbon monoxide *J. Therm. Anal. Calorim.* **100** 133–9

[8] Rasheed MA, Rao PLS, Boruah A, Hasan SZ 2015 Geochemical Characterization of Coals Using Proximate and Ultimate Analysis of Tadkeshwar Coals , *Geosciences*.**5** 113–9.

[9] Mayangsari W, Budi Prasetyo AB 2016 Proses Reduksi Selektif Bijih Nikel Limonit Menggunakan Zat Aditif CaSO4. *Metal LIPI* **1** 1–68

[10] Soedarsono J W, Kawigraha A, Sulamet-ariobimo R D, Asy M A, Yosi A and Putra E M 2013 The Influence of Coal and Reduction Process Parameters in Producing Iron Nugget*Adv. Mater. Res.* **789** 517–21

IOP Conf. Series: Earth and Environmental Science **105** (2017) 012068 doi:10.1088/1755-1315/105/1/012068

Effects of Na_2SO_4 addition on the selective reduction of limonite ore from Southeast Sulawesi

Rinanda Rahmat[1], Ahmad Maksum[1, 2], Delfiendra[3] and Johny Wahyuadi Soedarsono[1]

[1]Center of Minerals Processing and Corrosion Research, Department of Metallurgical and Materials Engineering, Universitas Indonesia, Depok 16424, Indonesia
[2]Department of Mechanical Engineering, Politeknik Negeri Jakarta, Depok 16425, Indonesia
[3]Center for Testing and Identification of Goods, Directorate General of Customs and Excise, Jakarta, Indonesia

Email: jwsono@metal.ui.ac.id

Abstract. In terms of its process, although only about 40% of the world's nickel reserves are classified as lateritic ore, nickel sulfide still dominates nickel laterite for its practicality. Because of the high demands on nickel, development and research on nickel mineral processing are needed to ensure that the demands are fulfilled from year to year. On this research, a study will be conducted on the effect of adding Na_2SO_4 to increase the level of nickel content in limonite by selective reduction process. The reduction process is done by heating limonite ore and coal as reducer from room temperature to 1000oC with temperature rising to 10oC/minute in mini furnace on varying addition of 0%, 5%, 10%, and 15% Na_2SO_4. After that, the reduction result is tested on XRD and AAS. It has been found that the nickel content has decreased, which is likely due to the excessive reduction time resulting in the magnetite being re-oxidized to hematite. Therefore, based on recovery calculations, the optimal addition of Na_2SO_4 is 5% by weight where the obtaining nickel content is 1.09%.

1. Introduction

Indonesia has become the sixth country to have the most nickel reserve for about 4.5 million ton. The composition of nickel in the world is divided in to nickel laterite for 60% and nickel sulfide for 40%. One of the nickel processes is pyrometallurgy process. In Indonesia, processing nickel itself still relies on nickel sulfide because the processing of nickel laterite is quite complex and cost more money [1,2]. Subsequently, nickel laterite process was developed with selective reduction by adding sulfur additives. In a study, it is mentioned that adding sodium sulfate in a selective reduction of limonite has increased the nickel content in ferronickel from 2.3% to 9.48%. After that, through the magnetic separation process, the nickel recovery has increased from 56.97% to 83.01%. This study shows that the sulfur in Na_2SO_4 reacts with Fe to form FeS (troilite) which acts as an activation agent to accelerate the melting phase and increase the displacement rate to gain more aggregation of ferronickel particles during the reduction process [3]. Based on the aspects above and previous study [4], this study will discuss the effect on adding Na_2SO_4 in the selective reduction process with limonite ore using the coal reducer.

2. Materials and method

2.1. Materials

Limonite ore samples used in the study carried out initial characterization by XRD test. Figure 1 provides the XRD test result of nickel ore.

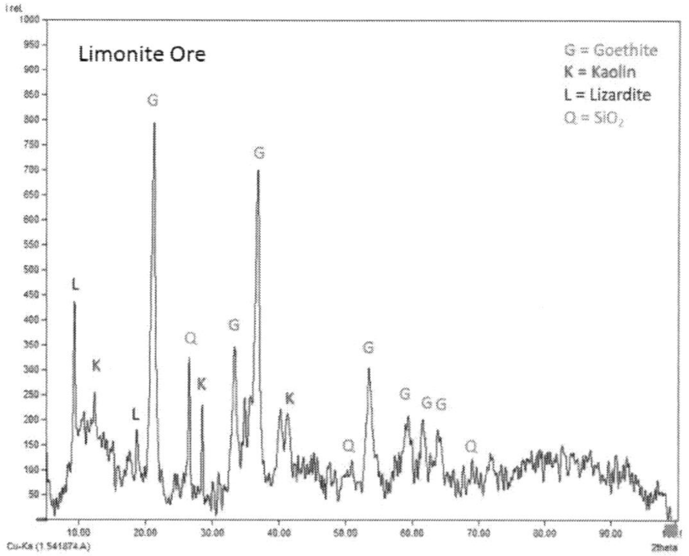

Figure 1. XRD pattern of nickel ore's initial sample.

While for the quantity test, the AAS test is performed to see the Fe and Ni contents in the limonite ore. The AAS test result can be seen in Table 1.

Beside the limonite, coal as a reductor was also tested with proximate and ultimate test to see the composition of the coal. The test results can be seen in Table 2.

Table 1. Atomic absorption spectroscopy analysis of iron, nickel, and cobalt content from limonite ore.

Element	Content (%)
Fe	41.46
Ni	1.16

2.2. Experiment

This selective reduction process takes place in a mini furnace. The limonite ore samples which has been processed into a fine powder (#300) were mixed with 6% of coal with the same mesh size and Na_2SO_4 with 0%, 5%, 10%, and 15% levels, and added by 3% of bentonite. After that, the samples were shaped into pellets with a small amount of water. The water content was removed by heat in the oven at 100°C for about 2 hours. After it dries, the samples can be put into a mini furnace. The reduction process of limonite ore is done in the range of 1000°C with constant temperature rise of 10°C /minute. According to the furnace procedure, the temperature is raised up to 60% of the target temperature and held for about 10 minutes, and then raised again until the target temperature that is 1000°C and held again for an hour. The cooling process is done inside the furnace. After the reduction process, AAS test was performed to determine the level of Fe and Ni on the limonite which has been under reduction. Also, XRD test is being held to analyze the compounds and nickel contents which were contained in the laterite ore.

Table 2. Proximate and Ultimate Analysis of Coal.

Parameter	Content (% adb)
Proximate Analysis	
Moisture in air dried sample	7.26
Ash	15.87
Volatile Matter	16.44
Fixed Carbon	60.43
Ultimate Analysis	
Carbon	64.02
Hydrogen	3.11
Nitrogen	0.60
Oxygen	16.32

3. Result and discussion

The XRD pattern of samples which has been under selective reduction on all Na_2SO_4 addition variations can be seen on Figure 2. It was shown that after undergoing reductions, the phases which present in the initial sample like lizardite, goethite, and kaolin was no longer seen. During the reduction process, goethite, and lizardite phase were facing dehydroxylation for the first time [5]. Dehydroxylation is a structural transformation or change that leads to the destruction of OH-group structures or the loss of water bonds in the crystalline of mineral. This process occurs naturally during reduction as a result of heat, and the next step is re-forming of the other crystalline structures.

From the XRD results above, there were four types of samples that contain hematite and magnetite compounds. This results showed that the reducer was not yet capable of reducing all hematite to become magnetite, wustite, or metallic iron. Furthermore, while SiO_2 was still in the reduction sample, the SiO_2 content will decrease as the temperature goes higher. This reduction was done at 1000°C where silica reduction starts at 1200°C. Regardless of the various temperature, the reduction temperature from SiO_2 could be decreased by the milling process [6]. It was predicted that the milling process of coal and nickel was done least from perfect because it was only done for 30 minutes. Moreover, the mix between Na_2SO_4 and bentonite was done manually and without the milling process. Consequently, the SiO_2 that were formed are still quite a lot.

The present of MgO in each reduced samples indicate that the MgO has not been fully reduced and reacts with other elements. Anyhow, some MgO reacted with other elements and formed a forsterite.

The formation of forsterite allowed oxide grains closure and prevented further reduction of nickel and iron. Also, sulfur had a major impact in this forsterite phase.

The presence of forsterite in XRD samples are possible due to the mixing of limonite, coal, and Na_2SO_4 were not yet formed as a well physical and chemical bond, thus making the samples less smooth and less reactive. As what has been mentioned above, only the mixing of limonite and coal used ball mill while the mixing of Na_2SO_4 was done manually. It was known that the ball milling was one of the mechanical activities which could improve the surface area, improve the surface reactivity, and reduce the particle size of powder [6]. As the mechanical was activated during the milling, the reduction reaction rate would increase and the sample would be optimized to 1000°C

The AAS test also aims to see the Fe and Ni contents after the reduction is done which later lay out the recovery calculation. The AAS samples results which have been reduced can be seen in Table 3.

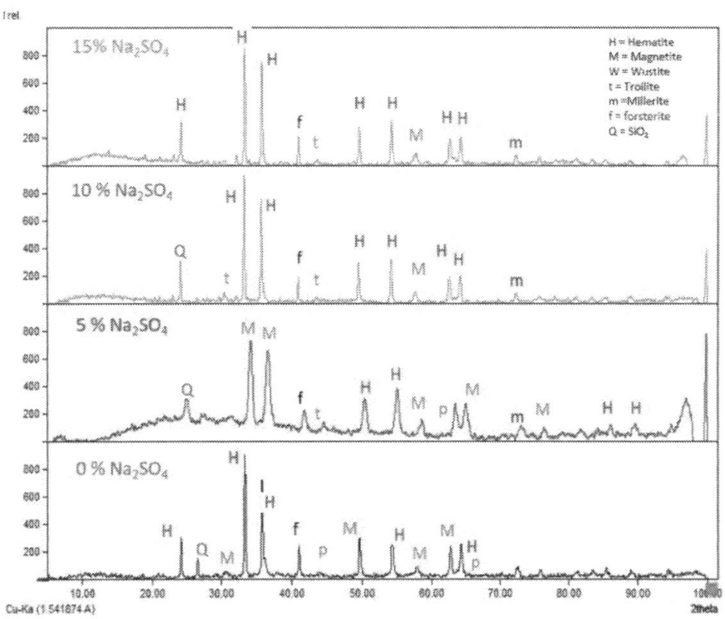

Figure 2. XRD pattern of initial nickel lateritic ore.

Table 3. Comparison of nickel and iron content of limonite between ore and after reduction samples.

Element	Ore	After reduction				Unit
		0% Na_2SO_4	5% Na_2SO_4	10% Na_2SO_4	15% Na_2SO_4	
Fe	41.46	8.38	16.34	13.49	7.64	% w/w
Ni	1.16	0.32	1.09	1.00	0.67	% w/w

The comparison data can be seen in Table 3. From the table above, it can be seen that after being reduced, there is a content decreaseboth in Fe and Ni. This can happen because of the time and duration during the reduction which have an important role in the reduction process [7]. Using the reduction time including the cooling phase for about 5 hours will have an effect on the level or concentrate of the results in reduction. This decrease also can be caused due to the weary reducer that has been used quite long. Moreover, the long the reduction goes, there will be an increase in CO_2 concentrate that cause the limit in gas-solid reaction to be filled by CO_2 which later can obstruct the reduction process [7].

The occurrence of this decline can also due to the wustite peak intensity weakened when the reaction time is extended. This extension can increase the nickel and iron accessions, yet decrease the contents of both materials. When the reduction time is shortened, the wustite reduction rate becomes slower and the hematite will block the diffusion from NiO resulting nickel with higher contents [8].

Even so, if it was seen from the four parameter that add the sulfur contents, the sulfur can actually capable of giving better nickel contents beside without an addition of Na_2SO_4.

The comparison of nickel and iron content can be seen from this graph at Figure 3 below.

IOP Conf. Series: Earth and Environmental Science **105** (2017) 012068 doi:10.1088/1755-1315/105/1/012068

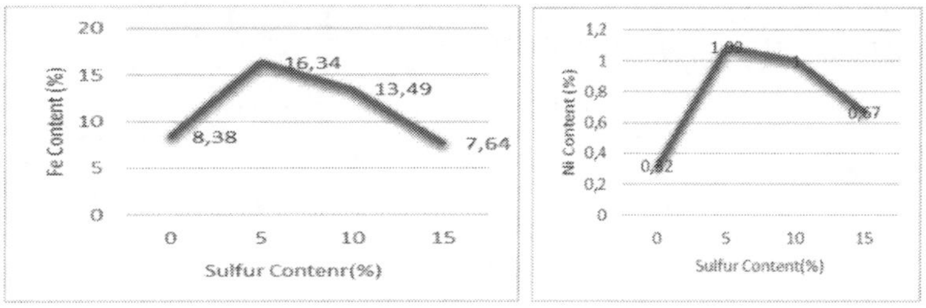

Figure 3. Comparison of nickel and iron content of limonite between after reduction samples with variation of sulfur addition.

Recovery calculation shows that by adding Na_2SO_4 for 5%can gives the maximum results compare to other vary additions. The recovery result can be seen in Table 4 and Table 5.

Table 4. Nickel recovery.

Sample	Concentrate weight, C (gr)	Ni Concentrate, c (%)	Ore weight, F (gr)	Ni ore, f (%)	Ni recovery
0% Na_2SO_4	12.6581	0.32	20.6067	1.16	16.9%
5% Na_2SO_4	13.0191	1.09	21.6377	1.16	56.56%
10% Na_2SO_4	11.2490	1.00	20.0248	1.16	48.42%
15% Na_2SO_4	12.4028	0.67	20.3553	1.16	23.61%

Table 5. Iron recovery.

Sample	Concentra te weight, C (gr)	Fe Concentrat e,	Ore weight, F (gr)	Fe ore, f (%)	Fe recovery
0% Na_2SO_4	12.6581	8.38	20.6067	41.46	12.41%
5% Na_2SO_4	13.0191	16.18	21.6377	41.46	23.48%
10% Na_2SO_4	11.2490	13.49	20.0248	41.46	18.28%
15% Na_2SO_4	12.4028	7.64	20.3553	41.46	11.22%

From the Figure 4 below, it can be seen the comparison of recovery of nickel and iron after reduction by variation of Na_2SO_4.

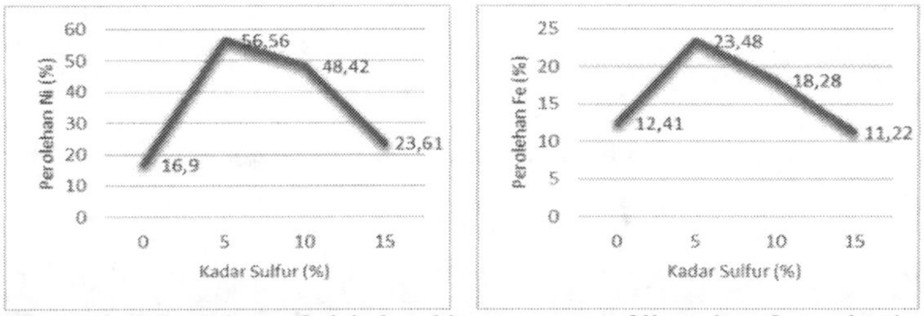

Figure 4. Comparison of nickel and iron recovey of limonite after reduction samples with variation of sulfur addition.

4. Conclusion

Adding extra sulfur in reduction process can increase the nickel and iron contents. It is mentioned before that the initial sample has 1.16% nickel. After the reduction for additional 0%, 5%, 10%, and 15% Na2SO4, each nickel contents obtained 0.32%, 1.09%, 1.00%, and 0.67%. Meanwhile, the iron contents in initial sample is 41,46% and after being reduced with additional 0%, 5%, 10% and 15% Na_2SO_4, the iron contents gain 8.38%, 16.34%, 13.49%, 7.64%.

Altough limonite selective reduction is supposed to be able to increase the nickel content. However, this study shows that the selective reduction decreasing the nickel and iron contents compared to samples before they undergo reduction. One of the factors that may affect the result is the duration during reduction.

But, the nickel and iron value after the reduction tends to be better with sulfur addition. For nickel, when it is added 0% of Na_2SO_4, the value becomes 16.9%. As for the addition of 5%, 10% and 15%, the nickel obtains 55.56%, 48.42% and 23.61% respectively.

The fact that the contents and recovery have a decrease when sulfur was added for more than 5%, shows that adding sulfur has an optimal value for each reductions.

Based on XRD test result, there is no ferronickel in each samples. This may be caused of the imperfect reduction of hematite into magnetite, wustite, and metallic iron. As for iron oxide and nickel oxide, they react to sulfur and form NiS and FeS.

5. Acknowledgement

Special thanks are due to Mr. Agus Budi Prasetyo (Research Centre for Metallurgy, Indonesian Institute of Sciences - LIPI, Indonesia) for helpful supports. The authors also wish to thank the Ministry of Research Technology and Higher Education and Directorate of Research and Community Engagement, Universitas Indonesia, under PITTA Research Grants Contract No. 791/UN2.R3.1/ HKP05.00/2017 for the financial support of this research.

6. References

[1] Mudd G M 2010 Global trends and environmental issues in nickel mining: Sulfides versus laterites *Ore Geol. Rev.* **38** 9–26

[2] U.S. Geological Survey 2017 *Mineral Commodity Summaries 2017* vol 1

[3] Li G, Shi T, Rao M, Jiang T and Zhang Y 2012 Beneficiation of nickeliferous laterite by reduction roasting in the presence of sodium sulfate *Miner. Eng.* **32** 19–26

[4] Soedarsono J W, Astini V, Fazri F, Kawigraha A, Sulamet-ariobimo R D, Rustandi A and Tjahyono S 2014 Effect of Carbon Content in Direct Reduction Process of Limonite Iron Oxide to Produce Pig Iron Substitute for Thin Wall Ductile Iron Process *Adv. Mater. Res.* **887– 888** 281–6

[5] Rao M, Li G, Zhang X, Luo J, Peng Z and Jiang T 2016 Reductive roasting of nickel laterite ore with sodium sulphate for Fe-Ni production . Part II : Phase transformation and grain growth *Sep. Sci. Technol.* **6395**

[6] Zhao Z, Zhang Y, Chen X, Chen A and Huo G 2009 Effect of mechanical activation on the leaching kinetics of pyrrhotite *Hydrometallurgy* **99** 105–8

[7] Li B, Wang H and Wei Y 2011 The reduction of nickel from low-grade nickel laterite ore using a solid-state deoxidisation method **24** 1556–62

[8] Lu J, Liu S, Shangguan J, Du W, Pan F and Yang S 2013 The effect of sodium sulphate on the hydrogen reduction process of nickel laterite ore *Miner. Eng.* **49** 154–64

IOP Conf. Series: Earth and Environmental Science **105** (2017) 012069 doi:10.1088/1755-1315/105/1/012069

Characteristics of calcium carbonate solution flowing in pipe

Yanuar[1], Pamitran A S[1], Sealtial Mau[2], Kurniawan T Waskito[1] and Winda Wulandari[2]

[1]Department of Mechanical Engineering, Faculty of Engineering, Universitas Indonesia, Kampus Baru UI Depok, 16424, Indonesia;
[2]Graduate Student Department of Mechanical Engineering, Faculty of Engineering, Universitas Indonesia, Kampus Baru UI Depok, 16424, Indonesia

E-mail: yanuar@eng.ui.ac.id

Abstract. Issue of energy is important because energy is one of economic foundation to support a nation. Indonesia have abundant resources of renewable energy and it can be used as friendly energy source. Innovations continue to be developed to find the right method about good quality and cheap of energy sources that is widely available in nature. Calcium carbonate powder of shellfish are the one of the environmentally friendly materials that are available in nature. In this study, calcium carbonate are used in aqueous solution of water-ethylene glycol fluid with a ratio of 40:60. The purpose of this study is to determine characteristics and hydrodynamic behavior of non-Newtonian calcium carbonate solution that is flowed in a small pipe. Concentration of powder was from varied 100 ppm, 300 ppm and 500 ppm on the base fluid, respectively. The test was performed on a circular pipe with 4 mm of inner diameter, horizontally. Pressure drop was determined between two different points in which there were high pressure and low pressure taps by different data acquisition. Changes of friction factor value become a parameter of the indicator to drag reduction. The highest drag reduction in transition flow is about 26.3 % for the 300 ppm concentration with Reynolds number is around 3000.

1. Introduction

Existence of a nation is depended by need for energy to support various productivity processes. Absorption and release of heat are phenomenon that occurs in the application of thermodynamics. Car cooling system; the engine releases heat and then coolant fluid absorbs the heat for cooling. This process is inseparable from friction effects along the circulated fluid flowed. Friction factor is a resistance that can cause energy loss. This phenomenon needs to be examined to find the right method of handling, so that the resistance can be reduced. Until now, there are two methods to reduce of friction, i.e. active control method and passive control method. Active control is a method of adding substances to the base fluid such as surfactants, fibers, biopolymer and nanoparticles [1-4]. Passive control is a method to form a cross-section of the geometry, hence it can reduce frictional resistance as the cross section in the form of a spiral, ellipses, squares and others [1,5,6].The phenomenon of reduction of friction resistance is called drag reduction (DR).

Toms was a first researcher that used polymers to increase DR. The phenomenon of DR using polymers continues to be developed by N. Yusuf et al, who investigated use of polymers in oil-water fluid flows. In the study, Yusuf determined DR up to 60%. The use of polymer was proven that it could increase DR in piping systems, in the Trans-Alaska oil [7]. Research is already done by Pouranfard et al, proved to reveal one of the reasons for the occurrence of DR. On this research, nanoparticles is potential to coat the contact area between the working fluid and the pipe surface that could potentially occurring slip condition [8].

The use of materials as a trigger for the occurrence of DR should be developed. Not only it-has the potential to produce similar effects, but also environmentally friendly and inexpensive. The experimental investigation on fluid flow with adding biopolymer were developed to produce DR. Bacterial cellulose when is dissolved into the working fluid and flowed through high flow Reynolds regime is potentially influence of DR. Satoshi Ogata et.al, managed to reveal the value of DR as result of bacterial cellulose by 11% [9]. A different study that conducted by Yanuar et al, examining the DR caused by guar gum. Guar gum biopolymer can produce the DR by 30% in the spiral pipe measuring P/Di 7.0 [10]. DR Research on piping systems should be focused on the use of environmentally friendly and inexpensive materials. On this study, calcium carbonate particle will be used as mixed substances in base fluids to trigger the occurrence of DR. The base fluid which used is pure water and ethylene glycol (40:60). The purpose of this study was to determine characteristics and hydrodynamics behavior of non-Newtonian calcium carbonate solution dispersed in pipe.

2. Research methods

2.1. Calcium carbonate solution preparation

The first step of the study was nanofluids preparation. Water and ethylene-glycol were used as solution with ratio of 40:60. The solution was mixed with $CaCO_3$ powder with concentrations which varied from 100 ppm, 300 ppm, to 500 ppm, respectively. The working fluid was stirred on 2000 rpm by hand mixer for 30 minutes. Working fluid-was measured by it mass, to obtain the value density of each solution. Measurements were carried out at 27 °C of temperature. Thus, several values were obtained as shown in table 1.

Table 1. Density of working fluids.

Pure water	100 ppm of working fluid	300 ppm of working fluid	500 ppm of working fluid
996.45 kg/m^3	1263.91 kg/m^3	1264.69 kg/m^3	1265.37 kg/m^3

2.2. Research procedure

The main focus of this study was to observe the characteristics of $CaCO_3$ nanofluids which flowed through circular pipe with variations of Reynolds numbers. The set-up of experimental tools was designed on the horizontal position to obtain experimental data such as flow rate and pressure drop. A circular pipe with 4 mm of inner diameter and 800 mm of length was used. The length of inlet pipe to achieve fully develop flow was 1000 mm. Working fluid was flowed through a pipe by a pump and flow rate. The flow was measured using by flow meter. Pressure drop was measured with differential pressure application DAQ adapter for emant 300 (DAQ component). The different pressure tools were placed on the high pressure and low pressure, separately. The different pressure value data was recorded by computer. Set-up experimental is showed in figure 1.

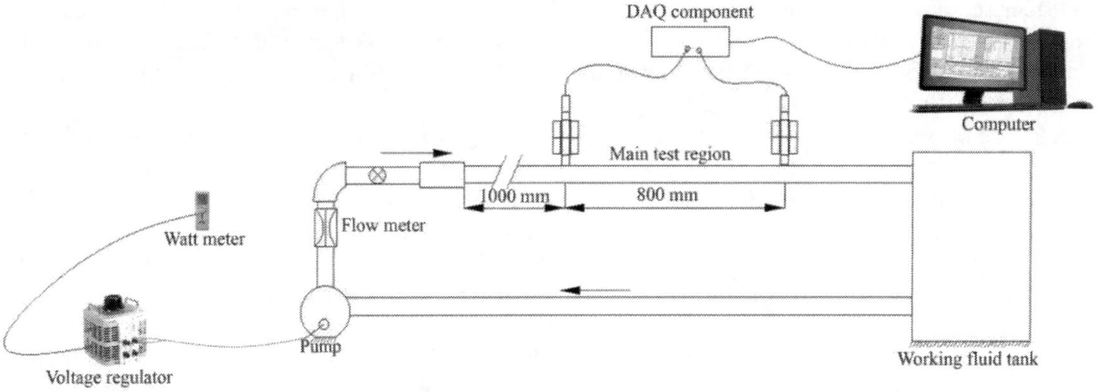

Figure 1. Experimental Setup.

First step after the experimental set-up was to validate in order to fully comply with testing standards on the piping system. The test with pure water and other data is processed using the relevant equations for Newtonian fluid, to know the correctness of horizontal designed tools. The calculation of validation data would be compared with the Hagen-Poiseuille equation for laminar flow and Blasius equation for turbulent flow. The computed data in this test would be calculated to obtain the friction factor correlation and the Reynolds number using the Fanning equation for Newtonian flow.

3. Method and rheological model

3.1. Methodology calculation

In this research, equations are used as the tools to get the results that accordance with the objectives of this research. Working fluids type of this study was non-Newtonian fluid, the equations are also adjusted for non-Newtonian fluid flow in circular pipe. Equation of generalize Reynolds number (*Re'*) for circular pipe as follows:

$$Re' = \frac{\rho_{nf} D^n V^{2-n}}{K 8^{n-1}} \quad (1)$$

For non-Newtonian fluid flow in the circular pipe, friction factor of laminar flow can be predicted through the following equation:

$$\frac{\Delta P}{L} = \frac{2 f \rho V^2}{D_h} \quad (2)$$

And to predict the friction factor in turbulent flow is used the equation as follow:

$$\frac{1}{\sqrt{f}} = \frac{40}{n^{0.75}} . log_{10}\left[R_e'(f_{turb})^{1-(1/2)}\right] - \frac{0.4}{n^{1.2}} \quad (3)$$

Where, ρ_{nf} is density of CaCO3 solution and V is velocity.

The experimental data on the first processed is determined by kind of fluids through the power law index value, n and consistency coefficient, K. The same relation of Newtonian fluid for non-Newtonian calculation which can be rheological model as follows:

$$\tau_w = \mu_a \gamma_w \tau_w = \mu_a \gamma_w \quad (4)$$

Where, μ_a is the apparent viscosity, and γ_w is the wall shear rate.

$$f(\tau_w) = \gamma_w = \left[\left(\frac{3n+1}{4n}\right)\frac{8u}{D}\right] \quad (5)$$

and,

$$\mu_a = K\gamma_w{}^{n-1} \tag{6}$$

The parameter of γ_w is the nominal shear rate. For fully developed laminar pressure drop, the shear stress, τ_w is computed as follows:

$$\tau_w = \frac{D\Delta P}{4L} \tag{7}$$

Where D is diameter, ΔP is pressure drop, L is the length and f is friction factor.

3.2. Rheological model

The each value of the shear rate and shear stress calculations from experimental data are plotted into the logarithmic graph. The value of power law index and coefficient consistency can be known through plotting the graphic line in figure 2.

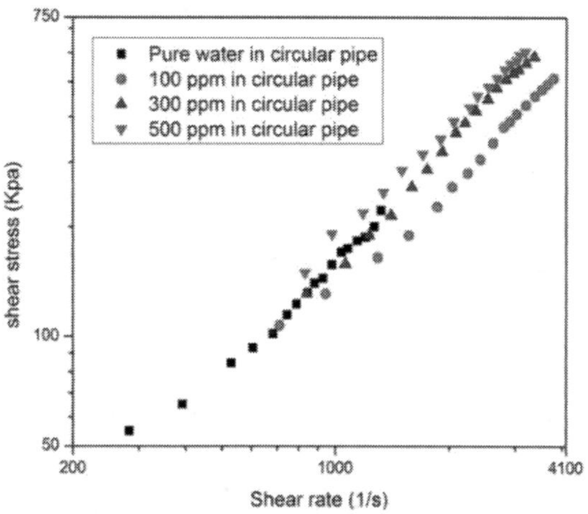

Figure 2. Graph of shear stress versus shear rate

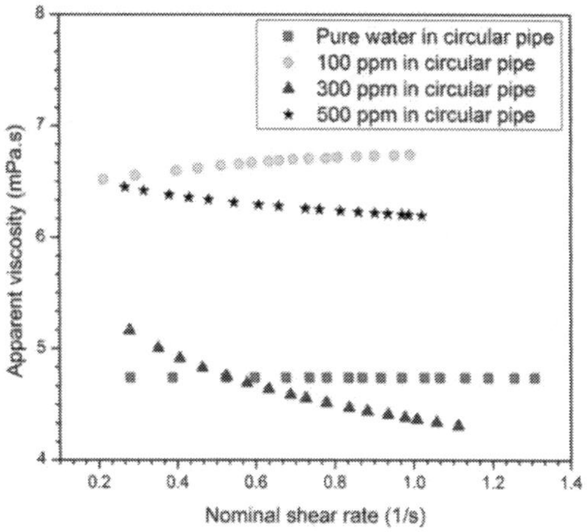

Figure 3. Apparent viscosity of nanofluids versus nominal shear rate.

On figure 2, it can be seen that maximum of shear stress and shear rate of pure water is lower than solution of 100 ppm, 300 ppm and 500 ppm. On figure 2, it also can be determined the value of n and K as in table 2.

Table 2. Power low power index values and consistency coefficients for solution

Power law parameters	100 ppm	300 ppm	500 ppm
n	1.02	0.87	0.97
K	6.71	7.03	6.20

The values of n and K at different functions is showed. The value of n can be used to determine the flow type of working fluid. For $n = 1$ is Newtonian fluid, for $n > 1$ is dilatant flow and for $n < 1$ is pseudoplastic fluid. For $n = 1$, the increase of viscosity value is constant to the shear rate. For $n > 1$, the viscosity value increases slowly in the low shear rate but at the high shear rate, it will increase of shear rate sharply. For $n < 1$, the viscosity value tends to decrease when shear rate is increased. Figure 3, can be seen that the value of K indicates the intensity of the apparent viscosity. The higher value of K, will indicate that generated shear stress will increase when fluids if flowed.

Figure 3 shows effect the viscosity of solution at a certain velocity. Through the graph, the base fluid has a constant trend line or fluid viscosity is not affected by shear rate. For the working fluid with the addition of particles, there are variations of viscosity value at each point, so this phenomenon is known as apparent viscosity. Working fluid with concentration 100 ppm shows behavior of dilatant fluid, where the trend line describes increasing viscosity with increasing shear rate. Working fluid with concentration of 300 ppm and 500 ppm shows behavior of pseudoplastic fluid type where the higher the shear rate value, the lower apparent viscosity value. The higher apparent viscosity value in the preliminary data of the working fluid of 100 and 500 ppm is influenced by the value of K which is lower than 100 ppm. From the graph, it can be assumed that higher the value of K will determine lower apparent viscosity value.

4. Result and discuss

4.1. Pressure drop investigation

The test apparatus which is flowed by working fluid is deemed to be standard, after its validated process. This research was focused on the pipe area with length of 800 mm and at both points: high pressure and low pressure points is mounted pressure measurement. The pressure tool to get the value of pressure drop and a flow meter on the up-stream to get the value of flow rate, flow meter is used to accommodate the flow of fluid on the up-stream per 60 minutes. Measurements were continuously recorded during the tests performed for each working fluid. In figure 4, it can be seen the value of pressure drop versus flow rate.

IOP Conf. Series: Earth and Environmental Science **105** (2017) 012069 doi:10.1088/1755-1315/105/1/012069

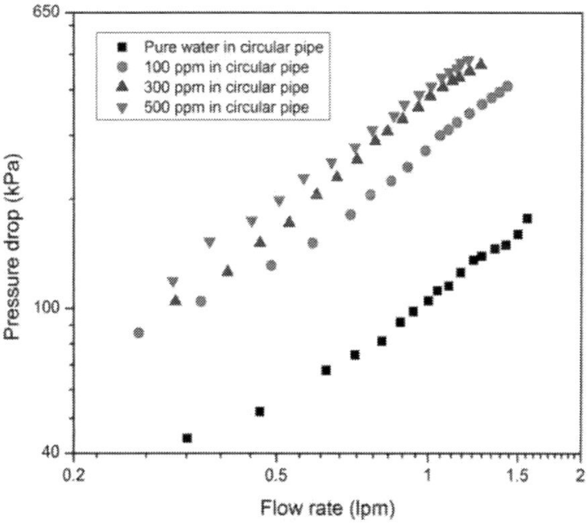

Figure 4. Comparison of pressure drop and flow rate.

Figure 4 shows the ratio of pressure drop and flow rate. In the graph, it was showed that pressure drop value of CaCO3 nanoparticles concentration is higher from water. It also could be said that the increasing pressure drop was influenced by powder of CaCO3, and so did the flow rate. The trend of graph corresponded to other references that applied nanoparticles [11,12]. Through this observation, the use of powder at high concentrations was ineffective, because the higher concentration will higher pressure drop will be generated.

4.2. Friction factor

Further experimental data are calculated using the relevant formulas for non-Newtonian fluid flow in relation with friction factor. The study was conducted at 27 °C temperature of room.

Figure 5 shows comparison of friction factor and generalize Reynolds number. In the graph, data ranges around laminar flow to the turbulent flow region. In figure, there are some points occurring delay in the transition flow. Where, in calculation of experimental data for the friction factor value is not fit on the trend line Hagen-Poiseuille or Blasius. The friction factor value in the transition region is caused by delay fluid flow that will enter to the turbulent region while the value of Re' number has reached 4000. Delaying flow phenomenon maybe influenced by Brownian motion of powder. This is seen in the graph that has a trend line transition flow that occurs below the value of Blasius equation.

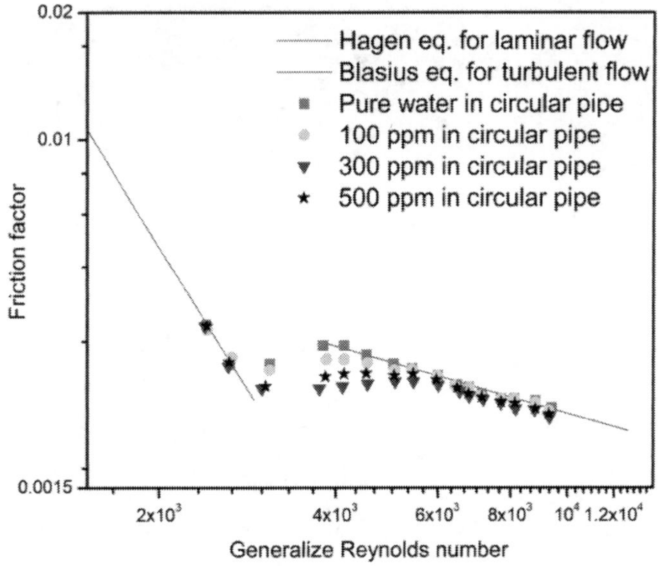

Figure 5. Comparison of friction factor with generalize Reynolds number.

Pseudo-random motion powder in solution for working fluid with the lowest coefficient of friction at concentration 300 ppm then followed by 500 ppm and 100 ppm which happened at transition regime. This value is influenced by the value of power law index on each working fluid. The *n* value, tend to near 1, where it will tend to linear which is similar to the trend line of Newtonian fluid.

Figure 6 shows the percentage of the DR value of the working fluid with an increase in *Re'* value. In the picture, it can be seen that there is a difference in value of the DR that is quite extreme percentage on the value of *Re'* laminar and turbulent.

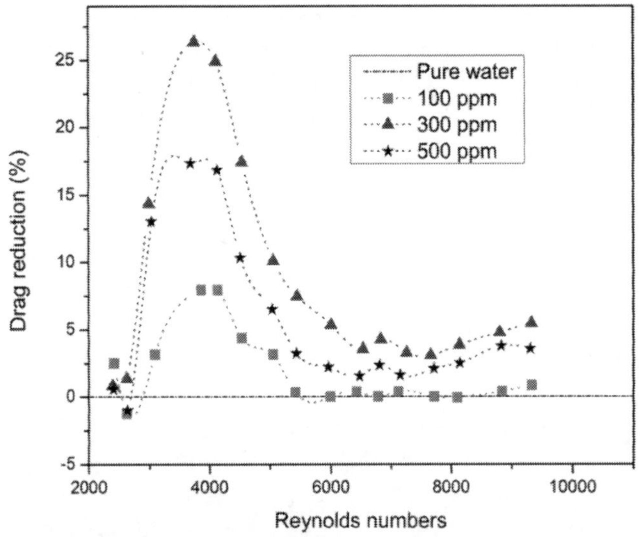

Figure 6. Comparison of DR values with generalize Reynolds number.

At the transition flow, DR is 26.3 % with 300 ppm concentration of nanoparticles and the *Re'* is around 3000. The percentage of the DR value is an interpretation of friction values in figure 5. In transition flow, maximum Brownian interaction of powder which pseudo-random motion is reduced frictional resistance in the boundary layer.

5. Conclusion

The present study was conducted to determine the hydrodynamic flow characteristics of calcium carbonate solution to increase flow efficiency in pipe. The effect of solution into friction factor value is varied. The highest DR in transition region is 26.3 % at 300 ppm and *Re'* is around 3000. Increase of Pressure drop when depends on the increasing concentration of the CaCO3 solution. Adding more concentration can cause increase pressure drop.

6. References

[1] Watanabe K, 2015 Drag Reduction by Fluid Slip *American Journal of Science and Technology* **2** pp 106-110

[2] Karami H and D Mowla 2012 Investigation of the effects of various parameters on pressure drop reduction in crude oil pipelines by drag reducing agents *Journal of Non-Newtonian Fluid Mechanics* **177** p37-45

[3] Gunawan, M Baqi, and S Fathern as 2013 Drag Reduction of Bamboo and Abaca Fiber Suspensions in Circular Pipe *Applied Mechanics and Materials Trans Tech Publ.*

[4] Tamano S, et al 2015 Drag reduction and degradation of nonionic surfactants with organic acid solutions in turbulent pipe flow *Journal of Non-Newtonian Fluid Mechanics* **215** p1-7

[5] Watanabe K, et al 1984 Flow in a Spiral Tube 1st Report Velocity Distribution and Pressure Drop *Bulletin of JSME* **27 (228)** p1105-1111

[6] Escudier M, A Nickson and R Poole 2009 Turbulent flow of viscoelastic shear-thinning liquids through a rectangular duct Quantification of turbulence anisotropy *Journal of Non-Newtonian Fluid Mechanics* **160 (1)** p2-10

[7] Joseph N, et al 2012 Experimental study on the effect of drag reducing polymer on flow patterns and drag reduction in a horizontal oil-water flow *International Journal of Heat and Fluid Flow* **37** p74-80

[8] Pouranfard A, D Mowl, and F Esmaeilzadeh 2014 An experimental study of drag reduction by nanofluids through a horizontal pipe turbulent flow of a Newtonian liquid *Journal of Industrial and Engineering Chemistry* **20 (2)** p633-637

[9] Ogata S, T Numakawa, and T Kubo 2011 Drag reduction of bacterial cellulose suspensions *Advances in Mechanical Engineering* **3**: p528373

[10] Yanuar, Gunawan, and M Baqi 2012 Characteristics of Drag Reduction by Guar Gum in Spiral Pipes *Journal of Technology* **58 (2)**

[11] Kumar S, G S Sokhal, and J Singh, Effect Of Cuo-Distilled Water Based Nanofluids On Heat Transfer And Pressure Drop Characteristics Characteristics *Int. Journal of Engineering Research and Applications* p2248-9622

[12] Kristiawan B and S Kamal 2015A Modified Approach for Power Law rheological Titania Nanofluids Flow Behavior in a Circular Conduit *Journal of Nanofluids* **4 (2)**: p187-195

IOP Conf. Series: Earth and Environmental Science **105** (2017) 012070 doi:10.1088/1755-1315/105/1/012070

Nutrient recovery from anaerobic digester liquid effluent using ornamental aquatic macrophytes

W Ediviani, C R Priadi and S S Moersidik
Environmental Engineering Program, Faculty of Engineering, University of Indonesia, Depok, Indonesia

Abstract. Anaerobic digestion (AD) which treat food waste is a waste-to-energy method that produces liquid effluent. This by-product, known as digestate, contains high nutrients that could be recovered using ornamental aquatic macrophytes in a constructed wetland system. This study investigates the capacity of nutrient recovery of *Canna indica*, *Iris pseudocarus*, and *Typha latifolia* from liquid digestate, together improving the quality of AD effluent. Constructed wetland with *T. latifolia* effectively removed TSS and COD to meet the wastewater quality standards (TSS = 71 mg/L, COD = 56.735 mg/L). *C. indica* removed up to 72% N as the highest N removal efficiency, and recovered most of N, even though it still needs longer detention time to meet the standard. *I. pseudocarus* removed up to 98% P yet the average TP level in the plant was slightly above *T. latifolia*. The result shows that nutrient recovery using constructed wetland improves the effluent quality within short operation period, meanwhile *C. indica* and *I. pseudocarus* as ornamental aquatic macrophytes also added the aesthetic value to the environment.

1. Introduction

Anaerobic digestion (AD) is a well-known method in waste-to-energy (WTE) system to treat organic waste, such as food waste, which also act as the problem solver of food waste generation. Researchers has been doing this method for reducing the rate of waste generation and producing energy. Meanwhile the effluent generated at the end of the process known as digestate is a by-product that needs to be treated [1]. Digestate which has two forms, solid (fibre) and liquid (liquor), contains high nutrient that could act as a pollutant when no post-treatment applied [2]. Due to the high nutrient content of digestate, it must be treated before disposing to prevent unwanted matter on the environment. Solid fraction/ fiber digestate has 8.8—13.0 kg/t DS of total nitrogen (TN), as for the liquid fraction/ liquor digestate has 4,000—6,000 mg/l of TN, 7,500—10,000 mg/l of biological oxygen demand (BOD), 90,000—115,000 mg/l of chemical oxygen demand (COD), and 5,000—63,000 mg/L of total suspended solid (TSS) content [2, 3].

The characteristic of digestate made it as a valuable nutrient source, yet not many technologies are implementable in developed countries. There are chemical and physical treatment could be applied towards digestate, such as alkaline stabilisation (chemical), thermal hydrolysis (physical), or centrifuge thickening (physical). But for chemical treatment, it has potential for releasing odour, could be unsuitable for all soil types and requires high chemical. As for the physical treatment, it requires high temperature, pressure, energy, yet it also still needs to dispose the liquor [4].

One other way to treat digestate is by biological treatment, using the nutrient recovery method, which not only could recover nutrients from digestate by plant uptake, but also creates a cleaner environment by making a friendly post treatment product (low nutrient, BOD and COD content). Nutrient recovery of digestate that has been done was only up to composting (biofertilizer), mostly on solid form, but it was not good enough to compete with synthetic fertilizers unless it is integrated with an agriculture industry which use AD technology as well [2, 4]. Plants, particularly aquatic macrophytes, has a reputation in nutrient removal system, such as constructed wetland. Aquatic macrophytes are type of plants which has a macro form, and lives in a wet ecosystem [5]. The vegetation that are quite popular in here is *Canna indica*, *Iris pseudocarus*, and *Typha latifolia*. Each of these ornamental aquatic macrophytes recovered the nutrients from liquor state as nutrient uptake which will be the source of their development.

Horizontal subsurface flow was the chosen type of the constructed wetland used in the research [6]. This system was the best option as the constructing location was near the campus which could prevent smell from the influent (since it flows in the subsurface), near from water and energy source, and fairly manageable. The site was also for adding the aesthetic view.

C. indica is one of the most popular ornamental aquatic macrophyte used in the wastewater treatment in a wetland system. As for treating domestic wastewater, *C. indica* could be used as an efficient nutrient removal up to 84% for nitrogen removal and 92% for phosphorus removal, which could create a healthier environment at the end of the treatment [7]. The selection of the vegetation was based on literature that mentioned emergent type is the most productive one from the other aquatic macrophyte plants. The other reason was also for the availability in the country, price and aesthetic [8]. With those capabilities and advantages, *C. indica* could be more likely used as a nutrient recovery agent for substances such as liquid products of organic waste treatment which has similarities with domestic wastewater; high nutrient content. This study focused on how ornamental aquatic macrophytes could recover nutrient and contribute in improving the effluent quality as a post-treatment for liquid digestate. *C. indica* was compared to two other ornamental aquatic macrophtyes, *I. pseudocarus* and *T. latifolia*, which are also common to be used as aesthetic plants.

2. Method

2.1. Operation and sampling method

The operation flow started from the AD reactor to the influent storage, and the influent later will be distributed to three beds (Figure 1). The operational time was done in 10 weeks; acclimatization (A) stage in 7 weeks and feeding (F) stage in 3 weeks. Both stages (A and F) started during the morning time (10 AM); flowing the influent inside the bed. Sampling was done three times a week (every two days) after the acclimatization stage.

Sampling was done in every stage in a certain period range. The A stage effluent sampling was done 3 times for each bed, and the F stage was done 2 times for each bed (HRT = 11 days) [9]. As for the plants sampling was done 2 times; once after the A stage was done, and once at the end of the operation time. Each sample was kept in the freezer until it was sent to the lab for checking the parameters.

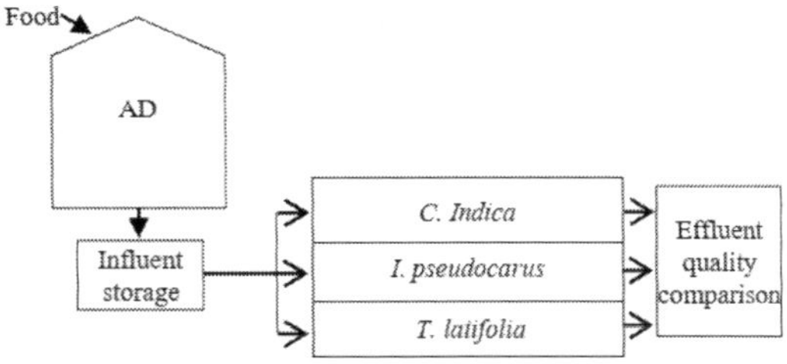

Figure 1. Operation flow

2.2. Influent characteristic, CW design and laboratory analysis

The constructed wetland influent was from an anaerobic digester which treat food waste [10, 11]. Effluent from the digester were pretreated by dilution (4x) to provide an influent with a BOD_5 characteristics $\pm$ 400 mg/L. This characteristic must be reach for the survival of the macrophyte itself [10].

Table 1. Characteristics of the influent

Periode	TSS (mg/L)	COD (mg/L)	BOD_5 (mg/L)	TN (mg/L)	TP (mg/L)
A1	2393.75	2253.165	420	617.219	305.563
A2	-	2626.583	275	658.808	401.890
A3	-	-	255	580.828	388.856
F1	6718.75	5158.228	530	669.948	37.943
F2	-	5039.558	520	649.154	3.983

A = acclimatization; F = feeding

As for the land used for each bed is 170 × 80 × 70 cm (length : width : depth), with planting area 120 × 60 cm (length : width). *C, indica*, *I. pseudocarus* and *T. latifolia* were the experimented object in the study. Sediment for the bed are as followed: top layer: clay (20 cm), middle layer: sandy loam (15 cm), and bottom layer: coarse sand (15 cm). In the inlet and outlet of the bed were used gravel (each side 15 cm)

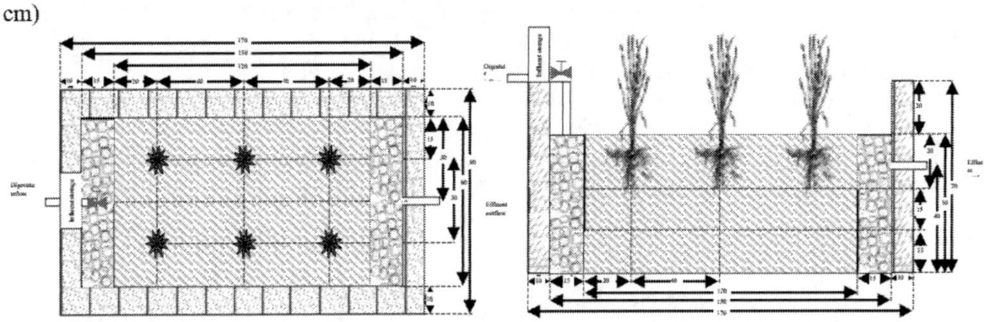

Figure 2. Top (left) and side (right) view of the HSSF CW design (scale in centimeter)

The parameters that were analyzed in this research are total suspended solid (TSS), chemical oxygen demand (COD), biological oxygen demand (BOD_5), total nitrogen (TN), and total phosphorus (TP). TSS content were tested on the initial influent flow and the last effluent sampling. BOD_5 was observed only on the influent. COD was check on all samples (influent and effluent), while TN, and TP were checked on influent, effluent, and plants. Those parameters will later be analysed to know the removal rate of COD, N, P, and nutrient uptake by each plant as the act of recovering nutrients.

2.3. Data analysis

The data which were gained from the study will be analysed later on by doing these steps.

1. Calculating each parameter removal (TSS, COD, and nutrient) by each plant using an equation.

$$\%\text{parameter removal}=\left(\frac{\text{parameter}_{\text{input}}\text{-parameter}_{\text{output}}}{\text{parameter}_{\text{input}}}\right)\text{mg/L x }100\% \tag{1}$$

2. Calculating the nutrient (N and P) uptake by each plant.

$$\text{Nutrient uptake (g/m}^2) = \left(\text{Nutrient}_F(\%)\text{x}\frac{\text{biomass F (g)}}{\text{planting area (m}^2)}\right) - \left(\text{Nutrient}_A(\%)\text{x}\frac{\text{biomass A (g)}}{\text{planting area (m}^2)}\right) \tag{2}$$

3. Result and Discussion

3.1. TSS removal

TSS content was check twice, at the end of A stage and the beginning of F stage. In table 2, the highest removal efficiency was shown by *T. latifolia* with 98.65% and the least was by *C. indica* with 95.74%.

Table 2. TSS content and removal in each effluent

Content in bed effluent (mg/L)			Removal (%)		
C. indica	*I. pseudocarus*	*T. latifolia*	*C. indica*	*I. pseudocarus*	*T. latifolia*
286	128	91	95.74%	98.09%	98.65%

C. indica had the least TSS removal efficiency than the other plants while showing a good growth performance (lots of shoots and leaves). *I. pseudocarus* growth was not as good as the other species (withered) during the A stage, yet has better growth development in F stage which made the TSS content lower than the A stage. However, each plant passed effluent quality for aquaculture, farms, plants irrigation and other purposes with the limitation of TSS $\leq$400 mg/L [13].

3.2. COD removal

Table 3. COD content and removal in each effluent

Periode	Content in bed effluent (mg/L)			Removal (%)		
	C. indica	*I. pseudocarus*	*T. latifolia*	*C. indica*	*I. pseudocarus*	*T. latifolia*
A1	750.000	1531.650	105.700	66.71	32.02	95.31
A2	118.350	629.110	99.370	95.49	76.05	96.22
A3	-	-	-	-	-	-
F1	235.440	755.700	20.797	95.44	85.35	99.60
F2	2.201	16.887	1.071	99.96	99.66	99.98

The COD measurement was done on the A1, A2, F1, and F2 stages. Based on the result in table 3, the lowest COD content was conducted by *T. latifolia* bed with the average 56.735 mg/L, also with the highest COD removal efficiency = 97.78%. Regarding to the regulation, only the bed planted with *T. latifolia* could produce effluent with the quality near two the 3nd class of wastewater classification (50 mg/L), while *C. indica* still need extra time to reduce the COD content to reach the 4th class wastewater (100 mg/L) [13].

3.3. N removal

Table 4. N total content and removal in each effluent

Periode	Content in bed effluent (mg/L)			Removal (%)		
	C. indica	*I. pseudocarus*	*T. latifolia*	*C. indica*	*I. pseudocarus*	*T. latifolia*
A1	194.740	223.555	235.735	68.45	63.78	61.81
A2	177.213	223.407	241.528	73.10	66.09	63.34
A3	1204.668	222.070	187.759	-	61.77	67.67
F1	158.201	192.215	257.272	76.39	71.31	61.60
F2	189.393	237.963	165.182	70.82	63.34	74.55

Table 4 shows that the effluent from every bed has lower N total content than the influent in table 1. This shows that N removal process has occurred in the constructed wetland system by each macrophyte. Looking at the result itself, the removal efficiency hasn't reach 90%, which is normal since the operation time was quite short for the constructed wetland to conduct a stable performance. However, there was an anomaly result on period A3. Seems like the result was inaccurate due to the sampling process since the other samples show a quite similar result. The efficiency calculation shows that *C. indica* could remove nitrogen up to 72.19%, the highest N removal efficiency among the other macrophytes in the study. Even though the influent's nitrogen content wasn't as high as the literature mentioned, but *C. indica* has the performance to improve the quality of the effluent. *I. pseudocarus* and *T. latifolia* shows similar efficiency performance, lower than *C. indica*. However, it still shows improvement to the effluent quality. The regulation state that wastewater that could be reuse for plant irrigation must meet 20 mg N/L, while the effluent has 20x N content and need further treatment.

3.4. P removal

Table 5. P total content and removal in each effluent

	Efluen Bed CW (mg/L)			Removal (%)		
	C. indica	*I. pseudocarus*	*T. latifolia*	*C. indica*	*I. pseudocarus*	*T. latifolia*
A1	1.100	4.953	0.810	99.64	98.38	99.73
A2	0.781	3.504	0.897	99.81	99.13	99.78
A3	55.467	2.809	2.375	85.74	99.28	99.39
F1	1.071	1.535	20.797	97.18	95.95	45.19
F2	2.201	16.887	1.071	44.74	-	73.11

The lowest average of P total content was generated by *T. latifolia* bed = 5.190 mg/L, yet *I. pseudocarus* shows the highest removal efficiency = 98.18% with the average P total content = 5.938 mg/L. While TSS, COD, and N removal were inclined to *T. latifolia* and *C. indica* positive results, P removal was inclined to *I. pseudocarus*. Due to the wastewater quality standards, *T. latifolia* dan *I. pseudocarus* almost reach the peak level (5 mg/L), while *C. indica* still contains 2x than what must be determined.

3.5. N and P uptake

C. indica shows a significant result for the N uptake, and has similar P uptake with *T. latifolia*. *C. indica* N uptake reached 166.39 g/m^2, which is 4.25x higher than *I. pseudocarus*, and 5.37x higher than *T. latifolia*. The gap in N uptake is exceptionally large, knowing that the initial biomass of each plants is not quite different. As for the P uptake, *T. latifolia* shows the most significant result by uptaking 16.13 g/m^2 P during the operation.

Table 6. N and P uptake by each macrophyte

Periode	TN (mg/L)			TP (mg/L)		
	C. indica	*I. pseudocarus*	*T. latifolia*	*C. indica*	*I. pseudocarus*	*T. latifolia*
A	28.53	20.36	25.08	0.78	0.62	0.74
F	194.92	59.51	56.04	15.61	6.63	16.87
Total uptake	166.39	39.15	30.96	14.82	6.02	16.13

This result shows on the plant growth/ development, which *C. indica* has the fastest leaves and flower growth, while *I. pseudocarus* has weak roots yet it kept flowering every once in 2 weeks. Morphology of *T. latifolia* showed the nutrient content as mentioned in the table; lack of TN content since the stem color was brown-yellowish, yet the shoots kept growing around the parent plant (increase in TP content).

3.6. Environment quality improvement

The data shows significant differences between influent and effluent quality performed by each plant. The TSS, COD, and TP content reduced with quite high efficiency removal percentage, while for TN it shows above the middle range removal efficiency. Following the regulation for wastewater quality standards, the TSS and COD of the AD effluent already meet the requirement, yet for TN and TP are almost there. The system does require dilution indeed for the plant growth reason (BOD content) [12], this must be solved by other pre-treatment to use less clean water, bearable enough for plants to load, yet suitable to be source of nutrients. However, within the short period, it was able to show a significant content level decrease of each parameter.

4. Conclusion

Nutrient recovery by ornamental aquatic macrophyte in a constructed wetland system could be a solution for treating the end-of-waste of AD to improve the environment quality. TSS content in the effluent from each bed has meet the requirement of the wastewater quality (TSS = $\leq$400 mg/L); *C. indica* = 181 and mg/L, *I. pseudocarus* = 223 mg/L, and *T. latifolia* = 71 mg/L. For COD level, *I. pseudocarus* and *C. indica* are beyond the standard (733.337 and 276.498 mg/L, respectively), only *T.*

latifolia (56.735 mg/L) who meets the quality standard (≤100 mg/L). *C. indica* removed N up to 72.19% from digestate, while *T. latifolia* and *I. pseudocarus* removed 65.79% and 65.26%, respectively. TN removal, it hasn't meet the standard yet, but the efficiency was above the middle range, and even though *I. pseudocarus* has the highest P removal efficiency, the average TP content is still higher than *T. latifolia*.

This study showed that all of the experimented plants could improve the effluent quality by recovering nutrients from liquid digestate, but the best performance goes to *C. indica*. Overall, the bed of *T. latifolia* performed better than *C. indica* and *I. pseudocarus*, yet *C. indica* and *I. pseudocarus* recovered more nutrients and have better morphology than *T. latifolia*. In terms to gain both improving water quality and adding the aesthetic value to the environment, further studies need to be done. This study shows that nutrient recovery method using constructed wetland does improve the effluent quality within the short operation period (10 weeks), yet for having better quality (less COD and TN level), longer period is recommended to have stable performance by the wetland. *C. indica* and *I. pseudocarus* as ornamental aquatic macrophytes also added the aesthetic value to the environment.

5. Acknowledgment
This study is funded by the grant of Green FT Universitas Indonesia and *Publikasi Internasional Terindeks untuk Tugas Akhir Mahasiswa* (PITTA) 2017 with number of grant contract 756/UN2.R3.1/HKP.05.00/2017. The influent material was supplied from Septiana Kurnianingsih and Dwita Fitriani Wijayanti research, influent preparation was helped by Mislan, and location of the CW was provided by Faculty of Engineering, University of Indonesia.

References

[1] Saveyn H 2012 EC *'End-of-Waste' Criteria for Biodegradable Waste Subject to Biological Treatment (Compost and Digestate)* (Rennes: Joint Research Centre, Institute for Prospective Technological Studies) p 39
[2] Anaerobic Digestion & Bioresources Association 2013 *Digestate* **3** chapter 6 p 81–92
[3] Ediviani W 2017 *Nutrient Recovery Potential Analysis from Digestate for Cultivating Ornamental Macro-phyte using Horizontal Subsurface Constructed Wetland Method* (Depok: Universitas Indonesia)
[4] Frischmann P 2012 *Enhancement and Treatment of Digestates from Anaerobic Digestion* (United Kingdom: Pell Frischmann Consultants Ltd) p 122
[5] Dhir B 2013 *Phytoremediation: Role of Aquatic Plants in Environmental Clean-up* (India: Springer) p 111
[6] Environment Protection Agency 1993 *Subsurface flow constructed wetlands for wastewater treatment: a technology assessment* (USA: EPA) p 87
[7] Calheiros CSC Bessa VS et al 2015 Constructed wetland with a polyculture of ornamental plants for wastewater treatment at a rural tourism facility *Ecological Engineering* **79** pp 1–7
[8] Hasan MR Chakrabarti R 2009 Use of algae and aquatic macrophytes as feed in small-scale aquaculture: a review (Rome: *FAO Fisheries and Aquaculture Technical Paper*) No. 531 p 123
[9] Ghosh D Gopal B 2010 Effect of hydraulic retention time on the treatment of secondary effluent in a subsurface flow constructed wetland *Ecological Engineering* **36** pp 1044–1051
[10] Kurnianingsih S Priadi CR 2017 Pilot scale anaerobic digestion of food waste process performance under various mixing intensities *The 1st International Conference on Science, Mathematics, Environment and Education 2017 Solo* presented on 16—17 September 2017 (ID: 3108) p 10
[11] Wijayanti DF Suwartha N Priadi CR 2017 Effect of addition of fat oil and grease (fog) towards performance of dry anaerobic digestion food waste reactor *The 2nd International Tropical Renewable Energy Conference 2017 Bali* presented on 3—4 Oktober 2017 (ID: 12732) p 9
[12] Tousignant E 1999 *Guidance Manual for The Design, Construction and Operations of Constructed Wetlands for Rural Applications in Ontario* (Ontario: Stantec Consulting Ltd) p 61
[13] Peraturan Pemerintah Republik Indonesia Nomor 82 Tahun 2001 Tentang Pengelolaan Kualitas Air Dan Pengendalian Pencemaran Air

Characteristics and effectiveness identification of restructured product of snakehead fish (*Channa Striata*) with transglutaminase enzyme addition

Setiadi and Sella Lametta

Department of Chemical Engineering, Faculty of Engineering, Universitas Indonesia

E-mail : setiadi@che.ui.ac.id

Abstract. The existence of fishery waste problem shows that the innovation of fishery processing not yet maximized in Indonesia. Both waste and fresh cuts of fish can be utilized by restructuring with transglutaminase enzyme addition as a crosslinking agent. This microbial transglutaminase (MTG) enzyme can alter protein functionalities by forming stable covalent bonds between residues of the amino acid of protein. Accordingly, this study focused on Snakehead Fish (*Channa striata*) is used as the protein source and variations of treatment conditions include incubation duration (2 days and 7 days) and the composition of enzyme and processed fish meat (0.0%, 0.5%, 1.0%) and. The results of this study showed the strong performance of crosslinking by MTG on processed fish meat. This is confirmed by an increase in texture profile parameters (hardness, cohesiveness, and elasticity), the detection of myosin changes, also wave number increment of C-N and an increase of intensity of C=O bonds. The highest effectivity value in the sample was achieved by variation of incubation duration of 7 days and the addition of 1.0% MTG enzyme with a value of 129,78% in hardness parameter in sample.

1. Introduction

Along with the increasing number of world population and the need for better nutrition source, the demand for fish which has a significant nutritional value continue to increase every year. Indonesia with a large population is fortunate enough to have abundant fishery resources with the position as the fifth largest fish producer according to FAO statistics. Indonesia's fishery products in public waters covering an area of 54 million hectares with a production potential of 0.9 million tons a year. The potential value of Indonesian marine and fishery products reaches 3000 trillion per year, however there was only about 225 trillion or about 7.5% that has been utilized. The government-driven potential must be supported by all stakeholders, including in terms of innovative processing of all fishery products [1].

One of the innovation of fishery product processing is chemical binding method which is done by bonding agent with various kinds of treatment. One of the most widely used is transglutaminase enzyme [4]. The use of transglutaminase enzymes that is specifically produced by microbes or known as microbial transglutaminase (MTG) species is expected to increase the value of meat scraps from previous product processing as well as to innovate the fresh fish processing. MTG can produce fish meat with reconstruction of a new and chewier structure because of its working principle as an agent forming of cross-linked covalent peptides between protein surfaces in enzyme-added fish tissues.

Previous studies [2-6] have shown that each species of protein source will have a different response to MTG and can produce various products (polymers) that vary in morphological, rheological, and physiochemical properties. In general, by adding enzymes in the re-structuring of fish meat, meat processing can be done more easily and the yield of marketable fishery products can be maximized.

2. Material and Methods

2.1. Materials
The materials used in the study are snakehead fish (*Channa striata*), Microbial Transglutaminase enzyme (MTG), water and alcohol.

2.2. Preparation Tools and Materials
Preparing the necessary tools in the experiment involves the provision of materials and tools, as well as the sterilization of tools used such as scales and gloves with alcohol. The transglutaminase enzyme was provided by a manufacturer in Spain that has been available since November 2016. Before being used, the enzyme was stored in a sterile and vacuum condition of a refrigerator at about -16^{0}C.

2.3. Preparation of Fish Meat and Enzymes
After preparing the required tools and ensuring the availability of the ingredients, fish meat was cleaned by cutting the meat carefully off the bones and unused body parts such as head and skin before washing. There are two types of treatment for fish meat based on their use for different analysis test.

- Treatment I: Cut fish meat into dice with dimensions of about 3 cm x 5 cm for texture profile samples
- Treatment II: Puree the fish meat and separate into 10 grams each as the sample unit.

2.4. Fish Meat Processing with Enzymes
On sample of texture profile analyzing, the enzyme was applied evenly on one surface of the meat, then stack one piece of other meat on top. As for other tests, the enzyme with adjusted dosage was mixed evenly through mixing. After that, the samples were wrapped up with plastic wrap until it was airtight and covered all the parts perfectly. All samples then are labeled before being stored and incubated in a chiller with an adjusted duration. Specifically, for the texture profile test, samples were made for three duplicates.

2.5. Product Analysis
The tests taken were Acidity Test Level, Texture Test with Texture Profile Analyzer, Detection of Functional Groups/Secondary Structure with Fourier Transform Infra-Red, Electrophoresis Profile Test with Sodium Dodecyl Sulphate-Polyacrilamide Gel Electrophoresis.

3. Result and Discussion
In the initial processing of fish, fish meat that has been cut off from other parts or called fillet washed with water as washing can affect the degree of acidity of fish. The washing stage was done for removing the natural components of fish, such as water-soluble proteins, blood and other components, which can affect the decay processes (lipid oxidation and microorganisms) during the storage process at low temperatures [4]. Table 1 shows 15 samples used in this study with various MTG addition and given incubation duration.

Table 1. Prepared Samples

Analysis Types	% Enzyme	Duration (days)	Samples
Texture Profile	0.0; 0.5; 1.0	2 and 7	6
FTIR*	0.0	7	1
	1.0	2 and 7	2
SDS-PAGE*	0.0; 0.5; 1.0	2 and 7	6

*all samples have passed the acidity test

Figure 1. One of incubated sample

After being incubated during given time to react, a description of restrutured samples are shown on Figure 1. The chopped and processed meat that has been incubated with MTG addition turned into a larger and solid meat.

3.1. Analysis of Acidity Level (pH)

To analyze the acidity level of the sample, the pH meter measurements with digital pH meters were performed on prepared samples beforehand. All incubated samples were thawed before the measurement. The results of pH values are shown in Table 2.

Table 2. The Result of pH measurements

Incubation Duration (days)	Enzyme Concentration (w/w)	pH
7	1.0%	6.1
7	1.0%	5.8
2	1.0%	6.3
2	0.0%	6.4

The optimum pH of MTG was ranged from 5 to 8, but the enzymatic activity is still processed at pH 4 or 9 [7]. Thus, all data show that the performance of this enzyme reaches optimum under this study conditions. The increased pH value as the MTG addition shows the results of crosslinking reactions in sample proteins, which chemically produce ammonia molecules [8]. Consequently, the content of more alkaline can affect the pH value of meat.

3.2. Texture Profile Analysis

Texture profile analysis was performed to find out the characteristics of fish meat texture that have been modified with MTG. The measurement conditions have been also adjusted according to a certain research reference [9]. Texture profile analysis was done on fish products, both with MTG enzyme or without MTG enzyme which used as control addition. The test was carried out at room temperature about 20-22°C [10,11].

Table 3. Sample Variations in Texture Profile Analysis

Incubation Duration (days)	% Enzyme	Parameter		
		Hardness (N)	Cohesiveness (-)	Elasticity (%)
	1	39.02	0.178	66.06
2	0.5	26.25	0.275	66.12
	0	16.98	0.205	61.87
	1	75.59	0.354	62.39
7	0.5	59.82	0.266	57.59
	0	55.14	0.195	55.59

The results in Table 3 briefly showed the changes on texture profile of samples after MTG was added and incubated in 2 or 7 days. These condition was developed from several previous researches on fish [3-5,7,10,12].

MTG allows the myofibril protein to dissolve and enable to induce the formulation of protein crosslinking. This has a major effect on changes in texture profile of various parameters in which each addition of MTG improves binding, hardness, cohesiveness, chewing and meat elasticity [11-12]. All values are compared and expressed on Figure 2.

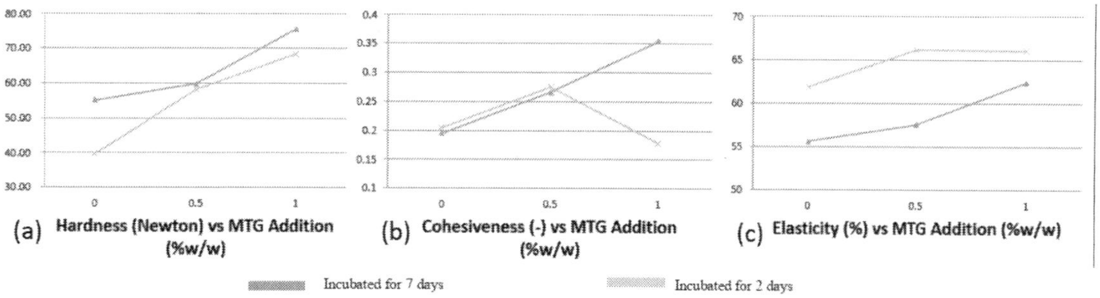

Figure 2. Results of Profile Texture Analysis

3.3. Level of Hardness

The level of hardness is the maximum force required to suppress the sample [11] and has units of force such as Newton (N) or gram-force (gf). The change of hardness value in snakehead fish only ranged from 16.98 to 75.59 N as expressed in Figure 2(a). The effect of MTG addition on samples that was incubated for 7 days shows higher hardness values. This supports various studies on other fish species such as silver goldfish [2], White Gulati Fish [4], Siamese catfish [5] and European hake and rainbow trout [11].

3.4. Level of Cohesiveness

Cohesiveness can be defined as how far a sample can undergo deformation before dehiscent [11]. To express the cohesiveness, no dimension number used. As for this study, the cohesiveness value is very low in all samples, which varied from 0.178 to 0.354 and somehow shows no significant changes od cohesiveness in each MTG addition (Figure 2(b)). In general, this shows a significant difference based on the incubation duration. However, based on different concentrations of MTG used in each sample, the values tend not to be as significant as those obtained by other studies [8,10,12].

3.5. Level of Elasticity

Measurement of springiness or elasticity values has the goal of determining a sample's ability to return to its original state after first pressure [11]. This elasticity level is expressed in a percentage that ranged from ~55% to ~66%. In Figure 2(c), it was shown that elasticity level in snakehead fish presented a significant change in both variations of incubation duration generally. For incubated samples of 2 days, the range of elasticity values ranged from 61.87% to 66.12%. While on the duration of 7 days, the value ranged from 55.59% to 62.39%. Both results showed that more MTG addition gave a higher value.

3.6. Analysis of Enzyme Effectiveness on Sample

As a whole, an increase in the value of hardness, elasticity and cohesiveness observed in this study can explain that MTG enzymes may result in protein crosslinking of fish. This affects the formation of larger protein polymer molecules between fish meat particles [2]. The formulae used in the calculation of the percentage of effectiveness and the results (Table 4) are as follows:

Table 4. Result of Samples Effectiveness

Incubation Duration (days)	% Enzyme	Parameter		
		Hardness (N)	Cohesiveness (-)	Elasticity (%)
2	1	2.10	(13.17)	6.77
	0.5	46.10	36.41	6.87
7	1	37.09	81.54	12.24
	0.5	8.49	34.15	3.60

$$\% \, Effectivity = \frac{(Sample\ parameter\ values\ with\ MTG\ addition)}{(Sample\ parameter\ values\ without\ MTG\ addition)} x \ 100\% \quad (1)$$

The effectiveness analysis was done based on the variation of the concentration of MTG (0.5% and 1,0%) and the incubation duration (2 and 7 days). In the term of incubation duration almost all samples with longer time showed more effective results. The highest effectiveness value was achieved by 7 days-incubated sample with MTG addition of 1.0% with 129,78% of hardness value.

3.7. Functional Group Analysis

In general, based on the results of this study, it was found that each group of atoms in the sample showed changes in the vibrational band with almost the same wave value. To verify the formation of crosslinking by MTG, there are two concerned waves. The preferred wave was assigned to Amide I vibrational mode, which ranges from 1600 cm^{-1} to 1700 cm^{-1} that involves mainly C=O stretching [11]. This mode will show an increase in the band intensity as the addition of MTG applied on samples. Figure 3 shows the result of functional group analysis results, which (a) shows sample with 1,0% MTG addition with 7 days-incubation duration; (b) shows sample with 1,0% MTG addition with 2 days-incubation duration; and (c) shows sample with 0,0% MTG addition with 2 days-incubation duration.

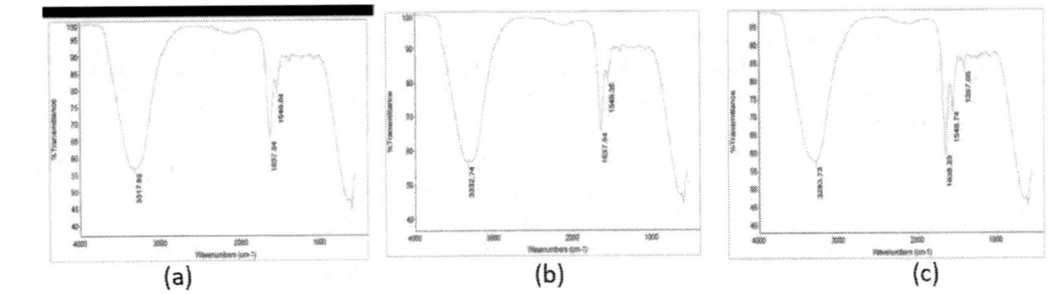

(a) (b) (c)

Figure 3. Result of Functional Group Analysis

As in Figure 3 show the corresponding result of wave intensity ~1637 cm^{-1} in (c) has a higher band intensity than (b). In addition, it was analyzed that the formation of new C-N bonds caused a decrease in the number of detected waves at ~1500 cm^{-1}. However, different results were obtained when comparing Figure 3(b) and (c), where samples with enzymes addition had slightly higher wave values than those without enzyme. Figure 3(a) and (b) was to compare the difference of incubation duration of samples, however did not show any significant result.

Therefore, each dominant peak that previously discussed is found in all sample variations. However, some of the results of the analysis show similar conclusions with previous studies [11] and concluded if FTIR test is useful to detect changes on protein binding as the result of MTG addition.

3.8. Analysis of Electrophoretic Profile with Sodium Dodecyl Sulphate-Polyacrilamide Gel Electrophoresis

The protein bands found in the performance analysis of MTG enzymes were in the 45 to more than 200 kDa band range. This result shown in Figure 4 is coresponding to the molecular weights of myosin respectively (200 kDa). There 3 lanes provided for samples with each different incubation duration, in which each lane was filled with different MTG addition whileLane 6 was used as protein marker,

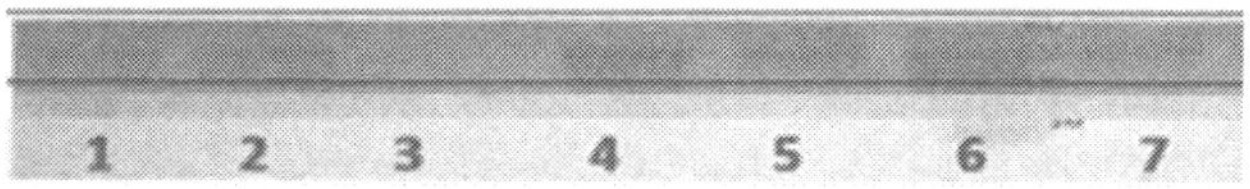

Figure 4. The Result of SDS-PAGE
Note: Lane 1-3 (2 days), Lane 6 (Marker), Lane 4,5,7 (7days)

In each samples shown in figure 4, the intensity of the myosin band was not clearly different. MHC (myosin heavy chain) molecule is crosslinked by MTG which ultimately complicates the separation of the molecule during gel analysis. The covalent bonds that occur cannot be disrupted by the electrophoresis reagent during analysis, so the detection becomes lessened [2]. This respect was an explanation of different result of SDS-PAGE based on its incubation duration, where samples with longer period have the decreasing protein band density.

4. Conclusion

The addition of microbial transglutaminase (MTG) enzyme in snakehead fish causes cross-linking of myosin proteins as seen from the results of functional group analysis with increasing C-N wave value and wave intensity C=O. Physically, the changes were also observed on the changes of final restructed samples. Also, the addition of MTG concentration and longer incubation duration were directly proportional to the increase in texture profile parameters (hardness, cohesiveness, and elasticity) of the sample and inversely proportional to the intensity of the protein band on the electrophoretic profile. The highest effectiveness value was achieved with variation of incubation duration for 7 days and addition of 1.0% MTG with effectiveness value of 129,78% in hardness parameter.

5. Reference

[1] Widodo, Johanes, Suadi 2008 *Pengelolaan Sumberdaya Perikanan Laut* (Yogyakarta: Gadjah Mada University Press)

[2] Tellez-Luis, Simon J, Jose´ A, Rocio M, Manuel V 2002 Low-salt restructured fish products using microbial transglutaminase as binding agent *J. Sci. Food and Agr.* **82** pp 953-959

[3] Helena M, José C, A. Javier B 2010 Use of microbial transglutaminase and sodium alginate in the preparation of restructured fish models using cold gelation *J. Inno.Food Sci. Technol.* **11** pp 394-400

[4] Gonçalves, Alex A, Marcelo G 2010 Restructured Fish Product from White Croacker (Micropogonias furnieri) Mince Using Microbial Transglutaminase *Brazilian Arch. Biotech. Technol.* **53** pp 987-995

[5] Sarika K, Manjusha L, Mithlesh K, Nagalakshmi K, Venkateshwarlu G 2013 Textural quality oxidative stability of restructured pangasius mince: effect of protein substrates mediated by transglutaminase *J Food Sci. Technol.* **52** pp 351-358

[6] Tzikas Z, Soultos N, Ambrosiadis I, Lazaridou A, Georgakis S 2015 Production of low-salt restructured Mediterranean horse mackerel (Trachurus mediterraneus) using microbial transglutaminase/caseinate system. *J. Hellenic Vet Med Soc* **3** pp 147-159.

[7] Jiang S, Yin L 2001 Application of Transglutaminase in Seafood and Meat Processing *J. Fish Sci.* **28** pp 151-162

[8] Buchert, Johanna, et al. 2010 Crosslinking Food Proteins for Improved Functionality *Rev. J. Food Sci. Technol.* **1** pp 113-138

[9] Huidobro R, Miguel E, Blázquez B, Onega E 2005 A comparison between two methods (Warner–Bratzler and texture profile analysis) for testing either raw meat or cooked meat *J. Meat Sci.* **4** pp 527–536

[10] Helena M, Javier B, Caroline P 2010 Evaluation of some physico-chemical properties of restructured trout and hake mince during cold gelation and chilled storage *Food Chem.* **120** pp 410-417

[11] Helena M, Cambero M, Ordonez J, L de la Hoz, Carmona P 2008 Raman spectroscopy study of the structural effect of microbial transglutaminase on meat systems and its relationship with textural characteristics *Food Chem.* **109** pp 25-32

[12] Min, Green B 2008 Use of Microbial Transglutaminase and Nonmeat Proteins to Improve Functional Properties of Low NaCl, Phosphate-Free Patties Made from Channel Catfish Belly Flap Meat *J. Food Sci.* **73** pp 218-226

Enhanced dielectric properties of Nanocrystalline $Ba_{(1-x)}Sr_xTiO_3$ (x=0 and 0.3) ceramics

E Yustanti[1], M A E Hafizah[2] and A Manaf[2]

[1]Department of Metallurgy, Faculty of Engineering, University of Sultan Ageng Tirtayasa, Cilegon Banten 42435, Indonesia
[2]Department of Physics, FMIPA Universitas Indonesia, Depok 16424, Indonesia

Email: azwar@ui.ac.id

Abstract. In this paper, we report the enhancement of the dielectric constant of barium titanate (BTO) through partial substitution of strontium to barium and crystallite size reduction through a high power ultrasonic treatment. The sample under investigation is $Ba_{1-x}Sr_xTiO_3$ or BST with x = 0 and 0.3 compositions prepared through mechanical alloying of $SrCO_3$, $BaCO_3$ and TiO_2 precursors as the feedstock. All mechanically alloyed samples were crystalline powders with a single phase as confirmed by an x-ray diffractometer (XRD). The mechanically alloyed materials were consisted of multi-crystallite particles as confirmed both by XRD data analysis and particle size evaluation. Subject to an additional ultrasonic treatment, the multi-crystallite particles were fragmented toward mono-crystallite particles with the mean crystallite size about 52 nm after 12 hours irradiation. Even a smaller mean crystallite size (18 nm) with a narrower crystallites distribution then that of BTO was obtained in $Ba_{1-x}Sr_xTiO_3$ with x = 0.3. Such a narrow crystallite size distribution with a small mean crystallite size has superior dielectric constant over those of BTO and doped BTO with a large mean crystallite size. The highest dielectric constant of 3000 was obtained at a frequency of 273 Hz in doped BTO after 12 hours ultrasonic irradiation. The value is 12 times higher than those of BTO and doped BTO with a large crystallite size.

1. Introduction

Barium titanate is a ferroelectric material, which widely used in many applications [1]. For the last few years, research on ferroelectric material have been developing rapidly due to the exponential increase of ferroelectric based materials that required by electronic products producer which should always be able to provide new innovations through improvement in material performances [2]. The characteristics and performance of BTO are influenced by its crystal and materials structure. When refer to previous reports [3,4], it is well known that BTO and Sr substituted BTO or BST possesses a perovskite structure with two different crystal structures, tetragonal and cubic respectively. The BST of both structures could produce unique electrical properties. It was reported that the BST with dopant substituted $x \leq 0.3$ possessed ferroelectric properties with tetragonal crystal structure, while the BST with $x \geq 0.3$ possessed para electric with a cubic crystal structure [4]. Structural modification of BTO like in BST could changes the intrinsic properties of BTO that possesses a better respond to the influence of external field like the changes of electric field, temperature, pressure and electromagnetic wave. Further improvement in properties could be obtained through refining the crystallite size toward nanostructure. There is a large surface to volume ratio in a nanocrystalline system, which expected to provide the enhancement in properties. Hence, BST based devices should be able to meet the current need of various electronic components with a better performance.

Various synthesis methods to prepare the BST based material have been introduced [5-9]. A few to mention here: sol gel, wet chemical, sonogel, citrate precursor, mechanical alloying and mechanochemical methods. However, among the all mentioned methods, the mechanical alloying has been the one of the most common and the most used because of the following reasons: it is a simple processing method, economically reliable, environmentally friendly and adaptable for a large production capacity [10,11]. The synthesis of strontium titanate based material generated through mechanical alloyed generally have micrometer particle size [12], thus a further treatment in material with nanostructure is required. In spite of various preparation techniques to produce nanometer scale material is available, in the current study, we have selected a mechanical alloying method to prepare the BST crystalline powders coupled with high power ultrasonic irradiation treatment in order to obtain nanoscale powders. It was demonstrated in our previous works [13,14] that crystalline powders with nanometres scale have successfully synthesized through a successive treatment ultrasonically to the mechanically alloyed powders [15,16]. The aim of this study is to increase the dielectric constant of barium titanate (BTO) through partial substitution of Sr^{+2} ion to Ba^{+2} during synthesis of the $Ba_{(1-x)}Sr_xTiO_3$ (BST) by mechanical alloying method coupled with a high frequency of ultrasonic irradiation treatment for further reduction their crystallite sizes.

2. Methodology/experimental

The $Ba_{(1-x)}Sr_xTiO_3$ with x=0 and 0.3 were prepared through stoichiometry mixing of the standard research grade $BaCO_3$, $SrCO_3$ and TiO_2 precursors of Sigma Aldrich with at least 99% purity level. All precursors were wet-ball milled using alcohol as dispersed media in a planetary ball mill apparatus for 60 hours at a 160 RPM speed. The ratio of ball mill:precursors was 10:1. The mechanical milling process to the precursors has produced heavily deformed powder materials. A 2 wt.% of Poly Vinyl Acetate (PVAc) diluted in a demineralized water were then added to the powder materials for wetting all the particles and let the paste in an oven for drying. The dried particles are poured into a cylindrical die and 10 tons of compacting were employed to make the green compact samples with 25 mm in diameter and 3 mm thickness. The green compact samples were sintered at a temperature of 1200°C for 4 hours. The characterization of polycrystalline samples was performed by X-Ray Diffractometer (XRD) PANalytical Empyrean and analyzed using a High Score Plus (HSP) software. The sintered BST pellet was then hand grounded and re-milled in a ball mill apparatus for 20 hours. The crystalline milled powders were dispersed in aquadest medium, an addition of 0.1% non-ionic surfactant. Ultrasonically irradiated treatments in Qsonica sonicator Q700, operated at 20 kHz with a transducer of 60 μm amplitude for 3, 6, 9 and 12 hours. Particle size characterization of powders was performed by means of particle size analyzers (PSA). The distribution of crystallite size in ultrasonically irradiated powders was calculated from refined XRD employing Whole Powder Pattern Modeling (WPPM) method [17,18].

3. Results and discussion

The diffraction pattern of a polycrystalline sample of x=0 and 0.3 compositions in the 2 theta range of 20-90^0 was previously reported along with phase analysis from the pattern [1]. It was shown that for x=0, the calculated lattice constants of tetragonal structure a and c is respectively 3.99603 Å and 4.02638 Å. While for x=0.3, the crystal structure changes to cubic with the lattice constants of a and c is 3.97334 Å. Obviously, the lattice changes are due to the presence of a partial Sr^{+2} ion substitution for Ba^{+2} in BST in which the ionic radius of Sr^{2+} is smaller than that of Ba^{2+} [19].

Figure 1 shows the crystallite size distribution of doped and un-doped BTO, along with their respective cumulative curves. The curve indicates that doped sample (x=0.3) has crystallite sizes <140 nm, which were smaller than that of un-doped (x = 0) sample (<350 nm). Apparently, inhomogeneity of crystallite size in the two samples is the same which indicated by nearly an equal breadth of their crystallite size distribution. However, the mean crystallite size of the two samples for doped and un-doped sample is 19 nm and 52 nm which showing a significant difference.

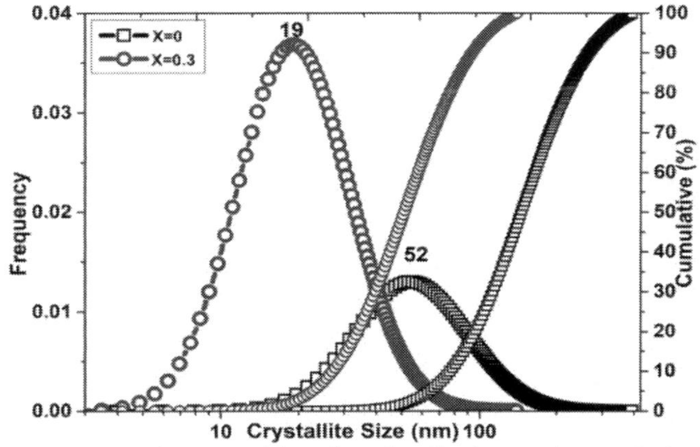

Figure 1. The crystallite size distribution and cumulative curve of mechanically milled crystalline $Ba_{(1-x)}Sr_xTiO_3$ powders.

The crystalite size distribution of particles undergone ultrasonic irradiation at a various irradiation time is compared in figure 2. It shows no significant change in their mean crystallite size, which is in the range 52-64 nm. All treated powders showed almost similar crystallite size distribution.

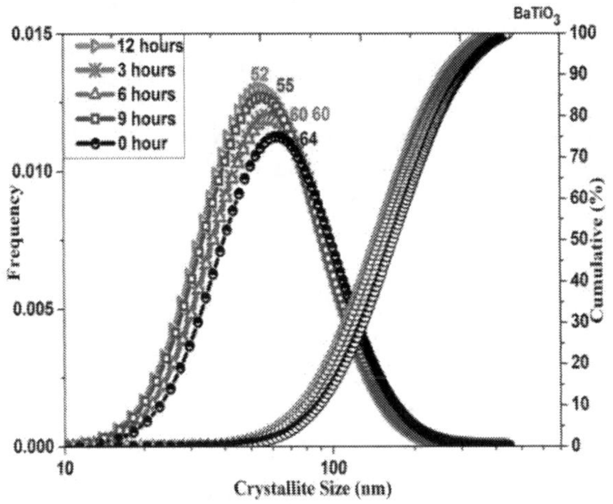

Figure 2. Crystallite distribution and cumulative curves for un-doped BTO.

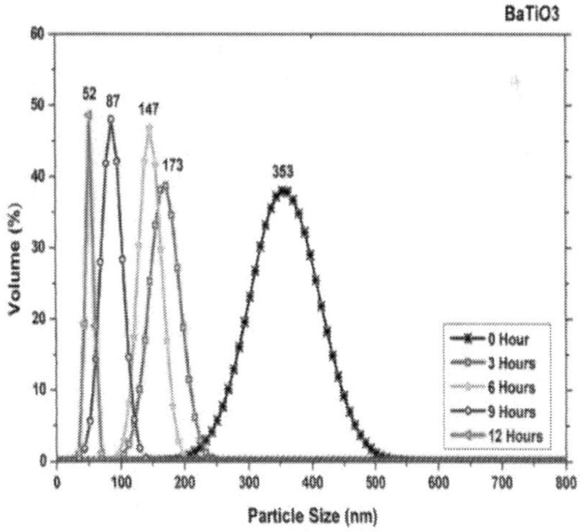

Figure 3. Particle size distribution of ultrasonically irradiated powders for un-doped BTO.

In contrary to the crystallites, the ultrasonic treated particles undergone progressive reduction in particle size as the irradiation time was extended from 0 to 12 hours. It is demonstrated in figure 3, the mean particle size of 353 nm in untreated powders was significantly reduced to 52 nm after 12 hours irradiation time. Reduction in particle sizes of mechanically powders was also shown in our previous report [20].

A similar data analysis for doped BTO is shown in figure 4. Trend of crystallite size reduction in doped BTO apparently similar to that of un-doped BTO. Before ultrasonic treatment, the particles have a mean crystallite size 38 nm. After 12 hours irradiation time, the mean crystallite size of ultrasonic treated particles reduced to 19 nm. It is clear that the treatment has brought devastating effect to the crystallites which fragmented the crystals to the smaller size. Hence, severe effects must also be experienced by the ultrasonically irradiated particles. It is shown in figure 5 which compares plots of the particle size distribution of doped-BTO before and after ultrasonic irradiation.

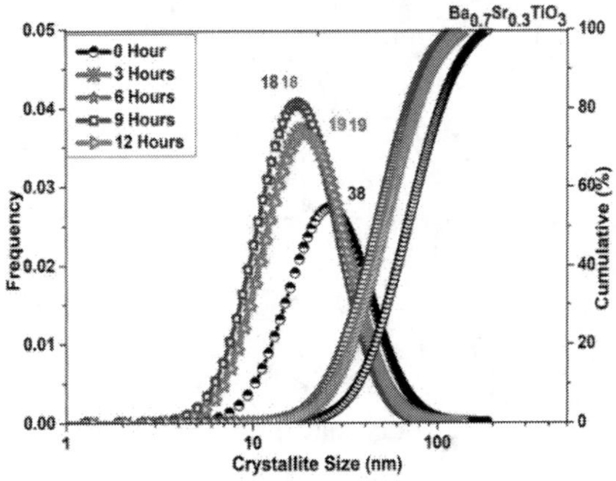

Figure 4. Crystallite distribution and cumulative curves for doped BTO.

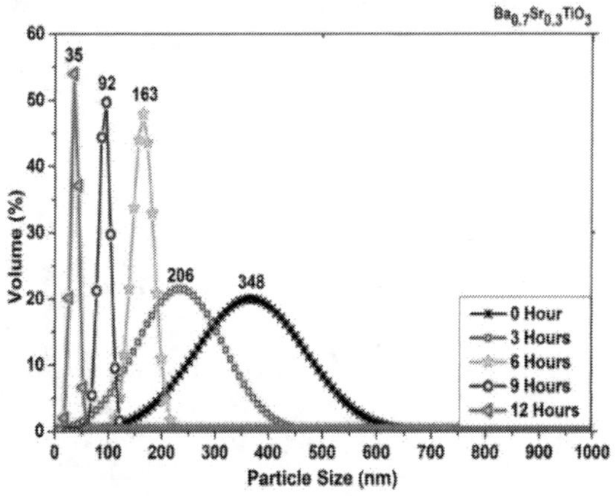

Figure 5. Particle size distribution of ultrasonically irradiated powders for doped BTO.

Prior to the irradiation, the particles are in the range size 100-650 nm. The particle sizes distribution is then gradually changed to the distributions with a much narrower particle size variation after ultrasonic treatment. Plots of particle sizes as shown in figure 5 indicated that the mean size of ultrasonic treated particles is progressively reduced. The longer irradiation time is applied, the smaller the mean size of particles. After the doped-BTO particles experienced ultrasonic treatment for 12 hours, the ultra-fine particles with the mean size of 35 nm was obtained. Plots in figure 5 assures that the increase of ultrasonic irradiation time is not only changing the particle size become smaller but also shows the narrower particle size distribution, which means that the particle size become uniform. It is again showing that the mechanical alloying method coupled with a high power ultrasonic destruction succeeded to produce doped BTO with nanometer scale in crystallite and particle sizes.

Figure 6 is the dielectric constant of the sample made of 12 hours ultrasonically treated particles (BS3T U12) compared with those of ultrasonic treated free (BS3T) and doped free (BTO) samples. The two last samples have an almost similar value of dielectric constant at least in the frequency range 273-1194 Hz. At frequencies lower than 273 Hz, an increase in the dielectric constant of the two samples is observed. At a frequency 203 Hz, the dielectric constant of BS3T sample is more than double of BTO. It was reported that strontium substituted BTO or $Ba_{1-x}Sr_xTiO_3$ at x=0.3 led to a crystal structure change from tetragonal to cubic, marked with the maximum value of dielectric constant in the whole composition. However, larger values of dielectric constant were obtained from a BS3T U12 sample when compared with the former two in the whole frequency range. At a frequency 203 Hz, the dielectric constant value is 3250 which more than six times larger than that of BTO and about 2.6 times larger than that of BS3T. Hence, the crystal structure and fine crystallite size of BS3T U12 are responsible for the increase in dielectric constant.

In the previous report [21], it was shown that the dielectric constant of the BST (x=0.03) with no additional treatment ultrasonically about 18.55 at a frequency 1 kHz. In another report, the dielectric constant of $Ba_{0.2}Sr_{0.8}TiO_3$ was 3,870 at a frequency 10 Hz [22]. The improvement of dielectric constant was shown due to the mean crystallite size of material in a nanometer scale that facilitates the grain exchange interaction of electric dipole moments on the surface. The fraction of interacting area will increase the decrease in crystallite size [23].

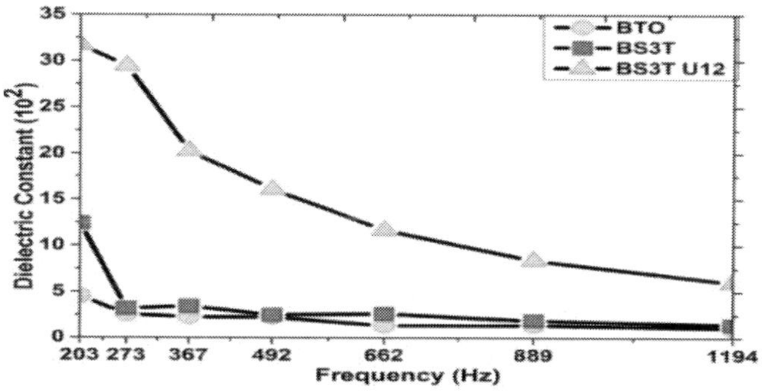

Figure 6. Dielectric constant of BTO-BS3T-BS3T U12.

Materials with fine crystallites size discourage the formation of multi-domain. Single domain grains are formed instead; this resulted in the increase in the dielectric constant of material [24]. The fine crystals with homogeneous single domain microstructure enables the stable and uniform ferroelectric $BaTiO_3$. External strain in grains has also shown a great influenced on the domain structure and nanoparticles is an additional parameter which could influence the domain energy [4].

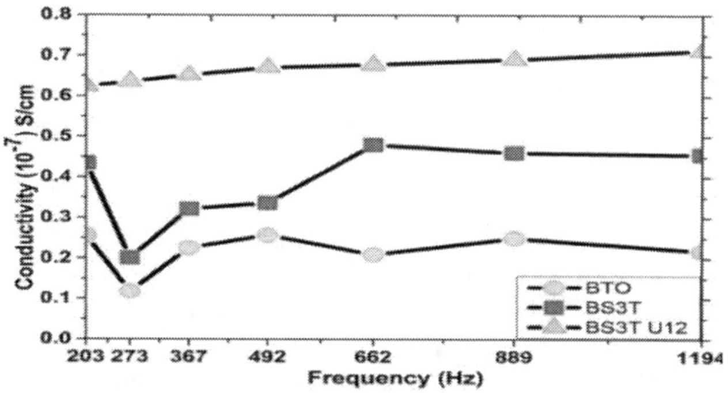

Figure 7. Electrical conductivity of BTO-BS3T-BS3T U12 in the frequency range 203-1194 Hz

Figure 7 shows electrical conductivity BTO-BS3T-BS3T U12 of 203-1194 Hz frequency, it can be seen that in general the electrical conductivity in the BTO-BS3T-BS3T U12 has relatively increased although fluctuate. Figure 7 shows electrical conductivity in the BTO-BS3T-BS3T U12 increased $2.5x10^{-8}$; $4.6x10^{-8}$; $6.9x10^{-8}$ S/cm respectively (f=889 Hz). The AC conductivity increases with increase in the frequency, which may be due to the strong hopping mechanism of ions [25].

4. Conclusion

A single phase of $Ba_{(1-x)}Sr_xTiO_3$ at x=0; 0.3 has successfully synthesized through mechanical alloying method coupled with an additional treatment ultrasonically. The treatment has proven effective to produce monocrystallite particles. Irradiation time to obtain such particles is 12 hrs which led to mono crystallite particles with the mean crystallite size 19 nm in the BS3T-U12 sample. The latter sample has an improved dielectric constant in the whole frequency range 203-1194 Hz over BTO and BS3T samples. A slight distorted cubic structure coupled with a fine mean crystallite size responsible for the improvement. The reduction in particle sizes also increased the electrical conductivity from $2.5x10^{-8}$ S/cm (BTO) to $6.9x10^{-8}$ S/cm (BS3T U12) at a 889 Hz frequency.

5. Acknowledgements

The authors gratefully acknowledge the support of Physics Department Universitas Indonesia for the research facilities. Research work is partially funded by the Universitas Indonesia under the research grant KOPIT with contract number: 1993/UN2.R12/HKP.05.00/2016. We also thank PT. DKSH Indonesia for an access Particle Size Analyzer Malvern Zetanano ZS.

6. References

[1] Ioachim A, Toacsan M I, Banciu M G and Nedelcu L 2007 *J. Solid State Chem.* **35** 513–20
[2] Wee F H, Malek F, Sreekantan S, Al-Amani A U, Ghani F and You K Y 2011 *J. Prog Electromagn Res.* **121** 181–213
[3] Wodecka-Dus B, Lisinska-Czekaj A, Orkisz T, Adamczyk M, Osinska K and Kozielski L 2007 *J. Materials-Science-Poland.* **25(3)** 791–9
[4] Vijatovic M, Bobie J D and Stojanovic B D 2008 *J. Sci Sinter.* **40** 235–244
[5] Ries A, Simões A Z, Cilense M, Zaghete M A, Varela J A 2003 *J. Mater Charact.* **50(2–3)** 217–21
[6] Quan J, Wang Y, Ping W, Tu L, Lai H and Chan W 2005 *J. Solid State Chem.* **178** 279–84
[7] Jiang W, Gong X, Chen Z, Hu Y, Zhang X and Gong X 2007 *J. Ultrason Sonochem.* **14** 208–12
[8] Mao C, Dong X and Zeng T 2007 *J. Mater Lett.* **61(8–9)** 1633–6
[9] Miclea C, Tanasoiu V, Spanulescu I, Gheorghiu A and Amarande L 2007 *Rom J. Inf Sci Technol.* **10(4)** 335–45
[10] Suryanarayana C 2001 *J. Prog Mater Sci* **46** 1–84
[11] Gomez-Yanez C, Cruz-aquino E, Cruz-rivera J J and Linares-miranda R 2007 *J. Alloys Compd.* **434–435** 806–8
[12] Yu Y, Wang X and Yao X 2013 *J. Ceram Int* **39(SUPPL.1)** S335–9
[13] Yustanti E, Hafizah M A E and Manaf A 2017 *Surfactant-assisted synthesis of* $Ba_{0.7}Sr_{0.3}TiO_3$ *nanoparticles by mechanical alloying and ultrasonic irradiation* in: international conference on engineering, science and nanotechnology 2016 ICESNANO 2016 (Solo-Indonesia, American Institute of Physics Conf. Proc.) pp 30119-1-4
[14] Yustanti E, Hafizah M A E and Manaf A 2016 *Synthesis of strontium substituted barium titanate nanoparticles by mechanical alloying and high power ultrasonication destruction* in: The 3rd International Conference on Advanced Materials Science and Technology ICAMST 2015 (Semarang-Indonesia, American Institute of Physics Conf. Proc.) pp 20102-1-6
[15] Dang F, Kato K, Imai H, Wada S, Haneda H and Kuwabara M. 2010 *J. Ultrason Sonochem.* **17** 310–4
[16] Markovic S, Mitric M, Starcevic G and Uskokovic D 2008 *J. Ultrason Sonochem.* **15** 16–20
[17] Leoni M, Confente T and Scardi P 2006 *Zeitschrift fur Krist Suppl.* **1(23)** 249–54
[18] Scardi P and Leoni M 2002 *J. Acta Crystallogr Sect A Found Crystallogr.* **58(2)** 190–200
[19] Cheng X and Shen M 2007 *J. Solid State Commun.***141(11)** 587–90
[20] Ali F, Reinert L, Leveque J M, Duclaux L, Muller F, Saeed S, Shah S S 2014 *J. Ultrason Sonochem.* **21(3)** 1002–9
[21] Suwarni, Zaidah A, Supriyanto A, Jamaluddin A and Iriani Y 2015 *J. Fis dan Apl.* **11(3)** 99–102.
[22] Zaidah A, Suwarni, Supriyanto A, Jamaluddin A and Iriani Y 2015 *J. Fis dan Apl.* **11(2)** 82–5
[23] Suastiyanti D, Hardiyanto M and Wijaya M 2014 *Karakteristik elektrik nanopartikel* $BaTiO_3$ *untuk aplikasi material multiferroic* in: Prosiding Seminar Nasional Sains dan Pendidikan Sains IX (Salatiga-Indonesia, Faculty of Mathematic and Natural Science Universitas Kristen Satya Wacana) pp 35–42
[24] Sunendar B P, Agustina M, Wibowo A and Oktavia E 2010 *J. Berita Selulosa* **45(2)** pp 64–9
[25] Hays A, Marsh CP, Franks R, Alvarado J. 2006 Int Refrig Air Cond Conf. Paper 829 Purdue USA pp R132, 1-7.

Visible light absorption and photosensitizing characteristics of natural yellow 3 extracted from *Curcuma Longa L.* for Dye-Sensitized solar cell

N Sofyan[1,2], F W Situmorang[1], A Ridhova[1], A H Yuwono[1,2] and A Udhiarto[3]

[1] Department of Metallurgical and Materials Engineering, Faculty of Engineering, Universitas Indonesia, Depok 16424, Indonesia
[2] Tropical renewable Energy Center, Faculty of Engineering, Universitas Indonesia, Depok 16424, Indonesia
[3] Department of Electrical Engineering, Faculty of Engineering, Universitas Indonesia, Depok 16424, Indonesia

E-mail : nofrijon.sofyan@ui.ac.id

Abstract. Curcumin has known to have a long intense wavelength absorption in the visible region ranging from 420-580 nm. Because of that, curcumin is a promising material for use in dye sensitized solar cells. In this work, curcumin was extracted from turmeric (*Curcuma longa L.*) through a simple extraction technique using different organic solvents, i.e. acetone, methanol, and ethanol. The results from the extraction were applied as a sensitizer in dye sensitized solar cell applications. Suitable solvent for extraction that could give high efficiency in DSSC device were analysed. Characterization of the dye was performed using Fourier transform infrared (FTIR) and ultra violet-visible (UV-Vis), while the performance of DSSC was analysed through a simple current and potential different (I-V) curve analyser. The infrared results showed that all of the turmeric extracted dyes have the same tendency with the characteristic of curcumin. UV-Vis absorbance characteristics of the curcumin natural dye extracted using acetone, ethanol and methanol has the same tendency with plateau-like absorbance peak at around 490, 492, and 505 nm, respectively. The maximum power conversion efficiency (PCE) is found to be 7.88% obtained from the solvent of ethanol whereas the PCE obtained from other natural dyes extracted using acetone and methanol are found to be low, 2.23% and 1.25%, respectively. This finding is convincing in terms of efficiency in which the value from the dye extracted using ethanol has higher efficiency than that of the commercial organic dye RK1 with a PCE of 4.02%.

1. Introduction

Dye sensitized solar cells (DSSCs), a third generation of solar cells, have attracted considerable attention due to their environmental friendliness, non-toxicity and low cost of production [1–3]. The principle of DSSCs is based on the photovoltaic effect that converts the sun-light photon into electrical energy through photo-electrochemical principle of band gap semiconductor and sensitization of the dye solution [4]. A DSSC is generally composed of conducting electrode, electrolyte, sensitizer and a counter electrode. Conducting electrode is usually in the forms of fluorine doped tin oxide (FTO) or indium doped tin oxide (ITO), semiconductor layers such as TiO_2, ZnO, CdSe, CdS, WO_3, Fe_2O_3, SnO_2, Nb_2O_5, and Ta_2O_5 [5]. The electrolyte is needed in process of reduction and oxidation reaction such as triiodide solution. The sensitizer dye will act as photon energy absorber such as anthocyanin, chlorophyll, carotenoid, and flavonoid that can so easy to be attained. The counter electrode will function to collect the electrons and at the same time to catalyse the reduction of electrolyte. Considering its function, the sensitizer dye is indeed one of the important elements that plays a significant role to enhance the performance of DSSCs in which it absorbs light spectrum and thus works as an anchoring group to improve the injection of electron into conduction band of TiO_2.

Some of inorganic compound such as porphyrins, platinum complex, phthalocyanines, and fluorescent dye have been used as a sensitizer. However, their availability is rather limited and difficult to be synthesized, making these compounds expensive in price. In order to overcome this problem, natural dyes are the best solution. They are not only abundance and easy to be attained, but also cheap, easily synthesized, biodegradable, and environmentally friendly. Interestingly, the use of curcumin derived dyes as sensitizers in dye sensitized solar cells has not attracted significant research attention [1]. In this paper, curcumin is used as a sensitizer. Curcumin can be extracted from turmeric or *Curcuma longa Linn.*

Turmeric is a rhizomatous herbaceous perennial plan (Curcuma *longa Linn*) of the ginger family. It is widely used as a spice, colouring agent in food processing industries, household medicine and insect repellent [6]. Moreover, under in vitro and in vivo studies, it has been confirmed that turmeric extracts have powerful biological activities, such as anti-inflammatory [7], antibacterial [8], antidepressant [9], antidiabetic [10], antitumor [11], immunomodulatory [12] and gastroprotective properties [13]. Turmeric or *Curcuma longa* is cultivated extensively in south and southeast tropical Asia, such as Bangladesh, China, Taiwan, Pakistan, Sri Lanka, Myanmmar, and Indonesia. Turmeric or *Curcuma longa* is not only abundance but also easy to be attained. In 2014, Indonesia produced about 112.088.181 Kg of turmeric to be used as a spice and colouring agent in food processing industries. The bio-active polyphenol component of turmeric is curcumin, also known as C.I. 75300 or Natural Yellow 3. The scientific chemical name of this active component is known as (1E, 6E)-1, 7-bis (4-hydroxy-3-methoxyphenyl)-1, 6 heptadiene-3, 5-dione [14, 15].

Curcumin is not easily dissolve in water. For this reason, the solvent used to extract the curcumin from the turmeric would be an important consideration. Previous investigation has used various types of solvents such as ethanol and acetonitrile. The result showed that ethanol has an efficiency of around 0.60% and acetonitrile 0.41% [1, 16]. In the current investigation, three different solvents, *i.e.* ethanol, methanol, and acetone were used to extract the curcumin or Natural Yellow 3 from the turmeric. The dyes resulted from the extraction were investigated in terms of their absorption spectra, functional groups, and their performance in the fabricated DSSC device. The results are presented and discussed detail in this article.

2. Experimental Setup

The natural material *Curcuma longa L.* was purchased from traditional market in Jakarta. The chemicals were titanium dioxide (TiO_2) nanoparticle (P-25 Degussa), platinum paste (Platisol, Solaronix), electrolyte (Iodolyte, Solaronix), ethanol 99% and methanol 99% (Merck) acetone and fluorine-doped tin oxide glass (FTO, Solaronix).

One kg of fresh turmeric was thoroughly washed with tap water and dried under ambient condition. The dried turmeric was crushed in a blender and air dried at room temperature. The crushed dried turmeric was grinded into fine powders and further was subjected to separation through a vibrating sieving machine. The fine turmeric powders were collected and put in a seal bottle. Curcumin extraction was carried out by mixing the fine turmeric powder in each of acetone, ethanol, and methanol. All of the mixtures have the same ratio of turmeric powder and solvent of 1:10 w/v. The mixture was stirred for 24 hours at room temperature. The extracted dye solutions were filtered and stored in a dark glass bottles for further characterization and DSSC device fabrication. The extracted dye solution was characterized using attenuated reflectance Fourier transform infrared (ATR FTIR, PerkinElmer Spectrum 2) for functional groups and ultra violet-visible (UV-Vis, PerkinElmer Lambda

25) for the dye absorption characteristic. As a comparison, 2.5 mg commercial organic dye (Sensidizer RK1, Solaronix) dissolved in 100 mL ethanol was also examined.

The electrode was prepared by adding one gr of TiO_2 nanoparticle P25 into 10 ml ethanol and stirred. Two drops of TRITON X-100 (Sigma Aldrich) were added into the paste and stirred again until a homogeneous phase was obtained. This white paste was then deposited onto fluorine doped tin oxide (FTO) glass using doctor blade method. For this purpose, the Scotch tape was used to adjust the thickness of the TiO_2 paste deposit with an area of 1 cm by 1 cm. The paste was dried at 450°C for one hour and cool down before soaking it in the curcumin dye solution for 6 hours and air dried. As a comparison, the commercial RK1 dye (Sensidizer RK1, Solaronix) dissolved in ethanol was used. The counter electrode was prepared using platinum paste deposited on an FTO glass. After the deposition, counter electrode was air dried before sintering it at 450 °C for 1 hour.

A double tape was used as an adhesive and to seal both electrode glasses. It also functions as a spacer to avoid the electric contact between the anode and the cathode. Further, electrolyte solution was dropped in the gap between the anode and the cathode. Finally, a binder clip was used to attach the electrodes together. Cell activity was then measured using a simple ammeter whereas the cell performance was examined using a Semiconductor Parameter Analyzer (SPA, Agilent 4155A) with a standard illumination of about 100 mW/cm^2.

3. Results and Discussion

Fourier transform infrared (FTIR) characterization was carried out to examine the functional group characteristic contained in the turmeric extracted dye. These specific functional groups are expected to be adsorbed onto TiO_2 layer in order to function properly. The dyes spectra were obtained at wavenumber 4000 – 400 cm^{-1} and the results are given in figure 1. The characteristic of pure commercial RK1 dye is given for a comparison.

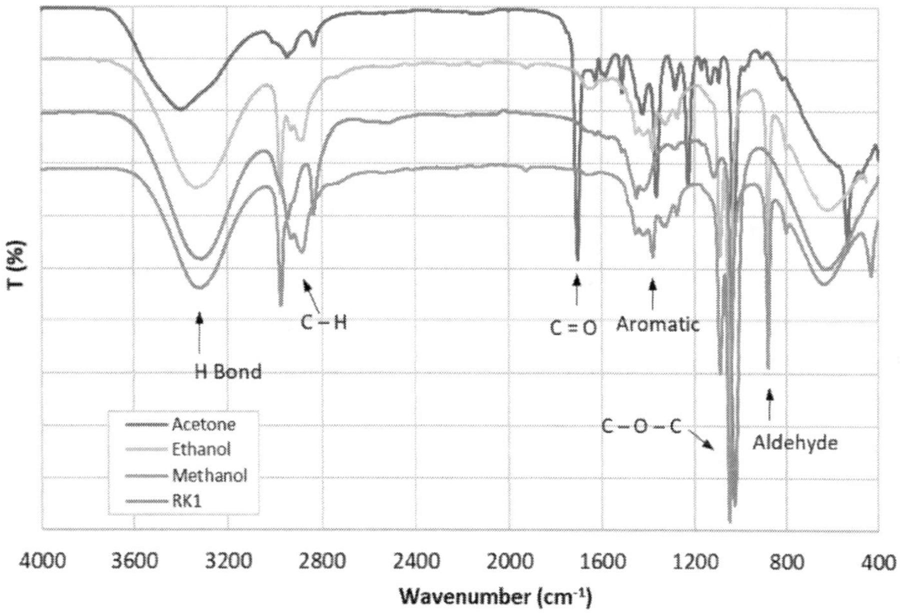

Figure 1. Fourier transform infrared transmittance characteristics of the commercial dye and the turmeric dye extracted using various solvents at wavenumber 4000 – 400 cm^{-1}

As seen in figure 1, except for the dye extracted using acetone, all of the spectra show the same tendency with that of the RK1 dye. In the case of the dye extracted using acetone, the absorption intensities of C = O and aromatic bands are higher than that of the other dyes. The absorption intensity of C – H band, however, is shorter than that of the other dyes. Nonetheless, all of the spectra show the appearance of hydrogen bonded OH stretching bands ranging from 3250 to 3450 cm^{-1}. The bands around 2878 to 2973 representing C-H groups stretching [17] are also present. The band at 1640 cm^{-1} indicates C=O stretching vibration. This band, as has been mentioned previously, except for the natural dye extracted using acetone, all are small. The small peak at 1450 cm^{-1} belongs to C – N, whereas the bands at 1423-1274 cm^{-1} belong to aromatic compound. Further, the band at 1045 belongs to C – O, while the stretching vibration of C–O–C esters is found at 1046 cm^{-1}. Due to inherent properties of the organic base natural dye, the aldehydes are also found in the wavelength between 880 cm^{-1} and 800 cm^{-1}.

The presence of carboxyl and hydroxyl group in the natural dye is expected to trigger the interaction with the semiconductor, in this instant to strongly bind onto TiO_2 surface [18]. The interaction between TiO_2 nanoparticle and functional groups of the dye would then will drive the electron transfer from the dye molecules to the conduction band of the semiconductor TiO_2 [19]. UV-Vis spectroscopy was used to examine the absorption spectrum of the dye in the visible range at wavelengths from 400 – 700 nm and the results are given in figure 2.

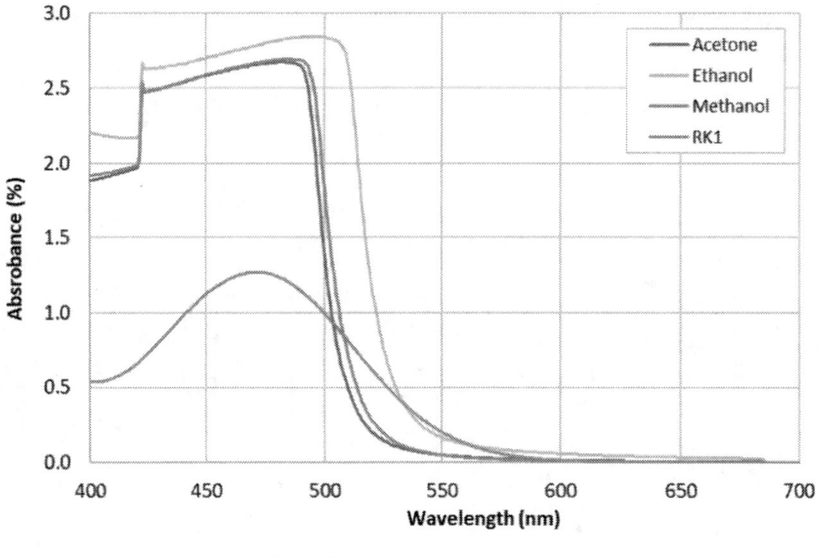

Figure 2. Ultraviolet-visible absorbance characteristics of the commercial RK1 dye and the curcumin natural dye extracted using various solvents at wavelength from 400 – 700 nm

As seen in figure 2, UV-Vis absorbance characteristics of the RK1 commercial dye has a specific absorbance peak at around 474 nm, whereas the curcumin natural dye extracted using acetone, ethanol and methanol has the same tendency with plateau-like absorbance peak at around 490, 492, and 505 nm, respectively. This absorption peak indicates at which the presence of the yellow 3 dye that would absorb high number of photon energy from the visible light source. It can also be noted from the spectra that the absorption characteristic of the dye has a broader absorption wavelength compared to that of the synthetic RK1 dye.

As a sensitizer in the DSSC device, the dye will establish bonds between the oxygen site of TiO_2 semiconductor and the functional groups of the natural dye such as carbonyl (C = O) and or hydroxyl (O-H) groups. Photocurrent activity could then be observed when an incident photon is absorbed by the natural dye during the service. In this instant, any successful bonding between TiO_2 semiconductor and the natural dye will promote electrons from the highest occupied molecular orbital (HOMO) to the lowest unoccupied molecular orbital (LUMO) state [21]. Photocurrent-voltage activity of the extracted dye on TiO_2 nanoparticle assembled in DSSC device was examined through a J-V curve characteristic using a semiconductor parameter analyzer and the results are given in figure 3.

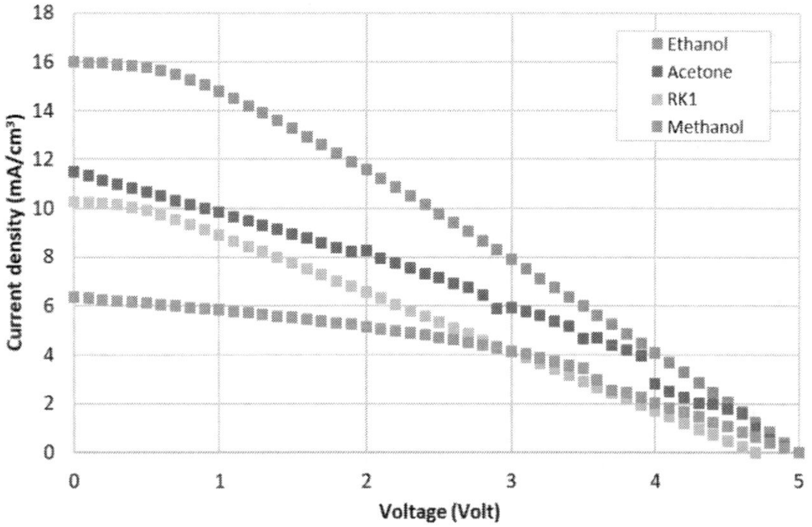

Figure 3. Photocurrent characteristic of the DSSC device sensitized using the commercial dye RK1 and the curcumin natural dyes extracted with various solvents

This J-V curve characteristic was then analyzed to obtain the power conversion efficiency [20]. Based on this analysis, the maximum power conversion efficiency (PCE) was found to be 7.88% obtained from the solvent of ethanol. The PCE obtained from other natural dyes extracted using acetone and methanol are found to be low as seen detail in table 1. This finding is convincing in terms of the PCE in which the value from the dye extracted using ethanol has higher efficiency than that of the commercial organic dye RK1 (4.02%) [22].

Table 1. Power conversion efficiency (PCE) of the DSSC device sensitized the curcumin natural dyes extracted with various solvents and the commercial dye RK1 as a comparison

	Acetone	Ethanol	Methanol	RK1*
V_{oc} (volt)	5.00	5.00	5.00	4.80
I_{oc} (mA)	11.49	16.01	6.37	10.28
V_{max} (volt)	0.20	0.50	0.20	0.40
I_{max} (mA)	11.16	15.76	6.27	10.06
FF	0.04	0.10	0.04	0.08
PCE (%)	2.23	7.88	1.25	4.02

*Source: Ref [22]

4. Conclusion

The results from infrared characterization showed that all of the turmeric extracted dyes indicate to have the same tendency with the characteristic of curcumin. All of the results from UV-Vis examination showed that the dyes have a visible light absorption with a plateau-like at a wavelength of 490, 492, and 505 nm, respectively. The maximum power conversion efficiency (PCE) is found to be 7.88% obtained from the solvent of ethanol. The PCE obtained from other natural dyes extracted using acetone and methanol are found to be low, 2.23% and 1.25%, respectively. This finding is convincing in terms of efficiency in which the value from the dye extracted using ethanol has higher efficiency than that of the commercial organic dye RK1 with a PCE of 4.02%.

5. Acknowledgment

The authors would like to express their appreciation for the funding from the Directorate of Research and Community Services (DRPM), Universitas Indonesia, through Hibah PITTA No. 823/UN2.R3.1/ HKP.05.00/2017.

6. References

[1] Kim H, Kim D, Karthick S N, Hemalatha K V, Justin R C, Ok S, and Choe Y 2013 *Int. J. Electrochem. Sci.* **8** 8320
[2] Al-Bat'hi S A M, Alaei I, and Sopyan I, 2013 *Int. J. Renewable Energy Research* **3** 1 138
[3] Aohan Y 2011 *Coating Titanium Dioxide and Solar Cell* (Saimaa University of Applied Sciences, Imatra)
[4] O'Regan B and Grätzel M 1991 *Nature* **353** 737
[5] Wang W-N, Soulis J, Yang Y and Biswas P 2014 *Aerosol and Air Quality Research* **14 533**
[6] Ammon H T P, Wahl M A 1991 *Planta Med.* **57** 1
[7] Jurenka J S 2009 *Altern. Med. Rev.* **14** 141
[8] De R, Kundu P, Swarnakar S, Ramamurthy T, Chowdhury A, Nair G B, and Mukhopadhyay A 2009 *Antimicrob. Aagents Ch.* **53** 1592
[9] Kulkarni S K, Dhir A, Akula K K 2009 *Scientific World J.* **9** 1233
[10] Wickenberg J, Ingemansson S L, Hlebowicz J 2010 *Nutr. J.* **9** 1
[11] Wilken R, Veena M S, Wang M B, Srivatsan E S 2011 *Mol. Cancer* **10** 1
[12] Rogers N M, Kireta S, Coates P T H 2010 *Clin. Exp. Immunol.* **162** 460
[13] Kim D C, Kim S H, Choi B H, Baek N I, Kim D, Kim M J, Kim K T 2005 *Biol. Pharm. Bull.* **28** 2220
[14] Zihlman J 2006 *Laboratory Chemical and Analytic Reagents* (Fluka: Switizerland)
[15] Sorachoti K 2006 *Crude Dyes Extracted from Plants and Monascus Rice Cultures as Sensitizers in Solid-State Dye sensitized Solar Cells* (Kasetsart University, Thailand)
[16] Jasim K E, Cassidy S, Henary F Z and Dakhel A A 2017 *J. Energy and Power Eng.* **11** 409
[17] Pereira Jr V A, de Arruda I N Q, and Stefani R 2015 *Food Hydrocolloids* **43** 180
[18] Al-Alwani M A M, Mohamad A B, Kadhum A A H and Ludin N A 2015 *Spectrochim. Acta Mol. Biomol. Spectrosc.* **138** 130
[19] Narayan M R 2012 *Renewable and Sustainable Energy Rev* **16** 208
[20] Fernando J M R C and Senadeera G K R 2008 *Current Sci.* **95** 663
[21] Maiaugree W, Lowpa S, Towannang M, Rutphonsan P, Tangtrakarn A, Pimanpang S, Maiaugree P, Ratchapolthavisin N, Sang-aroon W, Jarernboon W, Amornkitbamrung V 2016 *Sci. Rep.* **5** 15230
[22] Sofyan N, Ridhova A, Yuwono A H, Udhiarto A, Fabrication of solar cells with TiO_2 nanoparticles sensitized using natural dye extracted from mangosteen pericarps, *The 15th Int. Conf. on Quality in Research* (*QiR*), Bali, July 24 – 27, 2017.

A preliminary investigation on visualization of oscillating heat pipe with non-destructive test

Adi Winarta[1], Nandy Putra[1], Iwan Sumirat[2], Fahrurrozi[2] and Bharoto[2]

[1]Applied Heat Transfer Research Group, Department of Mechanical Engineering Universitas Indonesia, Kampus Baru Depok, 16424
[2]Centre of Science and Technology of Advanced Materials, National Nuclear Energy Agency of Indonesia (BATAN), Indonesia

E-mail : nandyputra@eng.ui.ac.id

Abstract. Oscillating Heat Pipe (OHP) is a passive heat exchanger operating on two phase flow principle. As one of the newest members of the heat pipe family, OHP has a complicated operating mechanism of working fluid flow. OHP is a potential device, which can be used for saving energy and cost. Although, OHP is a promising two-phase heat transfer device with excellent performance, simple structure and low cost but its operational characteristic is still not clear yet and debatable. As one of the non-destructive test, neutron radiography uses neutron beam techniques to penetrate heavy materials while beam are absorbed by lighter materials. This method is very suitable for use in visualization technique of the two-phase flow of heat transfer device with hydrogen-based fluids. In this paper, the visualization method is coupled with thermocouple temperature measurement. Aluminium OHP with 12 turns has been manufactured and tested. The dimension of aluminium OHP is made fit with the neutron beam size. DI Water with filing ratio 35 % has been chosen as the working fluid for the close loop OHP. Three level of heating power was applied to the evaporator section of OHP. The result showed that this preliminary study could reveal the temperature fluctuation of the thermocouple data. The condensation and evaporation also can be observed adequately with this method.

1. Introduction

Oscillating Heat Pipe (OHP) is one of the recent types of heat pipe family which has found by Akachi in the earlier 1990 [1]. Besides the high possibility of electronic cooling application for next thermal management device, OHP also has a good prospect in the heat recovery technology [2–4], solar energy [5] and much more [6].

OHP is made from a capillary tube which then bent forming many turns as snake shape model. It's structure typically either as a closed or open loop configuration. At one end of that meandering tube is called evaporator and the opposite turns called the condenser. The evaporator and condenser section typically connected by adiabatic part. The use of capillary tube is the most important characteristics of OHP [6]. It is characterized using the Bond Number (Bo) of working fluid properties. Which is $Bo=(\rho l-\rho v)gD2/4\sigma$, where ρl and ρv are liquid and vapor density respectively, g is gravity, D is tube diameter and σ is a surface tension. If Bo number is lower than 4 then it could be said that the surface tension force is dominant than the gravitational force. As a result, if a working fluid was injected into that capillary tube, it will lead to the formation of liquid and vapor slugs inside [7]. Then, if a sufficient heat is applied to the evaporator sections surface then the working fluid will start to move in such random motions. The movement of working fluid will deliver the heat from the evaporator to the condenser. Oscillating Heat Pipe (OHP) is one of the recent types of heat pipe family which has found by Akachi in the earlier 1990 [1]. In addition to the high possibility at electronic cooling application, OHP also has good prospect in the heat recovery technology [2–4], solar energy [5] and many more [6].

The performance of OHP depends on several multivariate parameters such as geometry, working fluid properties, heat load, inclination angle, filling ratio, number of turns and gravitational forces [6]. Therefore, the performance of OHP is very difficult to predict with simple correlation due to many parameter influences within. A proper visualization study should be taken in order to learn more deeply about the thermo-hydrodynamics process within the OHP.

Most of the visualization study used a visible light imaging method with a transparent tube made from Quartz glass or Pyrex [7–8]. However, in reality, OHP applications use the tube from conductor material (metal). Therefore, the visualization observation using transparent tube has not provided an actual flow phenomenon inside. This is because glass and metal having the poor properties in terms of heat transfer (heat transfer coefficient and thermal conductivity). They also have a different surface roughness from metal, which obviously affects the working fluid flow. Moreover, the observation of transient temperature in glass tube will be more difficult when compared with metal tube (aluminium/copper). Thus, the thermo-hydrodynamics behavior of working fluid between glass and metal material tube clearly will have a distinction.

The visualization of OHP has been studied experimentally by several researchers, focused on either visual observation of flow patterns and various operation parameters related with thermo-hydrodynamics of OHP [7–8]. Actually, the flow patterns inside the mini-channel are quite complex to fully describe precisely. There several types of fluid movements have been shown from transparent tube visualizations literature. Among others are small oscillation, intermittent oscillatory flow, bulk oscillation, permanent oscillatory flow, circulation, and dry-out or device failure [8–11]. Also, it stated there are several flow patterns, as follows; slug flow, semi-annular flow, a combination of slug and annular flow, transition semi-annular to fully annular flow, fully annular flow and fully vaporized flow (i.e. dry-out) [9–12]. Some of the literature stated that bubbly flow as the other term of a dispersed bubble. Basically, if a bubble has a diameter much less than the inner diameter of the tube then it can be said bubbly flow or dispersed bubble [8,11]. The dispersed bubble mostly only visible with transparent tube (Glass/Pyrex) visualization experiments. However, this is the most advantage of transparent tube visualization.

Neutron radiography is a visual observation method that utilizes the penetration of radiation energy to investigate the inside part of an optically opaque object. This method is the non−destructive test, which is provided by nuclear energy technology. The application of this method at heat pipe visualization study is still very rare, therefore there is not much literature discussing the uses of this method on heat pipe visualization study.

Takenaka et al. used neutron radiography to observe the flow boiling characteristics of cryogenic fluids [13]. By utilizing the difference of significant attenuation coefficient between nitrogen as cryogenic fluid with the aluminum heat exchanger. The result is used as a reference tool for next design of the device. Asano et al. have conducted visualizing and measuring the flow of refrigerant in the capillary and distributor pipes within a compression-type refrigerator [14]. They studied two-phase flow behavior on both channels and distributors using neutron radiography visualization methods. By taking pictures at 30 fps with a CCD camera and performing image processing, the phase distribution of the heat exchangers can be well shown. Sugimoto, et al. evaluated the behavior of working fluids on a Variable Conductance Heat Pipe (VCHP) with neutron radiography [15]. Videos with 30 frames per second were taken with Electron Bombardment Charged Coupled Device (EB-CCD) camera. The visualization results reveal the benefits of mounting thin plates to inhibit fluid blockage in the heat pipe arch. Also, the effect of the working fluid filling-ratio and the absence of evaporation in the downward inclination of heat pipe can be predicted. Using the same method, Putra et al. showed the captured image when boiling occurred inside the heat pipe with a sintered wick and wickless (thermo-syphon) [16]. Borgmeyer et al. studied the capability of heat transfer and fluid flow in copper Oscillating Heat Pipe (OHP) [17]. A number of thermocouples are also placed at a number of points on the surface of the pipe wall to measure the OHP temperature distribution. Borgmeyer found that neutron radiography could help to explain the physical phenomena and two-phase flow characteristics of the working fluid in the OHP [17]. Research studies of OHP visualization using neutron imaging have been done by several researchers. Based on the literature, all the OHP material tested using copper capillary tube [17–21]. All previous researchers have also used cooling water block at condenser section, which obviously blocked the visual observation.

The aim of this study is to observe the motion of working fluid inside the OHP with radiography neutron method. For this reason, the OHP structure was built with aluminium capillary tube. The OHP experimental set was also designed to improve neutron transparency, by using airflow to conduct cooling at the condenser section. This design was set to visualize the whole part of OHP structure. In order to find the clear explanation of the real phenomenon inside the tube. The visualization and temperature data were captured at the time which synchronized before. The result of this study could provide more information about the thermo-hydrodynamic of working fluid inside the OHP.

2. Methodology

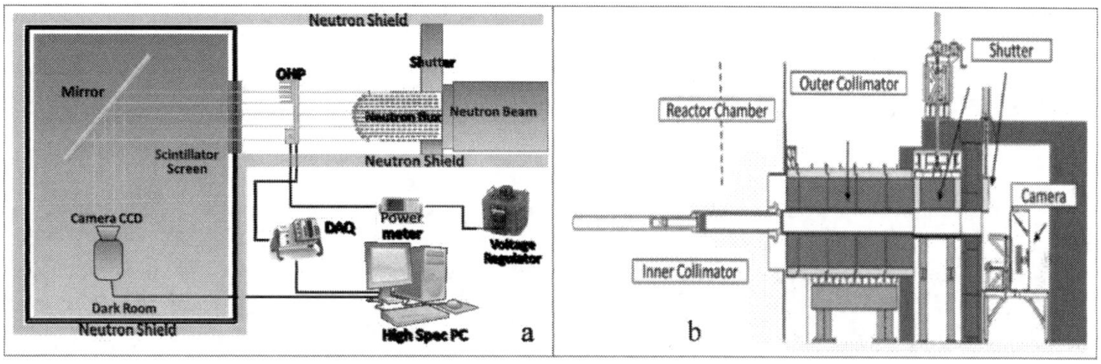

Figure 1. (a) The experimental setup of neutron radiography visualization test, (b) Neutron Radiography research facility RN1

Some explanation of this visualization method has been clearly written by Putra et al. [16]. For a simple explanation, when the object (OHP) is exposed to the beam of a neutron, each of material elements will be scattered and absorbed regard to the thickness and material properties. The lambert-beer law of attenuation explained the relation of neutron attenuation with property, thickness, and density of elements [18].

To perform the test, OHP was made from an aluminum capillary tube with an outer and inner diameter of 3 and 2 mm, respectively. The camera size of neutron imaging system determined the dimension OHP. Therefore, a closed loop OHP with an overall dimension of 13×14.5 cm2 was design and manufactured for the visualization test (Figure 2a). A pair of a sandwich structure made from aluminum block was constructed to provide uniform heat flux at the evaporator section. It has the overall dimension of 160×20×42 mm3. The contact surface between the round shape of aluminum tube and the block was improved with the semicircular groove. In order to minimize the thermal contact between them, thermal compound (Omegaterm "201") is used. Two cartridge heaters (Omega CSH-400) were inserted into the block through fit holes. The heater's power was controlled by a voltage regulator and measured by a power meter. The airflow cooling for condenser was chosen to get the clear picture of flow visualization image. Heat sink with centrifugal blower was used to cool the condenser sections. Semicircular grove is machining on the backside of aluminum heat sink to achieve better contact surface with the tube's shape. Thermal paste was used to minimize the thermal contact resistance. The wall temperatures of aluminum OHP are measured with type K thermocouples. Each of thermocouples are calibrated first before attached at OHP wall surface. The maximum error each of them was ± 0.5 °C. A data acquisition system was used to collect all the temperature data. The thermocouple positions showed at the figure 2c. All data of OHP visualization is tested at bottom heating mode. CCD Camera and data acquisition software were collected by a high specification desktop PC. The OHP was not encased in glass wool insulation to maximize the visualization result. Although this consideration setup would result in greater heat losses from the evaporator.

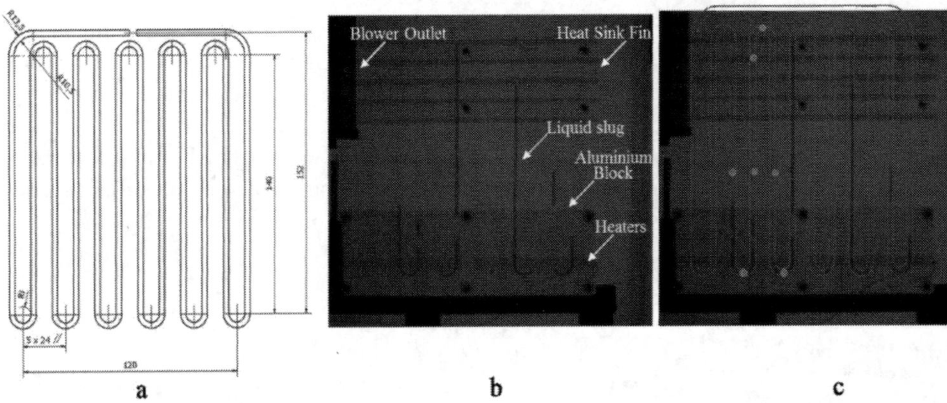

Figure 2. The OHP prototype (a) schematic of OHP, (b) Neutron radiography image of the OHP, (c) Image after marked with graphic editor software with the thermocouple positions.

Figure 1a and 1b showed the whole experimental setup of neutron radiography. The size of the OHP was dictated by the camera size of the neutron imaging system (see figure 2a). This test was conducted at neutron radiography facility, RN1, under the supervision of the Centre of Science and Technology of Advanced Materials (PSTBM), National Nuclear Energy Agency of Indonesia (BATAN). Figure 2b showed the picture of neutron radiography before the heat input is applied to the evaporator section. From the picture can be seen that the liquid slug inside the channels appeared as darkness image at the OHP loop. With Image J software, the filling ratio of OHP was confirmed at 35% (based on the ratio of liquid length and total loop of OHP). All the videos were collected at 15 fps (frame per second) rate and extracted frame by frame using video editor software. Vector graphic editor software is then used to mark the OHP structure at the extracted images from video data of visualization (figure 2c).

3. Result and Discussion

Figure 3 shows the transient temperature of evaporator, adiabatic and condenser section at OHP. Each turn of OHP numbered from left to right, to identify the channel clearly. Figure 4 shows the captured images from the start-up in time sequence. At initial condition (stage I), no heat supply was delivered to the evaporator. Stage I condition is represented very clear by figure 2b. This picture was analyzed using the image-j software to confirm the filling ratio of the OHP. After the initial condition was captured then 18 Watt of power input was supplied to the evaporator section through the cartridge heater. The evaporator's temperature then began to increase (Stage II). The captured image at second stage was showed no significant different with previous image (figure 2b). No motion of working fluid was observed at this stage. At stage III, the heat supply was increased into 40 Watt. Evaporator temperature increased more sharply than previous heat supply. However, there is no sign of working fluid movement at all. At stage IV, the heat supplied was increase to 85 Watt. Temperature evaporator still shows a sharp increase for about 200 seconds. The vapor plug may have expanded at this stage but it's hard to recognize the clearly picture. At #2 turn, the working fluid moved very fast (see figure

4) and the adiabatic temperature showed high amplitude fluctuations. The corresponding pictures of start-up showed in figure 4. The 4a and 4b frames showed the fast movement of liquid slug due to liquid evaporation at the evaporator section.

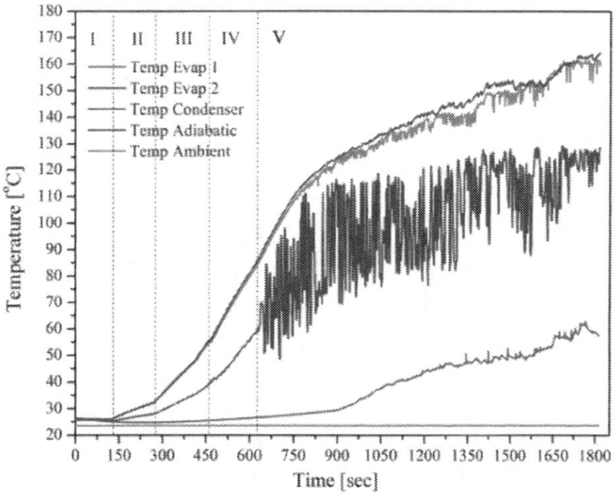

Figure 3. The transient temperature of OHP.

Figure 4a showed the sequence pictures of the instantaneous moment before the working fluid suddenly move very fast due to bubble expansion. At figure 4b, when the heat flux is high enough, the liquid slug at #2 turn was broken into two parts due to evaporation. Those two liquid slugs then quickly expelled with opposite direction. One of the liquid slug at right channel of #2 turn was merged with upper liquid slug and move together to condenser section. All liquid slugs on the channels seem moved with a different velocity. Next frames showed that only a few channels moved with significant distance (figure 4c and 4d). Interestingly, the condensation of vapor plug was observed at #2 turn left channel (figure 4c). In their report, Wilson et al. could not find this bubble collapse using the same visualization method [18].

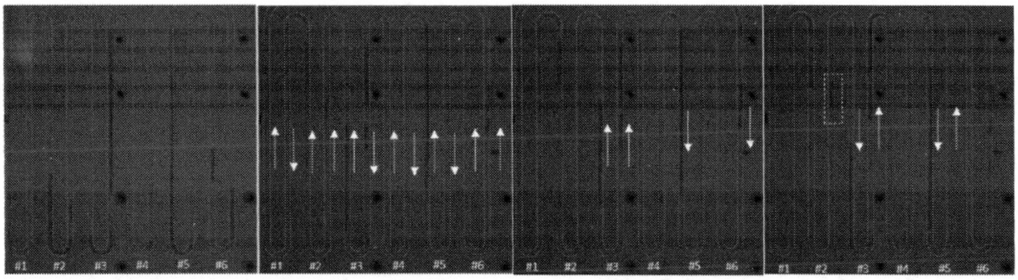

a) t= 641 s **b)** t= 641.066 s **c)** t= 641.2 s **d)** t= 642.46 s

Figure 4. Fluid movement at start-up frames.

Increasing the heat load to 100 Watt (stage V) still increased the evaporator temperature. Although the adiabatic section already shows the rapid fluctuation of working fluid (figure.3). After about 770 seconds then the evaporator temperature shows the fluctuations with lower amplitude.

Unlikely with the visualization using the transparent tube, this visualization cannot observe the detail of bubble and liquid slug boundaries (figure 5a). Thus, it very difficult to measure the contact angle from the liquid meniscus. And also, the dispersed bubble and bubble growth from nucleation cannot be observed, as shown in Figure 5b and 5c. Due to the non-uniform of liquid and vapor distribution, the velocity of fluid each of channel was extremely different (figure 5c). The liquid at #5 turn evaporated and then expelled the remaining slug with high velocity which results to blur frame pictures. However, the right channel of #6 turn show only little movement of liquid slug. Blurring images happened when the velocity of the liquid slug was higher enough to capture by only 15 fps of video camera adjustment. Even though the picture of moving liquid slug at evaporation sections was still blur due to inadequate of fps video. But more realistic phenomena could be observed from the related image and temperature data.

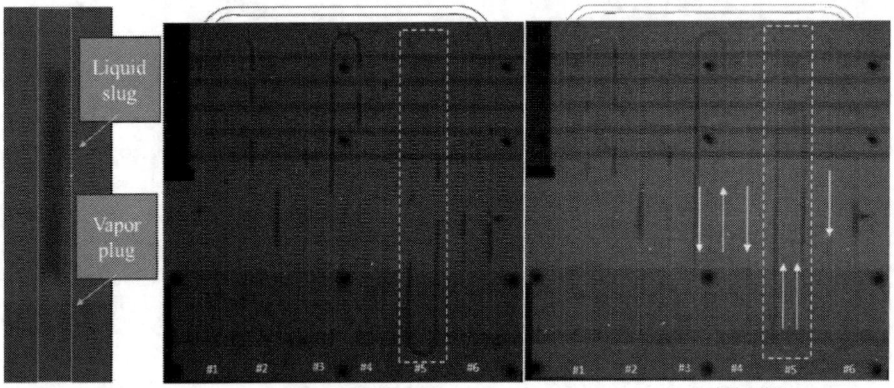

Figure 5. (a) Liquid slug and vapour plugs, (b) Instantaneous frames before evaporation, (c) After evaporation

From the fluid images and temperatures data, it can be confirmed that the fluctuations temperature at the adiabatic and evaporator were occurred due to the motion of liquid slug and vapor plug. If the slug was passing the thermocouple surface point then the significant temperature drop will occur at adiabatic sections. The temperature increased occurs when hot slug passes the point measurement surface again. The up and down motions of liquid slug caused temperature fluctuations in the adiabatic section (see figure 3). These results seem contradicted with already reported by Xu et al [22]. In summary, the temperature data coupled with related fluid flow movements will provide a better approach to thermo-hydrodynamic behaviour analysis in oscillating heat pipe.

4. Conclusion

Preliminary study on visualization fluid flow through neutron radiography has been conducted with aluminium OHP. From the resulted images, the motions of fluid flow within the capillary tube could observe well. The collapsing of vapor bubble was observed at condenser section. The evaporation of liquid slug was captured adequately also. The detailed bubble information, like nucleation, dispersed bubble growth and contact angle of the bubble meniscus could not observe clearly due to the small dimension of the OHP tube. Higher diameter tube with appropriate working fluid may solve this problem. The temperature data and fluid flow pictures could explain more accurate about the thermo-hydrodynamic of the working fluid. The fluctuations of the transient temperature at evaporator and adiabatic are one of the cases which could be explained well by this visualization method. The higher quality of video capture should be taken in the next study to maximize the quality of pictures. Thus, the deeply analysed could have been done from the result data.

5. Acknowledgment

The authors would like to thank DRPM Universitas Indonesia for funding this research through PUPT research scheme.

6. Reference

[1] Akachi H 1990 Structure of heat pipe. US Patent 4, 921, 041

[2] Meena P, Rittidech S and Poomsa-Ad N 2007 Application of close-loop oscillating heat-pipe with check valves (CLOHP/CV) air preheater for reduced relative-humidity in drying systems *J. Appl. Ener.* **84** 553
[3] Rittidech S, Dangeton W and Soponronnarit S 2005 Close-ended oscillating heat pipe (CEOHP) air-preheater for energy thrift in a dryer *J.Appl. Ener.* **81** 198
[4] Winarta A, Putra N and Bakhtiar F 2017 Thermal performance of oscillating heat pipe with ethanol/methanol for heat recovery application design *J.Ijaseit* **7** 1268
[5] Kargarsharifabad H, Mamouri S.J, Shafii M and Rahni M.T 2013 Experimental investigation of the effect of using closed-loop pulsating heat pipe on the performance of a flat plate solar collector *J. Ren. Sust. Ener.* **5** 013106
[6] Han X, Wang X, Zheng H, Xu X and Chen G 2016 Review of the development of pulsating heat pipe for heat dissipation *J. Ren. And Sust. Ener. Rev* **59** 692
[7] Tong BY, Wong T and Choi K 2001 Closed-loop pulsating heat pipe *J.ATE.* **21** 1845
[8] Xu J, Li Y and Wong T 2005 High speed flow visualization of a closed loop pulsating heat pipe *J. Int.J.Heat and Mass Trf.* **48** 3338
[9] Karthikeyan V, Ramachandran K, Pillai B and Solomon A.B 2015 Understanding thermo-fluidic characteristics of a glass tube closed loop pulsating heat pipe: flow patterns and fluid oscillations *J. Heat and Mass Trf.* **51** 1669
[10] Khandekar S, Charoensawan M, Groll M and Terdtoon 2003 Closed loop pulsating heat pipes Part B: visualization and semi-empirical modelling *J. ATE* **23** 2021
[11] Goshayeshi H.R, Chaer I 2016 Experimental study and flow visualization of Fe2O3/kerosene in glass oscillating heat pipes, *J. ATE* **103** 1213
[12] Li QM, Zou J, Yang Z, Duan YY, Wang BX 2011 Visualization of two-phase flows in nanofluid oscillating heat pipes *J. Heat Trf.* **133** 052901
[13] Takenaka N, Fujii T, Akagawa K, Ono A, Sonoda K, Nishizaki K, Asano H. 1990 Application of neutron radiography to visualization of multiphase flows *J. Flow and Instr.* **1** 149
[14] Asano H, Takenaka N, Fujii T, Shibata Y, Ebisu T, Matsubayashi M. 1999 Visualization and measurement of refrigerant flow in compression-type refrigerator by neutron radiography *J.N.Ins.Meth.in Phy.Res. Sec. A* **424** 98
[15] Sugimoto K, Asano H, Murakawa H, Takenaka N, Nagayasu T, Ipposhi S. 2011 Evaluation of liquid behavior in a Variable Conductance Heat Pipe by neutron radiography *J.N.Ins.Meth.in Phy.Res. Sec. A* **651** 264
[16] Putra N, Ramadhan RS, Septiadi WN 2015 Visualization of the boiling phenomenon inside a heat pipe using neutron radiography *J. ETFS* **66** 13
[17] Borgmeyer B, Wilson C, Winholtz RA, Ma HB, Jacobson D, Hussey D 2010 Heat transport capability and fluid flow neutron radiography of three-dimensional oscillating heat pipes *J. Heat Trf.* **132** 061502
[18] Wilson C, Borgmeyer B, Winholtz RA, Ma HB, Jacobson DL, Hussey DS, Arif M. 2008 Visual observation of oscillating heat pipes using neutron radiography *J.Therm. and H Trf.* **22** 366
[19] Wilson C, Borgmeyer B, Winholtz RA, Ma HB, Jacobson D, Hussey D. 2011 Thermal and visual observation of water and acetone oscillating heat pipes *J. H. Trf.* **133** 061502
[20] Thompson SM, Ma HB, Wilson C. 2011 Investigation of a flat-plate oscillating heat pipe with Tesla-type check valves *J.ETFS* **35** 1265
[21] Monroe JG, Thompson SM, Aspin ZS, Jacobson DL, Hussey DS. Neutron 2014 Neutron Imaging of an Unbalanced Flat-Plate Oscillating Heat Pipe *In52nd AIAA SciTec*
[22] Xu JL, Zhang XM 2005 Start-up and steady thermal oscillation of a pulsating heat pipe *J. H M Trf.* **41** 685

IOP Conf. Series: Earth and Environmental Science **105** (2017) 012075 doi:10.1088/1755-1315/105/1/012075

The effect of antioxidant additives on the growth of deposits on the use of biodiesel fuel (B100) at certain temperatures

B Sugiarto[1], M T Suryantoro , S Yubaidah[2] and M I Attharik[3]

[1]Departement of Mechanical Engineering, University of Indonesia, 16424, Indonesia
[2]BT2MP-BPPT, Area Puspiptek, Serpong, Muncul, Setu, South Tangerang, Banten 15310, Indonesia
[3]Departement of Mechanical Engineering, University of Indonesia, 16424, Indonesia

E-mail : bangsugi@lemtek-ui.com, muchammad.taufiq@bppt.go.id, lhamattharik@gmail.com

Abstract. Indonesia and several tropical countries produce a lot of energy sources including palm oil, coconut, nyamplung, rubber seeds, kosambi and various other plants. The source can be converted into biodiesel or bioethanol. This is reinforced by Regulation of The Minister of Energy and Mineral Resources, the government issued Indonesia's National Energy Policy under Regulation of The Minister of Energy and Mineral Resources of The R.I. No. 12/2015. This regulation formalized the promotion of biofuels in Indonesia, for both ethanol and biodiesel, and established a 30% biofuel in national energy consumption mandate by 2025. However, consideration of the use of biodiesel fuel is the formation of deposits in the engine or combustion chamber. In this study, the formation of a deposit of biodiesel fuel was carried out by comparison of biodiesel B100-NA without additive with biodiesel variation plus antioxidant additives such as B100 + BHT, and B100 + PG performed by the method of fuel droplet to heat plate to know the characteristics and mechanism of deposit formation On each fuel variation. The research was conducted by the deposition process and evaporation of Diesel fuel which was repeatedly carried out on a hot plate. The plate is heated with temperature variations in the enclosed space so that the conditions are close to the real condition of the engine. This test uses hot room temperature test rig. Use of antioxidant additives to inhibit oxidation in biodiesel is expected to keep the acid number low and increase oxidative stability, which will help prevent excessive deposits in the combustion chamber. This study aims to find the optimal temperature for the growth of the deposit can be controlled.

1. Introduction

Indonesia and several tropical countries produce a lot of energy sources including palm oil, coconut, nyamplung, rubber seeds, kosambi and various other plants. The source can be converted into biodiesel or bioethanol. This is reinforced by Regulation of The Minister of Energy and Mineral Resources, the government issued Indonesia's National Energy Policy under Regulation of The Minister of Energy and Mineral Resources of The R.I. No. 12/2015. This regulation formalized the promotion of biofuels in Indonesia, for both ethanol and biodiesel, and established a 30% biofuel in national energy consumption mandate by 2025 [1].

Nevertheless, the use of biodiesel with a greater percentage for applications in diesel engines still leaves some problems related to the occurrence of biodiesel degradation and the formation of deposits when used as fuel on the engine. Since rolling out the use of biodiesel fuel in engines, there has been an increase in reporting on the formation of deposits in both the combustion chamber and the injectors [2-4]. To date, studies on the mechanism of formation of biodiesel deposits are still small [3, 5, 6]. The ease of biodiesel for degradation is one of the factors that allegedly increases the formation of biodiesel deposits [3]. The addition of antioxidant additives has been shown to increase the stability value of biodiesel tested through accelerated tests in standard 110°C [7, 8], but the effect of antioxidant use on deposit formation has not been widely reported. The formation of deposits on the engine is a fairly complex phenomenon and is highly dependent on combinations of various parameters such as fuel, material surfaces, temperature, pressure, combustion conditions etc. [9].

To address this challenge, further studies and comprehensive research activities with a focus on fuels' characteristics and their deposit formation capability in different engine parts are strongly required. It is worth quoting that research on deposit formation in injector and combustion chamber basically includes measuring the thickness of the deposits in the engine componentd in response to the application of test and reference fuels [10]. Deposits can contain various materials and residues which gradually grow or accumulate on critical parts of internal combustion engines [6, 11]. Therefore, it is essential to analyze the chemical and physical attributes of deposits in order to figure out the causes. The analysis of micro-deposit structure in this research will be conducted with some analytical tools such as: FT-IR for fingerprint and SEM composition, TEM analysis for micro and nanostructures and EDS to know the physical chemical of the deposit [12].

2. System Description

2.1. Hot Room Temperature Test Rig

To obtain the appropriate deposit sample with the test fuel was tested using hot chamber method and different temperature plate. Each fuel sample is tested under the same conditions.

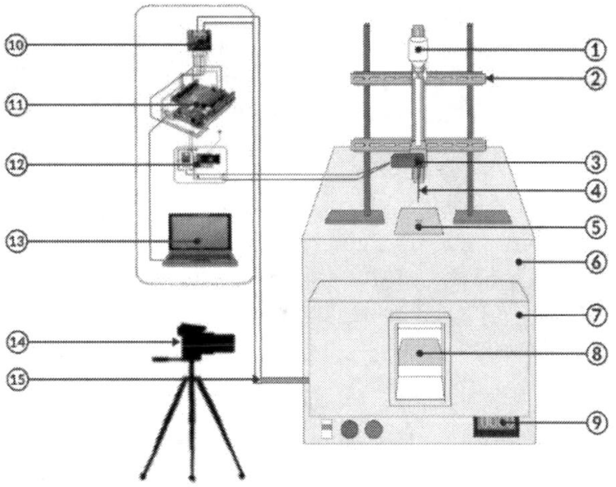

Figure 1. Hot Room Temperature Test Rig method.

This test rig is designed to test both droplets and spray of injectors. For droplet test, the droplet of fuel is dropped on the heat plate cooled in accordance with the temperature of the engine plate/component, while the room temperature is assumed to be the temperature of the combustion chamber. The temperature of the combustion chamber is varied according to the temperature of the combustion chamber.

Table 1. Caption of Hot Room Temperature Test Rig method.

Number	Caption of Hot Room Temperature Test Rig method
1	Fuel tank / container
2	Buffer fuel container
3	Solenoid Valve Shako PU220AR02-24V
4	Needle
5	The droplet entry holes into the furnace
6	Furnace with heater capacity up to 1200°C
7	Furnace door
8	Drip plate / heat plate (AISI 304)
9	Furnace temperature control
10	K type thermocouple module (Max6675)
11	Arduino Uno R3 Microcontroller
12	Power Supply 24 V DC & Relay Solenoid
13	Laptop, temperature data display & microcontroller programming
14	Video camera
15	K type thermocouple

Research on the formation of deposits in the use of biodeisel as alternative fuel is planned for 1 year. The formation of Pm/soot and Nox emissions is very important to be used as material for analysis of research results. Research in this time period will present the influence of chemical structure, biodiesel to the deposit formation.

2.2. Fuel Testing

2.2.1. Deposition process (Multi Droplet). The deposition process begins with the adjustment of the Hot Room Temperature Test Rig test device, then performs a mass measurement for the drop plate. Set the Hot Room Temperature Test Rig to the temperature per 2000 first drop and then the drop plate is inserted into the combustion chamber and depleted with fuel. After 2000 the first drop, the drop plate is removed and the mass measured again to obtain the mass of the deposit. Testing is done until the droplets reach 6000 and repeated with temperature variations.

2.2.2. Evaporation process (Single droplet). The evaporation process begins with setting the Hot Room Temperature Test, then set the temperature for the first temperature of 250ºC. Drop plates are inserted into the furnace and sprayed with fuel once. The evaporation process is recorded using a video camera and the evaporation time is calculated from the video that has been taken. Testing is done up to 350ºC and repeated with fuel variations.

2.3. Antioxidant Additives

Additives are the content added to a certain amount of fuel to alter the fuel properties of the fuel. Biodiesel has a low oxidation stability so antioxidant additives are needed. Several studies have proven that the addition of antioxidant additives to biodiesel can inhibit the biodiesel degradation process significantly [8]. Antioxidant additives used in this study are two, namely BHT and PG. Butylated Hydroxytoluene (BHT) has the $C_{15}H_{24}O$ molecular formula. BHT has a characteristic, which is white crystalline and is widely used because it is relatively cheap, has a molecular weight of 220, has a boiling point of 265°C, a melting point of 69.7°C and slightly smelly [13]. Propyl Gallate (PG) has the molecular formula $C_{10}H_{12}O_5$. Propyl errors ($C_{10}H_{12}O_6$) have heat-sensitive characteristics, decompose at its melting point of 148°C, can form a color complex with metal ions, resulting in low antioxidant capability. Three hydroxyl groups make PG a highly reactive antioxidant additive [13].

3. Result and discussion

In this study there are 3 variations of biodiesel sample and given different treatment for each sample. Here is a list of biodiesel samples and treatments.

Table 2. Biodiesel Variation and Treatment.

Variations of Biodiesel	Additives	Treatment
B100-NA	No Additives	Heated for 4.5 hours with temperature 80°C and container opened for 12 days)
B100-BHT	BHT (1000 ppm)	Heated for 4.5 hours with temperature 80°C and container opened for 12 days)
B100-PG	PG (1000 ppm)	Heated for 4.5 hours with temperature 80°C and container opened for 12 days)

Figure 2. Three variations of biodiesel sample (B100-Na, B100-BHT, and B100-PG)

In the three variations of biodiesel samples, B100-Na, B100-BHT, and B100-PG were performed to find the acid number, peroxide number, and iodine number. This test was conducted by IPB Integrated Chemistry Laboratory, Bogor. This test is conducted to determine the difference in quality of each of the three variations of biodiesel. Where expected differences in the quality of each variation and the influence of additives can inhibit the gradation of quality or oxidation of biodiesel fuel. Below is the test result data of three variations of biodiesel.

Table 3. Fuel properties of the B100 and B100 fuels have been added additives

Fuel	Test Temperature [°C]	Dynamic Viscosity [mPa.s]	Kinematic Viscosity [mm²/s]	Density [g/cm³]	Acid Numbers [mg.KOH/ mg.fat]	Iodine number [g/100g]	Peroxide Numbers [meq/kg]
B100	40	3.8732	4.5234	0.8563	0.47	46.32	15.17
B100 NA	40	3.7903	4.4263	0.8563	0.50	44.42	11.33
B100 + BHT	40	3.8801	4.5317	0.8562	0.42	47.62	11.34
B100 + BHA	40	3.8801	4.5307	0.8564	0.34	44.45	13.57
B100 + PG	40	3.8309	4.4731	0.8564	0.44	46.26	10.13

The above parameters have their respective standards and the following are the available standards to determine whether the biodiesel is still unoxidized or oxidized. The following table is the standard oxidation of biodiesel.

Table 4. Biodiesel Oxidation Standards

Standard Type	Value	Unit
Acid number (EN 14214)	Max 0.5	Mg KOH/g fat
Peroxide number (EN 14214)	Max 10	meq/Kg
Iodine number (EN 14214)	Max 120	g/100g

IOP Conf. Series: Earth and Environmental Science **105** (2017) 012075 doi:10.1088/1755-1315/105/1/012075

When viewed from existing data, the additive affects the oxidation velocity and is proven to slow the oxidation rate of biodiesel in the sample.

3.1. Test Rig Condition

The arrangement on the test rig is made in such a way as to represent conditions as in the real engine, including the room temperature in the test rig that resembles the temperature of the combustion chamber in the real engine and the temperature of the plate representing the temperature at certain places within the real engine. This can be seen in table 5.

Table 5. Single droplet yield data (evaporation).

Room Temperature (°C)	Plate Temperature (°C)	Cooler Box Temperatur (°C)	Mass Flow Rate (L/m)
500	250	217	28.2
500	300	271	11.76
500	350	326	6.9

3.2. Evaporation Process (Single droplet)

In the test of evaporation time (single droplet), tested at temperatures of 250°C, 300°C, and 350°C. The evaporation process of fuel on the heat plate is observed evaporation character and evaporation time for each fuel tested at each temperature. The evaporation time of the fuel, the temperature of the transition boiling regime and the appearance of fire are obtained from this process.

Table 6. Single droplet yield data (evaporation)

Temperature	Variasi Biodiesel		
[°C]	B100-NA	B100-BHT	B100-PG
250	42.77	34.80	52.29
300	3.69	20.48	17.90
350	2.47	9.20	11.49

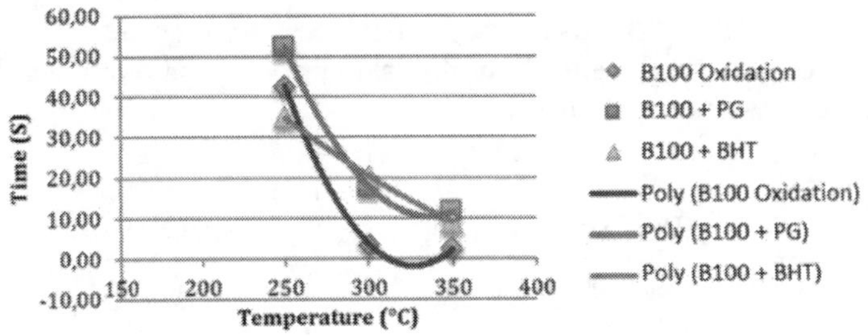

Figure 3. Evaporation time of one drop B100 Oxidation, B100 + PG, and B100 + BHT at each temperature

At a temperature of 250°C, B100-BHT takes the fastest time to evaporate perfectly compared to B100-Na and B100-PG. While at temperatures of 300°C and 350°C, the B100-PG takes a much faster time to evaporate perfectly than the B100-BHT and B100-Na burn perfectly at 300°C and 350°C.

The evaporation time affects the area of the deposit and the mass of the resulting deposit. Depositing deposits occur when the fuel has not fully evaporated but has been re-depleted with a fuel droplet (overlapping condition). The area produced will also become wider or narrow due to plate wetting that is affected by the evaporation time of the fuel. While in the gray zone is the zone where the fuel has undergone a trasition boiling regime. Transition boiling regime does not occur on B100-Na, B100-PG and B100-BHT fuels at temperature variations of 250°C, 300°C, and 350°C.

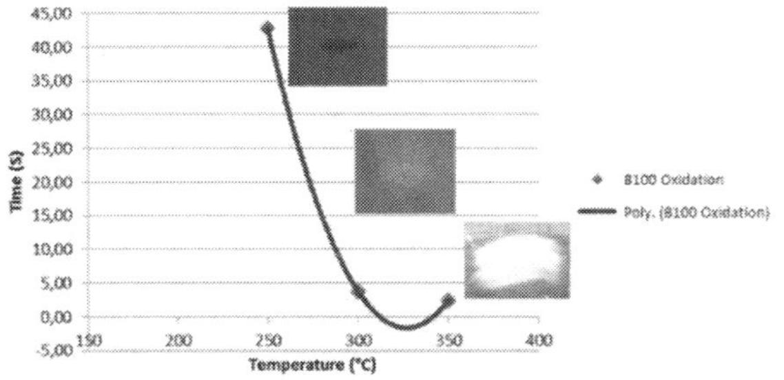

Figure 4. Time evaporation of one drop B100 Oxidized at each temperature On B100 is xidized.

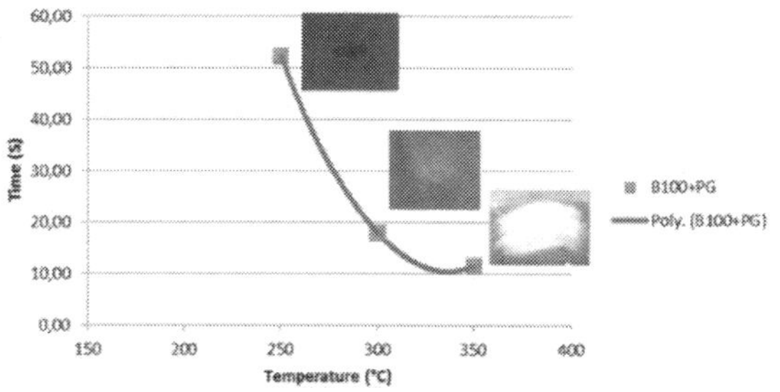

Figure 5. Evaporation time of one drop B100 + PG at each temperature.

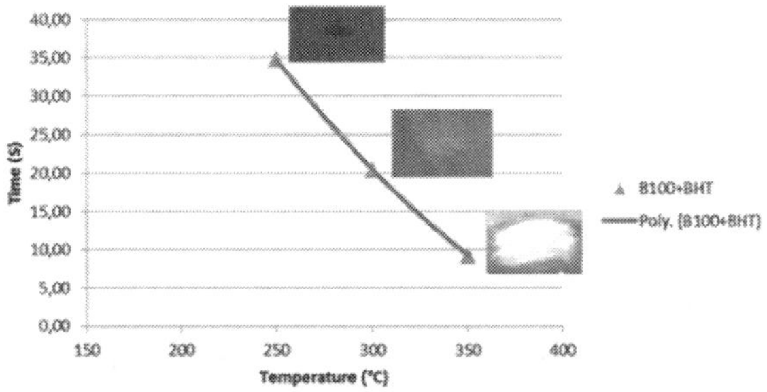

Figure 6. Evaporation time of one drop B100 + BHT at each temperature.

3.3. Deposition Process (Multi droplet)

The fuel deposition process on the hot plate produces the character and growth of the deposit at each temperature. The observable deposit character is the structure and contour of the deposit surface. Deposit growth can be obtained by measuring the mass of the deposit (the amount of the deposit). The results of the deposition process are as follows.

Table 7. The amount of deposit generated from the multidroplet testing of B0 and B100 as much as 6,000 drops (per 2,000 drops).

a. Temperature 250°C

Number of Droplets	Deposit Mass [mg]		*Total Relative Deposition Mass* [10^{-3}]	
	B0	**B100**	**B0**	**B100**
0	0	0	0	0
2000	21.6	24.9	1.27	1.74
4000	78	46.6	4.58	3.79
6000	82.9	113.4	4.87	5.63
8000	97.4	172	5.72	10.22
10000	119.7	190.3	7.03	12.48

b. Temperature 300°C

Number of Droplets	Deposit Mass [mg]		*Total Relative Deposition Mass* [10^{-3}]	
	B0	**B100**	**B0**	**B100**
0	0	0	0	0
2000	5.5	4	0.32	0.23
4000	10.1	5.1	0.59	0.3
6000	16.5	41.2	0.97	2.41
8000	22.7	62.1	1.33	3.63
10000	16.4	109.1	0.96	6.37

4. Conclusion

In this paper, forming of deposits from biodiesel with different variation, that are B100-NA, B100+BHT, and B100+PG will be observed with hot surface plate methode to find their characteristics and growth mechanisms of deposits from every biodiesel variation. The main conclusions are summarized as follows:

(1) Evaporation time B0 is faster than B100 because MEP at B0 occurs earlier or at a temperature lower than B100.
(2) The evaporation time affects the area of deposit area and the amount of deposit formed. The faster the evaporation time, the smaller the area of deposit produced.
(3) Temperatures of 250°C will result in the physical wet deposits on B100 and dry physical deposits at B0. A temperature of 300°C will result in a physical deposit that begins to dry on the B100 and a physical deposit that resembles the crust at B0. Temperatures approaching MEP will result in a physical deposit that resembles a crust while temperatures that move away from MEP will result in a wet physical deposit.
(4) Temperatures of 250°C produce contour-surface contours that have a notably different height between B0 and B100 but the intensity of the hills and valleys on the B100 deposit is less and the rougher areas.
(5) Temperatures of 300°C produce contour-surface contours that have a contrasting height difference. The elevation of the B100 deposit contour is almost 4 times that of B0, but the spread of the buildup becomes narrower resulting in a mounting deposit.
(6) The deposit structure B0 is tapered while the B100 deposit structure is spherical. The closer the temperature to the MEP point the more tapered the structure of the deposit B0. The closer the temperature to the MEP point the smaller the sphere of the B100 deposit structure.

5. Acknowledgments

The author would like to thank Mr. Taufik Suryantoro and Mrs. S. Yubaidah who provides access to the use of lab in BTMP, and to M Ilham Attharik, Rahmat Choirul D., Tezar Pratama And Timotius Halomoan as a team of researchers who participated in the research, so this research can be implemented well.

6. References

[1] Wirawan S S and Tambunan A H 2006 The current status and prospects of biodiesel development in Indonesia: A review *Third Asia Biomass Workshop*
[2] Liaquat A et al. 2013 Impact of palm biodiesel blend on injector deposit formation *Applied energy* **111**: p. 882-893
[3] Omori T et al. 2011 Biodiesel deposit formation mechanism and improvement of Fuel Injection Equipment (FIE) *SAE Technical Paper*
[4] Bouilly J et al. 2012 Biodiesel stability and its effects on diesel fuel injection equipment *SAE Technical Paper*
[5] Lin Y-S and Lin H-P 2011 Spray characteristics of emulsified castor biodiesel on engine emissions and deposit formation *Renewable energy* **36**(12) pp. 3507-3516
[6] Arifin Y B M 2009 Diesel and Bio-diesel Fuel Deposits on a Hot wall Surface
[7] Tang H et al. 2008 The effect of natural and synthetic antioxidants on the oxidative stability of biodiesel *Journal of the American Oil Chemists' Society* **85**(4) pp. 373-382
[8] Fattah I R et al. 2014 Effect of antioxidants on oxidation stability of biodiesel derived from vegetable and animal based feedstocks *Renewable and Sustainable Energy Reviews* **30** pp. 356-370
[9] Arifin Y M and Arai M 2010 The effect of hot surface temperature on diesel fuel deposit formation *Fuel* **89**(5) pp. 934-942

[10] Suryantoro MT, Sugiarto B and Mulyadi F 2016 Growth and characterization of deposits in the combustion chamber of a diesel engine fueled with B50 and Indonesian biodiesel fuel (IBF) *Biofuel Research* **3**(4) pp. 521-527
[11] Güralp O et al. 2006 Characterizing the effect of combustion chamber deposits on a gasoline HCCI engine *SAE Technical Paper*
[12] Suryantoro M T et al. 2016 Deposit characterization of a diesel engine combustion chamber by droplets at hot chamber temperature: Effect of temperature on evaporation time and deposit structure *Internasional Journal of Technology* **7**(8) pp. 1372-1380
[13] Buck L and Axel R 1991 A novel multigene family may encode odorant receptors: a molecular basis for odor recognition *Cell* **65**(1) pp. 175-187

Effect H_2SO_4 and Zymomonas mobilis concentration to bioethanol production by *bintaro* fruit (*cerbera manghas*) substrate

R E Santoso, N Fitriana, M Carolina, F C Dewi and Sumari

Universitas Negeri Malang, Semarang Street No. 5, Malang, Indonesia, 65145

E-mail : someonecalledega@gmail.com

Abstract. *Bintaro* fruit (*Cerbera manghas*) contains 36.945 of cellulose and 38% of lignin which is potential as a source of raw material for making bioethanol. The purpose of this research is to know the effectiveness of formula in producing bioethanol by the process of bintaro. The methods used in this research are pretreatment, delignification, hydrolysis, fermentation, and distillation. Pretreatment was performed by subcooling the substrate at 80^0C followed by milling it up to 60 mesh. Delignification was performed by soaking the substrate in 100^0C of 1N NaOH and 1 barr for 30 minutes. Hydrolysis was carried out with sulfuric acid catalyst (H2SO4) variations of 5.5%, 6.0%, and 6.5% at 120^0C for 60 minutes. Fermentation with Zymomonas in variation of 1%, 3%, and 5% at room temperature for 3 days. The fermented filtrate was distilled at 73^0C to obtain ethanol and tested the levels of ethanol with chromatography gas. The results showed that the highest levels of ethanol is 9.977% resulted from 6.5% of sulfuric acid hydrolysis and 5% of *Zymomonas mobilis* fermentation. The high levels of ethanol is supported by qualitative test result of the presence of reducing sugars with fehling yielding of 7002 ppm red brick tested quantitatively by nelson somogyi method. In conclusion, the level of bioethanol produced is directly proportional to the increased concentration of sulfuric acid and *Zymomonas mobilis*. The conclusion is levels of bioethanol produced is directly proportional to the increase in the concentration of sulfuric acid and *Zymomonas mobilis* is used.

1. Introduction

Bintaro fruit (*Cerbera manghas*) is a mangrove plant that grows in abundance in tropical areas of Indonesia, and it has not been used maximally. Bintaro fruit is a fleshy stone fruit with a coconut-like lignocellulose fiber. The lignocellulose concentration found in *Bintaro* fruit consisted of 36.945% cellulose and 38% lignin [1]. The cellulose content of *bintaro* fruit is potential to serve as a substrate for the production of bioethanol through a series of processes starting from pretreatment, delignification, hydrolysis, fermentation, and distillation.

Research on the hydrolysis of cellulose into glucose was once performed by [2] using ultra-high temperatures and vapor pressure explosions with a percentage of 67.7% glucose yield at 60 atm for 1 minute), but the method is less practical. Cellulose hydrolysis can also be carried out with acids by [3] using a heterogeneous catalyst and cellulose pretreatment with a precipitation method resulting in 50% glucose. This method is done at high temperatures (140°C) so it is not practical enough.

Alkaline delignification (NaOH 1N) at 100°C, pressure of 1 bar, and 30 minutes of holding time was able to remove lignin in the lignocellulose arrangement so that hydrolysis was optimal [4]. [5] in his research to pretreatment swelling NaOH and HCl regeration facilitate the process of depolymerizing cellulose into glucose. This is due to the amorphous hydrogel formed so that the structure is tenuous and easier to be hydrolyzed. Based on these facts, NaOH can be used for the preparation of cellulose from bintaro fruit. While the process of cellulose hydrolysis using HCl need to be replaced with H_2SO_4 because waste of HCl solution is difficult to degrade.

The highest glucose obtained by cellulose hydrolysis process from alfukate seeds using H_2SO_4 6.0% at 120°C for 60 minutes in the autoclave [6]. The concentration of H_2SO_4 used is not specific and in the process does not use delignification. Therefore, in this research, the specification of H_2SO_4 concentration is used to obtain optimal condition.

[7] has fermented using *Saccharomyces cereviceae* from cotton waste through pretreatment process first. *Saccharomyces cereviceae* mushroom in this study was only able to produce 0.900g/L of ethanol. Other research on bioethanol is reported by [8] that fermentation can be performed using *Z. mobilis* bacteria and capable of producing a larger bioethanol than ethanol paper sludge for 48 hours of incubation. The research of bintaro fruit processing into bioethanol has been done by [9] with enzymatic hydrolysis and fermentation methods using *Saccharomyces cereviceae*. The resulting bioethanol is relatively small at only 0.538% and is not proportional to the price of enzyme catalysts for relatively expensive hydrolysis. Therefore, *Zymomonas mobilis* bacteria potentially produce bioethanol from bintaro fruit with higher levels. The objectives of this study were (1) to know the effect of variation of concentration of sulfuric acid (H_2SO_4) solution to the glucose produced in cellulose hydrolysis of bintaro fruit, and (2) to know the effect of variation of *Zymomonas mobilis* concentration on bioethanol content of glucose fermentation process.

2. Methodology

2.1. Pretreatment
The initial stage of the research included soaked bintaro fruit weighing and cutting into small pieces. Subsequently, a physical pretreatment was done through drying at 80°C for three days, continued by dried bintaro fruit weighing and its water content measurement, ended with milling to obtain 40-60 mesh bintaro fruit powders.

2.2. Delignification
Substrate (bintaro fruit powders) immersion was done in 1L 1N NaOH at 100°C and 1 bar pressure for 30-minutes holding time in an autoclave in accordance with the research conducted previously by [4]. On the completion of the delignification process, the substrate was rinsed for several times using distilled water in order to neutralize the pH of the solution. Then, the process was followed by substrate filtration using filter paper to be then drained in the oven at 60°C for three days [10].

2.3. Hydrolysis
After drained in the oven during the delignification process, the substrate underwent a hydrolysis with sulfuric acid (H_2SO_4) catalyst varied to 5.5%, 6.0%, and 6.5% made by diluting 98% concentrated sulfuric acid in distilled water. 10.0g of the substrate was weighed and put into 3 Erlenmeyer flasks containing 100.0mL of sulfuric acid with the variations of 5.5%, 6.0%, and 6.5%. Next, the Erlenmeyer flasks were put into an autoclave to be then heated at 120°C for 60 minutes [9]. After 60 minutes of the hydrolysis process, the Erlenmeyer flasks were taken from the autoclave, and they were cooled down by being soaked in water. Afterwards, a filtration using filter paper was done to obtain the filtrate.

2.4. Fermentation
The pH of the filtrate needed to be increased until it reached 4.5 by adding 5 M NaOH solution via a universal indicator [11]. *Zymomonas mobilis* bacterium was then added into the 5.5%, 6.0%, and 6.5% filtrate with variations of 1%, 3%, and 5%. The addition of *Zymomonas mobilis* bacterium into 90.1mL filtrate was done in 0.901mL, 2.703mL, and 4.505mL respectively. The holding time for the fermentation

IOP Conf. Series: Earth and Environmental Science **105** (2017) 012076 doi:10.1088/1755-1315/105/1/012076

using *Zymomonas mobilis* bacterium was three days at room temperature [12]. After the 3-day fermentation, the filtrate was heated at 50°C to kill the *Zymomonas mobilis* bacterium.

2.5. Distillation
Distillation was done after the bacterium in the filtrate was eliminated. Heating filtrate at temperature was managed to be 74°C.

3. Research Findings and Discussion

3.1. Identification of Water Content in Substrate
The analysis results of water content of observed bintaro fruit substrate in reddish green color are presented below.

Table 1. Identification of water content.

(a) No	(b) Bintaro Fruit Weight	
	(c) Fresh Weight (g)	(d) Dry Weight (g)
(e) 1	(f) 160.146	(g) 41.336
(h) 2	(i) 158.375	(j) 40.660
(k) 3	(l) 160.630	(m) 41.448
(n) Average	(o) 159.717	(p) 41.148

Based on the above data, the average fresh weight of the bintaro fruit was 159.717g and the average dry weight after baked for three days was 41.148g. The water content in the fresh bintaro fruit was 74.24 %, which was obtained from the following equation.

3.2. Filtrate Visualization by Fehling Test
The visualization of hydrolysate using Fehling test is given below.

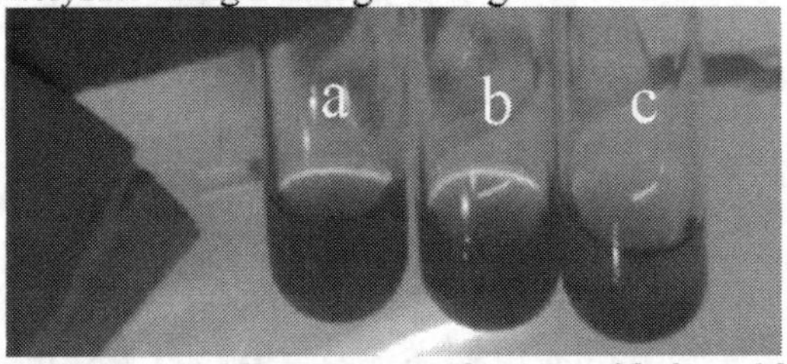

Figure 1. Fehling test results on sulfuric acid concentration of (a) 5.5%; (b) 6.0%; (c) 6.5%.

Figure 1 shows the occurrence of yellow or red cuprous oxide sediment, which indicated the presence of reducing sugar in the hydrolysate. Out of the three used concentrations, there was not any significant different red color among the 5.5%, 6.0%, and 6.5% samples. The sediment indicated that the polysaccharide conversion or breakdown process becoming reducing sugar was successful. The fehling test observation revealed that a high sulfuric acid concentration used at a high temperature would increase the production of the reducing sugar concentration in a red color.

3.3. Reducing Sugar Concentration by Nelson Somogyi Test
Figure 2 shows that the higher sulfuric acid concentration, the higher the produced glucose concentration. In this case, the sulfuric acid serves as a catalyst that is aimed to accelerate the hydrolysis reaction. According to [13], a high number of catalysts shall result in a quick hydrolysis reaction. [14] revealed that the increasing acid concentration in hydrolysis process could also lead to a degradation of glucose and other sugar compounds becoming hydroxyl methyl furfural (HMF) compound and furfural compound, which eventually forms formic acid. However, as seen from the graph above, the acid concentration that increased up to 6.5% had not caused the glucose and other

sugar compounds to degrade into hydroxyl methyl furfural (HMF) compound. Therefore, it can be concluded that 6.5% sugar concentration is still open for enhancement.

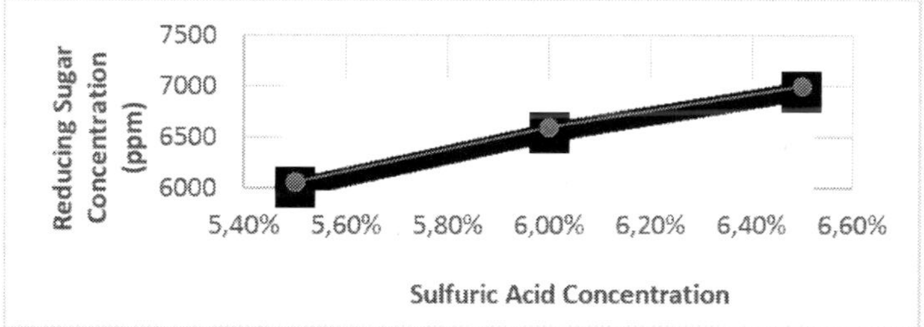

Figure 2. Comparison of H_2SO_4 Concentration and Reducing Sugar Concentration.

The above graph presents that there was 7002ppm reducing sugar concentration resulted from 6.5% sulfuric acid concentration. The higher the sulfuric acid concentration, the more the number of H+ bonding the hydroxyl group on the polysaccharide, which results in the release of interchain bonding that forms monomers, especially in the shape of monosaccharide. The increase of the sulfuric acid concentration lead to the breakdown of the polysaccharide became a high number of monosaccharide and disaccharide. The monosaccharide that was formed to reduce the cupric oxide would be detected by Uv-Vis after reacted to arsenomolybdate and formed dark blue-colored molybdenum. Such sample was then used for fermentation.

3.4. Gas Chromatography Test Results

The data analysis from the gas chromatography is presented in table 2 below.

Table 2. gas chromatography test results.

No	Name	Weight		Area		Ethanol	
		Sample	ACN	Ethanol	ACN	Ratio	Content (%)
1.	5.5% H_2SO_4 + 1% *Z. mobilis*	0.0198	0.2076	13,264.81	6,696,686.82	0.002	2.284
2.	5.5% H_2SO_4 + 3% *Z. mobilis*	0.1036	0.2229	184,499.16	8,114,377.92	0.025	5.380
3.	5.5% H_2SO_4 + 5% *Z. mobilis*	0.4933	0.2367	1,273,772.80	8,438,722.22	0.166	7.980
4.	6.0% H_2SO_4 + 1% *Z. mobilis*	0.4966	0.2407	514,893.53	8,819,420.15	0.064	3.112
5.	6.0% H_2SO_4 + 3% *Z. mobilis*	0.0989	0.2074	54,714.89	1,870,529.14	0.032	6.746
6.	6.0% H_2SO_4 + 5% *Z. mobilis*	0.4827	0.2394	1,382,787.81	8,087,229.88	0.188	9.326
7.	6.5% H_2SO_4 + 1% *Z. mobilis*	0.0424	0.0079	1,049,193.92	4,719,776.43	0.244	4.555
8.	6.5% H_2SO_4 + 3% *Z. mobilis*	0.3438	0.2397	1,117,349.23	9,580,964.72	0.128	8.942
9.	6.5% H_2SO_4 + 5% *Z. mobilis*	0.0499	0.2395	150,034.92	7,937,617.42	0.021	9.977

The gas chromatography test results were analyzed using double anova preceded by data normality test via Kolmogorov - Smirnov normality test. The normality test showed that the obtained data had a normal distribution with a p value of 0.137. The data with normal contribution was then statistically tested using double ANOVA, and the results presented that the increase of H_2SO_4 concentration and the use of *Zymomonas mobilis* affect the obtained amount of bioethanol. It showed that the increase of the sulfuric acid concentration and *Zymomonas mobilis* is directly proportional to the increasing production of the ethanol concentration. The high concentration of sulfuric acid will increase the capability to produce glucose from cellulose. The high concentration of glucose functions as the main source of nutrition for the microorganism, which resulted in more glucose to be converted into ethanol. From this research, it is known that 6.5% H_2SO_4 and 5% *Zymomonas mobilis* are the most efficient concentrations to produce bioethanol.

The ANOVA test on the interaction between the increase of H_2SO_4 concentration and *Zymomonas mobilis* used did not affect the production of bioethanol. It showed that the increase of the H_2SO_4 concentration and *Zymomonas mobilis* are not directly proportional to the production of bioethanol. Such ANOVA analysis results can be seen from the increase of 3 – 5% concentration (sample 8 and 9) which was relatively low, namely 1.035% while the 1-3 % concentration (sample 7 and 8) had 4.387% difference. The difference is such ethanol increase was influenced by the *Zymomonas mobilis* activity in the exponential phase (the 3rd up to the 24th hour), which initially utilized the glucose to grow and multiply itself, which is in a good agreement with the research conducted by [15] and by [16]. Therefore, it can be concluded that on the 3–5% concentration increase, the phase of the microorganism growth developed and they started to look for a food source (glucose) so that the death phase became faster and resulted in the low improvement of the ethanol concentration in sample 8 and 9.

The increase of acid concentration in the hydrolysis process also could cause a degradation of glucose and other sugar compounds become hydroxyl methyl furfural (HMF) and furfural, which shall form formic acid [14]. The first cause of the high concentration of ethanol produced in this research was the absence of sterilization on the hydrolysate that was going to be fermented via autoclave, since it would form an inhibiting compound that prevents the performance of the microorganism during the fermentation [17]. Second, the microorganism used in this research has better performance than *Saccharomyces cerevisiae* in producing bioethanol. Such fact is in line with the research conducted by [18] and [19] stating that the concentration of ethanol results from the *Zymomonas mobilis* fermentation is higher than that from the Saccharomyces cerevisiae fermentation.

3.5. Comparison of Research Result

This research is the second research to process the bintaro fruit into bioethanol. Previously [9] have conducted similar research. Therefore, the following is a detailed discussion of the comparative research results in terms of economic aspects, effectiveness, and efficiency of research.

Table 3. Comparative research in terms of effectiveness of glucose being converted.

Substrate Concentration	Hydrolyzate	Fermentation Glucose Level (ppm)			Ethanol Yield
		Before	After	Converted	
100 g/L	No	8604,9	1870,185	6734,185	0,538%

(source: [9])

Substrate Concentration	Hydrolyzate	Fermentation Glucose Level (ppm)			Ethanol Yield
		Before	After	Converted	
100 g/L	No	7002	109,9	6892,1	0,62%

Table 4. Comparison of research in terms of efficiency of research implementation

No.	Substance of Study	Iman and Handoko, 2011	Santoso et al., 2017
1	Preparation	3 days	1 days
	Delignification	-	30 minutes
	Hidrolysis	3 days	1 hours
	Fermentation	3 days	3 days
	Distillation	2 hours	2 hours
TOTAL (Time)		9 days dan 2 hours	4 days and 3,5 hours

Table 5. Comparative research in terms of economic value of research implementation

No.	Substance of Study	Iman and Handoko, 2011	Santoso et al., 2017
1	Delignification	-	4 g (80,- IDR)
	Hidrolysis	5 mL enzim *cellusoft* L (75.000 IDR)	6,63 mL H_2SO_4 (104,- IDR)
	Fermentation	5 g *S. cereviceae* (1.863,65 IDR)	4,5 mL *Z. mobilis* (1.750,- IDR)
TOTAL (Value of Substrate Processing)10 g/100 mL		76.863,65 IDR	1.934 IDR

Holistic conclusion obtained from the above three comparisons is this research is superior compared to [9]. It also indicates that this research is very feasible to be implemented massively to process the source of cellulose substrate in Indonesia to make it bioethanol.

4. Conclusion

From the research findings, it is known that bintaro fruit has 74.24% water content. Under the Nelson-Somogyi quantitative test, the identification of the highest reducing sugar concentration was 7002ppm in 6.5% sulfuric acid variation. Gas chromatography showed that the highest ethanol concentration was obtained from the 6.5% variation of sulfuric acid and 5% *Zymomonas mobilis* with a percentage of 9.977%. The results of this study also revealed a directly proportional relationship between the increase of the sulfuric acid concentration as a catalyst and the *Zymomonas mobilis* microorganism in the production of ethanol. The ethanol yield produced from 1 kg of fresh bintaro fruit was 9.0 mL/kg of fruits. The results of comparative observation with previous research indicate that this research has advantages in terms of effectiveness, efficiency, and economic value of implementation formula.

The suggestions are to develop the current research by increasing the sulfuric acid concentration for the hydrolysis process and to specify the *Zymomonas mobilis* content in the range of 3-5%.

5. References

[1] Chang L C, Gills J J, Bhat K P, Luyengi L, Farnsworth N R, Pezzuto J M and Kinghorn A D 2000 Activity Guided Isolation of Constituents of Cerbera manghas with Antiproliferative and Antiestrogenic Activities *Bioorganic and Medical Chemistry Letters* **10 (21)** pp. 2431–2434

[2] Sasaki C, Keisuke S, Chikako A and Yoshitoshi N 2012 Direct Hydrolysis of Cellulose to Glucose Using Ultra-High Temperature and Presure Steam Explosion *Carbohydrate Polymers Journal* **89** pp. 298-301

[3] Morales-delaRosa S, Campos-Martin J M and Gierro J L G 2017 Chemical Hydrolysis of Cellulose into Fermentable Sugars through Ionic Liquids and Antisolvent Pretreatments using Heterogenous Catalysts *J. Catalyst Today* **33**

[4] Maryana R, Ma"rifatun D, Wheni S and Rizal A 2014 Alkaline Pretreatment on Sugarcane Bagasse for Bioethanol Production. *Energy Procedia* **47** pp. 250-254

[5] Sun B, Peng G, Duan L, Xu A and Li X 2015 Pretreatment by NaOH Swelling and then HCl regeneration to Enhance the Acid Hydrolysis of Cellulose to Glucose *Journal of Bioresource Technology* **196** pp. 454-458

[6] Muin R, Lestari D, Sari T W 2014 Pengaruh Konsentrasi Asam Sulfat dan Waktu Fermentasi terhadap Kadar Bioetanol yang Dihasilkan dari Biji Alpukat. *J. of Chemical Engineering* **4 (20)**

[7] Nikolic S, Lazic V, Veljovic D and Mojovic L 2017 Production of Bioethanol from Pre-treated Cotton Fabrics and Waste Cotto Materials *Carbohydrate Polymers Journal* **164** pp. 136-144

[8] Yamashita, Y., Kurosumi, A., Sasaki, C., and Nakamura, Y.2008. Ethanol production from paper sludge by immobilized Zymomonas mobili *Biochemical Engineering Journal* **4 (3)** pp. 314-319

[9] Iman G and Handoko T 2011 Processing of Bintaro Fruit as a Source of Bioethanol and Activated Carbon *Proceedings of National Seminar on Chemical Engineering* ISSN 1693.

[10] Cheng D, Jiang S and Zhang Q 2013 Effect of hydrothermal treatment with different aqueous solutions on the mold resistance of Moso bamboo with chemical and FTIR analysis *Bio Resources* **8 (1)** pp. 371-382

[11] Yang S et al. 2009 Trancriptonic and Metabolomic Profiling of Zymomonas mobilis During Aerobic Fermentations *Journal of Bioedical Central Genomics* **10 (34)**

[12] Altintas M M, Eddy C K, Zhang M, McMillan J D, Kompala D S 2006 Kinetic modeling to optimize pentose fermentation in Zymomonas mobilis *Biotechnology and Bioengineering* **94** pp. 273–295

[13] Haviz M and Maryadi 2012 *Effect of Acid Concentration and Yeast Types on Hydrolysis and Fermentation Process of Bioethanol Preparation from Newspaper* Thesis Proceeding FT UnSri

[14] Taherzedah and Karimi 2007 Enzyme-Based Ethanol. University of Boras. Sweden: *Bioresources* **2 (4)** pp. 707-738

[15] Prescot S C and Dunn C G 1981 *Industrial Microbiology* New York. Mc. Graw-Hill Book Co.Ltd.

[16] Muslibah S and Herumurti W 2011 *Effect of pH and Zymomonas mobilis Concentration for Ethanol Production of Citrus Fruit Waste* Thesis Proceeding 2010/2011 FTSP ITS

[17] Yu Y, Lou X, Wu H 2008 Some Recent Advances in Hydrolysis of Biomass in Hot Compressed Water and Its Comparisons with Other Hydrolysis Methods *Energy Fuels* **22 (1)** p. 50

[18] Sivasakthivelan P, Saranraj P and Sivasakthi S 2014 Production of Ethanol by Zymomonas mobilis and Saccharomyces cerevisiae Using Sunflower Head Wastes - A Comparative Study *International Journal of Microbiological Research* **5 (3)** pp. 208-216

[19] Sulfahri, Nurhatika S. Nurhidayati T 2011 *The Ability of Zymomonas Mobilis Bacteria in Producing Ethanol from Algae Spirogyra* Thesis Proceeding 2010/2011 FMIPA ITS

IOP Conf. Series: Earth and Environmental Science **105** (2017) 012077 doi:10.1088/1755-1315/105/1/012077

SPEKTRA fast and smart software for renewable energy management

D Arengga[1], W Agustin[2], Y Rahmawati[3], S Sendari[4] and A N Afandi[5]
Department of Electrical Engineering, Faculty of Engineering, Universitas Negeri Malang, Jalan Semarang No. 5, Malang, Jawa Timur, Indonesia

E-mail : [1]arengga_oxcd@yahoo.com, [2]whynaagustin@gmail.com, [3]yrahmawati72@gmail.com, [4]siti.sendari.ft@um.ac.id, [5]an.afandi@um.ac.id

Abstract. Nowadays, many countries prefer to use renewable energy and increase the capacity of power generations. One method to increase the capacity power uses a hybrid system. The aim of this paper is to introduce a software for simulating a hybrid power generation included renewable energy, which combines solar and wind power generations. This software is named as SPEKTRA. In order to represent the actual system, there are some editable variables to show exchange of solar and wind power outputs. This simulation is used to increase the understanding of the power generation in the hybrid system for users and to estimate the spending cost of the power generation on hybrid system. The SPEKTRA is developed using Visual Basic. Functional and developed simulation is analysed using the black box method. the result show that the output graph is similar to the theoretical review.

1. Introduction

Currently, the usage and diversification of renewable energy to decrease dependency on fossil fuels become a major issue in many countries; the goal is to maintain the sustainability of energy. The renewable energy studies on diversification of energy, integration opportunities, challenges, and implications of using biomass, geothermal, hydropower, solar resources, wind-powered system and ocean technologies for generation technology improvement, electric system operational constraint, and electricity demand [1]. Indonesia has a large new and renewable-energy potential as explained by the Director General of Electricity and Energy Conservation at the Focus Group Discussion on the Supply and Demand of New and Renewable Resources [2]. However, the utilization of renewable and alternative energy is very low compared to the existing potential. The alternative energy is still limited to electrical power generator, where total of electrical energy is about 25.3TWh or 15% of total electrical resources in 2010 [3]. On the other hand, the energy consumption continues to rise along with economic growth.

Electricity is one of important points affecting to the economy of the countries. Electricity is the main energy source for housing, industrial sector, and public facilities. Electrical consumption growths in line with the population growth and regional development, so the needs should be balanced with the ability of electricity production [4]. Furthermore, the usage of power energy should be monitored and controlled [5]. Nowadays, increasing renewable energy as alternate energy production has been increased, using geothermal power plants (1.3 GW), hydropower (5.1 GW), while solar, wind, landfill, mini-hydro, biomass with a total capacity of 96MW [6]. Considering the potential usage of renewable energy, the developed simulation is proposed to be used for educational purposes such as represent the actual system of hybrid system.

The aim of this paper is to describe a simulation consisted of software called SPEKTRA. The software part of SPEKTRA will calculate the most potential renewable energy to be used according to the environment condition towards the capability of electricity result. There are some editable variables to show the exchange of solar and wind power outputs and it is also accomplished completed with financial review. The purposes of this simulation are to increase the understanding of making a power plant with hybrid system planning. This software was developed using visual basic because there are many features available as a data processor and has a great user interface so that users can easily learn independently. Including in this software, there are learning materials, the way data retrieval based on the parameters provided, see the parameter output, and the concept of management power plants renewable energy wind power and solar to review the calculation of the cost of the investment. The simulation will be able to give user an experience to understand the real condition without use the real power plant generator. It used to keep the users safely and minimalizing the accident while the users analyse the environment potential for sustaining the renewable energy.

2. Software Development of SPEKTRA

The proposed system is developed using Systems Development Life Cycle (SDLC) method consisted of planning, analysing, designing, implementing, and maintaining processes [7]. The simulation is used to show the real effect of the parameters inserted by user. These parameters represent the environmental conditions to apply the hybrid system. Meanwhile, the software section is used to calculate and analyse the electricity result which will be transmitted to the consumer. Based on the result shown in the software section, users could plan the most suitable renewable energy to be used and estimate the power generation on hybrid system. The detail of the proposed system is described on figure 1.

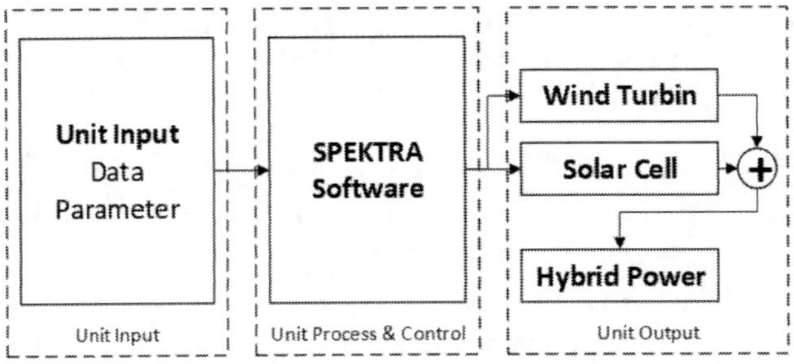

Figure 1. Block Diagram of SPEKTRA.

The software of SPEKTRA is developed using Visual Basic. Here, the software application is a desktop program which can be used easily using the computer. The software SPEKTRA is used to know the calculation of electricity result based on the parameter inserted by user which change the solar and wind power output and it also accomplished with financial review. Each power plant panel such as solar power generation, wind power generation, and hybrid generation have the different input of parameters to be analysed.

The calculation process of the software simulation is divided into three kinds of power plants. Those are solar power generation, wind power generation, and hybrid system included the wind and solar power generation. Input parameters of each power plant are shown in figure 2.

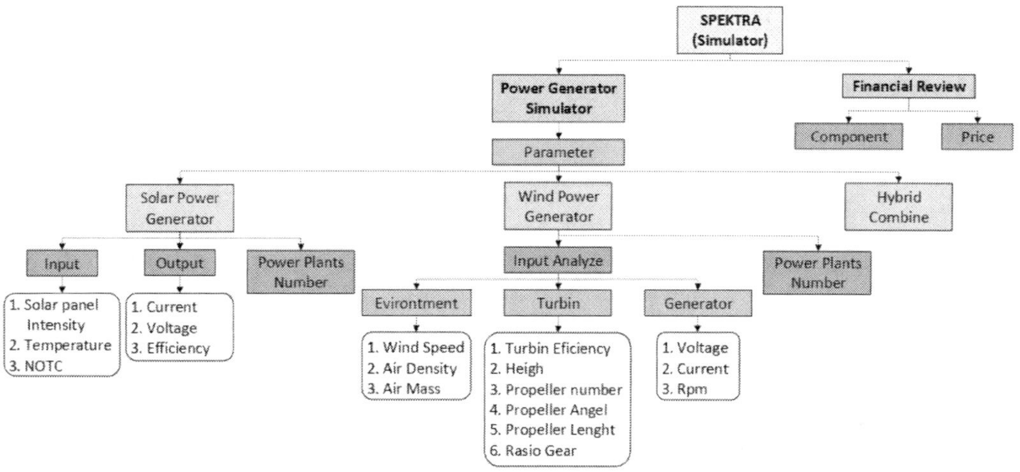

Figure 2. Hierarchy Concept of Software SPEKTRA.

2.1. Solar Power Generation

Solar power generation is a renewable energy generation utilizing the solar power resource. In order to calculate the solar power plant result based on the parameter input by user, the software simulation is developed using the formula (1). While to know the maximum power result, the software simulation is developed using the formula (2). Each solar panel has the threshold of temperature to work properly. If the temperature of the sun or environment condition is higher than the threshold, the power result will be reduced automatically by NOTC (Normal Operation Temperature Coefficient) percent set by user.

$$P = V \cdot I \tag{1}$$

$$P\max = Lux\max \tag{2}$$

2.2. Wind Power Generation

Wind power generation is a renewable energy generation utilizing the wind power resource. In order to calculate the wind result based in the parameter input by user, the software simulation is developed using following formula i.e. (3) is used to calculate the speed of the wind on the height represented by z, (4) is used to calculate power density of the wind parameter, and (5) is used to calculate the final power result created by wind power generator. The percentage of power result is equal with power generator calculated by formula (6) which is also equal to Rpm (rotation per minute) percent calculated by formula (7).

$$v = vref \ln(\frac{z}{z0}) \Big/ \ln(\frac{Zref}{z0}) \tag{3}$$

$$WPD = \frac{P}{A} = \frac{1}{2} p \cdot v^3 \tag{4}$$

$$Pm = Cp(\lambda \cdot \beta) \frac{\rho \cdot A}{2} v^3 \tag{5}$$

$$Pgen = V \cdot I \cdot Rpm\% \tag{6}$$

$$Rpm = P = V \cdot I \tag{7}$$

2.3. Hybrid Power Generation

Hybrid power generation is a combination of solar and wind power generation. The power result of the hybrid power plant is a total value between both of power plant as shown by formula (8). Contrasted to the previous power generation as a stand-alone power plant, the hybrid system allows the solar and wind power plants. The Advantage of the hybrid power plant is having stability, continuity, reliability more than the stand-alone power plant. A hybrid power plant can decide the type of power plants that produce maximum electricity in accordance with the environmental conditions.

$$Phybrid = Psolar + Pwind \tag{8}$$

3. Result and Testing of SPEKTRA

In the testing stage, the functionality of simulation are tested using Black Box testing method [8]. The experiment is done to ensure the simulation, which will show the output according to the actual system on the field based on environment parameter inserted by user. The test is carried out to avoid any errors when the system is executed.

3.1. Description of Software SPEKTRA

SPEKTRA, the developed software has a section of the parameter block that enables users to better understand how to perform simulation analysis. In this software, there are blocks of input parameters, blocks of output parameters, graphical display and circuit drawing tutorial for implementation. In the mater block the inputs of solar power plants contain input columns Voltage (V), Flow (A), Number of solar panels (Unit), NOCT (Normal operation Coefficient temperature), Solar panel power, Sun Light Intensity (Lux), temperature (oC). The outputs determine a voltage, current and power generated by the solar power plant. Wind turbine input parameters include columns covering area type, air mass (kg / m), air density, turbine coefficient, turbine height (m), turbine sweep area, (m), vane angle, generator voltage (V), Current Generator (I), Rpm speed minimum-maximum of generator and number of generators. Here the Figure 3 shows the specification of software simulation.

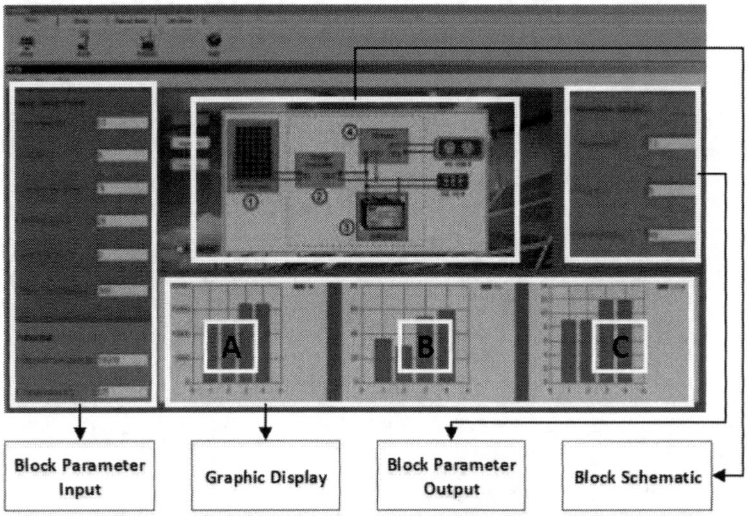

Figure 3. The Display and Specification of SPEKTRA.

3.2. Black Box Testing

The test has been done to software simulation. in the testing stage, the simulation was tested using the same parameter for each type of power plant. The same parameter was inserted in the software simulation to show the result calculation and the effect physically. The testing result confirmed that all parts of software simulation are running well. The testing of calculation function in the software simulation is done for each different power plant generator. The result follows to show that the calculation is equal with the theoretical calculation.

3.2.1. Solar Power Generation Testing. In the testing, the usage of power capacity is 100 Wp with variable data usage shown in table 1 and the Sun temperature usage is 25 degrees Celsius. In this calculation, the power peaks are getting out from the equation the ideal performance of the solar panel toward the sun light intensity in absorbency on solar panels.

Table 1. Solar Power Generation Result.

No	Solar Panel (Wp)	Sun Light (Lux)	Temperature (°C)	P Out (Watt)	P Theoretical Calculation (Watt)
1	100	51500	25	44.32	44.32
2	100	62300	25	53.61	53.61
3	100	94000	25	80.90	80.90

3.2.2. Wind Power Generation Testing. In the testing, the specification usage of power plant capacity is 3KW with the mechanics of A (cross-sectional area) is 5 meters, the angle of slope of the propeller blades is 30 degrees, and the number of propeller blades is 3 units. On the 15 meter height above ground level at the level of rudeness/roughness Beach area has wind speeds ranging from 2-25 Km/h. Based on the result, cut on minimum rpm generator and is cut off is the maximum rpm generator. The wind power generation result is shown in table 2.

Table 2. Wind Power Generation Result.

No	Sweep Area Turbine (m)	air velocity (m/s)	height (m)	P Out (Watt)	P Theoretical Calculation (Watt)
1	5	3	15	79.65	79.65
2	5	5	15	368.75	368.75
3	5	7	15	1011.85	1011.85

3.2.3. Hybrid Power Generation Testing. The estimation of hybrid system uses the previous result based on the solar and wind power generation. The combination result is shown in table 3.

Table 3. Hybrid Power Generation Result.

No	Solar Power (Watt)	Wind Power (Watt)	Hybrid (Watt)	P Theoretical Calculation (Watt)
1	44.32	79.65	123.97	123.97
2	53.61	368.75	422.36	422.36
3	80.90	1011.85	1092.75	1092.75

3.3. Financial Review

In order to know the financial review, users have to select the appropriate set of electronic materials displayed on the software based on the type of the selected power plant. Followed by entering the required amount and the price in accordance with catalog purchases. The system will display the calculation estimates the funds needed in the development of renewable electricity generation. Here Figure 4 shows the estimation of building cost of developing power plants.

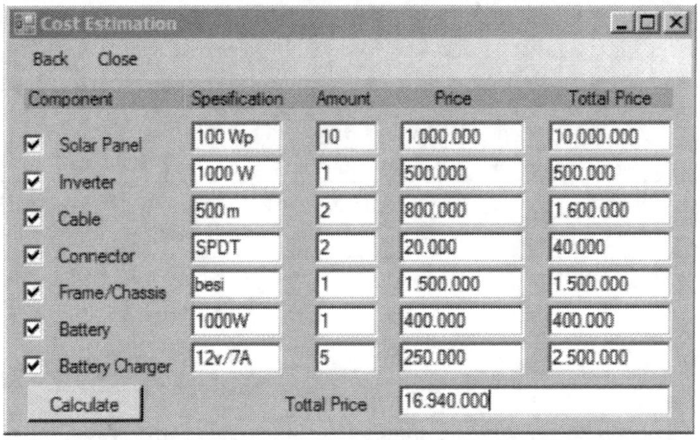

Figure 4. Display of Financial Review of SPEKTRA

4. Conclusion

In this paper, the simulation system has developed which aim to increase the understanding of power generation in hybrid system for users and to estimate power generation on hybrid system using the software simulation. The presence of this simulation will give the users an experience to know the actual system of renewable power generation utilizing the solar and wind resources. The testing has been done with functionality test using black box method. The testing result shows that the system worked properly, confirming that software of simulation and its financial review can work well and ready to be implemented.

5. Acknowledgments

We want to express our gratitude to Ministry Technology & Higher Education and Department of Electrical Engineering Faculty of Engineering Universitas Negeri Malang, which supported funds for this project.

6. References

[1] Mai T, Sandor D, Wiser R, and Schneider T 2012 Renewable Electricity Futures Study: Executive Summary NREL/TP-6A20-52409-ES. Golden, CO: *National Renewable Energy Laboratory*

[2] Indonesia's Renewable Energy Potential 2008 http://www.esdm.go.id/news-archives/general/49-general/1963-indonesias-renewable-energy-potential.html

[3] Iskandar M A 2012 Planning of Efficiency and Elasticity Energy *2012*, Center of Energy Technology, the Agency of Assessment and Application Technology (Balai Besar Teknologi Energi, Badan Pengkajian dan Penerapan Teknologi-BPPT), www.bppt.go.id.

[4] The Ministry of Energy and Mineral Resources 2014 *Book of Electricity Statistics* Number 28-2015

[5] Sendari S, Herwanto H W, Rahmawati Y, Putranto D M, Wibowo F S, Rachman I, Matsumoto T 2017 Developing Control and Monitor System with Recording of Operational Expense of Electrical Devices *31st International Conference on Advanced Information Networking and Application Workshop* (WAINA) pp 315-320

[6] Sugiyono A, Anindhita, Wahid L M A, Adiarso 2016 *Indonesia Energy Outlook 2016*

[7] Blanchard B S, Fabrycky W J 2006 System Engineering and Analysis *(4th Edition)* New Jersey: Prentice Hall

[8] Nidhra S, and Dondeti 2012 Black Box and White Box Testing Techniques – A Literature Review 2 *International Journal of Embedded Systems and Appications* (IJESA) pp 29-50If.

IOP Conf. Series: Earth and Environmental Science **105** (2017) 012078 doi:10.1088/1755-1315/105/1/012078

Internet-based monitoring and warning system of methane gas generated in garbage center

S Sendari[1], Y Rahmawati[1], W Kamdi[1], H Elmunsyah[1], W Agustin[1], N C Hadi[1], F S Wibowo[1], T Matsumoto[2] and I Rahman[2]

[1]Department of Electrical Engineering, Faculty of Engineering, Universitas Negeri Malang, Jalan Semarang No. 5, Malang, Jawa Timur, Indonesia, 65145
[2]Department of Live and Environment Engineering, Faculty of Environment, the University of Kitakyushu, 1-1 Hibikino, Wakamatsu, Kitakyushu, Fukuoka, 808-0135

E-mail : siti.sendari.ft@um.ac.id

Abstract. This paper introduces a system which can monitor the environment of the garbage center in a city. Comparing with developed countries, they have technologies to process garbage, while those technologies are still difficult to be implemented in developing countries. In Indonesia, garbage are mostly placed in a wide-open area called landfill, where methane gas could be produced by a decomposition process. The gas has a characteristic that it can be burned easily. However, it is potentially to be used as an alternative energy for gas stoves. Considering this characteristic, the gas should be managed appropriately for both of distributing and monitoring gas leakage. Since the production of methane gas is plentiful enough, the gas can be distributed among the households surrounding the garbage center. Monitoring of methane gas has to be done continuously and real time during producing and distributing, without generating pollutions to the environment. Here, the monitoring system has been developed using a micro controller equipped with a methane gas sensor, i.e., TGS 2611. The data of monitoring will be distributed using an Internet-based system, where a Wi-Fi module is attached to the micro controller to send and communicate with the web-based server. Furthermore, information of monitoring the gas can be accessed by using android application. The advantage of the proposed system is that the user can monitor, record, and get the information of gas level. The system can also be used as an early-warning system when the gas exceeds the limit and when a danger situation occurs. The proposed system was tested using the black box method to evaluate the functionality. Finally, the system is used to optimize the Waste Management program through the utilization of methane gas. Hence, the utilization of methane gas monitoring system is tested to the user, whether they can use it easily.

1. Introduction

Garbage could be the most problem of cities in Indonesia. Quantity of municipal solid waste (MSW) has been increasing year by year. Ministry of Environment and Forestry (LHK) reported that garbage tonnage in Indonesia could be 64 million tons per year in 2017 [1]. In order to reduce the capacity, there are some strategies used, such as landfill, combustion, recycling, recovery, reducing and compositing. Comparing with developed countries, where they have those technologies to process

[1] Author to whom any correspondence should be addressed

garbage, those technologies are still difficult to be implemented in developing countries. In Indonesia, there are some problems need to be addressed, such as high MSW generation, low quality waste management, limited final disposal sites, and financial [2]. According to the Central Bureau of Statistics, 84.51 % of MSW in Indonesia was disposed whithout shorting process [3]. Furthermore, the garbage are collected at the garbage center, which are mostly placed in a wide open area called waste landfill. From some research results, there are some advantages and disadvantages of surrounding waste landfill. An organic carbon in the waste landfill produces methane gas (CH4) and carbon dioxide(CO2) decomposed by microorganism. Waste landfills have been known as one of major sources of greenhouse gas (GHG) and as the responsible for 20% increasing global warming [4,5]. Furthermore, These waste landfill sites could be major source of land, air, ground and surface water pollutions and very harmful for communities who reside surround landfill sites. Improving the awareness of communities is the most important thing to reduce the impact of waste disposal pollutions. Management of landfill waste site is an important parameter to be implemented as a practical model to resolve the problems regarding pollution and hazardous health problems. Here, introducing control and routine preventive measures should be used for reducing the levels of various kinds of infections [6].

Reducing the negative effect of producing gas at waste landfill could be important part to reduce a potential conflict with communities surrounding the garbage center. CH4 is one of major GHG sources. However, it has a potential to be used for alternating energy for gas stove or also electricity. However, these applications require information about composition and quantity of biogas produced [7]. The aim of this paper is to describe the developed system, which can help communities to monitor, record, and get the information of gas level surrounding garbage center. Since CH4 production is not homogeneous, various representative gas monitoring sites is necessary to be developed [8]. Here, the monitoring system has been developed using a micro controller equipped with a methane gas sensor, i.e., TGS 2611. Thus, monitoring systems can provide the gas level to determine the action needed to control or decide actions considering situations at landfills and disposal sites [9]. The system can also be used as an early-warning system when the gas exceeds the limit and when a danger situation occurs. Methane concentration measurement results are evaluated according to the limit values. If the measurement results exceed the limit values, warning message is sent to the officers. While the previous work monitoring system is developed as web-based computer application [10], this research develops a monitoring system which can work using Wi-Fi communication. The data of monitoring will be distributed using an internet-based system, where a Wi-Fi module is attached to the micro controller to send and communicate with the web-based server. Furthermore, information of monitoring the gas can be accessed by using android application.

At the rest, this paper is organized as follows. Section 2 presents the situation of waste landfill, which is observed the potential CH4 produced at the landfill sites. In Section 3 describes the proposed system of CH4 monitoring system. Section 4 discusses the observation of CH4 monitoring result. The last, Section 5 present the conclusion.

2. Waste Landfill Situations

One of the biggest garbage center at East Java is the Klotok's landfill, which was established in 1992, and it has been already 25 years old. The area of this landfill is about 2.5 ha, which is used to collect garbage surrounding Kediri city with the capacity until to 576.35 m3. Here, the garbages are collected at the open area at arround 1 ha, with the height is designed up to 10 meters for each zone. Figure 1 shows Klotok's landfill area from satellite. The landfill is devided into two areas, i.e., old area, which is marked with red color, while the newer area is marked with green color. The newer area has been already 3 years old (called as active zone), while the older area of waste landfill has been 25 years old (called as inactive zone).

In this research, the observation was done using sampling method, i.e., 3 area sampling, which are selected from Klotok's landfill performed on 1.5 meter and 3 meter depth below the surface of landfill. This gas was monitored using pipes as shown in figure 2. The material of pipes use the high density landfill polyethylene (HDPE) with 4 inch of diameter. Here, CH4 flows through the pipes, which are connected to the developed CH4 Monitoring System. Then, the system will monitor, record and give the information of gas level.

Figure 1. The Picture of Klotok's Landfill

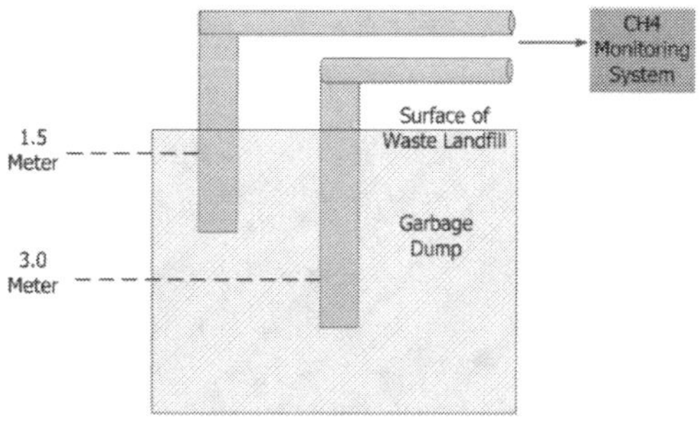

Figure 2. The picture of Pipe's Installation

3. CH4 Monitoring System

The CH4 Gas Monitoring Systems should be planned, designed and applied appropriately for monitoring and controlling gas at the waste landfill [11]. In order to develop CH4 Monitoring System, the developing system was done using System Development Life Cycle (SLDC) [9]. The developed CH4 monitoring system is designed to have abilities to monitor, record and give the information about CH4 gas level. Here, the developed system can also be used as an early warning system, which sends alert when the CH4 gas exceed the limit. The users can monitor the gas level using web-based computer application and android application. Figure 3 shows the developed monitoring system. The system consists of CH4 gas sensor, unit processing, wireless module data communication, unit display, buzzer, data storage, and data base.

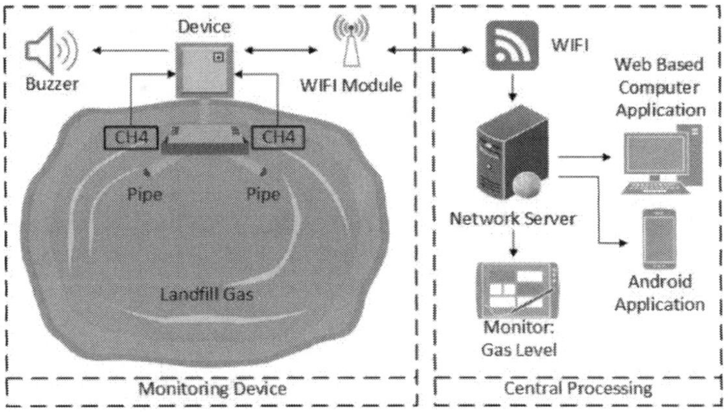

Figure 3. The Proposed of CH4 Monitoring System

Here, the developed system is equipped with WIFI module to be used for data communication, which is connected to the serial communication on digital pins of unit processing, so the system sends the data the data bse system via WIFI connection. The developed system reads the level of CH4 using TGS2611, CH4 gas sensor. Before CH4 gas sensor is used, it should be calibrated so system can send the data precisely [12]. The data is recorded to the storage unit and sent to the database system. Network server updates web based service for users, which can access using computer application or Android application. If the measurement of CH4 level exceeds the limit values, a warning alert is sent to the officers using buzzer, and network server sends a warning notification to users via the computer or android applications. The block diagram of detailed proposed system is shown in figure 4.

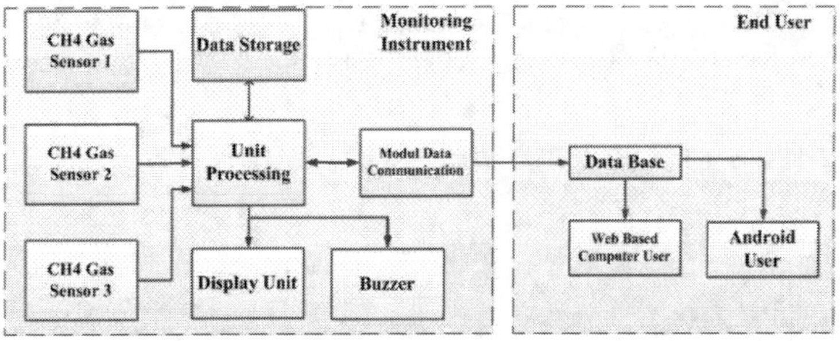

Figure 4. Block Diagram of CH4 Monitoring System

Here, the web-based computer and Android applications have some functions, i.e., (1) show the data as readable graph by real time; (2) report the data as document; and (3) give the notification if CH4 level exceeds the limit. The developed system works as the flowchart shown in figure 5 while the figure 6 will shown the flowchart of monitoring system at the network server.

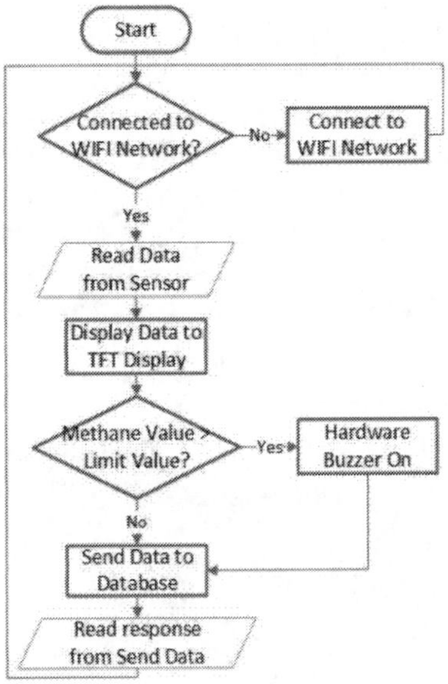

Figure 5. The Flowchart of CH4 Monitoring Device

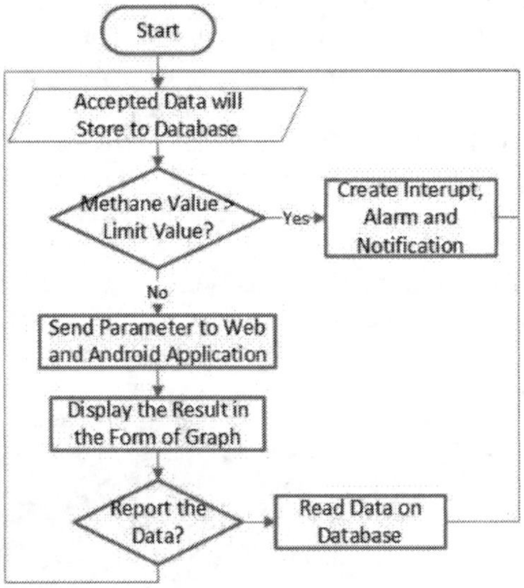

Figure 6. The Flowchart of monitoring system at the network server

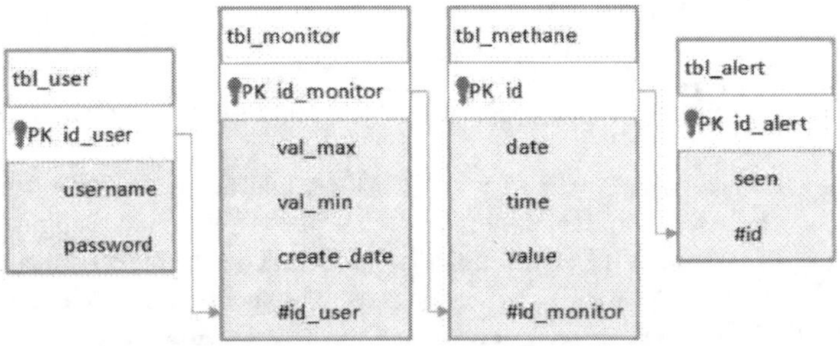

Figure 7. Data base of CH4 Monitoring System

The database system has important role on developing this system. Here, the database consists of tbl_user, tbl_methane, tbl_alert. The tbl_user is used to build a login system for the application. The

IOP Conf. Series: Earth and Environmental Science **105** (2017) 012078 doi:10.1088/1755-1315/105/1/012078

tbl_monitor is used to save the maximum and minimum values of CH4 gas every day, while tbl_methane is used to save all records of CH4 values as real time. The tbl_alert is used to save data when the measurement results exceed the limit values. These tables are connected user Entity Relationship (ER) diagram as shown in figure 7.

4. Experiment Result

The experiment is done to measure CH4 Gas of the newer landfill area compared with the older area. Each landfill devide into three terrace with 3 point area of measurement, where the gas level is compared based on depth of each pipes and time of sampling. Here, two kind of pipes are used, i.e., 1.5 meter and 3.0 meter. Table 1 shows the data comparison considering the depth of each pipes in older waste landfill.

Table 1. Data comparison based on depth of each pipes in older waste landfill

	Pipe 1 (% Vol)				Pipe 2 (% Vol)				Pipe 3 (% Vol)			
Depth	**1.5 meter**		**3 meter**		**1,5 meter**		**3 meter**		**1,5meter**		**3 meter**	
Day	**M**	**E**	**M**	**E**	**M**	**E**	**M**	**E**	**M**	**E**	**M**	**E**
1	0	0	2.86	2.78	59.91	59.10	52.88	61.72	0	0	1.04	0.78
2	0	0	4.95	8.06	63.59	61.92	59.88	65.60	0	0	4.03	4.84
3	0	0	2.84	4.03	64.19	66.34	66.10	65.58	0	0	1.10	3.28
4	0	0	2.57	2.22	66.47	61.18	60.10	61.62	0	0	10.66	4.50
5	0	0	2.06	4.60	61.36	60.42	61.01	61.16	0	0	10.00	11.18
6	0	0	3.99	4.70	58.86	61.05	62.64	64.21	0	0	15.57	14.19
7	0	0	2.68	4.58	60.62	63.01	67.62	69.86	0	0	1.84	3.05
Average	0	0	3.14	4.42	62.29	61.86	61.46	64.25	0	0	6.32	5.97
Dev	0	0	0.99	1.87	2.52	2.32	4.81	3.11	0	0	5.75	4.85

Notes: M: Morning E: Evening : CH4 Peak : CH4 Min : The waste dumping process

Table 2. Data comparison based on depth of each pipes in newer waste landfill

	Pipe 1 (% Vol)		Pipe 2 (% Vol)		Pipe 3 (% Vol)	
Day	**Morning**	**Evening**	**Morning**	**Evening**	**Morning**	**Evening**
1	57.37	58.75	46.28	63.16	22.40	31.89
2	64.90	64.68	67.72	68.13	3.59	5.22
3	65.96	67.84	65.39	66.04	2.90	9.62
4	61.76	63.66	65.00	66.19	50.38	50.27
5	63.20	64.87	55.20	64.22	10.25	42.03
6	63.56	64.82	59.38	64.32	7.72	14.94
7	61.66	61.88	61.18	62.16	7.46	17.94
Average	62.63	63.79	60.02	64.89	14.96	24.56
Dev	2.79	2.84	7.38	2.03	16.91	17.10

Notes: M: Morning E: Evening : CH4 Peak : CH4 Min

According to table 1, the pipe 1 and 3 are still produce CH4 gas if the depth is 3 meters in the morning and evening. While, the pipe 2 can produce CH4 gas when at the depth 1.5 and 3 meters. It caould be happen because the decomposition process of organic material by microbes still occurs normaly at the area of pipe 2, while the other areas can not produce CH4 sufficiently, because this waste landfill has been already over 20 years (not active zone).

Table 2 shows comparison data according the depth of each pipes in newer waste landfill zone. Here, CH4 gas procuction of pipe 1 and 2 are higher than that of pipe 3, so it has potential to harvest natural gas resources in this area. Meanwhile, pipe 3 there are a different level for observation point. This is influenced by age factor of waste dump. The highest CH4 gas content is at pipe 1 with a heap ages of ± 3 month.

5. Conclusions

Based on the aim of this paper, we conclude that the developed system works properly to monitor, record and give the information about CH4 gas distribution in Klotok's landfill. The proposed system has indicates the newer area of waste landfill has more potential of CH4 Gas than the older area. Hence, this paper is expected to be a initial research for mapping the CH4 gas in other area, considering the potential of natural gas is needed to be optimized.

6. Acknowlegment

We want to express out gratitude to Ministry Technology & Higher Education, which supported funds for this project.

7. References

[1] Alvionitasari R 2016 Indonesia Produce 64 Million Tons of Waste Annually: Ministry (Available from: http://en.tempo.co/read/news/2016/02/21/206746875/Indonesia-Produce-64-Million-Tons-of-Waste-Annually-Ministry)

[2] Kardono 2007 Integrated Solid Waste Management in Indonesia Proc. of International Symposium on EcoTopia Science 2007 (Nagoya, Japan, 23-25 2007) pp. 629-633

[3] Statistic Indonesia 2017 Percentage of Households by Province and Sorting Garbage Treatment, 2013-2014 (https://www.bps.go.id/linkTabelStatis/view/id/1360)

[4] Indonesia 2003 Jakarta Solid Waste Management System Improvent Project (https://www.jica.go.jp/english/our_work/evaluation/oda_loan/post/2003/pdf/2-14_full.pdf)

[5] Singh S, Anunay G, Rohit G, Shivangi G, Vipul V 2016 Greenhouse Gas Emissions from Landfills: A Case of NCT of Delhi, India *J Climatol Weather Forecast.* **4(1)** pp1–6.

[6] Maheswari R, Gupta S, Das K 2015 Impact of Landfill Waste on Health: An Overview. IOSR *Journal of Environmental Science, Toxicollogy and Food Technology.* **1(4)** pp17–23

[7] Ahmad Md I, Godara N, Tanweer S M 2015 Implementation and Optimizing Methane Content in Biogas for Production of Electricity, *International Journal of Engineering Research & Technology (IJERT).* **4** Issue 06 pp1326-1329

[8] Gonzalez C, Buenrostro O, Marquez L, Hernadez C, Moreno E, Robles F 2011 Effect of Solid Wastes Composition and Confinement Time on Methane Production in a Dump, *Journal of Environmental Protection.* **2** pp 1310-16

[9] Young G K, Centano A M 2009 Continuous Monitoring of Structures for Landfill Gas Intrusion (Available from: http://www.calrecycle.ca.gov/publications/Documents/Facilities/2009014.pdf)

[10] Sendari S, Lestari D, Rahmawati Y, Prabowo A 2016 Developing prototype of Web-Based Home Controlling for Teacher Trainer, *Proc of IEEE Region 10 Sysmposium TENSYMP 2016*

[11] Blanchard B S, Fabrycky W 2006 Systems Engineering and Analysis (4th Edition) (New Jersey: Prentice Hall)

[12] Sendari S, Rahmawati Y 2014 Instrumentation of Carbon Monoxide to Identify Traffic Jams *Proc. Of Conference on Applied Electromagnetic Technology (AEMT)* pp 183-188

IOP Conf. Series: Earth and Environmental Science **105** (2017) 012079 doi:10.1088/1755-1315/105/1/012079

Developing a simulator of renewable energy as a learning media of energy conversion

Y Rahmawati[1], A N Afandi[1], D Arengga[1], S Sendari[1], W Agustin[1], T Matsumoto[2] and I Rahman[2]

[1]Department of Electrical Engineering, Faculty of Engineering, Universitas Negeri Malang, Jalan Semarang No. 5, Malang, Jawa Timur, Indonesia
[2]Department of Live and Environment Engineering, Faculty of Environment, the University of Kitakyushu, 1-1 Hibikino, Wakamatsu, Kitakyusu, Fukuoka, 808-0135

E-mail : yrahmwati72@gmail.com; siti.sendari.ft@um.ac.id

Abstract. A renewable energy is a new topic in education in Indonesia, especially for vocational education. The problem is the teaching trainer as a learning media is still not available at Universitas Negeri Malang while developing a trainer is expensive. This paper introduces a renewable-energy simulator, which can show the process of converting energy using hybrid solar cell and wind power systems. In case of the solar cell system, the simulator shows the process of converting the sunlight beam to energy production. Furthermore, the simulator shows the watt peak (WP) of the daily solar beam. In the case of the wind-power system, the simulator shows the capacity of power generator considering the size of rotor, wind speed, and the type of generator. The unique point of this simulator is that the hybrid systems of solar cell and wind power systems are demonstrated. While solar cell can't be effective if the sunlight beam is not available, it can be supported by the wind-power system, which is available for 24 hours, but it is depending on the speed of wind. Thus, implementation of this simulator can help students easier to understand and optimise the development of power generation using renewable energy.

1. Introduction

Nowadays, the use of new and renewable energy is a major issue in many countries, the purpose is to maintain of sustainable energy. One of the country who has an enormous potential for renewable energy is Indonesia. Indonesia has a large new and renewable energy potential, which includes 450 MW of mini/micro hydro power, 50 GW of Biomass, 4,80 KWh/m2/day solar power, 3-6 M/sec wind power, and 3 GW of nuclear energy [1]. This new and renewable energy potential is explained by the Director General of Electricity and Energy Conservation at the Focus Group Discussion on the Supply and Demand of New and Renewable Resources held by Pusdatin ESDM [2]. Indonesia's government have been prioritizing the development of the renewable energy sector to increase the low supply and to meet the growth in consumption of electrical energy. The government's ambitious plan to raise renewable share to 23-25 percent by 2025 from just 6 percent now would expand access to electricity, support industrial growth, and support the country's less developed regions [3].

[1] To whom any correspondence should be addressed

Education also has a crucial role in the development of the renewable energy industry [4]. Renewable energy is seen by many as part of the appropriate response to these concerns and some national Governments have put programs in place to support the wider use of sustainable energy systems in education vocational education [5]. The education purposes are to prepare engineers, scientist, and energy planners to research and produce sustainable energy generation systems. The problem is that the teaching trainer as a learning media is still not available at Universitas Negeri Malang while developing a trainer is expensive. The aim of this paper is to introduce a renewable-energy simulator, which can show the process of converting energy using hybrid solar cell and wind power systems. A renewable hybrid energy system consists of two or more energy sources [6,7]. The Simulator is capable to simulate and demonstrate the development of renewable energy sourced to the solar and wind power. Solar PV has the highest resource potential in Indonesia at 532.6 GW, explained by the high solar resource across Indonesia, which is in the range of 4.0-6.9 kWh/m2 and averages 4.8 kWh/m2 [8]. However, the solar power system will not able to produce the electricity when there is no solar source for example at the time of evening. The unique point of this simulator is the hybrid systems of the solar cell and wind power systems can be demonstrated. While solar cell can't be effective if the sunlight beam is not available, it can be supported by the wind-power system, which is available for 24 hours, but it is depending on the speed of the wind and another condition.

In the case of the solar cell system, the simulator shows the process of converting the sunlight beam to energy production. Furthermore, the simulator shows the watt peak (WP) of the daily solar beam. In the case of the wind-power system, the simulator shows the capacity of power generator considering the size of the rotor, wind speed, and the type of generator. Simulator usage will help the user easier to understand the deference result of each type of power plant generation based on environment condition input by user. This will construct the user comprehension by simulator demonstration, thus the user can do observation without using the real equipment.

2. Simulation of Renewable Energy Development

In this section, the simulator of renewable energy is introduced. The simulator is developed using ADDIE model consists of analysing, designing, developing, implementing and evaluating [9]. The simulator has the ability to demonstrate and calculate power result of development for three kind of power generation i.e. 1) solar power generation, 2) wind power generation, and 3) hybrid power generation. Solar power generation simulator is developed to show how solar panels harvest light then converting it into electricity [10]. Wind power generation simulator is developed to show how the wind turns the blades, which spin a shaft, which connects to a generator and makes electricity [11]. Hence, hybrid power generation simulator is developed to show the combination of solar and wind power generation. The detailed simulator device is described as follows.

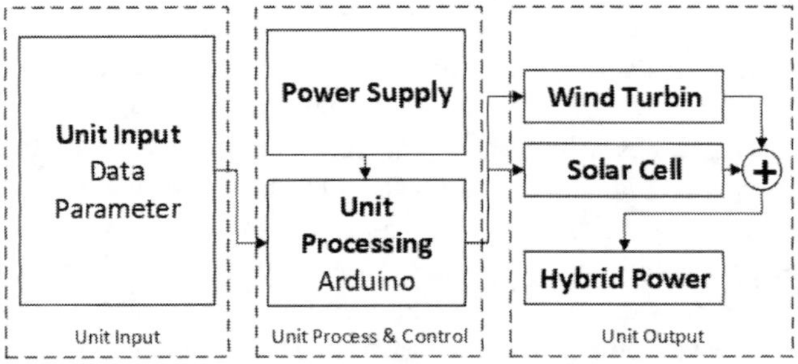

Figure 1. Block Diagram of Simulator Device

2.1. Simulator Device

Simulator device is built in a box of 80 mm height and width 120 x 40 mm. Here, the simulator device is controlled by arduino as unit processing and controlling. The simulator requires some input parameter by user indicate environment condition to start the simulation of development power generation. Solar and wind power generation have different input of variable depending on power resources to result the electricity. Solar power generation require three kind of input variable i.e., solar intensity (lux), solar power (Wp), and number of solar panel while wind power generation need the following input of variable i.e. density (Kg), velocity (m/s), and length of turbine (m). This input is used to calculate the power generation considering kind of power generation. The power result is shown in display monitor while the demonstration is shown till the electricity distribution process into costumer. The block diagram of detailed simulator device is shown in figure 1.

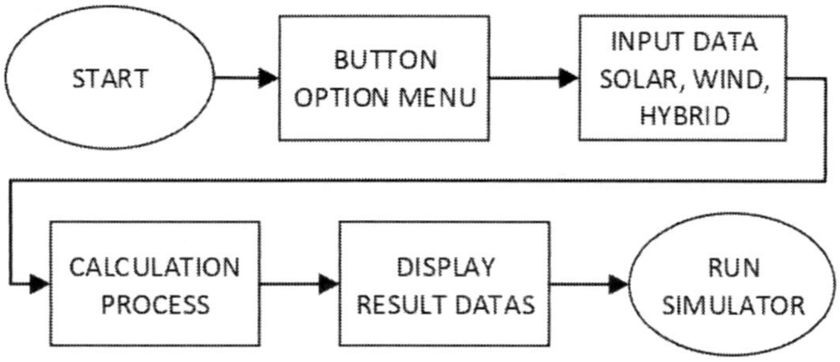

Figure 2. Flowchart of Simulator Device

Based on the simulator demonstration user are capable to analyse the result using hybrid power generation and stand-alone power generation. This simulator is designed to show the real condition of power plant development result based on the power generation usage. Figure 2 shows the flowchart of simulator device usage. Firstly, the user has to choose the type of power plant generator. Then input the variable of parameter needed by the chosen power plant generator based on the system guide. If the user chooses the solar power generation, the hardware simulator will not activate the wind power generation and also the reverse of it. The hardware simulator will show the power output and demonstrate the real condition of the actual system, display the environment effect and the electrical result transmitted to the consumer. Here, figure 3 shows the specification of the simulator device.

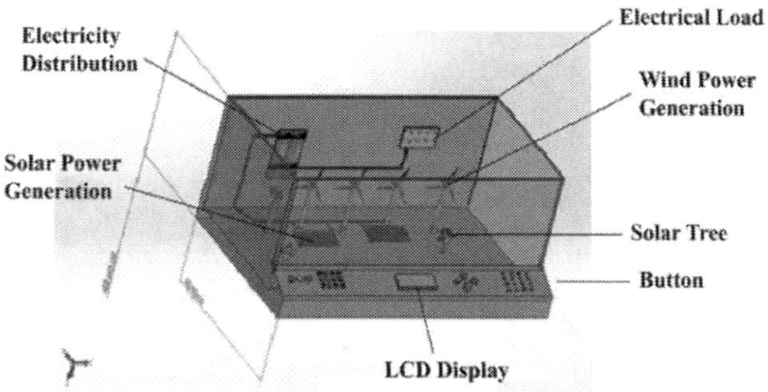

Figure 3. Specification of Simulator Device

2.2. Implementation
This learning media will be implemented into subject of new and renewable energy (code NTEL649) Department of Electrical Engineering, Faculty of Engineering, Universitas Negeri Malang. The Simulator is designed for the following basic competence i.e. 1) to clarify the working principles of new and renewable energy conservation in terms of technical, economic, financial, socio-cultural and environmental aspects and with a description of competence; 2) clarify the principle of solar energy conversion; 3) clarify the wind energy conversion principle; 4) clarify the working principle of a

hybrid generating system; and 5) predict the prospects for the provision of renewable energy in a decentralized and sustainable manner.

As shown in figure 4, the concept using simulator device will construct the students understanding to analyse environment potential before developing the power generation of renewable energy. Analysis result are capable to demonstrate using the simulator device. Thus, students have experience to study the effect and result of power generation process using miniatur equipment. Based on the result, the students will have the ability to optimise the power generation development considering potential environment and applying the energy management.

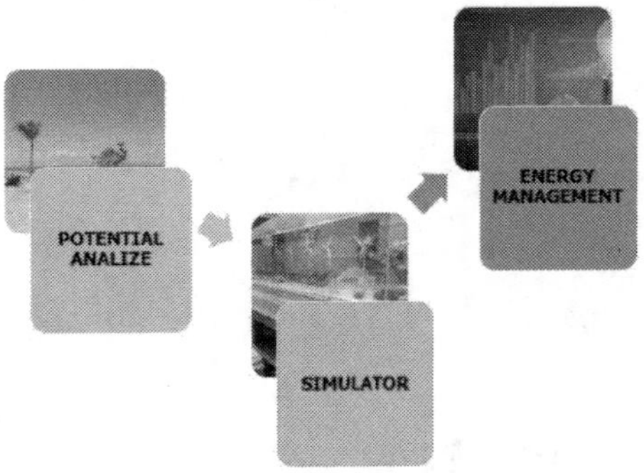

Figure 4. Learning concept using simulator device

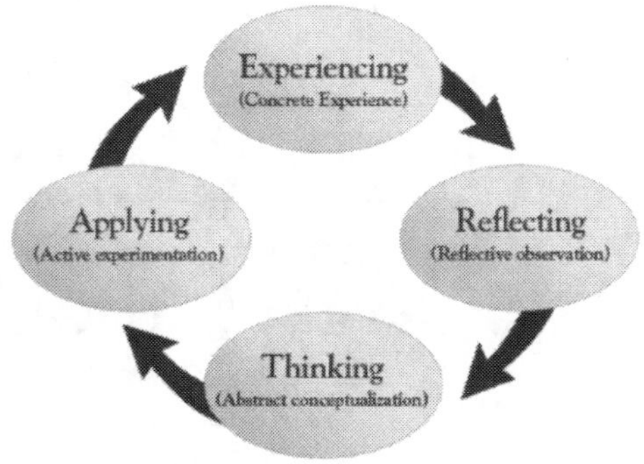

Figure 5. Kolb's Experiential Learning Cycle

Figure 5 shows the implementation of learning media use Kolb's Experiental Learning [12]. Simulator implementation on the learning process allows the students to get an experience to develop power generation of renewable energy. Thus the students can observe the results obtained by simulator. It has the ability to construct the student's thinking to plan the development of power generation considering the environmental conditions to produce optimal power.

Based on the implementation of the simulator which used as learning media of energy conservation, this simulator is equipped with job sheet as the guide the user when using the simulator. The jobsheet consists of 1) purpose and competence of learning, 2) theoretical concept of solar, wind, hybrid power generation, 3) job steps description, 4) analysis sheets, 5) questions of evaluation, and 6) simulator maintenance and repair guide.

3. Simulator of Renewable Energy Utilization

This section shows the result of power generation by the simulation process considering the parameter input manually inserted by user.

3.1. Solar Power Generation

Indonesia as an equator area has the maximum value of Sun's light i.e. 98000-116200 lux [13]. Based on the output power results of the measurement at maximum power solar panel from a voltage and electric current is proportional to the intensity of sunlight and temperature on the solar panels on the solar panels.

$$P_{Solar} = \frac{solar\_absorbed}{\max solar\_Intensity \cdot WpSolar} \tag{1}$$

The formula used in the simulator is taken from the value of the middle light solar maximum i.e. 107000 lux, so the equation obtained by the formula (1). Figure 6 and 7 shows the input and output process of solar power generation. Parameter usage of this attempt are 100 Wp of P max and 8920 for solar intensity. Based on the input, the calculation result is 12,14 (Watt).

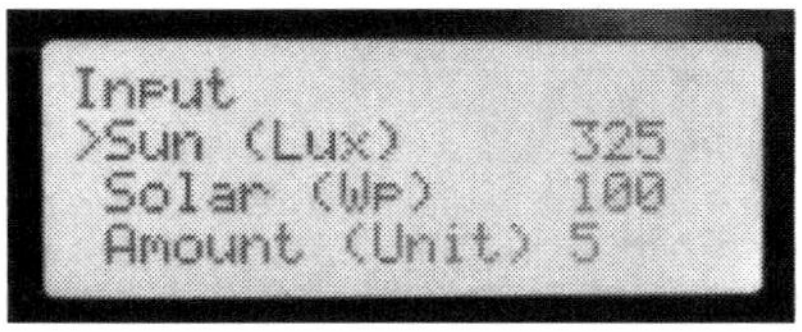

Figure 6. Solar power generation input

Figure 7. Solar power generation output

3.2. Wind Power generation

The wind power generation with horizontal axis used formulas (2). P (Watt) is power output, ρ (Kg/m^3) is power density, v (m/s) is velocity, and A (m) is the length of turbine. Figure 8 and 9 shows the input and output process of wind power generation.

$$P_{wind} = \frac{1}{2} \cdot \rho \cdot v^3 \cdot A \tag{2}$$

Parameter usage of this attemp are 5 Kg/m^3 for density, Wp 20 m/s for velocity, and 5 m^2 for length of turbine. Based on the input, the power calculation result is 1696 (Watt).

Figure 8. Wind power generation input

Figure 9. Wind power generation output

3.3. Wind Power generation

The hybrid power generation is generated based on the previous solar and wind power generation by sum the solar and wind power result as shown in formula (3). It combines the solar and wind power generation at the same time. The result of hybrid power generation is 12,14 + 1696 = 1708,14 (Watt) as shown in Figure 10.

$$P_{hybrid} = P_{solar} + P_{wind} \tag{3}$$

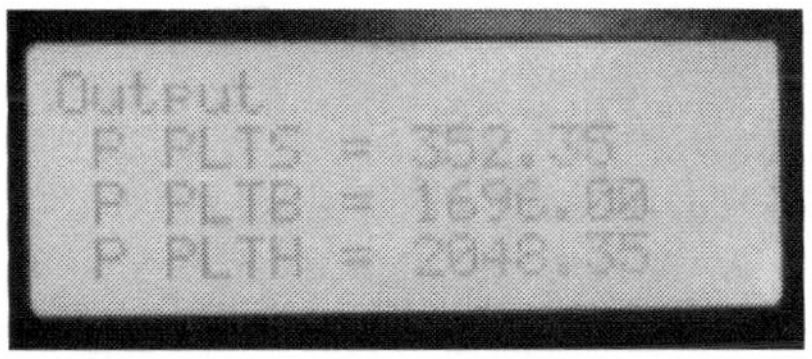

Figure 10. Hybrid power generation output

4. Conclusions

In this paper, a simulator device has developed using ADDIE method. Based on the evaluation process, the simulator has ability to demonstrate a power generation development into a miniature equipment. Simulator implementation are capable to minimize the time, avoid an accident, and reduce the expense during the education process. The presence of this simulator gives a basic experience for the students before apply it into the real development of power generation. The advantages of the simulator are spending low cost, modelling the real condition by insert the variable of parameter influence the power generation. The testing result shows that the simulator work properly. The implementation as a learning media has been confirmed by the expert judgement.

5. Acknowledgments

We want to express our gratitude to Ministry Technology & Higher Education, which supported funds for this project.

6. References

[1] Anon 2015 *Indonesia Energy Policy, Laws and Regulation Handbook Volume 1 Strategic Information and Basic Laws* (Washington DC: International Business Publications)

[2] Indonesia's Renewable Energy Potential 2008 (Availbale from: http://www.esdm.go.id/news-archives/general/49-general/1963-indonesias-renewable-energy-potential.html)

[3] Susantono B 2015 A Renewable Future for Indonesia (Available from: http://www.thejakartapost.com/news/2015/11/16/a-renewable-future-indonesia.html)

[4] Jennings P J 1997 Renewable Energy Education: an Essential Foundation for Market Proceedings of the 35th ANZSES Conference Canberra pp 33-35

[5] Jennings P J 2009 New Directions in Renewable Energy Education (Available from: https://www.researchgate.net/publication/222524310_New_directions_in_renewable_energy_education)

[6] Sendari S, Lestari D, Rahmawati D and Prabowo A 2016 Developing Prototype of Web-based Home Controlling for Teacher Training. *Proc. of IEEE Region 10 Symposium, Tensymp* pp 99-103.

[7] Pragya N, Nemab R K, and Rangnekara Saroj 2009 A Current and Future State of Art Development of Hybrid Energy System Using Wind and PV-Solar: A Review *Renewable and Sustainable Energy Reviews* **13** pp 2096-2103

[8] Veldhuis A J and Reinders A H M E 2013 Reviewing the Potential and Cost-Effectivess of Grid-Connected Solar PV in Indonesia On A Provincial Level *Renewable and Sustainable Energy Reviews* **27** pp 315–324

[9] Branch R M 2009 Instructional Design: The ADDIE Approach (USA: Springer)

[10] Baguley R and McDonald C 2016 Appliance Science: How solar panels convert light into electricity (Available from: https://www.cnet.com/news/appliance-science-how-solar-panels-convert-light-into-electricity/)

[11] Energy Efficiency and Renewable Energy 2016 How Do Wind Turbines Work? (Available from: https://energy.gov/eere/wind/how-do-wind-turbines-work)

[12] Kolb D A 1984 *Experiental Learning Experience as a Source of Learning and Development* (New Jersey: Prentice Hall)

[13] Asy'ari H, Jatmiko, and Angga 2012 Intensitas Cahaya Matahari terhadap Daya Keluaran Panel Sel Surya Simposium Nasional RAPI XI FT UMS pp 52-57

IOP Conf. Series: Earth and Environmental Science **105** (2017) 012080 doi:10.1088/1755-1315/105/1/012080

Tropical X-Y transformation design for building courtyard integrated photovoltaic

FX Teddy Badai Samodra

Department of Architecture, Institut Teknologi Sepuluh Nopember, Surabaya, Indonesia

*Email: fxteddybs@arch.its.ac.id

Abstract. The tropical region has a richness of solar radiation with its consequences on building thermal comfort. For that exception, the photovoltaic hold a role in generating eco-friendly energy also. Two-dimensional geographical location, recognized as X-Y for translating latitude (X) and altitude (Y) are analyzed for determining the optimum integrated thermal comfort of building and energy generating. This research utilizes open space, courtyard, which not only provides a slimming body of the tropical building to reduce deepness of cross ventilation work but also develops potential green energy by a simultaneous method in making opening area. All were done by *Ansys Fluent* and *Ecotect simulation* method in order to give a recommendation for each location. The result highlighted that higher width and length different ratio for East-West orientation courtyard, more energy can be generated. It has a capability in reducing more than 10% of energy consumption with maintaining indoor temperature by providing air. Comparing latitude (X), altitude location (Y) shows the higher effect on providing thermal comfort and energy efficiency because the different tropical location has less difference on solar distribution.

1. Introduction

The abundant solar radiation in the tropical region has consequences on building thermal comfort. For that exception, the photovoltaic hold a role in generating eco-friendly energy. Building surfaces can then be varied towards these sensitive orientations to generate optimized alternatives. Surfaces of optimized alternatives can have different photovoltaic (PV) integration options [1].

The well-designed courtyards may represent a valid option for building. The definition for optimum courtyard ratio should which allows the form to receive minimum solar radiation for the tropical regions [2]. The solar radiation is directly proportional to the atmospheric temperature while it is inversely proportional to the relative humidity. Moreover, wind speed has little influence on solar radiation [3].

Integrated PV increases the building's indoor air temperature by about 4°C, when compared to a building of the same size without PV integrated [4]. Therefore, the embedded closed ventilated courtyards achieve indoor thermal comfort and avoid excessive humidity in hot-humid climates and it needs a staggered form courtyard with V-shaped roofs [5]. This roof model could be integrated for PV. In this research, two-dimensional geographical location, recognized as X-Y for translating latitude (X) and altitude (Y) are analyzed for determining the optimum integrated thermal comfort of building and energy generating.

IOP Conf. Series: Earth and Environmental Science **105** (2017) 012080 doi:10.1088/1755-1315/105/1/012080

2. Method

This research utilizes open space, courtyard, which not only provides a slimming body of the tropical building to reduce the deepness of cross ventilation work but also develops potential green energy by a simultaneous method in making opening area. All were done by *Ansys Fluent* and *Ecotect* simulation method in order to give a recommendation for each location.

The lowland locations of the research are Singapore, Bangkok (Thailand), and Hanoi (Vietnam) as representative of 0-8°, 8-16°, 16-23.5°, respectively. For the highland tropical area, Bogota (Colombia), Santa Rosa de Copan (Honduras), and Mexico City (Mexico) are selected. Three difference tropical latitude locations are analyzed by <100 m above mean sea level (MSL) and more than 1000 m MLS for altitude (see figure 1). The X-Y locations indicate solar radiation distribution as consequences for X latitude difference (figure 2). The Y, altitude, determines tropical climate elements changes, with temperature as the most influence indicator. However, in this study, solar radiation will be the focus of discussion related to the photovoltaic potency.

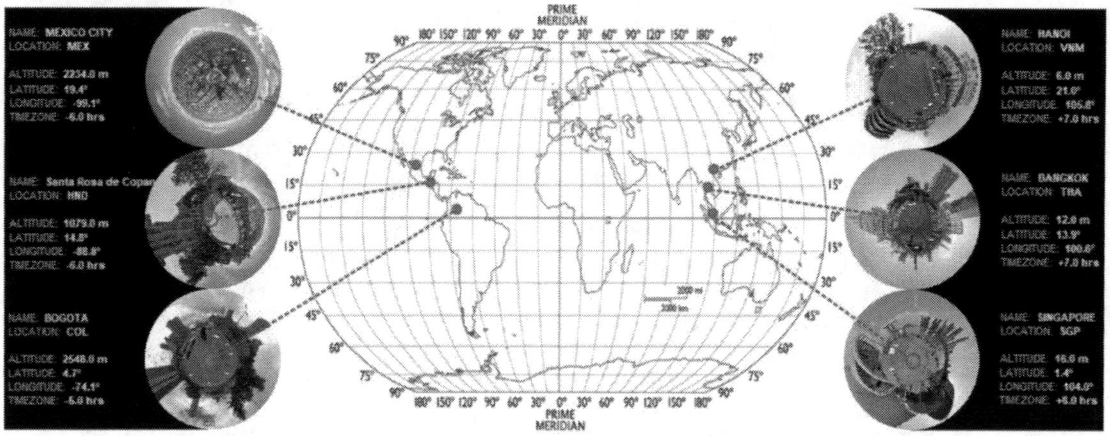

Figure 1. Research locations

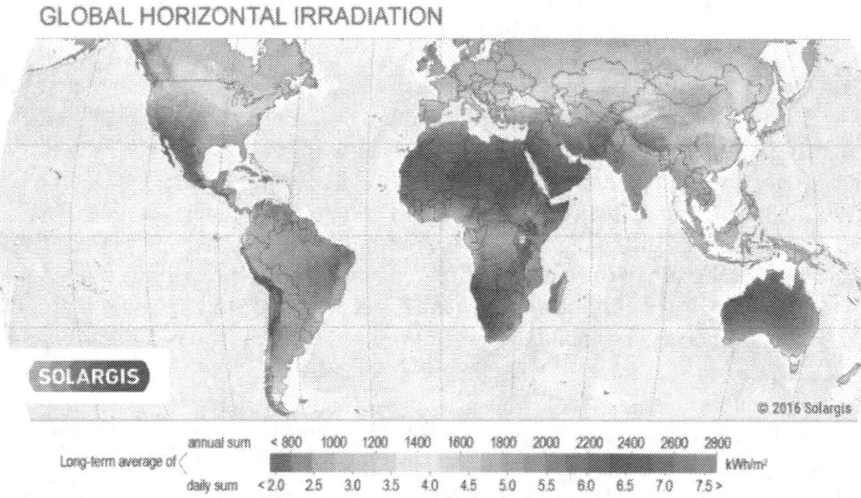

Figure 2. Global irradiance distribution
(source: www.solargis.com)

3. Results and Discussions

3.1. Courtyard Ratio

Tropical climate, both in north and south hemispheres receives solar penetration in various condition. In general, the lower latitude position gets a higher impact on its environment, hotter condition-higher Mean Radiant Temperature (MRT) even it is very depended on the micro or site condition of the courtyards.

For providing physiological cooling, wind acceleration should be designed for thermal comfort. Figure 3. evaluates air movement simulation using *Ansys Fluent Program* for critical courtyard (1:1 ratio for building width to height/length). Meanwhile, as shown in figure 4., the *Ecotect analysis* compares the various MRT resulted by the various ratios. It illustrates that the lowest ratio, 1:1, has potential in reducing heat accumulation into the building and increasing solar radiation for generating eco-friendly energy all at once. Therefore, this model is selected for PV integrated placement.

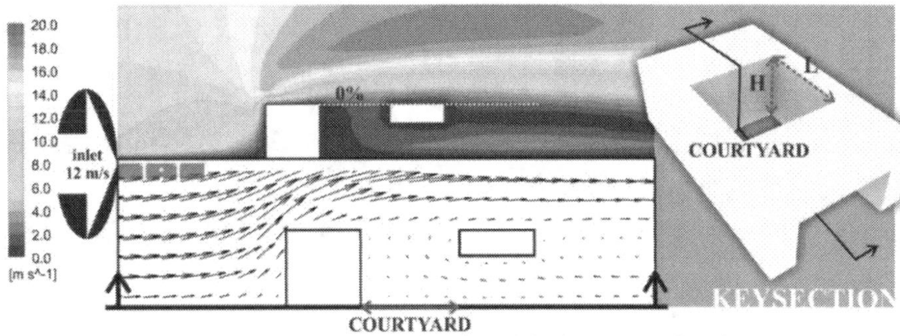

Figure 3. Air velocity for critical courtyard patio

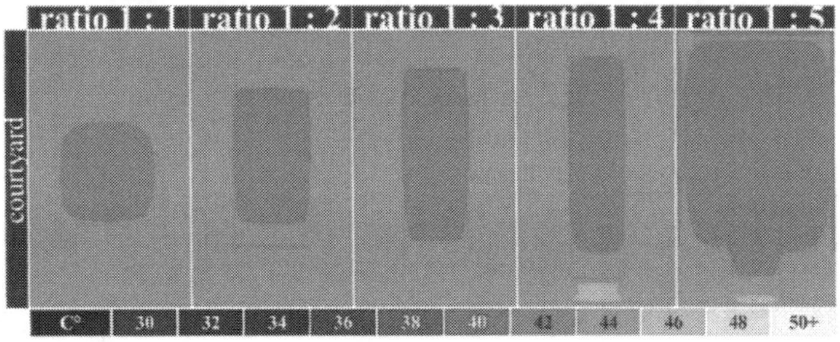

Figure 4. Mean Radiant Temperature of courtyard by different ratios

3.2. Total Radiation on X-Y Transformations

This section discusses the contribution of the surface (indicating for PV tilts) in various X and Y locations. As described by Figure 5., the results show that the 30° and 45° are the worst PV tilts where 90° reaches as the best. 90° means the perpendicular angle of PV surface from the incoming solar radiation or 0° for elevation. However, this angle should be avoided because it reduces the courtyard space. Thus, 0° and 60° could be decided as the most effective PV tilt based on technical and architectural reasons.

Discussing the difference between lowland and highland of total radiation, because of the closer distance to the main source of solar energy, highland should be the potential place in receiving solar radiation. However, in actual condition, for low, mid, high tropical location, it could not indicate the significant trend for both X and Y locations. It could be affected by site condition or local climate. In general, the X-Y transformation does not work for differing the recommendation of building courtyard

IOP Conf. Series: Earth and Environmental Science **105** (2017) 012080 doi:10.1088/1755-1315/105/1/012080

integrated PV (BCIPv). As research finding by Burg et al [6], the photovoltaic performance is more determined by global albedo impact, regional atmospheric heat islands, and locally heated surfaces. It also applied for the orientation variable. With the similar distribution, East (90°) and West (270°) have the same performance. Northern or Southern hemispheres do not affect on solar radiation distribution difference. North (0°) and South (180°) orientations could not imply the solar power source variation. For a while, it concludes that divergence on latitude for the tropical region, significant climate element dissimilarity, altitude, and building or courtyard orientation have no relationships for tropical regenerating energy.

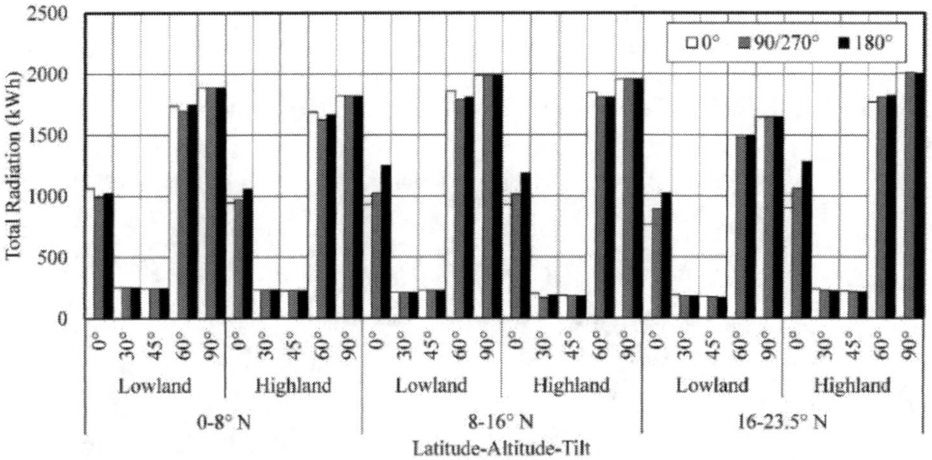

Figure 5. Total radiation on X and Y locations for various surface tilts

The optimum BCIPv produces potential solar energy by receiving in any location criteria but it could be deceiving trend that lowland more fluctuated result than highland (Figure 6. and 7.). In mid tropical latitude, it receives the highest when 16-23° is the most potential highland area. The problem of the courtyard as PV location is that inner shading occurs but it has disadvantages on utilizing the integrated inner wall or roof for beneficial limiting space for accelerating the wind. The demerits of the potential shading for thermal comfort could be anticipated by its average of total radiation for PV. With more 1000 kWh annual received solar radiation, the BCIPv ensures the eco-friendly power equipment and the problem of shading means nothing,

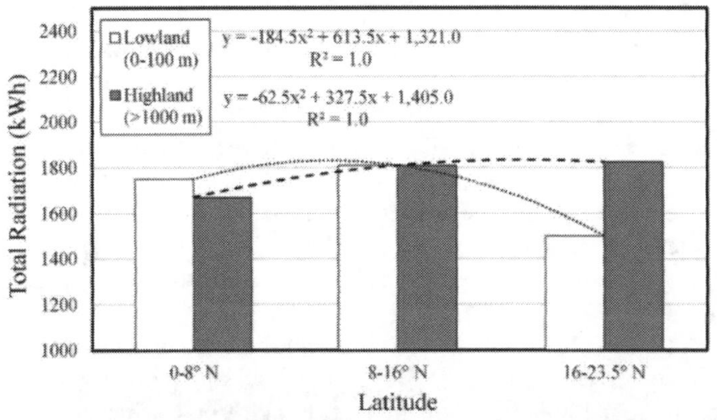

Figure 6. Total radiation for 60° of surface tilt

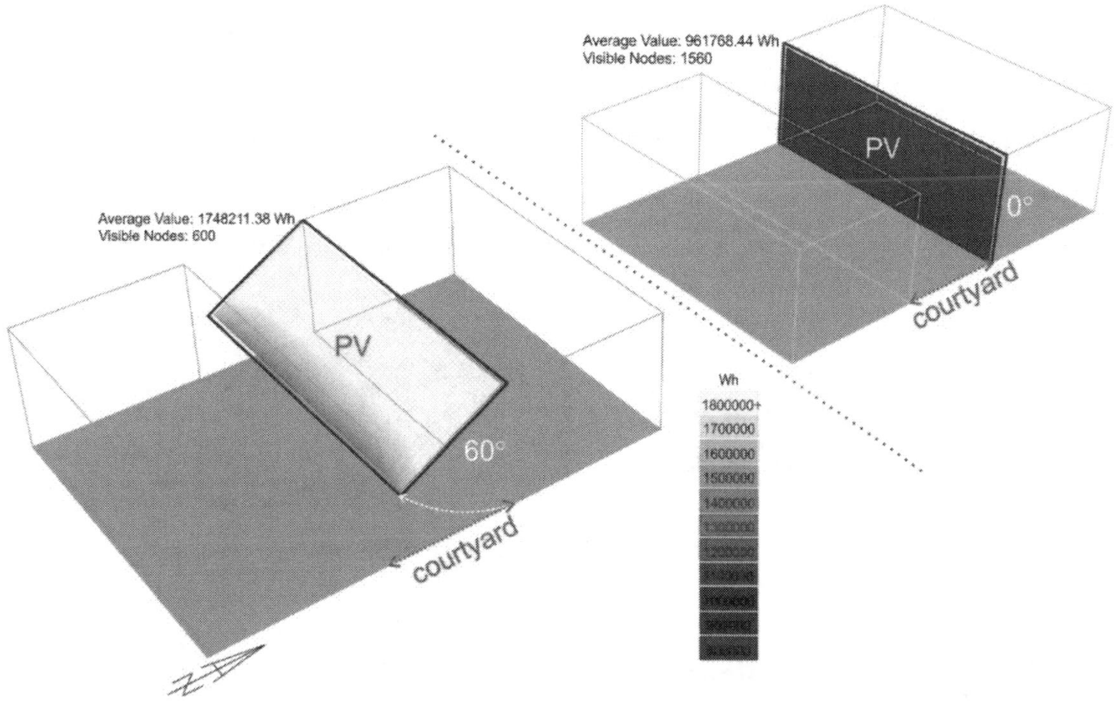

Figure 7. Optimum tilts for PV

3.3. Generating Solar Energy

The stereographic diagram in Figure 8. would be the next assurance for optimism of producing energy. In fact, the shadow happens in limited time in the morning or evening when solar energy has less effective than in the noon. As a result, Equation 1. proves the potential producing energy by BCIPv. In the case of 24 m² of panel area, 15% of solar panel yield, 0.75 PR and variation of 1750 kWh, 1810 kWh, and 1501 kWh of annual average irradiation on 60 tilted panels result in 4722 kWh, 4884 kWh, and 4050 Energy annually for 0-8°, 8-16°, and 16-23.5 location, respectively. Comparing to 8845 kWh annual Electric power consumption (kWh per capita) for Singapore [6], it has a capability in substituting more than 50%. When the declination angle is decreased, the result of energy rating difference increases [7]. In general, shading effects have an influence on the seasonal variation of produced power as much as ambient temperature, building orientation, and PV module's inclination angle. However, the tropical locations result in constant potential in generating energy for all orientations.

$$E = A \cdot r \cdot H \cdot PR \tag{1}$$

where:

E = Energy (kWh)
A = Total solar panel Area (m²)
r = solar panel yield (%)
H = Annual average irradiation on tilted panels (shadings not included)
PR = Performance ratio, coefficient for losses (range between 0.9 and 0.5, default value = 0.75)

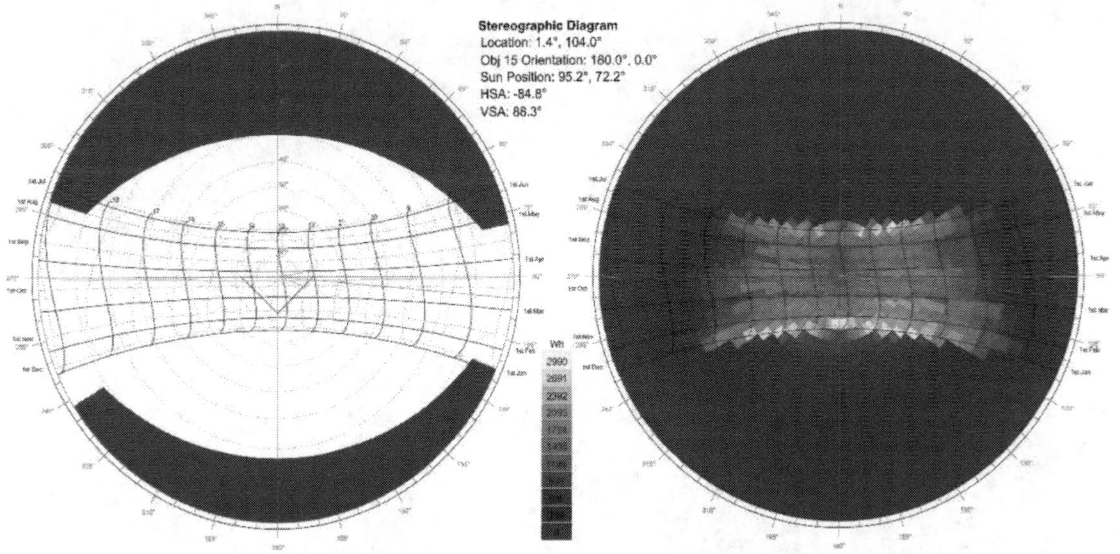

Figure 8. Stereographic diagram of 60° of PV tilt

4. Conclusions

Regarding providing better thermal comfort, the result highlighted that higher width and length different ratio for East-West orientation courtyard, more energy can be generated. In addition to Eicker et al [7] that the solar electric has a capability in reducing primary energy consumption by 21–70%, depending on the location, this study found more than 50% efficiency of energy consumption with maintaining indoor temperature by providing air movement. Comparing latitude (X), altitude location (Y) shows the higher effect on providing thermal comfort and energy efficiency because the different tropical location has less difference on solar distribution. In line with Karthick et al [8], the integrated modules should reduce the cooling load by minimizing the heat gain by the room around the courtyard.

5. Acknowledgments

This research is part of *Penelitian Doktor Baru* No. 881/PKS/ITS/2017 funded by *Dana Non PNBP ITS* 2017. The authors gratefully acknowledge this financial and technical supports.

6. References

[1] Youssef AMA, Zhai Z, Reffat RM 2015 *Building Simulation* **8** 353–366.
[2] Muhaisen AS 2006 *Building and Environment* **41** 1731-1741.
[3] Yazdani MG, Salam MA, Rahman QM 2014 *International Journal of Sustainable Energy* 1-14.
[4] Akata MAE, Njomo D, Mempouo B 2015 *Future Cities and Environment* **1**(1) 1-10.
[5] Kubota T, Zakaria MA, Abe S, Toe DHC 2017 *Building and Environment* **112** 115-131
[6] Burg BR, Ruch P, Paredes S, Michel B 2017 *Solar Energy* **147** 399-405.
[7] Eke R, Demircan C 2015 *Solar Energy* **122** 48-57.
[8] The World Bank Data, 1964-2014, Electric power consumption.
[9] Eicker U, Pietruschka D, Schmitt A, Haag M 2015 *Solar Energy* **118** 243–255.
[10] Karthick A, Murugavel KK, Kalaivani L, Babu US 2017 *Advances in Building Energy Research* 1-17.

IOP Conf. Series: Earth and Environmental Science **105** (2017) 012081 doi:10.1088/1755-1315/105/1/012081

Review: Home energy management system in a Smart Grid scheme to improve reliability of power systems

Hartono BS, Sri Paryanto Mursid and Sapto Prajogo

Department of Energy Conversion Engineering Politeknik Negeri Bandung Indonesia

Email: hartono@esi-labs.com, sp_mursid@yahoo.co.id, saptoprajogo@gmail.com

Abstract Home energy management system (HEMS) concept rises from the development of smart homes that build interaction between users with their home appliances in order to operate automatically, multi-functionally, adaptably and efficiently. In line with technological developments and published regulations related to environmental issues, smart home applications evolve into HEMS applications which are not only to provide ease and convenience, but also to monitor and to make efficient energy use at home, thereby reducing peak power quantity and electricity bill. Smart grid is an intelligent power grid starting from its generation, transmission and distribution. It combines computing technology, artificial intelligence and communications technology which creates a smarter power system and is able to produce better power quality and lower generation cost. In the smart grid scheme, by means of HEMS applications, consumers can participate in improving the quality of power systems. This study will discuss about the development of HEMS in associated with smart grid technology particularly the role of HEMS application with its DSM (Demand Side Managment) and PEV (Plug-in Electric Vehicles) programs in the smart grid scheme to improve quality of power systems. Several studies have shown that the contribution of HEMS to the smart grid system can improve the power losses and voltage profile.

1. Introduction

The concept of home energy management system (HEMS) rises from smart home development which has been developing from 1990 [1]. This application was initially intended to facilitate users in operating and managing some household appliances so that they can operate automatically and optimally. In addition, the users can achieve a more comfortable room or environment condition which suits their preference [2]. In the smart home application, interaction between users with their home appliances will be formed thereby the devices can be operated and programmed interactively either directly or remotely. Therefore, these appliances can operate automatically, multi-functionally, adaptably and efficiently [3]. To be able to do that, smart home application is completed with several technologies like sensors, controllers, communications and programming languages.

In line with the development of technology, the emergence of environmental problems and regulations related to these problems encourage the smart home applications not only to provide comfort and convenience, but also to achieve energy efficiency. Afterwards, smart home applications evolved by developing home energy monitoring capabilities to achieve savings. Based on the results of the study, by monitoring the use of energy, psychologically users will change the behavior of the use of household appliances and it enables to save energy up to 30% [4]. The use of smart meters and AMI (Advanced Metering Infraestruture) assists in observing the use of energy. By providing feedback on the amount of costs that must be spent on energy consumption, users will more consider for their energy use, if they object with it. There are several information feedback techniques related to this smart metering technology described on [4].

In line with its function related to energy management, there appears derivative of smart home application--application of home energy management system (HEMS)--which manages household energy usage, thus reducing peak power and electricity bill. HEMS application is equipped with capability of data collection, data proccesing, data representation, and preparation of interactive operation with user to be able to perform energy management. Therefore, it can provide feedback in real time, daily, weekly or monthly.

The development of HEMS technology can be grouped into several categories: interface technology, smart hardware and software platform [5]. Some studies related to the development of HEMS are conducted from sensor technology [6], Communication devices [7], storage energy device [8], smart meter [4], smart home appliances [5] and the HEMS application itself. In brief, the HEMS development map is described on [9-11].

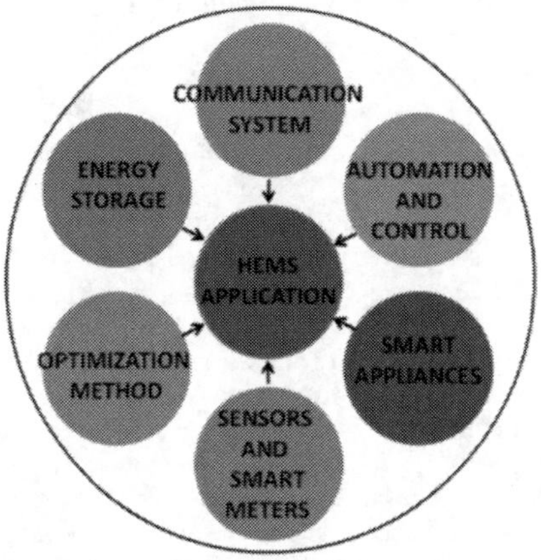

Figure 1. HEMS Technology Development

The increasing demand for electrical energy encourages the development of electric power technology, especially in large cities where the rapid population growth as well as the growth of the industry. The Increasing capacity demand of electrical energy, leading to the building of an electric power distribution becomes more complicated thus distributed generation as an alternative generation. Smart grid is a smart electrical network starting from the generation, transmission and distribution [12].

Smart grid technology combines the power system with computing technology, artificial intelligence and communications which make the power system smarter in order to produce better and cost effective electrical power [13]. One of the capabilities offered by smart grid technology is the ability of smart metering that is able to provide information about the amount of household power demand and self healing technology when there is a problem on the network either due to technical problems, sabotage or natural disasters, so as to increase Network reliability [14].

In the smart grid technology, the concept of generation changes from centralized to distributed and made it possible for consumers to become producers. With the support of computing and communications technology, consumers are allowed to participate/contribute in power system operation schemes through HEMS applications, as one of the entities of the smart grid system at the consumer level. In smart grid schemes, through HEMS applications, consumers can participate in improving the reliability of power systems. The presence of electric vehicle in the smart grid scheme connected in the HEMS application can participate effectively in helping to balance supply and demand by valley filling and peak shaving household loads [15].

This paper will review the development of HEMS and its relation with smart grid technology, particularly the role of HEMS in the smart grid scheme to improve the reliability of power systems.

This review is structured as follows. In Section II, we present HEMS position in smart grid system. In Section III, we review the power system infrastructure in smart grid concept, and last Section reviews the consumer contribution in improving power system reliability.

2. HEMS in Smart Grid

In the framework of smart grid, HEMS application is developed not only to manage energy use in households but also to conduct the management of energy supply, either from the provider of electrical energy or the own-generated one such as alternative energy sources through solar or wind power plant. The HEMS configuration consists of household loads that are divided into scheduled and unscheduled loads, energy storage, alternative energy sources, connections to grid, electric cars and HEMS control systems which is supported by communication technology and smart meter [16].

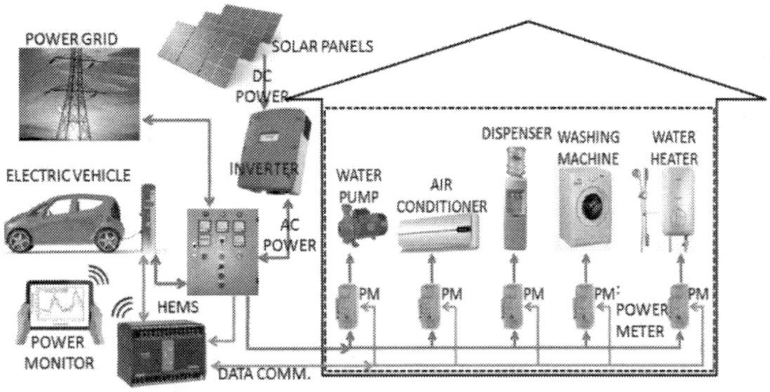

Figure 2. configuration of HEMS system

HEMS application performs managing operating time of the load, especially for scheduled loads so that minimum operating costs of electricity usage will be obtained. In HEMS the output power is generated from an alternative energy source rather than used to supply the household load; therefore, it can also be stored in energy storage and electric cars or supplied to grid. The presence of electric cars is not only intended to promote the use of environmentally friendly vehicles but also intended as energy storage; in addition, the vehicle itself can be used to contribute and maintain the reliability of power generated from the power grid.

In some countries, peak load rates and feed in tariffs are being applied. This led to research aimed at minimizing the costs to be paid for the electricity use. In general, the study related to the development of HEMS can be grouped into: problems of scheduling, automation and control, communication aspect and HEMS system [9]. Research development related to scheduling issues can be divided into modeling and optimization studies. The modeling study on schedule raises the problem of load modeling, pricing scheme, load forecasting and comfort level [17].

Optimization study uses heuristic and non heuristic approaches. Some heuristic algorithms developed for scheduling optimization problems include PSO, GA, taboo search and neural network. While for non heuristic optimization, among others, using LP and NLP [18-20]. The study on communications technology is aimed at discovering appropriate topology and communication media for use in HEMS. Several studies raised the use of sensor networks through wireless technologies as a communication medium such as zigbee technology. The communication architecture in HEMS is called home area network (HAN) used to gather information from each HEMS component beginning from household appliances, energy sources up to energy storage systems [21]. Comparison of some communication technologies on HAN are studied in [22] and some wireless network based on IEEE Standards as shown in table 1 [7].

The development of measurement methods on HEMS is directed to smart measurement. With the development of communications technology and micro-based processing, measurement devices have become smarter. Some studies related to smart measurements, are as follow [23] :

- Remote reading

- Bidirectional communication
- Support of advanced tariff systems and billing applications
- Remote energy supply control.

Table 1. Wireless Network Based on IEEE Standards.

Protocols / Specifications	ZigBee Over IEEE 802.15.4	WirelessHART Over IEEE 802.15.4	MiWi Over IEEE 802.15.4	Isa100.11 Over IEEE 802.15.4	Bluetooth (IEEE 802.15.1)	Wi-Fi IEEE 802.11
ISM Bands	2.4 GHz/915 MHz (USA)/868 MHz (EU)				2.4 GHz	5 GHz
Network Topology	Star, Peer-to-Peer and Mesh	Star, Peer-to-Peer and Mesh	Star, Peer-to-Peer	Star, Peer-to-Peer and Mesh	Star, Peer-to-Peer	Star, Peer-to-Peer
Modulation Scheme	BPSK (868-915 MHz) Q-QPSK (2.4 GHz)	O-QPSK (2.4 GHz)	FSK/OOK	O-QPSK (2.4 GHz)	GFSK/DQPSK 8DPSK (optional)	BPSK, QPSK, COFDM, CCK, M-QAM
Nominal Rate	250 kbps (2.4 GHz) 40 kbps (915 MHz) 20 kbps (868 MHz)				1 Mbps (v1.2/v4.0) 3 Mbps (v2.0) 24 Mbps (v3.0)	11-65-450 (IEEE 802.11 n) Mbps
Range (meters)	10–300	100	20–50	100–200	10	10–100
Application Areas	Demand Response, remote control and automation in residential and commercial buildings	Industrial Control, building control the sensory data conveying temperature, pressure or speed	AMR metering, consumer, electronics, home, automotive, industrial, automation, toys business and medical applications	Industrial and control market	Wireless connectivity between personal devices such as headphones, medical, sport & fitness, mobile phones or laptops	Wireless LAN connectivity, broadband Internet access
Advantages	Low Power consumption, several application profiles (home automation, smart energy) and topology flexibility	Communication Security, reliability and Environment with wired HART infrastructure	Flexible, cost-effective platform	Low energy consumption devices, Robustness in the presence of infrastructure, flexible and communication security	Speed and flexibility	Speed and flexibility

3. Power system infrastructure in smart grid

In a traditional power system, centralized generating stations generate bulk power which is supplied to the consumers through a one-way transmission and distribution system called the grid. The traditional bulk generation isa supplied power and then distributed to the consumers; at the same time, the integration of distributed generation is done at the bulk as well as distribution and customer levels [24]. As defined by the European Regulators' Group for Electricity and Gas (ERGEG), "*a smart grid is an electricity network that can efficiently integrate the behavior and actions of all users connected to it like generators, consumers and those that do both—in order to ensure an economically efficient, sustainable power system with low losses and high levels of quality and security of supply and safety*" [25]. The comparison of the traditional power grid and the smart grid can be seen in table 2 [24].

Today the operation of electric power generation system is required to fulfill the consumers power demand. However, it must consider other aspects like the environment, output power quality and finance. To be able to compete in free electric trade era, generating systems must be able to operate efficiently in order to reduce the cost of generation. To comply with these requirements, the generation system requires generating equipment modernization with the latest technology so it can operate more reliably, efficiently, safely and with better power quality [26].

The power system infrastructure in the smart grid is supported by communication and data processing technology. It allows the generation side to control a more planned power generation due to load forecast to be more measurable, so that the generation can be more cost effective. Overall, power system infrastructure in smart grid can be divided into: Smart infrastructure system, Smart management system and Smart protection system [14].

Table 2. Comparison of Traditional Grid and Smart Grid [24]

Traditional grid	Smart grid
Customer calls when the power goes out	Utility knows when power is out and restores it automatically
Utility meets peak demand	Utility suppresses peak demand, thus lowering the cost
Difficult to manage high wind and solar penetration	Utility can manage distributed energy resources safely
10%+ power loss in T&D	Utility reduces power loss by 2+%, and reduce emissions and electricity bills of customers
All centralized control	Local control provides distributed generation

In the smart grid it is not only the information but also the power which can be exchanged in two ways. Meanwhile in conventional power system of generator system the energy is distributed through transmission system and being distributed to consumers through distribution network. In smart grid power systemconsists of distributed generation. In smart grid scheme consumers can also act as electric producers to produce electric power and supply it to the network.

Smart infrastructure includes the smart energy subsystem, the smart information subsystem, and the smart communication subsystem. Smart energy sub system handles mechanism of generation, distribution and consumption of electrical energy. Smart information subsystem is responsible for handling information exchange related to measurement, monitoring and management of smart grid system operation. Meanwhile, the smart communication subsystem handles communication infrastructure issues for data exchange between smart grid systems, devices and applications associated with smart grid. The smart management system is done by utilizing smart infrastructure facilities to achieve goals related to energy efficiency improvements, supply and demand balance, emissions control, reduced operating costs, and utility maximization. The smart protection system is the subsystem in smart grid that provides advanced grid reliability analysis, failure protection, and security and privacy protection services.

The distributed generation on the smart grid brings out the concept of managing multiple distributed generation in a cluster of power systems in smaller scale called microgrid. Power generation in microgrid concept can work either in the cluster or independently and not connected in a larger network or utility grid. This concept provides power grids more reliability, high penetration of renewable sources, self-healing, active load control, and improved efficiencies [27]. A study from the International Energy Agency pointed out that a power system based on a large number of reliable small distributed generations can operate with the same reliability and a lower capacity margin than a system of equally reliable large generators [28].

In smart grid, plug in electric vehicles (PEV) are not only intended to reduce pollution from vehicle exhaust gases, but also to play a role in power system operation, through the concept of vehicle to grid (V2G) and grid to vehicle (G2V). In the concept of V2G, electric vehicles act as energy storage that can supply power to the grid when parked and connected to grid [29]. While on G2V, electric vehicles act as energy storage that can contribute to handling power fluctuations through coordinated charging of EVs that can improve power losses and voltage deviations by flattening out peak power [30]. When electric vehicles are being parked at home, HEMS application handled power transaction of electric vehicles to grid. On the network side, the smart grid concept is aimed at improving network reliability through several integrated control systems consisting of smart control centers, smart power transmission networks, and smart substations. By adding sensors, communications and control to the existing network systems, then the network system can be smarter and able to do self-healing [31].

In the distribution level, the smart grid system application is conducted through smart metering in every consumer's houses integrated with HEMS and smart grid application, thereby from the side of the power system management there can be an analysis and planing of power system operation. From the side of consumers with HEMS application it is possible to control power usage and power source plan in the house either from solar or wind, in order to get cheaper and more efficient energy use [32].

4. HEMS Application Improving Power System Reliability

Smart grid technology has the potential to substantially improve electricity service by increasing reliability, reducing environmental impacts, and decreasing costs. In the electricity industry, reliability is often measured differently at the distribution level (retail service) and at the bulk transmission/ (bulk) generation level [33]. The reliability issues in modern power grids becomes more challenging. Factors contributing to the challenges include: aggravated grid congestion, transfers over longer distances, increasing volatility and reduced reliability margins and the grid is being operated at its "edge" [34].

In the smart grid system, the home load as one of the entities in the electrical power system can also participate in maintaining the stability of the operation and improving the power quality of the power system network. Through HEMS applications customers can participate in maintaining or improving the reliability of power systems. One of the programs on a smart grid system aimed at improving network reliability is Demand Side Management (DSM). Through the DSM program customers can make arrangements of the use of loads that can impact on the reliability of power systems. There are several categories of DSM programs, among others [35]:

- Energy Efficiency (EE).
- Demand Response (DR).
- Time of Use (ToU).
- Spinning Reserve (SR).

Energy efficiency (EE) is aimed at reducing energy consumption by improving the load so that the required energy consumption can be reduced because it becomes more efficient. Energy efficiency programs at the household level through HEMS applications can contribute to increased efficiency at the network level, this can occur due to the large number of household electrical customers [36]. EE activities, one of them, is done by installing electricity-efficient equipment. There are several electricity saving targets for household expenses, ranging from lighting, air conditioning, washing machines and others. Energy savings are also done by changing costumer's behavior in using household appliances [37].

Demand response, As defined by U. S. Department of Energy (DOE) is : Consumers conduct activities to change the operating pattern of electrical energy consumption, in response to the application of tariff provisions and incentives for the use of electrical energy. Consumers change the pattern of electric energy consumption from normal usage to the usage pattern by paying attention to the different electric prices based on the operating time or incentive value provided, in order to encourage lower electricity usage when peak load time or the reliability of the power system is threatened [38].

Included in Demand Response (DR) among others are Time of Use (ToU) and Spinning Reserve (SR). ToU is used for Non Dispatchable loads while SR is used for Dispatchable loads. In some countries tariffs for the use of electric power are differentiated for some time, especially at peak load times, where electricity rates are more expensive than out of peak load [39]. With time-based pricing schemes, application development is to implement Time of Use (TOU) programs such as Power shifting, load shifting and load shedding. Power shifting is done by moving an alternative energy source, such as from the sun, to be used when the electricity price of the network is more expensive [40]. Load shifting is done by moving the load operating time to another time outside the peak load. While load shedding is done by load termination or by not activating load during peak load. Beside that, it can be done through priority load mechanism in which peak load for less priority load will be off.

Spinning Reserve is meant to face disturbances and contingencies so that the power system remains operating optimally at the frequency of 59.97 Hz-60.03 Hz. Spinning Reserve is divided into regulating reserves and contingency reserves. Regulating reserves are to handle normal operational disturbances in the system. Contingency reserves (also referred to as reliability response) handle supply contingencies such as loss of generation [41].

PEV (Plug-in Electric Vehicles) in the smart grid scheme is to improve the power quality of power systems. Several studies have shown that the contribution of HEMS-PEV to the smart grid system can improve the power losses and voltage profile, one of the studies showed that voltage fluctuations resulting from large disturbances can be reduced up to five times [42]. This study proposes automatically controlled vehicle group (ACVG) as basic control units used in power system model. PEVs connected to the same substation are aggregated together to form ACVG. Increasing number of EV chargers will impact to value of current total harmonic distortion (THD).Transformer lifespan will decrease depending on the battery charging type and concentrated charging of electric vehicles [43].

5. Conclusion

The contribution of HEMS application to the smart grid scheme in improving the power system reliability is conducted through demand side management (DSM) program. DSM implements one of them with demand response and improved system power reliability. Penetration of electric vehicles in the smart grid scheme can also improve the reliability of power systems through V2G and G2V. When the electric vehicle is at home then HEMS application will handle the mechanism of V2G or G2V. Contribution of HEMS to the smart grid system can improve the power losses and voltage profile; one of the studies showed that voltage fluctuations resulting from large disturbances can be reduced up to five times.

6. Acknowledgments

This work was financially supported by Ministry of Research, Technology and Higher Education of the Republic of Indonesia through grant of Penelitian Unggulan Perguruan Tinggi 2017

7. References

[1] Gabriele L, Salvatore C and Erica L 2016 A Review of systems and technologies for smart homes and smart grids *Energies* **9(348)** pp 1-33

[2] Muhammad RA, MBI Reaz and MA Mohd Ali 2012 A Review of smart homes – past, present, and future future *IEEE Transactions on Systems Man and Cybernetics Part C (Applications and Reviews)* · **42 (6)** pp1190-203

[3] Quynh L, Hoang B N and Tony B 2012 Smart homes for older people: positive aging in a digital world *Future Internet* **4** pp 607-17

[4] Darby S 2006 the effectiveness of feedback on energy consumption: a review for DEFRA of the literature on metering billing and direct displays *Environmental Change Institute, University of Oxford*

[5] Karlin B, Ford R, Sanguinetti A, Squiers C, Gannon J, Rajukumar M and Donnelly KA 2015 Characterization and potential of home energy management (hem) technology *Pacific Gas and Electric Report*

[6] Ding D, Cooper RA, Pasquina PF and Fici-Pasquina L 2011 Sensor technology for smart homes *Maturitas* **69** pp 131–6

[7] Tiago DPM, Radu G, Eduardo MGR, João COM and João PSC 2015 Smart home communication technologies and applications: wireless protocol assessment for home area network resources *Energies* **8** pp 7279-311

[8] Pascual J, Sanchis P and Marroyo L 2014 Implementation and control of a residential electrothermal microgrid based on renewable energies, a hybrid storage system and demand side management *Energies* **7** pp 210–37

[9] Ameena SAS, Mohammed HA and Magdy MAS 2014 Smart home activities: a literature review *Electric Power Components and Systems* **42** pp 294-305

[10] Giovanni P, Mario C, Antonio R and Jiahu Q 2017 Home energy management *Energies* **10(382)**

[11] Khan AM, N Javaid, S Razzaq, RD Khan and M Ilahi 2013 Home energy management

systems in future smart grids *Basic. Appl. Sci. Res.* **3(3)** pp1224-31

[12] Zhong F, Parag K, Sedat G, Costas E, Georgios K, Mahesh S, Ziming Z, Sangarapillai L and Woon HC 2011 Smart grid communications: overview of research challenges, solutions, and standardization activities *IEEE Communications Surveys & Tutorials* **15(1)**

[13] Smart Grid Task Force 2016 Outlook for increased adoption of smart grid technologies in ADB energy sector operations *ADB Sustainable Development Working Paper Series* **42**

[14] X Fang, Satyajayant M, Guoliang X and Dejun Y 2011 Smart grid – the new and improved power grid: a survey *IEEE Communications Surveys & Tutorials* **14(4)** pp 944-80

[15] Farid K, Mohamed B, Yassine A and Abdelaziz K 2015 Integrated energy management of a plug-in electric vehicle in residential distribution systems with renewables *IEEE 24th International Symposium on Industrial Electronics (ISIE)* pp 717-22

[16] B Zhou, Wentao L, KW Chan, Y Cao,Yonghong K, XiLiu and X Wang 2016 Smart home energy management systems: concept, configurations, and scheduling strategies *Renewable and Sustainable Energy Reviews* **61** pp 30-40

[17] Pengwei D and Ning L 2011 Appliance commitment for household load scheduling *IEEE Transactions On Smart Grid* **2(2)** pp 411-9

[18] Yuto Y, Tomoaki K, Michihiro K, Tatsuo N and Akinori S 2015 Model predictive control of ev storage battery with hems based on particle swarm optimization *IEEE Innovative Smart Grid Technologies - Asia* pp 1-5

[19] Pervez HS, Nursyarizal BMN, Perumal N, Irraivan E and T Ibrahim 2014 A Review on optimized control systems for building energy and comfort management of smart sustainable buildings *Renewable and Sustainable Energy Reviews* **34** pp 409–29

[20] Hussain SM, A Basit, ZA Khan, U Qasim and N Javaid 2015 A review on demand response: pricing optimization and appliance scheduling *Procedia Computer Science* **52** pp 843-50

[21] Jongbae K, Jinsung B, Daebeom J, Myeong-in C, Byeongkwan K and Sehyun P 2015 An IoT-based home energy management system over dynamic home area networks *International Journal of Distributed Sensor Networks* **2015** pp 1-15

[22] Anzar M, Nadeem J and Sohail R 2015 A review of wireless communications for smart grid *Renewable and Sustainable Energy Reviews* **41** pp 248–60

[23] Noelia UP, Luis H, David de la V and Itziar A 2016 Review, state of the art and trends review of smart metering in electricity grids *Applied Sciences* **6(68) pp** 1-24

[24] MS Thomas and JD. Mc Donald 2015 Power system scada and smart grids *CRC Press Taylor & Francis Group 6000 Broken Sound Parkway NW, Suite 300 Boca Raton*

[25] European Regulators' Group for Electricity and Gas. Annex 3 2009 Evaluation of responses—ERGEG conclusions paper on Smart grids *Technical Report for Council of European Energy Regulators ASBL Brussels Belgium*

[26] Lucian ID, Mihail A and Dorin B 2015 Smart grid economic dispatch *Procedia Technology* **22** pp 740-45

[27] RH Lasseter 2011 Smart distribution: coupled microgrids *Proceedings of the IEEE* **99(6)** pp 1074–82

[28] International Energy Agency 2002 Distributed generation in liberalised electricity markets *OECD/IEA*

[29] W Kempton, V Udo, K Huber, K Komara, S Letendre, S Baker, D Brunner and N Pearre 2009 A test of vehicle-to-grid (V2G) for energy storage and frequency regulation in the PJM system *Mid-Atlantic Grid Interactive Cars Consortium*

[30] K Clement, E Haesen and J Driesen 2009 Coordinated charging of multiple plug-in hybrid electric vehicles in residential distribution grids *Power Systems Conference and Exposition IEEE/ PES* 1–7.

[31] Bose 2010 Smart transmission grid applications and their supporting infrastructure *IEEE Transcations on Smart Grid* **1(1)** pp 11–19

[32] Mahmood, MN Ullah, S Razzaq, A Basit, U Mustafa, M Naeem and N Javaid 2014 A New scheme for demand side management in future smart grid networks *Procedia Computer Science* **32** pp 477–84

[33] Paul H, Jason V and Brian T 2014 Smart grid: reliability, security, and reslieincy *University of Vermont/MITRE*

[34] Khosrow M and Ranjit K 2010 Smart Grid - A Reliability Perspective *IEEE Transactions on Smart Grid* **1(1)** pp 57-64

[35] Peter P and Dietmar D 2011 Demand side management: demand response,intelligent energy systems, and smart loads *IEEE Transactions On Industrial Informatics* **7(3)** pp 381-88

[36] Ana RB and Jos´e S 2011 Towards efficient energy management: defining hems and smart grid objectives *International Journal on Advances in Telecommunications* **4(3&4)** 249-63

[37] E Gilbert, R Maslowski, S Schare and K Cooney 2010 Impacts of smart grid technologies on residential energy efficiency *ACEEE Summer Study on Energy Efficiency in Buildings*

[38] AEIC Load Research Committee 2009 Demand response measurement & verification, applications for load research *Association of Edison Illuminating Companies (AEIC)*

[39] Mahmood I K, S Razzaq, Z Najam, NA Khan, MA Rehman and N JavaidHome 2014 Appliances coordination scheme for energy management (hacs4em) using wireless sensor networks in smart grids. *5th International Conference on Ambient Systems, Networks and Technologies Procedia Computer Science* **32** pp 469-76

[40] Zhaoguang P, Hongbin S and Qinglai G 2013 TOU-based optimal energy management for smart home *4th IEEE PES Innovative Smart Grid Technologies* pp 1-5

[41] A AlMajali, Arun V and Clifford N 2017 Resilience evaluation of demand response as spinning reserve under cyber-physical threats *Electronics* **6(2)** pp 1-14

[42] Andrej G, Mirko T, Juergen K and Ljupco K 2014 Improving power grid transient stability by plug-in electric vehicles *New Journal of Physics* **16(115011)** pp 1-14

[43] Ryan L, Luther D and Edwin L 2011 A Survey of PEV impacts on electric utilities *Innovative Smart Grid Technologies (ISGT) IEEE PES*

IOP Conf. Series: Earth and Environmental Science **105** (2017) 012082 doi:10.1088/1755-1315/105/1/012082

The effect of Pineapple Leaf Fiber (PALF) incorporation into Polyethylene Terephthalate (PET) on FTIR, morphology and wetting properties

F M Abdul Aziz[1a)], S N Surip[1], N N Bonnia[1] and K A Sekak[1]

[1]Faculty of Applied Sciences, Universiti Teknologi MARA,40450 Shah Alam, Malaysia.

Email: fatimahmuyassarah@gmail.com

Abstract. Pineapple Leaf Fibers (PALF) which is rich in cellulose, relatively inexpensive and abundantly available has the potential for polymer reinforcement. In this study, Polyethylene Terephthalatate (PET) was added with PALF and electro-spinned. The resulting mats were compared with PET neat electrospun. The samples later were examined by Scanning Electron Microscope (SEM), Fourier Transform Infrared (FTIR) and contact angle (CA). Briefly, SEM results indicated that the present of fibers led to a tendency of lower average fiber diameter compared to the PET neat. Two distinct fiber networks with intersecting fibers were observed in PALF/PET. One networks probably corresponds to PET and the others to PALF. FTIR analysis shows the intensity peak represent carbonyl at ~3400cm^{-1} and ester at ~1100 $^{cm-1}$ decreased. It is suggested interaction occurred between lone pair of oxygen in the group with hydrogen group in PALF. New peak were observed at 3400 $^{cm-1}$ in PALF/PET that indicated present of hydrogen bonding as well as its hydrophilic tendancy. The contact angle of PET signify high average value 156^0 that comes with hydrophobic properties compared to PALF/PET with average value 16^0 with more hydrophilic properties.

1. Introduction

Nanofiber materials have many potential applications due to its large surface area-to-volume, small pore size and high porosity properties [1]. In the last few decades, natural fibers especially lignocellulosic fibers have attract researcher's attention due to its great mechanical properties as well as the positive contribution to the environment. Through the years, natural fibers have become highly demand in market and often use as reinforcing fibers for composite [2]. This study focuses on the potential of lignocellulosic fibers as replacement for current petroleum-based fibers.

Among these fibers, pineapple leaf was one of the potentially unexplored fibers that could offer more to the industry. Pineapple plant abundantly available especially in tropical country thus provides unlimited supply. A few studies were conduct to study mechanical properties of PALF and it shows high tensile strength and young modulus value [3-5]. This is due to high cellulose content that help the leaf to maintain its turgidity thus give great properties to PALF [6]. However, even there were papers already been published on PALF, there were only few proceed it to the PALF nanofibers size.

Various ways have been developed to produce high properties of nanofibers such as drawing, template synthesis, phase separation, self-assembly and electrospinning [7-11]. Electrospinning which is straightforward and versatile method with low cost of production has made it used widely to produce nanofibers with lower diameter [11]. Recently, a few studies has successfully produced an electrospun mats with high mechanical properties after addition of natural fibers [12-13]. Selection of a suitable solvent system for the polymers plays a big role to ensure electrospinning success [14]. Studies has been conducted on TFA and found it could favors the formation of continues nanofibers [10]

2. Experimental procedures

2.1. Materials
Pineapple Leaf fibers were received from local Malaysian company in fibers form. Polyethylene Terephthalate were obtained from Aldrich and received in granular form. The chemicals which are trifluoroacetic acid, cyclohexane and ethanol all were supplied by Aldrich Merck.

2.2. Soxhlet method (Dewaxing of PALF)
PALF were first dewax using soxhlet instrument to remove wax and terpenes thus increase its surface interaction with others chemicals. 10 g of PALF were put into the thimble and rinse with 200ml of ethanol:cyclohexane 1:1 ratio for 6 hours. Then, the remaining sample were rinse using distilled water and put in the oven with temperature 40 ^{0}C for 12 hours until the sample were completely dried.

2.3. Dissolution of PALF/PET
Dried PALF and PET were weight accordingly to table 1. The samples were dilute in the Trifluoroacetic acid (TFA) and stirred for 12 hours until homogenous solution were obtained

Table 1: Concentration of PALF

Sample	Conc PET (gmL^{-1})	PET (g)	TFA (mL)	Conc PALF (gmL^{-1})	PALF (g)	TFA (mL)	Total TFA (mL)
PET $_{neat}$	0.2	1	5.0	-	-	-	5.0
PET/PALF	0.2	0.5	2.5	0.02	0.06	3	5.5

2.4. Electrospinning
Electrospinning set up were mainly consists of power supply, syringe pump, syringe, needle and collector (figure 1).

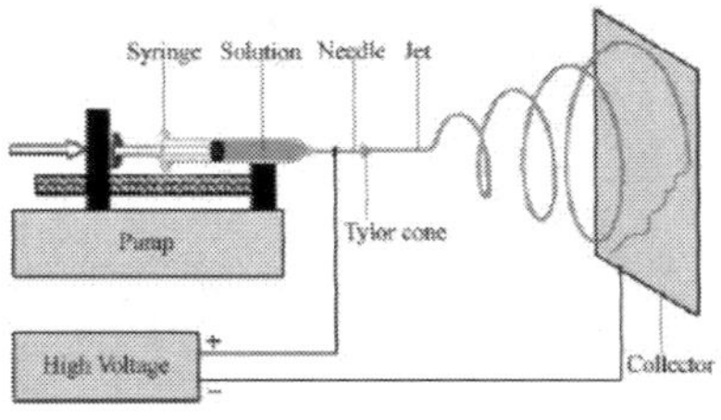

Figure 1. Electrospinning set-up

2.5. Characterization

2.5.1. FTIR. FTIR spectroscopy was carried out in order to study the interaction that occurred in TFA and PALF. This kind of interaction can be proven through frequency shifts, changes in band intensity, and shape of the FTIR spectra. The spectra were recorded by FTIR (Thermo Fisher Scientific Nicolet iS 10) equipped with Attenuated Total Reflectance (ATR) in the transmittance mode over a frequency range 500-4000 cm-1 with 4 cm-1 resolution.

2.5.2. FESEM. The surface morphology and size distribution were determined by FESEM.

2.5.3. Contact Angle (CA). The wettability of PET and PET/PALF electrospun was analyzed through a digital camera positioned on a static contact angle analyzer then analyzed with Image J 1.34S.

3. Results and Discussion

3.1. FTIR analysis

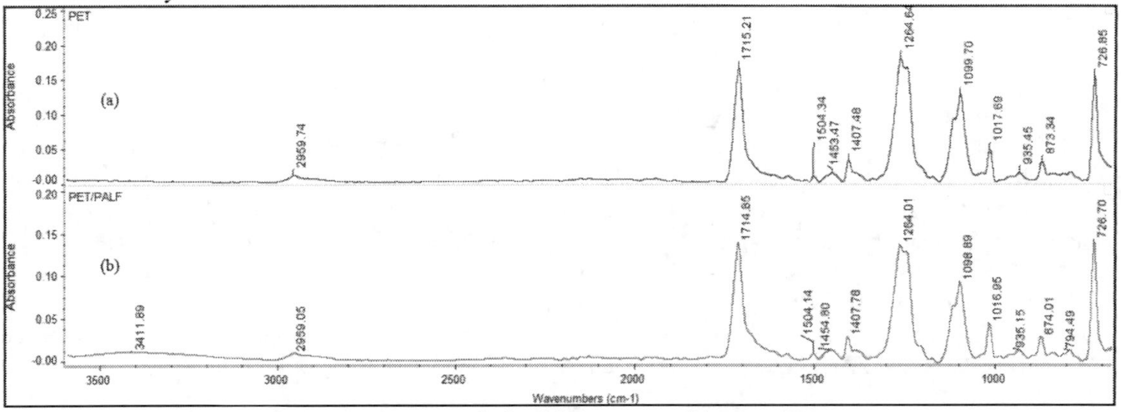

Figure 2. FTIR result for (a) $PALF_{neat}$, (b)PALF/TFA

Figure 2 showed FTIR spectra of (a)PET and (b)PET with PALF addition (PET/PALF) (b) at (600-3500) cm^{-1} region. For both PET and PET/PALF sample, peaks approximately at 2900 cm^{-1} were observed. This is due to the vibration of CH2, CH3 and CH bond in the PET structure. At peak ~1715 cm^{-1}, stretching of C=O bond were appeared indicated the conjugated ester group in the chemical structure of PET as well as acetyl group that present in the hemicellulose. C=C stretching that attribute to vibration of benzene and aromatic lignin of the skeletal vibration were observed in both sample at peak ~1050cm^{-1} wavelength. At peak ~725 cm^{-1}, peak signifies the present of C-H bond from glucose ring of the cellulose were observed. However, most of the peak shows a decreased intensity after PALF addition thus suggested there were interactions occurred between PET and PALF.

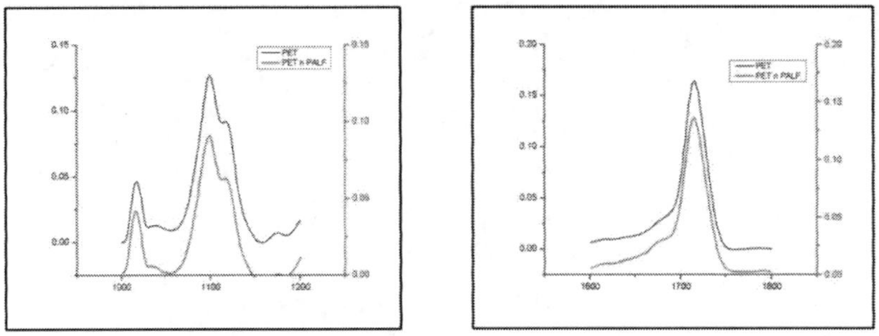

(a) (b)

Figure 3. Intensity of (a)C=O bond spectra and (b)C-C-O spectra for PET and PET/PALF

Based on few paper written on lignocellulosic fibers [14,15] a peak usually presented at ~1595 cm^{-1} represent the lignin however were absent at PET/PALF spectrum. This could indicate the reducing amount of lignin in the sample. It is suggested that this is due to the addition of TFA that plays role on disrupted the covalent bond that hold the cellulose (lignin) as well eliminates it. Peak at ~1715cm^{-1} and ~1099cm^{-1} shows a decreasing intensity as shown in figure 3. This indicated there is interaction between oxygen in the C=O and C-C-O group of PET with hydrogen bond of PALF. It is supported by the appearance of broad peak at 3400 cm^{-1} representing the hydroxyl group resulted from the interaction of PET/PALF contributed to the hydrophilic tendency of the fiber. Figure 4 shows the possible reactions occurred during PET and PALF interaction.

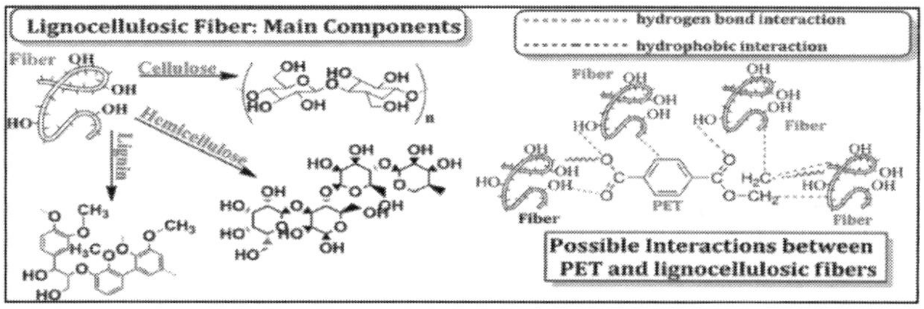

Figure 4. Main components of a lignocellulosic fiber and their reactive sites susceptible for interaction (*R.P.O Santos et al., 2015)*

3.2. FESEM analysis

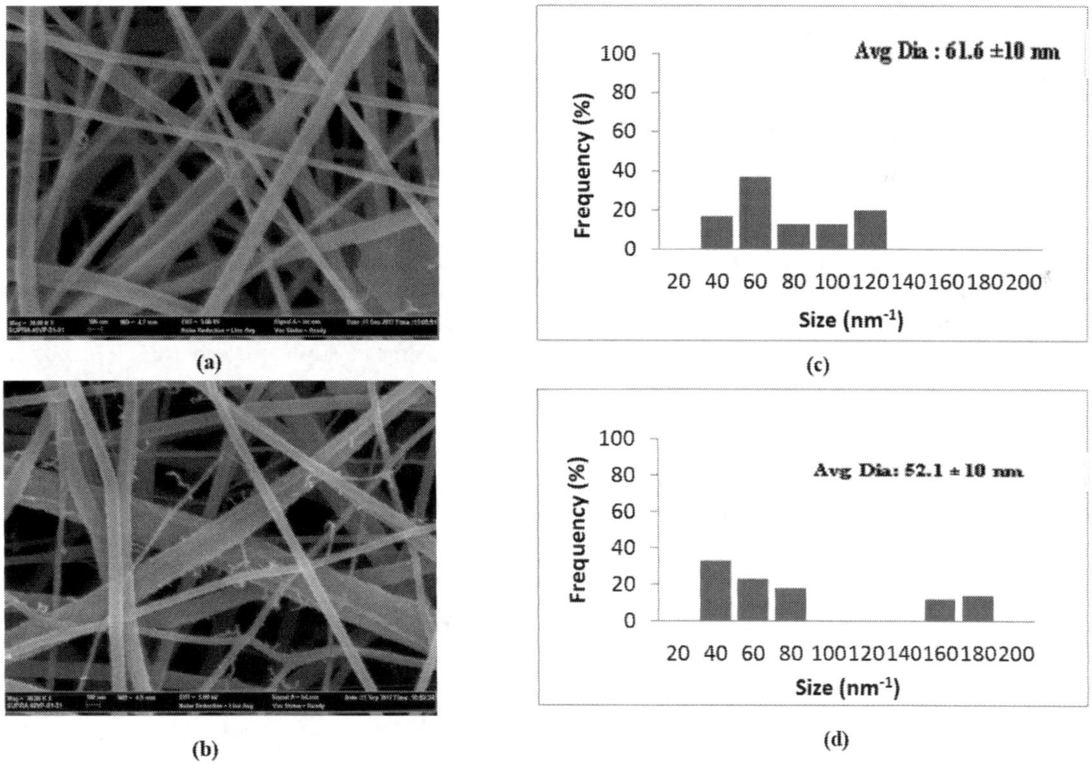

Figure 5. FESEM result for (a) PET (b) PET/PALF with magnification of X15K and frequency percentage size distribution for (c) PET (d) PET/PALF

FESEM images of electrospun nanofibrous structures of (a) PET and (b) PET/PALF shown in Figure 5 revealed that nano-scale of spider web structures were successfully created via electrospinning method. The fibers were found randomly orientated with interconnected pores in between to form a 3D scaffold. PET sample shows diameter range between 40-120 nm with average 61.6 ±10 nm while PET/PALF shows lower fiber average diameter which is 52.1 ±10 nm. with uniform size while PET/PALF shows diameter one with thicker fibers range between 160-180 nm while others shows thinner fibers range between 40-80 nm with average 52.1 ±10 nm.

FESEM results indicated that the present of fibers led to a tendency of lower average fiber diameter however has random diameter distribution. Two different fiber networks with intersecting fibers were observed in PALF/PET. It is suggested that one networks with bigger diameter probably corresponds to PET and the others with smaller diameter correspond to PALF. Diameter with thicker fibers range between 160-180 nm while the thinner fibers range between 40-80 nm. Thicker diameter might due to the increasing solution concentration after adding PALF. As shows in fig 5 (b), a thin fibers coexited among the electrospun fibers were present. It is suggested that a reaction has occured during addition of PALF into PET.

3.3. Contact Angle (CA) analysis

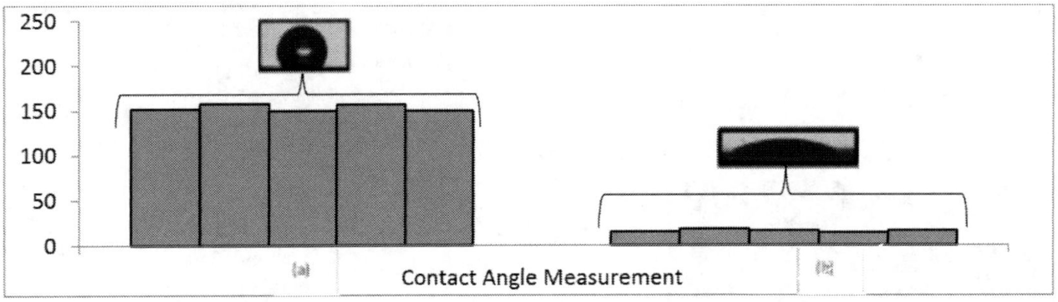

Figure 6: Contact Angle result for (a) PALF (b) PALF/TFA

Figure 6 shows contact angle results obtained from PET and PET/PALF electrospun mats. The contact angle of PET signify high average value 156^0 that comes with hydrophobic properties however decrease drastically after added with PALF with average value 16^0 thus made it more hydrophilic properties. This change indicates the presence of polar chemical groups onto the PET surface, increasing the surface energy of polymer and thereby decreasing in surface contact angle. Thus, it suggested PALF were successfully electro-spinned with PET polymer and PALF favored the changed of PET hydrophobic properties.

4. Conclusion

As the conclusion, PALF were successfully electro-spun together with PET and produced lower average finer diameter compared to PET electro-spun alone. Besides, contact angle result shows the ability to controls the hydrophilic/hydrophobic materials properties by tuning the PET/PALF ratio thus made it useful for many applications.

5. Acknowledgement

The authors would like to acknowledge University Technology Mara for providing grant to carry out this research under Nanocomposite Material & Industrial Application Research Group (600-IRMI/DANA 5/3/REI (0001/2016)).

6. References

[1] Huang Z M, Zhang Y Z, Kotaki M and Ramakrishna S 2003 *Composite Science and Technology* **63** pp 2223-2253
[2] Kaewpirom S and Worrarat C 2014 *Fibers and Polymers* **15** pp 1469-1477
[3] Khengketkit N and Amornsakchai T 2014 *Materials and Design* **55** pp 292-299
[4] Neto A R S, Araujo M A M, Souza F V D, Mattoso L H C and Marconcini J M 2013 *Industrial Crops and Products* **43** 529-537

[5] Dewi L U, Bhagawan S and Thomas S 1997 *Polymer Composite* (Wiley Inter-Science)
[6] Asim M, Abdan K, Jawaid M, Nasir M, Dashtizadeh Z, Ishak M R, Hoque M E 2015 *Int. Journal of Polymer Science* ID 950567
[7] Bajakova J, Chaloupek J, Lukáš D and Lacarin M 2011 *Proceedings of the 3rd International Conference on Nanotechnology-Smart Materials (NANOCON'11).*
[8] Wang Y, Zheng M, Lu H, Feng S, Ji G and Cao J 2010 *Nanoscale research letters* p **5** 913.
[9] Huang W, Wang M J, Liu C L, You J, Chen S C, Wang Y Z, and Liu Y 2014 *Journal of Materials Chemistry A* **22** pp 8416-8424
[10] Suvannasara P, Praphairaksit N and Muangsin N 2014 *RSC Advances* **102** pp 58664-58673
[11] Veleirinho B, Rei M F and Lopes J A 2007 *Journal of Polymer Science* (Wiley Inter-Science)
[12] Li G, Zhao Y, Lv M, Shi Y and Cao D 2013 *Colloids and surfaces A :Physicochem. Eng. Aspects* **436** pp 417-424
[13] Ahn Y, Lee S H, Kim H J, Yang Y H, Hong J H and Kim Y H 2012 *Carbohydrate Polymers* **88** pp 395-398
[14] Santos R P O, Rodrigues B V M, Ramires E C, Ruvolo A C and Frollini E 2015 *Industrial Crops and Products* **72** pp 69-76
[15] Chirayil CJ, Joy J, Mathew L, Mozetic M, Koetz J, Thomas S 2014 *Industrial Crops and Products* **59** pp 27-34

IOP Conf. Series: Earth and Environmental Science **105** (2017) 012083 doi:10.1088/1755-1315/105/1/012083

Hydrophilic and hydrophobic characteristics of dry peat

L R Perdana, N G Ratnasari, M L Ramadhan, P Palamba, Nasruddin, Y S Nugroho[1]

Department of Mechanical Engineering, Universitas Indonesia, Kampus UI Depok, 16424, Indonesia

E-mail: yulianto@eng.ui.ac.id

Abstract. Indonesia is a country with the largest peatland in Southeast Asia. These vast peatland is spread throughout the country. In the last 2 decades, Indonesia encounter land fires and now becomes an annual land fire event. Later, observed that the land fire occurred because dry season in Indonesia causing the peat becomes dried enough to have hydrophobic characteristics and easily burned. This fire phenomena on peatland is determined by some factor such as, physical properties, organic content, oxygen concentration, etc. Peat has hydrophilic and hydrophobic properties as physical properties when treated differently. Initially raw peat has hydrophilic properties. However, in certain low level of moisture content, peat becomes hydrophobic. To examine this phenomenon, mechanical understanding based experiment was done. For hydrophilicity, the experiment will be using Shimadzu MOC63u Moisture Balance Analyzer. The peat sample which are used are Indonesian natural peat. The wet peat was dried by using moisture balance equipment with various time (0.5 till 6 hours with interval 0.5h) at 100°C. and loading peat into a tube with 2.15 cm in inner diameter and 5.5 cm long where both end are covered by wire mesh. Afterwards, the tube was being dipped into water in a certain depth. As for hydrophobicity, the experiment was done with the same method as hydrophilicity by using some of the dried peat. From the experiment, the mass loss, adsorption time, volume shrinkage and adsorption capacity and as well as hydrophobicity of peat will be shown in the result.

1. Introduction

The vast territory makes Indonesia an archipelago. Within Southeast Asia, Indonesia has a best estimate of 149,056 km2 of peatland [1]. By having a lot of islands, means a lot of soil types around Indonesia. Peat is a heterogeneous mixture of more or less decomposed plants (humus) material that has accumulated in a water-saturated environment in the absence of oxygen. It is also known as the "forgotten fossil fuel" [2] as it is able to store a large amount of carbon. The peat condition will be determined by the environmental condition. Basically, in normal environmental, peat soil tends to have water retention trait. Although at some certain condition where the climate is hot, that water retention trait would weakened due to the change of its characteristics and structure. This state is called hydrophobicity, and where the strong water retention state is called hydrophilicity.

Peat works as water storage and also carbon storage. It has high water retention ability to store water. However, peat soil can only provide a few spaces for plant's growth compared to mineral soil. The decomposition of peat soil affect water retention ability [3], and the decomposed peat could retain

[1] To whom any correspondence should be addressed.

water twice till six time than its dry weight. Soil water has three different types classified from physical classification, they are: (1) gravitational water, water that move freely while drained due to gravitational force and do not absorbed by the micro pores, (2) capillary water, water that absorbed to the micro pores and become moisture content in peat, (3) hygroscopic water, this water forms very thin films around soil particles and it is hard to remove just by normal heat treatment.

Hydrophobicity is a state of peat that is unable to retain water or water repellency. Most peat is dominated by wooden plants which have hydrophobic trait [4]. This hydrophobic trait may cause problems in agricultural industries. Hydrophobicity makes peat easy to burn at high temperature. While in hydrophobic state which unable to retain water, makes it harder to be extinguished.

2. Materials and Methods

2.1. Sampling and methods

This study analyzed the characteristics of peat taken from Papua, Indonesia as in previous work [5, 6]. The peat was sealed to preserve the initial condition until it reached the laboratory and until it was ready to be used as sample. To have the initial condition as similar as possible will resulting in better data acquired. By sealing the peat, the less change occurred in peat condition. For lab experiment, the sample was done by inserting a pack of peat into a holed aluminum with the dimension of 2.15 cm in diameter. Afterwards, the sample would be put into Shimadzu MOC63u Moisture Balance Analyzer to had its moisture content controlled. After that, the sample would be sunk into water and let it absorb water and soaked for 30 minutes and drained for 12 hours. Lastly, the sunken sample was taken to be weighed to see the difference in mass that will give the result whether the sample was in hydrophilic or hydrophobic state.

2.2. Hydrophilicity observation

Hydrophilicity of the peat determined when the peat still able to absorb water to a certain point or called wettability. To proof that, the experiment was done by observing the mass gained after sinking the controlled moisture content peat for a certain period of time. This simple method was conducted to obtain a series of data with controlled moisture content peat sample ranging from wet to dry. Ten samples went through this experiment with ten different time taken using Shimadzu MOC63u Moisture Balance Analyzer at 100 C with real time data acquisition. The difference in time taken will result in different moisture content and later will change the characteristics of peat although slightly and also the hydrophilicity of the peat. The variable of the experiment can be seen in table 1**Table *1***.

Table 1. Experimental setup

Sample	**Drying time (hours)**									
Papuan peat	0.5	1	1.5	2	2.5	3	3.5	4	5	6

2.3. Hydrophobicity observation

Peat has a unique trait that perhaps only peat has. When reach a certain moisture contain, hydrophilic trait of peat which absorb water weakened and it becomes hydrophobic which decline water. The probability of hydrophobicity determined by the low moisture contain of peat condition. By using the same method as to observe hydrophilicity with the difference in moisture content, the hydrophobicity of peat can be observed. There are other advanced methods to observe hydrophobicity such as, Fourier Transform Infrared (FTIR) spectroscopy method and water drop penetration time method (WDPT) method [7].

2.4. Moisture content observation
Moisture content determine the condition of peat. Generally, high moisture content will result in peat still in hydrophilic state. On the other hand, while at low moisture content which change the dryness of peat, will result in hydrophobic state. The critical limit of moisture content was defined as a point where hydrophobicity phenomenon manifested [8]. The occurrence of hydrophilicity of peat was hypothetically estimated around 60-80% of moisture content. Thus, that makes the occurrence for hydrophobicity has lower percentage of moisture contain which go around 40-10%.

3. Results and Discussion
From the experiment, it can be seen that the more time taken for drying, the lower the moisture content became. While low in moisture content, means that the weaker the peat soil can retain water. The weaker the peat soil retain water the more hydrophobic peat become.

Mass loss phenomenon occurred because the moisture content evaporated. From figure 1 below, the percentage of mass loss after the drying process and mass gain after re-wetting process can be seen. The dryness of the peat affects the ability to re-adsorb water. The dried peat will not be able to fully return to the initial mass, even after re-wetting process. The weakened ability to adsorb water caused by the damaged structure of peat while in drying process and also causing irreversible drying. Following the drying time process, the water which can be adsorbed decrease. For sample no. 10 from table 1 (6 hours drying time), the mass of dried peat was 1.863 gram (10.3% of the initial mass) and after re-wetting process the mass increased to 2.1 gram (11.5% of the initial mass). It proved that dried peat cannot fully return to its initial mass. Thus, the longer the drying process, the more mass loss occurred. Meanwhile in re-wetting process, the mass gained from re-wetting was almost constant. However, at longer drying time which means low moisture content, the mass gained decreased caused by hydrophobicity.

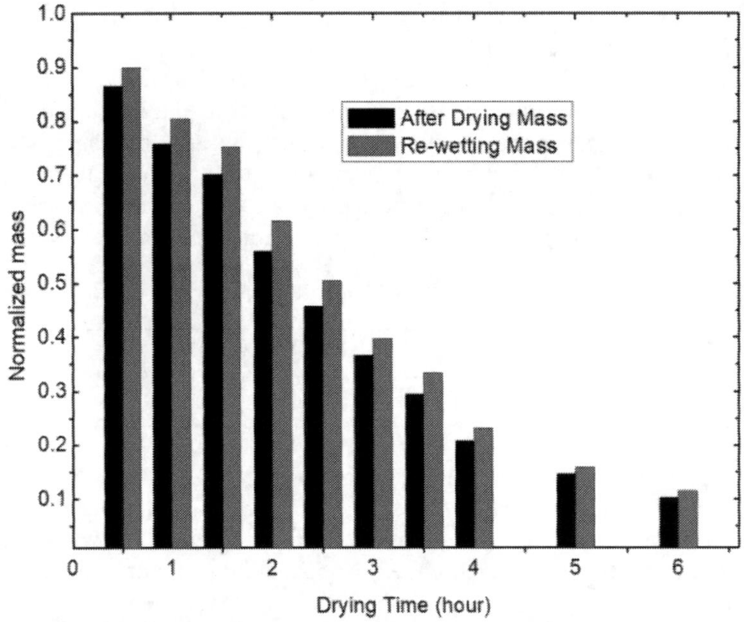

Figure 1. Mass measurement of after drying and after re-wetting at various drying time.

The ratio of hydrophilic and hydrophobic of peat as shown in figure 2**Figure** *2*, gave the information the hydrophilicity and also hydrophobicity of the peat. When the ratio almost reached 1, that means the hydrophobicity of the peat is lower, which also means the peat is hydrophilic. On the other hand, when the ratio almost reached 0, that means it becomes hydrophobic.

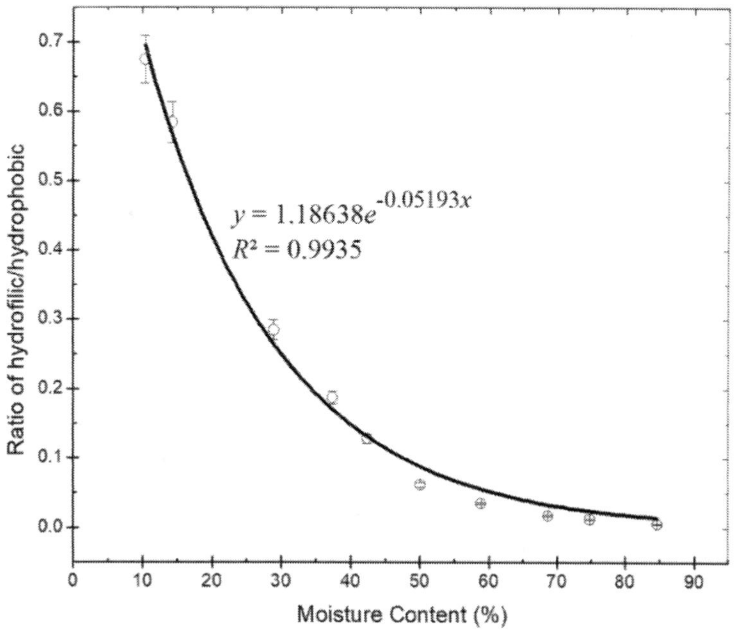

Figure 2. Hydrophilic and hydrophobic ratio of peat sample

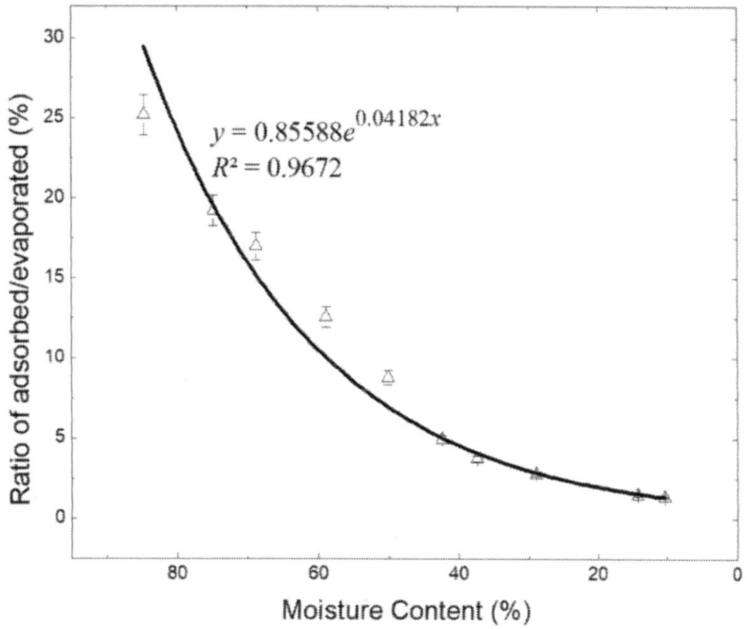

Figure 3. Ratio of adsorption with evaporation

In this experiment, adsorption and evaporation occurred where peat was heated and sunk into the water. To see the tendency from hydrophilicity and hydrophobicity, it can be done by comparing the amount of water adsorbed and evaporated while in drying and re-wetting process as shown in figure 3**Figure** *3*. Following the time taken in drying process, the ratio of water which is adsorbed and evaporated decrease. At 85% of moisture content (0.5 hours drying time) after re-wetting process, the ratio of adsorbed and evaporated water is 25.2% which is only ¼ can be re-adsorbed by the peat from evaporated moisture content. The drier the peat, the lower the ratio becomes.

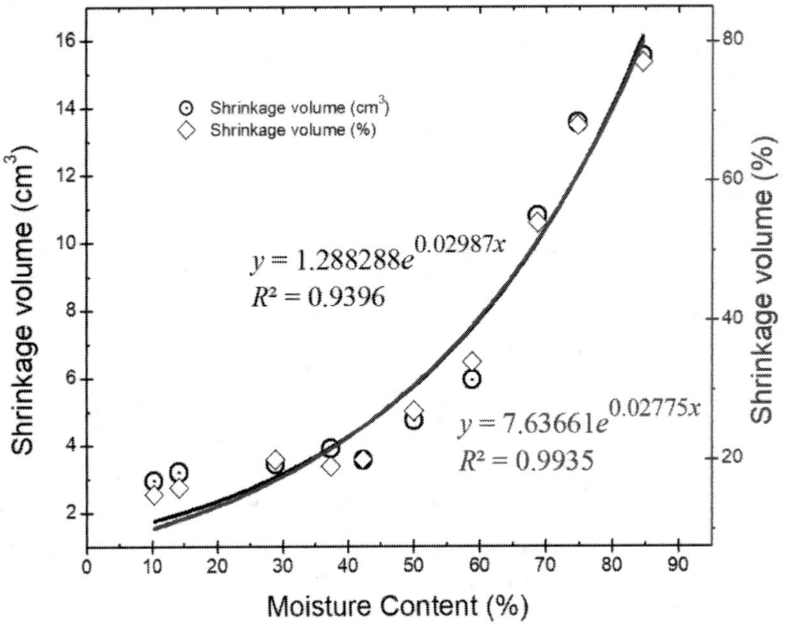

Figure 4. Volume shrinkage of after drying and re-wetting peat samples

Shrinkage analysis was done to observe the shrinking volume as the peat dried. It was also done to observe the damage and macro structural changes of peat. As seen in figure 4, the loss of moisture content affects the peat volume from drying process. At 75% moisture content (1 hour drying time), the remainder volume (shrinkage volume) is 68% from initial volume. Meanwhile, at 10% moisture content (6 hours drying time), the remainder volume is only at 15% of the initial volume. Thus, the more time taken for drying process, the smaller the remainder volume becomes.

4. Conclusion

Peat is a heterogeneous mixture of decomposed organic matters. Dried peat will become hydrophobic. To reach hydrophobic, the shrinkage volume and density and also mass loss will occur in the process. Shrinkage volume and mass loss will occur in drying process, while the water in the peat evaporated leaving the peat drying. While the volume shrinking and the mass loosing, the peat become more packed thus resulting in increased density. The denser the peat sample, the smaller the sample become.

Dried peat with low moisture content becomes hydrophobic. There was a critical point where peat started to change from hydrophilic to hydrophobic. Thus, this experiment indicates that the peat that undergoes heat treatment tends to be hydrophobic. Meanwhile from this experiment, it was known that moisture content played an important role to define peat as hydrophilic and hydrophobic. However, hydrophobic peat could still slightly adsorb water but not necessarily mean hydrophilic.

5. Acknowledgement
The authors would like to thank Directorate for Research and Public Services (DRPM) Universitas Indonesia for providing the International Indexed Publications for Student Final Project grant (PITTA) 2017 program, Grant no. 857/UN2.R3.1/HKP.05.00/2017.

6. References
[1] Ritung S, Wahyunto S, Nugroho S, Hikmatullah S and Tafakresnanto C 2011 *Indonesian Center for Agroclimate Res and Dev, Indonesian Agency for Agr. Res. and Dev., Ministry of Agr., Jakarta, Indonesia*
[2] Peat: The Forgotten Fossil Fuel 2011
[3] Rieley J, Ahmad-Shah A and Brady M 1996 *Tropical Lowland Peatlands of Southeast Asia* 17-54
[4] Valat B, Jouany C and Riviere L M 1991 *Soil Science* **152** 2 100-07
[5] Ramadhan M L, Palamba P, Imran F A, Kosasih E A and Nugroho Y S 2017 *Fire Safety Journal*
[6] Palamba P, Ramadhan M, Imran F, Kosasih E and Nugroho Y 2017 *AIP Publishing* **1826** 020017
[7] Utami S N H, Maas A, Radjagukguk B and Purwanto B H 2009 *Jurnal Tanah Tropika*
[8] Dekker L W, Doerr S H, Oostindie K, Ziogas A K and Ritsema C J 2001 *Soil Science Society of America Journal* **65** 6 1667-74

IOP Conf. Series: Earth and Environmental Science **105** (2017) 012084 doi:10.1088/1755-1315/105/1/012084

Effect of temperature and type of dispersant on treating oil spills

A A Yudiana, N Widiaksana, Y S Nugroho and M Wibowo

Fire Safety Engineering Research Group, Department of Mechanical Engineering, Universitas Indonesia.

E-mail: yulianto.nugroho@ui.ac.id

Abstract. Dispersant is a chemical surfactant which can break down an oil from the water surface. Therefore, dispersant can be used to treating oil spills. The effectiveness of dispersant as oil emulsifier can be influenced by the temperature of the environment and concentration of oil and dispersant. This paper shows the influence of temperature variations, dispersant types and layers of sea water on oil spills by testing the light absorbance using spectrophotometer UV-Vis. The results of the experiments show that the biggest light absorbance occurs on the water surface at temperatures of 26° C with area absorbing of 82.15 abs. Meanwhile, the smallest light absorbance is at temperatures of 16° C with wide absorbance 25.72 abs on the type of dispersant *soluble*. Wide light absorbance total greatest namely 26° C temperatures on the type of *soluble* with wide absorbance total 133.49 on three points of layers of water.

1. Introduction

Indonesia is a maritime country that has thousand of islands with a very wide sea territories. Indonesia has area reached 3.977 miles between the Indian Ocean and Pacific Ocean with land area of 1.922.570 km^2 and the area of the sea 3.257.483 km^2 [1]. It makes Indonesia have natural resources such as oil and gas. The industry activities of oil and gas can become a source of pollution at the sea such as oil spills. When the oil spilled at the sea, oil will spread on the surface of the water which formed a thin layer called *oil slick* [2]. Oil spills can be treated by giving a dispersant so it can breaks into the size of water column. Giving a dispersant on oil spills can be influenced by various factors such as the concentration of oil, the energy to mix the dispersant with the oil such as waves, the temperature of the environment, types of surfactant's molecule and oil ties to the water of the sea. The spread of oil spills was also influenced by the temperature of the water of the sea and the properties of oil itself. Temperatures have an important role to salinity and the stability of emulsion on oil. The sea water with low temperature increase oil and dispersant viscosity. When the oil becomes more viscose, the energy needed to mix the oil and dispersant increased [3]. At high temperatures, solubility of dispersant in sea water will tend to increase so that the temperature rise of water can reduce the viscosity and speed up the process of oil dispersion so that the *oil slick* on the surface of the water of the sea will gone. It can be use to determine the level of effectiveness of dispersant in different temperatures where testing the clarity of the water can be tested with *spectrophotometer* UV-Vis. Dispersant is a method of treating oil spills that is often used by the company in dealing with pollution at the sea, so the testing in the laboratory scale is expected to help the ability dispersion relatively better on the crude oil and processed oil.

IOP Conf. Series: Earth and Environmental Science **105** (2017) 012084 doi:10.1088/1755-1315/105/1/012084

2. Literature Review

Dispersant is a chemical substances mixture of emulsion substances (surfactant) and solvent so it can break down oil into small particles so that the surface of the oil layer became larger and easy to dispersion in the water. By adding a dispersant, the viscosity between oil and water is reduced so the waves can break a layer of oil into a small droplets [2]. At some of oil spills at the sea, the oil will dispersed naturally into water column. It releases them due to the viscosity grade on each of the different oil where oil with low viscosity more easily to dispersed naturally compared with oil with high viscosity [4].

Effectivity of dispersant can be reviewed by knowing by cmparing the light absorbance and wavelength from testing with spectrophotometer UV-Vis. Area under the curve light absorbance vs. wavelength, where samples of the experiment from the wavelengths done on the ratio of 340-400nm using trapezoidal rule calculation, based on the formula (1) [5]:

$$\text{Area} = \frac{(\text{Abs}_{340}+\text{Abs}_{370})30}{2} + \frac{(\text{Abs}_{370}+\text{Abs}_{400})30}{2} \tag{1}$$

Where the Abs_{340}, Abs_{370} and Abs_{400} are the absorbances measured at wavelength 340, 370, and 400 nm, respectively. Area can show the effectiveness of dispersant

A spectrophotometer can measure the intensity of light as a function of the color spectrum (light waves). An important feature of spectrophotometer namely spectral bandwidth (range of color spectrum that can be transmitted through samples tested), the percentage of the sample transmission and range of distric from absorption of samples. Using the appliance spectrophotometer, spectrophotometry measurements involving electronic energy is big enough on the molecule analyzed, so spectrophotometer UV-Vis used more for quantitative analysis compared qualitative research. The spectrum UV-Vis very useful for measuring quantitatively.

3. Methodology and Materials

The first phase of this experiment is looking for a literature that is derived from the theory, research and journals that have been there before. Samples are testing using spectroscopy to obtain data required to analyzed. The type of appliance spectrophotometry used a spectrophotometer UV-VIS in wavelength 340 nm, 370 nm and 400 nm. The data is in the form of temperatures, light absorbance, the speed of the dispersion, and other data. If all data needed has been achieved then the analysis can be started by connecting the data into a parameter. The outward from this research is expected to be a reference in the use of dispersant on the process of treating oil spills at the sea.

3.1. Sample Preparation

(a) (b) (c)

Figure 1. (a) MESLU-*Crude Oil* Dispersant. (b) NEO-CHEM M-405, (b) MAXI CLEAN-2.

The preparation of the samples that will be tested in this research is based on the needs of the data retrieval and restrictions that determined. The sample used in this research are dispersant and oil. The election of the type of oil for a sample of the experiment using MESLU-crude oil. Dispersant used as samples are OSD MAXI-CLEAN 2 and the OSD NEO-CHEM M-405.

3.2. Experimental Set-Up

Dispersant and crude oil taken using pipette and then poured into the beaker glass which filled with a sea water. Then the mixture diluted with magnetic stirrer which functions as the energy that work on the system as a replacement for the waves by set the frequency of the vibrations on the low frequency in order to produce a homogeneous mixture. After the oil has been dispersed, setting the temperature of each of the beaker glass done with the oven and cooler bag to achieve the temperature target.

After that, sample from the mixture taken using a syringe at the top, middle and bottom layer and then placed in a cuvette. The use of syringes aimed to sampling in each layer more precise so that the data obtained has a high accuracy. Cuvette placed into the spectrophotometer UV-Visfor testing of light absorbance. On this experiment, data taken directly by using the acquisition of data that is connected to the computer. The data then recorded in accordance with temperature and a specified time.

4. Results and Discussion

4.1. Graph of Light Absorbance and Temperatures at Layers of Sea Water (Dispersant A)

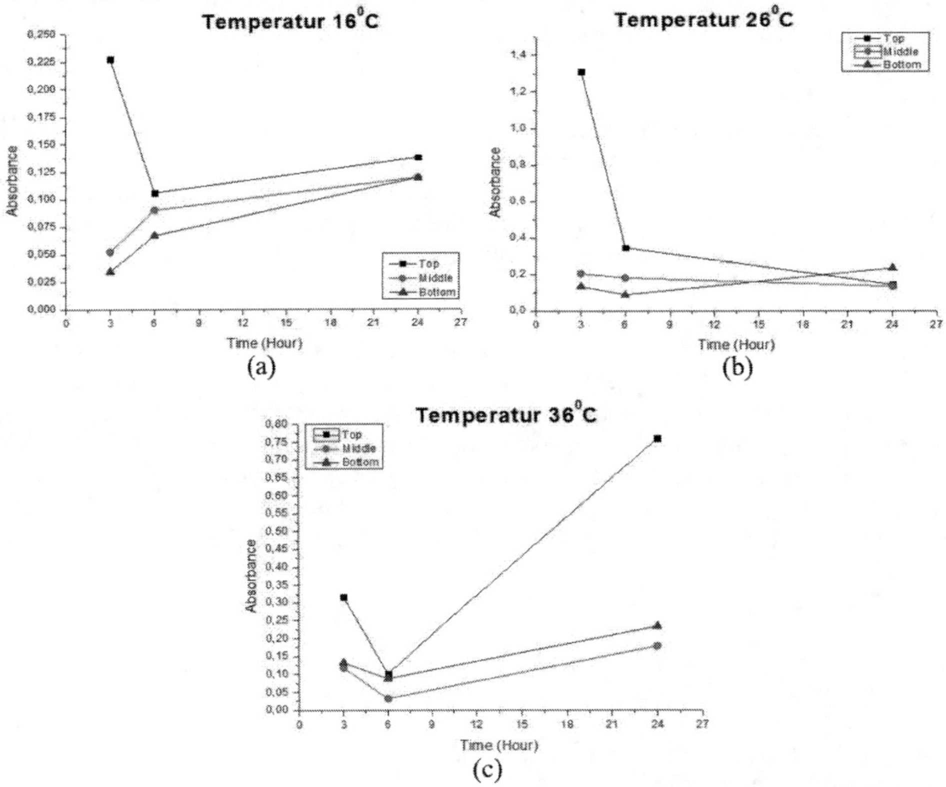

Figure 2. Dispersant A at Temperature (a) 16°C (b) 26°C and (c) 36°C.

Figure 2 shows each layer of water in each of the temperature variations have different absorbance level of light, therefore it can be proved that the temperature can affect the level of light absorbance on each layer of water. In this analysis, light absorbance will be divided into 3 layers, namely top layer, middle layer and bottom layer.

Emulsion occurred in a period of one to two days. In the analysis of light absorbance on a layer of water, it can be seen that the trend that occurred in the middle layer and bottom layers are same. The graph of light absorbance by dispersant soluble in water at temperature 16°C shows that the light absorbance of water layer shows value <0.25 abs by shooting UV with the wavelength of 340 nm.

IOP Conf. Series: Earth and Environmental Science **105** (2017) 012084 doi:10.1088/1755-1315/105/1/012084

The light absorbance on the bottom layer shows the value of 0.034 abs on the first three hours and increase until oil dispersed for 24 hours. For light absorbance on the middle layer has higher level compared with the bottom layer which is 0.052 abs and increase until oil dispersed for 24 hours. The top layer have different trends on the first six hours when oil dispersed. On the graph shows reduced light absorbance and increase until oil dispersed for 24 hours. The value of the light absorbance on the top layer is 0.227 abs, where the value has the difference about 0.2 abs with the bottom layer of water. But the value of light absorbance on the top layer after oil dispersed for 6 hours have values that are almost the same as the middle layer. It shows the stability of the emulsion is very low on each layer of water.

At temperatures 26^0 C, light absorbance on each layer has a different trends, where the layers of the surface of the water has a higher value compared with the middle layer and bottom layer about 1.0 abs, therfore it can be seen that the emulsion which happens have a good stability

Oil dispersing at temperature 36° C have the same trends in each layer of water. In the first three hours on the top layer has the difference value of absorbance about 0.2 abs with middle layer and bottom layer of water. The trend of the graph on the middle layer of water at temperatures 36° C have the same characteristics with middle layer at temperatures 26° C, where the value of light absorbance is lower than the top layer and the bottom of the water. This happened because of the higher temperatures increase from dispersant solubility in water and temperatures of oil spills, reducing the viscosity of oil. Decline in oil viscosity resulted in an increase in the effectiveness of the performance of dispersant.

4.2. Graph of Light Absorbance and Temperatures at Layers of Sea Water (Dispersant B)

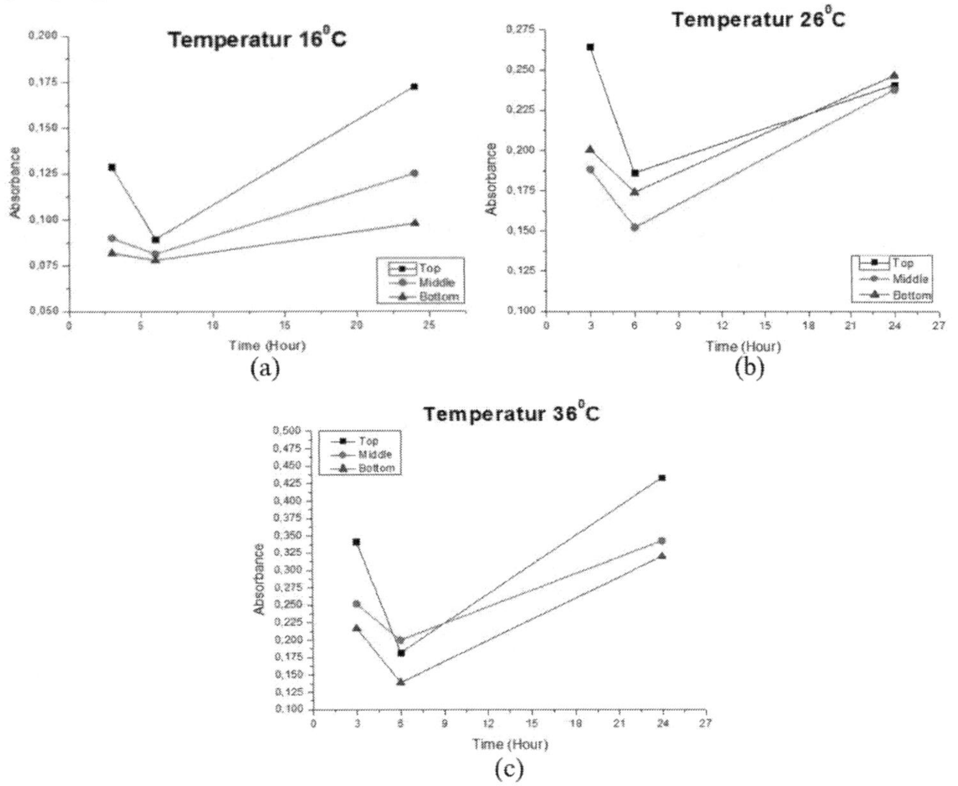

Figure 3. Dispersant B at Temperature (a) 16° C (b) 26° C and (c) 36° C

Dispersing of crude oil with dispersant insoluble in water at temperature 16° C shows the same trends in each layer of water. Based on the figure 3, the graphs shows a significant difference in the first three hours dispersing. The top layer of water has the light absorbance value 0.1229 abs, where the value of the absorbance have differences about 0.04 abs with middle layer and bottom layer of water. Changes in the level of turbidity in the upper layers of the water does not appear on the figure 4, but the results of the testing with spectrophotometer UV-VIS indicates that there is a change in the value of absorbance on each layer. After dispersing for three hours, the of value light absorbance on each layer up until dispersing for 24 hours having the result in the range of 0.07 - 0.1 abs.

The same thing happened on dispersing crude oil at temperature 26° C. The difference of light absorbance in the top layers with the middle layer and bottom layer about 0.06 abs. The level of turbidity decline at the time of dispersing 24 hours. In the first three hours of dispersing, top layers the water has the value of the light absorbance of 0.2645abs.

On dispersing crude oil at temperature 36° C, the value of the light absorbance on the top layer of the water on the first three hours is 0.3422 abs, where there are differences value oflight absorbance about 0.1 abs with the middle layer and bottom layer of water.

4.3. Area of Light Absorbance of Dispersed Oil

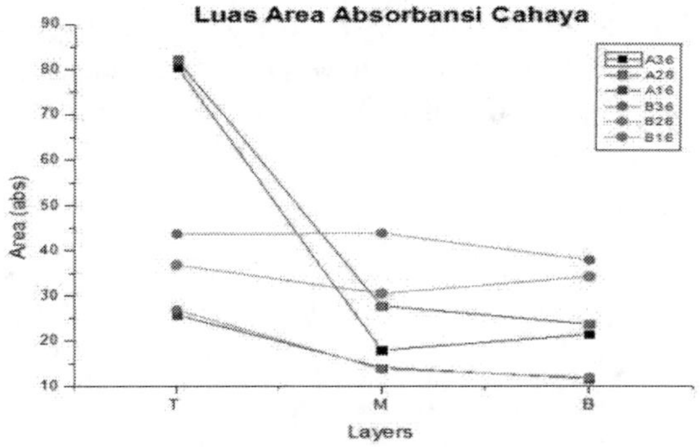

Figure 4. Area of Light absorbance

In this eresearch, the experiment done to correlate the phenomenon of light absorbance with the concentration of oil. As can bee seen on figure 4, each layer has a different area on each temperatures. Based on the graph, the top layer of dispersant soluble in water at temperatures 26° C have the biggest area of light absorbance 82.15 abs, whereas the smallest area of light absorbance at temperature 16^0 C with value 25.72 abs.

Table 1. Area Total of Light Absorbance of Dispersed Oil

Dispersant Types	Absorbance		
	36° C	**26° C**	**16° C**
A	119,84	133,49	51,42
B	125,32	101,54	52,55

Table 1 is a data from the area total on each temperatures of each type of dispersant. The largest area total of light absorbance is at temperatures 26° C with type of dispersant soluble in water with value 133.49 abs. This can happened due to the stability of the emulsifier substances, therefore the process of emulsion that occurs on when the oil dispersed more effective. The value of the area total of light absorbance which is not much different occur at temperature 36° C, both on the type of dispersant insoluble and soluble in the water with absorbance area each are 125.32 abs and abs 119.84. The high temperatures will reduce the viscosity of dispersant and crude oil to increase the performance of dispersant as the substance of oil emulsifier.

5. Conclusions

Oil dispersed caused by emulsification by emulsifier substances such as dispersant. Temperature can affect the effectiveness of dispersant. Area of light absorbance can represent in the determination of the effectiveness of dispersant. Light absorbance on each layer of the water on the oil dispersed has a different value. This is caused by the process of emulsion by emulsifier substances such as dispersant that mixed into crude oil. Area of light absorbance can be a reference to measure the effectiveness performance of dispersant. The larger an area then the effectiveness of the performance of dispersant getting better due to the level of turbidity is lower. The top layer of dispersant soluble in water at temperature 26° C have the biggest area of light absorbance about 82.15 abs, but the smallest area of light absorbance occurs at temperature 16° C by 25.72 abs. Largest area total of light absorbance found at the temperature of 26° C with types of dispersant soluble in water by 133.49 abs while the smallest area total area of light absorbance found at temperature 16° C with types of dispersant soluble in water by 51,42 abs. Based on the experiment that has been done, dispersant soluble in water at temperatures 26° C has the most effectiveness. But the temperature of 36° C still have a high enough of absorbance value, therefore it can be said dispersant can work effectively on the temperature range of 26° C - 36° C.

6. Acknowledgment

This research is funded by PITTA from Universitas Indonesia. Thank you are delivered to Department of Mechanical Engineering - Universitas Indonesia, Mechanical Engineering program study, and Fire Safety engineering group Universitas Indonesia.

7. References

[1] *National Geograhic*. Panjang Garis Pantai. http://nationalgeographic.co.id/berita/2013/10/terbaru-panjang-garis-pantai-indonesia-capai-99000-kilometer (accessed on November 19, 2016)

[2] Pancadewi Y 1990 Depok. Pengaruh Variasi Komposisi Dispersan dan Unsur Nitrogen Terhadap Proses Biodegredasi Tumpahan Minyak Bumi.

[3] Clayton J R Payner J R and Farlow J S 1993 *Oil Spill Dispersants. Mechanism of Action and Laboratory Test* CRC Press, Inc, Boca Raton, Florida p. 113

[4] The International Tanker Owner Pollution Federation Limited (ITOPF). The Use of Chemical Dispersant To Treat Oil Spills. United Kingdom.

[5] Desphande N, Chandrasekar S, Sorial G *Dispersant Effectiveness On Oil Spills – Impact of Environmental Factors* Cincinnati, Ohio.

IOP Conf. Series: Earth and Environmental Science **105** (2017) 012085 doi:10.1088/1755-1315/105/1/012085

Experimental study of thermal radiation from jet flame

P Satrio, R Adityo, R Agung and Y S Nugroho

Fire Safety Engineering Research Group, Department of Mechanical Engineering, Universitas Indonesia.

E-mail: yulianto.nugroho@ui.ac.id

Abstract. This paper presents studies on the influence of the nozzle tilt angle on the on the thermal radiation generated by diffusion jet flame. Analysis was limited to the nozzle tilt angle, distance and height of the heat flux measurement, and nozzle diameter. The studies were conducted for three different nozzle tilt angles of 0o, 45o and 90o, three different distances 10 cm, 15 cm and 20 cm, six different heights 10 cm, 20 cm, 30 cm, 40 cm, 50 cm and 60 cm, and two different nozzle diameters 17 mm and 11 mm. The thermal radiation intensity increased with the distance and height of measurements. The variation of horizontal and vertical jet flame was found to follow a linear equation. The studies also included investigation of flame shapes influence on the thermal radiation. The obtained results are presented in a graphical and tabular form. Based on the analysis of results the conclusion are established, important not only for the theoretical considerations but also in practice, especially in the context of immediate ignition on Liquefied Petroleum Gas as fuel for household leakage.

1. Introduction

Liquefied Petroleum Gas (LPG) is commonly used for households in Indonesia. Not only is LPG a relatively clean, safe and cost-effective fuel for households, its large-scale adoption also reduces the government budgets on subsidies for traditional kerosene consumption. LPG program in Indonesia has been effective for medium and higher income households in suburban areas [1]. High pressure gas inside storage tank heightens the risk rate especially in tubes which are closer to people. Since official using of LPG as the main sources of cooking fuel, several disastrous incidents have frequently occurred [2].

A. Palacios et al. were performed experiments of propane jet flame in still air [3]. The flame height up to 10 m were obtained using different exit diameters (10mm-43.1mm). The experiments were recorded with video and thermographic camera. Flame height and lift-off distance were determined by using visible and infrared images. Flame heights were expressed as function of mass flow rate, exit diameter, Froude number and Reynolds number. The results showed that the flame height is increased with the increase of orifice diameter and the fuel mass flow rate.

Mercedes Gomez-Mares et al. were performed experiment of axial temperature distribution in vertical jet fire of propane [4]. Thermocouples were used to measure temperature along the jet fire centerline and the flame contours were determined from infrared images. The results showed that temperature along the jet fire centerline increased from the bottom of the flame, then reached a maximum value and decreased again at the top zone. The temperature inside the flame were measured around 1600 K at 90% of the flame length. However, it was decided to use the temperature of 1500 K for comparing the data of lengths.

A. Palacios et al. were performed experiments of large outdoor jet fire. Geometric and thermal data were used to estimate the thermal radiation intensity from the flames towards targets located at diverse distances [5]. The experiments were studied the vertical turbulent sonic and subsonic exit velocity propane jet fires. The flame length up to 10.3m released in still air. Infrared images processing was used to analyze the temperatures of the flame surface and the surface emissive power of the flame. Solid flame model in both one-zone and multiple-zone configurations were applied to estimate thermal radiation intensity. The flame was considered as a cylinder defined by the 800 K isotherm. Experimental and predicted thermal radiation intensity values were compared.

K. Zhou et al. were performed predictions of heat flux radiation from jet fire using line source model calculation [6]. This method describing the flame emissive power and subsequent heat flux radiated from a horizontal propane jet fire. The results were evaluated through a testing against experimental fire data and comparison against other models. Comparison of model predictions against data collected in the experiment and predictions from the point source model and multipoint source model gives encouraging results relative to the validity of model system.

This experiment studies the effect different tilt angle of the nozzle against thermal radiation. Heat Flux Sensor was used to measure thermal radiation from jet fire. A tool was designed in such a way so the nozzle could be rotating in 360^0. In this experiment, the radiometer used is useful to measure the radiation received by an object and its effect on the distance and the angle of view of the heat source. The results then will be compared with the result of the theoretical thermal radiation flux calculations using line source model.

2. Line Source Model

There are four kinds of semi-empirical model that have been developed to calculate the thermal radiation from jet fires. Firstly, the point source model that commonly used to predict the far field radiation of both vertical and horizontal jet fires [7]. The second one is multipoint model and solid flame model which was used to calculate the near field radiation from a vertical methane jet fire and horizontal hydrogen jet fires [8]. The third one is cylindrical flame radiation model which was used for vertical and horizontal propane jet fires [9]. The last one is line source model which was recently developed to calculate the radiant heat flux from both small and large scale vertical jet fires [10]. The result using line source model was compared with experimental data measurement. It shows a better prediction than the other three models. Therefore, the line source model is developed to verify the thermal radiation measurement in this paper. The radiant heat flux received by a point using line source model is expressed by:

$$\dot{q}" = \int_S^H \frac{\tau E\prime}{4\pi R^2} \cos\theta dx \quad (1)$$

Where $\dot{q}"$ is the radiant heat flux received by a point, R is the length of connecting line between the target and any point at the jet flame centreline, θ is the angle between the connecting line and the target surface normal orientation, S and H is the lift-off distance and flame length of jet fires from the nozzle to the top of the fire.

The emissive power per line length (EPPL) is the emissive power of a certain point in the centreline. EPPL is proportional to the square of the flame radius [10]. Therefore, the EPPL can be expressed by:

$$E' = E'_0 . (b/b_0)^2 \quad (2)$$

Where is b and b0 are the flame radius and the maximum flame radius, respectively. The integral of EPPL over the full flame length result the total radiant heat output. The maximum EPPL ($E'_0 .$)can be expressed by

$$E'_0 . = \chi_R \dot{Q} / \int_S^H (b/b_0)^2 dx \quad (3)$$

Where χ_R and $\dot{Q}$ are the radiative fraction and the heat release rate, respectively.

IOP Conf. Series: Earth and Environmental Science **105** (2017) 012085 doi:10.1088/1755-1315/105/1/012085

3. Experimental Method

The total experimental program consisted of 36 individual tests, where a test is defined as obtaining data at all distances for each burner diameter setting of a single burner tilt angle. For example, the heat flux data for one test consisted of the measurements taken at each of the three horizontal measurement distances for each of the burner tilt angle settings. Each test was performed in 1 minute when the temperature read on the control system stable and constant. In this test, the data were collected with LabView 2016. Figure 1 outlines the research methodology of heat flux measurement test.

The nozzle was able to be rotated to provide a different field of view to the heat flux gauges. Three different burner angles were tested. Firstly, the burner angle could be set at 0°, meaning that the sides of the burner were parallel with the face of the heat flux gauges. The burner could then be set at 45° and 90^0. These three positions are shown in figure 1. The nozzle tilt angle was varied to be able to see the difference of the result of the heat flux obtained from the sensor.

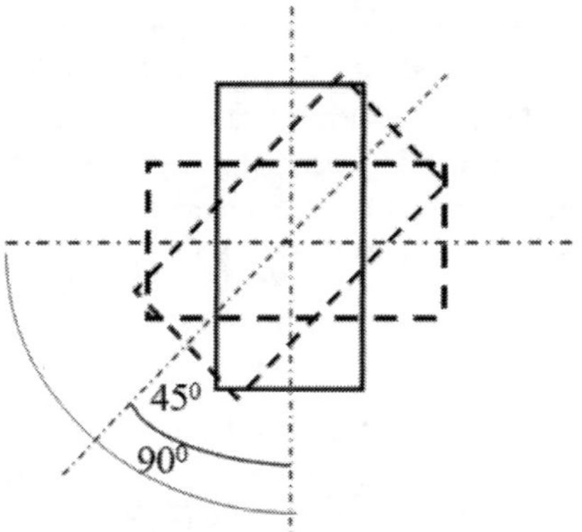

Figure 1. Relative angles of burner positions used in experiments.

The distance of the sensor with the jet flame centerline will be varied into three distances, which is as far as 10 cm, 15 cm, and 20 cm from the jet flame centerline. In this data measurement, the distance will be combined with all the variations in this test. The data variation could be collected only by rotating the gear so that the variation of sensor distance is the most basic variation in this test.

The sensor height is the length from the nozzle of the jet flame to the sensor. In this variation, the height will be varied in six heights, which is as far as 10 cm, 20 cm, 30 cm, 40 cm, 50 cm, and 60 cm. For the 0^0 orientations, the data was obtained by two sensors with the same distance (from left and right direction). The average of the value will be used in the data collecting. Nozzle diameter is the length of the gas burner. In this experiment, nozzle diameter was varied in two sizes, which is 11 mm and 17 mm. For this variable, the nozzle diameter will be combined with all the variations in this test. This experiment was carried out using welding equipment which mounted to the modified table so the nozzle can be rotated 360^0. Fig shows the table that are used in this experiment.

4. Results and Discussion

Figure 2 shows the comparison between the line source model prediction and the experimental data on different measurement heights. Fig a and b shows the result for the vertical jet flames with 17 mm and 11 mm nozzle diameter respectively. The results are in good agreement with the radiant heat flux prediction. Therefore, the line source model seems to be a good representative to calculate the radiant heat flux of the vertical LPG jet flames.

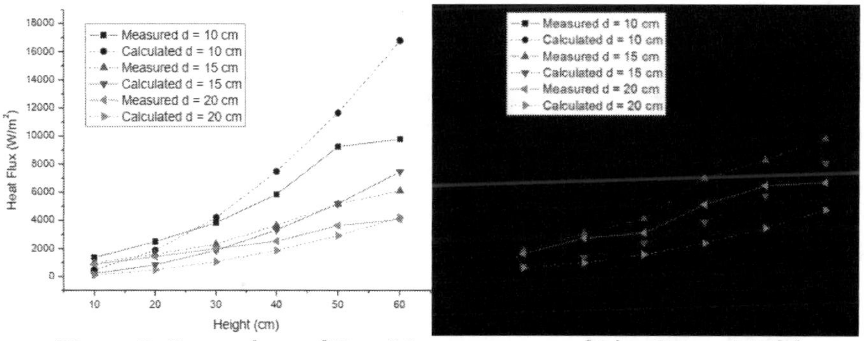

Figure 2. Comparison of Data Measurements and Line Source Model

The heat flux received by a target decreases as the distance from the flame surface increases. It is because the thermal radiation emitted by the jet flame spreads in all directions. Figure 3 shows the heat flux reaching a vertical surface (average values obtained for one minute), measured by heat flux sensors located at two different places with the same distance from the flame axis. It can be seen that the heat flux increases as the distance from the flames increases. This means that the hazards created by this thermal radiation will higher in short distance from the flame surface.

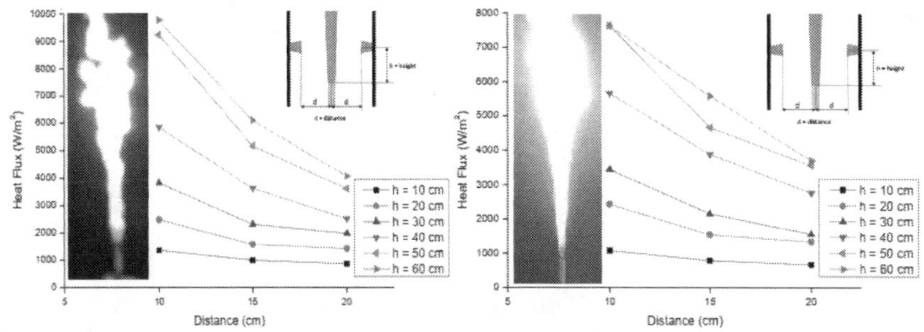

Figure 3. Variation in feasurement distance, height, and nozzle diameter for 90^0 tilt angle

Figure 4 presents the result of radiant heat flux measurement with 45^0 and 90^0 nozzle tilt angles. The hat flux received by a target are fluctuated as the distance from the flame surface increases. It is because flame shape of the flame was not steady. The distances of flame surface to the heat flux sensor are very affect the measured heat flux, the closer flame surface to the sensors, the higher heat flux received [11].

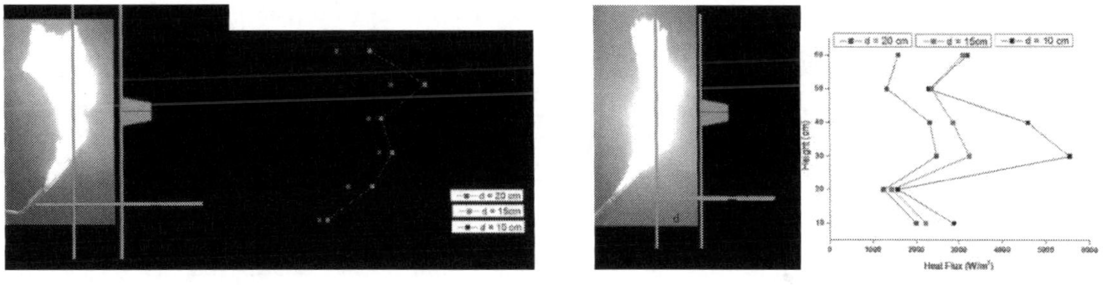

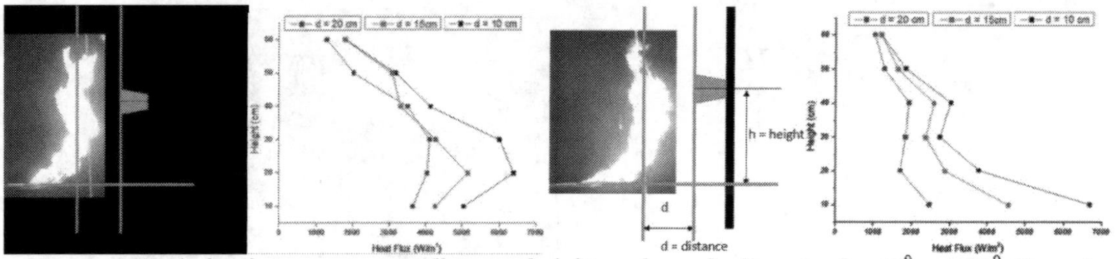

Figure 4. Variation in measurement distance, height, and nozzle diameter for 45^0 and 90^0 tilt angle

5. Conclusion

The experimental results obtained with different nozzle tilt angle jet flames have shown that the heat flux profiles of the flames depend on the fuel release. If the nozzle tilt angle is set at 0^0, the heat flux received by a target increase as the height of measurements increase. Heat flux trends in experimental data from 45^0 and 90^0 nozzle tilt angles has been fluctuated depends on flame shape. Although heat flux decreases quickly with the increase of distance from flames, at short distances it can be very high, which would could cause serious hazards. The experimental results also showed the validity of line source model calculation to predict thermal radiation from jet flame. Heat flux from vertical jet flames was used to compare with the line source model calculation. A comparison of the values thus obtained with the experimental results and the use of a line source model are in good agreement.

6. Acknowledgment

We acknowledge financial support provided by PITTA. The authors would like to thank the financial support provided by Universitas Indonesia through the PITTA 2017 funding scheme under Grant no. 857/UN2.R3.1/HKP.05.00/2017 managed by the Directorate for Research and Public Services (DRPM) Universitas Indonesia.

7. References

[1] Kities R, Mulder P and Rietveld P 2014 Energy poverty reduction by fuel switching. Impact evaluation of the LPG conversion program in Indonesia *Energy Policy* **66** pp. 436–449

[2] Addai E K, Tulashie S K, Annan J and Yeboah I 2016 Trend of fire outbreaks in Ghana and ways to prevent these incidents *Saf. Health Work* **7** 4 pp. 284–292

[3] Palacios A, Muñoz M and Casal J 2009 Jet fires: An experimental study of the main geometrical features of the flame in subsonic and sonic regimes *AIChE J* **55** 1 pp. 256–263

[4] Gómez-Mares M, Muñoz M and Casal J 2009 Axial temperature distribution in vertical jet fires *J. Hazard. Mater.* **172** 1 pp. 54–60

[5] Palacios A, Muñoz M, Darbra R M and Casal J 2012 Thermal radiation from vertical jet fires *Fire Saf. J.* **51** pp. 93–101

[6] Zhou K, Liu J and Jiang J 2016 Prediction of radiant heat flux from horizontal propane jet fire *Appl. Therm. Eng.* **106** pp. 634–639

[7] Lowesmith B J, Hankinson G, Acton M R and Chamberlain G 2007 An overview of the nature of hydrocarbon jet fire hazards in the oil and gas industry and a simplified approach to assessing the hazards *Process Saf. Environ. Prot.* **85** 3 pp. 207–220

[8] Hankinson G and Lowesmith B J 2012 A consideration of methods of determining the radiative characteristics of jet fires *Combust. Flame* **159** 3 pp. 1165–1177

[9] Gopalaswami N, Liu Y, Laboureur D M, Zhang B and Mannan M S 2016 Experimental study on propane jet fire hazards: Comparison of main geometrical features with empirical models *J. Loss Prev. Process Ind.* **41** pp. 365–375

[10] Zhou K and Jiang J 2015 Thermal radiation from vertical turbulent jet flame: line source model *J. Heat Transfer* **138** 4 p. 42701

[11] Satrio P 2017 Analysis of influence of jet flame tilt angle on the thermal radiation *Universitas Indonesia*

Analysis of effectiveness of oil spill recovery using disc-type oil skimmer at laboratory scale

N Widiaksana, A A Yudiana and Y S Nugroho

Fire Safety Engineering Research Group, Department of Mechanical Engineering, Universitas Indonesia.

E-mail: yulianto.nugroho@ui.ac.id,

Abstract. During the past 40 years, there have been 14 major accidents of oil spill in Indonesia, mostly as results of ship collisions. There are several methods in handling oil spill accident, in which the most effective methods are using mechanical oil skimmer with disc plate. The effectiveness of the oil skimmer on handling oil spills is influenced by various factors, such as the depth of the disk submerged or the disk surface area dipped into the oil spill, the area of the wiper sweep, the thickness of the oil on the disk surface, and the rotation speed of the disk. This paper presents the results of laboratory scale experiment of mechanical oil skimmer. The oil samples used are diesel engine oil with 15W-40 SAE and the water sample used is sea water. The experiment was conducted with variation of 27 mm and 55 mm of disc submersion depth or equal to dipped surface area of 31.35 cm^2 and 88,832 cm^2. The duration of the test for 3 data is 5 minutes. Based on theoretical calculations, the increase of rotation speed of the disk, the result of spill transport will also be higher. This is proved by the experimental results. The lifting process of oil spill is more effective with low rotation speed, because the result of oil spill transported will be more dominated by oil than water. In this test, the higher the rotation speed of the disk, the higher the water produced.

1.Introduction

Indonesia as a producer of Oil and Gas, also has the potential to contribute to environmental pollution in the form of oil spills as a result of oil and gas transportation process to and from Indonesia. According to data from the Ministry of ESDM, there are at least 14 large oil spill cases occurring in Indonesian waters since 1975 to 2009. One of the recorded accident was montera oil spills on Timor sea at 2009 which record 27.500 tons oil losses. Based on South East Asia oil trading lines, Indonesian water area in Malacca Strait, Lombok Strait, and Makassar strait are highly threatened by potential occurrence of sea pollution as a result of spilled oil by tanker or vessel collision. This condition also supported by high shipping activities from Japanese vessel for about 90% and Chinese vessel for about 70-90 % [1].

However, due to numerous accident of water pollution due to oil spill occurrence, counter action should be taken. Methods to counter oil spill are various, there are: Burning, Sinking, Dispersing, Sorbing, Biodegradation, Gelling, and Skimming [2]. Skimming method is known for its environmental friendly features, in which this method works mechanically to recover spilled oil on polluted area using its skimmer devices. There are two types of skimmer, Oleophilic Skimmer and Non-Oleophilic Skimmer, with each of them posing its own advantages and disadvantages [3].

This research is intended to study on working effectiveness of oil disc skimmer by conducting research of this custom-made oil spill recovery device in laboratory scale.

2. Methodology

2.1. Data Gathering
In order to test disc type oil skimmer working effectiveness, several methods will be applied, those test variables are depth of disc in liquid, area (slope of wiper) of wiped zone on disc, and disc rotation speed. These methods are shown in table 1. Each test will have 3 sets with each set consume 5 minutes for every take. Type of oil that will be used are Diesel Oil Lubricant with SAE 15W-40 produce by Fastron, and will be pour into sea water that will be placed into a vessel with certain amounts.

Table 1. Data Variables

Data Variables	Values
Wiped Area/Wiper Slope Area	18 degree/139 cm2
	37 degree / 90.5 cm2
Depth of Disc	27 mm
	55 mm
Disc Rotation Speed	Low Speed (5-20 RPM)
	Medium Speed (20-40 RPM)
	High Speed (40-55 RPM)

2.2. Test Schematic

2.2.1. Experiment Devices. Component of experiment devices consist of this following item:

1. Disc Type Oil Skimmer
   As shown by figure 1, this is the assembly of disc type oil skimmer that will be used in this experiment, moreover this device also considered as the main item for this experiment.

Figure 1. Disc Type Oil Skimmer [4].

   In this experiment, the dimension of water tub is 320 mm x 215 mm x 150 mm. This equipment will be filled with certain amount of water in order to duplicate the actual condition where oil spill occurs, also this is intended to get easier way to measure water and oil ratio.
2. Measuring Cylinder
   This equipment will be used to measure mixture of oil and water recovered by oil skimmer.
3. *Digital Tachometer*
   Works to measure rotation speed at disc of the oil skimmer during data gathering.
4. Diesel Engine Lubricant (*Fastron SAE 15W-40)*

This lubricant used as replacement of crude oil. In every set of test, the amount of oil poured into water vessel is 200 ml.

2.2.2. Procedure of Test Preparation. Before experiment being conducted, there are few things that have to be set up. These procedures are intended to make data gathering save and not consuming too much time. The procedure are as follow: First, we have to fill water tub with mixture of water and oil using ratio off 7500 ml : 200 ml and 9000 ml : 200 ml. next we have to calibrate every measuring devices before we start the experiment. Ex: Digital Tachometer and volt regulator. Do not forget to use body protection, such as: hand gloves, shoes, and mouth cover. Place water tub below working desk and ensure that disc have been dipped into water. Prepare measuring cylinder as collecting vessel for recovered oil spill.

2.2.3 Procedure of experiment. After completing procedure written on subsubsection 2.2.2, experiment proceeded into data gathering followed by this instruction: First, Set-up the rotation according to the variation of the desired rotation. Then, pour the seawater and oil into the water bath with a predetermined ratio. Finally, conducting testing with variations that have been determined before. For data gathering, we have to recording rounds of discs every minute. The experiment was conducted for 5 minutes in one experimental variation. Recording the results of oil spill haul contained in measuring cups. Repeat the instruction for desired variations.

3. Results and Discussion

3.1. Theoretical Calculation

Theoretical calculation will be used to identify amount of oil can be recover using disc type oil skimmer. This calculation consists of several components, depth/area of dipped disc, disc rotation, test duration, and oil thickness. Thus, such components can be expressed in form of following formula:

$$Amount\ of\ recovered\ oil\ spill = Area\ x\ rotation\ speed\ x\ time\ x\ oil\ thickness \quad (1)$$

Where:
Area : depth/area of dipped disc
Rotation speed : Disc rotation (RPM)
Time : Duration taken during each test
Thickness : Oil thickness (cm)

According to equation (1), area described as surface of disc that dipped into water solution and when the disc rotates several amounts of oil stick to the surface, then this area filled with oil can be wipe using wiper devices. There are three type rotation speeds: low speed, medium speed, and high speed. Duration of test of each session takes 5 minutes. We assumed oil thickness at surface of the disc will be around 0,5 mm. The result of theoretical calculation is shown in Table 2.

Table 2. Result of theoretical analysis

Area (cm2)	Rotation Speed (RPM)	Total recovered oil (ml)
31.5	20	157.6
	40	315
	55	433
89	20	444.2
	40	888.8
	55	1221

According to table 2 above, we can conclude when higher rotation speed applied will result in greater number of oil being recovered. This also applies when surface area is greater. Therefore, theoretically this would produce higher efficiency.

3.2 Result of Devices Experiment

3.2.1. Oil Spill Recovery for 27 mm disc depth. In this part of experiment, there are 2 identified surface areas that used. The first one is using 90.5 cm^2 or 37° of wiper angle of slope. The second variation is 139 cm^2 or 18° of wiper angle of slope. Based on those two variables, we produce this result:

3.2.1.1. Surface area of 9046,46 mm^2/ wiper angle of slope 37°

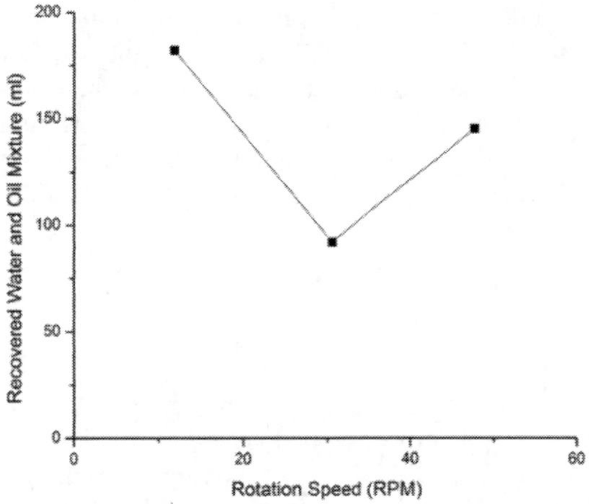

Figure 2. Result of oil spill recovered using surface area of 90.5 cm^2/ wiper angle of slope 37° with 3 variations of speed [4]

Based on the result of the test, there are significant decreases in number of oil recovered when medium speed being applied. According to figure 2, when low rotation speed (12 RPM) being applied number of oil being recovered is 182 ml, this is the highest amount if we compare to medium speed rotation (30.5 RPM) and high-speed rotation (48 RPM) which each of them produce 92 ml and 145 ml respectively.

3.2.1.2. Surface area of 139 cm^2/ wiper angle of slope 18°

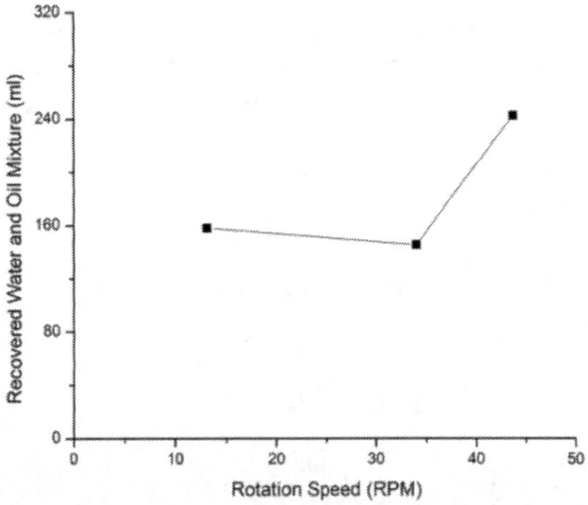

Figure 3. Result of oil spill recovered using surface area of 139 cm^2/ wiper angle of slope 18° with 3 variations of speed [4]

As shown in figure 3, when medium speed applied on disc depth of 27 mm there will be a decrease of recovered oil for wiper slope of 18° with recorded amount of 145 ml, this condition is similar to application of 37°. The decrease of recovered oil will be cause by several factors such as wiper error which lead to ineffectiveness of separating oil and water from the disc. If we compare to theoretical analysis, then we can conclude that there is a significant different result of the character from recovered oil amount based on type of rotational speed that used, since theoretical analysis implies that whenever the speed goes higher may increase amount of oil being recovered. On the other hand, based on experimental test the increasing of recovered oil only occurs when rotational speed is low.

3.2.2. Oil Spill Recovery for 55 mm disc depth. In this part of experiment, there are 2 identified surface area that used. The first one is using 90.5 cm^2 or 37° of wiper angle of slope. The second variation is 139 cm^2 or 18° of wiper angle of slope. Based on those two variables, we produce this result:

3.2.2.1. Surface area of 90.5 cm^2/ wiper angle of slope 37°

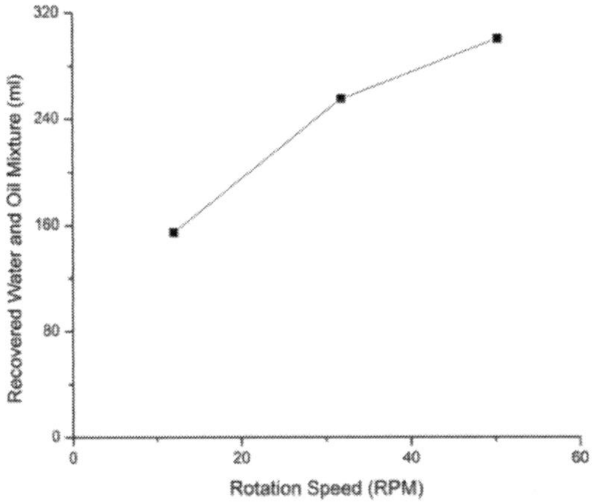

Figure 4. Result of oil spill recovered using surface area of 90.5 cm^2/ wiper angle of slope 37° with 3 variations of speed [4]

Based on the result of the test, there are significant increases in number of oil recovered when variety of speed being applied for 55 mm disc depth. In this variation, the characteristic of recovered oil is increasing gradually, parallel to the increasing of rotational speed. According to figure 4, when low rotation speed (12 RPM) being applied number of oil being recovered is 155 ml, for medium rotation speed (32 RPM) number of oil being recovered is 255 ml, and for high rotation speed (50 RPM) number of oil being recovered is 301 ml.

3.2.2.2. Surface area of 139 cm^2/ wiper angle of slope

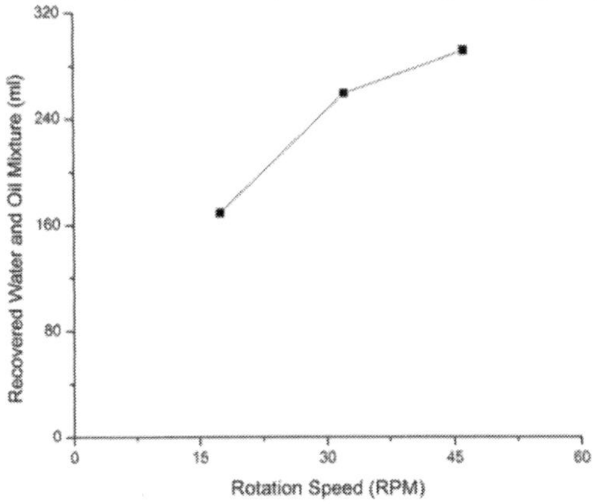

Figure 5. Result of oil spill recovered using surface area of 139 cm^2/ wiper angle of slope 18° with 3 variations of speed [4]

Based on the result of the test, there are significant increases in number of oil recovered when variety of speed being applied for 55 mm disc depth. In this variation, the characteristic of recovered oil is increasing gradually, parallel to the increasing of rotational speed. According to figure 5, when low rotation speed (17.5 RPM) being applied number of oil being recovered is 169 ml, for medium rotation speed (32 RPM) number of oil being recovered is 259 ml, and for high rotation speed (46 RPM) number of oil being recovered is 292 ml.

4. Conclusions

When 27 mm depth being applied, will produce area of 31.5 cm^2 and for 55 mm depth being applied will produce area of 89 cm^2. Based on theoretical calculation, if we apply larger area will produce higher amount oil being recovered. If higher rotation speed being applied, the quality of recovered solution will mostly consist with water. This condition can be proofed only with oil skimmer experiment test. Based on experiment data, slope angle of wiper or wiped area will not affect the amount of recovered oil. Since, when rotating oil will not covering every wiped area as an effect of tangential speed, so the oil tends to gather at the tip of the disc. In order to gain higher amount of recovered oil, it is necessary to consider the depth of disc on the water. If possible, the depth of dipped disc should reach its own radius [4].

5. Acknowledgment

This research is funded by PITTA from Universitas Indonesia. Thank you are delivered to Department of Mechanical Engineering - Universitas Indonesia, Mechanical Engineering program study, and Fire Safety engineering group Universitas Indonesia.

6. References

[1] Tumpahan Minyak Risiko Tinggi di Laut Masih Terus Mengintai *Kompas* 2017
[2] Clark R B 1986 *Marine Pollution* United States: Oxford University Press New York
[3] The International Tanker Owners Pollution Federation Limited 2012 *Use of Skimmers in Oil Pollution Respone* Impact PR & Design Limited, Canterbury, UK
[4] Widiaksana N 2017 Analisis efektivitas pengangkutan tumpahan minyak menggunakan oil skimmer tipe piringan pada skala laboratorium *Universitas Indonesia*

Exergoeconomic analysis and optimization of a combined double flash – binary cycle for Ulubelu geothermal power plant in Indonesia

A Rachmat, Nasruddin, A S Wibowo and A Surachman

Mechanical Engineering Department, Faculty of Engineering, Universitas Indonesia Kampus UI Depok, 16424, Indonesia

Email: nasruddin@eng.ui.ac.id

Abstract. In this paper, the combination of double-flash and binary cycles for Ulubelu geothermal power plant is proposed and optimized by using the Matlab software. This proposed system uses real data and properties of brine exploited from the Ulubelu geothermal well in Indonesia and four working fluid candidates, namely n-Pentane, R141b, R123 and R245fa are used in binary cycles. Optimization using a multi-objective genetic algorithm with an exergoenomic approach is applied to find out the proposed system performance from both thermodynamic and economic point of view. In the optimization procedure, the exergy efficiency and total specific cost of output power become objective functions while the first flash pressure, second flash pressure and Organic Rankine Cycle (ORC) turbine inlet temperature are selected as constraints. The system performance proposed in this paper is compared with the performance of the existing system. The results show that n-pentane is the best working fluid where multi-objective optimization indicates that the system can generate 63.54 MW of power with thermal and exergy efficiencies of 17.59% and 65.26% and specific cost of 1.7049 USD / GJ at the selected optimal design point. Compared to the existing system, there is a significant improvement in performance both from thermodynamic performance and economic performance.

1. Introduction

In recent years, renewable energy sources have been the focus of researchers because in addition to being environmentally friendly, these energy sources can also be quickly restored naturally so they will not run out if managed properly. Renewable energy source is one alternative solution to environmental problems caused by fossil energy. One promising renewable energy source because of its consistency and reliability is geothermal energy [1].

Based on geological condition, geothermal energy source can be divided into vapor-dominated and liquid-dominated. Vapor-dominated generally using dry steam system, while liquid-dominated mostly using flash system [2-4]. Organic Rankine Cycle (ORC) systems are also commonly used to harness the energy contained in liquid-dominated to generate electricity [2, 5-7]. The flash system is advantageous for higher temperature sources while the ORC system can utilize energy from sources that have lower temperatures [2, 8].

Several studies have been conducted related to double flash system and ORC system. Agung [9] proposed a double-flash system for Dieng GPP and compared it with existing single flash systems and concluded that double-flash system power output and exergy efficiency were 29.155 MW and 44.04%, respectively. Dipippo, in 2008, reported that the double-flash plant increased power output by 15-25% compared to a single flash plant under the same geothermal fluid conditions [9, 10]. Shokati et al [3] analyzed and compared the double flash and single flash / ORC cycles for geothermal power plants and reported that the dual flash cycle has a lower power cost.

Ulubelu is one of the geothermal power plant in Indonesia located in Lampung. Ulubelu has a liquid-dominated reservoir and uses a single flash system. Ulubelu geothermal power plant has a temperature and pressure of 179° C and 9.8 bar. With a single flash system, after passing through the separator, the brine coming out of the Ulubelu geothermal power plant is injected back into the reservoir after the precipitation process. Brine still has a mass flow rate, temperature and pressure respectively of 650 tons / hour, 171.48° C and 8.21 bar. In other words, this brine still contains the energy that can be extracted. One way that can be done to extract the remaining energy is by the addition of flash separator (double flash) and ORC.

This study aims to design and propose new configurations of combined flash-binary cycles that efficiently utilize the energy of available heat sources in Ulubelu geothermal field as well as analyze their performance by applying exergoeconomic analyzes. The performance of the proposed system is assessed and compared with the existing cycle from thermodynamic and economic point of view. Matlab software is used to simulate existing cycles and proposed cycles based on design data and actual data. In addition, this software is also used to perform multi objective optimization by using multi-objective genetic algorithm with an exergoenomic approach is applied to find out the proposed system performance from both thermodynamic and economic point of view.

2. System description and modelling

2.1. System description and assumptions

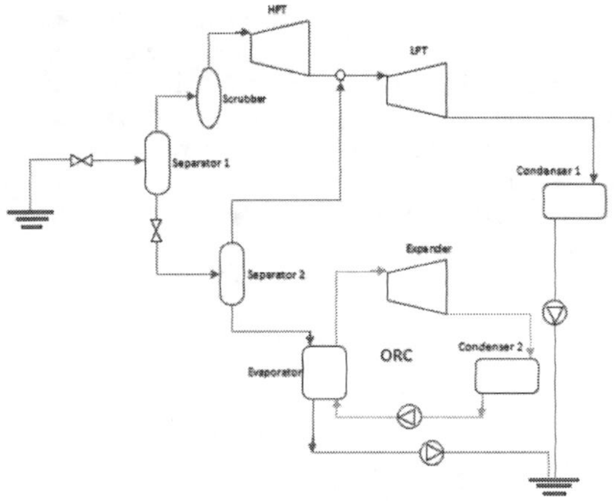

Figure 1. Schematic diagram of the proposed combined flash-binary cycle

Figure 1 shows the schematic diagram of the proposed combined flash-binary cycle. This system is designed by using real data such as pressure, temperature and mass flow rate in Ulubelu geothermal well. This cycle starts from a pressurized steam-brine mixture and high temperature flows through the throttling valve, then separated into two streams namely vapor and liquid brine in separator 1. The vapor is then flowed directly to the HPT by passing the scrubber to ensure the purity of the vapor. And the liquid brine flowed into the second separator after throttled. The steam that is still in separator 2 is flowed to the LPT after it is mixed with steam coming out of the HPT and then it is injected back to the injection well after it is condensed at condenser 1. Liquid brine from separator 2 which still has temperatures above 100° C, is passed to the evaporator in the binary cycle (ORC) before injection to the injection well. The binary cycle utilizes the thermal energy of the liquid brine coming out of separator 2, which is wasted when not in use. In binary cycles, four working fluids were used in this study to optimize the performance of this combined flash-binary system. The fluids are selected based on available heat sources [11]. The fluids include n-pentane, R141b, R123 and R245fa. Some of the assumptions used are similar to those of Aali et al [2].

2.2. Thermodynamic Analysis

To simulate and evaluate the thermodynamic performance of the proposed system, a computer program was developed by using Matlab software, while the thermodynamic properties of working fluid were calculated using Refprop software. The performance of GPP thermodynamics can be evaluated with thermal efficiency and exergy efficiency. The thermal efficiency and exergy efficiency for the system under consideration is defined as [12, 13]:

$$\eta_{th} = \frac{\dot{W}_{net}}{\dot{Q}_{geo}} = \frac{\dot{W}_{net}}{\dot{m}_1(h_1 - h_0)} \tag{1}$$

$$\eta_{ex} = \frac{\dot{W}_{net}}{\dot{E}_{geo}} = \frac{\dot{W}_{net}}{\dot{E}_1} \tag{2}$$

$$\dot{W}_{net} = \dot{W}_{HPT} + \dot{W}_{LPT} + \dot{W}_{Expander} - \dot{W}_{ORC\_Pump} - \dot{W}_{Pump1} - \dot{W}_{Pump2} \tag{3}$$

2.3. Exergoeconomic Analysis

The specific exergy cost theory methodologies (SPECO) have been widely used to investigate thermodynamic cycles in exergoeconomic analyzes. To perform an exergoeconomic analysis using the SPECO method, several equations are used, among them the cost balances equation as follows [2, 14]. For a specific total cost can be calculated by [15]:

$$c_{p\_total} = \sum_{i=1}^{n_p} c_{P_i} = \frac{\sum_{i=1}^{n_k} \dot{Z}_k + \sum_{i=1}^{n_f} \dot{C}_{f_i}}{\sum_{i=1}^{n_p} \dot{E}_{P_i}} \tag{4}$$

3. Results and discussion

The objective function in this study is to maximize exergy efficiency and minimize total specific cost. Some real data and assumptions used as inputs in simulations are shown by table 1 and table 2. The parameters chosen as the contraints are first flash pressure (P_2), second flash pressure (P_7) and ORC turbine inlet temperature (T_{18}), with details:

$$800 \leq P_2\ (\text{kPa}) \leq 980$$
$$300 \leq P_7\ (\text{kPa}) \leq 700$$
$$100 \leq T_{18}\ (^{o}\text{C}) \leq 130$$

IOP Conf. Series: Earth and Environmental Science **105** (2017) 012087 doi:10.1088/1755-1315/105/1/012087

Table 1. Real properties of geofluid available from Ulubelu geothermal wells

Parameter	Unit	Value
Pressure	kPa	980
Temperature	°C	179
Mass flow rate	kg/s	271.94

Table 2. Input data assumed for the system simulation

Parameter	Symbol	Value
Ambient pressure	P_0 (kPa)	100
Ambient temperature	T_0 (°C)	25
Turbine isentropic efficiency	η_T (%)	85
Annual plant operation hours	τ (hr)	7446
Interest rate	i_r (%)	10
Plant economic life	n (year)	20

3.1. Results of multi-objective optimization

As shown earlier, four ORC fluids namely n-pentane, R141b, R123 and R245fa were selected to maximize exergy efficiency and minimize total specific cost of the system. Optimization is done by using Matlab software and the results are shown in figure 2a-d.

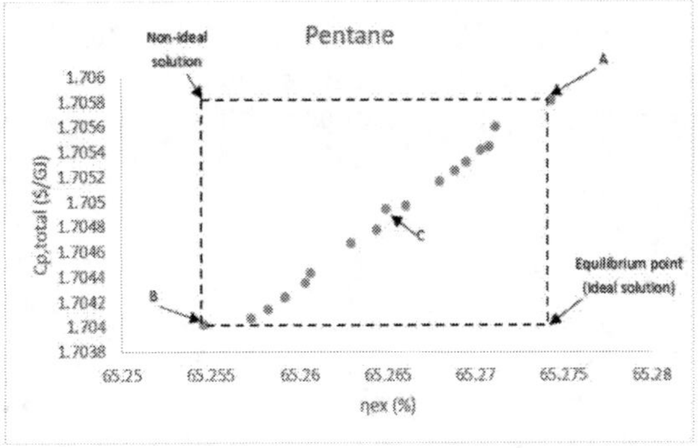

Figure 2a. Result of multi-objective optimization for n-pentane

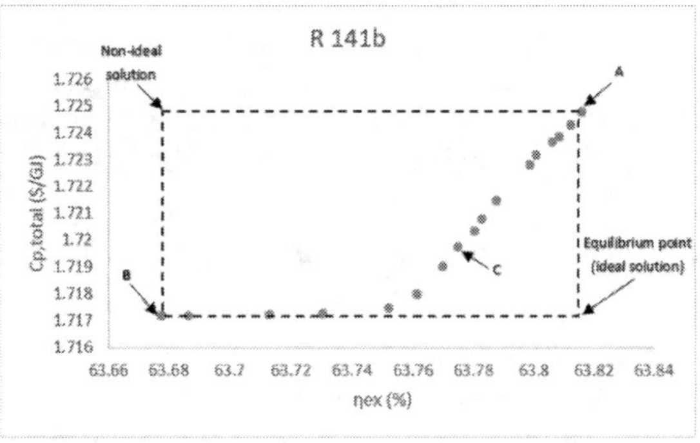

Figure 2b. Result of multi-objective optimization for R141b

IOP Conf. Series: Earth and Environmental Science **105** (2017) 012087 doi:10.1088/1755-1315/105/1/012087

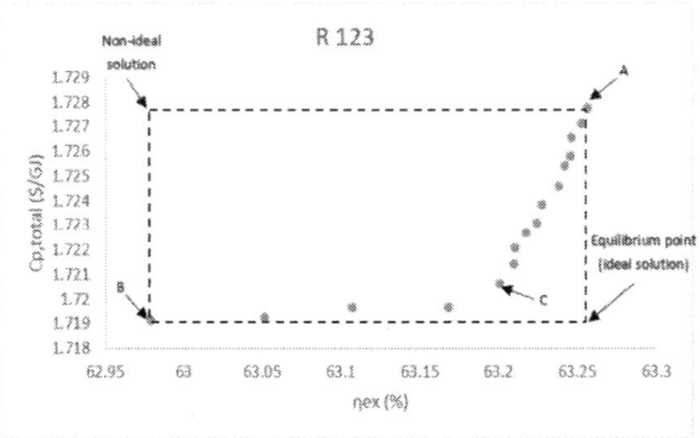

Figure 2c. Result of multi-objective optimization for R123

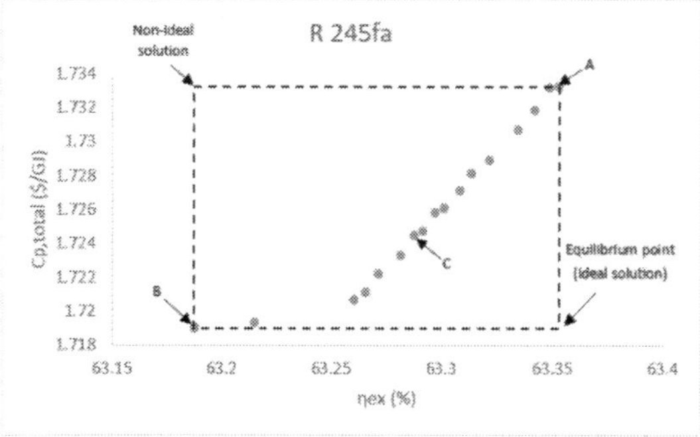

Figure 2d. Result of multi-objective optimization for R245fa

Figure 2a-d shows the Pareto frontier solution for the combined combined binary flash cycle with four working fluids namely n-pentane (figure 2a), R141b (figure 2b), R123 (figure 2c) and R245fa (figure 2d) as ORC working fluid. Each point of the Pareto frontier is the optimal solution for the two objective functions under consideration: the exergy efficiency and the total specific cost. The best thermodynamic performance for n-pentane (figure 2a) is achieved at point A where the exergy efficiency is 65.274% with the worst economic performance of 1.706 USD/GJ for the specific cost; Meanwhile, the best economic performance is achieved at point B where the specific cost of 1.704 USD/GJ with the lowest exergy efficiency of 65.255 %. Similarly, in other working fluids, point A shows the best thermodynamic performance where there is a maximum value of exergy efficiency while point B shows the best economic performance where there is a minimum value of specific cost. The points A and B values of each working fluid (figure 2a-d) are shown by table 3.

From table 3, it can be seen that from all four working fluids considered, n-pentane becomes the best thermodynamic performance fluid because it produces the highest value of exergy efficiency. Similarly, from the economic side, n-pentane is a working fluid with the best economic performance because it requires the lowest cost.

Table 3. Value of the best thermodynamic and economic performance for each ORC working fluid.

Working Fluids	The best thermodynamic performance (point A)		The best economic performance (point B)	
	η_{ex} (%)	c_{p_total} (USD/GJ)	η_{ex} (%)	c_{p_total} (USD/GJ)
n-pentane	65.274	1.706	65.255	1.704
R141b	63.816	1.725	63.678	1.717
R123	63.257	1.728	62.979	1.719
R245fa	63.353	1.733	63.188	1.719

To determine the point at which all objective functions have the most optimum value requires an approach where the nearest point at the Pareto frontier with the ideal solution (point C) is usually defined as the final optimal point [16]. Using this approach for the system under consideration, the final optimal point for each working fluid can be found in Pareto as shown in Figure 2a-d where the exergy efficiency and specific cost at the final optimum point (point C) for each working fluid can be seen in table 4.

Table 4. Performance comparison of the proposed cycle in this study with existing system at Ulubelu geothermal power plant.

No	Parameters	Unit	Single flash	Combine flash-binary (point C)			
				n-pentane	R141b	R123	R245fa
1	W_{net}	MW	52.38	63.54	61.99	61.55	61.36
2	η_{th}	%	14.73	17.59	17.18	17.04	17.04
3	η_{ex}	%	54.88	65.26	63.78	63.2	63.29
4	c_{p_total}	USD/GJ	1.938	1.7049	1.7197	1.7206	1.7245

Table 4 shows that n-pentane working fluid has a superior value compared to the other three working fluids, which results in an exergy efficiency of 65.26% and a specific cost of 1.7049 USD/GJ. When compared with the existing system (single flash), there is an increase in power of 11.16 MW. The thermal and exergy efficiency of the proposed combined flash-binary cycle is significantly higher than the existing system. In addition, the specific cost is also much lower than the existing system.

4. Conclusions

In this study, a new configuration of combined flash-binary cycle is proposed to utilize the available heat source energy in Ulubelu geothermal field more efficiently. To determine the proposed system performance from a thermodynamic and economic point of view, an exergeconomic analysis was performed with exergy efficiency and specific costs as an objective function of this analysis. The results show that n-pentane is the best working fluid where multi objective optimization indicates that the system can generate 63.54 MW of power with thermal and exergy efficiencies of 17.59% and 65.26% and specific cost of 1.7049 USD/GJ at the selected optimal design point. Compared to the existing system, there is a significant improvement in performance both from thermodynamic performance and economic performance.

5. Acknowledgment

The authors gratefully acknowledge to DRPM Universitas Indonesia for supporting this research by PITTA research grant 2017.

6. References

[1] Zare V 2015 A comparative exergoeconomic analysis of different ORC configurations for binary geothermal power plants *J. Energy Conversion and Management* **105** pp. 127-138

[2] Aali A, Pourmahmoud N and Zare V 2017 Exergoeconomic analysis and multi-objective optimization of a novel combined flash-binary cycle for Sabalan geothermal power plant in Iran *J. Energy Conversion and Management* **143** pp. 377-390

[3] Shokati N, Ranjbar F and Yari M 2015 Comparative and parametric study of double flash and single flash/ORC combined cycles based on exergoeconomic criteria *J. Applied Thermal Engineering* **91** pp. 479-495

[4] Sarr J-AR and Mathieu-Potvin F 2015 Improvement of Double-Flash geothermal power plant design: A comparison of six interstage heating processes *J. Geothermics* **54** pp. 82-95

[5] Guzović Z, Rašković P and Blatarić Z 2014 The comparision of a basic and a dual-pressure ORC (Organic Rankine Cycle): Geothermal Power Plant Velika Ciglena case study *Energy* **76** pp. 175-186

[6] Astolfi M et al. 2014 Binary ORC (Organic Rankine Cycles) power plants for the exploitation of medium–low temperature geothermal sources–Part B: Techno-economic optimization *J. Energy* **66** pp. 435-446
[7] Walraven D, Laenen B and D'haeseleer W 2013 Comparison of thermodynamic cycles for power production from low-temperature geothermal heat sources *J. Energy Conversion and Management* **66** pp. 220-233
[8] Coskun A, Bolatturk A and Kanoglu M 2014 Thermodynamic and economic analysis and optimization of power cycles for a medium temperature geothermal resource *J. Energy conversion and management* **78** pp. 39-49
[9] Pambudi NA et al. 2015 Performance improvement of a single-flash geothermal power plant in Dieng, Indonesia, upon conversion to a double-flash system using thermodynamic analysis *J. Renewable Energy* **80** pp. 424-431
[10] DiPippo R 2012 *Geothermal power plants: principles, applications, case studies and environmental impact* (Oxford: Butterworth-Heinemann)
[11] Zare V 2016 Exergoeconomic analysis with reliability and availability considerations of a nuclear energy-based combined cycle power plant *J. Energy* **96** pp. 187-196
[12] Jalilinasrabady S et al. 2012 Flash cycle optimization of Sabalan geothermal power plant employing exergy concept *J. Geothermics* **43** pp. 75-82
[13] Fiaschi D et al. 2014 An innovative ORC power plant layout for heat and power generation from medium-to low-temperature geothermal resources *J. Energy Conversion and Management* **88** pp. 883-893
[14] Bejan A and Tsatsaronis G 1996 *Thermal design and optimization* (New York: Wiley)
[15] Soltani S et al. 2013 A comparative exergoeconomic analysis of two biomass and co-firing combined power plants *J. Energy Conversion and Management* **76** pp. 83-91
[16] Mahmoudi S, Salehi S and Yari M 2017 Three-objective optimization of a novel triple-effect absorption heat transformer combined with a water desalination system *J. Energy Conversion and Management* **138** pp. 131-147

IOP Conf. Series: Earth and Environmental Science **105** (2017) 012088 doi:10.1088/1755-1315/105/1/012088

Hourly solar radiation in Depok, West Java, Indonesia (106.7942 Longitude, -6.4025 Latitude)

Nasruddin, M Arif Budiyanto and M Hanafi Lubis

Department of Mechanical Engineering, Faculty of Engineering, Universitas Indonesia, Depok, Indonesia

Email: nasruddin@eng.ui.ac.id, arif@eng.ui.ac.id, muhammad.hanafi41@ui.ac.id

Abstract. The objective of this paper is to predict hourly solar radiation in Faculty of Engineering, Universitas Indonesia, Depok, West Java, Indonesia (106.7942 longitude, -6.4025 latitude) as studied site. The calculation is based on several empirical models and compared to the measured data from BMKG. The comparison of theoretical data and measured data in order to know the influence of solar radiation parameters like solar irradiance condition, angstrom turbidity coefficient, and climate condition used to calculate the solar radiation at studied site. In order to evaluate the performance of solar radiation models, a statistical test was performed by using Mean Bias Error and Root Mean Square Error. The results indicate the best performance of solar radiation models that applicable in Indonesia. Therefore, Benjamin et al. model provides good performances for estimation of the hourly solar radiation.

1. Introduction

Indonesia as studied site has many potentials of solar radiation because it is located in the tropical area. Indonesia has several areas with the great solar energy potential from east until west region along the equator. The solar energy potential is considerable and it is estimated in 4.8 kWh/m^2 in all region [1]. The estimation of global solar radiation is essential for utilization the solar energy, design wherever appropriate observation missing, the values of solar radiation in clear skies are useful for determining the maximum performance heating and photovoltaic as well as for the design of air conditioning equipment in buildings or for determination of thermal load their solar installation.

The estimation of solar radiation has discussed in several literatures but there has not been discussing in Indonesia. The differences between several literature and this paper are the selected days, the studied site, and the parameters on the studied site. Analyzing of solar radiation intensity start with calculate the solar radiation intensity based on several empirical models, then compare the prediction data with the measurement data from BMKG, and evaluate the statically data use Mean Bias Error and Root Mean Square Error. The aim of this study is to know the solar energy power potential in Indonesia and to estimate the hourly solar radiation intensities in Depok, Indonesia by comparing solar radiation models with International Agency for Meteorology Climatology and Geophysics (BMKG) data.

2. Methodology

2.1. Date and Location Studied

The research based on recorded data from local meteorological station in Faculty of Engineering, Universitas Indonesia, Depok, Indonesia (106.7942 longitude, -6.4025 latitude) on February 20th 2014.

The measured global solar radiation shown in figure 1. The chosen of studied site based on the location of our university which has normal solar intensity and rain fall so it is good to be reference of solar radiation calculation in some places in Indonesia.

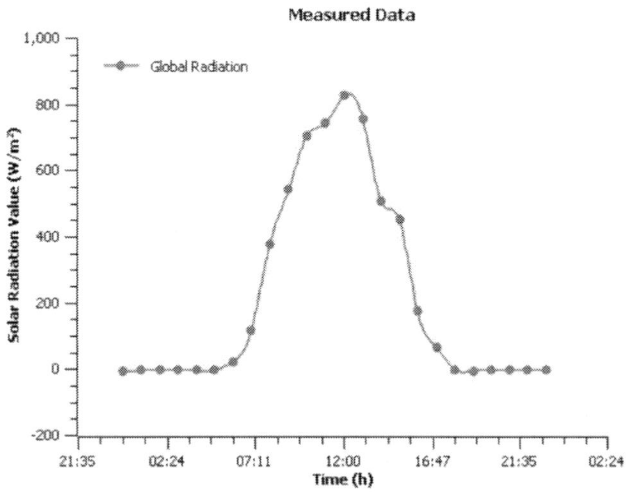

Figure 1. The graphic of solar radiation from local meteorological station

2.2. Solar Radiation Models

The research is about comparison the hourly solar radiation between the recorded solar radiations and solar radiation models. After comparing the prediction and measured data, we can conclude which model will be able to use in Indonesia based on meteorological parameters. The meteorological parameters include Earth-Sun distance (C_t), solar declination (δ), and hour angle (ω) and geographical parameters such as sunshine duration, mean ambient temperature, maximum and minimum ambient temperature, relative humidity, cloud cover, precipitation, latitude, longitude, altitude, and extraterrestrial radiation. The general data that we need to calculate the solar radiation is

$$C_t = 1 + 0.034.\cos(j - 2) \quad (1)$$

$$\delta = 23.45 \sin(0.986(j + 284)) \quad (2)$$

$$\omega = 15\,(12 - T_{sv}) \quad (3)$$

In the calculation, j is the day number of the year, ranging from 1-365 days on 1 January until 31 December. T_{sv} (h) is the true solar time at a given position.

2.3.1. Bird and Hulstrom Model. The Bird and Hulstrom model [2] , the direct solar radiation (W/m^2) as in (4) and the diffuse solar radiation(W/m^2) as in (5) are calculated using the following equation:

$$I = \cos\theta_z \,.\, I_n \quad (4)$$

$$D = D_r \,.\, D_a \,.\, D_m \quad (5)$$

The θ_z (degrees) is the zenith angle and the horizontal diffuse irradiance $D(W/m^2)$ at ground level is a combination of three individual components corresponding to the Rayleigh scattering after the first pass through the atmosphere D_r (W/m^2); the aerosols scattering and the multiple-reflection process between the ground and sky.

2.3.2. Model 2 ASHRAE. As recommended by ASHRAE [3], global solar radiation *(G)*, beam solar radiation in the direction of rays (I_n), and diffuse solar radiation (D) on the horizontal surface on a clear day are calculated using as the following equation;

$$G = I_n cos\Theta_z + D \tag{6}$$

$$I_n = A.exp(-B/\,cos\Theta_z) \tag{7}$$

$$D = C\,.\,I_n \tag{8}$$

A, B, and C values according to ASHRAE Model for solar irradiance calculation is provided on ASHRAE table. A (W/$\boldsymbol{m^2}$) is direct solar radiation values, B and C is dimensionless values that all parameters depends on date in a year.

2.3.3. Cooper Model. The solar radiation arriving on a horizontal surface under cloudless skies has two components; the direct component I (W/m^2), and the diffuse component D (W/m^2), forming the global solar radiation flux G (W/m^2). The direct solar radiation flux does not take into account the absorbed and the reflected portions by the atmospheric constituents (water vapor, aerosols, ozone, etc.), it depends only on the Sun direction. The diffuse solar radiation flux comes from all space and it has not of a preferred condition. The empirical formulas for this model [4] are;

$$I = I_O\,.\,sin\,h \tag{9}$$

$$D = 125\,.\,(sin(h))^{0.4} \tag{10}$$

2.3.4. Perin Brichambaut Model. In 1975 Perin Brichambaut [5] proposed a model which is designed to find direct, diffuse, and global radiation with relating to the astronomical parameters. The relationships between those formulas is the atmospheric condition:

$$I = R.exp(\frac{-A}{B.sin(h+1)}) \tag{11}$$

$$D = 125C.\ (sin(h))^{0.4} \tag{12}$$

Table 1. R, A, B, and C values according to Perin Brichambaut Model

Atmospheric condition	R (W/m^2)	A	B	C
Clear skies	1210	1.67	3.9	0.67
Normal condition	1230	1.61	3.1	0.47
Industrial zones	1260	2.33	4	0.45

2.3.5. Capderou Model. The Capderou [6] model uses the Linke turbidity factor to calculate the direct and diffuse solar radiation components received on horizontal plane. Several turbidity factors provided on table 3 and table 4. The factors value to determine the solar radiation for clear skies. The calculation of direct, diffuse, and global solar radiation flux is

$$I = I_O\,.\,C_t\,.\exp(-T_L\,.\ (0.9 + \frac{9.4\sin(h)}{T_1})^{-1})\,.\sin(h) \tag{13}$$

$$D = I_O\,.\,C_t\,.\exp(-1 + \log(\sin(h))) + a - \sqrt{a^2 - b^2} \tag{14}$$

Table 2. Angstrom Turbidity Coefficient and

Atmospheric condition	β_A	α_A
Rural site	0.05	0.11
Urban site	0.10	0.22
Industrial site	0.30	0.66

Table 3. Turbidity factor correction depending of climatic condition

Climatic conditions	Sky very pure	Normal condition	Sky very polluted
A	0.91	0.88	0.87
B	0.43	0.26	0.17

2.3.6. Chourad et al. Model Chourad et al. [7] model proposed empirical formulas which are based on the disturbing factor evaluations depending on the atmospheric conditions and astronomical parameters. The direct solar radiation by this model does not take into account the absorbed and reflected solar flux by the atmospheric components (water vapor, aerosols, ozone, etc). The diffuse solar radiation concern to all space and has no privilege area. The calculation of direct, diffuse, and global solar radition as follows;

$$I= I_o. C_t . A_1 . \exp(-\frac{A_2}{\sin(h)}). sin(h) \quad (15)$$

$$D= I_o. C_t. \left|0.2771 - A_1 . \exp(-\frac{A_2}{\sin(h)})\right| \sin(h) \quad (16)$$

$$G = 0.2771 I_o . C_t . A_1. \sin(h) + 0.706 I_o . C_t . A_1 . \sin(h) \exp(-\frac{A_2}{\sin(h)}) \quad (17)$$

Table 4. Turbidity factors depending according to Chourad Model

Climatic Condition	Sky very pure	Normal conditions	Sky very polluted
A_1	0.87	0.88	0.91
A_2	0.17	0.26	0.43

2.3.7. Tannaka model. Tannaka et al. [8] proposed the direct and the diffuse solar radiation on horizontal surface and they are calculated by the following equation;

$$I = I_O . \sin(h) . \tau_o^{\frac{1}{\sin h}} \quad (18)$$

$$D = \frac{1}{2} . I_O . \frac{\sin h . \left(1 - \tau_o^{\frac{1}{\sin h}}\right)}{1 - 1.4 \ln \tau_o} \quad (19)$$

2.3.8. Campbell model. The Campbell model [9] developed a computation of solar radiation components on a horizontal surface requires estimates of flux densities for at least three radiation streams; direct irradiance on a surface perpendicular to the beam (I_n), diffuse sky irradiance on a horizontal plane *(D)*, reflected radiation from the ground that supposed null for an horizontal plane and the total irradiance (G) on the horizontal (direct plus diffuse). All the components are evaluated as;

$$I = I_n. sin(h) \quad (20)$$

$$D= 0.3 I_o (1 - \tau_o{}^{ma}) sin(h) \quad (21)$$

2.3.9. El Mghouchi Y, et al. model. In 2014, El Mghouchi model [10] based in the the turbidity atmospheric factor Γ (dimensionless) for clear skies. This factor varies depending on the day number of the year. The direct, diffuse, and global solar radiation fluxes on an horizontal surface according to this model are given by the following formulas:

$$I = I_O \,.\, C_t \,.\, \Gamma \,.\, \exp\left(-\frac{0.13}{\sin(h)}\right) .\sin(h) \tag{22}$$

$$D = 120\Gamma \,.\, \exp\left(-\frac{1}{(0.4511+\sin(h))}\right) \tag{23}$$

2.3.10. Benjamin et al. model. The Benjamin et.al. [11] model proposed the direct and the diffuse solar radiation fluxes in a horizontal surface which depends on the direct and diffuse atmospheric transmittances are given by the following formulas

$$I = I_o . sin(h). \, \tau_{dir} \tag{24}$$

$$D = I_o . C_t . sin(h). \, \tau_{dif} \tag{25}$$

The calculation for global solar radiation is the sum of direct and diffuse solar radiation fluxes.

2.4. Data Analyze Methods

In order to evaluate the calculation result from the developed models, the performance of the prediction data assed from the following well established statically test of Mean Bias Error (MBE) and Root Mean Square Error (RMSE). These indicators are calculated as:

$$MBE = \frac{1}{K} \sum \left(G^i_{estimated} - G^i_{measured}\right) \tag{26}$$

$$\text{RMS}E = \left(\frac{1}{K} \sum \left(G^i_{estimated} - G^i_{measured}\right)^2\right)^{1/2} \tag{27}$$

3. Results and discussion

The estimation of hourly solar radiation has calculated using several empirical models from literature and several parameters provide in each model with the location in Depok, Indonesia (106.7942 longitude, -6.4025 latitude) and on February 20th 2014 as selected days. Figure 3 shows the comparison data between the empirical models and the measurement data from BMKG in studied site and selected days. The measurement data is lower than the several empirical models, it is because the empirical models have several parameters that constant and able to every location studied site. Indonesia has several parameters that will impact to the calculation like solar irradiance, atmospheric condition, and climatic condition. The chosen of the best performance models based on the closest graphic model to graphic measured data. It indicates that the solar radiation prediction of the model will be able to use in the studied site and date. Figure 3 shows several solar radiation models is close to measured data like Perin Brichambaut model, Chourad Model, and Benjamin model. After comparing the prediction data to measurement data, we should calculate the errors of each model with statistical test by MBE and RMSE in order to know the performance of each prediction model.

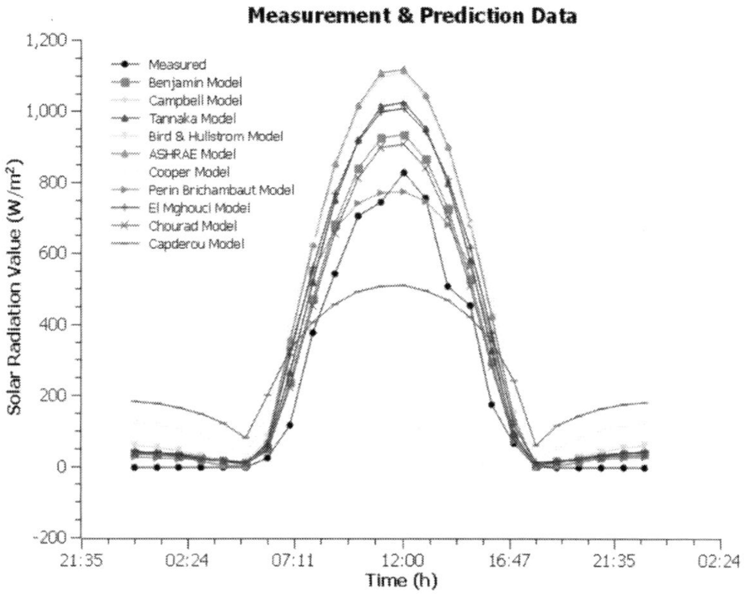

Figure 2. The Comparison Graph

Table 5 shows the results of statically indicators of all the empirical models and we can conclude that Tannaka model (Model 7), El. Mghouchi model (Model 9), and Benjamin model (Model 10) has lower MBE and RMSE than the other models. The lowest of MBE and RMSE will defines as the best comparing the statically errors data and it is applicable to estimate the solar radiation intensity. Model 10 that proposed by Benjamin *et al.* [11] is the lowest one with the MBE and RMSE value is 55 (W/m^2) and 273.6 (W/m^2), respectively.

Table 5. Results of Statically Indicators

Model	MBE (W/m^2)	RMSE (W/m^2)
Model 1	102.6234	502.7499
Model 2	116.6215	571.3263
Model 3	120.138	588.5534
Model 4	142.1679	696.4776
Model 5	138.6846	679.4132
Model 6	332.417	1628.504
Model 7	62.65497	306.9454
Model 8	106.3244	520.8813
Model 9	88.87161	435.3802
Model 10	55.84997	273.6079

4. Conclusion

The study of the solar radiation is the starting point of any investigation for a new energy. Several parameters of solar radiation calculation are very important of measuring the direct, diffuse, and global solar radiation. The measured data based on meteorological station since 2014 until 2016 collected every hour. The comparison between all the empirical models and measurement data from

IOP Conf. Series: Earth and Environmental Science **105** (2017) 012088 doi:10.1088/1755-1315/105/1/012088

local meteorological station has calculated and has assessed by the statistical indicators. Benjamin *et al.* [11] model are the best performance for the estimation of hourly solar radiation intensity because they provides small statistical errors than the other empirical models.

5. Acknowledgments

Authors would like to express our gratitude to Directorate Research and Community Engagement (DRPM) Universitas Indonesia in the providing of publication funding by PITTA 2017. The authors also grateful to the Indonesian Agency for Meteorology, Climatology, and Geophysics (BMKG) website (http://dataonline.bmkg.go.id) for providing temperature data set.

6. Reference

[1] Rahardjo I, Fitriana I 2005 *Analysis of the Potential of Solar Power Generation in Indonesia* Agency for Technology Assessment and Application (BPPT) Jakarta

[2] Bird R E, Hulstrom R I 1981 A simplified clear sky model for direct and diffuse insolation on horizontal surfaces *Golden, CO: Solar Energy Research Institute* SERI/TR-642-761

[3] ASHRAE 1985 *Handbook of fundamentals American Society of Heating, Refrigeration, and Air-Conditioning Engineers* Atlanta, Georgia

[4] Cooper P I 1969 The absorption of radiation in solar stills *Solar Energy* **12** pp 333-46

[5] Perin de de brichambaut C 1975 Estimation des Ressources Energétiques en France *Chaiers de l'A.F.E.D.E.S* **1**

[6] Capderou M, (Ed) 1985 Atlas solaire de l'Algérie énergétique *Tome 2 V 2, OPU Algern*: 2075-10-85 Alger pp 140-399

[7] Saïghi M 2002 Nouveau modéle de transfert hydrique dans le systéme sol- plante- atmosphere *Thése de Doctorat* Alger: USTHB

[8] Tannaka S, Takeda H, Tsuchiya T, Iwata T, Michijin T and Akimoto T 2006 Latest architectural environment engineering 4th edition *Inoue Shoin* pp 95-98

[9] Campbell G S, Norman J M 1998 *An introduction to environmental biophysics 2nd edition* (New York : Springer)

[10] El Mghouchi, El Bouardi A, Choulli Z and Ajzoul T 2014 New model to estimate and evaluate the solar radiation *Int J Sustain Built Environ* **3(2)** pp 225-34

[11] Liu B Y H, Jordan R C 1960 The interrelationship and characteristic distribution of direct, diffuse, and total solar radiation *Solar Energy* **4 (3)** pp 1-19

Prediction of ignition delay time of Liquefied Gas for Vehicle (LGV)

Y Muharam, F M Nur, T D Vollmer and A P Ramadhania

Department of Chemical Engineering, University of Indonesia, Depok 16424, Indonesia

E-mail: muharam@che.ui.ac.id

Abstract. Ignition delay time of liquefied petroleum for vehicle (LGV) fuel was predicted in this research. The method was using a detailed kinetic model of oxidation and combustion of propane/n-butane mixture. The validation of the model used experimental data in shock tubes in the temperature range from 1300 K to 1500 K and the pressure range from 2.5 bar to 10 bar for pure propane and pure n-butane. Calculated results of ignition delay time using the model have good agreement with experimental results indicating that the model may be used for LGV fuel. Simulation results of the model of LGV fuel show that the ignition delay time is faster when the n-butane fraction in the fuel is higher. The switchover of the ignition delay time order among different compositions of the fuels takes place at the initial temperature of 1200 K for the equivalence ratios of 1 and the initial pressure of 10 bar.

1. Introduction

Liquefied gas for vehicle (LGV) is liquefied petroleum gas (LPG) used as a fuel in internal combustion engines in vehicles as well as in stationary applications such as generators. It is a mixture of propane and butane. It is a green fuel since it reduces CO_2 exhaust emissions. It has an octane rating (RON) that is between 93.5 and 109.4 [1] and an energy content (higher heating value—HHV) that is between 25.5 MJ/L (for pure propane) and 28.7 MJ/L (for pure butane) depending upon the actual fuel composition.

One of combustion characteristics in internal combustion engines is autoignition. In diesel engine autoignition of a fuel is desired when the fuel injected into combustion chamber. The autoignition is caused by the high temperature which the air achieves when it greatly compressed. There is a period between the start of injection and the start of combustion, which is called delay time. In diesel engines, the delay time consists of physical delay wherein atomization, vaporization and mixing of air-fuel occur, and of chemical delay attributed to pre-combustion reactions [2]. The latter is caller ignition delay time. In spark-ignition engines which use spark plugs to ignite an air-fuel mixture autoignition is unwanted. The autoignition occurs because the increases in temperature and pressure of the unburned end gas cause it to ignite before it has been consumed by the propagating flame. This raises audible knocking. This malfunction is a serious limitation to the power developed and could lead to engine damage [3].

Research on the oxidation and combustion of LPG has been carried out by Kim et al. in a rapid compression machine [4]. It used pure butane and propane/butane mixture as the fuel. It investigated the ignition delay times of the fuel in the temperature ranging from 720 K to 900 K, the pressure range of 16 bar – 18 bar and equivalence ration of 1. The autoignition of LPG in spark-ignition engines was

IOP Conf. Series: Earth and Environmental Science **105** (2017) 012089 doi:10.1088/1755-1315/105/1/012089

investigated by Morganti et al. [5]. The experiments examined four species which are the primary constituents of LPG, i.e. propane, propylene (propene), n-butane and iso-butane.

Experimental and detailed kinetic study of the gas-phase oxidation of LPG (propane/n-butane/iso-butane) were performed in a jet-stirred reactor by Dagaut and Ali [6] at 1 bar, over the temperature range of 950 K – 1450 K and for equivalence ratios of 0.25, 0.5, 1, 1.5, 2 and 4. Mole fraction profiles as a function of temperature were obtained for molecular species. These experimental data were used to validate the detailed kinetic reaction mechanism consisting of 112 species and 827 reactions. Overall, the model is in very good agreement with the experimental data.

The recent research is intended to predict the ignition delay time of LGV using a kinetic model of detailed reaction mechanism of the fuel. The effects of propane/butane ratio, initial temperature, initial pressure and equivalence ratio on the ignition delay time were studied.

2. Method

Liquefied gas for vehicle considered in the research consists of propane and n-butane. The detailed reaction mechanism of LGV was derived from the detailed reaction mechanism for large aliphatic hydrocarbons developed by Muharam [7]. The validation was performed by comparing the calculated results of ignition delay time with the experimental data of ignition delay time for pure propane and pure n-butane. The data was obtained from Burcat and Radhakrishnan using a shock tube in the temperature range from 1300 K to 1500 K and the pressure range from 2.5 bar to 10 bar [8].

3. Modelling

In chemistry, a reaction mechanism is the step by step sequence of elementary reactions by which overall chemical change occurs [9]. If there are many elementary reactions, the net amount of a reactant consumed or a product produced will arise from the addition of the contributions of each elementary reaction. In general, species will take part in the reaction as a reactant and as a product on the other. The rate of production and consumption is a net result of all elementary reactions.

For a mechanism consisting of R elementary reactions of S species, which is given by

$$\sum_{s=1}^{S} v_{rs}^{(e)} A_s \xrightarrow{k_r} \sum_{s=1}^{S} v_{rs}^{(p)} A_s \tag{1}$$

where r =1, 2, …, R, $v_{rs}^{(e)}$ and $v_{rs}^{(p)}$ indicate the stoichiometric coefficients of reactants and products, respectively, the rate of formation of a species i is obtained from the addition of the whole rate equations where the species i is involved,

$$\left(\frac{\partial c_i}{\partial t}\right)_{\text{chem}} = \sum_{r=1}^{R} k_r \left(v_{ri}^{(p)} - v_{ri}^{(e)}\right) \prod_{s=1}^{S} c_s^{v_{rs}^{(e)}} \tag{2}$$

where i = 1, 2, ..., S.

The rate laws for a reaction mechanism consisting of R reactions among S species can be written as a system of first order ordinary differential equations,

$$\frac{dc_i}{dt} = F_i\left(c_1, \ldots, c_S, k_1, \ldots, k_R\right) \qquad i = 1, 2, \ldots, S \tag{3}$$

with initial conditions

$$c_i = c_i^0 \text{ at } t = t_0 \tag{4}$$

For many elementary reactions, a change in the parameters of the system (the rate coefficients) has nearly no effect on the time-dependent solution. Even if the reaction is included explicitly in the mechanism, it does not need a highly accurate rate coefficient. On the other hand, for a few of the elementary reactions, changes in the rate coefficients have really large effects on the outcome of the system. Accordingly, accurate rate coefficients are demanded. These few important reaction steps are

IOP Conf. Series: Earth and Environmental Science **105** (2017) 012089 doi:10.1088/1755-1315/105/1/012089

rate-determining steps or rate-limiting steps. The dependence of the solution c_i on the parameters k is called sensitivity and sensitivity analysis is a procedure to validate the kinetic model of detailed reaction mechanism.

4. Validation

The detailed reaction mechanism for the oxidation and combustion of propane/n-butane mixture consists of 155 species involving in 689 elementary reactions. Each elementary reaction is equipped with Arrhenius parameters which are pre-exponential factor (A), temperature exponent (n) and activation energy (E_a). Figure 1 and Figures 2 show the experimental and calculated ignition delay times of propane and n-butane. In general, the calculated ignition delay times are in good agreement with the experimental ones.

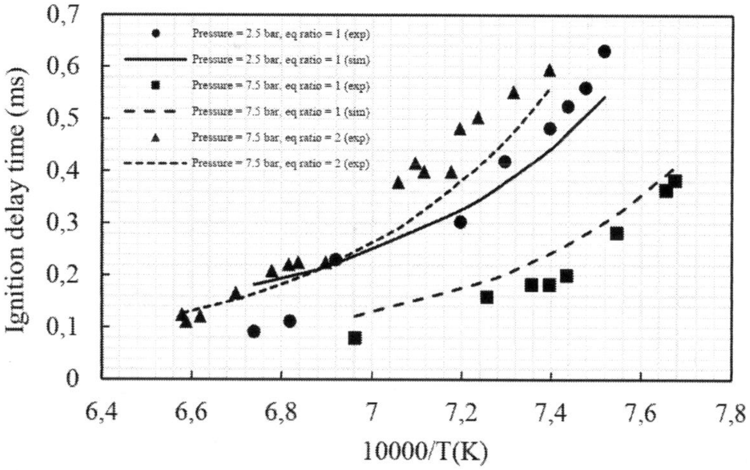

Figure 1. Experimental and calculated ignition delay times of propane.

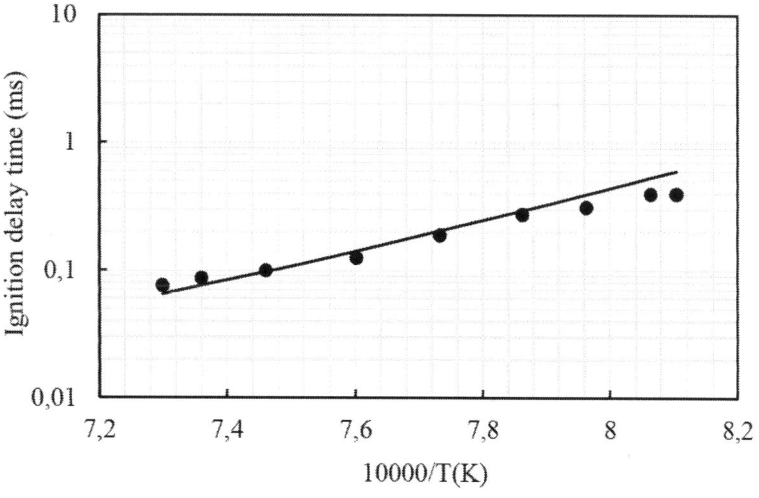

Figure 2. Experimental and calculated ignition delay times of n-butane.

5. Simulation

The valid model was then used to predict the ignition delay times in various fuel compositions, initial temperatures, initial pressures and equivalence ratios.

5.1. Fuel composition and initial temperature
Variations in fuel compositions were performed for the equivalence ratios of 1 and the initial pressure of 10 bar. The fuel compositions are: 0% propane and 100% n-butane (LGV-1), 20% propane and 80% n-butane (LGV-2), 40% propane and 60% n-butane (LGV-3), 50% propane and 50 % n-butane (LGV-4), 80% propane and 20% n-butane (LGV-5) and 100% propane and 0% n-butane (LGV-6). The fuels are burned with oxygen from the air so that there is nitrogen in the mixture as diluent.

Figure 2 illustrates the ignition delay times for various fuel compositions in the temperature range from 1100 K to 1500 K. It can be seen in the figure that the ignition delay time is faster when the n-butane fraction in LGV is higher. LGV-1, which is pure n-butane is the fastest, while LGV-6, which is the pure propane is the slowest. This is because carbon-carbon bonds and carbon-hydrogen bonds in longer hydrocarbon molecules, such as n-butane in this case, are weaker than the same bonds in propane molecules, making them easier to break off by radical attacks.

The figure shows an interesting phenomenon, i.e. there is a switchover of the ignition delay time order among different compositions of the fuels. When the temperature decreases to below 1200 K, LGV-1, LGV-6 and other LGVs switch their position. The ignition delay time of LGV-6 becomes the fastest, and that of LGV-1 becomes the slowest. This occurs because in the temperature range below 1200 K, oxidation reactions begin to play their role in ignition delay times.

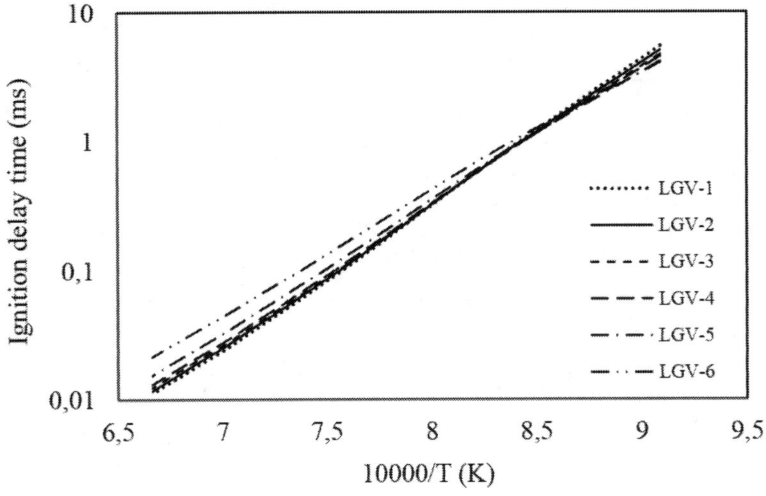

Figure 3. Ignition delay times for various fuel compositions and initial temperatures.

5.2. Initial pressure
Variations in initial pressure were carried out for the equivalence ratio of 1 and the fuel composition of 50% propane and 50% n-butane. The simulated initial pressures are 2 bar, 10 bar, and 18 bar.

Figure 4 shows the relationship between the ignition delay time and the initial pressure. It can be seen in the figure that the fuel with the initial pressure of 2 bar has the slowest ignition delay time. When the initial pressure increases to 10 bar, the ignition delay becomes faster. Similarly, when the initial pressure increases to 18 bar, the fuel becomes more combustible. This phenomenon is caused by the involvement of third bodies in a number of chemical reactions. Third body which is a collection of species other than the species expressed in a reaction equation provides energy to the collisions between the reactants so as to speed up the reaction.

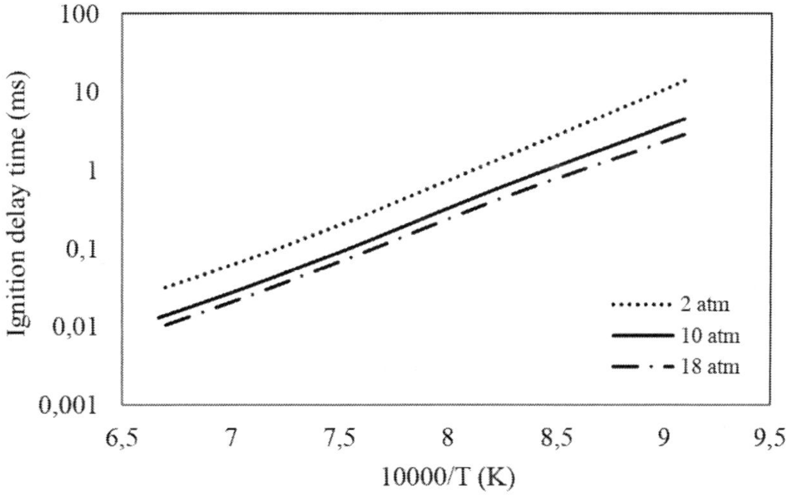

Figure 4. Ignition delay times at various initial pressures.

5.3. Equivalence ratio

Equivalence ratio varies at the initial pressure of 10 bar and the fuel composition of 50% propane and 50% n-butane. The three selected variants represent the mixture types, which are 0.5 (fuel lean), 1 (stoichiometric), and 2 (fuel rich).

The relationship between the ignition delay time and the equivalence ratio is exhibited in Figure 5. It can be seen in the figure that the fastest ignition delay time occurs for fuel lean mixture. This happens because in fuel lean mixture, although the reaction is slow, excess oxygen is able to initiate chain reactions by abstracting more H atoms from more fuel molecules so that the formation of radical pool is faster than that in stoichiometric mixture or fuel rich mixture. As a result, the ignition delay time is faster.

An interesting phenomenon takes place when the temperature decreases to below 1100 K where the fastest ignition delay time occurs for fuel rich mixture. This happens because at the temperatures below 1100 K, chain branching reactions are the decompositions of stable peroxide molecules. The formation of these molecules is preceded by the reaction of oxygen addition to radicals. Indeed, fuel rich mixture provides more fuel, and more fuel provides more alkyl radicals to be oxidized.

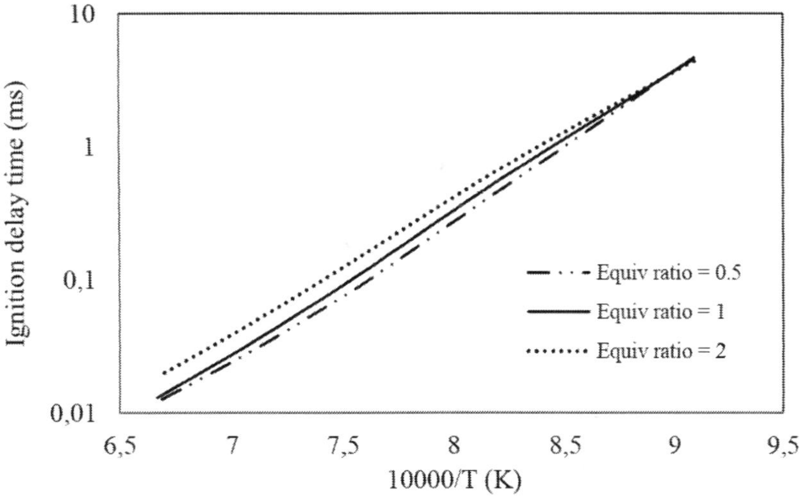

Figure 5. Ignition delay times at various equivalence ratios.

6. Conclusion

Calculated results of ignition delay time using the model have good agreement with experimental results indicating that the model may be used for LGV fuel. Simulation results of the model of LGV fuel show that the ignition delay time is faster when the n-butane fraction in the fuel is higher. The switchover of the ignition delay time order among different compositions of the fuels takes place at the initial temperature of 1200 K for the equivalence ratios of 1 and the initial pressure of 10 bar.

7. Acknowledgement

We express our gratitude to the University of Indonesia, which funded this research through the scheme of Hibah Publikasi Internasional Terindeks untuk Tugas Akhir Mahasiswa No 863/UN2.R3.1/HKP.05.00/2017.

8. References

[1] Morganti K J, Foong T M, Brear M J, da Silva G, Yang Y and Dryer F L 2013 Fuel 198 797
[2] Lakshminarayanan P A, Aghav Y V 2010 Ignition delay in a diesel engine *Modelling Diesel Combustion* (Dordrecht: Springer)
[3] Bradley D, Morley C 1997 Autoignition in spark-ignition engines *Low-Temperature Combustion and Autoignition* vol 35 ed M J Pilling (Amsterdam: Elsevier) chapter 7 pp 661-760
[4] Kim H, Lim Y, Min K and Lee D 2002 *KSME Int J* **16** p 1127
[5] Morganti K J, Brear M J, da Silva G, Yang Y and Dryer F L 2014 *Proc. Combust. Inst.* **36** p 2933
[6] Dagaut P, Ali K H 2003 *Fuel* **82(5)** p 475
[7] Muharam Y and Warnatz J 2007 *Chem. Phys.* **9 (31)** p 4218
[8] Burcat A and Radhakrishnan K 1985 *Combust. Flame* **60 (2)** p 157
[9] March J 1985 *Advanced Organic Chemistry: Reactions, Mechanisms, and Structure* 3rd ed. (New York: Wiley)

Numerical study on natural circulation characteristics in FASSIP-02 experimental facility using RELAP5 code

A R Antariksawan[1*], S Widodo[1], M Juarsa[1], D Haryanto[1], M H Kusuma[1,2] and N Putra[2]

[1]Center for Nuclear Reactor Technology and Safety, National Nuclear Energy Agency of Indonesia (BATAN), Building 80 Puspiptek Area Serpong, Tangerang Selatan 15310, Indonesia
[2]Heat Transfer Laboratory, Department of Mechanical Engineering, Universitas Indonesia, Kampus UI, Depok 16424, Indonesia

*E-mail: anhar@batan.go.id

Abstract. As in many other energy generating system, in the nuclear power plant, the use of natural circulation principle for the safety system is increasingly considered. The natural circulation could play an important role in providing emergency cooling of the nuclear reactor. The reliability of the use of natural circulation in the nuclear power plant should be demonstrated in order to assure the level of its safety. FASSIP-02 is a large scale test facility to study the natural circulation in the safety system of a nuclear power plant. To study the characteristics of the natural circulation and to help validating the design of FASSIP-02, a numerical study using RELAP5 code is undertaken. Based on the existing design of FASSIP-02, the numerical simulation is done with two variables, i.e. the heat flux and the pipe diameter. The effect of heat transfer surface area for dissipating the heat is also studied. The results show the natural circulation established in the FASSIP-02. The characteristics of the natural circulation with the values of several important parameters such as temperature, mass flow rate and pressure in the loop are obtained. The RELAP 5 calculations have provided the results that could be used to support the design and future operation of FASSIP-02.

1. Introduction

In the natural circulation, the fluid circulates as the result solely of buoyancy and gravity forces. There is no fluid moving machinery needed. The natural circulation system consists of a heat source, a heat sink and a loop piping system. The heat source is located lower than the heat sink allowing heated fluid flowing upward by buoyancy force. Contrarily, cooled fluid moves downward by gravitational force due to higher density. Such phenomena could be utilized to remove the heat from one hotter location to another cooler point while creating a cyclic process.

The use and existence of natural circulation based system could be found in many renewable energy generation system, such as solar energy based system [1] and geothermal system [2]. In the nuclear power plant (NPP), the implementation of natural circulation is increasingly considered as one of the passive safety system [3]. The passive safety system is a safety system that does not need an electrical sources, such as gravitation in the natural circulation. The need of the passive safety system is much more important in the case of station blackout (SBO) accident as in the Fukushima Daiichi NPP where all electrical sources to run the system were lost. In the case of SBO, the natural circulation is to be used for removing the heat from the nuclear reactor to the outside heat sink. Several designs of the small modular type NPP, which are under design, have included the passive safety system based on the natural circulation [4,5]. In the nuclear research reactor, the passive system and the natural circulation plays also in important role to keep its safety, such shown in [6]. In order to assure its reliability, system using natural circulation concept should show their stability in terms of continuous availability and performance for removing the heat. Then, many researches are still being performed to study such aspects, as for example in [7].

The Centre for Nuclear Reactor Technology and Safety (PTKRN), National Nuclear Energy Agency of Indonesia (BATAN) has since several years ago started the programme of the research and development on the use of natural circulation for nuclear reactor safety system. Mulya et al. started the study using a simple natural circulation test facility named NC-QUEEN [8]. Based on that study, a large scale rectangular single phase natural circulation loop called FASSIP-01 was built. Several studies have been done to assess the characteristic of the natural circulation in this facility [9-11].

On the other hand, in order to study more relevant geometry of an NPP, another facility FASSIP-02 is being designed. The geometry of this facility is designed to represent the passive residual heat removal system in a NPP design, though it is not intended to scale the real facility. This paper describes the preliminary results of the study on the characteristics of FASSIP-02 facility design. The study uses the RELAP5 computer code to simulate the existing FASSIP-02 design. The objective is to study the effect of the two important variables, which are the electrical heater power (heat flux) and the piping diameter on the characteristics of the natural circulation. Those characteristics studied are the evolution of the temperature in the loop, the water mass flow rate established in the loop and the heat removed to the cooling tank. As this study is conducted prior to the facility construction, the results of this study are expected to be considered in the finalization of the facility and the experimental design.

2. Methodology

2.1. RELAP5 computer code

RELAP5 computer code is one dimensional thermal-hydraulic code which is developed to study loss of coolant accident and transient in a light water reactor [12]. The code solves the mass continuity, momentum and energy conservation equations for each control volume to obtain thermal-hydraulic parameters such as temperature, pressure, void fraction etc. The system is divided into several thermal-hydraulic control volumes, named single volume, pipe or branch. The control volumes is connected to other using junction. To account the heat transfer, RELAP5 has a model of heat structure, which could exchange the heat with the fluid in its adjacent.

2.2. FASSIP-02 model and nodalization

Figure 1 depicts the scheme of FASSIP-02 test facility. The test facility consists of heater tank, the open cooling tank and the piping system. The pipe delivering hot water from the heater tank is called hot leg and the one flowing the cooler water from cooling tank is named cold leg. An electrical heater is placed in the bottom of the heater tank to heat the water. The hot water is cooled when it passes the tube which is submerged in the water placed in the open cooling tank. The submerged tube serves as a heat exchanger. The cooling tank is only cooled through natural circulation by transferring its heat to the air. All piping system is made of stainless steel, except the heat exchanger tube is from copper. The piping loop is equipped with the expansion tank to accommodate the hot water expansion and keep the loop pressure to the atmospheric pressure approximately.

Figure 2 shows the model RELAP5 for FASSIP-02 test facility. The system composes of the heater, the hot leg, the heat exchanger, the cold leg, the cooling tank and the expansion tank. The expansion tank and the cooling tank are connected to the atmosphere. All piping is insulated with an adiabatic boundary condition between the pipe wall and the environment.

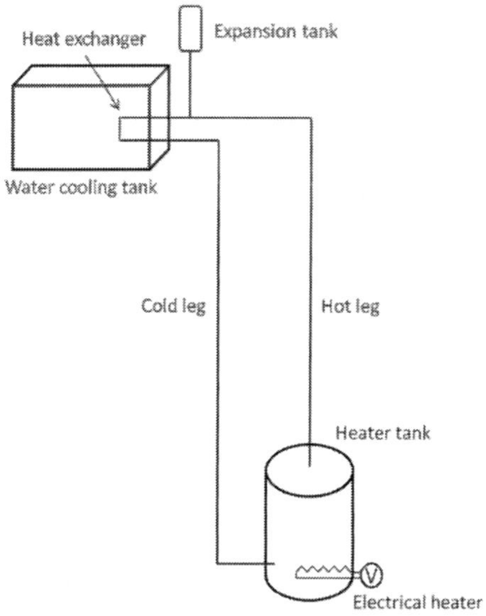

Figure 1. Scheme of FASSIP-02 facility.

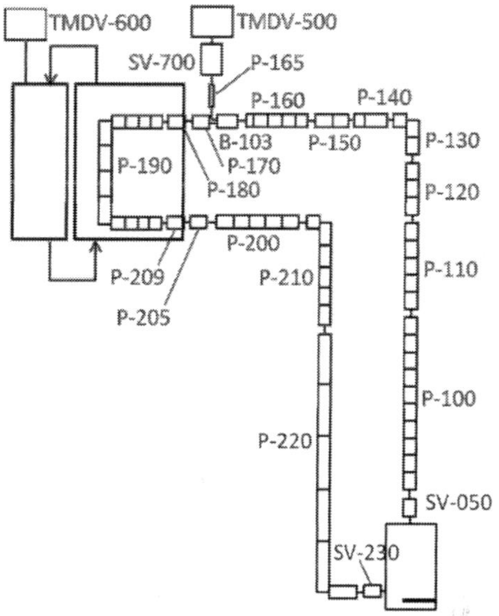

Figure 2. Nodalization of FASSIP-02 facility (P: pipe, B: branch, SV: single volume, TMDV: time dependent volume).

2.3. Calculation

The calculation with the pipe diameter of 2 inch (NPS), the electrical heater power of 10 kW and four variations of the heat transfer surface area of the submerged heat exchanger are done as the base calculation. The heat transfer surface area is an important parameter to be determined because it controls the heat removal from the natural circulation loop to the cooling tank. For that purpose, the heat exchanger is modelled as a bundle of vertical 1 inch pipe with 1 m length, which is composed of a number of pipes. There are four variation of the number of pipes, i.e. one pipe (case 1), ten pipes (case 2), twenty five pipes (case 3) and one hundred pipes (case 4). Based on the obtained results, the simulation to assess the effect of the pipe diameter and the electrical heater power is conducted while the heat transfer surface area of the heat exchanger is kept constant. Three pipe diameters, i.e. 1 inch, 2 inch and 3 inch, and three variations of electrical heater power of 2 kW, 5 kW and 10 kW are simulated. The initial and boundary conditions are identical for all calculation

The calculation was done for the duration of approx. 55.5 hours of problem time to simulate the long term characteristic of the circulation. As the concern is only for about the duration of the planned experiment, the discussion will be focused for the first 10 hours of the calculation time period

3. Results and Discussion

3.1. General characteristics

Table 1 shows the results of the base calculation. The case 1 to 4 are to simulate the variation of the heat transfer surface area. While the case 5 is similar to case 4 except with higher heater power. On the other hand, figure 3 shows the temperature of water at the horizontal hot leg region for four cases.

All calculation shows an instability at the beginning of calculation, but then the water temperature increases steadily with time. This increase is caused by the increase of the water temperature in the cooling tank, while the heat flux input is constant. The gradient temperatures in all cases are similar. However, in the case 1, the saturation temperature is reached earlier than other cases due to smaller total heat capacity, which is corresponding to smaller number of the pipe of the heat exchanger. As it could be seen from the figure 3, after 5 hours the water temperature of case 1 is the highest after which it increases with the same rate as three other cases. From the table 1, it also could be seen that after 10 hours, the water temperature in the case 1 has reached about 80.4 °C, while in the case 4 the

IOP Conf. Series: Earth and Environmental Science **105** (2017) 012090 doi:10.1088/1755-1315/105/1/012090

temperature is 52.2 °C. On other hand, with comparing the case 5 and case 4, the increase of the heat from the electrical heater resulted in faster achievement of the saturation temperature and the higher rate of the water temperature increase.

Table 1. Base case calculation.

Case number	Pipe diameter (inch)	Heater power (kW)	Number of pipe in the heat exchanger	Time to reach the saturation temperature (h)[a]	Temperature at t = 10 hours (°C)[a]
Case 1			1	35.8	80.4
Case 2		10	10	> 50	60.5
Case 3	2		25	> 50	54.7
Case 4			100	> 50	52.2
Case 5		15	100	36.0	59.7

[a] temperature at horizontal hot leg

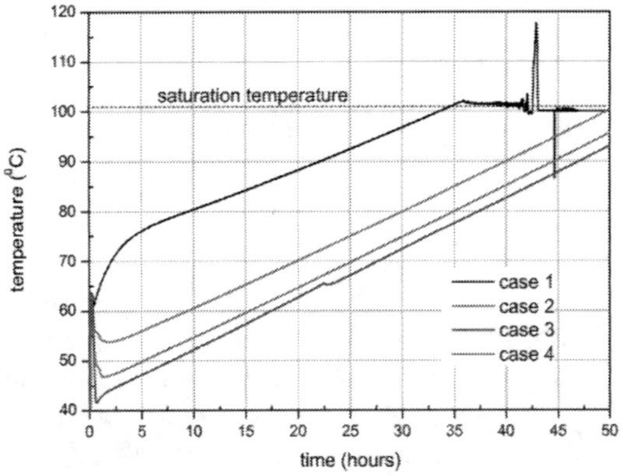

Figure 3. Evolution of the water temperature in hot leg for four cases.

The calculation results also indicate that increasing the heat transfer surface area did not increase the heat removed to the cooling tank. This is because the increase number of pipe of the heat exchanger, the velocity of the water flows in the pipe diminishes. As consequence, the overall heat transfer coefficient may unchange. The result might be different if it uses finned tube to increase heat transfer surface area, while the number of tube is not increased too much. The bundle of 25 tubes will be used for the next calculations because of it represents a reasonable number of pipe and provides good results.

3.2. Effect of the pipe diameter

To assess the effect of the pipe diameter, three diameters are simulated, i.e. 1 inch, 2 inch and 3 inch. In all cases, the heat exchanger is similar, which is a bundle of 25 pipes of 1 inch of diameter and 1 m of length. While, the electrical heater power is 10 kW. Table 2 shows the matrix of the calculation and the main results.

The results show that with smaller diameter size, the water was heated faster because of having smaller water mass. On the other hand, the water mass flow rate is almost similar for three cases. However, the bigger size of pipe tends to result a higher mass flow rate. This is due to lower friction loss for bigger diameter. The results from [13] indicated the same trend of the effect of the diameter.

3.3. Effect of the heat flux

Table 3 shows the simulation matrix and the main results to assess the effect of the heat flux transferred from the electrical heater to the water. The pipe diameter of 1 and 2 inch are chosen to be simulated because these sizes are the most likely used for loop of the FASSIP-02 facility. For each pipe size, the heat flux is varied, i.e. 2 kW, 5 kW and 10 kW. Figure 4 and 5 show the simulation results of the effect of heat flux on the temperature increase and the water mass flow rate for 1 and 2 inch pipe diameter, respectively.

Table 2. Matrix and main results of simulation for the effect of the pipe diameter.

Case number	Heater power (kW)	Number of pipe in the heat exchanger	Nominal diameter of the pipe (inch)	Temperature at t = 10 hours (°C)[a]	Mass flow rate in the loop (kg/s)[b]
Case 6			1	100.[c]	0.019 (0.05[d])
Case 7	10	25	2	54.6	0.19
Case 8			3	50.1	0.20

[a] temperature at horizontal hot leg
[b] average value from t=0 s to t=10 hours
[c] saturation temperature is reached at approx. 2.4 hours
[d] the average mass flow before the saturation temperature reached at about 10,000 s

Table 3. Matrix and results of simulation for the effect of electric power heater.

Case number	Pipe diameter (inch)	Number of pipe in the heat exchanger	Electrical heater power (kW)	Temperature at t = 10 hours (°C)[a]	Percentage of heat flux transferred to the cooling tank (%)
Case 9			2	95.6	0.4[b]
Case 10	1		5	saturated	0.4[c]
Case 11		25	10	saturated	0.4[c]
Case 12			2	36.7	94.9[b]
Case 13	2		5	44.6	94.4[b]
Case 14			10	54.6	94.5[b]

[a] temperature at horizontal hot leg
[b] average value for the period t = 3 to 10 hours (when the heat flux relatively steady)
[c] average heat flux for the period prior to the saturation condition.

It is obvious that the higher heat flux, the faster temperature increase for all diameter of pipe. For 1 inch pipe, for 5 and 10 kW heat flux, the water saturation temperature is reached at about 2.4 h and 4.8 h, respectively (see figure 4). For 2 inch pipe, the hot leg temperature reached about 54 °C, still far below saturation temperature, in the case of heat flux of 10 kW (see figure 5). As it could be seen from the figure 4, the water mass flow rate tends to be unstable at the beginning of the saturation.

While, the heat flux transferred to the cooling tank is quite small, approx. 0.4%, of the heat flux from the heater in case of 1 inch pipe and about 95% for 2 inch pipe. The heat transfer is small in case of 1 inch pipe might be due to very low mass flow rate. The heat received from the electrical heater is mostly used to increase the water temperature inside the pipe and also losses to the pipe wall.

On the other hand, the mass flow rate increases with the heat flux and the pipe diameter as well. In case of 1 inch pipe, the mass flow rate is low (about 0.03 to 0.05 kg/s) causing the heat flux transferred is small. In case of 2 inch pipe, the mass flow rate for 3 different heat fluxes is about 0.1 to 0.2 kg/s (see figure 5). Again these results agree well with the results from [13].

4. Conclusion

As expected, the characteristic of FASSIP-02 shows the establishment of a natural circulation. The smaller diameter of the pipe resulted smaller mass flow rate, heat flux transferred to the heat sink and faster achieving the water saturation temperature. The higher heat flux applied will cause faster water temperature increase and higher mass flow rate. If, the 1 inch pipe will be used, the power should be lower than 2 kW to avoid two phase flow in the loop. The 2 inch pipe provides more flexibility in terms of heat flux variation.

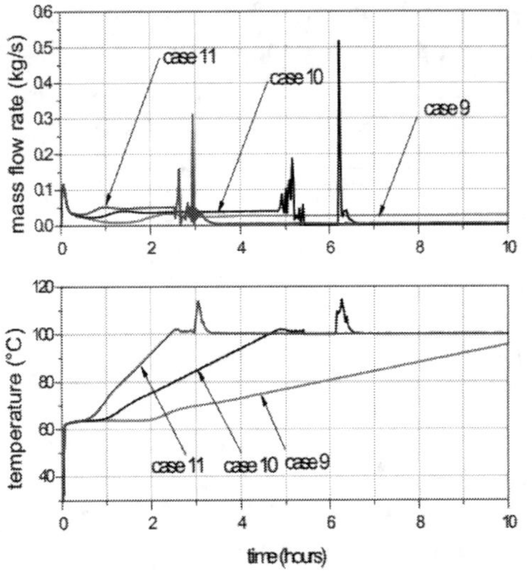

Figure 4. Mass flow rate and the water temperature for different heat flux for 1 inch pipe.

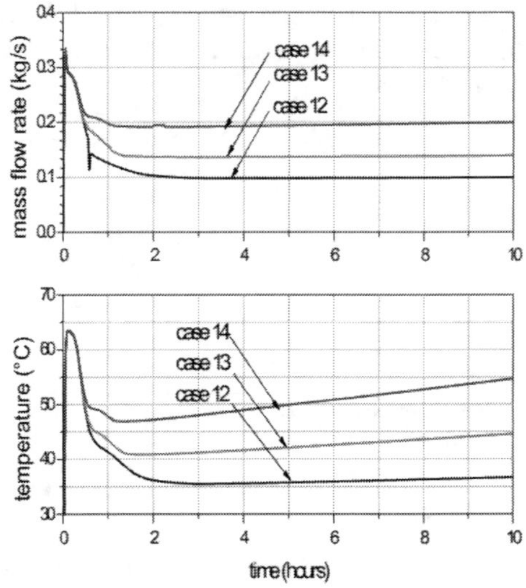

Figure 5. Mass flow rate and the water temperature for different heat flux for 2 inch pipe.

5. Acknowledgments

Authors acknowledge the support from Centre for Nuclear Reactor Technology and Safety (PTKRN), BATAN where the study is undertaken. Authors also wish to thank to the Ministry of Research, Technology and Higher Education of Indonesia for funding this work through INSINAS Program under the contract No.: 02/INS-2/PPK/E/E4/2017.

6. References

[1] Riahi A and Taherian H 2011 *Thermal Science* **15** (2) pp 447-56
[2] Faizal M, Bouazza A and Singh R M 2015 *Proc. World Geothermal Congress 2015 (Melbourne, Australia, 19-25 April 2015*)
[3] Xing J, Song D and Wu Y 2016 *Engineering* **2**(1) pp 79–87
[4] Nawaz H, Ilyas M, Ahmad M and Aydogan F 2016 *Annals of Nucl. Energy* **98** pp 191–9
[5] Karol M, John T and Zhao J 2015 *Renewable and Sustainaible Energy Reviews* **44** pp 643–56
[6] Antariksawan A R, Umar E, Widodo S, Juarsa M and Kusuma M H 2017 *Int. Journal of Technology* **8** (4)
[7] Chung Y J, Park H S Lee W J and Kim K K 2015 *Annals of Nuclear Energy* **77** pp 402-7
[8] Juarsa M, Purba J H, Kusuma H M, Setiadipura T and Widodo S 2014 Atom Indonesia **3** pp 141–7
[9] Juarsa M, Girano, Heru G B, Haryanto D and Prasetyo J 2016 *Prosiding Seminar Nasional Teknologi Energi Nuklir (BATAM, Indonesia, 4-5 August 2016)* pp 673–82
[10] Susyadi, Widodo S and Juarsa M 2016 *Prosiding Seminar Nasional Teknologi Energi Nuklir (BATAM, Indonesia, 4-5 August 2016)* pp 657–63
[11] Giarno, Heru G B, Prasetio J, and Juarsa M 2016 *Prosiding Seminar Nasional Teknologi Energi Nuklir (BATAM, Indonesia, 4-5 August 2016)*
[12] Mesina G.L. 2016 *Nuclear Science and Engineering* **182** pp 5-9
[13] Vijayan P K, Nayak K A, Saha D and Gartia M R 2008 Sci. Tech of Nucl. Installation **2008** pp 1-17

IOP Conf. Series: Earth and Environmental Science **105** (2017) 012091 doi:10.1088/1755-1315/105/1/012091

Estimation of natural circulation flow based on temperature in the FASSIP-02 large-scale test loop facility

M Juarsa[1,*], A R Antariksawan[1], M H Kusuma[1], D Haryanto[1] and N Putra[2]

[1]Center for Nuclear Reactor Technology and Safety, National Nuclear Energy Agency of Indonesia (BATAN), Building 80 Puspiptek Area Serpong, Tangerang Selatan 15310, Indonesia
[2]Heat Transfer Laboratory, Department of Mechanical Engineering, Universitas Indonesia, Kampus UI, Depok 16424, Indonesia

*E-mail: juars@batan.go.id

Abstract. In the future, the use of passive cooling system without electrical power input become highly considered to be implemented in the design of nuclear reactor safety system. Moreover, the lessons learned from the Fukushima-Daiichi accident encourage the incorporation of such passive cooling system. The passive safety cooling system is aimed to cope with both design bases as well as severe accidents. The natural circulation is one example parameter of the passive cooling system features being widely considered. However, some of the parameters involving in natural circulation are still being investigated, in particular the problem of flow instability. BATAN is designing a large-scale test loop facility, named FASSIP-02 loop, with a height of 11 meters between a heat source and a cooling source. The FASSIP-02 loop is designed to simulate the residual heat in the reference reactor. To ensure the validity of the design prior to the construction, it is necessary to make an analytical calculation using some correlations to estimate the natural circulation flow rate. The calculation method is performed by using the derived two principle correlations of the buoyancy force in the heating section and the gravitational force on the cooling section. The water temperatures in the heating section are varied at 50°C, 60°C, 70°C, 80°C and 90°C with the water temperature in the cooling tank pool are varied at 10°C, 20°C, 30°C, 40°C and 50°C. The calculation results can estimate the water temperature and volume are required. In addition, it could be used to determine the experimental matrix that is based on the FASSIP-02 loop design.

1. Introduction

The severe accident at Fukushima Dai-ichi nuclear power station has became a lesson learn to involving the passive system in nuclear safety system. The involving of passive system can reduce the recurrent of accident in nuclear power station. Due to the removal of heat generation after reactor shutdown, especially in station blackout case, the use of passive cooling system becomes highly considered to be implemented in the design of nuclear reactor safety system.

The additional of passive safety system features in advanced reactors are involved to prevent the accident occur. These passive safety system features are expected to enhance the nuclear reactor safety and prevent the release of radiation to the environment [1, 2].

The studies on using of passive safety system, such as passive residual heat removal system (PRHSR), in nuclear power station have done by many researchers using experiment and simulation.

Xia et.al has conducted the calculation analysis using RELAP5 on the natural circulation characteristics of passive residual heat removal system for integrated pressurized water reactor on accident condition [3]. Mangal et.al has investigated the natural circulation behavior using RELAP5 based on their experiment test section data [4]. Zou et.al was done the assessment of PRHRS cooling capability in core under design basis accidents and beyond design basis accidents [5]. Their results showed that the PRHRS has good performance to remove heat generation from primary coolant system under accident condition.

Specific investigation of thermal-hydraulics characteristics on passive system have been conducted by many researchers. Yan et.al studied the flow characteristics in two-phase passive cooling system loop of System-integrated Modular Advanced Reactor (SMART) design. The combination both experiment and simulation was conducted to know it characteristics. The characteristics obtained were used for standard design approval of SMART to verify its design bases, design tools, and analysis methodology [6].

We also have been conducted the previous investigation on heat pipe capability to remove the heat generation in nuclear spent fuel storage pool. Experiment and RELAP5 simulation were used to know the characteristics and thermal performance of heat pipe as passive cooling system under station blackout accident. The results showed that heat pipe has good cooling capability and can be proposed as passive cooling system in nuclear spent fuel storage pool [7, 8]. In another investigation, as a preliminary research, previous research has been done using small test loop called NC-QUEEN for transient condition [9]

we also have been constructed the FASSIP-01 loop research facility in order to investigate natural circulation flow based on the rectangular loop with medium size which is placed vertically. The HSS (heat sink system) loop used as cooling system to the cooler thank in the rectangular loop. The geometrical size of the rectangular loop is 3.5 m width and 6 m high, which was placed in a vertical position. from the investigation it's also founded the thermal characteristics of natural circulation flow in rectangular loop [10].

From the previous studies, there were no investigation which applied the large-scale test loop facility, named FASSIP-02 loop, with a height of 11 meters between a heat source and a cooling source. The FASSIP-02 loop is designed to simulate the residual heat in the reference reactor. To ensure the validity of the design prior to the construction, it is necessary to make an analytical calculation using some correlations to estimate the natural circulation flow rate. The calculation method is performed by using the derived two principle correlations of the buoyancy force in the heating section and the gravitational force on the cooling section.

2. Methodology

2.1. Design of FASSIP-02 Loop

The FASSIP-02 loop as a Large-Scale Test Loop Facility has been design to fulfil the needed of investigation on natural circulation phenomenon in large scale. The facility was named FASSIP-02 (the extension of Fasilitas Simulasi Sistem Pasif) unit 02. In the framework of understanding and mastering of the basic engineering to apply the natural circulation characteristics into passive cooling system for advanced nuclear reactors, it need to be made and carried out research using experimental facilities for the simulation of a passive cooling system. Figure 1 shows the design of FASSIP-02 loop.

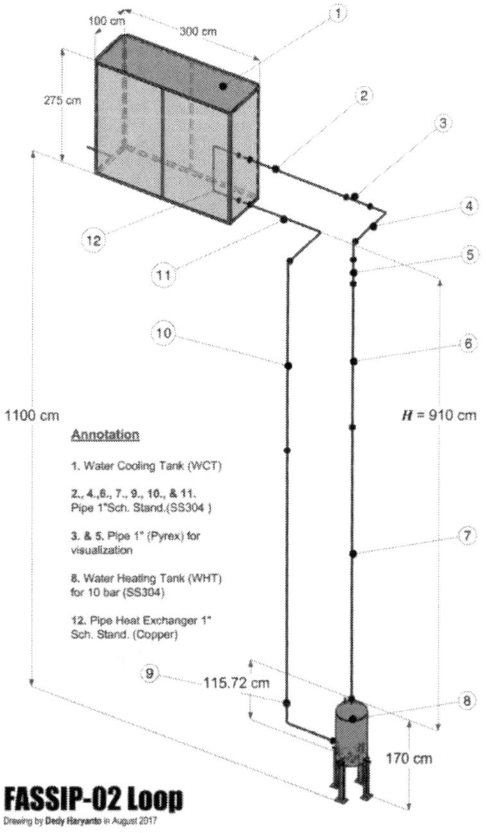

Figure 1. Design of FASSIP-02 Loop

Table 1. FASSIP-02 loop geometry data

Item	Geometry	Size/Material
Water Cooler Tank	shape	1 m x 3 m x 2.75 m (carbon steel, t= 8 mm)
Water Heating Tank	cylinder	Dia. 24 inch, Sch.40, h=1 m (cup=15.72 cm) (stainless steel 304)
Pipe	cylinder	Dia. 1 inch, Sch.20 (stainless steel 3040
Heat exchanger	cylinder	Dia. 1 inch, t=3 mm, (Cooper)
Heater (4 pieces)	-	Length= 30 cm, P=5 kW (total 20 kW)
Height differences	-	910 cm
Total height (*H*)	-	1100 cm
Total loss coef. (*K*)	-	12.71
Total pipe length (*L*)	-	43.17 m

There are several parameters that affect the natural circulation flow, where thermal hydraulics parameter is fluid temperature (T) and the geometrical parameters are height differences between heater and cooler (H), total loop length (L) and pipe diameter (D). The loss frictional of pipe defined base on elbow, tee, enlargements and contractions with K constantan. Then, then the design of FASSIP-02 loop was designed base on components and equipments related to thermal hydraulics and geometrical parameters. Design was done using 3D drawing software where figure 1 shows a heat source is a water heating tank and a heat sink tank (water cooling tank) which can simulate the temperature difference between hot area (bellow, to make density of water decreased) and cold area (upper, to make the density of water increased). Table 1 shows the geometry of FASSIP-02 loop.

2.2. Analytical

The concept of passive systems is the system which circulating the cooling fluid inside a loop without any intervention from outside forces and based on natural laws. The phenomenon of this fluid flow rate in the passive system called natural circulation, which occur due to differences in fluid density in hot and cold region [11]. Effect of fluid density changes in hot regions will lead to buoyancy force and the effect of fluid density changes in cold will cause the gravitational force. Thus, the implementation of passive systems in a nuclear power plant can be used for normal and non-normal condition (condition accident, transient) [12]. Figure 4 shows the simple concept of natural circulation based on buoyancy force and retarding frictional force with cold region in the top and hot region in the bottom of rectangular loop [13-16]. In the natural circulation loop, there is a hot part which is always placed at the bottom and the cold part placed at the top of a loop. In steady state condition, a phenomenon that occurs can be described by equation (1) as buoyancy force come from hot part (left side eq. 1) and in cold part is retarding frictional force (right side eq.1).

$$-g\oint \rho dz = \frac{R_h \dot{m}^2}{2\rho} \tag{1}$$

With the general hydro dynamics resistance is R_h [m^{-4}] as in equation (2)

$$R_h = \sum_{i=1}^{N}\left(\frac{f \times L}{D} + K\right)\frac{1}{A^2} \tag{2}$$

Then, equation (2) could be reform like,

$$R_h A^2 = \left(\frac{f \times L}{D} + K\right) \tag{3}$$

After equation (1) was integrated become,

$$\begin{aligned} g(\rho_h - \rho_c)H &= \frac{R_h \dot{m}^2}{2\rho} \\ \dot{m}^2 &= \frac{2g\rho(\rho_h - \rho_c)H}{R_h} \end{aligned} \tag{4}$$

The flow rate derivates using pipe frictional coefficient for laminar flow.

$$f = \frac{64\mu}{D\rho v} \tag{5}$$

Then, equation (5) substituted into equation (4).The new equation for hydrodynamics resistance is

$$R_h = \frac{64\mu L + D^2 \rho v K}{D^2 \rho v A^2} \tag{6}$$

Then, equation (6) substituted by the equation (2) and $\dot{m} = \rho v A$, finally

$$v = \frac{-64\mu L + \sqrt{(64\mu L)^2 + 8gHK\rho(\rho_h - \rho_c)D^4}}{2D^2 \rho K} \tag{7}$$

IOP Conf. Series: Earth and Environmental Science **105** (2017) 012091 doi:10.1088/1755-1315/105/1/012091

Equation (7) is natural circulation flow rate inside pipe rectangular loop derivate from to natural laws. It is mean that, all parameters consisting of total loop length (L), height differences (H) between heater and cooler, pipe diameter (D), temperature (T) and constantan K have influenced into natural circulation flow velocity. For estimation, the water temperatures in the water heating tank are varied at 70°C, 75°C, 80°C, 85°C and 90°C with the water temperature in the water cooling tank at 10°C, 20°C, 30°C, 40°C and 50°C. The calculation results obtained will be used as a basis for construction and commissioning test using FASSIP-02 loop.

3. Result and Discussion

Using water temperature simulation as described in above, the using the correlation for water density $\rho(T) = 1004.789042 - 0.046283(1.8T + 32) - 7.9738\times10^{-4}(1.8T + 32)^2$ and dynamics viscosity of water $\mu(T) = \exp\left[\frac{(A + CT)}{(1 + BT + DT^2)}\right]$ finally obtained natural flow rate curve to temperature difference in hot part and cold par as shown in following figure 2 and figure 3.

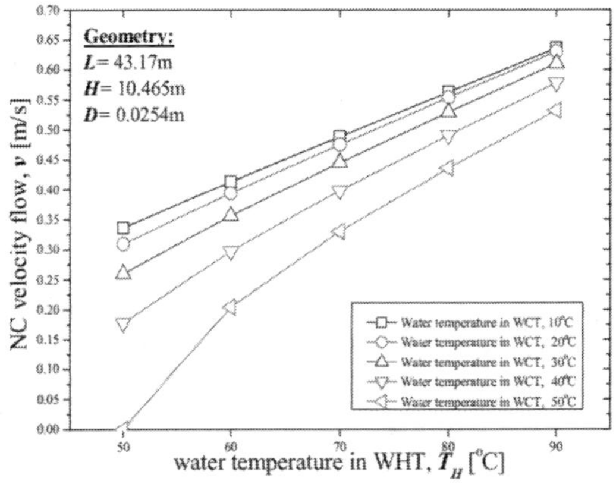

Figure 2. NC velocity for water temperature in WHT (hot part)

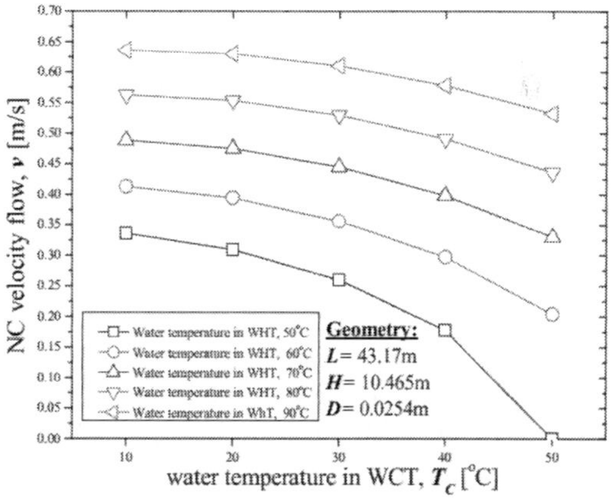

Figure 3. NC velocity for water temperature in WCT (cold part)

Figure 2 shows the shows the trend of NC flow velocity for each water temperature variation in WHT. The velocity of natural circulation (NC) was increased for each water temperature variation in hot part. Then, Figure 3 shows that the velocity of NC was decreased for each water temperature variation in cold part. It is shown that, the differences of temperature was increased causing the velocity flow of NC was increased as shown in Figure 4.

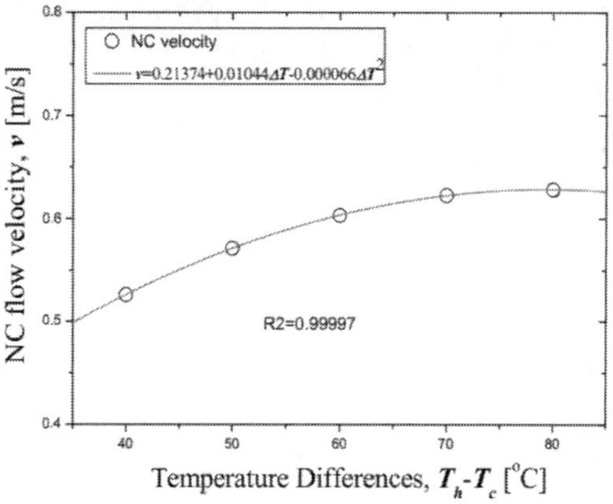

Figure 4. NC velocity for water temperature differences between hor part and cold part

4. Conclusion
Base on estimation with water temperatures in the heating section are varied at 50°C, 60°C, 70°C, 80°C and 90°C with the water temperature in the cooling tank pool are varied at 10°C, 20°C, 30°C, 40°C and 50°C can be concluded that the higher NC velocity flow is 0.628 m/s. The lower one is 0.1765 m/s fow water temperature in WHT is 50°C and in WCT is 40°C.

5. Acknowledgments
Authors wish to thanks to the Ministry of Research, Technology and Higher Education of Indonesia for funding this work through INSINAS Program under the contract No.: 02/INS-2/PPK/E/E4/2017. Authors also acknowledge the support from Centre for Nuclear Reactor Technology and Safety (PTKRN), BATAN where the study is undertaken.

6. References

[1] Chang M H, Sim S K and Lee D J 2000 SMART behavior under over-pressurizing accident condition *Nuclear Engineering and Design* **199**(1) pp 187-196

[2] Park H S et al. 2017 Contribution of thermal–hydraulic validation tests to the standard design approval of SMART *Nuclear Engineering and Technology*

[3] Xia G, Peng M and Dua X 2014 Calculation analysis on the natural circulation of a passive residual heat removal system for IPWR *Annals of Nuclear Energy* **72** pp 189-197

[4] Hou X, Sun Z and Lei W 2017 Capability of RELAP5 code to simulate the thermal-hydraulic characteristics of open natural circulation *Annals of Nuclear Energy* **109** pp 612-625

[5] Zou J et al. 2014 Assessment of passive residual heat removal system cooling capacity *Progress in Nuclear Energy* **70** pp 159-166

[6] Yan X, Fan G and Sun Z 2017 Study on flow characteristics in an open two-phase natural circulation loop *Annals of Nuclear Energy* **104** pp 291-300

[7] Kusuma M H et al. 2017 Investigation of the Thermal Performance of a Vertical Two-Phase Closed Thermosyphon as a Passive Cooling System for a Nuclear Reactor Spent Fuel Storage Pool *Nuclear Engineering and Technology* **49**(3) pp 476-483

[8] Kusuma M H et al. 2017 Simulation of Wickless-Heat Pipe as Passive Cooling System in Nuclear Spent Fuel Pool Using RELAP5/MOD3. 2. *International Journal on Advanced Science, Engineering and Information Technology* **7**(3) pp 836-842

[9] Juarsa M et al. 2014 Preliminary Study on Mass Flow Rate in Passive Cooling Experimental Simulation During Transient Using NC-Queen Apparatus *Atom Indonesia* **40**(3) pp141-147

[10] Juarsa M et al. 2016 *Passive System Simulation Facility (FASSIP) Loop for Natural Circulation Study* in Seminar Nasional Teknologi Energi Nuklir
[11] Jain V et al. 2010 Experimental investigation on the flow instability behavior of a multi-channel boiling natural circulation loop at low-pressures *Experimental Thermal and Fluid Science* **34**(6) pp 776-787
[12] Kozmenkov Y, Rohde U and Manera A 2012 Validation of the RELAP5 code for the modeling of flashing-induced instabilities under natural-circulation conditions using experimental data from the CIRCUS test facility *Nuclear Engineering and Design* **243** pp 168-175
[13] Vijayan P K, Austregesilo H and Teschendorff V Simulation of the unstable oscillatory behavior of single-phase natural circulation with repetitive flow reversals in a rectangular loop using the computer code athlet *Nuclear Engineering and Design* **155**(3) pp 623-641
[14] Greif R 1988 Natural circulation loops *Journal of Heat Transfer* **110**(4b) pp 1243-1258
[15] Misale M, Devia F and Garibaldi P 2005 Some considerations on the interaction between the fluid and wall tube during experiments in a single-phase natural circulation loops *IASME Trans*. **9**(2) pp 1717-1722
[16] Misale M et al. 2007 Experiments in a single-phase natural circulation mini-loop *Experimental Thermal and Fluid Science* **31**(8) pp 1111-1120

IOP Conf. Series: Earth and Environmental Science **105** (2017) 012092 doi:10.1088/1755-1315/105/1/012092

The effect of processing methods on the improvement of tensile properties of Random Ramie Fiber – Reinforced Tapioca Starch Biocomposites

H Judawisastra, Y S Syamsiar and Mardiyati

Materials Science and Engineering Research Groups, Faculty of Mechanical and Aerospace Engineering, Institut Teknologi Bandung, Ganesha 10, Bandung 40132.

Email: hermawan.judawisastra@material.itb.ac.id

Abstract. The disposal of Glass Fiber-Reinforced Plastic (GFRP) composites is becoming an environmental issue. To overcome the problem, biocomposites made of ramie fibers and tapioca starch might become an alternative to GFRP composites, due to their renewable properties, environmental friendly and easily degraded by nature. However, mechanical properties of fiber-reinforced starch based biocomposites are quite low due to high void content and poor fiber-matrix interface. In this research, the effect of the processing methods on tensile properties of tapioca starch - reinforced random ramie fiber biocomposites were studied. Ramie fiber reinforced tapioca starch biocomposites were fabricated by means of solution casting and compression molding technique. Random ramie fiber-reinforced tapioca starch biocomposites were succesfully made with the highest tensile strength and modulus elasticity of biocomposites were 18 MPa and 959 MPa respectively. The use of 1% NaOH alkalization process on ramie fiber increased tensile strength and modulus elasticity of biocomposites 64% and 54% respectively. Application of 120 °C compression temperature, increased tensile strength and modulus elasticity of biocomposites 18% and 74% respectively. Homogenous dispersion of ultimate fiber increased the tensile strength up to 91%. However, the addition of glycerol decreased the tensile strength and modulus elasticity for 46% and 39% respectively.

1. Introduction

Glass Fiber-Reinforced Plastic (GFRP) composites have been increasingly used in many applications in major industries. It is estimated that the production volume of GFRP in the European countries in 2015 increased by 2.5% to 1,069 megatons [1]. And the demand for GFRP will continue to grow rapidly worldwide. However, the growing production of GFRP caused the increasing of GFRP wastes. The GFRP wastes are commonly disposed by means landfilling and incineration which may affect the environment. To overcome this problem, the development of biocomposites derived from renewable resources are required. Ramie fibers and cassava starch have a great opportunity as an alternative to subtitute the glass fibers and the synthetic polymers on producing biocomposites.

Ramie is commonly found locally in Indonesia and belongs to the bast fiber. Ramie fibers have several advantages such as high grow rate over other bast fiber [2], high specific strength and modulus, and high cellulose content [3]. The tensile strength, modulus elasticity, and elongation of ramie fibers have been reported to be in the range of 400 – 938 MPa, 61 – 128 GPa, and 1.2 – 3.8 %, respectively [3]. As a result, ramie fibers have been widely used as reinforcing fibers for polymer composites [3-5].

Starch is a potential raw material for the production of biodegradable plastic and have been used to fabricate biocomposites due to renewable, biodegradable, and sustainable [6]. Cassava is one of the major resources for starches [7], and Indonesia is one of the largest producer of cassava with production of 24 million tonnes in 2014 [8]. The use of cassava starch as the matrix for the biocomposites have been reported in the past decades [9-14]. However, the tensile properties of fiber-reinforced starch based biocomposites are quite low, due to high void content and poor fiber-matrix interface [9-14]. Therefore, the main purpose of this study is to improve the tensile properties of biocomposites with the various processing methods. The effect of alkalization, compression temperature application, glycerol content, and fiber dipersion distribution variation were investigated.

2. Experimental

2.1. Materials

The ramie fibers used in this study were obtained from Pesantren Darussalam Garut, Jawa Barat (Indonesia). The native casava starch was purcased from PT. Budi Acid Jaya Tbk (Indonesia). The glycerol was used as plasticizer, was supplied by PT. Brataco Bandung (Indonesia). The polyvinyl alcohol (PVA) was supplied by Central Kimia Bandung (Indonesia).

2.2. Preparation of Ramie Fibers

Native-ramie fibers were washed with demineralized water and dried at room temperature for 48 h. The amounts of native-ramie fibers, were kept to be used as reinforcement for the biocomposites preparation. Another portion of native-ramie fibers were prepared for alkali treatment. Native-ramie fiber were refluxed in 1 wt% NaOH solution, with ratio 1:100 for 2 h at 98 °C. Last, the ramie fibers (untreated and treated fiber) were then cut into 10 mm lengths.

2.3. Preparation of Starch-Based Biocomposites

Tapioca starch-based biocomposites with 15 wt% ramie fiber were prepared by solution casting and compression molding technique. PVA 29 wt% was dissolved in demineralized water and stirred on magnetic stirer. The mixture was then heated to 85 °C. Thereafter, the mixture was cooled down to 55 °C. At 55 °C cassava starch, without and with addition of 20 wt% glycerol, were added under constant stirring, and was then heated to 70 °C. Ramie fibers (untreated and treated fiber) were spread randomly into the molds. The fiber distribution was varied and coded as D0, D1 and D2 reflect the unhomogeneous until homogeneous fiber dispersion respectively. After that, the resin mixture was removed from the heat and pouring it on the surface of ramie fibers and then dried at room temperature for 72 h. Other biocomposite specimens were fabricated by means of hot compression molding and were heated at 120 °C for 1 h. Table 1 shows the sample of biocomposites with the various processing methods.

Tabel 1. Sample of Biocomposites.

Wihout Compression	With 120°C Compresion Temperature
KMNG	K120MNG
KANG	K120ANG
KMG	K120MG
KAG	K120AG

N: Without alkalization NG: Without addition of glycerol
A: With alkalization G: With glycerol content

IOP Conf. Series: Earth and Environmental Science **105** (2017) 012092 doi:10.1088/1755-1315/105/1/012092

2.4. Tensile Test

Tensile test were carried out following ASTM D-3039. Tensile properties was measured using Tensilon RTF-1310, equipped with a 1 kN load cell at a crosshead speed of 5 mm/min. Three specimens were tested for every sample. The fractured surface of biocomposites were observed by optical microscope.

3. Results and discussion

3.1. Effect of Alkalization

Figure 1 shows the effect of alkalization on tensile properties of the biocomposites. It can be seen that alkalization improves the tensile strength and modulus elasticity of the biocomposites, with the maximum improvement by 64% and 54% respectively. Alkalization increased the fiber surface roughness and the number of free hydroxyl groups in the fiber surface, resulting in better mechanical interlocking and chemical bonding which leads to an improvement in fiber-matrix interface [15-17]. The improvement in fiber-matrix interface results in improvement the tensile properties of biocomposites.

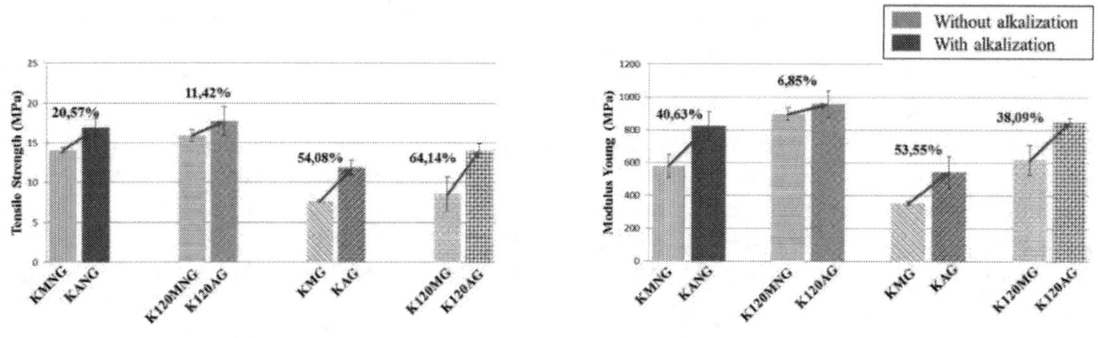

(a) (b)

Figure 1. The effects of alkalization on tensile properties of the biocomposites: (a) tensile strength and (b) modulus elasticity.

3.2. Effect of Compression Temperature

Figure 2 shows the effect of compression temperature on tensile properties of the biocomposites. The tensile strength and modulus elasticity increased after the application of compression temperature at 120 °C, with the improvement by 18% and 74% respectively. This improvement was mainly due to the decrease of void content on the biocomposites (see Figure 2.a).

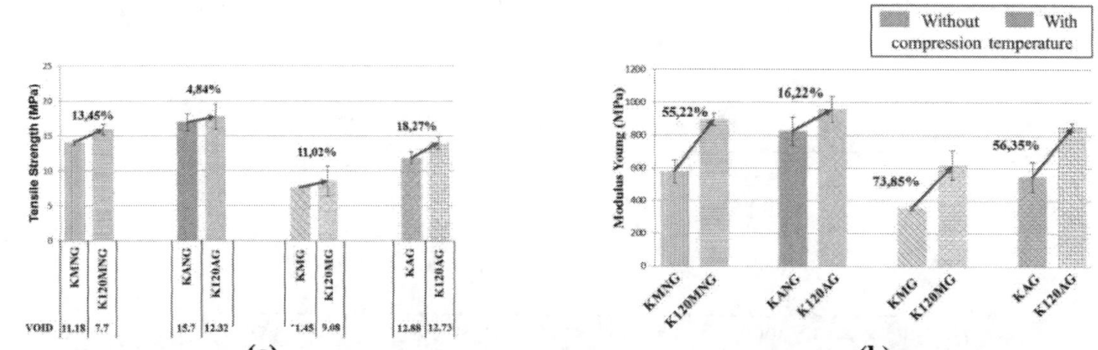

(a) (b)

Figure 2. The effects of compression temperature on tensile properties of the biocomposites: (a) tensile strength and void content and (b) modulus elasticity.

3.3. Effect of Fiber Dispersion

Figure 3. shows the effect of fiber dispersion on tensile properties of biocomposites. The tensile strength and modulus elasticity increased significantly with homogeneous fiber dispersion. The highest tensile strength (17.76 MPa) was obtained from the biocomposites made with 120 °C compression temperature using homogeneous ultimate fiber distribution, K120ANG-D2, see Figure 3.b. Homogeneous distribution of the ultimate fiber made the matrix easily penetrated into the fiber mats and resulting in better fiber-matrix interface. The tensile strength of ultimate fibers used in D2 sample which is much higher than fiber bundles (938 MPa vs 238 MPa [3,14]) used in D0 and D1 samples contributed to increase the tensile properties of biocomposites.

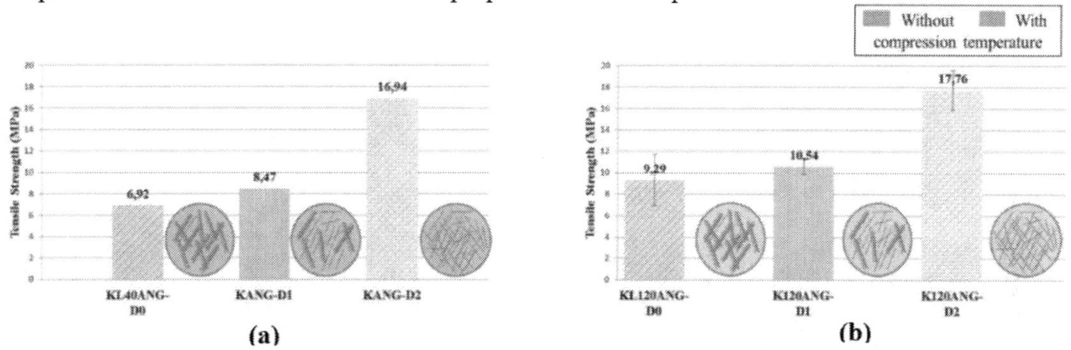

Figure 3. The effect of fiber dispersion on tensile strength of biocomposites: (a) without and (b) with 120 °C compression temperature.

3.4. Effect of Glycerol Content

The addition of glycerol content decreased the tensile strength and modulus elasticity for 46% and 39% respectively as shown in Figure 4. It shows that the addition of glycerol increased the void content on biocomposites (see Figure 4.a), resulting in lower tensile properties of biocomposites.

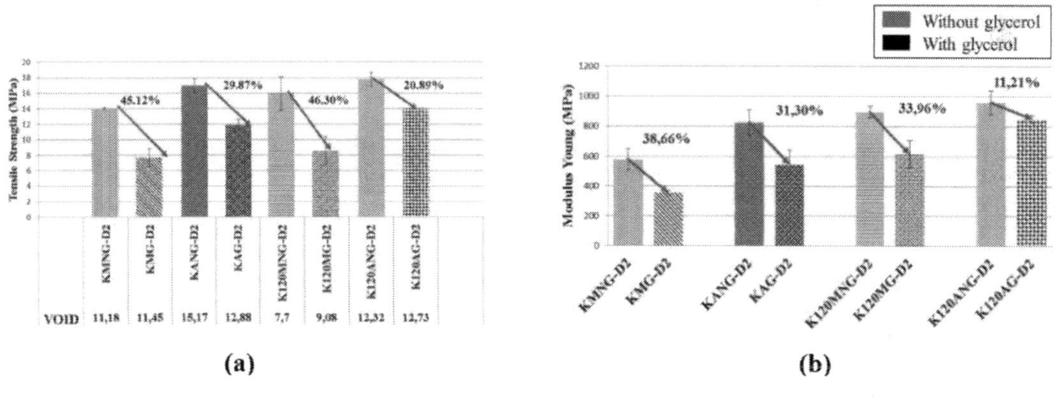

Figure 4. The effects of glycerol content on tensile properties of biocomposites: (a) tensile strength and void content and (b) modulus elasticity.

3.5. Fracture Surface of Biocomposites

Figure 5 shows the fracture surface of biocomposites. It can be seen that alkalization and high 120 °C compression temperature (see Figure 5.a and 5.b) decreased the existence of fibers pull out due to the improvement of fiber-matrix interface. These results confirm the improvement of tensile properties on biocomposites due to the alkalization and 120 °C compression temperature. On the other hand, the addition of glycerol content (see Figure 5.c) increased the number of fibers pull out due to deterioration of fiber-matrix interface, resulting in lower tensile properties of biocomposites.

IOP Conf. Series: Earth and Environmental Science **105** (2017) 012092 doi:10.1088/1755-1315/105/1/012092

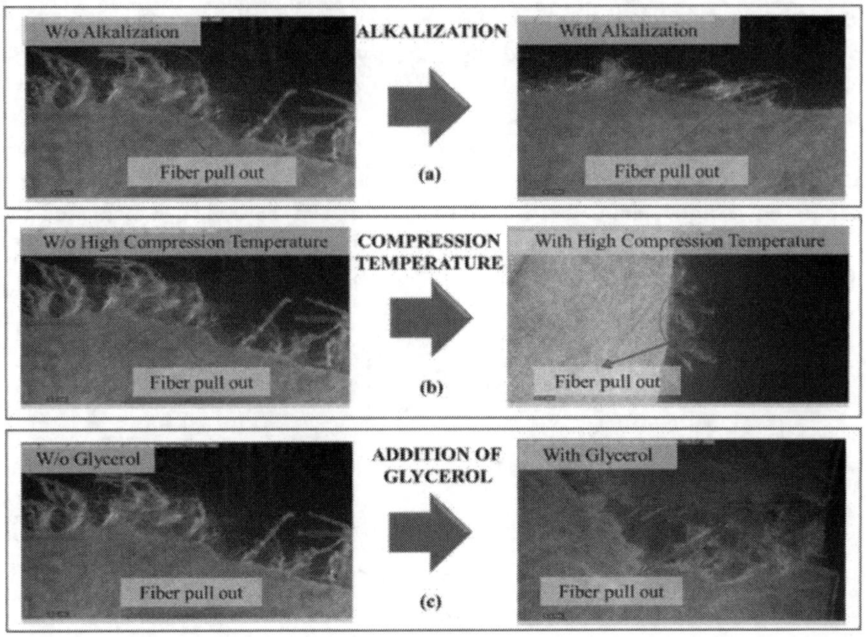

Figure 5. The fracture surface of biocomposites: (a) effect of alkalization, (b) effect of compression temperature, and (c) effect of addition of glycerol content.

3.6. Comparison of Tensile Strength Fiber Reinforced Based Starch Biocomposites

Figure 6 shows the comparison of tensile strength fiber reinforced based starch biocomposites. It can be seen, in Figure 6.a that the best processing method to obtain biocomposites with the highest tensile strength was by using homogeneous fiber distribution without glycerol addition. The tensile strength of random ramie fiber-tapioca starch developed in this research is higher compared to the literature results as shown in Figure 6.b [9-14].

(a) **(b)**

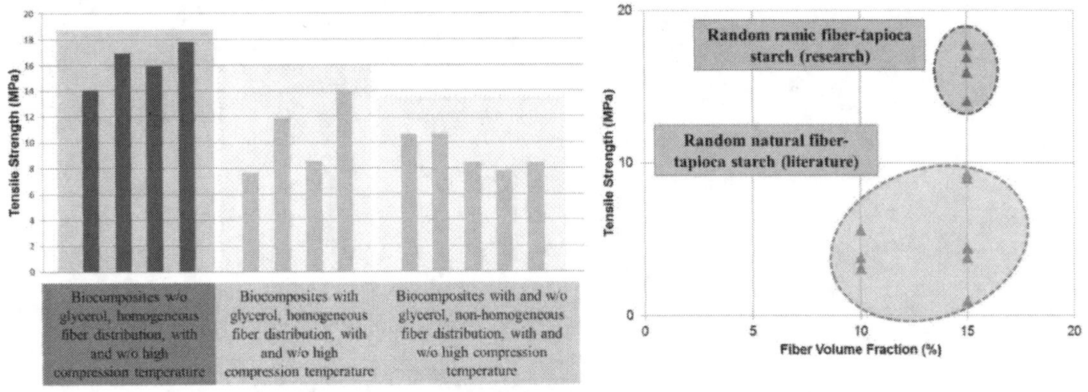

Figure 6. Comparison of Tensile Strength Fiber Reinforced Based Starch Biocomposites (a) with different processing methods and (b) with literatures [9-14].

4. Conclusions

The highest of tensile strength and modulus elasticity of biocomposites developed in this research were 17.76 MPa and 959 MPa respectively. The use of 1% NaOH alkalization process on ramie fiber increased tensile strength and modulus elasticity of biocomposites 64% and 54% respectively. The application of 120 °C compression temperature, increased tensile strength and modulus elasticity of biocomposites 18% and 74% respectively. Homogenous dispersion of ultimate fiber distribution increased the tensile strength up to 91%. The addition of glycerol decreased the tensile strength and modulus elasticity for 46% and 39% respectively. The results indicate that renewable resources such as ramie fibers and cassava starch, have potential ability to replace glass fibers and syntetic polymers as the reinforcement and the matrix for the biocomposites application.

References

[1] EuCIA-European Composites Industry Association 2016 *Composites Market Report 2015: Market developments, trends, challenges and opportunities (The European GRP Market/The Global CRP Market*)
[2] Mitra S et al 2014 *Ramie: The Strongest Bast Fiber of Nature* (India: Technical Bulletin) p 8
[3] Mohanty A K et al 2005 *Naturak Fibers, Biopolymers, and Biocomposites* (Boca Raton: CRC Press)
[4] Lodha P and Netravali A N 2002 *J. Mater. Sci.* **37** pp 3657-3665
[5] Nam S and Netravali A N 2004 *J. Adhes. Sci. Technol.* **18** pp 1063-1076
[6] Mo X Z, Zhong Y X, Liang C H Q, Yu S J 2010 *Advc. Matl Research* **87-88** pp 439-444
[7] Halley P J and Averous LR 2014 *Starch Polymers From Genetic Engineering to Green Applications* (Elsevier)
[8] FAO-Food and Agriculture Organization of The United Nations 2015 *Food Outlook: Biannual Report on Global Food Market*
[9] Guadalupe M, et al 2011 *Carb. Polymers* **86** pp 1712-1722
[10] Kaewtatip K and Thongmee J 2012 *Matl. and Design* **40** pp 314-318
[11] Raabe J, et al 2015 *J. Nanomatl.* **2015**
[12] Marsyahyo E, Astuti S, and Ruwana I 2011 *Advance in Composite Materials-Analysis of Natural and Man Made Materials* **11** pp 297-309
[13] Zanariah S Y et al 2013 *Carb. Polymers* **92** pp 2299-2305
[14] Judawisastra H, Virginia L, and Mardiyati 2015 *Prosiding Seminar Nasional* ISBN: 978-602-73461-0-9
[15] Mwaikambo L Y and Ansell M P 2002 *J. of Applied Polymers Science* **84** pp 2222-2234
[16] Yan L, Chouw N, and Yuan X 2012 *J. of Reinforced Plastic and Composites* **31** pp 425-437
[17] Wattanakornsiri A and Tongnunui S 2014 *J. Sci. Technol* **36(2)** pp 149-161

The use of Silica from beach sand as catalyst in Magnesium based hydrides for Hydrogen storage materials

Zulkarnain Jalil[1,*], Adi Rahwanto[1], Hizir Sofyan[2], Mustafa Usman[3] and Erfan Handoko[4]

[1]Department of Physics, Faculty of Mathematics and Natural Sciences, Syiah Kuala University, Banda Aceh 23111, Indonesia
[2]Department of Statistics, Faculty of Mathematics and Natural Sciences, Syiah Kuala University, Banda Aceh 23111, Indonesia
[3]Department of Agribusiness, Faculty of Agriculture, Syiah Kuala University, Banda Aceh 23111, Indonesia
[4]Department of Physics, Faculty of Mathematics and Natural Sciences, Jakarta State University, Jakarta, Indonesia

Email: zjalil@unsyiah.ac.id

Abstract. Magnesium (Mg) as one of the potential candidate material for absorbing hydrogen, because theoretically has the ability to absorb large quantities of hydrogen 7,6 wt%. However, the kinetic reaction of Mg is very slow. Its takes 60 minutes to adsorp hydrogen with the operating temperature (adsorb/desorb) high of 350^{0}C. Therefore, in this study discusses the hydrogen storage materials based on MgH_2-SiO_2. The purpose of this study to improve desorption temperature of hydrogen storage system based on MgH_2. The main material is MgH_2 combined with inserting SiO_2 catalyst was successfully extracted from quartz sand with coprecipitation method. As for the material preparation process is done with the technique of mechanical alloying. Milling MgH_2+SiO_2 samples using a ball milling for 5 hours, with the ratio of ball to powder 10:1 and a speed of 400 rpm. By variation of the catalyst insertion of 1 wt%, 3 wt%, and 5 wt% of SiO_2. The results of XRD measurement known that the sample was reduced to scale nanocrystal. Phase arising from the result of XRD observation are MgH_2 phase as the main phase, and SiO_2 phase as a minor phase. DSC testing results show that the lowest desorption temperature obtained on the sample with the addition of inserts weight of the catalyst 5 wt% SiO_2 has milling for 5 hours which is equal 307,11^{0}C.

1. Introduction

Magnesium-based hydrogen storage alloy is one of the most attractive hydrogen storage materials for fuel cell powered vehicle application. But, high desorption temperature, slow kinetics, however, limits its practical application. Several types of material (generally light-metal metal element) is believed to have the ability to absorb large amounts of hydrogen.

One of these is magnesium (Mg), which is considered a potential candidate for the hydrogen storage material. Magnesium, theoretically, have the ability to absorb large quantities of hydrogen (7.6 wt%). This amount exceeds the maximum limit of International Energy Agency (IEA) targeted i.e. amounting to 5 wt% and able to work at temperature below 100°C [2].

However, high working temperature, slow reaction kinetics and difficult activation limit the practical application of Mg-based hydrides.Some effort has been done to improve the properties of

Mg-based hydrides. For example the reduction of grain size of materials up to the nanoscale using the mechanical alloying technique, forming a composite material, and adding a specific catalyst, such as metals and metal oxides [3-5].

Furthermore, Norton *et al* [6] have been reported on the effectiveness of metal oxide type of silica (SiO_2) that can be used as material for hydrogen storage. In this study, we have used the natural silica (SiO_2) which was extracted from natural beach sand and catalyzed in MgH_2. The sample was prepared using a planetary high energy ball milling technique, as an effort to improve the properties of reaction kinetics of magnesium-based hydrogen storage materials.

2. Materials and Method

The main material MgH_2 (99.99%, size 50 μm) was purchased from Sigma Aldrich and quartz sand was collected from Calang Beach in Aceh Jaya District. Then, hydrochloric acid (HCL) and ammonium hydroxide (NH_4OH) solution and aquadest were used for silica extraction. Then, the extraction process of silica was done, which at the early stage of beach sand dissolved with HCL, and stirred with magnetic stirrer at 700°C until the solution is formed. The solution is then filtered with a filter paper. The solution is then deposited with NH_4OH and subsequently washed with aquadest. The precipitated silica was washed and then dried with an oven at a temperature of 70°C for 5 hours. The next process was the mixing of MgH_2 with SiO_2 (1wt%, 3wt% and 5wt%) using a planetary ball milling (Fritsch, P6) with ball to powder ratio (BPR) is 10:1 at a speed of 400 rpm for 5 hours. A small amount of samples were transferred into other containers in every variation of SiO_2 for further characterization. In order to find out the composition phase, the samples were carried out using XRD (Shimadzu D6000, Cu-Kα radiation λ = 1.54060 Å). Morphological observation was obtained by SEM (Philips, XL30). Furthermore, the thermal properties was observed by DSC (Shimadzu, D50).

3. Results and Discussion

Before the extraction process using coprecipation method, this quartz sand has been characterized by XRF. The chemical composition as stated in Table 1 below.

Table 1. XRF results of Calang natural beach sand

No.	Compound	Concentration (%)
1	SiO_2	88.78
2	CaO	3.7
3	Fe_2O_3	2.21
4	SrO	2.12
5	Al_2O_3	1.2
6	MoO_3	1.2
7	Yb_2O_3	0.25
8	TiO_2	0.24
9	NiO	0.15
10	BaO	0.09
11	MnO	0.078
12	CuO	0.056

From table 1 it can be seen that the compounds contained in quartz sand such as SiO_2 (%), CaO (%), Fe_2O_3 (%) and others. However, the highest percentage of SiO_2 compound content was 88.78% while the lowest percentage was CuO of 0.056%.

The extraction process of quartz sand is carried out to obtain high purity silica using coprecipitation method. Corrosive methods include extraction methods of inorganic compounds such as quartz sand, which can be extracted using low temperature. This extraction process will bring down a solute down to the desired sediment. In this method, treatment has been done. One of the initial treatments was to immerse the quartz sand in HCl 2M solution for 24 hours. With the soaking treatment of the HCl 2M solution, it was able to reduce the impurities in sand before proceeding in the extraction process.

Then, based on observations by XRD (figure 1), indicated that there are no changes from the 3 variations of SiO_2. X-ray diffraction pattern in figure 1 shows that almost a third of the curve have the same diffraction pattern due to milling in the same time of 5 hours, only in variations on the insertion of silica (SiO_2).

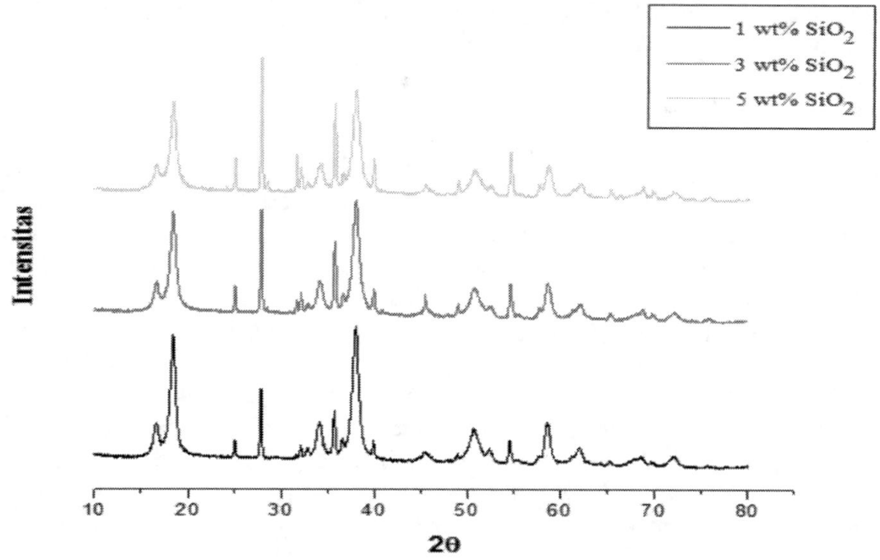

Figure 1. XRD pattern of the sample MgH_2-SiO_2

From the XRD result we can calculate the crystallite size in order to check whether the particle already decrease or not. Diffraction pattern can be calculated from the size of the sample using the Scherrer method [7],

$$D = (0.9\lambda)/B \cos \theta \quad (1)$$

where :

D : Crystallite size (Å)

λ : Wavelenght (λ= 1.54060 Å).

θ : The Bragg angle.

B : Line broadening at half the maximum intensity, FWHM (rad)

The results of the calculation of the size of the crystals is summarized in Table 2 is a result of calculations for each field phase MgH_2 + wt% SiO_2.

Table 2. Crystallite size of MgH_2 + SiO_2 in each variation of the SiO_2.

Catalyst	**Crystalline Plane (hkl)**	**Parameter**		
		2θ (°)	**FWHM (°)**	**Crystallite size (nm)**
1 wt%	0 2 1	37,9130	0,82480	10,75
3 wt%	0 2 1	37,9143	0,85420	10,38
5 wt%	1 1 0	27,8840	0,18540	46,60

In table 2 it can be seen that with the addition of catalysts the crystal size is getting smaller. The size of the crystals on the addition of the variation of the catalyst weight may change with the same milling time due to silica (SiO_2) as a catalyst having a hardening effect that will accelerate the process of breaking MgH_2 powder. Although SiO_2 is only used in very small quantities. This indicates that the grain size is smaller as the addition of the catalyst.

However, it does not apply to the addition of catalyst 5 wt% SiO_2. There is the agglomeration phenomena during the milling process. This agglomeration occurs due to the puncture and pressure given by the balls against the powder which causes the particles to join each other. Agglomeration is part of the mechanical alloying process. However, this excessive agglomeration can reduce the adsorption and desorption properties as it reduces the surface area of magnesium contact with hydrogen.

Thermal testing using the Differential Scanning Calometry (DSC) tool, to see changes in temperature and time function in a sample that occurs during the heating process. Thermal testing using DSC was carried out for a MgH_2- SiO_2 sample which has been milled for 5 hours with variation of SiO_2 catalysts of 1, 3, and 5 wt% can be seen in Figure 2.

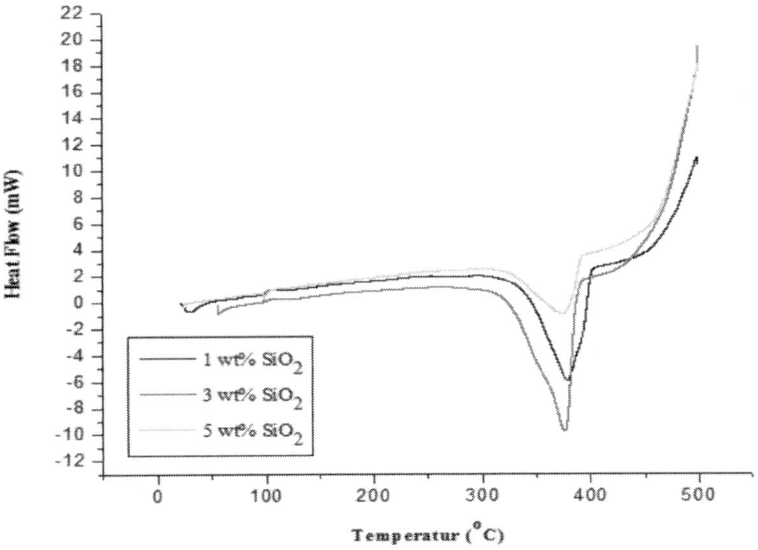

Figure 2. DSC of MgH2 - wt% SiO2 curve on the variation of catalyst for 5 hours of milling time.

Table 3. DSC results of MgH_2 with variation of SiO_2

No	Samples	Milling time (hours)	T_{onset} (°C)
1	MgH_2 + 1 wt % SiO_2	5	321,85
2	MgH_2 + 3 wt % SiO_2	5	313,86
3	MgH_2 + 5 wt % SiO_2	5	307,11

Based on figure 2. indicates that when the sample was added 1 wt% and 3 wt% SiO_2 the peak of the chart was sharp, it indicates that the reaction process when the hydrogen heating was released slower than MgH_2. Whereas the addition of SiO_2 catalysts as much as 5 wt% indicates that the peak of the chart is seen more shallow means that the reaction process when heated is so fast. This proves that the SiO_2 catalyst insertion has an effect on the MgH_2 desorption temperature. SiO_2 as a catalyst has an important role that can lower the activation energy of a reaction so it can accelerate the reaction.

Compared to previous research results with 2 wt% SiO_2 milling insertion, desorption temperature was at 350^0 C for 57.29 minutes [8]. Therefore, this research has improved certain characteristics of MgH_2 with the addition of SiO_2 catalyst, which was extracted from beach sand. The characteristic that can be improved is the temperature of the desorption and its timing. With the addition of small amount of catalysts [9,10], in this case as 1, 3, and 5 wt% SiO_2, has been able to lower the desorption temperature and its desorption time is shorter. This result corresponds to the pure MgH_2 before the addition of catalysts were in the range of 409^0 C and when given a catalyst, the desorption temperature tended to decrease.

4. Conclussion

The natural crystalline type catalyst of SiO_2 was succesfully produced from local beach sand and milled with MgH_2. This work has improved certain characteristics of MgH_2 with the addition of natural SiO_2 catalyst. The characteristic that can be improved is the temperature of the desorption and its timing. With the addition of small amount of catalysts as 1, 3, and 5 wt% SiO_2, has been able to lower the desorption temperature and its desorption time is shorter.

5. Acknowledgments

Authors would like to thanks Indah Anggraeni, S.Si and Syarifah Fathmiyah, S.Si (Materials Physics Lab, Syiah Kuala University) for the assistance in DSC and XRD sample preparation.

6. References

[1] Schlapbach and Zuttel 2001 Hydrogen storage materials for mobile applications *Nature* **414**

[2] Zuttel A, Borgschulte A and Schlapbach L 2008 *Hydrogen a Future Energy Carrier* Wiley VCH Verlag Weinheim

[3] Oelerich W, Klassen T and Bormann R 2001 Mg-based hydrogen storage materials with improved hydrogen sorption *Mater. Trans.* **42** (8) pp 1588-1592

[4] Zaluska A, Zaluski L and Ström-Olsen J O 1999 Nanocrystalline magnesium for hydrogen storage *Journal of Alloys and Compounds* **288** pp 217-225

[5] Ichikawa T, Hanada N, Isobe S, Leng H and Fujii H 2005 Composite materials based on light elements for hydrogen storage *Materials Transactions* **46** pp 1-14

[6] Norton M G, Mcleroy D N, Corti G and Miller M A 2009 Silica Nanosprings: A Novel Nanostructured Material for Hydrogen Storage *Proc. Clean Technology Conference*, Houston

[7] Barreto G P et al. 2013 Microwave Assisted Synthesis of ZnO Nanoparticles: Effect of Precursor Reagents, Temperature, Irradiation Time, and Additives on Nano-ZnO Morphology Development *Journal of Materials* Volume **2013** pp 1-11

[8] Mustanir, Nur S, Jalil Z and Fauzi 2011 Pengaruh Sisipan Katalis SiO_2 dalam MgH_2 yang Disintesis Melalui Rute Mechanical Alloying *J. Sains MIPA* **17** pp 43-47

[9] Jalil Z, Rahwanto A, Mulana F and Mustanir 2016 Desorption Temperature Characteristic of Mg-based Hydrides Catalyzed by Nano-SiO_2 Prepared by High Energy Ball Milling *International Journal of Technology* **7** (8) pp 1301-1306

[10] Jalil Z, Rahwanto A, Handoko E and Mustanir 2017 The role of nano-Ni catalyst in MgH_2 obtained by reactive mechanical milling method for solid hydrogen storage application *AIP Conference Proceedings* **1826** 020002

Micro-Bubble drag reduction with triangle bow and stern configuration using porous media on self propelled barge model

Yanuar, K T Waskito, B A Rahmat, A Y Perdana and B D Candra

[1]Department of Mechanical Engineering, University of Indonesia, Jakarta 16424, Indonesia

E-mail: yanuar@eng.ui.ac.id

Abstract. Drag reduction on ship to decrease fuel consumption and achieve higher speeds has became the topic since the last decades. One of the most atractive idea is micro-bubble, in order to reduce frictional resistance. The purpose of this research is to identify micro-bubble drag reduction effect on Self Propelled Barge Model at 1:28 scale. The influences of air injection ratio and speed will also be investigated. Five injection ratio were used in this experiments; 0,2; 0,3; 0,4; 0,5; and 0,6. Experiments were done in the towing tank, member of ITTC. Micro-bubble was injected on bow and midship using two configurations; triangle stern and triangle bow. The results show that triangle stern is the best configuration with 5% - 40% drag reduction at 0,2; 0,3; and 0,4 injection ratio.

1. Introduction

The resistance of a moving ship is a drag force that is opposite to the direction of the ship motion as it travels through the fluid at certain rate of viscosity. Drag on the ship can be categorized into frictional drag and residual drag. Drag reduction is divided into two methods: active and passive. The active is a method that adds substances continuously to the body of the ship, while the passive method using modification of the ship body in order to reduce resistances continuously.

Micro-Bubble Drag Reduction (MBDR) is an active method by injecting micro-sized air to the bottom of the ship. The use of micro-bubble is the most promising active methods to reduce frictional resistance. The first research on micro-bubble was done by McCormick and Bhattacraryya (1973) using electrolysis method. Madavan, *et al* (1985) injects the micro-bubble into the boundary layer layer using porous media in the water tunnel, resulting in 15-80% drag reduction.

Kodama, *et al* (2001) used tanker to determine the optimum location of air injection. They concluded that the most effective injection is at the bow of the vessel. Tanker and barge vessels have moderate L/B ratio, therefore the area covered by micro-bubble can become wider. This type of vessel also has a flat and wide bottom surface, making it suitable for micro-bubbles. Meanwhile, according to research of Yanuar, *et al* (2012) on the Fast Patrol Boat concluded that the best injection location is at 5 cm behind the midship. Another experiments, Kawabuchi, *et al* (2011) on container vessel injects micro-bubble on the bow and stern using CFD simulation, resulting up to 60% drag reduction.

In this research Self Propelled Barge Model was used with L=56m and scaled to ship model at 1:28 scale. Air compressor was used to inject air into porous media. The influence of speed and air injection ratio variations will be investigated. The effect of porous media location will also be analyzed by comparing two different sets up. The final result will be the most effective configuration and air injection ratio at certain speed.

2. Experimental Setup

The experiments were done in Indonesia Hydrodynamic Laboratory (IHL), Surabaya, Indonesia. This towing tank is a member of International Towing Tank Conference (ITTC). The Specifications of this towing tank are presented in figure 1.

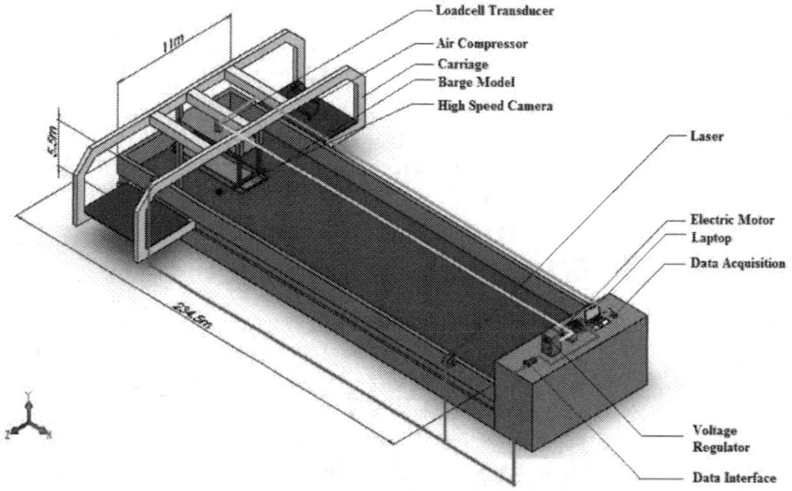

Figure 1. Experimental Set Up

A 2 m Self Propelled Barge Model has been constructed with fiber glass for experimental tests. The hull form of this model was straightforward, box shaped in the stern region and streamed in the bow, can be seen in figure 2.

Figure 2. Scaled SPB Model at 2 m Length

Air injection was undertaken at two positions along the length of the model, bow and stern regions with length ratio. The length to position ratio for injection was 0,35L to the bow and 0,025L to the stern. The injector was located in triangled positions shown in figure 3 and 4. The first configuration was triangle bow, consisted of one injector at the bow and two injectors at the stern. Thus, totally in one configuration consisted of three injectors. Another configuration was reciprocal with two injectors at bow and one at stern. Further detailed specifications of ship model can be seen in table 1.

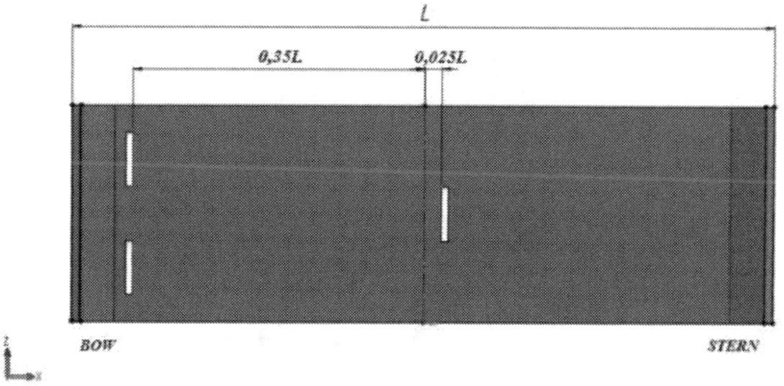

Figure 3. Triangle Stern Configuration

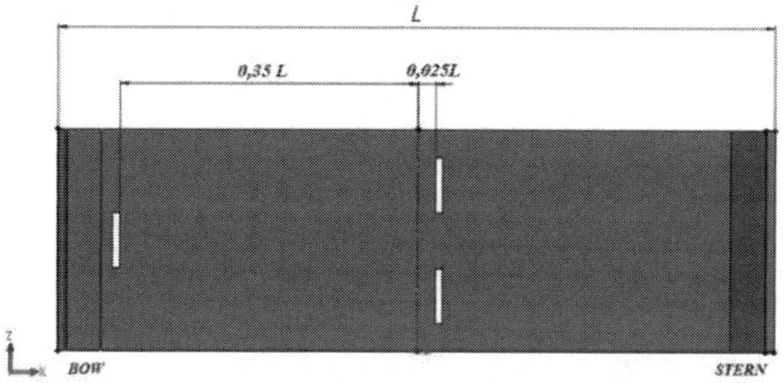

Figure 4. Triangle Bow Configuration

Table 1. Model Specification

Unit	Dimension
Length	2 m
Width	0.52 m
Height	0.12 m
Draft	0.08 m

In order to form bubbles around the hull of the model, air was injected through a plate with array holes (porous plate). The width and length of this plate was 20 mm and 100 mm respectively. The injection plates were covered with holes along the width and length, which is shown in figure 5.

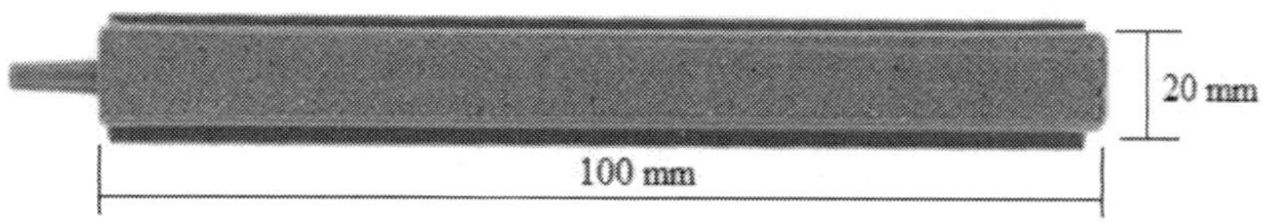

Figure 5. Porous Plate

To set and control the air injection rate, a flow meter was used. This flow meter shows the rate of air at Litre per Minute (LPM). The maximum rate could be measured at 100 LPM, which is shown in Figure 6. Thus, by using flow meter, air injection rate could be controlled at each step.

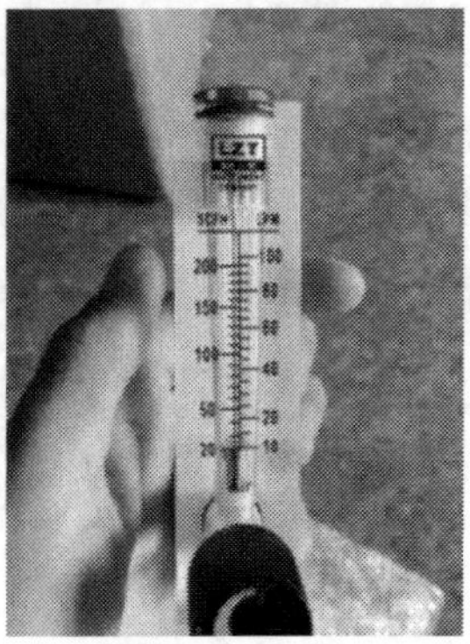

Figure 6. Air Flow Meter

In order to produce the injected air, an air compressor was used with 120 LPM maximum injection rate. This air compressor was connected to injectors using 6,35mm rubber tube. A divison manifold was used to create equal flow rate at each injector. Thus, the air injection schematic was created and shown in figure 7.

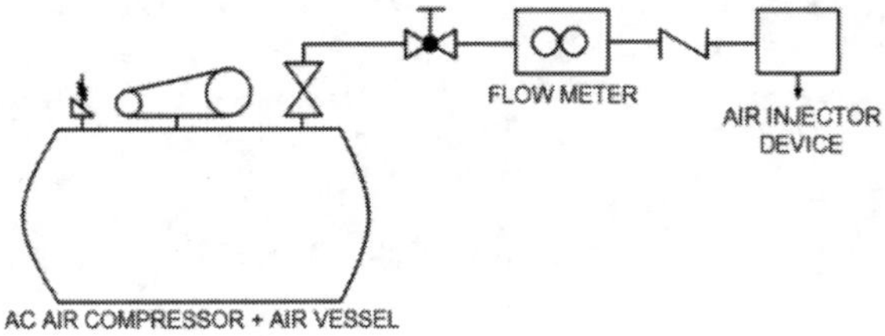

Figure 7. Air Injection Schematic

To get varied data, 10 steps of speed were used and shown in table 2. These steps were based on the actual ships Froude Number at 0,15. Then, five air injection ratio were also used to determine optimum rate for Self Propelled Barge Model; 0,2; 0,3; 0,4; 0,5; and 0,6.

Table 2. Variation of Speeds

V (ms^{-1})	Froude Number
0,5	0,11
0,56	0,13
0,67	0,15
0,83	0,19
1	0,23
1,11	0,25
1,25	0,28
1,3	0,29
1,33	0,3
1,38	0,31

3. Test Analysis

Froude simialirity law is followed in the extrapolation of resistance test result. According to it, the resistance of ship can be divided into two components and one of them is proportional to the frictional resistance of a flat plate at the same length and wetted surface area, towed at the same speed.

In this experiments, air injection ratio was used to determine optimum injection for Self Propelled Barge Model. According to Sayyaadi and Nematollahi (2013), the air injection (Q_a) injected into porous media can be calculated as:

$$\alpha = \frac{Q_\alpha}{Q_w} \tag{1}$$

With air injection ratio (α) determined at 5 steps; 0,2; 0,3; 0,4; 0,5; and 0,6. Then, Q_w is water flowrate within boundary layer (Sayyaadi and Nematollahi, 2013), can be evaluated as:

$$Q_w = 0,293.\ L^{0,8}.\upsilon^{0,2}.\ V^{0,8}.\ W \tag{2}$$

where L is the length of ship (m); υ is kinematic viscosity of the fluid (m^2s^{-1}); V is speed of ship (ms^{-1}); and W is the width of ship (m).

The first raw data obtained from the towing tank experiments were total resistances of ship (R_T), then it can be extrapolated into resistance coefficient (C_T) with the following equation:

$$C_T = \frac{R_T}{0,5.\rho.V^2.S} \tag{3}$$

where ρ is fluid density (kgm^{-3}) and S is wetted surface area (m^2). From the MAXSURF software, the wetted surface area can be determined; 1,1292 m^2. From the results above, then the results can be compared as:

$$DR(\%) = \left(\frac{C_T - C_{TO}}{C_{TO}}\right)100\% \tag{4}$$

where $DR(\%)$ is drag reduction percentage and C_{TO} is resistance coefficient without injection.

4. Results and Discussion

4.1. Total Resistance (R_T) vs Froude Number (Fr)

From the figure 8, it is shown that each configuration has the same trend, which is the total resistance of ship increases (R_T) as the speed increases. This corresponds to the resistance equation which the speed function is directly proportional to the resistance of the ship. When the injection process occurs, the triangle stern configuration tends to be stable, lower than without injection. Furthermore, in Figure 8 it can be seen as well that the total resistance of triangle stern configuration is lower than triangle bow. This is because in the triangle stern configuration consisted of two porous media located on the bow of the ship. The air that exits through the two porous media is more enveloped under the surface of the ship than the triangle bow configuration. Therefore, the ship resistance lowered, because the wet area is smaller.

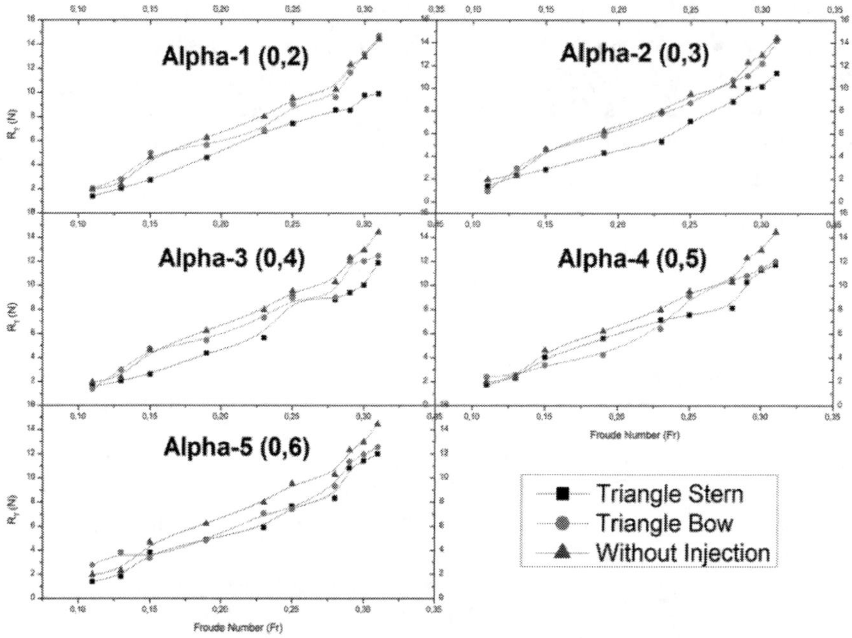

Figure 8. Total Resistance (N) vs Froude Number

4.2. Total Resistance Coefficient (C_T) vs Froude Number (Fr)

The next step is breakdown the total resistance into coefficient using equation 3. Wetted surface area (S) can be calculated precisely using naval software; MAXSURF and resulted a value of 1,1292 m^2. The graph trend in figure 9 has the same characteristics as the graph in figure 8 because each component has a comparable value.

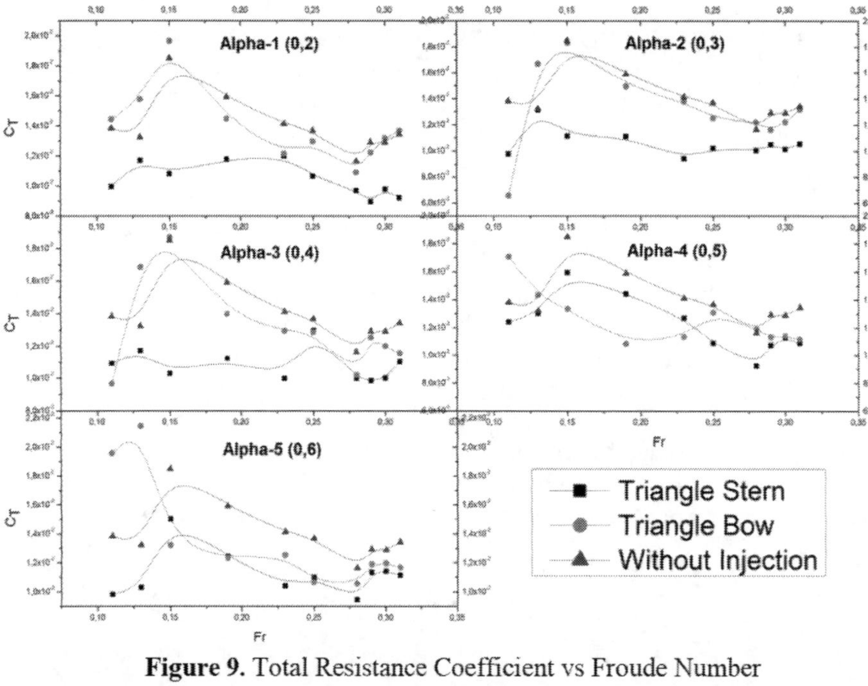

Figure 9. Total Resistance Coefficient vs Froude Number

In triangle stern configuration the overall graph is under the ship without injection and tends to be lower than triangle bow configuration graph. Whereas, in the position of triangle bow, on the first three Froude (Fr) numbers, the graph tends to be higher than ship without injection. This phenomenon can be explained through image visualization. In figure 10 and figure 11, can be seen the image comparison at Alpha-2 and Fr-3 on each configuration. In the triangle bow, there are micro-bubble built up from the front to the left, resulted in the formation of larger air bubbles. This larger bubbles can be detrimental if the bubble bursts and creates wake rake. This phenomenon is called pilling up effect (Sayyaadi and Nematollahi, 2013). From this phenomenon it can be concluded that at lower speeds micro-bubble injection in triangle bow configuration is not effective. While at higher speeds, micro-bubble is effective in reducing drag in both configurations. It can be shown on the graph that the total injection coefficient (C_T) is lower than without injection. Based on the five graphs can also be concluded that the triangle stern configuration is more effective than triangle bow configuration.

Figure 10. Triangle Bow at Alpha-2 (0,3) and Fr-3 (0,15)

Figure 11. Triangle Stern at Alpha-2 (0,3) and Fr-3 (0,15)

4.3. Drag Reduction (DR)

Based on analysis of the total resistance coefficient (C_T) vs Froude number (Fr) and also on the total resistance (R_T) vs Speed (V), the most effective drag reduction (DR) occurs in triangle stern configuration. Therefore, the drag reduction graph shown in figure 12 is the triangle stern configuration, to find the most effective air injection (α) ratio.

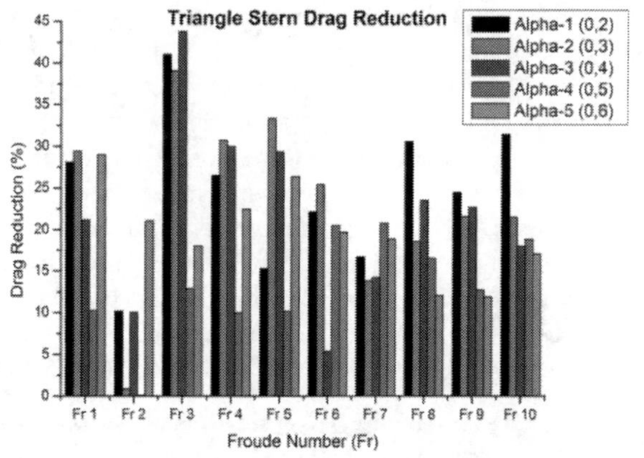

Figure 12. Drag Reduction on Triangle Stern Configuration

Based on the drag reduction graph in figure 12 it can be inferred that at low velocities from Fr-1 to Fr-5 the dominant air injection ratio in reducing drag is at 0.2-0.4, with a 10-43% reduction range. At low speed, the higher injection coefficient in some conditions were less effective in reducing the resistance. This is due to the pilling up effect (Sayyaadi and Nematollahi, 2013), where in higher injection coefficient at low speeds there is an accumulation of air on the bottom of the ship due to the low velocity of the vessel. The air moved slowly toward the stern of the vessel causes the formation of larger air bubbles. If this bubble is not broken, then it can be beneficial, because the wetted surface area is much reduced. However, in the event of a split on the bubble, it will cause a wake rake that created drags to the ship. This phenomenon showed in figure 13 and 14. In figure 13, micro-bubble did not coalesce and flows well behind the ship.

Figure 13. Triangle Stern at Alpha-2 (0,3) and Fr-3 (0,15)

However, in figure 14 and 15 there is a pilling up effect phenomenon which the micro-bubble joined into larger bubble in the midship but became small bubble in the stern, indicates that bubble burst has occured. But, this phenomenon did not occur at a ratio of 0.2; 0.3; And 0.4. At this ratio, the micro-bubble did not creating larger bubbles, flows smoothly to the stern of the ship. Therefore, it can be concluded that the effective injection ratio at low speed is at 0.2; 0.3; and 0.4.

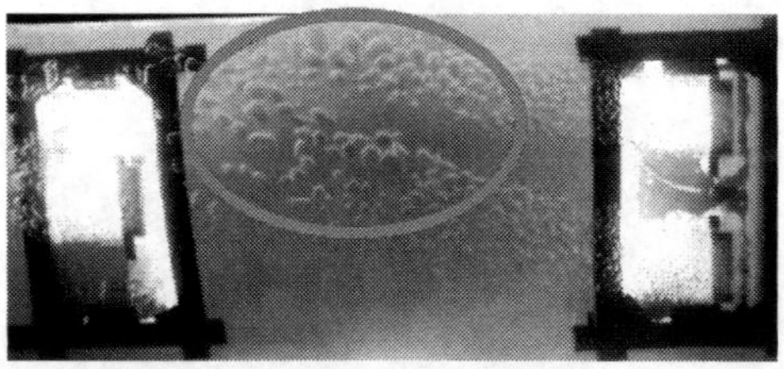

Figure 14. Triangle Stern at Alpha-4 (0,5) and Fr-3 (0,15)

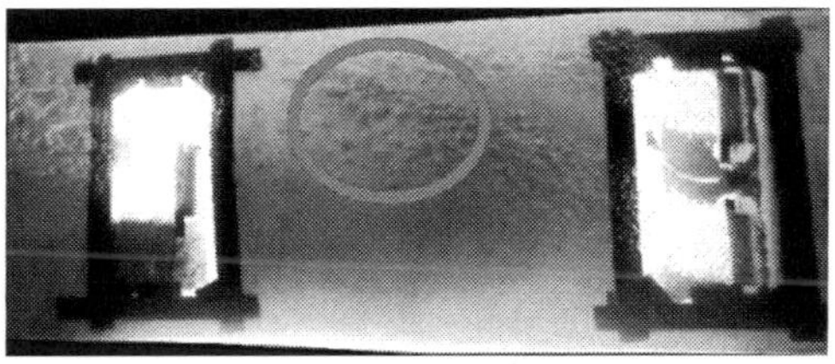

Figure 15. Triangle Stern at Alpha-5 (0,6) and Fr-3 (0,15)

At higher speeds from Fr-6 to Fr-10 there is no graphic trend occured. The increase in speed is not influential on the reduction of drags. The reduction percentage ranged from 15-30% with the highest reduction at Fr-8 and Fr-10 velocities in Alpha-1 injection coefficient (0.2). Meanwhile, at Fr-8 in the highest coefficient, drag reduction drops dramatically. This can be explained through image visualization. In figure 16 it appears that micro-bubbles were moving well toward the stern of the vessel, thus reducing the drags which occured better than Alpha-5 (0.6) in figure 17.

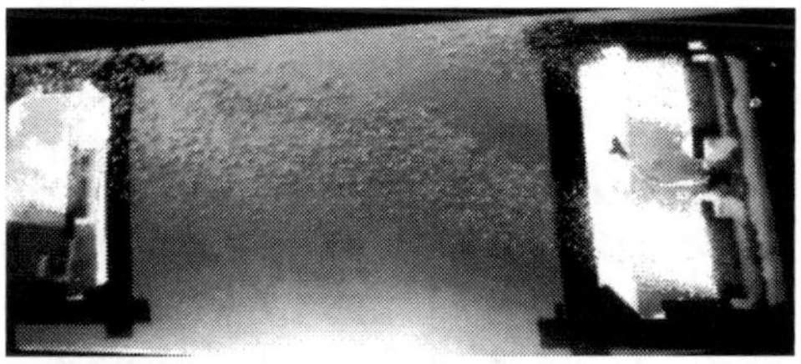

Figure 16. Triangle Stern at Alpha-1 (0,2) and Fr-8 (0,29)

Figure 17. Triangle Stern at Alpha-5 (0,6) and Fr-8 (0,29)

It appears the micro-bubble did not flow towards the stern of the ship properly, but goes straight to the side of the ship (leaking), thus drag reduction is lowered. Therefore, it can be concluded that at higher speeds the given injection ratio tends to provide drag reduction not much different at each step, thus the five ratios are effective to be used.

5. Conclusion

The experiment has been performed with and without injection of micro-bubble. Porous media was used to produce micro-bubble along the hull of model, with two configurations; triangle stern and triangle bow. By using 10 steps increment of speed 0.50 m/s; 0.56 m/s; 0.67 m/s; 0.83 m/s; 1.00 m/s; 1.11 m/s; 1.25 m/s;1.30 m/s; 1.33 m/s; 1.38 m/s. and 5 steps of air injection ratio, 0,2; 0,3; 0,4; 0,5; 0,6, this research yield the following conclusion:

1. The use of micro-bubble using porous media with triangle stern and bow configuration can reduce drag on ship model at 5 - 40%.
2. Triangle stern configuration is more effective than triangle bow, hence total resistance graphic shown in Figure 8 shows resistance in triangle stern lower than triangle bow.
3. The most effective air injection ratio at lower speeds were 0,2; 0,3; dan 0,4 while at higher

IOP Conf. Series: Earth and Environmental Science **105** (2017) 012094 doi:10.1088/1755-1315/105/1/012094

speeds all coefficients gave not so much different results. Therefore, the most effective air injection ratio for all conditions were 0,2; 0,3; dan 0,4.

6. References

[1] Harvald S A 1983 *Resistance and propulsion of ships* (New York: Wiley)
[2] Kodama Y, Kakugawa A and Takahashi T 2002 A full-scale experiment on micro-bubbles for skin friction reduction using Seiun-Maru, pt. 1: the prepatory study *J. Soc. Navy Architects japan* **192** pp. 1-13
[3] Kurniawan 2008 *Influence of Micro-bubble to Drag Reduction at Ship Model* Mechanical Engineering Department Universitas Indonesia
[4] ITTC 2002 *Testing and extrapolation methods in resistance towing tank test. Recommended procedures and guidelines*
[5] Lattore R, Miller A and Philips R 2003 Ship hull drag reduction using bottom air injection *Ocean Eng.* **30** pp 161-176
[6] Madavan N K, Deutsch S and Merkle C L 1983 Reduction of turbulent skin friction by micro bubbles *Phys Fluids* **27** pp 356-363
[7] Kawabuchi, Mizokami and Kodan 2011 CFD Predictions of Bubbly Flow around an Energy-saving Ship with Mitsubishi Air Lubrication System *Mitsubishi Heavy Industries Technical Review* **48** (1)
[8] Munson B R, Young D F and Okiishi T 2003 *Fluid Mechanics 4th Ed* Erlangga publishing
[9] NASA 2016 Global Climate Change *Fact Sheet: Global Warming* USA
[10] Sayyaadi H and Nematollahi M 2013 Determination of optimum injection flow rate to achieve maximum micro bubble drag reduction in ships; an experimental approach *Scientia Iranica.Transaction B, Mechanical Engineering* **20** (3) pp. 535-541
[11] Yanuar, Gunawan, Sunaryo and Jamaluddin A 2012 Micro-bubble Drag Reduction on a High Speed Vessel *JMSA* pp. 301-304

Selecting and testing of wind turbine blades of the local-wood growing fastly on local wind characteristics

Adi Surjosatyo[1], Bagus Rizky Dewantoro, Budiman R Saragih, Filbert Nainggolan, Wahyu Dwianto[2] and Teguh Darmawan

[1]Departement of Mechanical Engineering, University of Indonesia, Kampus Baru UI Depok, Depok 16424, Indonesia

[2]Indonesian Institute of Science Research Centre for Biomaterials

Email : adisur@eng.ui.ac.id

Abstract. The total potential of wind energy in Indonesia reaches 144 GW. This wind energy potential is dominated by small wind speeds ranging from 4-6 m / s and about 113 GW. This is the reason why the development of small-scale wind powerplant is more feasible to be developed in Indonesia. The characteristics of local wind and geographical conditions of Indonesia which is an archipelagic country are fundamental in designing wind power and utilizing of local materials. This study aims to get local fast-growing wood of wind turbines blades in Indonesia, which has the best physical and mechanical properties of wood, and is tested computationally on certain airfoils and shows a uniform distribution of stress. The types of wood are tested in this study were Sengon (Albizia chinensis), Jabon (Antocephalus cadamba), and Balsa (Ochroma Rowlee grandiflorum). The test covers the basic properties of the wood, physical and mechanical properties. The test results showed that Jabon have density 0.34 was the best compared to Sengon (0.24) and balsa (0.18). The best dimensional stability (T/R Ratios) are Jabon (1.97), sengon (2.19) and balsa (2.84), respectively. With the highest density result and the lowest dimensional stability value, mechanical testing of Jabon wood shows the best mechanical properties. The simulation result using Q-Blade software was obtained airfoil NACA 4415 most suitable for local wind characteristics, and got the low value of stress and distribution color uniformly on each wood.

1. Introduction

The development of wind power generations as part of renewable energy utilization has been implemented over the past few years, but the use of technology is not the same for every place, such as Indonesia which is a tropical area with wind energy potential of 113 GW dominated by wind speed 4-6 m/s. Therefore, the development of small-scale wind power generation is more feasible to be developed in tropical regions such as Indonesia, as lack of access to electricity can have a significant impact on education, economics, health, and gender equality [1-3]. The potential distribution of wind speed and geographical condition of Indonesia which is an archipelago country must be accompanied by the utilization of local materials as raw material for wind turbine blade manufacture.

F. Mahmuddin [4] performs computational computations based on the theory of blade element momentum (BEM) to find the performance of blade rotor [4], which is BEM theory based on momentum theory and blade element [5]. The comparison of the lift coefficient values of airfoils was tested using Q-Blade software [6], this is done to improve the performance of the blade rotor. Agung and Warits in

IOP Conf. Series: Earth and Environmental Science **105** (2017) 012095 doi:10.1088/1755-1315/105/1/012095

their essay conduct wind turbine evaluation using Q-Blade to seen similarity between blade characteristics and make blade design improvement with wind characteristic in Indonesia costal area [7, 8]. Blade is the key of wind power plants that convert wind kinetic energy into electrical energy in generator [9]. the optimal chord dimension can be made based on several parameters tailored to the design, such as wind speed, number of blade, tip speed ratio, and angle of attack, and if operated at low wind speeds it is possible to perform aerodynamic optimization of the rotor blades which are the most important part of the wind turbine. [10-12].

This study aims to get local fast growing wood wind turbines blades in Indonesia, which has the best physical and mechanical properties of wood, and is tested computationally on certain airfoils and shows a uniform distribution of stress. The types of wood are tested in this study were Sengon (Albizia chinensis), Jabon (Antocephalus cadamba), and Balsa (Ochroma Rowlee grandiflorum), which is a kind of fast growing wood in indonesia. The test covers the basic properties of the wood, physical and mechanical properties.

2. Methodology

2.1. Wood Properties Testing Set Up

A testing of physical properties, and set up of mechanical testing which are discussed in this paper is shown in figure 1 and figure 2.

Figure 1. Physical Testing of Wood Properties

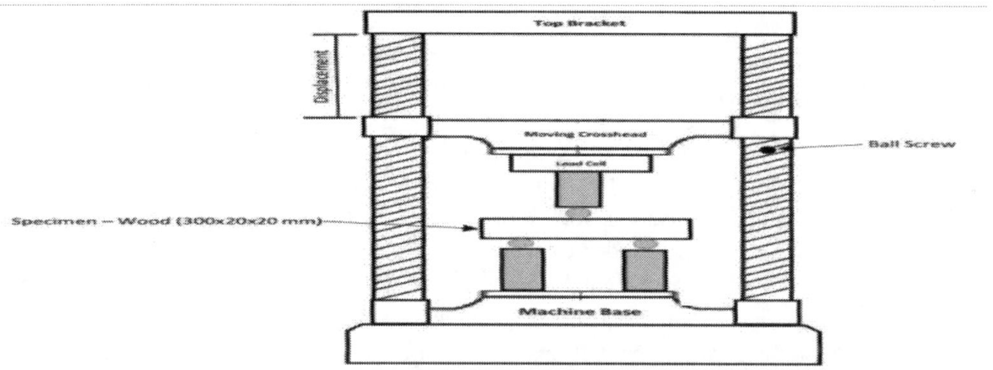

Figure 2. Mechanical Experimental Set Up

Testing of physical and mechanical properties of wood was carried out with ASTM 2359-69, ASTM D 143-52, and British Standard testing method (BS 373-1957) [13], to obtain values of wood density and modulus elasticity.

IOP Conf. Series: Earth and Environmental Science **105** (2017) 012095 doi:10.1088/1755-1315/105/1/012095

2.2. Wind Data Record

Local wind data record on the roof top of Lipi Pusinov's Building is shown in figure 3.

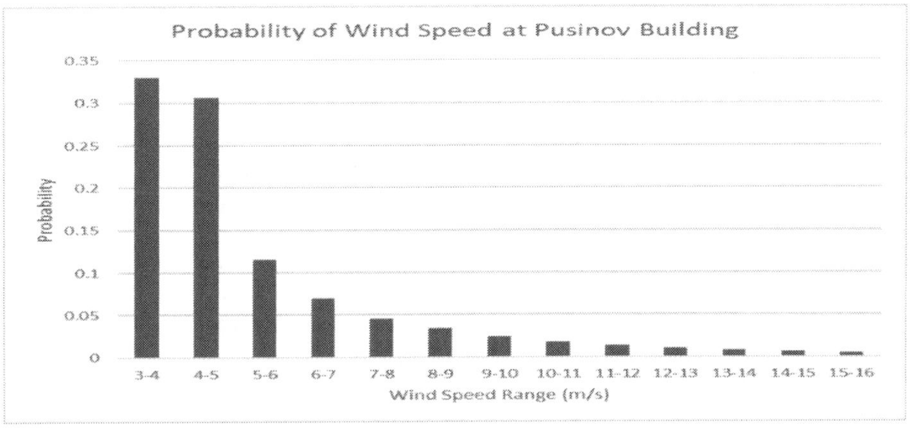

Figure 3. Wind Speed Probability Chart

2.3. Computational Fluid Dynamic Analysis

The computational simulation conducted to obtain the blade design and performance by using Q-Blade software. In this paper several theories on wind power plants used as a calculation, in order to get the performance of rotor blade in accordance with the characteristics of local wind.

The generated wind power can be calculated by [14] :

$$Pw = \frac{1}{2} A_r \rho V^3 \tag{1}$$

Which A_r is a rotor diameter, ρ is the air density and V is the wind speed, this is done in consideration in designing the blade later. We can determine the of lift and drag coefficient of the airfoil by the formula [15] :

$$C_L(\alpha) = L/\left(\frac{1}{2}\rho w^2 l\right) \tag{2}$$

$$C_D(\alpha) = D/\left(\frac{1}{2}\rho w^2 l\right) \tag{3}$$

with w as the relative speed of the wind and l is the chord width, the lift and drag coefficients are functions of angle of attack (α).

The Reynold number is determined by the equation below by taking into account the local characteristic of wind which is dominated by wind speeds range about 4 to 6 m / s.

$$Re = \frac{wl}{v} \tag{4}$$

With v kinematic viscocity (1.5111E-5 $m^2/_s$) dan l chord width, Reynold number will be used as parameter calculation of coefficient lift and drag.

3. Test Result

3.1. Result of Physical and Mechanical Test

The physical testing result of wood properties was carried out with ASTM 2359-69, ASTM D 143-52, and British Standard testing method (BS 373-1957) [13], to obtain values of wood density and modulus elasticity. The result of the physical testing is shown in table 1.

Table 1. Density values Jabon, Sengon, dan Balsa

No.	Wood Type	Density Value	Unit
1	Sengon	0.24	g/cm^3
2	Jabon	0.34	g/cm^3
3	Balsa	0.18	g/cm^3

From table 1, Jabon wood has better density value among another local wood with value 0.34. The greater the wood density value the more stable the wood when used as a blade.

The result of this mechanical testing shows that Jabon wood has the highest elasticity value, as shown in Table 2.

Table 2. Elasticity Value Jabon, Sengon, dan Balsa

No.	Wood Type	MOE	Unit
1	Sengon	4331.8915	mPa
2	Jabon	4615.5684	mPa
3	Balsa	2107.9982	mPa

3.2. Result of Computer Fluid Dynamic Simulation

3.2.1. Blade Performance Simulation. From the Figure 3, is determined blade that works at low wind speeds. In this paper, a computational simulation is performed for airfoil NACA 4415 and SG 6042. By using equation (4) and with the chord width 97.1953 mm, it can be obtained the Reynold Number Re 22515, as shown table 3.

Table 3. Computation Input data

Parameters	Unit	Parameters	Unit
Airfoil (a)	NACA 4415	Chord	97.1953
Airfoil (b)	SG 6042	Rotational Speed	216.68 rpm
Radius	925.5 mm	Wind Speed	3.5 m/s
Reynold Number	22515		

All of data is simulated using Q-Blade software to obtain the Coefficient Lift (Cl) and the TSR and CP ratio of airfoil NACA 4415 and SG 6042, as shown in the figure 4.

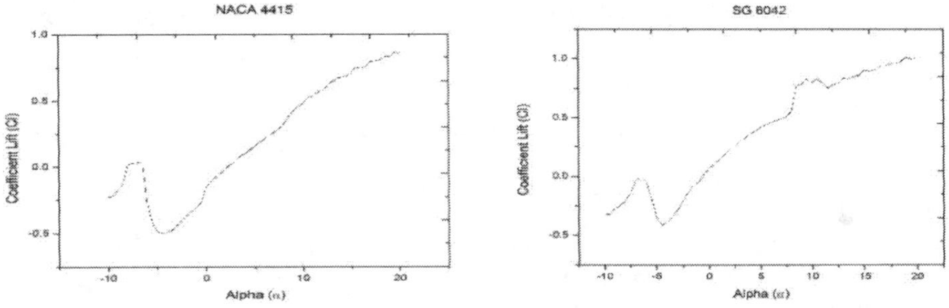

a. NACA 4415 b. SG 6042

Figure 4. Comparison of Coefficient Lift

In figure 4, the curve changes of Coefficient Lift vs Alpha on NACA 4415 has more constant increase compared to SG 6042, it is mean that the stability of lift values of blade can be maintained on the variation of the existing angle of attack.

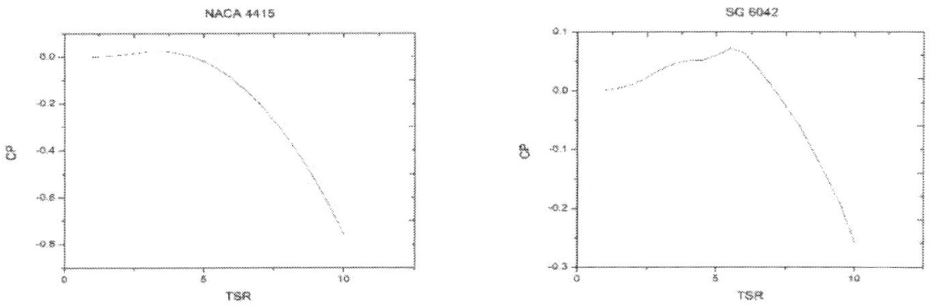

a.NACA 4415 b. SG 6042

Figure 5. Comparison chart of CP and TSR values

The change in TSR (tip speed ratio) values occurring in NACA 4415 is more stable than SG 6042, as shown as Figure 5, which can be concluded that NACA 4415 has a better TSR stability than SG 6042.

3.2.2. Blade Stress Distribution Simulation. The stress distribution simulation as part of computational simulation was conducted of three types of wood using airfoil which has the best performance that is NACA 4415. The Figure 6 shows that the stress distribution in each type of wood is different, Balsa has a better stress distribution value than Jabon and Sengon, this happens because Balsa has a lowest density value. The strongest of stress distribution of each wood can be indicated with the red colour on blade surface. The smaller the red color on the blade surface area the better the pressure distribution on the blade surface.

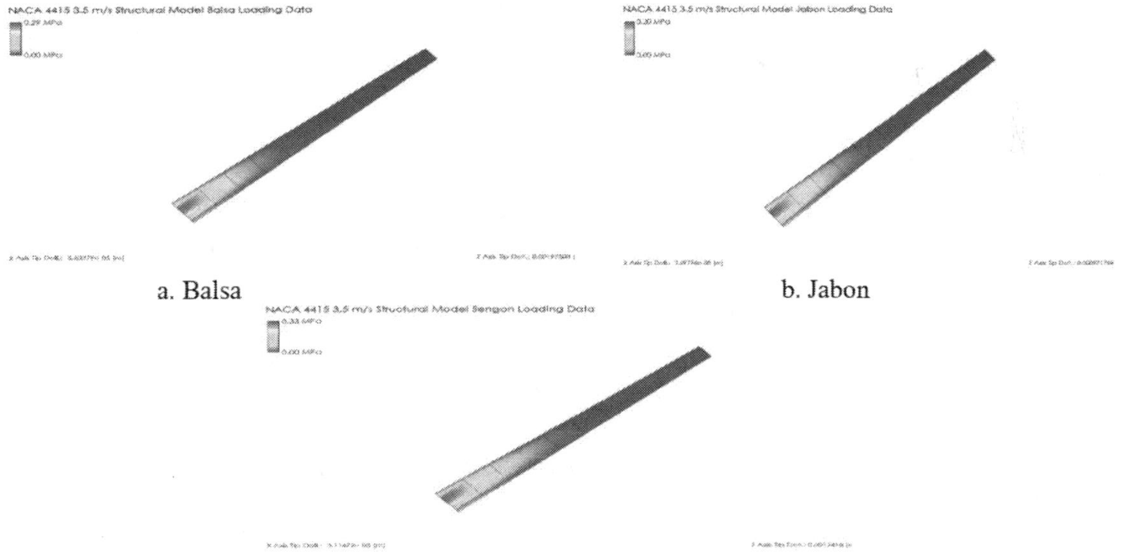

a. Balsa b. Jabon

c. Sengon

Figure 6. Stress Distribution Appearance of each wood

The maximum stress of each wood can be seen in Table 4, where the stress distribution in each type of wood is different, Balsa has a better stress distribution value than Jabon and Sengon, that is 0.29 mPa. This result happens because Balsa has a lowest density value.

Table 4. Maximum Stress on each wood

No.	Wood Type	Max Stress	Unit
1	Sengon	0.29	mPa
2	Jabon	0.30	mPa
3	Balsa	0.33	mPa

4. Conclusion
From the test results of physical and mechanical properties can be concluded that Jabon wood is better as a local wood material for wind turbine blades by considering the stability dimensions and mechanical strength when compared with Sengon and Balsa wood. Similarly, the type of airfoil used can be concluded that the airfoil NACA 4415 has a better lift and drag coefficient curve than SG 6042 at local wind speed, and for the graphical TSR and CP of NACA 4415 also has better stability form SG 6042.

5. Acknowledgement
The authors would like to thanks to Indonesian Institute of Science Research Centre for Biomaterials for providing facilities of physical and mechanical testing.

6. Reference
[1] Mettam, G. and L. Adams 1999 How to prepare an electronic version of your article Introduction to the Electronic Age *E-Publishing Inc.* New York p. 281-304.
[2] Rodrigues, R.M., et al 2016 Development of Guidelines for the Construction of Wind Turbines Using Scrap Material. Procedia Engineering 159: p. 292-299.
[3] Strunk Jr W 1979 White EB The elements of style New York: Macmillan.
[4] Mahmuddin, F 2017 Rotor Blade Performance Analysis with Blade Element Momentum Theory *Energy Procedia* 105: p. 1123-1129.
[5] Glauert H 1935 Airplane propellers in Aerodynamic theory Springer p. 169-360.
[6] Marten, D. and J. Wendler 2013 QBlade Guidelines. Ver. 0.6, Technical University of (TU Berlin) Berlin, Germany.
[7] Agung, Adi Surjosatyo 2015 Evaluasi Desain Blade Kincir Angin di Muara Gembong Essay Depok, University Indonesia.
[8] Warits, Adi Surjosatyo 2016 Desain Micro Scale Hotizontal Axis Wind Turbine Blade untuk Karakteristik Angin di Muara Gembong Bekasi Essay Depok, University Indonesia.
[9] Baoqing, X. and T. De 2012 Simulation and Test of the Blade Models' Output Characteristics of Wind Turbine *Energy Procedia* 17: p. 1201-1208.
[10] Ahmed, M.R., et al. 2011 Experimental and numerical studies on a low Reynolds number airfoil for wind turbine blades *J. of Fluid Science and Technology* 6(3): p. 357-371.
[11] Bai, C., et al. 2013 Design of 10 kW horizontal-axis wind turbine (HAWT) blade and aerodynamic investigation using numerical simulation *Proc. Engineering* 67: p. 279-287.
[12] Habali, S. and I. Saleh 2000 Local design, testing and manufacturing of small mixed airfoil wind turbine blades of glass fiber reinforced plastics part I: design of the blade and root. Energy conversion and management 41(3): p. 249-280.
[13] Standard, B.B., Standard 373–1957 methods of testing small clear specimens of timber London, UK: British Standards Institution.
[14] Bai, C.-J. and W.-C. Wang 2016 Review of computational and experimental approaches to analysis of aerodynamic performance in horizontal-axis wind turbines (HAWTs) *Renewable and Sustainable Energy Reviews* 63: p. 506-519.
[15] Dixon, S.L. and C. Hall 2013 Fluid mechanics and thermodynamics of turbomachinery Butterworth-Heinemann.

Formulating smart integrated workspace concept to improve energy efficiency

M A Berawi[1], P Miraj[2,3], F Naomi[1]

[1] Civil Engineering Department, Faculty of Engineering, Universitas Indonesia, Depok, Indonesia 16424
[2] Civil Engineering Department, Faculty of Engineering, Universitas Pancasila, Jakarta, Indonesia 12640
[3] Center for Sustainable Infrastructure Development (CSID), Universitas Indonesia, Depok, Indonesia 16424

Corresponding author: maberawi@eng.ui.ac.id

Abstract. The building is one of construction sector that consumes high energy and contributes to the greenhouse gas (GHG) emission. Many researchers and academics attempt to reduce the negative impact of the building by introducing alternative technology through the internet of things (IoT) and building management system (BMS). The research aims to examine components related to the workspace of building and evaluate the cost and energy usage. The study combining quantitative and qualitative approaches through energy simulation, life-cycle cost approach, and in-depth interview. The results show three components from lighting, cooling, and office equipment that can be quantified. Every component consists of equipment to tailor the concept of a smart integrated workspace. The energy per square meter from proposed concept generates 48.36 kWh per square meter or lower than the current design. Overall it requires 203,586,240 rupiahs or 18% more efficient to the current layout.

1. Introduction

Buildings consume forty percent of the world's energy production and contribute fifty percent to the carbon dioxide (CO2) emission [1]. Public buildings identified as the most sector that submits forty percent of global greenhouse gas (GHG) emission during its total life cycle [2]. Many experts estimate CO2 emitter will increase around 42% in 2035 worldwide. The high consumption of buildings not only occurs in developed countries such as China and European Union countries [3], but also in developing countries such as Indonesia [4].

In 2014, office buildings in Indonesia spent 15.2 million GWH to power the electricity and required 250 kWh per square meter [5]. The figure shows a higher consumption usage compared to other neighboring countries such as Malaysia and Singapore with an average about 100 kWh [6]. As the energy consumption and alternative production arise exponentially during the last decades, many researchers propose the new concept to combat energy issues. The key point from users in the building suggests that efficiency in energy shall not reduce the comfort, safety and buildings flexibility [7]. The smart office is a concept that use internet of things (IoT) to support the work environment behavior and in the longer term, increase the user's productivity [8].

IoT is part of architectural programming that manages and control buildings automatically through building management system (BMS) with an integrated workspace. It operates sensors to actuators from many building components such as temperature, noise, humidity, lighting and many other aspects of building [9]. Despite its advantages, owners and developers, particularly in developing countries, hesitate to invest the concept of the construction development. The concept most likely is rejected when the technology generates more cost and compromise the length of the payback period. Green building concept also shares a similar experience in Indonesia, where it took years to be implemented by the developers and project owners [10]. The research is then attempting to examine the most potential components in the workspace and to evaluate its cost for implementation in the project. The output expected to generate more research about the smart office concept and propose an alternative concept for the related stakeholders in office buildings.

2. Methodology

The research combines quantitative and qualitative method to achieve the targeted objectives. Using both method argued as solution to increase the reliability of the research output [11, 12]. The quantitative approach will be using energy simulation and life-cycle cost approach. Components in energy simulation follow GreenMark Interior V1.1 and generate a comparison between the baseline and proposed case study expressed by KWh. The life cycle cost evaluates the proposed case study based on its initial, operation and maintenance cost. The qualitative approach uses a desk study and in-depth interview to generate the concept of smart workspace design [13]. Desk study follows a benchmarking process and a literature review from many related resources such as journals, books, public records, policies and government reports [14, 15]. The in-depth interview is used to gain input from experts related to the case study.

The respondents categorized into practitioners and academics. Practitioners shall have experience in green building or smart building and hold green professional or LEED AP certificates. Academics should have research expertise in the green building or smart building in the past three years. The semi-structured questionnaire is used for the research instrument and conducted for about 15 – 30 minutes.

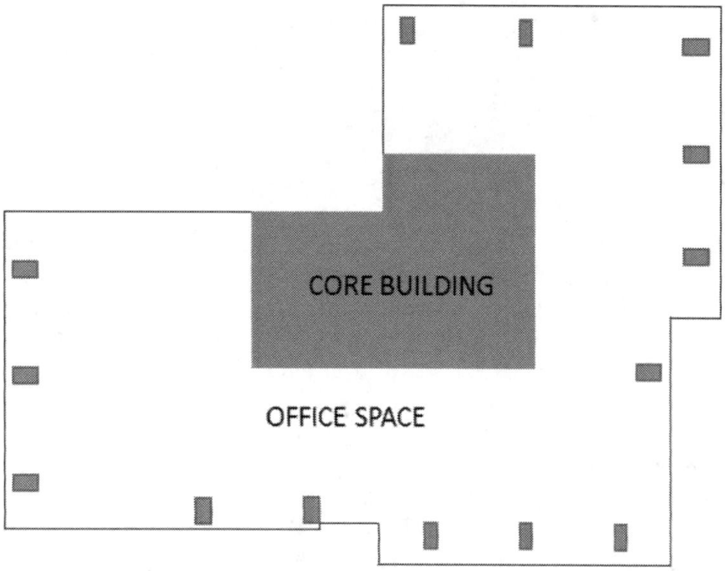

Figure 1. The building footprint as a case study.

The research uses a building located in the central business district in Jakarta as a case study. It has 39 floors and six basements with gross floor area about 90,828 square meters. The building has several functions such as tenants, function hall, retail, mechanical and electrical rooms, building management and security offices. Only the 34th floor with a net area about 1,356 square meters to be simulated in the research. The forecasted headcount is about 116 persons. The floor of the case study can be seen in figure 1.

3. Result and Discussion

3.1. Workspace Components in Building

Workspace-related to four factors including behavior, lighting, cooling, and office equipment. Building management that represents the behavior component should able to monitor the energy usage and inform to the tenants. Besides, the tenant also received real-time billing, which is longer term reduce the electricity consumption. Real-time monitoring shows load demand, distribution, power consumption per floor zone and many others. The changes from manual to automatic and centralized approach required a transformation of management, so that operated smoothly and managed thoroughly.

On the other hand, current lighting system use manual on/off and scheduling system. The alternative concept proposes a combination of scheduling and sensors of on/off system based on digital addressable lighting interface (DALI). Proposed concept offers LED light with specification as about 3,700 lm for the luminous flux and power about 40.5 W. It will be supported by the recess mountable 360-degree multifunction sensor, a sensor that combines motion detector (PIR) and ambient light level detection (PE). PE sensor aims to reduce energy usage by measuring daylight level and adjust along with the system to user's expected light level. Overall the building needs 97 points of the lamp with two gateway controllers. It acts to communicate the loop/group of lighting for on/off from the scheduling system or the user panel. The coverage area of sensors for the lighting system summarizes as figure 2.

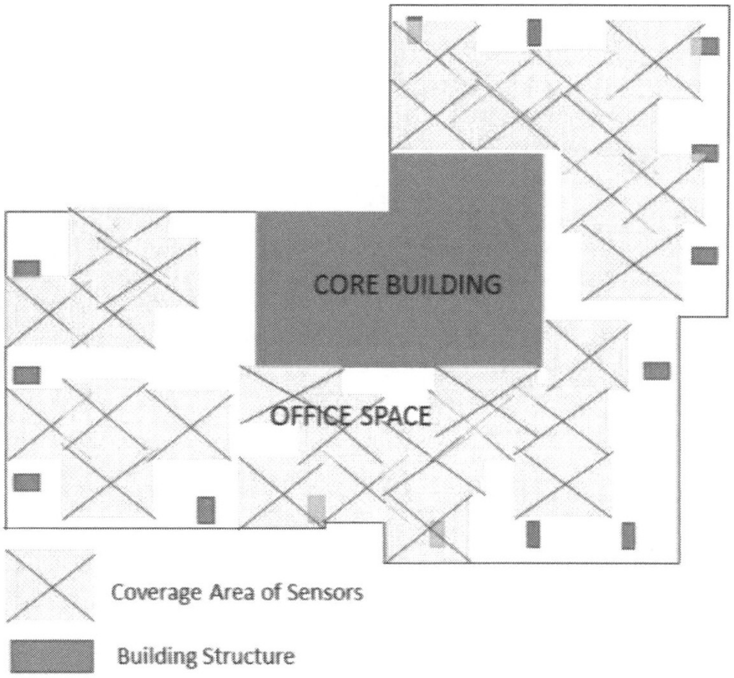

Figure 2. The coverage area of sensors for the lighting system.

Furthermore, cooling integrates variable air volume (VAV) with the building management system. The system maintained the air supply at a constant temperature and proven to reduce energy consumption by 20 – 30% [16]. However, the primary data shows that energy saving to VAV is varied on the user density at the space and the fan speed. When the speed is 100%, the power generates 18.5 kW. On the contrary, a 17% of fan speed produce a 0.1 kW of power.

In term of office equipment such as printing devices to the multifunction devices, it should pass the energy star standard or typical electricity consumption (TEC). While, LED display follows the needs of meeting room size from small, medium, large to the conference room.

3.2. Investment and Operational Cost

The cost in the smart integrated workspace that combines lighting cooling and office equipment consists of investment and operational cost. Due to qualitative result, behavior aspect is excluded from the cost evaluation. Site visit generates the quantity of each equipment. Unit price follows Indonesia's market price for the equipment. Office equipment is the largest contributor to investment cost with total 81.92%. Followed by lighting equipment and cooling equipment respectively. The initial cost for the concept design is summarized in table 1.

The operational cost is calculated by considers energy conversion from kWh to the electricity cost per year. Primary data generates the current annual kWh of each indicator from lighting, cooling and office equipment. The interior lighting has annual usage about 863,759 kWh to accommodate 65,630 square meter office area. Meanwhile, cooling and office equipment have larger usage about 57.13% and 90.91% compared to the lighting. Thus, the annual usage for the three indicators is about 13.16 kWh per square meter, 20.7 kWh per square meter, 25.2 kWh per square meter respectively. The tariff from Indonesia's state-owned enterprises in energy is about 1,472 rupiahs. Overall the annual operating cost of the components is about 248,634,100.

3.3. Current vs. Proposed Concept

The proposed smart integrated workspace evaluated energy efficiency and compared them from current condition. Lighting in proposed design minimizing electricity usage while maintaining lighting level through lighting control. The lighting scheduled to shut down and turn on automatically when users surpass the maximum usage. It supported by sensors and daylight harvesting concept. The quantification considers lighting spot and lamp power. Assuming the building operates for five days a week with eleven hours daily operation, the total electricity is about 10.42 kWh per square meter. Moreover, primary data shows that cooling system in the building require 4,631,146,303 Btu for 65,630 square meters with the level of temperature about 24° C. Assuming the VAV system provides 20% of efficiency compared to the other system, the proposed design produce 16.54 kWh per square meter. Last, the energy star certified office equipment generates 21.4 kWh per square meter.

Table 1. The initial cost of the smart integrated workspace design.

Equipment	Unit Price (Rp.)	Quantity (Unit)	Cost (Rp.)
Lighting			
LED lighting	2,226,031	184	409,589,704
Multi – function sensor	2,833,380	53	150,169,140
Gateway controller	1,713,513	4	6,854,052
Signal/ballast controller	12,566,011	1	12,566,011
Cooling			
VAV box	8,000,000	34	272,000,000
Temperature sensor	790,000	34	26,860,000
Office equipment (Star certified)			
Laptop	13,340,646	114	1,520,833,644
Multi – function devices	146,740,000	15	2,201,100,000
LED display			
Type 1	2,361,180	6	14,167,080
Type 2	6,923,460	6	41,540,760
Type 3	24,612,300	7	172,286,100
Voice over internet protocol	1,336,800	22	29,409,600
		TOTAL	4,857,376,091

Table 2. Energy and Cost Comparison between Current and Proposed Design.

Equipment	Current		Proposed Design	
	kWh/m^2	Unit price (Rp.)	kWh/m^2	Unit price (Rp.)
Lighting	13.16	30,470	10.42	24,346
Cooling	20.7	37,094	16.54	31,500
Office equipment (Star certified)	25.2	19,371	21.4	15,338
Total	59.06	86,935	48.36	71,184
Annual Operating cost		248,634,100		203,586,240

The proposed concept shows a lower usage in all building components. Current design produces a total of 86,935 rupiahs per square meter while proposed concept offers 71,184 rupiahs per square meter. Overall, less annual cost about 18.12% or equal to 203,586,240 rupiahs is generated. The comparison of current design with the smart integrated workspace concept shown in table 2.

4. Conclusion

Buildings consume high electricity and contribute to the increased figure of greenhouse gas (GHG) emission. Many researchers attempt to improve the power efficiency by using technology through the internet of things (IoT) and building management system (BMS) and introducing compact layout to accommodate energy saving.

The case study identifies four energy-related components includes behavior, lighting, cooling, and office equipment. However, the behavior is excluded from further analysis due to its categorization as a qualitative component. Every component consists of equipment to tailor the concept of a smart integrated workspace. Lighting has LED lighting, multifunction sensor, a gateway controller, and signal/ballast controller. On the other hand, cooling consist of two equipment from VAV box and temperature sensor. Last, office equipment comprises of the laptop, multi-function devices, LED display and voice over internet protocol. All of them have to meet energy star certification for further evaluation.

Lighting requires 579,178,907 rupiah or equal to 11.92% from the total initial cost. While, cooling needs a 298,860,000 rupiah (6.15%) and office equipment involve 3,979,337,184 rupiahs. In total, the additional cost to implement the concept is about 4.8 billion rupiahs. The energy per square meter from smart integrated workspace generates 48.36 kWh per square meter or lower than the current design. Overall it requires 203,586,240 rupiahs or 18% more efficient to the current design.

5. References

[1] Liu H and Lin B 2016 *Ecological Indicators*. 67 68-77
[2] Yan H, Shen Q, Fan L C, Wang Y and Zhang L 2010 *Build Environ.* 45(4) 949-95
[3] Ma J J, Du G, Zhang Z K, Wang P X and Xie B C 2017 *Build Environ.* 117 36-48
[4] Berawi M A, Miraj P, Berawi A R B, Silvia and Darmawan F 2016 *IJTech.* 7(8) 1445-1454
[5] Howes, S Davies R 2014 *Bulletin of Indonesian Economic Studies* 50(2) 157-183
[6] Hasan, M H, Mahlia T M I and Nur H 2012 *Renew Sust Energ Rev.* 16(4) 2316-2328
[7] Teich T, Wolf S, Neumann T, Junghans S and Franke S 2014 *Procedia Eng.* 69 597-602
[8] Furdik K, Lukac G, Sabol T and Kostelnik P 2013 *Int. J. of Computer Networks and Communications Security* **1**(6) 216-224
[9] Park S, Park S, Byun J and Park S 2016 *Int. J. of Distributed Sensor Networks* **12**(8) 1550147716660895
[10] Berawi M A, Susantono B, Abdul-Rahman H, Mustika S, Sesmiwati and Rahman Z 2013 *IJTech.* 1 45-55
[11] Creswell J W 2013 *Research design: Qualitative, quantitative, and mixed methods approaches* (Sage publications)
[12] Karim S B A, Rahman H A, Berawi M A, and Jaapar 2007 *A review on the issues and strategies of stakeholder management in the construction industry. In Meeting and Conference on Management in Construction and Researchers Association (MICRA)*

IOP Conf. Series: Earth and Environmental Science **105** (2017) 012096 doi:10.1088/1755-1315/105/1/012096

[13] Woodhead R M and Berawi M A 2008 *Int. J. of Management Practice.* **3**(1) 1-19
[14] Berawi M A and Woodhead R M 2008 *Value World,* 31(2) 4-7
[15] Berawi M A, Zagloel T Y, Miraj P and Mulyanto H 2017 *Procedia Eng.* 171 265-273
[16] Engdahl F and Johansson D 2004 *Energ Buildings*. 36(3) 205-218

Increasing energy efficiency by considering building's life cycle

M A Berawi[1], A Sihombing[1], P Miraj[2,3] and R Windrayani[1]

[1] Civil Engineering Department, Faculty of Engineering, Universitas Indonesia, Depok, Indonesia 16424

[2] Civil Engineering Department, Faculty of Engineering, Universitas Pancasila, Jakarta, Indonesia 12640

[3] Center for Sustainable Infrastructure Development (CSID), Universitas Indonesia, Depok, Indonesia 16424

Corresponding author: maberawi@eng.ui.ac.id

Abstract. Greenhouse gas and carbon emission have a considerable impact on the people behavior on constructing new buildings. One of the attempt to combat environmental issues is by implementing renewable energy such as photovoltaics (PV) into the buildings. However, a high cost from sustainability concept makes building owners or investors hesitate to use the device. The research aims to evaluate the solar energy implementation by taking university library building as the case study. Due to space and cost aspects, the analysis only considers lighting system. The life cycle cost is used to generate the best alternatives to the building. The results show three potential options; alternative one use public electricity from state-owned enterprises in the energy sector with a fluorescent lamp, alternative two use PV with a fluorescent lamp and alternative three use PV with LED lamp. Alternative three recommends as the most potential implementation on the building as it generates lower initial cost, moderate cost of operation and maintenance as well as fair salvage value.

1. Introduction

Environmental issues become a significant topic during the last decades due to the increased of world's carbon emissions and greenhouse gas effect [1, 2]. Reference [3] estimates an increasing percentage about 23.85% volume of carbon emissions during 2000 to 2013. Many researchers and academics believed that human activities significantly contribute to the global warming [4, 5]. The building sector is one of the largest contributors by fifty percent to the world's carbon dioxide (CO2) emission [3].

Office in Indonesia, for instance, needs a 15.2 million GWH to power the electricity or equal to 250 kWh per square meter. It is relatively higher than neighboring countries or any other developed countries with only 100 kWh [3]. Besides high energy consumption, many buildings still use non-renewable energy to meet the energy needs. On the other hand, solar power has the potential for development as it is environmentally friendly, produce insignificant pollution, and available during most of the time in developing countries such as Indonesia [6, 7].

Universitas Indonesia as one of the oldest and the largest university in Indonesia attempt to be a leader in implementing the sustainable concept in its building. The university built the main library that aims as a meeting point for students, lecturer, and many others to conduct research, discussion, and knowledge dissemination. The design attempt partial concept of sustainability from structural, vegetation, roof to the cooling system. High investment argues as one of the most significant factors that hinder the use of alternative energy in the building. Components are mostly imported from other countries [8]. Thus, the renewable energy device has a low impact on the financial and economic aspects of the building. Considering those problems, the research will evaluate the solar energy implementation to the university's library building. The result recommends the best alternative towards energy efficiency in the public construction sector.

2. Methodology

The research uses a life-cycle cost (LCC) to evaluate energy usage. The LCC measurement is an economic evaluation method aims to determine the economic impact of alternative building designs and building systems expressed by nominal value or currency [9, 10, 11]. The factors in LCC categorized into the initial cost (purchase, acquisition, cost construction), fuel cost, operational, maintenance, and repair cost, replacement cost, residual value or salvage value or disposal cost, finance charges (loan interest payment and non-monetary benefit).

The measurement of Life Cycle Cost system is as follows [12]:

$$LCC = I + Repl - Res + E + W + OM \& R + O$$

Where,
LCC = Total cost
I = investment cost
Repl = replacement cost
Res = residual cost
E = energy cost
W = water cost
OM & R = operational and maintenance cost, residual
O = other costs

The case study was conducted in 2011 and involved a series of interviews with general contractors in Indonesia.

3. Result and Discussion

The preliminary calculation considers the power needs of Universitas Indonesia's Library Building. Interviews from general contractors show that it reached three mW. The number considers huge accommodated only by the photovoltaics (PV) regarding space and cost. Thus, the study will evaluate the lighting system to substitute for the PV system.

PV calculated by multiplied power over the duration of use. It is assumed lighting standby for about eight hours from nine in the morning to five in the afternoon. The PV considers watt/peak measurement for analysis. Three alternatives are generated for analysis, firstly is the condition where all lighting will use the public electricity from state-owned enterprise in energy. Secondly, all lighting use PV system by assuming 50% of the lighting is on at daylight. Thirdly, the lighting is replaced by LED system to reduce the energy consumption of the building.

The building has six floors with a different capacity of lighting on each floor. The first floor has 50,900 watts. Meanwhile other has 69,745 watts, 41,724 watts, 40,501 watts, 6,589 watts and 6,645 watts respectively. Thus, the total power for building lighting is about 216,304 watts or 216 kW. As the lighting only used 50%, the daily usage per day is about 108 kW. The government of Indonesia issued electricity tariff per month about 800 rupiahs per kWh for usage past the peak load time and a coefficient of 1.5 multiplied by 800 rupiahs for usage in peak load time. The amount of usage past the peak load time is 40%, or equal to 10,368 kWh/month, Peak Load Time about 15,552 kWh/month. The total electricity bill per month estimated for 26,956,800 rupiahs.

The cost of maintenance, repair, and operation is 14,176,817 rupiahs per year. The cost of maintenance and repair for solar panels and inverters is about 4,828,404 rupiahs per annum including personnel, utility, land lease and labor management. Assumption for LCC calculation considers the inflation (4.66%), discount rate (9.25%), reinvestment rate (4.66%). Escalation factors also involved in the analysis of routine annual operation and maintenance, major repair/replacements, utility electric rate escalation, demand rate escalation, natural gas price escalation, photovoltaic degradation factor per year. All rate for escalation factors is about 4.66%.

3.1. Alternative 1

Option 1 use public electricity from state-owned enterprises in the energy sector. The lighting of Universitas Indonesia's public library uses the fluorescent lamp type. Investment costs consist of design costs, construction, and installation costs, material and equipment costs, indirect costs and contingency costs.

The investment cost is about 14,176,817,392 rupiahs; this cost includes construction and installation as well as the cost of material and equipment. The design cost is not taken into account as includes on the planning stage. Indirect cost assumed 5% from the investment cost as about 708,840,869 rupiahs, and contingency fee is 3% or equal to 425,304,521 rupiahs. Overall the total cost of investment and capital replacement in alternative 1 is about 15,310,962,783 rupiahs.

Annual expense on the project includes energy costs, operational costs, maintenance and repair and residual value and reinvestment. Energy cost derived from the use of electricity that requires 10,800,728,304 rupiahs per year. Maintenance and repair costs delivered from lamp usage and electricity installation both needs 3,344,938,167 rupiahs and 473,349,807 respectively.

Based on the interview, fluorescent lamp last for 5,000 hours. Lamp purchasing is about eleven times during twenty years of operation by assuming it used for eight hours per day. Thus, reinvestment for the lamp is about 195,901,100 rupiahs. Last, the salvage value calculated 10% from the investment period. Overall, the investment cost for the alternative 1 is about 15.31 billion rupiahs with 14.62 billion of operation and maintenance costs and 1.53 billion of salvage value.

3.2. Alternative 2

Option 2 combines PV with fluorescent lamp type for the analysis. An in-depth interview with local suppliers generates PV price and capacity for each panel of PV. One square meter of PV panel produces a 160-watt peak with a cost about US$4 or equal to 8,561 rupiahs. Compared to the regular calculation of electricity, the PV estimates energy usage for the building. The calculation will determine the required PV panel for the building. The construction and installation are similar to the previous alternative as about 14,176,817,392 rupiahs. Meanwhile, the material and equipment are much higher than option 1. The PV requires 31,247,756,183 rupiahs with tripled times of indirect costs and contingency. The three components lead to a higher investment cost about 49,058,539,461 rupiahs or three times greater than the alternative one about 15,310,962,783 rupiahs.

Unlike the alternative one that used public electricity, in the option 2, energy cost is no longer involved due to replace by the PV system. Maintenance and repair cost should be generated from primary data in Indonesia by taking into account the differences in weather, location, installation, and location. However, due to data limitations, a benchmarking from Electric Power Research Institute (EPRI) is used. They calculate rule of thumb for maintenance and repair cost for about US$ 47 per month or equal to 4,828,404 per year.

The electricity and lamp have similar cost with alternative 1. Maintenance and repair cost produce slight higher cost with 3,979,503,573 rupiahs. On the other hand, option 2 generates ten times greater salvage value compared to the previous option with 15,698,732,628 rupiahs over 1,531,096,278. Consequently, the investment cost for the option 2 is about 49.06 billion rupiahs with 3.98 billion of operation and maintenance costs and 15.70 billion of salvage value.

3.3. Alternative 3

Alternative 3 replace the fluorescent lamp with Light Emitter Diode (LED) lamp. It argues can reduce the energy usage nearly 50%. The replacement shall reduce the building energy usage from 108,152 watts per day to 62,818 watts per day. The PV investment costs from this replacement lowering the cost into 38,335,154,422 rupiahs. The construction and installation are similar with alternative 1 and alternative 2 for about 14,176,817,392 rupiahs. On the other hand, the initial cost of PV, as well as indirect and contingency costs are lower than alternative two but still higher than alternative 1. PV is half the cost from alternative 2. Meanwhile, indirect and contingency is also half the cost of alternative two but twice the cost of alternative 1.

Despite generate lower investment cost compared to the alternative 2, the maintenance and repair cost in the alternative 3 three times higher than alternative 1 and alternative 2. It because the LED lamp requires more maintenance and repair cost about 9,698,641,328 rupiahs or the highest cost among other alternatives. In general, the investment cost for the option 3 is about 38.33 billion rupiahs with 10.33 billion of operation and maintenance costs and 12.27 billion of salvage value.

3.4. Comparison of Each Alternative

Initial cost for PV systems is higher than using public electricity. As a comparison, the initial cost for both alternative 2 and alternative 3 is almost triple times from the alternative one that uses public electricity. However, the PV generates lower operation and maintenance cost and higher salvage value by ten times greater. On the other hand, LED initial cost over fluorescents lamp shows a little gap. However, LED needs higher maintenance and repair almost seven times over fluorescents lamp. The analysis recommends alternative three as the most potential implementation to the building. It generates lower initial cost, a moderate cost of operation and maintenance as well as fair salvage value for the end of building's life cycle. The comparison among three alternatives can be seen in figure 1.

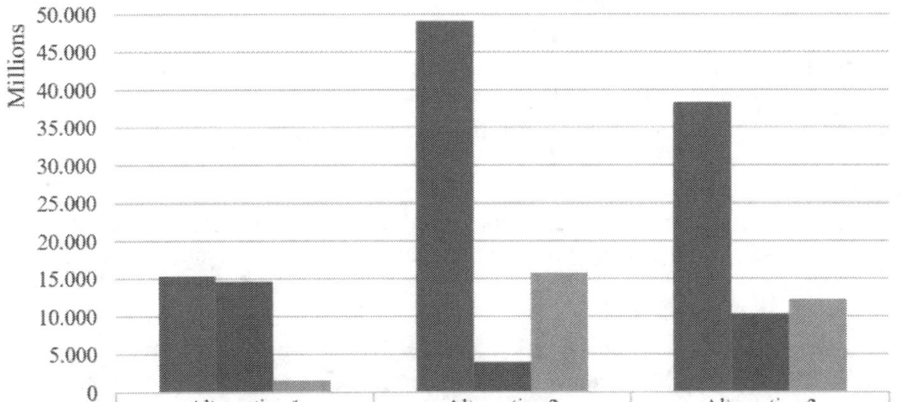

	Alternative 1	Alternative 2	Alternative 3
■ Initial Cost	15.310.962.783	49.058.539.461	38.335.154.422
■ O&M Cost	14.619.016.278	3.979.503.573	10.333.206.734
■ Salvage Value	1.531.096.278	15.698.732.628	12.267.249.415

Figure 1. A comparison of LCC for each alternative.

From the analysis, the initial cost for PV is relatively high in Indonesia about the US $ 10 per watt. Meanwhile, other countries have much lower price such as the United States, and Europe is about the US $ 3-4 per watt, India to US$ 2.80 and China about the US $ 2.50. Indonesia has a high price due to the country ability that unable to produce PV panels in the domestic market. Other countries attempt to reduce the high initial cost by implementing strategic policy related to renewable energy such as tax incentives, tax credits, and subsidies on the use of renewable energy. Similar policy and regulation should be further evaluated related to energy in building in Indonesia. It is expected that renewable energy contributes to reduce world's gas emissions and provide clean energy for better living.

4. Conclusion

Buildings sector recommends using renewable energy to combat global carbon emissions and greenhouse gas effect. Universitas Indonesia library building attempt to operate sustainability concept by implementing photovoltaics (PV) to the building. However, due to space and cost aspects, only the lighting system may be evaluated.

The case study compares three alternatives; alternative one use public electricity from state-owned enterprises in the energy sector with a fluorescent lamp, alternative two use PV with a fluorescent lamp and alternative three use PV with LED lamp. The result shows that the lowest initial cost is alternative one followed by alternative 3 and alternative 2. Regarding the lowest operation and maintenance cost is alternative two followed by alternative 1 and alternative 3. On the other hand, salvage value for alternative 2 is the highest compared to alternative 1 and alternative 3. The result recommends alternative three as the most potential implementation on the building by considering lower initial cost, moderate cost of operation and maintenance as well as adequate salvage value.

5. References

[1] Weingaertner C and Moberg Å 2014 *Sustainable Development* **22**(2) 122-133
[2] Shi Q, Zuo J, Huang R, Huang J and Pullen S 2013 *Habitat Int.* **40** 1-8
[3] World Bank. 2017 CO2 emissions (metric tons per capita), available at http://data.worldbank.org/indicator/EN.ATM.CO2E.PC?end=2013&start=2000&view=chart
[4] Berawi M A, Miraj P, Berawi A R B, Silvia and Darmawan F 2016 *IJTech.* 7(8) 1445-1454
[5] Darko A, Chan A P, Owusu-Manu D G, Ameyaw E E 2017 *J. Clean Prod.* **145** 386-394
[6] Aman M M, Solangi KH, Hossain M S, Badarudin A, Jasmon G B, Mokhlis H, Bakar A H A, Kazi S N 2015 *Renew Sust. Energ Rev.* **41** 1190-1204
[7] Tasri A, Susilawati A 2014 *Sust. Energy Technol. and Assessments* **7** 34-44
[8] Effendi P 2014 Contribution Matters 3 67
[9] Woodhead R M, Berawi M A 2008. *Int. J. of Management Practice.* **3**(1) 1-19
[10] Berawi M A, Woodhead R M 2008. *Value World* **31**(2) 4-7
[11] Karim S B A, Rahman H A, Berawi M A and Jaapar 2007 A review on the issues and strategies of stakeholder management in the construction industry *In Meeting and Conference on Management in Construction and Researchers Association (MICRA)*
[12] Fuller S 2010 Life-cycle cost analysis (LCCA) National Institute of Building Sciences, An Authoritative Source of Innovative Solutions for the Built Environment

MgH_2-SiC based hydrogen storage material prepared by reactive mechanical alloying method

Zulkarnain Jalil[1], Adi Rahwanto[1], Hasan Akhyar[2], Razali Razali[3] and Erfan Handoko[4]

[1]Department of Physics, Faculty of Mathematics and Natural Sciences, Syiah Kuala University, Banda Aceh 23111, Indonesia
[2]Department of Mechanical Engineering, Faculty of Engineering, Syiah Kuala University, Banda Aceh 23111, Indonesia
[3]Laboratory of Veterinary Public Health, Faculty of Veterinary Medicine, Syiah Kuala University, Banda Aceh 23111, Indonesia
[4]Department of Physics, Faculty of Mathematics and Natural Sciences, Jakarta State University, Jakarta, Indonesia

Corresponding Author: zjalil@unsyiah.ac.id

Abstract. Hydrogen storage in metal hydrides, compared to conventional methods, is regarded as one of the best solutions due to the higher volumetric storage capacity and safety. Magnesium and magnesium-based alloys are promising candidates for hydrogen storage. The hydrogen storage capacity of magnesium in the form of MgH_2 amounts to 7.6 wt%. Unfortunately, Mg has a high thermodynamic stability and therefore, relatively slow desorption kinetics, which are the major drawbacks for the application as a hydrogen storage material. Various techniques are developed to improve the sorption characteristics by accelerating the aforesaid processes. In this work we success to synthesis and investigate the catalytic effect of SiC and Ni (in nanostructure scale) on MgH_2 using reactive mechanical alloying method in 10 bar H_2. At first step, using SiC catalyst the sorption properties can be improved. The most promising step by using double catalysts of SiC and Ni (MgH_2-5wt%SiC-5wt%Ni) which absorp 5.7 wt% hydrogen and at the same time decrease the desorption temperature to 250°C. Compared to T onset of pure MgH_2 -which desorp at 380° C. To the best of our knowledge, this is the best result so far for MgH_2-SiC system.

1. Introduction

Presently, metal hydrides are the most promising materials for hydrogen storage, especially the lightweight metal alloy hydrides. The research is focused now on the new lightweight materials for hydrogen storage. Magnesium (Mg) is considered to be the most promising base material; it forms MgH_2 which corresponds to a hydrogen content of 7.6 wt% [1, 2]. Some efforts have been done in order to enhance the hydrogen absorption rate and to overcome the problems connected with the high absorption/desorption temperatures by alloying Mg/MgH_2 with other metals.

Some studies are devoted to the catalytic effect of 3d-transition metals, especially the Ni metal [3-7]. Recently, Eskandarany et al. [7] have showed a new approach for catalyzing MgH_2 with nano-Ni by high energy reactive ball milling under 50 bar of hydrogen atmosphere. They found that the sorption kinetics was examined at 275 °C within 8 minutes and produced ~5 wt.% H_2 for 25 h of milling time.

In this work we make attention on the effect dual catalyst of nano-Ni and SiC on the hydrogen storage properties of MgH_2 and prepared by mechanical milling in "mild" condition with relatively low hydrogen pressure (10 bar) for 2 hours.

2. Materials and Method

The pure MgH_2 powder (Sigma Aldrich, 99.99%, 50 µm) and Ni nanopowder (Sigma Aldrich, 99.9%, 90 nm) and SiC (Hongwu Nano, 50 nm, 99%) were used. Hardened steel crucibles of 9.5 mm in diameter containing 13 balls were filled; the total amount of powder was 5 g with a ball to powder weight ratio of 10:1. The Fritsch P6 planetary mill apparatus attached with a home-made reactive milling vial was used with a rotation speed of 200 rpm and with the milling times of 2h in 10 bar of hydrogen atmosphere.

The X-ray powder diffraction analysis after different milling was carried out by Philips PW3710 diffractometer using Co-K_α radiation, operating at 40 kV and 40 mA. The microstructural characterization were investigated by using SEM (JEOL JSM-5310LV). The hydrogen storage properties were investigated by gravimetric method in a wide temperature and pressure range for absorption (1-10 bar, 50-300 °C) and desorption (250-350 °C, 50 mbar-1 bar).

3. Results and Discussion

Figure 1 shows the evolution of X ray diffraction patterns for MgH_2-5wt% SiC-5wt% Ni material as a function of milling time and intensity. At the time of mixing the material is still in the pre-milling phase it is clear that the MgH_2 phase appears as the main phase at the angle of 2θ = 32.75°, 42.05°, 61.66°, 68.24°, 77.73°, 81.85°, 83.23°, 90.93°, 101,39°.

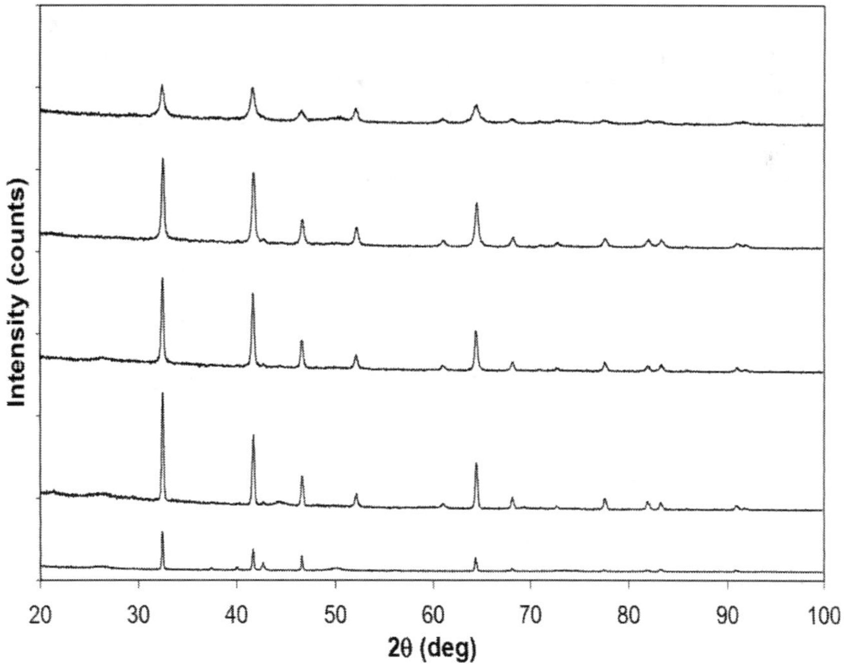

Figure 1. XRD pattern of the samples MgH_2-5wt%SiC-5wt%Ni

When milling lasts for 15-45 minutes, the form of diffraction pattern has not yet shown a significant change. Here it is known that crystal size is still microcrystalline. Different situations are shown on milling for 10 hours, where diffraction peaks begin to widen. It can be mentioned that at a time of 10 hours of milling, material begins to reduce grain size. Interestingly note that the smoothing process lasts relatively short (10 hours). Compared to non-reactive processes, it looks less effective in terms of production time.

In this study directly used commercial MgH_2 material. Referring to the literature, the use of MgH_2 material directly in the study, compared to pure Mg, would be able to obtain a nanocrystal structured material with a higher surface area which would affect the kinetics during the hydriding process [8]. Particle size particle size distribution of MgH_2-5wt% SiC-5wt% Ni (10 hours RMA) observed with particle size analyzer (PSA) indicates that grain size decreases as smoothness increases. From the measurement results it is known that the average size of particles is 0.6-0.28 μm, smoother than without the miling process. Reported by Varin et al [9] that increasingly decreasing particles will form an increasingly larger surface area so the atoms of H will easily recombine. This will have a strong effect on the change in the properties of MgH_2 uptake.

Revezs et al [10] uses the convolutional model of multiple whole profile fitting procedure (CMWP) to analyze the XRD peak profiles. Obtained information that fine grain has a large contribution to the absorption and desorption of MgH_2 material.

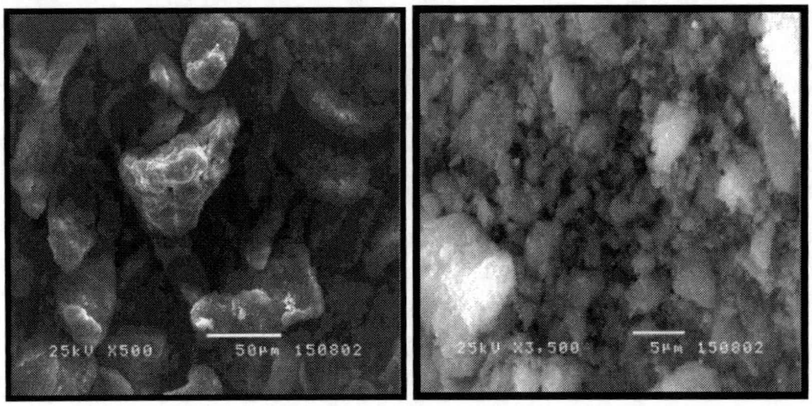

Figure 2. SEM images of the samples MgH_2-5wt%SiC-5wt%Ni

The surface of the powder are irregular, as a result of the fracturing during the milling process. The SEM image after 2 hours of milling shown the finer powder. It seem the powder was already in nanocrystal scale. If we compare with XRD, this is consistent with the structural analysis by XRD. After calculated by Scherrer method [13], the particles were reaches around tens nanometer after 2 h of reactive milling. The silicon carbide in the composites helps to break the magnesium particles into smaller sizes due to its hardness. addition of SiC can help to generate smaller MgH2 nanocrystals and it was found that the MgH_2 peaks in the XRD pattern became very broadened.

It appears that the pure MgH_2 (T onset) temperature of the pure MgH_2 is around 380° C. This result corresponds to reference which indicates that the high temperature MgH_2 operation. Then, the attempt to lower the onset was done by adding 5wt% SiC to the sample MgH_2-5wt% SiC. The onset showed a decrease compared to MgH_2, where it was decomposed at 350° C. Referring to Ranjbar et al [12] report, the result obtained indicates the suitability.

Furthermore, the addition of SiC and Ni nanoparticle catalysts (MgH_2-5wt% SiC-5wt% Ni) showed a significant decrease in temperature of desorption, whereas T onset could be reduced to 250° C. Suspected that the addition of nano-structured Ni catalysts could affect material performance based on MgH_2-SiC.

The observed attraction is the presence of Ni nanoparticle catalysts that affect the decrease of MgH_2 desorption temperature. Specifically, Xiea et al [11] studied the kinetics of MgH_2 desorption that increased dramatically by nanoparticle doping. However, the desorption kinetics do not increase as the Ni catalysts exceed a certain value.

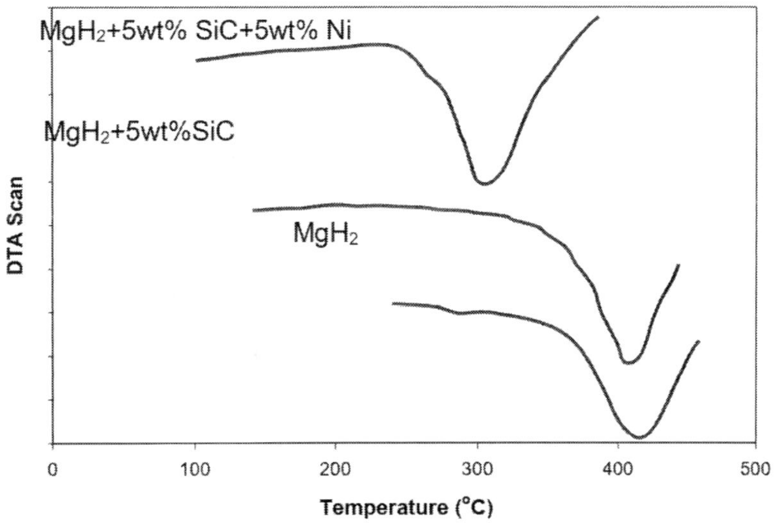

Figure 3. DTA scan of MgH_2 -5wt%SiC-5wt%Ni nano RMA 10 hours

Table 1. DTA scan of the MgH2 samples

Materials	T onset (°C)	Flow Rate (ml/min)
MgH2+5wt% SiC+5wt% Ni	250	20
MgH2+5wt%SiC	350	20
MgH2 , pure	380	20

By comparing the structure, particle size and configuration of the catalyst obtained, it can be demonstrated that the desorption kinetics can be improved. mainly due to the combination of atomic hydrogen acceleration on MgH_2 surface by Ni nanoparticles. Xiea believes that the size of the catalyst particles is beneficial to increase the catalytic effect and at the same time hydrogen storage capacity can still be maintained.

The amount of Ni catalyst in the material is also considered. When the amount of catalyst is too large it will lower the capacity of the hydrogen that can be absorbed. According to Xiea, the use of up to 25% catalysts can lower the onset of MgH_2. but if usage exceeds 25% then the onset T does not change, while H_2 capacity decreases. On the other hand, minimize Ni catalyst particle size also affects MgH_2 performance.

4. Conclussion

Preparation and structural characterization of magnesium-based material, in this case MgH_2 with addition of 5wt% of SiC and Ni double catalysts have been successfully applied using reactive mechanical alloying (RMA) preparation using hydrogen at 10 bar pressure. XRD observation results show that the MgH_2-5wt% SiC-5wt% Ni nanoparticles coated for 10 hours form the main phase of MgH_2. The addition of SiC and Ni nanoparticle catalysts (MgH_2-5wt% SiC-5wt% Ni) showed a significant decrease in temperature of desorption, where T_{onset} was reduced to 250° C. This result succeeded in repairing a pure MgH_2 Tonset which reached 380° C. Although operational temperature was still relative high mass applications, reactive mechanical alloying techniques and the addition of double catalysts (SiC, Ni) to MgH_2 material promising for magnesium-based hydrogen storage R & D materials.

5. Acknowledgments
Authors would like to thanks Syarifah Fathmiyah, S.Si (Materials Physics Lab, Syiah Kuala University) for the assistance in XRD sample preparation.

6. References
[1] Schlapbach and Zuettel 2001 Hydrogen storage materials for mobile applications *Nature* **414**
[2] Sandrock G and Thomas G 2001 The IEA/DOC/SNL on-line hydride databases *Appl. Phys. A* **72** pp 153–155
[3] Imamura H, Maruta Y and Tsuchiya S 1987 Magnesium-based hydrogen-storage materials: Solvated Mg-Ni by metal vapour deposition *J. Less Common Met.* **135**
[4] Song M Y, Kwon S N, Bae J S and Hong S H 2009 Hydrogen-storage properties of melt spun Mg-23.5 wt%Ni milled with nano Nb_2O_5 *Journal of Alloys and Compounds* **478** pp 501-506
[5] Yuan H, An Y, Xu G and Chen C 2004 Hydriding behavior of magnesium-based hydrogen storage alloy modified by mechanical ball-milling *Material Chem. Phys.* **83** pp 340
[6] Huot J, Pelletier J F, Liang G, Sutton M and Schulz R 2002 Structure of nanocomposite metal hydrides *Journal of Alloys and Compounds* **330-332** pp 727-731
[7] El-Eskandarany M S, Shaban E, Ali N, Aldakheel F and Alkandary A 2016 In-situ catalyzation approach for enhancing the hydrogenation/dehydrogenation kinetics of MgH_2 powders with Ni particles *Sci. Rep.* **6** 37335
[8] Bobet J L, Akiba E, Nakamura Y and Darriet B 2000 Study of Mg-M (M = Co, Ni, and Fe) mixture elaborated by reactive mechanical alloying *Journal of Hydrogen Energy* **25** pp 987-996
[9] Oelerich W, Klassen T and Bormann R 2001 Comparison of the catalytic effects of V, V_2O_5, VN, and VC on the hydrogen sorption of nanocrystalline Mg. *Journal of Alloys and Compound* **322** L5-9
[10] Oelerich W, Klassen T and R Bormann 2001 Metal oxides as catalysts for improved hydrogen sorption in nanocrystalline Mg-based materials *Journal of Alloys and Compounds* **315** pp 237-242
[11] Jalil Z, Rahwanto A, Mulana F and Mustanir 2016 Desorption Temperature Characteristic of Mg-based Hydrides Catalyzed by Nano-SiO_2 Prepared by High Energy Ball Milling *Int. J. Techol.* **7**(8) 1301-1306
[12] Jalil Z, Rahwanto A, Handoko E and Mustanir 2017 The role of nano-Ni catalyst in MgH_2 obtained by reactive mechanical milling method for solid hydrogen storage application AIP Conference Proceedings 1826, 020002

Thermodynamic analysis and multi objective optimization of kalina and absorption cycle for power and cooling driven by lahendong geothermal source

Nasruddin[1], Ratna Monasari, Bagus Rizky Dewantoro, Mochammad Ilham Attharik, Agung Satrio Wibowo and Arief Surachman
[1]Departemen of Mechanical Engineering, University of Indonesia, 16424, Indonesia

E-mail : nasruddin@eng.ui.ac.id

Abstract. Kalina cycle is a new concept in thermodynamics that convert heat energy into mechanical energy. The heat source in this system comes from the Lahendong geothermal source. The cycle is using a mixture of solution of two liquids with different boiling points for the working fluid. Water and ammonia are the most widely used combinations with a 30% - 70% ratio in this study. With the combined benefits of comparison ratios, the Kalina cycle is able to produce better exergy efficiency and exergoeconomoic compared to conventional Rankine cycle. Kalina cycle works on thermal efficiency around 40% - 60%. The objective of this study is to find optimization in exergy efficiency and exergoeconomoic based on Kalina cycle applied in Lahendong geothermal source used MATLAB-EES. The results showed that the optimal ammonia-water mixture occurred at 130.0626 °C, 2184.791 kPa, and 51.77% of basic solution can yield exergy value 2358.88517 W and use the cost 16994.9715 \$/year.

1. Introduction

Indonesia is one of the world's largest geothermal energy sources. With the increasing demand for electricity occurs in Indonesia each year requires additional capacity to produce electricity in accordance with national electricity consumption. Although Indonesia is one of the largest geothermal energy producers, its development is relatively slow and has many difficulties. Over the span of 20 years, Indonesia is only able to develop at least 807 MW or about 4 percent of the 20,000 MW of availability geothermal source [1].

Geothermal energy has its own challenges in its development. Some of the factors that affect the inhibition of geothermal energy development in Indonesia are: the process of exploration resource was expensive and there is no market for resources other than as electricity producers. The cost locating and verifying a new source for geothermal power projects is high, this phase is similar to the discovery of oil fields or coal mines. Geothermal energy is specific and must be converted into electricity around the energy source [2].

In 1984, Alexander Kalina discovered a power plant cycle using geothermal power with low to medium temperatures as a thermodynamic cycle [3]. The kalina cycle uses a mixture of water – ammonia as its working fluid. Water-ammonia are widely used as working fluids in the cycles because of their excellent thermodynamics to transfer heat. The kalina cycle has the potential to efficiently convert energy from the

IOP Conf. Series: Earth and Environmental Science **105** (2017) 012099 doi:10.1088/1755-1315/105/1/012099

source of thermal energy. At the low and medium temperature of geothermal energy, the utilization of the Kalina cycle produces a satisfactory thermodynamic performance.

In addition to kalina cycle, The cooling absorption cycle with water-ammonia is also expected to provide an important role in heat source recovery. Some researchers conducted energy and exergy analyze for cooling the absorption of water and ammonia mixtures and found the optimum temperature of the cultivated sources to maximize the coefficient of performance (COP) [4, 5].

Currently, the merger between the cooling and power cycle is becoming a widely used technology for energy efficient utilization. The reason to use kalina cycle is for improving the efficiency of absorption cooling system and power systems. In the kalina cycle, a low concentration of ammonia solution is extracted to the separator at high temperature. This study aims to utilize the heat contained in a low concentration of water-ammonia solution. Thus, the cooling absorption cycle inserted to the kalina cycle before the high temperature recuperator to produce simultaneous power and cooling output. The research in this paper is the first discusses the optimization multi objective from exergy destruction and cost in combined kalina and absorption cycle.

2. Description System

The kalina cycle in this study consist of several components part, namely: HRSG, Separator, Turbine, Generator, High Temperature Recuperator (HTR), Low Temperature Recuperator (LTR), Condenser, Pump, Throttle Valve, and Refrigeration Cycle Absorption. in this refrigeration section, the component parts are pre-cooler, condenser, SHX, Distillation Tower, throttle valve, pump, and evaporator. The schematic diagram as shown in figure 1.

After being heated at evaporator where heat is extracted from geothermal well, an ammonia-water mixture flows towards the separator and then separates into ammonia-rich vapor and ammonia-poor solution in separator. Ammonia-rich vapor with high content of 96% and mixture of ammonia with low levels of 49%. This ammonia-rich vapor will flow into the turbine and the rotate the generator, while the ammonia-low solution which still has a relatively high temperature and mass flow rate is used as a heat source to heat distillation tower in the absorption refrigerant cycle in state, 40% ammonia solution is cooled back to the HTR and flowed to throttle valve 1 for reduced pressure.

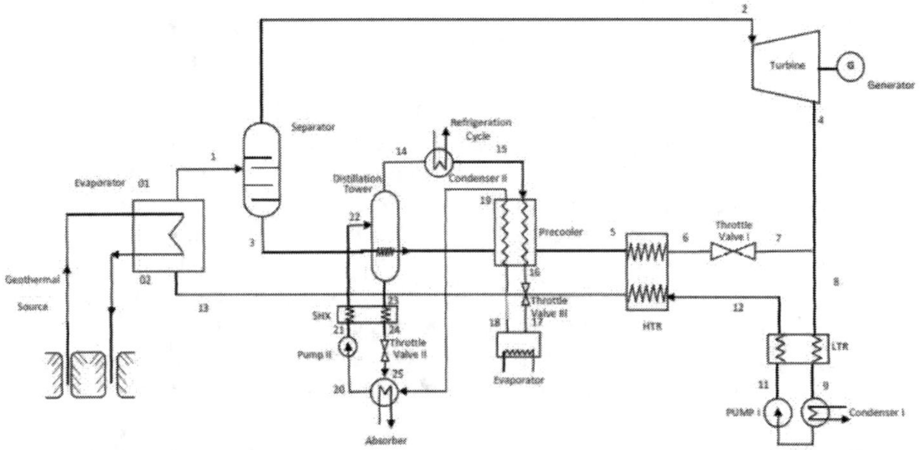

Figure 1. Schematic System of Kalina Cycle

After being used at turbine and absorption refrigerant cycle, ammonia-rich vapor and ammonia-poor solution will be re-mixed at state 7 to state 8 in the ammonia-water solution phase. The ammonia-water solution is then cooled by the LTR and condenser 1, temperature and pressure of ammonia-water solution will reduce on state 10. Then ammonia-water solution pressure will be raised by the pump 1. The compressed ammonia-water solution will absorb heat in the LTR and HTR and will return to state 13 and back to evaporator again to repeat the cycle.

In the absorption refrigerant cycle, the strong ammonia solution is pumped through SHX to distillation tower. The strong solution is heated by ammonia-poor solution in distillation tower and purified into refrigerant with high concentration [6]. High concentration refrigerant is condensed in condenser II, temperature and pressure at state 17 drops significantly after passes through the pre-cooler and throttle valve 3 to drops the pressure and temperature. Then back to pre-cooler and into absorber. The ammonia weak solution at distillation tower is cooled with cooling water at SHX flowed to the absorber and mixed with a strong ammonia solution. From absorber this mixed is pumped to distillation tower. This process occurs continuously (looping) in the cycle of refrigerant absorption.

3. Exergoeconomics Analysis

Exergoeconomics in which the concepts of exergy and economic have been combined is a relatively new area in engineering. In exergoeconomic analysis, the cost of thermodynamic inefficiencies is integrated in the cost of products of the system [7, 8]. Exergoeconomics reveals the cost formation process in energy systems which enables designer to design cost effective systems [9]. In economic analysis the prediction of the capital investment cost is important. In this respect, consulting with cost engineers or using vendor quotations is probably the most accurate method. In consulting with cost engineers to determine the new purchasedequipment costs after each design modification, the necessary thermodynamic data is submitted to the cost engineer. But generally the cost functions available in the literature are used for determining the capital investment cost of components in exergoeconomic analysis of a system. The cost values calculated from the cost functions are in agreement with the corresponding values provided by the cost engineer. It should be taken into account that in exergoeconomic analysis, if first, the values obtained for the capital cost of the components through the intended cost function are very different from the actual purchased-equipment costs in different sizes of components and second, inflation rate, interest rate and other economic parameters vary extensively during the lifetime of the system, the results obtained from exergoeconomic analysis will be very different from the practical results certainly. Perhaps the fundamental limitation of the theory of exergetic cost, as it was originally formulated, consisted of defining the productive structure in relation to the same flows and components presented in the physical structure. The resulting difficulties lie mainly in the adequate treatment of the dissipative units and of the residues of the system. Another difficulty is to identify the useful products of the system units as co-products or products-sub products. In recent years, many exergoeconomic studies have been conducted for different energy systems [10-18].

3.1. Energy and exergy analysis

The Kalina-based CCP cycle will be analyzed from energy and exergy perspective. Energy analysis based on the first law of thermodynamic discusses the energy conversion in quantity, but it is incapable of indicating the work potential or quality of different energy forms in relation to a given ambient condition [19]. Exergy method not only demonstrates the amount of usable work potential, or exergy, supplied as input to a system, but also reveals the system irreversibility distribution among different components and pinpoints the one contributing most to overall system inefficiency.

3.2. Validation for energy and exergy analysis
To verify the validity of exergy analysis, the available data in literature serves as contrast to the simulation results. The Kalina cycle and absorption refrigeration cycle are validated by data in Ref. [21] and Ref. [8], respectively. The Kalina cycle has the similar schematic diagram as illustrated in figure 1. When refrigeration cycle isn't attached to Kalina cycle, state 3 and state 5 are the same point. Therefore, all node numbers of Kalina cycle is in conformity with those in figure. 1 and state 3 and state 5 share the same thermodynamic parameters.

3.3. Economics and cost analysis
For exergoeconomic analysis different methodologies have been developed in recent years, among which the specific exergy costing theory (SPECO) has been widely applied to investigate thermodynamic cycles. This method establishes a rational technique to assess the cost rates of exergy streams by which the specific cost of each exergy stream (including the system products) can be evaluated. To perform an exergoeconomic analysis using the SPECO method, a cost balance equation along with required number of auxiliary relations, are applied to each system component [22-26].

In evaluating and optimizing energy system's performance, minimization of the specific cost of output power is an important objective [11, 27]. In the present study, this parameter is selected as the objective function on which the system performance is optimized. All the capital investment, operation and maintenance and fuel costs are included in this parameter [11].

4. Result and Discussion
In this section the process optimization on Kalina cycle driven by Lahendong geothermal reservoar will be displayed. The operation parameters of multi objective function and ranges of key thermodynamic parameters are listed at table 1.

Table 1. Operation parameters for optimization

Term	Value
Population size	100
Generation	100
Ranges of expander inlet pressure	1600 – 4000 kPa
Ranges of expander inlet temperature	130-180 °C
Ranges of concentration of basic solution	50-80%

Table 1 shows the values of thermodynamic parameters conditioned on the turbine's capability. Then the results of optimization Kalina cycle are shown in table 2.

Table 2. Optimization results

Term	Value
Expander inlet pressure	2184.791 kPa
Turbine inlet temperature	130.0626 °C
Concentration of basic solution	51.7 %
Exergy destruction at turbine	2.35 kW
Cost	16994.9715 $

In this study, two variables are exergy destruction and cost as consideration of multi objective optimization. Fig.2 shows the proposed pareto front solution for ammonia-water mixture as the working fluid of Kalina combine cycle. Each point in this pareto distribution illustrates relationship between two objective functions. When exergy values rise from 1.7 kW to 2 kW, there is a decrease amount of the cost. The optimum thermodynamic performance occurs at 2.3 kW exergy value with a cost value 16.9k $. The best economic performance occurs at point B where the exergy value is 3.1 kW with lowest cost 16 kW.it seen that's point A and B shown each objective function for cost and exergy value. The ideal solution in this multi-objective problem is to illustrate all points on the pareto distribution, then the final optimal point is determined.

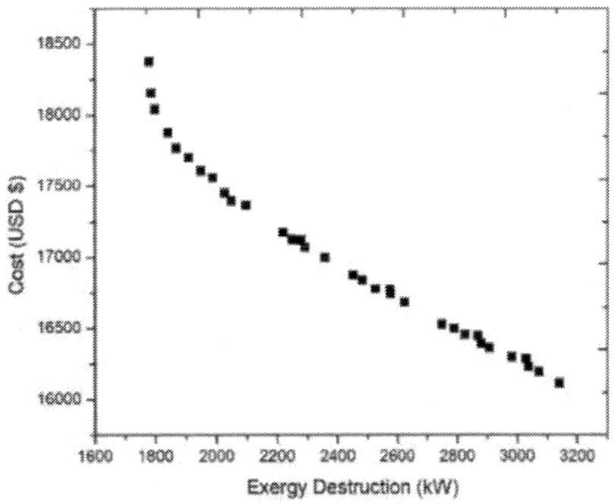

Figure 2. Distribution of the Pareto optimal solutions for Exergy Destruction and Specific Cost.

5. Conclusion

In this study, the analysis for kalina-based ccp cycles heated by Lahendong geothermal reservoirs was analyzed, considering the differences temperature and pressure on the wellhead. MATLAB-EES is used to obtain optimum value. Thermodynamic and exergoecenomic models were developed to analyze the performance of such systems. Multi-objective optimization is calculated to determine the optimum value of thermodynamic and exergoeconomic value of a system using optimization parameters of turbine, turbine inlet pressure, turbine inlet temperature, concentration of ammonia-water basic solution. The results showed that the optimal ammonia-water mixture occurred at 130.0626 °C, 2184.791 kPa, and 51.77% of basic solution can yield exergy value 2358.88517 W and use the cost 16994.9715 $/year.

6. Acknowlegment

The authors gratefully acknowledge to DRPM Universitas Indonesia for supporting this research by PITTA research grant 2017.

7. Reference

[1] Ibrahim R F and A Fauzi 2005 The Progress of Geothermal Energy Resources Activities in Indonesia in *Proceedings*
[2] Nasruddin et al 2016 Potential of geothermal energy for electricity generation in Indonesia *Renewable and Sustainable Energy Reviews* 53: p. 733-740
[3] Nasruddin et al 2009 Energy and exergy analysis of kalina cycle system (KCS) 34 with mass fraction ammonia-water mixture variation *J. Mechanical Sci. Technol.* **23**(7): p. 1871-1876

[4] Ziegler F and Alefeld G 1987 Coefficient of performance of multistage absorption cycles *Int. J. of Refrigeration* **10**(5): p. 285-295
[5] Chen J 1998 The coefficient of performance of a multi-temperature-level absorption heat transformer at maximum specific heating load *J. Phys. D: Applied Physics* **31**(22): p. 3316
[6] Cao L et al 2017 Thermodynamic analysis of a Kalina-based combined cooling and power cycle driven by low-grade heat source *Applied Thermal Engineering* **111**: p. 8-19
[7] Tsatsaronis G 2007 Definitions and nomenclature in exergy analysis and exergoeconomics *Energy* **32**(4): p. 249-253
[8] Abusoglu A and Kanoglu M 2009 Exergoeconomic analysis and optimization of combined heat and power production: A review *Renewable and Sustainable Energy Reviews* **13**(9): p. 2295-2308.
[9] Zare V et al 2012 Thermoeconomic analysis and optimization of an ammonia–water power/cooling cogeneration cycle *Energy* **47**(1): p. 271-283
[10] Lian Z, Chua K, and Chou S 2010 A thermoeconomic analysis of biomass energy for trigeneration *Applied Energy* 87(1): p. 84-95
[11] Soltani S et al 2013 A comparative exergoeconomic analysis of two biomass and co-firing combined power plants *Energy Conversion and Management* **76**: p. 83-91
[12] Dong R, Yu Y and Zhang Z 2014 Simultaneous optimization of integrated heat, mass and pressure exchange network using exergoeconomic method *Applied Energy* **136**: p. 1098-1109
[13] Mohammadkhani F et al 2014 Exergoeconomic assessment and parametric study of a Gas Turbine-Modular Helium Reactor combined with two Organic Rankine Cycles *Energy* **65**: p. 533-543
[14] Ahmadi P, Dincer I and Rosen M A 2012 Exergo-environmental analysis of an integrated organic Rankine cycle for trigeneration *Energy Conversion and Management* **64**: p. 447-453
[15] Shokati N, Ranjbar F and Yari M A 2014 comparative analysis of rankine and absorption power cycles from exergoeconomic viewpoint *Energy Conversion and management* **88**: p. 657-668
[16] Ahmadi P, Dincer I and Rosen M A 2011 Exergy, exergoeconomic and environmental analyses and evolutionary algorithm based multi-objective optimization of combined cycle power plants *Energy* **36**(10): p. 5886-5898
[17] Kostowski W J and Usón S 2013 Thermoeconomic assessment of a natural gas expansion system integrated with a co-generation unit *Applied energy* **101**: p. 58-66
[18] Lee Y D et al 2014 Exergetic and exergoeconomic evaluation of a solid-oxide fuel-cell-based combined heat and power generation system *Energy Conversion and Management* **85**: p. 154-164
[19] Kotas T 1985 *The exergy method of thermal plant analysis, Anchor Brendon ltd.* Tiptree Essex, Great Britain
[20] Yari M 2010 Exergetic analysis of various types of geothermal power plants *Renewable Energy* **35**(1): p. 112-121
[21] Zare V, Mahmoudi S and Yari M 2015 On the exergoeconomic assessment of employing Kalina cycle for GT-MHR waste heat utilization *Energy Conversion and Management* **90**: p. 364-374
[22] Herold K E, Radermacher R and Klein S A 2016 Absorption chillers and heat pumps CRC press
[23] Misra R et al 2003 Thermoeconomic optimization of a single effect water/LiBr vapour absorption refrigeration system *Int. J. of Refrigeration* **26**(2): p. 158-169
[24] Sholahudin S and Giannetti N 2016 Optimization of a cascade refrigeration system using refrigerant C 3 H 8 in high temperature circuits (HTC) and a mixture of C 2 H 6/CO 2 in low temperature circuits (LTC) *Applied Thermal Engineering* **104**: p. 96-103
[25] Smith R 2005 *Chemical process: design and integration* John Wiley & Sons
[26] Bejan A and Tsatsaronis G 1996 *Thermal design and optimization* John Wiley & Sons
[27] Zare V 2016 Exergoeconomic analysis with reliability and availability considerations of a nuclear energy-based combined cycle power plant *Energy* **96**: p. 187-196

IOP Conf. Series: Earth and Environmental Science **105** (2017) 012100 doi:10.1088/1755-1315/105/1/012100

Sodium percarbonate addition as electrolyte and buffer to produce electricity economically using industrial-tempeh-wastewater based microbial fuel cell

M A H V Kurnia[1] , T S Utami[1] and R Arbianti[1]
[1] Chemical Engineering Department, Faculty of Engineering, University of Indonesia, Depok, 16424, Indonesia

E-mail: vinci.kurnia@gmail.com

Abstract. Microbial Fuel Cell (MFC) is an energy conversion system used by bacteria to generate electricity from organic wastes. Currently, MFC electricity is still small, so it complemented by electrolytes. Previous research shows the addition of potassium persulfate can increase electric voltage 10-fold, but this is less economical so it is necessary to find alternative electrolyte. Sodium percarbonate is a cheap electrolyte which has buffering ability. Therefore, a performance study of single-chamber-MFC using sodium percarbonate electrolyte and or without buffer was conducted by measuring electricity production and tempe wastewater treatment quality (BOD and COD). This result was compared with the results of MFC with potassium persulfate with and without a buffer in the same procedure. In MFC containing sodium percarbonate, charge reversal occurred, with average power 0.04 mW/m^2, only 1.25% average power of MFC containing potassium persulfate. These results increased by buffer addition. Other results show that MFC containing sodium percarbonate will degrade more than 40% COD, greater than MFC with potassium persulfate, but still not efficient because Coloumb Efficiency was only in the range of 10^{-6} %. Nevertheless, this system can produce 63% BOD_5 reduction when a buffer was not added.

1. Introduction

Indonesia's energy consumption is increasing from year to year but, until 2015, only 88.30% of the regions in Indonesia can enjoy the power grid [1]. Electrical energy supply system that can be a solution is an energy conversion system that utilizes renewable resources, such as solar, wind, water, biomass and so on, but efforts need to be made to reduce investment costs by exploring the possibility of power plant mass production. The reduction of investment costs will have a direct impact on production costs [2].

Microbial Fuel Cell (MFC) is an energy conversion system used by bacteria to generate electricity from organic wastes [3]. MFC provides new opportunities for sustainable energy production from biodegradable fuels and can reduce substrate compounds [4]. The principle of the MFC system is that bacteria in the vessel produce electrons then transfer to the anode and conduct to the cathode connected by the conductivity device to produce a power that can run a machinery [5]. When bacteria degrade the substrate, electricity will be produced from MFC, so bacterial growth will be directly proportional to the increase in electricity production. Exclusively for waste substrates, bacterial growth will increase when electricity is produced and a decrease in COD (Chemical Oxygen Demand) [6].

The problem of MFC is the resulting voltage is still very small, so it cannot be applied immediately. To handle this, earlier researchers added electrolytes, such as potassium permanganate [7] and potassium persulfate [8], into the MFC compartment to speed up the electron transfer process that occurs in the anode. A study conducted by Citrasari [9] found that addition of potassium persulfate in MFC was able to increase the electrical voltage 10-fold. However, potassium persulfate price is still expensive, preventing it from being applied on a large scale. Therefore, we use cheap electrolyte, namely sodium percarbonate ($2Na_2CO_3 \cdot 3H_2O_2$) or washing soda which has natural buffering ability due to bicarbonate ion (HCO^{3-}). Power density produced by MFC with sodium percarbonate electrolyte is 9.6 W/m3 so it is worth paired with other electrolytic power density [10].

In this research, we will examine the effect of electrolyte addition by sodium percarbonate compared with potassium persulfate addition on MFC performance, which includes electricity production and COD/BOD_5 reduction. This research purpose is to produce an MFC system without buffer, low capital cost, and can be applied to turn on home appliances.

2. Methods

2.1. Microbial Fuel Cell (MFC) Preparation

MFC preparation comprises of electrode preparation and reactor preparation. The electrodes used in this MFC study include anodes and cathodes made of graphite. This electrode was immersed in 0.1 M HCl for 1 day then rinsed and immersed again with NaOH with the same concentration and immersion duration before used. Preparation of the MFC reactor involves preparing a tubular acrylic-based reactor. The anode that used is elongated cylindrical and installed at one end of the bioreactor, while the cathode is a square-shaped water-cathode. The reactor used is a batch reactor which the top can be opened and then locked with a bolt.

2.2. Substrate Mixture Preparation

The substrate mixture used in this study consisted of incubated liquid tempeh wastewater with/without buffers, electrolytes, and Gram-negative isolates. The prepared electrolytes were sodium percarbonate ($2Na_2CO_3 \cdot 3H_2O_2$) and potassium persulfate ($K_2S_2O_8$), each 0.03 M content with 225 mL volume. The buffer used was PBS (Phosphate Buffer Solution) 0.1 M pH 7 of 50 mL. The substrate used in this research was the residue of soybean cooking water in tempeh-making process in industrial tempeh manufacture, which the tempeh boiling ratio is 2:3 (w gr/v mL water) [11]. This wastewater is grown in mixed culture by incubating in a closed container for 3 days. 1 mL of this wastewater was then incubated in MacConkey agar for an inverted petri dish position for 24 h at 30°C to obtain Gram-negative isolates. Subsequently, the isolate was incubated in 50 mL of Luria Bertani (LB) medium for 24 hours and 1 mL of isolate was added to the substrate by separation using centrifugation at 400 rpm for 20 min. There are 4 mixture variants for this research which can be seen in table 1.

Table 1. 500 mL substrate mixture variant.

	225 mL sodium percarbonate 0,03 M	225 mL potassium persulfate 0,03 M
225 mL tempeh wastewater + 50 mL buffer	Sodium Percarbonate Buffer (SPB)	Potassium Persulfate Buffer (PPB)
275 mL tempeh wastewater	Sodium Percarbonate (SP)	Potassium Persulfate (PP)

2.3. MFC Experiment

MFC voltage measurements are performed using a digital multimeter connected directly to the CPU/PC. Multimeter used in this study is APPA 109N with data logger feature which is set to record data every 10 seconds. The multimeter is connected to the electrode with the power cable and then given the resistance to make the system into closed circuit with 820-ohm resistor. The experiment was carried out with the reactor condition being turned upside down. The results obtained are electric voltage data (V) in mV which is then converted to power density (P = V.I/A) in mW. A is the surface area of the anode with unit m^2. I is the electric current obtained from the equation I = V/R.

2.4. COD, Coulomb Efficiency (EC), and BOD_5 Analysis

Measurement of COD and BOD_5 were performed on wastewater mixtures in MFC before and after MFC was operated. COD measurements were performed using the spectrophotometric method with 2 mL sample reacted with Reagan HACH Cat. 2125925. Coulomb Efficiency (CE) calculated by using COD removal and current generated by MFC.

3. Result and discussion

3.1. Power Output of Electrolyte Mixture Variation

Overall, the average power density generated by wastewater (WW) + potassium persulfate buffer (PPB) is 15 times the average electrical density. This increase in electricity production is due to the buffer reducing the ohmic resistance [12] and facilitating ion transfer through the anode biofilm [13]. Buffers also inhibit the decrease in pH and increase the conductivity of solutions that affect the production of electricity and wastewater treatment efficiency [14].

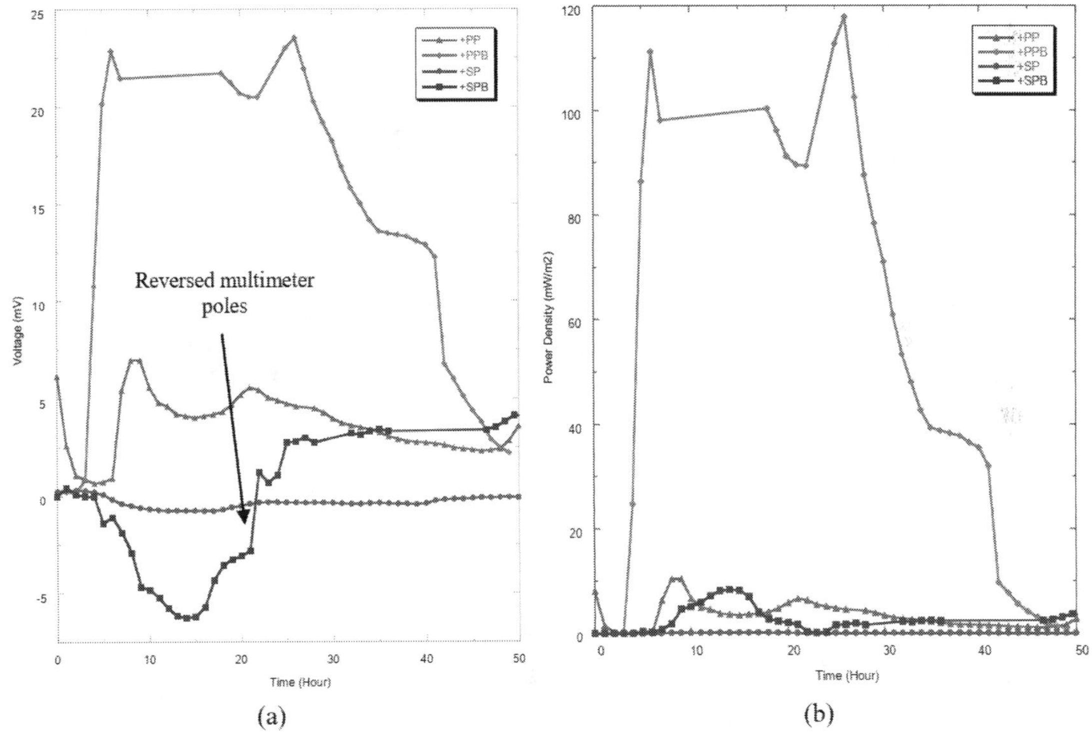

Figure 1. Profile of (a) electric voltage and (b) power density for each variation

The interesting thing to be gained from this research is the charge reversal on MFC using Sodium Percarbonate (SP), which is confirmed by reversing polar at the end of multimeter at 22 hours in MFC WW + Sodium Percarbonate Buffer (SPB) (figure 1(a)). This phenomenon has occurred in stacked MFC [15] One of the things that might explain the phenomenon of charge reversal in this research is the basic environment of wastewater due to pKa sodium percarbonate value is 10.3 (Sigma-Aldrich). This makes oxygen easy to dissolve, making oxygen in MFC is overflow where oxygen come from the decomposition of hydrogen peroxide by mixed culture bacteria [16] and diffusion from the environment. This condition encourages bacteria to stick to the surface of the MFC lid to absorb

IOP Conf. Series: Earth and Environmental Science **105** (2017) 012100 doi:10.1088/1755-1315/105/1/012100

oxygen that can produce larger ATPs [17]. The remaining oxygen will tend to do a reduction reaction because of the broadness of the anode surface contact with the wastewater.

The results showed very high average electrical density difference between MFC added by potassium persulfate (PP) (3.32 mW/m^2) with MFC added by SP electrolyte (0.04 mW/m^2). This occurs because the resistance of the flow of protons and electrons due to the oxygen present in MFC tends to be used for aerobic bacteria internal respiration and the presence of electrolyte resistance. This can be handled with the addition of a buffer that shows an increase in electrical density as much as 70-fold in SP electrolyte (0.04 mW/m^2 to 2.87 mW/m^2).

3.2. COD and BOD$_5$ Removal Measurement

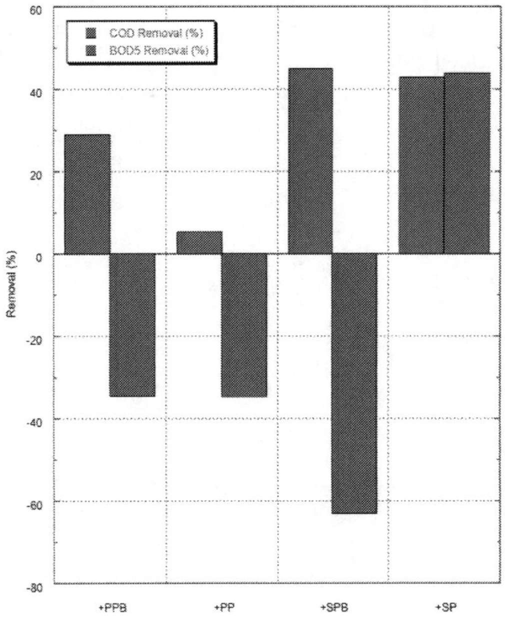

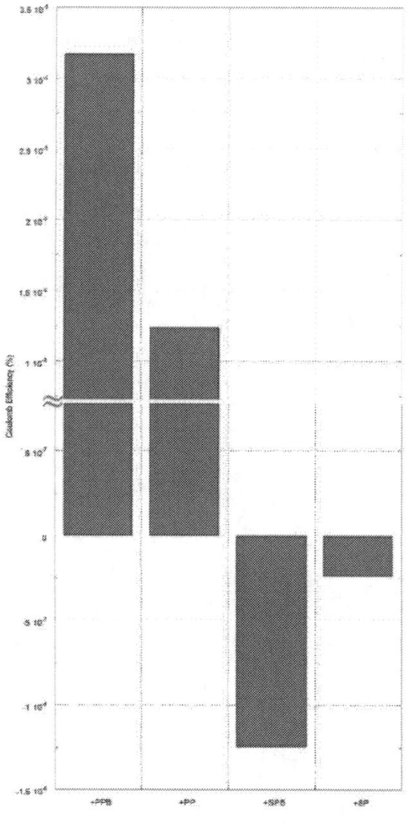

Figure 2. MFC's COD & BOD$_5$ removal measurement with different wastewater mixtures.

Figure 3. MFC's Coulomb Efficiency analysis measurement with different wastewater mixtures

From the measurement of the COD removal (figure 2), it was found that there was a decrease of COD of 28.9% in MFC added with PPB electrolyte, which was better than COD removal in MFC added with PP electrolyte. This is due to an increase in the conductivity of the solution [14] which is an important factor in improving the performance of wastewater treatment by MFC [18]. However, this is still low compared to MFC added with SP electrolyte, which exceeded 40% removal. This was because increasing the pH of waste after added sodium percarbonate (SP) makes aerobic bacteria grow more so that the conversion of nutrients in wastewater is greater.

The results of the BOD$_5$ measurements indicate an increase in oxygen in MFC in 3 of the 4 variations of this study, which indicates that bacteria are still actively growing. In the MFC added with sodium percarbonate there was a decrease in BOD$_5$ by 43.87% but increased by 63.05% when the buffer was performed. In the WW+SP variant, the basic wastewater environment making almost all mixed culture in wastewater respire aerobically so oxygen in wastewater decreased. This distinct with MFC WW+SPB which the pH was close to 7 (neutral), making a small proportion of mixed culture bacteria which still performs anaerobic respiration and H_2O_2 decomposition, resulting in the greater rate of decomposition and rate of oxygen diffusion to MFC compared to the growth rate of aerobic bacteria.

3.3. Coulomb Efficiency Analysis

Results of Coulomb Efficiency analysis (figure 3) show that MFC system with mixed waste with PP electrolyte is the most efficient variant compared to MFC system with SP electrolyte which has negative value due to the charge reversal phenomenon. The efficiency of Coulomb in MFC with SP electrolyte is very low, only within the range of 10^{-6} %. This shows the tendency of mixed culture bacteria not to release electrons and protons outside the cell and the ineffective hydrogen peroxide electrolyte which has been decomposed by the enzyme catalase in mixed culture bacteria [16].

4. Conclusion

The use of sodium percarbonate in single chamber MFC generates charge reversal electricity with average of power density is 0.04 mW/m^2, only 1.25% of MFC with an addition of potassium persulfate so it is necessary to add buffers to increase the electrical output. COD removal results showed that MFC with sodium percarbonate addition is a better wastewater treatment system than MFC with potassium persulfate addition, although the conversion of nutrients to electricity production is still not efficient. The BOD_5 removal analysis showed that MFC added by sodium percarbonate without buffer was able to reduce the BOD pollutant present in the wastewater. Thus, single chamber MFC with sodium percarbonate addition without buffer would be better applied as an economical natural waste treatment system without electricity production. Therefore, the use of sodium percarbonate as an MFC electrolyte needs to be further investigated, especially its optimum concentration.

5. Acknowledgments

The author is grateful for the financial support of Chemical Engineering Department University of Indonesia and author's family.

6. References

[1] Directorate General of Electricity of the Indonesian Ministry of Energy and Mineral Resources 2016 *Kebijakan Pemerintah dalam Pembangunan Infrastruktur Penyediaan Tenaga Listrik* (Jakarta: Institute for Essential Service Reform) p 9
[2] Indonesian Institute of Sciences 2004 *Pengembangan Energi Terbarukan Sebagai Energi Aditif di Indonesia* (Jakarta: Indonesian Institute of Science)
[3] Kurnianingsih N 2009 *Bakteri Microbial Fuel Cell* (Bandung: Alpen Steel)
[4] Rabaey K and Verstraete W 2005 *Trends Biotechnol.* **23** 291-8
[5] Logan B E 2008 *Microbial Fuel Cells* (New Jersey: John Wiley & Sons)
[6] Moon H, Chang, I S, Jang J K and Kim B H 2006. *Biochemical Engineering Journal* **27** 59-65
[7] You S, Zhao Q, Zhang J, Jiang J and Zhao S 2006 *Journal of Power Sources* **162** 1409–15
[8] Li J, Fu Q, Liao Q, Tian X, Ye D and Zhu X 2009 *Journal of Power Sources* **194** 269–74
[9] Citrasari A E 2014 *Pengaruh Jenis Elektrolit dan Penambahan Jumlah Bakteri Konsorsium dari Limbah Cair Tempe terhadap Produksi Listrik pada Microbial Fuel Cell* (Depok: Engineering Faculty University of Indonesia)
[10] Forrestal C, Huang Z and Ren Z J 2014 *Bioresource Technology* **172** 429-32
[11] Nout M J R and Rombouts F M 1990 *Journal of Applied Bacteriology* **69** 609-33
[12] Min B, Román Ó B and Angelidaki I 2008 *Biotechnol. Lett.* **30** 1213-8
[13] Torres C I, Marcus A K and Rittmann B E 2008 *Biotechnol. Bioeng.* **100** 872-81

IOP Conf. Series: Earth and Environmental Science **105** (2017) 012100 doi:10.1088/1755-1315/105/1/012100

[14] Huang L P and Logan B E 2008 *Appl. Microbiol. Biotechnol.* **80** 349–55
[15] Oh S E and Logan B E 2007 *Journal of Power Source* **167** 11-7
[16] Tortura G J, Funke B R and Case C L 2016 *Microbiology: An Introduction* 12th ed (London: Pearson Education)
[17] Karp J 2009 *The Three Metabolic Energy Systems* (*IDEA Ftiness Journal* vol 6 issue 2) (San Diego, CA: IDEA Health & Fitness Association)
[18] Feng Y J, Wang X, Logan B E and Lee H 2008 *Appl. Microbiol. Biotechnol.* **78** 873–80

Utilization of sesame seed sprout biomass as lipase source to hydrolyse palm oil

S Lalang Jati[1], A Nadia Tuada[1], U T Surya[1] and H Heri[1,2]
[1]Chemical Engineering, Faculty of Engineering, University of Indonesia, Depok 16424, Indonesia

E-mail: heri.hermansyah@ui.ac.id

Abstract. Lipase is known enzyme for its ability to hydrolyse triglycerides into free fatty acid. There are so much sources od lipase such as plants, microbes, and animal. But, lipase from plant seems interesting owing to its huge amount than microbes' lipase and less harm than animal's lipase. Many experts have proved that vegetable lipase is a potential biocatalyst so that lipase from sesame seed and sesame sprout was investigated in this research. Lipase which was obtained from each sources were then poured with 0.3 g PVA and 40% palm oil in demineral water. The lipase-palm oil solutions were then titrated with 0.05 M NaOH obtaining hydrolysis percentage. Based on the research, sesame seed supernatant produces 1.36 mmol FFA with 39% hydrolysis. Sesame sprout extract produces 1.37 mmol FFA with 39.14% hydrolysis. Those samples also contain lauric acid, miristic acid, palmitic acid, oleic acid and stearic acid which detected by GC instrument.
Keywords: Hydrolysis, Lipase Biocatalyst, Palm oil, Sesame Seed Sprout, Triglycerides

1. Introduction
Lipase is an enzyme which takes important group of enzymes in biodiesel production (processing fatty acid alkyl esters) [5]. It has different characteristic from chemical catalyst such as eco-friendly, high specificity, and stereoselective. In addition, can be found in many living creatures such as animlas, plants, and microbes [2]. So there is high potency of replacing the usage of chemical catalyst with biocatalyst since they both have good performance.

Commercial lipase is now available but the price is way more expensive than recoverable chemical catalyst [8]. This reaearch comes up with that problem and attempt to find economic method of producing lipase. Previous experiments focus on extracting lipase from sap papaya, beans sprout, and rice peel waste. In this research, sesame sprout was used to find its hydrolysing performance towards palm oil. Sesame seed (*Sesamum indicum L.*) known for its nutritious content and great source of essential oil (51.1%).

One benefit of utilising catalyst is adding palm oil value. Nowadays, palm oil is still restricted as a raw material of cooking oil or oleochemical product. Meanwhile, palm oil contains fatty acid that can be hydrolised into energy. This research tried to find another biocatalyst to help maximizing the esterification process.

---

[2] To whom any correspondence should be addressed.

IOP Conf. Series: Earth and Environmental Science **105** (2017) 012101 doi:10.1088/1755-1315/105/1/012101

2. Material and Methods

The research was conducted at Bioprocess Laboratory, Chemical Engineering Department, Engineering Faculty, University of Indonesia, Depok. Analysing sesame sprout content with HPLC in Balai Besar Pascapanen, Cimanggu, Bogor. Fresh sesame sprout and palm oil were purchased from local market around Jakarta. Polyvinyl alcohol (PVA) was purchased from Bratachem.

2.1 Preparation sesame seed into sprout

Washed sesame seed was divided into two group, the one which directly used and the other one to be cultivated into sprout. Sesame seed which prepared to be cultivated was then soaked into fungicide solution (1ml/l demineral water) for 10 minutes. Soaked sesame seed was removed from the fungicide solution, sowed onto sterile sand at 30^{0}C for six days [1].

2.2 Obtaining sesame extract

Both germinated and ungerminated seed were then washed with demineral water and homogenized for 10 minutes in 0.15 M phosphate buffer (5 ml/g fresh weight) containing 0.6 M sucrose, 1 mM EDTA, 10 mM KCl, 1 mM $MgCl_2$. Phosphate buffer was adjusted to pH 7.5 with KOH. Those solutions were centrifuged for 30 minutes at 5,000 rpm. Supernatant from centrifugation process and sesame sprout extract were used in hydrolysis activity test.

2.3 Hydrolysis reaction

Supernatant from sesame seed and sesame sprout extract were soluted into 4 ml phosphate buffer (0.1 M, pH 7.5) separatedly. Each solution was then stirred at 800 rpm for 30 minutes to dissolve enzyme content. After that, poured 10 ml palm oil in 15 ml demineral water and 0.3 g PVA into each enzyme solution and shook in a shakerbath at 150 rpm, 37^{0}C. Hydrolysed solutions were then filtered by Whatmann paper filter then continued by centrifugation to separated oil and water content. Lipase activity was determined by continuous titration of FFA released by lipolysis of emulsified oils. 2 ml centrifuged oil content was tested with 3 droplets of pp indicator and titrated with 0.05 M NaOH.

2.4 Free fatty acid analysis

Oil sample was weighed 0.1-0.5 g to be poured 2 ml NaOH-methanol 50% and heated up at 80^{0}C for 15 minutes. When the solution cool enough, added 2 ml BF_3 solution and heated up again at 80^{0}C for 45 minutes. Let the solution cool before added 1-2 ml Hexane and shook gently until homogen. Replaced the hexane solution before injected into GC instrument.

3. Results and Discussion

3.1 Hydrolysis reaction using sesame sprout's enzyme

The experiment shows that hydrolysis result from sesame seed and sesame sprout has only slight different. It means that both sesame seed and sesame sprout are potential as lipase source to hydrolize palm oil. Compared to candida rugosa and rhizopus oryzae, hydrolysation percent were not the highest, but sesame is the best lipase source from an economic consideration.

Beside that, both sesame seed and sesame sprout does not need long preparation to obtain its lipase. Unlike obtaining lipase from fungus, its process is relatively simple and easy. One of advantages to be considered to produce a commercial lipase from sesame seed or sesame sprout.

Here is the data result from 8 hours hydrolysis in pH 7, 37^{0}C and 150 rpm.

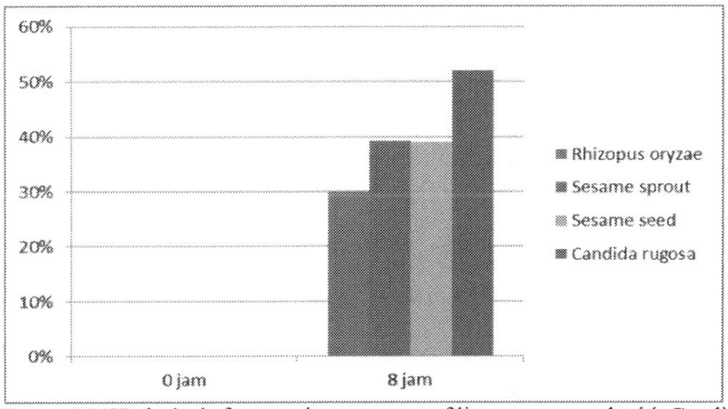

Figure 1 %Hydrolysis from various sources of lipase compared with Candida Rugosa (pH: 7,0; 37^{0}C; shaking rate$_{hydrolysis}$ 150rpm)

After 8 hours of hydrolysing, sesame seed could hydrolyse 39% palm oil and sesame sprout could hydrolyse 39.14% palm oil. Candida rugosa as the highest hydrolysation agent could hydroysze 52% olive oil and rhizopus oryzae could only hydrolyse 30% palm oil.

3.2 Content analysis using GC instrument

3.2.1 Pure Palm Oil. The data below shows that palm oil contains free fatty acid naturally. Those fatty acids were detected by GC instrument with various concentration for each fatty acid as seen below.

Table 1. Free fatty acids content in pure palm oil

Pure Oil		
Oil Concentration (mmol/100 gr)		
	0 hour	8 hour
Lauric Acid	0	5,580
Miristic Acid	0	0,644
Palmitic Acid	0	5,074
Oleic Acid	0	1,816
Stearic Acid	0	7,253

3.2.2 Sesame Sprout and Sesame Seed Supernatant. The data below shows free fatty acid concentration in sesame sprout and sesame supernatant. Those fatty acids were detected by GC instrument with various concentration for each fatty acid as seen below.

Table 2. Free fatty acids content in Sesame sprout

Sesame sprout		
Concentration (mmol/100 gr)		
	0 hour	8 hour
Lauric Acid	0	2,575
Miristic Acid	0	3,399
Palmitic Acid	0	3,066
Oleic Acid	0	3,362
Stearic Acid	0	4,032

Table 3. Free fatty acids content in Sesame supernatant

	Supernatant	
	Concentration (mmol/100 gr)	
	0 hour	8 hour
Lauric Acid	0	3,515
Miristic Acid	0	1,407
Palmitic Acid	0	0,000
Oleic Acid	0	8,606
Stearic Acid	0	5,419

The tables above show different contents of free fatty acids in sesame sprout and sesame supernatant after hydrolysis reaction was done. Sesame sprout seems having the average concentration for each fatty acids. In the other hand, sesame supernatant could only produce lauric acid, a little miristic acid, high oleic acid, and stearic acid while palmitic acid were not detected by GC instrument.

4. Conclusion

Enzyme activity during hydrolysis reaction showed good result. Both sesame seed and sesame sprout broke triglycerides into free fatty acid with relatively same amount (1.36 mmol and 1.37 mmol). Prosentase hydrolysis for each method is as follows 39% and 39.14%. From GC analysis, sesame seed formed oleic acid as the highest fatty acid continued by stearic acid, lauric acid, miristic acid, and the absence of palmitic acid (8.606, 5.419, 3.515, and 1.407 mmol/gr). On the contrary sesame sprout formed stearic acid, oleic acid, lauric acid, miristic acid, palmitic acid as muc as 4.032, 3.362, 2.575, 3.399, 3.066 mmol/100gr. From the data, lipase content in plants is relatively same with lipase content in microbes. But the interesting point in producing lipase from plant is its huge source in nature than microbes.

5. Acknowledgement

Author acknowledge the research grant supported by DRPM Universitas Indonesia and funded from Riset Insinas award that has resulted this article.

6. References

[1] Ambarwati and Susilo 2004 *Pengaruh Penambahan Biji Wijen (Sesanum indicum) dan Kecambah Jagung (Zea mays) terhadap Kadar Protein Susu Kedelai*

[2] Houde A, Kademi A and Leblanc D 2004 Lipases and Their Industrial Applications: An Overview *Applied Biochemistry and Biotechnology* **118**(1-3), pp.155-170.

[3] Jachmanian I and Mukherjee K D 1996 Esterification And Interesterification Reaction Catalyzed By Acetone Powder From Germinating Rapeseed *J. Am. Oil Chem Soc.* **73**(11): 1527-1532.

[4] Marno Septhian 2008 *Interesterifikasi Minyak Kelapa Sawit Dengan Metil Asetat Menggunakan Biokatalis Untuk Memproduksi Biodiesel* Fakultas Teknik UI, Depok.

[5] Ribeiro B, Castro A, Coelho M and Freire D 2011 Production and Use of Lipases in Bioenergy: A Review from the Feedstocks to Biodiesel Production *Enzyme Research* pp.1-16.

[7] Ritonga M Yusuf 2004 *Pengaruh Bilangan Asam Terhadap Hidrolisa Minyak Kelapa Sawit* Fakultas Teknik, Universitas Sumatera Utara

[8] Praswati P D K Wulan, Rejoso M Titis and H Heri 2011 *Hidrolisis minyak zaitun menggunakan lipase Rhyzopus oryzae yang diimobilisasi melalui metode Adsorpsi* Departemen Teknik Kimia

[9] Suhendra L et al 2006 *Aktivitas Hidrolisis dan Esterifikasi Lipase Ekstrak Kecambah Biji Wijen* Yogyakarta: Jurusan Teknologi Pangan dan Hasil Pertanian Universitas Gadjah Mada

Experimental study on horizontal axis wind turbine with splitted winglets

J Pratilastiarso[1], S Nugroho[1] , E Tridianto[1] and R I Syifa[1]

[1] Power Plant Engineering, Electronic Engineering Polytechnic Institute of Surabaya, Surabaya, Indonesia, 60111

Email : rizalibnu95@gmail.com

Abstract
Human needs increasing over years, energy is needed to fulfill the daily life. In the other hands, source of energy such as coal is decrease and become less and less. There is a renewable energy that can made from wind source. Wind energy can be obtained by using some tools such as wind turbine. Wind turbine is a device that used to convert wind energy into electricity. Wind turbine has developed to many variation. One of variation that added to the wind turbine especially on their blades is splitted winglet. This additional variation is an adaptation from recent aircraft. The advantages of using this typical of winglets on aircraft is increasing the lift coefficient and decreasing drag coefficient. The function of splitted winglet on wind turbine blades is to minimize the backflow on the tip of blades. This research expectation is turbine will get better performance. Comparing the turbine with the blade without winglets, with winglets and with splitted winglets. Turbine with the blades without winglets gives best performance among all the turbine variations. The air flow hit the turbine at 6 m/s, the performance of turbine without winglets give Cp about 5,2E-1. With the same velocity of airflow, the turbine with winglets and splitted winglets give Cp about 4,7E-1 and 2,1E-1.

1. Introduction

Nowadays, most of power plant that produce electricity from coal which is its source is decreasing over the years. Following by human needs of electricity to fulfill their life is increasing every time. One of solution that can solve this problem is by using renewable energy such as wind energy. Wind energy can be converted to make an electricity by using wind turbine. There are so many kind of wind turbine. From the axis of rotation, there are horizontal axis wind turbine and vertical axis wind turbine. Horizontal axis wind turbine give some advantages such as for using of area, fees, and noise. The turbine performance affected by external and internal factor. The external factor come from the ambiance condition including weather. The internal factor that affected in turbine performance come from turbine component such as blades, tail, generator, and its transmissions. Blades make the rotation of turbine to make an electricity. This device can make the turbine's performance become better by adding some variation. One of the variation that can added into the blades is splitted winglet.

Splitted winglet is the innovation from blended winglet that added by another winglet in the against direction. This additional variation is adapted from the recent aircraft. The purpose of adding this variation is to enhance the performance so that there are minimalize the back flow on the blade. From the research [1] the wings or aircraft with splitted winglet has better performance than wings of aircraft using blended winglet or single winglets.

This study is comparing wind turbine with blade with splitted winglets, fixed winglets, and no winglets. The performance of wind turbine can be shown by value of coefficient of performance by their tip speed ratio.

2. Methodology

Geometry that used in this paper is rotor of wind turbine with three blades. Wind turbine rotor with three blades give better performance [2]. This three blades rotor show better performance with Cp about 5,2 when the TSR in 4 until 5. Cp is coefficient performance which show the performance of rotor wind turbine. TSR is tip speed ratio which show the ratio of speed on tip of blade to the velocity.

The blade with splitted winglets dimension adapted form blade with sigle winglets a with additional other winglets in the against direction. The splitted winglets ratio is h2 is 0.4 times h1 [3]. Where h1 is the main winglets and h2 is the side winglets. The position of h1 or main winglets is in the same direction with the air flow with span angle 70° . The h2 or side winglets position is against the main winglets with span angle 52°.

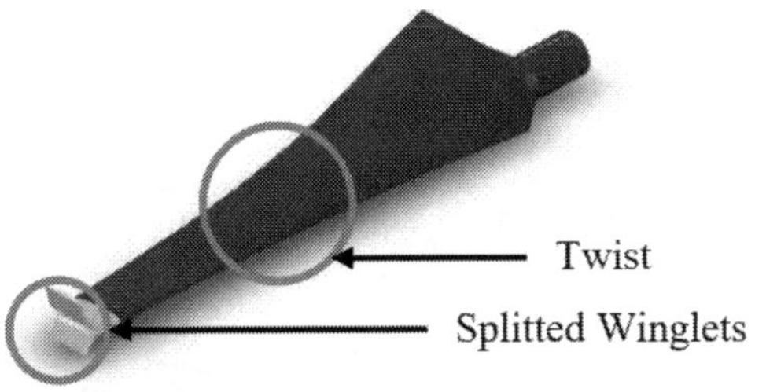

Figure 1. Blade with splitted winglets

Other additional method is using twist. Twist is the additional method in the segment of airfoil that construct blade with different angle of twist, dimension or chord length, and distribution of the segment. Twist give variable angle of attack through the length of the blade. Twist angle is an angle that given to the each segment of airfoil. Twist angle distribution is based on aerodynamic behavior of each individual airfoil at a certain angle of attack [4].

$$\beta(r) = \frac{2}{3} arctan \frac{R}{r\lambda} - \alpha_D \quad (1)$$

The airfoil variation is defined by the chord length. The chord length is the size of chprd in every segment of chord. The airfoil with different chord and twist angle is distributed based on the length of the blade [5]. The size of each blade is 12 cm and the hub radians is 3 cm.

$$c(r) = \frac{16\pi r}{BC_l} sin^2 \left(\frac{1}{3} tan^{-1}\left(\frac{R}{\lambda r}\right)\right) \quad (2)$$

So, the rotor diameter is 30 cm. From the equation above [5], The segment distribution is counted on excel and described in table below. In the table 1 from research [6] we know that longer segment position from the hub, smaller chord we get. The variation of twist angle is different in each segment.

The experiment of wind turbine is using wind tunnel. Wind tunnel is a testing tools for watching aerodynamic condition of instrument. The output of wind tunnel testing is torque and angular velocity. The rotor tested by turbine stand that using gear box that has ratio of 1:3,5.

Velocity that used on this study is about 0-7 m/s. The velocity is based on fan frequency that controlled on inverter. The consideration of using the velocity is because of the capability of turbine that influence the strength of the material that built the turbine.

Table 2.1 Detail Of Geometry Data Input

Segment	r/ (cm)	c/ (cm)	Twist β (°)
1	3,20	3,80	28,13
2	4,14	3,59	22,06
3	5,09	3,32	17,53
4	6,04	3,04	14,11
5	7,00	2,79	11,46
6	7,95	2,56	9,35
7	8,90	2,36	7,63
8	9,85	2,18	6,21
9	10,81	2,03	5,02
10	11,76	1,89	4,00
11	12,71	1,77	3,12
12	13,66	1,67	2,36
13	14,62	1,57	1,70
14	15,62	1,48	1,08
15	15,95	1,45	0,88

Figure 2. Rotor turbin angin saat pengujian

In this study also using variation of pitch angle. Pitch angle known by *pitch* control that is on the beginning part of blade. This variation purpose is to increase the performance by adjusting the pitch angle of blades in the fluctuate air flow condition [7] as shown on figure 3.

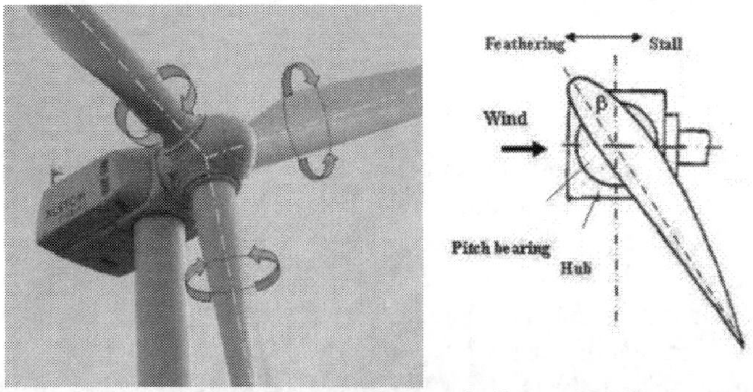

Figure 3. Pengaturan sudut *pitch* di pangkal *blade*
(Source : *Advances in Wind Power. InTech Janeza Trdine 9, 51000. Rijeka, Croatia.*)

Blade Element Momentum (BEM) is a simulation method by theory that used for comparing experimental product. The method works by dividing blade into some segment. BEM method calculate the change of performance of wind turbine in the variation of pitch blade angle in two dimension [8].

There is three vectors of velocity that affect the force that produce by airfoil in the segment of blade as shown on figure 2.

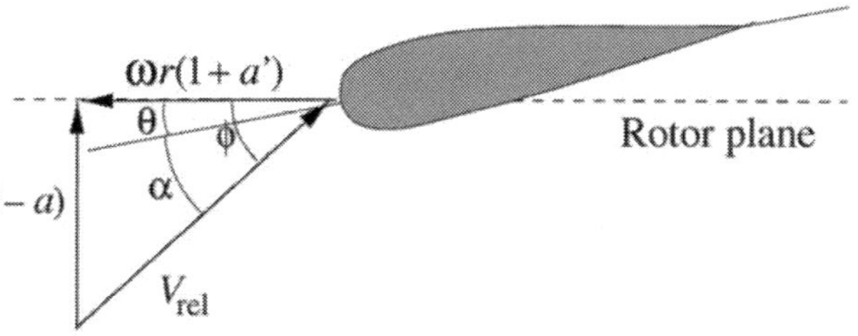

Figure 4. Triangle of velocity vectors on blade
(**Source**: *Martin O. L Hansen. 2008. Aerodynamics of Wind Turbines. Second Edition.*)

Relative velocity (V_{rel}) is the product of air velocity vector (V_o) with angular velocity (ωr), and ($\boldsymbol{\phi}$) that is inflow angle which is angle between relative velocity and angular velocity, inflow angle is sum of angle of attack ($\boldsymbol{\alpha}$) and pitch angle (θ).

Axis torque total is the sum of all $M_{i,i+1}$ contribution along one *blade* multiply by sum of *blade*.

$$M_{tot} = B\sum_{1}^{N-1} M_{i,i+1} \tag{3}$$

Sum of torque is multiply by angular velocity then power will calculated by equation below.

$$\mathrm{P} = \omega\, M_{tot} \tag{4}$$

3. Result and Discussion

The experiment result is about angular velocity and torque. This data is processed into tip speed ratio and coefficient of performance. Tip speed ratio is ratio on the tip of baldes In the free air flow. For the certain air velocity, TSR will influence the rotor speed. Lift typical of wind turbine has more higher TSR than drag typical of wind turbine.

The result of the processed data is shown in figure 5. The graph show experimental data of characteristic wind turbine in Cp to TSR. In the variation of 0 degree of pitch angle, the performance of wind turbine give the highest value of Cp among other variation. But the TSR that can make the turbine rotate is high. The TSR factor is the speed of air flow that affect the rotation of turbine and the torque. The peak of Cp not yet given in this variation. This condition happened because of this variation must be tested in the high speed of airflow. But in the variation of 10 degree of pitch angle. The turbine can rotate in the minimum air flow velocity.

Compared by BEM (Blade Element Momentum) that is calculation method. This method is calculated based on the TSR that produced by the experimental data calculation process. In this BEM, the graph that shown in figure 5, 0 degree variation give best performance among all variation on Cp. But this variation only works on high TSR condition. This phenomenon is occurred same as experimental process.

In the previous research [9] that compared between turbine without winglets, with fixed winglets, and with splitted winglets. The splitted winglets didn't get well performance between them. Turbine without winglets give best performance among them. As shown at the figure 6.

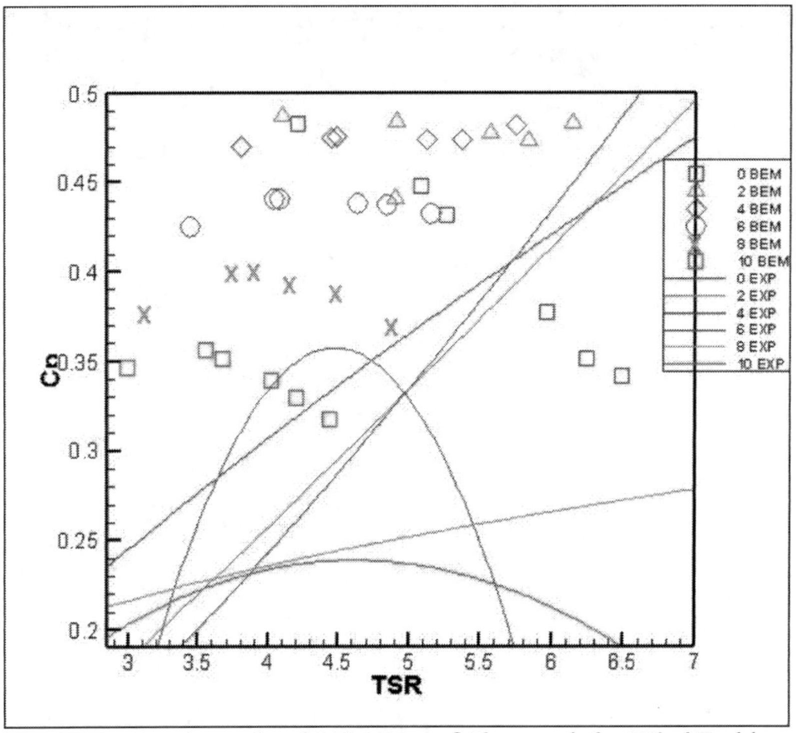

Figure 5. Experimental and BEM Data of Characteristics Wind Turbine Cp-TSR

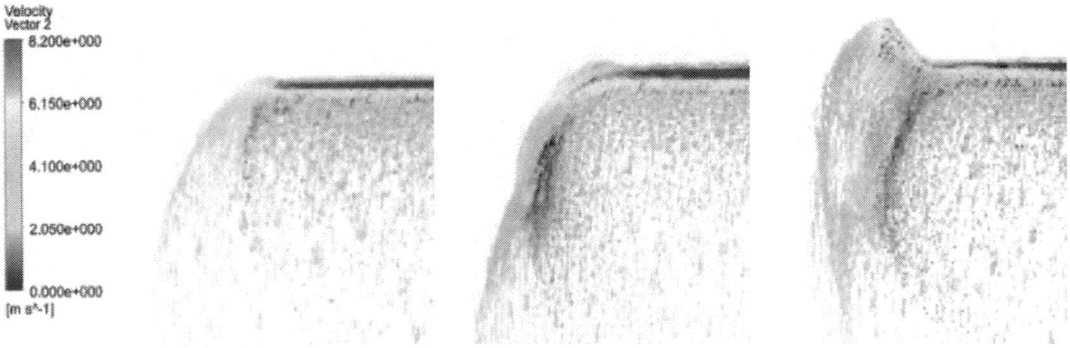

(a) (b) (c)

Figure 6. Visualisasi aliran kecepatan udara 5 m/s pada *blade* turbin angin (a) tanpa winglets (b) dengan *winglets* (c) dengan *splitted winglets*

In the blade a the vectors show minimum blue vectors than the vectors that happened in the blade b and c. This blue vectors show backflow air in the tip of blades. The backflow air is the air that move back and hit another air flow. The air flow that hit will decrease the velocity of its flow. So, in the tip of blades makes the lift force decrease and increase the drag force. This condition makes the torque lower. In the wind turbine smaller drag that happen small engine that required [8].

4. Conclusion

This research is successfully by doing experiment about horizontal axis wind turbine using wind tunnel testing. The additional of splitted winglets do not give better performance than blade without winglets.

The backflow on the tip of blade the torque decrease. But in some condition there are several conclusion that we get:

1. Performance of wind turbine with splitted winglets will increase when it used in high TSR with 0 degree of pitch angle.
2. Pitch angle 0 degree can't rotate the turbine with splitted winglets in minimum TSR.
3. 10 degree of pitch angle variation can be starting torque and changes of pitch angle also used for braking function.
4. Phenomenon that shown in CFD simulation show that the back flow in the tip of blade with splitted winglets that can cause the performance of turbine decrease.

5. Acknowledgement
Thank you to fluid laboratory-EEPIS for helping this research by providing data that needed in this research. Authors wish the data can be used for the next research.

6. References

[1] Sohail R, Helmut S, Abas A and George S 2015 Multi-Element Winglets: Multiobjective Optimization of Aerodynamic Shapes *Journal of Aircraft* **53** p 992-1000
[2] Tony B, David S, Nick J and Ervin B 2001 *Wind Energy Handbook* (John Willey & Sons Ltd., UK)
[3] Louis B 2012 *Split Blended Winglet* (Patent Application Publication, Seattle, WA, US)
[4] Rinaldo G, Diego M and Juancarlo R 2011 Winglets Design and Analysis for Wind Turbine *Florida International University Thesis*
[5] Soren G 2009 *Wind Turbines* (University College of Aarhus)
[6] Akhmad K 2016 *Experimental Study of Effects of blade pitch angle Against Horizontal Axis Wind Turbine Performance* (EEPIS Release)
[7] Benjamin B, Morten J, Matt K, Stephen B and Jakob S 2011 Wind Turbine Pitch Optimization *IEEE International Conference on Control Applications*
[8] Martin O L H 2008 *Aerodynamics of Wind Turbines* (Earthscan UK)
[9] Rizal I S and Setyo N 2017 Vortex Formation on Horizontal Axis Wind Turbine with Splitted winglets *International Conference on Computation in Science and Engineering, Bandung Institute of Technology, Indonesia*

Synthesis of biodiesel from palm oil with dimethyl carbonate and methanol as reagent variation using KOH and enzyme catalyst

Setiadi[1], Hamid and Fairuz Nawfal
[1]Bioprosess Engineering Study Program, Department of Chemical Engineering, Faculty of Engineering, Universitas Indonesia, Kampus Baru UI, Depok, Indonesia

E-mail: setiadi@che.ui.ac.id

Abstract. Biodiesel is a methyl ester transesterification process from the result of triglyceride with a methyl reagent source using a catalyst. One of novel reagent used is Dimethyl Carbonate (DMC) which can eliminate the formation of glycerol as the byproduct through the reaction resulting a compound of glycerol dicarbonate. The advantage of this compound has the same physical properties and to be a mixture in one phase liquid product with methyl ester. Lipase is a hydrolase enzyme that can catalyze the formation of fatty acid molecule from the triglyceride containing oil in the reaction of transesterification. The purpose of this research is to produce biodiesel product by observing the formation of one phase mixture of biodiesel product from the palm oilusing methanol and DMC as the reagent, KOH and lipase catalyst andthe variation of molar ratio of the oil. To obtain the best biodiesel quality examined by four characteristics: methyl ester (FAME) content, density, viscosity, and cloud point. The result of this research is that biodiesel can be produced by both reagent, and for biodiesel by DMC there was only one-phase biodiesel formed without impurities shown visually. For catalyst variation, the best biodiesel produced by methanol is biodiesel with KOH catalyst and has 98.2% ester content, 0.858 g/ml density, and 4.58 cSt viscosity. The best biodiesel produced by dimethyl carbonate is biodiesel with KOH catalyst and has 89.0% ester content, 0.883 g/ml density, and 4.91 cSt viscosity.

1. Introduction

Biodiesel is one of the most potential renewable fuel and has already been developed in this decade. It can be produced from Jatropha or palm oil that can be found easily in Indonesia because they were spread in many areas in this country. In a few years, biodiesel will be primary choice of fuel in Indonesia. To maintain the eco-friendly factor of biodiesel, in some research there were found a new type of catalyst used in the synthesis process called Lipase enzyme. Lipase could bring the reaction specifically towards the product without any disadvantageous side reaction. Lipase is a heterogen catalyst with the result that it required an easy separation treatment after being used**[1]**.

In the process of biodiesel synthesis, there are some drawbacks that need to be considered, for example the formation of by-product, glycerol. Glycerol is a side product of biodiesel production that formed by tryglyceride and alcohol reaction, while methyl ester is the main product. At the end of biodiesel production process, glycerol must be separated because it is an impurity that will decrease the biodiesel quality **[2]**. For this reason, researchers start to develop novel reagent called Dimethyl Carbonate (DMC) as alkyl reagent. This compound is known as *green reagent* and useful to eliminate glycerol as impurity in the biodiesel product**[2]**. And in this research, a new alternative of biodiesel production will be proposed. The material used will be palm oil, with novel reagent dimethyl carbonate as alcohol substitute and lipase enzyme as catalyst alternative in transesterification reaction.

2. Material and Methods

2.1. Palm Oil

Among the vegetable oils, palm oil is the most used oil especially in Indonesia. Palm oil contained triglyceride, a carboxylic acid with number of carbon from 6 to 30. Triglyceride is the largest component in palm oil, besides monoglyceride and diglyceride. Other advantage of palm oil is the content of saturated fatty acid, which is higher than other plant oil. Palmitic acid, one of the saturated fatty acid, has 47% content in the palm oil. And if the saturated fatty acid number is high, it will affect on cetane number [1].

Table 1. Fatty Acid Composition Comparison

Fatty Acid	Soybean Oil	Sunflower Oil	Corn Oil	Canola Oil	Coconut Oil	Palm Oil
C8					7	
C10					7	
C12					47	
C14					18	1
C16	11	6	11	4	9	45
C18	4	5	2	2	3	4
C20						
C22		1				
C16:1						
C18:1	23	29	28	61	6	39
C18:2	54	58	58	22	2	11

2.2. Dimethyl Carbonate

Dimethyl carbonate (DMC) is an organic compound, and categorized as carbonate esther $OC(OCH_3)_2$.DMC has liquid phase and colourless. DMC can be used as methylation agent, and can be used as solvent in biofuel production.In a normal biodiesel production with alcohol, there will be a glycerol as side product formed. Glycerol is an impurity for biodiesel and need to be separated with several processes[3]. This can be eliminated by DMC usage as solvent. Despite of producing glycerol, DMC reaction with triglyceride will form a compound called glycerol decarbonate. Glycerol decarbonate is a side product that has the same physical appearance as methyl ester, and biodiesel will visually seen as only one phase product[4].

2.3. Lipase Enzyme

Lipase enzyme (Triacylglycerol acylhydrolase) is one of the novel catalyst used in transesterification process [5]. Lipase is part of hydrolase enzyme that works in carboxylic esther bond. Lipase can be found inside animal, fungi, or in bactery. The general function of lipase is to hydrolize triglyceride and form diglyceride, monoglyceride, glycerol, and fatty acid. Because of this reason, lipase can be used in transesterification process[6].

2.4. Biodiesel Synthesis

2.4.1. Biodiesel with Methanol. For biodiesel synthesis with methanol as alkyl source, two types of catalyst were used as variation (KOH and Lipase). Mole ratio methanol/oil given is 1:6 and 1:12. For the synthesis with KOH catalyst, the first procedure was to heat up the oil until 80°C to vaporize water content, and then cool it down until 40°C. Then, the oil and methanol (which already dissolve the KOH about 0.8% oil mass) were mixed up in a beaker glass with magnetic stirrer for about 180 minutes at 60°C. After the product was formed, the methyl ester phase then separated from the glycerol by decantation and then washed by de-ionized water. Biodiesel was then heated up until 90°C to vaporize the water content inside it. For biodiesel reaction with lipase enzyme, the similar procedure were performed, but the difference step was only the lipase mass (5% oil mass) and the reaction time is about 300 minutes[1].

2.4.2. Biodiesel with DMC. For biodiesel with DMC as alkyl source, two types of catalyst will be used as variation (KOH and Lipase). Molar ratio between oil and DMC given is 1:6 and 1:12. The first procedure is to heat up the oil until 80°C to vaporize water content, and then cool it down until 40°C. Then, the oil and DMC (which already dissolve the KOH about 0.8% oil mass) were mixed up in a beaker glass with magnetic stirrer for about 180 minutes at 90°C.Biodiesel then heated up until 150°C to vaporize the water content and glycerol dicarbonate inside it.For biodiesel reaction with lipase, almost all the procedures are the same, the difference only the lipase mass (5% oil mass) and the reaction time is about 300 minutes at 70°C[3].

2.5. Methyl Ester Content

Biodiesel sample was analyzed using GC-FAME (*Gas Chromatography – Fatty Acid Methyl Ester*) CLARUS 680 and FTIR. This instrument analysis was used to determine the content of methyl ester in the product, and to analyze the fatty acid compounds in the biodiesel product. Like all other chromatographic techniques, a mobile and a stationary phase are required for this technique. The mobile phase is comprised of an inert gas. The stationary phase consists of a packed column in which the packing or solid support itself acts as stationary phase, or is coated with the liquid stationary phase.

2.6. Density Measurement

Density was measured using *automatic densitymeter*. Digital density meters based on the oscillating U-tube technique: The sample to be measured is filled into a U-shaped tube which is induced to vibrate. The eigenfrequency of the oscillation of the U-tube is influenced by the mass and therefore by the density of the sample. Biodiesel sample used in this test was about 20 ml. The densitymeter needs to be prepared until the operation temperature stabilized at 40°C. Sample of biodiesel was injected to the instrument apparatus for about three injections and the density can be obtained directly from the screen monitor.

2.7. *Viscosity Measurement*

The Viscosity of biodiesel was measured using Otswald Viscometer. Biodiesel was aspirated with a volumetric pipe until the first sign, and then the time for biodiesel to reach the second sign was measured using stopwatch[7]. Then, the viscosity can be determined using equation 2.1.

$$\mu = K \times t \qquad (2.1)$$

where:

μ = Kinematic Viscosity (centi stokes or cSt)
K = Oswald Viscometer Constant
t = Fluid Flow Time inside the Viscometer (s)

3. Result and discussion

3.1. Biodiesel with Methanol

After several steps of biodiesel synthesis with KOH catalyst, the product could be found in the form of two-phase liquid. The first phase layer visually appeared in lighter colour and expected to be methyl ester (biodiesel) and the second phase layer (below the first phase layer) seems to be product glycerol with more viscous. The existence of glycerol formation, it showed that the biodiesel synthesis through the transesterification reaction was proceeded smoothly according to its reaction theory.

In this reaction, methanol as the alkyl source along with KOH should break the binding chain between C-H and the esters, and then the C-H bond will connected to esters group from triglyceride compound. At the beginning, triglyceride will be converted into diglyceride and then methanol will start to attack the bond, and diglyceride will converted to monoglyceride. At last, methyl ester compound will be formed and hydroxy group from methanol will react with monoglyceride to form a glycerol [8].

A difference reaction mechanism is happening when lipase catalyst is used. Enzyme works with binding the carbon substance in triglyceride. This carbon substance then was binded by oxygen that exists on the active side of the enzyme. Later on, an Enzyme Substrate Complex (ES) will be formed.To form a methyl ester compound, methanol as methyl source will form a methoxy group by releasing hydrogen substance to active side of the enzyme and make a new group (N-H). The metoxy (CH_3O) will react with the Enzyme Substrate Complex and form Enzyme Alcohol Complex (E-AcA). At last, N-H group on enzyme active side will bind with oxygen and CH_2 compound in the E-AcA, and those bond will be released from the E-AcA so that enzyme structure will be shown the same asin the first form, and E-AcA will be a new compound called Fatty Acid Methyl Ester (FAME)[8].

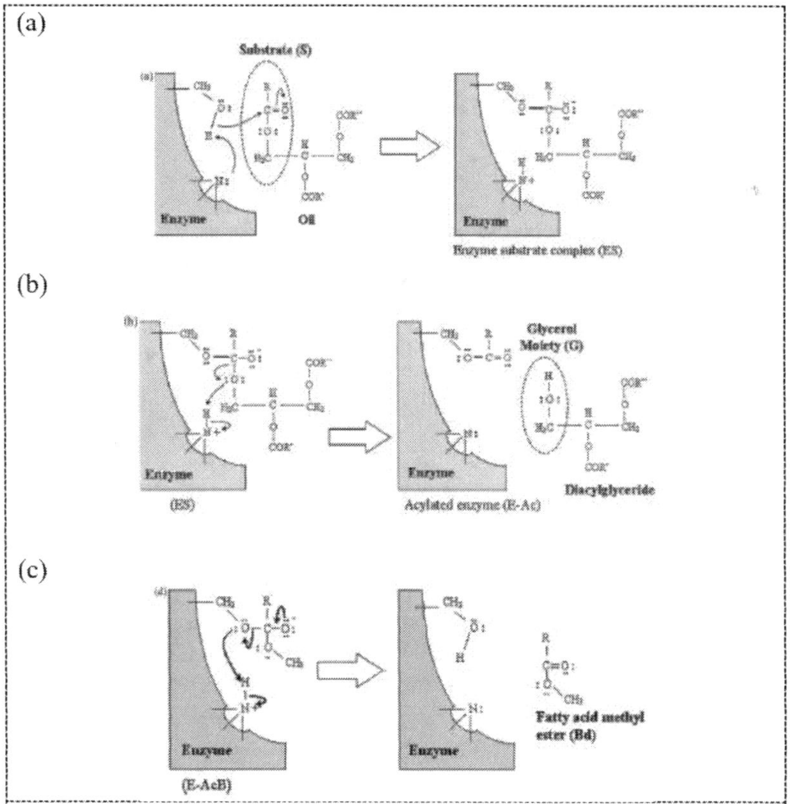

Figure 1. (a) Triglyceride Reaction with Enzyme; (b) Diglyceride and Monoglyceride Formation Mechanism; (c) FAME Formation by Enzyme from E-AcA

IOP Conf. Series: Earth and Environmental Science **105** (2017) 012103 doi:10.1088/1755-1315/105/1/012103

3.2. Biodiesel with Dimethyl Carbonate

In the experiment of biodiesel synthesis with DMC, it was visually found to be one phase product. There was no separated layer, unlike the product resulting from the trans-esterification using methanol. Almost all reaction mechanisms are the same, but this can be happened because of the C-H group from triglyceride will bind with ester group from DMC and will form glycerol di-carbonate compound [9]. Because both FAME and glycerol di-carbonate are having ester group inside their compound, then physically FAME and glycerol di-carbonate will not show any difference visually, and seems to be one-phase mixture. Glycerol is not formed in this reaction, on the other hand glycerol di-carbonate is found in the product. This can be proven by FTIR analyzer (Fig.2).

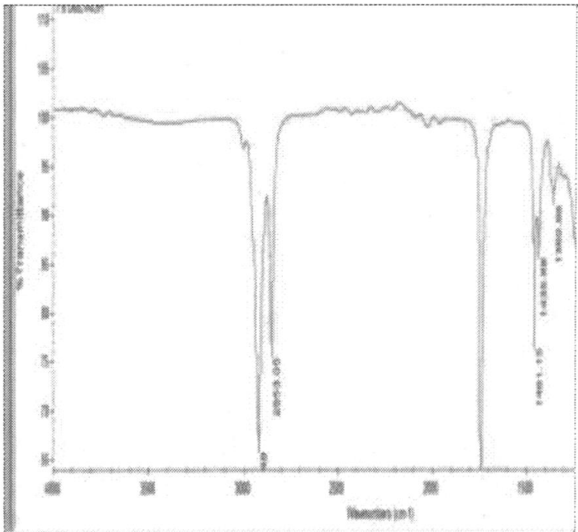

Figure 2. The Spectrum FTIR Analyzer Result shown a Compound with wavenumber 1790.1 cm^{-1}

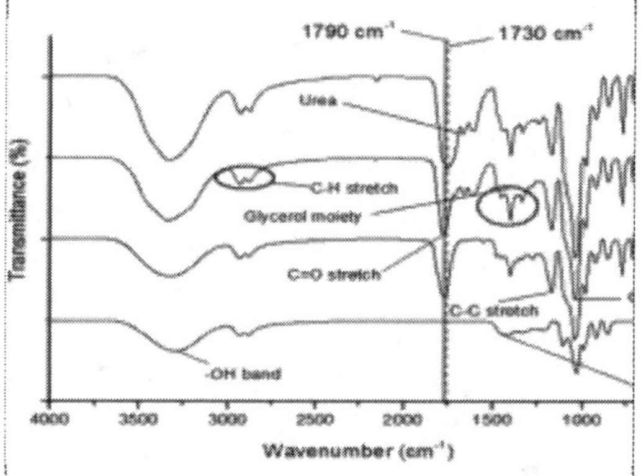

Figure 3. The Standard of FTIR Result of for of Glycerol Dicarbonate [10]

From this FTIR spectra, it can be concluded that a new compound is formed by reaction between triglyceride and DMC. The wavenumber was detected in 1790.15 cm^{-1} and based on former research, it was glycerol dicarbonate.

3.3. Biodiesel Characteristics

Based on these results on the data of density and viscosity characteristics, most of the biodiesel sample meet the requirement of standard biodiesel in Indonesia (SNI). But for methyl ester content (table 2), only enzyme biodiesel with molar ratio of 1:12 that seems to has a methyl ester compound. It happened for several reasons. Molar ratio 1:12 is the biggest molar ratio to be given in this research, so that the the number of oil and alkyl source are also in high value. Because of this reason, the reaction will occur better than the lower molar ratio. On the other hand, the number of enzyme given is getting higher while the reactant number is increased. Enzyme and substrate concentration are two important factors, that make the reaction become optimum. So in sample 1:12, the reaction is better than the other enzyme biodiesel sample.

Table 2. Characterization Result of Biodiesel Samples

Biodiesel Sample		Methyl Ester Content	Density	Viscosity
Methanol + KOH	1:6	97.5%	0.858	4.47
	1:12	98.2%	0.859	4.58
Methanol + Enzim	1:6	0.0%	0.887	5.75
	1:12	58.0%	0.858	4.48
DMC + KOH	1:6	89.7%	0.883	4.91
	1:12	81.8%	0.852	5.25
DMC + Enzim	1:6	0.0%	0.899	6.28

4. Conclusion

From this research, it can be concluded that transesterification reaction with methanol reagent could produce the compound of methyl ester and glycerol as side product. While the transesterification using dimethyl carbonate and KOH catalyst could resulted in glycerol dicarbonate compound as side product. The advantage of this compound is able to mix and dissolve each other with methyl esters product and so that the product would look in one-phase only. The characterization analysis shows that the transesterification with methanol, KOH catalyst and molar ratio 1:12 is the best biodiesel product achieved 98.2% methyl esther content, 0.858 g/ml density, and 4.58 cSt viscosity. For DMC biodiesel, the one with KOH catalyst and molar ratio 1:6 is the best biodiesel sample with 89.0% methyl esther content, 0.883 g/ml density, and 4.91 cSt viscosity.

5. Reference

[1] Lee K T and Ofori-Boateng C 2004 Sustainability of Biofuel Production from Palm Oil *Springer*

[2] Ilham Z and Saka S 2009 Dimethyl carbonate as potential reactant in non-catalytic biodiesel *Bioresource Technology* pp 1793-1796

[3] Fabbri D, Bevoni V, Notari M and Rivetti F 2006 Properties of Potential Biofuel Obtained from Soybean Oil by Transmethylation with Dimethyl Carbonate *J. Fuel Technol.* **86**(5-6)

[4] Kurle Y, Islam M and Benson T 2013 Process development and simulation of glycerol-free biofuel from canola oil and dimethyl carbonate *J. Fuel Processing Technology* pp 49-57

[5] Min J Y and Lee E Y 2011 Lipase-catalyzed simultaneous biosynthesis of biodiesel and glycerol carbonate from corn oil in dimethyl carbonate *Biotechnology Letter* **33** pp 1789-1796

[6] Al-Zuhair S, Ling W F and Jun S L 2007 Proposed kinetic mechanism of the production of biodiesel from pal oil using lipase *J. Process Biochemistry* **42**:951-960

[7] Knothe G and Steidley K 2006 Kinematic Viscosity of Biodiesel Fuel Components and Related Compounds *Fuel* **84**(9):1059-1065

[8] Caballero V, Bautista F, Campelo J, Luna D, Marinas J, Romero A, Hidalgo J, Luque R, Macario A and Giordano G 2009 Sustainable preparation of a novel glycerol-free biofuel by using pig pancreatic lipase: Partial 1,3-regiospecific alcoholysis of sunflower oil *Process Biochemistry* **44**(3):334-342

[9] Zhang L, Sheng B and Xin Z 2010 Kinetics of transesterification of palm oil and Dimethyl Carbonate for Biodiesel Production at The Catalysis of Heterogeneous Base Catalyst *J. Biortechnol.* **101**(21):8144-50

[10] Indran V P, Zuhaimi N S and Deraman M A 2014 An Accelerated Route of Glycerol Carbonate Formation from Glycerol using Waste Boiler Ash as Catalyst *RSC Publishing* **4**

IOP Conf. Series: Earth and Environmental Science **105** (2017) 012104 doi:10.1088/1755-1315/105/1/012104

Effect of medium and incubation time on production of AA, DHA and EPA from *Aspergillus oryzae* by solid state fermentation

N P Pinasthika, R Arbianti[1], T S Utami and H Hermansyah

[1]Bioprocess Engineering Laboratory, Department of Chemical Engineering, Faculty of Engineering, Universitas Indonesia, Depok, Indonesia 16424

Email: nadirapp@hotmail.com

Abstract.. Lately, artificial surfactant is more often used than natural surfactant for petroleum industry. The main advantages of using natural surfactant is because the continuous availability. Single Cell Oil, which utillizes one-cell microorganism, can be a solution, such as Aspergillus oryzae, to produce AA, DHA & EPA. A. oryzae was cultivated on Potato Dextrose Agar (PDA), Czapek Dox Agar (CDA) and Malt Extract Agar (MEA), and then the incubation times are 2,4,5,6 and 7 days in optimal medium. Lipids were extracted using ethanol and n-hexane. The characterization of lipid was done by gas chromatography (GC) method. The most optimal medium is CDA with a lipid concetration of 0,18 g. The best incubation time on CDA medium was 5 days with 0,216 g lipid content containing 0,123 g unsaturated fatty acid, 0,0613 g PUFA and 0,0615 g MUFA. The unsaturated fatty acid composition produced on the 5th day was 29.2% oleate; 29.1% linoleate and ± 0.046% EPA.

1. Introduction

Surfactants are widely used and have an enornmous number of their ability to influence the properties of surface and interface. The ability of surfactant can be used in petroleum industry at all stages [1]. Lately, artificial surfactant is more often used than natural surfactant. The main advantages of using natural surfactant is because the continuous availability. Nowadays, there are few experiments using microorganism that considered as efficient producer of surface active agent [2].

Fungi have the ability to grow in low pH, easy to handle, able to degrade complex carbon source and can grow in waste rapidly [3]. A fungus that is used in this research is *Aspergillus oryzae. Aspergillus oryzae* has about 18-57% of lipid content [4].

Solid state fermentation (SSF) method that uses solid medium gives similar condition as fungi habitat, where fungi is the main group that is usually cultured by SSF method[5]. SSF method has lower bacterial contamination level because of the absent of free-water and low pH level in the medium [5]. Solid mediums that are used in this research are Potato Dextrose Agar (PDA), Czapek Dox Agar (CDA) and Malt Extract Agar (MEA) because those medium are the most common medium to grow fungi. But, within those three mediums, it has never been proved which mediums is the best to produce AA, DHA and EPA.

Single cell oil technology has been used long time ago. According to Nisha et al. (2011), fungi *Mortierella alpina* was used to produce AA until 40% (w/w) [6]. *Mucor circinelloides* was also used for producing various type of fatty acid up to 50% (w/w) [7]. Fakas et al. (2009) used *Cunninghamelata enchinulata* and produced 40-60% of gamma-linolenic acid [8].

The other crucial factors in culturing *Aspergillus oryzae* is incubation time. Incubation time is time needed to control the growth of microorganism. Incubation time between fungi and other microorganism is different. The differences of incubation time between *Aspergillus oryzae* and other fungi raise a doubt, so this research will vary the incubation time to know the effect of incubation time on concentration profile of AA, DHA and EPA that produce by *Aspergillus oryzae*.

2. Method

2.1. Main Culture in Tilted Agar

Fungi *Aspergillus oryzae* that obtained from Indonesia Culture Collection (InaCC) grew on PDA and kept in ±4°C. Then, *A. oryzae* was regrowed on PDA by tilted agar method in test tubes. *A. oryzae* was incubated for three days in ±25°C.

2.2. Modification of the Medium

Three mediums that were used in this research is PDA (4 g potatoes infusion; 20 g dextrose; 15 g agar), CDA (30 g sucrose; 2 g $NaNO_3$; 1 g K_2HPO_4; 0.5 g $MgSO_4$; 0.5 g KCl; 0.01 g $FeSO_4$; 15 g agar) and MEA (30 g malt extract; 5 g mycological peptone; 15 g agar) manufactured by *HiMedia Laboratories Pvt. Ltd.* All mediums that were used during the experiment were modified. The ratio of C:N was modified into 30 : 1 [9]. Each medium was modified with certain carbon or nitrogen. The ratio of C:N of PDA was modified by adding $(NH_4)_2SO_4$, CDA was modified by adding sucrose and MEA was modified by adding glucose. All mediums was also supplemented with several compounds such as 7 g/L KH_2PO_4 and 2 g/L Na_2HPO_4 as buffer; 0.5 g/L $CaCl_2.2H_2O$; 0.1 g/L $FeSO_4.7H_2O$; 0.1 g/L $ZnSO_4.7H_2O$; 0.001 g/L $MnSO_4.2H_2O$; 0.005 g/L $CuSO_4.5H_2O$ as micronutrients [10].

2.3. Growth Curve on PDA

Cultures for growth curve were made on 25 ml of modified PDA each dish. The C:N ratio of PDA was modified into 30 : 1 and several micronutrients was added into PDA. Cultures for growth curve were made on ten petri dishes. Each petri dish was harvested every day for ten days and the dry cell weight (DCW) was measured. Cultures for growth curve were incubated in ±25°C.

2.4. Culture for Effect of Mediums

A. oryzae was cultured on three different mediums, which is PDA, CDA and MEA for five days according to the growth curve on PDA. Each petri dish was consists of 25 ml of each medium that had been modified. Cultures were incubated in ±25°C. These cultures were made in duplo. After five days of incubation, cultures were harvested and the lipid was extracted. The best medium was obtained based on the highest lipid productivity and would be used in the variation of incubation time.

2.5. Growth Curve on Highest-Lipid-Productivity-Medium

Based on the result of effect of medium, CDA was the medium with the highest lipid productivity. It needed a growth curve to determine the range of incubation time variation. Therefore ten cultures of 25 ml of modified CDA were made for a ten-days-growth-curve. Each petri dish was harvested every day for ten days. The cultures were incubated in ±25°C.

2.6. Culture for Effect of Incubation Time

Based on the growth curve on CDA, the range of incubation time that were going to use were 2nd (log phase), 4th (initial stationary phase), 5th & 6th (stationary phase) also 7th day (initial death phase). *A. oryzae* was cultured on 25 ml of CDA. These cultures were made in duplo. Two cultures were harvested and the lipid was extracted according to the range of incubation time.

2.7. Harvesting

Harvesting was done by autoclaving the culture in ±121°C for 15 minutes. After the medium melted down, the biomass could be taken and dried on filter paper for 24 hour in 60°C.

2.8. Lipid Extraction

Lipid extraction started by grinding dried biomass into powder. Then, the biomass powder was solved by ethanol 1:15 (w/v) and sonicated for three hour in 45°C according to Niawati et al. (2016). After it sonicated, the sampel was vortexed for five minutes then sentrifuged at 2000 rpm for 15 minutes [11,12] . Supernatant was dried over hot plate in <100°C until the odor of solvent was gone [12]. The fungus solid then solved by n-hexane with same ratio and sonicated for two hour in 60°C [13]. Next procedures were the same as the ethanol extraction.

2.9. GC Analysis

The composition of fatty acids, AA, DHA and EPA in lipids in incubation time variation were analyzed using Gas Chromatography (GC) SHIMADZU 2010 with the injector temperature was 250°C and the oven temperature was 400°C. The mobile phases for GC were UHP hydrogen gas, UHP helium and compressed air.

3. Results & Discussion

3.1. Growth Curve on PDA and CDA

The growth curves on PDA and CDA are shown on figure 1.

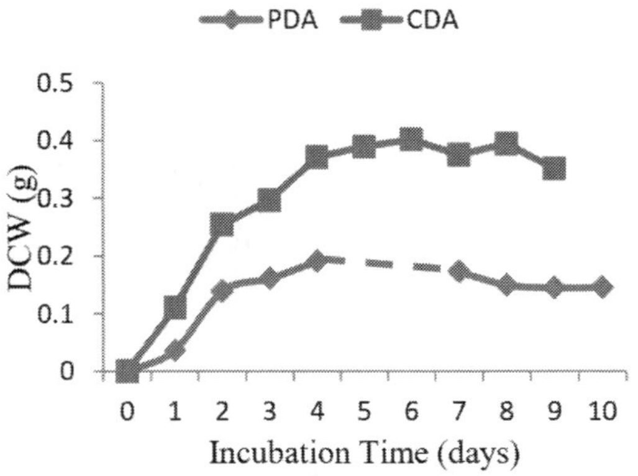

Figure 1 Growth Curve on PDA and CDA

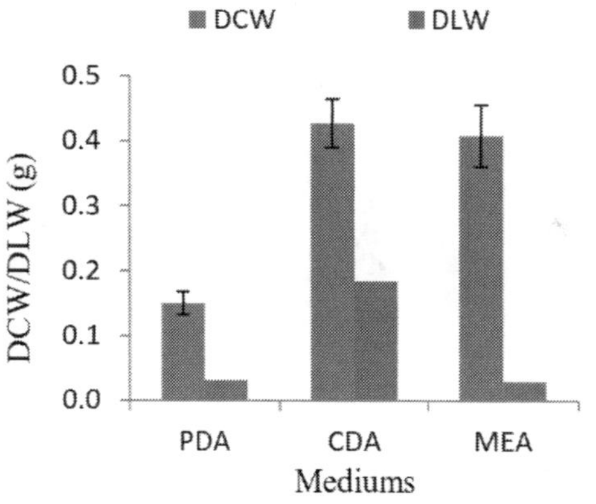

Figure 2 Effect of Medium on Dry Cell Weight and Dry Lipid Weight

According to the growth curve on PDA, lag phase and acceleration phase was started on day 0 until day 1. Exponential phase happened on day 1 to day 2. On day 2 until day 4, fungi were in decelaration phase. On day 4 until day 6 were considered as stationary phase. Accelerated death phase happened on day 6 until day 10. The optimal harvesting time based on this curve was day 5.

On the experiment of effect of medium, the medium with the highest lipid productivity is modified CDA. To determine the range of incubation time, growth curve on CDA was made. The incubation times that were chosen were in exponential phase, initial stationary phase, two dots in stationary phases and decelaration phase. According to the growth curve on CDA, those phases were on day 2, day 4, day 5, day 6 and day 7.

3.2. Effect of Medium on Dry Cell Weight and Dry Lipid Weight

The comparison of DCW and dry lipid weight (DLW) of each medium are shown on figure 2. DCW that were obtained from all mediums were ranged between 0,15-0,43 g per 25 ml medium with error value <5%. The DCW of MEA was also similar to the DCW of CDA. Based on figure 2, CDA produced the highest DLW.

According to figure 2, the highest DCW obtained at CDA because Aspergillus sp. can grow on all types of carbon source, and growth as well as the maximum biological activity occurs at the time of sucrose as sole carbon source [14]. The DCW on MEA is also similar to the DCW on CDA because MEA has been modified to get the ratio of C:N 30:1 by adding glucose into it. Pepton is organic nitrogen source, which source oforganic nitrogen provides vitamins, micronutrients and intermediate compounds needed by mould and can act as a stimulator and the essential precursor on the growth and ochratoxin production [15].

Medium CDA can produce highest DLW on figure 2 because the medium consists of simple sugar and pure synthetics. Mold will start forming lipids from the rest of the carbon on the medium after the nitrogen discharged. The remaining carbon in the medium that is available in CDA is simply sucrose. Meanwhile, in medium PDA and MEA, the amount of DLW produced by *A. oryzae* is much lower because the remaining carbon source in both medium is complex carbon compounds, causing fungus takes a longer time to break down the component that is used for the synthesis of cell and energy [16].

3.3. Effect of Incubation Time on Dry Cell Weight and Dry Lipid Weight

The comparison between DCW and DLW on several incubation times were shown below. According to figure 3, DCW on several incubation times ranged between 0,25-0,425 g/25 ml medium with error value <5%.

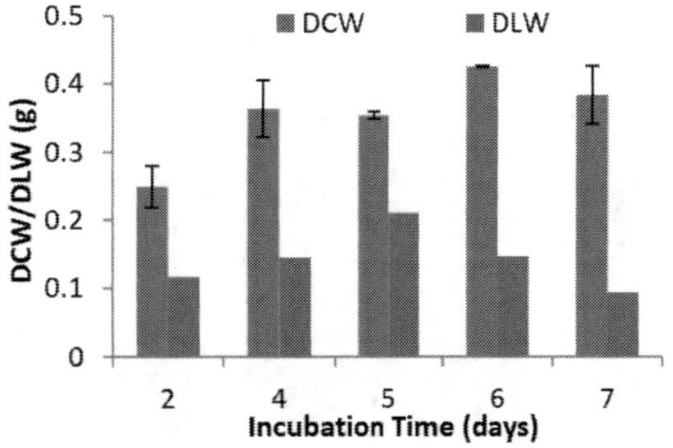

Figure 3 The Effect of Incubation Time on DCW and DLW

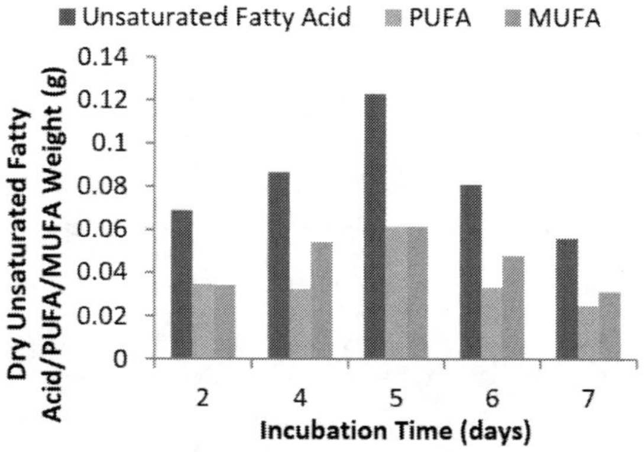

Figure 4 The Effect of Incubation Time on Dry Unsaturated Fatty Acid, PUFA and MUFA Weight

The highest DCW was obtained on day 6. The trend in DCW graph was inclining until day 6, and started to drop on day 7. The highest DLW was obtained on day 5. The trend in DLW graph was inclining until day 5. Dropping was found pretty drastic on day 6 and day 7.

Trends in figure 3 continue ascending until the day 6. The increase in DCW occurs because mold consuming carbon source and nitrogen source on the medium actively. On the 7th day, the DCW declined because of the lack of additional nutrients flow in the medium at that time period [17].

Trends in figure 3 from day 2 to 5 continue uphill because in the early phase of growth, there is sufficient nitrogen content in medium, therefore cells are more likely to reproduce it self rather than accumulate lipid. However, after certain cycles, microbes will start to confront the stress conditions that will trigger the accumulation of lipids [18]. The remaining carbon in the medium after a shortage of nitrogen will be assimilated by the cells and converted directly into lipids [19]. In the final phase of the growth cycle, the quantity of lipids will decrease because the mold cells will begin to use the stored lipids to meet energy needs for survival [18]. The results of this research are supported by statement of Dyal et al. (2005) that the amount of lipid is not always directly proportional to the DCW [20].

3.3.1. Analysis of Fatty Acid Composition

Analysis of lipid extracted from *A. oryzae* was done using GC. There are various fatty acid composition among several incubation times. The fatty acid compositions on each incubation time contain laurate (C12), myristate (C14), palmitate (C16-0), stearate (C18-0) as saturated fatty acid; 29,2-37,2% of oleic (C18-1) and 21,8-29,1% of linoleic (C18-2). Day 2 and day 4 also have 0,53% and 0,57% of linolenic (C18-3) respectively. AA, DHA and EPA appear to be not detected (n.d.).

Based on figure 4, unsaturated fatty acid that was produced ranged between 0,056-0,122 g. The highest unsaturated fatty acid weight was on day 5. The trend in unsaturated fatty acid graph is inclining until day 5 and starts to drop on day 6. The highest unsaturated fatty acid and PUFA, as our main target, were obtained on day 5. Unsaturated fatty acid concentrations, MUFA and PUFA in figure 4 continue ascending until day 5. Decreasing happens in the 6th day because of the lack of oxygen supply in medium. Fatty acid desaturation process is oxidative reaction that requires oxygen molecular [21].

4. CONCLUSION

Based on the results of the research, it can be concluded that the greatest concentration of lipids between PDA, CDA and MEA are 0.18 g obtained in medium CDA, therefore CDA is the best medium. The largest concentration of lipids between variations 2 days, 4 days, 5 days, 6 days and 7 days in medium CDA is 0.215 g on day 5. The best time variation is day 5 with unsaturated fatty acid content of 0,123 g, PUFA content of 0,0613 g and MUFA content of 0,0615 g. Unsaturated fatty acid composition are produced by Aspergillus oryzae on day 5: 29.2% oleic; 29.1% linoleic and ± 0.0467% EPA.

5. REFERENCES

[1] Schramm L L 2000 *Surfactant: Fundamentals and Application in the Petroleum Industry* (United Kingdom: Cambridge University)

[2] Holmberg K 2001 Natural Surfactant *Current Opinion in Colloid & Interface Science* **6** p 148-159

[3] Sumanti D M, Tjahjadi C, Herudiyanto M, Sukarti T 2005 Mempelajari Mekanisme Produksi Minyak Sel Tunggal dengan Sistem Fermentasi Padat Pada Media Onggok-Ampas Tahu dengan Menggunakan Kapang Aspergillus terreus *Jurnal Teknol. Industri Pangan* **16** no 1

[4] Mienda B S, Idi A, Umar A 2011 Microbiological Features of Solid State Fermentation and its Applications An overview *Research in Biotechnology* **2**(6) p 21–26.

[5] Muslimah, Utami T S, Arbianti, R 2015 Pengaruh Rasio Karbon/Nitrogen Berbasis Onggok dan Ampas Tahu untuk Produksi AA, DHA, EPA dari Aspergillus oryzae *Prosiding Seminar Nasional Teknik Kimia UNPAR* vol 1 (Bandung, November 19th 2015)

[6] Nisha A, Venkateswaran G 2011 Effect of Culture Variables on Mycellia Arachidonic Acid Production by Mortierella alpina *Food Bioprocess Technology* **4**(2) p 232-240

[7] Tang X, Zhang H, Chen H, Chen Y Q, Chen W, Song Y 2014 Effect of 20 Standard Amino Acids on the Growth, Total Fatty Acid Production and γ -Linolenic Acid Yield in Mucor circinelloides *Curr Microbiol* **69** p 899-908

[8] Fakas S, Papanikolaou S, Batsos A, Galiotou-Panayotou M, Mallouchos A, Aggelis G 2009 Evaluating renewable carbon sources as substrates for single cell oil production by Cunninghamella echinulata and Mortierella isabellina *Biomass and Bioenergy* **33**(4) p 573-580

[9] Bayizit, A A 2014. Fungal Lipids : The Biochemistry of Lipid Accumulation *J. Chem. Eng. Appl.* **5**(5) p 409–414.

[10] Weete J D, Shewmaker F, Gandhi S R 1998 γ -Linolenic Acid in Zygomycetous Fungi: Syzygites megalocarpus *Journal of American Oil Chemists Society* **75**(10) p 1367–1372.

[11] Niawati S, Utami T S, Arbianti, R, & Hermansyah H 2016 Pengaruh Suhu dan pH pada Medium Onggok dan Ampas Tahu dalam Produksi Asam Lemak Tak Jenuh dengan Menggunakan Aspergillus oryzae *Seminar Nasional Teknik Kimia Teknologi Oleo Petrokimia Indonesia 2016* (Pekanbaru, October 1st -2nd 2016)

[12] Sani N S, Racchmawati R., Mahfud 2012 Pengambilan Minyak Atsiri dari Melati dengan Metode Enfleurasi dan Ekstraksi Pelarut Menguap *Jurnal Teknik Pom ITS* **1**(1) p 1–4.

[13] Bhattacharyya P N, Jha D K 2011 Optimization of Cultural Conditions Affecting Growth and Improved Biactive Metabolite Production by A Subsurface Aspergillus strain TSF 146

[14] Hashem A, Abdallah E F, Al-obeed R S, Abdullah A 2015 Effect of Carbon , Nitrogen Sources and Water Activity on Growth and Ochratoxin Production of Aspergillus carbonarius (Bainier) Thom *Jundishapur Journal of Microbiology* **8**(2) p 1-8

[15] Gandjar I, Sjamsuridzal W, Oetari A 2006 *Mikologi: Dasar dan Terapan* (Jakarta: Yayasan Obor Indonesia)

[16] Ogbonna A I, Onyimba I A, Chuku A, Patience O, Ogbonna C I C, Onwuliri F C 2015 Growth assessment and amylase production by Aspergillus niger and A . terreus isolated from soils of Artemisia annua l . Plantation *European Journal of Biotcehnology and Bioscience* **3**(1) p 10–16.

[17] Gohel H, Ghosh S, Braganza V 2013 Yeast as a Viable and Prolonged Feedstock for Biodiesel Production. *Int. J. Renew. Energy Res.* **3**(1) p 126–131.

[18] Jadhav V V, Salunkhe D S, Bhadekar R K 2012 Effect of Alterations in Conventional Medium on Lipid Accumulation and Fatty Acid Content in Oleaginous Yeasts. *International Journal of Pharma and Bio Sciences* **3**(4) p 757–769.

[19] Aminah S, Mohtar W 2006 Medium Optimization for the Production of Lipidless Biomass By Cunninghamella sp . 2A1 Using Response Surface Methodology *Malaysian Journal of Microbiolody* **2**(1) p 40–45.

[20] Dyal S. D., Bouzidi L, Narine S 2005 Maximizing the Production of γ-linoleic Acid in *M. Rammanniana* as a Function of pH, Temperature and Carbon Source, Nitrogen Source, Metal Ions and Oil Supplementation *Food Res. Inter.* 38:815-829.

[21] Cook H W, & McMaster C R 2002 Chapter 7 Fatty acid desaturation and chain elongation in eukaryotes *Biochemistry of Lipids, Lipoproteins and Membranes* Vol 36 p 181–204.

Investigation of biomass gasification using indonesian clay as catalyst

SD Sumbogo Murti[1], Yuta SUDO[2], SUN Yan[2], Adiarso[1] and Reiji NODA[2]
[1]Center for Technology Energy Resources and Chemical Industry, BPPT, Energy Building 625, PUSPIPTEK Area, Serpong, South Tangerang, Banten 15314, Indonesia
[2]Chemical and Environmental Engineering, Graduate School of Engineering, Gunma University, 1-5-1, Tenjin-cho, Kiryu-shi, Gunma 376-8515, Japan

Email: sd.sumbogo@bppt.go.id

Abstract. Gasification is one of the technologies to process biomass as a renewable alternative energy source. Steam biomass gasification using various clays derived from Indonesia was carried out with laboratory scale fluidized bed reactor to evaluate the activities of clay as catalysts. At the same time to elucidated the relationship between material bed characteristics and gasification process activity. Tar capturing ability was compared for activated clay, silica sand and clay minerals derived from Indonesia by using a laboratory scale fluidized bed reactor. Even raw clay minerals were found to reduce heavy tar and water-soluble species emissions and increased carbonaceous materials deposited on the bed materials compared to activated clay. Activity of some Indonesian clay revealed high performance on the biomass gasification due to the amount of acid center.

1. Introduction

Indonesia is the largest producer of palm oil in the world [1]. Production of palm oil in Indonesia has recorded a phenomenal increase from 157, 000 tons in 1964 to 33.5 million tons in 2014 [2]. Production of crude palm oil (CPO) generates a large amount of residues (solid waste) which recently becomes an issue due to its environmental impact [3]. The process of fresh fruit bunches (FFBs) into CPO results in biomass wastes in large amount of empty fruit bunch fiber (EFBs) [4] As for the production of palm oil, almost same weight of EFB is left as waste [5]. These wastes can be used to generate power. From 1 tons of EFB is possible to produce power of 1.330 kWh [5]. However, the utilization rate of palm oil residues in Indonesia is still very low [6].

As the environmental issue such as global warming increases in severity, the promotion of energy conservation and the introduction of renewable energy has become a global challenge. Biomass is one of the renewable and potentially sustainable energy sources. It is a carbon neutral energy that does not increase CO2 in the atmosphere, and international research into biomass energy is currently moving forward. Biomass has been mentioned to be the fourth largest energy resource in the world, after coal, petroleum and natural gas [7]. Biomass gasification converts biomass into combustible gas that can be used for power and heat production. The product gas also can be upgraded to other valuable products such as FT liquids, SNG, hydrogen, or chemical.

The main technical obstacle of the biomass gasification is the efficient elimination of tar from the producer gas [8,9]. The objective of this research is to study tar reduction by using clay catalyst in the fluidized bed biomass gasification and to elucidate the relation between characteristics of bed material and activity of gasification process.

2. Material and Methods

2.1. Experimental setup and Bed material

In this research, steam gasification tests were carried out with a laboratory-scale fluidized bed reactor in a batch experiment to achieve these objectives. A comparison of various bed materials, such as Na-Bentonite and Ca-bentonite [10] derived from Indonesia with steam gasification was carried out. Silica sand and activated clay also carried out to compare these activities. A schematic diagram of the fluidized bed gasification reactor used in this research is shown in figure 1.

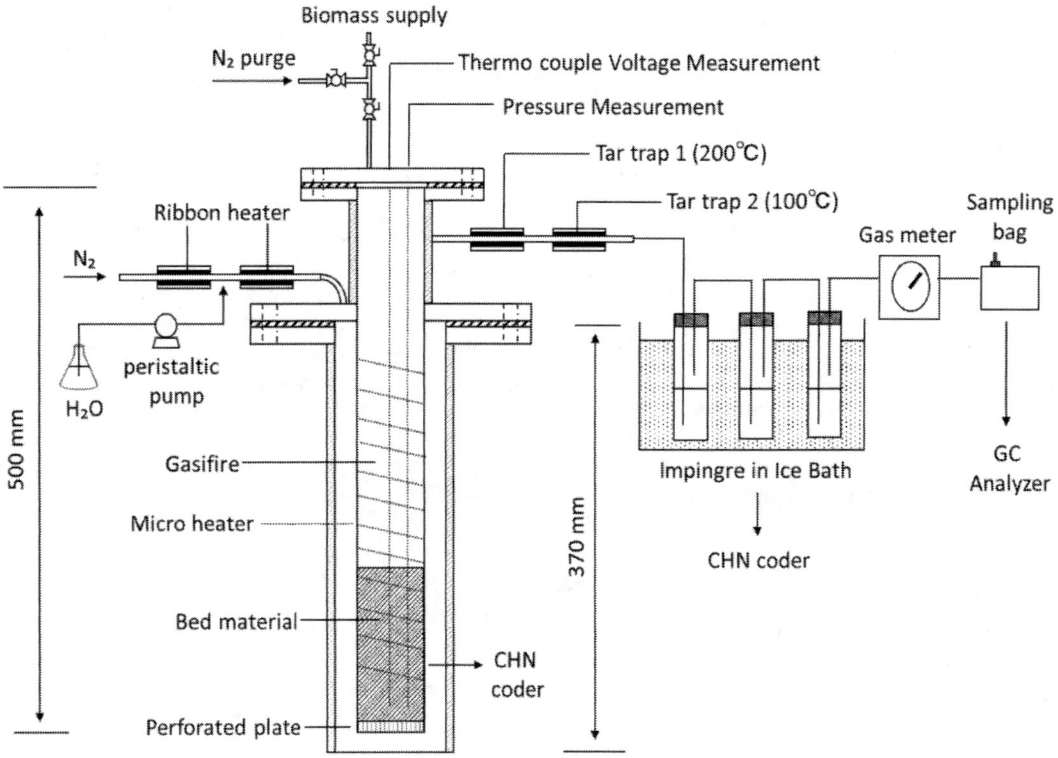

Figure 1. Experimental setup

Biomass sample is fed to the gasifier with feeding port pressured by nitrogen slightly. Products of synthetic gas in the reactor are introduced into a tar trap at 200°C and 100°C, to separate heavy tar. Small steel balls put inside the pipe can be used to trap the tar. Then, after tar trapping, products gas are flew through sampling line and quenched in three impingers cooled by ice bath, to separate water and water soluble organic compounds from gas products. The products of synthetic gas are collected into sampling bags. In experiments, cellulose is used as feedstock and clay derived from Indonesia as bed materials. Table 1 shows the basic chemical composition of cellulose. Table 2 shows information of clay as bed materials. There are 9 types of clay that have been collected from Java Island in Indonesia. Steam gasification of cellulose is performed in a fluidized bed gasifier at atmospheric pressure. Table 3 shows experimental conditions in this study.

Table 1. Ultimate analysis of cellulose as biomass

Components	Cellulose	
	Powder	Capsule
Ash (wt%-dry)	0	0
C (wt%-daf)	40.7	39.5
H (wt%-daf)	5.9	6.3
N (wt%-daf)	0.3	0.1

Table 2. Clay mineral derived from Indonesia as Bed material

No	Place of Origin	Note
1	Jampang (Banten)	
2	Pacitan (East Jawa)	Ca-Bentonite
3	Pacitan (East Jawa)	Na-Bentonite
4	Pacitan (East Jawa)	Na-Bentonite
5	Pacitan (East Jawa)	Ca-Bentonite
6	Bogor (West Jawa)	Ca-Bentonite
7	Blitar (East Jawa)	Ca-Bentonite
8	Blitar (East Jawa)	Na-Bentonite
9	Trenggalek, Tulungagung, Pacitan (East Jawa)	Ca-Bentonite

Table 3. Experimental Conditions

Condition	Unit	
Bed Material	-	Clay Samples
Biomass	-	Cellulose
Temperature	°C	650
Sample Weight	g	1
Gas Sampling Time	min	40
Bed Height	mm	80
Size of Bed Material	µm	75-150
U_0-U_{mf}	cm/s	2.5
Steam Concentration	%	85
Nitrogen Concentration	%	15 (Balance)

2.2. Analytical methods

After the experiment, product gas was analyzed by TCD Gas Chromatography (GC, 2014 SHIMADZU) after measured volume with gas meter and collected in sample bag to determine the carbon in each tar. Tar is trapped step wisely. The outlet pipe from the gasifier is covered by heater to trap tar condensable in the temperature more than 100°C [11]. The collected tar in each tar trap were burned in air to collects the combustion gas [12]. The condensed water that is collected by impingers is analyzed by CHN (LECO 628) analyzer to determine total carbon in solution. Char is separated from bed material by sieving with standard sieve of 150 µm opening. Total carbon remaining in the bed is determined by ultimate (CHN) analysis. After reaction process, the reactor was cleaned by nitrogen and cooled down the equipment.

2.3. Amount of acid center

The amount of acid center and its acidity were determined by the method developed by Benesi[13,14,15], where 0.1 g of bed material sample was put in 5ml of benzene and a predetermined quantity of butylamine-benzene solution. Four indicators benzalacetoph-enon (transparent to yellow): pKa=-5.6, dicinnamalacetone (yellow to red): pKa=-3.0, phenylazonaphthylamine (purple to yellow): pKa=+1.5, Methyl Red (yellow to red): pKa=+4.8) were applied to the titration by which amounts of butylamine adsorbed on acid centers were determined. If the addition of butylamine is less than the amount of acid center below pKa of the indicator applied, the color of the indicator does not change. Butylamine was added until each indicator changes its color to determine the amount acid center corresponding to the indicator.

3. Result and Discussion

3.1. Result of gasification

To confirm reproducibility two gasification runs were carried out in the same condition for 9 kinds of clay material. All of products could be recovered in the experiment, carbon recovery of the twice experiment were almost same, therefore reproducibility had been confirmed. As can be seen in figure 2, in the case of silica sand, carbon in the cellulose sample was gasified into around 35% gas, leaving 13% in residual solid (including char and carbonaceous materials trapped by particles), 11.7% in tar trap at 373K+473K (hereinafter, referred to as heavy tar), and 31.4% in water-soluble organics. The carbon fraction in each product are synthesis gas, in bed residue (char), trapped on bed materials (clay), heavy tar and water soluble. The total carbon recovery reached to around 90% for all sample. Figure 2 and table 4 show the carbon distribution of the fluidized gasification using Silica sand and activated clay. Markedly different things of this result are in trapped on bed material, heavy tar and water soluble. Result of silica sand which heavy tar is more than 10 times, water soluble is around 3 times compared with activated clay. In case of using silica sand, tar production is very high against activated clay. Amount of trapped on bed material using activated clay is around 3.5 times than silica sand. This result found to reduce heavy tar and water-soluble species emissions, and to increase trapped on bed materials than porous silica sand. Char, tar and carbonaceous materials in condensed water were decreased signification while the amount of carbonaceous materials in bed material was greatly increased by applying clay beds. Therefore, activated clay has catalytic activity. Activated clay showed significantly higher energy conversion to gas by gasification of captured carbonaceous materials.

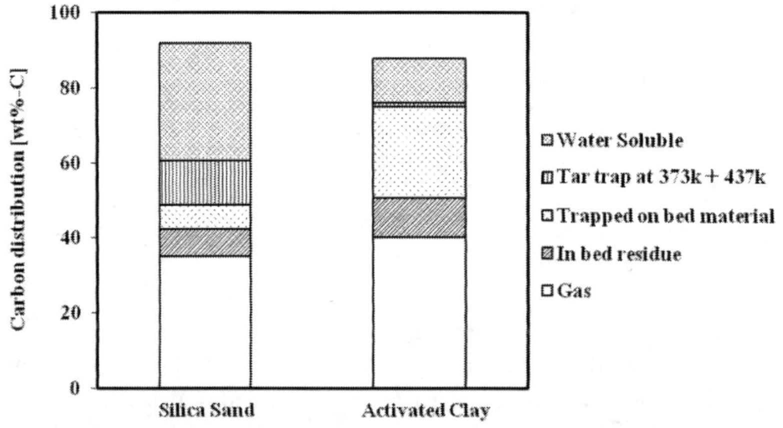

Figure 2. Result of Gasification using Silica sand and activated clay

Result of carbon distribution of the fluidized gasification using clay from Java Island in Indonesia is given in figure 3 and table 5. The carbon fraction in each product was resembles except for the case of silica sand. The different types of bentonite are each named for their respective dominant element, such as potassium (K), sodium (Na), calcium (Ca), and aluminium (Al) [16]. These clay samples can separate Ca-bentonite (No2, 5, 6, 7, 9.) and Na bentonite (No3, 4, 8.). Gas production of each sample is 35 to 41 wt%-C. A mount of tar production such as heavy tar and water soluble is smaller using Ca-bentonite. As showen on the table 5 that the best quality of the clay catalyst for biomass gasification process is the clay samples No. 2, type of Ca-bentonite. This is shown from the distribution of gasification products that produce relatively high product gas (syngas) and tar which can be captured by the highest catalyst / bed material compared to others catalyst types. Because one of the parameters for assessing the quality of the catalyst used in the gasification process is the ability of the catalyst to capture the tar, so the resulting syngas product is relatively more abundant and cleaner. A clay catalyst that can be selected based on the result of gasification is that has the ability to reduce emissions of heavy tar and water soluble species and can increased carbonaceous materials deposit on the bed material. Based on the results, Clay No. 2 and 6 produce very small tar of 0.88% and 0.68%, smaller than the tar produced by active clay. And increased carbonaceous materials deposited on the bed material to the value of 38.7% and 32% greater than using activated clay.

Table 4. Result of Gasification using Silica sand and activated clay

Products Yields	Silica	Activated
Gas	35.1	40
In bed residu	7.03	10.6
Trapped on bed material	6.6	24.3
Tar trap at 373k+437k	11.7	1.07
Water Soluble	31.4	11.7
Total Cabon	91.8	87.7
	wt%-C	

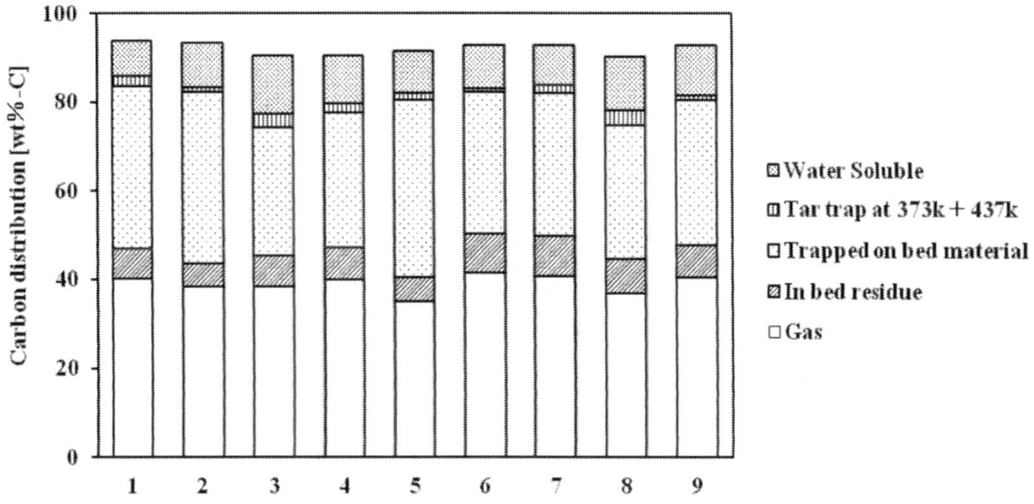

Figure 3. Result of Gasification using clay from Indonesia

Table 5. Result of Gasification using clay from Indonesia

Products Yields	1	2	3	4	5	6	7	8	9
Gas	40.2	38.5	38.5	40	35.2	41.6	40.8	36.8	40.5
In bed residu	6.6	5.1	7	7.1	5.4	8.6	8.9	7.7	7.3
Trapped on bed material	36.6	38.7	28.8	30.4	39.9	32	32.3	30.2	32.6
Tar trap at 373k+437k	2.5	0.9	3.1	2.2	1.5	0.7	1.66	3.5	1
Water Soluble	7.8	10.1	13	10.7	9.6	9.9	9.1	11.9	11.4
Total carbon	93.8	93.3	90.5	90.1	91.7	92.9	92.8	90.1	92.8
					Wt%-C				

3.2. Characterization of Clay

In this study, the amount of acid was analyzed to elucidate relation between characteristics of bed material and activity of gasification process. Fig.4 shows evaluation of acid center. Amounts of relatively strong acid are No.2, 4, 6, 9 (1.0, 0.87, 0.98, 0.72 mmol/g). On the other hand 7 and 8 are quite small (0.25, 0.22 mmol/g). No.1, 2, 3, 4, 6, 7, 9 are similar acid center (-3＜pKa＜+4.8). No.5, 8 are weak acid center (+1.5＜pKa＜+4.8) existed on the surface.

According to this result, amounts of strong acid are good conditions for gasification process. Because if using No.2, 6, 9 tar production such as heavy tar is smaller than others.

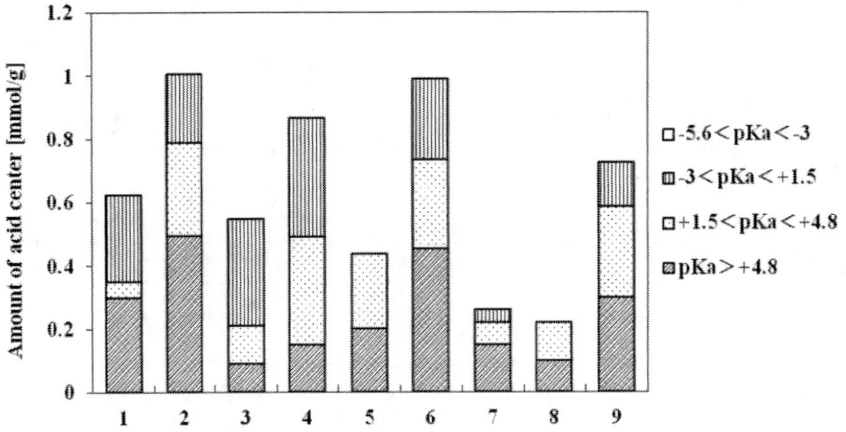

Figure. 4 Evaluation of Acid Center Conclusion

4. Conclusion

Biomass gasification using Indonesian clay as catalyst was conducted and succesed reducing tar production. In the case of using active clay as catalyst, the carbon in the cellulose sample was gasified producing about 40% gas, leaving 10.6% solids residual (including char and carbon trapped by particles), 1.07% tar and 11.7% organic soluble in water. Activity of Indonesian Clay No 2 and 6 showed the higher performance due higher of acid center.

5. Reference

[1] 2013 *Can Indonesia increase palm oil output without destroying its forest?* theguardian.com
[2] 2013 *Indonesia Palm Oil Production by Year* Indexmundi.com
[3] Rizky Fauzianto et al. 2014 Implementation of Bioenergy from Palm Oil Waste in Indonesia *Journal of Sustainable Development Studies* **5** No.1 100-115

[4] Yanni Sudiyani et al. 2013 Utilization of biomass waste empty fruit bunch fiber of palm oil for bioethanol production using pilot – scale unit *Energy Procedia* **32** 31-38

[5] Asia Biomass Energy Cooperation Promotion Office *Power Generation Potential EFB and Rice Husk in Indonesia* https://www.asiabiomass.jp/english/topics/0907_05.html

[6] Amzul Rifin et al. 2011 *The role of palm oil industry in Indonesian economy and its export competitiveness* Japan Univesity of Tokyo

[7] R.C. Saxena et al. 2009 Biomass-based energy fuel through biochemical routes: A review *Renewable and Sustainable Energy Reviews* **13** 167-178

[8] Shunsuke Nakamuraa et al. 2015 Development of Tar Removal Technologies for Biomass Gasification using the By-products Japan *Energy Procedia* **75** 208 – 213

[9] David Sutton et al. 2001 Review of literature on catalysts for biomass gasification *Fule Processing Technology* **73** 155-173

[10] Oliveira C. I. R. de et al. 2016 Characterization of bentonite clays from Cubati, Paraíba (Northeast of Brazil) **62** 272-277

[11] Atnaw S M et al. 2014 Study on Tar Generated from Downdraft Gasification of Oil Palm Fronds *Scientific World Journal* **2014(**2014) 497830

[12] Kamp W van de et al. 2005 Tar measurement standard for sampling and analysis of tars and particles in biomass gasification product gas *European Biomass Cenference and Exhibition*

[13] Benesi, H A 1957 Acidity of Catalyst Surfaces.II, Amine Titration Using Hammett Indicators *J.Phys. Chem* **61** 970-973

[14] Reiji NODA et al. 2009 Steam Gasification of Cellulose and Wood in a Fluidized Bed of Porous Clay Particles *Chemical Reaction Engineering* **42** 490-501

[15] Tulay ALEMDARO_GLU et al. 2009 Investigation of the Surface Acidity of a Bentonite Modified by Acid Activation and Thermal Treatment *Turk J Chem* **27** 675-681

[16] Christidis G E 1998 Comparative study of the mobility of major and trace elements during alteration of an andesite and a rhyolite to bentonite, in the islands of Milois and Kimolos, Aegean, Greece *Clays and Clay Minerals* **46**(4) 379-399

IOP Conf. Series: Earth and Environmental Science **105** (2017) 012106 doi:10.1088/1755-1315/105/1/012106

Measurement of thermal radiative heat transfer using a multi-axis heat flux sensor

Reisa Adityo[1], Rahmat Agung[1], Pujo Satrio[1] and Yulianto S. Nugroho[1]

[1] Fire Safety Engineering Research Group, Department of Mechanical Engineering, Universitas Indonesia, Depok, Jawa Barat 16424, Indonesia

E-mail : yulianto.nugroho@ui.ac.id

Abstract. Heat transfer always occurs in human life and can be perceived its benefits and losses. The heat transfer is divided into three according to transfers media, namely conduction, convection, and radiation. One example of cases where there are three processes of heat transfer is when there is a fire in a room. During a fire, the heat from the fire will travel into the surrounding environment and affect the objects around it with radiation, conduction, and convection. Heat that travels through radiation certainly does not require a medium of delivery because the heat is traveled in the form of electromagnetic waves and can travel to a very long distance. Distance, view factor, and placement of an object will affect the amount of radiation to be received from the heat of fire. Therefore, to examine the effect of distance, the viewing factor and the placement of objects against radiation, writer designs a multi-axis heat flux sensor. Instrument with multi-axis heat flux sensors or by the name of a Multi-Axis Radiometer will be designed which will then be tested to determine the results of the design. Instrument called Multi-Axis because the workbench of the instrument can be rotated 90°. Testing of the apparatus will include the heating temperature, the distance of the sensor to the heater, the sensor offset to the heater, the viewing angle of the sensor to the heater, and the orientation of the workbench. The results of testing distance and offset are validated by the theoretical calculations.

1. Introduction

The case where thermal radiation occurs is in the event of a fire in the room. In the event of fire, heat from fire will be transferred to the environment by conduction, convection, and radiation.Heat transfer in fires is an important factor in flame development, firespread, and fire suppression[1]. If the heat transfer in a fire is uncontrolled, it could make the fire become bigger and wider. When the heat of fire hit on a material surface by radiation, then there are three properties, some will be reflected, some will be absorbed, and some will be transmitted. The closer the distance to the object by the source of fire, then intensity of radiation will be greater. So that every object in the room has the same risk in the fire even if not directly exposed by fire[2].

In order to withstand the heat transfer and its effects is needed the appliance that can show the values from the effects of thermal radiation. The instrument is used to measure values is the Heat Flux Sensor. The Heat Flux Sensor will be mounted on a device called Radiometer. In this research, the use of radiometer is useful to measure the heat that is received by an object and the effect on the distance and the angle of view of the heat source. In addition to measuring heat from thermal radiation, the heat from thermal convection can also be measured with this instrument. The tests are performed to obtain data. Then, data will be compared with the result of the theoretical thermal radiation heat flux calculations as a reference for testing this instrument.

2. Theory

2.1. Radiation

Radiation is the transfer of energy through electromagnetic waves that can move through the empty space, or objects and transparent fluid[3]. The heat flux of radiation that is received a point is defined as the following

$$\dot{q}" = \varphi \varepsilon_e \sigma T_e^4 \tag{2.1}$$

Where the q" is the radiant heat flux (W/m2), φ is the configuration factor, ε_e is emissivity from the surface of the transmitter, σ isStefan-Boltzmann constant (5,67x10^{-8} W/m^{2}K^4), and T_e is the absolute temperature from the surface of the transmitter radiation (K).

When radiation emitted by a surface and accepted by a surface, the calculation of the heat flux becomes

$$\dot{q}" = \varphi \varepsilon \sigma (T_e^4 - T_r^4) \tag{2.2}$$

View factor or configuration factor is the value of how much radiation transmitter will be seen by the recipient surface.

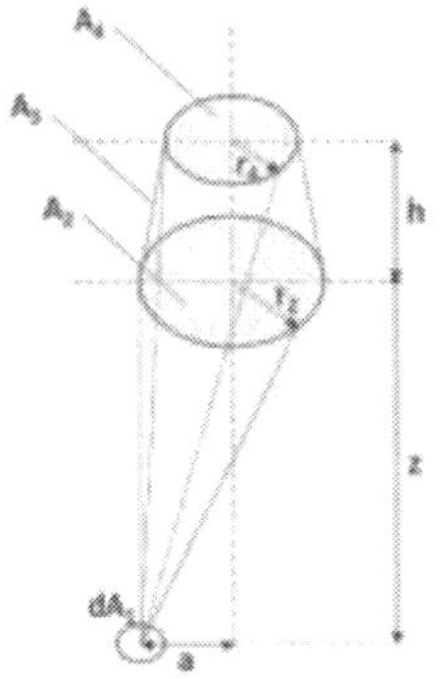

Figure 1. Radiation from one surface to another Surface.

From the picture above, from the position of dA_1, view factor is represented with equation $F_{d1\text{-}3}$,

$$F_{d1-3} = F_{d1-2} - F_{d1-4} \tag{2.3}$$

View factor $F_{d1\text{-}3}$ represent the fraction of radiation energy that leave the surface of the specimen dA1 to every direction until it reaches the area 3. View factor for the exchange between dA1 and the area 2 can be shown by the following equations.

$$F_{d1-2} = \frac{1}{2}\left(1 - \frac{1+H_2^2-R_2^2}{\sqrt{Z_2^2-4R_2^2}}\right) \tag{2.4}$$

where $H_2 = z/a$, $R_2 = r_2/a$ and $Z_2 = 1 + H_2^2 + R_2^2$. The exception on the common (2.4) occurs when dA_1 is located on the center line, where a = 0 then,

$$F_{d1-2} = \frac{r_2^2}{z^2+r_2^2} \tag{2.5}$$

Equation (2.2) also applies for $F_{d1\text{-}4}$ so $F_{d1\text{-}3}$ can be written to

IOP Conf. Series: Earth and Environmental Science **105** (2017) 012106 doi:10.1088/1755-1315/105/1/012106

$$F_{d1-3} = \frac{1}{2}\left[\left(1 - \frac{1+H_2^2-R_2^2}{\sqrt{Z_2^2-4R_2^2}}\right) - \left(1 - \frac{1+H_4^2-R_4^2}{\sqrt{Z_4^2-4R_4^2}}\right)\right] \quad (2.6)$$

And for a = 0,

$$F_{d1-3} = \frac{r_2^2}{z^2+r_2^2} - \frac{r_4^2}{(z+h)^2+r_4^2} \quad (2.7)$$

where $H_4 = (h+z)/a$, $R_4 = r_4/a$, and $Z_4 = 1 + H_4^2 + R_4^2$.

Equation (2.7) will be substituted into the heat flux equation (2.1) become,

$$q'' = F_{d1-3}\varepsilon\sigma T^4 \quad (2.8)$$

where T is the value of the average temperature from heating element[4].

2.2. Convection

In free convection fluid motion is due to buoyancy forces within the fluid, while in forced convection it is externally imposed. Buoyancy is due to the combined presence of a fluid density gradient and a body force that is proportional to density. In practice, the body force is usually gravitational but there also several ways in which a mass density gradient may arise in a fluid, but for the most common situation it is due to the presence of a temperature gradient [3]. In this research, the heater will be assumed as vertical plate and mostly buoyancy forces induce a free convection boundary layer in which the heated fluid rises vertically as shown below.

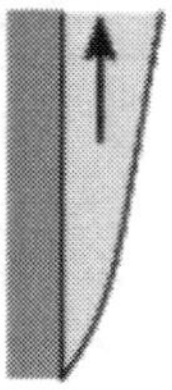

Figure 2. Free convection boundary layer on a vertical plate [3].

2.3. Radiometer

The Radiometer has thus schematic diagram as the figure 3 is a configuration to get the value of the heat radiation energy. The heat energy received by the detector will be related to the Stefan-Boltzmann constant with the view factor

$$R = \sigma F(T_s^4 - T_a^4) \quad (2.9)$$

Then,

$$F = \frac{R}{\sigma(T_s^4-T_a^4)} = \frac{R}{q_b} \quad (2.10)$$

View factor (F) will relate to the view factor (θ) from the heat source and the view factor can be stated in

$$F = sin^2\theta \quad (2.11)$$

So,

$$R = sin^2\theta \; x \; \sigma \; x \; (T_s^4 - T_a^4) = sin^2\theta \; q_b \quad (2.12)$$

Point of view (θ) can be different depends on the distance Radiometer (x) to a heat source such as figure 3.

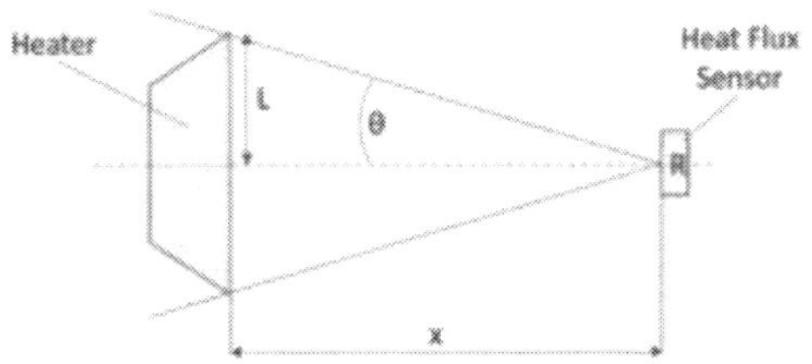

Figure 3. Schematic Point of View on Heat Sources.

P-355etoint of view (θ) can be defined to

$$\theta = \tan^{-1} L/X \tag{2.13}$$

where L is half of the length from surface of the heat source when it is square and the radius of the surface of the heat source when a circle[5].

The sensor used on the Radiometer is a thermal sensor to measure the heat flux. This sensor is usually called a heat flux sensor. This heat flux sensor uses the working principle of Gardon and Schmidt Boelter or have a working principle same as seebeck effect. The sensor is equipped with a black absorbent that is useful for capturing radiation. Inside the heat flux sensor there is a main sensor called thermopile. The thermopile will measure the temperature difference read on the front and back of the heat flux sensor[6]. The received temperature value will be converted to voltage so that it can be transmitted via cable and received in data collector.

Figure 4. Heat Flux Sensor[6].

Heat Flux (W/m^2) is obtained by the result of the voltage (U) produced divided by its sensitivity (S)

$$\phi = U/S \tag{2.14}$$

The sensitivity value is printed on the specification of the sensor. However, when the sensor temperature is below 10°C, then

$$\phi = \frac{U}{S} + \sigma(T + 273{,}15)^4 \tag{2.15}$$

Where σ is Stefan-Boltzmann Constant (5,67x10^{-8} W/m^{2}K^4).

3. Methodology

Radiometer that will be used for this research equipped with Conical Heater as heat source and will be done by several methods, such as heating temperature, sensors' distance to heater, sensors' offset to center point of heater, sensors' point of view to heater, and orientation of workbench. Each test is performed in 1 minute when the temperature read on the control system is stable and constant. In this test, the data will be read and collected through LabView 2016 and Microsoft Excel 2016 software. There will be at least three parameters of experiment, i.e. temperature of heat source, sensors' distance, and sensors' offset to center point of heat source. All of three parameters have many combinations for experiment. The results of experiment become experiment data of heat flux value from sensor readout. Furthermore, experiment data will be compared to radiative heat flux theoretical calculation in order to find the error value of the experiment data.

IOP Conf. Series: Earth and Environmental Science **105** (2017) 012106 doi:10.1088/1755-1315/105/1/012106

4. Results and discussion

4.1. Effect of sensors' distance

Testing the sensor distance from the heater at a distance of 10 cm, 20 cm, 30 cm, and 40 cm from the heater. The offset and point of view on this test is set at 0 cm and 0° with a flat work desk orientation. The control temperature used is 312.46 °C[7]. The data obtained from the test results are plotted so that the graph in figure 5. From the graphic results can be proved that the distance affects the amount of heat flux received by an object.

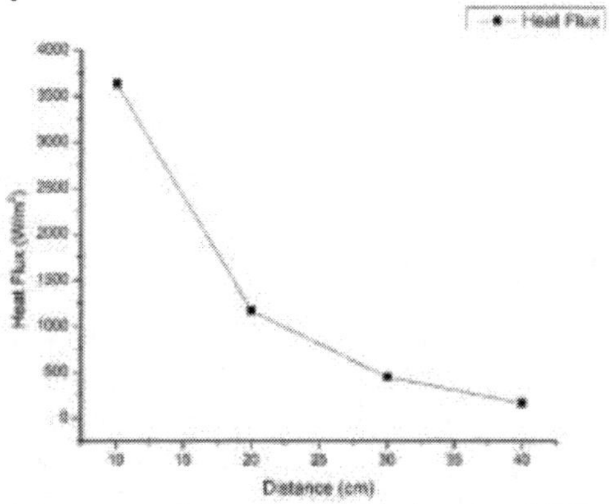

Figure 5. Distance vs Heat Flux Graph[8].

4.2. Effect of heating temperature and sensors' distance

Testing for variations in temperature and distance of sensor include all variations of the heating temperature and the sensor distance from the heater. The heating temperature are conducted at 273.74°C, 291,47°C, 312,46 °C, 331,16°C, and 351,11°C. For the sensors' distance are same as previous experiment explanation. The position of the offset and the sensor point of view on the heater in this test is 0 cm and 0° with the orientation of the flat work table. From the graph that has been plotted, it appears that the heat flux will increase with increasing temperature and the closer the sensor distance to the heater. At a distance of 10 cm from the heater, the heating flux value dominated and reads linearly with increasing temperature. While at a distance of 20 cm, 30 cm, and 40 cm is not completely linear because at the temperature control 330 °C to 350 °C increase heat flux is read only slightly. This can be caused by the distance of the sensor from the heater so that there is distraction from the environment. This distraction can occur because the Radiometer tool is not given a special cover. However, with the linear trend of graphs it can be proven that the distance and temperature affect each other from the magnitude of the heat flux.

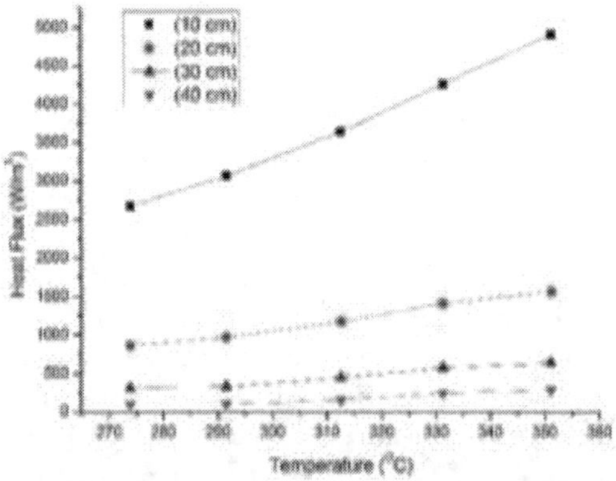

Figure 6. Temperature vs Heat Flux Graph[8].

4.3. Effect of offset and sensors' distance from heater

Variations are combined with the sensor distance from the heater to see the comparison between the distances. The offsets tested were -7 cm, -3.5 cm, 0 cm, 3.5 cm, and 7 cm. The negative value indicates that the sensor is on the left side of the heater and the positive value indicates the sensor is on the right of the heater. 7 cm is chosen as the largest offset value because the diameter of the heater is 16 cm so that the outer surface side of the heater is 7 cm. Distance variation in the test is all the distance variations, ie 10 cm, 20 cm, 30 cm, and 40 cm. The temperature used is 312.46°C with the sensor point of view 0° and the orientation of the flat work table. The graph of the test results is shown in figure 7.

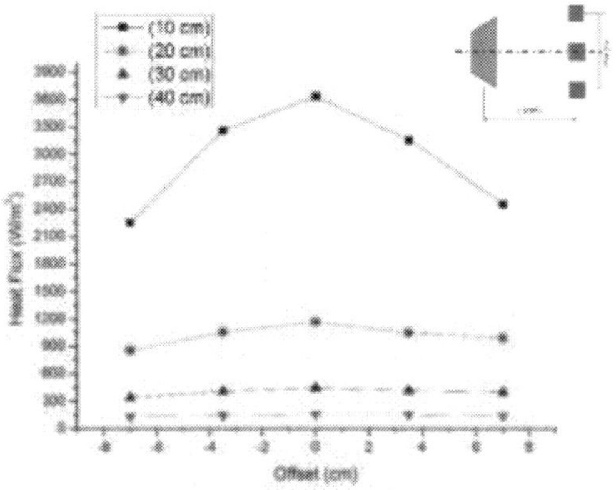

Figure 7. Sensors' Offset vs Heat Flux Graph[8].

4.4. Sensors' distance error

Theoretical calculation is done to compare and measure the error value of the experiment data. Theoretical calculation is done by equation (2.12) and T value are 273,74 °C, 291,47 °C, 312,46 °C, 331,16 °C, and 351,11 °C. When the T value of 273,74 °C, the error value at 10 cm is 48,54%, at 20 cm is 35,35%, at 30 cm is 1,58% which become the lowest overall error value, and at 40 cm is 42,99%. Then the T value is increased to 291,47 °C, the error value at 10 cm is 48,2%, at 20 cm is 32,12%, at 30 cm is 6,35%, and at 40 cm is 50,51%. After that, when the T value of 312,46 °C, the error value at 10 cm is 50,2%, at 20 cm is 36,5%, at 30 cm is 7,63%, and at 40 cm is 34,06%. Until T value 312,46 °C, the lowest error value still at distance of 30 cm. When the T value of 331,46 °C, the error value at 10 cm is 53,67 %, at 20 cm is 43,54 %, at 30 cm is 21,16 %, and at 40 cm is 11,4 %. Lastly, when the T value reach the highest temp on experiment of 351,11 °C the error value at 10 cm is 54,23 %, at 20 cm is 38,53 %, at 30 cm is 17,16 %, and at 40 cm is 8,76 %.The last two T value show that the lowest error value exists at distance of 40 cm.

4. Conclusions

The present work has allowed us to get to the following conclusions:

1. Radiative heat flux starts to dominate when the distance of sensor at 30 cm. In this case is proven by small error value which is the result of comparison between experiment data and radiative heat flux theoretical calculation data.
2. At the distance of 10 cm, there is assumption that convective heat flux is read by sensor. Big error value from experiment data and radiative heat flux theoretical calculation data shown if there could be an indicator to determine the deviation that occurs is a convective heat flux. Also, the convective boundary layer could be covered the area where the distance of sensor is 10 cm and make the data read by sensor includes convective heat flux.
3. Heating temperature is linear to heat flux. Hotter temperature of heat source makes the heat flux that is read by sensor also bigger.

4. Configuration factor will affect the amount of heat flux which is received. In this experiment configuration factor is represented by distance of sensor to heat source. Further sensor from heat source make the heat flux that is read smaller.

5. Acknowledgement

The authors would like to thank the financial support provided by Universitas Indonesia through the PITTA 2017 funding scheme under Grant no. 857/UN2.R3.1/HKP.05.00/2017 managed by the Directorate for Research and Public Services (DRPM) Universitas Indonesia.

6. Reference

[1] Makhviladze G M 2002 *Enclosure fire dynamics* **37** no. 1
[2] Buchanan A H 2001 *Structural Design for Fire Safety* (New York: Wiley)
[3] Incropera F P, Bergman T L, Lavine A S and DeWitt D P 2011 *Fundamentals of heat and mass transfer* (New York: Wiley)
[4] Wilson M T, Dlugogorski B Z and Kennedy E M 2003 uniformity of radiant heat fluxes in cone calorimeter *Fire Saf. Sci. 7* **7** 815–26
[5] King Fahd University *Radiation Heat Transfer Characteristics*
[6] Hukseflux, *Hukseflux Thermal Sensors USER MANUAL DR-SERIES PYRHELIOMETERS* pp. 1–47
[7] Agung R 2017 *Simulation and Examining Thermal Radiation Heat Flux Using Multi-Axis Heat Flux Sensor* (Depok: University of Indonesia)
[8] Adityo R 2017 *Designing and Examining Thermal Radiation Heat Flux Using Multi-Axis Heat Flux Sensor* (Depok: University of Indonesia)

IOP Conf. Series: Earth and Environmental Science **105** (2017) 012107 doi:10.1088/1755-1315/105/1/012107

Effect of nano Si addition on synthesized LTO for lithium battery anode

A Zulfia[1,3], **R Satria**[1], **B Priyono**[1] **and A Subhan**[2]

[1] Departemen Metalurgi & Material, Universitas Indonesia, Kampus UI, Depok, Jawa Barat 16424, Indonesia

[2] Pusat Penelitian Fisika, LIPI, Komplek PUSPIPTEK, Banten, Tangerang Selatan, 15314, Indonesia

E-mail : anne@metal.ui.ac.id

Abstract. $Li_4Ti_5O_{12}$ or LTO is one of many compounds that could be used as anode component in lithium battery. The most interesting aspect of using LTO as an anode is its long cycle life which is affected by its zero strain property during insertion and extraction of lithium ions. Despite its advantages, LTO still has problem in its capacity value which is limited to 175 mAh/g. Researchers have tried many methods to increase the capacity of LTO, such as making a composite from LTO host. In this composite, nano sized Si is used as additional element because its high theoretical capacity could increase the overall capacity of the LTO composite. In this research, LTO was synthesized by hydrothermal-mechanochemical methods before we mix it with nano Si in slurry making process. The mass variation of nano Si was 1%, 5%, and 10% in wt. XRD and SEM were used for material characterization. For the battery performance testing we used EIS, CV, and CD. This research will explain the effect of Si on the LTO/Si composite performance. From the testing, it is known that the highest capacity was obtained from LTO/Si-10% sample with 216.15 mAh/g, and able to retain 42.76% of its capacity at higher C-rate (4C). The results show that LTO/Si-10% could be used as an alternative for anode component.

1. Introduction

Nowadays, automotive sector is one of the most advancing technologies. Several innovation has been done by automotive industry, such as developing hybrid car which called by hybrid electric vehicle (HEV). In order to take advantage from this electric motor, we need a source of energy which comes from conventional li-ion battery with carbon-based anode. But, carbon-based anode has its disadvantages which is dendritic structure could be formed on the carbon and eventually compromising the safety factor. This is caused by dendritic structure could make short circuit easily happen when the operation voltage is identical to Li/Li^+[1]. The other disadvantage is the formation of solid electrolyte interface (SEI) which leads to loss in performance.

Based on those problems, the research on better anode material is still on going. One of the material that could be used as anode material is $Li_4Ti_5O_{12}$ or simplified as LTO, is considered to overcome the disadvantages from carbon-based anode. This compound has several advantages such as fast charging because of its superior surface area which is contributed from its nanostructure, a long

[3] To whom any correspondence should be addressed.

cycle life that is supported by its zero strain property, and better safety than carbon-based anode[2]. Despite all of that advantages, LTO still has its downsides. The concerning problems are its low conductivity with only 10^{-13}S/cm, low diffusion of Li^{+} ion (10^{-9}-10^{-13} cm^2/s), and low capacity (175 mAh/g). To overcome the problems, LTO needs modification by mixing it with other element such as Si which has high capacity (4200 mAh/g)[3]. The performance of synthesized LTO anode is expected to improve by mixing with Si.

2. Experimental procedure

The particle size of nano Si according to the specification handed by the manufacturer is 100 nm. In order to discover how much Si is needed to make an optimal mixture of LTO/Si as anode material, we used three variation of %wt Si. The variables consist of 1%wt, 5%wt, and 10%wt Si. For the LTO component, we synthesize it by using hydrothermal-mechanochemical method.

Before the synthesis process of LTO started, we are required to synthesize its precursor which is TiO_2. The process used to synthesize this TiO_2 is sol-gel method, with titanium tetra-n butoxide and ethanol as its starting material. The result of this process is clear TiO_2 gel and we crushed the dried gel in order to make it into powder form. When TiO_2 powder is formed, we calcined it at 300°C for 2 hours to reduce the water moisture, organic compounds, and improving crystallinity[4]. Calcination on the powder caused its color changed to light brown. The next step to make particle with broader surface area is using hydrothermal with temperature of 135°C for 15 hours. This hydrothermal process is followed by drying on hot plate to release the trapped water after the hydrothermal process and the synthesis of TiO_2 is finished.

Beside TiO_2, we also require commercial LiOH as the precursor for LTO. The process used to synthesize LTO is mechanochemical or ball milling. We used TiO_2, LiOH, and zirconia ball in milling machine and run it for 10 minutes. The product of this process is fine light pink LTO particle. This process is followed by sintering in 750°C for 3 hours in order to obtain the spinel phase of LTO and increasing the density of LTO powder[5,6]. After this sintering process, the LTO powder is ready to use.

The mixing process of LTO and Si is carried out in slurry making process in order to avoid the formation of silicon oxide. The slurry composition consists of active material, acetylene black as conductive agent, and PVDF as binder with ratio of 8:1:1. We also used 5 grams DMAC as solvent in this mixture. The active material composed by LTO and Si, with specific composition according to variable of %wt Si. Every sample's composition is in table 1. To make the slurry homogeneous, we made the slurry using magnetic bar to stir the mixture. After the slurry stirred, we used doctor blade to coat it on Cu foil and followed by drying for 2 hours. The anode material is ready to use.

Table 1. LTO/Si composite composition.

Material	Mass (gram)				
	LTO	Nano Si	Total Active Material	Acetylene Black	PVDF
LTO/Si 1%	1.98	0.02	2	0.25	0.25
LTO/Si 5%	1.9	0.1	2	0.25	0.25
LTO/Si 10%	1.8	0.2	2	0.25	0.25

After the synthesis process, then the composites are characterized. We used XRD to find out compounds contained in the synthesized LTO. Besides XRD, we also used SEM in order to understand the morphology of the powder. Meanwhile, to assess the performance of the battery we used cyclic voltammetry, electrochemical impedance spectroscopy, and charge-discharge testing. Specifically, cyclic voltammetry test to find out the working potential and the capacity. Electrochemical impedance spectroscopy or EIS test to compare the resistance of each variable. And charge-discharge test to get the information of retained capacity in particular C-rate.

3. Results and discussion

XRD pattern in figure 1 and figure 2 showed the pure Si sample and synthesized LTO respectively.It is found that there are few indications of SiO contained in the Si powder as showed in figure 1. After that, the first thing that must be checked is the compound contained in synthesized LTO. We used angle 10° up to 80° to find the existing compound. The obtained data is analyzed by X'pert High Score Plus application. From the application, it can be found that $Li_4Ti_5O_{12}$ compound is confirmed to be exist in the synthesized LTO powder. Beside the main compound, the trace of other compounds such as TiO_2 in rutile and anatase form are found as showed in figure 2. Those TiO_2 compounds are retained from the synthesis process. Actually, some previous research showed that rutile could give a negative effect on the battery performance because it can easily form Li_xTiO_2 when rutile exposed to lithiation. This Li_xTiO_2 will easily irreversibly transform to other form which lead to capacity loss[7].

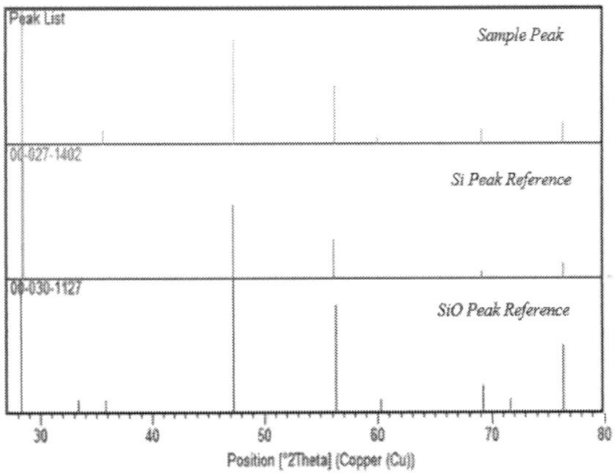

Figure 1. XRD Pattern for pure Si we used.

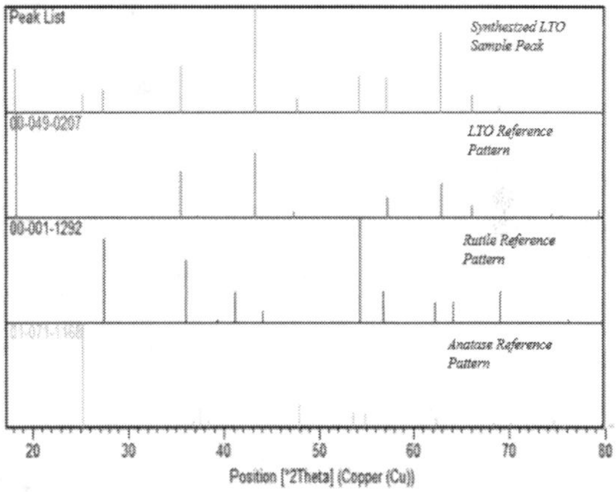

Figure 2. XRD pattern for synthesized LTO.

The observation of morphology of synthesized LTO was done by using SEM and compared to commercial LTO in figure 3 and figure 4 respectively. From the result, it is found that the morphology of the synthesized LTO and commercial LTO are generally different. The different aspect can be found in the homogeneity of the particle and also its particle size. In the synthesized LTO, the particles are less homogeneous and larger. Eventually from the previous study, larger particle could make the transfer range of lithium ion getting further and less surface area which could lead to slower charge-discharge time[8].

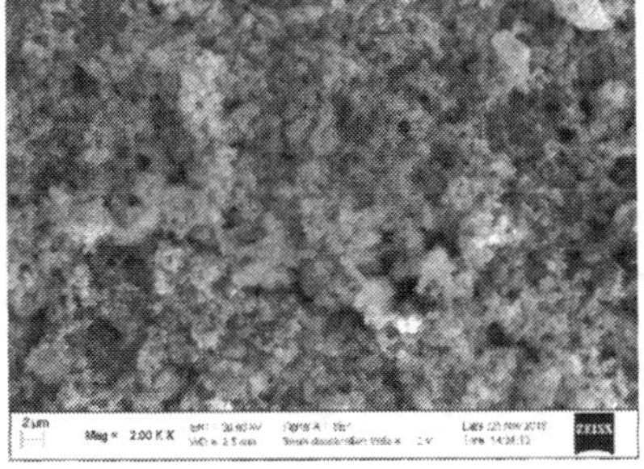

Figure 3. Commercial LTO SEM result.

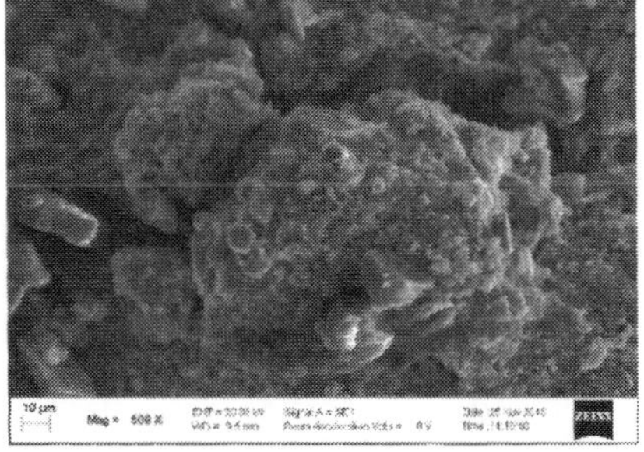

Figure 4. Synthesized LTO SEM result.

We also observed the morphology of LTO/Si composite after slurry making process at every composition. The results in figure 5 show the morphology of each composition are not too different and the particle sizes are observed in micro scale.

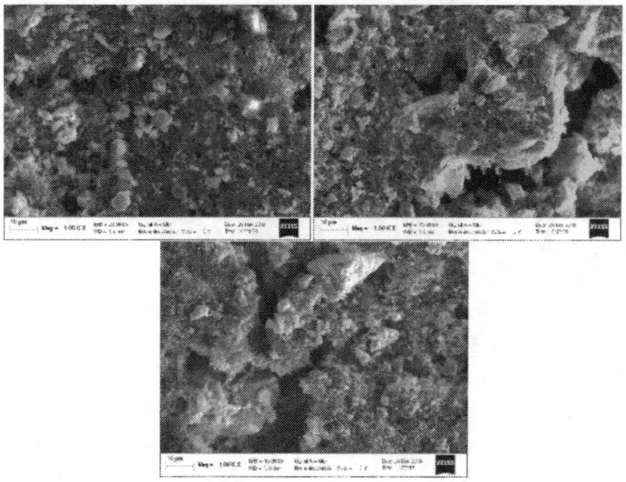

Figure 5. SEM result for LTO/Si-1% (top left), LTO/Si-5% (top right), and LTO/Si-10% (bottom).

From the EIS result in table 2, it is found that the lowest resistance to the highest resistance is in following order: LTO/Si-1%, LTO, LTO/Si-10%, and LTO/Si-5%. Generally, samples with higher Si content will have lower conductivity, which is supported by the fact that Si is one of the semiconductor or will have a good conductive property in high temperature. Beside that, the indication of SiO in the pure Si powder can affect the conductivity since SiO has the isolator property. The other reason of this result is the usage of PVDF binder, which is not the optimal binder for Si. This caused many gap exist on the surface and impedes the current [9].

Table 2. Rct values from all samples.

Sample	Rct (Ω)
LTO	10.6
LTO/Si-1%	4.165
LTO/Si-5%	120.61
LTO/Si-10%	88.425

We measured the working potential of every sample by using cyclic voltammetry test. From figure 6, there are two peaks which is consisted of cathodic and anodic peak at 1.703 V and 1.471 V respectively which means the average of those peak voltages or the working potential is 1.587 V. It also there is no significant difference from theoretical working potential.

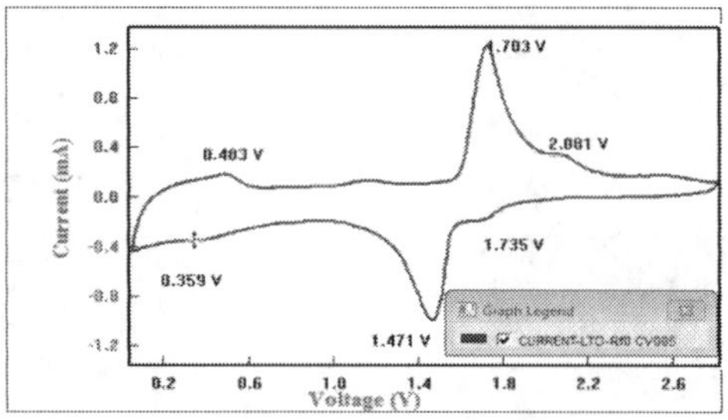

Figure 6. Cyclic voltammetry curve for synthesized LTO.

Meanwhile for the result of LTO/Si composite sampleson figure 7, it also shows that every sample also has two peaks around the same places. This means that LTO and Si do not disrupting each other's performance.

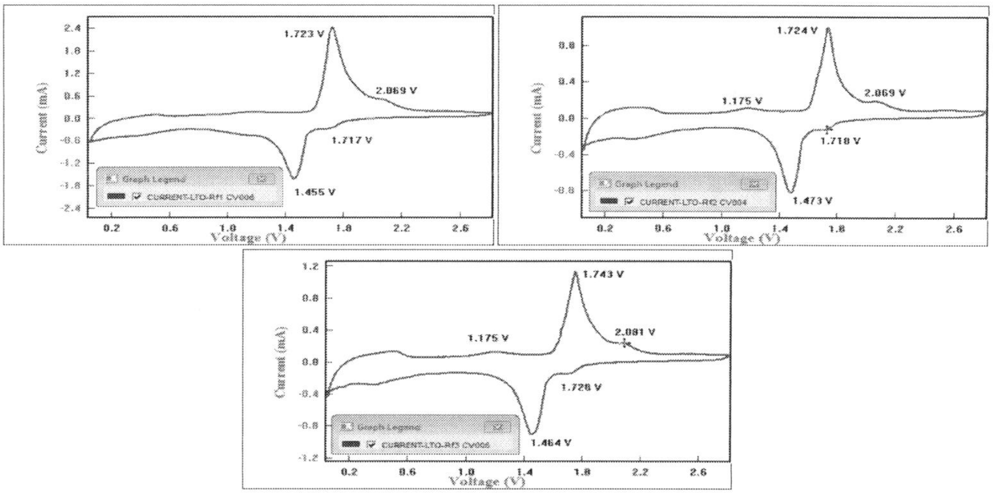

Figure 7. Cyclic voltammetry curve for LTO/Si-1% (top left), LTO/Si-5% (top right), and LTO/Si-10% (bottom).

Other aspect that we can study for the performance of battery is from its capacity, which is obtained from cyclic voltammetry testing and showed in figure 8. Our synthesized LTO without Si addition has capacity of 137.95 mAh/g, which is lower than the theoretical specific capacity of LTO (175 mAh/g) and the other commercial LTO sample which has 173.29 mAh/g. The effect of Si addition could be seen from the result, which generally has higher capacity with higher Si content. This is the result of the mixing process in composite, when the reinforce element has the better property then it could affect the mixture in good way. In this study, since Si has high theoretical capacity (4200 mAh/g), the mixture should be getting higher capacity than 137.95 mAh/g. The highest capacity that can be obtained from this study is the mixture of synthesized LTO and 10% wt Si with 216.15 mAh/g, which shows that Si can affect the overall capacity of the mixture.

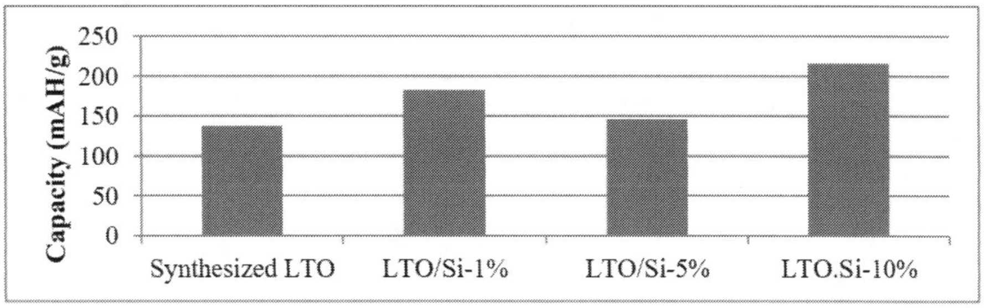

Figure 8. Capacity for every sample.

In charge-discharge test we will find out the capacity of the battery on certain C-rate and the results showed in figure 9. Generally the higher C-rate used, then the capacity on that C-rate is getting lower. From figure 9 it can be found that pure LTO and LTO/Si-1% has similar trend on the decreasing capacity with higher C-rate. But there's some exception for LTO/Si-5% and LTO/Si-10%. In higher C-rate, the LTO/Si-5% and LTO/Si-10% samples both has similar trend, which is they could still hold their capacity in higher value than LTO and LTO/Si-1%. Or in the other word, both LTO/Si-5% and LTO/Si-10% have better performance on higher C-rate. Previous study found that there are some complex effects from the lithiation process. We already know that some Si oxide have the isolator property. But from Chen, et al research, they found the complex effect of Si oxide [3]. The complex effects are the liberation of Si from lithiation process of SiO_x which will be the additional source of

Sifor higher capacity. Other product from lithiation of SiO_x is Li_ySiO_x which has the property to hold the volume expansion of Si[10]. The detail of Si addition will need further investigation to explain this complex effect. This complex effect can explain why the higher content of Si such as LTO/Si-5% and LTO/Si-10% have better performance in higher C-rate.

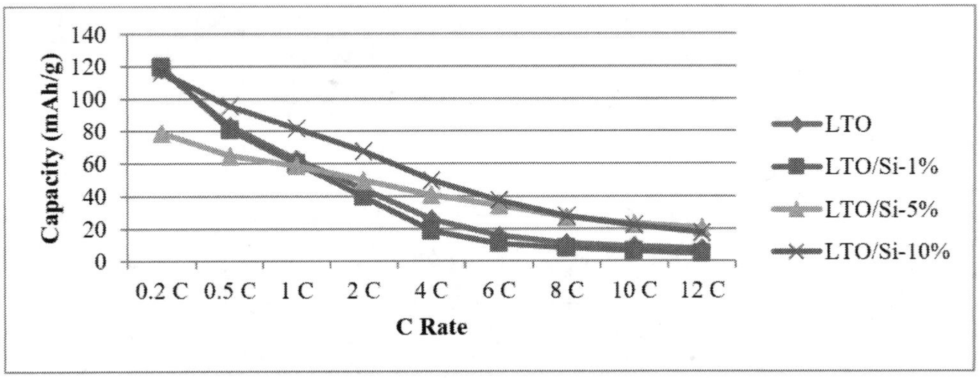

Figure 9. Specific capacity at particular C-rate.

Since HEV need C-rate around 4C, we also simplified the result from charge-discharge testing to see the retained capacity on table 3.

Table 3. Retained capacity at 4 C for every sample.

	LTO	**LTO/Si-1%**	**LTO/Si-5%**	**LTO/Si-10%**
Capacity Retained at 4 C	21.967 %	15.6354 %	51.584 %	42.7614 %

4. Conclusion

The addition of nano Si into synthesized LTO has been conducted by using slurry-making method. It's found that our synthesized LTO hasn't been made perfectly. But after we mixed it with nano Si, specifically LTO/Si-10%, our LTO/Si composite has been upgraded and has better performance in some aspect from its pure LTO counterpart. Although there are some complex effect from nanoSi addition into LTO host, this study indicates that nano Si additioncan be used to develop anode material further.

5. References

[1] Zhang Q, Liu Y, Lu H, Tang D, Ouyang C and Zhang L2016 *Electrochim Acta.* **189** 147-57
[2] Wang J, Zhao H, Wen Y, Xie J, Xia Q, Zhang T, Zeng Z and Du X 2013 *Electrochim Acta.* **113** 679-85
[3] Chen C, Agrawal R and Wang C 2015 *Nanomater.* **5** 1469-80
[4] Simonsen M E and Søgaard E G 2010 *J. Sol-Gel Sci.Technol.* **53** 485-97
[5] Zhang Z, Cao L, Huang J, Wang D, Meng Y and Cai Y 2013 *Electrochim Acta.* **88** 443-6
[6] Abbasian A R, Rahimipour M R and Hamnabard Z 2015*Procedia Mater. Sci.***11** 336-41
[7] Yang L, Li H, Liu J, Lu Y, Li S, Min J, Yan N, Men Z and Lei M 2016 *J. Alloys Compd.* **689** 812-9
[8] Daobin M, Chen Y, Wu B, Huang R, Jiang Y, Li L and Wu F 2016 *J. Alloys Compd.* **671** 157-63
[9] Hwang S S, Sohn M, Park H I, Choi J M, Cho C G and Kim H 2016 *Electrochim Acta* **211** 356-63
[10] Favors Z, Wang W, Bay H H, George A,Ozkan M and Ozkan C S 2014 *Sci.Rep.* **4** 4605

The impact analysis of postage multi-tariff system on gas transportation through the pipeline to increase gas price efficiency as the attraction of more environmentally friendly power plant substitution

A Bawono[1] and E Kusrini[1]

[1] Department of Chemical Engineering, Faculty of Engineering Universitas Indonesia, Kampus Baru UI, Depok 16424, Indonesia ; Corresponding Author: Tel.:62-85647166568, 62-21-7863516 Ext. 204; fax: 62-21-7863515;

E-mail: ajiagraningbawono@yahoo.com, ekusrini@che.ui.ac.id

Abstracy. The purpose of this study is to provide an analysis of the comparison and impact of changes in the model of Gas Transportation Tariff calculations through Pipes on the transportation of natural gas for fuel power plants in the region X. Which initially used a single tariff distance system that provides gas prices that are still not economically attractive, will be replaced with a more efficient multi-tariff postage stamp system. With the change of tariff calculation method, it will get the lower gas price and will inevitably increase the commercial appeal to be able to replace HSD (High-Speed Diesel) Generator which is expensive and cause more pollution. Initially, the substitution program is still less attractive because gas prices are still inefficient because the use of tariff calculation method is less precise. New tariffs are modeled by modifying cash flow and some variables with scenarios dividing tariffs into groups and re-modeling the volume of gas flowing in the pipeline through simulations using software and calculations. With the sensitivity test will also be done to provide the best scenarios to the efficiency of gas price formation. If gas prices are more efficient than fuel, it will further increase the attractiveness to accelerate fuel substitution of fuel with cleaner gas fuel and support Government programs. A comparison of the carbon decline will also be made to prove the substitution of diesel with gas to have a positive impact on the environment to provide an analysis of environmental issues. Finally, we propose that the application of a multi-tariff system will provide the best scenario for enhancing the efficiency of natural gas prices in the substitution of power generation from diesel into previously less attractive gas to be more efficient and more environmentally friendly.

1. Introduction

The Government of Indonesia is currently increasing the substitution program for the use of natural gas to replace fuel oil primarily for electricity generation and industry. From this program, people are expected to get fuel that is cleaner, safer, and cheaper [2]. One method for transporting the most common and most affordable gas from a source of production to a consumer center is through a pipeline. Provision of the infrastructure of natural gas pipelines is included in the type of general infrastructure activities such as electricity, telephone, and water business activities. The profit or income derived from the tariff payments is regulated by the Regulator [8] [10]. Profits are regulated through reasonable IRR (Internal Rate of Return) restrictions. Regulators or governments in some countries determine the maximum limits of IRR's fairness under the Cost of Capital or WACC (Weighted Average Cost of Capital) to determine the economics of return on investment [7] [11] [13]. The primary factor to support the program to increase the attractiveness of substitution of fuel oil source to natural gas for electricity and industrial needs, gas price should be lower and competitive. One important factor determining the price of natural gas through pipelines is the tariff for the transportation of natural gas. With a more efficient tariff, it will make gas prices cheaper and will help both electric and industrial companies to start switching to gas. It is worth noting that most of the tariffs for natural gas transportation through pipelines in Indonesia are determined by a single tariff distance system. Where tariffs are calculated by the number of pipelines passed by the gas. So that the actual gas volume in the same pipe is calculated by the sum of each pipe segment In the same pipe that is inefficient and causes the paid tariff to be more expensive, and it will reduce the attractiveness of the conversion of fuel oil to gas. Tariffs in Indonesia are also applied uniformly to all users of the pipe (shipper), so particular shipper that should get cheaper tariffs because its profit is limited to subsidies should pay the same tariff with profit-oriented shipper. To overcome the gas tariff to be more efficient it is necessary to study a tariff calculation with a new system. The proposed calculation system is postage stamp with a multi-tariff system. In this postage stamp system, several interconnected pipes are considered to be a single system. Multi-Tariff is the application of tariffs where in this case the transporter applies different tariff rates for each group of the shipper [9].

The purpose of this study is to provide an analysis of how the influence of the application of new model proposals in the calculation of tariffs for natural gas transportation through a pipeline that previously used. The method of distance single tariff system calculation compared with the new method of calculation of postage stamp multi-rate system. so that later will be obtained a proposed formulation can produce more efficient, fairer, more reasonable and cheaper gas prices as an attraction to encourage the conversion of cleaner, safer, and cheaper energy sources of fuel oil to gas.

2. Method

The research was conducted by simulating the case study of tariff calculation for natural gas transportation through the pipeline in open-access transmission pipe X consisting of four segments of the pipe section. Initially, the tariff for pipeline X was set with a single tariff distance system, so that the shipper transporting the gas to the end of the pipeline would pay four summed tariffs from each segment. Subsequent research is to change the system of tariff calculation which was initially using distance single tariff system was changed using multi tariff postage stamp system. Results obtained from both tariff calculation simulations, compared and analyzed.

The research stages started from the economic and technical data collection from open access transmission pipe X, then conducted literature study to enrich the reference source of the research. From the data obtained then done the simulation of calculation with a method of making free cash flow suitable for the system of tariff calculation which initially use single tariff distance system as well as for multi tariff postage stamp system. The results obtained will be analyzed by comparison and their impact on the pricing of gas. Furthermore, to support and prove that the use of gas as substitution of fuel oil for specific industry and electricity generation, will also make the comparison of the cost needed in the use of energy source. A comparison of the reduced impact of carbon emissions will also be made to prove that the use of gas will have a better effect on the environment [6].

3. Result and discussion

3.1. Scheme of calculation of tariff for natural gas transportation

The tariff calculation for the X transmission pipeline case study is conducted using several scenarios. In the first scenario, the tariff is calculated by the distance single tariff system model by calculating the tariff for each pipeline on the transmission pipeline X, so that there will be four tariff rates for the pipe. The equation model used is the general tariff equation as follows [7] [10] [12] :

$$tariff_n = \sum_{t=1}^{t} \frac{Cost\ of\ service}{Volume_t} = \sum_{t=1}^{t} \frac{(d)_t + (O\&M)_t + (A\&G)_t + (T)_t + (L)_t + (WACC \times NBA)_t}{V_t} \quad (1)$$

Where,

O&M : Operating and maintenance costs
A&G : Administration and general fee
d : Depreciation of CAPEX
T : Tax costs
L : Regulator levy
V : Volume of natural gas flowing
NBA : Asset base value (CAPEX)
WACC : Weighted Average Cost of Capital = IRR

Weighted Average Cost of Capital is formulated with the following equation [10] [12]:

$$WACC = CoE\frac{E}{E+D} + CoD\frac{D}{E+D} \quad (2)$$

Where CoE is the Cost of Equity, CoD is the Cost of Debt, E is Equity, and D is a Debt. The cost of Equity (CoE) and Cost of Debt (CoD) can be calculated by the following equation [4] [5]:

$$CoE = Rf + \beta(ICRP + BPMEM) \quad (3)$$

$$CoD = i \times (1 - T) \quad (4)$$

Where Rf is Risk-Free Rate, US Treasury Bond, Beta (β) is a measure of an investment portfolio fluctuation compared with the market (stock market), ICRP is Indonesia Country Risk Premium, BPMEM is based premium for the mature equity market, I is the interest rate, and T is Tax.

In a tariff calculation scenario with a distance system, tariffs are calculated for each pipeline. Economic and technical data such as investment value, operating and maintenance costs, and gas volume are calculated from each segment [3]. The total tariff for transmission pipeline X is the sum of each tariff per pipeline (see figure 1).

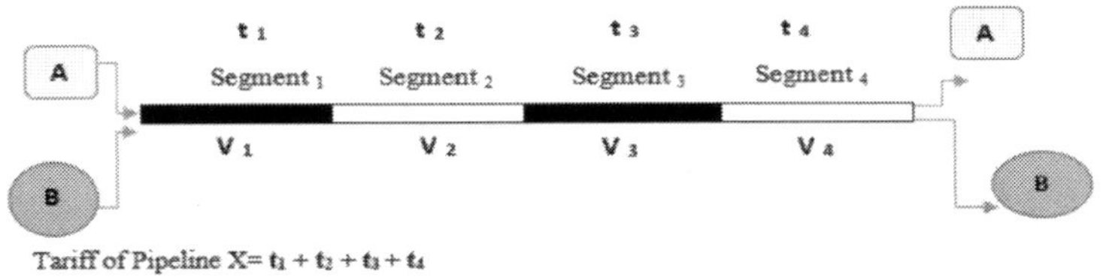

Figure 1. Distance system.

For the second scenario, the tariff is calculated using the postage stamp system model for single tariff as well as multi-tariff. For single postage stamp system tariffs are used equation (1). The difference is that the tariff for the postage stamp system is calculated by making several pipelines into one system. Economic and technical data such as investment value, operation, and maintenance cost are sums of each pipeline, while the volume used is the largest gas volume flowing in the X pipes (see figure 2).

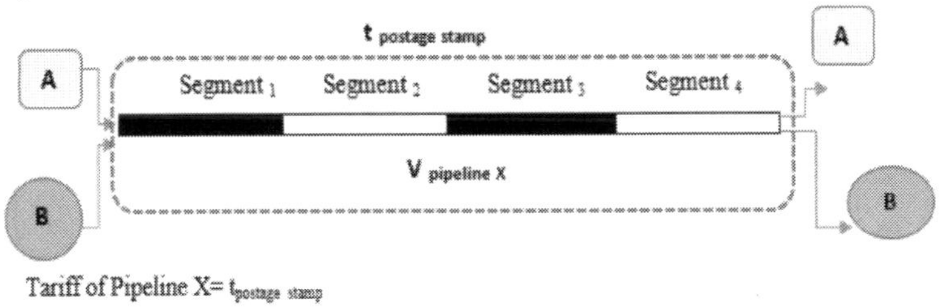

Figure 2. Postage stamp system.

The next scenario is the tariff calculation with multi tariff postage stamp system. The equation model used is the development of the previous calculation model that is as follows [1]:

$$multi-tariff_{(n)} = \sum_{i=1}^{t}\left(\frac{(d)_i + (O\&M)_i + (A\&G)_i + (T)_i + (L_{total})_i + (WACC \times NBA)_i}{V_1^2 + V_2^2 + V_3^2 + \ldots\ldots\ldots\ldots + V_{n-1}^2 + V_n^2}\right) \cdot (V_n)_i \quad (5)$$

Where n is the division of classes.

In the application of this multi-tariff system, the transporter (plumbing owner) applies different tariff rates for each shipper group, where specialized shippers requiring subsidies such as electricity generation and specialty industries will pay lower rates than profit-oriented shippers.

3.2. *Results and tariff calculation analysis*

After simulation of the calculation on the data transmission pipe open access X (see table 1) with various scenarios that have been made, then obtained the following results (see table 2 and table 3):

Table 1. Economic and technical data of transmission pipeline X

No	Pipeline Segment	specification		NBA (USD)	O&M (USD)	Volume (MMSCFD)
		Diameter (Inch)	Lenght (Km)			
1	Pipeline Segment 1	28	369,7	107.183.618	35.980.269	278
2	Pipeline Segment 2	28	52,63	15.258.518	5.122.172	278
3	Pipeline Segment 3	10	4,65	481.475	161.627	92
4	Pipeline Segment 4	24	3,42	849.881	285.298	92
	Total	90	430,4	123.773.492	41.549.366	

Parameter	Value
Depreciation Time and Inflation rate	10 Year and 1.31 %
Rf, BP MEM, ICRP, and β	2.14% , 5.70%, 3.24%, and 1.070
Coe and CoD	11.71% and 0%
D	0 %
E	100%
Tax and interest (i)	25% and 0
WACC	11.71%

Table 2. Calculation Result of single tariff system and tariff comparison

No	Pipeline Segment	Specification		Single Tarif (USD/Mscf)	
		Diameter (Inch)	Lenght (Km)	Distance System	Postage Stamp System
1	Pipeline Segment 1	28	369,7	0,584	**0,674**
2	Pipeline Segment 2	28	52,63	0,083	
3	Pipeline Segment 3	10	4,65	0,008	
4	Pipeline Segment 4	24	3,42	0,014	
			Total	**0,689**	

Table 3. Calculation Result of multi-tariff system and tariff comparison

Parameter	Gas Volume (MMScf/day)	Postage Stamp Tarif (USD/MScf)	
		Single Tarif	Multi Tarif
Category 1 (Power Generation or Special Industries)	92	0,674131239	0,401
Category 2 (Profit Industries)	186		0,809
Total Revenue (USD)		683.184.822	683.093.598
Δ Revenue			0,013%

From the results obtained can be seen that the tariffs generated by the method of calculating the single distance system tariffs produce the largest total tariff of 0.689 USD / MScf (see table 2). This condition happens because in this system the gas volume is calculated from each segment. The volume of gas flowing in a pipe consisting of several segments can be considered as one system (postage stamp). With the postage stamp system, the resulting tariff will be lower than the distance system which is 0.674 USD / MScf (see table 2). And to make the tariff more efficient again for the particular class, the multi-tariff system will get more efficient tariff that is 0.401 USD / MScf so that it will further increase cost savings and increase the attractiveness of substitution of the energy source of fuel oil to natural gas.

3.3. Fuel price calculation, price comparison for specific shipper energy source and carbon emisison
To prove that the use of natural gas through the pipeline can improve the efficiency of production cost, it is calculated the energy cost required to generate electricity by using High-Speed Diesel (HSD) compared to using natural gas flowed through pipeline X. Comparison of carbon emissions resulting from burning some The fuel will also be made so that it is known to affect the environment [6]. The results are as follows (see table 4 and table 5):

Table 4. Result of calculation and gas price comparison for pipe X with several scenarios.

Parameter	Production Gas Price (USD/MMBtu)	Transmission Tarif (USD/MMBtu)	Distribution Cost (USD/MMBtu)	Tax	Total Gas Price (USD/MMBtu)
Distance System	5,44	0,6887		0,0689	6,1976
Postage Stamp System					
Single Tarif	5,44	0,6741	0	0,0674	6,1815
Multi Tarif	5,44	0,4016	0	0,0402	5,8817

Table 5. Fuel price calculation, comparison and carbon emisison energy source.

Energy Source	Volume (MMScf/d)	Energy Requirement (MMBtu/d)	Price (USD/MMBtu)	Total Cost (USD/day)	CO2 emitted (pound per milion Btu)	Total CO2 emitted (pound/day)
High Speed Diesel (HSD)			26,00	2.392.000	161,3	14,8396
Natural Gas Distance	92	92.000	6,198	570.181		
Postage stamp					117	10,764
Single Tarif			6,182	568.702		
Multi Tarif			5,882	541.121		

It can be proven that the use of natural gas as an energy source will result in more efficient production costs compared to the use of fuel oil. Also, it can also be proven that the use of natural gas will produce carbon emissions as a result of lower combustion when compared with fuel oil, so it can be concluded that the use of natural gas as a substitution of fuel oil resources will have a good in economic and environmental benefits.

4. Conclusion

The use of natural gas as an alternative energy source for the substitution of fuel oil for the needs of power plants and other industrial needs will have a good impact. The application of multi-tariff postage stamp method in tariff calculation for natural gas transportation through the pipeline will increase efficiency and decrease natural gas price compared to single tariff method and single tariff postage stamp. With the use of natural gas will reduce production costs while increasing efficiency and more environmentally-friendly. Finally, it is expected to increase the attractiveness of more expensive and more expensive substitutes of fuel oil to cheaper and environmentally-friendly natural gas.

5. References

[1] Bawono A and Kusrini E 2017 *Tariff calculating model for natural gas transportation through open access transmission pipeline with multi tariff* (Depok: Universitas Indonesia)

[2] April L, Owen Z, Jeffrey L, et al 2012 Interactions, complementarities and tensions at the nexus of natural gas and renewable energy *The Electricity Journal* **25(10)** 38–48

[3] Chandra, Vivek, 2006 *Fundamentals of Natural Gas: An International Perspective* (USA: PennWell Corporation Tulsa Oklahoma) pp: 109-112

[4] Damodaran A 2012 *Equity Risk Premiums (ERP): Determinants, Estimation and Implications* Stern School of Business

[5] Damodaran A 2015 *Country Risk: Determinants, Measures and Implications – The 2015 Edition* Stern School of Business

[6] Energy Information Administration (EIA) 2017 *How much carbon dioxide is produced when different fuels are burned?* https://www.eia.gov/tools/faqs/faq.php?id=73&t=11

[7] Federal Energy Regulatory Commission (FERC) 1999 *Cost-of-Service Rates Manual* Washington, D.C. 20426 United States of America

[8] Hunt, Paul, 2014 *The Weighted Average Cost of Capital for Electricity and Gas Networks* Submission on the WACC to CMA Energy Market Investigation

[9] Png I P L and Cheng C W J 2001 *Pricing Policy* (Singapure: NUS School of Computing)

[10] Purwanto S W and Sommeng A N 2013 *The Regulation of Business Activities of Natural Gas Through Pipelines* (Jakarta: BPH Migas Indonesia) pp: 55-102

[11] Tapia J 2010 The 'Duty To Finance', The Cost of Capital and The Capital Structure of Regulated Utilities: Lessons from The UK *J. of Utilities Policy* **22** 8-21

[12] The Oil and Gas Downstrean Regulatory Agency of Republic of Indonesia, BPH Migas 2013 Rule Number 08 of 2013 Regarding *The Determination of Tariffs for the transportation of natural gas through pipes*

[13] Weijermars R 2011 Weighted average cost of retail gas (WACORG) highlights pricing effects in the us gas value chain: Do we need wellhead price-floor regulation to bail out the unconventional gas industry? *The Energy Policy J.* **39** 6291–6300

IOP Conf. Series: Earth and Environmental Science **105** (2017) 012110 doi:10.1088/1755-1315/105/1/012110

Utilization of waste heat from separation process of Ulubelu's geothermal power plant by implementing an Absorption Refrigeration System (ARS) to improve plant performance

Nasruddin[1] **and Agung Satrio Wibowo**[1]

[1] Departemen of Mechanical Engineering, University of Indonesia, 16424, Indonesia

E-mail : nasruddin@eng.ui.ac.id

Abstract. Utilization of geothermal energy for water-dominated reservoir usually involves a separation process that turns geothermal fluid mixtures into pure steam and brine water. This process also occurred on Ulubelu's geothermal power plant (GPP). As waste energy from the power generation process, the remaining heat energy in brine water is still high enough to run an absorption refrigeration system (ARS). This study proposed an integrated power generation and absorption system that operate side by side for a further cooling process. ARS will be employed to produce a lower temperature of cooling water from the GPP's cooling tower and then pass it to GPP's condenser. The lower temperature of cooling water will affect steam condensation process and the vacuum pressure of condenser, moreover, increase power production and exergy efficiency of Ulubelu's GPP. The improvement of exergy efficiency & production capacity will be observed along with the rise of the investment cost as form as the annual cost of the ARS. A Multi-objective optimization using genetic algorithm will be conducted to minimize exergy efficiency and the annual cost of ARS. The optimization will be conducted using MATLAB along with EES for work fluid properties database. The temperature of generator, absorber, condenser, and evaporator of ARS are used as decision variables. Finally, the effect of integrated system and optimum value for each decision variables are presented in this study.

1. Introduction

Increased demand for energy, decline of energy resources and environmental impact due to energy utilization have resulted to calls for a sustainable approach to development and management of energy resources [1]. Ulubelu is one of several water dominated reservoar in Indonesia [2]. The Ulubelu's geothermal fluid mixtures exploited from the ground need to be separated to extract steam from the liquid.

The extracted steam is then used to drive the turbine which coupled to a generator to produce electricity. The liquid separated from steam called brine water is considered a waste and in most cases is injected back to the ground. However, the waste liquid or brine water possesses a considerable amount of exergy, which can be utilized for other thermal applications such as absorption for cooling purposes. It is therefore clear that exergy efficiency of the geothermal power plant is lowered due to irreversibility during energy conversion process within the subsystems and exergy quality is lost.

In order to improve conversion efficiency of a geothermal power plant, cooling is essential for it increases the heat rejection, raises power output and increases heat to power conversion ratio [3]. Tesha [3] & Nyambane [4] conducted a simulation for integrating single effect absorption refrigeration system to produce a lower temperature of cooling water used in condenser to enchance the capacity for power generation. The single effect absorption has relatively low COP, therefore, multi effect absorption system with high temperature heat resource have now developed [5] but the research about combining multi effect absorption in geothermal power plant never conducted before.

Arora, et. al [6], Azhar, et. al [7], Ghani, et. al [8] performed energy and exergy of single effect and double effect absorption LiBr-Water system to maximize the COP and exergetic efficiency. Misra, et. al [9] conducted thermo economic optimization for double effect absorption to minimize overall product cost. Shirazi, et. al [10] performed multi-objective optimization of solar powered absorption chiller to find its optimum parameter for energy and economic objective function. But, the multi objective optimization for both objective function never conducted on integrated GPP and absorption refrigeration system. It is important to analyze economic component as the high efficiency of power plant need more cost and it worse for geothermal business. The power plant and absorption refrigeration system needs operational condition with high efficiency and low cost.

This study aims to utilize the heat contained in brine water that leaves from separator by implementing an absorption refrigeration system to improve Ulubelu GPP's performance. ARS will be employed to produce a lower temperature of cooling water from the GPP's cooling tower and then pass it to GPP's condenser. The lower temperature of cooling water will affect steam condensation process and the vacuum pressure of condenser, moreover, increase power production and exergy efficiency of Ulubelu's GPP. To be precise, the objectives are : to optimize the combination of absorption refrigeration cycle and power plant with genetic algorithm optimization from exergy desctruction and cost as objectives function with six parameters chosen as decision variables.

2. Description system

The GPP & Absorption refrigeration system in this study consist of several components, there are: separator, turbine, generator, condenser, cooling tower and absorption refrigeration. In this refrigeration system, the components are absorber, high desorber (D2), low desorber (D1), condenser, evaporator and solution heat exchanger (SHX 1 & SHX 2). The schematic diagram is shown in figure 1.

At point 1 in figure 1 the solution rich in refrigerant and a pump forces the liquid through solution heat exchanger 1 and 2 to high desorber (D2). The process incereases the temperature in the each SHX. The added thermal energy from brine water of separation process boils the refrigerant from the solution and separates the refrigerant and water. It results in refrigerant vapor (17) flows to condenser, where heat is rejected as the refrigerant condenses in condenser 1. The heat of condensation is utilized by low desorber to separate refrigerant vapor from the weak solution from D1. Secondary vapour is produced on D2 outlet, which the refrigerant vapour together flows to condenser 1. The total amount of liquid refrigerant leaving condenser 1 is sum of refrigerant vapour leaving D1 and D2. The condensed liquid then flows through expansion valve to the evaporator thus lowering its pressure. This change in pressure allows the evaporator temperature to be low enough for refrigerant to absorb heat from the water being cooled (27). the refrigerant evaporates and passes to the absorber where it gets absorbed by strong solution coming from D1 through heat exchanger SHX1. This process produced weak solution in the absorber and then it is pumped to the high desorber and the cycles completes [11].

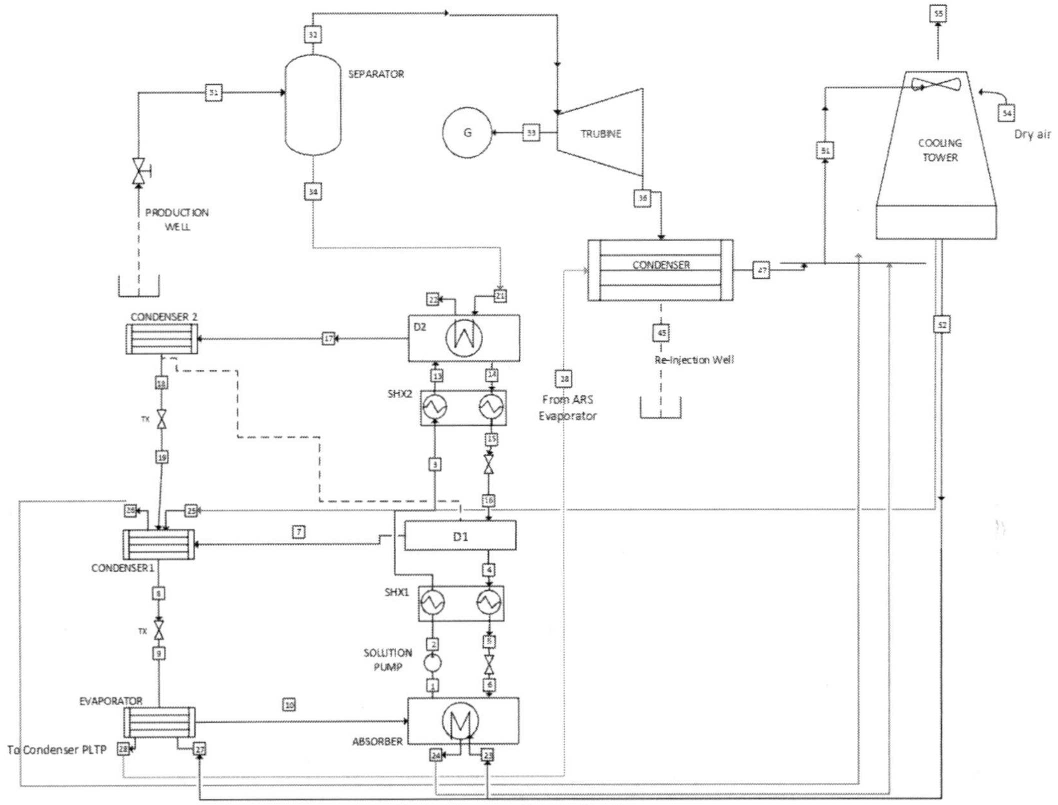

Figure 1. Schematic System of GPP and absorption refrigeration system.

The performance of the power plant is very dependent on the performance of the condenser to condense the steam from turbine output [12], the lower pressure on the condenser means more power to be produced by turbine, this is due to the large difference of enthalpy. This results in an entalhpy difference that reduces power generation results. Under these circumstances further cooling systems are required to lower the cooling water temperature. In this research, we are trying to use double effect absorption to reduce cooling water from cooling tower before it goes to power plant condenser for steam condensation process.

All the components in this system can be treated as control volume and the GPP-ARS is modeled on mass and energy conservation. Some assumptions are made as follows to simplify the mathematical model:

(1) Potential and kinetic energy change during the heat exchange and at all fluid streams are ignored.
(2) Heat exchanger is well insulated from the surroundings; that means heat exchange only occurs between hot and cold stream.
(3) The working fluid at absorber, condenser I and condenser II outlet is saturated liquid.
(4) The flows across the throttle valve are isenthalpic.

3. Mathematical models

Thermo economics which a combine concepts of both exergy and economic is a relatively new object in engineering. In exergoeconomic analysis, the cost of thermodynamic inefficiencies is integrated in the cost of products of the system [13-15]. Thermo economics reveals the cost formation process in energy systems which enables engineer to design cost effective systems [16].

3.1. Energy and exergy analysis

To simulate and analyse the thermodynamic performance of the proposed system, a code is developed in Engineering Equation Solver (EES) software. Each system threated as control volume which the mass and energy convservation as well as the principles of second thermodynamics law are applied.

The first law consists of mass, concentration and energy balance at each component of the combined system, while second law analysis dels with exergy destruction of the system components and exergetic efficiecny of the system [12].

Mass balance

$$\sum_{in} \dot{m} = \sum_{out} \dot{m} \tag{1}$$

Energy balance

$$\dot{Q} - \dot{W} = \sum_{out} \dot{m}.h - \sum_{in} \dot{m}.h \tag{2}$$

Exergy balance

$$\dot{X_{des}} = \sum_{out} \left(1 - \frac{T_{amb}}{T_j}\right) \dot{Q_J} - \sum_{in} \dot{m}.E_x - \sum_{out} \dot{m}.E_x \tag{3}$$

Exergy rate

$$\dot{E_{x,n}} = \dot{m}(h - h_o - T_o(s - s_o)) \tag{4}$$

In this research, double effect absorption using two different working fluids, water as refrigerant and lihtium bhromide as absorber. The initial design parameter for double effect absorption listed as follows : T_{21}=173.5°C, T_{evap}=5°C, T_{cond}=35°C, T_{des}=120°C, T_{abs}=32°C, nshx=0.7, T_{cw}=26°C

The requirement of the heat exchanger area for the absorption system was calculated using the log mean temperature difference (LMTD) method [17]. Since rate of heat transferred is a linear function of theoverall coefficient of heat transfer, the heat transfer area of the heat exchanger and the LMTD, for the heat transfer process could be derived from equation (8). The heat exchanger area then used to calculate the annual cost for this refrigeration system.

$$Q = U.A.LMTD \tag{5}$$

Table 1. *Overall coefficient of heat transfer* (U) value for each components [3].

Heat Exchanger	Type	U[kW/m^{2}C]
Absorber	Shell & tube	0.85
SHX	Plate	1.1
Desorber	Shell & tube	0.85
Condenser	Shell & tube	1.4
Evaporator	Shell & tube	1.5

3.2. Mathematical model validation for energy and exergy analysis

To verify the validity of the mathematical model, the available data in literature serves as contrast to the simulation results. The absorption refrigeration cycle are validated by data in Ref. [11]

3.3. Economics and cost analysis

The cost system in this research is considered as investement cost. Invenstment cost can be calculated by sum of each investation cost for each component of the system, scrubber, turbine, condenser, cooling tower, desorber, absorber, evaporator,and solution heat exchanger. The general equation for economic analysis [18-20]:

$$C_E = C_B(\frac{Q}{Q_B})^M \quad (6)$$

Table 2. Constant C_B and M value for each components.

Component	Market Price ($) $[C_B]$	M Constant
Scrubber	4920 USD (0.1 m^3)	0.53
Turbine	13 million USD (30MW)	0.6
Condenser	-	-
Cooling Tower	4430 USD (10 m^3/h)	0.63
Absorber	16500 USD (100 m^2)	0.6
Solution heat exchanger	12000 USD (100 m^2)	0.6
Desorber	17500 USD (100 m^2)	0.6
Condenser ARS	8000 (100 m^2)	0.6
Evaporator	16000 (100 m^2)	0.6

The capital recovery factor (CRF) equation can be written as

$$CRF = \frac{i(1+i)^n}{((1+i)^n)-1} \quad (7)$$

In evaluating and optimizing energy systems' performance, minimization of the annual cost of output power is an important objective [17]. In the present study, this parameter is selected as the objective function on which the system performance is optimized. All the capital investment included in this parameter which is defined as [21]:

$$C_{tot} = \left(C_{sep} + C_{tur} + C_{con} + C_{ARS}\right).CRF \quad (8)$$

4. Result and discussion

In this section the process optimization on GPP and Absorption refrigeration system will be displayed. The operation parameters of multi objective function and ranges of key thermodynamic parameters are listed at table 3.

Table 3. Operation parameters for optimization.

Term	Value
Population size	50
Ranges of throttle valve pressure	400 – 900 kPa
Ranges of absorber temperature	30-35 °C
Ranges of desorber temperature	120-140 °C
Ranges of condenser temperature	30-40 °C
Ranges of evaporator temperature	3-16 °C

Table 3 shows the range values of thermodynamic parameters used for optimization. Which has lower and upper bound. The results of multiobjective optimization are shown in table 4.

Table 4. Multi objective optimization results.

Term	Value
Throttle valve pressure	779.11 kPa
Absorber temperature	32.98 °C
Desorber temperature	121.13 °C
Condenser temperature	37.26 °C

Evaporator temperature	11.1 °C
Total exergy destruction	34257 W
Cost	73156.83 $

Multi-objective optimization needed to know what value of parameters that will give the optimum value for both objective function [6]. Figure 2 shown the graphic of multiobjective optimization as the result from MATLAB multiobjective optimization. The graphic called pareto front and every dot on that graphic represent optimum value. In order to find the optimum value with two objective function, multicriteria selection method used [21]. In this study, two variables are exergy destruction and cost as consideration of multi objective optimization. When exergy values rise from 3.4 MW to 3.45 MW, there is a decrease amount of the cost. The best economic performance occurs at point B where the exergy value is 3.44 MW with lowest cost 72243.6 $. While the maximum thermodynamics performance occurs at point A 3.4 MW with the highest annual cost 74608.9 $. It's clearly seen that multiobjective is needed to find the optimum value between exergy destruction and annual cost. Finally with TOPSIS method, the optimum thermodynamic performance occurs at 3.42 MW exergy value with a cost value 73156.8 $.

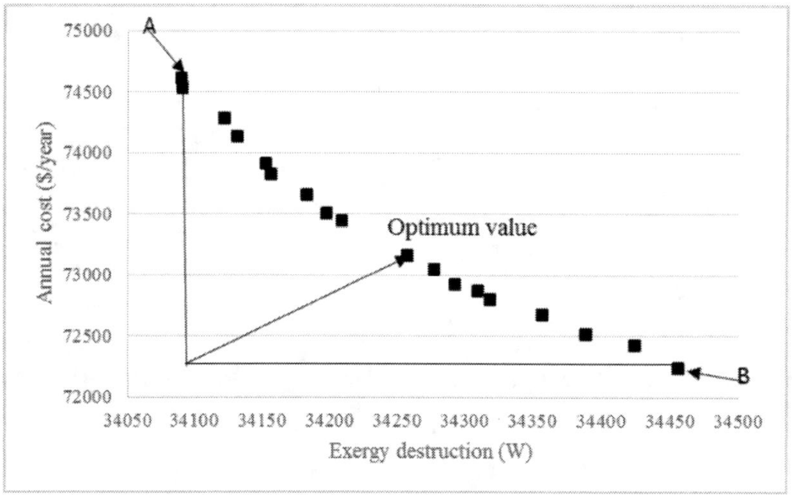

Figure 2. Distribution of the Pareto optimal solutions for Exergy Destruction and Specific Cost.

5. Conclusion

In this study, the thermodynamic & economic analysis for implementation of absorption refrigeration system to the Ulubelu GPP was simulated by using MATLAB-EES to obtain optimum value. Thermodynamic and exergoecenomic models were developed on EES to analyze the performance of such systems. Multi-objective optimization by MATLAB optimtool used to determine the optimum value of thermodynamic and exergoeconomic value of a system using 5 decision variables; throttle valve pressure, temperature of absorber, condenser, desorber and evaporator on absorption. The results showed that the optimum throttle valve pressure occurred at 779.11 kPa, temperature of absorber at 32.98°C, desorber at 121.13°C, condenser at 37.26 °C and evaporator at 11.1°C. with those optimum values, the system yield 34257 W of exergy destruction and annual cost of 73156.83 $/year.

6. References

[1] Nasruddin, Alhamid M I, Daud Y, Surachman A, Sugiyono A, Aditya H B, Mahlia T M I 2016 Potential of geothermal energy for electricity generation in Indonesia: A review *Renewable and Sustainable Energy Reviews* **53** 733-740

[2] Nugroho A P 2017 Geothermal power generation in Indonesia, a country within the ring of fire: Current status, future development and policy *In Renewable and Sustainable Energy Reviews*

[3] Tesha 2009 *Absorption Refrigeration System as an Integrated Condenser Cooling Unit in a Geothermal Power Plant* (Reykjavik, Iceland: United Nations University)

[4] Nyambane N 2015 *Hybridization of Cooling System of Olkaria II Geothermal Power Plant: Utilization of Energy and Exergy Concepts* (Nairobi: University of Nairobi)

[5] Wang X and Chua H T 2009 Absorption cooling: A review of lithium bromide-water chiller technologies *Recent Patents on Mechanical Engineering, Bentham Science Publishers Ltd.* **2** 193-213

[6] Arora A and Kaushik S C 2009 Theoretical analysis of LiBr/H2O absorption refrigeration systems *Int. J. Energy Res.* **33** 1321–40 doi:10.1002/er.1542.

[7] Azhar M D and Siddiqui M A 2017 *Energy and Exergy Analyses For Optimization of The Operating Temperatures in Double Effect Absorption Cycle* Int. Conf. on RAA. vol 109 pp 211-218

[8] Ghani M U and Zaman M 2016 *Thermodynamic Modelling and Optimization of Double Effect Series Flow Libr-H_2O Vapor Absorption Chiller* (Amsterdam: Elsevier)

[9] Misra R D, Sahoo P K and Gupta A 2005 Thermoeconomic evaluation and optimization of a double effect H_2O/LiBr vapour-absorption refrigeration system *Int. J. of Refrig* **28(3)** 331-43

[10] Shirazi A 2017 A Comprehensive, multi-objective optimization of solar-powered absorption chiller systems for air-conditioning applications *Energy Conversion and Management* **132** 281-306

[11] Herold K E R and Klein S A 1996 *Absorption Chillers and Heat Pumps* (Florida: CRC Press, Inc)

[12] DiPippo R 2016 *Geothermal Power Plants Principles, Applications, Case Studies and Environmental Impact 4th edition* (Kidlington, UK: Elsevier Ltd.)

[13] Ahmadi P, Dincer I and Rosen M A 2012 Exergo-environmental analysis of an integrated organic Rankine cycle for trigeneration *Energy Conversion and Management* **64** 447-453

[14] Tsatsaronis G 2007 Definitions and nomenclature in exergy analysis and exergoeconomics *Energy* **32**(4) 249-253

[15] Abusoglu A and M Kanoglu 2009 Exergoeconomic analysis and optimization of combined heat and power production: A review *Renewable and Sustainable Energy Reviews* **13(9)** 2295-2308

[16] Zare V, et al. 2012 Thermoeconomic analysis and optimization of an ammonia–water power/cooling cogeneration cycle *Energy* **47(1)** 271-283

[17] Bejan A, Tsatsaronis G and Moran M 1996 *Thermal Design and Optimization* (New York: John Wiley & Sons)

[18] Smith R M 2005 *Chemical Process: Design and Integration, 2nd ed* (New York: John Wiley & Sons)

[19] Estevez J 2012 *Geothermal Power Plant Projects in Central America: technical and financial feasibility assessment model* (Reykjavík, Iceland: University of Iceland)

[20] Misra R D 2003 Thermoeconomic optimization of a single effect water/Lihitum Bromide vapour absorption refrigeration system *Int. J. of Refrig.* **26(2)** 158-169

[21] Nasruddin, Sholahudin S, Giannetti N and Arnas 2016 Optimization of a cascade refrigeration system using refrigerant C_3H_8 in high temperature circuits (HTC) and a mixture of C_2H_6/CO_2 in low temperature circuits (LTC) *Applied Thermal Eng.* **104** 96-103

IOP Conf. Series: Earth and Environmental Science **105** (2017) 012111 doi:10.1088/1755-1315/105/1/012111

Comparative study of the monthly global solar radiation estimation data in Jakarta

Nasruddin[1], **Muhammad Arif Budiyanto**[1] **and Reyzando Nawara**[1]
[1] Department of Mechanical Engineering, Faculty of Engineering, Universitas Indonesia, Depok, Indonesia

E-mail: nasruddin@eng.ui.ac.id, arif@eng.ui.ac.id, reyzandonawara@ui.ac.id

Abstract. The objective of this paper is to get the most suitable model of monthly global solar radiation estimation in Jakarta which performs better than the others model. In this study, measured data of solar radiation were utilized to be compared with the empirical models from the literature and a new based-temperature model, to estimate the monthly intensities of global solar radiation in Jakarta, Indonesia (Latittude: 6.13° S, Longitude: 106.8° E). Global solar radiations were measured for two complete years, and the estimation of monthly solar radiation were calulated by four empirical models and also one linear regression model. The accuracy of the models was then analyzed in terms of some statistical indicators such as Mean Bias Error (MBE) and Root Mean Square Error (RMSE). The statistical analysis of result revealed that the empirical models have not the same accuracy, and it is concluded that there are the models that can be preferred for the estimation of monthly global solar radiation in Jakarta, Indonesia. The obtained results indicate that Allen is performing better than the others by obtaining maximum MBE of 0.52 and maximum RMSE of 1.82 . It also can be said that Allen is the most suitable model for estimating a monthly global solar radiation in Jakarta, Indonesia.

1. Introduction

Solar energy is the renewable radiation energy of the sun and it is fast becoming an alternative to other conventional sources of energy [1].There are several factors that affect the production of solar irradiance energy, such as the solar elevation. The amount of solar irradiance also varies depending on the time of the day and the season. The altitude of an areas also affect the amount value of solar irradiance. It has been found that solar irradiance increases by up to 4% for every 300 m increase in altitude [2]. Clouds also have a big impact on the amount of solar irradiance reaching the surface of earth. On average, clouds absorb or scatter about 20% of the incoming solar radiation.

Indonesia has several areas with a really great solar energy potential, especially those that located along equator line. Despite the potential of solar energy resources, the availability of its data is still limited. Regarding to that condition, there's been several studies to learn more about this solar radiation and to develop a new empirical models in order to estimate the amount of solar irradiance especially for the location that has limited measured data, such as at 1992, Graham L. Morrison from UNSW Sydney along side Sudjito create new model based on ambient temperature at 3 cities, Denpasar, Jakarta, and Kupang[3]. This empirical models are being used to estimate and predict the value of solar radiation using different meteorological data such as heat, sun duration period, and so on. Heat is one of the commonest indicators of solar energy radiation[4]. Therefore, temperature is a meterological indicator which can be used to quantify the size of solar radiation at a given location[5].

This study aims at using temperature data measured in Jakarta for a period of two years (2014-2015), as an independent variable to be calculated by four empirical models and one linear regression new models based on Javier Almorox research. The result from those models then being compared and analyzed statistically to know which one is the most suitable model to estimate the monthly solar radiation in Jakarta, Indonesia(Latittude: 6.13° S, Longitude: 106.8° E). The statistical analysis is performed to assess the accuracy of the proposed models by ranking the models with a purpose of finding the most suitable model.

2. Materials and methods

2.1. Characteristics of location under study

Jakarta is the capital city of Inonesia which located in DKI Jakrta province, with population 10187595 and 661.52 km^2area respectively. This city is located at 6.13° S latitude and 106.8° E longitude. Jakarta has an average monthly temperature varies from 24° C - 34° C approximately (**Table 1**). Jakarta is the largest city in Indonesia and classified as the industrial city. The region experiences a warm climate with a larger amount of rainfall in summer due to warm-wet flow from the tropics than in winter.

2.2. Instrumentation and data collection

The measurements were carried out for a period of two years (Januari, 2014-December 2015) to asses the solar radiation potential. The measured data then being calculated statistically to show which month reach the highest value of global solar radiation intensities. The measured data is shown on table 2 below. It can be observed that at 2014, the months August-Octobers show a high solar energy potential in terms of global solar radiation with September being the month where the mean global solar radiation reaches a maximum value of 16.62 MJ/m^2-day. While the minimum global solar radiation is observed in January and February due to the fact that in that moment is rainy season. The average minimum value of 7.21 MJ/m^2-day is observed in the month of January. For 2015, it can be observed that September reaches the highest value of average monthly global radiation of 15.49 MJ/m^2-day, while February is observed as the lowest value of monthly global solar radiation with value of 9.60 MJ/m^2-day.

Meteorological data comprising of maximum and minimum temperatures measured in Jakarta for two years period (2014-2015) were obtained from Indonesia Meterological Agency (BMKG). These maximum and minimum temperatures data of the study area were used as the materials to calculate the estimated monthly global solar radiation values by five empirical models.

Table. 1. Jakarta's Average Temperature.

	2014 Average Temperature		2015 Average Temperature	
Month	Tmin	Tmax	Tmin	Tmax
January	24.65	29.3	25.14	30.76
February	24.71	29.91	24.85	29.77
March	25.56	31.33	25.69	31.34
April	26.52	33.06	26.25	32.23
May	26.78	32.8	26.46	32.71
June	26.56	32.22	26.16	32.33
July	25.62	32.14	25.82	32.35
August	25.8	32.79	25.3	32.24
September	26.55	32.87	25.46	32.58
October	27.06	33.76	27.02	33.22

November	27.04	33.4	26.95	33.04
December	26.21	32.48	26.11	32.15

Table 2. Summary of statistical measured data.

						2014						
Months	Jan	Feb	Mar	Apr	May	Jun	Jul	Aug	Sep	Okt	Nov	Des
Statistics												
Mean	7.21	9.87	11.41	13.21	11.15	10.63	11.89	13.85	16.62	14.35	12.27	12.78
Median	7.31	10.42	11.13	12.45	10.65	10.27	11.25	13.39	15.44	13.03	11.50	12.33
Skewness	2.08	1.44	1.39	1.23	1.26	1.23	1.20	1.00	0.92	1.05	1.24	1.25
						2015						
Months	Jan	Feb	Mar	Apr	May	Jun	Jul	Aug	Sep	Okt	Nov	Des
Statistics												
Mean	10.76	9.60	12.60	12.26	12.46	12.21	13.07	15.32	15.49	15.48	12.66	12.65
Median	8.48	8.62	11.44	12.23	11.19	11.50	12.74	15.35	16.05	14.44	11.98	12.88
Skewness	1.60	1.74	1.37	-0.18	1.14	1.10	1.02	0.87	0.85	0.92	1.10	1.01

3. Empirical models

The monthly global solar radiation models are based on empirical correlations and depends on the most common meteorological parameters including cloud cover, ambient temperature, relative humidity, and also sunshine duration, which is the most widely used meteorological parameter to estimate monthly global solar radiation[6]. There are several methods to estimate and predict a monthly solar radiation, however since the temperature data available more in Indonesia's meteorological station, so the empirical correlations that was being used in this study is the temperature-based models. Temperature-based models for estimating global solar radiation utilize the direct affects of atmospheric transmissivity. These models assume that a high or low transmissitivity is due to an increase or decrease in air temperature. In this study, the estimation of the monthly global solar radiation calculated based on the temperature data measurement that available from BMKG shown in figure 1.

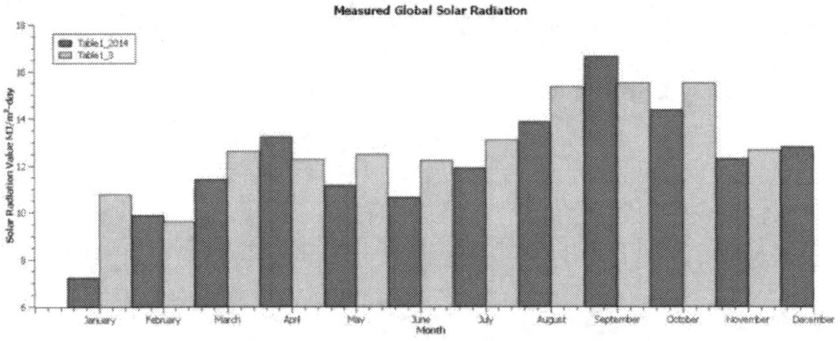

Figure 1. Measured Global Solar Radiation 2014 and 2015.

3.1. Model 1

Annadele et. al. proposed the estimation of solar radiation based on maximum and miniumum temperatures by take account the effect of decreased altitude and atmospheric thickness on global solar radiation by the suggestion of a correction factor. The model suggested by Annadele et. al. [7] :

$$G_D = A(1 + 2.7x10^{-5}Z)(T_{max} - T_{min})^{0.5}G_H \quad (1)$$

A is an empirical coefficient and the recommend value is 0.1072 in Jakarta. Z is the elevation in meter, T_{max} and T_{min} is the maximum and minimum of temperature in Celcius degree. Then G_H is extraterrestrial radiation which is a function of latitude and day of the year.

3.2 Model 2

Allen proposed a model of global solar radiation based on maximum and minimum temperature [8], as in:

$$G_D = A(T_{max} - T_{min})^B G_H \tag{2}$$

A and B are empirical coefficient, which the value in Jakarta for A is proposed to be 0.07768, while for B is to be 0.905. Z is the elevation in meter, T_{max} and T_{min} is the maximum and minimum of temperature in Celcius degree and also G_H is extraterrestrial radiation which is a function of latitude and day of the year[8].

3.3. Model 3

Bristow and Campbell proposed a simple model to estimate global solar radiation which is assumed to be exponential function of temperature difference as in [9]:

$$\mathrm{G_D} = \mathrm{A}[1 - \exp(-\mathrm{B}\Delta\mathrm{T}^{\mathrm{B}})\,\mathrm{G_H}] \tag{3}$$

$$B = 0.036\exp(-0.154\ \Delta\mathrm{T}) \tag{4}$$

A and C are empirical coeffecient which the value for A is 0.4895, and for C is to be 2.35.

3.4. Model 4

Hargreaves and Samani were the first to suggest that incident radiation could be evaluated from the difference between daily maximum and daily minimum temperature. The equation introduced by Hargreaves and Samani [10] :

$$\mathrm{H_c} = \mathrm{H_o} \times [\mathrm{Kr} \times (\mathrm{T_{max}} - \mathrm{T_{min}})^{\frac{1}{2}}] \tag{5}$$

where, Hc is the estimated solar radiation, Ho is the extraterrestrial solar radiation, Tmax is the maximum daily temperature, Tmin is the minimum daily temperature and Kr is an empirical coefficient that can be obtained by :

$$\mathrm{Kr} = \mathrm{A}(\mathrm{T_{max}} - \mathrm{T_{min}})^2 + \mathrm{B}(\mathrm{T_{max}} - \mathrm{T_{min}}) + \mathrm{C} \tag{6}$$

Kr is a function of the daily temperature range where, A; B; C are empirical coefficient

3.5. Model 5

A correlation analysis between variables provided by the meteorological station (daily maximum temperature and dailiy minimum temperature) and the extraterrestrial solar radiation (calculated for the model's application site) was performed. The general equation for new based-temperature model by linear regression :

$$\mathrm{H_c} = \mathrm{AT_{max}} + \mathrm{BT_{min}} + \mathrm{CH_o} + \mathrm{D} \tag{7}$$

Values for A is 0.62, B is -0.46, C is 0.42 and D is -7.07

4. Result and discussion

Four empirical models from literature and also one new based-temperature models by Jaview Almorox that was developed than being evaluated from its statistical point of view. The resulting values of monthly global solar radiation are presented in figure 2 and figure 3 below.

To evaluate the calculation result from the developed models above, the performance of those three models than being examined and analyzed by statistical test such as Mean Biass Error (MBE) and also Root Mean Square Error (RMSE). These two statistical indicators are being calculated by :

$$MBE = \frac{1}{K}\Sigma(\mathrm{G_p} - G_m) \quad (8)$$

$$RMSE = \sqrt{\frac{1}{K}\Sigma(\mathrm{G_p} - G_m)^2} \quad (9)$$

K is the total number of observations, $\mathrm{G_p}$ is the value of calculated global solar radiation, while G_m is the value of measured global solar radiation. The estimation value of monthly global solar radiation was being calculated based on temperature data measurement in Jakarta, Indonesia. From the figure 4 above we can see that that the value from all the models is still quite far from the measured data, but we can conclude that model proposed by Allen is the closest amongst two others models. Also from the table 3 below we can see that the model which proposed by Allen has the lowest value of MBE and RMSE. This result indicates a good fitting and the closest from the measured values.

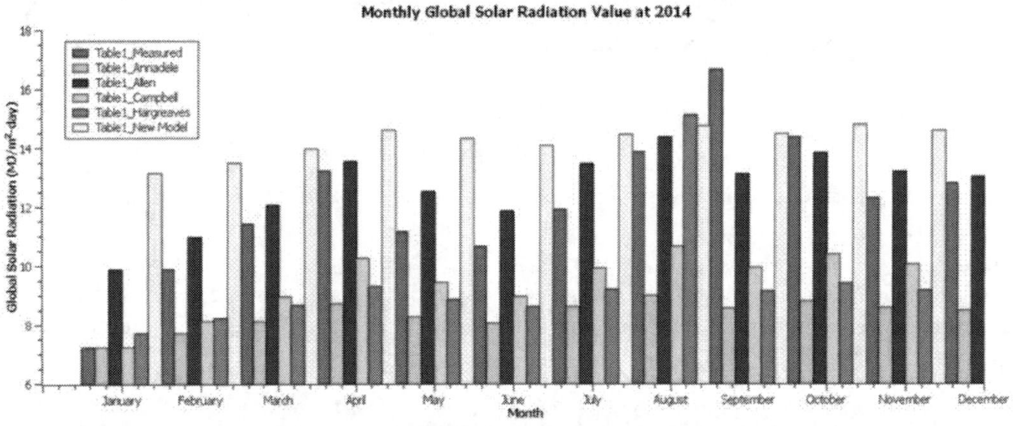

Figure 2. Monthly Global Solar Radiation Value at 2014 (measured and estimated).

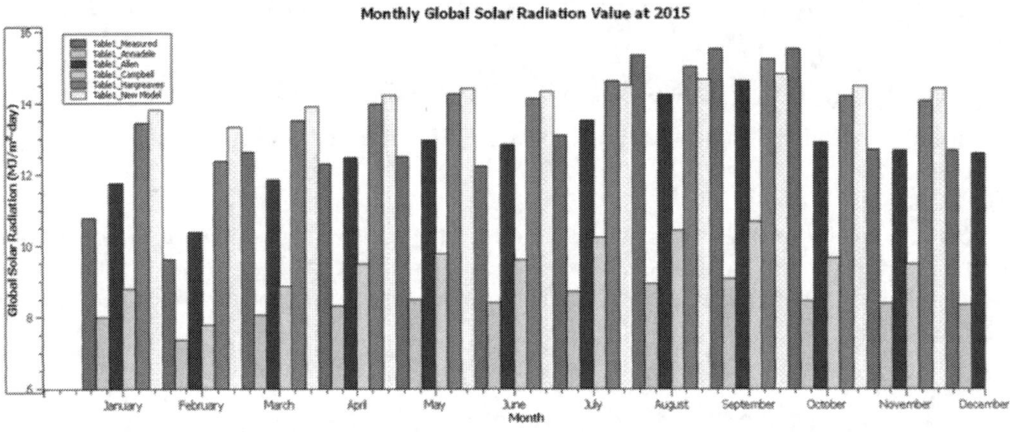

Figure 3. Monthly Global Solar Radiation Value at 2015 (measured and estimated).

Table 3. Statistical analysis of estimated solar global solar radiation.

			2014		
Models	Annadele	Allen	Campbell	Hargreaves	New Models
MBE	-3.774679276	0.52641898	-2.65418611	1.026145924	1.603394227
RMSE	13.07587258	1.82356884	9.19437039	3.554673753	5.55432053
			2015		
Models	Annadele	Allen	Campbell	Hargreaves	New Models
MBE	-4.524548075	-0.174433298	-3.382436419	5.686296492	1.526279762
RMSE	15.67349429	0.604254669	11.71710346	19.69790886	5.28718819

5. Conclusion

Solar energy is one of the resources that has a huge potential to fulfilling the demands of energy because of energy crisis, as the renewable energy. The estimation of monthly global solar radiation in Jakarta, Indonesia has been calculated by several empirical models from the literatures and one new based-temperature model. The calculation results from all models then being compared with the measured data and later being analyzed and assessed by the statistical methods. The best model regarding the lowest MBE and RMSE is Allen which provides the best performance for estimating global solar radiation and it can be said the most suitable model that can be used to estimate another areas in Indonesia.

6. Acknowledgments

Authors would like to express our gratitude to Directorate Research and Community Engagement (DRPM) Universitas Indonesia in the providing of publication funding by PITTA 2017.

7. References

[1] Okonkwo G N and Nwokoye A O C 2014 Estimating global solar radiation from temperature data in Minna location *European Scientific J.* **10(15)** 254-264
[2] Gueymard C A 2004 The sun ' s total and spectral irradiance for solar energy applications and solar radiation models *Solar Energy* **76** 423–453
[3] Morrison G L, Sydney U, Energy S T, Water S and Morrison G L 2015 Solar radiation data for Indonesia vol. December
[4] Dal A, Escobedo J F, Dal E and Cícero M 2014 Estimation of hourly , daily and monthly mean diffuse radiation based on MEO shadowring correction *Energy Procedia* **57** 1150–59
[5] For C *et al.* 2017 *Marlog* **6** 1–8
[6] Vindel J M, Polo J and Zarzalejo L F 2015 Journal of atmospheric and solar-terrestrial physics Modeling monthly mean variation of the solar global irradiation *J. Atmos. Solar-Terrestrial Phys.* **122** 108–118
[7] Allen R G 1997 Self-calibrating method for estimating solar radiation from air temperature *J. Hydrology Eng.* **2** 56–67
[8] Annandale J G, Jovanovic NZ, Benade´ N and Allen R G 2002 Software for missing data error analysis of Penman–Monteith reference evapotranspiration *Irrigation Sci.* **21** 57–67
[9] Bristow K L and Campbell G S 1984 On the relationship between incoming solar radiation and daily maximum and minimum temperature *Agricultural and Forest Meteorology* **31** 159–166
[10] Hargreaves G H and Samani Z A 1982 Estimating potential evapotranspiration *J. of Irrigation and Drainage Eng.* **108**(3) 225–230

The effect of centrifugation speed and Chitosan-Sodium Tripolyphosphate ratio toward the nanoencapsulation of Sambiloto *(Andrographis paniculata)* for the formulation of Hepatitis B drug

Devi Permata Sari[1], Tania Surya Utami[1], Rita Arbianti[1] and Heri Hermansyah[1]

[1]Department of Chemical Engineering, Faculty of Engineering, Universitas Indonesia, Depok, West Java 16424, Indonesia

E-mail: nana@che.ui.ac.id

Abstract. Hepatits B is a viral infection which attack the liver. One of the compound that can overcome and inhibits Hepatitis B is Andrographolide. The compound was derived from Sambiloto plants (*Andrographis paniculate).*Andrographolide compound works by inhibiting α-glucosidase which assists the secretion of Hepatitis B virus. The goal of this research is to make nanoencapsulation of sambiloto leaf extracts that was encapsulated in chitosan and STPP. The nanoencapsulation will increase the bioavailability of the body for the administered Andrographolide. The size of the resulting particle at a variation of centrifugal speed of 8.000 RPM with the concentration ratio of chitosan : STPP equals to 0.2%:0.1% (g/mL), was 68.3nm. The loading capacity of the nanoparticles is 67.20% and the encapsulation efficiency of the nanoparticles is 99.48%. The release profile has a cumulative release of 34.55% with slow release in gastric pH conditions and followed by a burst release in intestine pH conditions.

1. Introduction

One of the bad habits of society Indonesia is low in maintaining healthy. A less healthy causing the incidence of various health problems, such as the emergence of a wide range of dangerous diseases that can attack all of ages, from adult to Toddler-aged children.The cause of the occurrence of many diseases is due to the inability of the body against viruses, bacteria, as well as hazardous substances that are carcinogens. One of the dangerous diseases caused by virus is Hepatitis, which is comprised of Hepatitis A, B, C, D and E. Hepatitis is a disease characterized by inflammation that occurs in the liver organ [1]. Indonesia is a country with a high additional Hepatitis B is the second largest country in the South East Asian Region (SEAR) after Myanmar [2]. There are about 2 billion people in the world infected with the Hepatitis B virus and 240 million people of whom become people with Chronic Hepatitis B [2].

Wolrd Health Organization (WHO) states that the efforts that have been made to solve the Hepatitis B sufferers is to use adrenokortikosteroid, corticosteroids, and lamivudin, but these things have not been able to effectively solve the viral infections. Hepatitis B antivirals suchas drugs usage lamivudin already available and in development stage, but the drugs have not been evaluated for the treatment of acute hepatitis B [3].One of the compounds that can inhibit even the deadly hepatitis B virus replication is andrographolide. The source of the andrographolide found in Sambiloto (Andrographis paniculata). One of the andrographolide activity is can inhibits the α-glucosidase enzyme [4][5] which is a constituent of glikohidrolase enzymes i.e. enzymes that contribute in the formation of the viral envelope and is responsible for glikohidrolasi protein. The presence of inhibits glikohidrolase enzyme, then virus replication is not perfect and the virus will die because of the viral envelope formation will be blocked [6]

In this research, performed the extraction of sambiloto leaves with encapsulation by nano-sized particles that will be used is Chitosan. Chitosan has a fragile mechanical properties, so it must be stabilised by using sodium tripoliposphat (STPP) as a crosslinker, so when passing a very acidic pH, Chitosan is still not degraded and active substances not apart before reaching the target[7].

In this research was conducted on variation towards the concentration ratio Chitosan and STPP, because that affects the ratio of a particle may be nano-sized. Futhermore, it also conducted on variation of the centrifugation speed, to known how the appropriate centrifugation speed so that the particles that form can nano-sized. Then, controlled release by in vitro was done to know the performance of the extract of sambiloto has in nanoencapsulation as the hepatitis B drug.

2. Methods

Variations on this research is ratio of concentrations chitosan-STPP and centrifugation speed.Continued with particle distribution measurement by using particle size analyzer (PSA), determination of morphology using SEM and FTIR, quantitative test of andrographolide by using HPLC, and controlled drug release of nanoparticles in *in vitro*.

2.1. Extraction of Sambiloto and quantitative test of Andrographolide

Sambiloto leaves powder was dissolved in solvent ethanol 70%. Then extraction using sonicator for 50 minutes at frequency of 30 Hz and 25°C, continued by solvent evaporating using vacuum rotary evaporator. Calculating of andrographolide in sambiloto extract by using HPLC.

2.2. Nanoencapsulation

The selected method in the nanoencapsulation is ionic gelation method. Chitosan 0.2 g was dissolved in 100 mL acetic acid 1% (b/v), added 3 g of sambiloto extract, and dissolve 0.04 g 40 mL STPP into akuades then added 200 μL Tween 80 0.1% (v/v).Solution of STPP then added into chitosan-extract solution with flow speed of 0.75 ml/min accompanied with magnetic stirring for 2 hours. Then do size diminution using sonikator with 130 Watts and frequency of 30 Hz for 50 minutes and centrifugation for 15 minutes with a pause of 1 minute every 5 minutes. The last is drying using freeze dryer to a powder.

2.3. Controlled drug release

Controlled drug release was done on a synthetic fluid medium, i.e. in series in a seven hours. Synthetic fluid medium that is used in the two of kind of buffers that indicates the human digestive system, i.e. the Simulated Gastric Fluid (SGF) pH 1.2 which indicate gastric conditions and Simulated Intestinal Fluid (SIF) 7.4 pH which suggests intestinal conditions.

3. Result and discussion

3.1. Nanoencapsulation of Sambiloto extract

Method of extraction sonication produce yield 5.44% with active substance andrographolide 17.60%. The concentration ratio of Chitosan: STPP (g/mL) are made in different variations, i.e. 0.1% : 0.1% ; 0.1% : 0.15% ; 0.15% : 0.1% and 0.2% : 0.1%. Continued with the variation of centrifugation speed, i.e. 8,000 rpm, 10,000 rpm, 13,000 rpm and 15,000 rpm.

3.2. Particle size distribution of Chitosan:STPPRatio

Particle Size Analizer (PSA) test beginning with the sample dispersion in the aquademin with the addition of Dispersing Agent Coulter Type 1A. Here are the average particle size of the variation ratio of Chitosan: STPP obtained from the PSA analysis.

Table 1. Result of particle size distribution of chitosan:STPP ratio.

Concentration Ratio of chitosan : STPP (g/mL)	*Z-Average* (nm)	Particle Size Distribution (nm)
0,1%:0,1%	1917	90.78 ± 12.15
		125.2 ± 17.38
		232.7 ± 32.15
0,1%:0,15%	2569	206.6 ±27.87
		131.9 ± 18.32
		-
0,15%:0,1%	1454	144.1 ± 17.03
		269.2 ± 47.53
		62.68 ± 8.29
0,2%:0,1%	1424	170.5 ± 24.99
		90.41 ± 12.91
		1212 ± 153.2

Based on the table, that the higher concentration of chitosan than concentrations of STPP will earn average particle size (z-average) getting smaller. It is caused due to an increased of STPP concentration will enhance physical endurance of the particle, because of the increasing number of STPP will make ionic-crosslinking reaction by the positive force of chitosan more stable forming physically, but also increased nanoparticle diameter [8].Particle size distribution obtained from each variation has a relatively small size and reach a size of nanometer, but z-average or average particle size is still big even reach a size belongs to the micro.The Z-average size will only be comparable with the size measured by other techniques if the sample is monomodal, spherical or near-spherical in shape, monodisperse, and the sample is prepared in a suitable dispersant, as the Z-Average mean size can be sensitive to even small changes in the sample, e.g. the presence of a small proportion of aggregates[9].

3.3. Particle size distribution of Centrifugation Speed Variation

Based on the measurement of particle distribution, size obtained reach the nano is a variation of centrifugation speed 8,000 rpm i.e. 68.3 nm and 1663.3 nm at centrifugation speed 15,000 rpm. It is because at the high centrifugation speed will make particles collide and can cause clots flok. That is corresponds to the Budiyono (1999)[10] who explains that fluid movement due to stirring will be followed by microflok formed. Here are the average particle size of the centrifugation speed variation obtained from the PSA analysis.

Table 2. Result of particle size distribution of centrifugation speed variation.

Centrifugation Speed (rpm)	*Z-Average* (nm)	Particle Size Distribution (nm)
8.000	68.3	22.4 ± 3.2
		86.3 ± 27.3
10.000	1160.7	964.5 ± 161.8
		-
13.000	666.0	1.3 ± 0.3

15.000	1663.3	241.4 ± 78.1
		2780 ± 1619.4

According to Jong (2008)[11], the size of nanoparticles used as drug carriage has a range of <100 nm, so at the centrifugation speed 8,000 rpm can be said that the particles sucessfully formed with nano size.

3.4. Morphology determination using SEM

Morphology of the particles observed with the Scanning Electron Microscope (SEM). The resulting morphology is not smooth and has no pores on its surface, occurs the agglomeration, and the form is not spherical.

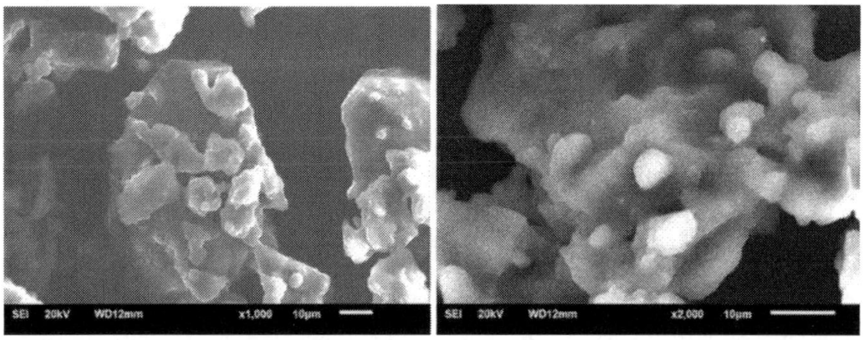

Figure 1. SEM result of nanoparticle

3.5. Loading capacity and encapsulation efficiency

Loading capacity shows the number of drugs contained in nanoparticles formed in this paper, the drug was active substance andrographolide in the crude extract of sambiloto. The results of the calculation of loading capacity from each variation can be seen below:

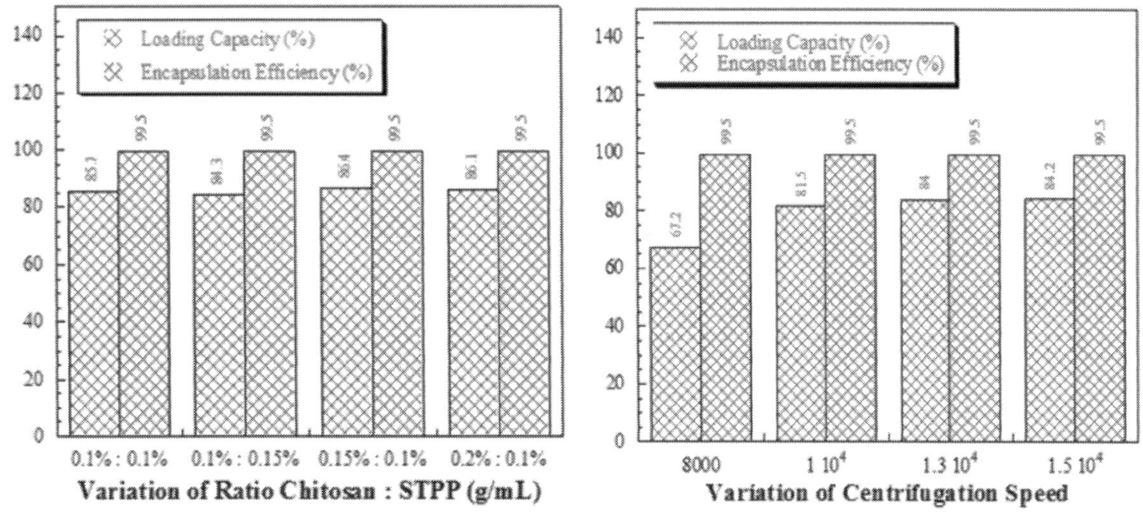

(a) (b)

Figure 2. (a) Concentration ratio of chitosan:STPP (b) Variation of centrifugation speed

In the figure (a) shows that with increasedof chitosan concentrations so the loading capacity and encapsulation efficiency will also increase, while the addition of STPP concentration will decrease of loading capacity and encapsulation efficiency. While in addition of the STPP concentrationwill decrease loading capacity because the droplets rapidly solidification occurs when the process of ionic gelation [12]. In figure (b) shows that increasing of the centrifugation speed will increase loading capacity, because with the increasing of the centrifugation speed, then small-sized particles will also precipitates (not only large-sized particles that precipitates), so the mass of nanoparticles after freeze drying will be more.

Encapsulation efficiency obtained ranged from 99.47%-99.51%, this result is appropriate with Chen *et al.* (2006)[13] who is state that good encapsulation efficiency is at least 80%, because it shows the processes that do not eliminate existing active substances. The high value of the encapsulation efficiency may also be evidenced by the results of morphology in FTIR in figure 2, because andrographolide contained in the sambiloto extract have been encapsulation with chitosan, as evidenced by the existence of andrographolide's functional group on nanoparticle.

3.6. Controlled drug release of Nanoparticle

The results showed that the cumulative release is 34.55% at the 7th hour. In figure 3 shows that at the 1st to 3rd hour which indicates gastric condition with the pH 1.2 longer than at pH 7.4 which indicates intestinal conditions shows profile release to be faster.

In figure 3 shows that at the 1st to 3rd hour in which indicates gastric condition is *slow release*. This is because in these conditions, the hydrogen bonds between molecules of Chitosan nanoparticles make getting stronger which causes diffusion into the moleculebecomes more difficult, and then followed the release of a drug that lasts longer[14]. The addition of surfactants such as tween 80 may increase the endurance of the particles on the acid condition in gastric.

Then on the intestinal conditions pH 7.4 shows the release profiles tend to be faster.That is called *burst release* i.e. mechanism of drug released actively, indicating the drug is at the surface of the nanoparticles so that chitosan having deprotonation on the surface when mixed with synthetic fluid medium [15].

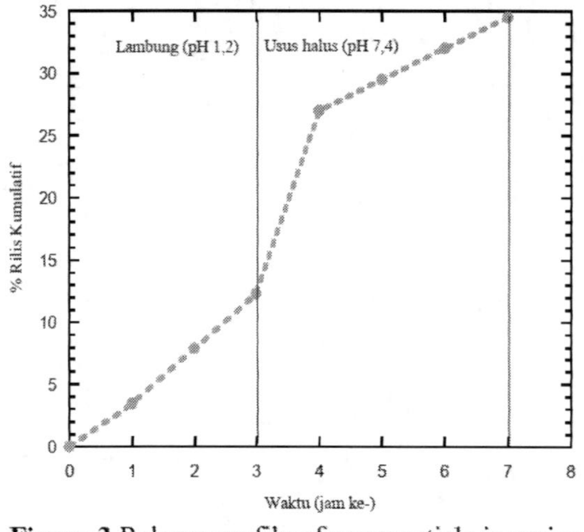

Figure 3. Release profile of nanoparticle in series

At the *Simulated Intestinal Fluid* (SIF), the concentration of STPP contributed to the polymer chain conformation more because deprotonation, so chitosan nanoparticle easy damaged [16]. Has been previously reported also that on media with pH above 6 ionization of amine reduced drastically [17]. This triggered the deprotonation of the amine chitosan and chitosan particles causing crosslink with STPP became unstable and start to degradation as reported on a similar research[18].

4. Conclusion

Based on the results of the research, it can be concluded that the results of measurements of the particles distribution have an average of 68.3 nm on a variation of the concentration ratio chitosan:STPP 0.2%:0.1% (g/mL) with the centrifugation speed 8,000 rpm. The variation has loading capacity and efficiency encapsulation each 67.20% and 99.48%, and cumulative release 34.55%.

5. References

[1] Misnadiarly A S 2008 *Mengenal Penyakit Liver* Mediakom Ed. XII (Jakarta: Badan Litbangkes)

[2] Kementrian Kesehatan Republik Indonesia 2014 Situasi dan Analisis Hepatitis. Pusat Data dan Informasi

[3] World Health Organization 2002 *Hepatitis B* Departement of Communicable Diseases Surveillance and Response

[4] Dai Gui-Fu *et al.* 2006 *Studies on the novel α-glucosidase inhibitory activity and structure–activity relationships for andrographolide analogues.* Bioorganic & Medicinal Chemistry Letters Vol 16

[5] Xu, Hai-Wei., Dai, Gui-Fu., Liu, Gai-Zhi., Wang, Jun-Feng., dan Liu, Hong-Min 2007 Synthesis of andrographolide derivatives: A new family of α-glucosidase inhibitors *Bioorganic & Medicinal Chemistry* **15** 4247-55

[6] Firdayani, et al. 2012 *Pengembangan Kandidat Senyawa Obat Turunan Naftokuinon sebagai Inhibitor Virus Hepatitis B* (Jakarta: Pusat Teknologi Farmasi dan Medika BPPT)

[7] Aini, D *et al.* 2015 *Uji Pelepasan Terkendali Ekstrak Daun Keji Beling (Strobilanthes crispus) untuk Sediaan Obat Antihiperkolesterolemia.* Prosiding Seminar Nasional Teknik Kimia UNPAR: 101-104

[8] Ko J A, Park H J, Hwang S J, Park J B and Lee J S 2002 Preparation and characterization of chitosan microparticles intended for controlled drug delivery *Int. J. of Pharmaceutics* **249** 165-174

[9] Malvern Instruments Limited 2011 *Dynamic Light Scattering Common Terms Defined.* [online] www.malvern.com/contact/

[10] Budiyono, *et al.* 1999 *Pengaruh Kecepatan dan Waktu Pengadukan pada Proses Pengendapan Limbah Radioaktif Cair yang Mengandung Aktinida dengan $FeCl_3$ dan $Al_2(SO_4)_3$* Prosiding Pertemuan dan Presentasi Ilmiah P3TM-Batan Yogyakarta

[11] Jong W H 2008 *Drug Delivery and Nanoparticles: Applications and Hazards. Int. J. of Nanomedicine* 133-149

[12] Mardliyati E, Muttaqien, Sjaikhurrizal E and Setyawati D R 2012 *Sintesis Nanopartikel Kitosan-Trypolyphosphate dengan Metode Gelasi Ionik: Pengaruh Konsentrasi dan Rasio Volume terhadap Karakteristik Partikel* Prosiding Pertemuan Ilmiah Ilmu Pengetahuan dan Teknologi Bahan (Serpong: BPPT)

[13] Chen, *et al.* 2006 Food protein-based materials as nutraceutical delivery systems *Trends in Food Science & Technol.* **17** 272-283

[14] Dhanasingh, et al 2010 Chitosan/Casein Microparticles: Preparation, Characterization and Drug Release Studies. *World Academy of Science, Eng. and Technology* **4** 08-23

[15] Siegel R A and Rathbone M J 2012 *Overview of Controlled Release Mechanisms. Advances in Delivery Science and Technology*

[16] Fauzia F 2014 *Mikropartikel Kitosan Tertaut Silang Tripolifosfat untuk Pelepasan Terkendali Senyawa Bioaktif Asetogenin dan Mangostin* Fakultas Teknik Universitas Indonesia

[17] Yu C Y, Yin B C, Zhang W, Cheng S X, Zhang X Z and Zhuo R X 2009 Composite microparticle drug delivery systems based on chitosan, alginate and pectin with improved pH-sensitive drug release property *Colloids Surf B Biointerfaces* **68** 245-9

[18] Konecsni K, Low N H and Nickerson M T 2012 Chitosan-tripolyphosphate submicron particles as the carrier of entrapped rutin *Food Chem* **134** 1775-9

Effect of pH, temperature and medium agitation rate in production of AA, DHA, EPA from *Aspergillus oryzae* with submerged fermentation

Astrid Miranti[1], **Rita Arbianti**[1] **and Tania Surya Utami**[1]

[1] Department of Chemical Engineering, Faculty of Engineering, Universitas Indonesia, Depok, West Java 16424, Indonesia

E-mail: nana@che.ui.ac.id

Abstract. There are several substances that needs to be fulfill to keep the brain cell growth such as AA, DHA and EPA. Fungi is one of the alternative source of omega 3, omega 6, omega 9 especially AA, DHA and EPA. This research variates operating condition that is suitable for the growth of *Aspergillus oryzae* in AA, DHA, and EPA fatty acid production with *Submerged Fermentation* using synthetic medium. *Aspergillus oryzae* cultivated in medium using glucose as carbon source and Ammonium sulfate and yeast extract as nitrogen source. The extraction method using ethanol and n-hexane as solvent. The result shows that optimum agitation rate for unsaturated fatty acid production of *Aspergillus oryzae* is 120 RPM, lipid yield 28,28% and unsaturated fatty acid content 50,36 % and EPA content 2,42 %. Optimum medium pH for PUFA production of *Aspergillus oryzae* is 6, lipid yield 22,35 % and unsaturated fatty acid content 45,5 %. optimum incubation temperature for unsaturated fatty acid production of *Aspergillus oryzae* is 25 °C, lipid yield 13,19 % and unsaturated fatty acid content 62,15 %. Unsaturated fatty acids produced from *Aspergillus oryzae* are oleic, linoleic, linolenic and EPA.

1. Introduction

Food problems became one of the biggest problems related to population growth. To guarantee the availability of foods for humans in large numbers, require large natural resources, both in terms of availability of land as well as the variety of food sources. According to economic experts who are members of the forum of Economic Intelligence Unit (EIU) in 2014, the food security index of Indonesia ranks 64 in the world, far below other ASEAN countries.

Indonesia food security index numbers are worrisome to be one of the causes of the poor condition of Indonesian public health. According to the World Health Organization (WHO) in 2013, the level of Indonesian health is one of the lowest in ASEAN. To meet nutrient intake on toddlers, there are some substances that have to be fulfilled for the process of development of brain cells mainly include carbohydrates, proteins, iron, vitamins and fatty acids such as DHA and EPA, which is divided into saturated fatty acids and unsaturated fatty acids.

Currently, omega 3 and omega 6 are derived from fish oil. However, the availability of fish in terms of food security are not proportional to the needsthat continue to increase. So the alternative source of fatty acids, especially the omega 3 and omega 6 besides fish oil is needed. Alternative sources of omega 3 and omega 6 besides fish oil should have several advantages both in terms of land use, the time it takes to produce fatty acids as well as cheaper cost [1]. A lot of research have been done on fatty acids fish oil replacement. Such research produces conclusion for using microorganisms as an alternative source of fatty acids [1]. The uses of microorganisms as an alternative source offatty acids also have an advantage as mentioned.

There are different types of microorganisms that can be used as an alternative source of fatty acids. These microorganisms have a high fat content, such as yeast, fungi, bacteria and mikroalga. In this study used microorganisms fungi (mold). Mold used in this study because the oleaginous microorganisms including mold, has various advantages such as easier handling, able to grow in low pH and can degradate a complex carbon source [2]. In this research the type of mold, Aspergillus oryzae is used which has a lipid content of approximately 18-57% [3].

In this research, molds were cultured in the synthetic medium with the ratio of C:N of 30:1 and submerged fermentation. Submerged fermentation method was chosen as a method of culture becausenutrients spread more evenly so that yield production of fatty acids expected to be greater. So, this research will find profile concentrations of unsaturated fatty acids from Aspergillus oryzae against variations in temperature incubation medium pH, as well as the rate of agitation of the medium in the production of unsaturated fatty acids, DHA and EPA

2. Methods

2.1. Culture stock in agar medium
Aspergillus oryzae cultured using PDA (*Potato Dextrose Agar*) medium containing 0.39 gram medium agar dissolved with 10 mL aquades.

2.2. Culture to make growth curve
Aspergillus oryzae cultured in medium for 10 days then harvested every day. Dry cell weight counted to make growth curve.

2.3. Culture in submerged fermentation
Previous cells cultured at medium agar transferred into liquid medium with glucose as the carbon source composition 4.5 g, ammonium sulfate and yeast extract 0.15 grams of micro nutrient sources, along with $(NH_4)_2SO_4$ as 0.25 g KH_2PO_4, 0.15 g K_2HPO_4, 0.05 grams, $MgSO_4.7H_2O$ 0.2 grams, $FeCl_3$ 0.015 grams into 150 ml aquades. Culture is carried out for 6 days with each variation rate of agitation, pH and temperature. Culture agitation rate variation of 70, 120 and 170 rpm. The ratio of C and N 30:1, pH 7 medium, incubation temperature 35 °Cmade during 144 hours (6 days). After obtained the rate of agitation with the optimum yield, a culture with pH variations variations medium conditions 5, 6 and 7, the ratio of C and N 30:1, incubation temperature 35 °C. Incubation time for 144 hours (6 days). After the pH and agitation rate obtained with optimum yield, a culture with temperature variations 25, 30 and 35 °C. Incubation time for 144 hours (6 days).

2.4. Cell harvesting
Cells harvested and dried with oven for 24 hours in temperature 60 °C. The cells then extracted with sonicator using ethanol as solvent for 3 hours. After that centrifuge 4000 rpm for 15 minutes. Then extracted again with n-hexane as solvent for 2 hours. After that centrifuge with 4000 rpm speed for 15 minutes to separate the lipid and biomass content.

2.5. Fatty acid composition analysis
Lipid extracted then analyzed using GC-MS in Balai Besar Industri Agro (BBIA) Bogor, Jawa Barat.

3. Result and discussion

3.1. Effect of agitation rate
The rate of agitation would affect the state of the mold as the impact both positive and negative. Positive impact caused by agitation is increasing the solubility of oxygen and nutrients distributed evenly or mold growth will increase the production of lipids [4]. But, like a double-edged blade, negatively impact the rate of agitation that is too high will damage the mycelium cells and tissues.

Therefore, this study tried to find out the relationship between the rate of agitation with the growth of mold.

In the figure it can be seen that on a medium with a higher rate of agitation, biomass growth mold tends to be faster, as seen on the medium with agitation 120 and 170 RPM compared to 70 RPM. From the graph we can say that optimum growth of the mold Aspergillus oryzae occurred at a range of 120 RPM agitation. On the agitation with 120 RPM mold growth is the highest with a biomass of 0.8391 g. This is because the medium mixed with good (perfectly mixed), so that the solubility of oxygen in high nutrient medium are mixed evenly. Another case with agitation 70 RPM which showed the lowest results, i.e. 0.3071 grams. It is thought to be caused by nutrients not distributed evenly in the medium.

3.2. Production of Lipids on the variation rate of Agitation

The influence of rate of agitation against the weight of lipid obtained there is in figure below. On the picture, seen that the resulting lipid ranged from 0.096 – 0.237 grams. The maximum amount of lipids formed at 120 RPM agitation. The abundance lipids produced at 120 RPM agitation is directly proportional to biomass mold that grows at a rate of agitation. It shows that the medium at a rate of 120 RPM agitation is the optimum medium for incubation mold either in terms of the growth of the biomass as well as the formation of lipids.

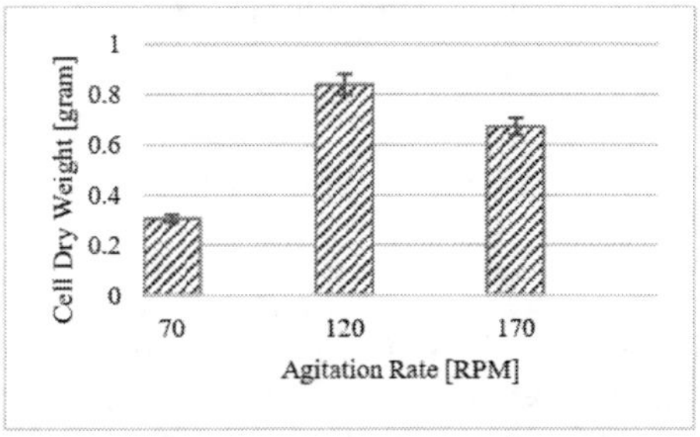

Figure 1. Effect of Agitation Rate to Cell Dry Weight.

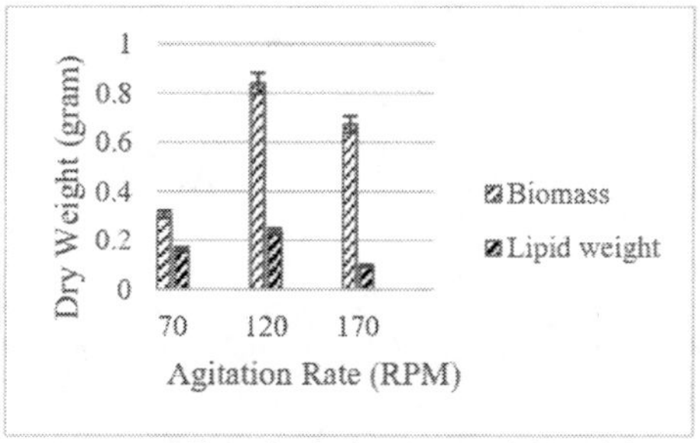

Figure 2. Effect of Agitation Rate to Lipid Weight.

The condition is the best agitation conditions where oxygen can be dissolved and the medium can be mixed well. While contact between Hypha can be maintained properly. Hypha quite contacted with each other and form a mycelium, later the mycelium does not suffer fragmentation and damage cells.

3.3. Saturated and unsaturated fatty acid analysis

Table 1. Fatty Acid Composition on Agitation Rate Variation.

Fatty Acid Composition	Agitation rate (RPM)		
	70	120	170
Saturated Fatty Acid (% w/w)			
Lauric (C12:0)	13,9	1,27	6,81
Miristic (C 14:0)	6,92	1,3	3,62
Palmitic (C 16:0)	34,3	19,7	21,9
Stearic (C 18:0)	6,93	22,9	22,8
Unsaturated Fatty Acid (% w/w)			
Oleic	31,6	24	17,7
Linoleic	6,31	25,9	20,5
Linolenic	$n.d^a$	0,46	$n.d^a$
AA	$n.d^a$	$n.d^a$	$n.d^a$

Saturated fatty acids resulted in variations of agitation 70, 120 and 170 RPM-including lauric, Palmitic, and stearic myristate. These fatty acids are produced with different compositions on any variation of the rate of agitation. The rate of 70 RPM agitation dominated Palmitic acid with percentage of 34.3%. The rate of 120 RPM agitation dominated stearic acid with percentage 22.9%. While the rate of 170 RPM agitation is also dominated by stearic acid with 22.8% percentage. The variation rate of agitation 70 RPM produces the largest saturated fatty acids with the largest percentage is a type of Palmitic (C16) and 34.3%. While saturated fatty acids produced in the smallest variations in the rate of 120 RPM agitation with the largest percentage of types of stearic (C20) namely 22.9%. Type of saturated fatty acids are produced depending on the success of elongase enzymes in adding chain lengths of fatty acids[5]. It showed variations 70 RPM elongase enzymes do not work in optimum. Elongase enzymes working the most optimum at 120 RPM agitation.

3.4. Effect of the pH of the incubation medium

The result of the growth of mold on the medium pH variations 5, 6 and 7 show the profile as shown in the figure below. At pH 5 obtained biomass mold of 0.98 g, in a medium with a pH obtained biomass mold of 1.2 grams. Whereas in the medium with a pH of 7, obtained biomass mold 0.3 grams. In General, on a medium with a slightly acidic pH of biomass produces a larger mold than a neutral pH. This indicates that the pH of the medium affects the growth of mold and mold can grow optimally at pH slightly acidic to neutral.

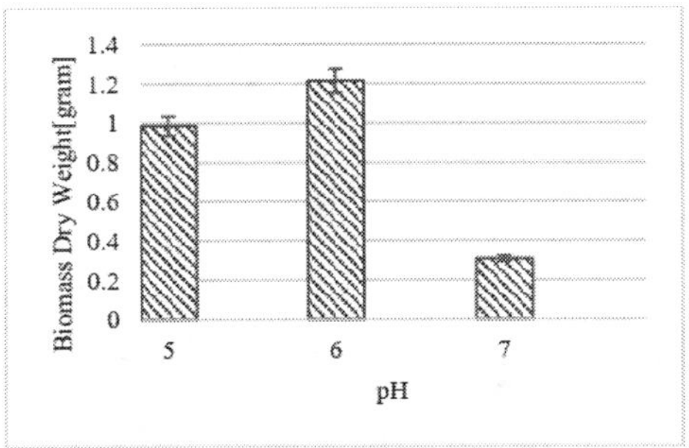

Figure 3. Effect of pH to Cell Dry Weight.

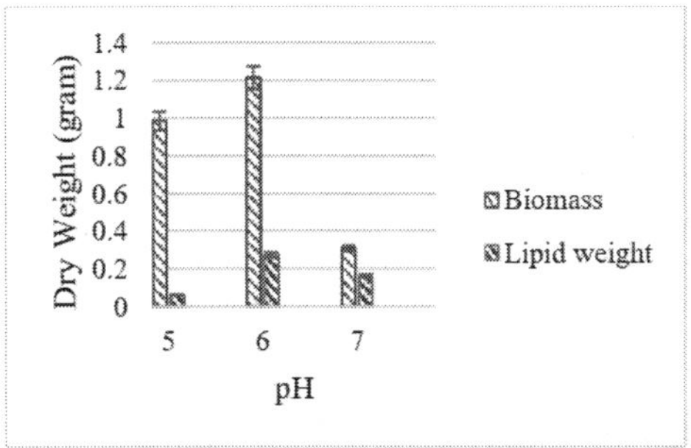

Figure 4. Effect of pH to Lipid Weight.

3.5. Production of lipids on the medium pH variation

The success of the accumulation of acetyl Coa to be transformed into a secondary metabolic products namely lipid depends on enzymes that synthesize lipids. The enzyme has a optimum pH to work in particular. Study on production of lipids on the mold known by using the weight of the lipids that formed during incubation.

On the picture, we can see that the resulting lipid range between 0.06-0.27 grams. The maximum amount of lipids are formed at pH 6. The abundance lipids produced at pH 6 is directly proportional to biomass mold, which grows at pH 6. It shows that the medium with a pH of 6 is the optimum medium for incubation and mold, both in terms of the growth of the biomass as well as the formation of lipids. Different results shown on a medium with a pH of 5. Mold growth well with the amount of biomass that is growing, though not as big a pH of 6. But the ammount of the biomass is not followed by the amount of lipid produced.

3.6. Saturated and unsaturated fatty acid analysis

Table 2. Fatty Acid Composition on pH Variation.

Fatty Acid Composition	pH		
	5	6	7
Saturated Fatty Acid (% w/w)			
Lauric (C12:0)	$n.d^a$	$n.d^a$	13,9

Miristic (C 14:0)	1,2	$n.d^a$	6,92
Palmitic (C 16:0)	21,4	54,5	34,3
Stearic (C 18:0)	22,1	$n.d^a$	6,93
Unsaturated fatty acid (% w/w)			
Oleic	27,7	45,5	31,6
Linoleic	21.8	$n.d^a$	6,31
Linolenic	0.39	$n.d^a$	$n.d^a$
AA	$n.d^a$	$n.d^a$	$n.d^a$
DHA	$n.d^a$	$n.d^a$	$n.d^a$
EPA	$n.d^a$	$n.d^a$	$n.d^a$

The highest concentration of saturated fatty acids found in the variations of the medium with a pH 7. This condition is an appropriate pH for enzyme-substrate complex formation. the corresponding medium pH will cause enzymatic activity can work well. On the formation of lipids by microorganisms especially mold, strongly influenced by the activity of the enzyme FAS (Fatty Acid Synthases). FAS Enzyme (Fatty Acid Synthases) can create a bond with NADPH well at pH 6-8.5 [6].

3.7. Effect of incubation temperature

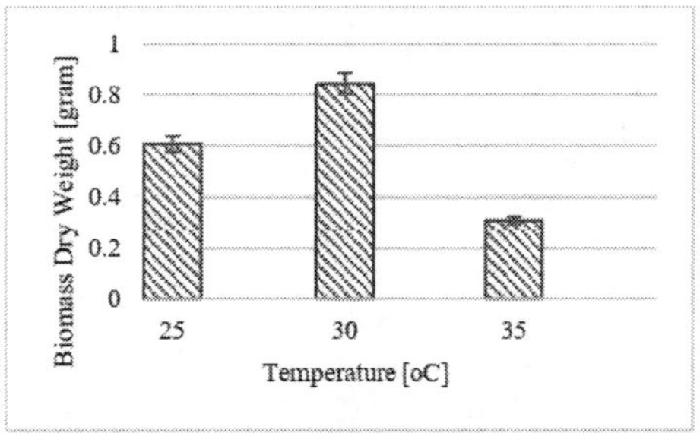

Figure 5. Effect of Temperature to Cell Dry Weight

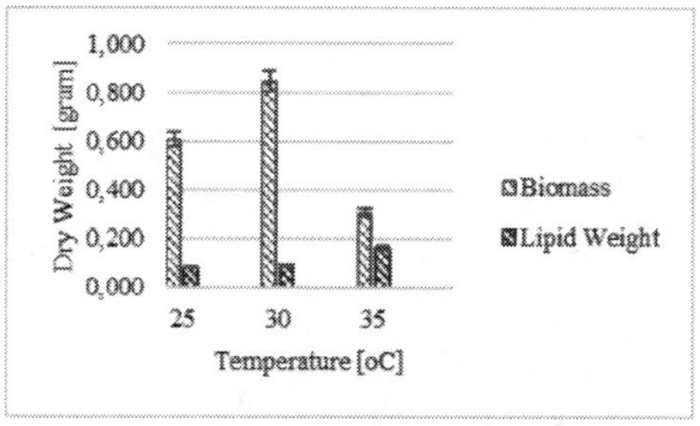

Figure 6. Effect of Temperature to Lipid Weight

The result of the growth of mould on temperature variation of 25 and 30 °C shows the amount of biomass that is higher than the temperature of 35 °C. This is in accordance with the theory that summed up by Fardiaz (1992) whereby the mould Aspergillus can grow well in the temperature range 25-30 °C [7]. This caused mold Aspergillus mold types with type mesofilik which is a species that fits on a medium temperature. The type of the mould Aspergillus will hampered its growth at temperatures below 18 °C [8].

3.8. Production of Lipid on Temperature Variation

In terms of the production of lipids, temperature variations show a slightly different profile with the acquisition of dry biomass. The lowest lipid production occurs at a temperature of 25 °C, while the highest at temperature 35 °C. The range of lipid production in incubation temperature variation ranged from 0.08 – 0.16 grams. The resulting lipid yield per dry weight biomass has a range between 13%-52%. At a temperature of 25 °C, the production of lipids are not too big, as well as biomass mold. This indicates that the temperature of 25 ° C is the temperature which is more suitable to mold growth than temperature 35 °C. A similar case is found at a temperature of 30 °C, the temperature at which the production of biomass is the highest. However, the magnitude of the biomass is not comparable to the lipids produced. Can be drawn the conclusion that the temperature 30 °C is the best temperature for growing of biomass mold, but not for the production of lipids on the mold.

3.9. Saturated and unsaturated fatty acid analysis

Table 3. Fatty Acid Composition on Temperature Variation

Fatty Acid Composition	Temperature (°C)		
	25	30	35
Saturated Fatty Acid (% w/w)			
Lauric (C12:0)	$n.d^a$	1,64	13,90
Miristic (C 14:0)	$n.d^a$	1,84	6,92
Palmitic (C 16:0)	18,03	34,36	34,30
Stearic (C 18:0)	19,81	16,48	6,93
Unsaturated Fatty Acid (% w/w)			
Oleic	21,13	19,87	31,60
Linoleic	41,03	23,57	6,31
Linolenic	$n.d^a$	1,58	$n.d^a$
AA	$n.d^a$	$n.d^a$	$n.d^a$
DHA	$n.d^a$	$n.d^a$	$n.d^a$
EPA	$n.d^a$	$n.d^a$	$n.d^a$

On the table, fatty acids produced from lipid is broken down into saturated fatty acids and unsaturated fatty acids. The percentage of saturated fatty acids are produced varies between 37.84% - 62.05%. Meanwhile, the percentage of unsaturated fatty acids that are generated vary between 37.91-62.16%. The highest saturated fatty acids obtained on incubation temperature 35 °C. Whereas, the highest unsaturated fatty acids are obtained at a temperature of 25 °C incubation.

The concentration of saturated fatty acids increase at the temperature variation range 25 – 35 °C. A different matter is shown by the acquisition of unsaturated fatty acids that decrease in temperature variation range 25 – 35 °C. The highest concentration of saturated fatty acids found in 35 °C temperature variations i.e. of 62.05%. It shows that saturated fatty acids dominated at a temperature of 35 °C. Concentrations of unsaturated fatty acids found on the highest variation in temperature 25 °C, namely in the amount of 62.16%. Thus, the unsaturated fatty acids are produced optimally at a temperature of 25 °C.

4. Conclusion

Optimum Agitation Rate for production of unsaturated fatty acids from Aspergillus oryzae is 120 RPM with yield lipid by 28.28% and yield unsaturated fatty acids of 50.36%. Optimum agitation Rate for production of EPA is 120 RPM with the composition of EPA obtained is 2.42%. Optimum pH medium for the production of unsaturated fatty acids from Aspergillus oryzae is pH 6 with yield lipid 22.35% and yield unsaturated fatty acids 45.5%. Optimum incubation Temperature for the production of unsaturated fatty acids from Aspergillus oryzae is 25 °C with yield lipid 13.19% and yield of unsaturated fatty acids 62.15%

5. Reference

[1] Bayizid, Arzu Akpinar, et al. 2014 Single Cell Oil (SCO) production by Fusarium species using cheese whey as a substrate *Mljekarstvo* **64(2)** 111-118

[2] Sumanti, D. T. C. H. M. S. T., 2003 *Mempelajari Mekanisme Produksi Minyak Sel Tunggal Dengan Sistem Fermentasi Padat Pada Media Onggok-Ampas Tahu Dengan Menggunakan Kapang Aspergillus Terreus* (Bandung: Universitas Padjajaran)

[3] Sawangkeaw R, Ngamprasetsith S 2013 A review of lipid-based biomasses as feedstocks for biofuels production *Renew. Sust Energy* pp 97-108

[4] Amanullah, A. et al. 2000 Agitation induced mycelial fragmentation of Aspergillus oryzae and Penicillium chrysogenum. *Biochem Eng.* pp 109-114.

[5] Certik M and Shimizu S 1999 Biosynthesis and regulation of microbial polyunsaturated fatty acid production *J. Bioscience Bioengineering* pp 1-14

[6] Yuan Z and Hammes G G 1984 Elementary steps in the reaction mechanism of fatty acid synthase, pH dependence of NADPH binding and isotope rate effect for beta ketoacyl reductase *The J. of Biological Chem.* pp 6748-51

[7] Fardiaz S 1992 *Mikrobiologi Pangan I* (Jakarta: Gramedia Pustaka Utama) *Science and Technology* pp 499-504

[8] Chavant L W 1981 Deviation from the usual relationship between the temperature, the growth rate, the fatty acid composition, and the lipid microviscosity of four different fungi. *Biochemical and Biophysical Research Comm.* **101** pp 912-20

IOP Conf. Series: Earth and Environmental Science **105** (2017) 012114 doi:10.1088/1755-1315/105/1/012114

Enchanced electrochemical performances of $Li_4Ti_5O_{12}$/Sn composites anode via sol-hydrothermal method for lithium ion batteries

A Z Syahrial[1,3], F Aldy[1], B Priyono[1] and A Subhan[2]
[1] Departemen Metalurgi & Material, Universitas Indonesia, Kampus UI, Depok, Jawa Barat 16424, Indonesia
[2] Pusat Penelitian Fisika, LIPI, Komplek PUSPIPTEK, Banten, Tangerang Selatan 15314, Indonesia

E-mail: anne@metal.ui.ac.id

Abstract. $Li_4Ti_5O_{12}$ (lithium titanate) were synthesized by sol-gel and hydrothermal method with LiOH as lithium ion source. $Li_4Ti_5O_{12}$/Sn composites anode were preparared by ball mill method with three of Sn variation. X-ray diffraction shows spinel, TiO_2, and Sn phases with anatase and rutile residue. The lowest electrolyte resistance obtained at the highest Sn value. The specific capacity of battery can be increased from addition of Sn by up to 258.6 mAh/g. Alloying and dealloying reaction of Li_xSn accomodate the increased specific capacity from charge/discharge. However, Thevolume expansion from Li_xSn leads to loss of capacity when the C rate increases. The efficient capacity at low and high charge-discharge rate obtained at the highest value of added Sn.

1. Introduction

The automotive industry is beginning to make progress towards renewable energy storage. Lithium-ion battery using graphite anode as its component is a conventional battery which often used in electric car. The lithium-ion battery is also considered one of the most promising candidates for the next generation as a large-scale power source in electric and hybrid cars with their high energy density and good cycle capabilities[1].

Spinel lithium titanate ($Li_4Ti_5O_{12}$, LTO) demonstrates a great potential as an alternative anode material that has great safety performance and stable face-centered cubic (fcc) with highly reversible crystalline structures (zero-strain material)[2]. LTO has high lithation voltage around 1.55 V vs. Li/Li^+, where this is higher than the reduction of the electrolyte solution, which may lead to the formation of a solid electrolyte intermediate (SEI) and lithium dendrite deposition on the surface of the anode material. However, the deficiencies in the LTO have, for example, almost half of the graphite theory capacity of 175 mAh/g, low Li ion diffusion capability, and low electrical conductivity[3].

Several ways have been developed to overcome the weaknesses of LTO. One in particular is by giving addition of metal Sn[4]. Sn has been found as an alternative anode material candidate with its

[3] To whom any correspondence should be addressed.

theoretical capacity of 994 mAh/g. The advantages of mixing $Li_4Ti_5O_{12}$/Sn using a solid-state method give a capacity of 480 mAh/g with an addition of 10 wt% Sn. Synthesized LTO using solid-state methods based on TiO_2 and Li_2CO_3. For example, Sn nano is obtained from chemical reduction process, so that Sn is more economical because Sn price of nanoscale is relatively high[5].

This paper will discuss on how to improve the quality of $Li_4Ti_5O_{12}$ by giving addition of Sn metal. $Li_4Ti_5O_{12}$were synthesized by using sol-gel and hydrothermal methods. LTO then mixed with Sn in variable wt% Sn using ball mill. From this variable, the author will discuss about the characteristic of LTO/Sn and the variation of added wt% Sn on its elemental, chemical, and electrochemical properties.

2. Experimental procedure

2.1. Synthesis of spinel $Li_4Ti_5O_{12}$ powder

Synthesis of $Li_4Ti_5O_{12}$ was prepared using 5 different main stages of sol-gel process, calcination, hydrothermal, mixing, and sintering. The first stage is a sol-gel process to produce xerogel TiO_2. It takes 2 main solution to get xerogel TiO_2 that is primary solution and secondary solution. The primary solution is a mixed solution of titanium tetrabutoxide ($Ti(OBu)_4$) with a pH 3 ethanol solution. The secondary solution is a mixed solution of pH 3 ethanol with aqueous water of H_2O. The sol-gel process is carried out for 2 hours over a magnetic stirrer by slowly adding a secondary solution into a primary solution. After the sol-gel hydrolysis and alcohol condensation process occurs from the solution, the precursor will further thicken and eventually solidify to form a clear gel (aging). The precursor is then subjected to an open air drying in the beaker glass for 24 hours, which is called the drying process[6]. The precusor then calcined at a temperature of 300°C for 2 hours in the furnace with pure O_2gas. The heat treatment in this phase causes the TiO_2 structure to change to anatase than to rutile or brookite due to lower anatase surface energy[7]. The hydrothermal process of TiO_2 anatase is to increase the surface area, reduce the particle, and increase the interaction between solids, expected that there's and increase interaction between TiO_2 and LiOH to form $Li_4Ti_5O_{12}$[8]. The hydrothermal process is carried out in the furnace at a temperature of 120°C for 15 hours. The precusor the ball-milled with zircon ball for 15 minutes. The precusor was sintered at 750°C[9] for 3 hours.

2.2. $Li_4Ti_5O_{12}$/Sn composites anode coin cell

$Li_4Ti_5O_{12}$/Sn composites were prepared using ball mill machine for 10 minutes with zirconia ball 1:10 ratio against the powder. The variation content of micro sized Sn particles on $Li_4Ti_5O_{12}$ were ball milled composed with 5%, 10%, and 15% Sn in table 1. Composites anode powder then mixed with DMAC solvent by slurry method using active material, acetylene black, and PVDF in 8:1:1 ratio, respectively. The samples then leveled by doctor blade on Cu foil and crimped with $LiPF_6$ electrolyte. The samples then subjected to EDS, XRD, and electrochemical performance studies.

Table 1. Comparison of Sn content of each variation on $Li_4Ti_5O_{12}$ powder.

	Active Material Sample $Li_4Ti_5O_{12}$/Sn (gr)			
	LTO+5%Sn	LTO+10%Sn	LTO+15%Sn	LTO Synthesis
$Li_4Ti_5O_{12}$	2,85	2.7	2.55	3
Sn	0.15	0.3	0.45	0

3. Results

3.1. Synthesis

Tests using X-Ray Diffraction (XRD) were performed to determine the phase formed and measure the crystallites in the $Li_4Ti_5O_{12}$/Sn sample. XRD result data is processed using X'pert High Score Plus program with Semi Quant percentage. In figure 1, the data generated from the X'pert High Score program display 3 phases formed from the $Li_4Ti_5O_{12}$ synthesis results. Those phases are the $Li_4Ti_5O_{12}$ spheres (JCPDS NO.49-0207), TiO_2 anatase (JCPDS NO. 21-1272), and TiO_2 rutile (JCPDS NO. 01-

1292)[10]. This indicates that the synthesis of $Li_4Ti_5O_{12}$is not fully successful and there are phases such as TiO2 anatase and rutile. The scores of the X'pert High Score program showed a fairly high fit of 71 ($Li_4Ti_5O_{12}$), 68 (rutile), and 53 (anatase). The presence of rutile and anatase indicates that when synthesis of Li4Ti5O12 on sintering process with 750°C temperature for 3 hours was not sufficient.

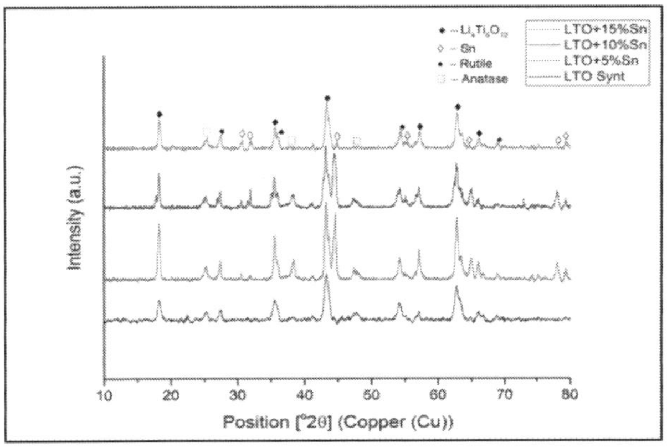

Figure 1. XRD sample test $Li_4Ti_5O_{12}$/Sn graphic.

There are five detected series in the sample which is Sn, Ti, C, O, and F. From the EDS test in figure 2, the purple color indicates that Sn is present in $Li_4Ti_5O_{12}$. The Ti element also indicates the presence of $Li_4Ti_5O_{12}$, means that the presence of spinel phases is dispersed in the sample. Li element cannot be detected by EDS testing due to low radiation energy characteristics. However, the authors can conclude that the Li element is adjacent to the Ti element. The element C as a yellow color is also spread evenly indicating that acetylene black as a conductive material where the path of the electron path[12] is scattered close to the elements Ti, O, and Sn.

3.2. *Electrochemical Performances*

The profile form of the EIS shows a semi-circle on its Nyquist plot. In the anode material of lithium ion batteries, the better pattern is the smaller semicircle. In figure 3, the entire half-circle line can be seen with Re and Rct. LTO+5%Sn and LTO+10%Sn has more Re value than LTO Synthesis and LTO+15%Sn. This indicates that when the coin cell is crimped, the emphasis is not the same on both samples. In table 2, the largest Rct is owned by LTO Synthesis among the samples with the addition of Sn. This shows that Sn influences the conductivity of the LTO/Sn sample, where Sn has good conductivity values. Conductivity is always inversely proportional to its resistance value. Sn metal has high conductivity value.

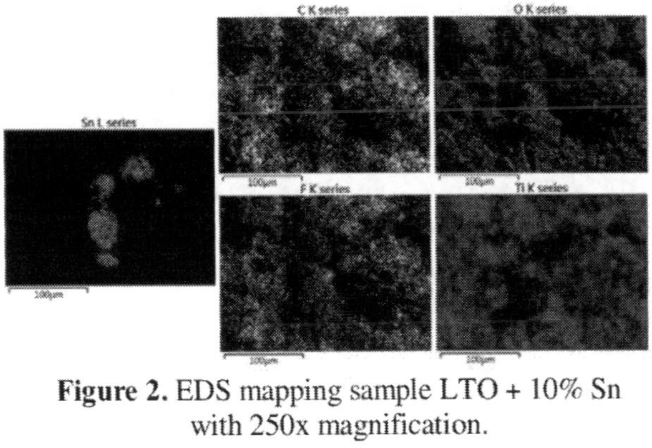

Figure 2. EDS mapping sample LTO + 10% Sn with 250x magnification.

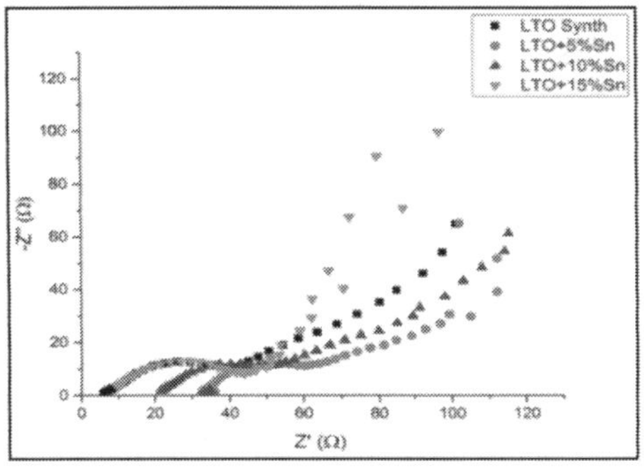

Figure 3. Semi-circle plot Nyquist.

From table 2 , LTO+5%Sn has the lowest Rct value and makes the greatest conductivity value. Re as an electrolyte constraint, reflects the electrical conductivity of electrolytes, separators, and electrodes[13], indicate the diffusion velocity of lithium ions from the LTO/Sn sample. Therefore, among the $Li_4Ti_5O_{12}$ samples with the addition of Sn, the LTO + 15% Sn sample with Re value of 9.55 Ω is estimated to have better lithium ion diffusion.There are different impedance values in LTO/Sn, LTO+5%Sn has a larger Rct value than other LTO/Sn. This may be due to the presence of rutile and anatase as shown in figure 1. Yang et al. (2016)[10], investigated that the presence of the rutile and anatase phases in $Li_4Ti_5O_{12}$ may increase its conductivity to Rct in the range of 45 Ω and have an increase in current density at low charge/discharge rates.

Table 2. Parameter impedance of sample LTO/Sn.

	Re (Ω)	Rct (Ω)
LTO Synthesis	5.87	35.13
LTO+5%Sn	35.7	27.40
LTO+10%Sn	32.9	27.90
LTO+15%Sn	9.55	34.35

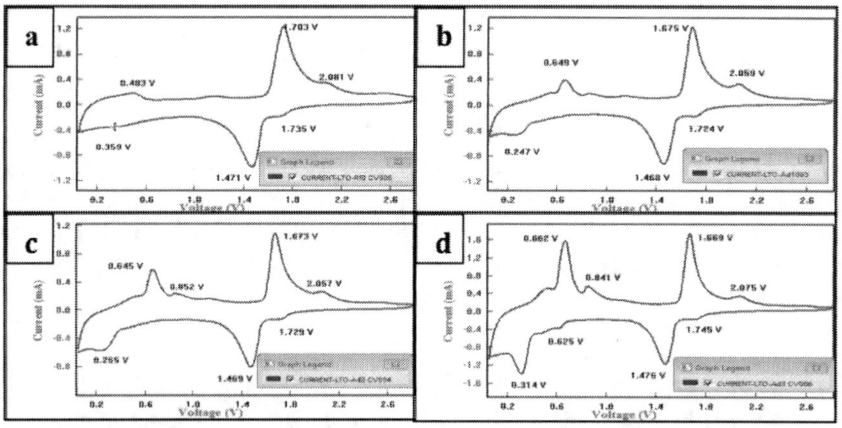

Figure 4. Cyclic voltammetry curve of sample **(a)** LTO Synthesis, **(b)** LTO + 5% Sn, **(c)** LTO + 10% Sn, and (d) LTO + 15% Sn.

Seen from figure 4, the peak at 1,703 V figure 5a is the working voltage of $Li_4Ti_5O_{12}$. $Li_4Ti_5O_{12}$ has a working voltage of ~ 1.5 volts[14], this means that synthesis of sol-gel and hydrothermal method has obtained a working voltage of ~1.5 volts. The cathodic peak at 2,081 V can be defined as the working voltage by anatase or rutile[10,15]. Figure 4b-d shows that there is a working stress on Sn metal at a peak of about 0.6 volts. The cathodic and anodic peaks of pure Sn metal have a voltage position of 0.58 and 0.43 V which is quite reversible[5]. These two main peaks constitute the alloying and dealloying abilities of Sn metal[14]. The process of alloying and dealloying can be written as follows.

$$Li_4Ti_5O_{12} + 3Li^+ + 3e^- \rightleftharpoons Li_{6-7}Ti_5O_{12} \ (reversible) \qquad (1)$$

$$xL^+ + +xe^- - +Sn \rightleftharpoons LixSn(0 \leq x \leq 4.4) \ \ (reversible) \qquad (2)$$

It is similar to LTO that has Li value of 4 to 7, while for Sn interacts with lithium ions can reach Li_xSn alloys up to 4.4. It is clearly seen that with the increase of Sn metal content on the $Li_4Ti_5O_{12}$ anode into a composite material, it will raise the current value from the peak of Sn. We can obtained the specific capacity by dividing the results with mass active materials which is 137.95, 152.6, 162.27, 258.6 mAh/g, with the increasing value of Sn on LTO/Sn. LTO Synthesis has a specific capacity value of 137.95 mAh/g. Theoretically, $Li_4Ti_5O_{12}$ has a capacity value of 150-160 mAhg[14]. Of the three samples, the LTO+15% Sn sample has the highest capacity of 258.6 mAh/g. This high capacity is the contribution of the theoretical nature of the Sn metal capacity of 994 mAh/g[16].

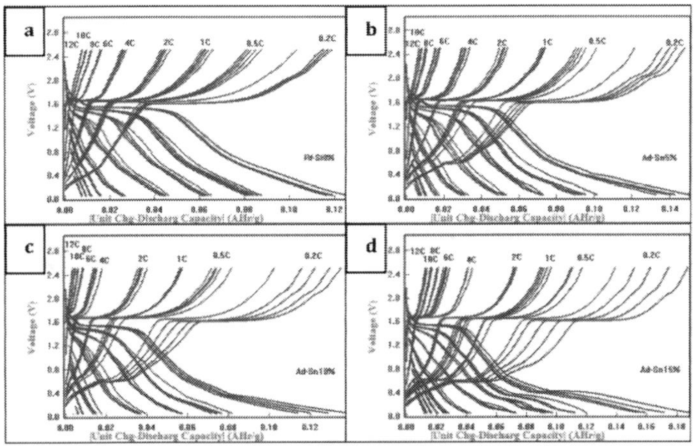

Figure 5. Sample charge / discharge curve **(a)** LTO Synthesis, **(b)** LTO + 5% Sn, **(c)** LTO + 10% Sn, and **(d)** LTO + 15% Sn.

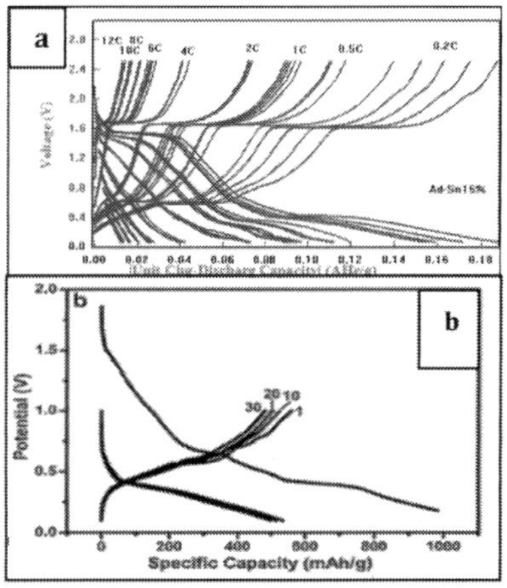

Figure 6. The charge-discharge curve (a) LTO + 15% Sn, and (b) Sn pure[5].

Figure 5 shows the relationship between charging-discharging and voltage on a given C. The LTO Synthesis sample in figure 5a has the characteristic of a straight line or plateau at a voltage of ~1.6 volts. The plateau voltage 1.5 and 0.6 V are the charge and discharge plateau characteristics of $Li_4Ti_5O_{12}$ spinel and are attributable to Ti^{4+}/Ti^{3+} transitions[5]. This plateau is the scene of the transition from the $Li_4Ti_5O_{12}$ spinel to the $Li_7Ti_5O_{12}$ rock salt. Looking from figure 6b, this is the charge-discharge graph of pure Sn. There is a characteristic charge-discharge curve for Sn that is in the range below 1.0 V, this is due to Li-Sn's multi-step alloying reaction when charging and responding to Li-Sn alloy to LiSn, Li7Sn3, and Li2Sn5[5]. From this comparison, seen in figure 6a, it has a plateau at voltage of about 0.5 V indicating the reversible properties of Li_xSn. Figure 5b-d also shows that with the addition of Sn level, the resulting plateau will be more elongated so as to increase its capacity.

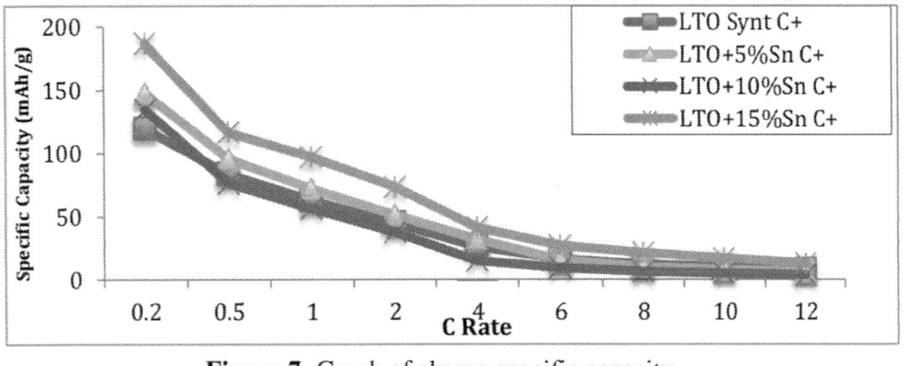

Figure 7. Graph of charge specific capacity.

Figure 7 shows the specific capacity relationships in the chargingprocess of the LTO/Sn sample. This data shows that the highest charging capacity at 0.2C and 12C LTO+15%Sn. While the lowest is owned by LTO+10%Sn. This maybe due to an overload of anatase and rutile. In the graph of discharge capacity to C rate. Discharging graph is almost identical as charging. Only the process of charging makes the resulting capacity slightly larger than the discharging process. At 0.2C, the resulting capacity of LTO Synt, LTO+5%Sn, LTO+10%Sn, and LTO+15% Sn respectively are 117.9, 142.1, 134.9, and 185.5 mAh/g. LTO+10%Sn has the lowest capacity. Sivashanmugam et al. (2011)[5], reported that with increasing levels of Sn on $Li_4Ti_5O_{12}$ will increase its capacity. This can be due to the unpredictable spread of anatase and rutile. Table 3 also illustrates the efficiency of the coulomb obtained from the ratio of charging to discharging capacity. Fullest potential can be obtained at 0.2 C. However, for the use of EV-HEV at 0.4 C, the capacity loss is considered to be extreme due to the Sn pulverization which has an expansion volume of up to 360% and increases the internal stresses[17] then continue into exhaustion at high C rate.

Table 3. Capacity loss from charge-discharge.

C Rate	Capacity Loss (%)							
	CLTO Syn	D LTO Syn	CLTO+ 5%Sn	DLTO+ 5%Sn	C LTO+ 10%Sn	DLTO+ 10%Sn	C LTO+ 15%Sn	D LTO+ 15%Sn
0.2	0	0	0	0	0	0	0	0
4	78.1	78.0	78.4	78.4	88.9	89.0	77.9	78.1
12	93.4	93.6	95.7	95.5	97.2	97.3	93.0	93.5

4. Disccusion

$Li_4Ti_5O_{12}$ has been successfully prepared by sol-gel process which produces xerogel TiO_2, calcination process, hydrothermal process, mechanochemical ball mill process with lithium ion source using LiOH powder, and sintering process produces brownish white sample. LTO + 15% Sn has the lowest resistance value for Re compared to LTO+10% Sn and LTO+5% Sn. There is a decrease in the value of Rct in all samples with the addition of Sn. Specific capacity increase with the increase of Sn vallue on cyclic voltammetry test up to capacity of 258.6 mAh/g on LTO+15% Sn. There is a reversible LTO and Li_xSn reaction in charge-discharge process with LTO+15% Sn resulting on having a better CD among other samples. Volume expansion due to Li_xSn phase changes. In conclusion, the addition of 15% Sn has better performance than 5% and 10% Sn.

5. Acknowledgment

The author would like to thank the Directorate Research and Public Services Universitas Indonesia for their financial support to this research under HIBAH PITTA 2017 with contract No: 736/UN2.R3.1/HKP.05.00/2017.

6. Reference

[1] Scrosati B and Garche J 2010 *J. Pow. Sources* **195(9)** 2419-30
[2] Etacheri V, Marom R, Elazari R, Salitra G and Aurbach D 2011*Energy Environ.*Sci.**4(9)** 3243-62
[3] Mosa J, Vélez J F, Lorite I, Arconada N and Aparicio M 2012 *J. Pow. Sources* **205** 491-4
[4] Kuo Y C and Lin J Y 2014 *Electrochim. Acta* **142** 43-50.
[5] Sivashanmugam A, Gopukumar S, Thirunakaran R, Nithya C and Prema S 2014 *Mat. Research Bull.***46(4)** 492-500
[6] Guglielmi M and Kickelbick G2014*Sol-Gel Nanocomposites*vol 1, ed A Martucci (New York: Springer) pp 2-4
[7] Hanaor D A and Sorrell C C 2011 *J. Mat. Sci.***46(4)** 855-74
[8] Yoshimura M and Byrappa K 2008 *J. Mat. Sci.***43(7)**2085-103
[9] Zhang Z, Cao L, Huang J, Wang D, Meng Y and Cai Y 2013 *Electrochim. Acta* **88**443-6
[10] Yang L, Li H, Liu J, Lu Y, Li S, Min J, Yan N, Men Z and Lei M 2016 *J. All. Comp.***689** 812-9
[11] Gu Y J, Guo Z and Liu H Q, 2014 *Electrochim. Acta* **123** 576-81
[12] Subhan A, Prihandoko B and Syahrial A Z 2011 *J. Ilmu Penge. Tehno.* Telaah **29** 27-35
[13] Yi T F, Yang S Y, Zhu Y R, Ye M F, Xie Y and Zhu R S 2014*Ceramic. Internat.***40(7)**9853-8
[14] Nitta N, Wu F, Lee J T and Yushin G 2015 *Mat. Today* **18(5)**252-264
[15] Liu S, Zhu K, Tian J, Zhang W, Bai S and Shan Z 2015 *J. Alloy. Comp* **639** 60-7

[16] Wen C J and Huggins R A 1981*J. Electrochem. Soc.***128(6)** 1181-7
[17] Tamura N, Ohshita R, FujimotoM, Kamino M and Fujitani S 2003 *J. Electrochem. Soc.* A **150(6)** 679-83

The influence of Titanium alloying and Co_3O_4 coating on the oxidation behavior of Fe-20Cr ferritic stainless steels for SOFC interconnects

A R Setiawan[1], A. Ramelan[2], R. Suratman[3] and A M Qizwini[4]
Research groups of Materials Engineering, Faculty of Mechanical and Aerospaces Engineering, Institut Teknologi Bandung,Indonesia

E-mail: [1]asep.ridwans@material.itb.ac.id, [2]ramelan@material.itb.ac.id, [3]rochim@material.itb.ac.id, [4]alfizqizwini@gmail.com

Abstract. Ferritic steel is a potential metal candidate for interconnect solid oxide fuel cell (SOFC) because its ability to form chromia layer at the surface during high temperature operation. However, chromia based oxide scales is reluctant to the evaporation of Cr from the oxide–gas interface. In this work, a combination of titanium alloying and application of Co_3O_4 spinel coating using dip coating method are carried out to improve the performance of ferritic steels as SOFC interconnect. The oxidation behavior of coated and uncoated Fe-20wt.%Cr alloys with different titanium contents: 0, 0.5, and 1 wt.% are studied as a function of time in air atmosphere. The samples were isothermally oxidized at 800°C for 24, 48, and 96 h in a box furnace. Preoxidation experiment were carried out on the alloys before the application of Co_3O_4 coating to improve its adherence. X-ray diffraction analysis (XRD) and Scanning Electron Microscope (SEM) were used for characterization of the prepared samples. The results shows that increasing the concentration of titanium in the alloys both for coated and uncoated samples caused the increased oxidation rates. The oxidation resistance of Co_3O_4 coated samples indicated from the weight change measurement was larger than that of the uncoated samples. Co_3O_4 coating is not effective for improving the oxidation resistance of samples.

1. Introduction

Fuel cell are the electrochemical device which capable to convert chemical reaction energy directly into electrical energy without involves combustion. One of the attractive fuel cell technology is solid oxide fuel cell (SOFC). This type of fuel cell has gain popularity due to its low emission and high efficiency. A single SOFC cell usually consist of solid ceramic electrolyte, an electrode (anode and cathode) and interconnect components. YSZ (Y_2O_3-stabilized ZrO_2) ceramics are typically used as an electrolyte for SOFC. The anode material is usually made from Ni-YSZ cermet, while the cathode is made from perovskite oxides, such as Sr-doped $LaMnO_3$ (LSM).

[1] To whom any correspondence should be addressed.

Interconnect is one of the main components of planar type SOFC. The electrical connection between anode in one cell with the cathode in the adjacent cell are provided by the interconnect component. It is also function as a fuel separator. Due to its severe working condition, interconnect component must satisfy several criteria, such as high electrical conductivity, oxidation resistivity, gas permeability, etc. In the past, those criteria are met by ceramics materials such as lanthanum chromate ($LaCrO_3$) [1]. These material are reported to have relatively high thermal stability and electrical conductivity in SOFC working condition at 1000 °C. However, ceramic interconnects are relatively difficult to fabricate and the cost is high. These factor become an obstacle for adoption in the industry, leading to the development of low cost metallic interconnects.

Recently, the operating temperature of solid oxide fuel cells (SOFC) are lowered from 1000 °C to 600-800 °C. The new condition has allowed the use of metallic material as interconnects. Among the various type of metal, ferritic steel is a potential candidate for SOFC interconnect because its ability to form continous chromia layer at the surface during high temperature operation. Ferritic steels also offer severals advantages for use as interconnect materials such as high mechanical strength, high thermal and electrical conductivity, CTE matching with other SOFC component and ease of fabrication. Despite these goods properties, these steels face many obstacle for adoption, such as unacceptably high oxidation rate, occurrence of buckling and spallation of the oxide scale when subjected to thermal cycling, and volatilization of high valence Cr-species in the form of CrO_3, or $Cr(OH)_2O_2$ which could lead to the degradation of overall cell performance [1,2]. The formation of oxide scale at the surface of ferritic steels will increase the contact resistance between electrodes and interconnect, reduce the SOFC cell efficiency.

Many studies have been done to improve the perfomance of ferritic steels as SOFC interconnect, from alloying design to surface treatments [1,3]. Several alloying element added to the ferritic steels such as Y, La, Ce [4] and Zr [5] are reported for improving oxidation resistance, scale adherence and conductivity. Small addition of element such as Nb, W, Ti, Mo to Fe-Cr alloys are effective to control the elemental diffusivities at the alloy grain boundaries by forming a laves type phase in air [6]. Nb, W and Mo addition to ferritic steels are also reported to improve the mechanical property, oxidation resistance, creep resistance and reduction of ASR [7,8]. Ti addition into ferritic steels was effective to reduce Cr evaporation [9-12]. However, if the amount of Ti is too high, oxidation rate and ASR will increase [10,12].

Protective coatings may also be utilize to improve the performance of ferritic steels. Several coating materials have been investigated extensively, such as perovskite ($LaSrCoO_3$, $LaSrCrO_3$, $LaSrMnO_3$) and spinel ($MnCo_2O_4$, $Mn_{1.5}Co_{1.5}O_4$, $(Cu,Mn)_3O_4$) coating materials [3,13]. The coating are not effective to prevent diffusion of chromium outward [13]. In the previous work [11,12], the author have been investigated the oxidation behavior of Ti containing alloys at 700 and 800 °C in air atmosphere. The results indicated that the presence of Ti in the alloy increase the oxidation rate of Fe-20Cr alloy. In this work, a combination of titanium alloying and application of Co_3O_4 spinel coating are carried out to improve the performance of Fe-20Cr during oxidation.

2. Experimental procedure

2.1. Sample preparation

Fe-20Cr alloy steels, with different titanium content; 0, 0.5 and 1 wt% were selected for the isothermal oxidation test at 800 oC. The chemical composition of chromium steels was analyzed using an Optical Emission Spectrometer (OES) ARL 3460. The results are shown in Table 1. Prior to oxidation, the specimens were cut into coupon with a size of 30 mm x 20 mm x 1 mm. After cutting, the surface of specimens were shot blasted using glass bead. Preoxidation experiment were carried out at 800 °C for 24 hour on the alloys before the application of Co_3O_4 coating to improve its adherence. Then, the specimens were vertically dip coated in Co_3O_4 slurries at room temperature. The slurries were prepared in a 250 ml beaker glass by adding 7 g Co_3O_4 powder, 20 g ethanol and 1 g polyvinil alcohol. The solution was stirred for one hour using magnetic stirrer at room temperature. The specimen were then withdrawn from the solution slowly with constant speed after an approximate time of 30 seconds. Finally, after coating, the specimen were held vertically and dried at room temperature for 3 hour.

Table 1. Chemical composition of specimens (wt%).

	Fe	Cr	Ti	Mn	Si	C	P	S	V
Fe-20Cr	Bal	20.01	0.01	0.12	0.12	0.01	0.04	0.01	0.01
Fe-20Cr-0,5Ti	Bal	19.89	0.47	0.12	0.12	0.01	0.04	0.01	0.01
Fe-20Cr-1Ti	Bal	19.69	0.98	0.12	0.11	0.01	0.04	0.01	0.01

2.2. Oxidation test

For the isothermal oxidation test, as coated and uncoated specimen were oxidized at temperature of 800 °C for up to 345.6 ks in a muffle furnace. The steel samples were hang using stainless steel wire inside an alumina crucible. The crucible were placed inside the furnace and heated for periods of 86.4, 172.8, 345.6 ks. When the oxidation was terminated, the furnace was cooled down to a room temperature. In order to evaluate oxidation resistance, mass gain of specimen was measured with electronic balance before and after the oxidation. The oxidation kinetics can be calculated with parabolic kinetics law as follows:

$$(\Delta m / A) = k_p t \qquad (1)$$

where $\Delta m/A$ is mass gain per unit area, k_p is the parabolic rate constant, t is the oxidation time.

2.3. Analysis

After oxidation, the crystal structure of the oxide scale was identified by X-ray diffractometer (XRD) with CuKa radiation. The microstructure of oxide scale was characterized using scanning electron microscopy (SEM) with energy dispersive X-ray analysis (EDX).

3. Result and discussion

3.1. Oxidation behavior

Figure 1 and figure 2 shows the mass gain per unit area as a function of oxidation time, for all uncoated and coated steels during oxidation at 800 °C in air atmosphere. The oxidation behavior for all specimen, uncoated and coated Fe-20Cr, Fe-20Cr-0.5Ti and Fe-20Cr-1Ti alloy were followed the classical parabolic relationship with time. For uncoated specimen, Fe-20Cr alloys has the lowest mass change, followed by Fe-20Cr-0.5Ti and Fe-20Cr-1Ti. For Co_3O_4 coated specimens, Fe-20Cr-0.5Ti has the lowest mass change, followed by Fe-20Cr, and Fe-20Cr-1Ti alloys. The presence of the Ti caused a significant changes in the oxidation resistance. The addition of titanium into the alloys increase the oxidation rate of Fe-20Cr alloy. Several researcher [4,14,15] reported that the presence of high amount of Ti in Cr_2O_3 matrix may generated an excess ionic defect and reduced the oxidation resistance of the alloys. High amount of Titanium may induce the creation of additional cation vacancies in the Cr_2O_3 matrix and subsequently promote fast diffusion of ion through Cr_2O_3 layer. Compared to the bare specimens, the oxidation resistance of Co_3O_4 coated samples indicated from the weight change measurement was larger. This indicated that Co_3O_4 layer are not protective as Cr_2O_3 in preventing oxygen inward diffusion into the substrate.

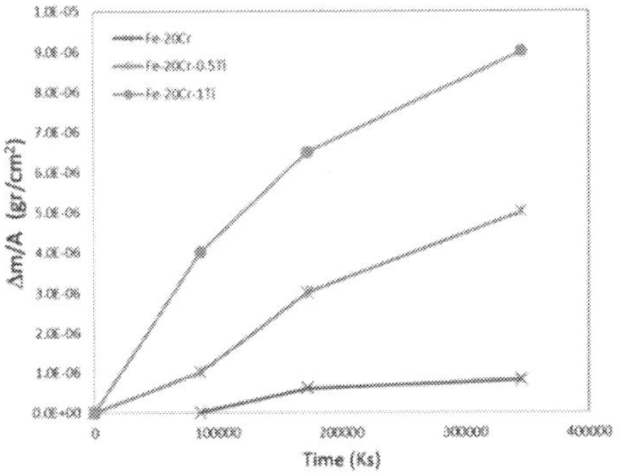

Figure 1. Oxidation behavior of Fe-20Cr with different titanium content : 0, 0.5 and 1 wt%.

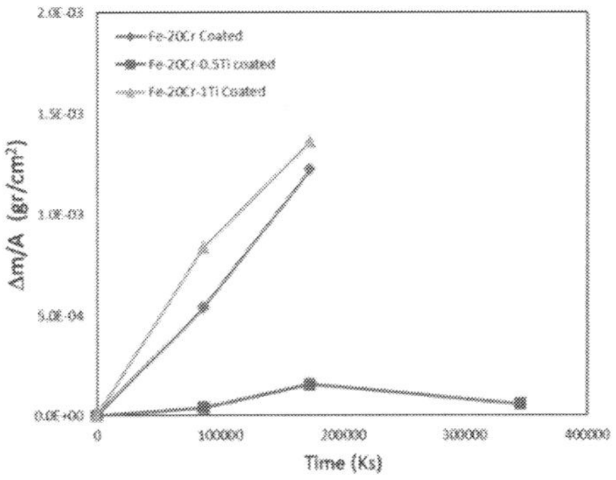

Figure 2. Oxidation behavior of Co3O4 coated Fe-20Cr with different Titanium content: 0, 0.5 and 1 wt%.

The calculated parabolic rate constant for uncoated and coated Fe-20Cr, Fe-20Cr-0.5Ti and Fe-20Cr-1Ti alloys are shown in Table 2. For uncoated alloys, the addition of Ti element into Fe-20Cr increase the parabolic rate constants (kp) value 2 order of magnitude. In the case of Co_3O_4 coated alloys, the addition of Ti element into Fe-20Cr increase the parabolic rate constants (kp) value 1 order of magnitude. Compared to the bare specimens, the parabolic rate constant of Co_3O_4 coated samples was larger, 6 order of magnitude.

Table 2. Parabolic rate constant.

Alloys	Parabolic Rate Constant, kp ($gr^2cm^{-4}s^{-1}$)	
	Uncoated	Co_3O_4 coated
Fe-20Cr	$2x10^{-18}$	$8.6x10^{-12}$
Fe-20Cr-0.5Ti	$7x10^{-17}$	$1.1x10^{-11}$
Fe-20Cr-1Ti	$2x10^{-16}$	$2.1x10^{-11}$

3.2. Phase identification

The XRD profile of the oxide scale formed on the uncoated specimen surface after 96 h oxidation at 800 °C is shown in figure 3.

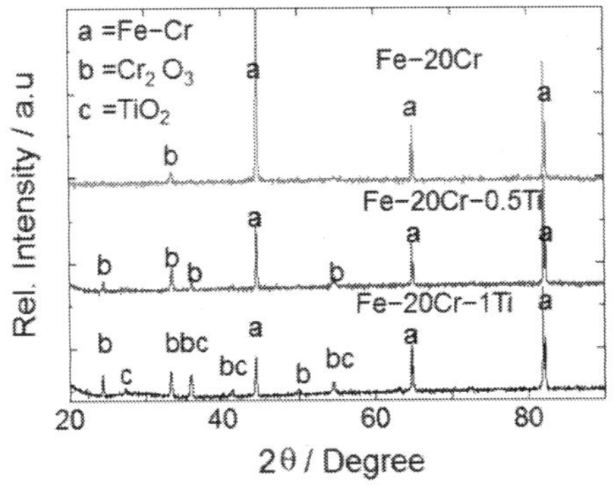

Figure 3. XRD Profile of uncoated Fe-20Cr, Fe-20Cr-0.5Ti and Fe-20Cr-1Ti alloy after oxidation at 800 °C for 96 hour.

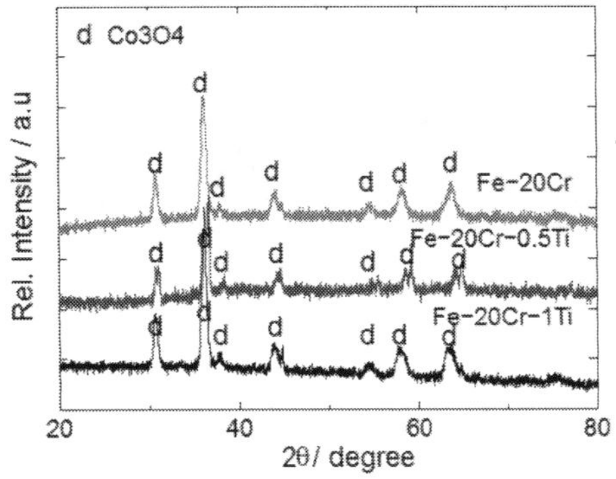

Figure 4. XRD Profile of Co_3O_4 coated Fe-20Cr, Fe-20Cr-0.5Ti and Fe-20Cr-1Ti alloy after oxidation at 800 °C for 96 hour.

Strong peaks of Cr_2O_3 are observed in all alloys. The presence of TiO_2 are detected on Fe-20Cr-0.5Ti and Fe-20Cr-1Ti alloy. The peak of alpha iron (Fe-Cr) in Ti containing alloys, decreases, indicating that the oxide scales was thicker so that X-ray penetration into the substrate are limited. The formation of TiO_2 and Cr_2O_3 in Fe-20Cr-0.5Ti and Fe-20Cr-1Ti alloy increases the mass gain of the alloys during oxidation.

The XRD profile of Co_3O_4 coated specimen after 96 hour oxidation at 800 °C are displayed in figure 4. Strong peaks of Co_3O_4 are observed in all alloy. Cr_2O_3 and TiO_2 phase were not detected on the XRD profile. This is due to very thick Co_3O_4 scales formed in the surface, so that the X-ray penetration into the substrate are limited.

3.3. Oxide morphology of Co_3O_4 coated specimen

Cross-section morphology of Co_3O_4 coated specimen after oxidation test was observed using Scanning electron microscope (SEM), as shown in figure 5. Thick oxide layer were found on the top surface, composed mainly of Co_3O_4. The oxide layers are porous. In Fe-20Cr-0.5Ti and Fe-20Cr-1Ti alloy, TiO_2 oxides were precipitated internally at the alloy/oxide interfaces. The results are similar with our previous work [11,12].

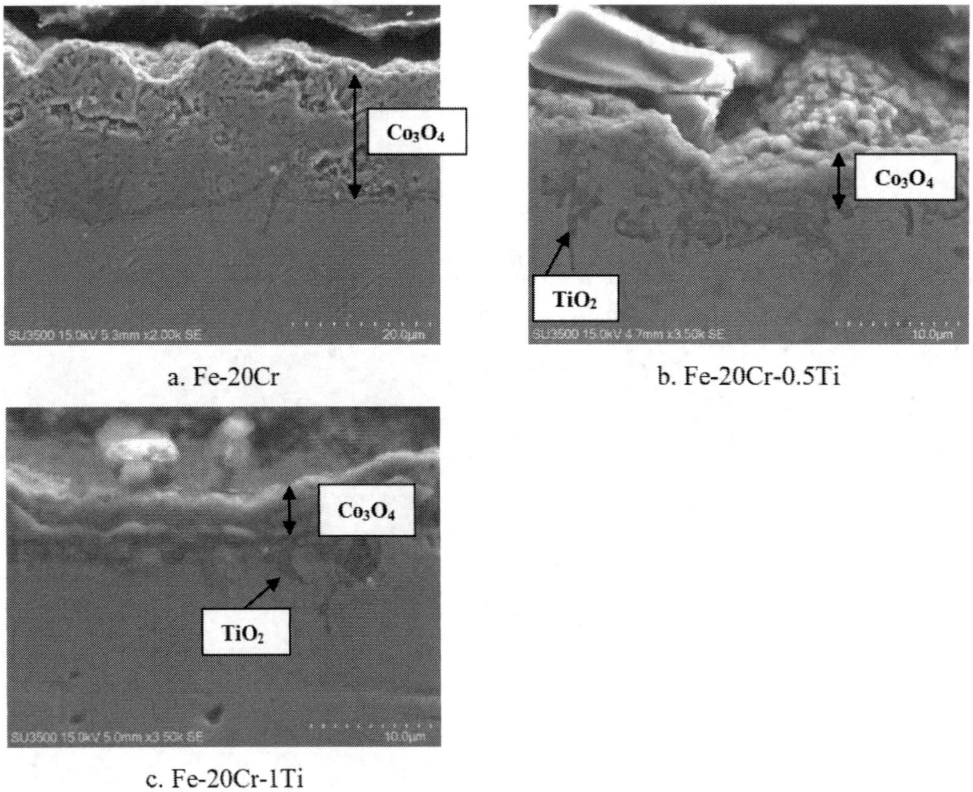

Figure 5. Cross section morphology of Co_3O_4 coated specimen after oxidation at 800 °C for 96 hour in air atmosphere.

The amount of Ti in Fe-20Cr alloy was not enough to form continous TiO_2 layers in the surface of alloys. The formation of TiO_2 oxide at the oxide/alloy interfaces indicated that the Co_3O_4 layer were not protective, so that oxygen from the atmosphere diffuse inward to react with Ti element in the substrate to form TiO_2 oxide. As a result, the oxidation behavior of Fe-20Cr-0.5Ti and Fe-20Cr-1Ti alloys increased significantly compared to the Fe-20Cr alloy, both for bare and coated alloy. Detail observation on the oxide morphology also indicated that the presence of Ti element in Fe-20Cr alloy may cause higher rate of spallation. High amount of Ti oxide in the interface formed the void and reduced the adhesion of the oxide to the substrate [16].

4. Conclusion

Investigation on the oxidation behavior of uncoated and Co_3O_4 coated Fe-20Cr, Fe-20Cr-0.5Ti and Fe-20Cr-1Ti alloys indicated that the alloy are still far from the application as SOFC interconnect. The oxidation test clearly indicated that Ti addition into Fe-20Cr alloy, both for coated and uncoated specimen, increases the oxidation rate of the alloy significantly. This is supported from SEM observation. In Ti containing alloys, large amount of Ti-oxide was dispersed in the oxide/metal interface and led to internal oxidation. The oxidation resistance of Co_3O_4 coated samples indicated from the weight change measurement was larger than that of the uncoated samples. Co_3O_4 layer is porous, and not effective for improving the oxidation resistance of samples

5. Reference

[1] Zhu W Z and Deevi S C 2003 *Mat. Sci. Eng.* A **348** 227
[2] Zhu W C and Deevi S C 2003 *Mater. Res. Bull.* **38** 957.
[3] Shaigan N, Qu W, Ivey D G and Chen W 2010 *J. Power. Sources.* **195** 1529
[4] Seo H S, Jin G, Jun J H, Kim D-H and Kim K Y 2008 *J. Power. Sources.* **178** 1
[5] Hua B, Pu J, Lu F, Zhang J, Chi B and Jian L 2010 *J. Power. Sources.* **195** 2782
[6] Wu J, S. Gemmen R, Manivannan A and Liu X 2011 *Int. J. Hydrogen. Energ.* **36** 4525
[7] Yun D W, Seo H S, Jun J H, Lee J M and Kim K Y 2012 *Int. J. Hydrogen. Energ.* **37** 10328
[8] Yun D W, Seo H S, Jun J H and Kim K Y 2013 *Int. J. Hydrogen. Energ.* **38** 1560
[9] Seo H S, Yun D W and Kim K Y 2012 *Int. J. Hydrogen. Energ.* **37** 16151
[10] Jablonski P D and Alman D E 2008 *J. Power. Sources.* **180** 433
[11] Setiawan A R and Artono T J 2016 *AIP. Conf. Proc.* **1711** 040003
[12] Ruhma Z, Setiawan A R, Ramelan A and Suratman R 2014 *Appl. Mech. Mater.* **660** 249
[13] Bi Z H, Zhu J H and Batey J L 2010 *J. Power. Sources.* **195** 3605
[14] Geng S and Zhu J 2006 *J. Power. Sources.* **160** 1009
[15] Holt A and Kofstad P 1999 *Solid State Ionics* **117** 21
[16] Safikhani A, Esmailian M, Tinatiseresht T and Darband G B 2016 *Int. J. Hydrogen. Energ.* **41** 6045–52

IOP Conf. Series: Earth and Environmental Science **105** (2017) 012116 doi:10.1088/1755-1315/105/1/012116

Preparation of activated carbon from babassu endocarpunder microwave radiation by physical activation

Maria de Fatima Salgado[1,2,3], Adekunle Moshood Abioye[3,4], Muhammad Mat Junoh[3], João Alberto Porto Santos[1] and Farid Nasri Ani[3,5]

[1] Department ofMathandPhysical, MaranhãoStateUniversity– CESC/UEMA., Praça Duque de Caxias S/N - Morro do Alecrim, 65.600-380. Caxias, Maranhao, Brazil.
[2] Programa de Pós-Graduação Eng. de Materiais do Instituto Federal do Piauí – Teresina - PPGEM / IFPI - Brasil.
[3] Faculty of Mechanical Engineering, Universiti Teknologi Malaysia, 81310 UTM Skudai, Johor, Malaysia.
[4] Department of Mechanical Engineering, Abubakar Tafawa Balewa University, Bauchi, Nigeria.

E-mail : farid@mail.fkm.utm.my

Abstract. Babassu endocarp were used to prepare activated carbons by physical activation via microwave radiation for the first time. The pyrolysis temperature was 600°C and the derived biochar were activated in CO_2atmosphereat 700, 750 and 800°C for 30 min. The material was characterized using scanning electron microscopy (SEM). The porous properties of the activated carbons obtained including the Brunauer-Emmett-Teller (BET) surface area, pore volume, and average pore diameter were determined by nitrogen adsorption isotherms at 77.32 K. The experimental results showed that most pores occurred during the activation predominantly as micropores. Endocarp babassu can be used as precursor to produce activated carbon with a rather well-developed porosity by pyrolysis and physical activation by two-steps with CO_2 activation via microwaves radiation. The activated carbon, with a low production cost, could be suitable for applications in gaseous pollutant adsorption, adsorb iodine, methylene blue, and residual chlorine.

1. Introduction

Babassu is considered the largest native oleifero resource in the world and one of the main forest products in Brazil [1]. The state of Maranhão is responsible for 94.5% of the whole national production of this native palm tree. The total production of babassu kernel in Brazil, was 83,917 tons in 2013 [2].The average composition of the babassu endocarp is 59%, summing up a residue of 707,300 tons in 2013. The majority of this rich mass lignocellulosic residue is improperly disposed, impacting negatively on the environment. The potential of babassu is numerous, however, in spite of its potential, only charcoal and oil have been produced in commercial scale [3]. The babassu nutshell showed high basic density and suitable lignin content for the sustainable production of bioenergy and

[5] To whom any correspondence should be addressed.

charcoal, and is technically capable of replacing coke in Brazilian steel plants [4]. This can contribute decisively to the economic development of extractive communities, who survive from the babassu nut collection by marketing a product with higher added-value. [5].

The babassu carbon study has prioritized its use to provide power to department such as: steel industry, metal recycling and household use. Considering that all bibliography consulted showed work produced in conventional oven, there is a blank that must be filled in this research area, as well as a need in minimize outlay and in find out others products. Thus, as example, the production of babassu activated carbon using microwave as power supply. The use of the residues in preparation or production of a higher valued material as an alternative transforming waste in raw material. Recently, many studies have reported the production of activated carbon (AC) from residues such as: Rice husk, [6]; Green Coconut Shell,[7];Babassu coconut,[8];Mukah coal, [9]; Residues of babassu, [10]; Coffee waste, [11];Oil palm, [12], etc. Many studies have discussed great potential of lignocellulosic as a renewable feedstock for preparation of Activated Carbon. [13, 14].

Activated carbon (AC) is a carbonaceous and porous material prepared from the carbonization and activation of organic substances, especially from vegetable source. It is widely used in pollutant adsorption in liquid and gas phases, as catalyst support, to purify various compounds, to treat effluents, etc. The adsorption capacity of AC depends on the nature of its precursor, size of its particles (granules or powder) and on the production process chosen. The AC quality is evaluated in terms of its physical properties of adsorption and superficial area [15]. The success of activated carbon among these areas is mainly due to its extensive surface area and diverse chemical functionalities that can be modified during and after its production process.

The use of lignocellulosic material to produce activated carbon has been of interest to numerous researchers since it represents a "green" alternative for producing the carbonaceous adsorbent used all over the world. Also by using waste biomass as an activated carbon precursors, we not only take advantage of the availability and low cost of these materials but also raise an alternative for the treatment and disposal of hundreds of tons of biomass waste generated worldwide, which in most cases leads to a negative environmental impact, [16].

Microwave irradiation has recently exhibited superior performance for some organic chemical transformations, when compared to conventional heating methods. This is related to the high efficiency of microwaves in rapidly heating media[17]. The microwave heating has huge potential as effective method in order to produce activated carbon with high quality, high carbon yield and porosity[18]. In addition, the pores of activated carbon produced via microwave heating are uniform and clean. The BET surface area of activated carbon is still high. The eco-friendly activated carbon prepared via microwave heating can save significant energy and time activation [19].

In this study, the microwave irradiation system was used to produce babassu endocarp activated carbon (BAC) using physical activation with carbon dioxide as an activating agent. The properties of the activated material, such as surface area, pore size and yield were analyzed for the first time for conditions presented.

2. Material and methods

2.1. Starting raw material

In this work, endocarp babassu was selected as the starting raw material. The babassu fruit presents an ovoid or ellipsoidal shape with average length of about 10.00 cm, diameter of 6.80 cm and a weight of 273.00mg, as shown in figure1. The fruit has 4 constituents: externally, it has the epicarp a thin peal that involves the mesocarp this is a starchy secondary layer, and the endocarp this is a hard wooden, which accommodation the kernels.

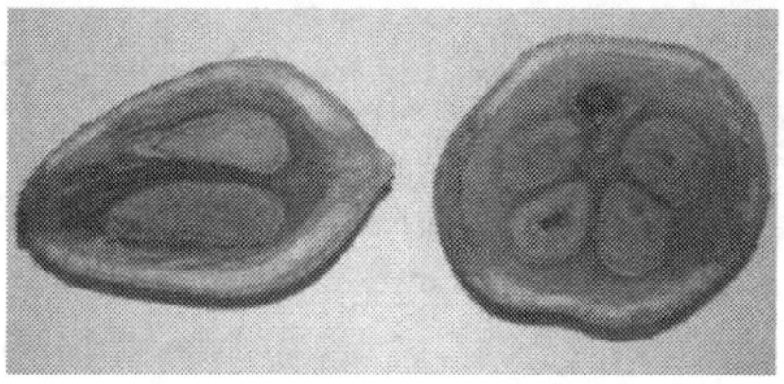

Figure 1. Babassu with 2 and 5 kernels.
Longitudinal and transversal section.

The babassu used in this workwere collected in the rural area of Caxias city, in the state of Maranhão, Brazil (04° 51' 32" S, 43° 21' 22" W) and comes from the extractive activity of local communities. Only the endocarp received and dried in the atmospheric ambient of the sun was used to make the activated carbons. The endocarp was first crushed to the size fraction of 5.0-8.0mm; the particle size plays an important role in microwave heating because the penetration depth becomes significant as the particle size increases[20]. 50 g of babassu endocarp were used in each experimental run. Pyrolysis of the pre-treated endocarp was performed in a system consisting of modified multimode domestic microwave, and commercial coconut-based activated carbon (CAC) used as microwave absorber in the pyrolysis process was obtained from Laju group of companies, Malaysia.

2.2. Carbonization process

Carbonization was carried out in a microwave system consisting of modified multimode domestic microwave (800W, 2500 MHz), in a glass quartz reactor with a length of 20 cm and an inner diameter of 5 cm with bottom wire mesh, which serves as screen support as well as distributor plate. N_2 and CO_2 gases, K-type thermocouple (0 – 1200°C range) were inserted into the bed region and data acquisition system was linked to personal computer for continuous monitoring and recording of process temperature[21-22]. For the pyrolysis of biochar, a single layer arrangement of precursor and coconut activated carbon (CAC) in the reactor was adopted because microwave heating requires the presence of microwave absorber to increase heat absorption. The size of CAC ranges from 1 mm – 1.8 mm. Thebabassu/CAC ratio was fixed at 1:0.5, irradiation time was 42min, with a heating rate of 50°C/minand 30 minutes of residence time. The final temperatures were 600°C. A significant increase in the mechanics of biocharis achieved by carbonization at a final temperature above 500°C [23]. The choice of residence time and carbonization temperature was obtained from previous study of biochar babassu characterization especially for the preparation of activated carbons. The experimental results showed that it was feasible to prepare biochar with sufficient densities and relatively high porosities for preparation of activated carbons the best retention time was 30 min. Endocarp babassu biochar depends on the pyrolysis conditions and indicated that high retention time might affect the biochar characteristics. The carbonization temperature of 600°C, i.e., the temperature at which the matter should be totally pyrolyzed, was possible by taking into account the thermo gravimetric curves of all raw materials used in this study. The microwave power of 800 W was selected for the pyrolysis. CAC was uniformly distributed on the wire mesh followed by babassu. N_2 was supplied at a flow rate of 400cm^3/min for about 5 min before commencement of experiments to ensure inert environment, and was reduced to 200cm^3/min during experiment so as to maintain the inert environment and to sweep out the vapor. After carbonization, the yield in biochar was calculated by dividing the dry mass of biochar by the dry mass of babassu endocarp multiplied by 100.

2.3. Activation process

The activation process was carried out in the same setup used for carbonization. All endocarp babassu biochar from the pyrolysis process were subsequently activated under the same conditions to prepare the final product. The activation temperature used were 700, 750 and 800°C.The temperature was raised at a rate of 50°C/min from room temperature to the desired activation temperature. During the heating process, the nitrogen flow rate was 200 cm^3/min. Once the final activation temperature was reached, the gas supply to the reactor was changed to carbon dioxide (99.998%).The CO_2 flow rate used for the activation process was 200 cm^3/min and continued for 30 min. After activation, $N_2$was used again as an inert flushing gas whilst the microwave was allowed to cool down to room temperature.

2.4. Characterizations of raw babassu,biocharand activated carbon
Proximate analysis was carried out by a thermogravimetric analyzer – TGA before and after carbonization. Ultimate analysis was used to determine the carbon, hydrogen, nitrogen and sulfur contents of babassu and biochar using an Elemental Analyzer (Vario Micro Cube). Scanning Electron Microscope – SEM was used to verify the presence of porosity on the raw material and biochar surfaces.

2.4.1. Proximate and ultimate analyses. The thermal degradation studies of the raw material endocarp babassuwere conducted on Shimadzu DTA-50 analyzer, with heating rate of 10°C/min from room temperature to 900°C under N_2 atmosphere (20 ml/min).Proximate analysis of Biochar precursor and activated carbon was performed in the Thermogravimetric analysis - TGA (Pyris 6-TGA) to determine the amounts of moisture, volatile matter, fixed carbon and ash content in activated carbons TGA was used to follow the thermal degradation profile of the material after activation. About 20mg of sample were heated from room temperature at 30°C to 100°C in nitrogen until complete dehydration was accomplished, followed by decomposition at 600°C for 7 min to determine the quantity of volatile matters. The atmosphere was then changed to be oxidizing. The sample was cooled to 700°C, 750°C and 800°C and maintained at this temperature until its weight remained unchanged. The weight loss during this period was due to the reaction of the fixed carbon with oxygen and the remaining residue was ash.

The quantification of the carbon (C), hydrogen (H), nitrogen (N) and sulfur (S) relative to raw material and biochar was performed in a universal brand Elemental Analyzer (Vario Micro Cube). The oxygen content was determined by difference. The molar ratios H/C, N/C and O/C were also determined.

2.4.2. Scanning Electron Microscopy – SEM. Scanning electron microscopy (SEM) analysis with Jeol VPSEM IT300 was carried out for babassu,biochar and activated carbon to study the development of surface morphology and porosity.

2.4.3. BET – Surface Area. The physical properties of the biochars and activated carbons relating to specific surface area and total pore volume were obtained by measuring their nitrogen adsorption-desorption isotherms at -196°C in a surface area porosimeter system (Micromeritics 3Flex 3.01). Brunauer-Emmett-Teller (BET) equation was used to calculate the BET surface area.

2.4.4. Statistic analysis. The three-sigma rule was used for statistic determination of the results which states that even for non-normally distributed variables, at least 98% of cases should fall within properly-calculated three-sigma intervals[24].

3. Results and discussion

3.1. Raw Material – Babassu
The proximate and ultimate analyses of raw material are listed in table 1.The babassu endocarp presents a high content of volatile matter, a high content of fixed carbon and a low ash content, which are favorable properties for the preparation of activated carbon. The density of the babassu gives it the waterproof suggesting high lign in value so that the carbonization yield will be high. All properties of the babassu endocarp show that it is good precursors for activated carbon.

IOP Conf. Series: Earth and Environmental Science **105** (2017) 012116 doi:10.1088/1755-1315/105/1/012116

Table 1. Main characteristic of the raw babassu.

Characteristics	Raw material
Proximate analysis (wt %)	
Volatiles	63.39
Fixed Carbon	26.03
Ash	1.98
Moisture	8.60
Ultimate Analysis (wt %)	
Carbon	44.02
Hydrogen	5.87
Nitrogen	0.32
Sulfur	0.91
Oxygen[1]	48.88
C/N	137.56
H/C	0.13
O/C	1.11

[1]by difference

According to TGA results, to babassu endocarp raw material, the first degradation stage is related to water loss. From 220°C starts the second degradation stage which corresponds to the decomposition of cellulose and hemicellulose molecules. This thermal event presents maximum degradation speed close 400°C with mass loss of 53.02%. The third stage with mass loss 10.37% represents ash and fixed carbon contents. How, [25] says, the removal of extractives and lignin compounds affect the thermal stability of the materials, probably the increasing porosity.

The results show good agreement with [26] which found for volatile materials (83.40%) and a fixed carbon content (15.16%) in the babassu endocarp, being this more suitable for burning and carbonization, in relation to the other constituents of the fruit. The hemicellulose decomposes at 200–260°C, cellulose at 240–350°C, and lignin at 280–500 °C[27]. Therefore, for the first peak at 280°C in the rate of weight loss curve in figure 2, this weight loss is mainly due to the combined decomposition of hemicellulose and cellulose. The decomposition of both cellulose and lignin contributes to the subsequent peak at 350 °C.

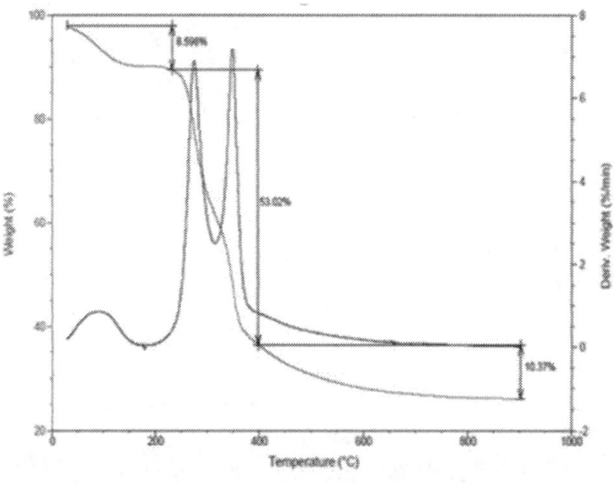

Figure 2. TGA curve for rawbabassu.

3.2. Temperature profiles

Pyrolysis temperature have significant effect on the characteristics of the activated carbons, [28]. During activation time, monitoring of process temperature was realized. The temperature ramp shown in figure 3, was obtained for all temperature from 700°C to 800°C. The feedback temperature control using microwave power is necessary for the precise temperature and the control in the reactor and thermocouples are the only available option[29]. It shows real time bed temperature profiles recorded during activation process of the AC. The cyclic ON/OFF of the magnetron in the multimode domestic microwave is responsible for the sinusoidal nature of the temperature profiles observed.

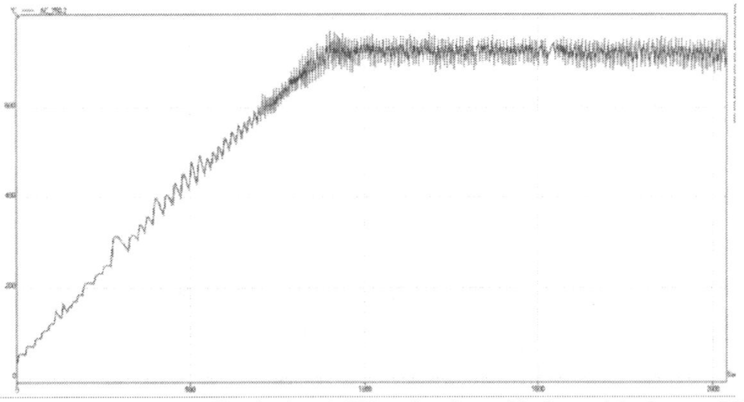

Figure 3. Temperature profile during MW activation of babassubiochar at 750 °C.

3.3. Thermogravimetric analysis of biochar and activated carbons

The Thermogravimetric analysis (TGA) profiles of the biochar and activated carbons are shown in figure 4. It is important indicator that all residues was burned and activated at 700°Cand have lower ash content. At 750 and 800°C the activated carbon have identical thermal properties.

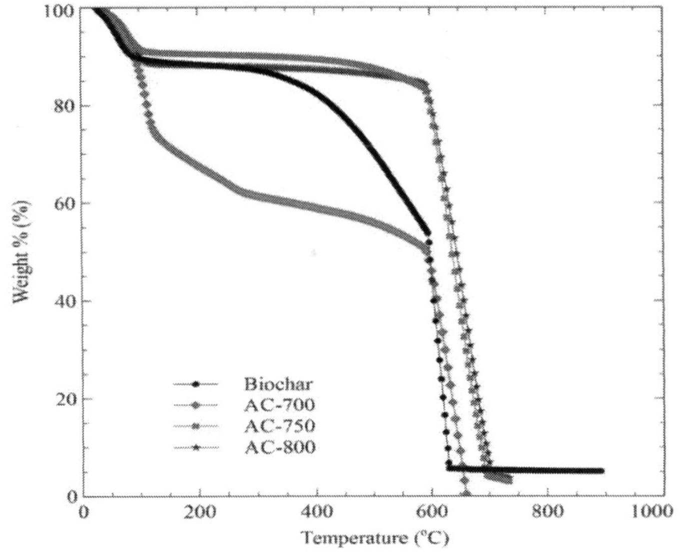

Figure 4. Thermogravimetric Analysis (TGA) of biochar and activated carbon.

The activated carbon produced under temperatures of 700, 750 and 800°C had its mass reduced in approximately 100°C (water lost), both showing begin of thermal decomposition around 340°C and stabilizing around 600°C, then, it can be assumed that the activated carbon produced, when compared to babassu biochar that originated it, have similar thermal stability . Another interesting aspect to be highlighted is regarding to the final mass. The babassu biochars resulted in 5.16% mass after stabilized, whereas the babassu in natural final mass was approximately 10.33%. AC_750°C biochar presented a residual mass around 2.8% and AC_800°C's was around 1.8%, whereas AC_700°C had no residual mass. The degradation process of lignocellulosic structures happened during all heating process according to [30–31]. It can be affirmed the material studied has an elevated lignin content in its solid matrix, once the amount of mass kept until temperatures of 400°C is considerably high. That condition is due to the fact that only the mesocarp of babassu coconut was used, i.e. it's most resistant part. Also should be emphasized the low level of ashes found in babassu activated carbon, which is an important property in choosing a good precursor for the activated carbon production. It is observed, as well, through the thermal analysis of activated carbon, a small amount of ashes after 700°C. These ashes suggest the presence of oxides or inorganic impurities formed after air heating. It is known that the presence of ashes can affect the chemical characteristics and the adsorptive behavior of activated carbon. This way, the low level of ashes found in the studied carbon is a desirable characteristic.

Table 2. Properties of Activated Carbons product at different temperature.

Properties	**Temperature (°C)**		
Proximate analysis (wt %)	**700**	**750**	**800**
Moisture	12.579	8.248	9.415
Volatiles	37.881	9.958	9.316
Fixed Carbon	62.119	78.926	79.509
Ash	-	2,867	1.760
Ultimate Analysis (wt %)			
Carbon	79.089	80.942	83.722
Hydrogen	2.279	2.120	1.623
Nitrogen	0.780	0.701	0.477
Sulfur	0.098	0.071	0.061
Oxygen[1]	17.753	16.166	14.118
C/N	101.376	115.496	175.655

[1]by difference

3.4. Yield of babassu endocarp biochar

The average yield for babassu endocarp biochars, as showed in figure 5, which passed through the pyrolysis process at 600°C with retention time 30 minutes, was 27, 66%. This value is similar to that obtained for the carbonization of babassunutshell25.8%[32] and shows the potential of fixed carbon production through the slow pyrolysis of babassu nut residues. The pyrolysis conditions are an important role in the development of the rudimentary pore structure in the biochar, through the release of volatiles from the carbon matrix. As this pore development in the biochar has a direct impact on the pore characteristics of the activated carbon the optimization of the pyrolysis parameters before activation is important, [33].

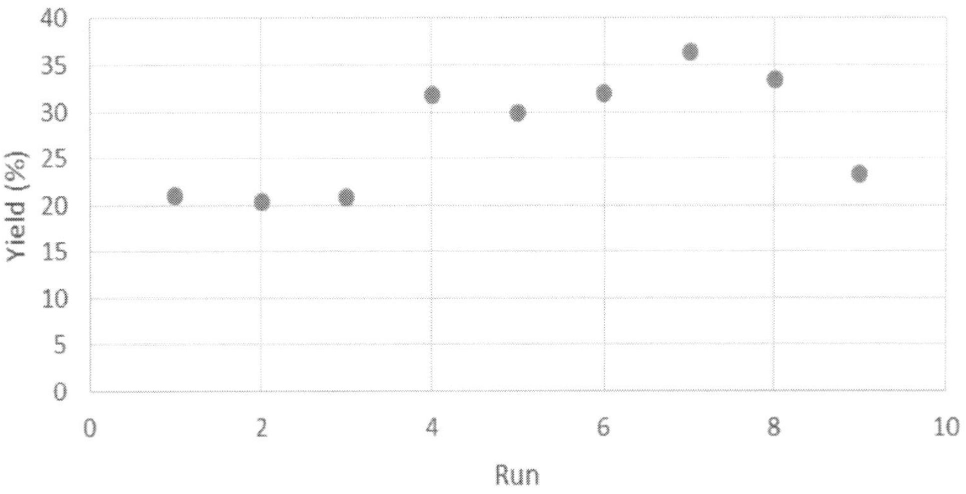

Figure 5. Yield of babassuendocarp biocharwith pyrolysis temperature at 600°C and retention time 30 minutes.

This highlights the carbonization potential of babassu endocarp, considering the quality and yield of charcoal which, combined with the high density of this biomass, provide a biochar that meets the specifications needed for production activated carbon. The statistic of activation carbon relative babassu endocarp biocharare shown in table 3. This statistical component aimed to evaluate the heat treatment reproducibility, the possible variations presented by the Microwave and the system as a whole. It can also demonstrate the reproducibility of activated carbon itself, thus providing information on the babassu biomass characteristics used in the activated carbon production.

Table 3. Yield Statistic Activate Carbon

Expected Percentage	Occurrence Interval	
68.27%	14.88816	17.02684
95.45%	13.81882	18.09618
99.73%	12.74947	19.16553

3.5. BET - surface area

BET surface areas of the AC pyrolyzed at 600°C and activated at 700, 750 and 800°C as shown in table 4, indicates an activated charcoal with high surface area and the pore size of all the activated carbons prepared in this study are mainly microporous.

Table 4. BET results.

Activation Temperature(°C)	BET surface area(m^2/g)	Total pore volume(cm^3/g)	Micropore volume(cm^3/g)	Pore Size(nm)
700	543.5550	0.224879	0.190095	1.65
750	480.7335	0.188455	0.173761	1.57
800	542.1852	0.226877	0.192316	1.67

The microporosity found in the activated carbon from the Babassu endocarp done in microwave agrees with the results obtained from babassu AC produced in conventional oven. The difference in the values obtained are probably due to different conditions of HTT and activation used by researchers and also due to the use of three components of the shell together i.e., epicarp, mesocarp and endocarp,[34-35].This development of microporosities in activated carbons prepared from oil palm stones points to potential applications in gas-phase adsorption for air pollution control,[36] and the granular activated carbons from babassu had capacities to adsorb iodine, methylene blue, and residual chlorine, [37]
The figure 6 show the adsorption and desorption isotherms of activated carbons produced at 700, 750 and 800°C. All the N_2adsorption isotherms at 77K of the activated carbons show similar shapes. In isotherm curves, most nitrogen uptake occurs at the initial steep part until a relative pressure of 0.1. This adsorption type is called a type I isotherm, which reveals that the activated carbons produced were a microporous material, [38-39].

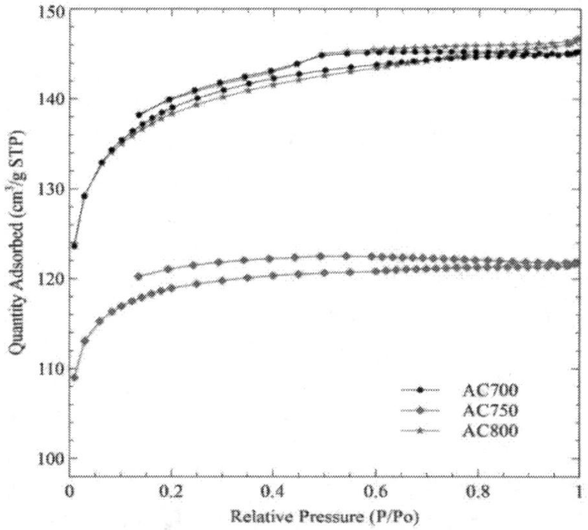

Figure 6. Adsorption–desorption isotherm of N_2 at 77 Kof the activated carbons at 700, 750 and 800°C.

3.6. SEM micrograph
The pore characteristics of the different activated carbonsstudied were analyzed using Scanning Electron Micrographs - SEM. Clear differences in porosityand particle shape were observed, and representative images areshown in figure 7. These images also show characteristics of babassu endocarp and the particlesthat formed during thebabassu endocarp pyrolysis and activation and their differentphysical characteristics.

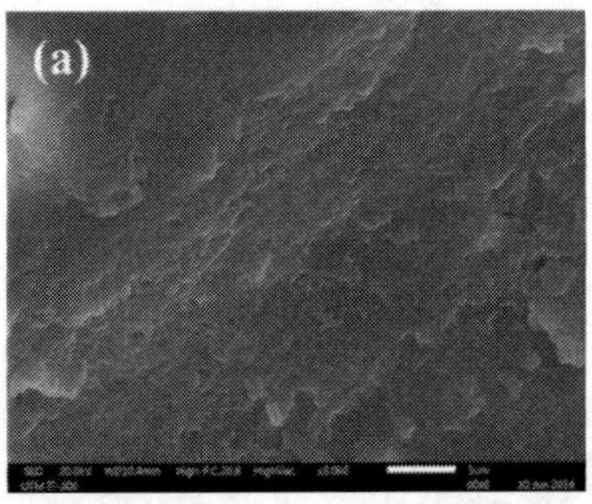

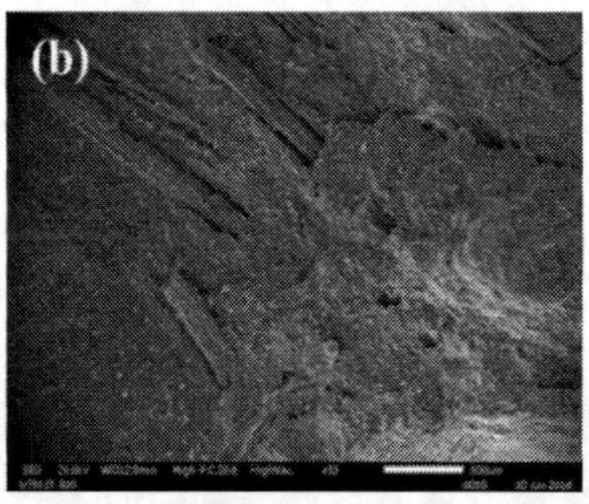

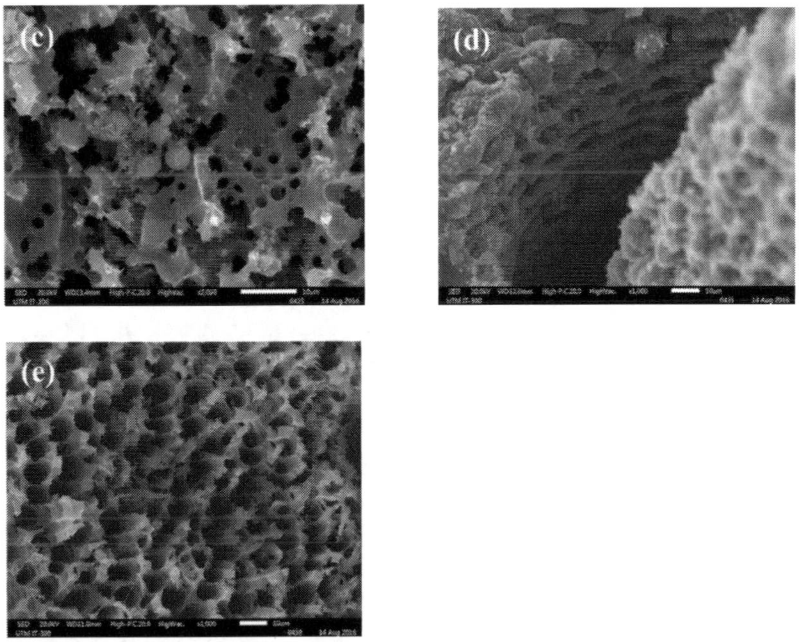

Figure 7. SEM images of(a) raw babassu endocarp, (b) biochar carbonized at 600°C, (c) AC-700, (d) AC-750,and (e) AC-800.

Scanning Electron Micrographs - SEM realized on raw babassu endocarp, babassu endocarp biochar spyrolyzed at 600°C with retention time 30 min and babassu endocarp activated Carbon indicatethat the initial structure of the raw material (figure 7a) shows a dense surface without pores. The pyrolysis creates a lotporosity (figure 7b). There are small "black spots" in lamellar structure of AC (figure 7c). These "black spots" are pores, perhaps micropores because pores are formed and destructed simultaneously during the activating period [40].The comparison of figure 7c, 7d and 7e indicates clearly that a regular porosity and a rather homogeneous surface was obtained by activation at 800°C. These indicate that there were many orderly pores over the surface, forming a system of advanced pore structures.

4. Conclusion

This research shows that endocarp babassu can be used as precursor to produce activated carbon with a rather well-developed porosityby pyrolysis and physical activation by two-steps with CO_2activation via microwaves radiation. The yield statistic of activated carbon is within the desirable. SEM examination of raw material, pyrolyzed endocarp babassu and produced activated carbons shows that the porosity increases gradually after pyrolysis and activation. Nitrogen adsorption shows that the activated carbons obtaine dare essentially microporous with an average BET surface area.The temperatures of 700 and 800°C produce AC with similar BET, and the temperature of 750°C produces AC with intermediate BET

In summary, the activated carbon with a low production cost, this carbon material could be suitable for applications in gaseous pollutant adsorption, adsorbent of iodine, methylene blue and residual chlorine.

5. Acknowledgements

The first authoracknowledges the UEMA, Brazil for supported post-doctoral stage and UTM, Malaysia for welcome. The authors are grateful to Cristiana Correa Rêgo, Cássio Salgado and Thaís Mendes for their scientific, technical support and Jack C. Wynkerfor the English revision.

6. References

[1] EMBRAPA - Empresa Brasileira de Pesquisa Agropecuária. Departamento de Orientação e Apoio à Programação de Pesquisa, Brasilia, D.F.Babaçu: Programa Nacional de Pesquisa. Brasilia. EMBRAPA-DDT, 1984.89p.CDD: 633.851

[2] Brasil 2014 Ministério do Planejamento, Orçamento e Gestão Instituto Brasileiro de Geografia e Estatistica/IBGE-ISSN 0103-8435 Prod. Extr. veg. e Silvic., Rio de Janeiro, **29** 1-56

[3] Mendonça C C, Lima A F, Silva G M, Barbosa C S, and Ferreira E J L 2014 Mapeamento de populações nativas, aspectos fenológicos e potencial de exploração de frutos de babaçu (*orbignyaphalerata*. arecaceae) na amazônia ocidental do Brasil. enciclopédia biosfera, Centro Científico Conhecer - Goiânia, **10** n.18 p. 2138

[4] Emmerich F G and Luengo C A 1996 Babassu Charcoal: A Sulfurless renewable thermo-reducing feedstock for steelmaking *Biomass and Bioenergy* **10**(1) 41-44

[5] Jonas O V, Claudio F L, Luiz C A Barbosa 2014 Analytical pyrolysis of the kernel and oil of babassu palm (Orbignyaphalerata) *J. of Analytical and Applied Pyrolysis* **107** 73–81

[6] Li *et al.* 2016 Preparation of activate carbon from Pyrolyzed Rice Husk by leaching out ash content after CO2 activation *BioResources* **11(2)** 3384-96

[7] Das D, Samal D P, Meikap B C 2015 Preparation of activated carbon from green coconut shell and its characterization *J. Chem. Eng. Process Technol* **6** 248

[8] Wandembruck M T, Mantovani D, Honório J, Bergamasco R, Vieira, Angélica M S, Vieira M F 2015 Capacidade adsortiva do carvão ativado de coco de *babaçu* frente à remoção de diuron em solução sintética p. 1155-64 . In: Anais do XXXVII Congresso Brasileiro de Sistemas Particulados - ENEMP 2015 [=BlucherEngineeringProceedings]. São Paulo: Blucher, 2015. .São Paulo: Blucher, 2015. ISSN 2359-1757, DOI 10.5151/ENEMP2015-PS-650

[9] Ani F N, Junoh M M, Muis Z A 2015 Granular-activated carbon from mukah coal using carbon dioxide activation **75:11** 123–127| www.jurnalteknologi.utm.my | eISSN 2180–3722

[10] Nascimento, Stéfani Caroline do, AnneliseKopp Alves, Produção de Carvão 2013 Ativado a partir de resíduos de processamento do ***BABAÇU*** (*Orbignyaphalerata*). XXV Salao de Iniciacao cientifica – UFRGS

[11] Brum S S, Bianchi M L, Silva V L, Gonçalves M, Guerreiro M C, Oliveira L C A 2008 Preparação e caracterização de carvão ativado produzido a partir de resíduos do beneficiamento do café *Quim. Nova* **31** No. 5 1048-52

[12] Jun'ichi Hayashi, Toshihide Horikawa, Isao Takedaa, KatsuhikoMuroyamaa and Farid Nasir Ani 2002 Preparing activated carbon from various nutshells by chemical activation with K_2CO_3 *Carbon 40* 2381–86

[13] Abioye A M and Ani F N 2015 Recent development in the production of activated carbon electrodes from agricultural waste biomass for supercapacitors: A review Renewable and Sustainable *Energy Reviews* **52** 1282–93

[14] Chowdhury Z Z, Hamid S B A, Das R, Hasan M R, Zain S M, Halid K, Uddin N 2013 Preparation of Carbonaceous Adsorbents from Lignocellulosic Biomass and Their use in removal of contaminants from aqueous solution. "Biomass carbon sorbents" *Bio Resources* **8(4)** 6523-55

[15] Reinoso R F, Marsh H 2006 *Activated Carbon*, 1st ed., (London: Elsevier)

[16] Thomas E. Rufford, Denisa Hulicova-Jurcakova and John Zhu *2014 Green carbon materials: advances and applications* (Singapore: Pan Stanford Publishing)

[17] Bassyouni F A, Abu-Bakr S M, Rehim M A 2012 Evolution of microwave irradiation and its application in green chemistry and biosciences *Res. Chem. Intermed* **38** 283–322

[18] Muthanna J A and Samar K T 2014 Optimization of microwave preparation conditions for activated carbon from *Albizialebbeck* seed pods for methylene blue dye adsorption *J. of Analytical and Applied Pyrolysis* **105** 199–208

[19] Abas F Z and Ani F N 2016 Characteristic of oil palm activated carbon produced from microwave and conventional heating *Applied Mechanics and Materials* **819** 606-611

[20] Salema A A and Ani F N 2011 Heating characteristics of biomass and carbonaceous materials under microwave radiation *IEEE First Conference on Clean Energy and Technology CET*. Doi. 978-1-4577-1354-5/111$26.002011

[21] Abioye A M and Ani F N 2015 The characteristics of oil palm shell biochar and activated carbon produced via microwave heating *Applied Mechanics and Materials* **695** 12-15

[22] Abubakar Z, Salema A A and Ani F N 2013 A new technique to pyrolysis biomass in a microwave system: Effect of stirrer speed *Bioresource Technology* **128** 578–585

[23] Assis M R, Brancheriau L A, Napoli A, Trugilho P F 2016 Factors affecting the Mechanics of carbonized wood: literature review *Wood Sci. Technol.* **50** 519–536

[24] Veronica Czitrom and Patrick D S 1997 Statistical Case Studies for Industrial Process Improvement. ISBN.: 0898713943,9780898713947 -514 pg. SIAM

[25] Maia C M B F M, Guitokuç C C M, Rauenç Daiana Signor 2014 Biochar from BABASSU residues chemical characterization and thermogravimetric analysis *17th Meeting of the International Humic Substances Society* Ioannina, Greece, 1-5 September, 2014

[26] Teixeira M A 2008 BABASSU – a new approach for an ancient Brazilian biomass *Biomass Bioenergy* **32(9)** 857–64

[27] Lua Aik Chong, Ting Yang, Jia Guo 2004 Effects of pyrolysis conditions on the properties of activated carbons prepared from pistachio-nut shells *J. Anal. Appl. Pyrolysis* **72** 279–287

[28] Lua AC, Fong Yow Lau a, JiaGuo 2006 Influence of pyrolysis conditions on pore development of oil-palm-shell activated carbons *J. Anal. Appl. Pyrolysis* **76** 96–102

[29] Hasan S W and Ani F N 2014 Review of Limiting Issues in Industrialization and Scale-up of Microwave-Assisted Activated Carbon Production *Ind. Eng. Chem. Res.* **53** 12185−91

[30] Hayashi J, Kazehaya A, Muroyama K E, Watkinson A P 2000 Preparation of activated carbon from lignin by chemical activation *Carbon* **38** 1873-78

[31] Nunes, Diego Luiz 2013 N972r Remoção de melanoidinas aplicando carvão ativado produzido a partir de torta de nabo forrageiro carbonizado em forno de microondas / Diego Luiz Nunes 127 f.: il

[32] Protasio, Thiago de Paula et al 2014 Mass and energy balance of the carbonization of BABASSU nutshell as affected by temperature *Pesq. agropec. bras., Brasília* **49(3)** 189-196

[33] Aik Chong Lua, Fong Yow Lau, Jia Guo 2006 Influence of pyrolysis conditions on pore development of oil-palm-shell activated carbons *J. Anal. Appl. Pyrolysis* **76** 96–102

[34] Liliane Schier de Lima a, Sueli Pércio Quináia a, Fabio Luiz Melquiades b, Gabriel E.V. de Biasi b, Jarem R. Garcia c,d 2014 Characterization of activated carbons from different sources and the simultaneous adsorption of Cu, Cr, and Zn from metallurgic effluent *Separation and Purification Technol.* **122** 421–430

[35] Freitas M M A and Figueiredo J L 2002 Preparation of activated carbons with controlled pore size - *Proceedings of the 6th International Symposium on the Characterization of Porous Solids (COPS-VI), Allicante, Spain*, ISBN 9780444512611

[36] Lua A C and Guo J 2000 Activated carbon prepared from oil palm stone by one-step CO activation for gaseous pollutant removal *Carbon* **38** 1089 –97

[37] Jaguaribel E F, Medeiros L L, Barreto M C S and Araujo L P 2005 The performance of activated carbons from sugarcane bagasse, babassu, and coconut shells in removing residual chlorine. **22** No. 01 pp. 41 - 47

[38] Su-Hwa Jung, Seung-Jin Oh, Gyung-Goo Choi and Joo-Sik Kim 2014 Production and characterization of microporous activated carbons and metallurgical bio-coke from waste shell biomass *J. of Analytical and Applied Pyrolysis* **109** 123–131

[39] Amphol A, Paitip T and Woranan N 2008 Preparation and characteristics of agricultural waste activated carbon by physical activation having micro- and mesopores *J. Anal. Appl. Pyrolysis* **82** 279–285

[40] Liu Q S, Zheng T, Wang P and Guo L 2010 Preparation and characterization of activated carbon from bamboo by microwave-induced phosphoric acid activation *Industrial Cropsand Products* **31** 233–238

Ecotect design simulation on existing building to enhance its energy efficiency

D Trisnawan[1]
Department of Architecture, Faculty of Engineering, Universitas Indonesia, UI Depok Campus, Depok 16424, Indonesia

E-mail: ditadesign1@gmail.com

Abstract. Ecotect Software has been used to calculate building's energy consumption by simulating its context within the environment. Mostly use by architects and building engineers to enhance their design advancements, the software is embedded to the main Autodesk CAD architecture, fully compatible with Autodesk REVIT. Research Studies on many existing buildings have been performed to evaluate its building's performance. It is related to its environment, especially on dealing with solar heat, its nature for day-lighting, natural airflow for ventilation, and its energy consumption for man-made systems such as Air Conditioning and Lighting.
This research is conducted to reach basic optimum efficiency of an existing building at UI Campus in Depok. After using ecotect software we presumed that some strategic upgrading design propositions should be done to minimize the daily energy consumptions on air conditioning and lighting. Sensitive to the annual solar path and air/wind direction, we could propose an alternative to maximize the use of natural resource to reduce the building's energy consumption.

1. Introduction

Human being builds to fulfil their needs of shelter. We make shelters for living and other activities. Architecture has evolved and changed tremendously nowadays, that we can easily simulate our buildings in advance. Using many software applications [1], we now can even simulate the level of comforts by making calculations towards the buildings' aspects in relations to nature. CAD (Computer Aided Design) technologies has come to a degree where we can measure and simulate our future built environment, by not only predicting its spatial forms, its spatial quality, and even further, its performance. The objectives of the writing are to deepen understandings the use of Ecotect Software on architecture, its green building concepts, and its applications for building performance evaluations, especially on an existing buildings. By simulating an existing building performance, we could then propose a better alternative solutions for renovations to achieve the maximum building's energy efficiency.

[1] Dita Trisnawan; To whom any correspondence should be addressed.

Figure 1. Existing building for ecotect calculations – Engineering Center building of Universitas Indonesia of Depok, Indonesia (source: FTUI)

2. Environment and Nature

Architecture is a compromised environment for man. People built to provide a more comfortable environment to live in. It is a necessary matter that people highly consider the nature as the main resource for comforts. But not until the discovery and intervention of AC (air conditioning) system and technology, architecture changed into a more industrial and technology dependent ones. AC system mimics the nature of air and climate and stabilizes the interior air conditions of buildings. They put a lot of tokens on general energy consumption.

3. Ecotect Simulation

3.1. Existing Condition

The Engineering Center (EC) building of the Faculty of Engineering Universitas Indonesia, were observed to find the performance and energy consumption flaws of the building. This effort was done as a pilot project within the faculty complex to enhance the green aspects of most buildings.

Energy consumption on existing structure like the EC Building, surprisingly has come to a state where it is necessary to be evaluated. Buildings with less consideration to green aspects, consume over too much energy comparing to buildings with consideration. On a quick observation of the building, some spots within floors were identified as uncomfortable as they contrary wished by design. These heat weaknesses then will be confirmed later after the Ecotect spatial thermal analysis, to find out on why the uncomfortableness happen.

3.2. Calculation Aspects

ECOTECT software is an environmental analysis tool that allows architects and designers to simulate building performance right in conceptual phase [2]. On building's energy efficiency efforts, research usually relates to the daylight factors, shadow parameters and thermal analysis to examine its performance level [3]. Those mentioned analysis leaned to a thermal simulation program named Ecotect by CAD-Autodesk for green building simulation. It is used to optimize solutions through modification of the existing condition, new design intervention to propose and making changes towards a better building's energy performance. These calculation aspects are the simulation tools for analysis to indicate possible optimized solution from this investigation.

Investigation was conducted to identify the issue/problems of the building. Façade details and materials were analyzed to see how they relate to the calculation aspects. Daily factors were simulated at first to map clearly the affected spaces within the building.

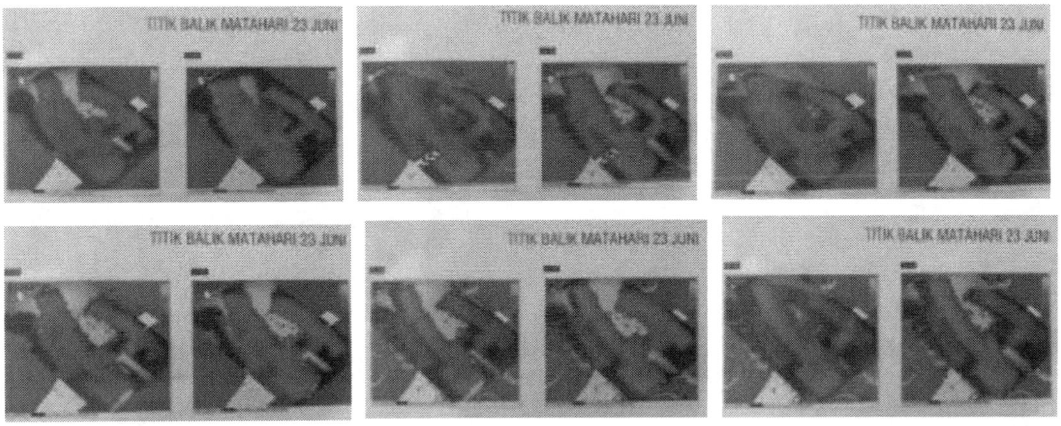

Figure 2. Samples of thermal analysis of the building on its ground floor – sun path turning points of June 23rd– (Source: Analysis results)

4. Research Method

Observative and quantitative method is used to measure the building performance based on several indicators. Ecotect Software application was done by first, recreating basic CAD drawings of the eexisting building. This drawings then generated into 3D Modelling with all building materials indicated. Solid and void parts are also clearly marked to simulate the sun-path effects to building.

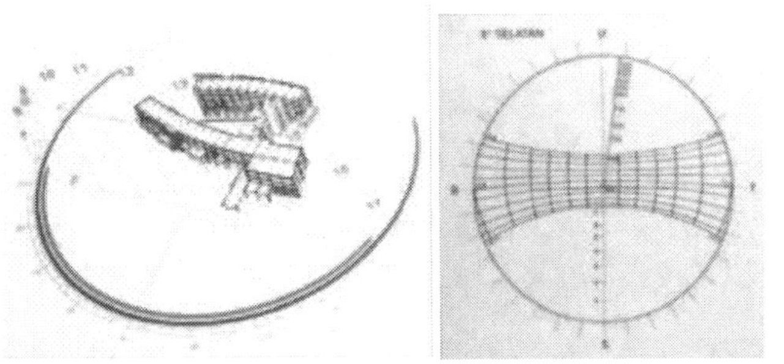

Figure 3. Sun path for thermal analysis of the building – normal turning points

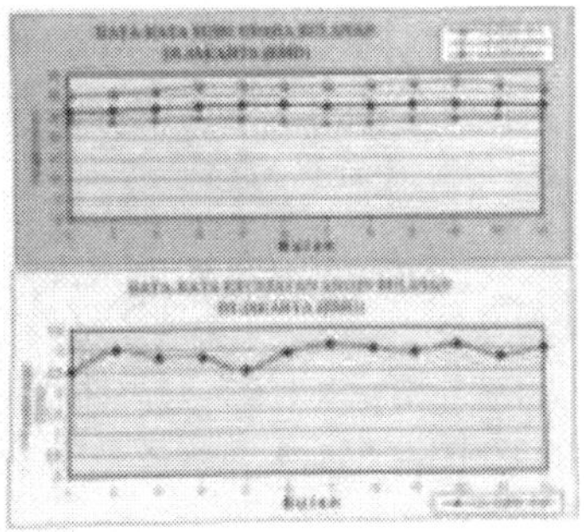

Figure 4. Average air temperature and wind speed of Jakarta/Depok area.

5. Research Results

Ecotect can simulate accurately the annual solar trace and positions (figure 2 and 3). It also provides accurate statistics of the building area's solar radiation intensity. According to the annual daily average solar radiation distribution, we can figure out solar radiation or day lighting [4] distribution of the building area and the shadings it generated. Shading analysis of floor samples based on sun-path applied on 3D model can be seen on figure 5 for sun's turning-points of June 23rd. On figure 6, the shadings generated for sun's turning-points of December 23rd.

Comparing the observation and the sun path on the area, we can assume that the building orientation was decided correctly by the architect to avoid maximum heated areas created. Sunshades are important aspect towards energy efficiency [5]. But yet, the curvaceous plan forms had created heated spots at some sun's position. These have been indicated after building's thermal analysis on figure 2. This building plan model is similar to courtyards building type [6], where the building footprint and plans can become a significant aspects for wind and solar movements that affecting the entity. The result analysis showed the facts that there are quite many heated spots which need to be fixed. Additional shadings or solid material board/wall should be applied to minimize the heated spots, especially along the inner corridors which face the afternoon sun directly. Even so these areas are affected by the robust sunrays, alternative model for shadings are always welcome to allow indirect sunlight to come in.

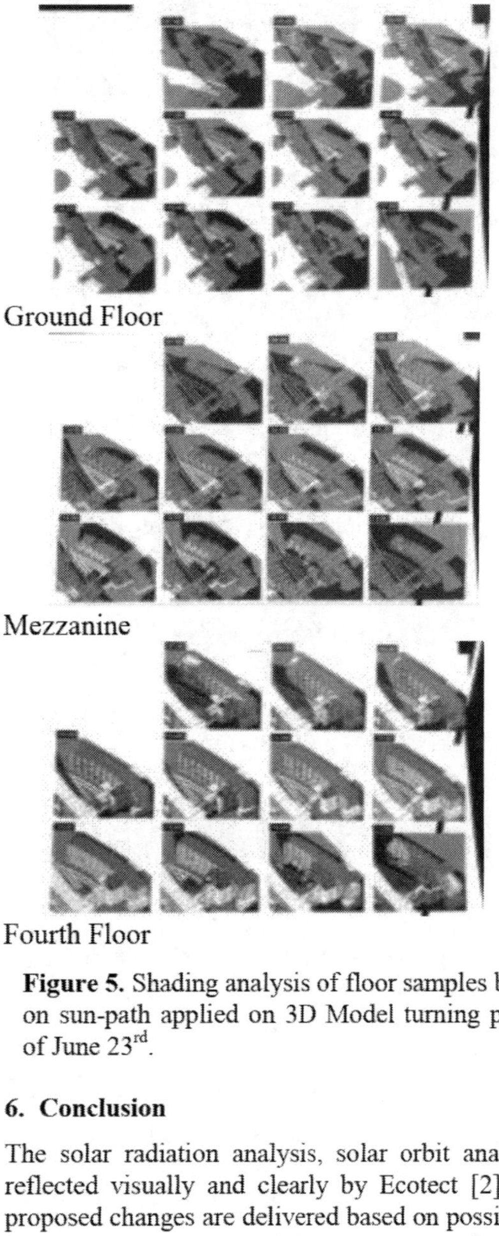

Figure 5. Shading analysis of floor samples based on sun-path applied on 3D Model turning points of June 23rd.

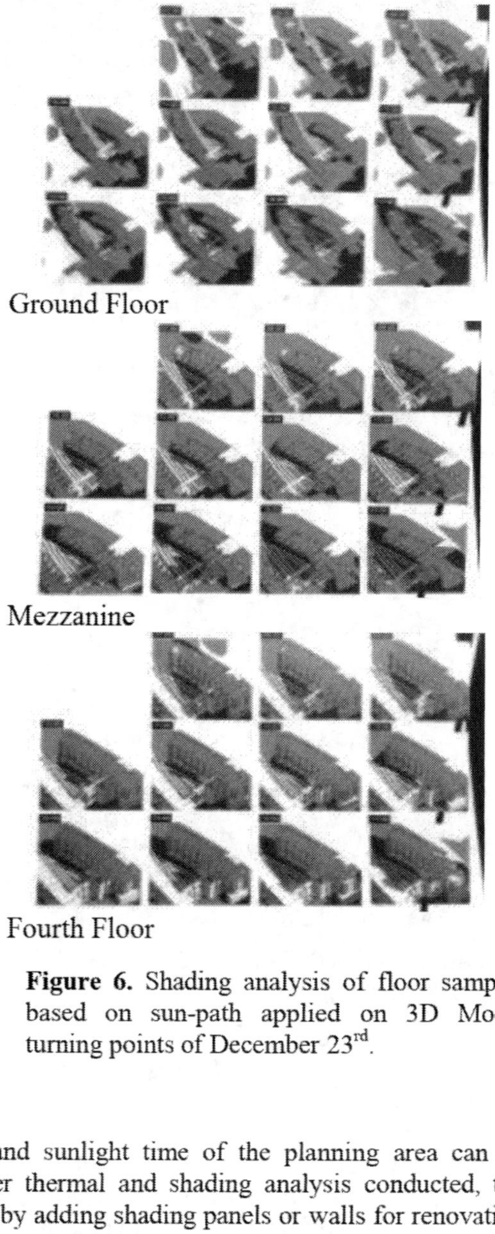

Figure 6. Shading analysis of floor samples based on sun-path applied on 3D Model turning points of December 23rd.

6. Conclusion

The solar radiation analysis, solar orbit analysis, and sunlight time of the planning area can be reflected visually and clearly by Ecotect [2]. After thermal and shading analysis conducted, the proposed changes are delivered based on possible fix by adding shading panels or walls for renovation in the future. Figure 7 and figure 8 are samples of proposed changes for consideration.

Building design performance evaluation provide overviews of conclusions with the design guideline from the investigated outcomes [7]. These results would maximize the energy efficiency for future use by considering the significant green building aspects for calculation. The building criteria for designing and building new environments should be based on the evaluation of existing ones, and modified when appropriate in the context of the design process. Rapid development in Indonesia and problems with the energy shortage, make ecological buildings and sustainable development have become the new focus of the society. It is very urgent to explore more ecological planning methods and strategies towards building performances. The Green Building Council of Indonesia has adopted strategies to evaluate aspects for improvement to the living comfort of buildings, which at the same time also reduce its energy consumption.

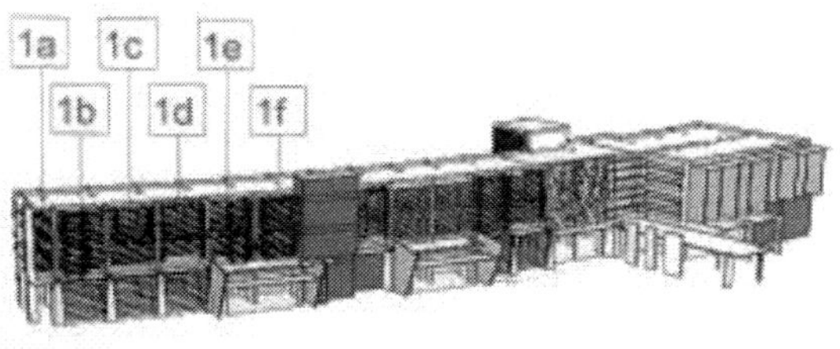

Figure 7. One proposed façade of the building. Northern Side.

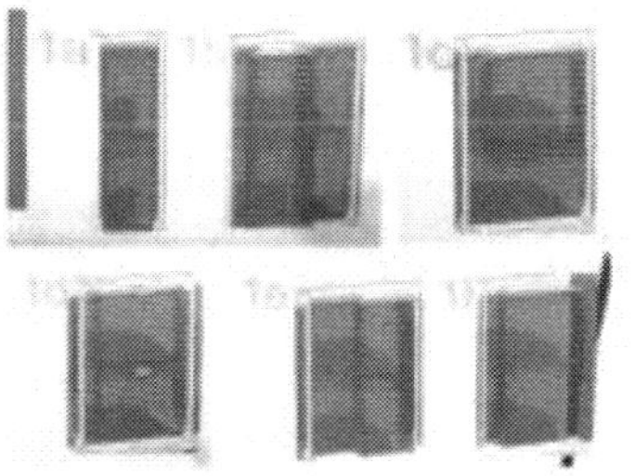

Figure 8. Sample details of the proposed Façade parts.

Architects and designer should give full consideration to various ecological energy-saving methods in the concept design by analyzing and then offer alternative designs. Software like Ecotect helps them to be sensitive to natural condition of a place where the building would be built. By considering nature as major potential source of energy, award winning designs would become more ecological, consume lower energy, and ultimately create better living environment. In the future, it is very significant to ensure that the building planning phase to be scientific to create a more comfortable, ecological, and energy-saving working/studying environment. As different place have different climates, temperature, and rainwater density, the global practice of green/ecological building and application, should also consider the local wisdom context. This is required to deepen understanding and define the local ecological knowledge for further research on this field.

7. References

[1] Beatriz Arranz, E. R.-U.-F.-S. 2014. Evaluation of Three Solar and Daylighting Control System Based on Calumen II, Ecotect, and Radiance Simulation Programmes to Obtain an Energy Efficient and Healthy Interior in the Experimental Building Prototype SD10. *Energy and Buildings* **83**, 225-236.

[2] Li Yang, B.-J. H. (2014). Application Research of ECOTECT in Residential Estate Planning. *Energy and Buildings* **72**, 195-202.

[3] Reza Fallahtafti, M. M. 2015. Optimisation of Building Shape and Orientation for Better Energy Efficient Architecture. *International Journal of Energy Sector Management* **9**(4), 593-618.

[4] Yuan Fang, S. C. 2017. *Optimization of Daylighting and Energy Performance Using Parametric Design, Simulation Modeling, and Genetic Algorithms.* Raleigh, North Carolina: North Carolina State University.

[5] Li Qing, G. R. 2017. University Library External Window Sunshade Device Simulation Analysis and Optimization Design. *Procedia Engineering*, 1196-1201.

[6] Namsu Jang, S. H. 2017. A Study on Courtyard Apartment Types in South Korea from the 1960s to 1970s. *Frontiers of Architectural Research 6*, 149-156.

[7] Wolfgang F.E. Preiser, J. C. 2014. *Assessing Building Performance.* Elsevier - Butterworth-Heinemann.

IOP Conf. Series: Earth and Environmental Science **105** (2017) 012118 doi:10.1088/1755-1315/105/1/012118

Estimating greenhouse gas emission level of a natural gas transmission pipeline from point A to B in West Java based on INGAA and IPCC guidelines

Cindy Dianita[1], Asep Handaya Saputra[1], M B P Almagistra[1]
[1]Department of Chemical Engineering, Universitas Indonesia, Kampus Baru UI, Depok 16424, Indonesia

E-mail: cindydianita@ui.ac.id

Abstract. When being transported by pipeline, natural gas is often emitted to the atmosphere, either for depressurization (venting emissions) or leak through the pipeline (fugitive emission). The emission level must be well estimated to provide relevant informations and recommendation to formulate strategies for reducing greenhouse gas (GHG) emission. Organizations such as INGAA (The Interstate Natural Gas Association of America) and IPCC (Intergovernmental Panel on Climate Change) provide GHG estimation guidelines which are adopted by many companies and countries. This study estimates the emission level of a natural gas pipeline in West Java using emission factors referring to INGAA and IPCC guidelines with flow rate variation. The result shows that the flow rate variation affects the total emission based on Tier 2 and Tier 3 INGAA as well as Tier 1 IPCC. It is also shown that fugitive emissions dominate the total emission of gas pipeline. However, the use of different methodologies and guidelines gives different emission level for the same pipeline. Different estimation results of emission level have been reviewed and national emission factors database for gas transmission is highly suggested to be developed.

1. Introduction

The "greenhouse effect" is the phenomenon where atmospheric gases absorb and trap the terrestrial radiation leaving the Earth's surface, causing a warming effect on earth [1]. Gases that trap heat in the atmosphere are called greenhouse gases (GHG). For emissions from oil and natural gas systems, CO, methane, and nitrous oxide are the main emissions released from combustion sources and natural gas processes [1]. The use of fossil fuel contributes in environmental problems due to their emission levels. However, natural gas can be considered as clear energy because it has lower emission level than other fossil fuel such as coal and petroleum. The low level of GHG emission of natural gas has become one of considerations of Indonesian Government in setting the target to optimize the consumption of natural gas in domestic market.

In Indonesia, the transportation of natural gas in Indonesia is still dominated by pipeline, both transmission and distribution pipeline networks. Although, natural gas is environmentally friendly fuel and cheaper than other fossil fuels but it still has risk to release GHG emission during the transporation process through pipeline. Methane, the main component of natural gas, is a more potent GHG gas than CO_2 as methane produces 21 times of trapped heat in the atmosphere when compared to CO_2 [2].

Several methane emissions studies have been reported as direct measurements from different natural gas supply chain facilities (exploration, production, gathering, processing,transmission, and distribution) with a variety of CH4 emission sources including fugitive, venting, and combustion emissions for objects in United States [3,4,5]. These studies basically estimate and calculate new emission factors based on direct measurements and mostly refer to The Environmental Protection Agency (EPA) Greenhouse Gas Inventory (GHGI) in comparing their calculation results. The other study in Russia was also done for Russia long distance gas transmission pipeline [6]. It was initiated by assessing 5 gas compression stations through direct sampling. The assessment result was then developed to extrapolate emission factors of Russian long distance gas transmission pipeline. However, these studies [3,4,5,6] had not assessed the variation of natural gas throughput and its contribution to emissions level, both fugitive and venting emissions.

This study highlights the estimation of emission level, both fugitive and venting, of an existing gas transmission pipeline from Point A to Point B in West Java (Indonesia). Fugitive emissions are defined as unintentional leaks from any activities of energy production and distribution such as leaks from pipe connection, valves, fittings, and compressors. While venting emissions are typically a deliberate action associated with plant activities, or are produced when emergency situations require or produce a rapid reduction in process pressures. Sources of vented emissions include venting from glycol dehydrator vents, pneumatic devices, an planned or non-routine maintenance venting or blowdown [1].

A methane emissions study of this gas pipeline was done previously by calculating fugitive emissions based on Tier 1 and Tier 2 of Interstate Natural Gas Association of America (INGAA) due to the limitation of data resources [7]. As the development of the previous study, the emission level in this study is estimated using emission factors from Interstate Natural Gas Association of America (INGAA) and Intergovernmental Panel on Climate Change (IPCC). Furthermore, emission level is also simulated using emission factors for Russian gas transmission pipeline system [6]. Flow rate variation is simulated as well in order to find correlation between the flow rate and emission level. Gas flow rate is varied according to actual condition of gas flow on site and initial design of gas flow.

2. Data and calculation

This study assesed the system of gas pipeline transmission from A to B as illustrated by Figure 1 with detail information on Table 1. In this study, the emission level is estimated by multiplying emission factors to activities data. The result of calculation is total emission consists of fugitive and venting emissions both for operation and design flowrate condition.

There are 3 tiers in estimating the emission of gas pipeline system by referring to INGAA guidelines. While emission level calculation based on IPCC guidelines is performed only for Tier 1 which is inline with National GHG Inventory Guidelines of Indonesia [8]. In this study, emission factors for Tier 1 IPCC are choosed for developing countries with economies in transition. Basically, IPCC has 3 levels of emission asessment , from Tier 1 to Tier 3. However, Tier 2 and Tier 3 approaches can be implemented in a country that has detailed production statistics and infrastructure data, based on a compilation of oil and gas companies reports, which then can be developed to calculate the national emission factor [9].

There are 2 types of Russian gas pipeline emission factors used in this study i.e emission factors assesed from five Russian gas compressor stations and from extrapolation for Russian long distance gas transmission pipeline. The result of both fugitive and venting emission is converted from CH_4 to CO_2 equivalent by multiplying CH_4 emission to 21.

3. Result and discussion

3.1. Emission level estimation as INGAA methodology

Generally, fugitive emissions are the biggest methane emission source account for transmission sector [1]. This phenomenon is also found in the calculation result of this study. As presented by table

2, total emissions significantly comes from fugitive emission. From calculation results of Tier 2 and Tier 3, ratio of fugitive emissions to total emissions are 62% and 59% respectively to total emission of gas pipeline with operation flow rate condition. The similar trend is also found when calculating the emision level with design flow rate condition, 54% and 57% respectively for Tier 2 and Tier 3. For Tier 1, fugitive emission contributes 67% to total emissions for both conditions of gas flow rate.

Fugitive emissions show higher value than venting emissions because of the higher emission factors. Fugitive emissions highly tend to occur because these emissions are difficult to be detected. Different from fugitive emissions, venting emissions have lower risk to occur as these emissions are emitted from emergency situation or over pressure condition in process which could be anticipated with several planned preventive actions.

Tier 1 approach only needs length of pipe as activity data without considering other process parameters. The result of Tier 1 estimation is used as basic or initial estimation and could be used as reference and recommendation for environmental assessment in order to obtain government permission.

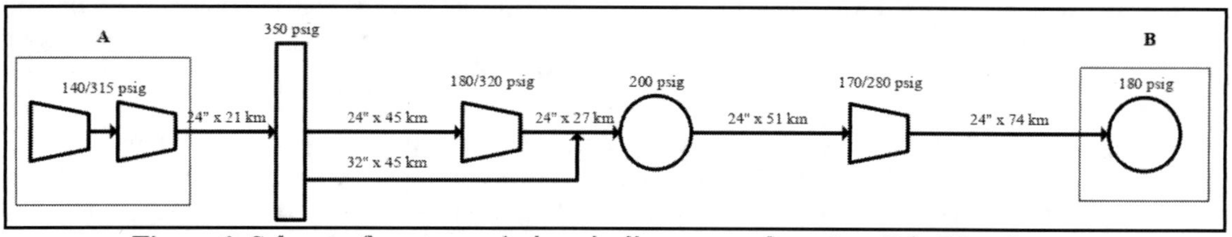

Figure 1. Scheme of gas transmission pipeline system from A to B in West Java.

Table 1. Activities data of gas transmission pipeline system from A to B in West Java.

Pipe length	218 km
Pipe material	Protected Steel
Pipe branch	18
Number of gas compression station	3
Number of gas compression station shop	3
Number of metering and regulator station (MRS)	1
Actual gas throughput	196,161.95 MMSCFY
Design gas throughput	551,150 MMSCFY
Type of MRS	Transmission Interconnected
Type of compressor	Centrifugal
Number of gas-operated pneumatics actuators	426
Number of continuous bleed pneumatics	399
Number of turbine valve operator	25
Number of pneumatic valve operator	2
Number of isolation valve operator	10
Number of MRS isolation valve	1
Number of gas compression control loop	3
Number of MRS control	1

For Tier 2, the calculation needs more data activity than Tier 1. Thus, the estimation result is expected to be more accurate. To perform Tier 2 estimation for fugitive emissions, several activity data are needed such as pipe length, the number of gas compressor station, as well as the number of metering and regulator station. Different activity data are used in order to calculate venting emission level. These emissions are estimated by venting emissions from glycol dehydrator which are calculated from annual gas flow rate or throughput and by venting emissions from pneumatic controller.

For Tier 3, the set of data activity is almost similar to Tier 2. However, Tier 3 is more detail as it includes the material of pipe and type of compressor for fugitive emissions. Moreover, Tier 3 also puts activity data of pipeline length, number of compressor station, and number of metering and regulator station both for fugitive and venting emissions estimation. For venting emissions, the type of pneumatic controller and actuator is considered as well as emissions from pneumatic driven isolation valves and control loops. In addition, pipe material is also considered in estimating emission level as it affects the pipe corrosion rate. Higher corrosion rate means higher risk of cracked pipe which gives possibility of natural gas to be emitted to atmosphere. Plastic is considered as a very good material for pipe and presented with lowest emission factor than cast iron, protected steel, and unprotected steel [1,3]. However, this material is not common for gas transmission pipeline due to its material properties [1].

Table 2 Estimation results of emission level of point A to B gas pipeline as INGAA Guidelines.

Tier	Fugitive Emission (Tonnes CO_2 equivalent/year)		Venting Emission (Tonnes CO_2 equivalent/year)		Total Emission (Tonnes CO_2 equivalent/year)	
	Actual gas flow rate	Design gas flow rate	Actual gas flow rate	Design gas flow rate	Actual gas flow rate	Design gas flow rate
Tier 1	6,371.10	6,371.10	3,162.46	3,162.46	9,533.56	9,533.56
Tier 2	36,131.43	36,131.43	21,863.18	30,184.07	57,994.61	66,315.50
Tier 3	84,869.24	84,869.24	56,715.86	65,036.74	141,585.10	149,905.98

The gas flow rate has effect on the venting emission both for Tier 2 and Tier 3 as these Tiers consider the vented emission from glycol dehydrator unit and quantified by the flowrate of gas processed in glycol dehydrator unit. Basically, there is no glycol dehydrator from Point A to Point B. However, before entering the transmission line, the natural gas is passed to glycol dehydrator in processing unit and this process is considered in Tier 2 and Tier 3. Glycol dehydrator is a unit to eliminate water content in natural gas with glycol as absorbent. In the gas dehydration process, glycol will absorb not only water but also green house gas such as CH_4 and CO_2 which will be released to atmosphere at the end of dehydration cycle process.

The factor of compression has significance contribution to the emission level according to Tier 2 and Tier 3 approaches in this study. The emission generated from compression station is more than 60% of total emission for Tier 2. While, the emission quantified by number of centrifugal compression contributes to more than 50% of total emission for Tier 3. Compressor is used to provide adequate pressure of the gas which is flowing in the pipeline to reach end users . Compressor is worked as the result of the driver, typically a gas or diesel engine or gas turbine. The driver of compressor is a potential leak source that emit GHG emissions, particularly CH_4 and CO_2.

Particularly in Tier 3, type of compressor is considered in estimating the emission level. There are 2 types of compressor in this methodology, reciprocating and centrifugal compressors with different emission factors. To reduce compressed natural gas being emitted to atmosphere, reciprocating type has flexible rings at the piston as a seal while centrifugal compressor has design with wet seal or dry seal . However, the design of wet seal centrigugal compressor leads more emission vented to atmosphere than dry seal centrifugal compressor and reciprocating compressor. Due to this reason, emission factors of centrifugal compressor are higher than reciprocating type.

The use of dry seal compressor actually has been applied by several oil and gas companies due to as it is more environmentally friendly but also more cost saving. Uptigrove et al. reported that dry seal compressor is more economic than wet seal type with $ 85,293 saving per year [10].

3.2. Emission level estimation as IPCC methodology

IPCC only provides emission factors for Tier 1 approach. Emission factors for Tier 2 and Tier 3 approaches are not available in IPCC guidelines because of the difference in technology, standard, and regulation in many countries. In this study, due to lack of resources to determine national specific emission factors and additional data needed for Tier 2 and Tier 3 approaches, thus only Tier 1 is applied in estimating the emission level. Tier 1 IPCC only considers the flow rate parameter to be multiplied to given emission factors. With the same object of pipeline, total emission for actual and design flow rate conditions are 15.035 and 42.243 tonnes CO_2 equivalent/year respectively . Both of these flow rate conditions contribute 79% of fugitive emissions to total emissions level.

When referring to Russian gas pipeline system, the highest emission factors come from compressor activities for both approaches [6]. In this estimation, the emission level is not affected by flow rate variation. Detail calculation results are presented on Table 3.

4. Review of different estimation results

Fundamentally, most of methodologies discussed above concern on almost similar activity data for emission factors such as properties of pipe, valve systems, gas compressor station, and component of gas compressor station. However, the use of different methodologies gives different results of emission level as presented on table 3.

Table 3. Estimation results of emission level of point A to B gas pipeline with different methodologies

Guidelines	Approaches	Total GHG Emissions (Actual gas flow rate) [Tonnes CO_2 equivalent/year]	Total GHG Emissions (Design gas flow rate) [Tonnes CO_2 equivalent/year]
INGAA	Tier 1	9,533.56	9,533.56
	Tier 2	57,994.61	66,315.50
	Tier 3	141,585.09	149,905.98
IPCC	Tier 1	15,035.01	42,243.40
Russian Pipeline System	Assessment of 5 gas compressor stations	488,627.80	488,627.80
	Extrapolation for Russian long distance pipeline system	3,044,255.19	3,044,255.19

Total emissions based on extrapolation for Russian long distance pipeline system gives highest result than other methods. This extrapolation basically is developed from direct assessment activities to collect the data. However, the geographic condition, infrastructure, technology, and equipment as objects of assessment are completely different from condition of gas transmission pipeline in this study. For example, the compressor power of Russian pipeline system is varied from 6 to 25 MW, while the compressor power in the pipeline system of this study is 2 to 3 MW.

Similar condition is found in INGAA guidelines as well. Although INGAA has become guidelines for many gas transportation companies in many countries but emission factors and activity data in the guideline were developed from direct sampling of gas transmission pipeline system located mostly in North America. The technology and operation condition are not exactly similar as pipeline system in Indonesia.

For IPCC guidelines, emission factors for Tier 1 are developed by considering geographic, economic, and technology conditions of group of countries. Thus, activity data and emission factors in IPCC are different for developed and developing countries. However, it needs to use higher level of "Tier" in order to get more accurate estimation. In IPCC guidelines, emission factors for Tier 2 and

Tier 3 are not provided due to the fact these data reflect actual condition of every country thus cannot be generalized.

5. Conclusion and recommendation

Some agencies provide guidelines to estimate the emission level of natural gas pipeline system with different emission factors and activity data. As the result, the emission estimation has various levels when using various guidelines and needs to be validated. For this validation purpose, the best method is direct sampling to the object. However, as it takes more time and cost, a rough estimation for validation purpose could be initiated by exploring supporting data related to gas losses in the gas flow system during the process of transportation. These losses are then considered as methane which is emitted to atmosphere.

In estimating emission level, there are some methodologies with different reliability level. Direct measurement is still the most accurate method but it has limitation as it needs more time, resources, and safety considerations. Practically, the use of existing emission factors is easier to be applied when related data and resources are not available. On the other hand, default emission factors mostly are not reliable because it generalizes conditions of assessed objects which are quite different from operation scenarios as the basis of developed emission factors. Thus, it is recommended that Indonesia has national database of emission factors based on source testing to accommodate actual object conditions. However, once these national emission factors have been established, continuous evaluation will be needed to keep the reliability of these emission factors particularly when conditions change on site.

6. References

[1] Interstate Natural Gas Association of America 2016 *Greenhouse Gas Emission Estimation Guidelines for Natural Gas Transmission and Storage* (Washington DC: Interstate Natural Gas Association of America) 6-80

[2] United States Environmental Protection Agency 2016 *Climate Change Indicators in the United States: Global Greenhouse Gas Emissions* p 1

[3] Lamb B, Edburg S, Ferrara T, Howard T, Harrison M, Kolb C et al. 2015 Direct measurements show decreasing methane emissions from natural gas local distribution systems in the United States *Environ. Sci. Technol.* **49** 5161

[4] Marchese A, Vaughn T, Zimmerle D, Martinez D, Williams L, Robinson A et al. 2015 Methane emissions from United States natural gas gathering and processing *Environ. Sci. Technol.* **49** 10718

[5] Brandt A, Heath G, Cooley D 2016 Methane leaks from natural gas systems follow extreme distributions *Environ. Sci. Technol.* **50** 12512

[6] Lechtenböhmer S, Dienst C, Fischedick M, Hanke T, Fernandez R., Robinson D , Kantamaneni R and Gillis B 2007 Tapping the leakages: Methane losses, mitigation options and policy issues for Russian long distance gas transmission pipelines *International Journal of Greenhouse Gas Control* **1** 387

[7] Dianita C and Saputra A.H 2016 Estimating Greenhouse Gas Emissions Level of A Natural Gas Pipeline – Case Study from A to B Point in West Java-Indonesia *MATEC Web of Conferences* **58** 1

[8] Kementerian Lingkungan Hidup Indonesia 2012 *Pedoman Penyelenggaraan Inventarisasi Gas Rumah Kaca Nasional* (Jakarta: Kementerian Lingkungan Hidup) pp 9–37

[9] Picard D, Ahmed A, Gjerald E, Nordrum S and Yesserkepova I 2006 *Fugitive Emissions* IPCC Guidelines for National Greenhouse Gas Inventories pp 4.32–4.70

[10] Uptigrove S, Harris T and Holzner D 1987 Economic justification of magnetic bearings and mechanical dry seals for centrifugal compressors *ASME. Turbo Expo: Power for Land, Sea, and Air* **3** 5–7

Modeling of turbine follow control with internal model control tuning method based on hysys

T B Satriyo[1], **H E G Prasetya**[1], **P D Permatasari**[1]
[1]Power Plant Engineering, Electronic Engineering Polytechnic Institute of Surabaya, Surabaya, Indonesia, 60111

Email : bimantara.box@gmail.com

Abstract. The power plant is one of the several industries whose product are used by consumers directly. The power plant requires an appropriate control strategy based on real-plant to maintain electricity production,. Turbine Follow Control (TFC) is one of the several control strategies for keeping electricity production. TFC is a control strategy that used to control throttle pressure based on Advanced Regulatory Control (ARC) concept. When there is a change of load on the generator, the error signal will be received by the indicator control. The error signal is used to respond fuel arrangement to maintain the main steam production. Furthermore, the changes of main steam production indicated by the pressure indicator control. It will be responded by the turbine governor valve to carry out a follow-up response in maintaining the main steam towards the turbine. The actual response is given in the form of the percentage of valve. This paper has been modeled TFC strategy using Internal Model Control (IMC) tuning method and First Order Plus Dead Time (FOPDT) approach for getting good control response based on HYSYS. In addition, IMC tuning response in this paper also depends on the ± 5% from setpoint test. From modeling of TFC based on HYSYS, the simulation shows that TFC is precise control strategy to handle the changes of load on the generator with integral absolute error value are 102,087.7 for +5% from setpoint test and 122,947.1 for -5% from setpoint test.

1. Introduction

Power plant products are used by consumers directly in every life sectors. In the process of maintaining electrical products, there are various efforts that can be done, such as performance evaluation of each unit operation, maintenance management, control strategy, and using high efficiency of unit operations. From these efforts for maintaining power production, the control strategy is the most precise effort considering simplification method for applying on the plant and cost of the operational process. At the power plant, there are several control strategies that are often used, namely boiler follow control (BFC), turbine follow control (TFC) and coordinated control (CC) [1]. These control strategy is used to maintain system reliability, mass, and energy balance, and set point value of several unit operations. Jacques Smuts, describes the concept of Advanced Regulatory Control (ARC) strategy as a way to control the operating units of boiler and turbine. ARC strategy classification is grouped by application. Turbine follows control (TFC) is part of controlling's application pressure valve (throttle pressure) [1]. Black and Veatch, explains the principle of turbine follow control strategy and provides the advantage and disadvantage of using TFC in power plant. TFC is used when load changes occur in the generator, the resulting error signal will be received by the controller indicator to respond to the fuel arrangements in order to maintain the main steam production. Furthermore, the

IOP Conf. Series: Earth and Environmental Science **105** (2017) 012119 doi:10.1088/1755-1315/105/1/012119

main steam production changes indicated by the pressure parameters will be responded by the turbine governor valve for further response in maintaining the main steam toward turbine. The actual response is given in the form of the percentage of a valve. Based on the control process, the structure of TFC scheme has a feedback control [2].

PI/PID control is assessed to produce a rapid system response [3]. Simple and easily understood algorithms are also the reasons that PI/PID control is often used [4]. Basically, PI/PID control needs to pay attention to the tuning process, i.e. the process of determining the characters K_c, τ_i, and τ_d in order to obtain the desired system response [5]. However, PI/PID control only is not sufficient to be applied to a power plant. Actually, PI/PID control is only used for linear conditions. Whereas given the complex system of the power plant, the resulting response tends to be nonlinear [6]. Therefore, in applying the design model of power plant control system, a true plant-based control strategy [7], which can be done by PID-based Internal Model Control (IMC) tuning method. To obtain PID tuning parameters, a mathematical approach is required based on real plant. Zhang Yao and Huang Chunqing simulate an IMC-PID control process through First Order Plus Dead Time (FOPDT) approach by considering filter (λ) [8]. The use of filters (λ) is considered capable of minimizing the integral absolute error (IAE) value for step response. In the simulation performed, filter (λ) has criteria referring to previous research. Rivera et al. [9] gives a value of $\lambda > 0.8\Theta$; $\lambda > 0.1\tau$. Chien and Freuhauf [10] gives $\Theta < \lambda < \tau$ values. While Sigurd Skogestad [11] gives a value of $\lambda = 1$.

This study will find a precise tuning method for turbine follow control (TFC) with real condition with the help of simulation program in the form of hsysys.

2. Modeling and Operating Conditions

The structure of TFC strategy is based on the feedback control mechanism with the output parameters in the form of a percentage of control valve openings. The control input on the TFC strategy consists of changes in load and changes in main-steam pressure. TFC structure is shown in figure 1(a).

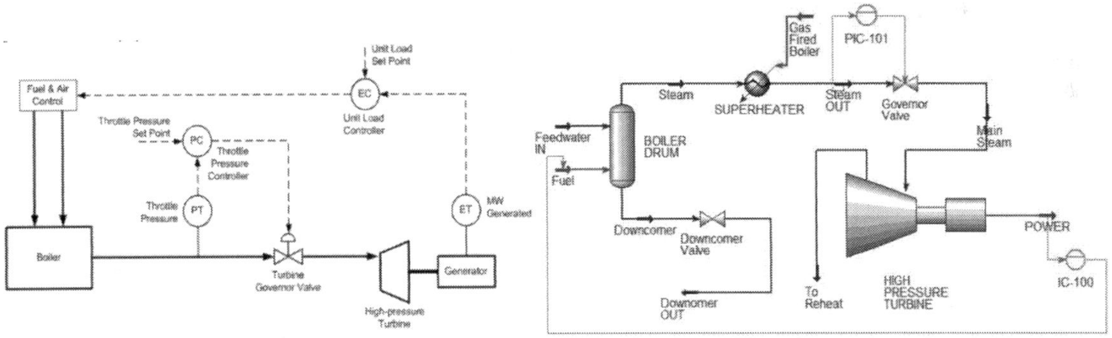

(a) (b)

Figure 1. (a) Turbine follow control structure (b) Process modeling with hysys in dynamic mode

In this paper, the modeling process is done by using HYSYS software. To obtain a modeling process that is appropriate to the actual plant conditions, P&ID and operational data scheme are required. The use of the P&ID scheme aims to represent the concept of the TFC strategy. P&ID modeling in HYSYS software is shown in figure 1(b). Whereas, the operational data required fulfilling the process simulation parameters in HYSYS, at least include mass flow, pressure or temperature that can be shown in table 1.

Table 1. Operating condition of power plant

Variable	Parameters	Unit	Value
Feedwater Outlet	Temperature	°C	352.82
	Pressure	bar	182.93
	Mass flow	kg/h	927,389.35
Main Steam	Temperature	°C	535.46
	Pressure	bar	162.62
	Mass flow	kg/h	976,027.92
Reheat Steam	Temperature	°C	346.07
	Pressure	bar	35.50
HP Turbine	Efficiency	%	82.57

3. Methodology

3.1. Open-Loop

The open-loop control system is a system whose output has no effect on the control action. In reality, on an open-loop control system is applied by manually opening the valve engine process. In the simulation process, the open-loop control system is used as a test condition. The open-loop test is used to find the transfer function of a process by First Order Plus Dead Time approach.

FOPDT is a mathematical model for obtaining a transfer function of a process which is then used to determine the PID tuning parameters of K_c, τ_i and τ_d [12] based on the basis of the tuning method used. The FOPDT mathematical model is an approach model that is widely used by industry as a method of controlling tuning approach.

To obtain the transfer function, it is approximated by the PRC equation of Cecil L. Smith [13] shown in the following equation (1), (2), and (3).

$$K = \frac{\Delta}{\delta} \tag{1}$$

$$\tau = 1{,}5\ (t_{63\%} - t_{28\%}) \tag{2}$$

$$\theta = t_{63\%} - \tau \tag{3}$$

where

K = Gain of steady-state
Δ = The magnitude of changes in the external variable
δ = The amount of change in input variables that affect the output variable
$t_{63\%}$ = Output response time reaches 63%
$t_{28\%}$ = Output response time reaches 28%
τ = Time constant
Θ = Dead time

3.2. Internal Model Control

The Internal model control PID (IMC-PID) is one of the controlling tuning methods that can only be achieved if the system contains parameters of the process to be controlled, either implicitly or explicitly [14]. IMC-PID can also be said as a method designed to control feedback from the output of a process.

The structure of the IMC-PID control method uses the transfer function of the process to locate the transfer function of the controller. Indirectly, the IMC-PID method is also an advanced process of determining the parameters Kc, τi, and τd.

Sigurd Skogestad formulates the stages of IMC-PID tuning as follows:

1. Modeling the system will produce the result characteristics of the consecutive equation
2. Create an open-loop system response by changing the controller mode: auto becomes manual. And change the set point value by ± 10% of the specified value
3. Using test signal in step signal. Because the step signal will produce an easy to observe system response
4. Using FOPDT method from open-loop response result to get parameters of values *Θ, τ, K,* $t_{63\%}$, and $t_{28\%}$
5. Specifies the control parameters of the values of *Kc, τi,* and *τd* with the value of the transfer function based on the IMC-PID Controller Tuning Rules table [13]. For the value of *Kc*, τ_i, and τ_d is given in equation (4), (5), and (6) below.

$$K_c = \frac{\tau_I}{K\,(3\lambda - 2\beta + \theta)} \qquad (4)$$

$$\tau_I = (\tau + 2\beta)\frac{\left(3\lambda^2 - \frac{\theta^2}{2} + 2\beta\theta - \beta^2\right)}{(3\lambda - 2\beta + \theta)} \qquad (5)$$

$$\tau_{Dc} = \frac{(2\tau\beta + \beta^2) - \frac{\left(\lambda^3 + \frac{\theta^3}{6} - \beta\theta^2 + \beta^2\theta\right)}{(3\lambda - 2\beta + \theta)}}{\tau_I} - \frac{\left(3\lambda^2 - \frac{\theta^2}{2} + 2\beta\theta - \beta^2\right)}{(3\lambda - 2\beta + \theta)} \qquad (6)$$

6. Performed analysis of control performance results.

3.3. Closed-Loop

The closed-loop control system is a system that measures the actual output of a process and compares it with the output set point. In a closed-loop system, the comparison of output condition with the desired condition (set point) can be analysed in qualitative and quantitative through the graph of system performance analysis as shown in three parameters below.

1. *Settling Time*
   Settling time is the time required by the response curve to exactly reach an absolute percentage value of 2% or 5% of the reference value (set point)
2. *Maximum Overshoot*
   Maximum overshoot (MO) is the maximum peak value of the response curve as measured from the beginning of the change. To determine the MO value can be expressed as a percentage through the following equation (7) approach.

   $$MO = \frac{c\,(t_p) - c(\sim)}{c(\sim)} \qquad (7)$$

3. *Integral Absolute Error* (IAE)
   Integral absolute error (IAE) is the sum of the error values of the resulting output response. The IAE criteria are preferred among ease industry practitioners in the measurement process. To calculate IAE value can be done by using the following equation (8).

   $$IAE = \int_0^{\infty} |SP(t) - CV(t)|\, dt \qquad (8)$$

IOP Conf. Series: Earth and Environmental Science **105** (2017) 012119 doi:10.1088/1755-1315/105/1/012119

4. Result and Discussion

At this stage, results and analysis of the TFC modeling process with IMC-PID tuning method will be presented.

4.1. Open-Loop Test Based on Control Valve Change

Open-loop simulation aims to find the transfer function process. The data is taken with operational valve change of 10% in 30 minutes duration on each laying of the controller. Based on the P&ID of the TFC strategy in figure 1., the controllers involved are the pressure indicator control (PIC) and the power indicator control (IC). The necessary data include an operation (OP), process variable (PV) and time in seconds. Furthermore, with FOPDT approach through PRC Cecil L. Smith equation (1), (2), (3) and considering open-loop graphic result, the value of parameters such as Θ, τ and K, for PIC and IC shown in table 2.

Table 2. FOPDT parameters on PIC and IC

Instrument	Parameter	Unit	Value	Instrument	Parameter	Unit	Value
	Θ	second	0		Θ	second	0
IC	τ	second	45.6024	IC	τ	second	39.7177
	K	-	16.551		*K*	-	1.590

4.2. Closed-Loop Test Based on Set Point Change

The analysis of closed-loop test refers to the use of known parameters such as Kc, τi, and τd to perform control actions based on ± 5% setpoint change. Analysis of IC and PIC control responses based on + 5% set point change are respectively shown in Figures 2.

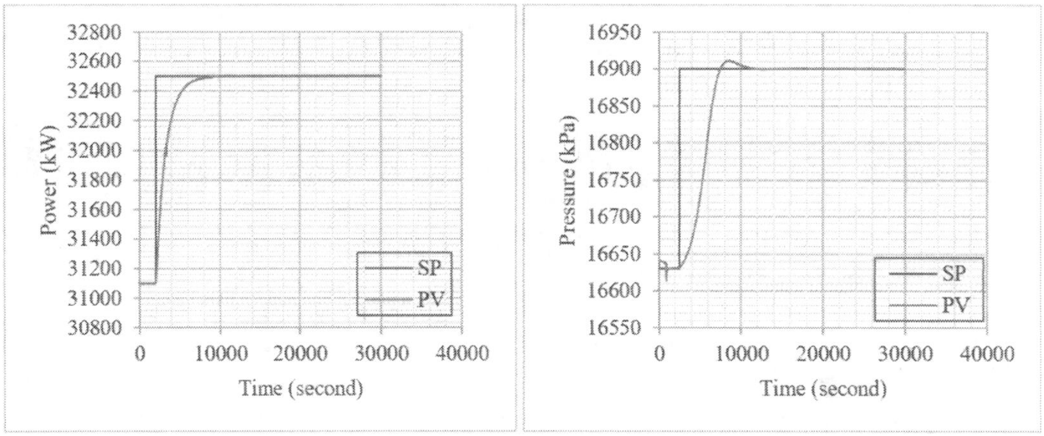

(a) (b)

Figure 2. (a) IC closed-loop test (b) PIC closed-loop test

Based on figure 2 (a) and (b), it can be seen the quality of the control response. Through ± 5% setpoint test, IC and PIC can meet the requested setpoint well. This is shown on the line of PV charts coinciding with the set point after the change. The process illustrates that the TFC strategy can meet set point changes. The control process also showed that the control scheme can't occur quickly. The control process in achieving the setpoint value had a certain time span. In quantitative response analysis, the performance of TFC control with IMC-PID tuning can be assessed by calculating settling time, maximum overshoot (MO) and integral absolute error (IAE). These three parameters of quantitative response analysis are shown in table 3.

Table 3. Closed-loop test based on set point change ±5%

Respond Analysis	IC		PIC	
	+5%	-5%	+5%	-5%
Settling Time	11,680	11,000	12,780	26,680
MO (%)	14.0	0.00	0.30	0.00
IAE	86,208.0	85,586.5	39,879.7	37,360.6

Based on the quantitative response analysis, the table showed every control strategy had a certain time span to achieve each set point. In each control instrument also showed that TFC can work quickly and accurately according to set point based on small of MO value.

5. Conclusions

The emphasis of this study is to the quantitative response of tuning TFC in the power plant as the variation of set point change ± 5%. For the results of modeling process above, the conclusions are summarized that TFC is precise control strategy to handle the changes of load on the generator.

Acknowledgments

The author's acknowledgment the support of PT UBJOM Pacitan and Power Plant Engineering of EEPIS for supplying material during all the phases of this project.

References

[1] Smuts J, 2010 Improving Boiler Stability Through Advanced Regulatory Control, OptiControls, Inc., Houston, Texas, ISA

[2] Black & Veatch 1996 Power Plant Engineering *Springer*, ISBN:0-412-064101-4, New York,)

[3] Karl J Å, Tore H 2004 Revisiting The Ziegler-Nichols Step Response Methode for PID Control *Journal of Process Control*, **14** 635-650.

[4] Willis M J 1999 *Proportional-Integral-Derivative Control.*

[5] Mochammad G *Desain Kontrol PID dengan Metoda Tuning Direct Synthesis untuk Pengaturan Kecepatan Motor DC* (ISSN 0853-8697, Jember, 2005)

[6] Lu S, Hogg B 2000 Dynamic Nonlinear Modeling of Power Plant by Physical Principles and Neural Networks, *International Journal of Electrical Power & Energy Systems*, **22** 67-78.

[7] Dale S, Thomas E, Duncan M, Francis D 1990 *Process Dynamics and Control* 3rd edition, pp. 212.

[8] Zhang Y, Huang C, 2014 IMC-PID Tuning Method for Stable FOPDT Processes with Stochastic Time Delay *IEEE: Navigation and Control Conference, China.*

[9] Rivera D, Morari M, Skogestad 1986 Internal Model Control. 4. PID Controller Design *Ind. Eng. Chem. Process Des. Dev.*, **25** (1) 252-265.

[10] Chien I L, Fruehauf P S 1990 Consider IMC Tuning to Improve Controller Performance *Chem. Eng. Prog.*, **86** 33-41.

[11] Skogestad 2003*Simple Analytic Rules for Model Reduction PID Controller Tuning* Elsevier.

[12] Bequette B W *Process Control: Modeling, Design, and Simulation* (Prentice Hall Professional, 2003)

[13] Smith C A, Armando C B 1985 *Principles and Practice of Automatic Process Control* John Wiley & Sons Inc.

[14] Totok R B, Hendrik E G P, Bayuaji, Nugroho, Soehartanto 2015 Design Plant-wide Control to Waste Heat Recovery Generation on Cement Indusrty Based HYSYS *Procedia Comput. Sci.*, **72** 170-177.

IOP Conf. Series: Earth and Environmental Science **105** (2017) 012121 doi:10.1088/1755-1315/105/1/012121

Optimizing the performance of $Li_4Ti_5O_{12}$/LTO by addition of silicon microparticle in half cell litium-ion battery anode

B Priyono[1], **P W Winowatan**[1], **A Z Syahrial**[1], **Faizah**[1], **A Subhan**[2]
[1]Departement of Metallurgical and Materials Engineering, Universitas Indonesia, 16424, Indonesia
[2]Research Center for Physics-LIPI, PUSPITEK, Tengerang, Banten, 15310, Indonesia
*Email : bambang.priyono@gmail.com

Abstract. The demand of lithium-ion battery (LIB) has been increased for high power application in transportation system. Thus, the current use of graphite as anode material needs to be replaced, due to formation of unwanted solid-electrolyte interphase (SEI) layer consuming intercalated Li+ that reduces the LIB performance and may cause ignition of the battery in high load usage. One of the candidates for anode material to replace graphite is lithium titanate (LTO), since the LTO does not form SEI and exhibits high-power with outstanding safety properties. This LTO compound was synthesized by mixing the TiO_2 xerogel of anatase phase and lithium carbonate (Li_2CO_3) as a source of lithium-ion followed by sintering at temperatures of 750°C to obtain the LTO with spinel crystalline phase. However, the LTO has the low theoretical capacity, i.e: 175 mAh/g, with real specific capacity obtained is at 114 mAh/g. To increase the LTO specific capacity, the addition of 10, 20 and 30 wt.% silicon microparticle which has theoretical capacity of 4200 mAh/g was conducted during preparation of the slurry anode mixture to minimize the formation of SiO_2. Anode sheet was made with Si/LTO and assembled into half-cell coin battery with lithium metal sheet as the counter electrode. Electro-impedance spectroscopy (EIS), Cyclic voltammetry (CV), and charge discharge (CD) testing were conducted to examine the battery performance. From EIS testing, the lowest impedance was obtained for the sample of 20 wt.% Si, while the highest impedance value obtained in 30 wt.% Si. The CV testing shows that the highest capacity at 141.1 mAh/g is achieved at the composition of 10 wt.% Si. Finally, from the CD testing, this Si/LTO anode could withstand the charge-discharge until 12 C and shows good stability until 100 cycles. From EIS and CV testing known that the optimum composition having the best performance is ranging from 10 wt.% to 20 wt.% Si. It is predicted that at higher composition, the pulverization of Si particle is occurred declining the performance of Si/LTO anode.

1. Introduction

Currently, the energy conservation and reducing harmful gas emission are the most crucial challenges in the transportation industries. The usage of Electric Vehicles (EVs) is one of the most promising solutions to this matter. However, the main obstacle to increase the application of Electric Vehicles (EVs) is in the battery for electrical energy storage system. The most notable advantage of using lithium ion battery is the high energy density and not to mention the operation temperature which can be said to have wide temperature range and long shelf life.

This advantage has made lithium-ion battery (LIB) replacing the conventional battery systems [1]. The lithium batteries exhibit promising capabilities for energy storage used in hybrid or electric vehicle (EVs). However, for EVs application the current use of graphite as anode material needs to be replaced, due to formation of unwanted solid-electrolyte interphase (SEI) layer consuming intercalated Li+ that reduces the LIB performance and may cause ignition of the battery in high load usage. One of the candidates for anode material to replace graphite is lithium titanate (LTO), since the LTO does not form SEI and exhibits high-power with outstanding safety properties [2]. However $Li_4Ti_5O_{12}$ also has a low both in the intrinsic conductivity (10^{-9} S.cm-1) and in its theoretical capacity (175 mAh/g) [3].

Recent researches regarding increasing the capacity, conductivity and rate capability have been conducted to find the types of optimization to overcome the issues. Nano-structuring, morphological optimization, addition or even the combination of them can boost the capacity and rate capability [4]. The addition of silicon (Si) which is a semiconductor element and has the highest specific capacity of ~4200 mAh g^{-1} is one of the promising strategies to increase the conductivity and capacity. Obviously, silicon gain an extensive attention towards other benefits such as the element relatively abundant and easier to obtain, arguably inexpensive, and slightly has the lithiation and de-lithiation potential that is higher than found in lithium, which can contribute to safe feature for the electrode [5].

In this research, the experiment was focused to synthesize of $Li_4Ti_5O_{12}$ (LTO) with the addition of silicon microparticle. The mixing of both active substances was carried out during slurry preparation to minimize the oxidation of Si element. Then, the slurry is coated to make an anode sheet and assembled into a Li-ion battery (LIB) half-cell with Li-metal sheet as the counter electrode. The influence of Si microparticle content (in weight %) in the LTO slurry mixture was observed using Electrochemical Impedance Spectroscopy (EIS), Cyclic Voltammetry (CV) and Charge-Discharge (CD) testing. The EIS is conducted to measure the conductivity and CV is performed to measure the specific capacity and working voltage. Then, the CD test is carried out to know the rate capability and columbic efficiency. To prove the stability of the present LTO/Si anode, the CD test is conducted until 100 cycles.

2. Materials and Methods

2.1. *$Li_4Ti_5O_{12}$ Powder Preparation and Characterization*

Methodology In the synthesis process of $Li_4Ti_5O_{12}$/Si (LTO/Si) was prepared through a sol-gel method, calcination, ball-milling, sintering and mixing. The first stage is the sol-gel method, where a primary solution consisting of ethanol pH 3 and titanium butoxide is mixed with secondary solution that contains ethanol and distilled water to form a clear gel using a magnetic stirrer. The obtained gel was dried in the ambient temperature for about 12 h to form TiO_2 xerogel. Calcination carried out at temperature of 300 °C for 2 h to obtain anatase TiO_2. Then, the high energy ball mill was used to mix anatase TiO2 with Li_2CO_3. Finally, the sintering process was then performed at 750 °C in 3 hour to form LTO spinel phase and ready to be used for the active electrode material. Then, to form the LTO/Si, the addition of silicon microparticle is done by mixing of the LTO and silicon microparticle during slurry preparation to minimize the oxidation of Si element.

The active materials were characterized using Scanning Electron Microscopy (SEM) and Energy Dispersive Spectroscopy (EDS) mappings to analyse the topography, microstructure and elements distributions. Further, Brunauer-Emmet-Teller (BET) Quantachrome is used to measure the surface area of the samples.

2.2. *Cell Assembly and Testing*

In this research, the half-cell coins battery were assembled and tested to observe the electrochemical properties and performance of LTO/Si compounds and use a lithium foil as counter electrode, and $LiPF_6$ 1M as an electrolyte. Further, DMAC, Acetylene Black (AB) and PVDF were used as solvent, conductive additive and binder, respectively. The compositions were prepared with fixed binder content of 10% wt.%, while the LTO and Si-microparticle contents were 10 wt.% for LTO/Si10-sr sample, 20 wt.% for LTO/Si20-sr sample, and 30 wt.% for LTO/Si30-sr sample. All of these material then mixed by magnetic stirrer at 300 rpm at 70°C for 2 h to obtain slurry.

The slurry was coated at copper foil current collector by electrode coater (doctor-blade) with 200 μm of wet thickness to form 100 μm of dry thickness. Coated Cu foil was then dried at doctor blade at 80 °C and cut to form coin shape. The assembly process was carried out in a glove box filled with high purity argon gas. Electrochemical performances were measured using a CR2032-type coin cell with the following devices: Wonatech instrument for Electrochemical Impedance Spectroscopy (EIS) and Cyclic Voltammetry (CV) and MTI Battery Testing instrument for Charge-Discharge(CD) .

3. Results and Discussion

3.1. Structure and Morphology of LTO

The examination of SEM is intended to observe the morphology of a material. The Figure 1 shown representative SEM image of LTO powder with different percentage silicon microparticle. The main problem is that the samples were agglomerated occurring which influence cycles efficiency and rapidly capacity decreased [6]. Distribution silicon and oxygen elements described in figure 2 and figure 3 via elements mapping of the following.

Figure 1. SEM image of LTO powder with different variables (a) LTO/Si10-sr, (b) LTO/Si20-sr, (c) LTO/Si30-sr

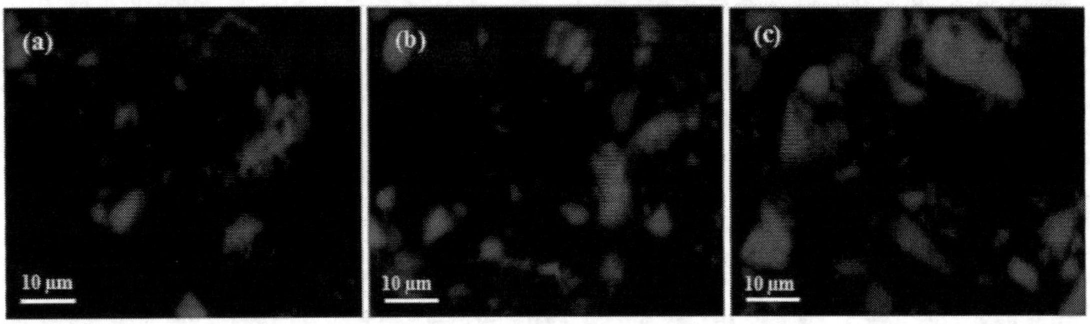

Figure 2. Silicon mapping in different samples (a) LTO/Si10-sr, (b) LTO/Si20-sr, (c) LTO/Si30-sr

Based on silicon element mapping in each samples as shown in Figure 2, it can be seen that silicon content increased in mapping can be investigated. With increased silicon content for each samples, it seems the increased agglomeration has occurred as shown from in Figure 2 (a) to (c). This may affect electrochemical performance of the present LTO/Si anode.

Figure 3. Oxygen mapping in different samples (a) LTO/Si10-sr, (b) LTO/Si20-sr, (c) LTO/Si30-sr

In figure 3, it is shown that oxygen distribution is relatively not the same as Si element mapping. It is proved that oxygen distribution is not related to the formation SiO_2. Thus, the effort to minimize the formation of SiO_2 conducted during mixing preparation of the slurry anode between LTO and Si element was successful.

Further, the surface area of the LTO/Si samples were measured and the result is listed in table 1.

Table 1. Surface area values LTO/Si microparticle powders.

Samples	Surface area (m^2/g)
LTO/Si10-sr	77.62
LTO/Si20-sr	95.30
LTO/Si30-sr	105.40

Table 1 shown result of the surface area for every sample has shown relatively large surface area. The highest surface area is LTO/Si30-sr. Surface area increased with added silicon microparticle. The influence of surface area of samples to the performance will be discussed later.

3.2. *EIS Characterization of LTO Electrodes*

In the application of battery, conductivity is important for battery performance. The conductivity of battery could be measure by testing the resistance of the battery by using Electrochemical Impedance Spectroscopy) can show the charge transfer resistance (Rct) electrode as seen in Figure 4 as follows.

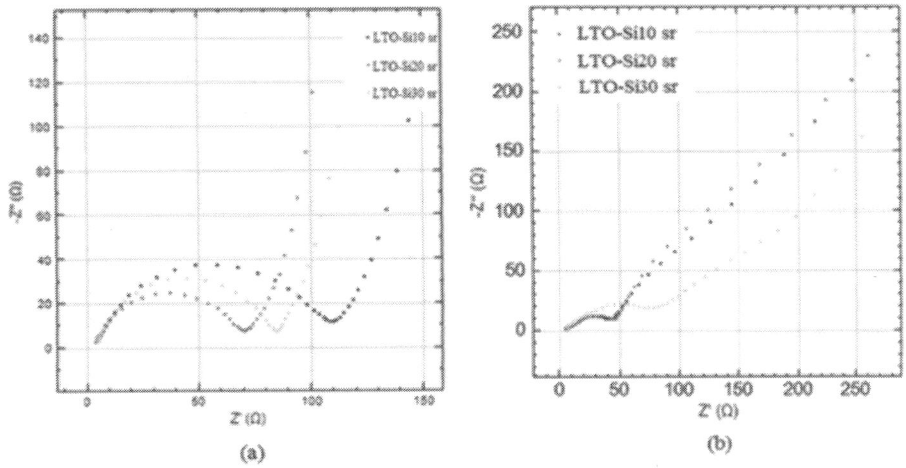

Figure 4. EIS semicircle curve of LTO electrode (a) pre-CV and (b) post-CV

As depicted in figure 4, the EIS test was conducted twice as shown on the above result. The first EIS was performed prior to cyclic voltammetry test (pre-CV) and the second EIS test was done after cyclic voltammetry (post-CV). The test result is presented in table 2.

Table 2. Impedance result of the samples

Sample	Pre-CV			Post-CV			Decreasing resistivity value
	Re (Ω)	Rct (Ω)	Rtotal (Ω)	Re (Ω)	Rct (Ω)	Rtotal (Ω)	(Ω)
LTO1/Si10-sr	4	107	111	4	41	45	66
LTO1/Si20-sr	4	68	72	4	41	45	27
LTO1/Si30-sr	4	84	88	4	76	80	8

Based on the result of EIS after cyclic voltammetry, it could be seen in Table 2 that the resistivity of all samples shifting to the left which mean that the conductivity is better than before the CV. It could be seen that the sample of LTO/Si20-sr has the lowest resistivity. But, the sample LTO1/Si10-sr has better decreasing resistivity value from pre-CV into post-CV resistivity value. It could mean that this composition has the best increased conductivity value during CV or lithiation-delithiation process. Further, for the sample of LTO/Si30-sr, the resistivity of pre-CV and post-CV values did not change significantly. It means that this composition has not good response to the lithiation-delithiation process compared to the other samples. This could be predicted as the result of pulverization of the Si particle that reducing the performance of this LTO/Si [7]. Addition of more Si contents in the mixture of LTO/Si, the reactivity of this compounds to lithiation-delithiation process decreases.

The influence of surface area of the samples has not played dominant role in this matter. Although the LTO/Si30-sr has the largest surface area, but reactivity to the lithiation-delithiation process is not the best amongst the samples.

3.3. Cyclic Voltammetry (CV) Characterization of LTO Electrodes

The use of cyclic voltammetry for battery performance is intended to measure energy density of a lithium-ion battery. Based on previous research, theoretically lithium titanate has working current of about 1.55 V with the lithium electrode having the capacity of 175 mAh/g [8]. As in this research the use of silicon microparticle in addition to the LTO, it is predictable that the curve of cyclic voltammetry would have two peaks of LTO and also the added element of silicon.

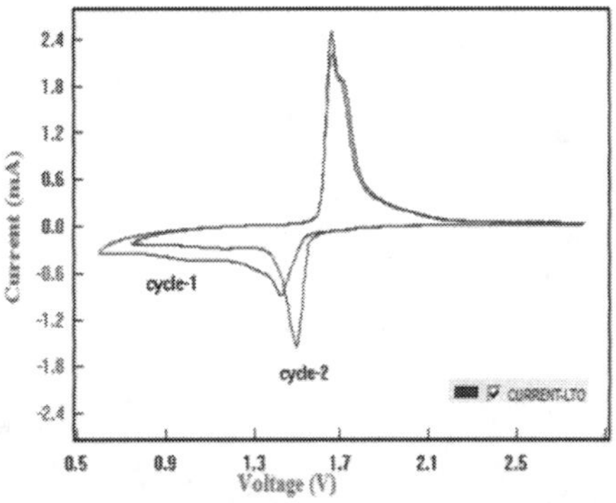

(a)

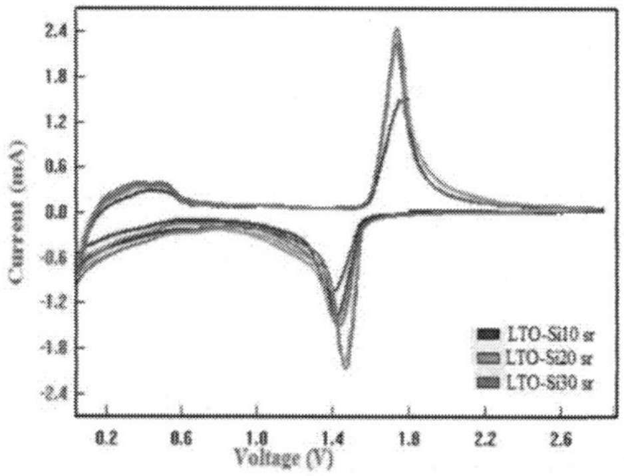

(b)

Figure 5. CV curve of half-cell, sweep rate 100 mVs^{-1}: (a) LTO and (b) LTO/Si

In figure 5, the highest peak is noted as the sample of LTO/Si20-sr and followed by LTO/Si10-sr. The sample of LTO/Si30-sr, which could be seen on the curve, having the lowest peak. It could be observed that there are anodic peak and cathodic peak in which indicating that the reaction is reversible. For silicon peak, the difference could be seen by comparing the LTO peak with the silicon peak. Silicon peak is not as high as the LTO peak and rather having a flat peak compared to LTO. However, it could be observe that in addition of silicon to the LTO has an impact to the LTO peak, in which the optimum is the silicon addition of 20 %.wt, in which the addition of silicon of 30% would lower the peak of LTO.

Table 3. Specific capacities of LTO/Si half-cell batteries

Samples	Specific capacity [mAh/g]
LTO	113.74
LTO/Si10-sr	141.08
LTO/Si20-sr	134.24
LTO/Si30-sr	102.66

Specific capacity in table 3 shows the sample of LTO/Si10-sr has the capacity of 141.08 mAh/g. For LTO/Si20-sr has 134.24 mAh/g and LTO/Si30-sr for 102.66 mAh/g. As seen in Figure 3, the trend of the specific capacity of every samples decrease with the addition of silicon concentration to the LTO. From Table 3, it shows that the addition of silicon does not always raise the specific capacity of the anode material as expected. One reason is the limited ability of the binder polyvinylidene fluoride (PVDF) used for the manufacture of a slurry containing the active ingredient of silicon. Binder type polyacrylic acid (PAA) may yield better performance [9].

3.4. Charge-Discharge (CD) characterization of LTO electrodes

In the test of Charge-Discharge, the examination is intended to investigate the capacity and performance of a battery. In this examination the battery will be given different current rate, in the case of this study, the current rates which used from C/5 up to 12C. Figure 6 shows the result of the charge-discharge capacity from C/5 to 12 C with the different variables of LTO/Si.

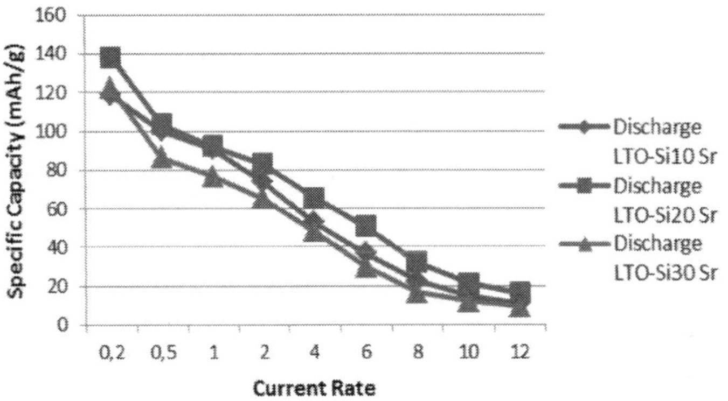

Figure 6. The discharge capacity of half-cell with current rate 0.2V to 12C

The discharge capacity of the battery showing an increase by the addition of silicon concentration. However, the LTO/Si30-sr which having the higher concentration of silicon, showing the decrement of the discharge capacity. The presents of silicon oxide which can be confirmed in the previous examination of SEM result has been a major drawback to the performance of the battery. In this case, the increase of the silicon concentration could increase the charge and discharge capacity which after reaching certain optimum point, the pulverization of Si microparticle would be inevitable and thus interfere with the properties and performance of the battery.

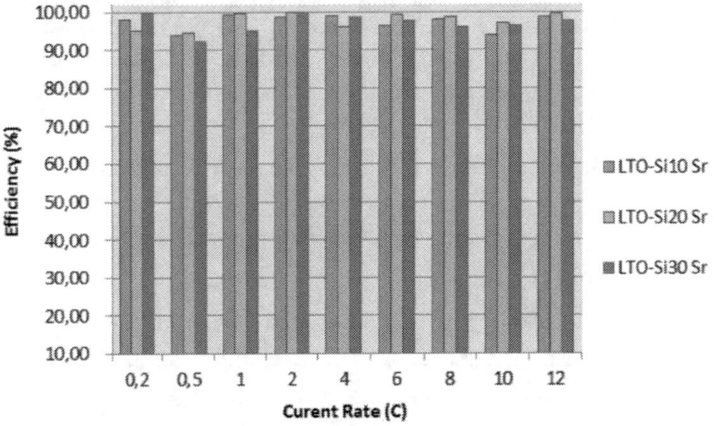

Figure 7. Columbic efficiency of half-cell during charge and discharge cycling

The coulombic efficiency which has been plotted into a graph in Figure 7 with the current rates, it can be observed that the efficiency of the battery samples reaching 90-100%. This would mean that the properties of LTO active material in spinel phase in which having zero-strain, where there is no change in structure during charge and discharge, has been proven to be confirmed.

Figure 7 shown that the value of the specific capacity of the sample composition LTO/Si20-sr generate the highest value for each current sample rate and followed for LTO/Si10-sr and the lowest value is for a sample LTO/Si30lr. This supports the above suggestion that the PVDF binder in this study has its limitations compared to the binder PAA (poly acrylic acid [9].

Furthermore, to determine the stability of the sample in the certain number of Charge-Discharge (CD) cycles, the CD testing then performed for one hundred (100) cycles with C-rate to charge is 0.5C while for discharge is at 1C. In Figure 8 is shown the value of specific capacity curve for 100 cycles of samples (a) LTO/Si10-sr, (b) LTO Si20-sr, and (c) LTO/Si30-sr.

The value of the CD test results for 100 cycle are presented in table 4 below.

Table 4. Specific capacity values in cycle 1 and cycle 100th of sample (a) LTO/Si10-sr, (b) LTO/Si20-sr, and (c) LTO Si30-sr

No.	cycles	LTO1 / Si10-slr		LTO1 / Si20-slr		LTO1 / Si30-slr	
		chg cap (mAh g)	Disch cap (mAh g)	Chg cap (mAh/g)	Disch cap (mAh/g)	Chg cap (mAh/g)	Disch cap (mAh/g)
1	number 1	109	103	106	96	88	80
2	100th	75	74	67	67	65	64
The decline in capacity		31%	28%	36%	30%	26%	20%

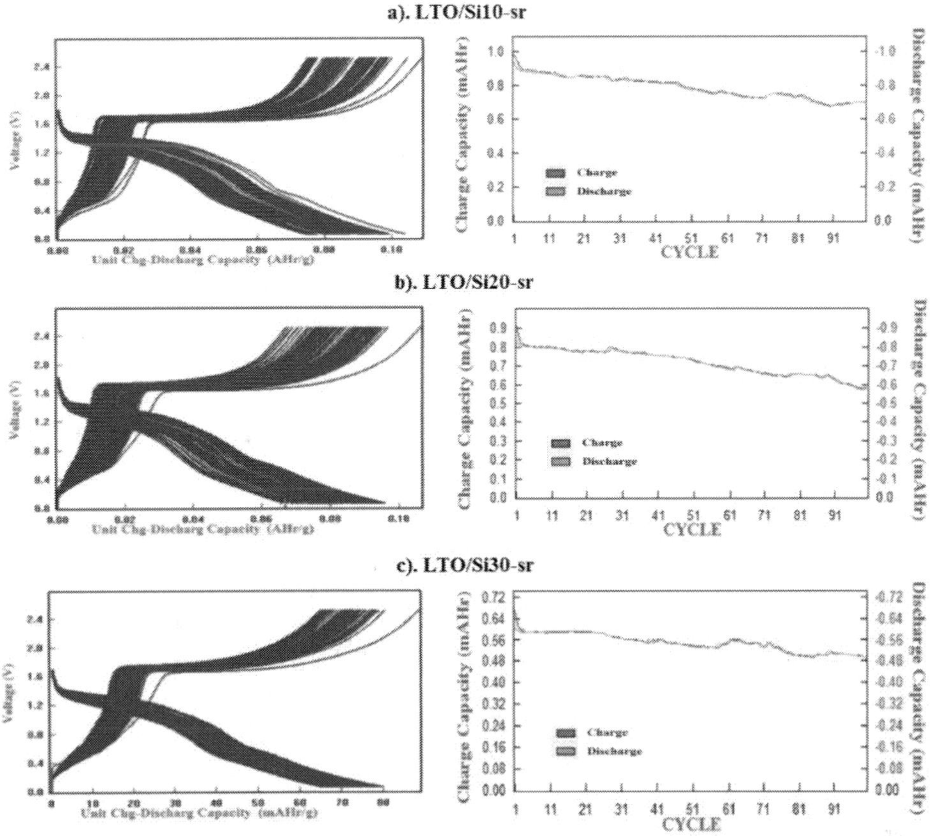

Figure 8. Specific capacity value curve for 100 cycles of samples: (a) LTO/Si10-sr, (b) LTO/Si20-sr, and (c) LTO Si30-sr

Figure 8 and Table 4 shows that all three samples have good stability and can last for 100 cycles. The decrease in the specific capacity of the battery cells occurred during charge-discharge process is due to the volume expansion of the silicon particles. The tension will cause cracking and collapse of silicon that causes loss of electrical contacts that will eventually eliminate its capacity [7].

The sample of LTO/Si30-sr is having a significant decrease compared with the sample of LTO/Si20-sr. This could indicate that pulverization as explained in previous section decreases the specific capacity of the battery. Nevertheless, if it is compared to the real capacity of LTO with specific capacity of 113.74 mAh/g, the addition of Si microparticle has increased the specific capacity of the LTO/Si compound.

4. Conclusion

$Li_4Ti_5O_{12}$/Si (LTO/Si) was successfully synthesized by a solid state reaction between xerogel TiO_2 and Li_2CO_3 for application in lithium ion batteries and mixing silicon microparticle during slurry preparation. Influence of surface area is not play major role in this LTO/Si compound. Influence of addition of silicon microparticle that has highest conductivity is at the silicon addition of 20 %.wt. The specific capacity of every samples decrease with the addition of silicon concentration to the LTO. The CV testing shows that the highest capacity at 141.1 mAh/g is achieved at the composition of 10 wt.% Si. The LTO/Si30-sr which having the higher concentration of silicon, showing the decrement of the discharge capacity. From EIS and CV testing known that the optimum composition having the best performance is ranging from 10 wt.% to 20 wt.% Si. Besides the limitation of PVDF binder capability, it is suggested that at higher composition, the pulverization of Si particle is occurred declining the performance of Si/LTO anode. Finally, all samples of LTO/Si shows a good stability which could withstand at 100 cycles charge and discharge test.

5. Acknowledgement

The authors would like to thank to the Direktorat Riset dan Pengabdian Masyarakat Universitas Indonesia (DRPM-UI) for the financial support under the grant of Riset HIBAH PITTA 2017 under contract No: 736/UN2.R3.1/HKP.05.00/2017.

6. References

[1] G. Lowe, M., Tokuoka, S., Tali Trigg, T. and Gereffi, 2010 "Lithium-ion Batteries for Hybrid and All-Electric Vehicles: the U.S. Value Chain," *Center on Globalization, Governance & Competitiveness.*, **1** 1–77.

[2] J. Shi, Y. Liang, L. Li, Y. Peng, and H. Yang, 2015 "Evaluation of the electrochemical characteristics of silicon/lithium titanate composite as anode material for lithium ion batteries," *Electrochimica Acta*, **155** 125–131,.

[3] S. Goriparti, E. Miele, F. De Angelis, E. Di, R. Proietti, and C. Capiglia, 2014 "Review on recent progress of nanostructured anode materials for Li-ion batteries," *Journal of Power Sources*, **257** 421–443.

[4] B. Liang, Y. Liu, and Y. Xu, 2014 "Silicon-based materials as high capacity anodes for next generation lithium ion batteries," *Journal of Power Sources*, **267** 469–490.

[5] Y. Nishi, R. J. Brodd, and A. Kozawa, 2014 "Lithium-Ion Batteries," in *Lithium-Ion Batteries*, 21–39.

[6] D. Mohanty, E. Hockaday, J. Li, D. K. Hensley, C. Daniel, and D. L. Wood, 2016. "Effect of electrode manufacturing defects on electrochemical performance of lithium-ion batteries: Cognizance of the battery failure sources," *Journal of Power Sources,* **312** 70–79.

[7] H. Wu and Y. Cui, 2012. "Designing nanostructured Si anodes for high energy lithium ion batteries," *Nano Today*, **7** (5) 414–429.

[8] J. Lu, C. Nan, Q. Peng, and Y. Li, 2012. "Single crystalline lithium titanate nanostructure with enhanced rate performance for lithium ion battery," *Journal of Power Sources*, **202** 246–252.

[9] T. D. Bogart, A. M. Chockla, and B. A. Korgel, 2013. "High capacity lithium ion battery anodes of silicon and germanium," *Current Opinion in Chemical Engineering*, **2** (3) 286–293,

IOP Conf. Series: Earth and Environmental Science **105** (2017) 012122 doi:10.1088/1755-1315/105/1/012122

Experimental Investigation on the Reduction of Catalyst Costs in the Polyethylene Pyrolysis Process

G K Upreti[1], S Awad*[1], G Burnens[1], C Kassargy[1], M TAZEROUT[1]
[1] GEPEA, UMR-6144, Département Systèmes Energétiques et Environnement, IMT Atlantique, 4 Rue Alfred Kastler, CS 20722, 44307, Nantes, France

E-mail: sary.awad@imt-atlantique.fr

Abstract. In our earlier works, the catalyst USY zeolite was tested for the pyrolysis of High density polyethylene (HDPE). It yielded 71% liquid oil that could be separated into diesel and gasoline-like fractions [3]. However, the high cost of this catalyst led to the present study where the effects of different low-cost catalysts on catalytic cracking of HDPE were investigated. GC-MS analysis were performed to check the carbon distribution of each sample and to make theoretical gasoline/diesel quantification in the pyrolysis liquids. Products where compared by studying carbon distribution, condensable range, branching degrees and reaction speeds. According to all those parameters, none of them outperformed USY zeolite. The Second Part has been conducted to observe the regeneration of USY Zeolite and monitor its performance for its economic value. The regeneration was done 10 times and the performance was monitored after every experiment with respect to % of heavier compound yields, carbon distribution range, char deposition range and gasoline and diesel yields. It was found that the regenerated catalyst loses its capability very slowly and even after 10 regenerations products characteristics were not sensibly altered.

1. Introduction

The constant development in the polymerization industries and processing machines are producing cost-effective production which are replacing the conventional materials with the plastics. Continuous innovation explains that plastics production has increased by an average of almost 10% every year on a global basis since 1950. The total global production of plastics has grown from around 1.3 million tonnes (MT) in 1950 to 322 MT in 2016 [1].

Due to the much economic cost of the plastic and the middle - class development the consumption is rising up in the last few years. This high consumption has led to the generation of more and more plastic waste. Plastic poses a greater threat to the environment and thereby it has become very difficult for its direct disposal. Depending upon its area of application, the service life range life of expectancy of plastics lies from 1 to 35 years. Plastic in municipal solid waste streams makes up only 7–9% of the weight of the total waste stream, by volume they may represent 20–30%. Of the organic waste stream, that is, after removal of glass, metals, etc., plastics are about 9%–12% by weight. In addition to the presence of plastics in municipal waste streams, many wastes collected from manufacturing or service industries may contain much higher proportions of plastics [1]. There is a great economical challenge for the management this plastic waste to recover its potential source of energy while processing it through the thermal degradation.

Energy recovery via pyrolysis does not require sorted or clean feedstock and it could be feasible at smaller scales than recycling. Pyrolysis is the thermal degradation of polymers under the effect of temperature under inert atmosphere. Pyrolysis transforms plastics into liquid and gaseous products, with high heating values, that could be used as fuels. Pyrolysis is done using two methods that is thermal pyrolysis and Catalytic Pyrolysis. In thermal Pyrolysis, the degradation is completed through free radical exchange mechanism which results in statistical distribution of wide range of carbon chain [4]. On the contrary, Catalytic pyrolysis degrades the plastic through carbocation exchange

IOP Conf. Series: Earth and Environmental Science **105** (2017) 012122 doi:10.1088/1755-1315/105/1/012122

mechanism. This carbocation exchange results in the β-Scission of the hydrocarbon chain which results in the bi-functional conversion of the long chain into shorter chains in consecutive reactions [5].

Polyethylene and polypropylene which represent the major part of plastic wastes are composed of long chains of hydrocarbons and their thermal degradation results in shorter carbon chain products. In order to enhance reaction kinetics and to select a range of carbon chain length distribution, catalysts are used (zeolites, activated carbon, metallic oxides...). Depending on its chemical and physical characteristics, yield and characteristics of products can be impacted.

The objective of this experimental study is to investigate the thermal degradation of plastic waste by pyrolysis using different catalysts, regenerating the catalyst and to find the most effective catalyst based on thermal cracking performances and the economic value. In this work, the initiation was done by selectivity of different materials and parameters used in the pyrolysis process and characterizing the catalysts based on their chemical and physical properties. Furthermore, testing of different catalysts in plastic waste pyrolysis process and their regenerative capabilities has been conducted using a pilot pyrolysis plant.

The final aim of this work is to find a way to lower the catalyst part price in the catalytic pyrolysis process. Cheaper catalysts were tested and compared according to their liquid fuel yield. Moreover, regeneration test campaign was done with USY Zeolite.

2. Materials and methodology

The Experiment was done using Batch type reactor of stainless steel. The reaction chamber was protected by Thermal resistance cover to avoid the reflux of the process. The Plastic used for the Experiment was High Density Polyethylene (HDPE). Catalysts were mixed with the plastics and introduced to the cold reactor. To create Inert Atmosphere, the nitrogen gas was purged for about 5-8 minutes. In condensing system, the direct flow of Tap water was continuously supplied throughout the whole experiment. The exhaust system was connected to a blower which was connected to general ventilation of the laboratory. The Catalysts that was used for conducting the experiments were CaO, Activated Carbon, Activated Carbon Sulphonated, Red Mud and Industrial Zeolite USY. The Red mud was from the origin of solid waste clay of Cameroon and the other catalysts were either taken or prepared in the laboratory.

The limit of the reactor temperature was controlled at 500°C with the heating rate of 21°C/min. Catalyst was added in 1:10 that is 20g of catalyst was taken for the pyrolysis of 200g of plastic. The liquid product was collected in a glass bottle with the measurement of volume in ml. The Time was recorded corresponding to the temperature and volume obtained at that particular time. Digitally, The Data of the process was registered using the software LabVIEW 7 Express.

The samples were analyzed by the machine PERKIN ELMER CLARUS 680 which is coupled with (GC/MS-FID). A column SLB - 5MS (30m x 0.250 mm; 0.25μm film thickness) was used to pass the helium at the flow rate of 1ml/min. The temperature commenced at 40°C for 10mins followed by 3°C/min to 170°C and then 10°C/min to 330°C, which was hold for 6mins. The peaks of the spectra was identified by NIST Library or attributed according to their molecular peak.

2.1 Catalyst regeneration

In this Study, the Regeneration was done for the USY Zeolite Catalyst. USY Zeolite has very high BET surface of about 617 m^2/g [2] and high Si/Al ratio which make it highly acidic in nature. The hypothesis has been made on the basis of its activity in different parameters. The experiments were all conducted at the same quantity ratio 1:10 and was controlled at same temperature limit of 550°C. The Images were taken at 3 different phases before the experiment, after experiment, and after regeneration which can be seen in figure 1.

The regeneration was done by taking the residue after the experiment in a ceramic bowl and heating it inside the oven with the presence of air. The oven configuration was set for an increasing rate of temperature from ambient temperature to 550°C for 47 mins and then holding at the same temperature of 550°C for the next 3 hrs. It was observed that in the regenerated catalyst, char is completely removed and visibly it is little pale yellow as compared to the white color of fresh catalyst. After each regeneration, the catalyst was weighted and the lost catalyst was completed according of the ratio of 1:10 with fresh catalyst.

Figure 1. Images of Catalyst (A) Before Experiment (B) Residue after Experiment (C) After Regeneration

3. Results and discussions

3.1 Gasoline and diesel yields

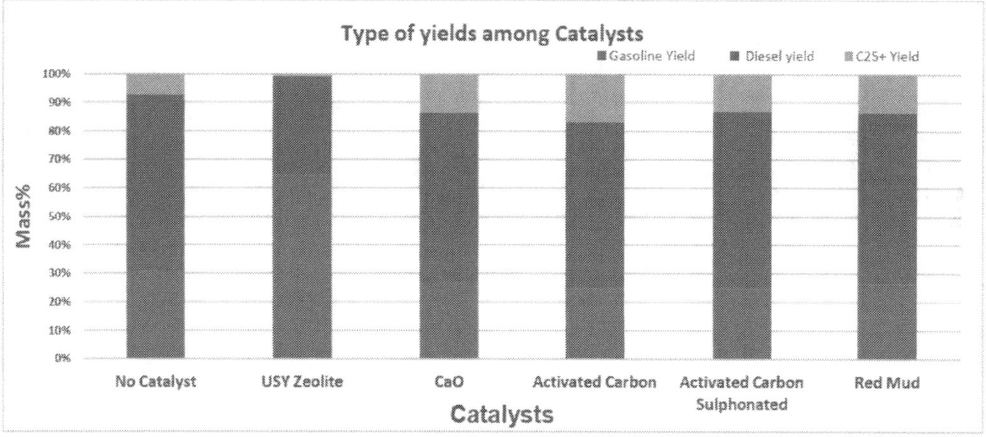

Figure 2. Comparison of Types of yield among the catalysts

The analyses of gasoline and diesel yields were monitored to find the maximum amount of different products that can be recovered after pyrolysis using different catalysts. The monitoring was done on the basis of hydrocarbon ranges which varies from C_5-C_{12} (Gasoline), C_8-C_{24} (Diesel) and C_{25+} (Other products) [6].

According to the Hydrocarbon ranges, Gasoline and Diesel share very small distribution of carbon range but this proportion can be separated further by distillation processes. It can be seen in figure 2, that the Maximum yield of gasoline is given by USY Zeolite and the rest of the catalyst has very low ranges of gasoline. This was due to the effect of internal structure that plays an important role in determining the cracking capability of catalyst. No catalyst also yielded a good amount gasoline and diesel but the quality of the fuel is very low. All the Catalyst except USY Zeolite are amorphous in internal structure that's why, it become difficult to crack long hydrocarbon chains. Whereas, the

Zeolite USY is crystalline in structure, which provides better cracking with higher rate of ion exchange activity.

3.2 Carbon distribution comparison

The carbon distribution shows the mass% and retention time of each type of compounds which are varied by the different catalyst that can be seen in figure 3. The graph with no catalyst has been considered as the reference while comparing with other catalyst. The shortening of carbon chains by the catalyst shows the active phase while cracking.

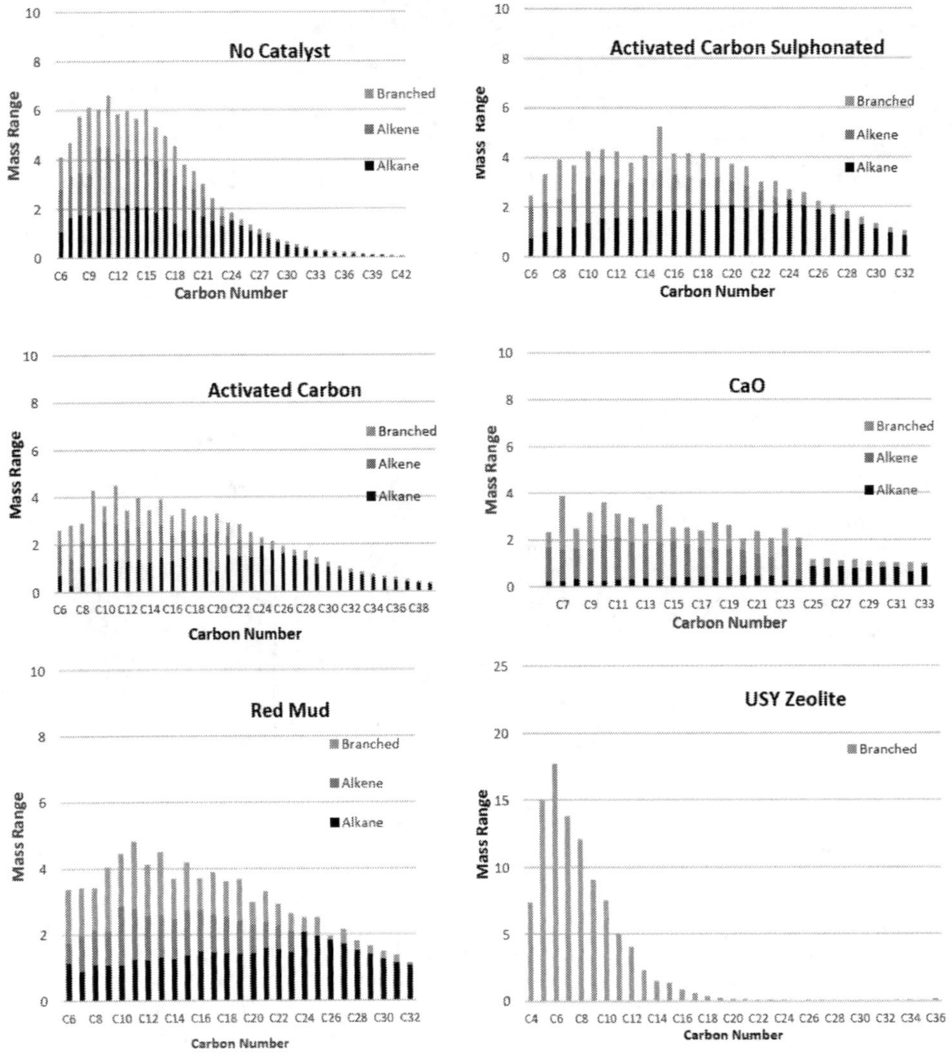

Figure 3. Carbon distribution comparison of catalysts

The phase of product yield in all the catalyst except USY Zeolite was obtained in wax phase while the product of USY Zeolite was in liquid phase. The wax formation in the products was due to the alternation in the catalytic cycle. It can be seen that the high % of the formation of heavier compounds in other catalysts results in the absorption of lighter compounds by the heavier compounds. On the other hand USY Zeolite produces high % of lighter compounds in comparison to others. This is due to the high presence of active acid sites and higher reaction degree mechanism of USY Zeolite.

3.3 Kinetics of different catalysts
The kinetics of catalysts were monitored by the consideration of the reference with the limit time in mins to the volume of 150 ml. The comparison was done with No catalyst which was our base point and then the order of kinetic was observed with catalysts. The variation was observed according to the increasing order that was **No Catalyst** (90mins) < **Activated Carbon Sulphonated** (68mins) < **CaO** (66mins) < **Red Mud** (65mins) < **Activated Carbon** (61mins) < **USY Zeolite** (31mins).

No catalyst surpasses the performance of Zeolite USY which is due to the absence of active bronsted acid sites in the other catalysts. This gives us a conclusion that the activity depends upon the crystalline structure and the acidic sites in the materials.

3.4 Regeneration of USY Zeolite
3.4.1 Relative difference in regenerations. The regeneration experiments were monitored by the relative difference in their carbon distribution. The experiments was conducted till 10 times but to observe the difference, the comparison was done between the 5th and 10th experiment. It can be seen from the figure 4 that the formation of heavier compounds are increasing at the 10th experiment from C_{18}. This might be due to the after effect of the increasing number of regenerations done which deactivates the acidic sites in the catalyst and destruct the internal structure. This makes the catalyst incapable to crack large hydrocarbons and thus produces heavier compounds.

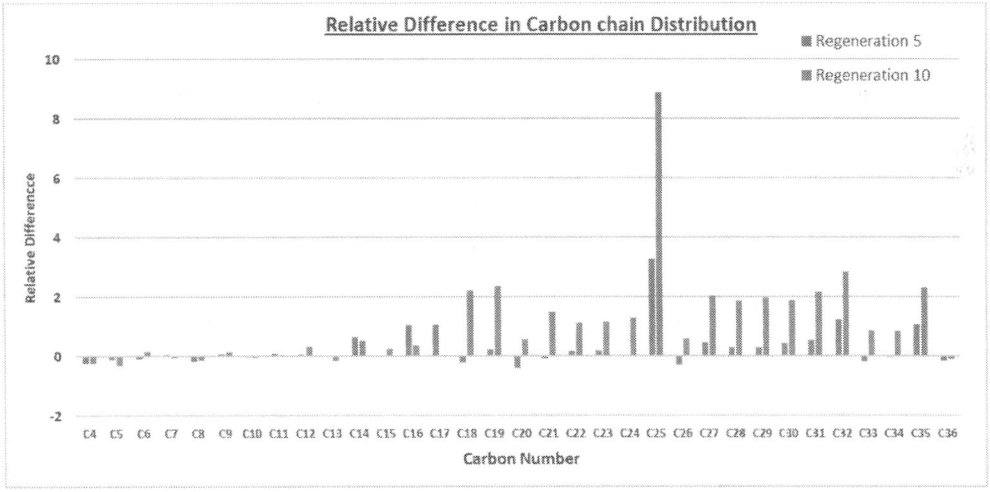

Figure 4. Difference between regenerations

It is also interesting to see that the formation of lighter compounds (till C_{17}) from the 5th and 10th experiment carries a very low relative difference. This gives a very strong indication that in both the experiments there was almost same percentage of formation of lighter compounds which follows the same carbon distribution range of USY Zeolite. It can also be noted that there is still the active presence of acidic sites in the catalyst till the 10th experiment.

3.4.2 Char deposition. According to the proximate analyses, the plastic contains the ash content and due to this, there is the formation of char in the residue of the pyrolysis process. While in the pyrolysis process, due to the polymerization reaction, there is no formation of the coke at the bottom of the reactor or less than 1% of the HDPE. The coke gets to stick to the internal structure of the surface of the catalyst. In the analyses it showed that, the mass% of char deposited has varied between 1.5% - 2.25 % of the total weight of the plastic. The activation state of the catalyst can be concluded from the char yield %. It can be concluded from that the catalyst shows little variation of char deposited throughout the experiments conducted and it is still in the active state.

4. Conclusion

It can be concluded from the experimental study that the catalytic cracking is far more advantageous than thermal cracking which is due the different type of cracking mechanisms. The variation in the activities of different catalyst are strongly dependent upon the presence of its acidic sites and the type of internal structure. In order to improve the performance of the catalyst, there can be treatment with acids to increase their overall strength.

The Pyrolysis process also yields the gases such as H_2, CH_4 etc. which can be further used for the energy production.

On the regeneration part of USY Zeolite, we observe that there is still presence of active acid site till the 10th experiment which shows that it is an economic method for the usage of the same catalyst. Industrial manufactured Zeolites like USY Zeolite are highly efficient and are very expensive but they can go through regeneration process to be economical while in continuous process of pyrolysis.

References

[1] Panda, A. K., Singh, R. K., & Mishra, D. K. (2010). Thermolysis of waste plastics to liquid fuel: a suitable method for plastic waste management and manufacture of value added products—a world prospective. *Renewable and Sustainable Energy Reviews*, *14*(1), 233-248

[2] Zheng, Q., Huo, L., Li, H., Mi, S., Li, X., Zhu, X., Deng, X. and Shen, B., 2017. Exploring structural features of USY zeolite in the catalytic cracking of Jatropha Curcas L. seed oil towards higher gasoline/diesel yield and lower CO 2 emission. *Fuel*, *202*, pp.563-571.

[3] A. C. Kassargy, S. Awad, and G. Burnens, "Experimental study of catalytic pyrolysis of polyethylene and polypropylene over USY zeolite and separation to gasoline and diesel- like fuels," J. Anal. Appl. Pyrolysis, 2017, https://doi.org/10.1016/j.jaap.2017.09.005

[4] Bockhorn, H., Hornung, A., & Hornung, U. (1999). Mechanisms and kinetics of thermal decomposition of plastics from isothermal and dynamic measurements. Journal of Analytical and Applied Pyrolysis, 50(2), 77-101.

[5] Martens, J. A., & Jacobs, P. A. (1990). Conceptual background for the conversion of hydrocarbons on heterogeneous acid catalysts (p. 52). Van Nostrand Reinhold: New York.

[6] http://www.caslab.com/Petroleum-Hydrocarbon-Ranges/

IOP Conf. Series: Earth and Environmental Science **105** (2017) 012123 doi:10.1088/1755-1315/105/1/012123

Utilization of super hydrophobic membrane contactor for NOx Absorption

S Kartohardjono[1], S Stephanie[1], Meylin[1], A O Dixon[1], Nelson Saksono[1]
[1]Process Intensification Laboratory, Department of Chemical Engineering, Universitas Indonesia, Kampus UI, Depok 16424, Indonesia

*E-mail: sutrasno@che.ui.ac.id

Abstract. NOx is produced from the reaction between nitrogen, oxygen and even hydrocarbons (during combustion), especially at high temperatures. Nitrogen oxide compounds are one of the harmful emissions that can result from fuel combustion. This research aims to find out the performance of super hydrophobic membrane contactors in the absorption process of N_2O from its mixture with air using acid solution and hydrogen peroxide. N_2O gas must be reduced from the exhaust gas especially to meet the regulations applicable to the environment due to the hazardous nature of the N_2O gases. The experimental results showed that the amount of NOx absorbed and the absorption efficiency increased with increasing absorbent flow rate in the membrane contactor. Meanwhile, the concentration of NOX in the outlet gas decreased with increasing absorbent flow rate.

1. Introduction

NOx is a term derived from oxide mono-nitrogen NO and NO_2 (nitric oxide and nitrogen dioxide) [1]. The amount of nitrogen oxide emitted into the atmosphere as air pollution can be significant in areas where vehicles are heavily congested, such as in big cities. NOx gas is formed whenever there is combustion in the presence of nitrogen - as in a vehicle engine, and can also be produced naturally by lightning [2]. The reduction of NOx in the exhaust gases such as those from boilers and the nitric acid industry is currently attracting much attention due to increasingly stringent environmental regulations. For this purpose, both dry and wet processing techniques are widely used and studied. In the case of the wet process, a major obstacle to absorption is the very low solubility of nitric oxide. To make the gas more soluble is by oxidizing NO to NO_2 in the gas phase with the help of ozone or chlorine dioxide. Oxidation may also be carried out in a washing liquid by adding an oxidizing agent such as potassium permanganate, sodium chlorite or sodium hypochlorite [3].

The gas absorption into the liquid can be carried out in the gas-liquid contactor on the hollow fiber membrane module. In recent years membrane contactors have attracted the attention of many researchers as an alternative technology for gas absorption process because it can overcome the weaknesses of conventional methods [4]. Hollow fiber membrane contactors have been widely used as gas-liquid contactors as they have large ratio between surface area for gas-liquid contact and volume of equipment [5]. In addition, gas absorption through membrane contactor also integrates separation and absorption processes to exploit the benefits of both processes [6]. The type of solvent and the membrane material is essential to maximize the efficiency of the process, wherein the membrane must be maintained so as not to be wetted by the solvent used, so that the membrane material used must be hydrophobic [7]. Based on research conducted by some previous researchers for gas absorption through membrane contactors, the hydrophobic property of the membrane is not enough. This membrane is still experiencing wetting at a certain period so that the mass transfer coefficient that occurs drastically due to membrane wetting [8, 9]. It is necessary that the membrane material is very resistant to not wetted by the absorbent solution for long periods, so that the advantage of gas absorption process through the membrane can be maximally exploited. The choice of membrane material is critical to the efficiency of the process [10]. To improve the hydrophobicity, The membrane can be incorporated with other material such as graphene nano sheets [11] or prepared as super-hydrophobic membrane [12]. This study aims to evaluate the performance super hydrophobic membrane contactor to absorb NOx, which will be represented by N_2O, using a mixture of nitric acid and hydrogen peroxide as absorbent.

2. Materials and Methods

The schematic diagram of N_2O absorption from its mixture with air using a membrane contactor is shown in Figure 1. The super hydrophobic hollow fiber membrane contactors used were supplied by PT GDP Filter Bandung. The feed gas (N_2O = 15000 ppm), and HNO_3 and H_2O were purchased from BOC Gases and Merck, respectively. The hollow fiber membrane contactors used sized of 6 cm and 34 cm in diameter and length, respectively, which consists of 2000 fibers. The fibers are polypropylene-based sized of 525 and 235 μm in outer and inner diameter, respectively. During the experiment, the absorbent (a mixture of HNO_3 and H_2O_2 1:1) was pumped through the lumen side of the membrane fibers, whilst the feed gas was sent to the shell side of the membrane contactor. The inlet and outlet gas compositions to and from membrane contactor were analyzed using gas chromatography Bruker Scion 436-GC. Meanwhile, the inlet and outlet gas flow rates to and from membrane contactor was measured using mass flow meter Sierra Top Trak Instruments. The amount of N_2O absorbed, N_2O_{abs}, by the absorbent and the absorption efficiency, $\%R$, from the experiment were calculated using equation (1) and equation (2), respectively.

$$N_2O_{abs} = (x_iF_i - x_oF_o)\frac{P}{RT} \quad (1)$$

$$\%R = \frac{N_2O_{Abs}}{x_iF_i}x100\% \quad (2)$$

where the x_i and F_i and x_o and F_o are N_2O concentration and flow rate in inlet and outlet membrane contactor, respectively. P, T and R, are pressure, temperature and gas constant, respectively. Meanwhile, Q_G and A_m are feed gas flow rate and membrane fibers surface area, respectively.

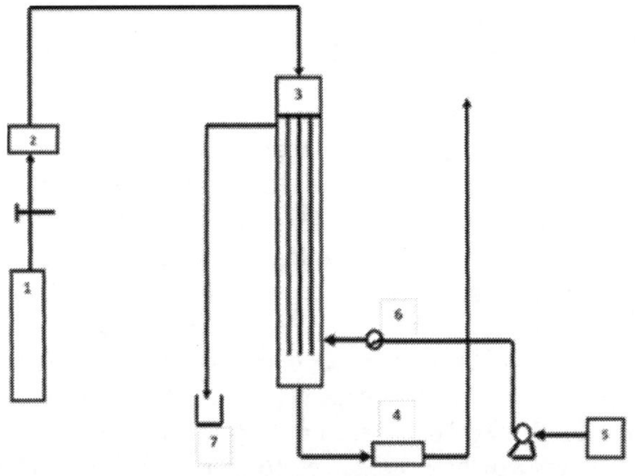

Figure 1. Schematic diagram of experimental set up:
1. Feed gas;
2. Mass flow meter;
3. Superhydrophobic hollow fiber membrane contactor;
4. Mass flow meter;
5. Absorbent reservoir;
6. Liquid flow meter

IOP Conf. Series: Earth and Environmental Science **105** (2017) 012123 doi:10.1088/1755-1315/105/1/012123

3. Results and Discussion

The transport mechanism of N_2O from the bulk gas in the feed side through the absorbent in the membrane contactor based on the experiment is following three steps:

- Firstly, the N_2O from the bulk gas is transferred to the fiber membrane surface in the gas phase
- The N_2O is then diffused to the other side of the fiber membrane surface in the liquid phase
- Finally, the N_2O is absorbed by the absorbent in the liquid phase through oxidation by H_2O_2 in the presence of HNO_3.

The amount of N_2O absorbed by the absorbent in the membrane contactor as a function of absorbent flow rate at various feed gas flow rates is shown in Figure 2. The N_2O absorbed increased with increasing absorbent flow rate due to more turbulence of the absorbent at higher flow rate. The turbulence in the liquid phase will reduce the resistance to the mass transfer in the absorbent side, which will lead to the increase in the amount of N_2O absorbed. Meanwhile, the increase in feed gas flow rate will increase the turbulence in the gas phase, that will reduce the mass transfer resistance in the gas phase, which also lead to the increase in the amount of N_2O absorbed.

The N_2O absorption efficiency increased with increasing the absorbent flow rate due to the increase in the amount of N_2O absorbed as shown in Figure 3. The absorption efficiency also increased with the increasing gas flow rate especially at the lower absorbent flow rate. The absorption efficiency at the higher absorbent flow rates (400 and 500 cc/min) are almost similar for feed gas flow rates of 100cc/min and 160cc/min, respectively. This is due to the N_2O absorption increment at absorbent flow rates of 400 cc/min and 500cc/min for feed gas flow rate of 160 cc/min was insignificant, so that the absorption efficiency almost constant. Meanwhile, the absorption efficiency for feed gas flow rate of 100cc/min increased in the range of absorbent flow rate observed in this study.

The N_2O concentration at outlet gas from the membrane contactor decreased with increasing the absorbent and the feed gas flow rates as shown in Figure 4. The increase in the absorbent flow rate will increase the amount of N_2O absorbed as shown in Figure 2, which lead to the decrease in the N_2O concentration at the outlet membrane contactor.

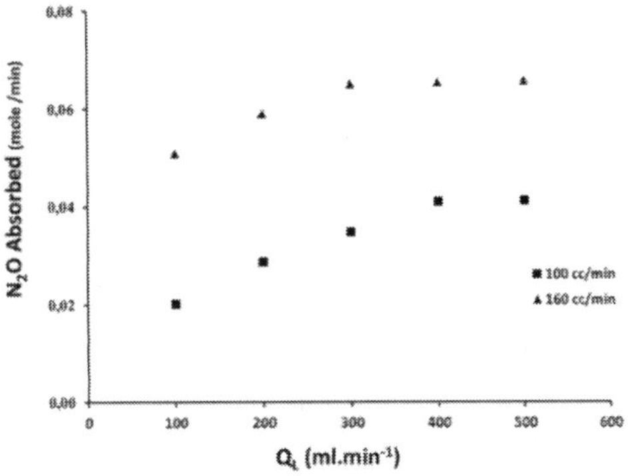

Figure 2. The amount of N_2O absorbed as a function of the absorbent flow rate, Q_L, at feed gas flow rates, Q_G, of 100 cc/min and 160 cc/min

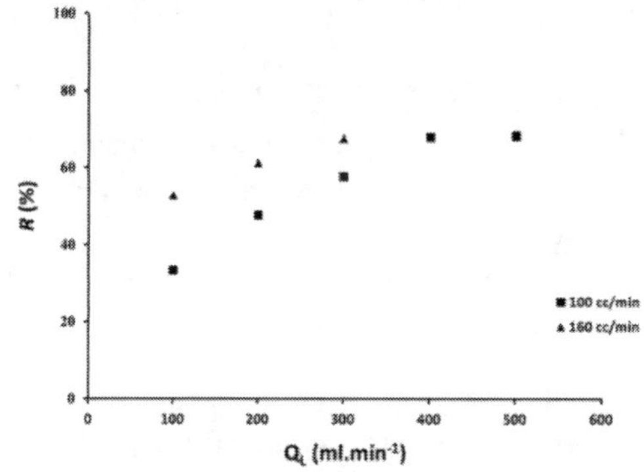

Figure 3. The N_2O absorption efficiency, %*R*, as a function of the absorbent flow rate, Q_L, at feed gas flow rates, Q_G, of 100 cc/min and 160 cc/min

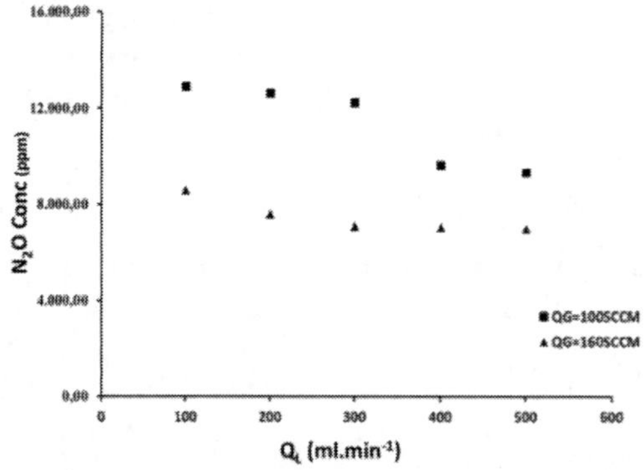

Figure 4. The N_2O concentration at the out let gas from the membrane contactor as a function of the absorbent flow rate, Q_L, at feed gas flow rates, Q_G, of 100 cc/min and 160 cc/min

4. Conclusion

Experiments have been conducted to see the effect of the absorbent and the feed gas flow rates on the N_2O absorbed, the N_2O absorption efficiency and the N_2O concentration at the outlet gas from the membrane contactor. The experimental results showed that the N_2O absorbed and the N_2O absorption efficiency increased with increasing the absorbent and the feed gas flow rates. Meanwhile, the N_2O concentration decreased with increasing the absorbent and the feed gas flow rates.

References

[1] H. Omidvarborna, A. Kumar, and D.-S. Kim, 2015 "NOx emissions from low-temperature combustion of biodiesel made of various feedstocks and blends," *Fuel Processing Technology,* **140** 113-18.

[2] K. Annamalai and I. K. Puri, 2007 *Combustion Science and Engineering* CRC Press.

[3] D. Thomas and J. Vanderschuren, 1996"The absorption-oxidation of NOx with hydrogen peroxide for the treatment of tail gases," *Chemical engineering science,* **51** 2649-54.

[4] J.-L. Li and B.-H. Chen, 2005 "Review of CO 2 absorption using chemical solvents in hollow fiber membrane contactors," *Separation and Purification Technology,* 41 109-22.

[5] F. Lipnizki and R. W. Field, 2001."Mass transfer performance for hollow fibre modules with shell-side axial feed flow: using an engineering approach to develop a framework," *Journal of Membrane Science,* **193** 195-208.

[6] Y. Lv, X. Yu, J. Jia, S.-T. Tu, J. Yan, and E. Dahlquist, 2012 "Fabrication and characterization of superhydrophobic polypropylene hollow fiber membranes for carbon dioxide absorption," *Applied Energy,* **90** 167-174.

[7] R. Faiz and M. Al-Marzouqi, 2010 "CO 2 removal from natural gas at high pressure using membrane contactors: model validation and membrane parametric studies," *Journal of Membrane Science,* **365** 232-241.

[8] W. Rongwong, R. Jiraratananon, and S. Atchariyawut, 2009 "Experimental study on membrane wetting in gas–liquid membrane contacting process for CO 2 absorption by single and mixed absorbents," *Separation and Purification Technology,* **69** 118-125.

[9] R. Wang, H. Zhang, P. Feron, and D. Liang, 2005 "Influence of membrane wetting on CO 2 capture in microporous hollow fiber membrane contactors," *Separation and Purification Technology,* **46** 33-40.

[10] N. Goyal, S. Suman, and S. Gupta, 2015 "Mathematical modeling of CO 2 separation from gaseous-mixture using a hollow-fiber membrane module: physical mechanism and influence of partial-wetting," *Journal of Membrane Science,* **474** 64-82.

[11] X. Wu, B. Zhao, L. Wang, Z. Zhang, H. Zhang, X. Zhao, *et al.*, 2016 "Hydrophobic PVDF/graphene hybrid membrane for CO 2 absorption in membrane contactor," *Journal of Membrane Science,* **520** 120-129.

[12] K.-J. Lu, J. Zuo, and T.-S. Chung, 2016 "Tri-bore PVDF hollow fibers with a super-hydrophobic coating for membrane distillation," *Journal of Membrane Science,* **514** 165-175.

Tapioca and tofu waste utilization to produce AA, DHA, and EPA

Ardita Rizky Putri Arcanggi[1], Prastiwi Arum[2], Ahmad Rafif[1], Adinda Eka[1] Kemala[1], Mustika Sari[1], Rita Arbianti[1], Tania Surya Utami[1], Heri Hermansyah[1]
[1]Departemen Teknik Kimia, Fakultas Teknik, Universitas Indonesia
[2]Departemen Farmasi, Fakultas Farmasi, Universitas Indonesia

Email : ardita.rizky@ui.ac.id, nana@che.ui.ac.id

Abstract. Human body has limited ability to synthesize unsaturated fatty acids such as AA, DHA, EPA. These fatty acids are essential for the body. This study initiated the efforts to produce unsaturated fatty acids of microorganisms, with low cost and high percentage. This study used tapioca and tofu waste as a medium for *Aspergillus*. In addition, this study analyzed the varying of carbon concentration to the maximum lipid production from *Aspergillus oryzae* and determine the optimum rate of agitation against lipid production from *Aspergillus oryzae*. The results shows that the optimum composition of AA, DHA, EPA are 0.18% (w / w), 0.33% (w / w), and 2.96% (w/w) respectively in concentration carbon of 9% (w/w) and agitation rate of 120 RPM.

1. Introduction

Sustainable Development Goals (SDGs) is to reduce the increasing malnutrition. According to the Basic Health Research Data (Riskesdas), malnutrition is characterized by an increase in malnutrition prevalence rate which increased to 19.6 percent in 2013. Malnutrition in children leads to growth and development disruption and impacts on the quality of future generations. One of the nutrient deficiency factors is caused by the lack of fatty acid intake in children under five years who are not optimally fulfilled, whereas the composition of human brain cells about 60% consists of fatty acids [1].

Alternative sources of fatty acid producer is microorganism. Fatty acid producing microorganisms consist of yeast, yeast, bacteria and microalgae. One of the best microorganism producing fatty acids is yeast where yeast has advantages such as the time required to produce relatively short and fatty acids that can produce is high [2]. Safety yeast for food so far is *Aspergillus oryzae*. *Aspergillus oryzae* used in the manufacture of soy sauce, tauco, and miso, so it is rated as a safe food grade [3]. *Aspergillus oryzae* is an easy-to-cultivate oleaginous microorganism that can survive in conditions with high levels of sugar or salt, energy saving, easy handling and easy use of technology that makes the *Aspergillus oryzae* potentially and highly prospective to be developed as an alternative source of AA, DHA, EPA producers [4]. *Aspergillus oryzae* does not require light in its growth, making it more energy efficient [5]. In addition, *Aspergillus oryzae* forms a filament in its infancy so it does not require filtration with a special membrane or special technology in its harvesting [6].

Factors which are affecting the growth of *Aspergillus oryzae*, among others, are the concentration of carbon source and the rate of agitation. According to Lopez et al. (2013), carbon concentrations can naturally regulate lipid production through catabolic repression processes. The presence of variations in

IOP Conf. Series: Earth and Environmental Science **105** (2017) 012124 doi:10.1088/1755-1315/105/1/012124

carbon concentrations determines the flexibility and ability of *Aspergillus oryzae* to consume food [7]. During this time, there has been no research in the variation of carbon concentration on the growth of *Aspergilus oryzae*. Therefore, further research is needed related to the relation of carbon concentration given to *Aspergillus oryzae* growth so it can be concluded whether the higher variation of carbon concentration determines the amount of lipid produced.

The rate of agitation plays an important role in the homogenization of nutrients in the fermentation medium. In general, the optimum agitation rate determines the transfusion of hyphae between the yeast affecting *Aspergillus oryzae* growth [8]. Low agitation rates result in slower growth and lower lipid rendement. Overly high agitation rates cause shear stress in the *Aspergillus oryzae* [9]. This is reinforced by Fadaly et al. (2009) that each type of *Aspergillus oryzae* has a large optimum agitation rate that varies depending on the type and magnitude of the reactor. Therefore, an optimal condition of carbon concentration and an appropriate agitation rate are required to produce an optimum unsaturated fatty acid.

In this study, the type of culture method used is submerged fermentation method using tapioca and tofu waste as a medium of fermentation. It aims to facilitate the manipulation of culture conditions so that more lipid rendement can be obtained. The researchers used industrial waste, i.e. tapioca starch and tofu waste as a fermentation medium for *Aspergillus oryzae*. In this research, two extraction methods were used to optimize the fatty acid yield, namely sonification extraction using ethanol and then followed by extraction using n-hexane. The ethanol and n-hexane extraction methods have several advantages such as safe and easy to apply in the food industry [10]. Lipids analyzed using GC (Gas Cromotography) to determine the composition of AA, DHA, EPA.

2. Experimental Methods

2.1. Tools and Materials

This study used gas chromarography, sentrifugator, spectrophotometer, autoclave, water bath shaker, laminar flow, and oven. The ingredients used for cultivation medium are tapioca waste, tofu waste, KH_2PO_4, Na_2HPO_4, $CaCl_2.2H_2O$, $FeSO_4.7\ H_2O$, and $CuSO_4.5\ H_2O$. The composition of carbon content using concentrated H_2SO_4 98%, $K_2Cr_2O_7$ 2N, and Glucose. The solvent used for the extraction is ethanol and n-hexane.

2.2. Research Variable

The variables that will be varied in this research are carbon concentration and agitation rate in *A. oryzae* fermentation medium. The carbon concentrations used in this study were 6% (w/w), 7% (w/w), 8% (w/w), 9% (w/w), and 10% (w/w), and the agitation rate used in this study were 70 rpm, 120 rpm, 170 rpm, 220 rpm. The variables that will be measured in this research are percentage of AA, DHA, and EPA content. The fixed variables in this study were temperatures in the range 25-30 °C, pH in the range of 4.5-6.5, and the incubation time to enter the stationary phase.

2.3. Preparation

Preparations include the manufacture of tapioca waste and tofu waste, testing of carbon content (Walkey and Black) and nitrogen (Kjeldahl) in tapioca waste and tofu waste, and sterilization of the tool.

2.4. Preparation of Stock Culture

Proliferation of yeast is done by culturing on PDA (Potato Dextrose Agar). Besides aiming to multiply the cells, culture on PDA medium also aims to increase the *Aspergillus oryzae* spores that will be used in liquid culture. Cultures on PDA medium were incubated for 3-4 days at 25-30 °C.

2.5. Liquid Medium Culture

The culture of the liquid medium was carried out 3 times, the first stage of the culture of liquid medium was carried out using variations of carbon concentration at concentrations of 6% (w/w),

7% (w/w), 8% (w/w), 9% (w/w), and 10% (w/w). The second stage of liquid medium culture is carried out using variation of agitation rate of 70 rpm, 120 rpm, 170 rpm, 220 rpm with carbon concentration is the optimal carbon concentration obtained in the previous stage. Liquid medium culture was carried out on a batch reactor using a medium of tapioca and tofu waste as a source of carbon and nitrogen, and also micronutrients added to the medium. Liquid culture is carried out at 25-30 °C, pH 4,5-6,5, with 6 days incubation time in water shaker bath.

2.6. Harvesting
Harvesting starts from filtration stage using filter paper, then weighing wet biomass. The biomass was then dried oven at 65 °C for 24 hours until a constant weight.

2.7. Extraction
The extraction process uses two solvents, ethanol and n-hexane. Fractionation method used is liquid-liquid partition. Partition separation techniques are based on differences in the degree of polarity of the solvent used. The separation technique involves two non-mixed solvents in a separating funnel, and then separates according to the partition coefficient [10]. Fractionation was carried out using a polar solvent using extraction with ethanol and continued for a non-polar solution using n-hexane.

According Ambarsari (2013), ethanol is a semipolar solvent that can dissolve polar and non-polar solutes. Extraction using ethanol was introduced as a simple extraction method because of its relatively easy processing and is often used in the process of sifting for microorganisms because it is food grade [11]. The definition of food grade is safe for use in producing foodstuffs. In this extraction process is also done simultaneously with the sonication process to optimize lipid production. Sonication is done for the process of breaking the cell wall. When the sonication process takes place, cavitation occurs due to the treatment of ultrasonic waves [12].

Furthermore, the nonpolar part was again extracted using n-hexane extract. Selection of n-hexane as solvent, since n-hexane is stable and volatile, selective in dissolving substances, extracting small amounts. Then the mixture was centrifuged to separate between the biomass and the ethanol-lipid mixture. Furthermore, the lipid and solvent mixtures are evaporated to separate the solvent from the lipids and regenerate the solvent. Lipid obtained in the form of dry lipid or concentrate which will then be tested. Then measure lipid mass to calculating the lipid yield.

2.8. Composition Analyisis of AA, DHA, EPA
The composition analysis of AA, DHA, and EPA conducted by using GC in the Agro Industry Center (BBIA), Bogor. The test method using GC is done by GC-IUPAC 2.301 7th Edition method.

3. Result and Discussion

3.1. Medium Waste Quality Analysis
The carbon analysis was performed using a UV VIS spectrometer with a wavelength of 651 nm. The following table is an analysis results of carbon and nitrogen content in the effluents carried out.

Tabel 1. Analysis of carbon content and waste nitrogen

Component	Tapioca Waste (% w/w)	Tofu Waste (% w/w)
Carbon	34.43	29.04
Nitrogen	0.12	5.2

The analysis of carbon and nitrogen content in tapioca waste and tofu waste shows that tapioca waste has a function as the highest carbon source compared to tofu wastes. The tofu waste basically has a fairly high carbon content, but the carbon in the tofu waste comes in part from the crude fiber and starch that are more difficult to degrade by microorganisms [13]. The tofu waste has the highest nitrogen content compared to the tapioca waste. Therefore, the tofu waste is known to act as a source of nitrogen.

3.2. The Growth of A. oryzae

The growth and proliferation of *Aspergillus oryzae* is characterized by increasing weight and number of cells [14]. The addition of dry weight of the cell during the fermentation process shows the growth of *Aspergillus oryzae* so that it can be made a *Aspergillus oryzae* growth profile. The addition of the dry weight of the cell is plotted on the Y axis, while the incubation time is plotted on the X axis (Fig. 1). The calculation of the dry weight of the cell is carried out to determine the number of cells present in the medium over a given period.

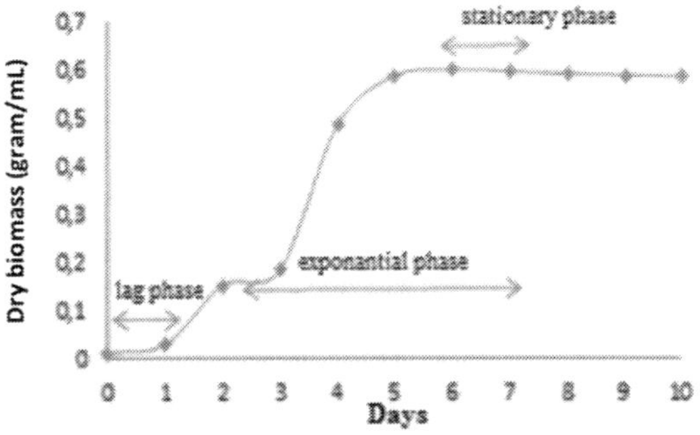

Figure 1. Growth curve of *Aspergillus oryzae*

Observations were made every 24 hour period. It aims to identify the growth curve of *Aspergilus oryzae*. According to Subrahamanyam et al., (2001) states that *Aspergillus oryzae* growth is not like a bacterium whose cell division is relatively rapid within a few hours. Early *Aspergillus oryzae* growth is transparent white, thin cotton-shaped, and floating - floating. This form is defined as a single spore or conidia that has grown into mycelium [14]. The mycelium continues to grow until it enters the stationary phase.

Figure 1 shows that *Aspergillus oryzae* passes through lag phase, log, and stationer. From the growth curve it can be seen that the first day there is a lag phase (or adaptation phase). Yeast began to grow hovering like fine white cotton. In this phase, the increase in the dry weight of the cells has not been significant because of the *Aspergillus oryzae* in the adaptation process so that the nutrients present in the medium have not been used for the growth process [14]. The exponential phase occurs on day one to day 5 is marked by a significant increase in cell dry weight.

The stationary phase begins on the 6th day. On the 6th and 7th day there is no significant biomass increase. The largest dry weight was obtained on day 6 so it can be proven that day 6 is the best time to harvest biomass. The same is also stated by research conducted by Sukma et al. (2010), rice bran oil fermented with *A. terreus* for 6 days resulted in an abundance of unsaturated fatty acids higher than the variation of incubation time for 5 days and 7 days.

On day 7 to day 10, *Aspergillus oryzae* begins to decrease dry weight. According Firdiana (2007), this is due to reduced nutrients and oxygen so that biochemical activity decreases. This causes many mycelia to tend to coagulate with each other, so at the time of sampling, many are not carried away.

According to Abu-Elreesh (2013), *Aspergillus oryzae* is harvested in the stationary phase yielding optimum fats. Abu-Elreesh (2013) using the DGB1 strain proves there is a relationship of lipid accumulation with *Aspergillus oryzae* growth. In the stationary phase there is maximum lipid accumulation and further decreases as the nutrients in the medium decrease, the accumulated lipids are used by the microorganisms for the metabolism [15].

According to Ratledge and Wynn (2003), lipid accumulation can occur when the nitrogen contained in the substrate has been used up for *Aspergillus oryzae* growth, then excess carbon is converted to triacylglyceride. After its optimum conditions, a decrease in lipid accumulation caused by the use of lipids by microorganisms to produce biomass. Therefore, the harvesting process is carried out at the stationary phase on the 6th day of the fermentation process.

3.3. Effect of Carbon Concentration on Production of AA, DHA, EPA

The process of catabolic repression to regulate the production of lipid microorganisms is influenced by the concentration of carbon sources [16]. To optimize the process of catabolic repression is still needed sources of nutrients such as mineral elements. *Aspergillus oryzae* growth depends on the concentration of carbon present in the medium. This is reinforced by Prihastuti and Yuliatun (2002) that higher carbon content provides a longer logarithmic phase than lower carbon content. The enzyme activity produced by oleaginous microorganisms is greater when the carbon and nitrogen sources in the substrate are sufficient to support the growth of stem cells and their metabolism, and can increase the ability of *Aspergillus oryzae* to produce lipids [17].

The yield of lipid with variation of carbon concentration in this study is presented in the following graph.

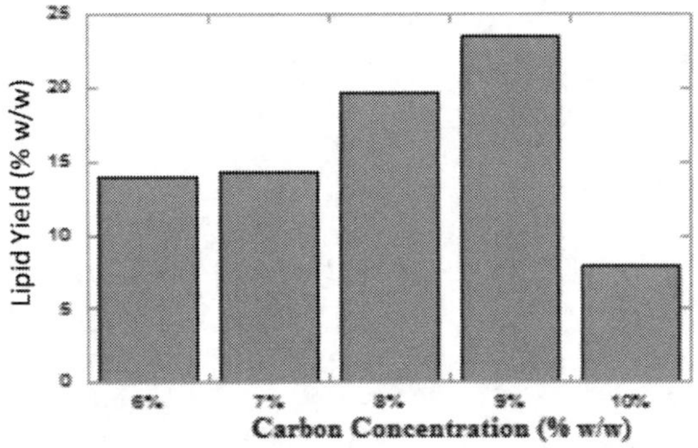

Figure 2. The yield of lipid biomass on the variation of carbon concentration

Based on Figure 2, the yield of lipid from variation of carbon concentration of 6% (w/w) was 14.01% (w/w). The lipid yield is relatively increased with an increase in carbon concentration of 7% (w/w), 8% (w/w), 9% (w w). The lipid yields were 14.29% (w/w), 19.6%, respectively, where the greatest lipid yield was obtained at a concentration of 9% (w w) of 23.59% (w/w). This increase occurs because *Aspergillus oryzae* are able to synthesize lipids by breaking carbon chains and releasing hydrogen atoms to produce oxidized compounds (pyruvic acid) (Sumanti et al., 2005). At 9% concentration (w/w), *Aspergillus oryzae* is able to optimize growth for the process of glycolysis where the resulting pyruvic acid enters the Krebs cycle in the mitochondria. This triggers the optimum lipid synthesis [18].

The lowest lipid yield was obtained at a concentration of 10% (w/w) of 7.96%. This is because the concentration of 10% is too concentrated so that the osmotic potential increase at substrate concentration becomes too high [18]. The high osmotic potential causes *Aspergillus oryzae* growth to be less optimum. According to Fadaly et al. (2009), less optimum *Aspergillus oryzae* growth is characterized by low biomass which has an effect on cell conversion to synthesize lipid to low. This is reinforced by Sumanti et al. (2005), the low yield lipid obtained because most of the carbon consumed is used for cell poliferation, whereas the carbon used in the lipid accumulation process becomes low.

Carbon concentrations of 6% (w/w) and 7% (w/w) had a significantly lower lipid yield of 14.01% and 14.29%, respectively. At a concentration of 8% (w/w) yielded a 19.6% lipid yield. Different carbon concentrations result in differences in cell growth in *Aspergillus oryzae*. As carbon concentration increases, cell growth and metabolism increase with increasing enzyme activity, thus affecting the ability of yeast to produce lipids [19]. In addition, the power of contact between nutrients with *Aspergillus oryzae* becomes intense thus increasing the resulting lipid.

The lipid yields obtained in this study had greater results than Putri, et al. (2016). The highest lipid yields produced by Putri, et al. (2016) is 8.2% (w/w). While in this research yield highest yield of lipid equal to 23.59% (w/w). This is due to the extraction factor. In a study conducted by Putri et al. (2016), the extraction carried out using only extraction using ethanol. In this study using fractionation extraction of ethanol and n-hexane which can optimize the amount of yield of lipid obtained.

The use of food grade solvents aims to produce safe lipids for consumption because the resulting lipids are designated as man materials [20]. The choice of n-hexane as a solvent because n-hexane is stable and volatile, selective in dissolving substances, extracting small amounts of solution and able to extract non-polar substances in sufficient quantities [21]. Based on the consideration of dry weight and yield of lipids produced for cultivation stage with variation of agitation used carbon concentration of 9% (w/w).

Lipid samples obtained from the extraction process. Furthermore, it was analyzed using GC-IUPAC 2.301 7th Edition method. The chromatogram of the lipid sample analysis results detected the fatty acid composition present in the lipid sample.

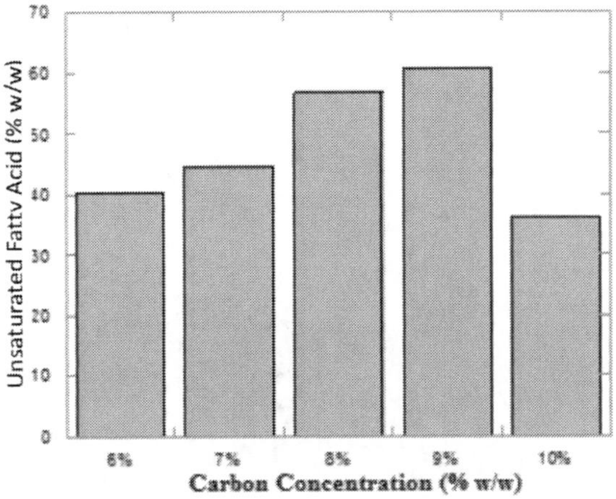

Figure 3. Percentage of unsaturated fatty acid of *A. Oryzae* with variation of carbon concentration

The largest unsaturated fatty acid composition was obtained at a concentration of 9% (w/w) in which the dry weight and high lipid yield affected the resulting fatty acid composition. In previous studies, the highest unsaturated fatty acids were obtained at a concentration of 6% [22]. This is due to an increase in concentration triggering nutrient transfer between medium and *Aspergillus oryzae* [23].

The maximum unsaturated fatty acid yields obtained in *A.oryzae* with RSM (RSM hydrolysate, molasses, and salt) mediums performed by Yangmin et al. (2015) is 8.55 mg/L or equivalent to 0.855 mg/100 ml of medium. While in Muniraj et al. (2013) yields a 3.5 gram/L lipid yield of 57.9% unsaturated fatty acid (w/w) or an unsaturated fatty acid of 202.65 mg/100 mL of medium. The study obtained 61.2% (w/w) unsaturated fatty acids under lipid conditions 122.44 mg/100mL medium. The unsaturated fatty acid yields produced by Muniraj et al. (2013) higher. This is because the medium used in the form of liquid waste potato containing starch is 31.2% (w/w).

The results of this study are higher than Yangmin et al. (2015) other than because the carbon source is also caused by the gradual extraction method using ethanol and n-hexane. The extraction method used can optimize the yield of lipid yield and the conversion of unsaturated fatty acid produced to be higher. The difference in the results is also due to differences in the medium used, Yangmin et al., (2015) using RSM medium whose main composition is cellulose, hemicellulose, and lignin with a tendency more difficult to be degraded by microorganisms.

AA, DHA, EPA are read only at a concentration of 9% (w/w) with a percentage of 0.18% (w/w), 0.33% (w/w), 2.96% (w/w) respectively. In a study conducted by Muslimah, et al. (2015), the composition of DHA is higher by 2.54% (w/w). Meanwhile, this study has a higher EPA than Muslimah, et al. (2015). This is due to the difference in concentration used by Muslimah by 2% (w/w).

Based on the results of this study, the concentration of carbon that can produce lipids by producing AA, DHA, and EPA is the optimal concentration of 9% (w/w). At this concentration, the yield of dry lipids and biomass has the highest percentage. Unsaturated fatty acids, other than PUFA and MUFA, the presence of AA, DHA, and EPA are considered in the selection of optimum concentrations. Therefore, Submerged Fermentation of the *Aspergillus oryzae* by using tapioca and tofu waste has the potential to be the source of production of AA, DHA, and EPA.

3.4. *Effect of Agitation Rate on Production of AA, DHA, EPA*

The agitation rate is able to influence the contact power between medium and nutrient [24]. In the medium of tapioca waste and tofu waste know the optimum agitation rate is required to produce high lipid yield on biomass. The optimum agitation rate can facilitate carbon storage in the form of a lipid reserve. According to Firdiana (2007), the agitation rate plays an important role in the homogenization of nutrients in the fermentation medium and the increased surface area of oxygen displacement in the fermentation medium. Agitation rates are too low resulting in slower growth and lower lipid yield. However, too high agitation rates cause shear stress in the *Aspergillus oryzae* [25].

The results obtained were linear with the yield of lipid yields obtained:

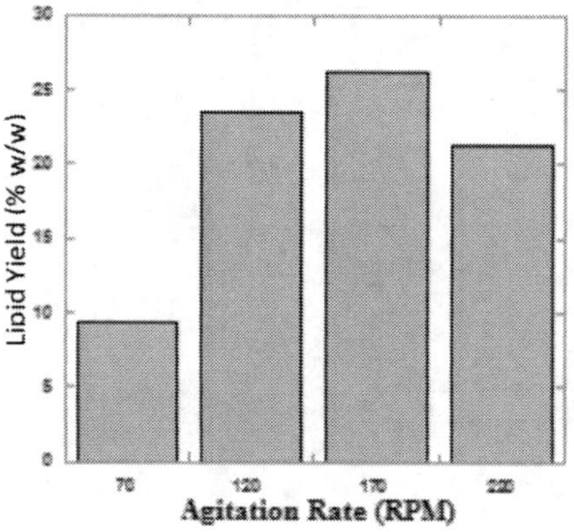

Figure 4. The percentage of lipid yields in variations in agitation rate

In Figure 4, agitation rate of 120 RPM and 170 RPM, the yield of lipid are 23.59% (w/w) and 26.1% (w/w), respectively. The resulting lipid yield has a correlation to dry weight. The amount of dry weight produced determines the amount of lipid obtained. The lipid yields obtained are quite high compared to other studies obtained by Hui et al. (2010) at 180 rpm agitation yields an 18% lipid yield (w / w). In the test of AA, DHA, EPA, GC-IUPAC 2,301 7th Edition was performed on the variation of agitation rate obtained in Figure 5.

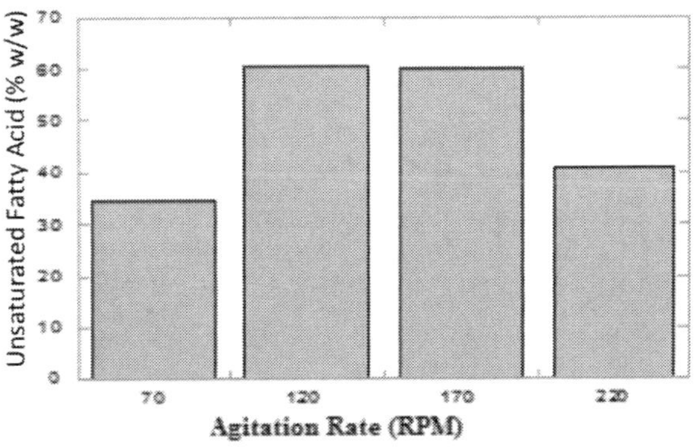

Figure 5. The percentage of unsaturated fatty acids of lipids *A. oryzae* with agitation rate variation

The results obtained at agitation rates of 120 RPM and 170 RPM have differences in unsaturated fatty acids that are not significantly different. The unsaturated fatty acids each obtained were 61.2% (w/w), while for the agitation rate of 170 RPM was 59.9% (w/w). These results indicate that the fungus is able to perform metabolism well with the agitation of 120 RPM and 170 RPM resulting in high unsaturated fatty acids. According [26], *Aspergillus oryzae* is able to transfer the hyphae well in the range of agitation of 120 RPM. Sari et al. (2015) obtained an unsaturated fatty acid of 59.42% (w/w). The percentage of unsaturated fatty acids obtained in this study was higher than that of Sari et al. (2015). This is because the medium of tapioca waste and tofu waste used in this study still contains nutrients that can help *Aspergillus oryzae* growth.

The agitation rate of 220 RPM has an unsaturated fatty acid content composition of 40.73% (w/w). According Purwanto et al. (2009), this is due to high contact between the mycelia causing *Aspergillus oryzae* can not do optimum growth. According Purwanto et al. (2009), at too high agitation rates the production of acetyl-KOA becomes unoptimum and results in low fatty acids being produced. The agitation rate of 70 rpm yields an unsaturated fatty acid of 34.84% (w/w). The low agitation rate makes the *Aspergillus oryzae* unable to perform the optimal metabolism because the nutrients in the medium are not dissolved properly.

AA, DHA, and EPA can be generated on agitation of 120 RPM and 170 RPM. At the agitation rate of 120 RPM AA, DHA, the resulting EPA was 0.18% (w/w), 0.33% (w/w), 2.96% (w/w). At the agitation rate of 170 RPM AA, DHA, and the resulting EPA was lower by 0.07% (w/w), 0.25% (w/w), and 1.51% (w/w). Meanwhile, the agitation rate of 220 RPM was detected only DHA and EPA of 0.1% (w/w) and 0.03% (w/w). In the study of Muslimah et al. (2015), AA, the resulting DHA is greater. Meanwhile, in this study, EPA contained higher.

Based on the results of this study, the agitation rate values with the highest unsaturated fatty acids as well as the composition of AA, DHA, and EPA optimum is 120 RPM. In this condition, *Aspergillus oryzaes* undergo optimal hyphae transfer and can carry out the homogenization of nutrients in the maximum fermentation medium. This is due to the large yield of lipid yield and biomass yield, as well as the magnitude of the composition of AA, DHA, and EPA.

4. Conclusion

Carbon concentration and agitation rate with optimum biomass yield and lipid yield, i.e. carbon concentration of 9% (w/w) and 120 RPM of 23.59% (w/w). The content of the largest unsaturated fatty acids was obtained in the variation of 9% carbon concentration and the agitation of 120 RPM of 61.2% (w/w) with the unsaturated fatty acid composition comprising 35.68% omega-9 fatty acid, 22.58% Omega-6 fats, and 4.02% omega-3 fatty acids.

5. References

[1] Abu-Elreesh, Gadallah dan Desouky Abd-El-Haleem. (2013). "An Effective Lipid-Producing *Fungal sp.* Strain DGB1 and its use for Biodiesel Production," *Academic Journals,* **12** 34.

[2] Akpinar-Bayizit, Arzu, Tulay Ozcan, Lutfiye Yilmaz-Ersan, Fikri Basoglu. 2014. "Single cell oil (SCO) production by *Fusarium species* using cheese whey as a substrate," *Mljekarstvo*, **64**(2) 111-118.

[3] Almatsier, S.2010.Prinsip Dasar Ilmu Gizi.Jakarta : Gramedia Pustaka Utama.

[4] Ambarsari, Mita. 2013 .Aktivitas Antibakteri Fraksi N-Heksan Ekstrak Etanol Daging Buah Sirsak (Annona Muricata L.) terhadap Pseudomonas Aeruginosa, Staphylococcus Aureus, Shigella Sonnei Dan Serta Bioautografinya. *Khazanah*, **6**(2), 1-11.

[5] Darah Ibrahim, Haritharan Weloosamy, and Sheh-Hong Lim. 2015. Effect of agitation speed on the morphology of *Aspergillus niger* HFD5A-1 hyphae and its pectinase production in submerged fermentation. **6**(3) 265–271.

[6] Fadaly, Husain., Naggar, Noura., Marwan. (2009). Single Cell Oil Production by an Oleaginous Yeast Strain in a low cost cultivation medium. Research Journal of Microbiology. ISSN 1816-4935

[7] Fardiaz, S. 1992. *Mikrobiologi Pangan I* (PAU Pangan dan Gizi IPB Bogor: P.T. Gramedia Pustaka Utama)

[8] Firdiana, Arie. 2007. "Pengaruh Variasi Kecepatan Agitasi Terhadap Produksi Epiglukan Oleh Epicoccum Nigrum Pada Media Molases," Universitas Jember.

[9] Frazier, W. C. dan D. C. Westhoff. 1998. Food Microbiology 4th ed. Mc GrawHill Book Co, Singapore.

[10] Hall, M.J. Ratledge. 1977. "Lipid accumulation in an oleaginous yeast (*Candida* 107) growing on glucose under various conditions in a one and two-stage continuous culture," *App. Environment. Microb*, **33** 577-584.

[11] Hendayana, Sumar. 2006 . *KIMIA PEMISAHAN Metode Kromatografi dan Elektroforensis Modern*. Bandung : PT. Remaja Rosdakarya.

[12] Indrati, R. 2008. "Pendidikan Konsumsi Pangan: Aspek Pengolahan dan Keamanan". Jakarta: Kencana PrenadaMedia Group.

[13] Ketaren, S. 1986. *Pengantar Teknologi Minyak dan Lemak Pangan*. Jakarta: Penerbit Universitas Indonesia.

[14] Kurtzman, C.P., M.J. Smiley, C.J. Robnett, and D.T. Wicklow. (1986). " DNA relatedness among wild and domesticated species of the *Aspergillus flavus* group," *Mycologia*, **78** 955-959.

[15] Kyle, D. J. and Ratledge, C. 1992. Industrial Application of Single Cell Oils.

[16] Lin, Hui, Qun Wang, Qi Shen, Junwei Ma, Jianrong Fu, dan Yuhua Zhao. (2004). "Engineering *Aspergillus oryzae* A-4 through the Chromosomal Insertion of Foreign Cellulase Expression Cassette to Improve Conversion of Cellulosic Biomassinto Lipids," *PLoS ONE Journal,* vol. 9(9).

[17] Logaraj, T.V. 2010. Studies on selected plants and microbes with special reference to polyunsaturated fatty acid. Thesis. Food Microbiology Department, University of Mysore

[18] Lopez, J. L., J. A. S. Perez, J. M. F. Sevilla, F. G. A. Fernandez, E. M. Grima. Y. Christi. (2003). Production of lovastatin by Aspergillus terreus : effect of the C:N ration. Enzyme and Microbial Technology **33** 270-277.

[19] Mamatha. 2009. Polyunsaturated fatty acids (PUFAs) of Mucor sp. with special reference to Gamma Linolenic Acid (GLA). Thesis. Food Microbiology Department, University of Mysore.

[20] Manpreet,S, Sawraj,S., Sachin,D., Pankaj,S.and Banerjee, U.C. 2005. "Influence of Process Parameters on the Production of Metabolites in Solid-State Fermentation," *Jounal of Microbiology*, **1**(2) 1-9.

[21] Meeuwse, Petra, Payman Akbari, Johannes Tramper, Arjen Rinzema. 2012. "Modeling growth, lipid accumulation and lipid turnover in submerged batch cultures of *Umbelopsis isabellina*," *Bioprocess Biosyst Eng*, **35** 591-603.

[22] Muniraj, Iniya Kumar, Liwen Xiao, Zhenhu Hu, Xinmin Zhan, dan Jianghong Shi. (2003). "Microbial lipid production from potato processing wastewater using oleaginous filamentous Kapang *Aspergillus oryzae*," *Elsevier Ltd*, 1-7.

[23] Muslimah, Sari, Arbianti, Utami. 2015. *Pengaruh Rasio Karbon/Nitrogen Berbasis Onggok da Ampas Tahu untuk Produksi AA, DHA, EPA dari Aspergillus oryzae.* Depok: Universitas Indonesia.

[24] Nakkeeran, E. 2009. Process Assessment for the Purification of Aspergillus carbonarius Polygalacturonase Produced by Submerged and Solid-State Fermentations. *International Journal Of Engineering And Technology*, **1**(4) 278-281.

[25] Nawangsari, R.T. 1996. "Penggunaan Berbagai Sumber Karbon dan Produksi Minyak Sel Tunggal Oleh Kapang *Mucor inaequisporus* M05II/4." Skripsi, , Yogyakarta: Fakultas Teknologi Hasil Pertanian UGM.

[26] Purwanto, Ibrahim, Sudrata, Ahmad. 2009. "Effect of Agitation Speed on Morphological Changes inAspergillus niger Hyphae During Production of Tannase". ISSN 1817-3128

[27] Putri, Septi, Arbianti, Utami. 2016. Optimasi PUFA dari *Aspergillus oryzae* berbasiskan Onggok dan Ampas tahu. Universitas Indonesia

Tariff calculating model for natural gas transportation through open access transmission pipeline with multi tariff system to increase gas usage as clean energy

A A Bawono[1], E Kusrini[1]

[1]Department of Chemical Engineering, Faculty of Engineering Universitas Indonesia, Kampus Baru UI, Depok 16424; Corresponding Author: Tel.:62-85647166568, 62-21-7863516 Ext. 204; fax: 62-21-7863515;

E-mail: ajiagraningbawono@yahoo.com

Abstract. The purpose of this study is to model the different tariffs on the transportation of natural gas through a pipeline for several shipper classes in an open access natural gas pipeline transmission line called a multi-tariff system. The determination of the natural gas tariff through the Open Access pipeline in Indonesia is mostly set using a distance system with the same value for all types of the shipper. The principle of justice is considered to be achieved from uniform tariffs for all the shipper, whereas in practice the separation of tariffs for each shipper group is currently essential. The proposed new tariff modeling is done by modifying the initial equations of the single-tariff model. Besides giving a new model proposal of tariff calculation for the multi-tariff system, in this research will also be modified free cash flow for new tariff calculation scheme and will be done the validity test on the formula. So that will get a multi-tariff calculation model which can become one of proposal material enter for Regulators to be able to set a fair tariff for the shipper without changing the economics of the transporter and can produce competitive gas prices to substitute clean energy sources for certain consumers such as electricity generation.

1. Introduction

Natural Gas is one of the clean energy sources that can be utilized for various needs, both for fuel power generation, industry, transportation and for household and commercial needs [1] [7]. The main problem in the regulation of the transportation of Natural Gas through the pipeline is the determination of tariff or cost of transportation. In determining the tariff, it must be accommodated to the interests of all parties, namely the transporter (the owner of the pipeline) and the shipper (the party using the pipeline) [9]. Determination of tariff for natural gas through the open access pipeline in Indonesia mostly determined using distance system with the same scale for all types of shipper on one segment of transmission pipeline, called as a single tariff [6], formulated by the following equation [4] [8]:

$$Tariff = \frac{depreciation + (O\,\&\,M) + (A\,\&\,G) + (Tax) + (Levy) + M\arg in}{Single\ Volume} \qquad (1)$$

The principle of fairness is deemed to be achieved from uniform tariffs for all such shipper, whereas in practice, in a pipe segment there can be a receiving point or a point of delivery of each different shipper [10]. The gas volume of each different shipper as well as the different gas utilization purposes (Industries, power plants, fertilizers, transportation, small and household customers) that require different policies for each tariff-related group that will be established. The purpose of this study is to model the different tariff calculations on the transportation of natural gas for several groups of the shipper in an open access pipeline transmission line called a multi-tariff system. So it can create a reasonable gas price without harming all parties and expected gas prices through a competitive pipeline compared to other more polluted fuel sources, which will encourage the substitution of gas usage as a clean energy source.

2. Method

The research procedure is done by the first step in the form of problem identification, then conducted literature study and data collection. To create a new modeling of tariff calculation for multi tariff system, the calculation of the formula of the general equation of single tariff calculation which has been determined by the previous regulator. The new equation is derived on the basis that the revenue derived from the single tariff system must be the same as the multi tariff system so that no party loses from the development of this scheme. With a multi tariff system, it is expected that each shipper pays a fair tariff according to the class determined based on the amount of natural gas volume that each shipper pours in a transmission pipe. To create a new modeling of tariff calculation for multi tariff system, the calculation of the formula of the general equation of single tariff calculation which has been determined by the previous regulator. The new equation is derived on the basis that the revenue derived from the single tariff system must be the same as the multi tariff system so that no party loses from the development of this scheme. With a multi tariff system, it is expected that each shipper pays a fair tariff according to the class determined based on the amount of natural gas volume that each shipper pours in a transmission pipe.

3. Result and Discussion

3.1. Proposed development of tariff calculation model for multi tariff system

By the general tariff equation, that tariff with the single system on a gas transmission pipe, formulated by the following equation [4] [9] [11]:

$$Tariff = \frac{Cost\ of\ Service}{Volume} \tag{2}$$

$$Tariff = \frac{(d)+(O\&M)+(A\&G)+(T)+(L)+(WACC \times NBA)}{V} \tag{3}$$

Where, *O&M* is operating and maintenance costs, *A&G* is administration and general fee, d is depreciation of CAPEX, *T* is Tax costs, *L* is Regulator levy, *V* is volume of natural gas flowing, *NBA* is Asset base value (CAPEX), and *WACC* is Weighted Average Cost of Capital = IRR.

Weighted Average Cost of Capital is formulated with the following equation [5] [8] [10]:

$$WACC = CoE\frac{E}{E+D} + CoD\frac{D}{E+D} \tag{4}$$

Where *CoE* is the Cost of Equity, *CoD* is the Cost of Debt, *E* is Equity, and *D* is a Debt. The cost of Equity (CoE) and Cost of Debt (CoD) can be calculated by the following equation [2] [3]:

$$CoE = Rf + \beta(ICRP + BPMEM) \tag{5}$$

$$CoD = i \times (1-T) \tag{6}$$

Where Rf is Risk-Free Rate, US Treasury Bond, Beta (β) is a measure of an investment portfolio fluctuation compared with the market (stock market), *ICRP* is Indonesia Country Risk Premium, *BPMEM* is based premium for the mature equity market, *I* is the interest rate, and *T* is Tax. From equation (3) we can derive the following equation:

IOP Conf. Series: Earth and Environmental Science **105** (2017) 012125 doi:10.1088/1755-1315/105/1/012125

$$tariff \times volume = (d) + (O\&M) + (A\&G) + (T) + (L) + (WACC \times NBA) \quad (7)$$

$$\text{Revenue} = tariff \times volume \quad (8)$$

$$\text{Revenue} = (d) + (O\&M) + (A\&G) + (T) + (L) + (WACC \times NBA) \quad (9)$$

In a multi-tariff system, the division of classes can be done based on the volume of natural gas entered by each group of shipper groups into the pipe. It can be modeled that each shipper class with each tariff group will enter the gas volume into the pipeline with the following equation:

Shipper Group 1 (large gas volume) with t_1 tariff, with volume = V_1 (10)
Shipper Group 2 (medium gas volume) with t_2 tariff, with volume = V_2 (11)
Shipper Group 3 (small gas volume) with t_3 tariff, with volume = V_3 (12)
Shipper Group n with t_n tariff, with volume = V_n (13)

Where n is the division of classes or group. Then the total existing gas volume flowing in the pipe can be formulated by the equation:

$$V_{total} = V_1 + V_2 + V_3 \ldots\ldots\ldots\ldots + V_{n-1} + V_n \quad (14)$$

If the single tariff is developed, it is divided into several classes according to the multi-tariff system with the limitation that the multi tariff system must maintain the existing condition. So that the transporter economy does not change, then it can be formulated that the sum of revenue from multi tariff must be equal to Income from the original tariff so that it can be described by equation model as follows:

$$\text{Revenue}_{multitarif} = \text{Revenue}_{existing} \quad (15)$$

$$(t_1 \times V_1) + (t_2 \times V_2) + (t_3 \times V_3) + \ldots\ldots\ldots\ldots + (t_{n-1} \times V_{n-1}) + (t_n \times V_n) = t_{existing} \times V_{total} \quad (16)$$

So by substituting equation (7) into equation (16), assuming the tariff is divided into three groups, it will be obtained the development of equation for the multi-tariff system to be:

$$(t_1 \times V_1) + (t_2 \times V_2) + (t_3 \times V_3) = (d) + (O\&M) + (A\&G) + (T) + (L) + (WACC \times NBA) \quad (17)$$

So by substituting equation (7) into equation (16), assuming the tariff is divided into three groups, it will be obtained the development of equation for the tariff calculation with multi-tariff system to be:

$t_1 = tariff\ of\ group$ 1 (shipper with large volume) (18)

$t_2 = at_1,\ tariff\ of\ group$ 2 (shipper with medium volume) (19)

$t_3 = bt_1,\ tariff\ of\ group$ 3 (shipper with small volume) (20)

$t_n = xt_1, tariff\ of\ group\ n$ (21)

Where,

$$a = \frac{V_2}{V_1},\ b = \frac{V_3}{V_1},,\ and\ x = \frac{V_n}{V_1} \quad (22)$$

Substituting equations (18), (19), and (20) into equation (17), we obtain the following equation:

$$(t_1 \times V_1) + (at_1 \times V_2) + (bt_1 \times V_3) = (d) + (O\&M) + (A\&G) + (T) + (L) + (WACC \times NBA) \quad (23)$$

Substituting equation (22) into equation (23), we obtain the following equation:

$$(t_1 \times V_1) + \left(\frac{V_2}{V_1} t_1 \times V_2\right) + \left(\frac{V_3}{V_1} t_1 \times V_3\right) = (d) + (O\&M) + (A\&G) + (T) + (L) + (WACC \times NBA) \quad (24)$$

$$(t_1 \times V_1) + \left(t_1 \times \frac{V_2^2}{V_1}\right) + \left(t_1 \times \frac{V_3^2}{V_1}\right) = (d) + (O\&M) + (A\&G) + (T) + (L) + (WACC \times NBA) \quad (25)$$

$$\left(V_1+\frac{V_2^2}{V_1}+\frac{V_3^2}{V_1}\right)\cdot(t_1)=(d)+(O\&M)+(A\&G)+(T)+(L)+(WACC\times NBA) \quad (26)$$

$$\cdot t_1=\left(\frac{(d)+(O\&M)+(A\&G)+(T)+(L)+(WACC\times NBA)}{V_1^2+V_2^2+V_3^2}\right)\cdot V_1 \quad (27)$$

The equation model for group-n is as follows:

$$t_n = xt_1$$

$$t_n = x\cdot\left(\frac{(d)+(O\&M)+(A\&G)+(T)+(L)+(WACC\times NBA)}{V_1^2+V_2^2+V_3^2+\dots\dots\dots\dots+V_{n-1}^2+V_n^2}\right)\cdot V_1 \quad (28)$$

$$t_n = \frac{V_n}{V_1}\cdot\left(\frac{(d)+(O\&M)+(A\&G)+(T)+(L)+(WACC\times NBA)}{V_1^2+V_2^2+V_3^2+\dots\dots\dots\dots+V_{n-1}^2+V_n^2}\right)\cdot V_1 \quad (29)$$

$$t_n = \left(\frac{(d)+(O\&M)+(A\&G)+(T)+(L)+(WACC\times NBA)}{V_1^2+V_2^2+V_3^2+\dots\dots\dots\dots+V_{n-1}^2+V_n^2}\right)\cdot V_n \quad (30)$$

If the single tariff is calculated for a period, then equation (3) will be as follows:

$$tariff = \sum_{i=1}^{t}\frac{(d)_i+(O\&M)_i+(A\&G)_i+(T)_i+(L)_i+(WACC\times NBA)_i}{V_i} \quad (31)$$

In the same way, the new equations for multi-tariffs if calculated within a particular time (t) are as follows:

$$multi-tariff_{(n)} = \sum_{i=1}^{t}\left(\frac{(d)_i+(O\&M)_i+(A\&G)_i+(T)_i+(L_{total})_i+(WACC\times NBA)_i}{V_1^2+V_2^2+V_3^2+\dots\dots\dots\dots+V_{n-1}^2+V_n^2}\right)\cdot(V_n)_i \quad (32)$$

3.2. Validation of proposed calculation model with manual simulation result of cash flow multi-tariff calculation

To prove that the development of the calculation model equation for the above multi tariff is valid and can produce the same result with the simulation result with cash flow manually, then the validation test is done. Validation is done by tariff case study for gas transportation through pipeline X previously calculated by single tariff system then modeled with the multi-tariff system using two method that is the simulation of cash flow development manually and through calculation with new equation model that has been derived. The results obtained are as follows (see table 1 and table 2). From the calculation result of the case study of pipe segment X and the result of comparison obtained, it can be concluded that the new equation model for multi-tariff calculation is valid because the multi-tariff calculation using the simulation of cash flow development compared with through the new equation model that has been derived produce a similar result.

Table 1. Economic and technical data of pipeline X

NBA and O&M	33,524,000 USD and 8,180,360 USD/year with escalation
Depreciation Time and Inflation rate	10 Year and 1.31 %
Rf, BP MEM, ICRP, and β	2.14% , 5.70%, 3.24%, and 1.070
Coe and CoD	11.71% and 0%
D and E	0 USD and 33,524,000 USD
Tax and interest (i)	25% and 0
WACC	11.71%
Gas Volume Total	150 MMScf/day
Category 1 (Shipper with large volume > 50 MMSCFD)	75 MMScf/day
Category 2 (Shipper with medium volume 10-50 MMSCFD)	45 MMScf/day
Category 3 (Shipper with small volume 0-10 MMSCFD)	30 MMScf/day

Table 2. Calculation Result, Validation and Tariff Comparison

Parameter	Single Tarif	Multi Tarif	
		Manual Cash Flow	New Model
Tarif (USD/MSCF)			
Category 1 (Shipper with large volume)	0.2789	0.3672	0.3672
Category 2 (Shipper with medium volume)		0.2203	0.2203
Category 3 (Shipper with small volume)		0.1469	0.1469
Total Revenue (USD)	152,796,993	152,802,798	152,802,798
Δ Revenue	0.004%		

From the results above can be seen that the tariff for category 3 is the cheapest tariff because it is a group that must receive a special rate because it concerns the public interest that must be subsidized. Tariff For other categories will be adjusted because it is a group based on profit orientation as high as possible. It can also be proven that the revenue earned between single tariffs and multi tariffs is not different. It can be proved that the result of deviation of revenue obtained from single tariff method compared with multi-tariff got a very small result, only by 0.004% (see table 2).

3.3. The result of comparison of tariff calculation with single tariff system with multi-tariff to produce more competitive gas price for particular group

In the pipeline X, there are several shipper classes, where there is a particular shipper group that should get a lower and fairer tariff than other shipper groups. The shipper of certain groups is channeling gas for special needs and subsidized so as to produce competitive prices and not burden the state budget and society widely. This particular shipper, among others, for the needs of electricity and fertilizer are arranged and subsidized profit because it has a broad impact on the community. Tariff is one of the components of gas price forming up to end user or consumer. The equation for calculating gas price through the pipeline to a consumer can be formulated by an equation as follows:

$$Gas\ \Pr ice = \Pr oduction\ \ \Pr ice + Transmission\ Tariff + Distribution\ Cost + Tax \qquad (33)$$

If previously with a single tariff method all shipper must pay the same tariff without differentiated class, so that particular group of subsidized shipper will pay the same tariff as the shipper which is a business entity with profit-oriented, so the principle of justice is not achieved. With cheaper rates, certain types of ships such as electricity and fertilizers will increase the attractiveness to switch to cleaner and cheaper gas energy sources than fuel oil. The results of calculation and comparison of gas price for certain class shipper before and after using multi tariff system are as follows (see table 3 and table 4):

Table 3. Result of calculation and gas price comparison for pipe X

Parameter	Production Gas Price (USD/MMBtu)	Transmission Tarif (USD/MMBtu)	Distribution Cost (USD/MMBtu)	Tax	Total Gas Price (USD/MMBtu)
Single Tarif	6.75	0.2789	0	0.028	7.0568
Multi Tarif	6.75	0.1469	0	0.015	6.9116

Table 4. Fuel price calculation and price comparison for specific shipper energy source

Energy Requirement (MMSCFD)	Energy Required (Assumed : 1 Scf=1000 Btu) (MMBtu/day)	High Speed Diesel		Natural Gas Pipeline With Single Tarif		Natural Gas Pipeline With Multi Tarif	
		HSD Price (USD/MMbtu)	Total Cost (USD/Day)	Gas Price (USD/MMbtu)	Total Cost (USD/Day)	Gas Price (USD/MMbtu)	Total Cost (USD/Day)
30	30,000	26	780,000	7.06	211,705	6.91	207,347
		Savings		568,295		572,653	

4. Conclusion

Modeling different tariff calculations on the transport of natural gas through pipes for several classes of the shipper in an open access pipeline transmission line called a multi-tariff system can create a fair gas price without harming all parties. Gas prices will be obtained through a competitive pipeline Compared to fuel oil sources, which will encourage the substitution of gas usage as a cleaner energy source.

5. Acknowledgement

This research was supported by The Oil and Gas Downstream Regulatory Agency of Republic of Indonesia, Ministry of Energy and Mineral Resources of the Republic of Indonesia, and Universitas Indonesia.

6. References

[1] April, L., Owen, Z., Jeffrey, L., et al 2012 Interactions, Complementarities and Tensions at The Nexus of Natural Gas and Renewable Energy *The Electricity Journal* **25** (10) 38–48.

[2] Damodaran, A. 2012 *Equity Risk Premiums (ERP): Determinants, Estimation and Implications* Stern School of Business

[3] Damodaran, A. 2015 Country Risk: Determinants, Measures and Implications – The 2015 Edition Stern School of Business, adamodar@stern.nyu.edu

[4] Federal Energy Regulatory Commission, FERC 1999 *Cost-of-Service Rates Manual* Washington, D.C. 20426 United States of America.

[5] Hunt, Paul, 2014 The Weighted Average Cost of Capital for Electricity and Gas Networks *Submission on the WACC to CMA Energy Market Investigation*

[6] I.P.L. Png & C.W.J. Cheng 2001 *Pricing Policy* NUS School of Computing

[7] Li, Gong, Tian, Jiao 2015 The Peak-Shaving Efficiency Analysis of Natural Gas Time-of-Use Pricing for Residential Consumers: Evidence from Multi-Agent Simulation *The Energy Journal* **96** 48-58.

[8] Purwanto, S.W and Sommeng, A.N 2013 The Regulation of Business Activities of Natural Gas Through Pipelines BPH Migas Indonesia 55-102.

[9] Siswanto, D 2011 The development of a model of tariff determination for the transportation of natural gas through transmission pipes to the monopoly system regulated in Indonesia *Doctoral Disertation Petroleum Engineering-Bandung Institute of Technology*

[10] The Oil and Gas Downstrean Regulatory Agency of Republic of Indonesia, BPH Migas 2013 Rule Number 08 of 2013 Regarding *The Determination of Tariffs for the transportation of natural gas through pipes*

IOP Conf. Series: Earth and Environmental Science **105** (2017) 012126 doi:10.1088/1755-1315/105/1/012126

Lipase production from *Bacillus subtilis* with submerged fermentation using waste cooking oil

M Suci[1,2], R Arbianti[1] and H Hermansyah[1,2]
[1]Department of Chemical Engineering, Faculty of Engineering, Universitas Indonesia, Depok West Java 16424, Indonesia

E-mail : maharani.suci1995@gmail.com; heri.hermansyah@ui.ac.id

Abstract. Bacterial lipase has been developed lately because of its advantage to produce with large scale. Culture of *Bacillus subtilis* was grown to produce lipase in Waste Cooking Oil (WCO) using submerged fermentation (SmF) method. The enzyme activity of the culture was improved by using different concentration of inoculum, substrate, nitrogen source, inducer, and Ca^{2+} ion at 30°C for 84h fermentation. Lipolytic activity of crude lipase was determined using titrimetric method with hydrolysis reaction. Maximum activity of lipase (4.96 U/mL) was found at 5% (v/v) inoculum, 4% (v/v) WCO, 0.5% (w/v) yeast extract, 0.25% (v/v) olive oil, and 10 mM Ca^{2+} that present in medium culture. Later, the crude lipase has been dried with spray dryer and resulting 17.33 gr of dry lipase powder per 500 mL crude lipase.

1. Introduction

Lipases (E.C. 3.1.1.3) have been known as one of enzyme which find usage in a wide array of industrial applications, such as detergent, food technology, textile, pharmacy, cosmetic, biodiesel, and agrochemical industrial [1]. Lipases are ubiquitous hydrolytic enzymes that catalyze the breakdown of fats and oils into free fatty acids, monoacylglycerols, diacylglycerols, and glycerol [2]. Among bacterial lipases, those from Bacillus shows interesting properties that make them potential for biotechnological applications. As cells have adapted to survive in such extreme climatic conditions, their enzymes and lipases in particular have evolved and exhibit exceptional properties [3]. *Bacillus subtilis*, *Bacillus pumilus*, *Bacillus licheniformis*, *Bacillus coagulans*, *Bacillus stearothermophilus*, and *Bacillus alcalophilus* are the most common bacterial lipases that being used [4]. Bacterial lipases are mostly extracellular and are produced by submerged fermentation (SmF) [5].

In the present paper, we describe the effect of inoculum concentration and nutritional factors for optimum production of *Bacillus subtilis* lipase. Nutritional factors were carried out by using different concentrations of carbon source, nitrogen source, inducer, and Ca^{2+} metal ion.

2. Materials and Methods

2.1. Chemicals

The chemicals were of reagent grade and purchased from commercial suppliers (Merck, Sigma-Aldrich). Waste cooking oil was collected from household waste with maximum usage is twice. The olive oil for enzyme assays and skim milk were supplied from the local market.

---

[2] Author to whom any correspondence should be addressed.

2.2. *Microorganism*

A lipase producing bacterial culture was purchased from InaCC - LIPI Cibinong. The culture was periodically sub-culture in Nutrient Agar (NA) containing agar 15 g/L, meat extract 1 g/L, peptone 5 g/L, sodium chloride 5 g/L, yeast extract 2 g/L with final pH 7.1±0.2 (25°C). The stock culture was stored at -4°C.

2.3. *Culture media and conditions*

Pre-culture was maintained on Luria Bertani broth medium. The stock culture was inoculated into 50 mL pre-culture medium and incubated at 30°C for 24 h. The cells were inoculated into 50 mL production medium where the concentration of inoculum, substrate, nitrogen source, inducer, and Ca^{2+} metal ion were used to find out the highest lipolytic activity. The standard operating conditions and the composition of production medium were referred to the previous study [6]: peptone, yeast extract, substrate in phosphate buffer (0.1 M pH 7.5) with temperature 30°C and stirring rate 150 rpm. The culture was grown in 100 mL Erlenmeyer flasks on shaker water bath for 84 h fermentation. The optimum condition of variation mentioned previously, will be the basic to scale up the lipase production.

2.4. *Enzyme assays*

The culture broth was centrifuged at 4000 rpm for 45 minutes. The supernatant was filtered and used as the crude enzyme for the estimation of lipase activity. Lipase activity was assayed by alkali titration using olive oil as substrate [7]. Olive oil (10% m/v) was emulsified with polyvinyl acetate (5% m/v) in 0.05 M phosphate buffer pH 7.5. A 2 mL sample of crude enzyme was added to 18 mL of this emulsion. After incubation in a shaker for 15 minutes at 37 °C and 150 rpm, the reaction was interrupted by the addition of 20 mL of ethanol. The solution was titrated with 0.05 M NaOH in the presence of phenolphthalein as indicator. Reaction blanks were run in the same way, but the sample was adding after the addition of the ethanol solution. The lipase activities were performed in duplicate. A unit lipase activity (U) was defined as the amount of enzyme that catalyzes the release of fatty acid per min under the conditions mentioned above.

2.5. *Protein analysis*

Protein concentration was measured spectrophotometrically at 750 nm by the Lowry method [8] using bovine serum albumin as protein standard.

2.6. *Dry lipase extract*

Skim milk (12% m/v) was added into 500 mL crude enzyme and stirred until homogenous. The mixture was dried using spray dryer (T_{in}=80°C, T_{out}=130°C) to obtain dry lipase powder from *B. subtilis* [9].

3. Results and discussion

3.1. *Growth and lipase production profile*

B. subtilis were grown at medium, which has been specified to find out the optimum incubation period. Maximum lipase activity of 4.72 U/mL was observed on 84 hours of production with the average optical density 2.452 at 600 nm. On the third day, decline in growth of *B. subtilis* and lipase activity was observed (figure 1). The decrease of *B. subtilis* growth may be due to over populated culture and fixed amount of nutrient with which the organism starts liberating proteolytic enzyme, enhancing self-consumption [10]. On the other hand, the decrease of lipase production at the later stage could be possibly due to pH inactivation, proteolysis, or both [11]. Maximum lipase activity on the 80 hours of fermentation also has been reported from *Bacillus sp.* [12]. Therefore, these conditions were employed for cultures used in future experimentation.

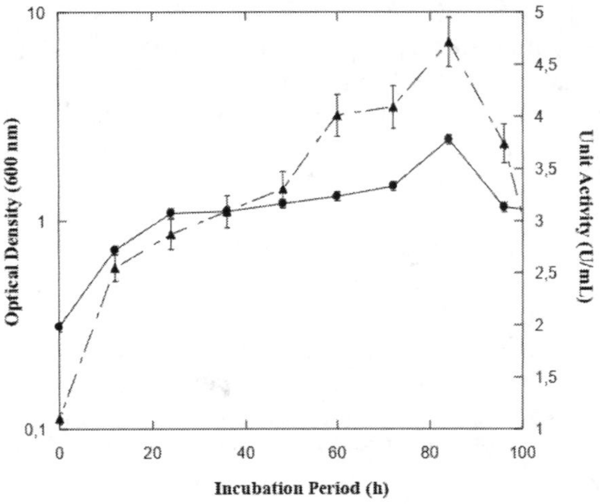

Figure 1. Lipase production and growth profile of Bacillus subtilis. The production was carried out at 30°C under shaking condition (150 rpm) in presence of 1% waste cooking oil, 0.3% yeast extract, 0.5% peptone, 0.25% olive oil, and 10 mM $CaCl_2.2H_2O$ in phosphate buffer 0.1 M pH 7.5. ●, cell growth; ▲, lipase activity.

3.2. Selection of inoculum concentration

The inoculum was added to the production medium in varying concentrations (1, 3, 5, 7 and 9%). The results showed that the optimum lipase production was observed in case of 5% inoculum with lipase activity about 1.83 U/mL (figure 2). Above 5% concentration, decline in lipase activity was observed. Beside due to over populated culture, this might be due to limited nutrient and dissolved oxygen to support better bacterial growth and lipase production [13]. Similar results were reported [14] that lipase production by *Bacillus sp*. FH5 was at optimum condition when concentration of inoculum (5%) was used.

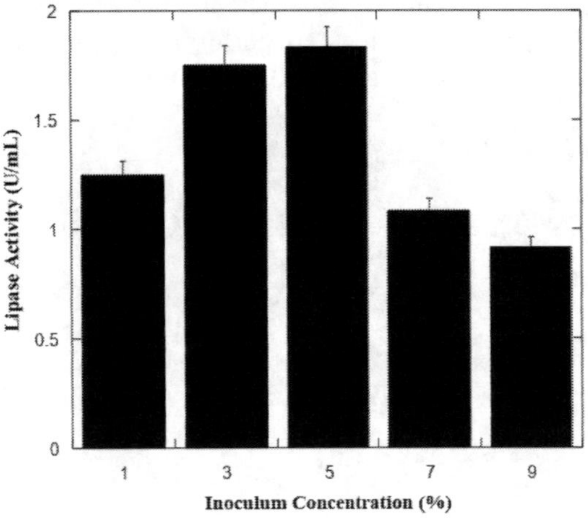

Figure 2. Effect of inoculum concentration on lipase production

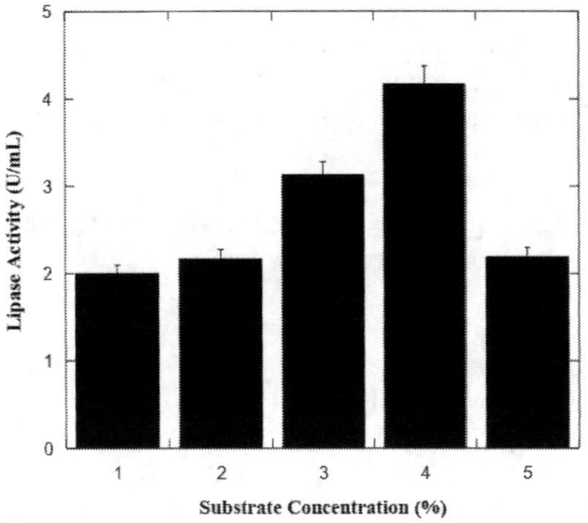

Figure 3. Effect of substrate concentration on lipase production

3.3. Selection of substrate concentration

The waste cooking oil (WCO) as the carbon source was added to the production medium in varying concentrations (1, 2, 3, 4 and 5%). The results showed that the optimum lipase production was observed in case of 4% substrate with lipase activity about 4.7 U/mL (figure 3). The increasing concentration of WCO increased the lipase activity, but it has also been observed that after 4% concentration, the lipase activity has decreasing. This may be attributed to the substrate inhibition (Feedback inhibition) [10]. Waste oil was theoretically beneficial for lipase yield [15]. However, results showed that a high concentration of oily substrate is unfavorable for lipase yield and its activity. This might be due to other carbon source or inducer present. On the previous study [16] found that the high concentration of oil usually decreased the transfer rate of oxygen to the fermentation medium, which reduces the specific growth rate and further reduces the yield of the products.

3.4. Selection of nitrogen source concentration

Yeast extract as the nitrogen source was added to the production medium in varying concentrations (0.1, 0.2, 0.3, 0.4 and 0.5%). The results showed that the optimum lipase production was observed in case of 0.5% (w/v) yeast extract with lipase activity about 4.75 U/mL (figure 4). The increasing concentration of yeast extract increased the lipase activity. Yeast extract, apart from acting as a nitrogen source also supplies vitamins and trace metals, thereby affecting the growth of the organism and thus increasing lipase production [17]. Yeast extract can act alone or in combination with other nitrogen sources. In the present study, yeast extract was added with 0.5% (w/v) peptone as organic nitrogen source.

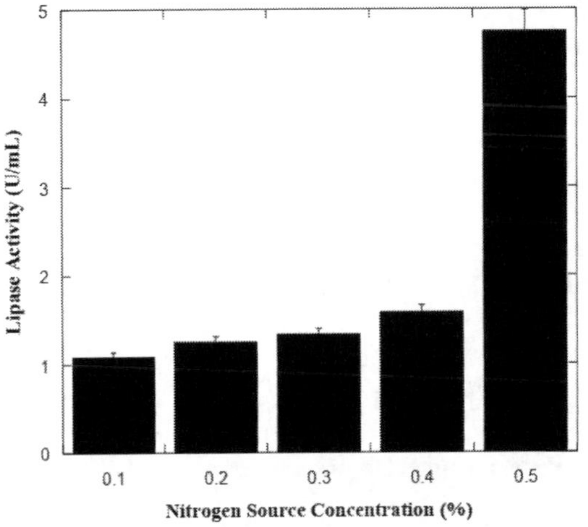

Figure 4. Effect of nitrogen source concentration on lipase production

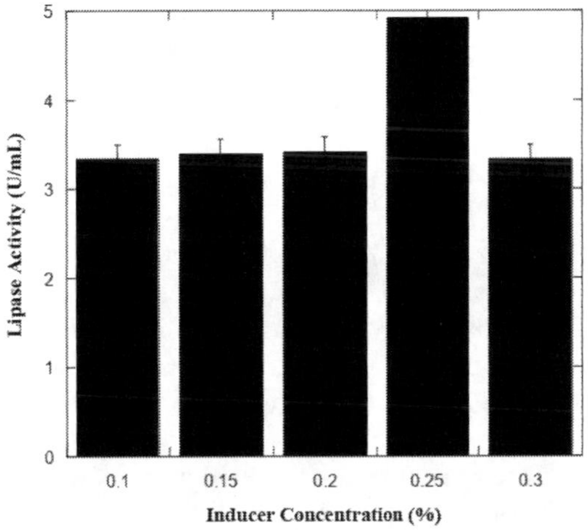

Figure 5. Effect of inducer concentration on lipase production

3.5. Selection of inducer concentration

Olive oil as the inducer was added to the production medium in varying concentrations (0.1, 0.15, 0.2, 0.25, and 0.3%). The results showed that the optimum lipase production was observed in case of 0.25% (v/v) olive oil with lipase activity about 4.92 U/mL (figure 5). The increasing of olive oil above 0.25% obtained the decline of lipase activity. Previous study [18] found that inducer from olive oil more effective to induce production of lipase. This might be due to lipid content in olive oil, tributyrin, which is structurally easily digested by microorganisms. However, the increasing of inducer was not proportional to the number of enzyme produced. This might be due to the ratio of carbon and nitrogen in the medium. Generally, the activity of intra and extracellular lipases increases with increasing lipid concentrations, although excessive levels in the growth medium may be cytotoxic [19].

3.6. Selection of Ca^{2+} metal ion concentration

$CaCl_2.2H_2O$ as the Ca^{2+} metal ion was added to the production medium in varying concentrations (10, 15, 20, 25, and 30 mM). The results showed that the optimum lipase production was observed in case of 10 mM $CaCl_2.2H_2O$ with lipase activity about 4.96 U/mL (figure 6). The increasing of metal ion Ca^{2+} obtained the decline of lipase activity. Several enzymes need metal ion to increase their activity [20]. However, those metal ions can act as an activator or inhibitor in a certain condition. On the previous study [21] found that the enzyme production was reasonably improved in the presence of Ca^{2+} which is significant in maintaining cell wall rigidity, stabilizing oligomeric proteins and covalently bounding protein peptidoglycan complexes in the outer membrane.

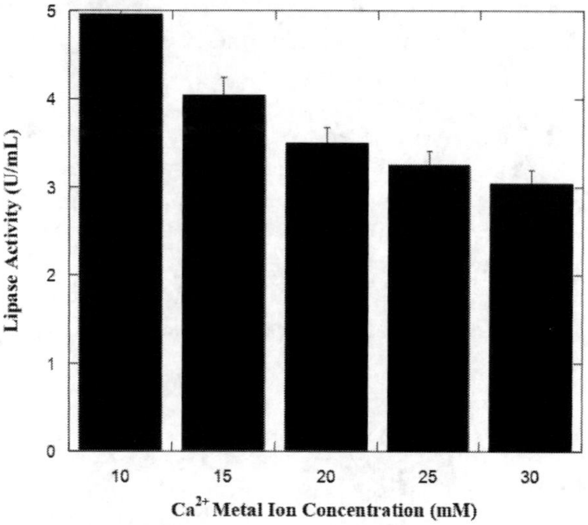

Figure 6. Effect of Ca^{2+} metal ion concentration on lipase production

3.7. Scale up production of enzyme

The production of lipase was scaled up from a flask (100 mL) to fermenter (20 L). The fermenter design is based on previous study. Pre-culture was grown in 1 L LB Broth medium and inoculated to 20 L production medium. The optimum condition from previous study was used with following composition: 5% (v/v) inoculum, 4% (v/v) WCO, 0.5% (w/v) yeast extract, 0.5% (w/v) peptone, 0.25% (v/v) olive oil, 10 mM $CaCl_2.2H_2O$ in phosphate buffer (0.1 M pH 7.5). The operating conditions are temperature 30°C and agitation rate 150 rpm during 84-hour fermentation. The culture broth was centrifuged at 4000 rpm for 45 minutes to obtain the supernatant. Furthermore, the supernatant was dried using spray dryer to get the dry lipase powder. Lipase activity was carried out with titrimetric and Lowry method through dry lipase enzyme and commercial enzyme (*Candida rugosa* lipase). The activity was 28 U/g solid and 694 U/g solid. While the specific activity was 1.44 U/mg protein and 12.5 U/mg protein respectively.

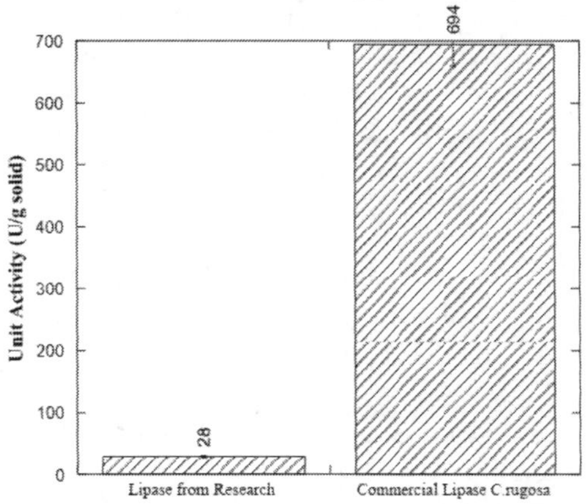

Figure 7. Comparison of lipase activity (U/g solid)

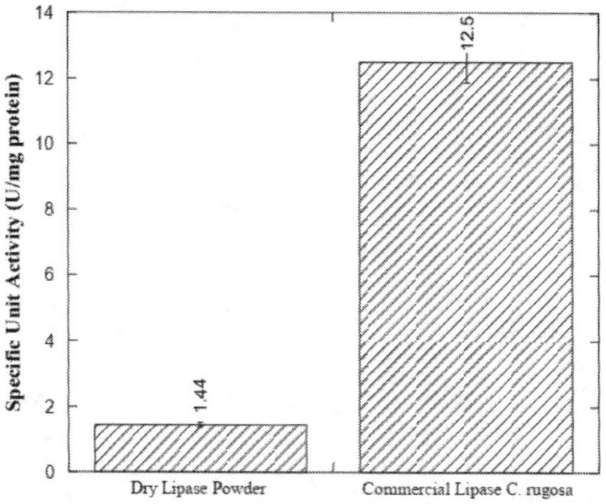

Figure 8. Comparison of lipase activity (U/mg protein)

4. Conclusion

The use of submerged fermentation for lipase production using waste cooking oil as substrate presented satisfactory results. A lipase *B. subtilis* was growth dependent as revealed by nutritional studies. The growth and lipase production are shown to be promoted by 5% (v/v) inoculum, 4% (v/v)

WCO, 0.5% (w/v) yeast extract, 0.5% (w/v) peptone, 0.25% (v/v) olive oil, 10 mM $CaCl_2.2H_2O$ in phosphate buffer (0.1 M pH 7.5) at 30°C under shaking condition (150 rpm). The maximum lipolytic activity was 4.96 U/mL in 84-hour fermentation. The activity in dry form decrease due to several factors but increase in specific activity. It is important to figure out the best purification method and immobilization to increase yield and activity of lipase.

5. Acknowledgement

The authors gratefully acknowledge KEMENRISTEK DIKTI, Universitas Indonesia, and USAID SHERA for supporting this research.

6. References

[1] Jaeger K, Dijkstra B, Reetz M 1999 *Ann. Rev. Microb.* **53** 315–51
[2] Dheeman D, Antony-Babu S, Frías J and Henehan G 2011 *J. Mol. Cat.* B **72** 256–62
[3] Guncheva M and Zhiryakova D 2011 *J. Mol Cat.* B **68** 1–21
[4] Treichel H, Oliveira D, Mazutti M A, Luccio M and Oliveira J V 2010 *Food and Bioprocess Technol.* **3** 182–96
[5] Gupta R, Gupta N and Rathi P 2004 *Appl. Microbiol. Biotechnol.* **64** 763–81
[6] Takaç S and Marul B 2008 *Journal Ind. Microbiol. Biotechnol.* **35** 1019–25
[7] Pinheiro T L F, Menoncin S, Domingues N M, Oliveira D, Treichel H, Luccio M and Freire D M G 2008 *Ciência e Tecnologia de Alimentos* **28** 444–50
[8] Lowry O H, Rosebrough N J, Farr A L and Randall R J 1951 *J. Biol. Chem* **193** 265
[9] Aliyah A N, Edelweiss E D, Sahlan M, Wijanarko A and Hermansyah H 2016 *Int. J. Technology* **7** 1392–1403
[10] Sarkar S, Sreekanth B, Kant S, Banerjee R and Bhattacharyya B C 1998 *Biop. Eng.* **19** 29–32
[11] Ahmed E H, Raghavendra T and Madamwar D 2010 *Appl. Biochem. Biotechnol.* **160** 2102–13
[12] Sugihara A, Tani T and Tominaga Y 1991 *J. Biochem.* **109** 211–16
[13] Shariff F M, Leow T C, Mukred A D, Salleh A B, Basri M and Rahman R N Z R A 2007 *J. Basic Microb.* **47** 406–12
[14] Hasan F, Shah A A and Hameed A 2006 *Annals of Microb.* **56** 247–52
[15] Xiaoyan L, Xinjun Y, Jinshun L V, Jiaxing X, Jun X, Zhen W, Tong Z and Yuanfang D 2017 *Food and Bioproducts Processing* **103** 86–94
[16] Ramani K, Boopathy R, Vidya C, Kennedy J L, Velan M, Sekaran G 2010 *Process Biochem.* **45** 986–92
[17] Gupta N, Sahai V and Gupta R 2007 *Process Biochem.* **42** 518–26
[18] R Sumathy, M Vijayalakshmi and M Deecaraman 2012 *Inter. J. Environ. Sci.* **3** 1072–78
[19] Zarevucka M 2012 *Olive Oil as Inductor of Microbial Lipase* (Institute of Organic Chemistry and Biochemistry: InTech Europe)
[20] Palmer T 1991 *Understanding Enzyme 3rd ed* (England: Ellis Horwood Limited)
[21] Devi A S, Devi K C and Rajendiran R 2012 *Inter. J. Biological, Biomol., Agricul., Food and Biotechnol. Eng.* **6** 840–45

IOP Conf. Series: Earth and Environmental Science **105** (2017) 012127 doi:10.1088/1755-1315/105/1/012127

Design and calculation of mechanical system for solar-powered electric boat

Sunaryo [1]**, Dendi Nurachman**[1]**, Pradhana S Imfianto**[1]
[1]Naval Architecture and Marine Engineering Study Program, Department of Mechanical Engineering, Universitas Indonesia Depok, 16424, Indonesia

E-mail: naryo@eng.ui.ac.id

Abstract. Indonesian government has a flagship program that is to promote the nation through the maritime sector and make Indonesia as Global Maritime Fulcrum. To achieve that, sea transportation is very important for the development of Indonesia. But sea transportation is also the highest contributor to the air pollution and global warming. In order to reduce the source of global warming Indonesian Government has a strategic plan to gradually replace the use of fossil energy with renewable energy. One of the renewable energy that can be used effectively in Indonesia is solar energy, because Indonesia located along the equator and has a high intensity of sunlight that means it has excellent solar energy availability. In attempt to contribute to the Government's energy strategic plan, the research is aimed to design a solar-powered electric boat with single pilot. The project is a further development of the solar boat created by the Universitas Indonesia team that participated in the International Solar-boat Challenge competition in the Netherlands. The paper is focused on the design and calculation of its mechanical system. The system is divided into propulsion system, steering system, and bilge pump system. The research is conducted based on literature study, discussion with experts, and surveys to the solar-powered electric boat had been designed previously for the race. Collected data are then used as input for the calculation and simulation of the vessel being designed. It is expected that the study could optimize the mechanical system of the solar-powered electric boat, and can be developed for recreational use in Indonesian waters.

1. Introduction

The Indonesian government has a flagship program that is to promote the nation through the maritime sector and make Indonesia as Global Maritime Fulcrum. As has been known that Indonesia is the largest archipelagic country in the world with more than 17,000 islands and 65% of the territory is covered by sea [1]. This make Indonesia has high potential in the maritime sector like fishery, tourism, shipping etc. Marine transport is one of the most important players to exploit the potential of maritime sector for the transportation and connectivity between islands in Indonesia.

In recent years, global warming effect is a major problem in almost all parts of the world including Indonesia. Transportation is the biggest contributor for this global warming effect [2]. Therefore trend towards using renewable and alternative energy sources on land has gathered momentum over the last decade or so as the general public and policy makers tackle the issues of pollution, energy security and climate change [3]. Indonesian government needs to replace the use of fossil fuels with renewable energy as an alternative to reduce this effect. One of the renewable energy that can be used in

Indonesia is solar energy, because Indonesia's geographical location along the equator, which has a high intensity of sunlight [4].

In attempt to contribute to Indonesia's strategy to replace the use fossil fuel to solar energy, this research's objective is to design a solar-powered electric boat with single pilot, which has a 6000 mm length, which is a further development of the solar boat created by Universitas Indonesia's team that participated in International Solar Boat Challenge competition held in the Netherlands in 2016.

The most important thing in building a solar powered boat is the efficiency of the electricity [5], to ensure the efficiency of the electric motor to reach the desired speed service and have a long cruising time. Mechanical system is one the most important thing to make an efficient solar-powered boat. This paper is focused onto the design and mechanical system of the boat, which is divided into 3 systems i.e. propulsion system; steering system; and pumping system.

2. Methodology and Results

The methodology being used to get the best electric-motor propulsion system is by analyzing the dimensions and specification of the first Universitas Indonesia's solar-powered boat, with the main particulars as follows: Ship type, Single pilot Monohull Semi-Trimaran; Maximum speed, 20-25 km/h (~13 knots); Length over all, 6 m; Width moulded, 2.3 m; Draft at full load, 0.12 m; Propulsion system, electric; Battery capacity, 48V. These data is used as an early reference to the design and calculation of the boat's mechanical systems, and then combined with literature study and discussion with experts so as to get more accurate results.

2.1 Propulsion System

In this method, the propulsion system being used for solar boat is determined. Propulsion is a system that will provide a force to push the ship to achieve the desired service speed. In determining the type propulsion, several factors need to be considered such as the dimensions of engine room, cooling system, motor mounting, and steering system. A PODSEM propulsion "POD Steerable Electrical Motor" is chosen. This propulsion combines outboard motor with a steering system as shown in figure 1. The reason for using this is because the engine room of the boat is considerably small i.e. 1000 mm long, 740 mm of width, and 330 mm of height. The propulsion system allows making flexible mounting between the motor and steering system, which will make better maneuverability, moreover this type of propulsion can make the cooling system much easier too.

After deciding the type of propulsion, next step is to calculate the amount of power required for the electric motor to get the desired speed of the vessel to reach 25 km/h. Based on Harvald's method [6] the effective power is obtained based on the following equation:

$$PE[W] = RT[N] x V\left[{}^{m}\!/_{s}\right] \quad (1)$$

$$PM[W] = {}^{PE[W]}\!\Big/_{\eta h x \eta p} \quad (2)$$

$$\eta h = {}^{(1-t)}\!\Big/_{(1-w)} \quad (3)$$

Where: *PE* is Effective Power; *RT* is Total Resistance; *V* is Speed of Vessel; ηh is hull efficiency; ηp is propulsion efficiency; *t* is thrust deduction factor; and *w* is wake fraction.

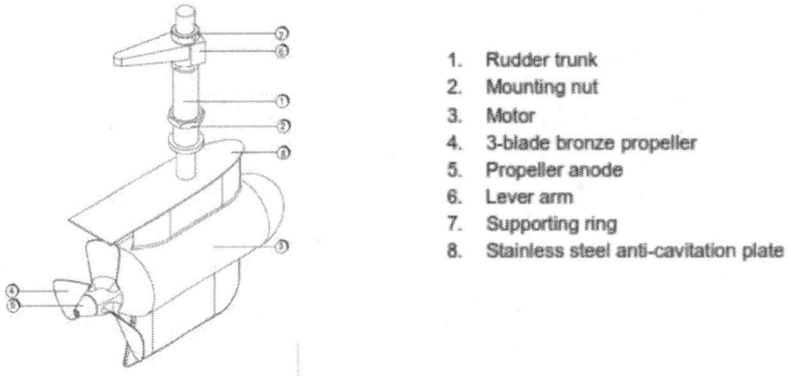

Figure 1. POD propeller

From the calculation, the results obtained are: that the value of RT is 600 (N) and V is 6.94 (m/s), then PE is 4164 (W). Furthermore, based on the calculation results the characteristics of the propulsion are obtained that: t = 0.3; w = 0.35; and ηh ≈ 1.07, while based on the available data chart for POD electrical motor the ηp is 92%. Thus the minimum power used for the motor is PM = 4230 (W).

Based on the obtained results the specifications of POD Drive Propulsion is shown in table 1, then the ship resistance was simulated using maxsurf resistance software by using slender body method. Parameter included in the simulation is 60% efficiency of hull resistance, where the results of this analysis are the relationship between motor power vs vessel velocity is presented in figure 2.

Table 1. Electric PODSEM Specification

Model	Voltage	Power	RPM	Current	Efficiency	Weight	Motor Type
UF 43	48 DC	4300 W	1550	96A	92%	27kg	Asynchronous

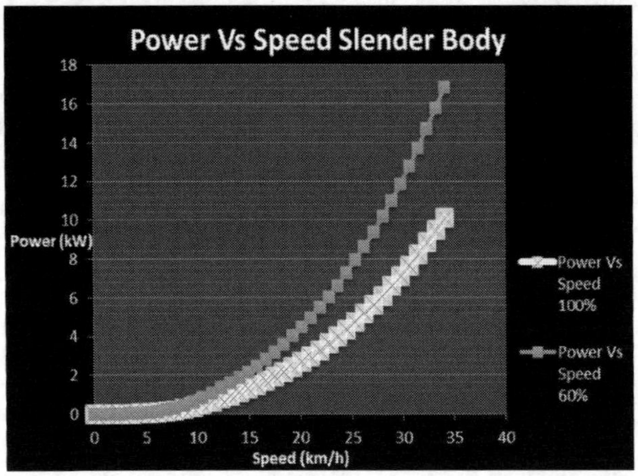

Figure 2. Power vs speed maxsurf resistance using slender body method

Using hull resistance data, the amount of power needed by each percentage used in throttle variations of the ship propulsion system was determined in order to match with the ship's operating characteristics [7]. This will follow the procedure of determining electrical current needed for motor by selected output power divided by electrical system voltage, as shown in Figure 3 and Table 2.

After determining the power, the required current of the motor, and the speed of the vessel for each variation of the throttle were obtained. The data were processed in accordance with energy consumption procedure [8], using 60 km sailing range as targeted distance and maximum PV charging at daylight condition. With data from figure 2 and figure 3 speeds to be used to fulfill 60 km sailing range was determined, in constant speed mode, as shown in table 3 and table 4.

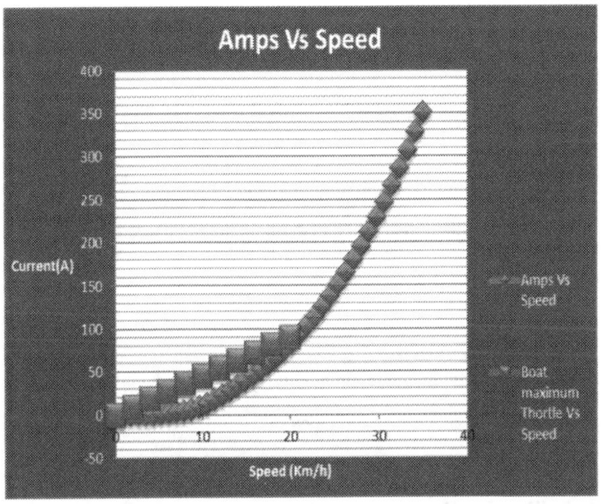

Figure 3. Amps vs speed

Table 2. Throttle variation

Throttle Percentage (%)	Current (A)	Power (kW)	Speed (km/h)
100	89.58	4300	20.67
90	80.63	3870	19.83
80	71.67	3440	18.93
75	67.19	3225	18.44
70	62.71	3010	17.96
60	53.75	2580	16.88
50	44.79	2150	15.71
40	35.83	1720	14.39
30	26.87	1290	12.96
25	22.39	1075	12.18
20	17.92	860	11.42
10	8.96	430	9.58

Table 3. Energy consumption

Speed (Km/h)	Input Current	Battery Capacity	Output Current (to motor) (A)	Throttle Percentage (%)	Battery Discharge Load(A)
	PV Panel (A)	Battery (80%) (Ah)			
20.67	21.67	26.4	89.58	100	67.92
19.83	21.67	26.4	80.63	90	58.96
18.93	21.67	26.4	71.67	80	49.99
18.44	21.67	26.4	67.19	75	45.52
17.96	21.67	26.4	62.71	70	41.04
16.88	21.67	26.4	53.75	60	32.08
15.71	21.67	26.4	44.79	50	23.13
14.39	21.67	26.4	35.83	40	14.17
12.96	21.67	26.4	26.88	30	5.21
12.18	21.67	26.4	22.395	25	0.73
11.42	21.67	26.4	17.92	20	-3.75
9.58	21.67	26.4	8.96	10	-12.71

Table 4. Energy consumption

Battery Draining Time Until 20%		Sailed Distance (S= Vxt) (Km)	Distance to Travel
Hours	Minutes		
0.39	23 minutes 32 seconds	8.03	51.97
0.45	27 minutes 26 seconds	8.88	51.12
0.53	32 minutes 8 seconds	10.00	50.00
0.58	35 minutes 19 seconds	10.69	49.31
0.64	38 minutes 59 seconds	11.55	48.45
0.82	49 minutes 37 seconds	13.89	46.11
1.14	1 Hour 8 minutes 27 seconds	17.93	42.07
1.86	1 Hour 51 minutes 12 seconds	26.82	33.18
5.07	5 Hours 4 minutes 2 seconds	65.69	-5.69
36.21	36 Hours 12 minutes	> 60 km	> 60 km
Battery Charging	Battery Charging	More than 60 km	>60 km
Battery Charging	Battery Charging	More than 60 km	>60 km

2.2 Steering System

The steering system to be used on solar boats is determined so that the boat can maneuver with a minimum turning angle of 40 degrees. In determining the type of steering system the compatibility with the propulsion system need to be considered, therefore the Hydraulic Steering System has been chosen, Hydraulic Steering System utilizes fluid oil from the steering wheel that passed through the cable to the actuator to move the piston, this piston movement is used to set the rudder direction on the shaft Pod propulsion by using the connecting rod [9]. Based on the trigonometric mathematical equations as shown in figure 4 and equation 4 [10].

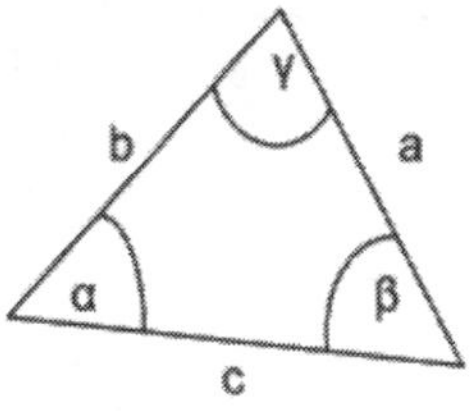

Figure 4. Trigonometric

$$\frac{a}{\sin\alpha} = \frac{b}{\sin\beta} = \frac{c}{\sin\gamma} \quad (4)$$

Since the shaft motion of the Hydraulic Actuator is only 8 cm to the right and 8 cm to the left, the required connecting rod length in this solar boat to connect between POD Propulsion with hydraulic steering actuator based on the sinus equation is obtained to be 12.4 cm. In addition, some other advantages of using a hydraulic steering system is that the force of pilot that need to turn the steering wheel and steer the ship is lighter, so that the use of space of engine room and cabin room can be reduced, and makes the mounting and maintenance much easier.

2.3 Bilge Pump System

The function of bilge pump is for disposing of water inside the vessel that come due to rain or leakage due to mounting Pod Propulsion Motor on the body of the vessel. The installation of bilge pump should be well designed to ensure the bilge pump system can be operated normally without disturbing other systems of the boat, including its capacity and operating processes. For this purpose formulas 5; 6; and 7 are being used.

$$Q_{in}\left(M^3/_{Hour}\right) < Q_{out}\left(M^3/_{Hour}\right) \tag{5}$$

$$Q_{in} = Q_{rain} + Q_{leakage} \tag{6}$$

$$Q = I\left(M/_{Hour}\right) x A\left(M^2\right) \tag{7}$$

Where: Q = Flow Rate; I = Rainfall Intensity; A = Cross Sectional Area.

Based on data from MDPI Journal [11] regarding rainfall in Indonesia, the rate of rainfall intensity in Indonesia is I ≈ 2.15 (M/Hour), the water flow rate of the rainfall that goes into the cabin of vessel with area A ≈ 0.48 (M^2) is Q rain ≈ 1.03 (M^3/Hour), it is also assumed that the incoming water flow rate due to leakage into the vessel is Q leakage ≈ 0.3 (M^3/Hour) so we can get Qin = 1.33 (M^3/Hour). Based on these results it is shown that the value of Q in <Q out. For this purpose Float Switch Automatic Bilge Pump with debit Q out = 1.36 (M3 / Hour) ≈ 360 GPH is chosen, and its specification as shown in table 5.

Table 5. Bilge pump specifications

Model	GPH	Amps@12V	Amps@13.6V	Fuse	Height	Width	Weight	Hose Ø
24	360	2.1	2.5	2.5	0.5”	3/8”	0.5 lbs	¾”

The Automatic Float Switch has the function to activate or deactivate the bilge pump when the water level inside the hull reaches a certain height so that it can save the energy used from the battery.

3. Conclusion

There are 3 sub system that have been determined i.e. propulsion system; steering system; batteries and bilge pump system.

PODSEM "POD Steerable Electrical Motor" type has been chosen for the propulsion system, because it has advantages such as save space of engine room, good cooling system, easy mounting, and flexible steering, moreover it can make better maneuverability. This particular boat will able to travel 60 km in constant speed at 30% throttle opening, which provide the boat with 12.96 km/h sailing speed and sailing time up to 5 hours 4 minutes and 2 seconds.

Hydraulic type steering system has been chosen for the Steering system, because it is easy in operation, easy installation, and does not require a large space.

Finally, in the Bilge Pump System Automatic Float Switch type was chosen in order to keep the efficiency of the battery because the pump can be on / off automatically based on the height of the water inside the vessel.

With this performance, this boat will be a perfect candidate to replace common type of water recreational vehicle in Indonesia.

4. Acknowledge

The Authors would like to thank Directorate of Research and Community Engagement Universitas Indonesia for providing fund for finalizing the research and publishing its results; PT Len Industri for supporting us with the propulsion system; Solar Boat Team Universitas Indonesia 2016 who had raced in the Netherlands that provide us with significant data and experiences to work on this research, and for every party that support the development of this research.

5. REFERENCES

[1] Wiryawan, B., et al. 1999. *Lampung Coastal Resources Atlas. Government of Lampung Province and Coastal Resources Management Project.* Coastal Resources Center, University of Rhode Island and Centre for Coastal and Marine Resources Studies, Bogor Agricultural University. Bandar Lampung. Indonesia

[2] Ministry of Environment. 2007. *Rencana Aksi Nasional dalam Menghadapi Perubahan Iklim [National Action Plan in Mitigating Climate Change].* Ministry of Environment Republic of Indonesia. Jakarta.

[3] Mustafa, M.M. 2014. Solar energy for river nile cruisers. *Portal of Scientific Journals of Croatia.* **65** (2).

[4] Handayani, N.A., et al . 2012. Potency of Solar Energy Applications in Indonesia. *International Journal of Renewable Energi Development,* 33 – 38.

[5] Gorter, T. 2015. Design considerations of a solar racing boat: propeller design parameters as a result of PV system power. *Proceedings of The 7th International Conference on Applied Energy,* 1901 - 06.

[6] Harvald S.A. (1983). *Resistance and Propulsion of Ships*. Department of Ocean Engineering, The techincal University of Denmark. *John Wiley and Sons,* 198 – 211.

[7] Kurniawan, A. 2016. A Review of Solar-Powered Boat Development. *Journal for Technology and Science,* **27** (1) 1 – 8.

[8] Spagnolo, G. S., et al. 2012. Solar-Electric Boat. *Journal of Transportation Technologies,* **2** 144 - 149.

[9] Xingrong, S., & Yuejin, C. 2013. Study on Hydrodynamic Performance of Podded Propulsor at Steering Conditions. *Proceedings of the Third International Symposium on Marine Propulsors*. 358 – 63.

[10] Islam, M. F. et al. 2009. Performance Charactecteristics of Static and Dynamic Azimuthing Podded Propulsor. *Proceedings of the First International Symposium on Marine Propulsors,* 482 - 492.

[11] Lee, H. S. 2015. General Rainfall Patterns in Indonesia and The Potential Impacts of Local Seas on Rainfall Intensity. *Basel: MDPI, Water* **7** (4) 1751 -68.

Synthesis and characteristic of Cu_2ZnSnS_4 thin film prepared by appropriate non-stoichiometry precursor

B Munir[1], **B E Prastyo**[1], **E Y Muslih**[2]
[1]Department of Metallurgy and Materials, University of Indonesia, Depok.
[2]Department of Mechanical Engineering, Trisakti University, Jakarta.

E-mail: bayu.eko@ui.ac.id

Abstract. Non-stoichiometry sulphur-containing precursors were studied to picture its connection to the formation of kesterite Cu_2ZnSnS_4 (CZTS) phase. The precursors were prepared from metal (Cu-Zn-Sn) chlorides with ethanol as solvent and 2-mercaptopropioninc acid as capping ligand. Mol ratio of S/Cu 4.4 was observed as the limit of non-stoichiometry. Confirmed by XRD characterization, the employment of sulphur in higher concentration resulted in the formation of secondary phases with a porous morphology. Further, a modification in holding time of annealing treatment was successfully improved the crystallinity of the CZTS film. The SEM micrograph was showed the formation of well-growth CZTS grains with size ~ 1.5 μm. However, porosity was still encountered at the surface film as indication of insufficient growth processes (heat treatment). The energy band characteristic of the film is 1.4 eV, determined by extrapolating the gradient line of $(\alpha h\upsilon)^2$ vs $h\upsilon$ (photon energy). The route presented in this paper offers alternative solution to synthesize CZTS semiconductor from ethanol-based precursor.

1. Introduction

Cu_2ZnSnS_4 (CZTS) has attracted much effort from scientists to develop low cost and highly efficient thin-film solar cells (TFSC) due to its internal properties: direct mechanism and wide-range absorption of sun spectra, high absorption coefficient ($\sim 10^4$ cm^{-1}) and tuneable band gap energy [1-3]. The latter property is potentially significance to the development of multijunction solar cells which involved certain adjustment on the band gap energy of absorber semiconductor. The energy band varies accordingly to metal composition and type of chalcogenide used in the preparation. CZTS firstly investigated as absorber semiconductor for photovoltaic by Ito and Nakazawa in 1988 and started get attention to develop TFSC in 1996 [1,2]. Further, CZTS-based solar cells intensively investigated by IBM research group led by Mitzi [1,4]. In 2014, Mitzi group has successfully attained 12.7% of energy conversion efficiency, a record in CZTS-based thin-film solar cells [5]. This efficiency is low enough compare to other active materials used in TFSC technologies such as CIGS, CIS and CdTe which have demonstrated cell efficiencies well beyond 14% for commercial module and around 20% for laboratory scale [6-8]. However, CZTS-based semiconductor has not yet reached mature production line due to shallow knowledge in this material; therefore, there seems to be still open room to improve in the direction of high-efficiency and cost-competitive way.

---

[1] Department of Metallurgy and Material 1st floor, University of Indonesia, Depok, 16424

The formation of chalcogenide quaternary system, Cu_2ZnSnS_4 (CZTS), involves complex competition between CZTS growth and secondary phases growth: binary phases (Cu_2S, SnS, ZnS, SnS_2) and terniary phases (Cu_2SnS_3) [4]. Attacking this issue remains to be the key solution to fill the performance gap between CZTS and CIGS/CIS/CdTe TF solar cells. SnS_2 is a n-type semiconductor with a band gap energy of 2.2 eV while ZnS is insulator in nature with wide band gap (3.54 eV) [1,9]. These binary phases may form a secondary diode inside absorbing semiconductor and reduce the active area to absorb sun spectra and thus generate low-density of electron-hole pairs. Cu_2S, SnS and other sulphide compounds have different electrical properties to CZTS phase. At the end, unwanted secondary phases can cause high photocarrier recombination and thus hamper solar cell performance [10-13].

In this study, we synthesized CZTS films using single-solution approach as precursor preparation and spin coating method as film deposition. Four elemental constituents, Cu-Zn-Sn-S, are directly included in the first stage by employing ethanol and 2-mercaptopropionic acid as solvent and capping ligand (stabilizer), respectively [14]. In previous research, this route has successfully been proven to fabricate CZTS phase without the use of sulfurization (annealing at sulfur-containing atmosphere) [13]. Different from common approach, 2-mercaptopropionic acid can efficiently form metal-sulfur complex ligand and induce the formation of binary sulphide phase before eventually form CZTS phase through heat treatment. These ligands mitigate the lost of elemental sulfur during high temperature annealing (±500 °C) and thus makes sulfurization no longer necessary. Figure 1 shows the XRD peak of CZTS kesterite phase processed from sulfurization route and without sulfurization [13]. The vanishingly difference in correspondence CZTS peak between both routes showed successful employment of 2-mercaptopropionic acid to eliminate sulfurization step. In this research, the attention is given to study non-stoichiometry sulfur-containing precursor on the formation of kesterite CZTS-phase and the growth of secondary phases.

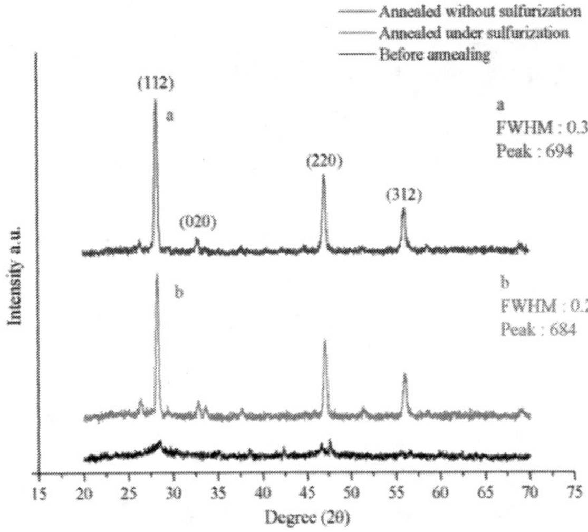

Figure 1. XRD data plot of as-deposited CZTS film, annealed with sulfurization, and annealed without sulfurization, respectively from bottom to top.

2. Experimental Setup

The synthesis of CZTS-films was carried out using chemical method followed by spin-coat deposition on top of soda lime glass substrate. The route is adapted from previous report; briefly, copper chloride ($CuCl_2$), zinc chloride ($ZnCl_2$) and tin chloride ($SnCl_2$) were dissolved at designed ratios [Cu/(Zn+Sn) = 0.8 and Zn/Sn = 1.2)] in ethanol and 2-mercaptopropionic acid under ambient environment and constant temperature at 80 °C [13]. After 15 minutes of agitation, thiourea was hot injected at different mol ratio to Cu and keep under agitation for another 10 minutes. After cooling down to ambient temperature, this solution then referred as precursor.

Soda lime glass (15 mm x 15 mm) is used as substrate for deposition stage. The glass firstly soaked in distilled water before ultrasonically cleaned in acetone and ethanol for 10 minutes, respectively. The precursor then deposited onto the glass using spin coater at 2000 rpm for 20 seconds followed by drying using hot plate at 150 °C for 12 minutes. These processes were repeated five times with the fifth (final) drying conducted at 170 °C for 60 minutes. The thus-obtained film was then annealed at 520 °C with a holding period of 60 minutes. The experimental images and final film are shown in figure 2.

(a) (b) (c) (d)

Figure 2. Experimental pictures of (a) Cu-Zn-Sn-ethanol solution, (b) addition of 2-mercaptopropionic acid, (c) addition of thiourea to form final CZTS precursor, and (d) as-deposited CZTS film.

3. Results and Discussion

Previous research, conducted by Si-Nae *et all.* (2014), has studied the use of thiourea as both stabilizer and source of sulphur [15]. Their work showed successful reaction between thiourea and metal chlorides to form complexes ligand. However, excess thiourea was required to obtain homogenous precursor solutions, and sulphurization was needed to induce CZTS phase. In this research, in favor of 2-mercaptopropionic acid, homogenous solution was obtained at stoichiometry of its constituents. Further, since the indication of unnecessary sulphurization step in our previous research, we developed precursor with excess-containing-thiourea to find optimum condition for the growth of CZTS phase in absence of sulphur atmosphere during heat treatment (annealing).

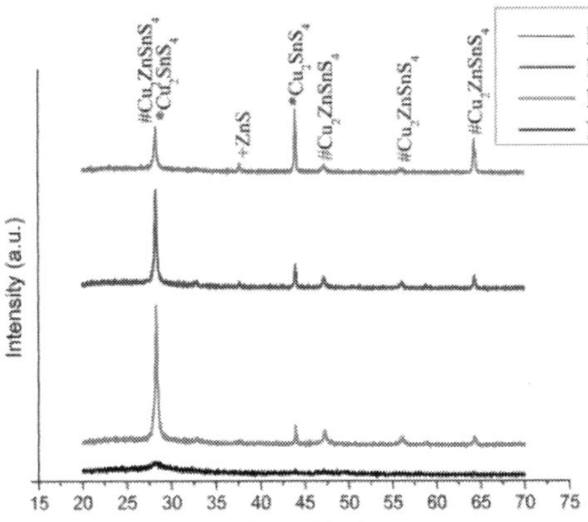

Figure 3. XRD signal of CZTS and secondary phases from the selected samples.

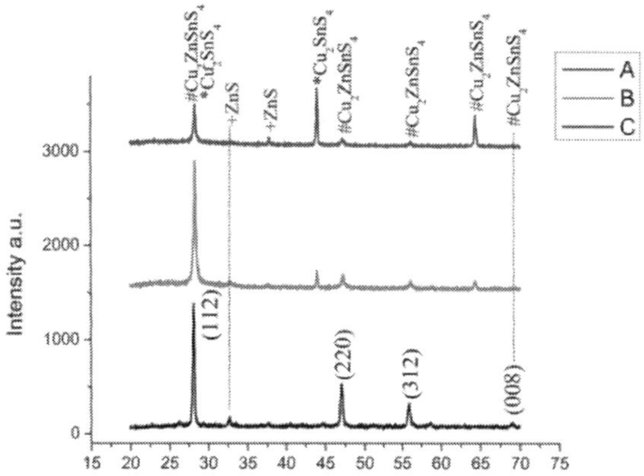

Figure 4. XRD spectra of CZTS film prepared from different condition: mol S/Cu:4.4 with annealing holding time of 30 minutes (A), 60 minutes (B), and mol S/Cu:6.4 annealed 60 minutes (C).

In order to understand the role of sulphur in the formation of CZTS phase, three samples with different mol ratio of sulphur to Cu were selected in the experiment: the ratio of 4.4, 5.4 and 6.4 were used, respectively. The crystallinity of CZTS phase, as first main consideration, is summarized in the form of XRD data plot at figure 3. The figure shows the existence of kesterite CZTS phase (JCPDS no. 26-0575) along with secondary phases, Cu_2SnS_3 (JCPDS no. 1-089-4714) and ZnS (JCPDS no. 5-0566), particularly at higher non-stoichiometry ratio of the precurcor chemicals (Mol S/Cu: 6.4). The formation of secondary phases at non-stoichiometry condition is also encountered in other researches [16-18]. Although the exact reaction mechanism of secondary phases is still unclear, recent research works suggest two responsible causes: the considerable loss of Zn, Sn and S during annealing and the excess of certain element, especially sulphur [15,19,20].

From the previous discussion, it is evidence that sulphur content in the precursor strongly affects the purity of CZTS phase. To support this conclusion, we managed to slightly change the sulphur content through thermal treatment, lowering holding time of the annealing from 60 minutes to 30 minutes. XRD measurements for this experiment is shown in figure 4 line C, with the main peaks indexed to the (112), (220), (312) planes of the kesterite structure. Minor peaks at 33.3° and 69.3° could also be observed for this sample, further confirm the formation of CZTS kesterite phase. Differ from the figure 4 line B, which represents XRD spectra of similar processed-sample apart from annealing holding time, relatively clear CZTS XRD peaks of sample C was observed with an increaces in the crystallinity (indicated from a sharper of the width and the height of the respective peaks).

Table 1. Chemical composition of respected CZTS film.

Ratio/ Sample	CZTS Film before annealing: Pre-dried	CZTS Film before annealing: dried	Annealed CZTS Film (Mol S/Cu): (4.4)*	(5.4)*	(6.4)*	(4.4)**
S/Metal	1,29	1,13	0,97	0,97	0,96	0,95
S/Cu	4,93	2,60	1,96	2,03	2,04	2,06
Cu/(Sn+Zn)	0,35	0,78	0,99	0,92	0,91	0,86
Zn/Sn	0,56	0,52	0,88	1,03	1,09	1,09

* annealing holding time 30 minutes.
** annealing holding time 60 minutes

Table 1 is EDS data representing the chemical composition of the CZTS films. To discuss the mechanism of changing peak intensity from the previous XRD analyses, the ratio of sulphur to metal, Cu/(Zn+Sn) and Zn/Sn are summarized in the table. Overall, the atomic ratio of S to metal and S to Cu share similar number for annealed CZTS film. However, the chemical composition at each certain temperature during annealing hides more significant information to picture the reaction path rather than the composition of final film. From table 1, the effect of sulphur concentration can be indirectly found by observing the ratio of Cu/(Zn+Sn) and Zn/Sn.

As identified by most of the reports, Cu poor (~0.8) and Zn rich (~1.2) hold the proper condition to fabricate high conductivity CZTS films, inferring the high purity of CZTS phase as secondary phases are naturally poor conductive compound [11-13,22]. In addition to final chemical ratio (Cu poor and Zn rich), the composition of the precursor also plays important role to the formation of CZTS and secondary phases during heat treatment [11]. Based on these considerations, and as suggested by EDS and XRD figures, 4.4 mol ratio of S to Cu is the limit of non-stoichiometry for CZTS ethanol-based precursor using 2-mercaptopropionic acid as capping ligand. Further, from the EDS information of 4.4* and 4.4** samples, lowering holding time of annealing treatment (from 60 minutes to 30 minutes) results in the closer composition to the desire Cu poor and Zn rich. Subsequently, confirmed by XRD spectra, 4.4** sample condition successfully formed relatively high and pure CZTS kesterite phase.

Figure 5 displays the surface micrograph of five samples, in correspond to different sulphur concentration of the precursor and holding time of the annealing treatment. Understanding the connection between preparation and final microstructure is crucial to minimize internal resistivity created by void, insulator phases, and grain boundary. Void and non-conductive compound have electrical barriers to hinder charge collection whereas grain boundary has supportive spaces for charge carriers (electron-hole pairs) to recombine [23-25]. However, specific to grain boundary, 1-2 μm is found to be the optimal size for kesterite CZTS rather than to reduce the number of grain boundary (increases grain growth). The grain boundary has previously been reported to provide a current path for minority carriers [22]. From figure 5, all of the annealed samples showed a densely packed structure except for figure 5e which showed a well-growth CZTS kesterite grains despite the porosity at the surface films. The densely packed structure of figure 5a, 5b and 5c, as confirmed from XRD spectra, might be attributed to the existence of the considerable amount of the secondary phases.

CZTS film prepare from 4.4 mol ratio of S to Cu (figure 5e), the highest crystallinity of CZTS kesterite phase, shows a uniform grain size around~ 1.5 μm. However, the growth still does not compensate the path (porosity) of gaseous phases from decomposed organic chemicals. Constraints on CZTS morphology are interestingly challenging as the kinetics of CZTS growth is affected by many condition: the chemicals, temperature and holding time of annealing as well as drying, and atmosphere during preparation and heat treatment. Another comprehensive study is needed to establish a robust environmentally-friendly route to synthesize well-characteristic of CZTS semiconductor, especially to manipulate the morphology of the film.

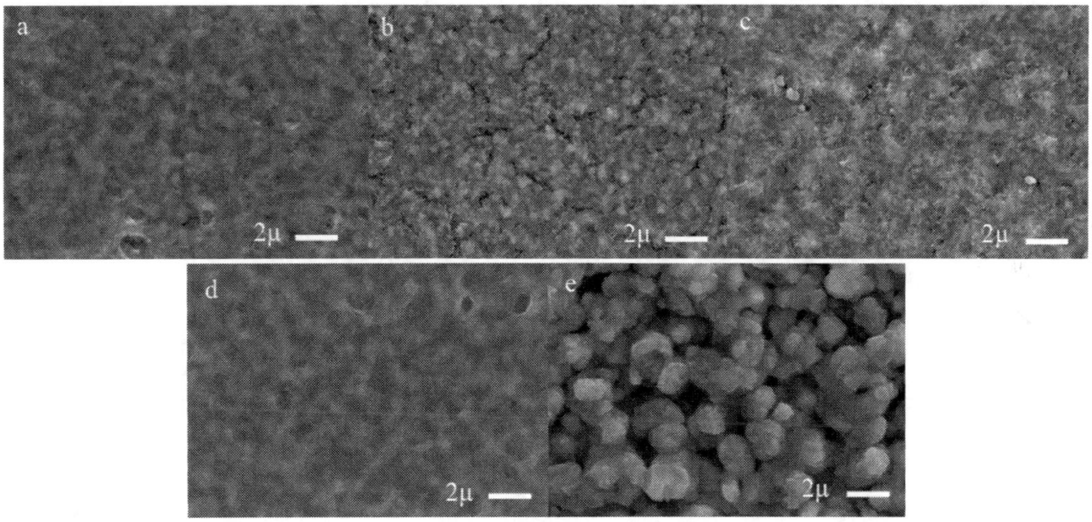

Figure 5. SEM micrographs of the surface morphology: Mol S/Cu (a) 4.4*, (b) 5.4*, (c) 6.4*, (d) as-deposited CZTS film and (e) 4.4**.

The final and main properties of semiconductor is the ability to absorb photon energy which represented by the band gap energy. The band energy was determined by extrapolating gradient line of $(\alpha h\upsilon)^2$ vs hυ (photon energy) curve to the based horizontal line of its axis as depicted in figure 6. It is evidence that the band energy fluctuates in correspond to chemical composition: particularly, the number of Cu/(Sn+Zn), Zn/Sn and the existence of secondary phases [26]. Cu_2SnS_3 (CTS), considered as secondary phase, is reported to has band gap energy in the range of 0.98 to 1.35 eV. CTS is p-type semiconductor with a similar crystal structure and band energy gap to CZTS semiconductor (~ 1.45 eV), and thus could absorb the same range of photon energy; however, it is reported to has a considerably lower efficiency than CZTS [27].

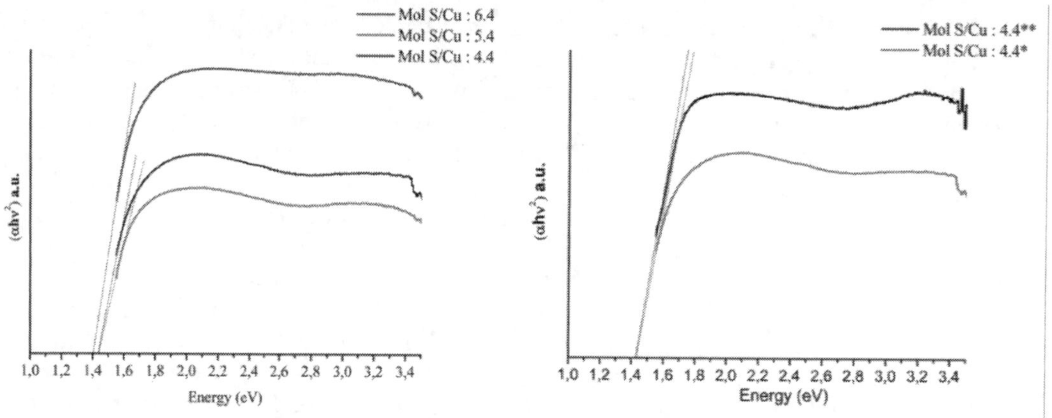

(a) (b)

Figure 6. $(\alpha h\upsilon)^2$ vs $h\upsilon$ (photon energy) curve to determine optical properties of CZTS films.

Figure 6 shows the energy gap of CZTS films slightly ranging from 1.4 eV to 1.42 eV. The lower energy gap, 1.4 eV, was observed for the sample prepared from the highest non-stoichiometry precursor. This condition is attributable to the existence of secondary phases, as discussed from the previous section. From figure 6 (b), both samples (4.4* and 4.4**) share similar band energy at around 1.42 eV. At this level, the figure indicated the vanishingly small influences of the crystallinity of CZTS to its energy gap; the chemical composition hold more significant part to the tunable band energy properties of CZTS. The highest device efficiency of CZTS-based solar cells, however, is found with the bandgap of ~ 1.12 eV prepared by exchange some part of sulphur with selenium (doping), forming CZTSSe matrix [12,28]. Meanwhile, other researches successfully tune the energy gap by modifying the cation matrix [4,5,12]

4. Conclusion

Non-stoichiometry sulphur-containing precursor has been observed to develop single solution approach to synthesize CZTS film. Atomic ratio of S to Cu 4.4 is found to be the limit of excess sulphur in the precursor. From this sample, kesterite CZTS phase has confirmed by XRD measurements with band gap energy ~ 1.42 eV. Further, technical modifications on the annealing treatment successfully reduce the existence of secondary phases and formed well-growth CZTS grains with size ~ 1.5 μm. Meanwhile, non-stoichiometry precursor with S/Cu higher than 4.4 demonstrated CZTS phase along with high concentration of secondary phases, particularly Cu_2SnS_3 and ZnS. All of the CZTS film prepared without the use of sulphur atmosphere during annealing treatment (sulfurization). This route offers the solution to fabricate CZTS semiconductor as low-cost and environmentally-friendly solar cells.

5. Acknowledgement

This work was fully funded by PUPT Research Grant from Universitas Indonesia through the Directorate of Research and Community Service Universitas Indonesia.

6. References

[1] Kentaro I and Nakazawa T 1992 *Japan J. Appl. Phys.* **27** 2094–97.

[2] Tanaka T, Nagatomo T, Kawasaki D, Nishio M, Guo Q, Wakahara A, Yoshida A, Ogawa H 2005 *J. Phys. Chem. Solids.* **66** 1978–81.

[3] Katagiri H, Saitoh K, Washio T, Shinohara H, Kurumadani T, Miyajima S 2001 *Sol. Energy Mater. Sol. Cells* **480-481** 499–502.

[4] Wang W, Winkler M T, Gunawan O, Gokmen T, Todorov T K, Zhu Y and Mitzi D B 2014 *Adv. Energy Mater.* **4** (1301465) 1–5.
[5] Mitzi D B, Gunawan O, Gokmen T, Todorov T K, Wang K, and Guha S 2011 *Sol. Energy Mater. Sol. Cells* **95** 1421–36.
[6] Green M A, Emery K, Hishikawa Y, Warta W, Dunlop E D 2013 *Progress in Photovoltaics: Research and Applications* **21** 1–11.
[7] Gessert T A, Wei S H, Ma J, Albin D S, Dhere R G, Duenov J N, Kuciauskas D, Kanevce A, Barnes T M, Burst J M, Rance W L, Reese M O and Moutinho H R 2013 *Sol. Energy Mater. Sol. Cells* **119** 149–155.
[8] Jackson P, Hariskos D, Wuerz R, Kiowski O, Bauer A, Friedlmeier T and Powalla M 2014 *Physics Status Solidi RRL* **9(1)** 28–31.
[9] Kumar M, Dubey A, Adhikari N, Venkatesan S and Qiao Q 2015 *Energy Environ. Sci.* **8** 3134–59.
[10] Schurr R, Holzing A, Jost S, Hock R, Voβ T, Schulze J, Kirbs A, Ennaoui A, Lux-Steiner M, Weber A, Kotschau I, Schock H-W 2009 *Thin Solid Films* **517** 2465.
[11] Munir B, Prastyo B E, Nurjaya D M, Muslih E Y and Alfauzan S K 2017 AIP Conference Proceedings **1788**, 030022 DOI: 10.1063/1.4968275.
[12] Mitzi D B, Gunawan O, Todorov T K, Wang K, Guha S 1992 *Sol. Energy Mater. Sol. Cells* **95** 1421–36.
[13] Munir B, Prastyo B E, Muslih E Y and Nurjaya D M 2016 *Int. J. of Tech.* **8** 1326–34.
[14] Sabyasachi B, Debasish M, Deepanwita G, Abhijit and Das K 2013 *J. Phys. Chem.* A **117** 1601–13.
[15] Si-Nae P, Shi-Joon S, Dae-Ho S, Dae-Hwan K, Mungunshagai G, Hyeonsik C and Jin-Kyu K 2014 *RSC Adv.* **4** 9118–25.
[16] Scragg J J, Watjen J T, Edoff M, Ericson T, Kubart T and Platzer-Bj¨orkman C 2012 *J. Am. Chem. Soc.* **134** 19330–33.
[17] Redinger A, Berg D M, Dale P J and Siebentritt S 2011 *J. Am.Chem. Phys. Soc.* **133** 1320–23.
[18] Scragg J J, Dale P J, Colombara D and Peter L M 2012 *Chem. Phys. Chem.* **13** 3035–46.
[19] Tiong V T, Zhang Y, Bell J and Wang H 2015 *RSC Adv.* **5** 20178–85.
[20] Zhenggang L, John C W, Lee K K, Zheng X, Wong T L H and Lam Y M 2014 *RSC Adv.* **4** 26888–94.
[21] Fernandes P A, Salom´e P M P and da Cunha A F 2011 *J. Alloys Comp.* **509** 7600–6.
[22] Yin L, Cheng G, Feng Y, Li Z, Yang C and Xiao X 2015 *RSC Adv.* **5** 40369–74.
[23] Werner M, Keller D, Haass S G, Gretener C, Bissig B, Fuchs P, Mattina F L, Erni R, Romanyuk Y E and Tiwari A N 2015 *ACS Appl. Mater. Interfaces* **7(22)** 12141–46.
[24] Buffiere M, Brammertz G, Sahayaraj S, Batuk M, Khelifi S, Mengin D, El Mel A, Arzel L, Hadermann J, Meuris M and Poortmans J 2015 *ACS Appl. Mater. Interfaces* **7(27)** 14690–98.
[25] Jie G, Chu J, Jiang J, Yan Y and Yang P 2015 *ACS Sustainable Chem. Eng.* **3(12)** 3043–52.
[26] Tanaka K, Fukui Y, Moritake N and Uchiki H 2011 *Sol. Energy Mater. Sol. Cells* **95(3)** 838–42.
[27] Vanalakar S A, Agawane G L, Kamble A S, Hong C W, Patil P S and Kim J H 2015 *Sol. Energy Mater. Sol. Cells* **138** 1-8.
[28] Rachmat A W, Woo S K, Eun S L, Munir B, Kyoo H K 2007 *Journal of Physics and Chemistry of Solids* **68 (10)** 1908–13.

IOP Conf. Series: Earth and Environmental Science **105** (2017) 012129 doi:10.1088/1755-1315/105/1/012129

Output voltage characteristic of heat pipe sink thermoelectric generator with exhaust heat utilization of motorcycles

Wayan Nata Septiadi[1,2,5], **Gemilang Ayu Iswari**[3], **Muhammad Ainur Rofiq**[3], **Boby Gitawan**[3], **Jeplind Morico Gugundo**[3] **and Chris Ave Duga Purba**[4]

[1] Department of Mechanical Engineering Udayana University, Kampus Bukit Jimbaran Badung-Bali, Indonesia

[2] Heat Transfer Laboratory Department of Mechanical Engineering Udayana University, Indonesia

[3] Bachelor Programme Student of Department of Mechanichal Engineering, Udayana University, Indonesia

[4] Bachelor Programme Student of Department of Electrical Engineering, Udayana University, Indonesia

E-mail: wayan.nata@gmail.com

Abstract. Nowadays the great challenge to energy engineers is how to develop new and renewable energy sources to reduce the use of fossil fuels. In this direction, the research focuses on the recycle of wasted heat energy on a motorcycle exhaust gas into electrical energy using the concept of seed back. Recycle energy using this concept on a motorcycle is done because the bike has the energy efficiency of ± 35 – 40 % while the rest were dumped unceremoniously in the form of heat energy that is issued through the exhaust. This research aims to know the characteristics of the output voltage heat pipe sink thermoelectric generator utilizing waste heat of the motorcycle. This research has been done on the measurement of the output voltage is generated by the thermoelectric module series TECI-12706B type SP1848-27145 SA as a heat pipe sink thermoelectric generator to produce electrical energy. This research was conducted to know the output voltage from 4 thermoelectric modules arranged in series or parallel. furthermore this thermoelectric generator will be the main energy source for SSUPER-BT devices. Stages of the research done by performing laboratory and road testing. In laboratory testing, simulation of thermoelectric generator made using aluminium rods contain Heater. two sides of the aluminium coupled to thermoelectric modules, heat pipe and CPU fan. in each experiment, the thermoelectric modules strung together in series or in parallel. The results showed on the TEGs with 4 thermoelectric modules in series were able to produce higher output voltage and current compared to TEGs with 2 thermoelectric modules with output voltage reaching 15.7 volts and 7.7 volts respectively. The TEGs circuit with 4 thermoelectric modules produces 72.6% greater power than TEGs with 2 thermoelectric modules.

1. Introduction

[5] To whom any correspondence should be addressed.

Development of renewable energy such as wind energy, geothermal, solar cell, ocean thermal energy conversation and others as power plants continue to seize the attention of energy engineers. The important consideration of research on the development of renewable energy is due to the desire to utilize renewable energy to replace the use of fossil energy as the main energy source.

The development of technology causes electricity be the one of the primary energy needs of the human being. Electricity is an energy that can be directly used in everyday life[1,2]. Electricity service providers in Indonesia such as PLN are still using non-renewable fossil energy such as coal to generate electricity to supply the country's electricity needs. In the process of electricity generation only about 34% is actually converted into electrical energy while the remaining energy is lost or wasted in the form of heat energy commonly called heat energy waste.

Waste heat energy is generated by machines that process energy. Waste heat energy also occurs in the combustion process in motorcycle fuel engines. The small efficiency of motorcycles and the concept of thermodynamic law that has the system's need to reject heat as a by-product of its operation causes only about 40% of the energy that can be used to drive the motor. The remaining energy is lost and simply thrown into the environment through the exhaust in the form of waste heat energy. This waste heat has a higher entropy than the original energy source.

This is the background of research on the reuse of waste heat energy in motorcycle into electricity with the help of thermoelectric generator. By using thermoelectric generators, waste heat generated from the heat of the motor cycle exhaust can be recycled into electrical energy, which can then be reused to another activities [3-5].

Electricity on the thermoelectric generator due to different temperatures between the heated and cold sides of the thermoelectric module by rapidly transferring heat [6-9]. The larger the temperature difference (ΔT) between the hot and cold sides, the greater the power generated. Low curves so thermoelectric has a weakness in the efficiency side that is classified as low as 10%. Factors affecting the low efficiency of the thermoelectric module are the overall unconverted heat and cooling system on the cold side of the imperfect thermoelectric module[10,11]. This study aims to find the Characterization Output Voltage Of Heatpipe Sink Thermoelectric Generator With Exhaust Heat Utilization Of Motorcycles

2. Methodology

2.1. Preparation and design of thermoelectric generator

Thermoelectric generators (TEGs) are designed using 4 thermoelectric modules TECI-12706B type SP1848-27145 SA dimensions 50 x 50 mm with 4 mm thickness in series in pairs between aluminium plate 120 x 120 mm, as shown in Figure 1. The TEGs system is used to capture the flow of heat flowing in the motorcycle exhaust channel, where the heat flowing in the drains is absorbed to the hot side of the thermoelectric through the heat absorber plate. The heat will be utilized to produce ΔT on the thermoelectric module. The cold side of the thermoelectric module will be fitted with a heat pipe sink functioning to lower the temperature of the cool side of the thermoelectric to obtain a larger ΔT. The amount of ΔT generated in the thermoelectric module is highly dependent on the heat of the exhaust and cooling by the heat pipe on the cold side of the thermoelectric module. Figure 2 is the design of the TEG system used to absorb heat in motorcycle exhaust channels.

IOP Conf. Series: Earth and Environmental Science **105** (2017) 012129 doi:10.1088/1755-1315/105/1/012129

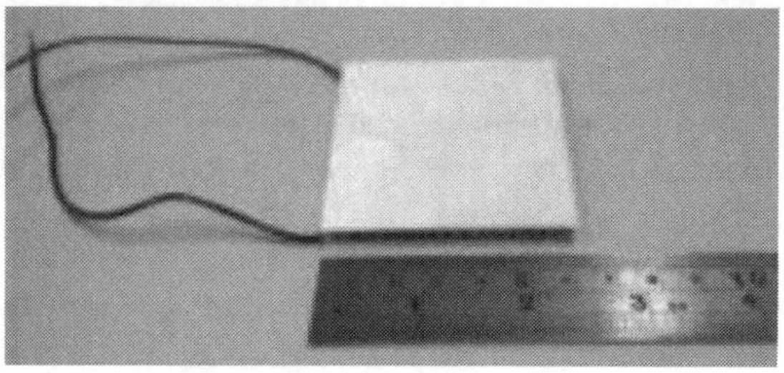

Figure 1. Thermoelectric Module.

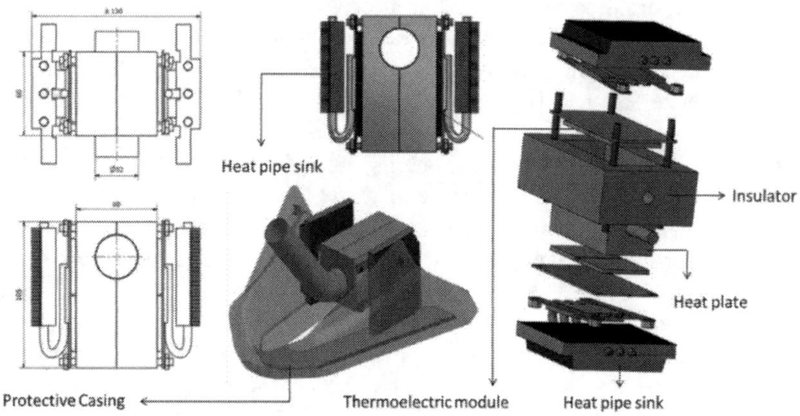

Figure 2. Design of TEGs system.

2.2. Experimental set-up

The exhaust heat in the motorcycle exhaust channel is simulated using a heating plate using a catheter heater with heat reaching 240°C. Figure 3 is a schematic of TEGs output voltage characterization test. Voltage regulators are used to regulating heat loading in simulated exhaust heat. TEGs mounted on the simulator plate are loaded 15.57 Watt up to 28.18 Watt to achieve exhaust gas temperature 80°C up to 240°C. The exhaust gas temperature on the simulator plate is varied from 80°C, 100°C, 120°C, 140°C, 160°C, 180°C, 200°C, 220°C and 240°C. Several K type thermocouples are mounted on the simulator plate, the thermal side of the thermoelectric module, the cold side of the thermoelectric module and on the heat pipe sink section. The thermocouple is connected to the NI 9213 and c-DAQ 9174 acquisition data. The voltage and current generated for each load are measured using the Avometer and clamp meter.

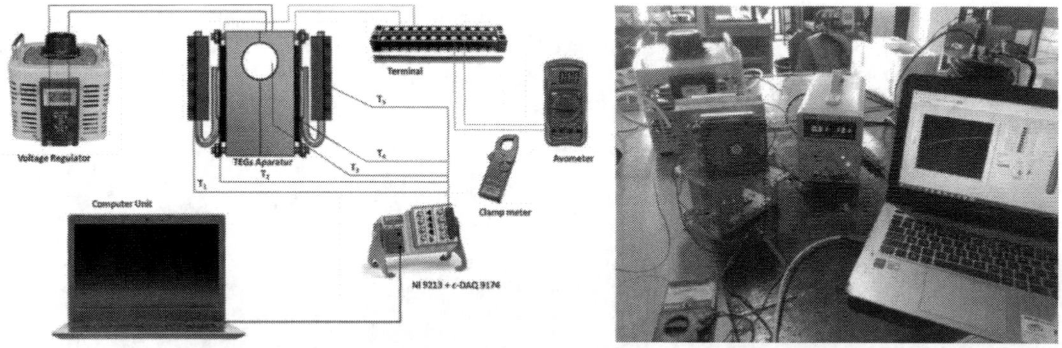

Figure 3. Schematic testing of output voltage characteristics of TEGs.

3. Result and Discussion

Tables 1 and 2 are voltage and current characteristics of TEGs that are coupled in series with 2 and 4 thermoelectric modules. From the table, it can be seen that in series circuit with two thermoelectric module, the highest voltage generated to 7.7 volts at a 77°C temperature different. In this circuit, the amount of maximum power that can be produced is 6.8 watts at a strong current of 884 mA. In the series circuit with four thermoelectric modules, the highest voltage is 15.7 volts which are achieved at a 126°C temperature different. In this circuit, the maximum power generated to 24.9 watts at a current of 1584 mA. The series with two thermoelectric modules produce 2 watts of power up to 6.8 watts in the temperature difference that range from 22°C to 77°C. The power of 2 watts to 24.9 watts in the 20 °C to 126 °C temperature different range is generated by series chains with 4 thermoelectric modules.

The temperature different that generated by series circuit with 2 and 4 thermoelectric module very influential to the magnitude of voltage and output current from TEGs, wherein circuit with 2 thermoelectric module, the temperature difference is not too big because in a certain time interval temperature hot side start propagating to the side cold so that the cold side of the temperature near the hot side temperature. This propagation is due to cooling on the cold side of the thermoelectric is less able to overcome the heat propagation from the hot side. Thermoelectric 2 circuits are also less able to manage exhaust gas temperatures so that thermoelectric efficiency is not able to overcome the exhaust gas temperature that affects the small delta temperature. The lower of temperatures different produces smaller voltages and stronger currents.

Tabel 1. Output voltage and current TEGs with 2 Thermoelectric Module.

Exhaust Gas Temperature (°C)	Thermoelectric module	Hot side Thermoelectric (°C)	Cool side Thermoelectric (°C)	Temperature Different ΔT (°C)	Voltage (Volt)	Current (mA)
80	2	68	46	22	3.6	558
100	2	88	61	27	3.8	569
120	2	102	74	28	4.1	602
140	2	122	84	38	4.5	633
160	2	148	93	55	5.3	679
180	2	163	101	62	6.2	724
200	2	184	112	72	6.8	783
220	2	201	124	77	7.4	841
240	2	222	145	77	7.7	884

Tabel 2. Output voltage and current TEGs with 4 Thermoelectric Module.

Exhaust Gas Temperature (°C)	Thermoelectric module	Hot side Thermoelectric (°C)	Cool side Thermoelectric (°C)	Temperature Different ΔT (°C)	Voltage (Volt)	Current (mA)
80	4	58	38	20	3.6	558
100	4	82	40	42	4.8	669
120	4	105	42	63	6.1	872
140	4	118	44	74	7.5	993
160	4	138	53	85	8.9	1109
180	4	158	61	97	10.6	1224
200	4	178	72	106	12.4	1343
220	4	198	84	114	13.8	1461
240	4	218	92	126	15.7	1584

Figure 4 was an output voltage and current distribution of TEGs with 2 thermoelectric modules assembled in series. The larger the temperature difference generated by the hot side with the cold side portion produces a higher voltage and current. Each of thermoelectric has the ability to produce temperature differently, but the temperature difference also depends on cold side cooling and efficiency of each thermoelectric module so that it will greatly affect the heat source temperatures. In a series 2 thermoelectric modules is only capable of producing a maximum temperature different of 77°C, which is smaller than the TGEs circuit composed of 4 thermoelectric modules. The maximum of temperature different at 126°C is capable of being generated by TEGs in circuits with 4 thermoelectric modules, as shown in figure 5. The larger of temperature different of TEGs with 2 thermoelectric modules result in greater output voltage and current TEGs also greater.

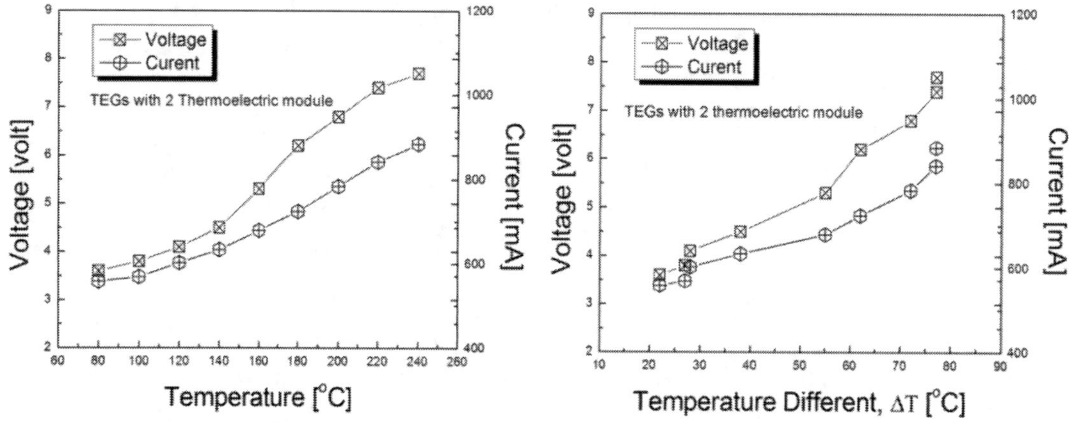

(a) (b)

Figure 4. Distribution of voltage and current with 2 thermoelectric module (a) at exhaust gas temperature (b) at the TEGs temperature different.

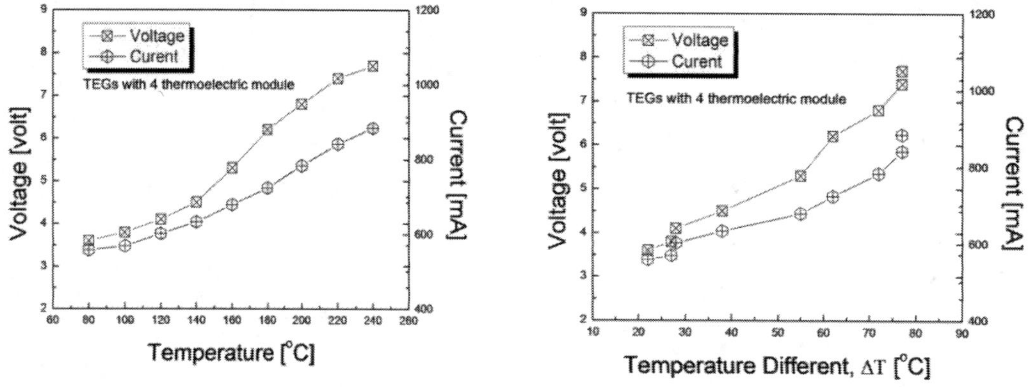

(a) (b)

Figure 5. Distribution of voltage and current with 4 thermoelectric module (a) at exhaust gas temperature (b) at the TEGs temperature different.

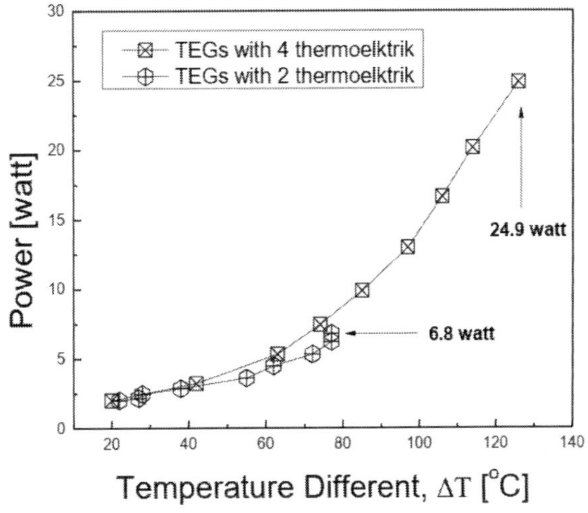

Figure 6. TEGs data in series circuit with 2 and 4 thermoelectric modules.

The magnitude of the output power generated by TEGs with 2 and 4 thermoelectric modules that are arranged in series are shown in figure 6. From the figure it appears that the temperature different and the amount of power generated by each of TEGs are also different. In a series of TEGs with 4 thermoelectric modules have better thermal management capability so the result of temperature different is greater than the circuit with 2 thermoelectric modules. This results in the maximum power capable of TEGs with 4 thermoelectric modules 72.6% greater than the power generated by TEGs with 2 thermoelectric modules, each producing maximum power of 24.9 watts and 6.8 watts.

4. Conclusion

From the research, it was found that TEGs with 4 thermoelectric modules in series were able to produce higher output voltage and current compared to TEGs with 2 thermoelectric modules with output voltage reaching 15.7 volts and 7.7 volts respectively. The TEGs circuit with 4 thermoelectric modules produces 72.6% greater power than TEGs with 2 thermoelectric modules.

5. Reference

[1] Kim S, Park S, Kim S, Rhi S H 2011 A thermoelectric generator using engine coolant for light-duty internal combustion Engine-Powered Vehicles *J. Electron. Mater.* **40** 812–816

[2] Hussain Q E, Brigham D R, Maranville C W 2009 Thermoelectric exhaust heat recovery for hybrid vehicles *SAE Int. J. Engines* **2** 1132–42

[3] Espinosa N, Lazard M, Aixala L and Scherrer H 2010 Modeling a thermoelectric generator applied to diesel automotive heat recovery *J. Electron. Mater.* **39** 1446–55

[4] Goncalves L M, Martins J, Antunes J, Rocha R, Brito F P 2010 Heat-pipe assisted thermoelectric generators for exhaust gas applications *Int. Mechanical Eng. Congress & Exposition* pp. 1387–96

[5] Orr B, Akbarzadeh A and Lappas P 2014 Predicting the performance of a car exhaust heat recovery system that utilises thermoelectric generators and heat pipes *Paper presented to SOLAR 2014*, Melbourne, Australia

[6] Fu J, Tang Y, Chen W, Ma Y and Zhu Z 2015 Research on engine exhaust energy recovery by a heat pipe exchanger with a semiconductor thermoelectric generator *Open Mech. Eng. J.* **9** 130–140

[7] Karri M A, Thacher E F, Helenbrook B T 2011 Exhaust energy conversion by thermoelectric generator: two case studies *Energy Convers. Manage.* **52** 1596–1611

[8] Liang X, Sun X, Tian H, Shu G, Wang Y and Wang X 2014 Comparison and parameter optimization of a two-stage thermoelectric generator using high temperature exhaust of internal combustion engine *Appl. Energy* **130** 190–9
[9] Rowe D M, Min G 1998 Evaluation of thermoelectric modules for power generation *J Power Sources* **73** 193–8
[10] Karami N and Moubayed N 2014 New modeling approach and validation of a thermoelectric generator *IEEE Int. Symposium on Industrial Electronics* pp. 586–91
[11] Liang G, Zhou J and Huang X 2011 Analytical model of parallel thermoelectric generator *Appl Energy* **88(12)** 5193–9

Institute of Physics
Dirac House, Temple Back
Bristol BS1 6BE UK

ISSN: 1755-1307
ISBN 978-1-5108-5686-8

CURRAN ASSOCIATES INC.
proceedings.com

22nd Conferencia Iberoamericana on Software Engineering (CIbSE 2019)

La Habana, Cuba
22-26 April 2019